# The Biology
# of Animal Viruses

*SECOND EDITION*

# THE BIOLOGY OF ANIMAL VIRUSES

## SECOND EDITION

**FRANK FENNER**

John Curtin School of
   Medical Research
Australian National University
Canberra, Australia

**B. R. McAUSLAN**

Roche Institute of Molecular Biology
Nutley, New Jersey

**C. A. MIMS**

Guy's Hospital Medical School
London, England

**J. SAMBROOK**

Cold Spring Harbor Laboratory for
   Quantitative Biology
Cold Spring Harbor, New York

**DAVID O. WHITE**

School of Microbiology
University of Melbourne
Victoria, Australia

ACADEMIC PRESS   New York   San Francisco   London   1974
*A Subsidiary of Harcourt Brace Jovanovich, Publishers*

ACADEMIC PRESS, INC.
111 Fifth Avenue, New York, New York 10003

*United Kingdom Edition published by*
ACADEMIC PRESS, INC. (LONDON) LTD.
24/28 Oval Road, London NW1

**Library of Congress Cataloging in Publication Data**

Main entry under title:

The Biology of animal viruses.

    Edition of 1968 by F. Fenner.
    Bibliography: p.
    1. Virology.    2. Virus diseases.    3. Virology—
Bibliography.  I.    Fenner, Frank John, Date
QR360.B488  1973        576'.6484        72-13614
ISBN 0–12–253040–3

# Contents

## 1  The Nature and Classification of Animal Viruses

## 2  Cultivation, Assay, and Analysis of Viruses

## 3   The Structure and Chemistry of the Virion: A Systematic Survey

## 4   Structure and Function of the Animal Cell

## 5   The Multiplication of DNA Viruses

## 6   The Multiplication of RNA Viruses

## 7   Viral Genetics

## 8   Interference and Interferon

## 9  Pathogenesis: The Spread of Viruses through the Body

## 10  Pathogenesis: The Immune Response

## 11  Pathogenesis: Nonimmunological Factors and Genetic Resistance

## 12  Persistent Infections

## 13 Viral Oncogenesis: DNA Viruses

## 14 Viral Oncogenesis: RNA Viruses

## 15 Prevention and Treatment of Viral Diseases

## 16  The Epidemiology of Viral Infections

## 17  Evolutionary Aspects of Viral Diseases

## Bibliography

# Preface

In the years that have elapsed since the first edition of this book was published there has been a very substantial growth in our knowledge of animal viruses, especially in the areas of molecular biology and tumor virology. In the Acknowledgments of the first edition the senior author, who was then the sole author, suggested that "to review critically as broad and as rapidly changing a field as animal virology is a bold, perhaps a foolhardy, enterprise." It is now an impossible enterprise, and this edition is the result of the collaborative efforts of five erstwhile colleagues from the Department of Microbiology, John Curtin School of Medical Research, who are now located in three continents: Australia, Europe, and North America. Each of us has taken prime responsibility for particular chapters, but all have read and criticized the whole text, with the senior author acting as editor where coordination was needed.

Several chapters of the first edition were almost completely rewritten, in two instances two chapters were consolidated into single chapters (now Chapters 3 and 7), two chapters (Chapters 5 and 8 of the first edition) were eliminated as separate chapters, and a new Chapter 2 was added for the benefit of scientists moving into animal virology from other fields.

We have endeavored to select references best suited to guide the newcomer to the appropriate literature, and have not always attempted to give due notice to those whose contributions have priority. This function is best fulfilled by histories or specialist review articles. We have provided article titles in the Bibliography as a valuable guide to readers wishing to consult original papers.

Because of the existence today of an International Committee on Nomenclature of Viruses that embraces all viruses and the realization that "animal" embraces both vertebrates and invertebrates, an apology is perhaps needed for retaining the words "animal viruses" in the title. We plead only that the term is widely used and was used in the first edition to designate the viruses of vertebrates.

FRANK FENNER
B. R. McAUSLAN
C. A. MIMS
J. SAMBROOK
DAVID O. WHITE

xi

# Preface to First Edition

During the last twenty years virology has developed into an independent science. It is now growing so rapidly that two new journals of virology (in English) were launched this year. Four major works on viruses of vertebrate animals have been published recently, which deal with the viral diseases of man (Horsfall and Tamm, 1965) and his domestic animals (Betts and York, 1967–1968), with techniques in virology (Maramorosch and Koprowski, 1967–1968), and provide an encyclopedic description of the viruses of vertebrates (Andrews, 1964). However, none of these books deals in a comprehensive way with the broader biological principles of animal virology, which is the aim of this two-volume work. It began as an attempt to revise Burnet's "Principles of Animal Virology," but changes in knowledge and emphasis made a revision impossible.

Since the publication of the second edition of Burnet's book in 1960, animal viruses have become major objects for study by molecular biologists, and the knowledge so gained has greatly clarified our understanding of the structure and composition of the virions and their interactions with animal cells. Volume I deals chiefly with the molecular biology of animal viruses and the cellular biology of viral infections. In Volume II a detailed description is given of viral infections at the level of organism, i.e., the pathogenesis of disease and the specific and non-specific reactions of vertebrates to viral infections, including those due to the oncogenic viruses. Finally, the ecology of animal viruses is discussed in relation to their spread through populations of vertebrates.

Only the viruses of warm-blooded vertebrates have been considered. One of the great simplifications of the last decade has been the realization that nearly all of the 500 viruses of warm-blooded vertebrates now recognized can be allocated to one of about a dozen groups. In each of these groups the viruses are very like each other in many properties, but often strikingly different from viruses of other groups. In the opening chapter of Volume I the basis for this classification and the properties of the major groups are discussed in detail, and in subsequent chapters different aspects of virology are discussed in terms of model viruses selected from some or all of these groups.

This work was written primarily for virologists and molecular biologists, including teachers, research workers, and graduate students in these fields. It should also prove useful to doctors and veterinarians interested in either clinical or public health aspects of the viral diseases of man or his domestic animals. Each chapter ends with a summary so that casual readers can rapidly orient themselves. In order to keep the text reasonably short no historical account is given of the development of virology; the reader is referred to the first chapter of Burnet's "Principles of Animal Virology" for such information.

The selection of references presented a problem. Rather than attempt to observe priority of discovery over so wide a field, the references were selected mainly for their potential value in providing the reader with an effective entry into the scientific literature of the subject under discussion. References to abstracts were retained only when no later paper on the topic could be found.

FRANK FENNER

# Acknowledgments

We are grateful to the following authors for their permission to use or modify figures or plates from their articles, and for their provision of original and sometimes previously unused photographs for the preparation of plates. Detailed sources of materials used are given in the legends:

Doctors A. C. Allison (Fig. 11–5); J. D. Almeida (Plates 3–7, 3–25); D. Baltimore (Fig. 6–3); A. R. Bellamy (Plate 6–4); F. L. Black (Fig. 16–4); S. S. Breese (Plate 3–8); F. Brown (Plates 3–4, 3–7, 3–12); F. M. Burnet (Figs. 11–2, 11–3); D. L. D. Caspar (Fig. 1–2); R. M. Chanock (Plate 3–17); P. W. Choppin (Plates 3–15, 6–2); R. W. Compans (Plates 3–6, 3–15); P. D. Cooper (Fig. 7–2); G. E. Cottral (Fig. 9–10); S. Dales (Plates 3–9, 3–23, 5–1, 5–3, 6–1, 6–3); J. E. Darnell (Fig. 3–5); N. J. Dimmock (Plate 2–8); A. W. Downie (Fig. 9–5); A. A. Ferris (Fig. 16–2); E. Fleissner (Tables 3–9, 3–10); E. A. Follett (Plate 3–1); H. Frank (Plate 3–22); A. J. Gibbs (Plate 3–16); M. Green (Fig. 5–5); P. M. Grimley (Plate 3–2); I. H. Holmes (Plate 3–14); R. E. Hope-Simpson (Fig. 12–2); K. Hummeler (Fig. 3–7); I. Jack (Plates 2–1, 2–2, 2–4, 2–5, 2–7); R. T. Johnson (Fig. 11–4); W. K. Joklik (Figs. 3–10, 6–4); Z. Kapikian (Plate 3–17); E. D. Kilbourne (Plate 2–3); W. G. Laver (Plates 3–3, 3–6, 3–15); F. Lehmann-Grube (Plate 3–18); H. D. Mayor (Plate 3–11); J. L. Melnick (Fig. 2–2); D. H. Moore (Plate 3–21); C. Morgan (Plate 6–1); B. Morris (Plates 9–1, 9–2); F. A. Murphy (Plate 3–19); J. Nagington (Plate 3–9); A. J. Nahmias (Fig. 17–1); G. J. V. Nossal (Fig. 10–3); P. L. Ogra (Fig. 10–5); N. Oker-Blom (Plate 3–19); H. G. Pereira (Plate 3–3); D. Peters (Plate 3–10); B. Roizman (Fig. 5–6, Plate 3–6); E. D. Sebring (Figs. 5–3, 5–4); W. Schäfer (Plate 3–22, Fig. 3–9); P. Sharp (Plate 3–2); E. Shelton (Plate 4–1); R. W. Simpson (Plates 3–13, 3–20); K. O. Smith (Plate 3–11); W. R. Sobey (Fig. 17–2); K. B. Tan (Plate 3–14); D. A. J. Tyrell (Fig. 9–3); D. W. Verwoerd (Plate 3–24); C. Wallis (Fig. 10–4); R. G. Webster (Plate 2–6); N. G. Wrigley (Plates 3–4, 3–5, 3–7, Fig. 3–2).

We also thank the publishers of the following journals and books for their permission to use photographs or figures that have previously appeared in the publications mentioned in individual captions and references:

*Annals of the New York Academy of Sciences* (Fig. 9–10)
*Archives für gesamte Virusforschung* (Fig. 11–5, Plate 3–7)

*Australasian Annals of Medicine* (Figs. 11–2, 11–3)
*Bacteriological Reviews* (Fig. 9–4)
*British Journal of Experimental Pathology* (Fig. 9–7)
*Center for Disease Control* (Fig. 16–3)
*Cold Spring Harbor Laboratory* (Figs. 1–2, 5–5)
*Journal of Cell Biology* (Plates 3–9, 5–1)
*Journal of Experimental Medicine* (Fig. 11–4)
*Journal of General Virology* (Plate 3–7, Figs. 3–2, 3–4)
*Journal of Hygiene* (Fig. 17–2)
*Journal of Immunology* (Plate 3–11)
*Journal of Molecular Biology* (Plates 3–3, 4–1, Fig. 3–5)
*Journal of Theoretical Biology* (Fig. 16–4)
*Journal of Ultrastructure Research* (Plate 3–10)
*Journal of Virology* (Plates 3–17, 3–24, 5–2, 6–1, Figs. 3–6, 5–3, 5–4)
*Lancet* (Fig. 9–6)
*Nature* (Fig. 10–3)
*Proceedings of the Royal Society of Medicine* (Fig. 12–2)
*Veterinary Record* (Fig. 9–5)
*Virology* (Plates 3–8, 3–9, 3–13, 3–14, 3–15, 3–19, 3–20, 3–23, 5–3, 6–1, 6–2, 6–3, Figs. 3–9, 3–10, 6–4)
"Atlas of Viruses" (Academic Press) (Plate 3–6)
"Fundamental Techniques in Virology" (Academic Press) (Plate 2–3)
"The Biology of Large RNA Viruses" (Academic Press) (Fig. 3–7)
"Myxomatosis" (Cambridge University Press) (Fig. 11–1)
"The Biochemistry of Viruses" (Marcel Dekker) (Fig. 5–6)
"Viruses Affecting Man and Animals" (W. H. Green) (Fig. 10–4)
"Microbiology" (Harper & Row) (Fig. 2–1)

All five authors acknowledge the help freely given by their scientific colleagues and secretaries; particular thanks are due to Mrs. M. Mahoney who typed and retyped the whole book. It is a pleasure to acknowledge the assistance of the staff of Academic Press in the production of the book and with the preparation of the figures.

FRANK FENNER
B. R. McAuslan
C. A. Mims
J. Sambrook
David O. White

CHAPTER 1

# The Nature and Classification
# of Animal Viruses

## INTRODUCTION

Virology began as a branch of pathology, the study of disease. At the end of the nineteenth century, when the microbial etiology of many infectious diseases had been established, pathologists recognized that there were a number of common infectious diseases of man and his domesticated animals for which neither a bacterium nor a protozoan could be incriminated as the causal agent. In 1898, Loeffler and Frosch demonstrated that the economically important disease of cattle, foot-and-mouth disease, could be transferred from one animal to another by material which could pass through a filter that retained the smallest bacteria. Following this discovery such diseases were tentatively ascribed to what were first called "ultramicroscopic filterable viruses," then "ultrafilterable viruses," and, ultimately, just "viruses."

Independently, the plant pathologist Beijerinck (1899) recognized that tobacco mosaic disease was caused, not by a conventional microorganism, but by what he called a "contagium vivum fluidum" (i.e., "contagious living fluid"). Many economically important diseases of domestic plants were subsequently found to be caused by viruses. Years later, the bacteriologists Twort (1915) and d'Hérelle (1917) recognized that bacteria also could be infected by viruses, for which d'Hérelle coined the name "bacteriophages." Insect viruses were not recognized as such until the 1940's (Bergold, 1958), and even more recently viruses have been recovered from fungi (Hollings, 1962), from blue-green algae (Schneider et al., 1964; review: Brown, 1972), from free-living mycoplasmas (Gourlay, 1971), and from protozoa (Diamond et al., 1972).

## THE NATURE OF VIRUSES

From the practical viewpoint of the plant pathologist and the public health worker, it is convenient to regard the viruses that cause disease as pathogenic microorganisms (Burnet, 1945). However, the question arose as to whether viruses, whatever their host, might have common properties that distinguished them from microorganisms. Lwoff has cogently argued (Lwoff, 1957; Lwoff and Tournier, 1966) that all viruses show some properties that distinguish them from any microorganism. Exceptions to some of Lwoff's generalizations have since been discovered, but two still apply: (a) unlike even the smallest micro-organisms (chlamydiae), viruses contain no functional ribosomes or other cel-lular organelles, and (b) in RNA viruses the whole of the genetic information is encoded in RNA, a situation unique in biology. Other distinctions apply to some but not all viruses, e.g., the isolated nucleic acid of viruses of several genera is infectious (i.e., the virus can be generated intracellularly from a single mole-cule of nucleic acid), and viruses of most genera contain either no virus-coded enzymes, or one or more enzymes that belong to particular classes (neuramini-dases and nucleic acid polymerases).

It is impossible to define viruses satisfactorily in a sentence or even a para-graph, bearing in mind both their intracellular states and the extracellular par-ticles or virions. Virions consist of a genome of either DNA or RNA enclosed within a protective coat of protein molecules, some of which may be associated with carbohydrates or lipids of cellular origin. In the vegetative state and as "provirus" (see Chapter 5), viruses may be reduced to their constituent genomes, and the simplest "viruses" may be transmitted from one host to another as naked molecules of nucleic acid, possibly associated with certain cellular compo-nents. At the other extreme, the largest animal viruses, e.g., the poxviruses and the leukoviruses, are relatively complex.

Lwoff's concept that "viruses are viruses" has had important theoretical and practical consequences; on the one hand, it emphasized their similarities irre-spective of the nature of the host (animal, plant or bacterium), and, on the other hand, it led to the possibility of freeing viruses from the rules of bacteriological nomenclature. However, the operational division of viruses made according to type of host continues to be used by the majority of virologists most of the time, and it is significant that the International Committee on Nomenclature of Viruses (ICNV), although dedicated to a universal classification, operates through Sub-committees on Bacterial, Invertebrate, Plant, and Vertebrate Viruses (Wildy, 1971).

## THE CHEMICAL COMPOSITION OF ANIMAL VIRUSES

The simpler viruses consist of nucleic acid and a few polypeptides specified by it. More complex viruses usually also contain lipids and carbohydrates; in the great majority of viral genera these chemical components are not specified by the viral genome but are derived from the cells in which the viruses multiply. In exceptional situations, cellular nucleic acids or polypeptides may be built into viral particles.

### Nucleic Acids

Viruses contain only a single species of nucleic acid, which may be DNA or RNA. Viral nucleic acid may be single- or double-stranded, the viral genome may consist of one or several molecules of nucleic acid, and if the genome consists of a single molecule this may be linear or have a circular configuration.

As yet, no animal viral nucleic acid has been found to be methylated, or to contain novel bases of the type encountered in bacterial viruses or mammalian transfer RNA's, but some virions contain oligonucleotides rich in adenylate, of unknown function. The base composition of DNA from animal viruses covers a far wider range than that of the vertebrates, for the guanine plus cytosine (G+C) content of different viruses varies from 35 to 74%, compared with 40 to 44% for all chordates. Indeed, the G+C content of the DNA of viruses of one genus (*Herpesvirus*) ranges from 46 to 74%.

The molecular weights of the DNA's of different animal viruses varies from just over 1 to about 200 million daltons; the range of molecular weights of viral RNA's is much less, from just over 2 to about 15 million daltons. The nucleic acid can be extracted from viral particles with detergents or phenol. The released molecules are often fragile but the isolated nucleic acid of viruses belonging to certain genera is infectious. In other cases, the isolated nucleic acid is not infectious even though it contains all the necessary genetic information, for its transcription depends upon a virion-associated transcriptase without which multiplication cannot proceed.

All DNA viruses have genomes that consist of a single molecule of nucleic acid, but the genomes of many RNA viruses consist of several different molecules, which are probably loosely linked together in the virion. In viruses whose genome consists of single-stranded nucleic acid, the viral nucleic acid is either the "positive" strand (in RNA viruses, equivalent to messenger RNA) or the "negative" (complementary) strand. Preparations of some viruses with genomes of single-stranded DNA consist of particles that contain either the positive or the complementary strand.

Viral preparations often contain some particles with an atypical content of nucleic acid. Host-cell DNA is found in some papovaviruses, and what appear to be cellular ribosomes in some arenaviruses. Several copies of the complete viral genome may be enclosed within a single particle (as in paramyxoviruses) or viral particles may be formed that contain no nucleic acid ("empty" particles) or that have an incomplete genome, lacking part of the nucleic acid that is needed for infectivity.

Terminal redundancy occurs in the DNA of some vertebrate viruses, but most sequences are unique. The largest viral genomes contain several hundred genes, while the smallest carry only sufficient information to code for about half a dozen proteins, most of which are structural proteins of the virion.

## Proteins

The major constituent of the virion is protein, whose primary role is to provide the viral nucleic acid with a protective coat. As predicted by Crick and Watson (1956), from a consideration of the limited amount of genetic information carried by viruses, the protein shells of the simpler viruses consist of repeating protein subunits. Sometimes the viral protein comprises only one sort of polypeptide chain, although, more commonly, there are two or three different polypeptides. The proteins on the surface of the virion have a special affinity for complementary receptors present on the surface of susceptible cells. They also contain the antigenic determinants that are responsible for the production of protective antibodies by the infected animal.

Viral polypeptides are quite large, with molecular weights in the range 10,000–150,000 daltons. The smaller polypeptides are often but not always internal, the larger ones often but not always external. There are no distinctive features about the amino acid composition of the structural polypeptides of the virion, except that those intimately associated with viral nucleic acid in the "core" of some icosahedral viruses are often relatively rich in arginine.

Viral envelopes usually originate from the cellular plasma membrane from which the original cellular proteins have been totally displaced by viral peplomers and a viral "membrane protein" (see Fig. 1–1). The peplomers consist of repeating units of one or two glycoproteins, the polypeptide moiety of which is virus-specified while the carbohydrate is added by cellular transferases. In many enveloped viruses, the inside of the viral envelope is lined by a viral protein called the membrane or matrix protein.

Not all structural viral proteins are primary gene products, since with many viruses the viral mRNA is translated into a large polypeptide that is enzymatically cleaved to yield two or more smaller virion proteins. Cleavage is often one of the terminal events in the assembly of the virion and it can occur *in situ* after most of the proteins are already in place.

Although most virion polypeptides have a structural role some have enzymatic activity. Many viruses contain a few molecules of an internal protein that functions as a transcriptase, one of the two kinds of peplomers in the envelope of myxoviruses has neuraminidase activity, and a variety of other enzymes are found in the virions of the larger, more complex viruses.

In addition to polypeptides that occur as part of the virion, a large part of the viral genome (most of it, with the large DNA viruses) codes for polypeptides that have a functional role during viral multiplication but are not incorporated into viral particles. Few of these "nonstructural viral proteins" have been characterized.

## Lipid and Carbohydrate

Except for the large and complex poxviruses, which constitute a special case, lipid and carbohydrate are found only in viral envelopes and are always of cellular origin. The lipids of viral envelopes are characteristic of the cell of origin, though minor differences between the viral envelope and the normal plasma membrane may be demonstrable. About 50 to 60% of the lipid is phospholipid and most of the remainder (20–30%) is cholesterol. Some of the viral carbohydrate occurs in the envelope as glycolipid characteristic of the cell of origin, but most of it is part of the glycoprotein peplomers that project from the viral envelope.

## THE STRUCTURE OF ANIMAL VIRUSES

During the 4 years that followed the introduction of negative staining for the electron microscopic study of viruses (Brenner and Horne, 1959), a general picture was obtained of the structure of representatives of most of the groups of animal viruses that were known at the time (review: Horne and Wildy, 1963). Three structural classes were distinguished: isometric particles, which were usually "naked" but in some groups were enclosed within a lipoprotein envelope; long tubular nucleoprotein structures, always (with viruses of vertebrates) surrounded by a lipoprotein envelope; and in a few groups, a more complex structure. Accepting a number of new terms defined by Lwoff *et al.* (1959a), Caspar and his colleagues analyzed the principles underlying the structure of simple viruses (review: Caspar, 1965). Their basic concepts remain valid, but subsequent work has rendered some of the original definitions ambiguous; where necessary these have been modified.

## Terminology

*Virion* (plural virions) is used as a synonym for "virus particle." The protein coat of an isometric particle or the elongated protein tube of viruses with helical symmetry is called the *capsid*. It may be "naked," or it may be enclosed within a lipoprotein *envelope* (peplos) which is derived from cellular membranes as the virus matures by budding. Where the capsids directly enclose the viral nucleic acid, as is usual with tubular capsids but less common with isometric capsids, the complex is called the *nucleocapsid*. With most isometric particles and in all complex virions, the capsid encloses another protein structure containing the viral genome, called the *core*.

Capsids consist of repeating units of one or a small number of protein molecules. Three levels of complexity can be distinguished. *Chemical units*, the ultimate gene products, are single polypeptides that may themselves constitute the *structural units*, or several polypeptides may form homo- or heteropolymers which constitute the structural units. The structural units, or groups of them, may be visualized in the electron micrographs as *morphological units*. Morphological units that form part of a capsid are called *capsomers*; those projecting from the envelope are the *peplomers* (sometimes called "spikes," an unsatisfactory term since they are never pointed and may, indeed, have knob-shaped ends).

The chemical units are sometimes held together by disulfide bonds to form the structural units, hence the practice of using reducing agents in polyacrylamide gel electrophoresis when analyzing viral proteins to determine their constituent polypeptides. The structural units are held together to form the capsid by noncovalent bonds, which may be polar (salt and hydrogen bonds) or nonpolar (van der Waals and hydrophobic bonds). The capsids of some viruses are readily disrupted in molar calcium or sodium chloride, suggesting electrovalent bonds between the structural units; others are unaffected by salt and can only be disrupted by detergents, suggesting that they are hydrophobically bonded.

## Isometric Viruses

It has been found that the isometric virus particles that have been adequately studied by X-ray diffraction and electron microscopy have capsids in which the capsomers are arranged with icosahedral symmetry. According to Caspar and

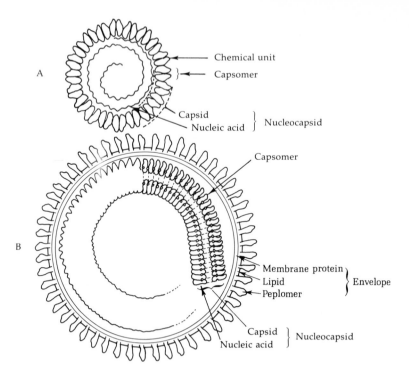

Fɪɢ. 1–1. *Schematic diagrams of the structure of a simple nonenveloped virion with an icosahedral capsid (A) and an enveloped virion with a tubular nucleocapsid with helical symmetry (B). The capsids consist of morphological subunits called capsomers, which are in turn composed of structural subunits that consist of one or more chemical subunits (polypeptide chains). Many icosahedral viruses have a "core" (not illustrated), which consists of protein(s) directly associated with the nucleic acid, inside the icosahedral capsid. In viruses of type B the envelope is a complex structure consisting of an inner virus-specified protein shell (membrane protein, made up of structural subunits), a lipid layer derived from cellular lipids, and one or more types of morphological subunits (peplomers), each of which consists of one or more virus-specified glycoproteins (modified from Caspar et al., 1962).*

Klug (1962), this occurs because the icosahedron is that polyhedron with cubic symmetry which, if constructed of identical subunits, would least distort the subunits or the bonds between them.

An icosahedron (Fig. 1–2) has 20 equilateral triangular faces, 12 vertices, where the corners of 5 triangles meet, and 30 edges, where the sides of adjacent pairs of triangles meet. It shows twofold symmetry about an axis through the center of each edge (Fig. 1–2A), threefold symmetry when rotated around an axis through the center of each triangular face (Fig. 1–2B), and fivefold symmetry about an axis through each vertex (Fig. 1–2C). Each triangular face may be thought of as containing, and being defined by, three asymmetric units (i.e., units that have no regular symmetry axes themselves) so that a minimum of sixty asymmetric units are required to construct an icosahedron.

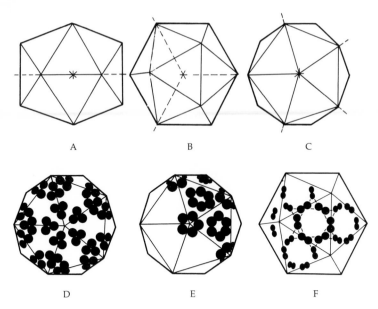

FIG. 1–2. *Features of icosahedral structure. Above. Regular icosahedron viewed along twofold (A), threefold (B), and fivefold (C) axes. Various clusterings of structural subunits give characteristic appearances of capsomers in electron micrographs. With T = 3 the structural subunits may be arranged as 20 T trimers (D), capsomers are then difficult to define, as in poliovirus; or they may be grouped as 12 pentamers and 20 hexamers (E) which form bulky capsomers as in Parvovirus, or as dimers on the faces and edges of the triangular facets (F), producing an appearance of a bulky capsomer on each face, as in Calicivirus.*

The triangular faces of an icosahedron can be subdivided into smaller identical equilateral triangles, to form a solid called an icosadeltahedron. Only certain subdivisions are possible; the number of new triangles per facet is called the triangulation number $(T)$, and $T = h^2 + hk + k^2$, where $h$ and $k$ are any pair of integers. When $h = k$ ($T = 3$, 12, 27, etc.), or when either $h$ or $k = 0$ ($T = 1$, 4, 9, 16, etc.), the triangles are arranged symmetrically on the underlying icosahedral face, but with other values for $h$ and $k$ (e.g., $h = 2$, $k = 1$ and $T = 7$) they are in a skew arrangement. A complete description then requires determination of the hand of the structure (right-*dextro* or left-*levo*). The hand of the icosahedral shells of some papilloma viruses has been investigated by Klug and Finch (1965) and Finch and Klug (1965), who concluded that the human papilloma virus (human wart virus) had a $T = 7d$ icosahedral surface lattice whereas rabbit papilloma virus had a $T = 7l$ lattice.

In an icosadeltahedron with a triangulation number of 3, each icosahedral face has 9 and the whole solid $20 \times 9 = 180$ asymmetric units. These structural units may differ in shape and clustering so that the morphological units (capsomers) visible by electron microscopy may differ greatly in viruses with the same triangulation number. There are three basic types of clustering pattern:

1. The three units defining each triangular face may cluster at the center of the triangle, forming trimer capsomers (Fig. 1–2D).

2. The structural units may cluster at the vertices of the triangles, so that where five triangles meet at the vertices of the icosahedron there are pentamer capsomers, and where six triangles meet on faces of the icosadeltahedron there are hexamer capsomers (Fig. 1–2E).

3. Pairs of structural units from adjacent triangles may cluster on the edges between the triangle to give dimer capsomers (Fig. 1–2F).

The pattern seen on the surface of the virion need not reflect the way in which the structural units are bonded together, and gives no clue as to whether the structural units are constituted by single chemical units or are homo- or heteropolymers of the chemical units. However, the number of structural units in each capsomer can be guessed at from the arrangement and size of the capsomers (Fig. 1–2).

All known animal viruses whose genome is DNA have isometric (or complex) capsids, as do all those whose genome is double-stranded RNA and the viruses of two major families (Picornaviridae and Togaviridae) whose genome consists of a single molecule of single-stranded RNA.

## Viruses with Tubular Nucleocapsids

Tobacco mosaic virus occupies a unique position in virology. Not only was it the agent whose "viral" nature was first appreciated (Beijerinck, 1899), and the first virus to be crystallized (Stanley, 1935), but more is known of its physical and chemical structure than any other virus (reviews: Caspar, 1965; Kaper, 1968). The virus particles are nonenveloped straight rods, which consist of 2100 repeating polypeptide (chemical) units, which are the structural units and also, without clustering, constitute the capsomers. These protein molecules are arranged in a helical manner so that except at the ends of the particles every capsomer is in a structurally equivalent position in relation to the long axis of the rods. Many plant viruses and a few bacteriophages have similar nonenveloped tubular virions, their capsomers being arranged in helices whose pitch is characteristic for the virus group. Such viruses are structurally defined by their length and width, the pitch of the helix, and the number of capsomers in each turn of the helix.

Tubular nucleocapsids are found in many groups of viruses of vertebrates, but only among those whose genome consists of single-stranded RNA. None of these occurs as "naked" virions; the flexuous helical tubes are always inside lipoprotein envelopes. The diameters of the nucleocapsids of several viruses have been measured, but in only a few cases is the length or the pitch of the helix known. The best studied example, the nucleocapsid of Sendai virus, a paramyxovirus, is a helix about 1 $\mu$m long and 20 nm wide, with a pitch of 5.0 nm (Finch and Gibbs, 1970). There are about 2400 hourglass-shaped structural units in the nucleocapsid, with either eleven or thirteen units per turn of the helix. The structural units are single polypeptides with a molecular weight of about 60,000 daltons, arranged with their long axes at an angle of about 60° to the long axis of the nucleocapsid, which therefore has a herringbone appearance in electron micrographs (see Plate 3–16). The contact surface between adjacent turns of the basic helix is conical, so that contact is maintained even when the nucleocapsid is sharply flexed, and the viral RNA is thus protected.

## Relationship of Nucleic Acid and Capsomers

Unlike cells, which contain several different species of nucleic acid that subserve different functions, the only nucleic acid in viruses, apart from small amounts of host tRNA in leukoviruses, is their genome. It may consist of either DNA or RNA, it may be single- or double-stranded, it may be linear or cyclic, and the genome may consist of one or several molecules of nucleic acid.

Although the only detailed studies have been made on a few plant and bacterial viruses (review: Tikchonenko, 1969), it is clear that the interaction between the viral nucleic acid and the capsomers is different in nucleocapsids with helical and icosahedral symmetry. In tobacco mosaic virus, there is a maximum regular interaction between the single strand of viral RNA and the protein subunits which form a protective coat around it. A similar relationship probably exists in most animal viruses with tubular nucleocapsids, but in some viruses (e.g., influenza virus) the integrity of the tubular structure is destroyed by treatment with RNase but not by proteases, suggesting a different relationship of RNA and protein. In icosahedral viruses, on the other hand, there can be no such regular relationship of the nucleic acid and each polypeptide subunit. In the simplest isometric viruses, the folding of the flexible single-stranded RNA may have some regularity in relation to the capsomers and their constituent chemical subunits. X-ray diffraction studies of turnip yellow mosaic virus (Klug et al., 1966), for example, show that a significant portion of the single RNA chain is deeply embedded within the protein shell, large segments being intimately associated with the 180 structural units, which as hexamers and pentamers make up the 32 capsomers. The presence of the RNA in and about these positions enhances the definition of 32 capsomers seen in electron micrographs (Finch and Klug, 1966). Except for some togaviruses, even the simple isometric viruses of vertebrates have a more complex structure than this, since they contain several virus-coded polypeptides. One or more of these polypeptides are known to be "internal" to the capsid and it is thought that these rather than the capsomers interact with the viral RNA. Reovirus particles have two concentric protein shells, each consisting of well-defined morphological units. The proteins of the larger DNA viruses are arranged in several layers, not all of which display symmetry. The internal proteins of many DNA viruses are highly basic and are thought to be bonded to the viral nucleic acid, constituting a core within the isometric capsid.

## Viral Envelopes

Although occasionally used in a more general way to refer to the outer viral coats of some complex viruses like the poxviruses (Mitchiner, 1969), we think that it is desirable to restrict the use of the term "envelope" to the outer lipoprotein coat of viruses that mature by budding through cellular membranes.

Enveloped viruses contain 20–30% of lipid, all of which is found in the envelope. Chemical analyses show that the lipid is derived from the cellular membranes through which the virus matures by budding, but all the polypeptides of viral envelopes are virus-specified. *Herpesvirus* is the only virus of vertebrates that matures by budding through the nuclear membrane, and its envelope contains several virus-specified glycoproteins. All other enveloped viruses bud through cytoplasmic membranes, and contain one or more different polypeptides. The Togaviridae have an isometric core to which a lipid layer is directly applied, and virus-specified glycoprotein peplomers project from this. All animal viruses with tubular nucleocapsids are enveloped, and in these the lipid layer from which glycoprotein peplomers project is probably applied to a protein shell (the membrane protein; see Fig. 1–1), which may be relatively rigid, as in *Rhabdovirus*, or readily distorted (as in the myxoviruses) so that in negatively stained electron micrographs the virions appear to be pleomorphic.

### Complex Virions

Viruses that have large genomes have a correspondingly complex structure. Apart from the undetermined nature of the "cores" of many of the isometric viruses (e.g., *Herpesvirus* and *Adenovirus*), the virions of the two largest animal viruses (*Poxvirus* and *Iridovirus*) have highly complex structures, which are described in the appropriate sections of Chapter 3. The RNA viruses that have the largest (single-stranded) genomes, those of the *Leukovirus* genus, also have a highly complex structure with an envelope enclosing an icosahedral capsid that, in turn, surrounds a tubular nucleocapsid.

## CLASSIFICATION AND NOMENCLATURE IN BIOLOGY

The aim of classification in biology is to make an ordered arrangement of a particular class of biological objects that will indicate their similarities and differences. Adoption of a system of classification also involves consideration of the nomenclature of the objects to be classified. Linnaeus introduced a latinized binomial nomenclature into biology 200 years ago, and phylogenetic classifications of animals and plants based on the theory of evolution have since been introduced. International Codes of Nomenclature with rigid sets of rules, and Judicial Commissions to pass judgement on proposed names, have been set up for the naming of plants and of animals. An International Code of Nomenclature of Bacteria and Viruses was approved in 1947 and has since been revised (Buchanan *et al.*, 1958). Although they are primarily concerned with nomenclature, all these Codes involve agreement upon a system of classification. Codes are based on "acceptances," i.e., beliefs we would like to justify but are unable to prove, the principal one being that we are able to arrange living things in an orderly system that is indicative of both rank in a hierarchy and phylogenetic relationships (Cowan, 1966). Classifications of animals and plants attempt to be scientific by deriving their taxa from a consideration of phylogenetic relatedness. More recently this approach has been reinforced by tests for genetic relatedness, i.e., the information content of the genetic material of the agents concerned. This has been tested by homology experiments with DNA's extracted from the cells of a variety of animals (McCarthy, 1969), and it is to be expected that the phylogenetic and the molecular biological approaches will eventually be combined.

The classification of bacteria into the same hierarchical pattern as that of plants and animals (phyla, subphyla, classes, orders, suborders, families, genera, and species) has led to a chaotic situation (Cowan, 1970). Some bacterial taxonomists are looking to numerical methods, readily exploited with the aid of electronic computers, for the solution of their problems (Sneath, 1964). Disadvantages of this approach are that the weighting of characters tends to be involuntary, and that pleiotropism may lead to some characters being scored more than once. Most virologists believe that certain characters of viruses, such as the type, amount, and conformation of the viral nucleic acid, are taxonomically more important than characters like host range or pathogenic potential.

Molecular biology provides an alternative to phylogenetic relationships for making a scientific classification of microorganisms, viz., by the determination of genetic relatedness, using both the genetic material and the polypeptides that it specifies (Mandel, 1969). There are two groups of agents, the mycoplasmas and the viruses, for which detailed "official" classifications are still in the process of formation. Because of their small genomes, they are particularly suitable for molecular taxonomy, i.e., classification based on the molecular weights and base ratios of their genomes, and on the results of nucleic acid hybridization experiments. Applied to mycoplasmas, this approach has disproved claims that these microorganisms were derived from certain bacterial species (Razin, 1969).

Nucleic acid hybridization experiments have now been performed with many different viruses; detailed references to the results obtained will be given in Chapter 3. In general, they have provided some useful data on relationships within genera and species, but not at higher taxonomic levels. With the methods used thus far many viruses now allocated to the same genus have shown little or no homology of their nucleic acids. Indeed, nucleic acid hybridization may be too critical a method to be useful except for the comparison of closely related viruses, and less exacting tests for the similarity of viral genomes may be more pertinent when considering different viral species. Bellett (1967a,b) analyzed the data available in 1966 on the molecular weights and base ratios of the nucleic acids of different viruses. His results on the "clustering" of the viruses of vertebrates are consistent with the genera proposed by the ICNV (Wildy, 1971). However, newer knowledge about the fragmented nature of the genomes of some RNA viruses and of the varied modes of their transcription and translation (Baltimore, 1971b) suggests that these data, where available, should be added to the parameters used by Bellett. The differences between the molecular weights of the DNA's of the DNA viruses of vertebrates are such that sophisticated analysis is not needed to define the currently accepted families and genera.

## PREVIOUS CLASSIFICATIONS OF VIRUSES

Until about 1950, little was known about viruses other than their pathogenic behavior. Most early proposals for viral classification were confined to either plant or animal viruses and were based mainly upon the symptomatology of diseases caused by them, which tended to classify the host responses rather than the viruses. Bawden (1941) made the pioneering suggestion that viral nomenclature and classification should be based upon properties of the virus particle.

In the early 1950's Bawden's approach was exploited by animal virologists (Andrewes, 1952), and viruses were allocated to groups which were usually given latinized names constructed from a chosen prefix plus the word "virus." Thus, myxovirus (Andrewes et al., 1955), poxvirus (Fenner and Burnet, 1957), herpesvirus (Andrewes, 1954), reovirus (Sabin, 1959), papovavirus (Melnick, 1962), picornavirus (International Enterovirus Study Group, 1963), and adenovirus (Pereira et al., 1963) groups were described. In the meantime, a classification using quite different criteria had been established by epidemiologists. Since they were so concerned with the transmission of infection, epidemiologists have used a classification based on the mode of transmission of disease; they have grouped viruses together as "respiratory viruses," "enteric viruses," or "arthropod-borne (arbo-) viruses." The last term, in particular, has been widely used, but it is generally agreed that this epidemiological classification, although useful, is in no sense taxonomic.

Concurrently with these suggestions relating to the viruses of vertebrates, Lwoff (1957) insisted upon the similarities between viruses, whatever their natural host, and the differences between viruses and all other biological entities. He was instrumental in arranging for the establishment of an international committee (Anon., 1965; Lwoff and Tournier, 1966) to discuss nomenclature. Its major proposal was to select "type species" upon which names for groups would be based. It also proposed a classification based on (a) the chemical nature of the nucleic acid, (b) the symmetry of the nucleocapsid (helical, cubical, or binal), (c) the presence or absence of an envelope, and (d) certain measurements: for helical viruses, the diameter of the nucleocapsid, for cubical viruses, the triangulation number and the number of capsomers.

The official International Committee on Nomenclature of Viruses (ICNV), which was set up at the Ninth International Congress for Microbiology in 1966, adopted the physicochemical criteria of Lwoff and Tournier, but rejected the detailed hierarchical classification. The more important nomenclatural proposals accepted by ICNV were: (a) an "effort should be made" toward a latinized binomial system of nomenclature, (b) the "law of priority" is unacceptable, (c) no taxon should be named from a person, and (d) anagrams, siglas, hybrids of names, and nonsense names should be prohibited.

Before describing the classification of animal viruses that we shall use throughout this book, it is appropriate to consider some of the problems of classification and nomenclature that have not yet been tackled by ICNV. One of the most important is the level of taxa that should be used. So far, only three families of animal viruses have been accepted (see below), but it is clear that large and heterogeneous groups currently classed as genera (e.g., *Poxvirus, Herpesvirus, Paramyxovirus, Leukovirus*; Wildy, 1971) should be regarded as families. Indeed, it would not be unreasonable to regard all the currently accepted isolated "genera" as families, some of which (e.g., *Adenovirus*) might at this stage contain only a single genus. The conventional physicochemical criteria [(a) nucleic acid: type, strandedness, fragmentation, and molecular weight; (b) virion: shape, size, and symmetry] are suitable for classification at this level of family/genus, perhaps assisted by the serological cross-reactivity of "group" antigens where these have been recognized.

At the other end of the nomenclatural spectrum, there is hopeless confusion in the ways in which the terms "species," "type," "subtype," and "strain" are used. For example, "types" of influenza virus exhibit no serological cross-reactivity and their nucleic acids do not hybridize; they should be regarded at least as distinct species. On the other hand, many alphaviruses and flaviviruses with distinct names, which exhibit extensive serological cross-reactivity, should perhaps be regarded as types within the same species. Serological cross-reactivity and nucleic acid hybridization tests are probably most useful for making comparisons at this "species" level.

## A CLASSIFICATION OF ANIMAL VIRUSES

The ICNV, working under the chairmanship of Professor P. Wildy, presented its first report at the Tenth International Congress for Microbiology in Mexico City in 1970, and has published valuable basic data on forty-three viral groups encompassing viruses of bacteria, invertebrates, plants, and vertebrates (Wildy, 1971). In spite of the problems referred to above, we shall follow the classification set out in the Report, amplifying it with proposals that have come forward since then, but maintaining accepted usages of the terms "type," "strain," etc.

### Families, Genera, and Species

Only two families were established by ICNV in 1970 (Wildy, 1971): Papovaviridae and Picornaviridae, and subsequently the family Togaviridae was defined. Most accepted "groups" of vertebrate viruses were given generic names. No species names were adopted by ICNV, although "type species" were designated for several of the genera.

### Cryptograms

In the descriptions of families and genera in this chapter the four terms of the cryptograms of Gibbs et al. (1966), as modified in Wildy (1971), are shown. The data refer to the infective viral particle (the virion). The first term of the cryptogram describes the type of the nucleic acid (R = RNA, D = DNA)/strandedness (1, 2 = single-, double-stranded). The second term describes the molecular weight of the nucleic acid (in millions)/the percentage of nucleic acid in the virion. Where the genome of infective particles consists of separate pieces occurring together in a single virion the symbol "$\Sigma$" indicates this fact and the figure gives the total molecular weight of the genome. The third term describes the outline of the virion/outline of nucleocapsid [S = essentially spherical; E =

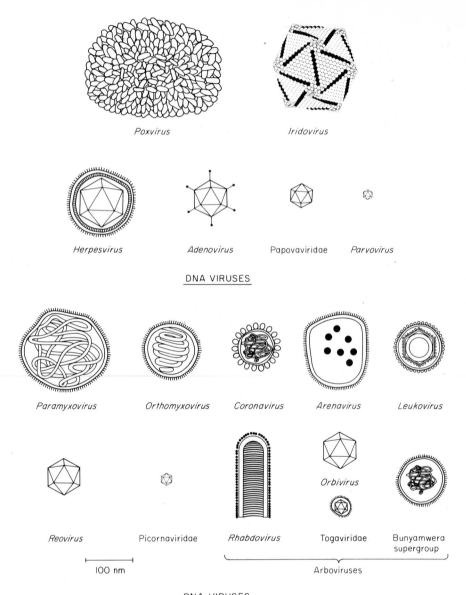

*Poxvirus*                                    *Iridovirus*

*Herpesvirus*          *Adenovirus*        *Papovaviridae*      *Parvovirus*

DNA VIRUSES

*Paramyxovirus*     *Orthomyxovirus*      *Coronavirus*        *Arenavirus*        *Leukovirus*

*Reovirus*           *Picornaviridae*       *Rhabdovirus*        *Orbivirus*        Bunyamwera
                                                                *Togaviridae*       supergroup

100 nm                                                  Arboviruses

RNA VIRUSES

Fig. 1–3. *Diagram illustrating the shapes and relative sizes of animal viruses of the major taxonomic groups (bar = 100 nm).*

elongated with parallel sides, ends not rounded; U = elongated with parallel sides, end(s) rounded; X = complex]. The fourth term describes the kinds of host infected (V = vertebrate; I = invertebrate)/the kinds of vector (O = spreads without a vector; Di = diptera; Ac = tick or mite; Si = flea). An asterisk indicates that a particular property is not known.

## SHORT DESCRIPTIONS OF THE MAJOR GROUPS OF DNA VIRUSES

TABLE 1–1

*Properties of the Virions of the Recognized Genera of DNA Animal Viruses*

| GENUS | GENOME[a] | | VIRION | | |
|---|---|---|---|---|---|
| | MOL. WT. ($\times 10^6$ DALTONS) | NATURE[b] | SHAPE[c] | SIZE (NM) | TRANS-CRIPTASE |
| *Papillomavirus*[d] | 5 | D, cyclic | Icosahedral (72) | 55 | — |
| *Polyomavirus*[d] | 3 | D, cyclic | Icosahedral (72) | 45 | — |
| *Adenovirus* | 20–29 | D, linear | Icosahedral (252) | 70–80 | — |
| *Herpesvirus* | 100–120 | D, linear | Icosahedral (162), enveloped | Envelope 150; capsid 100 | — |
| *Iridovirus* | 130–140 | D, linear | Icosahedral ($\sim$ 1500) ? enveloped[e] | Capsid, 190 | ? |
| *Poxvirus* | 160–200 | D, linear | Brick-shaped | 300 $\times$ 240 $\times$ 100 | + |
| *Parvovirus* | 1.2–1.8 | S, linear | Icosahedral (32) | 20 | — |

[a] Genome invariably a single molecule.
[b] D, double-stranded; S, single-stranded.
[c] Figure in parentheses indicates number of capsomers in icosahedral capsids.
[d] Members of family Papovaviridae.
[e] Insect iridoviruses, no envelope; vertebrate members probably enveloped.

### Family: Papovaviridae [D/2: 3–5/7–13: S/S: V/0, Di, Ac, Si]

The family Papovaviridae (sigla: Pa = papilloma; po = polyoma; va = vacuolating agent, SV40) encompasses two genera, *Polyomavirus* (poly = many; oma = tumor) and *Papillomavirus* (papilla =nipple; oma = tumor), which differ substantially in size and nucleic acid content of the virion (Table 1–2) but share many other properties.

TABLE 1–2

*Family: Papovaviridae[a] [D/2: 3–5/7–15: S/S: V/0, Di, Ac, Si]*

| Genus: | *Papillomavirus* | *Polyomavirus* |
|---|---|---|
| Size of virion: | 55 nm | 45 nm |
| Mol. wt. of DNA: | 5 $\times$ 10$^6$ daltons | 3 $\times$ 10$^6$ daltons |
| Type species: | Rabbit papilloma virus | Mouse polyoma virus |
| Other members: | Rabbit oral papilloma virus | Simian virus 40 |
| | Human papilloma virus | "K" virus |
| | Canine papilloma virus | Rabbit vacuolating virus |
| | Canine oral papilloma virus | Viruses of multifocal |
| | Bovine papilloma virus | leukoencephalopathy of man |

[a] Characteristics: single cyclic molecule of double-stranded DNA, 5 or 3 million daltons; icosahedral capsid 55 or 45 nm diameter, with 72 capsomers; no envelope; multiply in nucleus. Two genera: *Papillomavirus* and *Polyomavirus*.

An important property of many papovaviruses is their capacity of produce tumors. In nature, some produce single benign tumors (which may undergo malignant change) and are highly host specific; others may cause primary malignant tumors within a short period of their inoculation into newborn rodents.

### Genus: Adenovirus [D/2: 20–29/12–14: S/S: V/0]

The adenoviruses (adeno = gland) are nonenveloped icosahedral DNA viruses which multiply in the nuclei of infected cells, where they may produce a crystalline array of particles. Many serological types have been isolated from human sources. These have an antigen that is shared by all mammalian strains, but differs from the corresponding antigen of avian strains. Allocation to the genus is made primarily on the basis of the characteristic size and symmetry of the virion as seen in electron micrographs (icosahedron with 252 capsomers).

TABLE 1–3

*Genus: Adenovirus[a,b]*

| SUBGROUP[c] | RECOGNIZED SEROTYPES | MOL. WT. OF DNA (MILLION DALTONS) |
|---|---|---|
| Human | 33 | 20–25 |
| Simian | 12 | 22 |
| Avian | ? | 29 |

[a] Type species: human adenovirus type 1 [D/2: 23/13: S/S: V/0].

[b] Characteristics: single linear molecule of double-stranded DNA, 20 to 29 million daltons; icosahedral capsid 80 nm in diameter, with 252 capsomers and fibers projecting from the twelve vertices; no envelope; multiply in nucleus.

[c] Adenoviruses have been recovered from many other species: cow, pig, sheep, horse, mouse, opossum, and dog (including canine hepatitis virus).

Most adenoviruses are associated with respiratory infection and many such infections are characterized by prolonged latency. Some multiply in the intestinal tract and are recovered in feces. Many adenoviruses, from both mammalian and avian sources, produce malignant tumors when inoculated into newborn hamsters.

In the laboratory, stable hybrids have been produced between certain adenoviruses and the *Polyomavirus,* SV40 (see Chapter 7).

### Genus: Herpesvirus [D/2: 80–100/7: S/S: V/0]

The herpesviruses (herpes = creeping) are readily recognized by their morphology. Their icosahedral capsid is assembled in the nucleus and acquires an envelope as the virus matures by budding through the nuclear membrane.

Electron microscopic examination by negative staining of many previously unclassified viruses showed that several of them had large icosahedral capsids with 162 capsomers enclosed within lipoprotein envelopes, similar to the type species, herpes simplex virus. When examined further, such viruses were found to be DNA viruses that multiplied in the nucleus, and have now been included in the genus *Herpesvirus*. Table 1–4 shows some of the viruses now regarded as members of this genus; a more complete list is given by Andrewes and Pereira (1972). There is a group-specific antigen(s) associated with the nucleocapsids and demonstrable by immunodiffusion, and several type-specific antigens associated with the nucleocapsid and envelope. Some type-specific antigens cross-react (e.g., herpes simplex viruses type 1 and type 2 and B virus).

TABLE 1–4

*Genus: Herpesvirus*[a,b]

| VIRUS | NATURAL HOST | COMMENT |
|---|---|---|
| Herpes simplex type 1[c] | Man | |
| Herpes simplex type 2[c] | Man | Genital tract |
| Varicella-zoster | Man | Varicella and zoster are different manifestations of infection by one virus |
| Epstein-Barr | Man | Causes infectious mononucleosis: associated with Burkitt lymphoma and nasopharyngeal carcinoma |
| B virus[c] | Monkey | |
| Pseudorabies | Cow, pig | |
| Infectious bovine rhinotracheitis | Cow | |
| Equine abortion (equine herpes type 1) | Horse | |
| Infectious laryngotracheitis | Chicken | |
| Marek's disease | Chicken | Oncogenic in birds |
| Lucké | Frog | Produces adenocarcinomas in frogs |
| Cytomegaloviruses | Man, mouse guinea pig, etc. | Several related viruses, each host specific |

[a] Type species: herpes simplex virus type 1 [D/2: 100/7: S/S: V/0].
[b] Characteristics: single linear molecule of double-stranded DNA, about 100 million daltons; icosahedral capsid 100 nm in diameter, with 162 capsomers, enclosed by envelope 150 nm diameter; multiply in nucleus; mature by budding at nuclear membrane. Group-specific antigen(s) associated with nucleocapsid.
[c] Serologically related by cross-neutralization tests.

Different herpesviruses cause a wide variety of types of infectious diseases, some localized and some generalized, often with a vesicular rash. A feature of many herpesvirus infections is prolonged latency associated with one or more episodes of recurrent clinical disease.

### Genus: Iridovirus [D/2: 130–140/15: S/*: I,V/*]

This genus (irido = iridescent) was defined on the basis of several viruses of insects whose structure and nucleic acid content have been carefully studied

(Bellett, 1968; Wrigley, 1969). Several DNA viruses of vertebrates that are similar in morphology and certain other characteristics have been tentatively grouped with the genus *Iridovirus* (Table 1–5). Like poxviruses, but unlike other DNA viruses, iridoviruses multiply in the cytoplasm. Their DNA consists of a single linear molecule, with a molecular weight of about 130–140 million daltons, and the virion is a large and complex nonenveloped icosahedron, with an outer shell composed of about 1500 capsomers. The vertebrate "iridoviruses" may be enveloped. Several enzymes are found within mature virions.

TABLE 1–5

*Genus: Iridovirus[a,b]*

| Members of genus (from insects) | *Tipula* iridescent virus |
|---|---|
| | *Sericesthis* iridescent virus |
| | *Chilo* iridescent virus |
| | *Aedes* iridescent virus |
| Possible members of genus (from vertebrates) | African swine fever virus [c] |
| | Amphibian cytoplasmic DNA viruses, including frog virus 3 (FV3) [c] |
| | Gecko virus |
| | Lymphocystis virus of fish |

[a] Type species: *Tipula* iridescent virus [D/2: 126/15: S/*: I/*].
[b] Characteristics: single linear molecule of double-stranded DNA, 130–140 million daltons; complex structure, with outer icosahedral capsid 190 nm in diameter, with about 1500 capsomers; no true envelope.
[c] Iridoviruses of vertebrates may be enveloped.

The best studied of the vertebrate "iridoviruses" are some viruses of frogs, notably FV3 (review, Granoff, 1969); the most important economically is African swine fever virus (review, Hess, 1971).

### Genus: Poxvirus [D/2: 160–200/5–7: X/*: V/0, Di, Ac, Si]

The poxviruses (pock = pustule) are the largest animal viruses, and contain a larger amount of DNA (160–200 million daltons of double-stranded DNA) than any other virus. The structure of the brick-shaped virion is complex, consisting of a biconcave DNA-containing core surrounded by several membranes of viral origin. There is a poxvirus group antigen which is probably an internal component of the virion, and can be demonstrated by complement fixation or gel diffusion tests. Several enzymes, including a transcriptase, are found within mature virions. Multiplication occurs in the cytoplasm and the virions mature in cytoplasmic foci. Occasionally, the virion may be released within a loose membrane derived from the cytoplasmic membrane. This is not essential for infectivity, and must be distinguished from the envelope of viruses that mature by budding through cellular membranes.

The genus is divided into several subgenera (Table 1–6), and there are several poxviruses that have still to be classified. The properties outlined for the genus

TABLE 1–6

*Genus: Poxvirus*[a,b]

| SUBGENERA | | | | | OTHER POXVIRUSES NOT YET ALLOCATED |
|---|---|---|---|---|---|
| A. VACCINIA | B. ORF | C. SHEEP POX | D. BIRDPOX | E. MYXOMA | TO GENERA |
| Vaccinia | Orf | Sheep pox | Fowlpox | Myxoma | Swinepox |
| Cowpox | Bovine | Goatpox | Canarypox | Rabbit | Molluscum |
| Ectromelia | papular | Lumpy skin | Pigeonpox | fibroma | contagiosum |
| Monkeypox | stomatitis | disease | Turkeypox | Squirrel | Yaba monkey |
| Variola | Pseudocowpox | | | fibroma | tumor virus [c] |
| | (milkers' | | | Hare fibroma | Tana virus [c] |
| | nodes) | | | | Entomopox- |
| | Chamois | | | | virus [d] |
| | contagious | | | | |
| | ecthyma | | | | |

[a] Type species: vaccinia virus [D/2: 160/5: X/*: V/0].

[b] Characteristics: single linear molecule of double-stranded DNA, 160–200 million daltons; large brick-shaped virion measuring 300 × 240 × 100 nm; complex structure; multiply and mature in cytoplasm. Contain several enzymes including a transcriptase. Members of subgenera A to E share a group antigen; additional serological cross-reactivity within subgenera. Some members (subgenera B, C, and swinepox virus) differ in shape from vaccinia virions.

[c] Serologically related.

[d] Viruses of insects that show many resemblances to poxviruses of vertebrates (morphology, nucleic acid, enzymes).

apply to all the subgenera, except that the virions of members of the subgenera B and C (see Table 1–6), and swinepox virus, are narrower than those of other poxviruses, and virions of subgenus B (orf) have a distinctive surface structure. Species within each subgenus show a high degree of serological cross-reactivity by neutralization as well as complement fixation tests. Genetic recombination occurs within, but not between, subgenera; nongenetic reactivation (complementation) occurs between most poxviruses of vertebrates (Chapter 7).

Certain viruses that multiply in insects have many of the attributes of poxviruses and have been tentatively called entomopoxviruses (entomo = insect) (review: Bergoin and Dales, 1971).

Poxviruses cause diseases in man, domestic and wild mammals, and birds. These are sometimes associated with single or multiple benign tumors of the skin, but are more usually generalized infections, often with a widespread vesiculo-pustular rash. Several poxviruses are transmitted in nature by arthropods acting as mechanical vectors.

## Genus: Parvovirus [D/1: 1.2–1.8/35: S/S: V/0]

Parvoviruses (parvo = small) are unique among the DNA viruses of vertebrates in that their genome is a single molecule of single-stranded DNA. Two subgenera are recognized: several viruses of rodents which are "normal" infectious viruses (subgenus A), and the adenovirus-associated viruses, which are able to replicate only in cells concurrently infected with an adenovirus (subgenus B). In subgenus B the single strands of DNA found in a population of virions are complementary and anneal after extraction to form a double strand.

TABLE 1–7

*Genus: Parvovirus[a,b]*

|  | MEMBERS | SEROTYPES |
|---|---|---|
| Subgenus A | H-1 | Serotype 1 |
| (infectious viruses) | H-3 | |
| | RV | Serotype 2 |
| | X-14 | |
| | Minute virus of mice (MVM) | Serotype 3 |
| Subgenus B [c] | Adenovirus-associated | Four serotypes |
| (satellite viruses) | viruses | |
| Possible members of genus | Avian, porcine, bovine parvoviruses | |
| | Feline panleucopenia virus | |
| | Hemorrhagic encephalopathy virus (rat) | |
| | Mink enteritis virus | |

[a] Type species: latent rat virus (Kilham) [D/1: 1.8/34; S/S: V/0].
[b] Characteristics: single linear molecule of single-stranded DNA, 1.2 to 1.8 million daltons; icosahedral capsid about 20 nm in diameter; no envelope; multiply in nucleus.
[c] Viruses of subgenus B are defective; their multiplication depends upon concurrent infection of cells with an adenovirus.

The viruses of rodents cause acute fulminating disease when inoculated into newborn hamsters (Kilham, 1961). The adenovirus-associated viruses are not known to cause any symptoms.

## SHORT DESCRIPTIONS OF THE MAJOR GROUPS OF RNA VIRUSES

### Family: Picornaviridae [R/1: 2.5–2.8/30: S/S: V/0]

The picornavirus group (sigla: pico = small; rna = ribonucleic acid), which includes a very large number of viruses, was accepted by ICNV as a family, Picornaviridae, with three genera: *Enterovirus* (entero = intestine), *Rhinovirus* (rhino = nose), and *Calicivirus* (calici = cup). Newman *et al.* (1973) believe that this subdivision of the family is unduly restrictive; on the basis of particle density, base composition of the viral RNA's, and stability at various pHs they differentiate cardioviruses (cardio = heart) from *Enterovirus*, and foot-and-mouth disease virus and "equine rhinovirus" from the genus *Rhinovirus* (Table 1–9).

### Genus: *Enterovirus* [R/1: 2.6/30: S/S: V/0].
Enteroviruses have the family characteristics of the Picornaviridae. The particles are 20–30 nm in diameter, acid stable (pH 3), and have a buoyant density (in CsCl) of 1.34–1.35 g/cm$^3$. They are

## TABLE 1–8

*Properties of the Virions of the Recognized Genera of RNA Animal Viruses*

| GENUS | GENOME | | VIRION | | | | |
|---|---|---|---|---|---|---|---|
| | MOL. WT. ($\times 10^6$ DALTONS) | NATURE [a] | ENVELOPE | SHAPE [b] | SIZE (NM) | TRANSCRIPTASE | SYMMETRY OF NUCLEOCAPSID [c] |
| *Enterovirus* [d] | 2.6 | S,1 | – | Icosahedral | 20–30 | – | Icosahedral (20–30) [h] |
| *Rhinovirus* [d] | 2.6 | S,1 | – | Icosahedral | 20–30 | – | Icosahedral (20–30) [h] |
| *Calicivirus* [d] | 2.8 | S,1 | – | Icosahedral | 20–30 | – | Icosahedral (20–30) [h] |
| *Alphavirus* [e] | 4 | S,1 | + | Spherical | 50–60 | – | Icosahedral (30–40) |
| *Flavivirus* [e] | 4 | S,1 | + | Spherical | 40–50 | – | Icosahedral (20–30) |
| *Orthomyxovirus* | 5 | S,7 | + | Spherical | 80–120 | + | Helical (9) |
| *Paramyxovirus* | 7 | S,1 | + | Spherical | 100–300 | + | Helical (18) |
| *Coronavirus* | ? | S,? | + | Spherical | 80–120 | ? | Helical (9) |
| *Arenavirus* | ? 3.5 | S,?3 | + | Spherical | 85–120 | ? | Helical (?) |
| *Bunyamwera* [f] | ? | S,? | + | Spherical | 90–100 | ? | Helical (?) |
| *Leukovirus* | 10–12 | S,4 | + | Spherical | 100–120 | + (Reverse) | Helical (?) |
| *Rhabdovirus* | 4 | S,1 | + | Bullet-shaped | 175 × 70 | + | Helical (5) |
| *Reovirus* | 15 | D,10 | – | Icosahedral | 70–80 | + | Icosahedral (45) |
| *Orbivirus* [g] | 15 | D,10 | – | Icosahedral | 50–60 | + | Icosahedral (50–60) [h] |

[a] All molecules linear; S, single-stranded; D, double-stranded; number, number of molecules in genome.
[b] Some enveloped viruses are very pleomorphic (sometimes filamentous).
[c] Figure in brackets indicates diameter (nm) of nucleocapsids.
[d] Members of family Picornaviridae.
[e] Members of family Togaviridae.
[f] "Bunyamwera supergroup," morphologically and serologically related arboviruses.
[g] Name suggested by Borden *et al.* (1971).
[h] Nucleocapsid not distinguishable from virion.

TABLE 1–9

Family: Picornaviridae[a,b,c]

| GENUS | MEMBERS | ACID STABILITY | BUOYANT DENSITY IN CsCl (G/CM³) |
|---|---|---|---|
| Enterovirus | Human, including polio-viruses, coxsackieviruses and echoviruses<br>Bovine and porcine<br>Murine encephalomyelitis virus<br>Duck hepatitis virus<br>Nodamura virus [e] | Stable at pH 3 | 1.34–1.35 |
| Rhinovirus | Human rhinoviruses, > 90 serotypes<br>Bovine rhinoviruses | Labile at pH 3 | 1.38–1.43 |
| Calicivirus | Vesicular exanthem of swine virus group—several serotypes<br>Feline picornavirus | Labile at pH 3<br>variable at pH 5 | 1.37–1.38 |
| Cardiovirus [d] | EMC virus<br>Mengo virus<br>ME virus | Stable at pH 3 and pH 8 but unstable at pH 6 | 1.34 |
| Aphthovirus [d] | Foot-and-mouth disease virus (several serotypes) | Labile at pH 3 | 1.43 |
| Equine rhinovirus [d] | — | Labile at pH 3 | 1.45 |

[a] Type genus: Enterovirus.
[b] Type species: Poliovirus type 1 [R/1: 2.6/30: S/S: V/0].
[c] Characteristics: single linear molecule of single-stranded RNA, 2.6–2.8 million daltons; purified RNA is infectious; nonenveloped; capsid 20–30 nm in diameter, with cubic symmetry; multiply in cytoplasm.
[d] Not official genera, but groups distinguished from Enterovirus and Rhinovirus, respectively, by Newman et al. (1973) on basis of density in CsCl, base composition of RNA, and stability of virion at different pH's.
[e] Possibly an arbovirus.

primarily inhabitants of the intestines, and a large number of serotypes have been found in the feces of man and of various animals.

The enteroviruses of man have been subdivided into three major subgroups: poliovirus, three serotypes; echovirus (acronym: echo = enteric cytopathogenic human orphan), thirty-four serotypes; and coxsackievirus (Coxsackie = town in New York State), twenty-four serotypes of type A and six of type B. The polioviruses, which show some serological cross-reactivity, are distinguished by their capacity to paralyze humans. Coxsackieviruses were originally defined in terms of their capacity to multiply in infant mice, but subsequently some echoviruses were found to do the same. It has been recommended that all future enteroviruses that are discovered should be numbered sequentially from 68, irrespective of subgroups (Rosen *et al.*, 1970).

Most infections with enteroviruses are inapparent, a few are associated with gastrointestinal disorders, and some may cause generalized infections with rash, central nervous system involvement, including poliomyelitis and aseptic meningitis, or specific damage to the heart.

**Genus:** *Rhinovirus* **[R/1: 2.6–2.8/30: S/S: V/0].** The rhinoviruses resemble the enteroviruses in several characteristics but they are acid labile (pH 3) and have a buoyant density (in CsCl) of 1.38–1.43 g/cm³. Most have a low ceiling temperature of growth and are characteristically found in the upper respiratory tract of man and various animals. There are a large number of different serotypes of human rhinoviruses, and there are several serotypes of foot-and-mouth disease virus, which resemble rhinoviruses in some respects, but not in others (Table 1–9).

Most rhinoviruses cause mild localized infections of the upper respiratory tract, but foot-and-mouth disease virus causes a severe generalized disease with rash in cattle.

**Genus:** *Calicivirus* **[R/1: 2.8/20–30: S/S: V/0].** This genus differs substantially from the other genera of the Picornaviridae, both in its morphology and the chemical composition of the capsid, the outstanding difference being the distinctive "chunky" arrangement of the capsomers (see Plate 3–12). The other properties of the genus are those common to the Picornaviridae, with acid stability and a buoyant density intermediate between those of *Enterovirus* and *Rhinovirus*.

## Family: Togaviridae [R/1: 4/4–6: S/S: V,I/0,Di,Ac]

During the last quarter century intensive world-wide efforts have been made to recover viruses which would multiply in both arthropods and vertebrates, and some 200 different agents with these biological properties are now known. They have been called "arthropod-borne viruses," a name which was shortened to "arborviruses" and then (in order to avoid the connotation of "tree") to "arboviruses." The arboviruses have been defined, on epidemiological grounds (mode of transmission), as a group comparable to the "respiratory viruses." Arboviruses are viruses which, in nature, can infect arthropods that ingest infected vertebrate blood, can multiply in the arthropod tissues, and can then be transmitted by bite to susceptible vertebrates (World Health Organ., 1961).

For many years arboviruses have been recovered from vertebrate tissues and suspensions of arthropods by the intracerebral inoculation of mice, and advantage has been taken of certain chemical and physical properties found to be commonly associated with them to avoid confusion with murine picornaviruses. The property generally tested was sensitivity to lipid solvents. Many arboviruses have lipoprotein envelopes and their infectivity is destroyed by these reagents (Theiler, 1957; Casals, 1961). There was thus a tendency to equate sensitivity to lipid solvents with "arbovirus." During the last decade it has been recognized that the arbovirus group is quite heterogeneous in its physicochemical properties (see Table 16–3). Some members are not enveloped (*Orbivirus*, Nodamura virus), and those sensitive to lipid solvents belong to at least three major groups (Togaviridae, *Rhabdovirus*, and Bunyamwera supergroup).

This preamble has been necessary because in the past the term "arboviruses" has been regarded as applying particularly to viruses with the physicochemical properties of the group A and group B arboviruses. These viruses now form two genera (*Alphavirus* and *Flavivirus*) of the family Togaviridae (toga = cloak).

TABLE 1–10

*Family: Togaviridae*[a,b]

| GENUS | COMMENTS |
|---|---|
| *Alphavirus* | Type species: Sindbis virus [R/1: 4/5–7: S/S: V,I/Di]. All show serological cross-reactivity and all are mosquito-borne viruses. |
| | Members: Equine encephalitis viruses—Western, Eastern, and Venezuelan; Semliki Forest; Chikungunya; Sindbis; and thirteen other named viruses. |
| *Flavivirus* | Characteristic species: Dengue type 1 [R/1: 4/7: S/S: V,I/Di]. All show serological cross-reactivity, some are mosquito-borne and some are tick-borne viruses. |
| | Members: yellow fever, St. Louis encephalitis, Japanese encephalitis, dengue (four serotypes), West Nile, Murray Valley encephalitis, Russian tick-borne encephalitis and 27 other named viruses. |
| Possible members of family | Rubella virus |
| | Hog cholera virus and bovine mucosal disease virus (serologically related) |
| | Equine arteritis virus |

[a] Type genus: *Alphavirus*.
[b] Characteristics: single linear molecule of single-stranded RNA of molecular weight 4 million daltons, within a capsid of cubic symmetry, 20–40 nm in diameter, which is enclosed within a lipoprotein envelope 40–70 nm in diameter; multiply in cytoplasm and mature by budding from cytoplasmic (*Alphavirus*) or intracytoplasmic (*Flavivirus*) membranes; purified RNA is infectious.

**Genus:** *Alphavirus* **[R/1: 4/4–6: S/S: V,I/Di].**   The alphaviruses (alpha = Greek letter A), formerly known as the group A arboviruses, have the familial characteristics (Table 1–10) and show serological cross-reactivity by the hemagglutinin-inhibition test. The arthropod vectors are mosquitoes, but some alphaviruses may be transmitted congenitally by vertebrates. In nature, they usually cause inapparent infections of birds, reptiles, or mammals, but some can cause generalized infections associated with encephalitis in man and in other mammals.

**Genus:** *Flavivirus* **[R/1: 4/7–8: S/S: V,I/0, Di, Ac].**   This genus (flavi = yellow) comprises the group B arboviruses. All members show serological cross-reactivity. The arthropod vectors may be ticks or mosquitoes, and some of them may be transmitted by the ingestion of contaminated milk. They differ from the alphaviruses in that budding usually occurs into cytoplasmic vacuoles rather than from the plasma membrane.

Most cause inapparent infections in mammals and less commonly in birds, but generalized infections of man may occur with visceral symptomatology (e.g., yellow fever), rashes (e.g., dengue), or encephalitis (e.g., Japanese encephalitis).

**Other Possible Members of the Family Togaviridae.** On the basis of the physicochemical definition proposed, several other viruses that are not transmitted by arthropods should probably be included in this family. Generic names have not yet been proposed for these viruses, which include rubella and equine arteritis viruses, and the two serologically related viruses of hog cholera and bovine mucosal disease.

## Genus: Orthomyxovirus [R/1: Σ5/1: S/E: V/0]

In early classifications, some members of two very different genera, now distinguished from each other as *Orthomyxovirus* (ortho = correct; myxo = mucus) and *Paramyxovirus*, were grouped together as *Myxovirus* (Andrewes *et al.*, 1955). The common properties were an RNA genome, a tubular nucleocapsid, and a pleomorphic lipoprotein envelope that carried the properties of hemagglutination and enzymatic elution. The term "myxovirus" is now only used as a vernacular expression to encompass the viruses that have these properties (*viz.*, influenza, mumps, Newcastle disease, and parainfluenza viruses); it has no taxonomic status.

Type A influenza viruses have been recovered from a number of different species of animal (birds, horses, and swine) as well as man; types B and C are specifically human pathogens. They are an important cause of respiratory disease in man and other animals, and some of the avian influenza viruses may cause severe generalized infections.

TABLE 1–11

*Genus: Orthomyxovirus[a,b]*

| | |
|---|---|
| Influenza type A of man, swine, horse, fowl,[e] and other birds | Share type-specific nucleoprotein and membrane protein antigens |
| Influenza type B<br>Influenza type C[d] | Each type has distinctive nucleoprotein and membrane protein antigens: recovered only from man |

[a] Type species: Influenza A virus [R/1: Σ5/1: S/E: V/0].

[b] Characteristics: genome consists of seven separate pieces of single-stranded RNA, total molecular weight 5 million daltons; tubular nucleocapsid 6–9 nm diameter is type-specific antigen; lipoprotein envelope 80–120 nm in diameter contains strain-specific hemagglutinin and neuraminidase antigens; virion contains a transcriptase; multiply in nucleus and cytoplasm; mature by budding from the plasma membrane.

[e] Includes fowl plague virus.

[d] Few studies of physicochemical properties available; may not be an *Orthomyxovirus*.

## Genus: Paramyxovirus [R/1:  6–8/1:  S/E:  V/0]

In contrast to the orthomyxoviruses, the paramyxoviruses (para = alongside; myxo = mucus) are enveloped viruses whose RNA occurs as a single linear molecule with a molecular weight of about 7 million daltons (Table 1–12). The tubular nucleocapsid has a diameter of 18 nm and is about 1.0 $\mu$m long. It is enclosed within a pleomorphic lipoprotein envelope 150 nm or more in diameter; long filamentous forms with the same diameter also occur.

TABLE 1–12

*Genus: Paramyxovirus*[a,b]

|  | VIRUS | COMMENT |
|---|---|---|
| Accepted members | Mumps<br>Newcastle disease<br>Parainfluenza 1 (human and murine)<br>Parainfluenza 2 (human, simian and avian)<br>Parainfluenza 3 (human and bovine)<br>Parainfluenza 4<br>Other avian parainfluenza viruses | Contain neuraminidase and hemagglutinin |
| Possible members | Measles [e]<br>Distemper [e]<br>Rinderpest [e]<br>Pneumonia virus of mice<br>Respiratory syncytial virus | Lack neuraminidase |

[a] Type species: Newcastle disease virus [R/1: 7/1: S/E: V/0].

[b] Characteristics: single linear molecule of single-stranded RNA, 7 million daltons, within tubular nucleocapsid 18 nm in diameter; pleomorphic lipoprotein envelope 100–300 nm in diameter carries specific hemagglutinin and, among accepted members, neuraminidase peplomers; virion contains a transcriptase; multiply in cytoplasm; mature by budding from cytoplasmic or intracytoplasmic membranes.

[e] Serologically related to each other, but not to other paramyxoviruses.

Three serologically related viruses, those of measles, distemper, and rinderpest, have been tentatively allocated to the *Paramyxovirus* genus on the basis of the morphology of the virion and nucleocapsid; they do not have a neuraminidase; respiratory syncytial virus is different again.

Some paramyxoviruses cause localized infections of the respiratory tract and several produce severe generalized diseases; among the latter some are characteristically associated with skin rashes.

## Genus: Coronavirus [R/1:  */*:  S/E:  V/0]

The genus *Coronavirus* (corona = crown) comprises a small number of enveloped RNA viruses with a tubular nucleocapsid 9 nm in diameter. The genome consists of single-stranded RNA; its molecular weight has not been determined. The envelope carries characteristic pedunculated projections. Human strains cause common colds; in other animals coronaviruses infect the respiratory or alimentary tract, or may cause systemic disease.

TABLE 1–13

*Genus: Coronavirus[a]*

| |
| --- |
| Human respiratory coronaviruses |
| Mouse hepatitis viruses |
| Transmissible gastroenteritis of swine virus |
| Infectious avian bronchitis virus |
| Hemagglutinating encephalomyelitis virus of pigs |

[a] Type species: Avian infectious bronchitis virus [R/1: */*: S/E: V/0].

[b] Characteristics: genome consists of single-stranded RNA, molecular weight undetermined; tubular nucleocapsid 9 nm in diameter; lipoprotein envelope 80–120 nm in diameter with large pedunculated peplomers; multiply in cytoplasm and mature by budding into cytoplasmic vacuoles.

## Genus: Arenavirus [R/1: Σ3.5/*: S/*: V/0]

The genus *Arenavirus* (arena = sand) was defined in terms of the electron microscopic appearance of the virions in thin sections, and serological cross-reactivity (Rowe *et al.*, 1970a). The pleomorphic enveloped virions are 85–120 nm in diameter (sometimes larger), and have closely spaced peplomers. The structure of the nucleocapsid is unknown, but in thin sections the interior of the particle is seen to contain a variable number of electron-dense granules 20–30 nm in diameter, hence the name.

All members of the genus are associated with chronic inapparent infections of rodents; some cause acute generalized diseases in other hosts (e.g., Lassa fever virus in man).

TABLE 1–14

*Genus: Arenavirus[a,b]*

| |
| --- |
| Lymphocytic choriomeningitis virus (cosmopolitan), Lassa virus (Africa), "Tacaribe complex": Junin, Latino, Machupo, Parana, Pichinde, Pistillo, Tamiami, Tacaribe (Western Hemisphere) |

[a] Type species: Lymphocytic choriomeningitis virus of mice (LCM) [R/1: Σ3.5/* : S/* : V/0].

[b] Characteristics: single-stranded RNA probably in several pieces, total molecular weight 3.5 million daltons; lipoprotein envelope 85–300 nm in diameter; multiply in cytoplasm; mature by budding from plasma membrane. All members share a group-specific antigen. Envelope encloses "granules" 20–30 nm in diameter; some of these are cellular ribosomes.

## Bunyamwera Supergroup [R/1: */*: S/E: I,V/Di]

The Bunyamwera "supergroup" of arboviruses (Bunyamwera, a locality in Africa) was established by Casals (World Health Organ., 1967) to bring together a number of minor arbovirus groups linked by distant serological reactions between occasional "bridging" viruses. The subgroups of viruses included are shown in Table 1–15. All these viruses, numbering well over 100, are known or suspected to be arthropod-borne.

TABLE 1–15

*Bunyamwera Supergroup<sup>a</sup> [R/1: */*: S/E: I,V/Di]*

|  | VIRUSES |
|---|---|
| Serologically related members of Bunyamwera supergroup | Bunyamwera subgroup |
|  | Group C subgroup |
|  | Guama subgroup |
|  | Capim subgroup |
|  | Simbu subgroup |
|  | Bwamba subgroup |
|  | California subgroup |
|  | Patois subgroup |
|  | Tete subgroup |
|  | Koongol subgroup |
| Viruses serologically unrelated to Bunyamwera | Phlebotomus fever subgroup |
|  | Uukuniemi |
|  | Turlock |
|  | Rift Valley fever |

[a] Characteristics: genome consists of single-stranded RNA, possibly occurring as pieces, molecular weight undetermined; tubular nucleocapsid 12–15 nm in diameter, within lipoprotein envelope 90–100 nm in diameter. Genus will probably include the Bunyamwera supergroup and some serologically unrelated but morphologically similiar viruses. All multiply in and are transmitted by arthropods.

Morphologically, those that have been studied have enveloped roughly spherical virions 90–100 nm in diameter with a tubular nucleocapsid. Several other arboviruses serologically unrelated to those of the "supergroup" have a similar morphology (Table 1–15). Their genome consists of single-stranded RNA probably occurring in several pieces; its molecular weight has not been determined.

## Genus: Leukovirus [R/1: Σ10–13/2: S/E: V/0]

The outstanding characteristic of the genus *Leukovirus* (leuko = white) is that all members contain an RNA-dependent DNA polymerase ("reverse transcriptase"). The viruses contain three or four pieces of single-stranded RNA, with a total molecular weight of 10 to 12 million daltons, associated with a helical nucleocapsid, which is enclosed within a capsid with cubic symmetry.

This is, in turn, enclosed within a lipoprotein envelope about 100 nm in diameter, containing peplomers which confer the type specificity. Leukoviruses mature by budding from the plasma membrane.

TABLE 1–16

*Genus: Leukovirus[a,b]*

Subgenus A: leukosis-leukemia-sarcoma viruses
   Includes the well studied murine leukemia and avian leukosis viruses (C-type particles). Some strains produce sarcomas, some leukemia, others fail to transform cells or to induce neoplasia. Carried in the genome of normal cells as a DNA copy of viral genome. Rodent strains show serological cross-reactivity, but also have species- and type-specific antigens.
Subgenus B: mammary tumor virus
   Differs from other murine leukoviruses antigenically, in mode of maturation (A- and B-type particles), and in pathogenic potential (mammary adenocarcinoma).
Subgenus C: progressive pneumonia-visna viruses
   Associated with respiratory or demyelinating diseases of sheep. Will transform nonpermissive cells.
Subgenus D: foamy agents
   Cytopathogenic viruses causing inapparent infections in cats, monkeys, and cattle. Distinctive morphology. Viral antigen found in nucleus of infected cells, as well as in cytoplasm.

   [a] Type species: Rous sarcoma virus [R/1: Σ10–12/2: S/E: V/0].
   [b] Characteristics: Virion contains a virus-specified RNA-dependent DNA polymerase and other enzymes. Genome is a linear molecule of single-stranded RNA, 10–12 million daltons molecular weight, consisting of three to four linked pieces and probably associated with tubular nucleocapsid. Structure of virion is complex, the nucleocapsid being enclosed within a capsid of cubic symmetry, which is enclosed in an envelope that carries type-specific antigens. Virion also contains species-specific (e.g., feline or murine) and interspecies-specific (e.g., avian or rodent) antigens.

As Table 1–16 illustrates, the genus *Leukovirus* accepted by ICNV is clearly an inadequate taxon for the variety of viruses that now fulfill the physicochemical criteria set out above. The term "oncornaviruses" (Nowinski *et al.*, 1970) is also not suitable for the taxon as a whole or any subgroup of it, for not all the viruses conforming to the physicochemical specifications of the genus *Leukovirus* are tumor viruses, many "RNA tumor viruses" are not transforming (Temin, 1972), and in any case the leukemia-sarcoma viruses and the mammary tumor virus (both of which produce tumors) belong to different subgenera.

   In order not to prejudge a classification of the group, we shall adhere to the accepted generic name, but indicate four subgenera. Subgenus A includes the "C-type particle" viruses, some of which cause leukemia-sarcoma and are currently being subjected to intensive study. The mammary tumor virus, which is the best known representative of the subgenus B, differs from viruses of subgenus A in morphology and maturation ("B-type particles") and shows no serological cross-reactivity with the murine viruses of subgenus A. Subgenus C includes a group of serologically related viruses that cause slowly progressive diseases in sheep. They have all the physicochemical properties of leukoviruses. Although they do not cause neoplastic disease, they will transform cells that are nonpermissive for viral growth (Takemoto and Stone, 1971). The viruses of subgenus D (foamy agents) include a number of viruses of monkeys, cats, and cattle, that have no known pathogenic potential but have been frequently isolated from tumors (as "passenger viruses") or healthy animals. They have a different morphology from other leukoviruses (Clarke *et al.*, 1969) and produce an intranuclear antigen as well as cytoplasmic antigens in infected cells (Parks and Todaro, 1972), but they contain a reverse transcriptase and are much more resistant to UV irradiation than other RNA viruses.

### Genus: Rhabdovirus [R/1: 4/2: U/E: V,1/0, Di, Ac]

The rhabdoviruses (rhabdo = rod) are enveloped RNA viruses with single-stranded RNA of molecular weight 4 million daltons. The RNA is associated with a very regular double-helical nucleocapsid 5 nm in diameter, enclosed within a bullet-shaped shell that measures about 175 × 75 nm (Table 1–17).

TABLE 1–17

*Genus: Rhabdovirus*[a,b]

| MEMBERS | HOSTS |
|---|---|
| Viruses of vertebrates | |
|   Vesicular stomatitis–two serotypes [c] | Mammals and diptera |
|   Cocal [c] | |
|   Flanders-Hart Park | Birds and diptera |
|   Mt. Elgon bat | Bats and diptera |
|   Kern Canyon | Bats (invertebrate host not determined) |
|   Rabies [d] | Mammals |
|   Lagos bat [d] | |
|   Nigerian shrew [d] | |
|   Bovine ephemeral fever | Cattle, sheep, and diptera |
|   Hemorrhagic septicemia | Trout |
| Insect and plant viruses | |
|   Sigma virus | "$CO_2$ sensitivity" virus of *Drosophila* |
|   Potato yellow dwarf virus | These and several other viruses of plants have a morphology similar to vesicular stomatitis virus and multiply in leafhoppers as well as plants |
|   Lettuce necrotic yellow virus | |

[a] Type species: Vesicular stomatitis virus [R/1: 4/2: U/E: V,I/0, Di].

[b] Characteristics: bullet-shaped enveloped viruses measuring 175 × 70 nm and containing single-stranded RNA with molecular weight about 4 million daltons; virion contains a transcriptase; multiply in cytoplasm and mature by budding from the plasma membrane.

[c] Serologically related.

[d] Serologically related (Shope *et al.*, 1970).

Several arboviruses belong to this genus, which also includes rabies virus and the virus of hemorrhagic septicemia of trout. It has been claimed that rabies virus can be adapted to multiply in *Drosophila melanogaster* (Plus and Atanasiu, 1966).

Several viruses with a somewhat similar morphology cause diseases of insects and plants (Table 1–17, and see Table II of Howatson, 1970), but it may well turn out that these resemblances are superficial. Examination of the nature of their genomes and polypeptides is necessary before it can be confidently stated whether these viruses rightly belong to the genus *Rhabdovirus* or even to an enlarged family that might be called Rhabdoviridae.

### Genus: Reovirus [R/2: Σ15/15: S/S: V/O]

This genus was described and named by Sabin (1959) (acronym: reo = Respiratory Enteric Orphan). Their characteristic features are a genome that consists of ten pieces of double-stranded RNA with a total molecular weight of about 15 million daltons, enclosed in a double capsid (Table 1–18).

TABLE 1–18

*Genus: Reovirus*[a,b]

| |
|---|
| Members: three mammalian serotypes, five avian serotypes, bat, simian, and canine viruses |
| Possibly related: clover wound tumor virus, rice dwarf virus |

[a] Type species: reovirus type 1 [R/2: Σ15/15: S/S: V/0].
[b] Characteristics: double-stranded RNA, 15 million daltons, occurring as ten separate pieces; icosahedral outer capsid diameter 75–80 nm; icosahedral inner capsid 45 nm diameter; no envelope; virion contains a transcriptase; multiply in cytoplasm.

The mammalian serotypes share a common antigen, which differs from the group antigen of the avian serotypes.

Clover wound tumor virus, which multiplies in plants and leafhoppers, resembles the reoviruses of vertebrates morphologically and chemically but does not cross-react with them serologically.

### Genus: Orbivirus [R/2: Σ15/20: S/S: I,V/Ac, Di]

Bluetongue virus, an arbovirus, was found to resemble the reoviruses in some properties but not in others (review: Howell and Verwoerd, 1971). Subsequently, a large number of similar viruses have been recognized (Table 1–19) and the name *"Orbivirus"* (orbis = ring) was suggested for them (Borden *et al.*, 1971). Reoviruses and orbiviruses may eventually be grouped together in the same family for which the name "diplornavirus" has been suggested (Verwoerd, 1970). Apart from its "illegality" (according to ICNV Rules), the occurrence of other quite different viruses with genomes of double-stranded RNA (like some of the viruses of fungi and insects) cautions against ready acceptance of this term.

All members of the genus multiply in arthropods as well as vertebrates. Some of them (bluetongue and Colorado tick fever viruses) cause severe generalized diseases with viremia in some vertebrates.

TABLE 1-19

*Genus: Orbivirus[a,b,c]*

| SEROLOGICAL SUBGROUP | HOST | |
|---|---|---|
| | VERTEBRATE [d] | INVERTEBRATE |
| Bluetongue; sixteen serotypes | Mammals | Culicoides |
| African horse sickness; nine serotypes | Mammals | Culicoides |
| Kemerovo and six serologically related viruses | Mammals | Argasid and Ixodid ticks |
| Changuinola and four serologically related viruses | Mammals | Phlebotomus |
| Colorado tick fever | Mammals | Dermacentor |
| Epizootic hemorrhagic disease of deer | Mammals | Culicoides suspected |
| Palyam and serologically related viruses | (Mammals) | Mosquitoes and Culicoides |
| Eubenangee and one serologically related virus | (Mammals) | Mosquitoes |
| Corriparta and one serologically related virus | (Mammals, birds) | Mosquitoes |

[a] Type species: Bluetongue virus [R/2: Σ15/20: S/S: I,V/Di].

[b] Characteristics: double-stranded RNA, 15 million daltons, occurring as ten separate pieces; icosahedral capsid (single shell), outer diameter 50–60 nm; no envelope; virion contains a virus-specific transcriptase; multiply in cytoplasm of cells of vertebrates and several kinds of arthropods.

[c] Borden *et al.* (1971). Not an official generic name.

[d] Parentheses indicate experimental susceptibility or serological evidence, without isolation in nature from that source.


## UNCLASSIFIED VIRUSES

It is pleasing to note that several of the viruses listed as "unclassified" in the first edition of this book have now been allocated to genera (lymphocytic choriomeningitis of mice, *Arenavirus;* mouse hepatitis virus, *Coronavirus;* rubella virus, Togaviridae; visna virus, *Leukovirus;* and African swine fever virus, *Iridovirus*). A few unclassified viruses remain that warrant special mention here, such as the human hepatitis viruses, the agents of the subacute spongiform encephalopathies (scrapie, etc.), lactic dehydrogenase elevating virus (LDV), and the Marburg agent.

### Human Hepatitis Viruses

Experiments with human volunteers many years ago (Neefe *et al.*, 1945) and again more recently (Krugman *et al.*, 1967) have shown that the diseases commonly known as infective hepatitis and serum hepatitis are caused by two viruses that differ serologically, in their clinical expression, and in their usual routes of transmission. Because both can be transmitted orally it is better to use noncommital names for them, and "serum hepatitis" is now termed hepatitis B; infective hepatitis, hepatitis A. Study of these viruses has been greatly inhibited by the lack of susceptible laboratory animals (chimpanzees may get clinical hepatitis; marmosets and rhesus monkeys subclinical infection, while other laboratory animals are insusceptible), and the difficulty of obtaining reproducible cytopathic changes in cultured cells. The recognition in the sera of cases of serum hepatitis of lipoprotein particles of characteristic serological specificity, called "Australia antigen," hepatitis-associated antigen (HAA), and now hepatitis B antigen (HB-Ag), has led to a great expansion in studies on the incidence and pathogenesis of hepatitis B, but the actual virions have not yet been unequivocally demonstrated (see Chapter 3). Serologically unrelated particles of similar morphology have been reported to occur in feces from patients with hepatitis A (Cross et al., 1971).

## Agents of Subacute Spongiform Encephalopathies

Four diseases of similar nature, scrapie of sheep, transmissible encephalopathy of mink, and kuru and Creutzfeld-Jakob disease in man appear to be caused by similar agents, which differ from all known viruses by being nonimmunogenic. The causative agents are filtrable, highly heat-resistant, and highly resistant to ionizing radiation. It has been suggested that they may be small molecules of naked RNA, protected by being closely associated with cellular membranes (Diener, 1972a), but a definitive description of these agents is still awaited.

## Lactic Dehydrogenase Elevating Virus (LDV)

This virus, which occurs as an inapparent infection in many laboratory mice and as a contaminant of cells and viruses derived from or passaged through mice (review: Notkins, 1965), shows some resemblances to the togaviruses. It appears to have an isometric core and a lipoprotein envelope, and its RNA is infectious. However, the viral RNA is large, perhaps 5 million daltons (Darnell and Plagemann, 1972).

## The Marburg Agent

In Germany, in 1967, a small outbreak of a serious new disease occurred in laboratory workers who had handled the tissues of recently imported vervet monkeys (review: Siegert, 1972). The causative agent grows in cultured cells and kills guinea pigs. Studies with inhibitors suggest that it contains RNA; of known viruses it most closely resembles rhabdoviruses in structure but is much larger and more pleomorphic (Murphy et al., 1971b).

## THE ORIGINS OF VIRUSES

The foregoing account has shown how varied are the agents that we classify as viruses, for reasons based on their composition and their mode of intracellular replication. We can only speculate about their origins and relationships to each other, except in cases where the relationship is very close. It seems likely that different viruses belonging to any one genus, and in at least some cases, different genera allocated to a particular family, may be phylogenetically related. No useful suggestions can be made concerning the relationships between families or genera (except in some of the cases where genera have been allocated to the same family), a fact which underlines the undesirability at this stage of our knowledge of erecting any taxa at levels higher than the family.

Two suggestions have been made concerning the origin of viruses: (a) that they are the result of progressive parasitic degeneration of microorganisms (Green, 1935) and (b) that they have developed from components of the cells of their hosts (Andrewes, 1966; Luria and Darnell, 1967), or are indeed still a permanent part of the host's genome (Todaro and Huebner, 1972). With our present knowledge of the morphological and chemical complexity of the poxviruses, it is not difficult to envisage these agents as being the next degenerate step in the series: bacterium, rickettsia, chlamydia. Although they resemble bacteria in most important respects, rickettsia and chlamydiae are, like viruses, obligate intracellular parasites lacking the metabolic equipment for independent multiplication.

On the other hand, some DNA viruses could well have arisen from episomes, by the acquisition of genetic information specifying a protein coat. Even this may not be essential, if Diener's (1972b) observations on potato spindle tuber virus are confirmed and generalized. The two alternatives are not mutually exclusive; some viruses may have evolved from cellular organelles like chloroplasts or mitochondria, themselves probably derived from bacteria (Swift and Wolstenholme, 1969). It is difficult to see where most RNA viruses could have originated except from cellular RNA's.

Comparing the nearest neighbor nucleotide doublet frequencies of the nucleic acids of several large and small viruses, Subak-Sharpe (1969) noted that the patterns shown by small viruses with genomes of less than 5 million daltons (two enteroviruses, three parvoviruses, two polyoma viruses, and two papilloma viruses), closely resembled the pattern of mammalian DNA. On the other hand, the doublet frequency patterns of several viruses with large genomes (two herpesviruses and a poxvirus) differed strikingly from that of mammalian DNA. The doublet patterns of three adenoviruses resembled each other and showed a slight resemblance to the pattern of mammalian DNA, which could derive from some earlier natural fusion of genomes, like that recognized as a laboratory artifact with adenoviruses and SV40 (see Chapter 7). This evidence supports the notion that the small viruses may have originated from vertebrate cells whereas the herpesviruses, poxviruses, and probably the adenoviruses did not. These large viruses may have originated from the nucleic acid of cells of a different phylum, or as suggested earlier, by parasitic degeneration of microorganisms.

*CHAPTER 2*

# Cultivation, Assay, and Analysis of Viruses

Viruses are obligatory intracellular parasites and cannot replicate in any cell-free medium, no matter how complex. Some viruses are fastidious about the sorts of cells that they infect; for instance, some known human viruses have not yet been cultivated under laboratory conditions. Fortunately, however, most viruses can be grown in cultured cells, embryonated eggs, or laboratory animals; indeed, the cultivation of viruses in experimental animals, or better still in cultured cells, is an essential prerequisite for their detailed study.

In this chapter we shall describe in general terms the ways in which animal viruses are isolated and grown in the laboratory, how they are assayed, how they are purified and, finally, some of the methods used for their biochemical analysis. For technical details about these procedures the reader is referred to several excellent recent textbooks on methodology: Maramorosch and Koprowski (1967–1971), Habel and Salzman (1969), and Lennette and Schmidt (1969).

As a preliminary step it will be necessary to describe the ways in which animal cells are cultured, since the major differences between the laboratory procedures used in bacterial, animal, and plant virology depend upon the characteristics of the cells in which the viruses can be grown.

## CELL CULTURE

More than half a century has elapsed since human and animal cells were first grown *in vitro*. However, it is only since the advent of antibiotics that cell culture has become a routine procedure. Aseptic precautions are still essential, but the problems of contamination with bacteria, mycoplasmas, fungi, and yeasts are no longer insurmountable, and many kinds of animal cells can be cultivated *in vitro* for at least a few generations (Rose, 1970). Since 1949, when Enders, Weller, and Robbins reported that poliovirus could be grown in cultured nonneural cells with the production of recognizable histological changes, a large number of animal viruses have been grown in cultured cells, and hundreds of previously unknown viruses have been isolated and identified. The discovery of the adenoviruses, echoviruses, and rhinoviruses, for example, is directly attributable to the use of cultured cells, as is the revolution in the diagnosis of viral diseases and the development of poliomyelitis, measles, and rubella vaccines. Likewise, biochemical investigations of viral multiplication, that were impossible before the development of ways of synchronously infecting cultured cells, are now carried out in laboratories the world over.

## Methods of Cell Culture

Cells may be grown *in vitro* in several ways, listed below.

**Organ Culture.** Slices of organs, if carefully handled, maintain their original architecture and functions for several days or sometimes weeks *in vitro*. Such organ cultures (strictly tissue cultures) of respiratory epithelium have been used to study the histopathogenesis of infection by respiratory viruses; indeed, some respiratory viruses can only be grown outside their natural host by using organ cultures (see review by Hoorn and Tyrrell, 1969). Organ cultures of fetal intestinal epithelium are being used to study certain enteric viruses.

**Tissue Culture.** This term was originally applied to the cultivation *in vitro* of fragments of minced tissue in suspension, or "explants" of tissues embedded in clotted plasma. Subsequently, the term came to be associated with the *in vitro* culture of cells in general. "Tissue culture" in its original sense is now obsolete, and there is no logic in perpetuating the general use of the term; "cell culture" will be used throughout the text that follows.

**Cell Culture.** Tissue is dissociated into a suspension of single cells or small clumps by mechanical mincing followed by treatment with proteolytic enzymes. After the cells are washed and counted, they are diluted in medium and are permitted to settle on to the flat surface of a specially treated glass or plastic container. Most types of cells adhere quickly, and under optimal conditions they then divide about once a day until the surface is covered with a confluent monolayer of cells. Detailed accounts of technical methods used in cell culture can be found in Paul (1972) and Schmidt (1969a).

## Media

Cell culture has been greatly aided by the development of chemically defined media containing almost all the nutrients required for cell growth. The best known of these media, developed by Eagle (1959), is an isotonic solution of simple salts, glucose, vitamins, coenzymes, and amino acids, buffered to pH 7.4, and containing antibiotics to inhibit the growth of bacteria. Serum must be added to Eagle's medium to supply the cells with an additional factor(s), the nature of which is still undefined, but without which most cells will not grow. More recently, attempts have been made to experiment with a higher oxygen content in the gaseous environment and to substitute nonvolatile buffers for the conventional bicarbonate buffering system (Massie *et al.*, 1972). Such buffers, either phosphates or substituted sulfonic acids (HEPES, i.e. $N$-2-hydroxyethylpiperazine-$N'$-2-ethanesulfonic acid; or TES, i.e., N-tris (hydroxymethyl)methyl-2-aminomethanesulfonic acid), provide better pH control and reduce the requirement for incubators to be gassed with $CO_2$.

## Types of Cultured Cells

Some types of cells are capable of undergoing only a few divisions *in vitro* before dying out, others will survive for up to a hundred cell generations, and some can be propagated indefinitely. These differences, the nature of which is not fully understood, give us three main types of cultured cells.

**Primary Cell Cultures.** When cells are taken freshly from animals and placed in culture, the cultures consist of a variety of cell types, most of which are capable of very limited growth *in vitro*—perhaps five or ten divisions at most. However, they support the replication of a wide range of viruses, and primary cultures derived from monkey kidney, human embryonic kidney or amnion, and chicken and mouse embryos are commonly used for this purpose, both for laboratory experiments and vaccine production.

**Diploid Cell Strains.** These are cells of a single type that are capable of undergoing up to about 100 divisions *in vitro* before dying. They retain their original diploid chromosome number throughout (Hayflick, 1965). Diploid strains of fibroblasts established from human embryos are widely used in diagnostic virology and vaccine production and have some use in experimental studies. It should be kept in mind that certain aspects of the expression of the viral genome may require the use of cells that retain their normal state of differentiation.

**Continuous Cell Lines.** These are cells of a single type that are capable of indefinite propagation *in vitro*. Such immortal lines usually originate from cancers, or by "transformation" occurring in a diploid cell strain. Often they no longer bear any close resemblance to their cell of origin, as they have doubtless undergone many sequential mutations during their long history in culture. The most usual indication of these changes is that the cells are "dedifferentiated," i.e., they have lost the specialized morphology and biochemical abilities that they possessed as differentiated cells *in vivo*. For example, it is no longer possible to distinguish microscopically between the various epithelial cell lines arising from cells of ectodermal or endodermal origin, or between the "fibroblastic" cell lines arising from cells of mesodermal origin (Plate 2–1). Cells of continuous cell lines are often aneuploid in chromosome number, and may be tumorigenic.

Continuous cell lines such as HEp-2, HeLa, and KB, all derived from human carcinomas, support the growth of a number of viruses. These lines, and others derived from mice (L929,3T3) and hamsters (BHK-21), are widely used in experimental virology; there are now available a range of continuous cell lines derived from a variety of animals. The American Type Culture Collection holds deep-frozen samples of many cell lines, tested for freedom from extraneous viruses and mycoplasmas.

The great advantage of continuous cell lines over primary cell cultures is that they can be propagated indefinitely by subculturing the cells at regular intervals. Furthermore, they retain viability for many years when suspended in glycerol or dimethyl sulfoxide and stored at low temperatures ($-70°$ to $-196°C$). Since cells inevitably undergo changes during serial passage, it is standard laboratory practice to store at low temperature a large number of ampules of cells that are

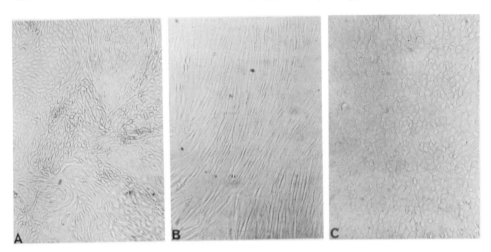

PLATE 2–1. *Types of cultured cells. Confluent monolayers of the three main types of cultured cell, as seen by low power light microscopy through the wall of the culture flask. (A) Primary monkey kidney epithelial cells. (B) Diploid strain of human fetal fibroblasts. (C) Continuous line of epithelial cells (HeLa). X 60 (Courtesy I. Jack.)*

satisfactory for the investigations being conducted, and to make up batches of cells for experimental use from such seed cultures.

Some continuous cell lines have been adapted to grow in suspension culture, i.e., as a suspension of single cells continuously stirred by a spinning magnet. Such "spinner cultures" are particularly useful for biochemical studies of viral multiplication and for the commercial production of some vaccines.

For the molecular biologist, continuous cell lines are obviously preferred for the study of macromolecular processes. Suspension cultures are often used, since they provide a large and uniform population of cells that can be synchronously infected with virus and from which samples can be rapidly processed. The use of large-scale "roller cultures," in which cells are attached as a monolayer in cylindrical bottles which are slowly rotated in a horizontal position, has improved the logistics of growing large quantities of cells that will grow only as monolayers or of viruses that prefer to grow in monolayer cultures.

## Cell Type and Viral Growth

Of paramount importance is the selection of cell lines that will support the optimal growth of the virus under study. Some viruses will multiply in almost any cell line; some cell lines are favorable for supporting the replication of many different types of viruses. On the other hand, many viruses are quite restricted in the kinds of cell in which they will multiply, although repeated blind passage may lead to adaptation.

Cultured cells can serve three main purposes: (a) primary isolation, in which emphasis is placed on high sensitivity and readily recognized cytopathic effects (CPE), (b) vaccine production, where the emphasis is placed on yield, and (c) basic biochemical research for which continuous cell lines, preferably growing as suspension cultures, are usually chosen. The same cells may be used in mono-layer cultures for infectivity assays, or sometimes a primary culture will be used for this purpose. Information on the cells used for the cultivation of human pathogens is given in Lennette and Schmidt (1969), and for all animal viruses in Andrewes and Pereira (1972). Many research workers have particular cell lines that they favor for the viruses that they are studying; Table 2–1 lists some of the kinds of cultured cells commonly used for biochemical experiments and plaque assay of model viruses of various genera.

## Recognition of Viral Growth in Cell Culture

The growth of many viruses in cell culture can be monitored by a number of biochemical procedures indicative of the intracellular increase in viral macro-molecules and virions, as described below (Chapters 5 and 6). In addition, there are cruder methods that are commonly used for diagnostic work, some of which also have an important place in research laboratories.

**Cytopathic Effects (CPE).**   Many but by no means all viruses kill the cells in which they multiply, so that infected cell monolayers gradually develop histo-logical evidence of cell damage, as newly formed virions spread to involve more and more cells in the culture. These changes are known as cytopathic effects (CPE); the responsible virus is said to be cytopathogenic. Most CPE can be readily observed in unfixed, unstained cell cultures, under low power of the light microscope, with the condenser racked down and the iris diaphragm partly closed to obtain the contrast required in looking at translucent cells. A trained observer can distinguish several types of CPE in living cultures (Plate 2–2 and Table 2–2), but fixation and staining of the cell monolayer is necessary in order to see de-tails such as inclusion bodies and syncytia. Fluorescent antibody staining, de-scribed below, is widely used to recognize viral antigens in such cultured cells.

Observation of CPE is an important tool for the diagnostic virologist, who is concerned with isolating viruses from infected animals or human patients. Some viruses multiply readily in cell culture on primary isolation; the time at which cytopathic changes first become detectable depends to some extent on the number of virions that the specimen contained, but, far more important, on the growth rate of the virus in question. Enteroviruses and herpes simplex virus, for example, which have a short latent period and a high yield, often show detectable CPE after 24 to 48 hours, destroying the monolayer completely within about 3 days. On the other hand, cytomegaloviruses, rubella, and some of the more slowly growing adenoviruses may not produce detectable CPE for several weeks. Since the cell cultures may have undergone nonspecific degen-eration during this period it may be necessary to subinoculate the cells and su-pernatant fluid from the infected culture on to fresh monolayers. CPE often appears soon after such "blind passage," either because this enhances the titer or selects variants adapted to grow better in the cultured cells.

**Hemadsorption.**   Cultured cells infected with orthomyxoviruses, paramyxo-viruses, and togaviruses, all of which bud from cytoplasmic membranes, acquire the ability to adsorb erythrocytes. The phenomenon, known as hemadsorption

## TABLE 2–1

### Cell Cultures Used for Cultivation of Animal Viruses Commonly Studied in the Laboratory[a]

| | |
|---|---|
| DNA viruses | |
| *Parvovirus* | |
| Subgenus A | Rat embryo |
| Subgenus B (AAV) | Human embryo kidney, KB (both coinfected with adenovirus) |
| Papovaviridae | |
| Polyoma virus | Mouse embryo, 3T3 |
| SV40 | African green monkey kidney, rabbit kidney, BSC-1 |
| *Adenovirus* | |
| Human | African green monkey kidney, human embryo kidney, WI-38, HeLa, HEp-2, KB |
| Avian | Chick embryo kidney |
| *Herpesvirus* | |
| Herpes simplex virus | African green monkey kidney, human embryo kidney, chick embryo, WI-38, HeLa, HEp-2 |
| Pseudorabies virus | Rabbit kidney, baby hamster kidney, RK13 |
| Cytomegalovirus | African green monkey kidney, human embryo kidney, rabbit kidney, WI-38 |
| *Iridovirus* | |
| Frog virus 3 | Chick embryo, baby hamster kidney, FHM |
| *Poxvirus* | |
| Vaccinia | Chick embryo, HeLa, L929, KB |
| RNA viruses | |
| Picornaviridae | |
| *Enterovirus* | African green monkey kidney, human amnion, WI-38, HeLa |
| *Rhinovirus* | African green monkey kidney, human embryo kidney, WI-38, HeLa, KB |
| *Aphthovirus* | Bovine embryo kidney, pig kidney, BHK |
| *Cardiovirus* | Mouse embryo, HeLa, L929 |
| Togaviridae | |
| *Alphavirus* | Chick embryo, mouse embryo, BHK, Vero, HeLa |
| *Flavivirus* | BHK, Vero, HeLa |
| Rubella virus | RK13, Vero, WI-38, BHK |
| *Orthomyxovirus* | |
| Influenza A virus | Chick embryo, calf kidney |
| *Paramyxovirus* | |
| Newcastle disease virus | Chick embryo, hamster kidney, HeLa, KB, Vero |
| *Rhabdovirus* | |
| Vesicular stomatitis virus | Chick embryo, BHK, HeLa, L929 |
| *Leukovirus* | |
| Avian (including Rous sarcoma virus) | Chick embryo |
| Murine | Mouse embryo, 3T3 |
| *Reovirus* | |
| Reovirus type 3 | Human kidney, WI-38, L929 |
| *Orbivirus* | |
| Bluetongue virus | Bovine embryo kidney, lamb kidney |

[a] Abbreviations: primary cultures, abbreviations not used; diploid strain: WI-38, human embryonic lung; heteroploid lines: BHK, baby hamster kidney cell line BHK-21; FHM, fat head minnow (fish); HeLa, HEp-2, and KB, human carcinoma, cervical, epidermoid, and nasopharyngeal, respectively; L929 and 3T3, lines of mouse fibroblasts; RK13, line of rabbit kidney cells; Vero and BSC1, lines of African green monkey kidney cells.

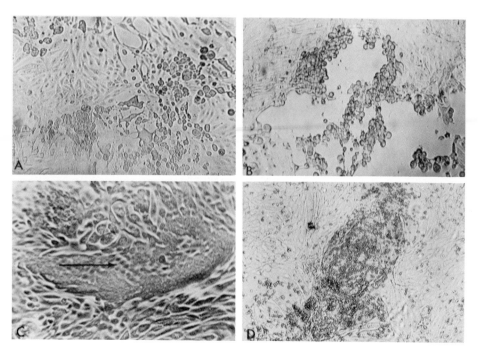

PLATE 2–2. *Cytopathic effects (CPE) produced in monolayers of cultured cells by different viruses. The cultures are shown as they would normally be viewed in the laboratory, unfixed and unstained (X 60). (A) Enterovirus, rapid, complete cell destruction. (B) Herpesvirus, focal areas of swollen rounded cells. (C) Paramyxovirus, focal areas of fused cells (syncytia: arrow indicates nuclei). (D) Hemadsorption. Erythrocytes adhere to those cells in the monolayer that are infected. The technique is applicable to any virus that causes a hemagglutinin to be incorporated into the cellular membrane. Most enveloped viruses that mature by budding from cytoplasmic membranes produce hemadsorption. (Courtesy I. Jack.)*

(Shelokov *et al.*, 1958), is due to the incorporation into the plasma membrane of newly synthesized viral protein that has an affinity for red blood cells (Plate 2–2D). Hemadsorption can be used to recognize infection with noncytopathogenic viruses, as well as the early stages of infection with cytocidal viruses.

**Interference.** The multiplication of one virus in a cell often inhibits the multiplication of another virus entering subsequently (Chapter 8). Rubella virus was first discovered by showing that infected monkey kidney cell cultures, showing no CPE, were nevertheless resistant to challenge with an unrelated echovirus (Parkman *et al.*, 1962). The phenomenon was also exploited for a time for the isolation of rhinoviruses. Although it is no longer used for either of these viruses, because cell lines have become available in which they produce CPE, interference is a useful technique when searching for new noncytopathogenic viruses and for the assay of nontransforming avian leukosis viruses (see Chapter 14).

TABLE 2–2

*Cytopathic Effects of Viruses in Cell Culture*

| CYTOPATHIC EFFECT | VIRUS |
|---|---|
| Pyknosis, shrinkage, cell destruction | Enteroviruses, poxviruses, reoviruses,[a] togaviruses,[a] rhinoviruses,[a] vesicular stomatitis virus |
| Aggregation | Adenoviruses |
| Cell fusion to give syncytia | Paramyxoviruses, herpesviruses |
| Minimal | Orthomyxoviruses, rabies virus, coronaviruses, leukoviruses, arenaviruses |

[a] Often produces incomplete cytopathic effect.

## EMBRYONATED EGGS

Prior to the 1950's, when cell culture really began to make an impact on virology, the standard host for the cultivation of many viruses was the embryonated hen's egg (developing chick embryo). The technique was devised by Goodpasture (Goodpasture *et al.*, 1932) and extensively developed by Burnet over the ensuing years (Beveridge and Burnet, 1946). Nearly all of the viruses that were known at that time can be grown in the cells of one or another of the embryonic membranes, namely the amnion, allantois, chorion, or yolk sac (Fig. 2–1).

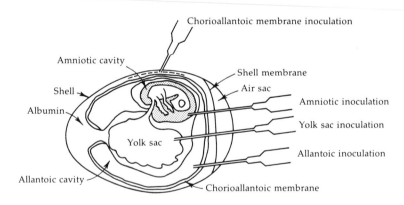

FIG. 2–1. *Routes of inoculation of the embryonated egg. Yolk sac inoculation is usually carried out with a 5–8-day-old embryos; amniotic and allantoic inoculation with 10-day-old embryos; chorioallantoic inoculation with 11- or 12-day-old embryos. (Modified from Davis et al., 1967.)*

Eggs are inoculated 5–14 days after fertilization, depending on the state of development of the membrane it is proposed to infect. A hole is drilled in the shell, and virus is injected into the fluid bathing the appropriate membrane. Following incubation for a further 2–5 days, viral growth can be recognized by one or more of the criteria listed in Table 2–3.

TABLE 2–3

*Growth of Viruses in Embryonated Eggs*

| MEMBRANE | VIRUSES | SIGNS OF GROWTH |
|---|---|---|
| Yolk sac | Herpes simplex | Death |
| Chorion | Herpes simplex ⎫ | |
| | Poxviruses ⎬ | Pocks |
| | Rous sarcoma ⎭ | |
| Allantois | Influenza | Hemagglutination |
| | Mumps and Newcastle disease | Death |
| | Avian adenovirus | Death |
| Amnion | Influenza | Hemagglutination |
| | Mumps | Death |

Embryonated eggs are rarely employed now for viral isolation, but the allantois produces such high yields of certain viruses, like the influenza viruses and avian adenoviruses, that this system is used both by research laboratories and for vaccine production.

## LABORATORY ANIMALS

Like embryonated eggs, laboratory animals have almost disappeared now from diagnostic laboratories, since cell cultures are so much simpler to handle and much more versatile, although suckling mice are still used in the isolation of some coxsackieviruses and in many arbovirus laboratories.

However, laboratory animals are still essential for many kinds of virological research. Primates are used to study a few human viruses, like the kuru agent and hepatitis viruses, that will not grow in other laboratory animals or in cultured cells. Hamsters are widely used in tumor virology, because they are highly susceptible to tumor production by oncogenic viruses and then yield valuable antisera. Experiments on pathogenic mechanisms and the role of the immune response can only be carried out with suitable laboratory animals, usually primates, hamsters, rabbits or mice. Finally, since serology looms large in much virological research, laboratory animals, usually rabbits, are extensively used for producing antisera. Blaškovič and Styk (1967) have reviewed methods employed for studying viruses in laboratory animals.

## ASSAY OF VIRAL INFECTIVITY

All scientific research depends upon reliable methods of measurement, and with viruses the property we are most obviously concerned with measuring is infectivity. The content of infectious viruses in a given suspension can be "titrated" by infecting cell cultures, chick embryos, or laboratory animals with dilutions of viral suspensions and then watching over the next few days for evidence of viral multiplication. Two types of infectivity assay should be distinguished: quantitative and quantal.

## Quantitative Assays

A familiar example of this type of assay is the bacterial colony count on an agar plate. Each viable organism multiplies to produce a discrete clone; the colony count therefore represents a direct estimate of the number of organisms originally plated. The parallel systems in virology are the counting of pocks on the chorion of the chick embryo or plaques on monolayers of cultured cells.

**Plaque Assays.** Dulbecco (1952; Dulbecco and Vogt, 1954) introduced to animal virology a modification of the bacteriophage plaque assay that is now used very widely for the quantitation of animal viruses. A viral suspension is added to a monolayer of cultured cells for an hour or so to allow the virions to attach to the cells, then the liquid medium is replaced with a solid gel, which ensures that the spread of progeny particles is restricted to the immediate vicinity of the originally infected cell. Hence, each infective particle gives rise to a localized focus of infected cells that becomes, after a few days, large enough to see with the naked eye (Plate 2–3).

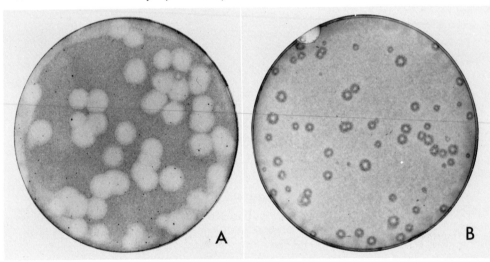

PLATE 2–3. *Plaques produced by influenza virus in monolayers of a continuous cell line derived from human conjunctival cells (Chang). Neglecting possible overlap of plaques or clumping of virions, each plaque is initiated by a single virion and yields a clone. (A) Normal plaques, seen as clear areas in monolayer stained with neutral red. (B) "Red" plaques, characteristic of certain strains of influenza virus, and some other viruses (see Chapter 9); demonstrated by using half strength neutral red in overlay medium. (From Kilbourne, 1969; courtesy Dr. E.D. Kilbourne.)*

Various materials have been used to form the gel with which the cell monolayer is overlaid shortly after inoculation. They include agar, methylcellulose, tragacanth, and starch gel. Agar has been most commonly employed, but suffers from the disadvantage that it contains sulfated polysaccharides that inhibit the growth of some viruses; however, this inhibitory substance can be neutralized by the addition of DEAE-dextran to the agar. Of course, the gel must also incorporate the usual nutrient medium required to maintain the cells in a viable condition.

After an incubation period of 2 days to 4 weeks, depending on the virus under study, the cell monolayers are stained with a vital dye, such as neutral red or tetrazolium. The living cells absorb the stain and the plaques appear as clear areas against a red background. Noncytocidal viruses can be titrated in a similar fashion, plaques being recognized by such techniques as hemadsorption, interference, or fluorescent antibody staining. Some viruses, e.g., herpesviruses and poxviruses, will produce plaques even in cell monolayers grown in liquid medium, because most of the newly formed virus remains cell-associated, so that plaques form by direct spread to adjacent cells through intercellular bridges.

There are a number of technical variations of Dulbecco's plaquing method (Cooper, 1967). For instance, virus may be allowed to attach to cells in suspension. The cells can then be permitted either to adhere to the glass and be overlaid with agar medium, or, alternatively, the cell suspension with virus attached may be embedded in agar medium directly. The latter method is suited only to viruses small enough to diffuse from cell to cell through the agar; the plaque that eventually results is then spherical.

Ordinarily, infection with a single virus particle is sufficient to form a plaque (Dulbecco and Vogt, 1954), so that the infectivity titer of the original viral suspension can be expressed in terms of "plaque-forming units" (pfu) per milliliter. The error will be minimal if the titer is determined from plates inoculated with a dilution of virus producing about 20 to 100 plaques per plate, depending on plaque size—enough to minimize errors attributable to the Poisson distribution (which governs the distribution of very small numbers of discrete particles), while avoiding overlap of adjacent plaques.

**Infectious Center Assays.** An important modification of the plaque assay is the infectious center count. Suspensions of cells that have been infected with virus are plated on a preformed cell monolayer during the latent period, before new virions are produced, and the monolayer is overlaid carefully with agar. The number of plaques that appear is a measure of the number of cells originally infected, so that it is a matter of simple calculation to determine the proportion of infected cells in the culture being assayed. The number of infected cells in a suspension or monolayer culture can also be assayed a few hours after inoculation of the culture by counting microscopically the cells that are positive for immunofluorescence, or in suitable cases, for hemadsorption (White et al., 1962).

**Transformation Assays.** Methods other than the plaque assay have been devised to quantitate the infectivity of those viruses that do not cause cell death. We have already referred to the use of hemadsorption or interference to assay some noncytocidal viruses; in addition, there are some oncogenic viruses that interact with some cultured cells in a cytocidal fashion, but "transform" other types of cell. Such cells are not killed, but their social behavior is changed so that they take on many of the properties of malignant cells. Compared with noninfected cells, transformed cells show little contact inhibition, so that they grow in an unrestrained fashion to produce a heaped-up "microtumor" that stands out conspicuously against the background of normal cells in the monolayer (review, Stoker and Macpherson, 1967). Like malignant cells excised from a tumor, the transformed cells also assume the ability to grow in sloppy suspension of agar or methocel (review by Macpherson, 1969). For certain oncogenic viruses like Rous sarcoma virus, transformation is the basic method for assaying viral infectivity (see review by Vogt, 1969a); for others, such as, the DNA tumor viruses, transformation is a relatively inefficient process and these viruses are usually assayed by conventional procedures.

**Pock Assays.** A much older assay, still occasionally used for the poxviruses, is the titration of viruses on the chorion of the chick embryo. Newly synthesized virus escaping from infected cells spreads mainly to adjacent cells, so that each infecting particle eventually gives rise to a localized lesion, known as a "pock." The nature of the pock is often highly characteristic of a particular group of viruses or even a particular mutant (Gemmell and Fenner, 1960).

## Quantal Assays

The second type of infectivity assay is not quantitative but quantal, i.e., it does not register the number of infectious virus particles in the inoculum, but only whether there are any at all. Being an all-or-none assay, it is not nearly as precise as a quantitative assay; accordingly, it is only used for viruses that do not form plaques. Serial dilutions of virus are inoculated into several replicate cell cultures, eggs, or animals. Adequate time is allowed for virus to multiply and spread to destroy the whole cell culture, or kill the animal, as the case may be. Hence, each host yields only a single piece of information, namely, whether or not it was infected by that particular dilution of virus. A more economical procedure can be used with viruses such as vaccinia that produce localized skin lesions in an animal like the rabbit. Twenty or thirty skin sites can be separately inoculated, each giving an all-or-none answer equivalent to the death or survival of a cell culture or a mouse. A good example of an accurate quantal assay is the titration of influenza viral infectivity in squares of allantois-on-shell maintained *in vitro* in plastic trays, the test being read by hemagglutination (Fazekas de St. Groth and White, 1958).

**Statistical Considerations.** Isaacs (1957) reviewed the results of infectivity titrations of a number of animal viruses, compared with particle counts carried out with the electron microscope. With most viruses one particle is sufficient to initiate infection ("one-hit kinetics"). The evidence for this is the Poissonian distribution of "takes" (infected mice or eggs, lesions in rabbit skin, etc.) when closely spaced dilutions near the end point are tested, or more conveniently the linear relationship between dose and plaque count (review: Dougherty, 1964). However, a number of situations have now been recognized in which the infectivity assay (usually a plaque count) follows two-hit rather than one-hit kinetics (Fig. 2–2) indicating that two different types of virus particle must

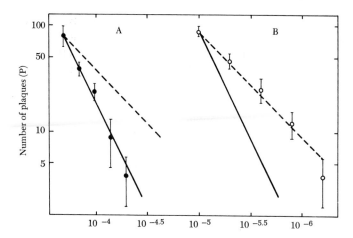

Fig. 2–2. *One- and two-hit kinetics with adenovirus 7 and defective hybrid (Ad7 and SV40) particles, plated on African green monkey kidney (AGMK) cells. Broken line, one-hit curve, i.e., the number of plaques is proportional to the first power of the concentration; solid line, two-hit curve, i.e., the number of plaques is proportional to the second power of the concentration. When a mixed population of Ad7 and (Ad7-SV40 def) yielded by AGMK cells is plated alone, plaques are produced only at low dilution, and with two-hit kinetics (A); to produce a plaque, cells must be infected by both a hybrid particle and an Ad7 virion (of which there are only a few in the mixture). If a high concentration of Ad7 is added to all plates, so that all cells receive an Ad7 virion as well as whatever hybrid particles are present, the mixed population plaques with one-hit kinetics (B). (Modified from Boeyé et al., 1966.)*

infect a single cell in order that one of them (or occasionally both) may replicate. The adenovirus-associated viruses, some of the SV40-adenovirus hybrids, and the murine sarcoma viruses (Hartley and Rowe, 1966) cannot multiply except in cells coinfected with a "helper" virus (review by Rowe, 1967). Assays of such defective viruses provide examples of two-hit kinetics (Fig. 2–2).

The results of typical quantal and quantitative infectivity titrations are given in Table 2–4. It will be seen that at high dilutions of the inoculum all the hosts

TABLE 2–4

*Comparison of Quantitative and Quantal Infectivity Titrations*

| VIRUS DILUTION | QUANTITATIVE ASSAY (PLAQUE COUNT)[a] | QUANTAL ASSAY (CPE) |
|---|---|---|
| $10^{-2}$ | C, C, C, C, C | + + + + + + + + + + |
| $10^{-3}$ | 50, 42, 54, 59, 45 | + + + + + + + + + + |
| $10^{-4}$ | 5, 7, 3, 6, 4 | + + − + + + + + + + |
| $10^{-5}$ | 0, 0, 1, 0, 1 | − + + − + − − + − + |
| $10^{-6}$ | 0, 0, 0, 0, 0 | − − − − − − + − − − |
| $10^{-7}$ | 0, 0, 0, 0, 0 | − − − − − − − − − − |

[a] C, confluent (uncountable).

remained uninfected, because they failed to receive even a single infectious unit. The "endpoint" of a quantal titration is taken to be that dilution of virus which infects 50% of the inoculated hosts; the infectivity titer of the original virus suspension is then expressed in terms of "50% infectious doses" ($ID_{50}$) per milliliter. At first sight, it may be thought that Table 2–4 shows some anomalous results, in that some hosts have become infected following inoculation with higher dilutions of virus than those that failed to infect others. This sort of finding is quite normal and explicable in terms of the Poisson distribution. At any given multiplicity ($m$) of virus there is a finite probability ($e^{-m}$) that a given aliquot of the inoculum will contain no virus particle. In the example shown, each 0.1 ml sample of the $10^{-5}$ dilution contained an average of one $ID_{50}$, i.e., 0.5 of an infectious unit. Hence, each 0.1 ml sample of the $10^{-6}$ dilution contained an average of 0.05 infectious units, i.e., about one host in twenty received one infectious particle; whereas, each 0.1 ml sample of the $10^{-4}$ dilution contained an average of five infectious units, i.e., $e^{-5}$ (about one in a hundred) of these hosts failed to receive an infectious particle at all. In practice, quantal assays rarely produce such a nicely balanced result as the example presented in Table 2–4, and statistical procedures must be used to calculate the endpoint of the titration.

**Efficiency of Plating.** In their pioneering quantitative studies of the multiplication of bacterial viruses, Ellis and Delbrück (1939) introduced the term efficiency of plating (EOP) to give a numerical value to the plaque count obtained with the same phage preparation using different susceptible bacteria as host organisms. The absolute EOP is defined as the plaque count relative to the total number of viral particles in the sample. Luria et al. (1951), using electron microscopic counts, showed that the absolute EOP of the T-even phages could approach unity. In animal virus systems, the absolute EOP is always low and often very low, ranging from $10^{-1}$ to $10^{-6}$. In other words, depending on the virus–cell combination under study, the proportion of virus particles in the suspension that score as infectious ranges from an upper limit of about 10 to as low as 0.0001%. This constitutes a major difference between bacteriophage and animal virus systems and has been a major cause of difficulty in animal virus genetics. Several factors contribute to the low plating efficiency of animal viruses. A certain proportion of the virions in any given suspension are noninfectious; most of these were once infectious but have been inactivated during growth or preparation, others are genetically defective in that they lack a complete genome or are completely "empty" (see Chapter 3). However, the major contribution to low EOP comes from the low susceptibility (i.e., high resistance) of the assay systems available, and a great deal of effort has been devoted to improving the efficiency of certain model systems so that genetic and biochemical studies can be performed.

A further complication is that at least some of the noninfectious particles can contribute to the yield, when they occur in cells in which an infectious particle is also multiplying. In experiments with two serotypes of poliovirus, for example, Ledinko and Hirst (1961) found that the number of doubly neutralizable poliovirions produced, the number of doubly infected cells, and the number of cells producing phenotypically mixed virus all exceeded theoretical expectations based on the infective virus input. Similarly, using phenotypic mixing of Newcastle disease virus as an index of mixedly infected cells, Granoff (1961) obtained evidence which suggested that the majority of virions that were nonplaque producers contributed something to the yield from doubly infected cells, which yielded phenotypically mixed virus and both parental genotypes.

## Infectious Nucleic Acids

Infectious nucleic acids can be extracted from a number of different viruses (Table 2–5; review, Pagano, 1970). Positive results are regularly obtained only with viruses whose genomes consist of a single molecule of nucleic acid and whose virions do not contain a transcriptase.

TABLE 2–5

*Viruses Yielding Infectious Nucleic Acids*

| | |
|---|---|
| DNA viruses | |
| *Parvovirus* | Positive results with adenovirus-associated virus subgenus only, in cells preinfected with helper adenovirus |
| Papovaviridae | Readily demonstrable; single-strand scission does not destroy infectivity |
| *Adenovirus* | Positive results claimed with certain simian and human adenoviruses |
| RNA viruses | |
| Picornaviridae | Readily demonstrable; both single-stranded viral RNA and double-stranded replicative form are infectious |
| Togaviridae | Readily demonstrable; infectious RNA is one of the distinguishing characteristics of the family |
| Lactic dehydrogenase virus | Readily demonstrable |

**Infectious DNA.** Most work on infectious viral DNA has been carried out with polyoma virus, which on extraction yields three different forms of DNA molecule (see Chapter 3). These are a closed circle double-stranded supercoiled form (sedimentation coefficient in neutral conditions, 20 S), a relaxed form (sedimentation coefficient, 16 S) resulting from at least one scission in one of the strands, and linear double-stranded molecules of varying length, which may result from double-stranded breakage of the viral nucleic acid or may be cellular DNA from pseudovirions (Winocour, 1969). Both the 20 and 16 S molecules are infectious. In some cases, extracted DNA can elicit other biological effects such as cell transformation, tumor formation in animals, or induction of new antigens. For example, Bourgaux *et al.* (1965) demonstrated transforming activity with polyoma virus DNA, the supercoiled double-helical form being at least ten times more efficient in transforming capacity than any other form.

For isolated viral DNA to be infectious, it is clear that one or other of the cellular DNA-dependent RNA polymerases (see Chapter 4) is able to transcribe the incoming viral DNA. It is not known why it has proved so difficult to demonstrate infectious DNA from human herpesviruses and adenoviruses. Possible explanations include the following: (a) difficulty in preparing nonfragmented DNA, an unlikely explanation with adenoviruses, (b) a topological requirement—uncoating may have to occur at a specific cellular site, (c) an internal viral protein associated with the DNA may perhaps serve to modify transcriptional patterns, and (d) these viruses may contain undetected polymerases. An unusual example of what is apparently an infectious DNA has been reported with an avian leukovirus. Certain strains of Rous sarcoma virus transform hamster cells but yield no virions, whereas transformed chick cells yield infectious particles (see Chapter 14). Hill and Hillova (1972) used a genetically marked mutant (temperature-sensitive and of unusual antigenicity) to transform hamster cells. DNA extracted from these, in turn, transformed chick cells which produced infectious particles having the same phenotype as the mutant of Rous sarcoma virus originally used.

**Infectious RNA.**   For isolated viral RNA to be infectious the essential requirements seem to be that the genome is a single piece of nucleic acid and that the viral RNA can itself act as messenger RNA. These conditions are fulfilled by viruses of two large families, Picornaviridae and Togaviridae.

Extensive investigations on the optimum conditions for the assay of infectious RNA have been carried out with poliovirus RNA, which is a linear molecule with a molecular weight of 2.6 million daltons. In one effective assay system (Koch and Bishop, 1968), suspended cells are exposed to the viral nucleic acid and then plated on a monolayer of indicator cells for an infectious center assay. The suspended cells may be sensitized by exposure to dimethyl sulfoxide or polycations such as DEAE-dextran, which may augment infectivity as much as 10,000-fold (Vaheri and Pagano, 1965). Both the single and double-stranded forms (see Chapter 6) of poliovirus RNA are infectious. Indeed, in the presence of polycations, the double-stranded RNA has a higher specific infectivity than single-stranded viral RNA, probably because it is resistant to cellular ribonucleases.

Inefficient as are assays for infectivity of poliovirus, the infectivity levels obtained with free viral RNA are usually several orders of magnitude lower, even though most of the infecting RNA seems to enter the cells, as judged by conversion from an RNase-sensitive to a RNase-resistant state after adsorption.

**Significance of Infectious Viral Nucleic Acids.**   The demonstration that some viral nucleic acids are infectious is of considerable theoretical and practical importance. More clearly than any other observation, it showed that the viral genome was the repository of all viral genetic information. In the laboratory, it has extended the host range of viruses; for example, poliovirus RNA will infect many cells and animals that are resistant to infection with poliovirions due to the absence of cellular receptors for the viral capsid (see Chapters 6 and 11). This demonstrates clearly that such "resistant" cells are in fact capable of supporting viral growth if a mechanism can be found for facilitating the entry of undegraded viral RNA into the cell. With a few viruses, notably SV40, the use of infectious DNA has been of great assistance in dissection of the functions of the genome. Persistent failure to demonstrate infectivity of the nucleic acids of some viruses led to the successful search for polymerases in their virions. Finally, infectious viral nucleic acids provide a convenient model for studying the more general problem of introducing functional nucleic acids into vertebrate cells. Although early attempts at mammalian gene transfer by isolated DNA (Szybalska and Szybalski, 1962) have not been repeated, the range of such possibilities was dramatically broadened by the claim that a metabolic lesion in a mammalian cell could be repaired by introducing the necessary gene via the DNA of a transducing bacteriophage (Merril et al., 1971).

## ASSAY OF OTHER PROPERTIES OF VIRIONS

### Hemagglutination

Many viruses contain in their outer coat virus-coded proteins capable of binding to erythrocytes (Table 2–6; reviews, Rosen, 1964, 1969). Such viruses can, therefore, bridge red blood cells to form a lattice. The phenomenon, known as hemagglutination, was first described by Hirst (1941), who then went on to analyze the mechanism of hemagglutination by influenza virus. In this case, the hemagglutinating protein (hemagglutinin) on the virion is a glycoprotein,

TABLE 2–6

*Hemagglutination by Viruses*

| VIRUS | | RBC |
|---|---|---|
| GROUP | SUBGROUPS OR SPECIES | |
| Parvovirus | Subgenus A: several species | Human, guinea pig, mouse, 4°C |
| | Subgenus B: AAV type 4 | Human, guinea pig, 4°C |
| Papovaviridae | Polyoma | Guinea pig, 4°C |
| Adenovirus | Most types | Monkey, rat, 37°C |
| Herpesvirus | | Nil |
| Iridovirus | | Nil |
| Poxvirus | Smallpox, vaccinia | Fowl (some birds only), 37°C |
| Picornaviridae | Coxsackie (some serotypes) | Human |
| | Echo (some serotypes) | Human |
| | *Rhinovirus* (some serotypes) | Sheep, 4°C |
| Togaviridae | *Alphavirus, Flavivirus,* rubella | Goose, pigeon, pH and temperature critical |
| Orthomyxovirus | Influenza types A and B | Fowl, human, guinea pig, 4°C |
| Paramyxovirus | Parainfluenza, mumps | Fowl, human, guinea pig, 4°C |
| | Measles | Monkey, 37°C |
| Rhabdovirus | Rabies | Goose, 4°C |
| Coronavirus | Human | Rat, mouse, fowl, 37°C |
| Arenavirus | | Nil |
| Bunyamwera supergroup | | Goose, pH critical |
| Leukovirus | | Nil |
| Reovirus | Types 1–3 | Human |
| Orbivirus | Bluetongue | Nil |

which occurs in the form of many short projections (see Plate 3–15). The virus will attach to any species of erythrocyte carrying complementary receptors, which are glycoproteins of a different sort. Hemagglutination by influenza and the paramyxoviruses, but not other viruses, is complicated by the fact that the virion also carries an enzyme, neuraminidase, which destroys the glycoprotein receptors on the erythrocyte surface and allows the virus to elute, unless the test is carried out at a temperature too low for the enzyme to act (4°C). About $10^7$ influenza virions are required to cause macroscopic agglutination of a convenient number of chick erythrocytes (conventionally 0.25 ml of a 1% suspension of red cells). Thus hemagglutination is not a sensitive indicator of the presence of small numbers of virions, but because of its simplicity it provides a very convenient assay if large amounts of virus are available. The viral suspension is diluted serially (usually in twofold steps) in a plastic tray using a calibrated wire loop, and erythrocytes are added to each well. Unclumped red blood cells settle to form a "button," whereas agglutinated cells form a "shield" (Plate 2–4).

## Direct Particle Counts

Negative staining is such a simple technique that with many viruses the number of particles in relatively crude suspensions can be counted directly in

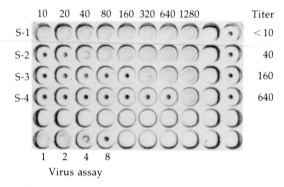

PLATE 2–4. *Hemagglutination by influenza virus, used for titrating antibodies to the viral hemagglutinin. (Titers are expressed as reciprocals of dilutions.) In the example illustrated, a cardiac invalid was immunized against the influenza strain, A2/Hongkong/68. Serum samples, S-1, S-2, S-3, and S-4 were taken, respectively, before immunization, 1 week after the first injection, 4 weeks after the first injection, and 4 weeks after the second injection. The sera were treated with periodate and heated at 56°C for 30 minutes to inactivate nonspecific inhibitors of hemagglutination, then diluted in twofold steps from 1/10 to 1/1280. Each cup then received four hemagglutinating (HA) units of influenza A2/Hongkong/68 virus, and a drop of red blood cells. Where enough antibody is present to coat the virions, hemagglutination has been inhibited, hence the erythrocytes settle to form a button on the bottom of the cup. On the other hand, where insufficient antibody is present, erythrocytes are agglutinated by virus and form a shield. The virus assay (bottom line) indicates that the viral hemagglutinin used gave partial agglutination (the endpoint) when diluted 1/4. Interpretation: The patient originally had no hemagglutinin-inhibiting antibodies. One injection of vaccine produced some antibody, the second injection provided a useful booster response. (Courtesy I. Jack.)*

the electron microscope. The virus can be mixed with a known concentration of latex particles to provide an easily recognizable marker (Luria *et al.*, 1951). Alternatively, the virions contained in a known volume can be sedimented in an ultracentrifuge directly onto an electron microscope grid (Sharp, 1965).

## Comparison of Different Assays

If a given preparation of virus particles were to be assayed by all of the methods described above, the "titer" would be different in every case. For example, an influenza virus suspension may provide the following data (tabulated below):

| METHOD | AMOUNT (PER MILLILITER) |
|---|---|
| (1) Direct electron microscope count | $10^{10}$ EM particles |
| (2) Quantal infectivity assay in eggs | $10^9$ egg $ID_{50}$ |
| (3) Quantitative infectivity assay by plaque formation | $10^8$ pfu |
| (4) Hemagglutination assay | $10^3$ HA units |

There are clearly great differences in the sensitivity of different methods of assay. Relative to an absolute standard (1), the EOPs are $10^{-1}$ (2), $10^{-2}$ (3), and $10^{-7}$ (4).

## PURIFICATION AND SEPARATION OF VIRIONS AND THEIR COMPONENTS

The expansion of our knowledge of the biology of animal viruses has occurred largely as a consequence of three new found technical skills: first, the ability to grow many types of animal cells *in vitro* has enabled us to produce in turn large quantities of many animal viruses; second, the development of physical methods of separation of macromolecules has given us the skill to purify to homogeneity large quantities of many sorts of viruses and their various constituent molecules; third, the availability of radioactive precursor compounds of high specific activity has allowed us to follow in detail many of the biochemical events that occur both in uninfected and in virus-infected cells. It is the marriage of these three disparate advances that has made many of the techniques, which were esoteric or impossible a few years ago, no more than matters of routine manipulation today.

## Purification

A few animal viruses can be purified by taking advantage of certain characteristic properties of their virions. For example, influenza viruses can be concentrated and purified by adsorbing the virus on to red blood cells at 4°C. The cells are then washed free of impurities and the temperature is raised to 37°C. The enzyme neuraminidase, which is an integral part of the virions, begins to act and destroys the viral receptors on the red blood cells so that the viral

particles are released. The red cells are then removed by low-speed centrifugation (Ada and Perry, 1954). However, for most animal viruses no such convenient
short cuts are available, and purified suspensions are obtained by the sequential
application of different fractionation techniques. Because the customary starting
materials are large volumes of medium, infected cells or body fluids, containing
small numbers of virions, the first step of purification often involves concentrating the viral particles by salting out with ammonium sulfate, precipitation
with polyethylene glycol or ethanol, or adsorption onto, and elution from, ion-
exchange resins. The host components that are different in size from the particular virus under study are usually separated by differential centrifugation;
low-speed centrifugation deposits large pieces of cell debris and subsequently
high-speed centrifugation is used to pellet the virus.

**Rate Zonal Centrifugation.**   Further purification is usually achieved by rate
zonal centrifugation through preformed density gradients (Fig. 2–3; review by
Mazzone, 1967). The solutes most commonly used in such gradients are buffered
sucrose and glycerol, although for those viruses that are unaffected by solutions
of high ionic strength, salts such as potassium bromide and tartrate are also
satisfactory. The gradients are formed directly in the centrifuge tubes by mixing
together solutions of different solute concentrations in such a way that there

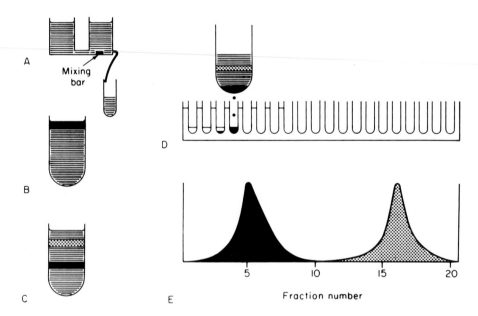

FIG. 2–3. *Rate zonal centrifugation. (A) Gradient is formed by mixing solutions (usually sucrose or glycerol) of different densities. (B) Sample is applied to top of preformed
gradient. (C) Under centrifugal force, the different components in the sample sediment
through the density gradient at different rates, forming bands with different sedimentation coefficients. (D) At the end of the centrifugation samples are collected and (E)
the position of the separate components is determined by optical density tracings or
radioisotope counting.*

is a continuous linear gradation of density from the top to the bottom of the final column of liquid. The sample of virus is layered onto the gradient and during centrifugation sediments as a band through the gradient at a rate (sedimentation coefficient, S) which is determined by the size and weight of the viral particles, and which can be predicted by classic centrifugation theory. The centrifugation is usually stopped when the band of virus has traveled one-half to two-thirds the length of the gradient, and samples are collected by piercing a hole through the bottom of the centrifuge tubes, or pumping a dense solute through the bottom of the tube and monitoring the effluent as it passes from the top. The viral band is located optically (with a spectrophotometer) or by radioactivity.

**Equilibrium Density (Isopycnic) Gradient Centrifugation.** Virions can be separated from other particles with similar sedimentation coefficients by isopycnic gradient centrifugation, which is based on differences in the buoyant densities of virions and contaminating particles (review, Brakke, 1967; Fig. 2–4). The

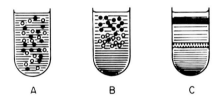

FIG. 2–4. *Equilibrium density gradient (isopycnic) centrifugation. (A) Samples under investigation are mixed with a solution of high density (usually an alkali metal salt like CsCl). (B) During prolonged high-speed centrifugation, the heavy metal atoms tend to sediment, eventually forming a density gradient. As this process occurs, the viral particles or molecules in the sample begin to collect at a position in the gradient where their buoyant density matches the density of the supporting medium. (C) Eventually sharp bands are formed, which are collected and plotted as shown in Fig. 2–3.*

virions are suspended in solutions, usually of alkali metal chlorides such as cesium chloride, which have high density and low viscosity. During high-speed centrifugation, the metal ions form a stable density gradient whose steepness is a reflection of the balance that is established between the tendency of the heavy metal ions to sediment through the solvent and their back-diffusion rate from more to less concentrated areas. Once the gradient is established the viral particles migrate to a position in the centrifuge tube where their density and that of the supporting medium are equal. When equilibrium is reached, the centrifuge is stopped, the bands of virus are collected, and the solute removed by dialysis or by passage through Sephadex columns.

**Criteria of Purity.** There are no absolute criteria of purity for viral preparations. Minimally one would require that the particles are homogeneous by electron microscopic examination, that further purification steps fail to remove any more components from the preparation without reducing the infectivity of the particles, and that when radioactively labeled uninfected cells are subjected to the same purification procedure no host components are found which will copurify with the viral particles. Although these criteria have been fulfilled for many types of viruses (e.g., polyoma viruses, adenoviruses, picornaviruses), there are others (e.g., RNA tumor viruses and coronaviruses) which have the same size and density as components of uninfected cells (Anderson *et al.*, 1966). Despite many attempts these viruses have not been purified successfully and further progress on their chemistry will require the development of new purification techniques or the refinement of those already in existence.

## Radioactive Labeling

Most of our knowledge of the biosynthetic events produced by infection has to be gained by using radioactive tracers. Radioactively labeled molecules enter all the intracellular pools and are used metabolically in exactly the same way as unlabeled molecules, but because they are radioactive their pathways and fates within the cell can be followed very easily. However, it should be borne in mind that we still know little about the intracellular compartmentalization of pools of metabolites.

The radioisotopes that are most commonly used in animal virology are $^3$H, $^{14}$C, $^{32}$P, $^{35}$S, and $^{125}$I or $^{131}$I, and they are usually monitored by scintillation spectroscopy. For most experimental purposes isotopes are not added to cells as free atoms but rather in precursor compounds that are used in the synthesis of specific cellular components. For example, labeled amino acids are used to follow protein synthesis, uridine for RNA, thymidine for DNA, glucosamine for glycoprotein, and choline for lipid synthesis. By using a mixture of precursors labeled with, for instance, $^3$H and $^{14}$C, it is possible to follow two metabolic pathways at once (double-labeling experiments).

The choice of precursor and its specific activity depends on the purpose of the experiment. When isotopic uridine is used as an RNA precursor, it is largely incorporated into RNA. However, it may also be converted to dUMP and dCMP and thus appear in DNA. To avoid this, [5-$^3$H]uridine should be employed since the conversion from dUMP to dTMP by reductive methylation at the 5 position eliminates the tritium. Thymidine is a specific precursor of DNA and by blocking the endogenous synthesis of dTMP with fluorodeoxyuridine (see Fig. 4–4), exogenous thymidine is incorporated into the DNA of most cells at an enhanced rate.

In general, care should be taken that changes in the rate of precursor uptake or incorporation are not merely a reflection of a change in the intracellular pool size rather than a change in the rate of macromolecular synthesis. Further, the level and activity of the exogenous isotope should not be rate limiting. With high specific activities of some precursors, in particular tritiated thymidine, cell functions may be damaged by irradiation effects. Provided attention is paid to these points, radioactive isotopes are very powerful tools which allow precise analysis of the effects of animal viruses on cells. They also allow us to prepare radioactive virus, a fact which has not only made possible much of the detailed analysis of the composition and architecture of the particles which is described in Chapter 3, but by allowing the application of microtechniques has relieved many scientists of much drudgery.

## ELECTRON MICROSCOPY OF VIRUSES

Several different electron microscopic techniques are routinely used to study the morphology of virus particles and their multiplication in infected cells. Negative staining, introduced into virology by Brenner and Horne (1959), has proved a very powerful method for revealing details of viral architecture (review, Horne, 1967). At neutral pH, phosphotungstic acid does not combine readily with proteins or nucleic acids, but after drying it forms a relatively uniform electron-opaque background in which small objects such as virions stand out in detail. Because the phosphotungstate enters interstices between adjacent macromolecules, this method, when used at high magnification, reveals fine details of viral structure (see Chapter 3).

Thin sectioning of viral pellets or more commonly of infected cells also reveals important information about viral structure and morphogenesis (review, Morgan and Rose, 1967). In order to locate specific viral proteins either within particles or in infected cells, auxiliary techniques are used, such as autoradiography (review by Granboulan, 1967), treatment of the thin sections with ferritin-labeled antibody (review, Breese and Hsu, 1971), or antibody labeled with horseradish peroxidase (review, Kurstak, 1971).

Finally Kleinschmidt's method (Kleinschmidt, 1968) of spreading nucleic acids and photographing the extended molecules after metal shadowing or better by dark field illumination is used increasingly for the study of viral nucleic acids. With double-stranded DNA molecules, a length of 1 $\mu$m corresponds to a molecular weight of about 2 million. A more recent extension of this technique is the direct visualization of DNA heteroduplexes to determine the degree of homology between the nucleic acids of related viruses (review by Davis et al., 1971).

## NUCLEIC ACID HYBRIDIZATION

Nucleic acid hybridization (reannealing) is the interaction of single-stranded polynucleotide chains with complementary base sequences to form double-stranded structures. Depending on the initial reagents, the products of the reaction can be RNA:RNA, RNA:DNA, or DNA:DNA duplexes, each of which can be measured by a variety of methods. There are several excellent reviews of the most commonly used experimental techniques (Raskas and Green, 1971; Bøvre et al., 1971; Gillespie, 1968; Kohne and Britten, 1971) and this short section will do nothing more than describe two basic methods and examples of their use in animal virology.

### Hybridization in Solution

Preparations of double-stranded nucleic acid are denatured by heating or alkali treatment and incubated under conditions which promote reannealing

of the polynucleotide strands (neutral pH, 25°C below the $T_m$). Because reassociation of the polynucleotide strands is a bimolecular reaction, the kinetics of
reannealing are second order, and the rate of the reaction depends on the concentration of the nucleic acid, its size, its complexity, and the salt concentration in
the solution (Wetmur and Davidson, 1968; Britten and Kohne, 1968). The formation of native nucleic acid can be assayed by hydroxylapatite chromatography
or by the use of nucleases which differentiate single- and double-stranded nucleic
acids. A good example of the use of liquid hybridization in animal virology is
the determination of the number of copies of viral DNA in transformed cells (see
review by Gelb and Martin, 1971). Small amounts of highly radioactive viral
DNA are allowed to reanneal in the presence of large amounts of DNA either
from virus-transformed or from control cells. Because the rate of reannealing of
the radioactive viral DNA is proportional to its concentration in the solution,
hybridization occurs faster in the presence of DNA from transformed cells than
in the presence of control cell DNA's. By following the kinetics of the annealing
of the radioactive DNA it is possible to determine directly the number of viral
copies in DNA extracted from transformed cells (see Chapter 13).

## Hybridization to DNA Immobilized on Filters

DNA is first denatured and immobilized on nitro-cellulose filters, which are
immersed in solutions of radioactive nucleic acid under conditions which favor
hybridization. If the solution contains any labeled nucleic acid sequences which
are complementary in base sequence to the DNA immobilized on the filter, then
reannealing will occur. The extent of the reaction can be determined by measuring the amount of radioactivity specifically absorbed to the filters. This basic
method has been used to detect the synthesis of virus-specific RNA in infected
cells. However, many other uses have been described and many variations of the
technique are available, for example, competition hybridization, in which filters
containing immobilized DNA are reacted first with an excess of unlabeled RNA
and then with smaller amounts of radioactive RNA. If the unlabeled RNA contains sequences which completely overlap those of the radioactive RNA then
no label will adsorb to the filter because all the available sites on the DNA will
have been saturated. If the sequences of the two RNA's are different then
hybridization of the labeled RNA will occur. By using experiments of this sort
it has been possible to determine which particular segments of viral genomes
are transcribed into RNA at different times after infection. Clearly hybridization
of nucleic acids is a powerful technique, and because of its simplicity and sensitivity, it has become an essential tool for the study of nucleic acid metabolism
in virus-infected and transformed cells.

## EXAMINATION OF AN UNKNOWN VIRUS

Suppose that we have isolated a new virus and have managed to produce
a suspension of purified particles. How can we classify the virus, and how do
we find out about its chemical composition? A lead may be provided by its past

history—the species of animal from which it was isolated and whether or not it was related to a disease. This information, in conjunction with that obtained by electron microscopic examination of stained, unstained, and sectioned particles, might be enough for us to make a preliminary identification. However, we may want to know more about the physical and chemical properties of the virus, as well as the nature and size of its genome. In most cases, this information can be obtained quite easily by applying tests of the following types.

## Size and Shape of the Viral Particle

The most commonly used method for measuring the size of virus particles is electron microscopy, using catalase crystals as calibration standards. The method is extremely rapid and simple and tells us not only about the size of the virions but also something of their shape and symmetry, with the minor reservation that the virus particles may undergo distortion during the fixation and staining processes. Independent methods for determining the diameter of the virion include filtration through membranes of known pore size, determination of sedimentation, and/or diffusion coefficients or light scattering.

## Nucleic Acid of the Viral Genome

Whether the virus contains DNA or RNA as its genetic material can best be determined by performing direct sugar analysis on the isolated nucleic acid, which is extracted from the virus with hot or cold phenol. However, several other tests are almost as good and much simpler, and are more sensitive when only small quantities of virus are available. For instance, if the virus is grown in the presence of [$^{14}$C]thymidine and [$^{3}$H]uridine, the isolated nucleic acid will contain only $^{14}$C counts if it is DNA and only $^{3}$H counts if it is RNA.

Whether the nucleic acid is single- or double-stranded can be determined by several physicochemical techniques, such as determination of the buoyant density, base analysis, and melting characteristics. However, the easiest method is probably by treatment of the nucleic acid with nucleases that are known to be specific for different types of nucleic acids. A serious caveat is that certain types of virus are known to exist (e.g., adenovirus-associated viruses) whose particles contain single strands of DNA but some of the particles contain one strand of DNA and others its complement. During extraction the isolated strands reanneal to form a double-stranded DNA duplex. With the adenovirus-associated viruses the state of the nucleic acid within the particles was determined by examining in the fluorescence microscope unfixed particles stained with acridine orange (see Chapter 3). Acridine orange binds to the nucleic acid within the particles and, when exposed to UV light, fluoresces green if the nucleic acid is double-stranded and orange if it is single-stranded.

Finally, an unreliable method that was used widely in the past was to determine whether 5′-bromodeoxyuridine (BUdR) depressed the yield of virus from infected cells. BUdR is a base analog of thymidine and, when it is incorporated into DNA in the place of the normal base, causes the DNA to be inactive. Hence, it was argued that if growth in BUdR depressed the yield of a virus, then the virus must contain DNA as its genetic material. It is now known that some types of RNA viruses (notably the RNA tumor viruses) require DNA synthesis soon after infection if they are to replicate, so that by the BUdR test these viruses score as DNA viruses.

The size of the molecule(s) of nucleic acid that makes up the viral genome can be determined in a number of ways. If the virus contains single- or double-stranded DNA, centrifugation methods are useful (reviews by Bauer and Vinograd, 1971; Szybalski and Szybalska, 1971), but the method of choice at present is to examine the nucleic acid directly in the electron microscope by Kleinschmidt's technique (review, Kleinschmidt, 1968). For double-stranded DNA, a length of 1 $\mu$m is equivalent to approximately 2 million daltons of nucleic acid. This method also tells us something about the topology of the DNA, for example, whether it is circular or linear, and if circular, whether it is superhelical or not.

For viruses containing RNA, the problem is rather more difficult. In many RNA preparations, there are severe problems of aggregation of molecules, on the one hand, and degradation due to nuclease action on the other; and there are no reliable methods of determining the length of RNA molecules by electron microscopy. However, there are excellent techniques for determination of molecular weights of RNA by electrophoresis through sieves of polyacrylamide or agarose (reviews by Adesnik, 1971; Dingman and Peacock, 1971) or by sedimentation in denaturing solvents such as 99% dimethyl sulfoxide (Strauss et al., 1968b). As long as adequate numbers of calibration markers are used, these methods are very accurate and reliable.

Once the molecular weight of the nucleic acid molecules extracted from the virus has been determined, it is necessary to decide how many of each class are present in each virus particle. Enough chemical data have accumulated to show that each virus particle of adenoviruses, polyoma viruses, poxviruses, herpesviruses, and picornaviruses contains only one molecule of nucleic acid, but with some of the RNA viruses, it is more difficult to be so certain. For instance, the genomes of reoviruses, leukoviruses, and influenza virus contain more than one piece of RNA (Chapter 3). However, we cannot say with assurance exactly how many pieces of RNA constitute the genome. In the absence of complete genetic maps, the only way to determine the composition of the genome of these particles is to try to fit together different combinations of RNA so that they add up to the amount of RNA that is found by chemical analyses of purified particles. In many cases this remains more a matter of guesswork than science.

## Proteins

Information about the size and subunit composition of the proteins that are present in viruses is most commonly determined by electrophoresis through polyacrylamide gels (review, Maizel, 1971). In this method, acrylamide is polymerized to form cylinders or slabs, and the mixture of proteins under investigation is added to the top of the gel. When an electric current is applied across the ends of the gel, the proteins in the mixture migrate through it as bands at speeds which are governed, on the one hand, by the net charge of the protein molecules in the buffer conditions used, and, on the other hand, by the pore size of the gel. The smaller protein molecules pass through the pores formed by the polyacrylamide cross-links more easily than the larger proteins, and so they migrate faster. By adjusting the acrylamide concentrations within the range 5–15%, gels with pore sizes suitable for the separation of protein molecules with a wide spectrum of molecular weights can easily be generated. The position of the protein bands is determined either by staining the fixed gel with dyes like Coomassie Brilliant Blue, or by autoradiography, or by cutting it into small segments and counting the amount of radioactivity in each fraction.

Two sets of conditions can be used for polyacrylamide gel electrophoresis. (a) *Denaturing conditions.* Before being placed on the gel the proteins are dissociated into their constituent polypeptide chains. The most commonly used conditions are heating to 80°C for a few minutes in the presence of reducing agents such as dithiothreitol or $\beta$-mercaptoethanol, and the anionic detergent sodium dodecyl sulfate (SDS). The detergent binds to the separated polypeptide chains and gives all of them an overall net negative charge. The polypeptides separate in the gel entirely on the basis of their size; all charge differences are eliminated. Given that certain precautions are taken, the method gives a catalog of all the polypeptides present in a particular viral preparation, and as long as the polypeptides are not derived from glycoproteins, which behave anomalously in the system, the method gives extremely accurate estimates of the molecular weights of the polypeptides under study. (b) *Nondenaturing conditions.* These are sometimes used in order to keep the proteins in a native state, so that the separation in the gels is not of constituent polypeptides but rather of entire proteins. Since the migration of proteins without bound SDS is a function not only of their size but also of their overall charge, the buffers are tailored to match the characteristics of the particular protein under study.

**Assignment of Proteins to Morphological Subunits.** In several cases, it has been possible to assign particular viral polypeptides to morphological subunits of viral particles. The details of the methods used vary from virus to virus, but the principle remains the same—the viral particles are gently disrupted into their morphological subunits, usually by treatment with detergents or dilute alkali, and the subunits are then separated from one another by immunoprecipitation, sedimentation velocity or isopycnic gradient centrifugation. It is then a simple matter to dissociate the subunits into their constituent polypeptide chains by heating in sodium dodecyl sulfate and $\beta$-mercaptoethanol, and to determine by polyacrylamide gel electrophoresis the precise composition of the morphological subunit in terms of the number of each species of constituent polypeptide and how they are bound together. The method cannot be applied universally, since it depends on successful disruption and separation of morphologically identifiable subunits, but at its best [see, for example, Rueckert's (1971) work with EMC (Chapter 3)] the technique provides an elegant and satisfying correlation between chemical and morphological subunits.

In those viruses where assignment of polypeptides to subunits is incomplete or impossible, the position of individual polypeptides in the particles can often be mapped by techniques such as radioactive labeling of the external proteins of the virion by iodination with [125]I and lactoperoxidase (Stanley and Haslam,

1971) or with ferritin or horse radish peroxidase antibody labeling and auto-radiography.

## Viral Envelopes

All viruses contain nucleic acid and protein and many contain nothing else, except for the ribose or deoxyribose associated with their nucleic acid. Others also contain lipids and carbohydrate. Except for the poxviruses, which are in an exceptional position, this lipid and carbohydrate is located in a viral envelope (see Chapters 1 and 3) that surrounds the genome and its associated proteins— such viruses are said to be enveloped. Whether or not a virus is enveloped is most easily and unambiguously determined by electron microscopic examination. Less direct methods that have been used in the past include determination of the sensitivity of the virus to lipid solvents such as ether or sodium deoxycholate and radioactive labeling with sugars such as fucose or glucosamine which are incorporated preferentially into membrane glycoproteins.

## Viral Enzymes

Neuraminidase was recognized as an integral component of the envelope of myxoviruses many years ago (review, Drzeniek, 1972). An even more impor-tant discovery was the demonstration that many viruses contain nucleic acid poly-merases (review, Baltimore, 1971b). Three types of polymerases have been found associated with viruses of different groups: DNA-dependent RNA polymerase, RNA-dependent DNA polymerase, and RNA-dependent RNA polymerase. All of them can be readily assayed by mixing viral particles with suitable labeled precursors and following conversion into nucleic acid. In a few cases, whole viral particles can be assayed directly, but usually it is necessary to open up the virions by removing their envelopes or protein shells with detergents or proteolytic enzymes.

Several other enzymes have been described as being associated with the virions of the more complex viruses (see Chapter 3). It is important to distin-guish enzymes that may be absorbed on the outside of the particle from those that are part of the virion. If they are truly virion-associated it is important to determine their location—are they in the envelope, within the outer part of the capsid, or within the viral core (see Kang and McAuslan, 1972)?

## MEASUREMENT OF ANTIBODIES AND ANTIGENS

Serological surveys of populations of vertebrates are widely used in epidemi-ological studies, and provide information about the distribution and spread of viruses in time and space. In diagnostic virology, antibodies of known specificity are used to identify unknown viruses that may be recovered from diseased or healthy individuals, by the use of one or several of the techniques set out below. Frequently, viral isolation is unsuccessful, but observation of a rising antibody titer to a particular virus between acute phase and convalescent specimens of serum is taken as presumptive evidence that the virus being tested was involved in the disease episode.

In laboratory research, serological methods provide the potential for recognizing different antigenic components of a virus both in infected cells and during purification procedures, and for measuring the concentration of such antigens. For this purpose, antisera against viral antigens are produced in suitable laboratory animals, usually rabbits (review, Horwitz and Scharff, 1969a). Further, the antibody response plays an important role in the pathogenesis of viral diseases (see Chapter 10), and methods for measuring different kinds of antibodies in various parts of the body are essential for the proper understanding of disease processes in intact animals.

Many different serological techniques are available (Table 2–7), each appropriate for particular purposes (see Casals, 1967; Lennette and Schmidt, 1969).

TABLE 2–7

*Serological Procedures Used in Virology*

| PROCEDURE | PRINCIPLE | SENSI-TIVITY [a] | SPECI-FICITY [b] |
|---|---|---|---|
| Virus neutralization | Antibody neutralizes infectivity | High | High |
| Hemagglutination inhibition | Antibody inhibits viral hemagglutination by coating the virus | High | High |
| Gel diffusion | Antibodies and soluble antigens diffuse toward one another through agar and produce visible lines of precipitate where homologous antigens and antibodies are present in optimal proportions | Low | High |
| Complement fixation | Antigen–antibody complex binds complement, which is thereafter unavailable for the lysis of sheep RBC by hemolysin | Low | Low |
| Immunofluorescence | Antibody (usually) or antigen can be "tagged" by conjugation to a dye (e.g., fluorescein) which fluoresces on excitation by ultraviolet light | Low [c] | Low |
| Radioimmunoassay | Antibody or antigen can be radioactively labeled and antibody–antigen reaction (by precipitation or gel diffusion) recognized by scintillation counting or autoradiography | High | High |

[a] The sensitivity of a serological test refers to its ability to detect small quantities of antibody.

[b] The specificity of a serological test refers to its ability to discriminate between serotypes within a viral group.

[c] "Low" in the sense that the test is negative if the serum is greatly diluted, but high in that small numbers of infected cells can be readily detected.

## Virus Neutralization

Certain antibodies interact with virions and neutralize their infectivity. The neutralization test is usually conducted as follows (Habel, 1969a). Serum is heated at 56°C for 30 minutes to destroy nonspecific inhibitors of viral infectivity. Dilutions of the serum are then mixed with a constant dose of virus, say 100

$TCID_{50}$. The mixtures are allowed to stand for a time, e.g., 60 minutes at ambient temperature, and then assayed for residual infectivity by inoculation into cultured cells, embryonated eggs, or laboratory animals. The endpoint of the titration is taken as the highest dilution of antiserum that inhibits the development of CPE in cultured cells (Plate 2–5), or the multiplication of virus in the animal used.

The interaction of virions and antibodies is to some extent reversible, so that the apparent titer of antiserum will be influenced by factors such as the time and temperature of incubation of the virus–antiserum mixture, the volume of fluid in which the test is conducted, the susceptibility of the cell type used, and the time at which the CPE is read (review, Fazekas de St Groth, 1962). Since all these factors may affect the final equilibrium between bound and free antibodies, strict comparisons of titers are only valid within a laboratory where a particular protocol has become standard. For example, all tests should be terminated when the "virus only" controls (no antiserum) first show complete CPE; otherwise, the apparent neutralizing titer of the serum drops the later the test is read, as the virus "breaks through" the higher dilutions of antiserum. There are many other ways of carrying out virus neutralization tests. The most important of these is the plaque reduction test, which measures the reduction in the num-

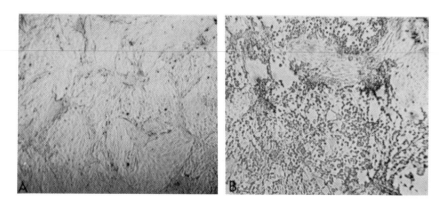

PLATE 2–5. *The virus neutralization test, a specific and sensitive measure of antiviral antibodies. "Acute" and "convalescent" sera from a patient with aseptic meningitis was heated at 56°C for 30 minutes to inactivate nonspecific inhibitors of infectivity, then serially diluted. Aliquots of each dilution were separately mixed with 100 $TCID_{50}$ of coxsackievirus type B1, B2, B3, B4, B5, or B6. Following brief standing, each mixture was inoculated into monolayers of monkey kidney cells, and incubated at 36°C for several days. All cultures were inspected daily for development of CPE in comparison with controls receiving virus only. Only two key cultures are shown (unstained, X 23). Tube A received a $10^{-2}$ dilution of convalescent serum plus $10^2$ $TCID_{50}$ of coxsackievirus B2. Tube B received a $10^{-2}$ dilution of convalescent serum plus $10^2$ $TCID_{50}$ of coxsackievirus B1, B3, B4, B5, or B6 (all giving the same picture, which corresponded also to that obtained when the "acute" serum was inoculated together with any one of the six serotypes). Interpretation: The meningitis was probably attributable to coxsackievirus B2. (Courtesy I. Jack.)*

ber of plaques produced by a given viral suspension in the presence of graded dilutions of antibody.

The virus neutralization test is the most sensitive and the most specific serological procedure available. Only antibodies directed against surface antigens of the virion, particularly those involved in adsorption to the host cell, will register in this test. These superficial antigens are those that have been most subject to the selective pressures of evolution (e.g., by antigenic drift, see Chapter 17), and they are therefore specific to the viral type or strain. Hence, neutralizing antibodies against a given serotype usually show little or no cross-reaction with other viruses within the same group.

## Hemagglutination Inhibition

Antibody can inhibit virus-mediated hemagglutination (HA) by blocking the particular antigens on the surface of the virion that are responsible for this phenomenon. The hemagglutination-inhibition (HI) test is conducted as follows (Rosen, 1969). The serum being tested is first treated to destroy nonspecific inhibitors of HA. For example, inhibitors of influenza virus HA are of two varieties: serum glycoproteins, which can be destroyed by periodate, trypsin, or the receptor-destroying enzyme of *Vibrio cholerae*; and heat-labile inhibitors, destroyed by heating at 56°C for 30 minutes. The treated sera are then diluted serially in a plastic tray using a calibrated wire loop. About four or five "agglutinating doses" of the virus being tested are added to each cup from a calibrated dropping pipette, and erythrocytes of the appropriate species are added in the same way. The tray is allowed to stand at a suitable temperature and pH for enough time for the erythrocytes to settle (about 45 minutes). The HI titer is taken as the highest dilution of serum inhibiting hemagglutination (Plate 2-4).

The HI test is highly sensitive and, except in the case of the togaviruses, highly specific. It measures only those antibodies that bind directly to viral hemagglutinin (i.e., to the projecting tip of the peplomers of most enveloped viruses), and possibly also those capable of attaching to other antigens so closely adjacent to the hemagglutinin that HA is inhibited by steric hindrance. Such superficial antigens are usually type specific. Moreover, the HI test is simple, inexpensive, and rapid. It is the serological procedure of choice for assaying antibodies to any virus that causes hemagglutination.

## Gel Diffusion

Antigen–antibody interactions can be detected by observing precipitation reactions in semisolid gels (reviews by Ouchterlony, 1964; Barron, 1971). Antigen and antibody are placed in adjacent wells cut in a thin layer of agar on a glass slide or Petri dish. The reactants diffuse through the agar at a rate inversely related to their molecular weights. Where antigen and antibody meet in "optimal proportions" a sharp line of precipitate forms in the agar. If several antigens and their corresponding antibodies are present, as in the case of most unpurified preparations of virus and of unabsorbed antiviral sera, each antigen–antibody complex forms a discrete line. If two qualitatively identical preparations of antigen (or antiserum) are placed in adjacent wells and allowed to diffuse toward a common well of antiserum (or antigen), equidistant from them both, the corresponding pairs of precipitin lines will join exactly. This is called the reaction of identity (Plate 2–6).

Gel diffusion is a powerful technique because it permits the simultaneous recognition of all the antibody (or antigen) specificities present in the test mate-

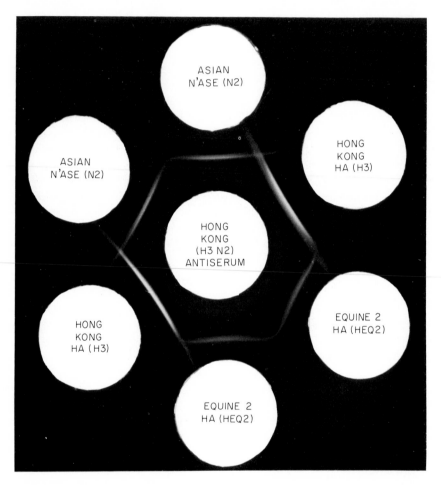

PLATE 2–6. *Gel diffusion test. Example illustrates its use to analyze relationships between envelope antigens of influenza A virus. Center well: antiserum to Hong Kong influenza virus (H3 N2). Peripheral wells, purified antigens: Hong Kong hemagglutinin (H3), Asian neuraminidase (N2), and Equine 2 hemagglutinin (HEq 2). Antiserum to Hong Kong virus contains antibodies to all the antigens tested. Note (1) Two pairs of antigens (N2 and HEq 2) each show fusion of precipitin lines ("reaction of identity"), (2) Neuraminidase (N2) and hemagglutinin (H3) show complete crossing over of precipitin lines ("reaction of complete nonidentity"), and (3) Equine (HEq 2) and Hong Kong (H3) hemagglutinins show partial fusion of lines ("reaction of partial identity") indicating serological cross-reactivity. (Courtesy Dr. R.G. Webster.)*

rial. Furthermore, it is capable of providing a definitive answer overnight or sometimes within a few hours. On the other hand, the method does not lend itself readily to quantitation, i.e., it gives no accurate idea of the "titer" of the unknown antibody or antigen. Moreover, rather high concentrations of both reagents are necessary to ensure the formation of a visible line. It should be appreciated that as a rule purified viruses cannot be used as antigen in gel diffusion, because most virions are too large to diffuse through the agar at a satisfactory rate; crude preparations, on the other hand, usually contain most of the antigens of the virion in "soluble" or small particulate form, although some antibodies will react only with assembled virions and not with isolated coat proteins. This can present something of a technical problem, because if a particular antigen is present in particles of different size, several precipitin lines may result. Many of these drawbacks are overcome by an important new development whereby gel diffusion is rendered much more sensitive and can be quantitated (Schild et al., 1972). Antigen, which can be whole virions but is preferably "soluble," is incorporated throughout the agar; the lower the antigen concentration the higher the sensitivity of the test. Antisera placed in wells diffuses out and forms a precipitate with the antigen. The test is quantitated by measuring the diameter of the circle of precipitate. Sensitivity is increased by staining with thiazine red after washing out unprecipitated antigen (if "soluble"). Even greater sensitivity can be achieved if the antigen is radioactively labeled with $^{125}$I, by autoradiography, or by counting a punched out circle of agar in a scintillation spectrometer.

## Complement Fixation

Antigen–antibody complexes will "fix" complement, and virus–antibody complexes are no exception to the rule (Schmidt, 1969b). Indeed, following attachment of antibody to enveloped viruses, the phospholipase that comprises an integral component of the complement system may actually puncture the viral envelope.

For the complement fixation (CF) test the acute and convalescent sera are heated (56°C, 30 minutes) to inactivate complement, then serially diluted. Usually small plastic trays are used, in order to conserve reagents. Two units of antigen (e.g., a crude preparation of live or inactivated virus) are then added to each serum dilution together with two units of complement, derived from guinea pig serum. The reagents are allowed to interact at 4°C overnight (or at ambient temperature for a shorter interval) to allow complement to become "fixed." Sheep erythrocytes, "sensitized" by the addition of rabbit antiserum against them ("hemolysin"), are then added and the trays are incubated at 37°C for about 45 minutes. In those cups where complement has been fixed by the virus–antibody complex, the hemolysin agglutinates the RBC but cannot lyse them; where complement is still available, the RBC are lysed (Plate 2–7).

To ensure that the test is working properly, controls must be set up to exclude the following:

1. The serum may be "anticomplementary," i.e., it may fix complement even in the absence of viral antigen. This is most commonly due to hemolysis or bac-

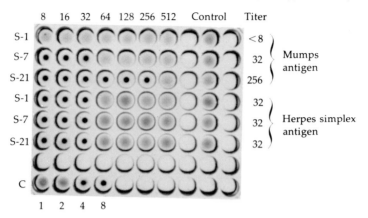

PLATE 2–7. *Complement fixation test. (Titers are expressed as reciprocals of dilutions.) The example illustrates the results of examination of three consecutive serum samples from a patient with aseptic meningitis complicated by herpes simplex infection of the lip (S-1, S-7, and S-21, serum taken, respectively, on day 1, 7, and 21 following admission to hospital). Following heating at 56°C for 30 minutes to inactivate complement, each serum was diluted in twofold steps from 1/8 to 1/512. A standard dose of antigen (inactivated mumps or herpes simplex virus) and complement (two hemolytic units) was added to each cup, and allowed to stand at 4°C overnight. Sheep erythrocytes, "sensitized" by addition of "hemolysin" (rabbit antiserum against sheep erythrocytes), were then added, and the tray incubated at 37°C for 45 minutes. Where complement has been fixed, there is no lysis, and the red blood cells have sedimented to the bottom of the cup. Titration of the complement used in the test is shown in the lowest row (C). Interpretation: (1) The rising titer of antibody against mumps antigen confirms the diagnosis of mumps meningitis. (2) The unchanged titer of antibody against herpes simplex antigen indicates that the herpes labialis represented a recrudescence of a previously existing infection. (Courtesy I. Jack.)*

terial contamination in serum that has been improperly collected or stored, or to the high lipid content of serum taken too soon after a fatty meal. The anticomplementary activity can sometimes be removed by heating, by absorption with kaolin, or by absorption with excess complement followed by heating.

2. The viral antigen preparation may be anticomplementary, i.e., it fixes complement even in the absence of antiviral serum. This does not commonly occur, except with mouse brain extracts (togaviruses). The offending lipids can be removed with acetone or fluorocarbon.

3. The viral antigen preparation may fix complement "nonspecifically" as a result of interaction between cellular antigens and anticellular antibodies in the serum. Such problems can be avoided by proper care in the selection of host species for the growth of virus and the preparation of antisera, respectively.

4. On the other hand, anticomplementary activity of serum may indicate the presence of immune complexes in the circulation, which may bind complement *in vivo*. Since such antigen–antibody complexes may play an important role in pathogenesis (see Chapter 10) it is necessary to differentiate their presence from nonspecific anticomplementary activity of the types just mentioned.

The crude preparations of virus (e.g., cell culture supernatants) employed as antigen in most CF tests contain besides virions, "soluble" antigens, corresponding to unassembled components of the virion and also virus-coded nonvirion proteins. Antibodies to all these antigens may be present in convalescent sera and will register by CF. Accordingly, the CF test, as usually performed, will demonstrate substantial "crossing" between serotypes with a given group of viruses if those serotypes share common group ("family") antigens (e.g., adenoviruses). In this sense, it is not a highly specific test, since it does not always allow one to discriminate between antibodies to different serotypes. It is also not a particularly sensitive test, in that antibody titers determined by CF are lower than those obtained in neutralization or HI tests on the same serum. Yet, this is the test of choice for the preliminary screening of a serum for antibody, when one has little idea of even the group to which the causative virus belongs. For example, a single CF test using any adenovirus serotype as antigen will detect serum antibodies provoked by any other adenovirus serotype. Once "group-specific" antibodies have been determined in this way, the neutralization or HI test can be invoked to identify "type-specific" antibodies, should that information be required.

## Immunofluorescence

Over 30 years ago, Coons (Coons et al., 1941) brought to virology a technique that has been of great benefit to research, and is now being extensively used as a diagnostic aid. "Fluorescent antibody" is specific immunoglobulin which has been "tagged" with a dye such as fluorescein or rhodamine that fluoresces on exposure to ultraviolet or blue light. The chemical process of conjugation can be accomplished as a simple laboratory routine or fluorescein-conjugated immunoglobulins are commercially available. There are three main variants of this technique (reviews, Fraser, 1969; Vogt, 1969b).

**Direct Immunofluorescence** (A–af).  Viral antigen (A) (e.g., in the form of an acetone-fixed, virus-infected cell monolayer on a cover slip) is exposed to fluorescein-tagged antiviral serum (af). Excess af is washed away and the cells are inspected microscopically using a powerful light source (xenon or mercury vapor lamp) from which all but the light of short wavelength has been filtered out. Additional filters in the eyepieces, in turn, absorb all the blue and ultraviolet light, so that the specimen appears black except for those areas from which fluorescein is emitting greenish-yellow light (Plate 2–8).

**Indirect Immunofluorescence** (A–a–gf).  This system, sometimes known as the "sandwich" technique, differs in that the antiviral antibody (a) is untagged, but fills the role of the "meat in the sandwich." It binds to antigen (A) and also to a fluorescein-conjugated antigammaglobulin (gf), which is subsequently added. For example, if the antiviral serum were a human convalescent serum, then it would be appropriate for g to be antibody globulin made in a goat or a rabbit by injecting normal human globulin. The indirect technique has the advantages of greater sensitivity, and more important, in diagnostic virology, of requiring

PLATE 2–8. *Indirect immunofluorescence test used for determining the site of as-sembly of components of influenza virus. Antibody against the NP (nucleocapsid) anti-gen shows nuclear accumulation at 4 hours after infection of chick cells. Guinea pig antiserum to NP antigen; fluorescein-conjugated rabbit anti-guinea pig IgG. (Courtesy Dr. N. Dimmock.)*

only a single tagged reagent, e.g., fluorescein-conjugated goat antihuman gam-maglobulin, with which to test the interaction between any antigen and the corresponding human antibody.

**Complement Immunofluorescence** (A–a–C–gf).   Here complement is bound to the antigen–antibody complex, then fluorescein-conjugated anticomplement immunoglobulin is added. The system is not without its difficulties, and is not used as extensively as the other two methods.

Immunofluorescence has two great advantages over other serological tech-niques. For the research worker, it has provided a method for determining which cells in an organ contain virus, and at another level, the location of particular viral antigens in the infected cell. Although problems of nonspecific fluorescence have delayed its widespread acceptance in the diagnostic laboratory, immuno-fluorescence has great potential for speeding up viral identification, whether it is used directly on biopsy or autopsy material, or after the growth of viruses in cell culture.

### Radioimmunoassay

The availability of carrier-free preparations of radioisotopes, especially of iodine ($^{125}$I or $^{131}$I), which can be readily coupled to the tyrosine residues of pro-teins, has led to the development of radioimmunoassays for analyzing either antigens or antibodies. These procedures are very sensitive, e.g., the technique of radioautography of cells labeled with radioactive [125]I antigen or antibody is over a thousand times more sensitive than is the use of fluorescent antigen or antibody. Radioimmunoassays are also very accurate and simple to perform. One of the two major reagents, either antigen or antibody, is labeled with isotope so that it can be followed by standard procedures, such as radioautography or counting in a scintillation spectrometer. There are three major requirements: (a) One of the reagents, antigen or antibody, should be available in pure form. (b) The extent of substitution of the protein with iodide should be kept low (less than two atoms of iodide per protein molecule of about 40,000 daltons), otherwise the serological properties of the protein may be modified. (c) The process of iodination should not change the serological properties.

Two principal methods for iodinating proteins are available. In the first, iodide added to the protein is oxidized by adding an oxidant, usually chloramine-T (Greenwood et al., 1963). This is efficient but suffers from the disadvantage that sometimes methionine residues in proteins are readily oxidized and this may change the serological properties of the protein. In the second method, iodide added to the protein is oxidized by incorporating an enzyme, lactoperoxidase, into the reaction mixture (Marchalonis, 1969). This technique is more specific and less damaging to the antigen but not quite as efficient at causing substitution of the protein by iodide.

Two commonly used types of radioimmunoassay are radioimmunoelectrophoresis and radioimmunodiffusion (Salzman and Moss, 1969; Yagi, 1971). In both procedures, use is made of the knowledge that globulins with antibody activity may react with antigen even after their precipitation by antisera previously prepared against the globulin. For example, if one wishes to know what class of antibody molecules which specifically react with a given antigen are present in serum, the following approach could be used. A sample of the serum is subjected in the normal way to immunoelectrophoresis in agar gels on glass slides. Antiglobulin serum is then diffused into the gel so that arcs of precipitation occur, corresponding to the different classes of antibody. The slides containing the agar and reagents are thoroughly washed and the radiolabeled antigen is diffused into the gel. The slides are again thoroughly washed, stained, and subjected to radioautography. Identification of radioactive arcs (and hence specific antibody) is made by superimposing the film bearing the autoradiograph over the stained slide.

While useful, this procedure does not readily allow a quantitative estimate of the amount of specific antibody (or antigen). This may be achieved in several ways (Horwitz and Scharff, 1969b), one of which is as follows (Wistar, 1968). Aliquots of [125]I-labeled antigen (or antibody) are mixed with serial dilutions of the antiserum (or antigen). An appropriate amount of antiglobulin antibody is added, the mixture incubated for 3 hours at $37°C$, and allowed to stand for 16 hours at $4°C$. The mixtures are then centrifuged so that complexes of antigen, antibody, and antiglobulin antibody, but not free antigen or antibody, are sedimented. The radioactivity in the washed sediments is then estimated in the scintillation spectrometer.

## CHAPTER 3

# The Structure and Chemistry of the Virion: A Systematic Survey

## INTRODUCTION

In Chapter 1 we outlined the principles underlying the structure of virus particles and described general features of their chemical composition, as a preliminary to describing the classification of animal viruses that is used in this book. Virions of the different genera differ greatly in size and shape (see Fig. 1–3) and in chemical composition. In this chapter, we present brief descriptions of the chemical composition and structure of representative viruses of currently accepted genera. Since the first edition of this book was published, the wide exploitation of polyacrylamide gel electrophoresis and molecular hybridization has yielded much new information about viral proteins and nucleic acids, which has made it possible to relate chemical composition and viral structure in a meaningful way. Accordingly, we now bring the two together in a single chapter.

Because of the large amount of information available this chapter is unavoidably a long one. Even so, some physicochemical features of viruses that are very important to special groups of virologists, e.g., antigenic structure and serology, with which medical and veterinary diagnostic virologists are vitally concerned, have been accorded brief treatment. Readers interested in these facets of virology are referred to appropriate textbooks and review articles on diagnostic virology.

## PAPOVAVIRIDAE

The two genera, *Polyomavirus* and *Papillomavirus*, that compose this family consist of viruses that share many structural and chemical properties (see Table 1–3) but differ in size.

The most important viruses of genus *Polyomavirus* are polyoma, which is a virus of mice (review, Gross, 1970) and simian virus 40 (SV40), which was discovered by Sweet and Hilleman (1960) in cultures of monkey kidney cells. Although they are antigenically distinct, both viruses are oncogenic for rodents and both cause transformation of cells in culture (see Chapter 13). The virions are about 45 nm in diameter, have a sedimentation constant of 240 S, and contain about $3 \times 10^6$ daltons of DNA, that is, about 12% of the particles by weight (Murakami *et al.*, 1968). The remaining 88% is protein; there are no detectable lipids or carbohydrates in the virions.

The genus *Papillomavirus*, which includes the Shope rabbit papilloma virus and the virus of human warts, is distinguished by larger virions, 55 nm in diameter, with a sedimentation coefficient of 300 S and a density in CsCl of 1.34 g/cm$^3$. They contain a nucleic acid molecule of $5 \times 10^6$ daltons molecular weight.

## Structure

Most unfractionated preparations of papovaviruses contain both "full" and "empty" spherical particles (Wildy *et al.*, 1960a; Plate 3–1) as well as a variety of particles of smaller diameter than normal virions, or of tubular structure (Finch and Klug, 1965; Mattern *et al.*, 1967). Full particles have a buoyant density in cesium chloride of 1.32 g/cm$^3$ and contain DNA and all the polypeptides; empty particles have a buoyant density of 1.29 and lack DNA and the three smallest polypeptides (Crawford *et al.*, 1962; Frearson and Crawford, 1972).

Klug and Finch (1968) made an exhaustive study of the capsid structure of the viruses of human warts (Klug and Finch, 1965), rabbit papilloma (Finch and Klug, 1965), and other papovaviruses including polyoma virus (Klug, 1965). They concluded that the capsid of Papovaviridae has a skew icosahedral structure with a triangulation number $T = 7$ and 72 capsomers. Human warts virus (Klug and Finch, 1965) and SV40 (Anderer *et al.*, 1968) have a right-handed surface lattice, rabbit papilloma virus a left-handed lattice. The capsomers appear to be squat hollow cylinders about 7.5 nm long in polyoma virus (Breese, 1964) and 10 nm long in papilloma virus (Howatson and Crawford, 1963).

## Serological Reactions

*Polyomavirus.* SV40 and polyoma virus show no serological cross-reactivity, but antigenic variants of polyoma virus (Hare, 1967) and SV40 (Ozer *et al.*, 1969)

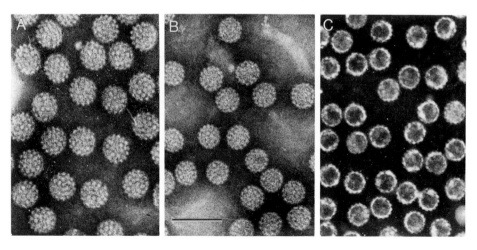

PLATE 3–1. *Papovaviridae. (bar = 100 nm). (A) Human wart virus (genus Papillomavirus), full particles. (B) Mouse polyoma virus (genus Polyomavirus), full particles. (C) Mouse polyoma virus, empty particles. (Courtesy Dr. E.A. Follett.)*

have been described. Some but not all isolates of a human polyoma virus obtained from cases of progressive multifocal leukoencephalopathy (see Chapter 12) cross-react with SV40.

Polyoma virus, but not SV40, agglutinates red blood cells of guinea pigs and other species at 4°C, by attaching to neuraminidase-susceptible receptors (Hartley *et al.*, 1959).

*Papillomavirus.* Canine, human, bovine, and rabbit papilloma viruses fail to cross-react in immunodiffusion tests (Le Bouvier *et al.*, 1966).

## Proteins

Both polyoma virus and SV40 particles contain six or seven different types of polypeptide chains (Table 3–1). Most of the protein (70–90%) of both viruses consists of a polypeptide with a molecular weight of approximately 45,000 daltons, with a blocked N-terminal amino acid (Murakami *et al.*, 1968). Tryptic peptide maps show that the major capsid proteins of SV40 and polyoma virus are not related and that the minor polypeptides of the particles, which have lower

TABLE 3–1

*Properties of Virions of Papovaviridae* [a]

| | Polyomavirus | | |
| | POLYOMA VIRUS | SV40 | *Papillomavirus* |
|---|---|---|---|
| Diameter (nm) | 45 | 45 | 55 |
| Symmetry | Icosahedral | Icosahedral | Icosahedral |
| Triangulation number | 7 Levo | 7 Dextro | 7 Levo (rabbit) |
| | | | 7 Dextro (human) |
| Structural subunits | 420 | 420 | 420 |
| Capsomers | 72 | 72 | 72 |
| DNA content (% w/w) | 12.3 | 12.5 | 10 |
| DNA, molecular weight (daltons) | $3 \times 10^6$ | $3 \times 10^6$ | $5 \times 10^6$ |
| Percentage G + C [b] | 48 — 49 | 41 | 48–49 (rabbit) |
| | | | 41 (human) |
| | | | 43 (canine) |
| | | | 45.5 (bovine) |
| Polypeptides,[c] molecular weight   1. | 47,000 | 46,000 | |
| (daltons)   2. | 36,000 | 42,000 | |
| 3. | 33,000 | 30,000 | |
| 4. | 23,000 | | |
| 5. | 15,000 | 15,000 | |
| 6. | 14,000 | 14,000 | |
| 7. | 12,000 | 12,000 | |

[a] Modified from Tooze (1973).

[b] The G + C contents of the various DNA's are from J. Smith *et al.* (1960) for polyoma virus; Crawford and Black (1964) for SV40; Watson and Littlefield (1960) for rabbit papilloma; and Crawford and Crawford (1963) for human, canine, and bovine papilloma viruses.

[c] The molecular weights of the polypeptides are those of Hirt (unpublished). Similar results have been obtained by Girard *et al.* (1970), Estes *et al.* (1971), Barban and Goor (1971), Roblin *et al.* (1971), and Hirt and Gesteland (1971).

molecular weights, are unrelated to the major capsid protein (B. Hirt and R. Gesteland, unpublished results). The exact location of each of the polypeptides within the virus particles is not known, but because the three smallest are relatively rich in basic amino acids, remain tightly attached to the viral DNA when the virions are dissociated in alkali (Anderer *et al.*, 1968; Estes *et al.*, 1971), and are not found in "empty" particles (Frearson and Crawford, 1972) it seems clear that they are internal. Even though these internal proteins are integral parts of the virion, two lines of evidence suggest that they are not coded by viral DNA. First, comparison of the tryptic fingerprints of the proteins of purified polyoma-virus and histone fractions of uninfected host cells shows correspondence of several peptides, and second, if cells are labeled with radioactive lysine before infection, the three small virus-associated polypeptides are labeled to a much greater extent than the capsid proteins (Frearson and Crawford, 1972). The most likely interpretation of these results is that proteins which are present in host cells before infection are incorporated directly into virus particles. Presumably the other three or four polypeptides found in the virions are virus-coded capsid proteins.

Limited information is available about the proteins of papillomaviruses, and the size of the polypeptides that make up the virions is unknown. The amino acid composition of Shope rabbit papilloma virus has been determined (Knight, 1950); there is an acetylated N-terminal amino acid at one end and a C-terminal threo-nine at the other of what presumably is the main coat protein (Kass and Knight, 1965).

## Nucleic Acid

*Polyomavirus.* Although polyoma virus and SV40 are similar morphologi-cally, their DNA genomes have different base compositions [polyoma virus: 48% G + C (J. Smith *et al.*, 1960); SV40: 41% G + C (Crawford and Black, 1964)] and do not hybridize with one another (Winocour, 1965). However, the DNA mole-cules extracted from the viruses have virtually identical physical properties. In any preparation of polyoma virus or SV40 DNA, most of the molecules are double-stranded, circular and superhelical, with a molecular weight of about $3 \times 10^6$ daltons (see Fig. 3–1 and Plate 3–2). These closed circular molecules are known as component I. Because each strand of the double helix is closed upon itself and is base-paired with its partner in the usual Watson-Crick fashion, the two strands are topologically joined and cannot be separated by conditions, such as high pH or temperature, which destroy hydrogen bonds. Consequently, component I DNA shows sedimentation properties which were at first rather puzzling, but for which satisfying explanations are now available (reviews by Vinograd and Lebowitz, 1966; Crawford, 1969, and see Fig. 3–1). In neutral conditions, the DNA molecules have a sedimentation coefficient of about 21 S; at pH 12.5 or higher, all the base pairs are disrupted without the strands sepa-rating and the molecules collapse into dense supercoils, which sediment at 53 S (Vinograd *et al.*, 1965).

SV40 and polyoma virus DNA's contain about 5000 base pairs. As there are about ten base pairs per turn of the Watson-Crick helix, there must be about

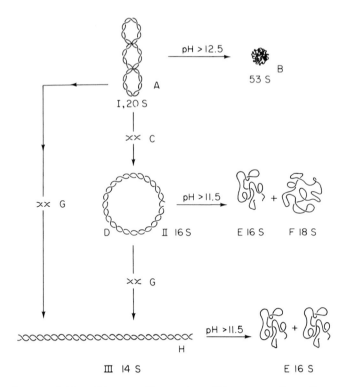

Fig. 3–1. *Papovaviridae. Diagram illustrating different configurations of the DNA molecule of polyoma virus and the sedimentation coefficients associated with the different forms. (A) Component I (20 S), a twisted cyclic double-stranded molecule, shown here with three right-handed twists (actually 15–20), is the form found in the virion. At pH >12.5, it assumes a different configuration (B), with a sedimentation coefficient of 53 S. A break in one strand (C) converts component I to component II [(D) 16 S] by undoing the twists. At pH >11.5 component II dissociates into two single-stranded molecules: (E) 16 S, linear and (F) 18 S, cyclic. A break (G) in the second strand, or in both strands of (A), produces component III [(H) 14 S], a linear double-stranded molecule that dissociates into two linear single-stranded molecules [(E) 16 S] at pH >11.5. (Modified from Vinograd and Liebowitz, 1966.)*

500 turns in component I DNA. For some reason as yet unknown, but presumably connected with the mechanism of replication of the DNA, the component I DNA from both polyoma virus and SV40 contains fewer turns than usual at the moment of completion of the last phosphodiester bond. Consequently, the molecules are under strain, which is relieved by building 15–20 right-handed superhelical turns into each molecule (Vinograd *et al.,* 1965; Bauer and Vinograd, 1968). When a single phosphodiester bond in one of the strands of component I DNA is broken, the superhelical turns are lost immediately, because free rotation around the phosphodiester bond opposite the break is possible and the deficiency of Watson-Crick turns can be rapidly corrected. Some "nicked" molecules (component II) are always found in preparations of viral DNA. These two components of viral DNA can be separated by two methods. First, component II

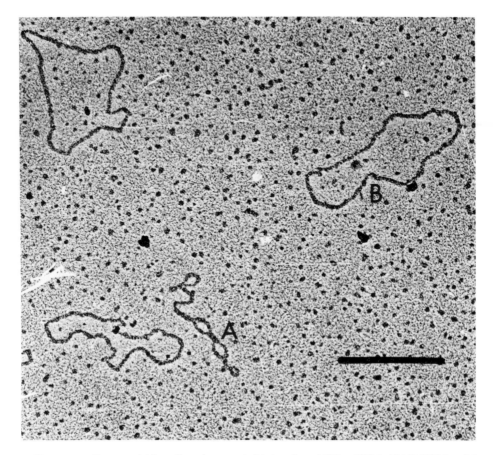

PLATE 3–2. *Papovaviridae. (bar = 0.5μm). Molecules of SV40 DNA. SV40 DNA exists in two major forms. When it is isolated from the virus particles most of the DNA occurs in the configuration shown in A—as double-stranded closed circular molecules containing superhelical twists. If one of the DNA strands is broken, the superhelical twists are relieved and the molecule assumes a relaxed circular configuration (B). (Courtesy Dr. P. Sharp.)*

is less compact than component I, and it sediments more slowly than the closed circular form in neutral conditions (see Fig 3–1), so that the two forms can be resolved by sedimentation velocity centrifugation. Second, both components I and II bind intercalating dyes such as ethidium bromide or propidium diiodide; because of its closed circular structure, component I cannot bind as many molecules of the dyes per base pair as component II, and as a consequence the two forms of viral DNA have different buoyant densities when centrifuged to equilibrium in gradients of cesium chloride–ethidium bromide (Radloff *et al.*, 1967) or cesium chloride–propidium diiodide (Hudson *et al.*, 1969) (for preparative methods see review by Pagano and Hutchison, 1972). Like component I, component II DNA carries both infectivity and transforming ability (Pagano and Hutchison, 1972).

Finally, it has been known for some time (reviews, Crawford, 1969; Wino-cour, 1969) that many preparations of polyoma virus DNA and some preparations of SV40 DNA, depending on the host cells used to grow the virus (Ritzi and Levine, 1970; Basilico and Burstin, 1971), contain linear pieces of DNA. The source of some of these molecules may be component I DNA which has suffered a double-strand break; the majority, however, are pieces of host DNA which also weigh about $3 \times 10^6$ daltons (Michel *et al.*, 1967; Trilling and Axelrod, 1972). It is not known how these fragments become excised from the host genome and packed into viral coats (pseudovirions).

*Papillomavirus.* The DNA's of papilloma viruses have molecular weights ranging from $4.0 \times 10^6$ to $5.3 \times 10^6$ daltons and make up 12% of the weight of the particles (Crawford, 1969). They have the same general structure as that of polyoma virus DNA's, i.e., they are double-stranded, superhelical, and circular and can be converted to more slowly sedimenting forms by "nicking" with DNase (review, Crawford, 1969). The DNA is infectious (Ito and Evans, 1961; Kass and Knight, 1965) and has a nearest neighbor pattern similar to that of host DNA (Subak-Sharpe *et al.*, 1966b; Morrison *et al.*, 1967). There is no homology detectable by hybridization between human and rabbit papilloma DNA's or between polyoma virus and rabbit papilloma DNA's (review by Crawford, 1969).

**Defective Virions of SV40 and Polyoma Virus.**   In addition to empty particles containing no DNA (Crawford *et al.*, 1962), preparations of polyoma viruses may contain particles with reduced amounts of DNA. When polyoma virus or SV40 is passed repeatedly at high multiplicities, the yield of infectious particles decreases markedly, whereas the total yield is only slightly reduced (Uchida *et al.*, 1966; Blackstein *et al.*, 1969). The defective particles are more heterogeneous in density than those of fully infectious virus, and their circular DNA is from 10 to 50% shorter than the DNA from infectious particles (Yoshike, 1968; Thorne *et al.*, 1968). Even though the defective particles are not infectious, they retain some biological activity since they can induce the formation of virus-specific antigens in infected cells and can cause tumors when injected into susceptible animals (Sauer *et al.*, 1967; Uchida and Watanabe, 1968).

The mechanism by which part of the viral genome is lost is unknown, but it may be the result of a process which causes not only deletions of viral DNA, but also other large-scale changes in viral DNA structure. The evidence for such a process comes from heteroduplex mapping and hybridization experiments; when the DNA extracted from SV40 grown at high multiplicity is examined by heteroduplex mapping, it is clear that as well as deletions, many inversions and substitutions have occurred (Tai *et al.*, 1972). While some of the substitutions may reflect rearrangements within the viral genome, others are due to recombination events between viral and host DNA which result in the incorporation of host sequences into the viral genome (Lavi and Winocour, 1972). Virus stocks grown at low multiplicities do not show these changes. The meaning of all this is not clear. Certainly, it is enough to shake our confidence in the defined nature of SV40 DNA and causes us to be more fastidious in the ways we use to grow virus. However, knowledge of the existence of the phenomenon may provide us with ways to explain how these two small viruses cause malignant transformation, one manifestation of which is the permanent integration of viral genomes into host DNA.

## ADENOVIRUS

This genus includes many species from human, simian, bovine, and other mammalian sources, and several from birds (review, Norrby, 1971). All have a very characteristic icosahedral structure. They lack envelopes, and contain no lipid or glycoprotein.

### Structure

The capsid is about 70 nm across, with 252 capsomers of approximately spherical shape ($T = 25$) (see Plate 3–3). Six capsomers form the 42 nm long edges of the equilateral triangular faces, and these are shared by neighboring faces.

The 240 capsomers with six neighbors, that occupy the faces and edges of each triangular facet, are designated *hexons* (Ginsberg *et al.*, 1966). Electron microscopic examination of isolated hexons (Plate 3–4) suggests that they are hollow prolate ellipsoids, with an inner diameter of 2.5 nm, an outer diameter of 8 to 9.5 nm., and a length of 8.8 to 11.5 nm (Pettersson *et al.*, 1967). Purified hexon protein has been crystallized (Pereira *et al.*, 1968; Plate 3–4). X-ray diffraction studies of crystallized hexon protein of adenovirus type 2 (Franklin *et al.*, 1971) and type 5 (Cornick *et al.*, 1971) show that the crystals belong to the cubic system and that there are three crystallographic units per hexon, each with a molecular weight of 110,000–120,000 daltons (review, Franklin *et al.*, 1972).

The twelve capsomers at the vertices of the icosahedron, each with five neighbors (*pentons*), are structurally complex, consisting of a round head (the *penton base*) embedded in the capsid, with a long rod 2 nm in diameter (the *fiber*) projecting from it (Plate 3–5). Isolated penton base subunits appear to have a globular structure [outer diameter, 8 nm; with a pentagonal outline when seen end-on (Pettersson and Höglund, 1969)].

The fibers of different human adenovirus serotypes vary considerably in length, measurements ranging from 10 to 31 nm being recorded for eleven different serotypes (Norrby, 1969). Serotypes belonging to the same subgroups appear to have fibers of equal length (Table 3–2). All have a "knob" about 4 nm in diameter at the outer end. On the other hand, the avian adenovirus (CELO virus) has a characteristic double fiber structure, with two knobbed fibers 42.5 and 8.5 nm long, respectively (Laver *et al.*, 1971). The fiber protein of adenovirus type 5 has been crystallized (Mautner and Pereira, 1971), forming needle-shaped crystallites with fine structure visible down to 3.5 nm in electron micrographs (Plate 3–5). Optical diffraction studies indicate a highly ordered structure.

Numerous methods have been employed to disrupt the adenovirion sufficiently gently to release groups of capsomers and cores relatively undamaged.

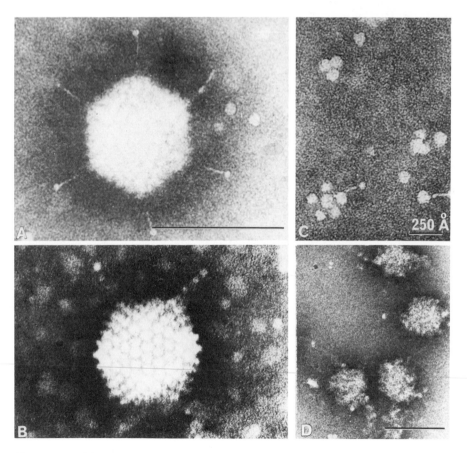

PLATE 3–3. *Adenovirus. Negatively stained preparations of adenovirus type 5 (black bars = 100 nm). (A) Virion showing the fibers projecting from the vertices of the icosahedron. (B) Virion showing the icosahedral array of capsomers. The capsomers at the vertices are surrounded by five nearest neighbors, all the others by six. (C) Clusters of capsomers from disrupted virion. Penton base with attached fiber is surrounded by a group of five hexons. (D) Purified cores released from acetone-disrupted virions with some contaminating groups of nine hexons and single capsomers. (A, B, and C from Valentine and Pereira, 1965; D from Laver et al., 1968, courtesy Drs. Pereira and Laver.)*

Pentons may be released by dialyzing at pH 6.0 to 6.6 (Laver *et al.*, 1969), then hexons by freezing and thawing (Prage *et al.*, 1970). Pyridine (10%), or heating at 56°C in the presence or absence of deoxycholate (Russell *et al.*, 1971), tends to release hexons in groups of nine from the faces of the icosahedron, and occasionally groups of five hexons encircling a solitary penton from one of the vertices of the particle. Continued treatment with any of these agents, or with 5 M urea, formamide, acetone, or at pH 10.5, liberates all the surface capsomers

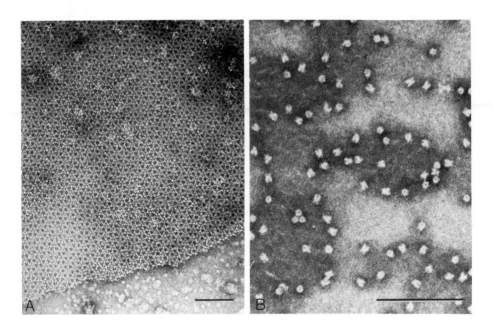

PLATE 3–4. *Adenovirus. Hexons of adenovirus type 5 (bars = 100 nm). (A) "Two-dimensional" crystal (electron micrograph by Mr. J.V. Heather). (B) Individual hexons, seen both in end-view and side-view (courtesy Dr. N.G. Wrigley).*

leaving the viral core (Plate 3–3), which consists of DNA plus two internal proteins, one of which (core protein 2) has a very unusual amino acid composition (see below). Russell *et al.* (1971) designate the complex of DNA and the two core proteins as the *viral core* and the complex of DNA and core protein 2 as the *inner nucleoprotein*.

## Serological Reactions

The antigenic characteristics of the adenovirus capsid proteins have been described in detail by Norrby (1969, 1971) and Wadell (1970). All large polypeptides carry several antigenic determinants, and corresponding polypeptides from two distinct viral serotypes may share long amino acid sequences yet differ in an exposed region of antigenic importance. So it is not surprising to find that sophisticated serological techniques can detect group-specific, type-specific, and even "subgroup-specific" determinants on the hexon polypeptide, group- and subgroup-specific determinants on the penton-base, and subgroup- and type-specific determinants on the fiber. Type-specific neutralizing antibodies presumably attach to fibers and hexons.

Most adenoviruses agglutinate red blood cells, those of the rat or monkey being the most generally useful (review, Norrby, 1971). Rosen (1960) separated human adenoviruses on the basis of their hemagglutinating properties; the sub-

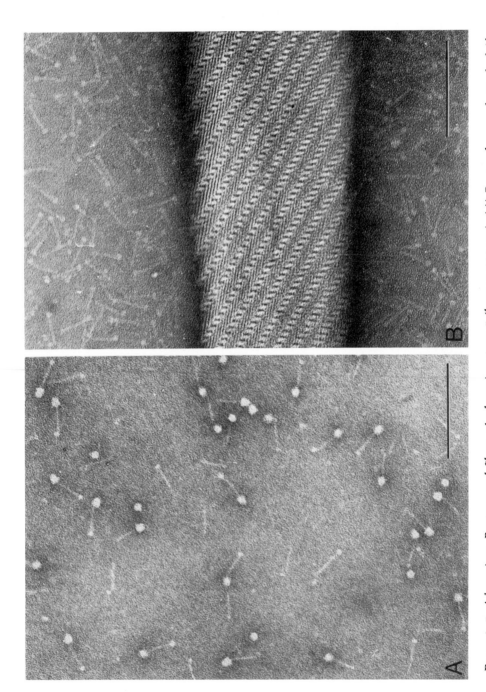

Plate 3–5. *Adenovirus. Pentons and fibers of adenovirus type 5 (bars = 100 nm). (A) Penton base and attached fiber. (B) Crystal of type 5 fiber protein and associated fibers (courtesy Dr. N.G. Wrigley).*

TABLE 3–2

*Classifications of Some Human Adenoviruses Based on T Antigens; Correlation with Other Properties* [a]

| | | | TRANSFOR- | HEMAGGLUTINATION [c] | | | FIBER |
| | | ONCO- | MATION | MONKEY | RAT | G + C | LENGTH |
| SEROTYPES | T ANTIGEN | GENICITY [b] | *in vitro* [b] | CELLS | CELLS | (%) | (NM) |
|---|---|---|---|---|---|---|---|
| 12, 18, 31 | A | High | + | — | ± | 48–49 | ? |
| 3, 4, 7, 11, 14, 16, 21 | B | Low | + | + | — | 49–52 | 9–11 |
| 1, 2, 5, 6 | C | Nil | + | — | ± | 57–59 | 25–30 |
| 9, 10, 13, 15, 17, 19, 26 | D | Nil | + | — | + | 58–59 | 12–19 |
| 8, 22, 23, 24, 27, 29, 30 | ? | Nil | + | — | + | 57–60 | ? |

[a] Cytopathology in permissive cells also shows a general correlation with these groupings.
[b] See Chapter 13.
[c] +, Complete; ±, partial.

groups thus defined shared several other biological properties, and in spite of some exceptions provide the basis of a useful classification (Norrby, 1971). Type-specific hemagglutination-inhibiting antibodies must attach to fibers because only the fibers adsorb to red blood cells. Aggregates of two or more isolated pentons will agglutinate erythrocytes by virtue of the hemagglutinin at the tip of the fibers. A single penton or a single fiber will hemadsorb but not hemagglutinate because it is univalent; such "incomplete hemagglutinin" can be converted into "complete hemagglutinin" by artificial aggregation using antibody against another viral serotype, which will aggregate pentons by binding to group-specific antigenic determinants in the penton base.

Nonstructural antigens (T antigens) are constantly present in transformed cells (see Chapter 13) and appear early and fleetingly during cytolytic infections. Cross-reactivity of T antigens allows subdivision of the human adenoviruses into subgroups of different oncogenicity (Table 3–2) which correspond in general with the subgroups defined by hemagglutinating properties.

## Proteins

Most of the detailed studies on the fine structure and protein composition of adenovirions were originally made on types 2 and 5 but the findings have in general been confirmed with other adenoviruses. The hexon, penton base, and fiber are composed of three distinct polypeptides with molecular weights of 120,000, 70,000, and 62,000 daltons, respectively, while two or three additional arginine-rich (but tryptophan-containing) proteins of molecular weight 46,000, 24,000, and 22,000 daltons are located in an inner core that makes up 20% of the mass of the virion (Maizel *et al.*, 1968b). Since each hexon has a molecular weight of about 350,000 daltons as determined by sedimentation coefficient in zonal centrifugation, it was concluded that the capsomer is a polymer constructed of three identical molecules of molecular weight 120,000 daltons. This has recently been elegantly confirmed by X-ray diffraction studies of crystallized hexon protein (review by Franklin *et al.*, 1972). The two or three other polypeptides often found in trace amounts in association with hexons (Maizel *et al.*, 1968b; Laver, 1970) are considered by Pereira and Skehel (1971), but not by Everitt *et al.* (1973), to be products of proteolytic cleavage of the hexon polypeptide.

The penton antigen causes reversible clumping and detachment of cells from glass (Rowe *et al.*, 1958; Valentine and Pereira, 1965). Purified fiber protein has a striking inhibitory effect on cellular synthesis of RNA, DNA, and protein, in both adenovirus infected and uninfected cells (Levine and Ginsberg, 1967).

Russell *et al.* (1971) have examined the cores obtained by heating virions in deoxycholate. They find that "core protein 1" (molecular weight, 46,000 daltons) can be quite readily removed from the "inner nucleoprotein," consisting of DNA complexed with "core protein 2," which with their serotype, 5, runs as a single band of molecular weight 22,000 daltons. Core protein 2 has a very unusual amino acid composition, being extremely rich in arginine (22%), alanine (19%), and glycine (11%), while lacking phenylalanine and having only a single tryptophan residue and an N-terminal alanine (Laver, 1970; Prage and Pettersson, 1971; Russell *et al.*, 1971). Taken together with its ready solubility in distilled water and in acid, these properties are reminiscent of those of the F3 class of arginine-rich histones, though neither histones nor protamines are usually as large.

Endonuclease activity has been identified in several adenovirus serotypes (Burlingham *et al.*, 1971) but is not generally associated with the penton base as previously thought (Burlingham and Doerfler, 1972; J. Williams, unpublished). The enzyme cleaves any native DNA, especially in $G + C$-rich regions.

## Nucleic Acid

Detailed studies on adenovirus nucleic acids (reviews by Green, 1969, 1970b) have been stimulated by the oncogenic potential of these viruses. All mammalian adenoviruses thus far examined contain a single molecule of linear double-stranded DNA of molecular weight 20–25 million daltons, representing 11.6–13.5% of the mass of the virion. The $G + C$ content varies from 48% to nearly 60%. Among human adenoviruses the more highly oncogenic have a lower $G + C$ content (Piña and Green, 1965); the reverse is true for simian adenoviruses (Goodheart, 1971).

DNA–DNA annealing experiments demonstrated that all human and two simian adenoviruses share 10–25% of their nucleotide sequences, but pairs of serotypes within each subgroup show over 80% homology (Green, 1969, 1970b). Electron microscopic studies of heteroduplex formation gave similar results but showed much closer homologies than 10–25% in inter-subgroup comparisons (Sharp, 1973).

Seeking an analogy with temperate bacteriophages that might help to explain the mechanism of integration of adenovirus DNA's, several workers looked without success for evidence of classic circular permutation and terminal redundancy ("sticky ends") in the DNA's of several human adenoviruses (Green *et al.*, 1967; Green, 1970b; Doerfler *et al.*, 1972). However, Garon *et al.* (1972) and Wolfson and Dressler (1972) have now reported that melted adenovirus DNA forms single-stranded circles of the anti-parallel ("panhandle") type, which they interpret to mean that the base sequence at one end of each strand is an inverted terminal repetition of the sequence at the other end.

The DNA of the avian adenovirus, CELO virus, is substantially larger than that of any mammalian adenoviruses, with a molecular weight of 29 million daltons and a G + C content of 54% (Laver *et al.*, 1971). Denaturation studies revealed that mature CELO virus DNA molecules, like the DNA's of human adenoviruses, contain regions of differing base composition, but that they have neither complementary single-chain nor duplex terminal repetitions. Further, there are no regular single-strand interruptions in the molecule and the sequences are unique and not permuted (Younghusband and Bellett, 1971).

Burnett and Harrington (1968a, b) reported that the isolated DNA of the simian adenovirus SA7 had both transforming and infectious capacities. Early results with the DNA from human adenoviruses were negative, but Russell *et al.* (1971) have shown that the "inner nucleoprotein" of adenovirus 5 is infectious. More recently, Nicholson and McAllister (1972) reported that highly purified adenovirus 1 DNA is infectious.

## HERPESVIRUS

Members of this genus have been recognized primarily by the architecture of the virion: the herpesviruses are large enveloped DNA viruses 150–170 nm in diameter, with an icosahedral capsid about 100 nm in diameter. The genus is a large one that includes viruses native to vertebrates of all orders that have been examined (see Chapter 17), but no herpesviruses have been found in insects, plants, or bacteria. The description that follows will be based almost entirely on herpes simplex virus type 1, the best studied member of the genus (review by Roizman *et al.*, 1973).

### Structure

Herpesvirions have a complex and unique structure. There is an outer lipoprotein envelope largely derived from altered host–cell nuclear membrane, and a less certainly defined inner membrane, which surrounds an icosahedral capsid with 162 capsomers. Within this there are two much less well-defined "capsids" surrounding a nucleoprotein core that has no clearly visible structural symmetry.

The capsomers of the outer icosahedral capsid ($T = 16$) are arranged as pentamers and hexamers (Plate 3–6). Individually, the capsomers are elongated hollow prisms that in cross-section appear to be hexagonal or (at the vertices) pentagonal, and measure $9.5 \times 12.5$ nm, with an axial hole 4 nm in diameter (Wildy *et al.*, 1960b).

Within this outer capsid, which is the structure seen in standard negatively stained preparations, there appear to be two other capsids, that can be designated

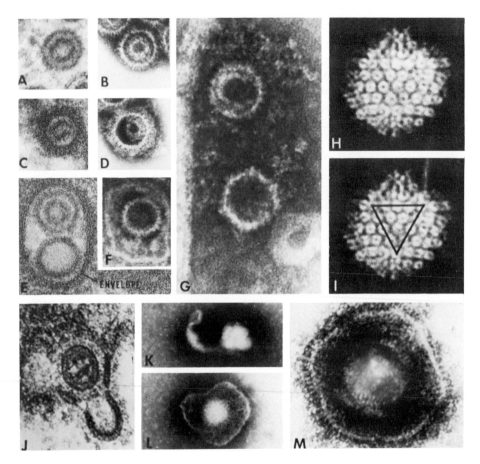

PLATE 3–6. *Herpesvirus. (A–F) From Burkitt lymphoma. (G) Lucké adenocarcinoma of the frog. (H–M) Herpes simplex virus. (A and B) Particles consisting of a core and three concentric capsids. (C and D) The same as (A) and (B) but coated by an amorphous material (inner membrane). (E and F) Enveloped nucleocapsids. (G) Naked nucleocapsids from Lucké adenocarcinoma, the upper nucleocapsid showing details of an internal structure. (H and I) Naked nucleocapsid of herpes simplex virus with a triangular face of the icosahedron outlined. (J) Thin section of a particle, apparently coated with an inner membrane, in the process of acquiring its outer envelope from the nuclear membrane. (K and L) Enveloped particles at the same magnification, one with an intact envelope impermeable to negative stain (K) and one into which the stain has penetrated (L). (M) Enveloped particle penetrated by stain and showing details of the envelope. A, C, E, and J, thin sections. Remainder are negatively stained whole mounts. (From Roizman and Spear, 1973, courtesy Dr. B. Roizman.)*

"middle" and "inner" capsids. The middle capsid has been most clearly demonstrated with a frog herpesvirus (Stackpole and Mizell, 1968), and has an outer diameter of 75 nm and an inner diameter of 45 nm. It encloses the inner capsid, 45 nm in diameter, which in turn encloses a nucleoprotein core 25 nm in diameter (Spring *et al.*, 1968). The multiple shells are best seen in thin sections

(Plate 3–6). Furlong *et al.* (1972) suggest that the core is a toroid that contains DNA wound around the cylindrical structure with a spacing of 4–5 nm. The polyamines that it contains (see below) are sufficient to neutralize 40–60% of the phosphate in the DNA; there appears to be only one structural protein in the core (Gibson and Roizman, 1973).

Unlike the many enveloped RNA viruses whose nucleocapsids develop in the cytoplasm and mature by budding through cytoplasmic membranes, herpesvirus capsids develop in the nucleus and mature as they bud through the nuclear membrane (review, Darlington and Moss, 1969). In fully intact virions that have been recently released from cells, the envelope forms a tight sheath around the capsid (Plate 3–6), but in aged preparations it often becomes larger and irregular, and separates from the capsid. Two concentric membranes can sometimes be distinguished (Plate 3–6). The outer envelope appears to be about 20 nm thick and shows a repeating unit structure sometimes visible as peplomers (Spring *et al.*, 1968). There is some evidence for a lipid-containing electron-lucent inner membrane which Roizman (1969a) suggests may be required for the capsid to acquire an affinity for the nuclear membrane and thus get its outer envelope. Under some conditions, nonenveloped particles occur in the cytoplasm, usually late in infection and probably because of breaks in the nuclear membrane. Such particles may become enveloped at cytoplasmic membranes or through the plasma membrane (Darlington and Moss, 1969).

## Serological Reactions

Virions containing as many proteins as the herpesviruses clearly give rise to a variety of different antibodies in infected animals. Immunodiffusion tests reveal that there is a herpesvirus group antigen associated with the nucleocapsids. This was demonstrated in direct comparisons of herpesviruses from man (herpes simplex virus, EB virus, cytomegalovirus), chicken (Marek's disease virus), and frog (Lucké virus) (Kirkwood *et al.*, 1972). The group antigen probably accounts for reports of cross-reactivity in complement fixation and gel diffusion tests with several other herpesviruses, e.g., herpes simplex and varicella viruses (Schmidt *et al.*, 1969) and between equine rhinopneumonitis and infectious bovine rhinotracheitis viruses. Further studies may reveal subgroup specificies; some such tool will be needed if the present confusing state of classification and nomenclature of herpesviruses (Roizman and de-Thé, 1972) is to be resolved.

Neutralization tests, which depend upon the binding of antibodies to envelope antigens, reveal only closer relationships. By such tests herpes simplex types 1 and 2 are closely related and both are related to B virus (Watson *et al.*, 1967). Nevertheless, herpes simplex viruses types 1 and 2 share only eight of the twelve or so antigens that have been detected by gel diffusion (Wildy, 1972).

## Proteins

It has proved very difficult to purify herpesviruses satisfactorily because of their site of maturation and the nature of the virion.

TABLE 3–3

*The Major Polypeptides of Herpes Simplex Virus* [a]

| DESIGNATION | MOLECULAR WEIGHT (DALTONS) | PERCENTAGE OF VIRION PROTEIN | GLYCO-PROTEIN | PROBABLE STRUCTURAL ROLE |
|---|---|---|---|---|
| II | 110,000 | 25 | — | Main capsomer consists of 9 molecules of II |
| III | 106,000 | 12 | + | Major peplomer of outer envelope |
| IV | 83,000 | 11 | + ⎫ | |
| V | 69,000 | 11 | + ⎬ | Inner envelope |
| VI | 60,000 | 5 | — | Capsid (minor component) |
| VII | 40,000 | 6 | — | Arginine-rich, in core |
| VIII | 32,000 | 2 | — | Capsid (minor component) |

[a] After Becker and Olshevsky (1972).

Becker and Olshevsky (1971, 1972) noted eight main polypeptides in their gels of enveloped herpes simplex virus (Table 3–3). These accounted for over three-quarters of the total virion protein; there were fourteen additional minor components. Three of the major polypeptides, all glycoproteins, were missing from naked virions; two arginine-rich core proteins were absent from virus grown under conditions of arginine starvation. Robinson and Watson (1971) resolved at least eight polypeptides in nonenveloped capsids of highly purified HSV-1. Using a different protocol for purification, Spear and Roizman (1972) reported no less than twenty-four proteins in the whole virion, of which one-half were glycoproteins in the envelope. In view of the problems associated with purification of herpesviruses and the possibility that some "viral polypeptides" may be products of degradation, aggregation, or incomplete glycosylation, confirmation is needed before acceptance of this high figure, particularly for the number of envelope glycoproteins. Nevertheless, contrary to earlier beliefs (Watson and Wildy, 1963), it is unlikely that there are any host proteins in purified virions, although the membranes of infected cells contain both host and viral proteins (Heine *et al.*, 1972).

Herpes simplex virus shows a high minimal requirement for arginine (Y. Becker *et al.*, 1967), which Gibson and Roizman (1971) ascribe to the presence of the polyamines spermine and spermidine in the virion. The spermine is located in the nucleocapsid (probably associated with DNA) while spermidine is found in the envelope. It is not known whether cell- or virus-coded enzymes are responsible for the synthesis of polyamines.

## Lipids

Purified enveloped virions of HSV-1 contain over 20% lipid, all of which is located in the envelope. It is likely that as in other enveloped viruses the lipids are derived from host cell membranes; with HSV-1, the likely origin of such lipids appears to be the inner lamella of the nuclear membrane, but direct studies have yet to be made.

## Nucleic Acid

Herpesviruses contain about 100 million daltons of double-stranded DNA. Direct length measurements in the electron microscope gave a value for HSV-1 of 101 million daltons (Becker *et al.*, 1968), while velocity sedimentation produced a virtually identical figure of 99 million (Kieff *et al.*, 1971). Calculation of the kinetics of reannealing between single strands of fragmented DNA that had been subjected to shearing before denaturation gave a molecular weight of 95 million daltons or slightly more, and revealed that there were no repetitive nucleotide sequences (Frenkel and Roizman, 1971). When HSV-1 DNA was denatured and sedimented in alkaline sucrose gradients, several discrete bands of single-stranded material were obtained, suggesting that the DNA was double-stranded and linear with no cross-linking but with breaks at specific points in the individual strands (Kieff *et al.*, 1971). All the breaks appear to be in one strand, and Frenkel and Roizman (1972b) suggest that HSV-1 DNA molecules exist in four forms, each consisting of one intact strand and one characteristically fragmented strand. Single strand breaks have also been reported to occur in the DNA extracted from Marek's disease virus (L. Lee *et al.*, 1971). One HSV-1 mutant (MP) appears to be a deletion mutant that contains $4 \times 10^6$ daltons less DNA than wild type (Bachenheimer *et al.*, 1972b).

Little or no genetic homology has been found between antigenically unrelated herpesviruses (Bachenheimer *et al.*, 1972a). Even with the two closely related herpes simplex viruses, HSV-1 and HSV-2, good base-pair matching was observed for somewhat less than one-half the molecule, the remainder showing little if any homology (Kieff *et al.*, 1972).

Molecular weight estimates of other herpesvirus DNA's range from 60 (a figure of doubtful reliability) to 120 million daltons for the DNA of Marek's disease virus (L. Lee *et al.*, 1971). There are large differences in the G + C content (Plummer *et al.*, 1969; Roizman and Spear, 1971), figures ranging from 46% for Marek's disease virus to 68 and 74% for herpes simplex and pseudorabies viruses, the highest G + C contents of any vertebrate viruses. There has been one report (Lando and Ryhiner, 1969), as yet unconfirmed, that infectious DNA can be extracted from herpes simplex virus.

## IRIDOVIRUS

This genus comprises a small number of insect viruses (review, Bellett, 1968) and is included here because several otherwise unclassified viruses of vertebrates [African swine fever virus (review, Hess, 1971) and the polyhedral cytoplasmic DNA viruses of amphibia (review, Granoff, 1969)] are chemically and morphologically rather similar. It is likely that eventually the iridoviruses of vertebrates will be recognized as a separate genus within an enlarged family, or even as a separate family.

## Structure

Elegant studies that have been carried out on the structure of the outer capsid of insect iridoviruses (Wrigley, 1969, 1970) will be summarized here as they provide a background for what may be expected with the vertebrate iridoviruses.

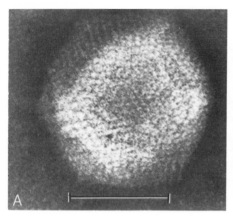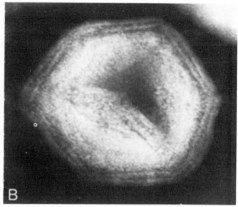

PLATE 3–7. *Iridovirus. Negatively stained virions. (bar = 100 nm). (A) Tipula iridescent virus, after treatment with "Afrin," showing capsomers (Wrigley, 1970). (B) Mature virion of African swine fever virus, from a "cell-spread" preparation, showing outer membranes within which capsomers are just visible (Almeida et al., 1967). (Courtesy Drs. Wrigley and Almeida.)*

The outer shells of *Sericesthis* iridescent virus (SIV) and *Tipula* iridescent virus (TIV) consist of morphological subunits closely packed about 7 nm apart and in hexagonal array (Plate 3–7 and Fig. 3–2). The particles of SIV have a diameter of about 190 nm and an icosahedral edge length of $86.0 \pm 2.7$ nm. By ingenious use of the Goldberg diagram, which categorizes the ways in which spheres can be packed onto the surface of an icosahedron, Wrigley deduced that the outer shell was composed of 1472 subunits ($T = 146$) (Fig. 3–2). There is a core 90 nm in diameter consisting of DNA and protein.

In thin sections of infected cells, both the polyhedral cytoplasmic DNA viruses of amphibia (e.g., FV3) (Darlington *et al.*, 1966) and African swine fever virus (ASFV) (Breese and De Boer, 1966) are seen as regular hexagons about 130 (FV3) or 180 nm (ASFV) across, with two well-defined membranes (Plate 3–8) and a central dense core about 75 nm in diameter. These viruses acquire an envelope as they leave the cell; it has not been determined yet whether this is a true envelope containing virus-specified proteins or merely a pseudoenvelope derived from the plasma membrane of the cell.

The only evidence in the vertebrate iridoviruses of a complex capsomer structure like that found in the insect iridoviruses comes from electron micrographs of "cell-spread" preparations of ASFV, in which somewhat broken particles were found that showed that the outer shell consisted of very large numbers of subunits (Plate 3–7), an appearance closely resembling cell-spread preparations of TIV (Almeida *et al.*, 1967). It is perhaps relevant that elucidation of the capsid structure of the insect iridescent viruses by Wrigley was not possible until the fortuitous discovery that treatment of particles with a nasal decongestant (Afrin) revealed their surface structure. This approach has yet to be applied to the vertebrate iridoviruses.

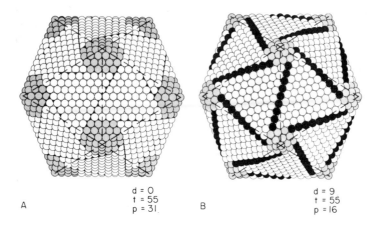

d = 0
t = 55
p = 31

d = 9
t = 55
p = 16

A

B

FIG. 3–2. *Iridovirus. Diagrams illustrating possible structures for the outer capsid of insect iridescent viruses. Trisymmetrons (t) shown in white, disymmetrons (d) in black, pentasymmetrons (p) in gray. (A) Model with 1472 capsomers. (B) Model with 1562 capsomers (Wrigley, 1969, 1970).*

The central core can be released by treatment of purified virions with detergent; cores of FV3 had a density in cesium chloride of 1.365 g/cm³, compared with 1.275 g/cm³ for intact particles (Aubertin *et al.*, 1971).

### Serological Reactions

There is extensive cross-reactivity between different insect iridoviruses (Bellett, 1968), but none between them and the vertebrate iridoviruses, and none have yet been detected between the different vertebrate iridoviruses (Granoff, 1969).

### Proteins

Only preliminary investigations have been made of the proteins of iridoviruses. As expected from their large genome, the virions contain many polypeptides. FV3 contains at least sixteen polypeptides, five of which are firmly associated with the DNA-containing cores that remain after treatment with detergent (Tan and McAuslan, 1971). Furthermore, FV3 has been shown to contain several enzymes. A nucleotide phosphohydrolase (ATPase) is associated with the viral core (Aubertin *et al.*, 1971), as is a DNase active at pH 5.0 (Kang and McAuslan, 1972). Protein kinase, a DNase active at pH 7.5, and a ribonuclease are located within the capsid but outside the core (Kang and McAuslan, 1972).

### Nucleic Acid

The large double-stranded DNA molecules of three iridescent viruses of insects have been analyzed by Bellett and Inman (1967). Comprising 12–16%

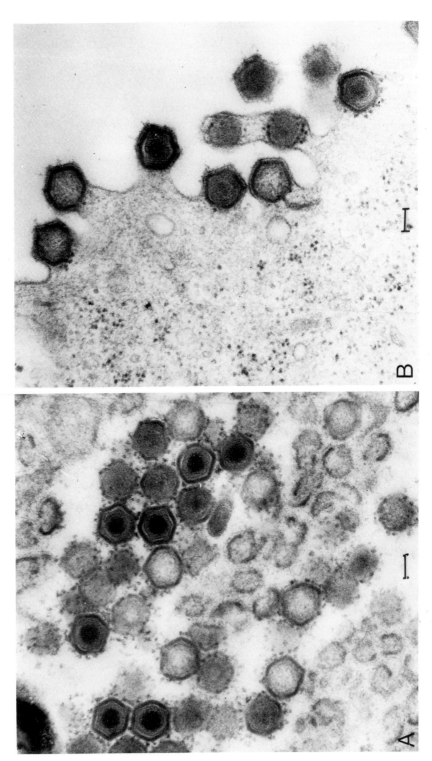

PLATE 3–8. Thin section of pig kidney cell infected with African swine fever virus (bars = 100 nm). (A) Developing and mature virions in cytoplasmic "factory" area. (B) Mature virions being released by budding from plasma membrane. (From Breese and De Boer, 1966, courtesy Dr. S.S. Breese, Jr.)

of the virion mass, the molecular weight has been calculated at 140 million daltons, which is slightly larger than the molecular length of 59 $\mu$m would suggest. A buoyant density of 1.69 and $T_m$ of 82°C indicate a G + C content of around 30% (review, Bellett, 1968).

By contrast, the DNA of FV3, which is 70–80 $\mu$m long and hence has a molecular weight of 137–157 million daltons, has a much higher G + C content of around 58% (Granoff, 1969). W. Smith and McAuslan (1969) estimated the molecular weight to be somewhat lower (130 million daltons). In experiments conducted with RNA's synthesized *in vitro* on viral DNA templates, the DNA of FV3 did not show any homology with DNA's of the insect iridoviruses (Bellett and Fenner, 1968).

## Lipids

The insect iridoviruses contain no essential lipids, although occasionally particles yielded by cultured insect cells (but not in intact insects) have an "envelope" that appears to be acquired from the plasma membrane (Bellett, 1968). On the other hand, both ASFV and FV3 were estimated to contain 14% lipid (Smith and McAuslan, 1969), but no precise measurements of overall chemical composition have been made with purified preparations.

## POXVIRUS

This large genus, which probably merits a higher taxonomic status, includes several subgenera that infect vertebrates (see Table 1–6) and a subgenus that infects insects (*Entomopoxvirus*, review by Bergoin and Dales, 1971). The type species, vaccinia virus, being large, stable, simple to grow in quantity, and fairly readily purified, was well characterized in crude chemical terms long before most other animal viruses, but it is so complex that we still do not know the structural and antigenic role of most of the proteins in the virion.

## Structure

The basic structure of the different species in each subgenus is similar, and viruses of several subgenera, vaccinia, birdpox, and myxoma, for example, are superficially identical in both thin sections and negatively stained preparations. Virions of the orf and sheep pox subgenera are slightly narrower in relation to their length.

**Vaccinia Virus (Subgenus A).** The structure of the virion of vaccinia is illustrated diagrammatically in Fig 3–3. Thin sections and negatively stained electron micrographs show that it consists of a well-defined central core containing the viral DNA, on either side of which there is an oval mass called the lateral body. The core and lateral bodies are enclosed within another well-defined surface membrane, which has a characteristic ridged surface structure (Plate 3–9). Two interconvertible appearances, termed "C" and "M" forms, are seen in negatively stained preparations, depending upon the penetration of the phosphotungstate (Westwood *et al.*, 1964).

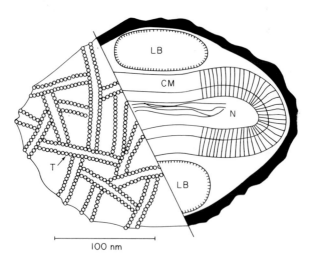

FIG. 3–3. *Poxvirus. Diagram illustrating the structure of vaccinia virus. Left-hand side, surface structure; right-hand side, in section. The viral DNA (N = nucleoid) is contained within the core, probably associated with protein, but the structural arrangement is not known. The core has a complex outer membrane (CM) 9 nm thick, with a regular subunit structure. In the virion, the core assumes a dumbbell-shape because of the large lateral bodies (LB) which are, in turn, enclosed in a protein shell 22 nm thick, the outer part of which is seen by freeze-etching (left-hand side) to consist of irregularly arranged tubules (T), which in turn consist of small globular units.*

The ultrastructure of the cores has been studied by Easterbrook (1966), who obtained them by treating a purified suspension of mature virions with mercaptoethanol and a nonionic detergent. Within the virion, the core is an oval biconcave disc, dumbbell-shaped in cross section, but after release the concavities disappear and the core adopts a rounded brick shape about the same size as a mature virion. Its wall is composed of an inner smooth membrane about 5 nm thick and an outer layer of regularly arranged short cylindrical subunits about 10 nm long and 5 nm in diameter. When the complete virions are negatively stained at alkaline pH (Peters and Müller, 1963) or when sections are stained with uranyl chloride or indium chloride, the core can be seen to contain two or three broad cylindrical elements which may be connected in a tight S-shape. The arrangement of this continuous folded structure that is thought to form the backbone for the viral DNA is seen best in sections of fowlpox virus fixed at pH 3 (Plate 3–10; Hyde and Peters, 1971).

The concavities of the core within the intact virion accommodate the "lateral bodies," which are seen by freeze-etching to be well-defined oval structures (Medzon and Bauer, 1970). On treatment of suspensions of virions with different detergents the lateral bodies may remain attached either to the core (Easterbrook, 1966) or to the outer membrane (Medzon and Bauer, 1970).

The outer membrane contains all the phospholipid of the virion (Dales and Mosbach, 1968), which unlike phospholipids of enveloped virions is synthesized *de novo* rather than being derived from cellular membranes. The structure of

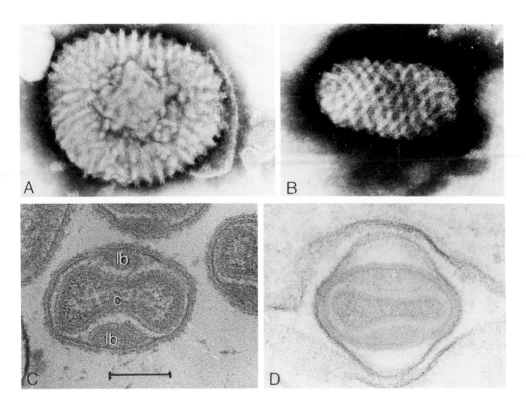

PLATE 3–9. *Poxvirus (bar = 100 nm). (A) Negatively stained vaccinia virion, show-ing surface structure of rodlets or tubules. (B) Negatively stained orf virion, showing characteristic surface structure of poxviruses of subgenus B. (C) Thin section of vac-cinia virion in its narrow aspect, showing the biconcave core (c) and the two lateral bodies (lb). (D) Thin section of mature extracellular vaccinia virion lying between two cells. The virion is enclosed by a single membrane originating from the inner membrane of the cisterna. (A and D, from Dales, 1963a; B, from Nagington et al., 1964; C, from Pogo and Dales, 1969, courtesy Drs. Dales and Nagington.)*

the outer membrane has been studied by thin sectioning, negative staining, and freeze-etching (Medzon and Bauer, 1970). Negative staining reveals a number of randomly arranged "tubules" (Plate 3–9), which are seen by freeze-etching to consist of double ridges composed of spherical subunits about 5 nm in diameter. The double ridge ("tubule") is 15 nm wide, with a center-to-center spacing of the ridges of about 10 nm. The subunits are spaced at 6–7.5 nm centers along the ridges which vary in size but are commonly about 70–100 nm long.

**Orf Virus (Subgenus B).** The virions of orf (contagious pustular dermatitis) and related viruses (milkers' node and bovine papular stomatitis viruses) have a distinctive appearance (Nagington et al., 1964). The particles measure 260 × 160 nm, compared with 300 × 240 nm for vaccinia (Nagington and Horne, 1962). Sections show that the core is biconcave and that lateral bodies are present (Büttner et al., 1964). Treatment of virions with alkali and selective staining of sectioned material demonstrates an internal structure within the core very like that produced by similar treatment of vaccinia virus (Büttner et al., 1964). The

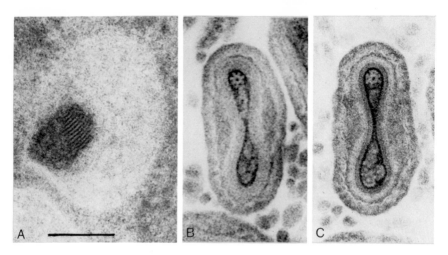

PLATE 3–10. *Poxvirus. Fowlpox virus: structure of the internal component as seen in thin sections after fixation at pH 3 (bar = 100 nm). (A) Immature particle showing nucleoid striation (5 nm center-to-center). (B and C) Mature virions showing substructure of nucleoid. (Hyde and Peters, 1971, courtesy Professor D. Peters.)*

most distinctive feature of orf virions is the threadlike structure which makes up their outer surface; a single long thread is wound around the rest of the particle as a rather loose spiral (Plate 3–9).

**Other Subgenera.** The virions of members of the myxoma and birdpox subgenera, and those of Yaba monkey tumor poxvirus and molluscum contagiosum virus, are morphologically indistinguishable from vaccinia virions.

### Serological Reactions

The antigens of poxviruses or poxvirus-infected cells can be analyzed by all the known serological techniques. Neutralization and cross-protection tests form the basis of the division of the family into five subgenera and several ungrouped species (see Table 1–6); there is substantial cross-reactivity in these tests between the species in each subgenus. In addition, precipitin tests show that most poxviruses, to whatever subgenus they belong, contain a common antigen that is an internal component of the virion (Woodroofe and Fenner, 1962).

Members of the vaccinia subgenus produce a hemagglutinin that is a lipoprotein separable from the virion, and hemagglutinin-negative mutants of vaccinia virus have been reported (Fenner, 1958). Hemagglutinins of different members of the subgenus cross-react.

### Proteins

Vaccinia virions contain a large number of structural proteins and several enzymes. Their location in the virion has not been accurately determined but data is accumulating that indicates which polypeptides are on the surface and which are associated with the cores.

**Structural Proteins.** Improved analytical methods have increased the number of polypeptides that could be distinguished in disrupted vaccinia virions from seventeen (Holowczak and Joklik, 1967a) to thirty (Sarov and Joklik, 1972a; Table 3–4).

The majority (17) of the thirty polypeptides are associated with the viral cores; two of them (VP4a and VP4b) account for 50% of the total protein of the core, and the cores themselves constitute one-half the mass of the virion. Five polypeptides are located on or near the surface of the intact virion; they account for 20% of the virion protein.

Two components (VP6a and VP6b), comprising 9% of the virion protein, are glycopolypeptides. When complete uncoating is blocked by infecting cells with vaccinia virus in the presence of cycloheximide (see Chapter 5), subviral components accumulate that resemble cores but contain, in addition, the viral glycopolypeptides (Sarov and Joklik, 1972b). These constitute a layer of surface tubules (measuring 10 × 5 nm) on the outside of the core. The failure of con-canavalin A to agglutinate vaccinia virions (Zarling and Tevethia, 1971) suggests the absence of surface glycoproteins. According to Holowczak (1970) and Garon and Moss (1971) the carbohydrate component of the vaccinia glycoproteins (unlike those of enveloped RNA viruses—see below) does not vary according to the cell in which the virus is grown, but consists solely of glucosamine. Since vaccinia virions develop in sites remote from the cytoplasmic smooth membranes where cellular glycosyltransferases are located it is possible that the carbohydrate as well as the polypeptide moieties of the vaccinia glycoproteins are virus-specified.

**Viral Enzymes.** After discovering that transcription of mRNA from "early" viral genes could occur inside intact poxvirus cores *in vivo* or *in vitro*, Kates and McAuslan (1967a, b) demonstrated that such cores contain a DNA-dependent RNA polymerase. Several other enzymes have now been identified in purified cores from a wide variety of poxviruses. As well as vaccinia, the simian poxvirus Yaba and entomopoxviruses (Bergoin and Dales, 1971) have been shown to contain four enzymes in their cores; transcriptase, a $Mg^{2+}$-dependent nucleo-tide phosphohydrolase, an exonuclease active on single-stranded DNA with an optimum pH of 5.0, and an endonuclease active on single-stranded DNA with an optimum pH of 7.5 (Schwartz and Dales, 1971; Aubertin and McAuslan, 1972).

## Lipid

The outer membrane of vaccinia virus, which appears to consist of a random arrangement of "tubules" in negatively stained preparations, is a lipoprotein in which lecithin is the predominant phospholipid (Dales and Mosbach, 1968). Unlike all enveloped viruses, which acquire their envelopes by budding through altered cellular membranes, the outer membrane of vaccinia virus is synthesized *de novo* in the cytoplasmic viral "factories." It nevertheless has the same tri-laminar structure as cellular membranes, when seen in thin sections, with the tubules ("spicules") applied on its outer surface, as an integral part of the

## TABLE 3–4

### The Properties of Vaccinia Virus Polypeptides [a]

| POLYPEPTIDE | MOL. WT. | PERCENT-AGE[b] | SURFACE | OUTER REGION NOT ON SURFACE | CORE | COMMENT |
|---|---|---|---|---|---|---|
| VP 1a | 200,000– | } ~0.5 | — | — | + | |
| 1b | 250,000[c] | | — | — | + | |
| 1c | 152,500 | } 2.4 | — | — | + | |
| 1d | 145,000 | | — | — | + | |
| 2a | 94,000 | } 2.2 | — | — | + | |
| 2b | 90,000 | | — | — | + | |
| 2c | 84,000 | 0.3 | — | — | + | |
| 3a | 78,500 | } 2.0 | — | — | + | |
| 3b | 73,500 | | — | — | + | |
| 3c | 70,000 | 0.7 | — | + | — | |
| 4a | 63,000 | 13.8 | — | — | + | Principal virion polypeptide |
| 4b | 58,500 | 11.1 | — | — | + | components |
| 4c | 56,000 | } 2.4 | + | — | — | |
| 4d | 54,000 | | — | — | + | |
| 5a | 51,000 | | — | — | + | |
| 5b | 47,500 | } 3.4 | — | + | — | |
| 5c | 46,000 | | — | + | — | |
| 5P | 46,750 | <0.5 | — | — | + | Phosphoprotein? |
| 6a | 41,000 | } 9.1 | — | + | — | |
| 6b | 39,000 | | — | + | — | Glycopolypeptides |
| 7P | 31,500 | <0.1 | — | + | — | Phosphoprotein? |
| 7a | 31,000 | 6.5 | + | — | — | |
| 7b | 27,000 | } 1.0 | — | — | + | |
| 7c | 26,000 | | — | — | + | |
| 8 | 23,000 | 7.0 | — | — | + | |
| 9a | 18,500 | } 7.3 | — | — | + | |
| 9b | 17,000 | | — | + | — | Not precipitated by ETOH[e] |
| 9P | 17,000 | <0.1 | — | — | + | Phosphoprotein? |
| 10a | 16,000 | } 10.5 | + | — | — | Not precipitated by ETOH |
| 10b | 14,500 | | — | + | — | |
| 11a | ~11,800 | 1.5 | — | — | — | Phosphoprotein? |
| 11b | ~11,000 | 11.4 | — | + | — | |
| 12 | ~8,000 | 6.6 | + | — | — | Not precipitated by ETOH |
| Total[d] 30 | 2 × 10⁶ | 99.0 | 5 | 8 | 17 | |

[a] From Sarov and Joklik (1972a).

[b] Calculated from autoradiograms of gels in which polypeptides labeled with $^{14}$C-amino acids had been electrophoresed.

[c] Estimated values.

[d] Excluding 5P, 7P, and 9P.

[e] ETOH, ethyl alcohol.

membrane (Dales and Mosbach, 1968). Unlike the lipid components of all other viruses, which are derived from the host cell, the lipid of vaccinia is distinctive, having, for example, a ratio of stearic to oleic acid of 1.0 (compared with 0.5 for cellular lipid).

## Nucleic Acid

Poxvirus DNA is the largest nucleic acid molecule in any vertebrate virus, or indeed in any known virus. Comprising 5% of the virion, vaccinia DNA was calculated by Joklik (1962) to have a total molecular weight of 160 million daltons. Measurements by electron microscopy of the contour length of vaccinia DNA molecules have confirmed that each virion does indeed contain a single molecule of double-stranded linear DNA up to 80 nm long, i.e., with a molecular weight of over 150 million daltons (Becker and Sarov, 1968). Berns and Silverman (1970) postulate that the duplex is cross-linked since the two strands do not separate completely on alkaline denaturation. The DNA of fibroma virus (subgenus E) is about the same size as that of vaccinia (Jacquemont *et al.*, 1972); that of fowlpox virus somewhat larger (100 $\mu$m long, 200 million daltons molecular weight; Gafford and Randall, 1970). The G + C content of poxvirus DNA (vaccinia: 35%, fibroma: 40%) is lower than that of any other group of vertebrate viruses.

Under conditions where viral mRNA is produced, cores of vaccinia virus synthesize poly A molecules about 150 nucleotides in length (Kates and Beeson, 1970b); these are said to hybridize with viral DNA (Kates, 1970), but the results need confirmation.

Poxvirus DNA is not infectious but subviral particles may be (Takahara and Schwerdt, 1967), a result that is not unexpected from the fact that the core of the virion carries a transcriptase.

## PARVOVIRUS

The smallest viruses of vertebrates, parvoviruses (reviews by Hoggan, 1970, 1971), are unique in that their genome consists of a single linear molecule of single-stranded DNA of molecular weight 1.2–1.8 million daltons. They are non-enveloped and contain no lipid and no glycoproteins. Some members of the genus (subgenus A) are "normal" infectious viruses; others [subgenus B, the adenovirus-associated viruses (AAV)] are "satellite" viruses that multiply only in the presence of helper adenovirus.

The parvoviruses are highly resistant to physical and chemical reagents, withstanding heating at 60°C for at least an hour and being unaffected by ether, chloroform, and anionic detergents (Toolan, 1968; Hoggan, 1970), and have a high buoyant density in CsCl (1.38–1.46 g/cm$^3$; Hoggan, 1971), a property that has proved valuable for separating AAV from adenoviruses.

## Structure

Parvovirus particles are very small, and their apparent diameters are affected by the method of staining (phosphotungstate or uranyl acetate) and of measure-

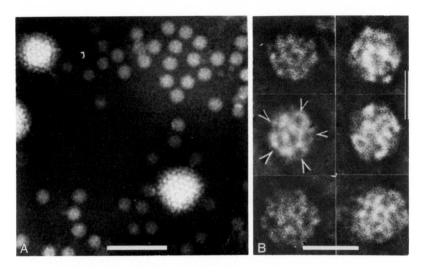

PLATE 3–11. *Parvovirus, subgenus B [Adenovirus-associated viruses (AAV)]. (A) AAV-4 and the simian adenovirus SV15 (bar = 100 nm). (Courtesy Dr. H. D. Mayor.) (B) Enlargement of AAV associated with human adenovirus type 4 (bar = 20 nm). (From K. Smith et al., 1966, courtesy Dr. K.O. Smith.)*

ment. Under comparable conditions several authors have found that some parvoviruses (notably Kilham rat virus, the serologically related H3 virus, and minute virus of mice) were smaller ($\sim$ 19 nm diameter) than others (Hoggan, 1971).

Determination of the symmetry and number of capsomers of these very small particles has been difficult. They are clearly polyhedral (Plate 3–11), but do not seem to have the "knobby" structure of ΦX174, described by Tromans and Horne (1961). Vasquez and Brailovsky (1965) and Karasaki (1966) believe that they are probably icosahedra of triangulation number $T = 3$, with 12 pentamers (at the vertices) and 10 $(T-1) = 20$ hexamers on the triangular facets, i.e., 32 capsomers, but this structure is not compatible with the chemical data (see below). K. Smith *et al.* (1966) believe that the capsid of AAV is best represented as a network of fibers with "holes" at each vertex, as was once proposed for reovirus (Vasquez and Tournier, 1962).

## Serological Reactions

Comprehensive tests for serological relatedness have not yet been performed on the many parvoviruses recently recovered from various domestic animals (rat, dog, cat, mink, pig, etc.), but several of the viruses from rodents cross-react in neutralization tests and do so to an even greater extent by fluorescent antibody tests (Hoggan, 1971). The four AAV serotypes can be distinguished by neutralization and fluorescent antibody tests. There is some cross-reactivity between AAV-2 and AAV-3, but not between AAV's and any of several infectious parvoviruses tested (Hoggan, 1971).

Many of the infectious parvoviruses (subgenus A), but only AAV-4 among the satellite parvoviruses (subgenus B), produce hemagglutination, often only with red cells from particular species, and sometimes only at 4° and not at 37°C (Toolan, 1968; Hoggan, 1971).

## Proteins

Using Kilham rat virus, AAV-2 and AAV-3, respectively, three independent groups of workers (Salzman and White, 1970; Rose et al., 1971; F. Johnson et al., 1971) have reported that parvoviruses contain three polypeptides, in a molecular ratio of about 10:1:1. There appear to be about 60 molecules per virion of the major polypeptide (molecular weight, 62,000–66,000 daltons), suggesting that the capsid may be composed of 12 pentamers, 20 trimers, or 60 monomers. The significance of the minor polypeptides is unknown; the aggregate molecular weight of the three polypeptides (189,000–238,000 daltons) exceeds the coding potential of the genome (estimated at 170,000 daltons).

Salzman (1971) claimed that there was a virion-associated DNA polymerase in RV grown in rat (nephroma) cells, but it is probably adventitiously adsorbed, as has been found with poxviruses (Tan and McAuslan, 1972). Virions of RV or the minute virus of mice grown in rat and mouse embryo primary cultures lack any polymerase activity (P. Tattersall, personal communication).

## Nucleic Acid

The first member of the group whose nucleic acid was carefully characterized was the minute virus of mice (MVM), which contains a molecule of single-stranded DNA of molecular weight 1.5–1.8 million daltons (Crawford, 1966; Crawford et al., 1969). Subsequently, rat virus (RV) and H-1 were also found to contain single-stranded DNA, and, in a comparative study, McGeoch et al. (1970) found all three DNA's to have molecular weights of 1.5 to 1.7 million daltons and similar base compositions. Nearest neighbor base sequence analyses indicated the close relationship between these three viruses of subgenus A. It was also noteworthy that the CpG doublet occurred very infrequently, as in other small DNA viruses and in mammalian cells (Subak-Sharpe, 1969).

The DNA of the satellite parvoviruses was originally thought to be a double-stranded molecule of molecular weight 3.6 million daltons, because the DNA extracted from purified suspensions showed base pairing and its melting curve was steep ($T_m$ 90°–93°C) (Rose et al., 1966). However, from a comparative study of AAV-1, the minute virus of mice, and the bacteriophage $\Phi$X174, Crawford et al. (1969) concluded that AAV's could not contain more than 1.8 million daltons of DNA, and they suggested that it was single-stranded, but that the two complementary single strands were encapsidated in different particles in approximately equal numbers, and that these single strands annealed spontaneously on extraction. Earlier experiments with acridine orange staining (Mayor and Melnick, 1966) also suggested that the virions contained single-stranded DNA.

Elegant studies by Rose et al. (1969) confirmed Crawford's prediction. By growing virus in the presence of BUdR, which substitutes for thymidine, they were able to separate two distinct populations of virions by equilibrium gradient centrifugation: "heavy" virions, containing the thymine-rich DNA strand, and "light" virions containing the thymine-poor strand. The DNA's extracted from the two classes of virion also banded separately, and on mixing they annealed to give double-stranded DNA (Rose and Koczot, 1971; Berns and Adler, 1972).

Molecular hybridization studies between the DNA's from the four AAV serotypes and the corresponding RNA's, synthesized *in vitro*, showed substantial homology between serotypes. None of the AAV-specific RNA's hybridized with DNA's from adenoviruses types 2 or 7 or SV15 (Rose *et al.*, 1968), or with DNA's from two rodent parvoviruses (Hoggan, 1971).

DNA has not been extracted from the infectious parvoviruses in an infectious form, but AAV DNA is infectious when plated on cells preinfected with helper adenovirus (Hoggan *et al.*, 1968).

## PICORNAVIRIDAE

This family comprises three accepted genera: *Enterovirus*, *Rhinovirus*, and *Calicivirus*. As pointed out in Chapter 1, some workers believe that a genus *Cardiovirus* should be distinguished from the enteroviruses, and that foot-and-mouth disease virus and possibly equine rhinovirus are sufficiently distinctive to be separated from the genus *Rhinovirus*.

All picornaviruses are nonenveloped and have an icosahedral capsid 25–40 nm in diameter. Detailed chemical and structural investigations have been carried out with poliovirus, ME virus (a "cardiovirus") (review, Rueckert, 1971), and with foot-and-mouth disease virus (Talbot and Brown, 1972). The structure and chemistry of all picornaviruses except members of the genus *Calicivirus* are very similar (Table 3–5), and all genera except *Calicivirus* will be considered together.

TABLE 3–5

*Some Physical Properties of Picornaviruses*

| GENUS | VIRUS | BUOYANT DENSITY IN CsCl | ACID STABILITY | DIAMETER (NM) | NO. OF CAPSOMERS | RNA (%) |
|---|---|---|---|---|---|---|
| *Enterovirus* | Poliovirus | 1.34 | Stable pH 3–10 | 27 | 60 | 29 |
| (*Cardiovirus*)[a] | ME virus | 1.34 | Stable pH 3–10 Labile pH 5–7 [b] | 24 | 60 | 31.5 |
| *Rhinovirus* | Rhinovirus 14 | 1.38–1.41 | Labile pH <7 | 23 | 60 | 30 |
| (*Aphthovirus*)[a] | Foot-and-mouth disease virus | 1.43 | Labile pH <7 | 23–25 | 60 | 31.5 |
| *Calicivirus* | Vesicular exanthema of swine virus | 1.37–1.38 | Labile pH <5 | 35–40 | 32 | 22 |

[a] Not accepted genera.
[b] In the presence of 0.1 M sodium chloride (Rueckert, 1971).

## Structure

The virions of *Enterovirus* and *Rhinovirus* appear to be rather uniform in size (Plate 3–12), with a diameter of 25 to 30 nm measured by electron microscopy of negatively stained particles (McFerran *et al.*, 1971). The probable icosahedral structure of the poliovirion was suggested by X-ray diffraction studies, which led to the further suggestion that the shells contained 60 $n$ identical or structurally equivalent protein subunits (Finch and Klug, 1959; Klug and Caspar, 1960). Something of the arrangement of these protein subunits (capsomers) can be seen in electron micrographs of preparations of infected cells (Horne and Nagington, 1959), but determination of details of the symmetry has been difficult because the capsomers of normal virions are not clearly defined by phosphotungstate.

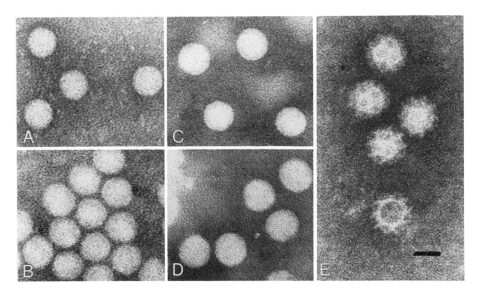

PLATE 3–12. *Picornaviridae (bar = 25 nm). (A) EMC virus (Cardiovirus). (B) Porcine enterovirus (genus Enterovirus); (C) Foot-and-mouth disease virus (Aphthovirus), (D) Human rhinovirus (genus Rhinovirus). (E) Vesicular exanthema virus of swine (genus Calicivirus). (Courtesy, Dr. F. Brown.)*

Talbot and Brown (1972) have compared the structure of foot-and-mouth disease virus (FMDV) with that of the cardiovirus, ME virus (Rueckert *et al.*, 1969; Dunker and Rueckert, 1971). Degradation at pH 6.5 or by heating yields large subunits, free RNA, and an insoluble precipitate. The large subunits of ME virus sediment at 14 S, and can be further dissociated by urea treatment into 5 S subunits (molecular weights 440,000 and 88,000 daltons, respectively). The 5 S subunits (the capsomers) are protein molecules consisting of one each of three nonidentical polypeptide chains (VP1, VP2, VP3), that are generated by cleavage of a precursor polypeptide (see Chapter 6). Whereas the large 14 S subunits of ME virus are pentamers, the large subunits of FMDV, which sediment at 12 rather than 14 S, appear to be trimers of similar monomeric subunits (Fig 3–4). The protein–protein bonds in FMDV break along the edges of the triangular faces whereas with ME virus the bonds break so as to yield pentamer subunits.

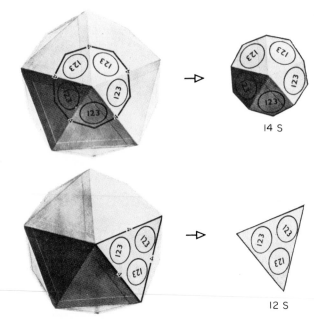

FIG. 3–4. *Picornaviridae. Diagram representing the dissociation of the icosahedral structure of ME virus (above) and FMDV (below) into 14 and 12 S subunits, respectively. These large subunits are pentamers (ME) or trimers (FMDV) of monomeric capsomers, each of which consists of equimolecular amounts of VP1, VP2, and VP3. VP4, the smallest structural protein, is thought to be located at the edges, as shown; other experiments show that the capsomers are so arranged that VP1 is at the vertices (of FMDV). (From Talbot and Brown, 1972.)*

Talbot and Brown (1972) suggest that the fourth structural protein of the virion (VP4) is located at the apposition of the faces of the icosahedron through which the twofold axes of symmetry pass (see Fig 1–2A and Fig. 3–4). A combination of trypsin treatment and electron microscopy of the antibody-binding site of FMDV suggests that VP1 is located at the vertices of the icosahedron.

**Serological Reactions**

Although both *Enterovirus* and *Rhinovirus* are large genera they contain very few subgroups that show serological cross-reactivity. An exception is poliovirus, whose three serotypes are distinguished by neutralization and cross-protection tests, but cross-react in complement-fixation tests. Cross-reactions between various echoviruses can also be recognized by complement fixation tests, but not by neutralization (Bussel *et al.*, 1962). Some but not all members of both *Enterovirus* (reviews: Rosen, 1964, 1969) and *Rhinovirus* (Stott and Killington, 1972) produce hemagglutination.

With poliovirus it has long been known that empty capsids, occuring naturally, or artificially produced by heating or UV irradiation, possess a capsid antigen ("C" or "H") quite distinct from that ("D" or "N") of complete virions (Hummeler *et al.*, 1962). It is now clear that antigen D, the binding site for neutralizing antibody, is in fact VP4, the smallest of the virion polypeptides, which is missing from both naturally and artificially produced empty capsids (Maizel *et al.*, 1967; Breindl, 1971). Attachment of enterovirus particles to receptors on cellular membranes alters their properties (see Chapter 6); the first step in this change appears to be dissociation of VP4 from the virion (Crowell and Philipson, 1971).

### Proteins

Maizel and Summers (1968) have identified four polypeptides (designated VP1, VP2, VP3, and VP4) in infectious poliovirions, with molecular weights of 35,000, 28,000, 24,000 and 6,000 daltons, respectively. Empty virions ("procapsids") lack VP2 and 4, but have instead another protein designated VP0 (molecular weight 41,000 daltons). In the presence of guanidine, viral RNA synthesis is inhibited and procapsids accumulate in the cell; on reversal of the block VP0 is enzymatically cleaved to yield VP2 and VP4, and RNA associates with the four capsid proteins to yield infectious virions (Jacobson *et al.*, 1970; see Chapter 6).

All other picornaviruses except the caliciviruses also have four virion polypeptides, with molecular weights very similar to those of the poliovirus structural proteins (review, Rueckert, 1971). The probable location of these polypeptides in the virion is indicated in Fig. 3–4.

### Nucleic Acid

Poliovirus RNA appears to be representative of that of all the picornaviruses. It consists of a single linear molecule of single-stranded RNA with a molecular weight of 2.6 million daltons (Granboulan and Girard, 1969; Tannock *et al.*, 1970). Repeated passage of poliovirus at high multiplicity eventually produces defective particles containing only 85% of the normal RNA complement (C. Cole *et al.*, 1971). A characteristic of the Picornaviridae is that RNA extracted from the virion is infectious, and double-stranded replicative form RNA extracted from infected cells is also infectious. Each molecule of poliovirus RNA contains at its 3' end one sequence of poly A that is 53–55 nucleotides long (Armstrong *et al.*, 1972), or longer according to recent estimates of Yogo and Wimmer (1972).

Hybridization experiments between viral RNA's and complementary strands obtained by denaturation of double-stranded replicative form RNA revealed about 30% cross-hybridization between RNA's of the three poliovirus serotypes (Young *et al.*, 1968).

All other picornaviruses that have been carefully examined (ME virus, FMDV, and rhinovirus) have yielded RNA's virtually indistinguishable in size and base composition from that of poliovirus (Medappa *et al.*, 1971; Wild and Brown, 1970).

### Calicivirus

The virions of the caliciviruses are larger than those of other picornaviruses (35–40 nm diameter, McFerran *et al.*, 1971), they are unstable at pH 3 but not at pH 7 (Oglesby *et al.*, 1971), they have a relatively lower RNA content (22% compared with 30% for *Enterovirus*), and the appearance after negative staining is strikingly different from that of other picornaviruses (Plate 3–12). As with the reoviruses, which also show a symmetrical pattern of dark pits surrounded by a lighter rim, there is no general agreement about the interpretation of this electron microscopic appearance in terms of capsomer symmetry. Almeida *et al.* (1968) suggest a 32-subunit structure.

The viral RNA is a single molecule of molecular weight 2.8 million daltons; the isolated RNA is infectious. So far no chemical analyses of the virion polypeptides have been reported.

## TOGAVIRIDAE

This large family is defined by two criteria (a) the morphology of the virion, which consists of a small nucleocapsid of cubic symmetry enclosed by a lipoprotein envelope, and (b) the viral RNA, which can be extracted in an infectious state. It includes two genera of arboviruses, *Alphavirus* and *Flavivirus*, and as possible members a number of viruses which are not arthropod-borne (Horzinek *et al.*, 1971; Table 1–10). The majority of studies on the morphology and chemistry of togaviruses has been carried out with two alphaviruses, Sindbis and Semliki Forest viruses.

### Structure

Electron micrographs of negatively stained fixed *Alphavirus* particles reveal uniform spherical particles 50–70 nm diameter with irregular surface projections up to 10 nm long (Simpson and Hauser, 1968; Plate 3–13). Sometimes penetration of the stain outlines a clear region between radii of 20 and 25 nm (Compans, 1971). Treatment of 60 nm diameter virions with a protease removes the fringe of irregular surface projections (which are 6.5 nm long) to leave regular spherical particles 45 nm in diameter, which on analysis are found to contain all the viral RNA and lipid and unchanged core protein, but no glycoproteins (Compans, 1971). On the other hand, treatment of virions with detergent yields nucleocapsids free of the lipid and glycoprotein of the envelope (Strauss *et al.*, 1968a), like those obtained from infected cells (Plate 3–14; Acheson and Tamm, 1970a). The RNA within such "cores" is accessible to ribonuclease (Kääriäinen and Soderlund, 1971).

By analogy with other "spherical" viruses, it is reasonable to assume that the nucleocapsid has cubic symmetry, but the detailed structure has not been

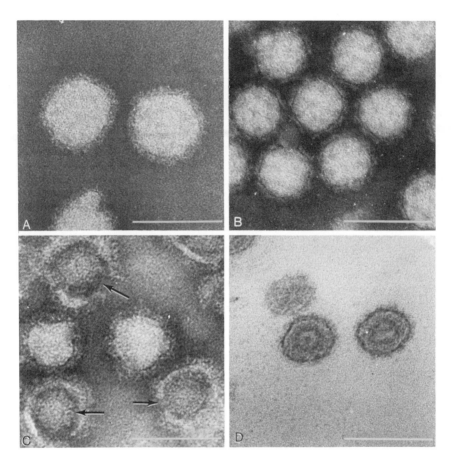

PLATE 3–13. *Togaviridae, Alphavirus (A, C, and D, Middleburg; B, Sindbis) (bar =
100 nm). (A) Negatively stained virions show poorly defined surface projections on envelope. (B) Virions prefixed with formaldehyde before negative staining. Sixfold symmetry is evident in these particles. (C) Damaged particles reveal an inner core (arrows).
(D) Thin section, showing the internal component cut in two different planes of section.
(From Simpson and Hauser, 1968, courtesy Dr. R.W. Simpson.)*

determined. In negatively stained particles there is occasionally a suggestion of
a symmetrical pattern (Plate 3–13); Horzinek and Mussgay (1969) suggest that
the nucleocapsid consists of 32 "hollow" polygonal capsomers arranged within
a $T = 3$ surface lattice.

X-ray diffraction studies (S. Harrison *et al.*, 1971) reveal the relationship
between the lipid and protein components of the Sindbis virion. Four nonoverlapping zones can be distinguished: RNA, core protein, lipid, and outer protein
(Fig. 3–5). The lipid occurs as a bilayer about 5 nm wide that is applied directly
to the surface of the nucleocapsid. The glycoprotein peplomers are attached to the
outer surface of this lipid bilayer but they do not penetrate deeply into it. Since
the core protein consists of about 400 molecules of a single polypeptide with a

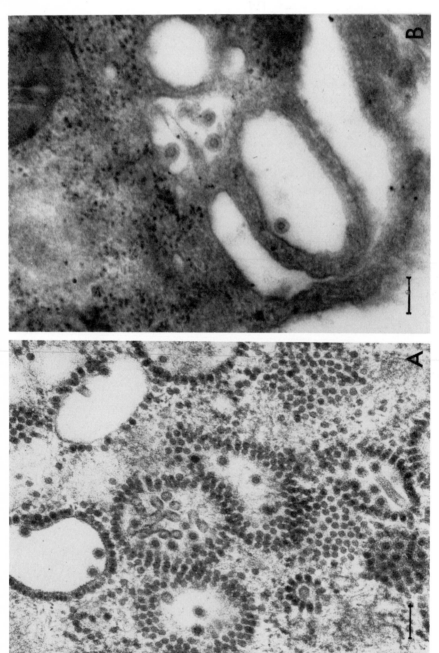

PLATE 3–14. *Togaviridae. Thin sections of infected cells (bars = 100 nm). (A) Semliki Forest virus nucleocapsids in the cytoplasm of infected chick cells. Nucleocapsids acquire an envelope by budding through the membranes of cytoplasmic vacuoles (courtesy, Dr. K.B. Tan). (B) Rubella virus budding into the vacuoles of BHK21 cells (from Holmes et al., 1969, courtesy Dr. I.H. Holmes.)*

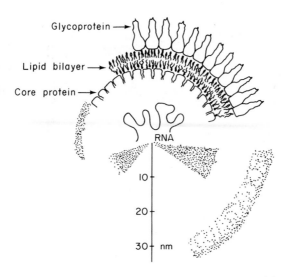

Fig. 3–5. *Alphavirus. Structure of Sindbis virions as determined by X-ray diffraction and electron microscopy of enveloped virions and of cores. The upper part shows the organization of protein, lipid, and RNA; the elongated glycoprotein subunits, the nucleocapsid subunits, and the lipid components of the bilayer are depicted schematically, and the apparent concentration of RNA near the particle center is indicated by the wavy line. The lower right-hand portion shows the deposition of negative stain, both in normal images of intact virus (below) and in stain filled particles (above). The left-hand portion shows the staining of cores: the usual negatively stained image (above) and the positively stained image (below). The scale at the bottom shows radial distances (in nm). (From S. Harrison et al., 1971.)*

molecular weight of 30,000 daltons (Strauss *et al.*, 1968a), the protein shell of the nucleocapsid must be at least 14 nm thick.

The flaviviruses are slightly smaller than alphaviruses [overall diameter about 50 nm (World Health Organ., 1967)], and detailed study of their structure has proved difficult. However, like alphaviruses, they appear to have a central "spherical" nucleocapsid surrounded by a lipid layer, then a layer of protein peplomers (Matsumura *et al.*, 1971).

The ungrouped togaviruses (Table 1–10) were first considered as possible members of this family because of their morphological and biological similarities to the accepted togaviruses (Holmes and Warburton, 1967; Horzinek *et al.*, 1971).

### Serological Reactions

The two named genera of togaviruses were defined by their serological reactions. *Alphavirus* represents the erstwhile "group A arboviruses," *Flavivirus* (named from its best known member, yellow fever virus) the "group B arboviruses." Viruses of both genera agglutinate red blood cells, the most susceptible cells being those of geese or newly hatched chicks. Optimum conditions of pH and temperature for performance of the hemagglutinin test vary. Some 20 viruses have been grouped into the genus *Alphavirus* and nearly 40 into the genus *Flavivirus* on the basis of cross-reactivity as detected by the hemagglutination inhibition test (Casals, 1971). Complement fixation and neutralization tests are more specific, but show some cross-reactions within each genus, as do cross-protection tests.

Rubella virus has hemagglutinating properties like those of alphaviruses (Holmes and Warburton, 1967) but shows no serological cross-reactivity with other togaviruses. Two ungrouped togaviruses that do not agglutinate red cells, hog cholera virus and bovine diarrhea virus, are serologically related to each other (Darbyshire, 1962).

### Proteins

Togaviruses contain very few species of protein. Early studies resolved only two proteins in alphaviruses, one with a molecular weight of 30,000 daltons that comprises the protein of the nucleocapsid, and an envelope glycoprotein of molecular weight 53,000 daltons (Strauss *et al.*, 1968a). Recently, M. Schlesinger *et al.* (1972) have identified a second minor glycoprotein of similar molecular weight. The nucleocapsid protein of Semliki Forest virus is rich in lysine, whereas the envelope glycoprotein is rich in various hydrophobic amino acids that may facilitate the attachment of peplomers to the underlying lipid (Acheson and Tamm, 1970b).

All investigators report that flaviviruses contain at least three proteins (Stollar, 1969; Shapiro *et al.*, 1971); Kunjin virus is said to contain four (Westaway and Reedman, 1969).

Rubella virus contains two or possibly three polypeptides: an arginine-rich nucleocapsid protein with a molecular weight of 35,000 daltons, and an envelope glycoprotein of molecular weight 45–50,000 daltons, which may possibly be a constituent of a larger glycoprotein (molecular weight 62,500 daltons ) (Vaheri and Hovi, 1972).

### Carbohydrate

The envelope glycoprotein of Sindbis virus has been analyzed for its carbohydrate content by molecular sieve separation of the glycopeptides released following digestion with pronase (Burge and Strauss, 1970). These glycopeptides differ from those of the envelope glycoprotein of vesicular stomatitis virus (a *Rhabdovirus*) grown in the same type of host cell only in respect of the number of sialic acid residues they contain. If sialic acid is removed by mild acid treatment, the Sindbis and vesicular stomatitis virus glycopeptides are indistinguishable by exclusion chromatography (Burge and Huang, 1970), tending to support the view that the composition of the carbohydrate sidechains is determined by the glycosyltransferases present in the host cell membranes. About one-third of the viral carbohydrate is in the glycolipid of the envelope.

## Lipid

Renkonen *et al.* (1971) have analyzed the lipids of Semliki Forest virus grown in two distinguishable clones of BHK cells. About 31% is neutral lipid (mainly cholesterol), 61% phospholipid (mainly phosphatidylcholine, phosphatidylethanolamine, phosphatidylserine, and sphingomyelin) and 8% glycolipid, probably mainly sialolactosylceramides. The authors proposed a model of envelope structure based on the fact that phospholipid and cholesterol molecules are present in equimolar numbers. Their observation that the composition of the viral lipids closely resembles that of the host cell plasma membrane is not totally incompatible with that of David (1971), who reports that the lipids of Sindbis and other enveloped viruses show less correlation with those of the host's plasma membrane than with those of the same virus grown in a different host. David postulates that the structure of the envelope glycoprotein plays a determinative role in selecting from a variety of host lipids those that have a particular affinity for the hydrophobic end of the peplomer. Since the peplomers do not penetrate the lipid layer that separates them from the protein of the nucleocapsid (S. Harrison *et al.*, 1971), the bonds linking them to lipid must be very firm.

## Nucleic Acid

Alphaviruses contain a single linear molecule of single-stranded RNA with a molecular weight of about 4 million daltons (Burge and Strauss, 1970). Although the extracted RNA is infectious, there were suggestions, based on its behavior in dimethyl sulfoxide, that the RNA consisted of a hydrogen-bonded aggregate of several smaller fragments (Dobos *et al.*, 1971). More recently, Arif and Faulkner (1972) have claimed that if adequate care is taken in its preparation, the RNA of Sindbis virus is indeed a single molecule with a molecular weight of 4 million daltons. However, in a comparative study, Boulton and Westaway (1972) reported that the *Flavivirus*, Kunjin virus, and Sindbis virus yielded RNA's of similar molecular weights (4.2 million daltons), but that either could be split into two equal halves of 2.1 million daltons by treatment with 8 $M$ urea. Like picornaviruses, togaviruses contain adenylate-rich oligonucleotides; Eastern equine encephalitis viral RNA contains a single poly A sequence about 70 nucleotides long (Armstrong *et al.*, 1972).

Rubella virus is the best studied member of the "unclassified" togaviruses. Its genome consists of a single-stranded RNA, with a molecular weight of about 3 million daltons, which is infectious (Hovi and Vaheri, 1970).

## ORTHOMYXOVIRUS

Detailed study is limited to a few strains of influenza A virus which, after adaptation to laboratory hosts, occurs as a roughly spherical virus about 100 nm in diameter. Strains recently isolated from man are usually long filaments rather than irregular spheres (Plate 3–15A).

The structure and chemistry of influenza viruses has been extensively reviewed; Hoyle (1968) provides a comprehensive analysis of earlier work while White (1973) and Laver (1973) give up-to-date summaries of present knowledge.

## Structure

The influenza virion consists of several double-helical ribonucleoprotein fragments enclosed within a lipoprotein envelope from which project two kinds of peplomers, the hemagglutinin and the neuraminidase.

The envelope has been investigated by electron spin resonance (ESR) spectroscopy of intact particles and particles from which the peplomers have been

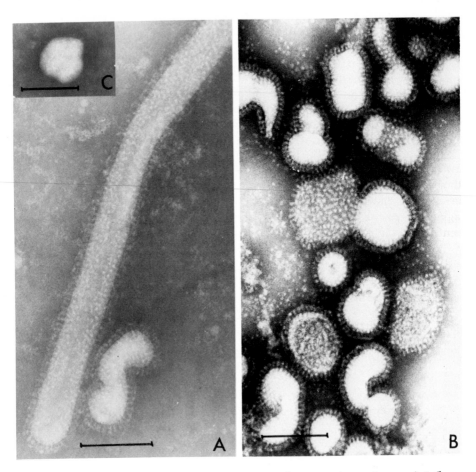

PLATE 3–15. Orthomyxovirus. Negatively stained preparations of type A influenza virus (bars = 100 nm). (A) Filamentous form, characteristic of strains recently isolated from man. (B) Pleomorphic "spherical" particles with well defined peplomers. (C) Particle lacking peplomers, obtained by bromelain treatment. (A from Choppin et al., 1961, courtesy Dr. P.W. Choppin; B, courtesy Dr. W.G. Laver; C, from Compans et al., 1970a, courtesy Dr. R.W. Compans.)

removed by treatment with chymotrypsin (Landsberger *et al.*, 1971). Removal of the peplomers had no effect on the ESR spectra, indicating that they are not involved in determining the organization of the lipid. Calculations suggest that the lipid of the virion is disposed as a single bilayer of cholesterol and phospholipid. The inside of the envelope is lined by a 6 nm layer of protein (M) which can be clearly visualized by electron microscopy beneath the 6 nm electron-lucent lipid layer in the smooth lipid-coated "cores" derived by stripping the peplomers from the virion with chymotrypsin or bromelain (Schulze, 1970, 1972; Compans *et al.*, 1970b; Plate 3–15). The M protein layer remains *in situ* when the lipid is removed from glutaraldehyde-fixed chymotrypsin cores with the nonionic detergent NP40 (Schulze, 1970, 1972).

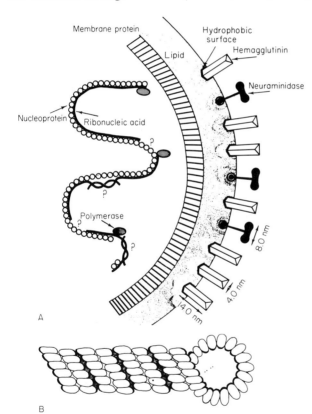

FIG. 3–6. *Orthomyxovirus. Schematic diagrams illustrating the structure of (A) the influenza virion (from Laver, 1973) and (B) one of the pieces of ribonucleoprotein. (From Compans et al., 1972a.)*

Two glycoprotein peplomers, the hemagglutinin and the neuraminidase, project from the envelope (Fig. 3–6). The numbers of peplomers per virion have been calculated by measurement of their distribution in electron micrographs, based on different spacings on spherical virions of various sizes. The figures for spherical virions 120 nm in diameter, with a center-to-center spacing between the peplomers of 7 nm (Nermut and Frank, 1971) was 738 (Tiffany and Blough, 1970), compared with 620 hemagglutinin and 118 neuraminidase molecules calculated on chemical grounds (Webster and Laver, 1971). The exact concordance is clearly coincidental, since the relative proportion of hemagglutinin and neuraminidase subunits varies greatly in different strains of influenza A virus (Webster et al., 1968), and a relatively small difference in the diameter of the virion would mean a substantial change in the number of peplomers. In electron micrographs, the peplomers often appear to be arranged in rows so that they form triangles and hexagons (Almeida and Waterson, 1967).

The hemagglutinin peplomers are rods about 14 nm long and 4 nm wide, with one hydrophobic and one hydrophilic end. The neuraminidase subunits, at least of Asian strains, are mushroom-shaped structures (Fig. 3–6) consisting of a head 8.5 × 5 nm and a stalk 10 nm long, with a knob 4 nm in diameter at its end (Laver and Valentine, 1969).

The only morphologically recognizable component inside the lipoprotein envelope is the ribonucleoprotein (RNP). Electron micrographs of partially ruptured virions occasionally reveal what appears to be a continuous coil of RNP occupying the whole of the interior of the particle (Almeida and Waterson, 1967). However, there is strong evidence that, like the viral RNA, the RNP occurs as several separate pieces, both in the virion and in the infected cell (Kingsbury and Webster, 1969; Duesberg, 1969; Compans et al., 1972a; Pons, 1971; Schulze, 1972). RNP extracted from influenza virions by detergent treatment occurs in three size classes that correspond to the size classes of RNA. Electron micrographs (Pons et al., 1969; Schulze, 1972; Compans et al., 1972a) indicate that the RNP is folded back upon itself in a hairpin-like manner and twisted into a double-helical structure with a single-stranded loop at one end (Fig. 3–6). The single helix is about 7 nm in diameter and the double helix 15 nm.

In most viruses that have tubular nucleocapsids, the RNA is protected by the protein shell. The RNA of isolated influenza virus RNP, however, is accessible to ribonuclease and hydroxylamine (Duesberg, 1969) suggesting that it may be on the outside. The RNA can be readily displaced from RNP by the anionic polymer, polyvinyl sulfate, leaving structures that are morphologically indistinguishable from the original (Pons et al., 1969).

### Serological Reactions

A vast amount of information has been accumulated about the serological reactions of influenza viruses (reviews, Hoyle, 1968; Webster and Laver, 1971). The essential features are that the type specificity (A, B, or C) resides in the internal antigens (NP and M; see Fig. 3–6 and Table 3–6), and is usually recognized by performing complement fixation tests with appropriate antisera, whereas the strain specificity resides in the peplomers, the hemagglutinin (HA), and neuraminidase (NA). Strains are usually distinguished by hemagglutination-inhibition or neutralization tests.

Hemagglutination by influenza viruses was first recognized by Hirst (1941), and is due to bridging of sensitive erythrocytes by HA peplomers of the virion. Isolated peplomers attach to red cells, but being monovalent do not agglutinate them unless the peplomers themselves aggregate by their hydrophobic ends (Laver and Valentine, 1969). Influenza virus spontaneously elutes from red cells at 37°C, because of the activity of the other viral peplomer, neuraminidase (NA), which digests the cellular receptors by cleaving the glycosidic bonds joining the keto group of neuraminic acid to D-galactose or D-galactosamine (Gott-schalk, 1957).

## Proteins

Polyacrylamide gel electrophoresis of disrupted purified influenza virions has shown that they contain six or seven proteins, plus one that is partially cleaved to form two of the others (Skehel and Schild, 1971; reviews by Laver 1973; White, 1973; Table 3–6).

TABLE 3–6

*Designations of Polypeptides of Influenza Virions* [a]

| DESIGNATION | APPROXIMATE MOL. WT. (DALTONS) | FUNCTION | REMARKS |
|---|---|---|---|
| $P_1$ $P_2$ | 81,000–94,000 | ? | Internal, nonglycosylated proteins present in small amounts, possibly the virion polymerase(s) |
| HA | 75,000–80,000 | Hemagglutinin | Glycoprotein. May be cleaved to two smaller polypeptides, $HA_1$ and $HA_2$, which are held together in the virion by disulfide bonds. Two HA molecules (or $HA_1 + HA_2$ complexes) form the hemagglutinin peplomer. Strain-specific antigen |
| NP | 53,000–60,000 | Nucleocapsid | Each capsomer is a monomer of NP. Type-specific antigen |
| NA | 55,000–70,000 | Neuraminidase | Glycoprotein. The neuraminidase may consist of two polypeptides in some strains, and $NA_1$ and $NA_2$ can then be used as designations. The active enzyme (the peplomer) is a tetramer of molecular weight 200,000–240,000 daltons. Strain-specific antigen |
| $HA_1$ | 50,000–60,000 | Hemagglutinin | See remarks on HA |
| $HA_2$ | 23,000–30,000 | Hemagglutinin | See remarks on HA |
| M | 21,000–27,000 | Major or "membrane" protein of viral envelope | Associated with the inner surface of the lipid layer of the envelope. Type-specific antigen |

[a] Modified from Kilbourne *et al.* (1972).

Bromelain- or chymotrypsin-treated cores contain two major proteins and one or two minor ones. One of the major proteins (NP), with a molecular weight of 53,000 to 60,000 daltons, comprises the subunits (capsomers) of the double-helical RNP (Fig. 3–6). The RNP is surrounded by a layer of protein of lower molecular weight (21–27,000 daltons), which is the most abundant molecule in the virion, comprising one-third of its total mass (White *et al.*, 1970). This polypeptide is known as the membrane protein (M), for it is intimately associated with the viral lipid and to all intents and purposes, represents the inner layer of the envelope (Compans *et al.*, 1970b; Schulze, 1970). Trace amounts of one or two high molecular weight proteins (81–94,000 daltons) (P$_1$, P$_2$) are also evident in viral cores and possibly comprise the polymerase.

The hemagglutinin peplomer is a dimer (molecular weight 150,000 daltons) of two noncovalently linked glycoproteins (HA, molecular weight 75–80,000 daltons), which are sometimes cleaved into two smaller molecules, HA$_1$ and HA$_2$, of molecular weights 50–60,000 and 20–30,000 daltons, respectively. Cleavage only occurs in HA molecules synthesized in certain cell types, and even then the two cleavage products remain associated as a single molecule covalently linked by disulfide bonds (Haslam et al., 1970a,b; Skehel and Schild, 1971; Laver, 1971; Stanley and Haslam, 1971; Lazarowitz et al., 1971). The larger glycopeptide HA$_1$ contains proportionately about four times as much glucosamine as the smaller one and several times as much proline suggesting that it has a much more tightly organized tertiary structure; HA$_2$ is more firmly associated with the lipid layer of the envelope and has a more open structure as revealed by iodination studies (Laver, 1971; Stanley and Haslam, 1971). The fact that antigenic drift occurs in both HA$_1$ and HA$_2$ (Laver and Webster, 1972) suggests that both may have some influence on the antigenic site at the tip of the peplomer which binds neutralizing antibody.

The other envelope protein is the viral neuraminidase (review, Drzeniek, 1972). The neuraminidase content of influenza virus varies markedly from strain to strain (5–15%). It is a strain-specific enzyme, has a pH optimum of 6 to 7 and some viral neuraminidases require divalent cations for maximal activity (Wilson and Rafelson, 1967). The active enzyme (molecular weight 220–250,000 daltons) has been variously reported to be composed of one or two different species of polypeptide (Webster, 1970; Haslam et al., 1970b; Skehel and Schild, 1971; Bucher and Kilbourne, 1972). It now seems likely that the enzyme is a tetramer of four glycoprotein molecules of molecular weight 60–65,000 (Lazdins et al., 1972).

## Carbohydrate

Apart from the ribose associated with the viral RNA, influenza virions contain glucosamine, mannose, galactose, and fucose, but no neuraminic acid (Ada and Gottschalk, 1956; Klenk et al., 1970b). The two peplomers that project from the envelope, HA and NA, are glycoproteins. About 20% of the mass of HA is carbohydrate (Laver, 1971), at least some of which is located near the free end of the HA subunits because "host-specific antigen," demonstrable by binding of anticellular antibody, is composed of the carbohydrate moiety of the HA and neuraminidase glycoproteins (Laver and Webster, 1966). Some of the viral carbohydrate is present as glycolipid in the envelope lipid that remains after the glycoprotein peplomers have been removed (Klenk and Choppin, 1970; Klenk et al., 1972).

## Lipids

Influenza virions, like other enveloped viruses, contains about 25% lipid, which is located on the outside of the membrane protein (M). Studies with purified plasma membranes of four different types of cultured cells and viruses grown in them has shown that, with few exceptions, the lipids of the host cell plasma membranes are quantitatively incorporated into the virions (Klenk *et al.*, 1972).

## Nucleic Acid

Influenza was the first virus of vertebrates that was shown to contain RNA rather than DNA (Ada and Perry, 1954). Because of the high rate of genetic recombination observed with influenza virus, Hirst (1962) suggested that the genome might consist of several distinct molecules. It is now clear that this is indeed the case. Acrylamide gel electrophoresis resolves the RNA into six or seven single-stranded molecules of molecular weights ranging from about 0.35 to 1.0 million daltons and totaling nearly 5 million daltons (Bishop *et al.*, 1971; Content and Duesberg, 1971). Some electron micrographs suggest that the several pieces may be linked together into a single molecule in the virion; RNA from three strains, spread in the presence or absence of urea or formamide, had a mean length corresponding to a molecular weight of 2.5 to 2.9 million daltons and a maximum length corresponding to about $5 \times 10^6$ (Li and Seto, 1971). However, the RNP, like the RNA, occurs in several size classes. The seven component pieces of RNA do not result from random breakage, because double-stranded pieces of similar length are synthesized in the infected cell (Duesberg, 1968a; Pons and Hirst, 1969) and there is some degree of specificity in the hybridization of these double-stranded molecules with the corresponding single-stranded viral molecules (Content and Duesberg, 1971).

Serial passage of concentrated suspensions of influenza virus usually leads to the production of an overwhelming preponderance of noninfectious but hemagglutinating particles, called "incomplete" virus (review by von Magnus, 1954). These contain all the viral components except the largest piece of viral RNA (Pons and Hirst, 1969) and the equivalent class of large nucleocapsid fragments (Kingsbury and Webster, 1969), and have a greatly reduced content of transcriptase (Chow and Simpson, 1971).

The RNA of influenza virus does not act as a messenger RNA in *in vitro* systems, and it hybridizes with RNA extracted from the polyribosomes of infected cells (Pons, 1972). Like all other viruses whose RNA is complementary to the messenger strand, the influenza virion contains a transcriptase (Chow and Simpson, 1971; Shekel, 1971). The enzyme, which is found in chymotrypsin-derived spikeless "cores" and in purified RNP (Compans and Caliguiri, 1972), has a temperature optimum of 28° to 32°C, a requirement for $Mg^{2+}$ and a preference for $Mn^{2+}$, is insensitive to actinomycin D, and catalyzes the *in vitro* transcription of small complementary molecules from the viral genome (Chow and Simpson, 1971; Bishop *et al.*, 1971).

## PARAMYXOVIRUS

This genus comprises a number of pleomorphic enveloped viruses that may occur as irregular spheres of sizes ranging from about 150 to nearly 300 nm in diameter, or as long filaments. Most chemical studies on paramyxoviruses have been made on three model viruses—Newcastle disease virus (NDV), simian virus 5 (SV5), and parainfluenza 1, strain Sendai. The findings have been remarkably consistent (review, Compans and Choppin, 1971) and will be considered together.

## Structure

In electron micrographs (Plate 3–16), the envelope resembles that of orthomyxoviruses although the peplomers are not as prominent. The detailed structure of the peplomers has not been determined, but Tiffany and Blough (1970) have made calculations of their probable number for particles of different dimensions, depending on the separation of the peplomers, which has been estimated as about 9 nm. Spherical particles 150 nm in diameter would then contain about ⌐50, those 300 nm in diameter about 3500 peplomers.

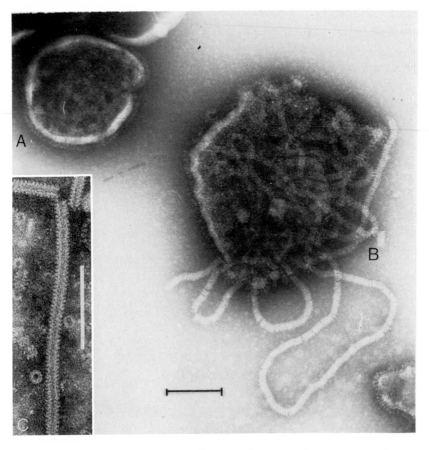

PLATE 3–16. *Paramyxovirus. Negatively stained virions of mumps virus (bars = 100 nm). (A) Intact virion; peplomers visible at lower edge. (B) Partially disrupted virion, showing nucleocapsid. (C) Enlargement of portion of nucleocapsid, in longitudinal and cross section. (Courtesy Dr. A.J. Gibbs.)*

The nucleocapsid can be readily extracted from paramyxoviruses as a single tube 1 $\mu$m long, composed of RNA and helically symmetrical repeating protein subunits (Hosaka et al., 1966). The nucleocapsids obtained from different species of *Paramyxovirus* vary somewhat in flexuousness and in the pitch of the single-start helix: 5 nm for Sendai virus, and 6 nm for measles and mumps viruses (Finch and Gibbs, 1970). They are constructed of between 2400 and 2800 identical protein subunits each of molecular weight about 60,000 daltons (Mountcastle et al., 1970), with an hour-glass shape and arranged in the helix at an angle of 60° to the long axis of the nucleocapsid. There are eleven or thirteen subunits in each turn of the helix (Finch and Gibbs, 1970). By examining nucleocapsids with a tilted stage, Compans et al. (1972b) determined that the sense of the helices of both SV5 and mumps virus was left-handed. Another less flexuous form of nucleocapsid, in which the subunits have molecular weights of 43,000 to 47,000 daltons, is found after treatment of infected cells by trypsin. Mountcastle et al. (1970) suggest that these smaller subunits may also be produced by proteolytic cleavage *in vivo*, and may account for the accumulation of nucleocapsid in cells persistently infected with paramyxoviruses.

The single linear molecule of single-stranded RNA, which constitutes the genome, is intimately associated with the protein subunits which protect it from degradation by ribonuclease.

As originally reported for parainfluenza type 1 by Hosaka et al. (1966), one nucleocapsid about 1 $\mu$m long is essential for infectivity of paramyxoviruses, but particles are often produced that contain several nucleocapsids enclosed within a single envelope. In a general way the size of the virion (which may vary greatly above the basic 150 nm diameter) reflects the number of nucleocapsids it contains.

### Serological Reactions

Like the orthomyxoviruses, the accepted members of the *Paramyxovirus* genus (see Table 1–12) agglutinate red blood cells by virtue of the HA peplomers and elute from them, since all possess a neuraminidase. All show some cross-reactivity in neutralization or complement fixation tests with at least one other member of the genus (Cook et al., 1959). Brostrom et al. (1971) found that the hemagglutinins and neuraminidases of four paramyxoviruses were immunologically distinct, except for cross-reactivity between the neuraminidases of Sendai virus and NDV.

Three other viruses that are classified as "possible" paramyxoviruses, measles, distemper, and rinderpest, exhibit a substantial degree of serological cross-reactivity (DeLay et al., 1965), but have distinctive host ranges. Measles virus causes hemagglutination, but none of the three viruses has a neuraminidase.

Other "possible" members of the genus, respiratory syncytial virus and pneumonia virus of mice, do not cross-react serologically with each other or any other paramyxoviruses.

### Proteins

At least five or six polypeptides have been resolved in acylamide gel electrophorograms of NDV, Sendai, and SV5 (Table 3–7; Mountcastle et al., 1971).

TABLE 3-7

*Major Structural Proteins of Paramyxoviruses
and Their Molecular Weights* [a,b]

| PROTEIN | SV5 | SENDAI | NDV |
|---|---|---|---|
| Nucleocapsid | 61,000 | 60,000 | 56,000 |
| Glycoprotein | 67,000 | 65,000 | 74,000 |
| Glycoprotein | 56,000 | 53,000 | 56,000 |
| Membrane protein | 41,000 | 38,000 | 41,000 |
| Other proteins | 76,000(?) [c] | 69,000 | 62,000(?) |
| | 50,000 | 58,000 | 53,000 |
| | | 46,000(?) | 46,000 |

[a] From Mountcastle *et al.* (1971).
[b] Molecular weight in daltons.
[c] Question mark indicates a protein whose status as a virion protein is uncertain.

As in *Orthomyxovirus*, the most conspicuous is the "membrane protein" (molecular weight 40,000 daltons) which surrounds the nucleocapsid. Glycoprotein peplomers project from the lipid of the envelope. Some of these peplomers are hemagglutinins, in all paramyxoviruses except respiratory syncytial virus (RSV); peplomers also have neuraminidase activity in the case of all paramyxoviruses except respiratory syncytial virus and the measles-distemper group. Scheid *et al.* (1972), while able to recover two glycoproteins from SV5, found that both hemagglutinating and neuraminidase functions were associated with the larger glycoprotein. Most paramyxoviruses also possess powerful hemolytic and cell fusing potential, but the precise mechanism is still uncertain (review by Poste, 1970).

Hosaka and Shimizu (1972) have performed interesting reconstitution experiments with membrane components extracted from Sendai virions solubilized with NP40. By mixing dialyzed preparations of Sendai virus lipid with isolated peplomers, they were able to produce artificial particles comprised of a membrane with morphologically identifiable peplomers projecting from both sides. Commercial preparations of phosphatidylethanolamine in conjunction with smaller amounts of cholesterol proved a satisfactory substitute for viral lipids. The reconstituted membranes had hemolysin activity, though this was lacking from lipid or peplomers alone. They concluded that the hemolysin characteristic of paramyxoviruses requires the presence of phospholipid (potentiated by cholesterol) to "activate" the glycoprotein peplomers.

The nucleocapsid of paramyxoviruses contains a transcriptase (Huang *et al.*, 1971; Robinson, 1971a). The enzyme activity is of a very much lower order than that found in the virions of vaccinia, vesicular stomatitis, or reovirus.

### Carbohydrate

The glycoproteins of SV5 contain the sugars glucosamine, fucose, mannose, and galactose, while the glycolipid of the envelope contains galactosamine, glucose, and galactose. As might be expected in a virus that carries neuraminidase, neither the glycoprotein nor the glycolipid contains neuraminic acid (Klenk *et al.*, 1970b). The glycolipid embodies the blood group and Forssman antigens characteristic of the cell species from which the virus budded (review, Choppin *et al.*, 1971).

## Lipids

The lipids of the envelope of paramyxoviruses, like those of the orthomyxoviruses, reflect the lipid composition of the plasma membranes of the cells in which they are grown (Klenk and Choppin, 1969), but some selection from these lipids does occur (Klenk and Choppin, 1970).

## Nucleic Acid

Paramyxovirus RNA occurs in the virion as a single molecule of molecular weight 6.5–7.5 million daltons (Duesberg and Robinson, 1965; reviews, Blair and Duesberg, 1970; Kingsbury, 1970). It cannot be dissociated into smaller fragments without breaking covalent bonds, and represents the longest known molecule of continuous single-stranded viral RNA. The finding that most virus-specific RNA synthesized in infected cells, including that associated with polyribosomes, hybridizes completely with RNA from virions (Bratt and Robinson, 1967) led to the conclusion that the parental RNA is "negative" (anti-message). It was, therefore, not unexpected to find that the virion contained a transcriptase (Huang *et al.*, 1971; Robinson, 1971a). Some virions contain RNA that is complementary to most of the virion RNA, for partial self-annealing of RNA extracted from NDV has been demonstrated (Robinson, 1970).

## CORONAVIRUS

This genus was established on the basis of the highly characteristic morphology of the virions of viruses recovered from several species of host animal (Anon., 1968; Table 1–13).

In negatively stained preparations, the virions are seen to be pleomorphic enveloped particles about 120 nm in diameter, with large petal-shaped peplomers 15–20 nm long and quite narrow at the base (Plate 3–17). Thin sections of cells supporting the multiplication of coronaviruses reveal that the peplomers project from a thick shell that consists of an outer double membrane and an inner layer (W. Becker *et al.*, 1967). There is no evidence of a spherical nucleocapsid inside this shell, nor has a tubular nucleocapsid been convincingly demonstrated, although thin sections of pelleted virus reveal threadlike structures, 7–8 nm in diameter, inside the particles, that could be so interpreted (Apostolov *et al.*, 1970).

The extreme fragility of coronaviruses has hampered chemical studies. Polyacrylamide gel electrophoresis reveals at least six different polypeptides, two or which appear to be glycopolypeptides associated with the peplomers (Hierholzer *et al.*, 1972). There is a host–cell antigen in the lipoprotein envelope, since treatment of virions with complement-containing antisera to the host cell in

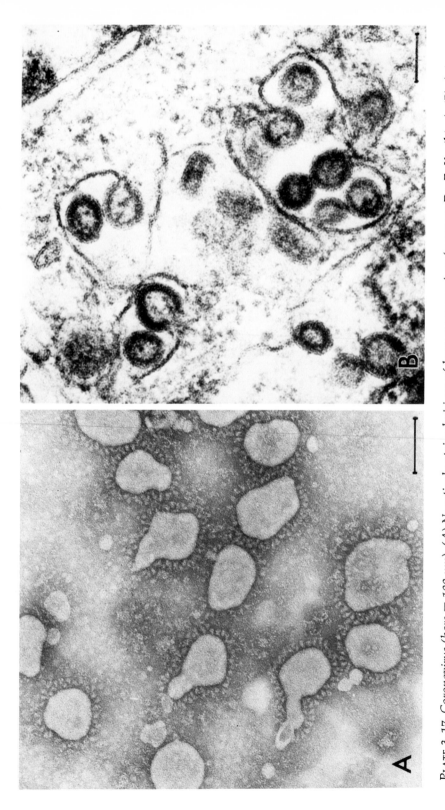

PLATE 3–17. *Coronavirus (bars = 100 nm). (A) Negatively stained virions of human coronavirus (courtesy Dr. Z. Kapikian). (B) Avian coronavirus (infectious bronchitis virus); thin section of chorioallantoic membrane showing mature virions in cytoplasmic vacuoles and virions budding into vacuoles. (From W. Becker et al., 1967, courtesy Dr. R.M. Chanock.)*

which they were grown results in the production of holes in the viral envelope like those produced in red cell membranes by immune hemolysis (Berry and Almeida, 1968).

The genome appears to consist of single-stranded RNA (Anon., 1968), but its size and structure have not been determined.

Several of the coronaviruses of mammals show serological cross-reactivity (Bradburne, 1970), but avian infectious bronchitis virus appears to be quite distinct. Human coronavirus agglutinates red blood cells of several species of animal (Kaye and Dowdle, 1969).

## ARENAVIRUS

This name was proposed by Rowe et al. (1970a), on the basis of the distinctive morphology of several apparently unrelated viruses when examined by thin section electron microscopy (Murphy et al., 1969), and the subsequent discovery that some of these agents were serologically related (Rowe et al., 1970c). The type species is lymphocytic choroiomeningitis virus (LCM), a virus long studied in experimental animals but whose structure and chemical composition have just begun to be investigated.

The buoyant density in sucrose is 1.18 g/cm$^3$, i.e., much the same as all the other enveloped RNA viruses (Pedersen, 1970). The virions are very pleomorphic, varying in diameter from 100 to over 300 nm (average 110–130 nm), with a heavy unit membrane envelope bearing closely spaced peplomers about 6 nm long (Plate 3–18; Murphy et al., 1970). They characteristically contain a variable number of electron-dense granules 20–25 nm in diameter, from which their name (arenosus=sandy) was derived (Plate 3–18).

The RNA of both LCM virus and another arenavirus, Pichinde (Carter et al., 1973), is single-stranded and can be resolved into four components by gel electrophoresis. Two of these comigrate with 18 S and 28 S ribosomal RNA, and it appears likely that the ribonuclease-sensitive "granules" visible inside the virion (Dalton et al., 1968; Murphy et al., 1970) are indeed ribosomes of host cell origin (Pedersen, 1971). The other two RNA species, presumably viral, have molecular weights of 1.1 and 2.1 million daltons.

Analysis of the structural proteins of Pichinde virus revealed four polypeptides, two of which were glycopolypeptides (Ramos et al., 1972).

## BUNYAMWERA SUPERGROUP

What was called the "Bunyamwera supergroup" (World Health Organ. 1967a) is the largest serogroup of arboviruses, with members distributed all over the world. Like the togaviruses, their infectivity is destroyed by lipid solvents (i.e., they have a lipoprotein envelope), but they are larger, with a diameter of 90–100 nm, and have a ragged closely adherent envelope with projections (Murphy et al., 1968a,b; I. Holmes, 1971). Their buoyant density in cesium chloride is 1.19 g/cm$^3$.

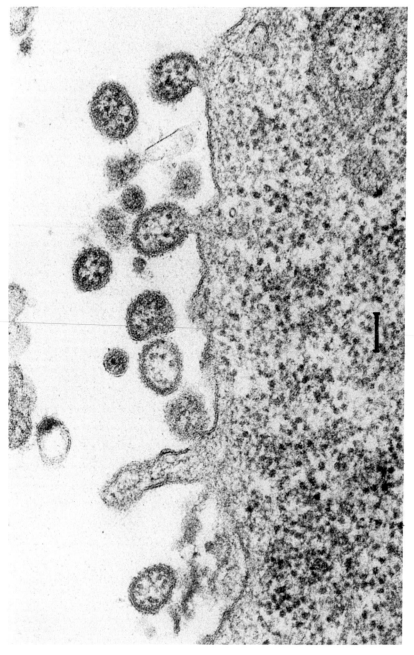

PLATE 3–18. *Arenavirus. Thin section of L cell infected with lymphocytic choriomeningitis virus (bar = 100 nm). Note the peplomers projecting from the envelope, the electron-dense particles within the free virions, and the large numbers of ribosomes in the cytoplasm just beneath the budding particles. (Courtesy Dr. F. Lehmann-Grube.)*

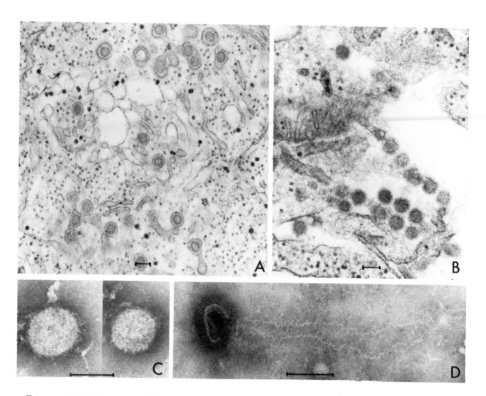

PLATE 3–19. *Viruses of Bunyamwera supergroup (bars = 100 nm). (A and B) Thin sections of mouse brain after infection with La Crosse virus. (A) Virions in Golgi vesicles. (B) Extracellular virions. (C) Negatively stained virions from cell culture-grown Guama virus. (D) Inkoo virus disrupted by storing at 4°C for 24 hours, showing released internal tubular component. (A, B, and C, courtesy Dr. F.A. Murphy; D, from von Bonsdorff et al., 1969, courtesy Professor N. Oker-Blom.)*

Finnish workers (von Bonsdorff *et al.*, 1969; Saikku *et al.*, 1970, 1971) have demonstrated that the core, unlike that of Togaviridae, with which they were at first confused, is not cubical, but is probably a long helical structure (Plate 3–19), which like the helical nucleocapsid of rhabdoviruses uncoils freely when released from the virion. In its uncoiled form, the nucleocapsid is 2.5 nm in diameter and at least 1 $\mu$m long. It has a buoyant density in CsCl of 1.31 g/cm$^3$ and is ribonuclease-resistant. The RNA was resolved into two bands by rate zonal contrifugation (21 and 27 S), while acrylamide gel electrophoresis revealed a single nucleocapsid protein of molecular weight 25,000 daltons and one or more other proteins in the range 65–75,000 daltons (Pettersson *et al.*, 1971), the latter being a glycoprotein (Lazdins and Holmes, 1973). By analogy with other viruses, it might be expected that the helical nucleocapsid is enclosed within a protein shell to which a lipid bilayer is closely applied, and from which virus-specified peplomers project; but no relevant investigations have yet been published.

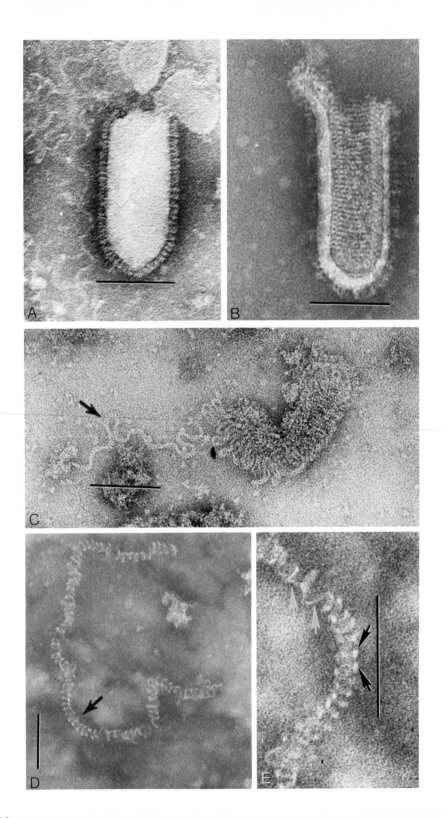

## RHABDOVIRUS

Structural and chemical studies have been carried out in some detail on two of the rhabdoviruses of vertebrates, vesicular stomatitis virus (VSV), a safe and convenient virus for laboratory studies, and rabies virus (reviews by Howatson, 1970; Hummeler, 1971). Both are enveloped bullet-shaped RNA viruses with a highly characteristic structure.

### Structure

Rhabdoviruses of vertebrates are rigid but fragile cylinders about 170 nm long and 70 nm in diameter, hemispherical at one end and usually planar at the end that buds off last. A lipoprotein envelope contains prominent peplomers (Plate 3–20). Wound inside this cylinder there is a nucleoprotein helix which, in negatively strained preparations, gives the interior a striated appearance.

The peplomers, which are 10 nm long and appear to consist of hollow knobs at the end of short stalks, are polymers of the glycoprotein G. They, and that protein, are selectively removed from virus particles by protease treatment (McSharry et al., 1971). The peplomers are responsible for the induction of neutralizing antibody to VSV, and constitute the hemagglutinin of rabies virus. After mixed infection of cells, VSV peplomers can be replaced in phenotypically mixed virions (see Chapter 7) by the peplomers of the paramyxovirus SV5, without affecting the overall morphology of the particle (McSharry et al., 1971). Since the other virion proteins are not affected by protease, except after pretreatment of particles with detergent to remove the lipid, it appears likely that in VSV, as in influenza virus, the lipid is external to the membrane protein "M" and the nucleocapsid protein "N" (Fig. 3–7).

The smooth envelope remaining after protease treatment consists of the viral lipid and a single protein "M" making up a clearly defined shell 65 nm in diameter. The space within this is fully occupied by the nucleocapsid, which appears to be a helix with thirty turns capped by four turns of diminishing diameter at the hemispherical end of the virion (Nakai and Howatson, 1968), giving the appearance in negative staining of a series of thirty-four parallel striations. The unwound extended nucleocapsid is 3.4–4.0 $\mu$m in length and consists of an RNA molecule of about 4 million daltons associated with about 1000 capsomers spaced 3.5 nm apart and consisting of the third major protein "N." Individual capsomers of the nucleoprotein of VSV measure $9 \times 3 \times 3$ nm (Table 3–8). The ribonucleoprotein produced by deoxycholate disruption of VSV

PLATE 3–20. *Rhabdovirus. Negatively stained preparations of vesicular stomatitis virus (bars = 100 nm). (A) Flattened particle showing peplomers at periphery. (B) Particle penetrated by stain and showing the characteristic cross-striations. (C) Unwinding nucleocapsid: a continuous 5 nm wide strand with repeating subunits. (D) Portion of nucleocapsid showing helical arrangement (arrow). (E) Higher magnification of (D). Black arrows indicate apparent rectangular units $4 \times 5.5$ nm. (From Simpson and Hauser, 1966, courtesy Dr. R.W. Simpson.)*

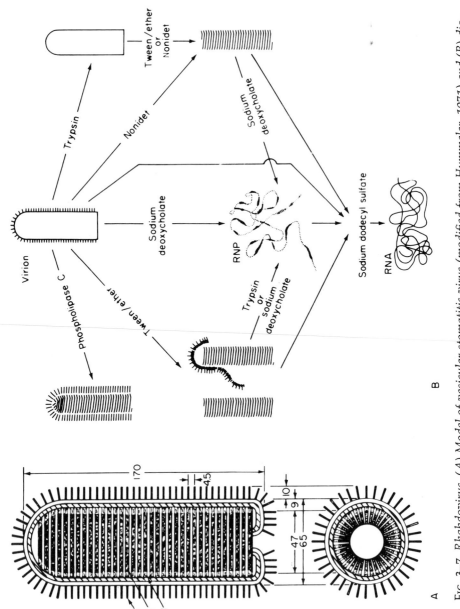

Fig. 3–7. Rhabdovirus. (A) Model of vesicular stomatitis virus (modified from Hummeler, 1971) and (B) diagram illustrating the structural relationships between different components of VSV (from Cartwright et al., 1970). G = glycoprotein peplomers; M = membrane protein; N = nucleoprotein.

TABLE 3–8

*The Proteins of Vesicular Stomatitis Virus* [a]

| PROTEIN | |
|---|---|
| L (associated with nucleocapsids) | 190,000 daltons |
| G (glycoprotein peplomers) | 69,000 daltons |
| M (membrane) | 29,000 daltons |
| NS | 40,000–45,000 daltons |
| N (nucleoprotein) | 50,000 daltons |
| Length of ribonucleoprotein | 3.5 $\mu$m |
| Number of subunits | 1000 |
| Size of subunits | 9 $\times$ 3 $\times$ 3 nm |

[a] Data from Hummeler (1971) and Wagner *et al.* (1972).

contains active transcriptase and is infectious (Cartwright *et al.*, 1970; Szilágyi and Uryvayev, 1973).

The structure and composition of rabies virus are very like those of VSV (Sokol *et al.*, 1969, 1971; György *et al.*, 1971).

Aberrant forms are frequently observed in the morphogenesis of both VSV and rabies virus. Long rods are found late in the growth cycle (Hummeler *et al.*, 1967), and noninfectious spherical particles called "T" (truncated) particles are produced in large numbers during serial undiluted passage of VSV (Cooper and Bellett, 1959). T particles have the same chemical composition as complete (B = bullet-shaped) particles but contain less nucleic acid (Huang *et al.*, 1966). They are not infectious but produce VSV-specific interference (see Chapter 8).

### Serological Reactions

Serological analysis of this large and heterogeneous group has not proceeded very far. Several small subgroups have been defined among the rhabdoviruses of vertebrates (Table 1–17). The nucleoprotein antigens of several VSV serotypes cross-react (Cartwright and Brown, 1972), but the Indiana and New Jersey serotypes are distinguishable by neutralization, which depends on antibody to the glycoprotein peplomers (Kang and Prevec, 1970). Two African viruses recovered from shrews and bats, respectively, cross-react with rabies virus in complement fixation and neutralization tests (Shope *et al.*, 1970).

Rabies virus agglutinates goose erythrocytes at 0°–4°C and pH 6.4 (Halonen *et al.*, 1968); hemagglutination has also been demonstrated with VSV and Kern Canyon virus.

### Proteins

Workers in the field (Wagner *et al.*, 1972) have recently suggested a standard alphabetical nomenclature for the viral polypeptides of rhabdoviruses (Table 3–8). Purified VSV virions (B or T) contain three major proteins: a glycoprotein (G) of molecular weight 69,000 daltons, which comprises the peplomers, the nucleoprotein (N) of molecular weight 50,000 daltons, and a nonglycosylated membrane protein (M) of molecular weight 29,000. There is also a high molecular weight protein L, which appears to be associated with nucleocapsids, and a fifth minor protein NS that is associated with proteins N and L in the nucleocapsids. The virion-associated transcriptase activity is associated with the ribonucleoprotein core which has been reported to contain two minor polypeptides in addition to N, L, and NS (Bishop and Roy, 1972).

Five proteins have been described in rabies virions, including an internal nucleocapsid protein, at least one membrane protein, and glycoprotein peplomers (Sokol et al., 1971). T particles lack transcriptase (Perrault and Holland, 1972a).

## Lipids

As with all enveloped viruses, the lipids of VSV are fundamentally those of the host cell's plasma membrane with minor variations in the proportions of phospholipids and neutral lipids; they have a higher content of phosphatidylethanolamine and sphingomyelin, and a lower amount of phosphatidylcholine (McSharry and Wagner, 1971). Similarly, the envelope of VSV grown in BHK cells contains the same glycolipid (a ganglioside containing neuraminic acid) as is present in the plasma membrane of that particular cell species (Klenk and Choppin, 1971).

## Nucleic Acid

The virion of VSV contains 3% RNA, 13% carbohydrate, 20% lipid, and 64% protein. On the basis of sedimentation data, Huang and Wagner (1966b) calculated the RNA content of the B particles of VSV to be about 4 million daltons, and that of the T particles to be only one-third of that figure. By direct length measurements under the electron microscope, Nakai and Howatson (1968) reached virtually the same answer (3.5 and 1.1 million daltons, respectively). It is now clear that T particles of different sizes may occur, containing RNA with a corresponding variety of lengths. Huang et al. (1970) and Mudd and Summers (1970b) confirmed that the total RNA content of infectious particles was around 3.8 million daltons and demonstrated that it is complementary to messenger RNA obtained from polyribosomes. Such a finding implies the presence of a virion-associated transcriptase, which has been demonstrated (Baltimore et al., 1970), and shown to transcribe the complete genome into small complementary mRNA molecules in vitro (Bishop and Roy, 1971).

## LEUKOVIRUS

The genus Leukovirus has been divided into four subgenera on the basis of biological and serological properties (see Table 1–16). The avian leukosis and the murine-feline leukemia/sarcoma viruses form subgenus A and the mouse mammary tumor viruses constitute subgenus B. Subgenera C and D, less securely classified in the same genus, comprise, respectively, visna of sheep and the foamy agents of cats, cattle, and primates. The features common to all these viruses are architectural similarity, the possession of a virion-associated RNA-dependent DNA polymerase (reverse transcriptase), and the presence of a characteristic specific antigen in every member of each of the respective sub-genera. All the leukoviruses have a very similar chemical composition and consist of about 60 to 70% protein, 20 to 30% lipid, 2 to 3% carbohydrate, about 1% RNA, and very small amounts of DNA. The chemistry of these viruses was reviewed by Duesberg (1970) to whom we refer the reader for early references.

## Structure

**A-, B-, and C-Type Particles.** Before describing the morphology of these complex virions it is well to understand the origin and meaning of terms like "C-type particle" that are commonly used in current literature. The designations were introduced by Bernhard (1958), after consideration of the morphological appearances seen with the electron microscope in thin sections of mouse mammary adenocarcinomas and mouse leukemia cells. Bernhard's diagram and a modern interpretation of the significance of these particle types is shown in Fig. 3–8. The essential difference between the morphological development of mouse leukemia virus (and all other viruses of *Leukovirus* subgenus A) and that of mouse mammary tumor virus (subgenus B) is that in the latter well developed spherical particles (intracytoplasmic A particles) form in the cytoplasm. These viruses (Plate 3–21) then mature by acquiring an envelope as they bud from cytoplasmic membranes to form what are called "enveloped A particles." The B particle, found only with mammary tumor virus, appears to be a degraded "enveloped A particle," in which the envelope has separated from the core. Similar changes occur in other enveloped viruses, e.g., herpes simplex virus.

The viruses of subgenus A are commonly called C-type particles. Unlike mammary tumor virus, no intracytoplasmic spherical particle can be distinguished, and the particle appears to assemble, in all its complexity (Fig. 3–9) as it buds from the cytoplasmic membrane (Plate 3–22).

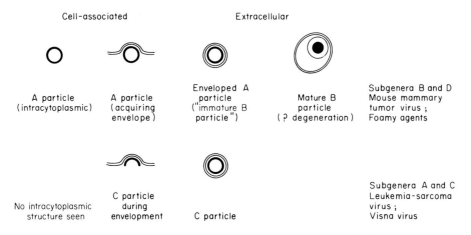

| Cell-associated | | Extracellular | | |
|---|---|---|---|---|
| A particle (intracytoplasmic) | A particle (acquiring envelope) | Enveloped A particle ("immature B particle") | Mature B particle (? degeneration) | Subgenera B and D Mouse mammary tumor virus ; Foamy agents |
| No intracytoplasmic structure seen | C particle during envelopment | C particle | | Subgenera A and C Leukemia-sarcoma virus ; Visna virus |

FIG. 3–8. *Leukovirus. Diagram illustrating morphogenesis of leukoviruses and the differences between A, B, and C type particles.*

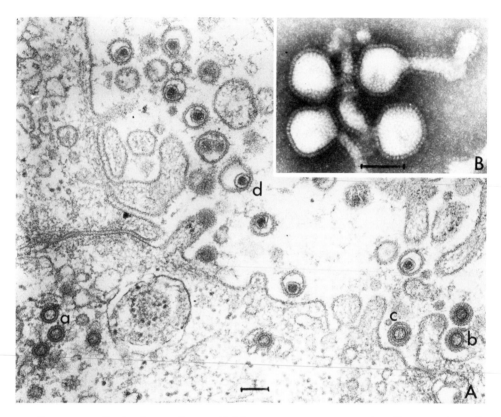

PLATE 3–21. *Leukovirus, subgenus B. Mouse mammary tumor virus (bars = 100 nm). (A) Thin section of mammary adenocarcinoma of mouse; a, intracellular A particle; b, budding A particle; c, enveloped A particle; d, B particle. (B) Negatively stained preparation, showing peplomers and "tails." (Courtesy Dr. D.H. Moore.)*

The morphological development of members of the other subgenera has not been investigated in much detail. Viruses of subgenus C (e.g., visna virus) resemble the leukemia/sarcoma viruses of subgenus A, i.e., they mature as C-type particles. The foamy viruses of subgenus D, on the other hand, develop as spherical particles in the cytoplasm, rather smaller than the mammary tumor virus (50 nm compared with 75 nm diameter), and then bud to form "enveloped A particles."

**Structure of** *Leukovirus;* **Subgenus A.**   Analysis of the structure of the leukoviruses has proved extremely difficult. In negatively stained preparations, they are pleomorphic and often "tailed" (as in Plate 3–21). This is an artifact because sections and properly fixed preparations (Plate 3–22) reveal that the particles are roughly spherical and about 100 nm in diameter. The lipoprotein envelope contains rather knobby peplomers, that have been identified as being glycoproteins (Rifkin and Compans, 1971). Freeze-drying, and negative staining with

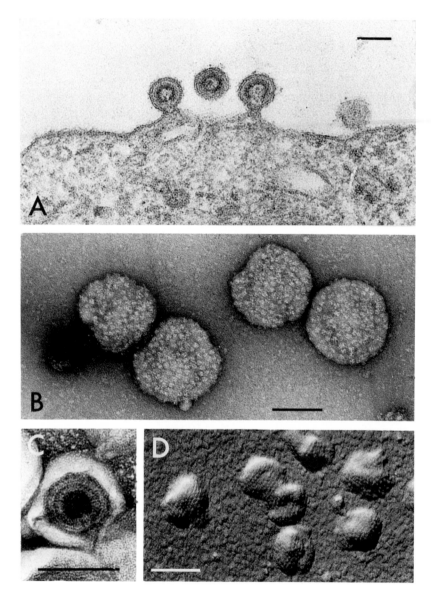

PLATE 3–22. *Leukovirus, subgenus A. Murine leukemia virus (bars = 100 nm). (A) Budding of virions from a cultured mouse embryo cell. (B) Virions negatively stained with uranyl acetate, showing peplomers on the surface. (C) Virion somewhat damaged and penetrated by uranyl acetate, so that the concentric arrangement of core, shell, and nucleoid becomes visible. (D) Cores isolated by ether treatment of virions, freeze dried and shadowed. The hexagonal arrangement of the subunits of the shell around the core is recognizable. (Courtesy Drs. H. Frank and W. Schäfer.)*

uranyl acetate, revealed clearly that the peplomers were globular structures about 8 nm in diameter that projected from a membrane (probably lipoprotein) about 8 nm thick (Fig. 3–9) (Nermut *et al.*, 1972).

The viral core that is enclosed within this envelope has a more complex structure than that of any other RNA virus, which is perhaps not surprising considering the amount of genetic information carried by leukoviruses. The best description is probably that of Nermut *et al.* (1972) (Fig. 3–9). The core appears

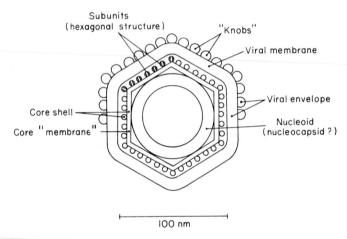

FIG. 3–9. *Leukovirus. Diagram illustrating the structure of the virion of mouse leukemia virus. (From Nermut et al., 1972.)*

to have cubic symmetry and to consist of two elements: (a) a layer of ringlike morphological subunits about 6 nm in diameter, forming a close hexagonal pattern, applied to (b) a core "membrane" about 3 nm thick. Within this there is a tubular structure that sometimes shows indications of helical symmetry. It is usually seen as a ring within the core "membrane," but sometimes appears to fill the core more completely. It is assumed that the viral nucleic acid is associated with this structure.

### Proteins and Serological Reactions

**Structural Proteins.** The polypeptides of several leukoviruses have been examined by polyacrylamide gel electrophoresis, by gel filtration in the presence of guanidine hydrochloride, and by their antigenic properties (review, Nowinski *et al.*, 1972). Although there are minor disagreements between different laboratories concerning the exact number of polypeptide chains and their locations within the particles, in general, the picture is satisfyingly cohesive. However, the whole subject is currently made unnecessarily difficult because of the failure to agree on a system of nomenclature for the polypeptides and antigens. Tables 3–9 and 3–10 show a comparison of the nomenclature and results obtained by different groups, with the avian and murine leukoviruses of subgenus A.

TABLE 3–9

*Comparison of Nomenclature for Proteins of Avian Tumor Viruses*

| | AUTHORS | | | | |
|---|---|---|---|---|---|
| FLEISSNER (1971) | DUESBERG *et al.* (1971) | BOLOGNESI AND BAUER (1970) | ALLEN *et al.* (1970) | HUNG *et al.* (1971) | APPROXIMATE MOLECULAR WEIGHT [a] |
| m1 | I | G [d] | | 1, 3 [f] | 70,000 |
| m2 | II | G [d] | | 2 | 32,000 |
| gs1 | RSV3 | 4 | gsa [e] | 4 | 27,000 |
| gs2 | (RSV3) [b] | 3 | | 5, 6 [g] | 19,000 |
| gs3 | RSV1 [c] | 1 [c] | gsb [e] | 8 [c] | 15,000 |
| gs4 | RSV2 | 2 | | 7 | 12,000 |
| p5 | (RSV1) [c] | (1) [c] | | (8) [c] | 10,000 |

[a] According to estimates by Fleissner in Tooze (1973).

[b] The protein corresponding to *gs2* appears as a shoulder on RSV3.

[c] Proteins corresponding to *gs3* and p5 appear superimposed in PAGE (Fleissner, 1971). The strong *gs* antigen reported at this mobility must thus be *gs3*.

[d] Protein G contains carbohydrate and has a molecular weight (ca. 115,000) compatible with a complex of $m_1$ and $m_2$, neither of which appear as distinct monomers in this system.

[e] Identified by molecular weight, *gs* antigenicity, and amino acid composition.

[f] Protein 1 contains carbohydrate and may represent an aggregate of protein 3 (=m1), since m1 aggregates strongly even in 6 M guanidine hydrochloride (Fleissner, 1971). For additional evidence that the smaller glycoprotein forms residual aggregates in SDS during polyacrylamide gel electrophoresis, see Fig. 5 of Duesberg *et al.* (1970) and Fig. 7 of Hung *et al.* (1971).

[g] Protein 6 as isolated in 4 M urea probably represents an aggregated form of protein 5 with a different pI. See Bolognesi and Bauer (1970) for evidence that a protein of this size in polyacrylamide gel electrophoresis aggregates strongly in the absence of reducing agents.

The purified virions of all leukoviruses contain seven or eight polypeptides. Two of these are glycoproteins, which are located as peplomers in the viral envelope (Rifkin and Compans, 1971), and carry a strain-specific antigenic activity which can be identified by neutralization tests, immunofluorescence of infected living cells, and viral interference. The nucleoid contains several polypeptides which carry group-specific antigenic activity, identified by immunodiffusion, complement fixation, and immunofluorescence of fixed cells. The group-specific antigenic activity is distributed among several of the polypeptides of the nucleoid and falls into two classes: (a) an activity that is specific to all leukoviruses isolated from one animal species, and (b) an activity which is common to viruses isolated from different species. All RNA tumor viruses carry group-specific (*gs*) antigens, but only the mammalian leukemia/sarcoma viruses possess an "interspecies" antigen. As can be seen from Tables 3–9 and 3–10, there is still a confusing nomenclature for the assignment of antigenic activities to individual polypeptides.

TABLE 3–10

Comparison of Nomenclature for Proteins of Mammalian (Murine) Tumor Viruses

| | AUTHORS | | | | | |
|---|---|---|---|---|---|---|
| NOWINSKI et al. (1972) | SHANMUGAM et al. (1972) | SCHÄFER et al. (1972) | DUESBERG et al. (1971) | OROSZLAN et al. (1971b) [e] | MORONI (1972) | APPROXIMATE MOLECULAR WEIGHTS [i] |
| m1 | (V) [d] | II$_\mathrm{v}$ [f] | I | | IV | 100,000 |
| m2 | (VI) [d] | II$_\mathrm{gs}$ [f] | II | | V | 70,000 |
| gs1, (gs3) [a] | IV | IV, V [a] | | 3 | III | 27–30,000 |
| p2 [b] | II [b] | I [g] | | 1 [b] | (II) [h] | 15,000 |
| p3 [b] | III [b] | III [g] | | 2 [b] | (II) [h] | 12,000 |
| p4 [c] | I [c] | (I) [c] | | (1) [c] | I [c] | 10,000 |

[a] The gs1 and gs3 antigens are on the same molecule. Antigens IV and V of Schäfer et al. (1972) correspond respectively, to gs1 and gs3 of Old's group. (Antigen V is also designated gs-interspecies by Schäfer's group.)

[b] Protein p2 behaves as though larger than p3 in gel filtration (in 6 M guanidine hydrochloride) and smaller than p3 in polyacrylamide gel electrophoresis in the presence of SDS.

[c] Proteins p2 and p4 have essentially identical mobilities in polyacrylamide gel electrophoresis (E. Fleissner; R. Nowinski; unpublished), hence they usually appear superimposed as a single species by this method. It is not clear whether the minor species (I) reported by Moroni (1972) and by Shanmugam et al. (1972) represents a separation of p4 under their conditions.

[d] Shanmugam et al. (1972) do not explicitly identify these proteins as glycoproteins; in addition, they present evidence of a host–cell origin for these species in their virus preparations. Therefore, these species are not equivalent to the viral glycoproteins.

[e] Polypeptide designations of Schäfer et al. (1972), which are not shown, are similar to those of Oroszlan et al. (1971b).

[f] The particular glycoproteins carrying these two antigens have not been identified.

[g] Provisional assignments of antigens to polypeptides (W. Schäfer, personal communication).

[h] Moroni did not resolve two proteins at this position.

[i] According to estimates by Fleissner in Tooze (1973).

**Enzymes.** A variety of enzyme activities have been found in the virions of leukoviruses including lactic dehydrogenase and hexokinase (Mizutani and Temin, 1971), ribonuclease (Rosenbergova et al., 1965), ribonuclease H (Mölling et al., 1971a, b), an endonuclease (Mizutani et al., 1970), and an exonuclease (Mizutani et al., 1971) directed against DNA, polynucleotide ligase (Mizutani et al., 1971; Hurwitz and Leis, 1972), ATPase (Mommaerts et al., 1952), nucleotide kinase and phosphatase (Mizutani and Temin, 1971), protein kinase (Strand and August, 1971), RNA methylase (Gantt et al., 1971), aminoacyl transfer RNA synthetases (Erikson and Erikson, 1972), and RNA-dependent DNA polymerase (Temin and Mizutani, 1970; Baltimore, 1970). Several of these enzymes are located on the virion envelope, e.g., ATPase, but many of them are associated with the nucleoid so that it is necessary to disrupt the virion envelope with neutral detergent in order to demonstrate them. There is no evidence that bears on whether these particle-associated enzymes are host- or virus-coded, but there seems little doubt that many of them are adsorbed onto viral nucleoids or envelopes during maturation or are present in vesicles which copurify with the virions. Although direct evidence is still lacking, it seems likely that at least some of the virion-associated enzymes play a role in transformation and viral growth. For instance, it is tempting to believe that the function of the RNA-dependent DNA polymerase is to synthesize proviral DNA and that the virion-associated ligase and the nucleases act together to integrate the proviral DNA into the host genome (see Chapter 14).

RNA-dependent DNA polymerase has been purified from a number of different leukoviruses, and its properties have been extensively studied in a number of laboratories (reviews by Gallo, 1971; Temin and Baltimore, 1972). The enzyme is located inside the envelope of the virion and it has been estimated that there are about ten enzyme molecules per virus particle (Kacian *et al.*, 1971). The purified enzyme from murine RNA tumor viruses has a molecular weight of about 90,000 daltons (Hurwitz and Leis, 1972). The enzyme from avian viruses is larger and consists of two polypeptide subunits of molecular weights 110,000 and 69,000 daltons, respectively. In addition to DNA polymerizing activity, the purified enzyme from avian viruses contains a ribonuclease H activity which degrades the RNA moiety from DNA:RNA hybrids (Mölling *et al.*, 1971b; Keller and Crouch, 1972).

The virion RNA-dependent DNA polymerase carries an antigenic activity that is distinct from all other antigens of the virus particles and its polypeptide constitution does not correspond to any of the major viral polypeptides, as would be expected from the minute quantities of enzyme present in the virus particles. The mammalian and avian enzymes are immunologically distinct from one another although there is some cross-reaction between the enzymes isolated from different mammalian subgenus A viruses. There is no cross-reaction between the enzyme from subgenus A viruses and that from the subgenus B mouse mammary tumor virus (Aaronson *et al.*, 1971b; Oroszlan *et al.*, 1971a; Scolnick *et al.*, 1972; Parks *et al.*, 1972; and see Chapter 17).

## Lipids

Almost all of the lipids in leukovirus particles are located in the envelope of the virions. The envelope is derived from the plasma membrane of the host cell and so it is hardly surprising that the overall lipid constitution of the viruses bears a close resemblance to that of isolated plasma membranes. However, some differences have been reported. For instance, the phospholipids of at least some strains of Rous sarcoma virus contain significantly more sphingomyelin and significantly less phosphatidylcholine than the plasma membranes of the host cells (Quigley *et al.*, 1971). Presumably the virions bud through particular areas of the cell membrane which are slightly different in composition from the majority of the membrane.

## Nucleic Acid

The nucleic acid of all of the RNA tumor viruses so far examined consists of at least five components listed below.

1. *A high molecular weight single-stranded RNA.* This sediments in non-denaturing conditions at 60 to 70 S (Robinson *et al.*, 1967). Although this RNA is presumed to be the viral genome, there is no direct evidence to support this belief because the isolated molecules are not infectious. The molecular weight of this RNA calculated from its sedimentation velocity is 9–12 million daltons and it seems to be made up of about 4 noncovalently joined subunits which can be dissociated from one another in denaturing conditions such as elevated temperature or 99% dimethyl sulfoxide (Duesberg, 1968). After denaturation, the RNA subunits sediment in sucrose gradients at 35 S (apparent molecular weight of about 3 million daltons) and at least some of them contain tracts of polyadenylic acid of molecular weight 60,000 daltons (Green and Cartas, 1972; Lai and Duesberg, 1972a). The exact number of subunits, their degree of homology with one another, and the way they are held together to form the 70 S RNA is unknown.

There may be a correlation between the biological activity of at least some of the RNA tumor viruses and the subunit structure of their 70 S RNAs. Duesberg and his collaborators have shown that different viruses of the avian leukosis subgroup yield two classes of subunits (*a* and *b*) which differ in electrophoretic mobility in polyacrylamide gels. The larger subunit (*a*) is found only in viruses which are capable of transforming fibroblasts (Duesberg and Vogt, 1970). Transforming viruses can spontaneously lose the type *a* subunit and the resulting mutants are no longer capable of causing transformation but are still capable of replication (Martin and Duesberg, 1972). The base composition of the 60–70 S RNA's of those leukoviruses that have been examined show no peculiarities.

2. *4 S RNA.* Avian leukosis viruses contain about 20 molecules of 4 S RNA (Bishop *et al.*, 1970a) at least some of which have the same electrophoretic mobility as the 4 S RNA's of the host cells and are methylated to the same extent (Erikson, 1969). They can be charged with amino acids (Travnicek, 1968), they hybridize to cellular DNA, and the hybridization is competed by cellular transfer RNA's (Baluda and Nayak, 1970). These experiments show that the RNA tumor viruses contain transfer RNA derived from host cells, presumably during viral maturation. The incorporation of transfer RNA's into virions seems to be selective because different species of transfer RNA's are not present in the same ratios in the virus particles as in the host cells (Travnicek, 1968; Erikson and Erikson, 1970; J. M. Taylor *et al.*, unpublished). Some of the 4 S RNA in the virus particles is attached to the 60–70 S RNA by hydrogen bonds (Erikson and Erikson, 1971) and recent evidence suggests that its function may be to act as primer during DNA synthesis from the viral RNA template (J. M. Taylor *et al.*, unpublished).

3. *5–7 S RNA.* Rous sarcoma virus has been shown to contain about 4 molecules of a single-stranded RNA with a molecular weight of about 80,000 daltons, a very high G + C content, and a sedimentation coefficient of 5–7 S which closely resembles the 5 S RNA found in ribosomes (Bishop *et al.* ,1970b).

4. *18 and 28 S RNA.* Many preparations of leukovirus RNA contain small amounts of 18 and 28 S ribosomal RNA (Bishop *et al.*, 1970b). Presumably this

RNA is derived from host ribosomes which are contaminants of the virion preparations.

5. *DNA*. Many of the RNA tumor viruses contain small amounts of DNA (Levinson *et al.*, 1970; Riman and Beaudreau, 1970; Rokutanda *et al.*, 1970; Biswal *et al.*, 1971). The DNA sediments at about 7 S and seems to be homologous in sequence to cellular DNA (Varmus *et al.*, 1971). Its function is unknown.

## REOVIRUS

Two genera of viruses of vertebrates, *Reovirus* and *Orbivirus*, comprise viruses with a genome consisting of several separate pieces of double-stranded RNA. They share a number of other properties: "spherical" shape, lack of an envelope, presence of a transcriptase in the virion, and cytoplasmic multiplication. They are distinguished morphologically by the obvious "double-capsid" structure of reoviruses, whereas orbiviruses appear to have a "single-capsid" structure, and biologically by the fact that all orbiviruses, but no reoviruses, multiply in arthropods as well as vertebrates.

### Structure

The virions of *Reovirus* are spherical particles 75–80 nm in diameter, consisting of an outer capsid and a second inner capsid (Dales *et al.*, 1965b), sometimes called the core (Plate 3–23). The capsomers of the outer shell have not yet been visualized clearly enough to determine their spatial arrangement. Two interpretations have been suggested: (a) that there are 92 hollow prismatic capsomers arranged on the surface of an icosahedron (Mayor *et al.*, 1965), or (b) that there are 180 solid capsomers positioned equidistantly from the center of the virion and arranged around 92 holes (Vasquez and Tournier, 1964; Amano *et al.*, 1971). Both these models require that the number of capsomers visible around the circumference of the particle (peripheral capsomers) should be a multiple of 6. However, careful measurements by Luftig *et al.* (1972) reveal that there are 20 peripheral capsomers, leading to a calculated total number of capsomers of about 120. The capsomers are 9 nm in diameter, with a center-to-center spacing of 10 nm.

The outer capsid consists of the proteins $\mu_2$, $\sigma_1$, and $\sigma_3$ (Zweerink *et al.*, 1971; Table 3–11). Astell *et al.* (1972) have reassembled virions *in vitro* by adding $\sigma_3$ to "subviral particles" isolated from infected cells; $\sigma_3$ seems, therefore, to be a major component of the outer shell. However, a careful study of the progressive degradation of the outer capsid by chymotrypsin indicated that $\mu_2$, and not $\sigma_3$, is essential to the stability of the capsid and the infectivity of the virion (Joklik, 1972). Chymotrypsin first removes $\sigma_3$ completely, leaving the virions fully infectious; then $\mu_2$ is degraded by stagewise cleavage and this is accompanied by a precipitous drop in infectivity, loss of the virion-associated oligonucleotides, and "activation" of the virion transcriptase; the minor protein, $\sigma_1$, is lost last,

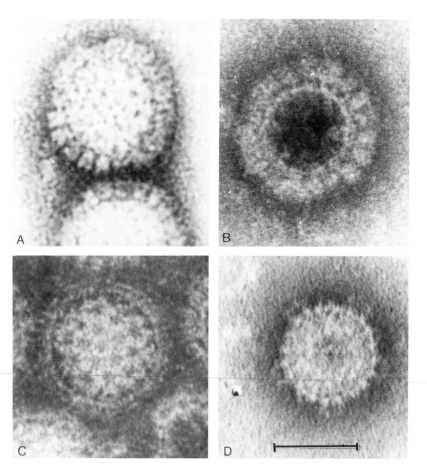

PLATE 3–23. *Reovirus. Electron micrographs of negatively stained preparations of reovirus type 3 (bar = 50 nm). (A) Untreated virion, with individual capsomers project-ing laterally. (B) Apparently empty particle, showing inner shell and peripheral capsom-ers. (C) After trypsin treatment the inner shell can be seen more clearly. (D) Inner shell almost devoid of outer capsomers, obtained by treatment of purified virions with sodium pyrophosphate. (From Dales et al., 1965b; courtesy of Dr. S. Dales.)*

the final product being the 45 nm diameter inner capsid, or "core" (Shatkin and Sipe, 1968b). Paradoxically, chymotrypsin in higher concentration removes only $\sigma_3$ and a 12,000 molecular weight segment of $\mu_2$, and neither activates the transcriptase nor decreases the infectivity of the virion. The properties of these "paravirions" resemble those of the "cores" with enhanced infectivity described by Spendlove *et al.* (1970).

The capsomers that make up the inner capsid (Plate 3–23) are smaller (about 4 nm in diameter) than those of the outer capsid, but the number of capsomers visible around the periphery is again 20 (Luftig *et al.*, 1972). In addition to approximately 120 capsomers, the inner capsids contain twelve "spikes" located

TABLE 3–11

*The Capsid Polypeptides of Reovirus* [a]

| POLYPEPTIDE | CLASS | LOCATION | MOLECULAR WEIGHT (DALTONS) |
|---|---|---|---|
| $\lambda_1$ | Primary gene product | Inner capsid | 155,000 |
| $\lambda_2$ | Primary gene product | Inner capsid | 140,000 |
| $\mu_1$ | Primary gene product | Inner capsid | 80,000 |
| $\mu_2$ | Derived from precursor | Outer capsid | 72,000 |
| $\sigma_1$ | Primary gene product | Outer capsid | 42,000 |
| $\sigma_2$ | Primary gene product | Inner capsid | 38,000 |
| $\sigma_3$ | Primary gene product | Outer capsid | 34,000 |

[a] Data from Zweerink *et al.* (1971).

on fivefold vertices. These have a diameter of 10 nm and they project about halfway to the outer surface of the outer capsid. They appear to be hollow at their distal end, leading to the suggestion that they may be "organelles" through which transcripts of the reovirus genome are released. Proteins $\lambda_1$, $\lambda_2$, $\mu_1$ and $\sigma_2$ are associated with the inner capsid (Table 3–11).

As with other icosahedral viruses, empty shells as well as full particles are found in preparations of reovirions. The two forms can be separated by density gradient centrifugation (Mayor *et al.*, 1965), i.e., "emptiness" is probably due to absence of the viral nucleic acid. Cores obtained from such empty particles contain the same proteins as those obtained from infectious virions. Treatment of the latter cores with sodium dodecyl sulfate releases viral RNA as a tightly associated mass of strands, which appear to be coiled either randomly or concentrically (Luftig *et al.*, 1972).

### Serological Reactions

The three mammalian serotypes cross-react by immunofluorescence and complement fixation, but can be distinguished by neutralization or hemagglutination-inhibition tests. There are five distinct serotypes of avian reovirus (Kawamura *et al.*, 1965). Group-specific antigens have been described for both mammalian (Sabin, 1959a) and avian strains (Kawamura and Tsubahara, 1966). Mammalian reoviruses agglutinate human erythrocytes (review, Rosen, 1969).

### Proteins

Reovirions contain seven major polypeptides that are located in the outer and inner capsids (Table 3–10). Like the viral RNA molecules (see below) they fall into three size classes (Loh and Shatkin, 1968). R. Smith *et al.* (1969) have calculated that these structural polypeptides have molecular weights corresponding almost exactly with the coding potential of the mRNA classes *l*, *m*, and *s* corresponding to the double-stranded RNA classes L, M, and S (Fig. 3–10). Not all the structural polypeptides of the virion are primary gene products; $\mu_1$, for example, is a precursor of $\mu_2$ (Zweerink *et al.*, 1971).

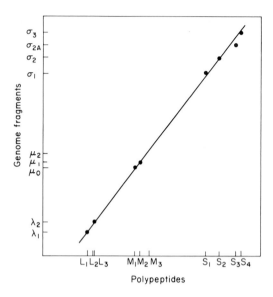

FIG. 3–10. *Reovirus. Relative electrophoretic migration rates of reovirus genome RNA segments and reovirus-specified polypeptides. (From Zweerink et al., 1971.)*

Two enzyme activities have been recognized as internal components of reo-virions, a transcriptase (Borsa and Graham, 1968; Shatkin and Sipe, 1968b) and nucleoside phosphohydrolase (Borsa *et al.*, 1970; Kapuler *et al.*, 1970). Neither activity has yet been related to any specific viral polypeptide.

The transcriptase transcribes each of the ten double-stranded RNA's into complementary mRNA, even while still *in situ* in the core (Skehel and Joklik, 1969; Banerjee and Shatkin, 1971). Electron microscopic studies suggest that each viral core contains multiple enzymatic sites from which the nascent mRNA molecules are extruded (Gillies *et al.*, 1971; see Plate 6–4).

## Nucleic Acid

Reovirus RNA is double-stranded by all the standard criteria, including ribonuclease resistance, a sharp thermal denaturation transition, base pairing ($A = U$, $C = G$) (Gomatos and Tamm, 1963), and a double helical structure with eleven nucleotides per turn by X-ray diffraction (Arnott *et al.*, 1967). A single molecule 7.7 $\mu$m long (corresponding to a molecular weight of 15 million daltons) can sometimes be visualized intact if great care is taken to disrupt the virion gently with sodium perchlorate (Dunnebacke and Kleinschmidt, 1967), but on extraction, the RNA is regularly found as ten fragments of three size classes (Bellamy and Joklik, 1967; Watanabe and Graham, 1967), $L_1$, $L_2$, and $L_3$ of about 2.7 million daltons, $M_1$, $M_2$, and $M_3$ of about 1.3 million daltons, and $S_1$, $S_2$, $S_3$, and $S_4$ of 0.6 to 0.8 million daltons (Shatkin *et al.*, 1968). These ten molecules do not arise by random breakage, or by breakage at specific weak points. Evidence for the existence of ten unique molecules in the virion itself is provided by an ingenious experiment in which the 3' terminal nucleoside resi-dues of reovirion RNA were labeled *in situ* by reduction with [³H]borohydride following periodate oxidation, and it was found that all ten pieces of the ex-

tracted RNA (i.e., 20 ends, not 2) were uniformly labeled (Millward and Graham, 1970). Furthermore, Banerjee and Shatkin (1971) examined the 5' termini of the ten fragments and discovered that they were all ppGpPyp, suggesting that each had been synthesized as a separate entity. The fact that each of the ten double-stranded molecules is transcribed *in vivo* or *in vitro* into a mRNA molecule of the same length and complementary base sequence (Bellamy and Joklik, 1967; Watanabe *et al.*, 1967) which in turn codes for a single protein (R. Smith *et al.*, 1969) indicates that each represents a single gene.

In addition to the 15 million daltons of double-stranded RNA, reovirions contain about 5–15 million daltons of single-stranded adenine-rich RNA consisting of heterogeneous small pieces (Bellamy and Joklik, 1967). About one-half of this material consists of poly A or, to be more precise, aligonucleotides containing 10–15 bases, mainly adenylate with a pAp 5' end (Shatkin and Sipe, 1968a; Stolzfus and Banerjee, 1972). The remainder consists of pieces 2–10 nucleotides long with ppGp at the 5' end (Stolzfus and Banerjee, 1972). Nichols *et al.* (1972a) postulate that both classes of oligonucleotide may arise as products of abortive transcription of the 3' ends of the viral genes by the virion transcriptase.

## ORBIVIRUS

This recently recognized genus (Borden *et al.*, 1971; Table 1–19) resembles *Reovirus* in a number of important characteristics but all members of the genus are arboviruses, and they differ structurally from *Reovirus*. The only member of the genus that has been studied in any detail is bluetongue virus (reviews, Verwoerd, 1970; Howell and Verwoerd, 1971).

### Structure

Detailed studies have been made of the morphology of bluetongue virus by Els and Verwoerd (1969), Bowne and Ritchie (1970), and Verwoerd *et al.* (1972); several other members of the genus have been examined by Murphy *et al.* (1971a).

Recent studies show that the intact virion is larger than was previously believed; an outer diffuse layer comprising two of the seven capsid polypeptides is lost during centrifugation in cesium chloride, leaving the smaller 55 nm particles with well defined capsomers (Verwoerd *et al.*, 1972; see Plate 3–24). Because of the small number of capsomers the triangulation number of the capsid is difficult to determine by observing the relationship between neighboring pentamers. Both Els and Verwoerd (1969) and Murphy *et al.* (1971a) believe the capsid to be icosahedral with 32 capsomers ($T = 3$). The capsomers are large (8–11 nm diameter), rather widely placed, and have a hollow cylindrical (doughnut) shape. Different members of the genus differ somewhat in size (65–80 nm in diameter, according to Murphy *et al.*, 1971a) but most are smaller than reoviruses (Plate 3–24). Occasionally "enveloped" virions are seen; these "pseudoenvelopes" appear to be derived from the unmodified cell membrane during release, and are not required for infectivity.

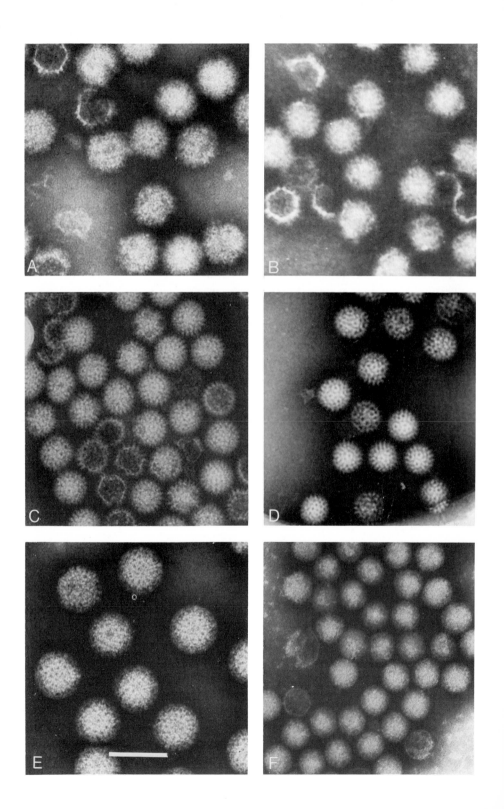

## Serological Reactions

There are sixteen serotypes of the bluetongue virus subgroup, and eight other serological subgroups have been recognized (Table 1–19).

## Proteins

Bluetongue virus contains seven structural polypeptides, five in the well defined capsid and two (molecular weights 110,000 and 61,000 daltons) in the diffuse outer layer that surrounds it. As with reovirus (Fig. 3–10) the molecular weights of the polypeptides correlate closely with the sizes of the genome segments (see below), on the assumption that each acts as a gene for one of the polypeptides (Verwoerd et al., 1972).

The virion transcriptase, which is activated by removal of the diffuse outer coat (just as reovirus transcriptase is activated by removal of the outer capsid), differs from the reovirus transcriptase in its dependence on $Mg^{2+}$ ions and its low optimum temperature of 28°C (Verwoerd and Huismans, 1972).

## Nucleic Acid

The genome of bluetongue virus consists of ten molecules of double-stranded RNA with a total molecular weight of 15 million daltons, in three size classes (Verwoerd et al., 1970). The molecular weights of the fragments differ from those of reovirus RNA's, and there is no cross-hybridization (Verwoerd, 1970).

## UNCLASSIFIED VIRUSES

As pointed out in Chapter 1, several viruses that have not yet been classified are important either in experimental investigations or in human medicine. Current knowledge of the structure and physicochemical properties of some of these agents is summarized here.

### Human Hepatitis Viruses

The historical background is set out in Chapter 1. In 1965, a significant advance in the search for a causative agent of human hepatitis was made with the recognition that a serum protein previously considered as a genetic marker in certain racial groups was in fact specifically associated with serum hepatitis (Blumberg et al., 1965). Because it was first discovered in the serum of an

PLATE 3–24. *Orbivirus. Virions of bluetongue (A–D) and reovirus (E and F) negatively stained with phosphotungstate (bar = 100 nm). (A) From sucrose gradient. (B) From CsCl gradient pH = 8. Note woolly appearance. (C) From CsCl gradient pH = 6 and (D) from CsCl gradient pH = 7. Note smaller size and well-defined surface structure. For comparison (E) reovirus, intact virions. (F) Reovirus cores derived by chymotrypsin treatment. (A–D, from Verwoerd et al., 1972; all illustrations courtesy Dr. D. W. Verwoerd.)*

Australian aboriginal it was originally called "Australia antigen" (Au), but has now been renamed hepatitis B antigen (HB-Ag); the antibody to it is called anti-HB-Ag. HB-Ag is found in very high concentrations in the sera of some persons for many years after they have been infected with hepatitis B virus.

Electron microscopic examination shows that HB-Ag consists of spheres and tubules 20 nm in diameter and comprised of subunits resembling capsomers (Plate 3–25). Chemically, most investigators believe that HB-Ag is devoid of nucleic acid, although Józwiak *et al.,* (1971) have reported that it consists of 70% protein, 25% lipid, and 5% RNA. The immunogenicity of HB-Ag is damaged by some proteases, especially after prior treatment with sodium dodecyl sulfate or diethyl ether, suggesting that it is a lipoprotein. The major lipid components are cholesterol, lecithin, and sphingomyelin, but the antigenic determinants reside in the protein moiety (Kim and Bissell, 1971). Acrylamide gel electrophoresis

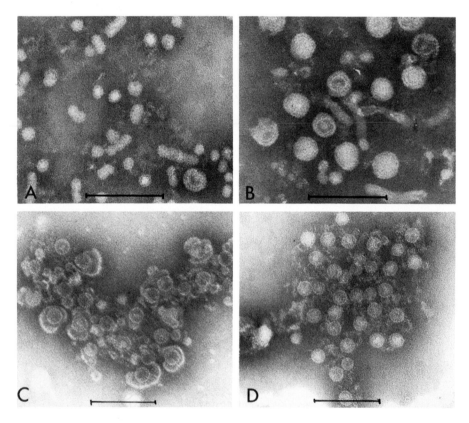

PLATE 3–25. *Human hepatitis virus (bars = 100 nm). (A) Hepatitis antigen B (HB-Ag: Australia antigen); spherical and tubular particles; single Dane particle at bottom right. (B) HB-Ag specimen with many Dane particles. (C) Specimen B treated with Tween 80. (D) Particles from liver homogenate of case of hepatitis; same diameter as internal sphere of (C). (Courtesy Dr. J.D. Almeida.)*

reveals two major polypeptides, with molecular weights of 26,000 and 32,000 daltons, and a minor polypeptide of molecular weight 40,000 daltons (Gerin et al., 1971).

HB-Ag is antigenically complex. All preparations share a common antigenic determinant, *a*; three minor determinants, *d, x,* and *y* (Le Bouvier, 1971) are distributed differently in different HB-Ag positive sera, which can be divided into three subgroups according to their possession of *d, y,* or neither.

Samples of blood or serum containing HB-Ag are usually infectious for man, but the 20 nm particles of antigen are not the virions. In some sera, the 20 nm HB-Ag particles are accompanied by larger spherical particles with a diameter of 42 nm (Dane et al., 1970; Plate 3–25), which appear to have a double shell. Study by immunoelectron microscopy of a serum sample that contained predominantly this larger particle (Almeida et al., 1971) and examination of sections (Huang, 1971) and homogenates of livers of acute hepatitis (Almeida et al., 1970), led to a tentative picture of the structure and morphogenesis of the virion (Almeida et al., 1971). The 42 nm particles can be disrupted by detergent treatment to release 27 nm spherical particles that are identical with the spherical particles found in the nuclei of liver cells (Plate 3–25) and in liver homogenates. When the disrupted particles, or the material obtained from liver homogenates, was treated with posthepatitis sera that contained no anti-HB-Ag the 27 nm particle specifically attached antibody molecules, visualized by electron microscopy. Some sera that contained anti-HB-Ag, on the other hand, agglutinated HB-Ag particles and tubules but did not attach to the 27 nm spheres. Tentatively, it is suggested that the virions are spherical particles with a diameter of 42 nm that consist of two protein shells. The inner one is assembled in the nucleus; the outer capsid is added later. The outer capsid consists of capsomers of a different antigenic specificity, which are made in excess, and often aggregate to form small HB-Ag particles.

Preliminary claims have been made that morphologically similar but serologically distinct particles may be found in fecal extracts but not serum from cases of infectious hepatitis (Cross et al., 1971).

## Lactic Dehydrogenase Virus (LDV)

This murine virus, originally and more correctly termed lactic dehydrogenase-elevating virus, has a distinctive morphology not matched by any other known virus (de-Thé and Notkins, 1965). In negatively stained preparations, most particles are elliptical in shape, 60–65 nm wide and 70–85 nm long; they appear somewhat smaller in sections. LDV appears to replicate exclusively in macrophages, sections of which reveal both oval and circular profiles. The latter show a core 26–29 nm in diameter, surrounded by several shells or layers, an inner dense shell 7–9 nm wide, and an outer layer of low density 5–7 nm wide limited by a thin outer membrane.

Extraction with phenol releases infectious RNA (Notkins, 1965), which appears to consist of a single molecule of single-stranded RNA with a molecular weight of about 5 million daltons (Darnell and Plagemann, 1972).

## Scrapie Virus

Agents that share the characteristics of scrapie virus (reviews, Gibbs and Gajdusek, 1971; Hunter, 1972) are known to cause four diseases, scrapie of sheep, mink encephalopathy, and kuru and Creutzfeld-Jakob disease in man (see Chapter 12). These "viruses" are nonimmunogenic, and can be distinguished from each other only by their pathological behavior (see Table 12–4).

TABLE 3–12

*Some Physical, Chemical, and Biological Properties of Scrapie* [a]

| TREATMENT | RESULTS |
| --- | --- |
| Size | |
|    Filtration | 20–30 nm |
|    Ionizing irradiation | 7–10 nm |
| Density cesium chloride | 1.32–1.34 $g/cm^3$ |
| pH (2.1–7.0) | Stable |
| Heat | |
|    80°C for 60 minutes | Negligible loss of titer |
|    100°C for 10 to 60 minutes | Substantial but incomplete loss |
|    80°C for 20 minutes after | Substantial loss |
|       fluorocarbon treatment | |
| UV irradiation (240, 250, 254, 265, | Negligible loss of titer |
|    280, 290, 315, and 330 nm) | |
| UV at 254 nm for 700 seconds | 1.6 log drop |
| Ether | Sensitive |
| Formalin, 0.5–18% | Resistant |
| β-Propiolactone, 1% | Slight loss of titer |
| Periodate, 0.01 M | Sensitive |
| Urea, 6–8 M | Sensitive |
| Phenol, 90% | Sensitive |
| Pepsin, trypsin, acetylethyleneamine | Resistant |
| Deoxyribonuclease, ribonuclease | Resistant |

[a] Data from Gibbs and Gajdusek (1971), which contains detailed references.

Reliable experiments reveal that the infectivity of scrapie is filterable. As assayed by infectivity, the scrapie agent passes through membranes of 43 but not 27 nm average pore diameter. High resistance to ionizing and ultraviolet irradiation (Alper *et al.*, 1966; Haig *et al.*, 1969) suggest that the nucleic acid has a very small target area, equivalent to $2 \times 10^5$ daltons, which is hardly enough to code for one small protein. Infectivity is highly resistant to heating, formalin, proteases, and nucleases but susceptible to ether, periodate, urea, and phenol (Table 3–12). However, it should be remembered that all these tests have been carried out with impure preparations, and heat resistance, for instance, is greatly reduced after fluorocarbon treatment. There seems to be a close association between the infectivity and cellular membrane fragments (Lampert *et al.*, 1971b; Kimberton *et al.*, 1971).

The radiation target size and the lack of immunogenicity are compatible with the notion that the scrapie agent is a free nucleic acid of up to $2 \times 10^5$ daltons molecular weight, that is normally closely associated with and protected by cellular membranes (Diener, 1972a; Hunter, 1972). There is a model for such an infectivity in the "viroid" of potato spindle tuber disease, which is a ribonucleic acid with a molecular weight of about 50,000 daltons that is protected by its close association with nuclear chromatin (review, Diener, 1972b).

## GENERAL OBSERVATIONS

This summary of current knowledge about the structure and chemical composition of the virions of animal viruses is based on observations made on representative viruses of the various genera. One advantage of a valid system of viral classification is that the findings should, in general, apply reasonably well to other species within these taxa.

Basic features of the structure of animal viruses have already been presented in Chapter 1 (Tables 1–1 and 1–8, and Fig. 1–3). Here we shall set out, largely in tabular form, some general aspects of their chemical composition and add to the generalizations already given in Chapter 1.

### Proteins

Present evidence suggests strongly that the majority of polypeptides associated with purified virions are virus-specified. Two kinds of exceptions are worth noting: (a) chance inclusion of minor proteins associated with cellular membranes in the envelopes of some enveloped viruses, e.g., ATPase in avian myeloblastosis virus particles grown in myeloblasts, but not in virions grown in cells that lack ATPase in their membranes (de-Thé et al., 1963); (b) cellular proteins that may be included in the cores of some icosahedral viruses, e.g., the basic proteins of the polyomaviruses.

Proteins in the virion are of four main kinds, capsid, core, envelope, and enzymatic (Tables 3–13 and 3–14). In most isometric viruses, several polypeptide molecules (of one or more kinds) combine to form the basic morphological units (capsomers) of the capsid, whereas the capsomers of all tubular nucleocapsids are monomers of a single polypeptide species. It is clear that the term "core" is employed rather too all embracingly, to mean different things by different workers. The core of reovirions is more correctly described as an inner capsid since it shows typical icosahedral symmetry; herpesvirions have several concentric capsids. The "core" of adenovirions appears to consist of histonelike proteins bound intimately to the DNA. That of the poxviruses is probably extremely complex and it is likely that further research will reveal that its seventeen polypeptides are distributed in several layers.

RNA animal viruses with tubular nucleocapsids have envelopes that contain two kinds of proteins, an inner nonglycosylated membrane or matrix protein, and projecting peplomers which are always glycoproteins. The carbohydrate

TABLE 3–13

*Proteins of the Virion*

| GENUS | PROTOTYPE VIRUS | TOTAL STRUCTURAL POLYPEPTIDES | ENVELOPE GLYCOPOLYPEPTIDES | ENVELOPE "MEMBRANE PROTEIN" | OUTER CAPSID | INNER CAPSIDS AND/OR CORE | HELICAL NUCLEOCAPSID |
|---|---|---|---|---|---|---|---|
| *Parvovirus* | AAV | 3 | 0 | 0 | ? | ? | 0 |
| *Papovaviridae* | Polyoma | 6[a] | 0 | 0 | ? | 3[a] | 0 |
| *Adenovirus* | Human 2 | 5–9 | 0 | 0 | 3–6 | 2–3 | 0 |
| *Herpesvirus* | HSV-1 | 12–24[b] | 3–12[b] | ? | ? | 2+ | 0 |
| *Iridovirus* | FV3 | >15 | 0 | 0 | ? | 5 | 0 |
| *Poxvirus* | Vaccinia | ~30 | 2[c] | 0 | ? | 17 | 0 |
| *Enterovirus* | Poliovirus 1 | 4 | 0 | 0 | 4 | 0 | 0 |
| *Rhinovirus* | FMDV | 4 | 0 | 0 | 4 | 0 | 0 |
| *Reovirus* | Reovirus 3 | 7 | 0 | 0 | 3 | 4 | 0 |
| *Orbivirus* | Bluetongue | 7 | 0 | 0 | 2[d] | 5[d] | 0 |
| *Orthomyxovirus* | Influenza A | 7 | 3–4 | 1 | 0 | 1–2[e] | 1 |
| *Paramyxovirus* | SV5 | 6 | 2 | 1 | 0 | ?[e] | 1 |
| *Togaviridae* | Sindbis | 3 | 2 | 0 | 1 | 0 | 0 |
| *Rhabdovirus* | Vesicular stomatitis | 7 | 1 | 1 | 0 | 4[e] | 1 |
| *Leukovirus* | Rous sarcoma | 7–8 | 2 | 1 | ? | ? | ? |
| *Bunyamwera* supergroup | Uukuniemi | 2+ | 1+ | ? | 0 | 0 | 1 |

[a] Three core polypeptides are of cellular origin.
[b] Different investigators give very different figures.
[c] Not in envelope.
[d] An icosahedral capsid composed of 5 polypeptides, surrounded by an amorphous layer containing 2 polypeptides.
[e] Minor polypeptides associated with the helical nucleocapsid.

TABLE 3-14

*Enzymes Associated with the Virion*

| GENUS | POLYMERASE: WITHIN CORE OR ASSOCIATED WITH NUCLEOCAPSID | OTHER | |
|---|---|---|---|
| | | NATURE | LOCATION |
| *Parvovirus* | — | — | |
| *Polyomavirus* | — | — | |
| *Adenovirus* | — | DNA endonuclease | Within capsid |
| *Herpesvirus* | — | Protein kinase | Within capsid |
| *Iridovirus* | ? | Nucleotide phosphohydrolase | Within core [a] |
| | | DNA endonuclease; SS, DS, pH 5 | |
| | | DNA endonuclease; SS, DS, pH 7.5 | Within capsid, |
| | | RNA endonuclease; SS, DS, pH 7.5 | outside core [a] |
| | | Protein kinase | |
| *Poxvirus* | DNA-RNA | Nucleotide phosphohydrolase | |
| | | DNA exonuclease; SS DNA, pH 5 | |
| | | DNA endonuclease; SS DNA, pH 7 | Within core |
| | | Protein kinase | |
| *Enterovirus* | — | — | |
| *Alphavirus* | — | — | |
| *Orthomyxovirus* | RNA-RNA | Neuraminidase | Peplomers in |
| *Paramyxovirus* | RNA-RNA | Neuraminidase | envelope |
| *Rhabdovirus* | RNA-RNA | ? Protein kinase | |
| | | Nucleotide phosphohydrolase | Adsorbed to envelope |
| *Leukovirus* | RNA-DNA | DNA exonuclease | |
| | | DNA endonuclease | Associated with core |
| | | DNA ligase, etc. [b] | |
| *Reovirus* | RNA-RNA | Nucleotide phosphohydrolase | Within core |

[a] Location difficult; core not well defined.
[b] Many others, listed on page 136.

151

moiety of envelope glycoproteins is characteristic of the cell of origin, but the glycoproteins of vaccinia virus, which are located just outside the core, are completely specified by the viral genome.

Many viruses contain enzymes as intrinsic parts of the virion. These are of two main types: (a) neuraminidase in the envelopes of myxoviruses, and (b) transcriptases and other enzymes concerned with nucleic acid synthesis and degradation, in many genera (Table 3–14). It is quite likely that some enzymes also have a structural role, e.g., the RNP-associated transcriptases of most RNA viruses. As might be expected, there are more viral enzymes in the larger and more complex viruses than in the small viruses but none have yet been unequivocally proved to be virus-specified.

## Lipids

Poxviruses contain lipid that is an intrinsic component of the virion, but is synthesized *de novo* in the cytoplasmic factories where these viruses multiply. In all other animal viruses, with the possible exception of the iridoviruses of vertebrates (which have been inadequately studied), lipids, if present (Table 3–15), occur only in the viral envelope, and are derived from the cellular membranes through which enveloped viruses mature by budding. Different viruses do, however, select somewhat different combinations of lipids from those available in the plasma membrane of their cell of origin. Viral glycolipids are derived directly from the cellular membrane and contain the blood group and Forssman antigens.

TABLE 3–15

*Chemical Composition of Representative Viruses*

| | | PERCENTAGE DRY WEIGHT | | | | |
|---|---|---|---|---|---|---|
| GENUS | EXAMPLE | DNA | RNA | PROTEIN | LIPID | CARBOHYDRATE |
| *Parvovirus* | AAV | 20 | — | 80 | — | — |
| *Polyomavirus* | SV40 | 12 | — | 88 | — | — |
| *Adenovirus* | Human | 12–13 | — | 87 | — | — |
| *Herpesvirus* | Herpes simplex 1 | 7–10 | — | 67 | 20 | 1.5 |
| *Iridovirus* | FV3 | 30 | — | 56 | 14 | — |
| *Poxvirus* | Vaccinia | 5 | — | 88 | 4 | 3 |
| *Enterovirus*[a] | Poliovirus 1 | — | 30 | 70 | — | — |
| *Rhinovirus*[a] | Rhinovirus 14 | — | 30 | 70 | — | — |
| *Calicivirus*[a] | Vesicular exanthem | — | 22 | 78 | — | — |
| *Alphavirus* | Semliki Forest | — | 7 | 62 | 24 | 6 |
| *Orthomyxovirus* | Influenza A | — | 1 | 70 | 23 | 6 |
| *Paramyxovirus* | SV5 | — | 1 | 73 | 20 | 6 |
| *Rhabdovirus* | Vesicular stomatitis | — | 3 | 64 | 20 | 13 |
| *Leukovirus* | Rous sarcoma | — | 2 | 60 | 30 | ?8 |
| *Reovirus* | Reovirus 3 | — | 15 | 85 | — | — |

[a] Family Picornaviridae.

## Nucleic Acids

The nucleic acids enclosed within virions, like the virion proteins, are in most cases viral and not of cellular origin. However, there are exceptions. Some *Polyomavirus* "pseudovirions" contain only host DNA, and in the same genus other virions carry sequences of host DNA covalently bonded to viral DNA. The complex virions of leukoviruses always include some low molecular weight cellular RNA's and often considerable quantities of cellular DNA, while arenaviruses regularly contain cellular ribosomes.

Viral nucleic acid takes many forms. In DNA viruses, the genome usually consists of a single molecule, linear or cyclic, which may contain single-stranded breaks (Table 3–16). The parvoviruses, however, contain single-stranded DNA and separate virions encapsidate + or − strands. Parvovirus DNA carries so few genes (about 3) that some members of the genus are defective, whereas the largest DNA viruses (poxviruses) carry well over 100 genes (see Chapter 7).

TABLE 3–16

*Properties of DNA's of Representative Animal Viruses*

| GENUS | EXAMPLE | AMOUNT [a] | NATURE [b] | INFEC-TIVITY | LENGTH ($\mu$m) | BUOYANT DENSITY (CsCl) | PERCENT-AGE G + C |
|---|---|---|---|---|---|---|---|
| *Parvovirus* | MVM | 1.8 | S,L | — | | | 48 |
| | AAV-1 | 1.5 | S,L [c] | + [d] | 1.5 | | 58 |
| *Polyomavirus* [e] | Polyoma | 3 | D,C | + | 1.5 | | 48 |
| *Papillomavirus* [e] | Rabbit papilloma | 5 | D,C | + | 2.5 | | 48 |
| *Adenovirus* | Human type 2 | 24 | D,L | ? | 12.6 | 1.716 | 57 |
| | Human type 12 | 22.5 | D,L | — | 12 | 1.708 | 49 |
| | Avian (CELO) | 29 | D,L | — | | 1.713 | 54 |
| *Herpesvirus* | HSV1 | 100 | D,L | — | 50 | 1.727 | 68 |
| | Pseudorabies | | | — | | 1.732 | 74 |
| | Marek's disease | 120 | D,L | — | | 1.705 | 46 |
| *Iridovirus* | FV-3 | 130 | D,L | — | 70–80 | 1.720 | 57 |
| *Poxvirus* | Vaccinia | 160 | D,L | — | 80 | 1.695 | 37 |

[a] Molecular weight in million daltons.
[b] S, single-stranded; D, double-stranded; L, linear; C, cyclic.
[c] Virions contain a single strand of either polarity (+ or −).
[d] In cells preinfected with helper adenovirus.
[e] Members of family Papovaviridae.

The RNA animal viruses present a very diverse picture in the nature of their RNA and in the ways in which the viral genetic information is eventually translated into polypeptides (Table 3–17, and see Chapter 6). Double-stranded RNA occurs in only two genera, in both of which the genome is fragmented. All other RNA viruses contain only single-stranded RNA, usually in the form of a single continuous molecule, but fragmented in the case of *Orthomyxovirus*. The RNA's extracted from viruses of the families Picornaviridae and Togaviridae are

TABLE 3–17

*Properties of RNA's of Representative Animal Viruses*

| GENUS | EXAMPLE | AMOUNT[a] | NATURE[b] | SENSE[c] | INFEC-TIVITY | BASE COMPOSITION (%) A | G | C | U |
|---|---|---|---|---|---|---|---|---|---|
| *Enterovirus*[d] | Poliovirus 1 | 2.6 | S,1 | + | + | 29 | 24 | 22 | 24 |
| (*Cardiovirus*)[d] | EMC | 2.7 | S,1 | + | + | 27 | 24 | 23 | 26 |
| *Rhinovirus*[d] | Rhinovirus 2 | 2.6 | S,1 | + | + | 34 | 20 | 20 | 26 |
| (*Aphthovirus*)[d] | FMDV | 2.8 | S,1 | + | + | 26 | 24 | 28 | 22 |
| *Alphavirus*[e] | Sindbis | 4 | S,1 | + | + | 30 | 26 | 25 | 20 |
| *Flavivirus*[e] | Dengue | 4 | S,1 | + | + | 31 | 26 | 22 | 21 |
| *Orthomyxovirus* | Influenza A | 4.8 | S,7 | − | − | 23 | 20 | 24 | 33 |
| *Paramyxovirus* | NDV | 7.5 | S,1 | − | − | 22 | 25 | 23 | 30 |
| | SV5 | 6.5 | S,1 | − | − | 28 | 21 | 21 | 30 |
| *Rhabdovirus* | Vesicular stomatitis | 4 | S,1 | − | − | 28 | 20 | 22 | 30 |
| | Rabies | 4.6 | S,1 | − | − | 26 | 21 | 23 | 29 |
| *Leukovirus* | Rous sarcoma | 10 | S,4 | − | − | 25 | 29 | 24 | 22 |
| | Rauscher leukemia | 12 | S,4 | − | − | 25 | 26 | 26 | 23 |
| *Reovirus* | Reovirus 3 | 15 | D,10 | − | − | 28 | 22 | 22 | 28 |
| *Orbivirus* | Bluetongue | 15 | D,10 | − | − | 28 | 22 | 21 | 29 |

[a] Molecular weight in million daltons.

[b] S, single-stranded; D, double-stranded; number, number of molecules in genome.

[c] +, indicates that the (single-stranded) viral RNA functions as mRNA; −, mRNA (or in *Leukovirus*, DNA) is transcribed from viral RNA.

[d] Members of family Picornaviridae.

[e] Members of family Togaviridae.

infectious, because they can be directly translated into polypeptides by ribosomes (i.e., the viral RNA functions as a messenger RNA). Among other RNA viruses, however, mRNA must first be transcribed by a transcriptase present in the core of the virion. Details of these processes will be discussed in Chapter 6.

## CHAPTER 4

# Structure and Function of the Animal Cell

## INTRODUCTION

Many of the differences between animal viruses and bacteriophages stem from the fact that the animal host is a complex and integrated assembly of millions of cells, whereas the bacterial cell is an isolated independent unit. Nevertheless, from the point of view of the central interest of virology, i.e., viral multiplication, the individual cell is the important unit, and something must be known of the structure and function of animal cells before viral multiplication and the effects of viruses on cells can be properly understood. A six-volume treatise has been written on "The Cell" (Brachet and Mirsky, 1959–1964) and Willmer's three-volume work "Cells and Tissues in Culture" (1965–1966) contains much information of direct interest to virologists. In this chapter, we will present a brief summary of the structure and function of a nonspecialized vertebrate cell, concentrating attention on aspects that may be important for the understanding of viral multiplication. In Chapter 9, features of some specialized cells and tissues which are of particular importance in the pathogenesis of viral infections will be described. Possible mechanisms of action of several metabolic inhibitors are described in terms of their effects on cellular and viral functions, as a background for understanding their use in studies of viral multiplication (Chapters 5 and 6).

## STRUCTURE OF AN UNDIFFERENTIATED ANIMAL CELL

Our knowledge of the ultrastructure of cells comes, for the most part, from electron microscopy. Cells in tissue culture, with which most studies of viral multiplication are carried out, usually appear undifferentiated in electron micrographs (Fig. 4–1).

Animal cells are bounded externally by the cytoplasmic (or plasma) membrane, within which is contained the cytoplasm and a number of cytoplasmic organelles, the mitochondria, ribosomes, Golgi apparatus, endoplasmic reticulum, and lysosomes being the most prominent. In the cells of higher organisms the genetic apparatus is concentrated in the nucleus, which also contains organelles called nucleoli.

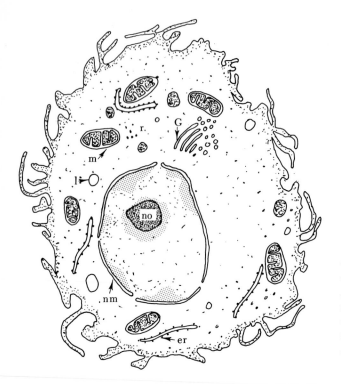

FIG. 4–1. *Diagram of an undifferentiated animal cell, showing various organelles. G, Golgi apparatus; er, endoplasmic reticulum; l, lysosome; m, mitochondrion; nm, nuclear membrane; no, nucleolus; r, ribosomes.*

## THE CYTOPLASMIC MEMBRANE

### Structure

The cytoplasmic membrane (plasma membrane) plays an important part in many aspects of cell behavior. Not only does it form the retaining sheath which encloses the cellular constituents, but also it is responsible for creating and maintaining concentration gradients of many different types of cellular metabolites. Furthermore, there is growing evidence that division and differentiation of cells are influenced by events which occur at the membrane, and most pertinent to this book, the membrane provides the first surface that a virus encounters when it infects a cell. Unlike some bacteriophages, which use enzymes to digest the rigid bacterial cell wall and a contractile protein to inject their genetic material into the cell, most animal viruses are taken into the cell by an active process of engulfment (Chapters 5 and 6). The specificity which determines whether or not viral attachment will occur resides on the outer surface of the cytoplasmic membrane, and movement of the membrane and the cytoplasm beneath it leads to the engulfment of some of the particles which do attach. For many viruses, the membrane is also the last barrier that they must cross when they leave the cell (or what remains of it) at the end of the infectious process. For others (for example, the lipid-containing RNA viruses), the plasma membrane is the site at which the final steps of viral synthesis occur, and modified membrane material becomes part of the virion as they mature.

The outer surface of mammalian cells consists of a "unit membrane" consisting of a phospholipid bilayer, the two monolayers of which are about 20 Å thick separated by a gap of about 35 A (review by Robertson, 1959). There are both charged and uncharged lipids in the bilayer. The charged lipids are arranged with their polar groups to the outside and their hydrophobic chains to the inside of the unit membrane, and it seems probable that there is an asymmetric distribution of the charged lipids between the two monolayers, with phosphatidylcholine and sphingomyelin largely in the outer layer, and phosphatidylserine and phosphatidylethanolamine largely in the inner layer (Bretscher, 1972). The neutral lipids (e.g., cholesterol) are probably evenly distributed.

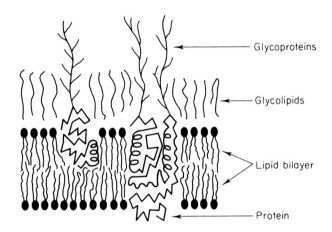

Glycoproteins

Glycolipids

Lipid bilayer

Protein

FIG. 4–2. *Lipid–globular protein mosaic model of membrane structure (modified from Singer and Nicolson, 1972).*

Much of the protein of the membrane is found in the gap between the lipid bilayer (Branton, 1969) in globular particles which can be seen in freeze-etched sections. Most probably these proteins protrude through both tiers of the bilayer. Complex oligosaccharides are often attached both to the lipids exposed on the outer cell surface, to form cerebrosides, gangliosides, and other complex glycolipids, and to the proteins, to form glycoproteins (see Fig. 4–2 and review by Singer and Nicolson, 1972). There is no reliable estimate of the number of different proteins found in the plasma membranes of cells in culture, or of their molecular weights and functions.

It is now clear that membranes are much more fluid structures than was believed at one time. The movement of lipid components has been followed by electron spin labeling and it has been found that there is an extremely rapid lateral diffusion of lipid molecules within both layers of the unit membrane (Hubbell and McConnell, 1968); the translocation of lipid from one monolayer to the other is much less rapid (Kornberg and McConnell, 1971). Furthermore, several types of experiments involving addition of specific antibodies or lectins to cells carrying the complementary surface receptors have shown that proteins in the membrane are also very mobile (Frye and Edidin, 1970; Taylor et al., 1971; Nicolson, 1971). Because radical differences in the distribution of protein molecules on cell surfaces can occur within minutes, it seems that the proteins in the membranes behave like icebergs gliding in a moving sea.

## Cell Movement and Contact Inhibition

Given this idea of fluid membranes, it is not surprising that the outer surface of cells shows great motility when seen in time-lapse cinematographs. Long villi may protrude from the cytoplasmic membrane and long indentations and pleats may occur in it. In sparse cultures, before a monolayer is formed, the cells themselves are quite mobile. Ruffles, which may reach a height of 5 $\mu$m, form continuously on the leading edge of the moving cell and travel backward toward the nucleus, and the cell moves across the glass by making intermittent contacts generated by the peristaltic movements of the cell membrane. The mechanism by which this movement occurs is unknown, but seems to be mediated by one or other of the classes of cytoplasmic fibers visible in cells of all types (see Goldman, 1971). When a ruffled membrane makes contact with a neighboring cell, adhesion results and the ruffles become stationary. If many contacts are made, large-scale movement of the cell ceases. This is called contact inhibition (Abercrombie and Ambrose, 1958). Contact between normal cells also seems to inhibit cellular division, so that normal cells grow to form a monolayer and then cease dividing or grow at a much reduced rate. Such contact inhibited cells have extremely low rates of macromolecular synthesis (Fried and Pitts, 1968) and most of their cytoplasmic polyribosomes disappear (Levine *et al.*, 1965). At one time it was thought that the growth of cells in culture was controlled completely by contact inhibition (Abercrombie and Heaysman, 1954), so that when the surfaces of two cells touched there was mutual inhibition of both movement and division. It is now clear that the situation is more complicated. First, it appears that inhibition of movement and division may operate through different mechanisms (Stoker and Rubin, 1967) and for this reason, at present, the term "contact inhibition" is generally restricted to arrest of movement; cessation of growth is called "density-dependent inhibition." Second, density-dependent inhibition can be reversed to some extent by restoration of alkaline pH in the culture medium (Ceccarini and Eagle, 1971) as well as by addition of growth factors in serum (Kruse and Miedema, 1965; Holley and Kiernan, 1968). What these factors are and how they act is unknown. In any case, neoplastic cells or cells transformed *in vitro* by tumor viruses require less of these factors to sustain growth than do normal cells (Dulbecco, 1970; Jainchill and Todaro, 1970). Such cells continue to divide at a rapid rate even at high cell densities, so that they form piled-up multilayers and reach saturation densities 10–25 times higher than normal cells. As we shall see in Chapter 13, it is this differential growth that forms the basis of the transformation assay for tumor viruses.

## Cytosis

Nutrients may enter cells by an active process which involves special enzymes, the permeases. In addition to this, the exposed surfaces of cells actively engulf droplets of liquid from their environment by a process called pinocytosis, and they may also ingest particles by a similar process which is called phago-cytosis. Pinocytosis may be an important stimulus to lysosome production, at least in macrophages (Cohn and Benson, 1965); phagocytosis plays a major role in the infection of cells by viruses. Pinocytosis and phagocytosis have much in common, and it is convenient to use the term "cytosis" to include both. Cytosis is temperature-dependent and does not occur at 2°C; it is inhibited reversibly by metabolic poisons like sodium fluoride (Sbarra and Karnovsky, 1959). In cytosis, pseudopodia form and fuse to enclose drops of medium or particles adhering to the cell surface. The vesicles thus formed separate from the membrane and often move across the cytoplasm, and lysosomes discharge their contents into these vesicles. In cells exposed to viral suspensions, the phagocytic vesicle may contain ingested virions, which often undergo at least the first stage of uncoating there.

## CYTOPLASMIC ORGANELLES

Electron microscopy reveals the presence of several types of membrane-bound structures within the cytoplasm. The functions of these components have been analyzed after separation by centrifugation of disrupted cells. When care is taken to protect the organelles from osmotic damage, four main fractions are obtained: the nuclear, mitochondrial, and microsomal fractions, and the supernatant.

### Mitochondria

Mitochondria are the largest cytoplasmic organelles; they consist of an enclosing membrane and an inner membrane which is thrown into folds to form a large internal surface. Biochemical studies on isolated mitochondria have shown that they contain a large number of enzymes, some adsorbed on the internal membrane and some in solution. The soluble enzymes are especially concerned with the oxidation of the breakdown products of fat and carbohydrate, while the adsorbed enzymes use the reducing power so obtained to regenerate adenosine triphosphate (ATP). Mitochondria are semiautonomous; they synthesize proteins and are capable of division. Each of them contains several molecules of DNA apparently identical to each other in base sequence (Borst, 1969; Clayton et al., 1970). Most of this DNA occurs as double-stranded, closed circular molecules with a molecular weight of about $10^7$ daltons (see reviews by Borst and Kroon, 1969; Nass, 1969a). However, dimers (Nass, 1969b), catenates (Clayton and Vinograd, 1967), and replicating forms (Robberson et al., 1972) have also been described. Mitochondria contain DNA-dependent RNA polymerase which transcribes both strands of the DNA symmetrically in vivo (Aloni and Attardi, 1971). The RNA complementary to one of the strands is rapidly degraded, and the stable RNA that remains seems to be composed mainly of 16 and 12 S species which are incorporated into mitochondrial ribosomes (Attardi and Ojala, 1971) and a family of 12 to 15 different 4 S molecules, some of which serve as transfer RNA's (Galper and Darnell, 1969; Nass and Buck, 1970).

Very little is known about protein synthesis in mitochondria; their ribosomes sediment at 60 S and are therefore much smaller than cytoplasmic or bacterial ribosomes (Attardi and Ojala, 1971). The source of the messenger RNA that they translate and the nature of the polypeptides that they synthesize are unknown. However, the amount of information in mitochondrial DNA is not enough to code for all the proteins and enzymes present in the organelles, so at least some of these proteins (for example, cytochrome c, the genes for which show Mendelian inheritance) must be coded in nuclear DNA. Whether these proteins are synthesized inside the mitochondria or are ferried from the cytoplasm is unclear.

Although the proper functioning of the mitochondrion may be needed to supply the energy required for viral multiplication, no direct association has ever been reported between these organelles and viruses.

### Lysosomes

Lysosomes are slightly smaller than mitochondria and consist of vesicles which contain a variety of hydrolytic enzymes, bounded by a single smooth lipoprotein membrane (review, de Duve, 1963). The lysosomal membranes appear to be formed by budding from the Golgi apparatus, particularly when pinocytic vesicles make contact with it (review, de Robertis et al., 1970). Lysosomes occur in most vertebrate cells but are particularly numerous in phagocytes, in which they constitute the granules familiar to hematologists (Hirsch and Cohn, 1964). Following phagocytosis, the lysosomes discharge their contents into the phagocytic vesicle and thus lead to the rapid digestion of its contents.

The lysosomal membrane effectively segregates digestive enzymes from the cytoplasm of the cell. Allison (1967a) has suggested that lysosomes may participate at three stages in the growth cycle of animal viruses (see Chapter 9). They may supply enzymes which uncoat the virion within the phagocytic vesicle. Lysosomal enzymes may also be involved in the breakdown of host–cell polynucleotides, which results in a markedly increased pool of acid-soluble nucleotides in virus-infected cells. Finally, they may play a part in cell degeneration, facilitating the liberation of those viruses which mature at an intracellular site rather than at the cell membrane, and producing the cytopathic effects (CPE) that are used as an index of viral growth in cultured cells (see Chapters 2 and 9).

### Golgi Apparatus

The Golgi complex consists of stacks of flattened vesicles situated at one pole of the cell and arranged so that one end of the stack is close to the nuclear envelope and the other near secretory vesicles. The complex can be isolated from cell homogenates by differential centrifugation (Morré, 1969). The function of the Golgi apparatus is not known precisely but it seems to be the site at which sugar residues are attached to proteins (see review by Neutra and Leblond, 1969). Because it is especially well developed in secretory cells, it is thought that the complex is involved in transporting proteins from the endoplasmic reticulum to the outside of the cell.

## Ribosomes

The mitochondria are the largest and most complex of the cytoplasmic organelles; the smallest are the ribosomes, but these are so numerous (about $10^7$ in a single HeLa cell) that up to 25% of the total production of macromolecules in the cell may be devoted to their manufacture. Ribosomes are oblate spheroids measuring about $25 \times 18$ nm with an approximate sedimentation coefficient of 80 S (see Spirin and Gavrilova, 1969). In negatively stained preparations examined with the electron microscope (Plate 4–1), it is clear that each

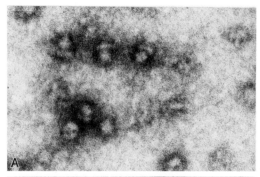
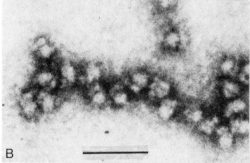

PLATE 4–1. *Ribosomes and polyribosomes from mouse cells, in negatively stained preparations (bar = 100 nm). (A) Monomers, showing the small and large subunits. (B) Large polyribosome. The subunit structure is visible in individual ribosomes, which are connected by a strand the axis of which runs through the small subunits (Shelton and Kuff, 1966, courtesy Dr. E. Shelton).*

ribosome consists of two dissimilar subunits (60 and 40 S), which separate from each other in the absence of magnesium ions or in solutions containing high concentrations of KCl (Girard *et al.*, 1965; Martin and Wool, 1968). Each subunit consists of a large number of different polypeptide chains, which have not yet been cataloged completely, as well as one or two molecules of RNA. The 60 S subunit contains $1.75 \times 10^6$ daltons of RNA, apparently in two pieces, which are hydrogen-bonded together. In nondenaturing conditions, this RNA sediments at 28 S; however, after heating or during centrifugation in dimethyl sulfoxide, a 7 S fragment is released (Pene *et al.*, 1968). The 40 S subunit contains one piece of RNA which sediments at 18 S and has a molecular weight of $0.70 \times 10^6$ daltons. In addition, ribosomes contain a piece of 5 S RNA whose location is unknown, but whose sequence has been determined (Forget and Weissman, 1969; Williamson and Brownlee, 1969). Ribosomal RNA's possess several distinguishing characteristics: they are high in G and C, a high proportion of the bases are methylated, and they contain "odd" bases such as pseudouridine (review, Maden, 1971).

Virtually all the steps in the biosynthesis of ribosomes take place in the nucleolus, for it is there that "ribosomal DNA" is situated (review, Birnstiel *et al.*, 1971). This DNA includes sequences complementary to 18 and 28 S ribosomal RNA species as well as "spacer" regions. The DNA is highly redundant and saturation hybridization experiments show that in HeLa cells it contains about 1100 copies of ribosomal sequences (Jeantur and Attardi, 1969). These are probably arranged in contiguous units in which the sequences for 18 and 28 S RNA's alternate (see Brown and Weber, 1968). The ribosomal DNA is transcribed probably by a distinct DNA-dependent RNA polymerase (Blatti *et al.*, 1970) into an RNA molecule with a sedimentation coefficient of 45 S and a molecular weight of $4.1 \times 10^6$ daltons. This RNA is a precursor of ribosomal RNA and is sequentially and specifically cleaved into several pieces, two of which are the final 18 and 28 S ribosomal RNA's (reviews by Perry, 1969; Darnell, 1968b; Attardi and Amaldi, 1970; Burdon, 1971; Maden, 1971). How this cleavage occurs and how the ribosomal sequences are preserved while the rest of the precursor is degraded is unknown.

The newly formed ribosomal RNA complexes with ribosomal proteins in the nucleolus, and fully formed 40 and 60 S ribosomal subunits enter the cytoplasm (Girard *et al.*, 1965; Perry, 1965). Although several structures intermediate in the assembly process have been isolated, the mechanism of ribosome morphogenesis remains obscure. The site of synthesis of the ribosomal proteins is unknown; there have been claims from time to time that they are made on "nuclear ribosomes," but it now seems more reasonable that these proteins are synthesized on conventional cytoplasmic ribosomes and are then transported back into the nucleolus (review by Maden, 1971).

Once in the cytoplasm, the newly synthesized 40 and 60 S particles enter a pool of ribosomal subunits, 80 S monosomes, and polyribosomes (Girard *et al.*, 1965; Joklik and Becker, 1965), the latter consisting of groups of ribosomes held together by a strand of messenger RNA (Plate 4–1). They constitute the "work bench" upon which polypeptides are synthesized. All viral polypeptides are probably synthesized on polyribosomes, and even for virions which are assembled in the nucleus (like adenoviruses), the viral proteins appear to be synthesized on polyribosomes in the cytoplasm (Thomas and Green, 1966).

## Endoplasmic Reticulum

In cells which manufacture and secrete large amounts of a particular sort of protein (e.g., plasma cells making antibody, and secretory cells of the pancreas and other glands), the ribosomes are associated with membranes, making up the "rough" endoplasmic reticulum. The rough endoplasmic reticulum is poorly developed in most cells used for the culture of viruses, but particle-covered membranes are never entirely absent, even in HeLa cells. The "microsome" fraction of cell homogenates consists mainly of isolated ribosomes and fragments of the endoplasmic reticulum; the polyribosomes are usually disrupted in such preparations by enzymatic destruction of the mRNA which holds the ribosomes together.

## THE NUCLEUS

In contrast to bacteria, the genetic apparatus of cells of higher organisms is segregated from the cytoplasm within a special cellular organelle, the nucleus. This is surrounded by a nuclear membrane and contains the chromosomes and one or more specialized organelles called nucleoli.

### The Nuclear Envelope

The nucleus of interphase cells is surrounded by an envelope consisting of two concentric unit membranes separated by a space of 10 to 15 nm. The envelope appears to be continuous with the membranes of the endoplasmic reticulum and is pierced at regular intervals by pores, where the parallel inner and outer membranes join around a gap of variable diameter. The pores account for about 10% of the surface area of the nuclear envelope (Watson, 1959). Negative staining techniques have shown that the pores are octagonal in shape and are surrounded by circular structures called annuli (Gall, 1967), which may act as diaphragms to control the passage of particles to and from the nuclei. The nuclear envelope disappears during mitosis by breaking up into small separate vesicles, but reforms late in telophase, presumably from the cytoplasmic membrane system (Barer *et al.*, 1959).

### The Chromosomes

There are about $3.8 \times 10^{12}$ daltons of DNA in diploid mammalian cells. Cultured cells usually contain slightly more DNA than this because they tend to be aneuploid. Whether the DNA exists as one piece or many and its arrangement in interphase nuclei are unknown (for discussion, see Dupraw, 1970). However, there is no doubt that most of the DNA is associated with a variety of basic proteins, known as protamines and histones (reviews by Elgin *et al.*, 1971; DeLange and Smith, 1971). Protamines are a very simple and homogeneous group of polypeptides with molecular weights around 4000 daltons; histones are larger and more diverse with molecular weights in the range 10–18,000 daltons. Both sorts of proteins are very rich in basic amino acids, such as arginine and lysine, and seem to attach to DNA by binding to one of the grooves of the double helix (review, DeLange and Smith, 1971). Histones and protamines undoubtedly control the expression of the genes coded in DNA in some manner (Paul and Gilmour, 1968), but the mechanism by which this regulation is achieved is still obscure.

The DNA–protein complex (chromatin) is not usually visible in the living interphase nucleus. Just before cell division, however, the chromatin condenses into chromosomes which can be seen in the light microscope. The number of chromosomes and their structure are highly characteristic features of each animal species (Hsu and Benirschke, 1967). The different chromosomes that make up the karyotype of an animal can now be unambiguously distinguished from one another by the following three features. First, each chromosome has a distinctive size. Second, every chromosome has a small constriction—the centromere—which forms the attachment point of the chromosome to the mitotic spindle. Because the position of the centromere is at a fixed site on each chromosome, it can be used as an easy landmark for classification purposes. Third, each chromosome has a characteristic banding pattern when stained with Giemsa after alkaline or heat denaturation of the DNA (Gall and Pardue, 1969; Patil et al., 1971), or with quinacrine mustard (Caspersson et al., 1970).

During interphase the chromosomes become so long and thin as to be virtually invisible, except in two unusual cases, the lampbrush and polytene chromosomes of certain oocytes and diptera, respectively. Lampbrush chromosomes have an axis from which a series of paired loops project laterally. Autoradiographical studies have demonstrated that there is active synthesis of RNA and protein in the loops (Izawa et al., 1963). Each loop follows its own pattern of activity, some being continuously active while others are active only at the beginning or end of oocyte growth. For a particular locus in the lampbrush chromosome of a newt, Gall and Callan (1962) found that RNA synthesis started at one end of the loop and progressed gradually toward the other end (review, Callan, 1963). Beautiful electron micrographs of transcription of lampbrush chromosome loops have been obtained by Miller and his colleagues (Miller et al., 1970).

Beermann and his colleagues (Beermann and Clever, 1964) have shown that "puffs" in the polytene chromosomes of various diptera are also engaged in transcription (Ashburner, 1970). One can conclude that "looping out" and "puffing" are manifestations of regional gene action and suggest that differential gene action during development is, at least in part, controlled through mechanisms that determine the specific coiling and uncoiling of regions of the chromosome.

## Nucleoli

Nucleoli are the only organelles easily visible in the nucleus of living cells. Usually there is one nucleolus for each haploid set of chromosomes, associated with specialized regions of certain chromosomes called nucleolar organizer regions. The nucleoli disappear during cell division and reappear at telophase by reforming around these organizer regions.

Nucleoli have no surrounding membrane, but they can be isolated and purified very easily (Penman, 1969). DNA–RNA hybridization studies using DNA from isolated nucleoli (McConkey and Hopkins, 1964; Steele, 1968) or fractionated chromosomes of HeLa cells (Huberman and Attardi, 1967) have shown that the nucleoli and nucleolar organizer regions contain the genes for 18 and 28 S ribosomal RNA. We have already seen that the nucleoli are the site at which these genes are transcribed and the RNA product is processed and assembled into ribosomes.

## THE CELL CYCLE

Cells in culture grow and divide until further synthesis and division are arrested by density-dependent inhibition or by unfavorable changes in the composition of the medium. The sequence of events affecting a cell emerging from a division up to the end of the following division is called the cell cycle. The time it takes a cell to progress from one mitosis to the next is called the cell cycle time, which for most types of cells in culture is 18–24 hours. The cell cycle consists of a long interphase, in the course of which the cell is metabolically active, and a short mitotic cycle, when cellular syntheses are inhibited but major cytological changes occur and the cell divides. DNA synthesis and chromosomal duplication occur before mitosis begins (see Fig. 4–3).

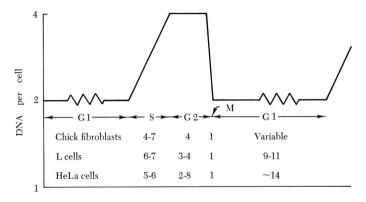

FIG. 4–3. *The four stages of the cell cycle and their duration (in hours) in three different types of cultured cell, and the amounts of DNA per cell at different stages of the cycle ($G_1$, first gap; S, synthesis; $G_2$, second gap; M, mitosis). (Data from Firket, 1965.)*

### The Mitotic Cycle

The duration of the mitotic cycle is variable even in neighboring cells in a single culture, but with most vertebrate cells it lasts between 1 and 2 hours. The occurrence and duration of mitosis are very sensitive to temperature, and cells of warm-blooded vertebrates do not divide at temperatures below 21° or above 44°C. The mitotic cycle has been arbitrarily divided into a number of phases, depending upon cytological appearances. At the end of the long *interphase* period, the mitotic cycle begins with *prophase*, during which the cell rounds up and the previously invisible chromosomes become shorter and thicker and can be seen distinctly within the nuclear membrane. Prophase ends suddenly with the disappearance of the nuclear membrane and the nucleoli. *Prometaphase* is a variable period during which the chromosomes move out of the former nuclear area and attach themselves one by one to the mitotic spindle. *Metaphase* proper begins when the chromosome group "freezes" into an equatorial plate. The most dramatic moment of the cycle is *anaphase*, when the two chromatids of each chromosome begin to draw apart. *Telophase* begins with the constriction of the cytoplasm in the area formerly occupied by the equatorial plate, and is prolonged into the phase of reconstruction when the nuclear membrane and nucleolus reappear, the chromosomes fade into the intermitotic chromatin, and the two daughter cells separate.

## Synthesis of Chromosomal DNA

At any time, the interphase population of cells consists of two groups, with very few intermediates: cells with the normal (usually diploid) amount of DNA and cells with twice this amount. Relatively few cells are found to be synthesizing DNA at any one time, i.e., the synthetic phase is short, relative to the whole cell cycle. Depending upon the behavior of the DNA the cell cycle can be divided into four periods (Howard and Pelc, 1952) (Fig. 4–3): $G_1$ (first gap), postmitotic stage, DNA stable; S (synthesis), DNA doubles in amount; $G_2$ (second gap), premitosis stage, DNA stable; M (mitosis), DNA quantitatively halved and distributed between daughter cells.

$G_1$ is variable in length (Sisken and Kinosita, 1961) and is probably responsible for variations found in the duration of the growth cycle in particular cells grown under different conditions. Nongrowing tissue, in which the idea of a cycle is meaningless, may be considered as remaining physiologically at this stage (Todaro *et al.*, 1965).

During the S period, DNA replication occurs by a semiconservative mechanism. The chromosomes seem to be duplicated in a specific order (Taylor, 1963), and the "satellite" DNA's situated near the centromere of the chromosomes (Pardue and Gall, 1970) are replicated last (Tobia *et al.*, 1970; Evans, 1964). DNA synthesis is initiated at many discrete sites along the chromosomes and appears to proceed in both directions from each initiation point. The length of the replication units varies from 7 to 30 $\mu$m, and the maximum rate of DNA synthesis is about 2.5 $\mu$m per minute, which is about 100–150 base pairs per minute (Huberman and Riggs, 1966). The enzymes that accomplish this synthesis are unknown, but presumably one or other of the DNA-dependent DNA polymerases recently described (Weissbach *et al.*, 1971) are involved.

Mitosis follows after $G_2$ and results in an exactly equal distribution of the chromosomes to the two daughter cells.

## Synthesis of Other Macromolecules

While nuclear DNA synthesis occurs during a well-defined period (S) in the growth cycle of dividing cells, synthesis of other macromolecules (RNA and protein) occurs throughout the cycle, except during mitosis. Investigations using metaphase-arrested cells show that in such cells there is a profound depression of synthesis of both RNA and protein (Prescott and Bender, 1962), and in HeLa cells Scharff and Robbins (1966) found that the polyribosomes disaggregate during metaphase. The interference with macromolecular synthesis in such cells also inhibits the replication of viruses which infect them, to the extent that viral replication is held up until mitosis is completed (Marcus and Robbins, 1963).

## Synchronization of Cell Division

Studies on biochemical events in cultured cells, and the possible effects of the stage of the growth cycle on infection or transformation by viruses, are greatly simplified if the growth cycles of all cells in a culture are synchronized. Many methods have been reported to give partial synchronization: chemical synchronization using drugs which interfere reversibly with DNA synthesis or mitosis, selective detachment of mitotic cells, and gravity sedimentation. Drugs such as colchicine or vinblastine can be used to arrest all cells in metaphase but the effect is irreversible; Colcemid has the same effect and can be reversed by washing the culture. For cells growing in monolayers the least damaging and most effective method depends upon the fact that cells in metaphase round up and attach only loosely to their supporting glass surface. This allows them to be removed from monolayers of nondividing cells by gentle washing (Pfeiffer and Tolmach, 1967), and for many hours after removal the descendant cells show a considerable amount of synchrony. However, the yield of cells is low, since only a small percentage of the total culture is undergoing mitosis at any given time. It can be increased very greatly by combining this method of obtaining mitotic cells with phasing of the cultures by the addition of high concentrations of thymidine, which inhibits DNA synthesis (Xeros, 1962). Excess thymidine (2 mM) is applied long enough (24 hours) to accumulate all the cells in the S phase; then the block is relieved for 8 hours, and all the cells finish DNA synthesis and pass to $G_2$. Excess thymidine is added again, and this time the cells all accumulate at the beginning of S (Puck, 1964). If this method is combined with the washing off of mitotic cells, up to 95% of the cells in a culture can be obtained in a highly synchronized condition.

Synchronous populations of suspension cultures can be obtained by layering cells on the top of columns of medium. The larger cells (those in S and $G_2$) fall through the medium most rapidly so that the cells remaining in the top few milliliters of the column are virtually all in $G_1$. These can be recovered easily and show synchronous growth for a generation or two (Shall and Mc-Clelland, 1971).

## MOLECULAR BIOLOGY OF ANIMAL CELLS

Biochemical studies of animal viruses are concerned largely with two problems: first, how the introduction into cells of the nucleic acids of cytocidal viruses leads to synthesis of viral nucleic acids and proteins and how these products are assembled into progeny virus particles; second, how infection with tumor viruses causes the changes in cell behavior that are responsible for malignancy. Although our understanding of these problems is still very far from being complete, there has been a remarkable expansion of our knowledge of both these processes during the past few years. This progress would have been impossible without understanding of the basic facts of molecular biology. An excellent description of the molecular biology of prokaryotes is available (Watson, 1970), and in this section we will describe briefly the mechanism of gene expression in higher cells, i.e., the process whereby the information coded in the chromosomal DNA is expressed as protein.

The sequence of amino acids in a particular protein is specified by the sequence of nucleotides in a particular segment of DNA. This is called the genetic code and it seems to be the same for all organisms. The transfer of genetic information from DNA to protein is effected through messenger RNA (mRNA). The nucleotide sequence is transcribed into messenger RNA, which is ferried to the cytoplasm where it becomes attached to ribosomes, the sites of protein synthesis. The nucleotides are read in groups of three (known as codons) in the 5' to 3' direction along the RNA, by transfer RNA molecules which carry amino acids. Proteins are synthesized in the N-terminal to C-terminal direction so that amino acids are added to the free carboxyl group of a growing polypeptide chain. At the end of the RNA message, the ribosome encounters one or more codons which signal chain termination. The newly synthesized polypepetide chain is released and assumes its correct three-dimensional configuration. This basic mechanism of protein synthesis is very similar in all living things.

## Chromosomal Organization

It is now clear that the chromosome of many viruses and bacteria consists of a single giant molecule of DNA. The chromosomes of higher organisms, on the other hand, are certainly more complex. In the first place, they contain thousands of times as much DNA (the genome of a human cell consists of $7 \times 10^9$ base pairs; that of *E. coli*, $4.5 \times 10^6$ base pairs; and that of bacteriophage lambda, $4.8 \times 10^4$ base pairs). Second, the genome of somatic mammalian cells is diploid, so that there are at least two copies of the vast majority of genes. Third, the DNA is regularly associated with basic proteins, the histones and protamines; and, finally, there are functional subunits within the chromosomes concerned with transcription, replication, and perhaps packing, of the DNA.

The one physical property that most clearly distinguishes the DNA of eukaryotes from that of simpler creatures is the presence of redundant sequences. Britten and Kohne (1968) observed that when DNA from eukaryotes is heat-denatured, a certain fraction of it renatures rapidly. This phenomenon means that some base sequences are repeated many times in the total genome; such redundant sequences account for about 35% of the human genome (Saunders *et al.*, 1972). The average size of the redundant sequences seems to be about 400 base pairs (Britten and Kohne, 1968), and they can be classified on the basis of their relative reassociation rates into DNA families that occur different numbers of times in the genome. The function and structure of some of these families is known. For instance, the genes for ribosomal RNA, 5 S RNA, and histones (Kedes and Birnstiel, 1971) are all reiterated. The chromosomal location of a few other families has been determined by *in situ* hybridization techniques (Pardue and Gall, 1970). However, by and large, we have little idea of the functions of most of the repetitive classes of DNA; although some of the redundant species are transcribed, it seems unlikely that the majority of highly repetitious DNA functions as template for RNA sequences found *in vivo*. Perhaps the untranscribed regions of DNA are important in housekeeping functions such as organizing the DNA into chromosomes, or attaching the chromosomes to spindle microtubules during mitosis.

Although a large proportion of the genome consists of repetitions of one sort or another, somewhat more than one-half of the DNA is found to be made up of sequences that occur only once in the total genome (Britten and Kohne, 1968). These unique sequences are transcribed *in vivo* (Gelderman *et al.*, 1971; Davidson and Hough, 1971) and presumably consist of structural genes that are represented only once in each haploid set of chromosomes (J. Bishop *et al.*, 1972).

## Transcription

The nuclei of eukaryotic cells contain two major and several minor DNA-dependent RNA polymerase activities which can be separated from each other by chromatography on DEAE-cellulose (Roeder and Rutter, 1969). These enzymes are different in so many properties that most people believe them to be unrelated proteins, presumably with different transcriptional roles *in vivo*. One of the enzymes is found almost exclusively in nucleoli (Roeder and Rutter, 1970) and is thought to be responsible for transcription of the ribosomal cistrons; in all probability, the primary product of this enzyme is the 45 S ribosomal precursor RNA. Most of the second major species of DNA-dependent RNA polymerase is found in the nucleoplasm, where presumably it is responsible for the production of the major species of nucleoplasmic RNA ("heterogeneous nuclear RNA"). This RNA consists of molecules which are heterogeneous in size and which sediment between 20 and 100 S (Warner *et al.*, 1966). Even though its existence has been known for 10 years (Harris, 1962), its function is still unclear. Most of this RNA is never transported out of the nucleus, but is broken down to acid-soluble nucleotides within a few hours of synthesis (Attardi *et al.*, 1966; Scherrer and Marcaud, 1965). Thus, heterogeneous nuclear RNA accounts for more than 75% of the total RNA synthesized in the cell (review, Darnell, 1968b), but because of its short half-life it comprises less than 1% of the total RNA present in the cell. The reason for this apparent wastage is unknown.

Although the evidence is not yet complete, it seems very likely that a small fraction of the heterogeneous nuclear RNA is transported out of the nucleus and functions as mRNA in the cytoplasm. The data on which this hypothesis is based are: (a) the base competition of heterogeneous nuclear RNA is very similar (40–45% G + C) to that of the total DNA genome and that of mRNA (Scherrer *et al.*, 1963); (b) both heterogeneous nuclear RNA and mRNA contain sequences in common (Lindberg and Darnell, 1970); and (c) pulse chase experiments (Tonegawa *et al.*, 1970) are consistent with the notion that cytoplasmic mRNA molecules are derived from large nuclear RNA precursors.

The mechanism which selects out the desired RNA sequences from the pool of heterogeneous nuclear RNA and transports them to the cytoplasm is unknown. However, there is growing interest in the sequences of polyadenylic acid (poly A), which often are found attached to the 3'-hydroxyl end of some species of heterogeneous nuclear messenger (S. Lee *et al.*, 1971; Darnell *et al.*, 1971), as well as some viral RNA's (Kates, 1970; Lai and Duesberg, 1972a). It is thought that these poly A sequences may play a role in post-transcriptional processing of RNA.

The enzyme responsible for synthesis of the other major species of RNA in mammalian cells, transfer RNA (tRNA), is unknown. There is some evidence that tRNA, like mRNA and ribosomal RNA (rRNA), is also derived from a larger precursor molecule (see Darnell, 1968b), but the details of the process are not yet clear.

## Messenger RNA and Translation

The absolute proof that a particular RNA is a messenger is that it directs the synthesis of a particular polypeptide in cell-free protein synthesizing systems. This proof has been achieved for messenger RNA's coding for $\alpha$ and $\beta$ globin chains (Lockhart and Lingrel, 1969; Mathews *et al.*, 1971); myosin (Heywood, 1969); immunoglobulin light chains (Starnezer and Huang, 1971), and lens $\alpha$ crystallin (Mathews *et al.*, 1972; Berns *et al.*, 1972), which have been shown to be translated in cell-free systems derived from reticulocytes or Krebs II ascites tumor cells. These systems also translate RNA's from viruses such as EMC (A. Smith *et al.*, 1970). Such stringent proof of messenger function of most putative messenger RNA preparations from mammalian cells is not possible to obtain. Consequently, mRNA is usually defined in operational terms, as a rapidly labeled species of cytoplasmic RNA which is associated with polysomes. This RNA sediments more slowly than heterogeneous nuclear RNA (Latham and Darnell, 1965), and its base composition, in contrast to rRNA or tRNA, is very similar to that of total cell DNA (review, Darnell, 1968b). It can be released from polysomes by EDTA or puromycin treatment (Penman *et al.*, 1968).

The first step in translation seems to be the binding of the smaller (40 S) ribosomal subunit to the mRNA near the 5' end (Joklik and Becker, 1965). This initial binding and the subsequent addition of the 60 S subunit to form a monosome seems to require the presence of GTP, a specialized transfer RNA (met-tRNA), and several additional factors (Heywood, 1969) which are now being identified in increasing numbers. As in *E. coli*, met-tRNA selects an AUG codon at or near the 5' end of the messenger RNA, hence the N-terminal amino acid of newly synthesized proteins is methionine (see review by Lucas-Lenard and Lipmann, 1971). Unlike the prokaryotic systems, however, there is no evidence that the N-terminal methionine is formylated. The ribosome ratchets along the message, laying down amino acids in the order specified by the triplets in the message. Additional protein factors as well as GTP are needed for this process of polypeptide elongation. As one ribisome leaves the 5' end of the RNA, another is added so that a polysome is formed. At the end of the message, the polypeptide chain is completed when the ribosome encounters a chain termination signal. At this point, a third sort of factor as well as GTP is required (Goldstein *et al.*, 1970), and in all probability the ribosome then dissociates into its subunits releasing the newly synthesized protein.

## INHIBITORS OF CELL FUNCTION AND VIRAL MULTIPLICATION

Chemical agents that specifically block the biosynthesis of macromolecules or otherwise interfere with intracellular organelles are among the most convenient and powerful probes available for understanding the molecular biology of cells and viruses. Almost every year new and useful inhibitors become available, usually as a by-product of the efforts of the pharmaceutical industries in searching for drugs that might cure cancer or viral infections. Frequently such drugs are found to be too toxic to provide a selective action against malignant tumors or viruses. However, if they can be shown to have a specific intracellular target, they constitute valuable tools with which cell biologists can elucidate the interrelations of macromolecules.

### Inhibitors of DNA Synthesis

As described earlier, drugs that inhibit DNA synthesis have been used to synchronize cell division or the onset of viral DNA synthesis in order to see how regulatory mechanisms are linked to DNA replication. A brief description of several such inhibitors is presented. As will be evident from their properties and mode of action, the selection of an appropriate inhibitor will depend on the cell system under study and the requirement for a reversible inhibition.

**Fluorodeoxyuridine.** The essential features of thymidine metabolism are presented in Fig. 4–4. Fluorodeoxyuridine (FUdR) is a reversible inhibitor of DNA synthesis effective at concentrations of about $10^{-6}$ M. FUdR becomes phosphorylated by cellular thymidine kinase thus becoming an analog of thymidylic acid (dTMP). In the phosphate form, FUdR inhibits thymidylate syn-

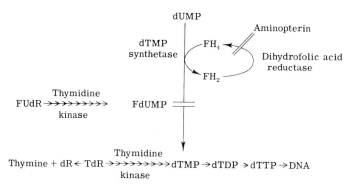

FIG. 4–4. *Pathway to thymidine metabolism. dR, deoxyribose; TdR, thymidine; dTMP, thymidylic acid; dUMP deoxyuridylic acid; FH₄ and FH₂, tetra- and dihydrofolic acid, respectively; dTTP, thymidine triphosphate; FUdR, fluorodeoxyuridine.*

thetase which normally catalyzes the endogenous synthesis of dTMP from dUMP. The intracellular pool of thymidine in most cultured animal cells is apparently too low to be a source of dTMP.

If a source of thymidine becomes available to a cell either from extensive degradation of DNA or from free thymidine in the medium, inhibition of the thymidylate synthetase reaction can be bypassed by direct phosphorylation of thymidine. When using FUdR, it is advisable to add at the same time an excess of deoxyuridine; if FUdR becomes degraded intracellularly to fluorouracil and deoxyribose, deoxyuridine will inhibit competitively the formation of fluorouridine and its incorporation into RNA.

Bromouracil deoxyribose (BUdR) or iodouracil deoxyribose (IUdR) are also phosphorylated by thymidine kinase, but in contrast to FUdR they are incorporated into DNA in place of thymidine after conversion to the corresponding triphosphates. This has provided the basis for the chemotherapy of herpesvirus infections (see Chapter 15).

**Aminopterin.** From Fig. 4–4, it is obvious that DNA synthesis could be prevented by blocking reductive methylation of dUMP via tetrahydrofolic reductase. Aminopterin is an analog of folic acid. Provided no exogenous supply of thymidine is available, it reversibly inhibits the reductase and thus, indirectly, the synthesis of DNA. The selectivity of aminopterin for inhibiting pyrimidine synthesis is achieved by simultaneous addition of adenosine and glycine (Hakala, 1957; Simon, 1961). Aminopterin is the inhibitor of choice to block DNA synthesis in cells that are deficient in thymidine kinase or where it is desired to promote the incorporation of a thymidine analog into DNA.

**Arabinosyl Nucleosides.** Cytosine arabinoside (ara-C) and adenosine arabinoside (ara-A) are as effective as FUdR or aminopterin in inhibiting mammalian DNA synthesis. The toxicity and inhibitory effects of ara-C can be overcome by simultaneous addition of deoxycytidine; however, the latter is not effective in reversing ara-C inhibitory effects once established. In contrast, the effects of ara-A are not prevented by normal deoxynucleosides such as deoxyadenosine. It should be noted that the efficacy of ara-C can be diminished by high intracellular levels of deoxycytidine deaminase which convert it to uridine arabinoside, which does not inhibit DNA synthesis.

Multiplication of most if not all DNA viruses in cell cultures is markedly inhibited by ara-C at concentrations as low as $10^{-6}$ M. Both ara-C and ara-A are converted *in vivo* to the corresponding triphosphates and as such, inhibit conversion of uridine to dCMP. Another, and probably primary, site of inhibition is DNA polymerase. Ara-C as the triphosphate inhibits polymerization of deoxynucleotides and to some extent is also incorporated into DNA.

For a comprehensive review of the chemistry and inhibitory effects of arabinosyl nucleosides, the reader is referred to the review of Cohen (1966) and to the papers of Furth and Cohen (1968) or Furlong and Graham (1971) for details of the mechanism of action.

## Inhibitors of RNA Synthesis

The central problems of molecular biology are those concerned with the control of gene expression. Any compound that will directly inhibit transcription of cell or viral genes has obvious value.

**Actinomycin D.** This polypeptide antibiotic is probably the only currently available compound that rapidly and specifically inhibits transcription. Its main drawback is that its action is not readily reversible by removal of the drug from the growth medium. *In vivo* and *in vitro* actinomycins interact only with helical deoxypolynucleotides that contain guanine (Reich, 1966). Based on X-ray studies, it is thought that actinomycin becomes located in the minor groove of helical DNA, binding to guanine residues by hydrogen bonds.

Whereas levels of the antibiotic of about 1 to 5 $\mu$g/ml are sufficient to block DNA-dependent RNA synthesis by over 95%, the replication of many single-stranded RNA viruses is unaffected (Reich *et al.*, 1962). This indicates that most cellular functions on which viral replication depends are not directly impaired by actinomycin. By judicious application of actinomycin D, one can enhance the incorporation of isotopic uridine into single-strand RNA-dependent viral RNA synthesis, or at another level it can be used to investigate the programming of DNA transcription and the stability of various messenger RNA species (Mc-Auslan, 1963; G. Tompkins *et al.*, 1969).

At relatively low levels of actinomycin D, labeling of 45, 28, and 18 S RNA species, but not the 4 S species, is inhibited in mammalian cells. At high concentrations, all RNA species are inhibited (Franklin, 1963; Perry, 1963). An additional effect of actinomycin D is to inhibit the transport of RNA from nucleus to cytoplasm (Levy, 1963). Reich and Goldberg (1964) present a useful review of actinomycin D and its effects.

**$\alpha$-Amanitine.** There are at least two or possibly three distinct RNA polymerase activities associated with the mammalian cell nucleus (Roeder and Rutter, 1970). RNA polymerase II is located in the nucleoplasm and appears to be responsible for the major fraction of heterogenous RNA synthesis in the uninfected cell.

$\alpha$-Amanitine, a cyclic polypeptide, is a potent and selective inhibitor of RNA polymerase II and can, therefore, be used to test the role of this enzyme in the multiplication of DNA viruses that replicate in the cell nucleus. $\alpha$-Amanitine does not readily enter cultured cells. However, isolated nuclei are permeable. Nuclei of adenovirus-infected cells synthesize adenovirus RNA *in vitro*; this synthesis is inhibited by $\alpha$-Amanitine, suggesting that adenovirus is either transcribed principally by RNA polymerase II of the host cell or an enzyme that resembles it closely (Price and Penman, 1972; Wallace and Kates, 1972).

## Inhibitors of Protein Synthesis

Inhibitors of protein synthesis comprise a very large and diverse group of compounds. The biosynthesis of a protein is the end result of many interrelated metabolic events and accordingly is subject to inhibition at many different steps. Pestka (1971) provides a comprehensive survey of many well known as well as new inhibitors of protein synthesis, and their mode of action. This outline will be confined to a few of these.

**Glutarimide Antibiotics.**   Cycloheximide, acetoxycycloheximide, and streptovitacin A, an hydroxylated cycloheximide, are glutarimide antibiotics that are potent reversible inhibitors of protein synthesis in animal cells effective both *in vivo* and *in vitro* (Ennis and Lubin, 1964; Wettstein *et al.*, 1964). Acetoxycycloheximide is perhaps the most potent of these on a molar basis (Ennis, 1968), although the relative efficiency of the congeners appears to vary depending on the cell system under study. Streptovitacin A does not appear to be as effective in L cells as cycloheximide, whereas in KB or HeLa cells protein synthesis is markedly inhibited by doses as low as 10 $\mu$g/ml. These three inhibitors have a similar site of action, namely inhibition of polysomal breakdown linked to protein synthesis, by preventing detachment of ribosomes from polysomes and the incorporation of amino acids into nascent polypeptide chains.

**Puromycin.**   Its structure suggests that puromycin could act as an analog of aminoacyl-adenyl tRNA. Like the latter it can serve as an acceptor of the nascent polypeptide chain of ribosome-bound peptidyl-tRNA (Rabinowitz and Fisher, 1962) and be incorporated into nascent chains, which are then released prematurely. A secondary effect is the breakdown of polysomes, an energy-dependent reaction that can be prevented by cycloheximide.

**Emetine.**   Emetine is the principal alkaloid of ipecacuanha. Although a potent inhibitor of protein synthesis in HeLa cells (Grollman, 1968), its effects, in contrast to those of cycloheximide and puromycin, are irreversible. In the presence of the drug, nascent polypeptide chains remain attached to the polyribosomes and amino acid incorporation is inhibited. There is some evidence that at high concentrations the synthesis of various RNA species may be differentially inhibited (Gilead and Becker, 1971).

**Pactamycin.**   The synthesis of protein is initiated near the 5' end of RNA and proceeds toward the 3' end. Pactamycin at very low concentrations acts as a selective inhibitor of protein synthesis by interfering with the firm attachment of initiator tRNA to the initiation complex. In poliovirus-infected cells, inhibition of the initiation of protein synthesis by pactamycin is followed by a transient period of protein synthesis during which the relative amounts of the species of protein synthesized should be a reflection of their position on the viral genome. Taber *et al.* (1971) made use of this point to determine the order in which the three major cleavage products of poliovirus capsid protein are synthesized (see Chapters 6 and 7).

### Inhibitors Affecting Intracellular Microfilaments

**Colchicine and Colcemid.**   Colchicine and Colcemid (*N*-deacetyl-*N*-methyl colchicine) are mitotic poisons that arrest cells in metaphase. Colchicine binds to the protein subunits of microtubules, the most widely distributed type of intracellular microfilament, and may therefore interfere with a wide variety of cellular functions. Colcemid is believed to prevent the centrioles from organizing the microtubules which are necessary for their migration to the poles, and in this way blocks mitosis at an early stage.

A valuable property of Colcemid is the ease with which its action can be readily reversed by removal of the drugs, allowing mitosis to proceed some 5 minutes later. Accordingly, it has been used to treat monolayer cultures to yield a population of cells that are synchronized for entry into the $G_1$ period (Stubblefield et al., 1967).

**Cytochalasin B.** This agent has a great diversity of effects including interference with cytokinesis, induction of extensive morphological changes in cultured cells, and dissociation of embryonic epithelia. Many of its effects are directly related to the disruption of microfilaments. Unlike colchicine, cytochalasin has not been proved to bind to microfilament precursors. It may also inhibit mucopolysaccharide synthesis, thus interfering with cellular adhesion (Sanger and Holtzer, 1972).

For the virologist, perhaps the most useful effect of cytochalasin is the enucleation of cells. Prescott and collaborators (Prescott et al., 1971) have been able to achieve up to 80% enucleation of L cells by treatment with cytochalasin, and thus demonstrate that replication of the DNA of poxvirus is independent of nuclear function.

### Other Viral Inhibitors

The compounds just described are those that have a fairly well defined mode of action and are in general use. Many other compounds inhibit replication of specific viruses, for example, rifampicin, isatin β-thiosemicarbazone, and guanidine. These compounds and some others are described in Chapter 15, together with a discussion of their potential value in chemotherapy.

CHAPTER 5

# The Multiplication of DNA Viruses

## INTRODUCTION

In Chapter 3 we described the chemistry and structure of virions. But the term "virus" encompasses much more than a rather dry description of virus particles. For virions are only the inert forms assumed by viruses to travel from one host cell to the next; they undergo no metabolism and they do not reproduce.

All dynamic events associated with viruses occur within host cells. This chapter and the next are concerned with these events most of which are facets of the process of viral multiplication. Because DNA and RNA viruses differ in important details of their mechanisms of multiplication, particularly with regard to transcription and replication of their genomes, we shall consider these two groups separately. It will be useful first to give a brief outline of the multiplication cycle as it applies to all viruses.

## THE MULTIPLICATION CYCLE

The study of the multiplication of animal viruses has been greatly influenced by earlier work with bacterial viruses, the concept of a single "growth" cycle being developed from studies of cytocidal infections with virulent bacteriophages. Ellis and Delbrück (1939) developed the "one-step growth" experiment by manipulating the conditions in a population of infected cells so that the behavior of the population as a whole could be used to test hypotheses about what was happening in any individual cell at a given time after infection. In essence, the one-step growth experiment, as it is used in animal virology, consists of the simultaneous infection of a population of cells with at least one infectious virus particle per cell, usually 10–100 to ensure that every cell is infected and that there is a high degree of synchrony of subsequent events in the infected cell population. After taking steps to inactivate the unabsorbed viral inoculum the culture is assayed at frequent intervals for the number of infectious virus particles formed, both within the infected cells and released into the medium (Fig. 5–1).

*Eclipse* is a term of historical interest, that figured prominently in discussions of viral multiplication 10 or more years ago. It refers to the fact that the infecting virion disappears and can no longer be demonstrated inside the cell during

176

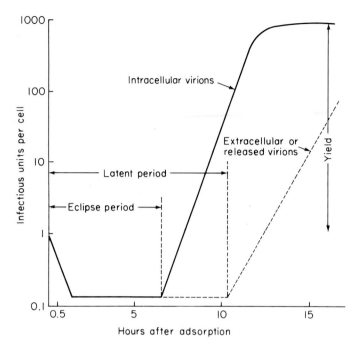

FIG. 5–1. *Generalized diagram of the multiplication cycle of an animal virus that is released from the cell late and incompletely. In other cases, virions mature as they are released and the curves for intracellular and extracellular infectivity correspond.*

the "eclipse period" of several hours that elapses before progeny particles appear. The term *latent period* refers to the interval between the disappearance of the inoculated virions and the release of new virions into the supernatant fluid. Unlike bacteriophages, most animal viruses are released one by one over an extended period, or else most of the virions remain cell-associated for the whole period of viral multiplication. Even when the plasma membrane is destroyed by viruses, animal cells do not "burst" like bacteria, but rather shrink and leave an intact mass of debris that may enclose most of the viral progeny. The *viral yield* is equivalent to what bacterial virologists called the "burst size." This can be measured in terms of the number of virus particles or of "infectious units" produced per cell. These two figures are not identical, for reasons discussed in Chapter 2.

## THE ESSENTIAL STEPS IN MULTIPLICATION

It is possible to dissect the viral multiplication cycle into several more or less sequential steps (although one process grades into the next, and after the first few hours several of the processes which follow are going on simultaneously): (a) attachment; (b) penetration, leading to or coincidental with (c) uncoating. This leads to (d) eclipse, the disappearance of infectious virions, during which a complex series of biosynthetic events occurs, viz., (e) transcription of mRNA from specific sequences of the parental viral DNA, (f) translation of this mRNA into virus-coded enzymes and other "early" proteins, (g) replication of the viral DNA, (h) transcription of further mRNA from progeny as well as parental DNA, and (i) translation of these "late" mRNA's into structural and other virus-coded proteins some of which are involved in regulatory functions. Finally the eclipse phase ends with (j) assembly and release of new virions.

## Attachment

Attachment of virions to the plasma membrane is electrostatic and follows the establishment of contact by collision between virions and cells. It is independent of temperature except in so far as this affects Brownian movement of virions and cells and thus the likelihood of their collision.

Attachment will occur only if there is a certain affinity between the cell surface and the virions, areas of the cell membrane showing this affinity being called viral receptors. Lack of receptor sites is the cause of the resistance of some cells and some organisms to infection by certain viruses (see Chapter 11); attachment is a necessary but not necessarily a sufficient condition to allow productive infection.

## Penetration

With bacterial viruses the coat of the virion remains outside the cell; only the nucleic acid penetrates (Hershey and Chase, 1952). This is probably an exceptional event with animal viruses. Opinions differ as to whether disruption of the virion occurs at the plasma membrane, or only after the engulfment of intact virions; there are probably important differences between viruses. Fazekas de St Groth (1948a) and, subsequently, Dales (Chardonnet and Dales, 1970b; review; Dales, 1965) have been the main proponents of the engulfment (viropexis) hypothesis; Morgan (see Miyamoto and Morgan, 1971) believes that enveloped viruses undergo a preliminary fusion of their envelope with the plasma membrane prior to release of the nucleocapsid into the cytoplasm. Both interpretations depend primarily upon electron microscopic evidence, and it is very difficult to distinguish the behavior of the small percentage of virions that actually initiate infection (usually much less than 10%, see Chapter 2) from the majority that do not.

## Uncoating

This term implies release of the infectious nucleic acid from the viral coat. Again, a variety of situations exists, ranging from release of the viral nucleic acid at the cell membrane (as appears to happen with enteroviruses, see Chapter 6) to the complex intracellular uncoating of poxviruses. Some viruses, for example reoviruses, may never undergo complete uncoating; the virion-associated transcriptase functions within viral cores that are maintained in a more or less intact state (see Chapter 6).

## Transcription

Poxviruses carry their own transcriptase but as far as we know those DNA viruses that multiply in the nucleus depend on cellular RNA polymerases. In both cases, transcription is regulated, by mechanisms that are unknown. "Early" genes may be defined as those segments of parental DNA which are transcribed when cytosine arabinoside is present to prevent DNA replication. "Late" genes are those whose transcription rate becomes significant after DNA starts to replicate. With poxviruses, as with some DNA bacteriophages, there is evidence that even the "early" mRNA species fall into two classes: "immediate early" and "delayed early," only the former being made in the presence of cycloheximide, which inhibits the synthesis of a protein that brings about the last stage of uncoating of the transcriptase-containing core of the virion.

Early transcripts usually represent only a minority of the viral genes. In the main, they code for "early" enzymes and other proteins (e.g., T antigens) of unknown function, whereas late mRNA's code for the structural proteins of the virion, but the distinction is not always clear-cut. In general, the transcripts, whether early or late, are large polycistronic molecules which are subsequently cleaved into shorter, monocistronic messengers. Polyadenylic acid (in stretches of up to 150 adenine residues) is then added to the 3' end of each mRNA molecule before it is transported out of the nucleus into the cytoplasm.

Not all mRNA is transcribed from the same strand of double-stranded viral DNA; early and late mRNA's may differ in this regard, and it doubtless represents a control mechanism we know nothing of. Furthermore, in the case of papovaviruses and adenoviruses, whose DNA may become integrated with that of the host cell, long sequences of cellular and viral DNA may be transcribed to give a continuous "hybrid" RNA molecule which is subsequently cleaved.

As will be discussed in detail in Chapter 6, the RNA viruses vary widely in the ways in which their genetic information is transcribed and translated. In some cases, the viral RNA is translated directly; in others, the viral genome is transcribed by a transcriptase carried in the virion.

## Translation

Monocistronic mRNA's are translated into proteins on cytoplasmic polyribosomes. There is no evidence with DNA viruses for the type of posttranslational cleavage of giant polypeptides that occurs with certain RNA viruses (Chapter 6), although minor cleavages may play a role in the final assembly of some structural proteins into the virion.

The "early" proteins, synthesized in the presence of inhibitors of DNA synthesis, include the T antigens detectable in both cytocidal and transforming infections by papovaviruses or adenoviruses. They also include many enzymes concerned in DNA replication. Some of these enzymes are virus-specified but in other cases (e.g., in Papovaviridae) they represent cellular enzymes which have been derepressed by infection.

Arginine-rich proteins are present in the core of most DNA viruses. These histonelike proteins may be closely associated with the viral DNA during its replication and/or transcription. Under conditions of arginine starvation assembly of new viral cores is inhibited.

## Replication of Viral DNA

Once an adequate concentration of the necessary enzymes has developed, viral DNA begins to replicate by standard Watson-Crick base-pairing. Cellular DNA polymerases may be involved but the larger DNA viruses may code for their own.

Evidence is accumulating that arginine-rich proteins may remain attached to viral DNA throughout replication. Association with membranes may also be crucial at certain stages.

Special problems attend the replication of the circular DNA of the papovaviruses. To overcome insuperable physical problems in twisting the superhelix, an endonuclease is induced which "nicks" one strand while a short region is replicated, then a ligase repairs the break again.

## Assembly and Release

In the case of the poxviruses and iridoviruses, both DNA replication and structural protein synthesis occur in a common cytoplasmic "factory," hence, there are minimal topological constraints on automatic assembly of virions when the concentration of the two "reactants" reaches an adequate level. On the other hand, herpesviruses, adenoviruses, and papovaviruses are assembled in the nucleus, after structural proteins, synthesized in the cytoplasm, have migrated back to the nucleus, where DNA replication and transcription have been occurring. Little is known about the factors that control these movements of molecules.

All DNA viruses except the smallest (*Parvovirus* and Papovaviridae) have a complex structure consisting of several concentric layers of protein. These are laid down stagewise, the basic "core" proteins being the first to associate with viral DNA and the capsomers of the outer capsid being added last. The herpesviruses and many of the RNA viruses (Chapter 6) acquire a lipoprotein envelope by budding through cellular membrane that has been modified by the incorporation of viral proteins (mainly glycoproteins).

The release of the first virion into the supernatant fluid technically marks the end of the latent period. Usually, virions are released one or a few at a time over many hours, and often remain cell-associated for long periods, sometimes in crystalline aggregates.

## Regulation of Viral and Cellular Macromolecular Synthesis

Reference has been made to the fact that both transcription and translation of the viral genome are subject to certain controls. Only a minority of viral cistrons are transcribed before DNA replicates, and even after the whole genome becomes available for transcription the various "late" genes are transcribed at different rates. Moreover, the half-life of these mRNA's varies, so affecting translation rates. More subtle mechanisms also regulate translation. For example, structural viral proteins may act as repressors of transcription or translation by binding to particular segments of viral DNA or mRNA. Little is known about these mechanisms but they are bound to attract increasing attention from virologists over the next few years.

Cellular macromolecular synthesis is also affected by viral infection. DNA, RNA, and protein synthesis may all be "shut down" to varying degrees by different viruses. The advantages of this to the virus are clear enough; if the virus carries all the genetic information it requires, there is a selective advantage in preventing the supply of cellular mRNA's to ribosomes, thus freeing them for the exclusive use of viral messengers. Moreover, interferon will be one of the cellular proteins whose synthesis is turned off.

Virus-coded proteins are responsible for the shutdown. Under artificial laboratory conditions of very high multiplicity, infection may cause a rapid suppression of synthesis of cellular macromolecules, even if the viral genome has been inactivated by irradiation; clearly, structural proteins of the virion are responsible. At low multiplicity, however, viral proteins must be synthesized in order to bring about these effects. Cellular transcription and translation seem to be separately suppressed; DNA synthesis is blocked as a consequence of the other two inhibitions.

Cell shutdown is rapid and almost complete within a few hours of infection by poxviruses and herpesviruses, slow in the case of adenoviruses, and does not occur with papovaviruses. In fact, papovaviruses actually stimulate the synthesis of cellular DNA; their DNA becomes integrated with that of the cell in productive cytocidal infections as well as in nonproductive transformation.

Table 5–1 summarizes the types of protein that may be specified by viruses.

TABLE 5–1

*Products of the Viral Genome*

1. Structural polypeptides of the virion
   (may also be enzymes or act as regulatory proteins)
2. Enzymes of the virion (for use in the next cell infected)
3. Enzymes, nonstructural, involved in DNA transcription or synthesis
4. T antigens and other proteins detectable serologically inside infected cell or in plasma membrane
5. Regulatory proteins suppressing cellular transcription or translation
6. Regulatory proteins suppressing expression of "early" viral genes

This brief outline of the basic principles of viral multiplication has been intended to do no more than provide the nonvirologist with sufficient background to appreciate the intricacies of what follows. Studies on the multiplication of bacterial viruses will not be referred to again, but since they provide useful background information on some aspects of the molecular biology of the multiplication of animal viruses, the reader's attention is drawn to the following reviews: Calendar, 1970; Echols, 1971; Hershey, 1971; Stent, 1971.

## A SYSTEMATIC SURVEY

The major groups of DNA viruses exhibit great diversity in structural complexity and DNA content (see Chapters 1 and 3), and in their mode of multiplication. Accordingly, there are great differences in their dependence upon host functions, and in the strategies they employ to initiate infection. The DNA of viruses of most genera (*Parvovirus*, Papovaviridae, *Adenovirus*, and *Herpesvirus*), is replicated in the nucleus, where the virions are finally assembled, while with other genera (*Poxvirus* and *Iridovirus*) the entire multiplication cycle occurs in the cytoplasm. The timing of events in the multiplication cycle varies characteristically with different genera; Fig. 5–2 illustrates approximate growth curves for the viral species commonly used for studies of viral multiplication. These are usually the species and strains that multiply most rapidly and to highest titer; other species of the same genera may have much more prolonged multiplication cycles.

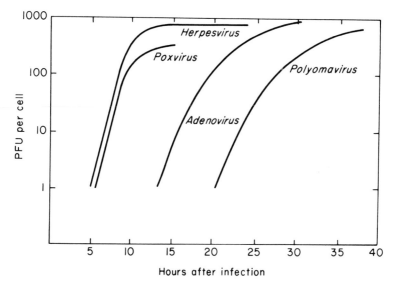

FIG. 5–2. *Idealized multiplication curves (total infectious virus per cell at intervals after infection at high multiplicity) of viruses representing the major genera of DNA viruses.*

The bulk of the information about the multiplication of viruses of a particular genus usually comes from data provided by one or two model systems chosen by investigators for the facility with which those systems can be manipulated to yield information. These are usually mutants that have been selected after many passages for their capacity to grow rapidly and to high titer in convenient lines of cultured cells, often malignant cells growing in suspension. The use of logarithmically growing cells rather than resting cells is also "unnatural" and may give misleading results on such matters as virus-induced enzymes and cell shutdown. Furthermore, to ensure synchrony of infection, such cultures are inoculated with very high multiplicities of virus such as are never encountered in nature. As a result the progeny may consist predominantly of defective (noninfectious) virions and the synthesis of cellular macromolecules is shut down far more rapidly than at low multiplicities. In these respects, the system favored by biochemists is quite artificial and may or may not give results that can be extrapolated to the behavior of a wild strain infecting differentiated tissues *in vivo*.

Two main types of association of DNA virus and cell can be recognized. The genome of the virus may become integrated into that of the host cell and replicate as part of the cellular DNA. In such cases, the outcome may be either (a) a cryptic "infection," if expression of the viral genome is completely repressed or (b) the morphological transformation of the cell and the expression of a limited number of viral genes (see Chapter 13). More commonly, the genome of the virus may replicate independently of the cellular DNA, leading to the production of new virions. If the virus is cytocidal, cellular functions are arrested and infected cells are destroyed. Some viruses affect cellular synthesis less severely, the extreme case being steady-state infections in which both viral and cell multiplication proceed (for example, rabbit fibroma in cultured rabbit kidney cells; see Chapter 12).

## PARVOVIRUS

Since 1968, when discussion of *Parvovirus* multiplication warranted no more than a brief mention (Fenner, 1968), there has been little progress because of the lack of suitable systems in which a high percentage of cells can be synchronously infected and from which high yields of virus can be obtained. During the last few years, however, it has become increasingly clear that the distinctive feature of the multiplication of these viruses is their dependence upon cell division (Tattersall, 1972); their pathological behavior in animals reflects this dependence (see Chapter 9).

### Subgenus A (Rat Virus and H-1 Virus)

The growth cycles of rat virus (RV) and H-1 have been followed by plaque assay. The latent period of RV is about 6 hours and the virus titer reaches a maximum at about 28 hours (Brailovsky, 1966). H-1 infection is slower, the latent period being about 10 hours and maximum titer reached after 50 hours (Al-lami et al., 1969).

Infection by RV has been studied in cultures of rat embryo under conditions where the physiological state of the cells could be manipulated (Tennant et al., 1969). Based on fluorescent-antibody staining, viral protein synthesis was detected 12–25 hours after infection, but only about 5% of infected stationary phase cells supported viral multiplication. When cells were stimulated to incorporate thymidine by the addition of fresh serum, and then infected 10 hours later, the number of cells synthesizing virus increased in direct proportion to the increase in the number of cells stimulated to take up thymidine. These data suggest that a host function must be expressed during S phase which directly influences the rate of viral synthesis. In support of this conclusion, it was shown that the capacity of rat embryo cells to synthesize RV was impaired by doses of X rays, UV irradiation or 5-fluorouracil that blocked cellular DNA synthesis. These treatments were selective for RV and did not impair the multiplication of other DNA viruses (e.g., *Herpesvirus*) that multiply in the cell nucleus (Tennant and Hand, 1970).

## Subgenus B (Adenovirus-Associated Viruses, AAV)

The quantitation of AAV is handicapped by the necessity for adenovirus coinfection, but a fluorescent focus assay and a neutralization test have been described (Ito *et al.*, 1967; Boucher *et al.*, 1970).

The nature of the function supplied by adenovirus to AAV is not known but the temporal relationship between replication of AAV and helper virus has been established. When cells were infected simultaneously with adenovirus and AAV the latent period of both was 12–16 hours. The latent period of adenovirus was not reduced by preinfection of the cells with AAV, but that of AAV could be reduced to about 4 hours if cells were preinfected with adenovirus (Parks *et al.*, 1967). AAV-1 is complemented by coinfection with an adenovirus *ts* mutant at the nonpermissive temperature (Ito and Suzuki, 1970). Since this mutant does not initiate DNA synthesis at this temperature (Suzuki and Shimojo, 1971), the AAV helper function is not due to one of the adenovirus structural proteins, which are "late" proteins (see below).

There is evidence that in some systems adenoviruses may act as helpers for the subgenus A (nondependent) parvoviruses. For example, infection of rat cells with adenovirus 12 and RV led to stimulation of RV synthesis (Chany and Brailovsky, 1967); and preinfection with adenovirus shortened the eclipse period of H-1 virus from 24 to 12 hours (Ledinko and Toolan, 1970). Observations of this sort led Toolan (1968) to suggest that all parvoviruses are defective in some way, and this seems to be borne out by the results of Tennant *et al.* (1969) showing RV dependence on transiently expressed cell functions. The nature of the cell function required for parvovirus multiplication is not known; it could be either a viral or cellular product or a replication site.

## PAPOVAVIRIDAE

The major interest of virologists in the papovaviruses has been focused on their oncogenic potential (see Chapter 13). As a background to understanding this, detailed studies have been made of the multiplication of two viruses of the genus *Polyomavirus*, namely polyoma virus and SV40, in productive infections of cultured cells. Lack of a satisfactory cell system has precluded studies of the multiplication of viruses of the genus *Papillomavirus*.

## The Multiplication Cycle

The biochemical study of polyoma virus had long been hampered by such difficulties as inability to infect productively a large proportion of cells in culture. Productive infection occurs only in mouse cells and much work has been done with mouse kidney primary cultures or secondary mouse embryo fibroblasts. The merits of these systems with respect to convenience and suitably low backgrounds of DNA synthesis have been discussed by Fried and Pitts (1968). Generally, productive infection has been studied in stationary phase cells where small increases in macromolecular synthesis can be detected over low backgrounds. The time course of polyoma DNA and virion antigen synthesis in these cell systems is similar; both commence between 12 and 24 hours after infection and continue throughout the cycle. Progeny virions appear about 36 hours after infection, reaching a maximum about 30 hours later.

SV40 replicates well in primary African green monkey kidney cells and in several established monkey cell lines. The kinetics of SV40 DNA synthesis are similar to those of polyoma virus. Progeny virions appear in about 24 hours, and maximum yields of up to 100 pfu per cell are obtained 3 days after infection, after which cell lysis may occur.

## Initiation of Infection

Very little information is available about attachment, penetration, and uncoating, except that polyoma virus infection is prevented by treatment of the cells with neuraminidase.

## Transcription

In productive infections, viral DNA may be integrated into the host genome or occur as free supercoiled molecules (see below); in virus-free transformed cells it is covalently linked to the host genome (see Chapter 13). The detection of virus-specific RNA is hampered by the continued production of large amounts of host RNA, but is possible by means of DNA–RNA hybridization techniques.

At least two patterns of transcription occur in lytically infected cells. The first pattern occurs early after infection and continues until the onset of viral DNA synthesis. The stable species of viral RNA present during this period correspond to about 30% of the sequences of the same strand of SV40 DNA that is copied *in vitro* by E. coli RNA polymerase. However, the molecular weight of this RNA is about 900,000 daltons (Weinberg et al., 1972a), considerably larger than 30% of the SV40 genome, suggesting that the early phase of SV40 transcription occurs from a temporarily integrated viral DNA molecule, so that the transcript contains SV40 sequences covalently linked to host RNA.

The second pattern of viral transcription occurs after viral DNA synthesis has begun. There is an increase of at least tenfold in the concentration of SV40 RNA sequences in the cell, and although "early" RNA sequences are still transcribed, much of the "late" RNA is synthesized from the strand of SV40 DNA that was silent during the early stages of infection. This "true" late RNA corresponds to about 70% of the sequences of that strand. So by late times after infection, 50% of the total sequences of SV40 DNA are represented in stable species of RNA (Khoury et al., 1972; Lindstrom and Dulbecco, 1972; Sambrook et al., 1972). The stable species seem to fall into two size classes of molecular weights 650,000 and 900,000 daltons, but it is uncertain whether the RNA molecules in these classes contain different sequences of SV40 or whether one is derived from the other (Weinberg et al., 1972a). In addition, RNA molecules of apparently very high molecular weight and which contain SV40 sequences can be found in the cell nucleus (Weinberg et al., 1972a). It has been suggested that these high molecular weight molecules may be precursors of the stable species of SV40 RNA present in the cytoplasm and that they may contain viral and host sequences joined by covalent bonds in both nucleus and cytoplasm. Many of the RNA molecules containing SV40 sequences also contain polyadenylic acid residues (Weinberg et al., 1972b).

## Translation

Polyoma virions contain six or seven polypeptides of which the three smallest, rich in basic amino acids, are probably host-specified (Chapter 3). Ninety percent of the proteins synthesized in infected cells are host proteins, the two major viral structural polypeptides (see Chapter 3) constituting most of the other 10%. The structural proteins are "late" proteins, since their synthesis depends upon prior viral DNA synthesis (Rapp et al., 1965).

Both polyoma and SV40 synthesize virus-specific T antigens. They can be detected by fluorescent-antibody tests in the nuclei of both transformed and productively infected cells (Habel, 1965; Black et al., 1963b), appearing 6–10 hours after infection and prior to the appearance of virion antigens. T antigen synthesis is not inhibited by cytosine arabinoside, i.e., transcription of the relevant mRNA occurs from the input viral genome. The function of T antigens in viral multiplication or in cell transformation is not known.

Other nonstructural polypeptides have been described (Anderson and Gesteland, 1972; Walter et al., 1972), but it has yet to be shown that they are not host enzymes elicited by viral infection.

## Replication of Viral DNA

The intact mature form (component I) of papovavirus DNA occurs in the form of a covalently closed circular duplex molecule (Chapter 3). Considerable effort has gone into experiments to identify and characterize the replicative intermediate of polyoma or SV40 DNA, the main approach being to pulse-label infected cells with radioactive thymidine, isolate intermediates by equilibrium centrifugation and characterize these by electron microscopy and physicochemical techniques.

The idea that papovavirus DNA replicated according to the Cairns model (Bourgaux et al., 1971; A. Levine et al., 1970) was complicated by the existence of circular oligomers and catenated molecules (Cuzin et al., 1970; Jaenisch and Levine, 1971). However, these may not be replicative forms at all, but merely the result of recombinational events between circular molecules. An important advance in clarifying the picture of replication of SV40 DNA was made by Sebring et al. (1971), who described a replicative intermediate (RI) that had not been reported previously and which is likely to be the essential intermediate form. While a number of replicative forms similar to those described by other workers were detected, Sebring et al. noted that the most frequently occurring molecules contained two branch points, three branches, no free ends, and a superhelical region in the unreplicated portion of the molecule (see Fig. 5–3). This form has also been described by Jaenisch et al. (1971).

When isotopically labeled RI molecules were isolated from sucrose gradients, Sebring et al. showed that in these molecules newly replicated SV40 DNA was not covalently attached to the parental DNA strands, and that the parental DNA strands were covalently closed. Figure 5–4 represents an interpretation of these results.

An important problem that has still to be solved in order to understand the replication of circular DNA molecules concerns the unwinding of the parental duplex. During replication of a covalently closed circular DNA molecule, unwinding would introduce superhelical turns into the molecule and as replication proceeded it would become increasingly difficult to unwind parental strands. The existence of a "swivel" was postulated in order to overcome this difficulty (Cairns, 1963). Since the parental strands of the RI remain covalently closed in alkali, Sebring et al. (1971) suggest that such molecules do not have a permanent swivel point in the unreplicated region. They envisage an intermittent swivel involving the alternate action of a nicking endonuclease and a ligase to seal the nick after each phase of DNA synthesis, so that replicative intermediates isolated from the cell at any given moment may not contain a swivel. Following such repair, for which an appropriate polynucleotide ligase is induced in SV40-infected cells (Sambrook and Shatkin, 1969), replication would not proceed until another nick was introduced. Of particular interest is the recent discovery of a protein isolated from mouse embryo cells that is capable of untwisting closed circular DNA's containing superhelical turns (Champoux and Dulbecco, 1972). The "enzyme" apparently acts by introducing a single-strand nick into the DNA, thus forming a DNA–enzyme complex that allows the DNA strands to rotate relative to the helical axis; the break is then resealed. This activity could be the postulated swivel.

## Integration of Host and Viral Genomes

When nonpermissive cells are infected and transformed by polyoma viruses, viral genomes are stably integrated into the host genome in such a way that they are transcribed and inherited as the transformed cells multiply (see Chapter 13). Until quite recently it was assumed that in permissive systems the replicating viral genomes remained independent of the host genome. It now appears that viral genomes may become integrated into the host genome during the multiplication of polyoma and SV40 in permissive cells, but it has yet to be established whether the recombinational events leading to integration and excision of viral DNA are necessary for viral multiplication.

During productive infection with polyoma virus, some linear host DNA can be found encapsidated in viral particles (Michel et al., 1967; Winocour, 1968). These so-called pseudovirions were also found after SV40 infection of primary

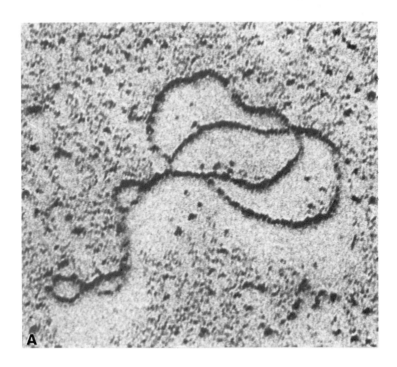

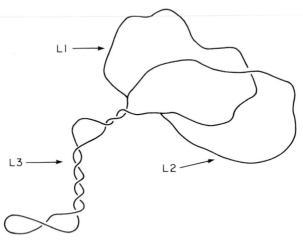

FIG. 5–3. *Papovaviridae.* (A) *Electron micrograph of a twisted SV40 DNA replicating molecule. In some regions the individual strands which comprise the superhelical branch can be seen. Magnification:* $1.5 \times 10^5$. *In an interpretive drawing of the molecule (B), the branches of the molecules measured are indicated. The two branches that were not superhelical were designated L1 and L2. The superhelical branch was designated L3. Most measurements were carried out on stained molecules. The duplex length of L3 was estimated by measuring its linear length and multiplying by 2. (From Sebring et al., 1971.)*

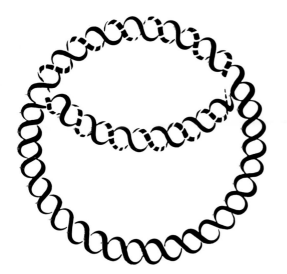

FIG. 5–4. *Papovaviridae. Diagrammatic representation of replicating SV40 DNA. The salient features of the molecule are: (1) both parental DNA strands (solid lines) are covalently closed, and (2) the two newly synthesized DNA strands (broken lines) are not covalently linked to the parental DNA nor are they linked together. (From Sebring et al., 1971.)*

African green monkey cells, but not with infection of BSC-1 cells (Levine and Teresky, 1970).

Investigating interactions between SV40 and permissive BSC-1 cells, Aloni *et al.* (1969) found that closed circular SV40 DNA could be hybridized to some extent with host cell DNA, from which they concluded that host DNA sequences had become covalently incorporated into some viral DNA molecules. This occurred only at high input multiplicities or when plaque purified virus was passaged undiluted (Lavi and Winocour, 1972; Tai *et al.*, 1972). By denaturing and then renaturing SV40 DNA from particles produced at high or low multiplicities, Tai *et al.* produced heteroduplex molecules with inhomogeneities arising from deletions or deletions with substitutions. The substitutions are probably due to host DNA arising from integration of SV40 DNA into the chromosome followed by excision of the SV40 DNA together with chromosomal DNA. At high multiplicities, such DNA's will replicate because complementation should provide any missing viral functions.

Additional evidence for integration of DNA during lytic infection has come from studies of SV40-infected Chinese hamster cells (Hirai *et al.*, 1971) and polyoma-infected mouse embryo cells (Ralph and Colter, 1972). Host cell DNA was substituted with a "heavy" analog of thymine prior to infection and subsequently the cells were infected with polyoma virus in the presence of an inhibitor of DNA synthesis. After separation of the lighter viral DNA, "heavy" host DNA could be hybridized with virus-specific RNA, from which the number of viral genome equivalents per cell (about eight in the polyoma system) could be estimated.

The evidence in both productive and transformation systems for integration of the viral DNA, and the induction of T antigens and of cell DNA synthesis (reviews, Dulbecco, 1969; Green, 1970a), suggest that there are fewer differences between productive infection and transformation than previously thought, but it is not yet clear whether integration is a necessary step in viral multiplication.

## Assembly and Release

All the papovaviruses are assembled in the nucleus, within which they may form crystalline arrays. Polyoma virus and SV40 develop in a rather similar manner in the nuclei of infected mouse embryo cells (Mattern *et al.*, 1966) and *Cercopithecus* monkey cells (Granboulan *et al.*, 1963), respectively. Mattern *et al.* (1966) studied the intracellular development of polyoma virus during a single-step growth cycle in mouse embryo cells. Unfortunately, the inoculum used contained membrane-bound aggregates as well as single virions, and the early stages after engulfment are difficult to interpret. By 16–20 hours after infection, bundles of densely staining filaments developed in the nucleus and later became very prominent. By 24 hours there were many filaments, scattered spherical particles, and small crystalline arrays of spherical particles in the nucleus; later, larger nuclear crystals were sometimes seen. From the twenty-eighth hour, virions, but not filaments, were found in the cytoplasm of cells whose virus-laden nuclei were degenerating. These progeny virions were arranged on various cellular membranes as monolayers of tightly packed particles. The strong affinity of virions released from the nucleus for cell membranes, from which they can be released by neuraminidase, provides an explanation for the value of this enzyme in the preparation of concentrated suspensions of virus (Crawford, 1962).

## Alterations of Cellular Metabolism

Productive infection of confluent cultures with polyoma virus or SV40 markedly stimulates the rate of cellular DNA synthesis (Dulbecco *et al.*, 1965; Weil *et al.*, 1965), the increase being dependent upon several factors including the initial cell state and the multiplicity of infection (Vogt *et al.*, 1966; Branton and Sheinin, 1968). The greatest stimulation of DNA synthesis is observed in confluent primary cells that normally possess low DNA synthetic ability. SV40 infection of CV-1 cells (Kit *et al.*, 1967b), but not of BSC-1 cells (Gershon *et al.*, 1966), induces DNA synthesis. Polyoma virus infection of exponentially growing mouse embryo cells did not induce DNA synthesis at high multiplicities but did so at lower multiplicities (Branton and Sheinin, 1968). Only part of the viral genome is involved in derepressing host DNA synthesis, since infectivity is inactivated five times more rapidly than the ability of polyoma virus or SV40 to induce DNA synthesis (Defendi and Jensen, 1967; Gershon *et al.*, 1966).

The stimulation of DNA replication by polyoma virus and SV40 involves all cellular DNA, including mitochondrial DNA, and is accompanied by the induction of synthesis of histones. When the replication of density-inhibited cultures of uninfected fibroblasts is stimulated by a change of medium the rate of nuclear acidic protein synthesis is immediately increased (Farber *et al.*, 1971). Rovera *et al.* (1972) have found that after infection of permissive or nonpermissive cells with SV40, synthesis of this same class of acidic proteins is induced within 3 hours of infection, i.e., about 9 hours before the synthesis of DNA and histones is induced. The analogy between the virus-induced change and the serum-induced change in nuclear acidic protein synthesis suggests a common pathway for induction of DNA synthesis by the different stimuli.

At least eight enzyme activities increase after infection of permissive cells by SV40 or polyoma virus (review, Green, 1970a). These enzymes, which include thymidine kinase and DNA polymerase, are all involved in deoxyribonucleotide conversions. As there is clearly insufficient viral genetic information to code for this number of enzymes, it seems likely that they represent derepressed host cell functions. This argument is supported by demonstrations that thymidine kinase was not elicited in kinase-negative cells transformed by or productively infected with polyoma virus (Littlefield and Basilico, 1966; Basilico et al., 1969). However, there are reports of differences in physicochemical properties between pre- and postinfection activities of thymidine kinase and DNA polymerase (Carp, 1967; Kit et al., 1967a). It remains to be established if these differences indicate modification of preexisting enzymes by viral infection, derepression of host genes for isoenzymes, or the coding of a new enzyme by the virus.

The synthesis of ribosomal or tRNA species is not altered appreciably by polyoma virus or SV40 infection (Oda and Dulbecco, 1968), but synthesis of cellular mRNA is stimulated about twofold (Benjamin, 1966; Oda and Dulbecco, 1968).

## ADENOVIRUS

Adenoviruses are well suited for virological studies because large amounts of the virions and their components can be readily obtained and purified. Much is known about their structural proteins and DNA (see Chapter 3), but little information is as yet available concerning the nonstructural proteins that probably account for two-thirds of the coding potential of their genome. The two most commonly used adenoviruses in laboratory studies with cultured cells are the human types 2 and 5 (reviews by Schlesinger, 1969; Green et al., 1970); most of the following information pertains to these.

### The Multiplication Cycle

High yields of adenovirus (up to $10^4$ pfu per cell) can be obtained from infected suspension cultures of KB cells or certain lines of HeLa cells. Plaque assays are carried out in KB cells or human embryo kidney monolayers. Assembly occurs in the nucleus, where large aggregates of virions are found; these are not released until the necrotic cell is disrupted.

The multiplication cycle extends over a prolonged period; the eclipse period varies with cell–virus system and multiplicity of infection but is usually between 13 and 18 hours (Fig. 5–5).

### Initiation of Infection

Attachment, as measured by mixing highly purified radioactive adenovirions with cells resuspended at high concentrations, can take place at $0°–4°C$, although the subsequent steps of penetration and eclipse are prevented until the temperature is raised to about $37°C$. The elongated fibers projecting from the vertices of the virion (see Chapter 3) probably constitute the points of contact between virus and the attachment sites on cell surfaces, since receptor sites can bind and be inactivated by purified adenovirus type 2 fiber antigen (Schlesinger, 1969). There are about $10^4$ adenovirus receptors per cell; studies of their selective destruction with different proteases indicates that they are different from picornavirus receptors (Philipson *et al.*, 1968).

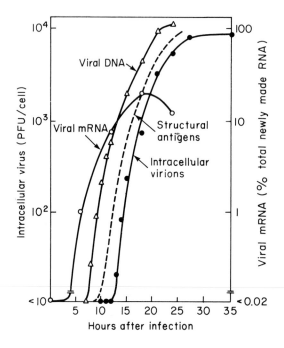

FIG. 5–5. *Adenovirus. Idealized representation of the multiplication of adenovirus type 5 in KB cells, illustrating the relationship between increases in viral mRNA (transcription), viral proteins (translation), replication of viral DNA, and intracellular appearance of infectious virions. (From Green et al., 1970.)*

Once adenovirus is attached to cells, it is rapidly taken into the cell either by engulfment into a phagocytic vacuole (Chardonnet and Dales, 1970a) or by direct penetration into the cytoplasm (Morgan *et al.*, 1969). Both mechanisms may operate, the apparent predominance of one over the other being a function of the cytotic activity of the cells used. Isotopically labeled adenovirus type 5 particles isolated soon after entering the cytoplasm have a buoyant density (1.35 $g/cm^3$) slightly greater than that of intact virions (1.34 $g/cm^3$), due to the loss of about 5% of the original virion protein (Sussenbach, 1967), probably the penton antigen from the vertices of the particle. This loss of protein, which may occur within 5 minutes of attachment, is not dependent upon synthesis of new enzymes (Lawrence and Ginsberg, 1967; Sussenbach, 1967) and is sufficient to eclipse viral infectivity.

Analysis by gradient centrifugation and electron microscopy of subcellular fractions of cells infected with radioactive adenovirus shows that the altered particles move into the cell nucleus (Chardonnet and Dales, 1972). Here they are rapidly converted to particles which lack more than two-thirds of the orginal virion protein and correspond to the viral "cores" produced *in vitro* by urea treatment (Lonberg-Holm and Philipson, 1969). Viral DNA is then released in a form virtually free of protein. The entire process from attachment to liberation of viral DNA in the nucleus is largely completed within 1 hour.

## Transcription

Adenovirus is a useful system for the study of viral gene transcription and regulation. The chemical and physical properties of the virus and its DNA are well known, milligram quantities of viral DNA are easily obtained, the virus synthesizes large amounts of mRNA, and the structural polypeptides of the virion have been characterized in great detail. Furthermore, gene transcription can be studied with either productively infected cells or virus-transformed cells. Nuclei from adenovirus-infected cells synthesize high molecular weight virus-specific RNA *in vitro*. Since synthesis is inhibited by $\alpha$-amanitine, it is likely that adenovirus DNA is transcribed principally by an activity which is identical to or resembles closely the RNA polymerase II of the host cell (Price and Penman, 1972).

Two distinct categories of viral mRNA can be distinguished: "early" RNA, made prior to DNA synthesis, and "late" RNA appearing in significant amounts only after viral DNA synthesis. Early and late messengers are qualitatively different as seen by hybridization competition tests; total adenovirus type 2 nuclear mRNA contains 58–59% G+C, while early cytoplasmic viral mRNA contains only 53%, indicating early selective transcription of a segment of the viral DNA with a relatively low G+C content (Parsons and Green, 1971; Parsons *et al.*, 1971; review, Green *et al.*, 1970).

Early virus-specific RNA, which may appear within 2 hours of infection, is homologous to about 8 to 20% of the viral genome and when this RNA is isolated from the cytoplasm and resolved by electrophoresis on polyacrylamide gels, two major species (23 and 16 S) can be resolved. If cells are infected in the presence of cycloheximide, there is a fivefold increase in the rate of early messenger synthesis and another viral mRNA species (27 S), normally present in minor quantities, becomes readily detectable. All early virus-specific RNA sequences are also detectable late (18 hours after infection). In adenovirus-transformed cells, the sequences transcribed are homologous to about 50% of the early sequences transcribed in productive infection (Green *et al.*, 1970). In close agreement with this figure, Raska and Strohl (1972) found 60% of early viral sequences transcribed in BHK cells undergoing abortive infection by adenovirus type 12. Neither in adenovirus-transformed cells nor in the abortive infection of BHK cells by adenovirus type 12 can late viral sequences be detected.

The virus-specific RNA sequences that are transcribed in cells transformed by adenovirus 2 are heterogeneous in size (4–45 S). The 4 S species hybridize with viral DNA but do not appear to be tRNA species (Kline *et al.*, 1972), and are said to be transcribed by polymerase I (amanitine-insensitive) and thus not a breakdown product of viral mRNA (Price and Penman, 1972). A single molecule of 45 S would represent 30–50% of the viral genome as a polycistronic viral RNA. These 45 S species may contain host transcripts, as in the case of SV40. Green *et al.* (1970) claim that adenovirus DNA can be separated into heavy

(H) and light (L) strands. Virtually all the viral genome is transcribed in productive infection; 80% of the transcripts are produced from L strands and 20% from H strands. Of the early messenger sequences, 40% are H strand transcripts and 60% L strand transcripts.

The late viral RNA species (appearing 18 hours after infection and dependent upon DNA synthesis) can be detected in the cytoplasm as six to eight species with sedimentation coefficients in the range 10–29 S, at least the smaller molecules being of the right size to serve as monocistronic messengers. Short (15-minute) pulse labeling of nuclear virus-specific RNA with [³H]uridine revealed four RNA species (36, 38, 40, and 43 S). Since these labeled species can be "chased" into the cytoplasm where 10–29 S species are detected and since viral RNA sequences present in the 36–43 S species are also found in cytoplasmic RNA by hybridization competition tests, at least some of the cytoplasmic virus-specific RNA molecules are derived by cleavage of high molecular weight precursors from the nucleus. The late mRNA species that specify structural proteins of the virion are relatively stable, declining with a functional half-life of about 6 hours in the presence of actinomycin D (White *et al.*, 1969).

Evidence is now rapidly accumulating to support the idea that after adenovirus RNA is produced as large molecules and then cleaved, polyadenylic acid is attached and the messenger molecules are transported to the cytoplasm (Philipson *et al.*, 1971; Raskas, 1971; Raskas and Okubo, 1971).

## Translation

The size of the adenovirus genome indicates that it could code about twenty-three polypeptides of average molecular weight 50,000 daltons (Table 7–1).

**Structural Proteins.**   The virion contains several structural proteins, with molecular weights varying between 22,000 and 120,000 daltons, the principal ones comprising the hexon (120,000), penton base (70,000), fiber (62,000), and core proteins (46,000 and 22,000) (see Chapter 3). Considerable quantities of these structural proteins ("soluble antigens") are produced during adenovirus infection, and the time of their appearance is dependent upon the multiplicity of infection. They are made in such excess that they accumulate as protein crystals in the nucleus at later stages of infection (Godman *et al.*, 1960).

The structural antigens are all "late" products of the genome, their synthesis being dependent on synthesis of viral DNA. The fiber and hexon antigens appear about 12 to 16 hours after infection, the penton base antigen about 2 hours later. Infectious virus increases rapidly in parallel with appearance of penton base, as all components are then available for assembly into complete virions. Free pentons, detected by direct hemagglutination, take an even longer period than penton base to appear. Russell *et al.* (1971) suggest that the penton bases, when produced, are not incorporated into free pentons but are preferentially assembled in the virions. Most of the structural polypeptides are not derived by posttranslational cleavage but are separately synthesized at different rates in molecular ratios that do not correspond with those in the virion itself

(White *et al.*, 1969). For example, the penton base and fiber polypeptides are made in considerable excess relative to their representation in the virion, and the polypeptides of the cores are produced in minimal amounts (Horwitz *et al.*, 1969), i.e., some control mechanism must modulate the transcription and/or translation of different mRNA species.

Highly purified adenovirions have an endonuclease activity associated with them (Burlingham *et al.*, 1971). This activity appears in productively infected cells at approximately the same time as the capsid proteins (Burlingham and Doerfler, 1972).

**Nonstructural Proteins.** Because of the delay in arrest of host protein synthesis in adenovirus-infected cells (see below), it has been difficult to detect nonstructural virus-coded proteins. A number of investigators were able to demonstrate that an antigen (called T antigen, see Chapter 13) immunologically distinct from capsid antigens is synthesized early in infection. One distinctive feature of this T antigen, which is detectable by gel diffusion or complement fixation with sera from hamsters bearing virus-induced tumors (Berman and Rowe, 1965), is that its biosynthesis is not affected by inhibiting viral DNA synthesis (Gilead and Ginsberg, 1968), i.e., it is specified by an "early" mRNA.

Using specific antisera prepared against the purified antigens, Russell *et al.* (1967) were able to follow the development of the antigens in adenovirus type 5 infected cells. They detected another early antigen, P antigen, using rabbit antiserum prepared against an extract of cells infected with adenovirus type 5 in the presence of an inhibitor of DNA synthesis. The P antigen appears before the capsid antigens; it is first detectable about 5 to 7 hours after infection and reaches a maximum level at about 12 hours. The P antigen is similar in several respects to the T antigen; it is an early antigen whose synthesis is not dependent on viral DNA synthesis, and it is relatively thermolabile. The function of T and P antigens is not established, although Russell *et al.* (1971) suggested that P antigen might be a viral internal protein.

In an attempt to detect early nonstructural polypeptides, Russell and Skehel (1972) have labeled infected cells with [$^{35}$S]methionine of high specific activity and examined the polypeptides by gel electrophoresis and autoradiography. Ten hours after infection with adenovirus type 5 they found five nonstructural "infected cell-specific" polypeptides (ICSP 1–5). They are probably virus-specified. ICSP 3, which is synthesized in the presence of cytosine arabinoside and is a major component of the P antigen, is detectable as early as 2.5 hours after infection; the others first appear at 10 to 13 hours. These polypeptides are not synthesized simultaneously with capsid polypeptides and ICSP's 1, 2, and 3, like the arginine-rich core polypeptides, are not synthesized later in infection. Russell and Skehel have estimated that if all the ICSP's are classified as separate viral gene products, then about 15% of the viral polypeptides can be classified as "early," which is in agreement with the estimates of Green *et al.* (see above) for the early transcription of the genome. Thus far there is no clear evidence in the literature that P antiserum reacts with antigens containing core polypeptide 2, although there is circumstantial evidence that core polypeptide

2 is a "late" component of P antigen. To date there have been no convincing data for the detection of adenovirus-induced enzymes early in infection and the biological role of these early infected cell-specific polypeptides is unknown.

### Replication of Viral DNA

Synthesis of adenovirus DNA in the cell nucleus precedes maturation of viral particles by about 6 to 7 hours and, as is usually the case for DNA viruses, DNA is synthesized in considerable excess (about tenfold) over that incorporated into virions. DNA synthesis continues for about 12 hours and results in an accumulation of viral DNA equivalent to the DNA content of the uninfected cell. The large pool of unencapsidated viral DNA forms part of the characteristic inclusion bodies of adenovirus-infected cells. Host DNA is not broken down and utilized for viral DNA synthesis.

Recent investigations have been directed toward understanding the intranuclear sites and forms of replicating adenovirus DNA. There is evidence that viral DNA can be extracted as a complex with cellular protein, possibly representing a replicative stage (Doerfler et al., 1972). Based on the nature of the association of the isolated DNA replication complex with membranelike material, it has been tentatively proposed that viral DNA attaches to the nuclear membrane, probably competing with cellular DNA for attachment sites (Pearson and Hanawalt, 1971). Although the present evidence is conflicting, it is likely that adenovirus DNA replicates as linear genome length molecules, one daughter strand displacing one parental strand as a single-stranded branch (Sussenbach et al., 1972; Bellett and Younghusband, 1973).

Newly replicated adenovirus type 2 DNA can be isolated from the nucleus of infected HeLa cells as a fairly homogenous complex with a sedimentation coefficient of 73 S (Wallace and Kates, 1972). This affords an opportunity to study the association of viral DNA with protein and the early steps in its maturation into virions.

### Assembly and Release

Electron microscopy reveals that assembly of adenovirions is confined to the nucleus, from which virions escape only when the nuclear membrane is disrupted, and large crystalline arrays of virions are often seen in the nuclei of cells infected with adenoviruses.

By studying the synthesis of individual polypeptides comprising the structural antigens, it has been possible to trace them from synthesis on cytoplasmic polyribosomes to incorporation into whole virions (Horwitz et al., 1969). Hexon polypeptides require 3–4 minutes to be synthesized and released from ribosomes, and 3–4 minutes later 80–90% of the hexons are assembled into complete capsomers. The polypeptides comprising the penton base and fiber are even more rapidly synthesized and released from ribosomes (2 minutes), but although some assembly of fiber and penton base into pentons was detected, Horwitz and co-workers found that many hours were required for completion of penton assembly, in accord with the findings of Russell et al. (1967), based on measurement of antigens. Radioactive core proteins appear in virions within 15 minutes of their synthesis and prior to the time that hexon polypeptides are detectable in assembled virions. Synthesis of individual polypeptides is not coordinated with structural requirements of the virus and the significance of overproduction of pentons and fibers is not understood, unless it is that being the least numerous molecules in the capsid, they need to be produced in excess so as to ensure an adequate pool.

Arginine is particularly important for the multiplication of several DNA viruses, and must be exogenously supplied to ensure the production of adeno-virions (Bonifas, 1967). In the absence of arginine, production of infectious adenovirus type 2 is inhibited, although T antigen, viral DNA, hexon, penton base, and fiber are synthesized. Upon restoration of arginine to the medium late in infection, new infectious virus becomes detectable quite rapidly. This phenomenon provides an opportunity to study experimentally some stages of virion assembly, although it is difficult to standardize conditions and results from different laboratories therefore vary. It was suggested that arginine deficiency prevents a late replicative step, such as the synthesis of a maturation factor, and that this factor might be an arginine-rich core component of the mature virion (Russell and Becker, 1968). Data on the proteins synthesized under conditions of arginine deprivation (Arg$^-$) have recently been compiled (Rouse and Schlesinger, 1972; Raska et al., 1972). All major viral structural polypeptides are made but in reduced amounts. Surprisingly, these are not used for virion assembly subsequent to restoration of arginine; Everett et al. (1971) found that capsid antigen titers, in particular hexon antigen, were considerably diminished under these conditions, but penton base and fiber antigen titers were less inhibited. Certain virus-specific antigens such as P antigen appeared to be reduced when assayed by immunofluorescence but not when detected by complement fixation tests (Russell and Becker, 1968). There are no precise data on whether viral structural antigens synthesized under Arg$^-$ conditions fail to migrate from cytoplasm to nucleus, as is found in herpesvirus-infected arginine-deprived cells.

The amount of adenovirus DNA synthesized under Arg$^-$ conditions is reduced about tenfold as is the synthesis of virus-specific RNA. There is no significant difference in the species of RNA sequences transcribed in Arg$^+$ or Arg$^-$ infected cells. The arginine-dependent step follows DNA synthesis, since prolonged inhibition of viral DNA synthesis in the presence of arginine is not followed by formation of infectious virus upon reversal of the block in the absence of arginine. The general conclusion is that some "maturation" factor is critically dependent upon the availability of an external arginine supply. Russell and Skehel (1972) note that synthesis of core polypeptide 2, an arginine-rich component, is inhibited under conditions of arginine deficiency and it might be that this polypeptide controls the rate of viral DNA replication as well as the aggregation of viral DNA and proteins into nucleocapsids.

A more complete understanding of the mechanisms involved in adenovirus assembly and maturation is likely to come from the sort of studies conducted by Winters and Russell (1971), who have been able to demonstrate assembly of infectious adenovirus from infected cellular extracts *in vitro*. Since it is now possible to separate the internal viral nucleoprotein containing only one very arginine-rich protein in association with viral DNA, it is feasible to test directly the role of the arginine-rich protein in assembly. However, it may well be that more than one maturation factor is involved and Winters and Russell point out that with crude extracts lipid-containing membranes might be of some significance, since nonionic detergents and fluorocarbon inhibit the process.

### Alterations of Cellular Metabolism

A characteristic of adenovirus infection is that despite the accumulation of viral DNA and protein beginning 6–10 hours after infection, the overall rates of synthesis of these macromolecules, as compared with uninfected suspension cultures, remains unchanged for about 40 hours. Total RNA synthesis is unaltered in rate for about 24 hours and decreases thereafter. However, using methods that distinguish cellular and viral synthesis, it was found that adenovirus inhibits cellular DNA synthesis, but that this may take about 10 to 14 hours (Ginsberg, 1969). Synchronized cells that are infected at the beginning of the S phase of DNA synthesis will proceed through S normally but will not enter mitosis or initiate a subsequent round of DNA synthesis. As a result, adenovirus DNA replication can be studied in isolation from cellular DNA synthesis (Hodge and Scharff, 1969). Measurement of host RNA synthesis by hybridization techniques demonstrates that transcription of the host genome may not be inhibited until about 15 hours after adenovirus type 5 infection.

Host protein synthesis declines in parallel with the rise in capsid protein synthesis, beginning at about 15 hours after infection. Fluorodeoxyuridine, which inhibits late viral protein synthesis, also completely prevents the shutdown of host protein synthesis, even at high multiplicity of infection (White *et al.*, 1969). The authors suggest that the inhibition of cell protein synthesis may be attributed to competition for ribosomes between host and viral mRNA's rather than to any virus-coded "early" protein. Fiber antigen, which in high concentrations can inhibit viral and cellular protein synthesis (Levine and Ginsberg, 1967), is not a selective cell shutdown protein, though it may be instrumental in the eventual death of the cell. There is, however, the interesting finding that some late viral protein prevents the association of heterologous viral mRNA with ribosomes in cells coinfected with adenovirus and poxvirus (Giorno and Kates, 1971).

### HERPESVIRUS

The herpesviruses have a large genome (100 million daltons) and are structurally complex (see Chapter 3). The capsid is assembled in the nucleus and maturation occurs as the nucleocapsid acquires an envelope at the nuclear membrane. Many herpesviruses cause persistent infections characterized by long periods of latency (see Chapter 12); others are integrated into the cellular genome and are oncogenic (see Chapter 13). Here we shall be concerned only with productive cytocidal infection. Two models have been extensively studied: herpes simplex type 1 (HSV-1) and pseudorabies virus (reviews by Roizman, 1969a, b; Kaplan, 1969).

## The Multiplication Cycle

The duration of the eclipse phase is about 8 hours. With HSV in HEp-2 cells, viral protein appears at 2 hours and reaches a maximum at about 8 hours, viral DNA has a similar time sequence, about an hour later, and free infectious virions appear at 10 hours and reach peak titers at 15 hours (Fig. 5–6; Roizman, 1969b).

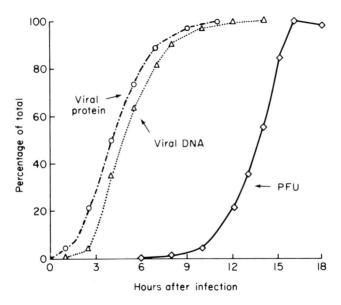

FIG. 5–6. *Herpesvirus. The synthesis of viral protein, viral DNA, and assembly of herpes simplex virus type 1 in infected HEp-2 cells. (From Roizman, 1969b.)*

## Initiation of Infection

There are conflicting views concerning the relative importance of cytosis and fusion. Hummeler *et al.* (1969), suggest that both kinds of penetration may occur but that engulfment of the enveloped virion may be the more important.

Controversy over the role of the envelope in the infection process has been reviewed at length by Darlington and Moss (1969). Recently, Abodeely *et al.* (1970) have shown that either enveloped or chemically deenveloped particles can be taken into cells by viropexis and initiate infection. The presence of the envelope probably facilitates adsorption to cells and increases the efficiency of plating, but is not intrinsically essential for infectivity (Spring and Roizman, 1968; Watson *et al.*, 1964). Virtually nothing is known of the initial processing of herpesvirus particles within cells.

## Transcription

Despite the size and complexity of the herpesvirion, there is no evidence that herpesvirus, like poxvirus, carries its own transcriptase. Virus-specific RNA is made first in the nucleus, perhaps using cellular RNA polymerase as adenovirus appears to do. This argument is supported by a report (Lando and Ryhiner, 1969), which still requires confirmation, that HSV-DNA is infectious.

Virus-specific RNA is detectable in the nucleus as molecules with sedimentation coefficients in the range 40–60 S (Wagner and Roizman, 1969; Wagner, 1972). After a lag of about 10 minutes, RNA hybridizable with viral DNA appears in the cytoplasm (Roizman et al., 1970) as molecules with sedimentation coefficients over the range 4–30 S, but predominantly in the range 10–20 S. The higher molecular weight viral RNA from the nucleus has been shown by molecular hybridization competition analyses to have some sequences in common with the cytoplasmic viral RNA (Wagner, 1972) and presumably the cytoplasmic species arise as a result of cleavage of the high molecular weight transcripts found in the nucleus. Herpesvirus-specific RNA's may be "early" or "late," i.e., transcribed prior to or after the onset of viral DNA synthesis; both classes exist as larger molecules in the nucleus than in the cytoplasm. Evidence for two classes of early viral RNA in HSV-infected cells has been provided by hybridizing unlabeled RNA from infected cells with labeled purified viral DNA (Roizman and Frenkel, 1972). Both early (2 hours after infection, prior to viral DNA synthesis) and late (8 hours after infection), they detected two classes of RNA differing in abundance. Competition hybridization tests demonstrate that the sequences in the most abundant early RNA are present in the most abundant late RNA. Roizman and Frenkel argue that the function of the most abundant classes of RNA is to specify certain structural proteins of the virion, but data in support of this are preliminary.

In cycloheximide-pretreated pseudorabiesvirus-infected cells, there is production of virus-specific RNA with a sedimentation coefficient of about 26 S. Unless the ratio of viral DNA to this RNA is kept high in hybridization tests, only a small proportion is hybridizable, suggesting that it results from repeated transcription of a small segment of the viral genome (Rakusanova et al., 1971). Competition hybridization tests and comparison of saturation characteristics of viral DNA with RNA synthesized by infected cells in the presence or absence of cycloheximide have led Rakusanova and colleagues to the conclusion that RNA synthesized before the onset of DNA synthesis in pseudorabies-infected cells ("early" mRNA) can be divided into two distinct classes, one of which does not require synthesis of proteins following infection whereas the other does. These two classes of RNA may be analogous to the "immediate early" and "delayed early" species of T4 phage-specified RNA (Salser et al., 1970).

In a series of papers, Subak-Sharpe and co-workers (Subak-Sharpe and Hay, 1965; Subak-Sharpe et al., 1966a) reported that herpes simplex virus could code for certain tRNA's. Although this claim has now been withdrawn (Bell et al., 1971), it did stimulate similar investigations with other animal viruses (see sections on Poxvirus and Rous sarcoma virus).

## Translation

During the first few hours of herpesvirus infection, the overall rate of protein synthesis declines, followed by a period of stimulation lasting about 5 hours before declining irreversibly. These phases correspond to the inhibition of cellular protein synthesis followed by a period of viral protein synthesis and are reflected in the sedimentation profiles of polyribosomes prepared from infected cells. Polydisperse host polysomes (sedimentation coefficient approximately 170 S in sucrose gradients) become disaggregated and are replaced starting about 2–3 hours postinfection by more rapidly sedimenting viral polysomes (270 S) (Sydiskis and Roizman, 1966, 1967).

More than twenty polypeptides have been detected in herpes-infected cells; at late times (8–9 hours after infection) only about thirteen viral polypeptides are labeled. In pulse-labeled cells infected with herpes simplex or pseudorabies-virus, labeled viral polypeptides first appear in the cytoplasm but migrate into the nucleus (Roizman, 1969b; Fujiwara and Kaplan, 1967). The transport of viral polypeptides from cytoplasm to nucleus, where they are assembled into capsid proteins, is comparatively slow, the larger polypeptides requiring 2–3 hours to be chased after pulse labeling, and there appears to be some selectivity since some polypeptides are restricted to the cytoplasm (Roizman, 1969b).

**Structural Proteins.** Determination of which polypeptides synthesized in infected cells were actually viral structural components has been hampered by the inherent difficulties in purifying herpesvirions free from cell material and by the problem of obtaining clean envelope-free virions (see Chapter 3). These problems have been partially resolved by improvement in techniques of purifying virions (Robinson and Watson, 1971), by comparing unenveloped virus obtained from infected nuclei with enveloped virus obtained from cell cytoplasm fractions, and by the use of a nonionic detergent to remove viral envelopes completely (Olshevsky and Becker, 1970). Eight major and some fourteen minor components have been recognized in enveloped herpes virions (see Chapter 3). Two of the core proteins are rich in arginine. If cell cultures are deprived of arginine, viral DNA and viral polypeptides are synthesized but the arginine-rich polypeptides are synthesized at a rate too low to support viral maturation (Y. Becker *et al.*, 1967). A further block in viral assembly is due to the failure of viral structural proteins to migrate into the nucleus in arginine-deprived cells (Mark and Kaplan, 1971). A contrasting but practically useful observation is that incubation of human Burkitt lymphoblasts in arginine-deficient medium results in the synthesis of Epstein-Barr viral antigens and virions in most of the cultured cells (Henle and Henle, 1968; Weinberg and Becker, 1969; see Chapter 13).

A large amount of confusing literature on the synthesis and cellular location of viral antigens, of which about twelve can be detected by gel diffusion tests (Watson *et al.*, 1966), has been summarized by Roizman (1969b). Precise data on "early" proteins (synthesized before viral DNA synthesis) are lacking, although it is becoming increasingly clear that unlike the situation with other DNA viruses, nearly all viral structural proteins are synthesized by input herpesvirus

DNA prior to its replication (Frenkel and Roizman, 1972a). The only viral proteins whose synthesis has been followed in detail are the glycoproteins, all of which are located in the envelope (see Chapter 3). Following herpesvirus infection, synthesis of host plasma membrane proteins ceases, and new glycoproteins accumulate in the smooth membranes (Heine *et al.*, 1972; Spear and Roizman, 1970; Keller *et al.*, 1970). The number and electrophoretic mobilities of these glycoproteins are genetically determined by the virus; for example, HSV strains MP and G differ not only in viral envelope glycoproteins but also in the smooth membrane glycoproteins elicited in cells they infect.

The new antigens that appear on the surface of infected cells are identical to those present on the surface of infectious virus particles (Heine *et al.*, 1972). Glycosylation of the viral proteins occurs after they are bound to smooth membranes. The enzymes responsible are probably preformed cellular enzymes since glycosylation is not blocked by concentrations of puromycin that inhibit viral protein synthesis. Further, there are differences in the glycosylation of membrane proteins following infection of different cells, such as Vero and HEp-2. Since the extent of glycosylation is probably influenced by the primary structure of viral envelope proteins, variations in the proteins specified by different strains of virus might lead to different glycoproteins in infected cell membranes, and thus, to use Roizman's (1970) phrase, to different "social behavior" of infected cells.

Proteins associated with the nuclear membrane at the time of infection do not become part of the virion nor are they lost from the membrane after infection (Ben-Porat and Kaplan, 1971). Therefore, viral proteins must be embedded, presumably prior to glycosylation, into newly forming nuclear membrane, in such a way that the nucleocapsid buds through locally altered areas and acquires only virus-specific proteins.

**Nonstructural Proteins.**   Assuming that each gene for a structural protein of herpes simplex virus is represented only once in the viral genome, it can be estimated from the sum of the molecular weights of the known viral structural proteins that only about 12 to 15% of the viral DNA is required to code for all the structural proteins (see Table 7–1), so that the genome carries a large amount of additional information. Looking to the analogous situation in phage T4, some of this information may specify enzymes, and there is a growing body of evidence to support the contention that HSV codes for several enzymes (Hay *et al.*, 1971).

Herpesviruses elicit increased activities of DNA polymerase, thymidine kinase, dCMP deaminase, deoxycytidine kinase, dTMP kinase, and DNase (reviews by Kit, 1968; Hay *et al.*, 1971). There is evidence that DNA polymerase, thymidine kinase, and the DNase induced by HSV infection are biochemically and immunologically distinct from the corresponding activities in uninfected cells. The virus induces thymidine kinase in kinaseless mutant cells and a corresponding viral mutant has been obtained which cannot produce thymidine kinase. Hay *et al.* have provided convincing data to support the idea that the virus-induced thymidine kinase and deoxycytidine kinase activities are expressions of one enzyme with two separate active sites. High concentrations of thymidine are known to block cellular DNA synthesis in cultured cells, probably due to feedback inhibition of ribonucleotide reductase (Morris and Fisher, 1963). Herpesvirus DNA replication is not blocked by such high thymidine levels; this could be due to the stimulation of ribonucleotide reductase activity in infected cells (Cohen, 1972). It is not established whether the induced reductase is virus-specified or a modification of a preexisting activity. There is considerable speculation but no concrete evidence about the role of virus-elicited enzyme activities in herpesvirus multiplication. The few studies on the regulation of synthesis of the induced enzyme activities (see Kaplan *et al.*, 1967) are complicated by leakage of intracellular protein late in infection.

## Replication of Viral DNA

Herpesvirus DNA synthesis takes place in the nucleus. Host DNA synthesis becomes progressively inhibited after infection, and viral DNA synthesis can be followed by virtue of its high G+C content (see Chapter 3), which permits separation of host from viral DNA by density gradient centrifugation. In HEp-2 cells infected with HSV, the bulk of viral DNA is made between 3 and 7 hours after infection (Roizman, 1969b); in cells infected with equine abortion virus DNA synthesis may extend almost throughout the reproductive cycle (O'Callaghan *et al.*, 1969). As one might expect, herpesvirus DNA has been shown to replicate semiconservatively (Kaplan, 1964). The initiation of viral DNA synthesis requires the early synthesis of proteins. Once initiated, viral DNA synthesis will continue in the absence of concomitant protein synthesis but at a reduced rate. This contrasts strikingly with the situation for poxvirus DNA synthesis (see below).

There have been suggestions that herpesvirus latency or the capacity to initiate infection may be related to mitotic activity in cells. In synchronized KB cells infected in suspension with herpes simplex virus, initiation of viral DNA synthesis, as well as production of new infectious virus, is independent of the phase of the mitotic cycle of the host (Cohen *et al.*, 1971b). A corresponding situation is found for the onset of vaccinia and adenovirus type 2 DNA synthesis. In sharp contrast is a report that initiation of synthesis of equine herpes virus DNA in the same cells is dependent upon some cellular function related to the S phase of the cell cycle (Lawrence, 1971).

## Assembly and Release

Withdrawal of viral DNA from the replicating pool into virions is a random but inefficient process; only 15–20% of the pseudorabies DNA pool is incorporated into subviral structures impermeable to DNase (Ben-Porat and Kaplan, 1963). The time required for viral proteins to be transported from the cytoplasm into the nucleus (Olshevsky *et al.*, 1967) is responsible for the delay in their entry into capsids, which is followed by a further delay of about 2 hours prior to their appearance in mature virions.

Morphological aspects of herpesvirus multiplication have been studied with the electron microscope. Depending upon the strains of virus and cell, non-enveloped capsids may appear singly or in small groups scattered through the nucleus, but if large numbers of particles are formed rapidly and synchronously they may form intranuclear viral crystals (Morgan et al., 1959; Epstein, 1962). Areas of fine granular material (about 10 nm in diameter), which may consist of capsomeric protein, are found in the cell nuclei (Watson et al., 1964). The so-called single membrane particles seen in thin sections are nonenveloped capsids, and the double membrane forms are the enveloped virions. The acquisition of the envelope occurs at the nuclear membrane (Morgan et al., 1959), or occasionally, in some types of cell, at the cytoplasmic membrane (Epstein and Holt, 1963).

Herpesviruses remain cell-associated for a prolonged period and the herpesvirus-infected cell maintains its physical integrity and certain functional activities. Enveloped virions are gradually shed from the surface or by egestion from the cytoplasmic vacuoles. Nonenveloped virions probably appear in the supernatant fluid only from the occasional cell which is disrupted before the normal process of "excretion" (and envelopment) of the virus can occur.

Infection of dog kidney cells from the MPdK⁻ strain of HSV or of adult mouse macrophages with the H4 strain leads to abortive infections (Roizman et al., 1969; Stevens and Cook, 1971a). Viral structural components are made in both systems and, in the case of infected dog kidney cells, naked nucleocapsids are produced but do not become enveloped. The virus-specified proteins essential for maturation have not been determined; however, such abortive systems may be one way to analyze the steps in viral maturation and may be relevant to an understanding of latency.

We see, then, that during herpesvirus multiplication, input viral DNA enters the nucleus, viral mRNA's are sent to cytoplasmic ribosomes, and viral structural proteins are transported back into the nucleus, where they are assembled into nucleocapsids which pass back into the cytoplasm through the specifically altered nuclear membrane, acquiring part of it as an envelope. An understanding of traffic control in this system, in particular the controls on selective macromolecular transport, present challenging problems.

## Alterations of Cellular Metabolism

Herpesviruses cause a gradual inhibition of cellular macromolecular syntheses. Inhibition of cellular DNA synthesis starts about 2 hours after infection and is complete by 7 hours. Since this inhibition is sensitive to puromycin, Ben-Porat and Kaplan (1965) suggested that specific virus-induced protein blocked cellular DNA synthesis by some undefined regulatory mechanism. Autoradiographic studies show that inhibition of host DNA synthesis is accompanied by extensive aggregation and displacement of chromatin to the nuclear membrane (Roizman, 1969). There is no convincing evidence that infection with herpesviruses causes degradation of host DNA, although chromosome breakage does occur (Waubke et al., 1968).

Inhibition of host RNA synthesis is evident soon after infection and the incorporation of uridine into nuclear RNA drops to less than 25% of uninfected controls by 7 hours after infection. During this time the nuclear 45 S host ribosomal RNA is made, although at a diminishing rate, but is not processed to the cytoplasmic 18 and 28 S species (Roizman *et al.*, 1970). In cycloheximide-treated infected cells, inhibition of cell-specific RNA synthesis does not occur, indicating that synthesis of a viral protein is involved; however, the processing of 45 S RNA is inhibited in cycloheximide-treated cells (Rakusanova *et al.*, 1971).

Overall synthesis of protein declines rapidly between 1 and 3 hours after infection, and is then enhanced due to the synthesis of viral proteins between 4 and 8 hours. Inhibition of host protein synthesis early in infection corresponds with a breakdown of cytoplasmic polyribosomes (Sydiskis and Roizman, 1966).

Different strains of herpes simplex and pseudorabies virus differ markedly in their effects on cell morphology. Some strains cause cell rounding but no cell fusion or adhesion; some cause polykaryocytosis and others produce effects intermediate between these extremes. Such modifications, including altered antigenicity (Watkins, 1964), no doubt reflect modification of cellular membranes by viral products (Keller *et al.*, 1970) as described above. Early in the multiplication cycle of herpes simplex virus (2 hours after infection), concanavalin A agglutinin sites appear on the cell surface (Tevethia *et al.*, 1972). This is a virus-determined function but whether or not the agglutinin sites correspond to a specific viral antigen is not known.

## IRIDOVIRUS

The "iridoviruses" of vertebrates, which are only tentatively allocated to this genus, have been little studied biochemically; the only work on their multiplication that is worth reporting here concerns frog virus (FV3) (review by Granoff, 1969). Work on African swine fever virus has been reviewed by Hess (1971).

### The Multiplication Cycle

FV3 multiplies in a variety of cells (see Table 2–1), but multiplication is temperature-sensitive and does not occur at temperatures higher than 28°–30°C. In BHK cells at 26°C, FV3 undergoes an eclipse period of about 5 hours, followed by rapid multiplication for the next 7–16 hours, reaching titers of up to 100 pfu per cell (Kucera and Granoff, 1968; McAuslan and Smith, 1968). FV3 matures within the cell cytoplasm, and inclusion bodies containing arrays of polyhedra can be seen in electron micrographs (Granoff, 1969). Late in infection, virus is enveloped by the plasma membrane as it buds through it.

Variants of BHK cells have been obtained in which there are wide ranges of rates of viral DNA synthesis and assembly of virions ranging from completely nonpermissive systems through systems in which viral DNA but little infectious virus is produced, to systems in which the virus matures but over an extended period (Vilaginès and McAuslan, 1970).

## Initiation of Infection

Other than its probable independence of prior protein synthesis little is known about the initiation of infection by FV3.

## Transcription

FV3-specific RNA synthesis begins in the cytoplasm about 2 to 3 hours after infection and rises in parallel with viral DNA synthesis. Late in infection RNA synthesis is dependent upon DNA synthesis, but if cells are infected in the presence of an inhibitor of protein synthesis, some viral RNA is synthesized (Gravell and Cromeans, 1971). The latter observation suggests that the virus either utilizes a preexisting RNA polymerase or brings its own as part of the virion, but no DNA-dependent RNA polymerase has yet been detected in FV3 virions.

## Translation

The synthesis of FV3 structural proteins (see Chapter 3) in infected cells has not yet been studied. A few investigations have been made of virus-induced enzyme activities. Conflicting data have been reported on DNA polymerase activity in infected cells (review, Granoff, 1969); it has yet to be established that the assays used were a valid measurement of DNA polymerase activity *per se*. Kang and McAuslan (1972) have found that FV3 induces a novel "late" activity, namely, a ribonuclease that cleaves double-stranded RNA endonucleolytically. It has not been established whether this is specific for double-stranded RNA or whether it also cleaves single-stranded RNA, and its function is unknown.

## Replication of Viral DNA

In a fully permissive system, such as minnow cells at 26°C, FV3 DNA synthesis commences in the cytoplasm 2–3 hours after infection and continues for about 12 hours. Initiation of DNA replication, but not DNA replication itself, is temperature-sensitive. Initiation cannot occur at temperatures over 33°, but if initiated at 28°, synthesis will continue when the temperature is raised to 34°C (Kucera, 1970).

The onset of viral DNA replication requires protein synthesis, but once underway, can proceed for some time in the absence of protein synthesis. There is some disagreement as to whether the proteins needed for DNA synthesis can accumulate to a significant extent under the direction of input DNA templates (McAuslan and Smith, 1968; Granoff, 1969).

## Assembly and Release

In the absence of data on the synthesis of the viral structural proteins, the only information available on assembly and release of the virions comes from electron microscope studies (Granoff, 1969). These show that maturation occurs entirely within the cytoplasm, and that late in infection virions may acquire an envelope as they bud through the plasma membrane.

The observation that nonencapsidated DNA had a high affinity for isolated nuclei allowed McAuslan and Smith (1968) to demonstrate that maturation of virions was much more sensitive to inhibitors of protein synthesis than was viral DNA synthesis.

### Alterations of Cellular Metabolism

In cells infected with FV3, regardless of whether they are permissive for viral multiplication, host DNA and RNA synthesis are rapidly and extensively inhibited and the nuclei become distorted and shrunken. These effects appear to be due to some structural component of the virions, since virus that has been extensively irradiated will produce them in nonpermissive cells, at temperatures nonpermissive for viral multiplication (Maes and Granoff, 1967; McAuslan and Smith, 1968; Guir et al., 1971). Host DNA synthesis is completely arrested within 1 or 2 hours of infection, and RNA even more rapidly; preliminary experiments suggest that this is probably due in part to the pronounced decrease in activity in RNA polymerase type II (Campadelli-Fiume et al., 1972). No detailed study of the inhibition of host protein synthesis in infected cells has been made.

## POXVIRUS

This genus includes the largest and most complex of all viruses and the genome not only codes for a large number of structural proteins (Chapter 3) and for enzymes that are not included in the virion, but also for several enzymatic activities that are assembled as integral components of the virion. Like the iridoviruses, poxviruses multiply in the cytoplasm. Almost all biochemical studies of poxviruses have been carried out with one variant or another of vaccinia virus, and the following account will focus on this species (reviews by Joklik, 1968; McAuslan, 1969a).

### The Multiplication Cycle

The rate of formation of viral DNA and infectious virus shown in Fig. 5–7 is typical of members of the vaccinia subgenus in a wide variety of cells in suspension culture. Viral yields of over 100 pfu or 1000–10,000 particles per cell are not uncommon and most virions remain cell-associated as late as 24 hours or more after infection. The timing of all events in the multiplication cycle of vaccinia virus is much more rapid than with other DNA viruses (see Fig. 5–2), but the multiplication cycles of poxviruses of other genera may be quite prolonged. For example, in BSC-1 cells under one-step growth conditions, Yaba tumor poxvirus DNA synthesis reaches a peak 20 hours after infection; infectious virus appears at 30 to 40 hours and reaches a maximum at about 70 to 80 hours (Yohn et al., 1970).

Regardless of differences in timing with different poxviruses, Fig. 5–8 summarizes our knowledge of the sequence of events in poxvirus multiplication, as determined by the use of different inhibitors.

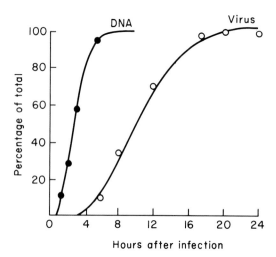

FIG. 5–7. *Poxvirus. Time course of synthesis of viral DNA and appearance of infectious vaccinia virus in infected HeLa cell spinner cultures.*

## Initiation of Infection

Following adsorption, a process which in this case does not involve specific cellular receptors, the entire virion is engulfed and transferred into phagocytic vacuoles. Data from both electron microscopy (Dales and Kajioka, 1964; Dales, 1965) and radiochemical studies with isotopically-labeled virus (Joklik, 1964a, b) have provided a fairly detailed picture of the fate of virus within cells.

Soon after it appears in the vacuoles, the outer viral membrane is degraded by cellular enzymes and the virus is converted to the subviral structure called the "core" (see Chapter 3, and Plate 5–1). This coincides with the dissolution of the membrane of the vacuole and the release of the core into the cytoplasm. Before it can replicate, the DNA must be released from the core by a second un-

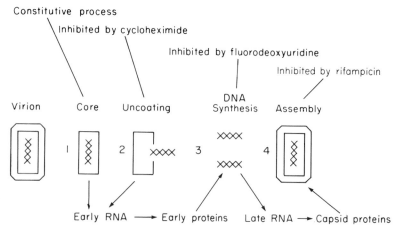

FIG. 5–8. *Poxvirus. Diagram illustrating the sequence of events in vaccinia virus multiplication.*

coating step. The mechanism of this second stage, which appears to involve disruption of cores and release of their components (Plate 5–1), is not completely understood. Experiments with inhibitors of RNA and protein synthesis showed that release of DNA in a form susceptible to exogenous DNase depended upon initiation of gene expression.

The early hypothesis, based on the assumption that a coated viral genome could not synthesize mRNA, was that a host gene for synthesis of an uncoating system was derepressed by infection (Joklik, 1964a, b). However, it is now known that the cores contain a DNA-dependent RNA polymerase that enables them to synthesize virus-specific mRNA *in vivo* or *in vitro* (Kates and McAuslan, 1967a, b). The prevailing idea is that if special proteins must be synthesized to complete the second stage of uncoating, their mRNA's could be transcribed from the viral genome while it is still within the core. If their experiments are taken at face value, the demonstration by Prescott *et al.* (1971), that L cells "enucleated" with cytochalasin could be infected with vaccinia virus would dispose finally of the idea that the host cell nucleus is required for poxvirus uncoating.

### Transcription

The existence of a transcriptase in poxvirions and the clear separation between host and viral transcription sites in the cell makes vaccinia virus one of the most useful systems with which to study the control of the expression of DNA viral genes. The temporal aspects of mRNA synthesis (review, McAuslan, 1969a) are summarized in Figs. 5–8 and 5–9.

The overall production of poxvirus mRNA's is clearly biphasic. The initial small burst of early mRNA synthesis resulting from transcription of cores is followed by a second wave representing mostly "late" mRNA sequences whose synthesis is dependent on viral DNA replication. Regulation of the rate of synthesis of early mRNA is governed by the uncoating process. Soon after penetration of the virion and removal of the outer protein coat, the core synthesizes and releases virus-specific cistron-sized RNA, primarily 10–14 S (Kates and McAuslan, 1967a). About 14% of the viral genome is transcribed from cores and extruded by a mechanism which specifically requires ATP (Kates and Beeson, 1970a). Following the completion of uncoating, other early genes may be expressed although the evidence for this is indirect and comes from studies on regulation of DNA polymerase induction (Kates and McAuslan, 1967b). If the second stage of uncoating depicted in Fig. 5–8 is blocked by cycloheximide, then the rate of early mRNA synthesized by cores is markedly enhanced for extended periods of time; on the other hand, if uncoating is allowed to proceed but initiation of DNA synthesis is prevented by fluorodeoxyuridine, there is normally a burst of early mRNA synthesis which rapidly declines (Fig. 5–8; McAuslan, 1969a).

Early mRNA synthesized *in vitro* or *in vivo* from viral cores contains poly A sequences about 180 nucleotides long; these are attached at the 3' end of viral RNA (Kates, 1970). Their function is unknown, but their existence is of particular interest in view of the idea that poly A sequences found in other mRNA's

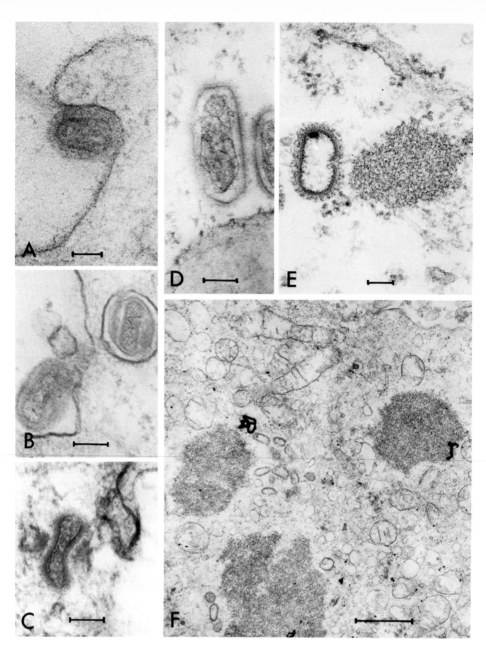

PLATE 5–1. *Attachment, engulfment, and uncoating of vaccinia virus in L cells; electron micrographs of thin sections.* [Bars = 100 nm, except in F, where bar = 1000 nm.] (A) Attachment and early stage of engulfment; (B) one virion attached to surface, another intact virion enclosed within a phagocytic vesicle; (C) intracellular particle undergoing disruption; the outer coat has disappeared and the lateral bodies are being detached from the core; (D) uncoated core, free in cytoplasm, showing densely staining filaments (? DNA) within; (E) empty shell of a core lying free in the cytoplasm with extruded DNA adjacent to it; (F) areas of viroplasm in cytoplasm of cell with associated autoradiographic label and empty "shells" of virions used for inoculum, 3 hours after inoculation. (All illustrations courtesy Dr. S. Dales; A, from Dales and Siminovitch, 1961; E, from Dales, 1965; remainder from Dales, 1963a.)

might be involved in the cleavage of mRNA's or their transport from nucleus to cytoplasm.

Early mRNA continues to be made late in the multiplication cycle; some of this at least could be from virions that do not completely uncoat, for only about 50% of the input virus ever becomes DNase-susceptible (Joklik, 1964a,b). Early and late mRNA sequences can be distinguished by molecular hybridization competition and by differences in sedimentation rates (Oda and Joklik, 1967); the sedimentation rate of mRNA made at late times is 16–23 S as compared to 10–14 S for early mRNA. Both size classes appear to contain poly A sequences.

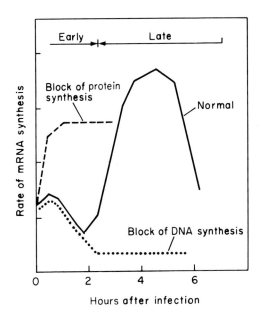

FIG. 5–9. *Poxvirus. Effect of inhibition of viral protein or DNA synthesis on the rate of vaccinia virus mRNA synthesis. (Data from McAuslan, 1969a.)*

Two experimental approaches have provided estimates of the metabolic stability of different poxvirus mRNA's. The first indications of metabolically stable messengers in poxvirus-infected cells came indirectly through studies on the regulation of early enzyme synthesis. The mRNA for thymidine kinase (McAuslan, 1963) was shown to have a very long half-life (at least 6 hours) as does the messenger for induced DNA polymerase (Jungwirth and Joklik, 1965), whereas other early mRNA's (for example, those for poxvirus DNase and a protein involved in DNA replication) are quite unstable (McAuslan and Kates, 1966; Kates and McAuslan, 1967c). Determination of the rate of decay of different classes of messenger in the presence of actinomycin D indicated that late mRNA is relatively unstable, decaying with a half-life of about 13 minutes, whereas early mRNA has a half-life of about 120 minutes (Sebring and Salzman, 1967). However, in a detailed study of the temporal pattern of poxvirus messenger synthesis Oda and Joklik (1967) found that in vaccinia-infected L cells, late messengers were just as stable as early messengers.

Parallel to what one finds with DNA–protein complexes isolated from phage-infected bacteria (Snyder and Geiduschek. 1968; Chesterton and Green, 1968), DNA complexes isolated from vaccinia-infected cells synthesize both "early" and "late" mRNA *in vitro* by the activity of an endogenous RNA polymerase (Dahl and Kates, 1970). Pertinent to an understanding of transcription control of these complexes are the observations of Obert *et al.* (1971) on the requirement for arginine in late mRNA transcription. As we have seen, multiplication of adenovirus and herpesvirus is inhibited by arginine deficiency, the lesion probably being the absence of arginine-rich core proteins that are required for assembly. Multiplication of vaccinia virus also requires arginine. If there is a deficit of it in the medium, early mRNA sequences are transcribed, vaccina DNA is replicated, but late mRNA sequences are not transcribed. Presumably arginine is required for synthesis of a new polymerase for late mRNA synthesis or perhaps for an arginine-rich protein, translated from early mRNA, which is important for maintaining newly replicated DNA in the appropriate physical state for initiation of transcription of late sequences.

Vaccinia virus stimulates infected cells to produce interferon. Following observations of the effectiveness of double-stranded RNA and the ineffectiveness of single-stranded RNA, or DNA, as inducers of interferon (see Chapter 8), Colby and Duesberg (1969) sought and found a vaccinia-directed double-stranded RNA. Ribonuclease-resistant RNA is synthesized *in vitro* by vaccinia virus cores and is similar in properties to that made *in vivo*. The double-stranded species is thought to arise as a result of overlap of convergent transcription from complementary DNA strands (Colby *et al.*, 1971).

Transfer RNA's made after vaccinia virus infection may be altered. Quantitative changes in arginyl-tRNA and phenylalanyl-tRNA have been detected (Clarkson and Runner, 1971). There is also a noticeable relative increase in the synthesis of methylated tRNA (Klagsbrun, 1971). The significance of these changes is unknown.

## Translation

As one might expect from the classes of mRNA produced, the proteins synthesized by poxviruses fall roughly into early and late classes but there is no simple grouping of either structural proteins or induced enzymes into one or other of these classes. Combining quantitative immunoprecipitation with autoradiography of gel diffusion plates, Salzman and Sebring (1967) showed that five viral structural antigens were synthesized before the onset of viral DNA synthesis, but synthesis of these ceased and other viral porteins appeared after viral DNA replication began. Viral antigens synthesized in the infected cell are divisible into two molecular weight classes; synthesis of the low molecular weight antigens begins early and ceases 4 hours after infection, when production of high molecular weight antigens begins (Wilcox and Cohen, 1967). A more definite picture of the programming of vaccinial protein synthesis comes from determination of the individual structural polypeptides and from the assay of virus-elicited enzyme activities (assuming that such activities are virus-specified).

**Structural Proteins.** Vaccinia virions are composed of at least thirty different polypeptides, of which seventeen (including the two principal virion polypeptides) have been assigned to viral cores and five are located near or on the virion surface (see Chapter 3). The coordination of their synthesis, which must control assembly, is somewhat unexpected. Two of the core polypeptides and one of the surface polypeptides are made in the absence of viral DNA synthesis, and are thus early proteins (Holowczak and Joklik, 1967b). A third core polypeptide, a major virion component which may represent a family of polypeptides of similar molecular weights, is made late, as are most of the lower molecular weight species. One or two polypeptides appear to be made throughout the multiplication cycle (Moss and Salzman, 1968; Holowczak and Joklik, 1967b). The structural polypeptides are not made in infected cells in the same proportion as they appear in virions and some of the lower molecular weight species aggregate into very large molecular weight structures soon after they are formed.

**Virus-Induced Enzymes.** Several enzyme activities are elicited by poxvirus infection (Table 5–2) and since the coding potential of the poxvirus genome (about 160 proteins of molecular weight 50,000 daltons; see Table 7–1) far exceeds the known structural proteins, it is not unreasonable to assume that vaccinia virus carries information for synthesizing enzymatic systems concerned with nucleic acid biosynthesis. Although based on circumstantial evidence, there are several compelling arguments in favor of induced enzymes being virus-coded. The evidence (reviews, Kit, 1968; McAuslan, 1969a) includes demonstrations that an increase in enzyme activity requires *de novo* protein synthesis, physicochemical differences between pre- and postinfection enzyme activities, and the induction of an enzyme in host cells deficient in that particular activity.

TABLE 5–2

*Enzymes Induced in Cells Infected with Vaccinia Virus*

| ACTIVITY | CLASS |
|---|---|
| Thymidine Kinase [a] | Early |
| Polynucleotide Ligase [b] | Early |
| DNA Polymerase [c] | Early |
| dAT-dependent Polymerase [f,g] | Early |
| Exo-DNase (DS DNA) [d] | Early |
| Endo-DNase (SS DNA) [c] | Early |
| Exo-DNase (SS DNA) [e] | Late |
| Virion Transcriptase [g] | Late (?) |

[a] McAuslan (1963).
[b] Sambrook and Shatkin (1969).
[c] Jungwirth and Joklik (1965)
[d] McAuslan and Kates (1966).
[e] McAuslan and Kates (1967).
[f] McAuslan (1971).
[g] Pitkanen et al. (1968).

Thymidine kinase is induced by both herpesvirus and vaccinia virus, but we know nothing about its role in natural viral infections other than that it is not essential for viral multiplication in cultured cells. The synthesis of this early enzyme is regulated in an interesting way, which provided the first clear demonstration that protein synthesis could be regulated at the translational level (Fig. 5–10; McAuslan, 1963). Thymidine kinase synthesis begins 1.5–2 hours after infection, proceeds for about 4 hours, and then ceases abruptly. Cessation of synthesis is controlled by the viral genome. Studies with actinomycin D and puromycin showed that the messenger for thymidine kinase was very stable and that synthesis of some protein was necessary to effect inhibition of enzyme synthesis. Poxvirus-induced DNA polymerase is controlled similarly, but not all early proteins are under the same regulatory mechanism. For example, two poxvirus-induced DNases (review, McAuslan, 1969a), the exonuclease (DS-specific) and the endonuclease (SS-specific), appear to be synthesized early (1.6–6 hours after infection) from mRNA's with a very short half-life, although their synthesis is terminated about the same time as thymidine kinase. One outer structural polypeptide, an early polypeptide, does not appear to be switched off but is synthesized throughout the multiplication cycle.

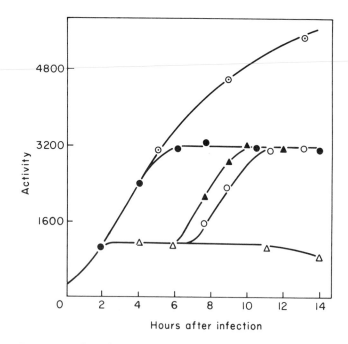

FIG. 5–10. Poxvirus. The effect of inhibitors on the establishment of repression of thymidine kinase synthesis. (●) Kinase activity in poxvirus-infected cells; (⊙) kinase activity after addition of actinomycin at 2 hours; (△) kinase activity after addition of puromycin at 2 hours; (▲) kinase activity after removal of puromycin at 5½ hours; (○) kinase activity after addition of actinomycin 30 minutes before removal of puromycin. (From McAuslan, 1963.)

The discovery of transcriptase within poxvirions (Kates and McAuslan, 1967b) led to the finding by different investigators that poxvirions contained several other enzyme activities including nucleotide phosphohydrolase (ATPase, Munyon et al., 1968), a single-strand specific exo-DNase, and a single-strand specific endo-DNase (Pogo and Dales, 1969; Aubertin and McAuslan, 1972). The virion exo-DNase is clearly a late poxvirus-induced activity that has been purified and characterized (McAuslan and Kates, 1966). Although the functions of these virion enzymes are unknown, they should be useful markers to follow assembly.

Late in the multiplication cycle, the RNA polymerase activity is found in a particulate form and probably represents new virions in some late stage of assembly (Pitkanen et al., 1968), since it transcribes only early mRNA in vitro. One might expect to see at early times a soluble form of this virion transcriptase but it has not yet been found. During such a search, a dAT-primed poly AU synthesizing activity was found as an early function with a very stable messenger (McAuslan, 1971). So far, it has not been possible to make this activity transcribe any template other than dAT. Thus there are at least two poxvirus-induced polymerase activities: the dAT-dependent early activity and the late, bound activity which synthesizes early mRNA in vitro.

**Glycoprotein Synthesis.** Extensive alterations in glycoprotein synthesis occur within 1 hour after vaccinia infection (Moss et al., 1971c; Garon and Moss, 1971). The new glycoproteins are virus-specific and are to a considerable extent associated with cellular membranes. Perhaps introduction of new glycoproteins into cell surfaces is responsible for the morphological changes that take place after infection with fibroma virus or those strains of vaccinia virus that cause polykaryocytosis.

The virion also contains two glycoproteins that have glucosamine as the only sugar; these are neither on the surface nor in the viral core (Chapter 3). Since the viral factories constitute large foci in the cytoplasm, distant from the cellular membranes where the glycosylating enzymes of the cell are found, it is possible that glycosylation is effected by virus-specified rather than cellular enzymes.

## Replication of Viral DNA

Poxvirus DNA is synthesized in the cell cytoplasm and can be initiated in most cells regardless of whether or not the cell is making DNA at the time of infection (Cairns, 1960). A convenient way to determine poxvirus DNA synthesis quantitatively is to measure the incorporation of [³H]thymidine with time into cytoplasmic acid-soluble material after separating the infected cell into nuclear and cytoplasmic fractions. As shown in Fig. 5–7, vaccinia virus DNA synthesis commences 1.5–2 hours after infection and is virtually complete by 6 hours, a time when new infectious virus is just appearing. In contrast, fowlpox DNA is synthesized between 12 and 48 hours after the infection of chick cells.

By the time viral DNA synthesis ceases, most of the DNA is still susceptible to exogenous DNase, so that synthesis is not arrested because DNA becomes encapsidated or because precursors are depleted, suggesting that some finer

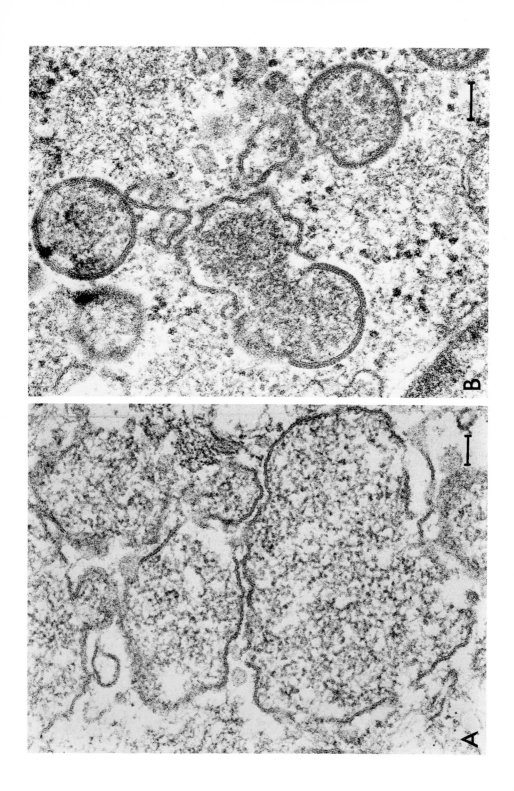

control must be involved. Continuous protein synthesis is required for viral DNA synthesis, so that inhibition of protein synthesis with cycloheximide abruptly inhibits DNA replication (Kates and McAuslan, 1967c). The essential protein is a stable early viral function distinct from known enzymes involved in DNA synthesis, and it accumulates when DNA synthesis is selectively blocked. The stoichiometric relationship between protein requirement and DNA replication suggests that the protein involved may be a nonenzymatic protein intimately associated with the viral DNA.

## Assembly and Release

Our understanding of the maturation of poxviruses is largely based on electron microscopic studies of infected cells, assisted more recently by the use of rifampicin. The most comprehensive study of correlated biological events and morphological appearances in poxvirus infections is due to Dales, who has investigated the replication of vaccinia virus in L cells (Dales, 1963a; Kajioka et al., 1964).

Initial transcription occurs from cores, and then the viral DNA is released and sets up what Cairns (1960) called a "factory" area (viroplasm) in the cytoplasm, where the viral components are synthesized and assembled. The viroplasm is first seen 2–3 hours after infection, as an area of dense granules and randomly oriented fine threads, which appears to have pushed aside the mitochondria and other cytoplasmic organelles. The dense material of the viroplasm then aggregates into clumps of dense filaments within and around which limiting membranes form. These membranes appear to develop as "caps" at one edge of the viroplasm and gradually grow to form a spherical surface which eventually completely surrounds the viroplasm to form an "immature particle." Subsequently, and to a differing extent in different preparations, small foci of the enclosed viroplasm appear to condense eccentrically within the immature particle. These electron-dense condensations are probably the viral DNA (Rosenkranz et al., 1966). The mature virion is slightly smaller than the immature particle just described, but it is much more complex structurally, with large lateral bodies and an outer double membrane on either side of the core (see Chapter 3).

**Rifampicin and Poxvirus Proteins.** Rifampicin (see Chapter 15) inhibits poxvirus multiplication at the level of assembly, but causes no obvious inhibition of poxvirus mRNA or protein synthesis (Moss et al., 1969; McAuslan, 1969b). The induction of the particulate transcriptase is, however, completely

---

PLATE 5–2. *Effects of rifampicin on maturation of vaccinia virus and reversal of effect after the removal of the inhibitor (bars = 100 nm). (A) HeLa cell treated with rifampicin and infected with vaccinia virus for 8 hours. Precursor membranes fail to form viral coat membranes. (B) Reversal 3 minutes after removal of rifampicin. Note continuity of viral coats with precursor membranes. (From Grimley et al., 1970, courtesy Dr. P.M. Grimley.)*

inhibited (McAuslan, 1969b) perhaps because of failure to assemble a structure to the minimal level necessary for active transcription.

Another effect of rifampicin involves polypeptide cleavage. One of the core polypeptides, with a molecular weight of 76,000 daltons, is derived by cleavage of a larger precursor (125,000 daltons), during the formation of the viral core. Rifampicin completely prevents the formation of this core polypeptide without inhibiting the synthesis of its precursor (Katz and Moss, 1970a, b).

The primary action of rifampicin on poxvirus morphogenesis occurs during the early stage of formation of the outer membrane. Electron-dense areas, presumably DNA, are formed in the presence of rifampicin. According to Pennington et al. (1970), the unique spicule-covered outer membranes of the immature forms are not seen in the presence of rifampicin. When rifampicin is removed, spicule-covered membranes begin to form around the electron-dense viroplasm. Single immature forms then appear, nucleation begins soon after, and development of lateral bodies and modelling of biconcave cores becomes evident 30 minutes after drug removal (Grimley et al., 1970). Some of these stages (see Moss et al., 1969) are shown in Plate 5–2.

**Inclusion Bodies.** Late in infection some, but not all, poxviruses produce A-type inclusions that are distinctly different from so called B-type inclusions which are the sites of viral multiplication (see Chapter 9). Protein recovered from A-type inclusions is antigenically distinct from the virion proteins (Ichihashi and Matsumoto, 1966). Depending on the viral species and strain, these A-type inclusions have distinctly different properties and may either contain virions or be completely free of them (Ichihashi et al., 1971; see Plate 5–3). In either case, the inclusion bodies are surrounded by clusters of polyribosomes.

## Alterations of Cellular Metabolism

In the commonly used HeLa cell–vaccinia virus system, at high input multiplicity, cellular DNA synthesis is rapidly inhibited by either active or UV-irradiated virus even in the presence of puromycin (Joklik and Becker, 1964; Jungwirth and Launer, 1968). Host DNA is usually not degraded and reutilized to any significant extent, although two recent papers provide claims to the contrary; Walen (1971) found that after infection with vaccinia virus, part of the cellular DNA was degraded and subsequently appeared in the viral "factories," and Oki et al. (1971) reported degradation of host DNA and incorporation of the products into vaccinia virions, although their system was abnormal in that the host cells were prelabeled with high specific activity [³H]thymidine.

Fowlpox has a more extended multiplication cycle, in which host DNA synthesis may proceed at a normal rate for the first 12 hours of infection (Gafford et al., 1972). With rabbit fibroma virus, which produces localized tumors in rabbits, host DNA synthesis in infected cell cultures may become arrested for long periods, and then begin again after the peak of viral maturation (Tompkins et al., 1969). Even in cells whose division is blocked by gamma irradiation, fibroma virus infection causes a burst in cellular DNA synthesis.

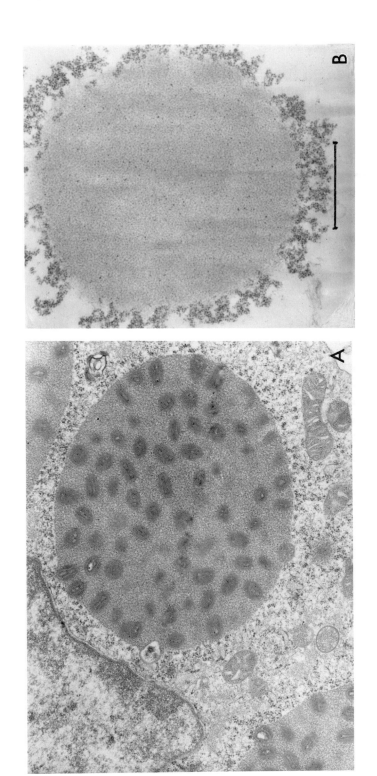

PLATE 5–3. *The A-type inclusion bodies of cowpox containing (A) and lacking (B) virions (bar = 1000 nm). (A) In situ in infected cell. (B) Inclusion body with attached polyribosomes after separation from cell. (From Ichihashi et al., 1971, courtesy Dr. S. Dales.)*

Inhibition of cellular RNA transcription in poxvirus-infected HeLa cells is a gradual process taking about 6 hours for completion, although transport of host mRNA from nucleus to cytoplasm is rapidly blocked (Salzman and Sebring, 1967), so that briefly labeled viral mRNA, which is made in the cytoplasm, can be readily separated from cellular RNA's (Oda and Joklik, 1967).

Inhibition of host protein synthesis by vaccinia virus is quite different from that found with adenovirus and herpesvirus, which depends on synthesis of new viral protein(s). Inhibition of HeLa cell protein synthesis begins within 20 minutes after infection and is virtually complete within 1–4 hours, depending on the virus multiplicity. Synthesis of vaccinia viral protein is not required; perhaps a constituent of the virion has a selective effect on host protein synthesis at the translational level (Moss, 1968).

## CHAPTER 6

# The Multiplication of RNA Viruses

## INTRODUCTION

The multiplication of RNA viruses differs from that of the DNA viruses we have just considered in several important respects. Unique among genetic systems, their genome is RNA, which is transcribed, translated, and replicated quite differently from DNA; indeed, the RNA viruses themselves vary widely in the mechanisms whereby these processes are accomplished. Compared with most DNA viruses, all are extremely limited in the amount of genetic information they carry (see Table 7–1), hence they code for relatively few enzymes, although all require at least one RNA-dependent RNA polymerase. Finally, most RNA viruses mature by budding from cytoplasmic membranes.

## THE ESSENTIAL STEPS IN MULTIPLICATION

The basic features of the viral multiplication cycle have already been described (Chapter 5). Before embarking upon a systematic description of the multiplication of RNA viruses we shall summarize those steps in the cycle in which RNA viruses differ fundamentally from DNA viruses.

### Transcription and Translation of Viral RNA

Baltimore (1971b) has done a great deal to clarify the thinking of virologists on the various ways in which transcription, translation, and replication of RNA genomes occurs in RNA viruses of different genera (Table 6–1). It will be noted that the RNA extracted from a virus is infectious only if (a) the genome does not occur in pieces, and (b) there is no virion transcriptase. The latter is a reflection of the polarity of the viral RNA. A transcriptase is only necessary when the viral RNA is "antimessage," so that a complementary mRNA must be transcribed. The reverse transcriptase of *Leukovirus* constitutes a special case.

To simplify discussion it is first necessary to explain the terminology we shall use for describing different species of virus-related RNA. Throughout this book we use the term cRNA to mean RNA complementary to RNA in the virion (vRNA), regardless of the polarity of the vRNA. In other words, the

TABLE 6–1

*Expression of Viral RNA* [a]

| FAMILY OR GENUS | VIRAL GENOME | | | | VIRION TRANSCRIPTASE | POSTTRANS-LATIONAL CLEAVAGE OF PROTEINS |
|---|---|---|---|---|---|---|
| | STRANDED-NESS | MOLECULES | MESSENGER FUNCTION | INFEC-TIVITY | | |
| Picornaviridae ⎫<br>Togaviridae ⎬ | SS | 1 | + | + | − | + |
| Paramyxovirus ⎫<br>Rhabdovirus ⎬ | SS | 1 | − | − | + | − |
| Orthomyxovirus | SS | 7 | − | − | + | − [b] |
| Leukovirus | SS | 4 | + | − | + (Reverse) | − |
| Reovirus ⎫<br>Orbivirus ⎬ | DS | 10 | − | − | + | − [b] |

[a] Modified from Baltimore (1971b).
[b] Minor cleavages do occur.

mRNA of picornaviruses is identical with vRNA (one may call it "vmRNA"), whereas the mRNA of rhabdoviruses is cRNA (or "cmRNA"). Hopefully, this terminology should lead to less confusion than either of the two alternatives that the reader may encounter in the literature: namely (a) the conventional practice of calling virion RNA "+" and its complement "−," or (b) Baltimore's more recent idea of calling mRNA "+" and its complement "−." Both ideas are logical, but the existence of two mutually contradictory schemes suggests that we should use neither; the terminology proposed describes unequivocally the nature of the RNA under discussion.

The alternative mechanisms whereby the genetic information encoded in vRNA can be translated into protein (Table 6–1) can be summarized as follows:

1. The single-stranded viral RNA (vRNA) molecule of picornaviruses and togaviruses can be viewed as a giant messenger RNA (mRNA) molecule containing meaningful genetic information ("sense") directly translatable by ribosomes into protein. The huge polypeptide that results is subsequently enzymatically cleaved into progressively shorter polypeptides.

2. The single-stranded RNA molecule of the paramyxoviruses and rhabdoviruses, on the other hand, is "antimessage"; RNA of complementary nucleotide sequence (cRNA) must first be transcribed and only this is seen by tRNA's and ribosomes as mRNA that can be translated into proteins. A virus-coded RNA-dependent RNA polymerase ("transcriptase"), carried in the helical ribonucleoprotein (RNP) nucleocapsid in these genera transcribes cRNA while still *in situ* in intracellular ribonucleoprotein. Some at least of these transcripts are full length; it is not certain whether or not any are shorter. However, whether by direct transcription or by cleavage of full length cRNA, short lengths of "monocistronic" cRNA arise, each of which has poly A added to its 3' end, and is separately translated into a separate polypeptide.

3. The seven (or more) separate single-stranded RNA molecules of the orthomyxoviruses are also "antimessage." A separate cRNA molecule, representing a monocistronic mRNA molecule, is transcribed from each one by a virion-associated transcriptase.

4. The single-stranded RNA of the leukoviruses is transcribed by a virion-associated RNA-dependent DNA polymerase ("reverse transcriptase") into an RNA:DNA hybrid, which in turn serves as a template for the synthesis of double-stranded DNA. Such virus-specific DNA can be integrated into cellular DNA leading either to a cryptic carrier state or to transformation of the cell (Chapter 14). It may, in turn, serve as a template for transcription of single-stranded mRNA and vRNA, which are identical in sense but differ in size.

5. The ten double-stranded RNA molecules of the reoviruses and orbiviruses are transcribed into ten distinct single-stranded mRNA molecules by a transcriptase present in the core of the virion; indeed, virions uncoated only to the core stage can produce mRNA for prolonged periods both *in vivo* and *in vitro*. After addition of poly A to the 3′ end, each mRNA is translated on ribosomes into a single identifiable protein.

The RNA viruses have evolved along very different paths to reach equally effective answers to the common problem of how to produce a number of separate proteins from a single set of genetic information, for compared with its bacterial counterpart, the mammalian protein-synthesizing apparatus does not seem to have the capacity to initiate translation of individual cistrons at multiple points along a polycistronic mRNA.

### Replication of Viral RNA

The replication of RNA is, as far as we know, something unique to RNA viruses; if cells had the capacity to generate RNA from RNA it would greatly complicate the control of the orderly flow of information from DNA→RNA→protein. The precise mechanism of viral RNA replication is therefore a matter of considerable interest. Though debated for several years, it is now agreed that viruses with a single-stranded genome replicate by a semiconservative process in which the parental (vRNA) molecule serves as a single-stranded template for the simultaneous transcription of several complementary (cRNA) strands (Spiegelman et al., 1968; Weissman et al., 1968). The "replicative intermediate" (RI) is, therefore, a partially double-stranded structure with single-stranded "tails" (review by Bishop and Levintow, 1971). The single-stranded cRNA progeny molecules peel off and in turn serve as templates for the synthesis of new vRNA (Fig. 6–2).

Virus-coded RNA-dependent RNA polymerases ("replicases") are required to catalyze this process. These enzymes tend to be specific for the RNA of their own viral species. In some cases, different polymerases are involved in the production of "−" and "+" strands, respectively, but the generality of this situation has yet to be established.

### Membranes and Budding

Cellular membranes play a dominant role throughout the multiplication of RNA viruses (review, Allison, 1971b). The plasma membrane is involved in attachment and penetration (see Chapter 5), translation of viral proteins occurs mainly on polyribosomes associated with membranes of the rough endoplasmic reticulum, glycosylation of viral glycopolypeptides takes place in various smooth cytoplasmic membranes, and RNA replication occurs in membrane-bounded "replication complexes." Moreover, since in most genera, the virions of RNA viruses are enveloped, cytoplasmic membranes (usually the plasma membrane but sometimes internal cytoplasmic membranes) play an important role in the assembly and release of virus particles by a process of budding. In the latter process, the sequence of events is as follows:

1. Viral glycoproteins enter the plasma membrane, displacing all of the existing cellular protein.

2. Viral "membrane protein" (a low molecular weight nonglycoprotein) forms a layer on the inside of the modified plasma membrane.

3. Viral nucleocapsid attaches to the membrane protein and the viral glycoproteins develop into visible projections (peplomers) on the outside of the plasma membrane.

4. Budding occurs; this is a sort of "reverse cytosis" in which there is an evagination of the altered area of plasma membrane, and the new virion, enclosing its nucleocapsid, is nipped off (Plate 6–2).

## Regulation of Viral Gene Expression

Very little is yet known of the mechanisms whereby viral RNA transcription, translation, and replication are controlled. In the Picornaviridae, the single vmRNA molecule is translated directly into a single polypeptide, and there is no control of transcription or translation. However, with the transcriptase-carrying viruses, in all of which monocistronic cmRNA's are synthesized, there is evidence that individual vRNA genes are transcribed at different rates and that the resulting cmRNA molecules are translated at different rates. The various viral proteins are not made in equimolar proportions, nor in numbers that are inversely proportional to their molecular weights. Moreover, the relative rates of synthesis of the several viral proteins fluctuate in a systematic fashion as the multiplication cycle proceeds, nonstructural proteins and sometimes "core" proteins tending to be most conspicuous early and capsid proteins later. In addition, there is good evidence that some regulatory mechanism must determine whether at any given phase of the multiplication cycle, mRNA molecules are utilized for translation into protein or for RNA replication or, in the case of vmRNA, for assembly into virions.

Since we know almost nothing about how such regulation is accomplished in animal cells, it may be useful to consider briefly the sorts of controls that have so far been found to operate with RNA phages (reviews by Calendar, 1970; Stavis and August, 1970; Kozak and Nathans, 1972; Sugiyama et al., 1972). Perhaps the outstanding finding has been that most RNA phage proteins are multifunctional. The f2 capsid protein, for example, also acts as a repressor of replicase synthesis; Qβ replicase inhibits protein synthesis by blocking the attachment of ribosomes to mRNA. There is also evidence that the f2 "maturation protein" may remain bound to vRNA during translation and perhaps throughout replication. Moreover, ribosomes from infected bacteria have an enhanced affinity for viral mRNA, and certain "initiation factors" associated with ribosomes select particular mRNA's for translation. The secondary and tertiary structure of vmRNA, which may change systematically during translation, play a major role in regulating the rate of synthesis of individual proteins from this polycistronic messenger. A virus-coded protein may also change the specificity of Qβ polymerase so that it transcribes strands of the opposite polarity. Such "sigma factors" (σ) have been explored in depth in DNA phage systems (Hershey, 1971).

Some recent observations on the regulation of expression of mammalian RNA viral genomes can be assessed in the light of this experience.

1. Arginine-rich "core" proteins may remain associated with viral RNA throughout transcription, and possibly throughout translation and replication. Though the evidence now available is minimal, there is sufficient indication from recent data obtained with three different viral genera (*Orthomyxovirus*, *Rhabdovirus*, and *Reovirus*) to suggest that the nucleoprotein of the nucleocapsid, which shows such a high affinity for RNA, plays a role in addition to that of mere support.

2. The virion transcriptase appears to be active when associated with such "nucleocapsids" whether they be in the form of the helical ribonucleoprotein of the orthomyxoviruses, paramyxoviruses, and rhabdoviruses, or the "core" of reoviruses. It is not yet clear whether these inner structural proteins of the virion are in fact the basic subunits of the active enzyme or whether the enzymatic activity resides in the minor proteins of the viral core.

3. A nonstructural protein of influenza virus is most abundant early in the multiplication cycle and accumulates rapidly in nucleoli and ribosomes. It may render ribosomes virus specific by binding viral mRNA and rejecting cellular mRNA, so shutting down host protein synthesis.

4. The capsid protein precursor of poliovirus may regulate the translation, replication, and assembly of the viral RNA.

5. The conformation of vRNA may be important in determining selective transcription of "early" genes by transcriptase *in situ* within viral cores, or there may be topological constraints on the transcription of certain genes until uncoating is complete.

### Regulation of Cellular Macromolecular Synthesis

Cellular macromolecular synthesis is abruptly suppressed by some RNA viruses and totally unaffected by others. At one end of the spectrum, picornaviruses code for proteins that separately shut down the transcription of cellular RNA and its translation into protein; DNA synthesis stops in consequence. Many of the paramyxoviruses, on the other hand, have virtually no effect on cellular metabolism and bud from cells for prolonged periods without affecting their continued growth and division; such persistent noncytocidal infections, or "carrier cultures", are discussed in Chapter 12. Leukoviruses are not only noncytocidal, but they may transform infected cells to malignancy following integration of virus-coded DNA into cellular chromosomes (Chapter 14). There is some evidence that the orthomyxoviruses may require the transcription of cellular DNA during the first part of their multiplication cycle; this contrasts sharply with the paramyxoviruses and picornaviruses which multiply quite normally in the presence of actinomycin D or even in enucleated cells.

## A SYSTEMATIC SURVEY

There are more genera of RNA viruses than of DNA viruses. Little is known of the multiplication process in some; these will be dealt with in summary fashion. A great deal of work has been carried out with model viruses of other genera; descriptions of their multiplication will be correspondingly detailed.

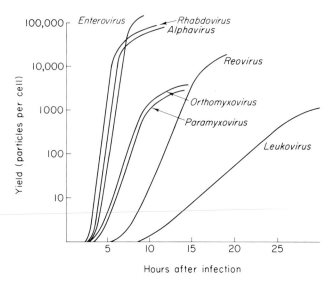

FIG. 6–1. *Idealized multiplication curves (total virus particles per cell at intervals after infection at high multiplicity) of viruses representing the major genera of RNA viruses (Enterovirus, poliovirus type 1; Alphavirus, Sindbis virus; Rhabdovirus, vesicular sto-matitis virus; Orthomyxovirus, influenza type A; Paramyxovirus, Newcastle disease virus; Reovirus, reovirus type 3; Leukovirus, RAV–1). Latent periods, multiplication rates, and final yields are all affected by species and strain of virus and by cell strain and the kind of culture; latent period and multiplication rate, but not final yield, by multiplicity of infection.*

The basic multiplication cycle of different RNA viruses varies substantially in both the length of the latent period and the viral yield. Figure 6–1 illustrates representative examples of the multiplication curves of model viruses of the best studied genera.

## PICORNAVIRIDAE

Poliovirus type 1, the type species of the genus *Enterovirus,* which is the type genus of the family Picornaviridae, has for many years served as the model for the detailed biochemical analysis of the multiplication of RNA viruses. As already pointed out, the mechanism of translation of the genome of Picornaviridae is by no means common among RNA viruses; but data on the replication of poliovirus RNA have wider applicability. In this section, we shall present the current state of knowledge about poliovirus multiplication in considerable detail in the hope that it will serve as a solid foundation upon which the discussion of less thoroughly explored genera may be built as the chapter develops. Relatively little will be said of other picornaviruses apart from occasional reference to pioneering work with the cardioviruses, ME, EMC, and mengovirus.

Several excellent reviews provide valuable guides to the earlier literature (Baltimore, 1969, 1971a, b; Baltimore et al., 1971; Bishop and Levintow, 1971; Cooper, 1969; Summers et al., 1971).

## The Multiplication Cycle

Poliovirus multiplies rapidly and to high yield in many human and simian cell lines, in monolayer or suspension cultures. The whole growth cycle is confined to the cytoplasm (Franklin and Baltimore, 1962) where viral RNA replication, protein synthesis, and virion assembly all occur in close association with membranes (Mosser et al., 1972). Actinomycin D has no inhibitory effect (Reich et al., 1962), nor indeed does enucleation of the cell (Crocker et al., 1964). Viral RNA can be separated from cellular RNA by taking the cytoplasmic material labeled in a short pulse of radioactive uridine in the presence of actinomycin, purifying it by chromatography or rate zonal centrifugation, and assaying it by infectivity. Viral proteins can be preferentially labeled with amino acids after host mRNA transcribed before the addition of actinomycin has been given time to decay in the presence of guanidine, a reversible inhibitor of viral multiplication (Summers et al., 1965). Recently cell-free systems have been employed to provide further insights into the expression of the viral genome (Rekosh, 1972).

## Initiation of Infection

Human picornaviruses multiply only in human or simian cells because only the latter carry the specific receptors to which these viruses adsorb (reviews by Holland and Hoyer, 1962; Holland, 1964). The receptor substance has been isolated, but not purified, from plasma membranes of susceptible cells and shown to contain protein closely associated with membrane lipid. Cells from human amnion, which are refractory to poliovirus infection in vivo, were found to lack the specific receptors on initial explantation in vitro, but such receptors appeared after a period in culture (Holland, 1961; Chany et al., 1966). Similarly, M. Taylor et al. (1971) have presented evidence that a bovine enterovirus selectively destroys certain malignant tumors of mice in vivo because the cancer cells bear receptors that are not exposed in other mouse tissues. The receptors for various enteroviruses appear to differ from one another as judged by cross-saturation experiments, differential susceptibility to blocking by anticellular sera and to proteases, and differences in regeneration time at 37°C (see Levitt and Crowell, 1967).

"Nonsusceptible" mammalian cells may be infected with purified enteroviral nucleic acid; though the "plating efficiency" of the RNA is extremely low, the progeny consist of perfectly normal infectious virions (Holland, 1964). The device of cell fusion by inactivated Sendai virus has also been used as a means of introducing poliovirus into cells lacking the specific receptor; normal multiplication of virus follows (Enders et al., 1967).

"Uncoating" of picornaviruses commences while the adsorbed virion is still outside the cell. Joklik and Darnell (1961) showed that about one-half of the poliovirions that had attached to cells at 4°C subsequently eluted into the supernatant medium when the temperature was raised to 37°C; eluted virions were found to have suffered a distortion of the capsid which rendered them unstable to high salt concentration and no longer capable of readsorption to susceptible cells. Mandel (1971a) analyzed the effects of this temperature-dependent process on the conformational state of the capsid; the physical properties of the virion are greatly changed, and it no longer binds neutralizing antibody efficiently. Crowell and Philipson (1971) demonstrated that eluted coxsackievirus B3 virions, which were noninfectious by virtue of being unable to reattach to HeLa cells, had specifically lost the polypeptide VP4; it seems probable, therefore, that VP4 is the first protein to be released in the uncoating process. This may suffice to weaken the capsid sufficiently to ensure its automatic dissolution on entering the cytoplasm.

Hall and Rueckert (1971) have demonstrated that the extracellular uncoating process proceeds even further with the cardioviruses. At 37°C more than one-half the adsorbed ME virions liberate their RNA into the supernatant fluid; the capsid remains temporarily attached to the plasma membrane then dissociates into 14 S pentamers. Extracellular uncoating occurred only at 37°C and could be inhibited by alkaline pH (8.5) or hypertonic salt (40 mM $MgCl_2$), conditions that have previously been found to stabilize ME virions against heat inactivation, to increase phagocytosis (Dales, 1965a), and to increase the plating efficiency of isolated viral RNA (review, Pagano, 1970). Obviously, release of RNA into the surrounding medium is not in itself relevant to the establishment of successful infection in nature, but the disruption of the capsid which makes RNA release possible is probably the first step in uncoating of the virion.

Precisely how the virion traverses the plasma membrane is not clear. Dales et al. (1965a) presented evidence for cytosis, but Dunnebacke et al. (1969) propose that adsorbed virus can "penetrate directly" into the cytoplasm. It is also not clear whether the removal of VP4 and/or the extrusion of RNA normally occurs before the virion enters the cytoplasm (Chan and Black, 1970) or after association with the membrane of a phagocytic vacuole (Mandel, 1967a, b, c).

## Translation

Once uncoated, the infecting viral RNA molecule attaches by its 5′ end to a 45 S ribosomal subunit and protein synthesis soon begins. No preliminary transcription of cRNA is necessary because the vRNA itself constitutes mRNA suitable for direct translation into protein. Formal proof of this fact has gradually accumulated over the years. Poliovirus-infected cells were found to contain unusually large polyribosomes with 20 to 40 (average 35) ribosomes distributed along the length of an mRNA of molecular weight and base composition indistinguishable from vRNA (Penman et al., 1963, 1964). Such polysomes synthesized proteins which were immunologically precipitable with poliovirus-specific antisera (Scharff et al., 1963; Warner et al., 1963). Ultimate proof has come with the demonstration that vRNA serves as a template for the in vitro synthesis of proteins with peptide maps resembling those of infected cells and including those of purified virions (Rekosh et al., 1970).

There are no punctuation points in the poliovirus RNA molecule. In this respect, it differs from the polycistronic messengers represented by certain bacterial virus RNA's. Ribosomes attaching to the 5' end progress down the whole length of the molecule until they encounter the sequence of fifty adenines near the 3' end. Hence a giant polypeptide with a molecular weight of 250,000 daltons is synthesized, corresponding to the whole poliovirus genome. Very rapidly, however, this "polyprotein" (Baltimore, 1971a) is cleaved by proteolytic enzymes in a systematic fashion.

Posttranslational cleavage was demonstrated by "pulse labeling" virus-infected cells with radioactive amino acids for a very brief period, then "chasing" with an excess of unlabeled amino acids to reveal quantitative movement of counts from large proteins into smaller and progressively smaller ones (Summers and Maizel, 1968; Holland and Kiehn, 1968; Jacobson and Baltimore, 1968b). Formal confirmation that the small proteins did indeed represent cleavage products of the short-lived high molecular weight precursors came with the demonstration that peptide maps of precursor and products could be equated (Jacobson et al., 1970). Recognition of the very large polypeptide corresponding to the whole poliovirus genome was not a simple matter because of its very short life. Jacobson and Baltimore (1968b) demonstrated its existence by allowing protein synthesis to proceed in the presence of amino acid analogs, or at a temperature of 43°C, or in the presence of diisopropyl fluorophosphate to inhibit cleavage (reviews by Baltimore, 1971a; Baltimore et al., 1971). Kiehn and Holland (1970), using another enterovirus, coxsackievirus B1, reported that the high molecular weight protein was readily detected early in the multiplication cycle, before large quantities of the cleavage enzyme(s), which they believed to be virus-induced, had accumulated in the cell. Then Roumiantzeff et al. (1971) found that much of the protein synthesized in vitro by membrane-associated polyribosomes extracted from poliovirus-infected HeLa cells was uncleaved.

There may be some ambiguity in the choice of cleavage loci by the putative proteolytic enzyme(s) (Cooper et al., 1970b). Usually, however, the precursor polypeptide is cleaved at specific sites and in a definite order to yield functional proteins that are now well characterized (reviews, Baltimore, 1971a; Baltimore et al., 1971). Figure 7–2 (Chapter 7) depicts the sequence of events. NCVP00 represents the "polyprotein" translated from the whole polioviral genome. While nascent on the ribosome it is cleaved in two places to yield NCVP1, NCVP2, and NCVPX (Jacobson and Baltimore, 1970). There is good evidence from genetic maps constructed from recombination experiments using temperature-sensitive mutants that NCVP2 is (or includes) the viral polymerase I, and more tentative speculation that NCVPX may represent polymerase II, or perhaps a sigma factor conferring viral specificity on a host polymerase (Cooper, 1969; Cooper *et al.*, 1971). NCVP1 clearly contains all the structural polypeptides of the viral capsid, for the second round of cleavage (which may sometimes occur before NCVP00 has been completed but is usually delayed for some minutes) converts it to VP0, VP1, and VP3, which comigrate in gel electrophoresis with the three structural proteins found in empty virions, and have almost identical tryptic peptide maps (Baltimore, 1971a). As will be discussed in detail below, the final cleavage of VP0 to yield VP2 and VP4 does not occur until 20–30 minutes later, coinciding in time with the assembly of RNA and protein into mature virions.

Cell-free systems have recently been developed to facilitate the detailed examination of protein synthesis (reviews by Lucas-Lenard and Lipmann, 1971; Phillips and Sydiskis, 1971). Several years ago, Warner *et al.* (1963) were able to demonstrate that poliovirus RNA could serve as a template for the *in vitro* translation by *E. coli* ribosomes of protein precipitable by antisera raised against poliovirions. Recently, Rekosh *et al.* (1970) have shown by peptide mapping that poliovirus RNA is translated as efficiently as f2 RNA by *E. coli* ribosomes, but the resulting proteins are much smaller than NCVP00. Initiation occurs at several sites in this bacterial *in vitro* system, each of the resulting proteins beginning with N-formyl methionine. This being so, such systems lose much of their appeal as an approach to the study of translation of animal viral RNA's. Fortunately, however, satisfactory mammalian cell-free systems are now being developed. As first demonstrated some years ago, EMC viral RNA is translated efficiently *in vitro* by Krebs' ascites ribosomes from the mouse (see Kerr and Martin, 1971). Ribosomes from rabbit reticulocytes were then shown to translate enteroviral RNA with fidelity to yield specific protein of high molecular weight (Mathews and Korner, 1970). Human HeLa cell ribosomes translate RNA from any of the cardioviruses almost as well as do ribosomes of murine origin (ascites or L cells) provided that certain soluble factors are supplied from mouse cells (Eggen and Shatkin, 1972).

The vagaries of mammalian cell-free protein-synthesizing systems are now sufficiently well understood and controlled for them to begin to be extensively used to examine the fine controls of the translation process, as well as the mechanism of action of interferon, virus-induced repressors of cellular protein synthesis, and "late" viral proteins that suppress the expression of "early" viral genes.

## Replication of Viral RNA

As soon as the infecting RNA molecule has been translated, a polymerase molecule becomes available to transcribe cRNA (−) from the vRNA (+) template. Kinetic studies show that RNA replication commences within ½ hour of infection and proceeds at an exponential rate until 10–20% of the final yield of RNA has been made by 3.5 hours, after which synthesis continues at a linear rate of about 2500 molecules per minute until a total of 250,000 molecules have been synthesized (Darnell *et al.*, 1967). Fully one-half of these eventually become encapsidated in virions—quite an efficient process. It takes less than 1 minute for a viral RNA molecule to be transcribed (100 nucleotides per second) but the doubling time of vRNA during the linear phase is about 15 minutes, indicating that chance factors and probably regulatory mechanisms we know little about determine whether new vRNA molecules are utilized for transcription, translation, or encapsidation.

From the outset of studies on the multiplication of RNA viruses, it was recognized that at least one novel enzyme must be involved: a polymerase capable of catalyzing the transcription of RNA from an RNA (rather than a DNA) template. It was assumed that no preexisting cellular enzyme would suffice and this intuitive assumption was confirmed by the demonstration that many RNA viruses could multiply normally in the presence of actinomycin D, which blocks transcription by DNA-dependent RNA polymerase (Reich *et al.*, 1962). Virus-induced RNA-dependent RNA polymerase ("RNA synthetase" or "replicase") was soon discovered in mengo- and poliovirus-infected cells (Baltimore and Franklin, 1963b). Crude extracts of cytoplasm stimulated poliovirus RNA synthesis *in vitro*, though the enzyme was somewhat unstable. Actinomycin D had no inhibitory effect on its synthesis or on its activity. Similar replicases have since been discovered for a wide range of RNA viruses but each tends to be specific for its own particular RNA, implying that viral RNA possesses near its 3' end a specific nucleotide sequence that serves as a recognition and binding site for its own polymerase.

The precise mechanism of replication of viral RNA was debated heatedly during the 1960's but has now been clarified, largely as a result of definitive studies with bacterial viruses of the f2 group (reviews, Weissmann *et al.*, 1968; Spiegelman *et al.*, 1968). In synopsis, vRNA (+) is the template for the synthesis of cRNA (−). Several cRNA molecules may be transcribed simultaneously by separate polymerase molecules; the whole structure is known as the "replicative intermediate" (RI). Released cRNA molecules in turn serve as templates for the simultaneous transcription of several vRNA strands. Different polymerases may be required for the two stages in this cyclic process. A double-helical form of viral RNA, known as the "replicative form" (RF), is an irrelevant end product or artifact. A schematic diagram of the replication of viral RNA is given in Fig. 6–2.

Poliovirus RI, first described by Baltimore and Girard (1966) and positively identified by Baltimore (1968b) and Bishop *et al.* (1969), possesses all the properties to be expected of the structure described by Weissmann *et al.* (1968) and Spiegelman *et al.* (1968). Table 6–2 lists these properties and compares the RI with single-stranded vRNA and with the double-stranded RF first described by Montagnier and Sanders (1963).

The RI (reviews, Bishop and Levintow, 1971; Montagnier, 1968) can be separated from mRNA, cRNA, and RF by such techniques as agarose gel filtration or benzoylated DEAE-cellulose chromatography. It consists of a single

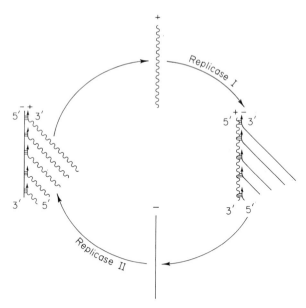

Fig. 6–2. *The replication of single-stranded RNA. (After Spiegelman et al., 1968.)*

complete strand (usually −, sometimes +) on which about six or seven complementary strands (+ or −) are growing simultaneously. Transcription begins
at the 3′ end of the template. Each progeny strand is hydrogen bonded to its
template over only a short sequence of base pairs at its growing end. The
polymerase molecule present at this growing end may play an important role in
holding the strands together. The 5′ end of each nascent strand quickly separates
from the template and floats free as a single-stranded "tail." These tails can be
artificially removed from the RI by ribonuclease I, leaving a partially double-
stranded structure consisting of one intact strand hydrogen bonded to several
shorter sequences, with gaps in between. Heating the RI produces a biphasic
hyperchromic shift; the single-stranded branches are gradually denatured at
relatively low temperatures, then hydrogen bonds are melted abruptly when the
$T_m$ of the double-stranded RF is reached. On analysis, most of the complete
single-stranded templates are found to be cRNA (i.e., −, as indicated by their
ability to hybridize with vRNA) while the short pieces of heterogenous lengths
are found to be mainly +. In other words, the synthesis of RNA is asymmetric,
+ strands being made at five to ten times the rate of − strands (Bishop and
Levintow, 1971).

It is not known whether some control mechanism is involved or whether the
preponderance of cRNA templates in RI is the inevitable consequence of the fact
that mRNA is siphoned off for translation and encapsidation. It is also not
known whether separate polymerases are required for the transcription of +
from −, and − from +, respectively, or whether a single enzyme performs both
functions, as occurs with Qβ phage (Mills *et al.*, 1966). Cooper *et al.* (1970a,
1971) have presented genetic evidence for the existence of two enzymes: poly-

TABLE 6–2

*Properties of Intermediates in the Replication of Enterovirus RNA*

| PROPERTY | VIRAL RNA (vRNA) | REPLICATIVE INTERMEDIATE (RI) | REPLICATIVE FORM (RF) |
|---|---|---|---|
| Electron microscopy (Kleinschmidt) | Single-stranded | Single-stranded with up to seven incomplete strands attached | Double-stranded |
| X-Ray diffraction | Single-stranded | Partially hydrogen bonded | Double helix |
| Infectivity | + | + | + |
| Nucleotide composition | $A \neq U; G \neq C$ | $A \neq U; G \neq C$ | $A = U; G = C$ |
| Susceptibility to ribonuclease I (at high ionic concentration) | Susceptible | Partially resistant | Resistant |
| Susceptibility to ribonuclease III | Resistant | Partially resistant | Susceptible |
| Melting curve (heat or DMSO) | Gradual | Biphasic | Sharp |
| Reannealing to self or to vRNA | — | + | + |
| Molecular weight | $2.6 \times 10^6$ | Heterogeneous | $5.2 \times 10^6$ |
| Migration in gel electrophoresis | Fast | Slow | Intermediate |
| Sedimentation in rate zonal centrifugation (in high salt) | 35 S | 18–70 S | 18 S |
| Buoyant density in equilibrium gradient centrifugation | 1.66 | 1.63–1.64 | 1.61 |
| Solubility in 1.5 M NaCl or 2 M LiCl | Insoluble | Insoluble | Soluble |

merase I, coded by the 3′ end of the genome, is considered to be the enzyme catalyzing the transcription of − from +; polymerase II, possibly coded by the short middle section of the genome, NCVPX, is postulated to transcribe + from − templates.

RI accumulates at an exponential rate during the first few hours of infection (Baltimore and Girard, 1966). The best evidence that the RI does represent the real intermediate in the replication of viral RNA comes from "pulse chase" experiments in which radioactivity rapidly entered RI then passed into single-stranded progeny molecules (Girard, 1969; McDonnell and Levintow, 1970).

RF with typical double helical structure accumulates in moderate amounts only late in infection (Baltimore and Girard, 1966) and is an irrelevant end product of viral RNA replication (Baltimore, 1968b; Girard, 1969). Furthermore, there is good evidence that much of the double-stranded material isolated from infected cells is an artifact resulting from the fact that single complementary strands anneal spontaneously *in vitro* following extraction (Weissman *et al.*, 1968). Oberg and Philipson (1971) were able to abolish the annealing artifact by adding diethylpyrocarbonate prior to phenol extraction of the RNA. Their work finally disposed of the suggestion that poliovirus RNA replication occurs by a conservative mechanism, i.e., by the transcription of single strands from a double-stranded template. The process is clearly semiconservative.

## Assembly

The viral capsid is composed of 60 NCVP1 molecules. Though each is cleaved enzymatically within 10 minutes to VP0, VP1 and VP3 (Summers and Maizel, 1968), all three cleavage products remain permanently associated by noncovalent bonds as a 5 S (100,000 molecular weight) "immature protomer" (Rueckert, 1971; see Chapter 3). Pulse-chase experiments demonstrate that these monomers polymerize in less than 5 minutes *in vivo* to form 14 S pentamers (Phillips *et al.*, 1968; Hall and Rueckert, 1971; see Fig. 6–3). Twelve such pen-

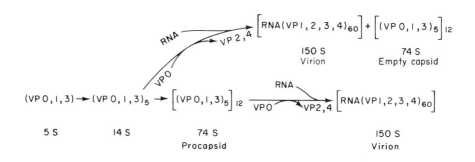

FIG. 6–3. *The assembly of poliovirions. Lower pathway envisages the 74 S "procapsid" as a virion precursor (from Baltimore, 1971a); the upper alternative views the 74 S "empty capsid" as a by-product (see text).*

tamers may aggregate to form 74 S empty capsids *in vivo* (Maizel *et al.*, 1967) or *in vitro* (Phillips *et al.*, 1968; Phillips, 1971). Jacobson and Baltimore (1968a) presented evidence that these "procapsids" are essential precursors in virion assembly. They maintain that at the time VP0 is cleaved to yield VP2 and VP4 (20–60 minutes after its synthesis) a viral RNA molecule enters the procapsid to form the complete virion. It is easier to envisage that RNA is able to associate with capsomers to form a virion only after the "immature protomers" have assumed a particular tertiary configuration as a result of cleavage of VP0 → VP2 + VP4. "Procapsids" would then represent experimental artifacts or unstable by-products of viral assembly.

The kinetics of virion assembly are presumably determined by the local concentration of the reactants in the vicinity of cytoplasmic membranes. Since assembly can occur spontaneously *in vitro* it appears that no special "maturation mechanism of action of interferon, virus-induced repressors of cellular protein protein" is required, apart from the proteolytic enzyme that cleaves VP0 into VP2 and VP4 immediately beforehand. As a result of the diversion of mRNA from transcription to translation which occurs from 3.5 hours onward, there is a substantial pool of capsid protein but not of vRNA. Hence virions are assembled at an exponential rate from RNA synthesized 0–15 minutes earlier and protein made an average of 20 to 60 minutes earlier. A single HeLa cell produces 150,000 poliovirions (Baltimore, 1969). These are readily visualized by electron microscopy, free in the cytoplasm, or occasionally as crystalline aggregate (Dales *et al.*, 1965a).

## Regulation of Viral Gene Expression

Every newly synthesized vRNA molecule is confronted with a triple choice: (a) to serve as a template for transcription of cRNA (replication), (b) to serve as a template for translation of protein, or (c) to associate with capsid proteins and form a virion. We know little about the factors that govern this choice, but they may well be dictated by considerations of topology.

A few examples will suffice to illustrate the problem. It was unexpected to find that polymerase I, which is presumably the first virus-coded protein required for the replication of the parental RNA molecule, is in fact synthesized last of all the viral proteins, being situated at the C terminal end of NCVP00. However, there may be great advantage in the polymerase molecule being completed in close apposition to the 3' end of the vRNA because that is the end to which the polymerase must bind to initiate the first round of RNA replication. Until at least one complementary RNA molecule has been transcribed by this replicase the exponential phase of viral multiplication cannot begin. Indeed, there must be only a small probability that the enzyme can successfully complete its 1-minute journey from 3' terminus to 5' without encountering a ribosome progressing slowly (12 minutes) in the opposite direction. Failure to accomplish this mission before the template is degraded by RNase could lead to an abortive infection, and is perhaps one of several reasons for the low "plating efficiency" of viruses in mammalian cells (see Chapter 2). The chance delays encountered by the infecting vRNA molecule in transcribing the first molecule of cRNA, or indeed in managing to attach to a 45 S ribosomal subunit to translate the first polymerase molecule, may contribute substantially to the well known phenomenon of "multiplicity-dependent delay" (Cairns, 1957).

The function of the long sequence of adenine residues (poly A) situated near the 3' end of poliovirus RNA and of other vmRNA's (Armstrong *et al.*, 1972), is unknown, but it may be involved in regulation.

It is probably significant that the quite separate processes of transcription and translation both occur in close association with membranes. A striking proliferation of cytoplasmic vesicles (smooth membrane cisternae) is clearly visible by electron microscopy between 3 and 7 hours after infection (Amako and Dales, 1967; Mosser *et al.*, 1972). By rate zonal and equilibrium gradient centrifiguration, it is possible to separate the sites of viral RNA replication from those of protein synthesis (Caliguiri and Tamm, 1970). RNA is synthesized in a "replication complex" (Penman *et al.*, 1964; Girard *et al.*, 1967) consisting of RI with about one-half dozen molecules of attached polymerase in association with smooth membrane. Viral proteins, on the other hand, are assembled mainly but not exclusively on membrane-bound polyribosomes, which can be clearly separated from RI and polymerase. It seems, therefore, that ribosomes do not attach to the 5' end of nascent mRNA while it is still being transcribed from a cRNA molecule, as happens in certain bacteria. Nevertheless, the fact that RNA engaged in transcription or translation must be attached to a membrane ensures that ribosomal subunits will find themselves in close apposition to mRNA molecules as the latter detach from the RI. In fact, Huang and Baltimore (1970a) have calculated that the average newly synthesized mRNA molecule has thirty-five ribosomes attached to it in little more than 10 minutes, i.e., the minimum time required for one ribosome to travel the length of the message. Similarly, localization of these processes to a restricted area on the surface of a membrane may enable capsid proteins to accumulate in sufficient concentration to stand a good chance of encountering a vRNA molecule and assembling automatically into a virion.

Once viral RNA replication is well underway and new vRNA templates are accumulating exponentially, some regulatory mechanism must operate to ensure that most of the vRNA molecules are deployed as templates for protein synthesis. That this is indeed so is apparent from the striking preponderance of mRNA over cRNA in the infected cell. Cooper et al. (1973) have pointed out that, since each virion contains 60 NCVP1 molecules for every molecule of vRNA, an efficient multiplication program would demand that the overall rate of viral protein synthesis exceed that of vRNA by a factor of 60. Yet six to seven molecules of cRNA are simultaneously transcribed from vRNA in less than 1 minute, while each of thirty-five ribosomes using a similar template simultaneously manufacture only thirty-five molecules of NCVP1 in 11 to 12 minutes. Cooper et al. (1973) postulate the existence of a regulatory protein, the "equestron," corresponding to the partially cleaved capsid protein precursor (VP0, 1, 3) and present an ingenious scheme to explain how this protein could control viral transcription, translation, and assembly.

## Alterations of Cellular Metabolism

Poliovirus and other picornaviruses produce substantial inhibition of host cell RNA, protein, and DNA synthesis. These specific effects which begin quite early in the viral multiplication cycle are not to be confused with the later decline of cellular and viral macromolecular synthesis that eventually accompanies the death of the cell in any cytocidal infection. Virus-induced cell "shutdown" ("shutoff," "cutoff") has been recognized for many years (Martin et al., 1961; Franklin and Baltimore, 1962; Zimmerman et al., 1963) but there is still uncertainty about the precise mechanisms involved (reviews, Darnell et al., 1967; Martin and Kerr, 1968). Direct comparison of results obtained in different systems is difficult because the speed and extent of the shutdown depends critically on the multiplicity of infection and the medium ("growth" or "maintenance") as well as on the strain of virus and cell involved. Early data that cellular protein synthesis declines at least as rapidly as cellular RNA, despite the fact that mammalian mRNA has a half-life of several hours, pointed to separate effects of viral infection on translation and transcription, respectively. The decline in cellular DNA synthesis is thought to be a secondary consequence of inhibition of host protein synthesis. Most of the available evidence suggests that newly synthesized virus-coded protein(s) are responsible for the shutdown of all these cellular macromolecules (Franklin and Baltimore, 1962; Penman and Summers, 1965), although it is puzzling that interferon has no effect on the phenomenon (Martin and Kerr, 1968).

**Inhibition of Cellular Protein Synthesis.**   Within an hour of picornavirus infection polyribosomes begin to dissociate and most of the ribosomes are freed
within 2 hours. Concomitantly with the rise in virus-coded protein synthesis,
polyribosomes reappear, but now they are uniformly large (20–40 ribosomes,
average 35) because vRNA, with a molecular weight of 2.6 million daltons, has
supplanted cellular mRNA of heterogeneous but generally lower molecular
weight (Penman *et al.*, 1963; Dales *et al.*, 1965a). By the time substantial
amounts of new viral protein become detectable the rate of cellular protein synthesis has dropped to 25 to 50% of normal, and it continues to fall further
throughout the remainder of the cycle (Franklin and Baltimore, 1962). Experiments with cycloheximide indicate that a newly synthesized virus-coded protein
is responsible for the shutdown (Baltimore and Franklin, 1963; Penman and
Summers, 1965); it acts by preventing the attachment of cellular mRNA to ribosomes (Leibowitz and Penman, 1971). If viral RNA replication is prevented by
guanidine or by infection with virus inactivated by ultraviolet irradiation or
hydroxylamine, shutdown of host protein synthesis is negligible at low multiplicity of infection but still occurs at high multiplicity, presumably because a
certain amount of the repressor protein is synthesized, albeit at a very slow rate,
off parental vRNA templates (Penman and Summers, 1965). Cooper *et al.* (1973)
postulate that the repressor is the "equestron," namely, the viral capsid protein
precursor NCVP1, at the stage when it has been cleaved to VP0, 1, and 3 but
before its final cleavage to VP1, 2, 3, and 4.

Ehrenfeld and Hunt (1971) have demonstrated that double-stranded RNA
(RF) inhibits polioviral protein synthesis when added to an *in vitro* system of
rabbit reticulocyte ribosomes. This phenomenon is not specific because synthetic
polynucleotides as well as double-stranded RNA from Q$\beta$ and reovirus are also
somewhat inhibitory.

**Inhibition of Cellular RNA Synthesis.**   Relatively little new information on
the mechanism of shutdown of cellular RNA synthesis has accumulated since
the phenomenon was first described by Franklin and Baltimore (1962). Virus-
coded protein synthesis seems to be necessary for shutdown of cellular RNA
transcription but the possibility has not been excluded that the phenomenon is
secondary to the shutdown of host protein synthesis. If one of the viral polymerases is formed by the substitution of a virus-coded $\sigma$ factor in a host-coded
transcriptase, so changing its template specificity from cellular to viral nucleic
acid, this could have the incidental effect of blocking the transcription of cellular
RNA.

Inhibition of transcription of cellular mRNA might be expected to enhance
the rate of viral protein synthesis by decreasing the competition for ribosomes
as existing mRNA decays. However, mammalian mRNA has such a long half-
life that the impact of shutdown of mRNA synthesis on the production of cell
protein must be reckoned as insignificant in comparison with the more rapid
direct effect of viral infection on translation. Moreover, the production of ribosomal RNA is inhibited at least as markedly as mRNA (Darnell *et al.*, 1967). It is
questionable, therefore, whether virus-induced suppression of cellular RNA synthesis plays an important role.

**Inhibition of Cellular DNA Synthesis.**   Cellular DNA replication is so absolutely dependent on RNA and protein synthesis that a decline in its production must inevitably follow a decline in the latter. Hand *et al.* (1971) have used autoradiography following a pulse of [³H]thymidine to examine the reduction of cellular DNA synthesis during infection by mengovirus, as well as by Newcastle disease virus and reovirus. They found a diminution in the number of active sites at which DNA replication was initiated but no reduction in the rate of chain elongation, nor was there any destruction of existing DNA, as occurs with some bacteriophages.

**Inhibition of Cell Division.**   Needless to say, virus-induced shutdown of host protein synthesis immediately blocks cell division, which is sensitive to inhibition of protein synthesis up to the moment that mitosis begins (Tobey *et al.*, 1965). If, however, mitosis has already commenced before host protein synthesis is suppressed, cell division preempts viral multiplication for the duration of mitosis (about 1 hour) and the viral multiplication cycle is extended for a corresponding period; arresting cell division in metaphase with chemical inhibitors also blocks viral multiplication (Marcus and Robbins, 1963).

**Cytopathic Effects and Cell Death.**   Clearly, the irreversible shutdown of cellular protein, RNA, and DNA synthesis that follows the synthesis of virus-coded protein (Franklin and Baltimore, 1962) must inevitably kill the cell. However, cytopathic effects become visible within a few hours of infection; the fact that this CPE is prevented by guanidine led Bablanian *et al.* (1965) to conclude that newly synthesized viral coat protein may be the cause. Allison and Sanderlin (1963) postulated that virus-induced CPE results from the release of destructive enzymes from damaged lysosomes (review, Allison, 1967). More recently, Blackman and Bubel (1969), using $\beta$-glucuronidase as an index, demonstrated that lysosomal enzymes are liberated shortly before virus is released and proteins begin to leak from infected cells.

### Defective Interfering (DI) Particles

As originally described by von Magnus (1954), serial transfer of influenza viruses at high multiplicity leads to the production of "incomplete" (noninfectious) virus in progressively greater amounts. A similar phenomenon is demonstrable with poliovirus although 16–18 consecutive passages at high multiplicity were necessary to produce a population of noninfectious (DI) particles (C. Cole *et al.*, 1971). These particles interfere with the multiplication of infectious virus but are only capable of reproduction themselves when complemented by infectious virions. In the absence of the "helper," they are competent to make viral proteins and defective viral RNA but no infectious virions. Analysis of the RNA of the major class of DI particles indicated that 13–15% of the viral RNA was missing from the 5' end of the molecule, hence an abbreviated version of the capsid protein precursor (NCVP1) was made.

### Foot-and-Mouth Disease Virus (FMDV)

Relatively little is known about the multiplication of human rhinoviruses, except that the polypeptides synthesized are very similar to those of poliovirus

(Medappa *et al.*, 1971). However, FMDV has attracted considerable attention because of the threat it poses to the livestock of countries that seek to prevent its entry by quarantine (review, Bachrach, 1968). Arlinghaus and Polatnick (1969) described two $Mg^{2+}$-dependent, actinomycin-resistant, RNA-dependent RNA polymerase activities which appeared to be compartmentalized in different regions of the cytoplasm of FMDV-infected BHK21 cells, one preferentially making single-stranded and the other double-stranded RNA. Cell protein synthesis is markedly depressed by FMDV infection and cell RNA slightly so. A temporary decline in the degree of methylation of tRNA has also been described (van de Woude *et al.*, 1970).

## TOGAVIRIDAE

It will be recalled that togaviruses are small enveloped icosahedra containing a single molecule of single-stranded RNA of molecular weight 4 million daltons and very few structural proteins: one (or sometimes two) in the "core" and two peplomer glycoproteins. The recent revival of interest in the multiplication of these viruses stems from the fact that their simple envelopes provide useful models for the study of membrane structure and glycoprotein synthesis.

Togaviruses are potent inducers of interferon and are very sensitive to its action, and much of our basic knowledge of togavirus multiplication was obtained in the early 1960's as a byproduct of investigations on interferon. More recently, Pfefferkorn and Burge (1968) have produced an extensive collection of *ts* mutants for complementation studies which have yielded useful information (see Chapter 7), while Burge and Strauss (1970) have made pioneering contributions on the incorporation of viral glycoproteins into membranes. Apart from some early studies on Western equine encephalitis virus (WEE), most of the recent work has been done with two other alphaviruses, Sindbis and Semliki Forest virus (SFV) which behave almost identically in most respects. The account that follows will be based on these viruses; subsequently, the behavior of viruses of the genus *Flavivirus* and the ungrouped togavirus, rubella virus, will be briefly described.

### The Multiplication Cycle

Togaviruses grow to high titer and are simply assayed by plaquing in cultured chick fibroblasts, BHK21, or Vero cells. The latent period is only 3 hours at high multiplicity with some alphaviruses but longer with the flaviviruses. Virions are produced by budding and reach maximum titers after 7 hours or more (Mécs *et al.*, 1967). The synthesis of viral macromolecules can be readily followed because these viruses grow at a normal or enhanced rate in the presence of actinomycin D (Taylor, 1965). The whole replication cycle is confined to the cytoplasm. Viral RNA is infectious (Wecker and Schonne, 1961) indicating that the genome is not divided and requires no accompanying transcriptase.

### Translation

The mechanism of entry of togaviruses into cells has not been studied, but the speed of multiplication suggests that the viral RNA is uncoated rapidly and is then translated into protein. As with picornaviruses, togavirus proteins are synthesized on membrane-bound cytoplasmic polyribosomes (Friedman, 1968). The structural viral proteins are detectable in infected cells 1.5 hours after infection (Wecker and Schonne, 1961); actinomycin pretreatment may be used to reduce the background of cell protein synthesis. Strauss *et al.* (1969) described up to a dozen additional electrophoretic peaks in Sindbis virus-infected cells which they believed may represent nonstructural virus-coded proteins or high molecular weight precursors. Evidence for cleavage was obtained by Burrell *et al.* (1970) using short pulse labels in an SFV-BHK21 system. Scheele and Pfefferkorn (1970) were not able to demonstrate posttranslational cleavage using pulse-chase experiments in this system, but described a *ts* mutant that produced large amounts of a high molecular weight protein instead of nucleocapsid protein at the nonpermissive temperature. More recently, Pfefferkorn and Boyle (1972a) reported that a high molecular weight protein accumulated in Sindbis virus-infected cells in the presence of an inhibitor of chymotrypsin, while Sefton and S. Schlesinger (unpublished) have data showing that a precursor of MW 130,000 is rapidly cleaved to yield the core protein and a 100,000 MW intermediate, which represents the precursor of the two envelope glycoproteins.

Burge and his colleagues have addressed themselves to the problem of how viral glycoproteins are synthesized. Most of their analyses were conducted before they discovered that there are two (not one) species of envelope glycoprotein in Sindbis virus (Schlesinger *et al.*, 1972), but many of their earlier conclusions are still valid. Virtually all the radioactive sugar label incorporated into acid-precipitable material between 3 and 11 hours after infection in Sindbis virus-infected cells was in viral glycoprotein in cytoplasmic membranes (Burge and Strauss, 1970). The glycopeptides extractable from these glycoproteins were, except for the number of terminal sialic acid residues, indistinguishable by exclusion chromatography from those extracted from vesicular stomatitis viral glycoproteins grown in the same cell line (Burge and Huang, 1970). Since separate transferases are required to add each individual sugar it is unlikely that togaviruses (or any other budding viruses) carry enough genes to code for them all. Evidence that the enzymes involved are cellular comes from the demonstration that sialyltransferase and fucosyltransferase are not altered in their specific activity or acceptor specificity as a result of Sindbis virus infection; moreover, sialyltransferase from uninfected cells was shown to be capable of transferring [$^3$H]sialic acid to Sindbis virus glycoprotein (Grimes and Burge, 1971).

David (1971) presented evidence to support the view that the composition of the hydrophobic end of the viral glycoprotein peplomers determines the selection of host lipids for the viral envelope, since the lipid composition of viruses varies with the virus as well as the host cell from which the virus buds (Renkonen *et al.*, 1971; see Chapter 3).

## Replication and Transcription of Viral RNA

The fact that togavirus RNA is infectious (Wecker and Schonne, 1961) is powerful evidence that the genome is a single molecule of mRNA, but this has never been formally proved by a demonstration that vRNA can serve as a template for *in vitro* synthesis of viral protein. Such an experiment, or even a description of the mRNA isolated from polyribosomes, would help to clarify the mystery that surrounds the role of the 26 S "interjacent RNA" (Sonnabend *et al.*, 1967), which is consistently found in infected cells in addition to vRNA (42–49 S), RF (16–23 S) and RI (polydisperse, 22–30 S) (Stern and Friedman, 1969; Cartwright and Burke, 1970; Sreevalsan, 1970a; Simmons and Strauss, 1972a, b). Because interjacent RNA is single-stranded, RNase-sensitive, and of the same polarity as vRNA but shares only one-third of its base sequences, Simmons and Strauss (1972a, b) postulate that this 26 S molecule and the 33–38 S species described by Levin and Friedman (1971) represent mRNA molecules corresponding to an incomplete segment of the vRNA. Any proteins translated from such large messengers would presumably need to be cleaved. By contrast, Eaton *et al.* (1972) have isolated from polysomes of alphavirus-infected cells a heterogeneous population of 16–18 S RNA's which contain a 4 S RNase-resistant A-rich region, but this has not been positively identified as mRNA either.

Stern and Friedman (1969) isolated a putative RI by chromatography on columns of benzoylated-DEAE cellulose; this heterogeneous RNA was partially RNase-resistant and was preferentially labeled in short radioactive pulses. More recently Simmons and Strauss (1972b) have described two species of RI, one producing complete vRNA molecules, the other producing shorter molecules of the same polarity. Treatment with RNase yielded three species of RF, corresponding in length to one-third, two-thirds, and 100% of the viral genome. Stollar *et al.* (1972) have isolated even shorter (12 S and 15 S) double-stranded RNA species late in the cycle from Sindbis virus-infected mammalian or avian (but not mosquito) cells. Sreevalsan (1970a) found that most of the RI, RF, and viral polymerase in Sindbis virus-infected cells was bound to cytoplasmic membranes. Parental vRNA from radioactively labeled virions also rapidly attached to membranes, leading Sreevalsan to the conclusion that intimate association with membranes is essential for viral RNA replication.

Martin and Sonnabend (1967) described a cytoplasmic membrane-associated polymerase appearing 2–3 hours after SFV infection, which was capable of catalyzing the *in vitro* synthesis of double- but not single-stranded viral RNA from endogenous intracellular substrate. They took this to be the transcriptase and postulated the existence of a second enzyme, the replicase. Pfefferkorn and Burge (1968) obtained evidence for two polymerase cistrons on the basis of complementation between their *ts* mutants of groups A and B (see Table 7–6). The preparations of Sreevalsan and Yin (1969) may have contained the replicase as well as the transcriptase since they synthesized single-stranded vRNA as well as RI and RF *in vitro*.

## Assembly and Release

Electron microscopic studies (Morgan *et al.*, 1961; Acheson and Tamm, 1967; Tan, 1970; Matsumura *et al.*, 1971) have clearly demonstrated that alphaviruses acquire an envelope in the process of budding from cytoplasmic membranes (see Plate 3–14). The first step in assembly is the association of vRNA with protein to form the nucleocapsid (Friedman, 1968). Nucleocapsids then migrate to the plasma membrane and budding occurs at a relatively constant rate of 200–1000 pfu per hour from those regions of the membrane into which viral peplomers have been inserted (Acheson and Tamm, 1967). Relatively few virions form by budding into cytoplasmic vacuoles; aggregation of excess nucleocapsids around such vesicles and into paracrystalline arrays (Morgan *et al.*, 1961) occurs only very late in the cycle when the release of virions into the supernatant fluid is almost complete (Acheson and Tamm, 1967). The extensive proliferation of cytoplasmic vacuoles that is also seen in thin sections may be connected with some necessary role for membranes in viral RNA replication and protein synthesis rather than with the budding process, because the phenomenon is equally conspicuous in poliovirus-infected cells.

Waite and Pfefferkorn (1970) discovered that budding can be completely arrested by reducing the ionic strength of the surrounding medium. On reversal of the block, virus that has accumulated intracellularly is suddenly released in a synchronous burst.

### Alterations of Cellular Metabolism

Togaviruses inflict little obvious damage on cultured invertebrate cells (see Weiss, 1971) but produce a moderate degree of CPE in vertebrate cell cultures. It is not surprising to find therefore that they cause a moderate suppression of host macromolecular synthesis in the latter but not the former.

In both Sindbis virus- and SFV-infected vertebrate cells, a decline in host protein synthesis is detectable by 3 hours and develops progressively until over 90% of the protein and glycoprotein being made toward the end of the cycle is viral (Friedman, 1968; Strauss *et al.*, 1969; Burge and Strauss, 1970). Cellular RNA synthesis does not slow significantly until quite late in the cycle (Taylor, 1965).

### Flavivirus

Flaviviruses multiply much more slowly than alphaviruses (Stollar *et al.*, 1967), and tend to bud into cytoplasmic vacuoles rather than from the plasma membrane (Filshie and Reháçek, 1968), but in most other respects follow a similar pattern. For example, with dengue type 2 virus, the best studied member of the genus, RNA synthesis becomes detectable about 6 hours after infection and the three major RNA species formed sediment at much the same rate as their alphavirus counterparts (Stollar *et al.*, 1967). Electron microscopy reveals a striking increase in the number of cytoplasmic vacuoles; nucleocapsids bud mainly into these vesicles (Matsumura *et al.*, 1971). Crystalline aggregates of virions accumulate inside the vacuoles late in infection of Vero cells, but this is not a prominent feature in BHK21 or KB cells, despite higher yields in the latter. The latent period was 12 hours in all three cell lines. Increasing the $Mg^{2+}$ concentration in the medium led to a sudden stimulation of release.

Trent *et al.* (1969) studied RNA synthesis by St. Louis encephalitis virus and found three major species: vRNA (43 S), RF (20 S), and an RNase-resistant 26 S material thought to be the RI. Westerway and Reedman (1969) presented tentative evidence that the proteins of Kunjin virus might be derived by posttranslational cleavage.

## Rubella Virus

Holmes *et al.* (1969) were the first to show that rubella virus closely resembles the prototype togaviruses in its morphogenesis. Sedwick and Sokol (1970) found essentially the same species of RNA in rubella virus-infected BHK21 cells as have been reported for other togaviruses: a 38 S single-stranded vRNA (which is infectious), a 20 S double-stranded RF, and a partially RNase-resistant RI. The proteins of the rubella virion resemble those of other togaviruses in number and molecular weight but their intracellular synthesis has yet to be studied.

## Growth of Togaviruses in Arthropod Cell Cultures

Arboviruses, by definition, multiply in both arthropods and vertebrates, despite the very considerable differences between the two in metabolism, including of course body temperature. The independent establishment of continuous lines of mosquito cells by Grace, Singh and Peleg presents virologists with the opportunity to compare the growth of togaviruses in invertebrate and vertebrate cells (reviewed by Weiss, 1971; Dalgarno and Kelly, 1973). Reháçek *et al.* (1971) and Buckley (1971) have shown that several flaviviruses and some alphaviruses grow well in mosquito cell lines. There are usually no cytopathic effects, although some flaviviruses lyse a few cells or cause others to fuse together, but the culture eventually recovers. Persistent infections usually develop with only a small proportion of the cells producing virus. SFV grows with essentially the same kinetics (latent period, rate of viral production, final yield) in cultured cells of the mosquito *Aedes albopictus* at 28° or 37°C as it does in vertebrate cells (Vero) at 37°C (Dalgarno and Kelly, 1973).

## ORTHOMYXOVIRUS

The influenza viruses differ in several important respects from the viruses we have been discussing so far. The genome occurs as 7 or more separate RNA molecules, each corresponding to a single cistron. A transcriptase associated with the ribonucleoprotein transcribes cRNA molecules which serve as messengers for translation into the seven known viral proteins. Mixed infection commonly leads to reassortment of the separate influenza genes giving rise to hybrid particles (see Chapter 7). Repeated passage of a single strain of virus at high multiplicity leads to the production of noninfectious particles lacking the largest of the genes ("incomplete virus" and the von Magnus phenomenon, discussed below). The multiplication of influenza virus is inhibited by actinomycin D, $\alpha$-amanitine, enucleation of the host cell, or preirradiation of the cell with ultraviolet light; the meaning of these observations is uncertain, but one interpretation is that transcription of some cellular DNA is essential for viral replication.

Various aspects of influenza virus multiplication have been reviewed by Hoyle (1968), Robinson and Duesberg (1968), Scholtissek et al. (1969), Blair and Duesberg (1970), Kingsbury (1970), Pons (1970), Compans and Choppin (1971), Shatkin (1971), and White (1973).

## The Multiplication Cycle

Until the 1950's much of what was known of the multiplication of animal viruses came from studies of influenza virus in the embryonated egg (reviews, Burnet, 1960; Hirst, 1965). The failure until recently to find cell lines that supported influenza virus multiplication delayed the application of modern biochemical methods, but some useful studies have been made with fowl plague virus in chick cells (Scholtissek et al., 1969) and other influenza virus strains in MDBK cells (Choppin, 1969). The latent period is 3–4 hours for influenza type A and somewhat longer for type B. Release by budding continues for many hours. Hemagglutination (Hirst, 1941) and hemadsorption (Shelokov et al., 1958) have proved useful shortcuts in monitoring the progress of infection.

## Initiation of Infection

The discovery and detailed description of the specific attachment of influenza virus to host cells via a glycoprotein–glycoprotein interaction was one of the pioneering achievements of the early days of animal virology (reviews, Hirst, 1965; Scholtissek et al., 1969). Following his discovery of hemagglutination in 1941, Hirst demonstrated that cells infectible by influenza virus also carried a receptor for the viral hemagglutinin and could be rendered resistant to infection by destroying those receptors with neuraminidase. Body fluids including respiratory mucus and serum were found to contain an inhibitor of hemagglutination and of infectivity, which competed with cellular receptors for virus but, like the cell receptors, could be destroyed by viral or bacterial neuraminidase. Gottschalk (review, Gottschalk et al., 1972) identified the cellular receptors and the soluble inhibitors as glycoproteins with $N$-acetylneuraminic acid (sialic acid) as the terminal sugar in the carbohydrate sidechains and demonstrated the removal of neuraminic acid with neuraminidase (sialidase).

There was a widespread belief that the viral neuraminidase must play a role in the initiation of infection, but Fazekas de St Groth (1948a) disproved this notion, and proposed that the host cell actively engulfed the influenza virion by a process he called "viropexis." Dales (review, Dales, 1965a) has supported this hypothesis; others (Hoyle et al., 1962; Duc-Nguyen et al., 1966; Morgan and Rose, 1968) have postulated that the envelope of the influenza virion, and by implication that of other enveloped viruses, fuses with the plasma membrane of the cell, liberating the nucleocapsid directly into the cytoplasm (Plate 6–1). Morgan and Rose (1968) demonstrated that fusion occurs very rapidly at 37°C (within about 5 minutes of adsorption at 4°C); they agree that phagocytosis (viropexis) does occur but believe that the phagocytosed virions are mostly destroyed by lysosomal enzymes. The matter is still unresolved; it is probable that virus enters in both ways, and possible that both are relevant.

Following penetration, but before the actinomycin D-sensitive step, there is a delay, the duration of which varies for each infecting virion strictly according to chance (White et al., 1965). The mean delay at a multiplicity of infection of $<1$ is 2.5 hours but can be reduced to zero at high multiplicity. This fact provides the rationale for using high multiplicity inocula to synchronize infection.

## Transcription

Complementary RNA begins to be transcribed as early as 15 minutes after infection and remains the dominant species of virus-specific RNA until vRNA takes over during the third hour (Scholtissek and Rott, 1970). Doubtless, the virion transcriptase is responsible for the early cRNA synthesis, and the enzyme may be active before the RNA is freed from the proteins of the core. The virion transcriptase (Chow and Simpson, 1971; Penhoet et al., 1971; Skehel, 1971) is known to be present in the ribonucleoprotein itself (D. Bishop et al., 1972; Compans and Caliguiri, 1973); NP40-treated cores actively produce abbreviated strands of cRNA in vitro (Bishop et al., 1971; Chow and Simpson, 1971; Skehel, 1971). An RNA-dependent RNA polymerase with the same properties (production of short lengths of cRNA in vitro, requirement for $Mg^{2+}$ and stimulation by $Mn^{2+}$, actinomycin-insensitivity) is synthesized in influenza-infected cells (Ho and Walters, 1966; Scholtissek, 1969). The concentration of this transcriptase builds up slowly from 1 hour after infection to reach a peak in the cytoplasm late in the cycle (Scholtissek and Rott, 1970). The cRNA almost certainly represents mRNA since it is found in association with polyribosomes (Pons, 1972). Final proof of whether each molecule represents a monocistronic message will have to await the in vitro synthesis of individual viral proteins from purified populations of each cRNA species.

## Translation

The viral structural proteins (see Table 3–6) are synthesized on cytoplasmic polyribosomes but rapidly migrate to other parts of the cell, the NP accumulating in the nucleus and the HA in cytoplasmic smooth membranes (White et al., 1970). Protein M also moves into smooth membranes (Lazarowitz et al., 1971; Compans, 1973).

The main nonstructural protein, NS, accumulates in the nucleolus within 5 minutes of its synthesis in the cytoplasm (Taylor et al., 1969, 1970; Lazarowitz et al., 1971), and is also found in association with ribosomes (Pons, 1972; Compans, 1972). It is possible that the two events are related since ribosomes are assembled in the nucleolus; NS may direct ribosomes to translate viral rather than cellular mRNA. Alternatively this protein, which is the most plentiful virus-induced macromolecule, may represent a subunit of a viral RNA polymerase (Taylor et al., 1970). NS almost certainly represents the basic polypeptide subunit of the type-specific nonstructural protein previously recognized in the nucleolus by immunofluorescence (Dimmock, 1969). Recently, Skehel (1972) has reported a second nonstructural polypeptide (molecular weight 11,000 daltons) on gel electrophorograms of infected cells.

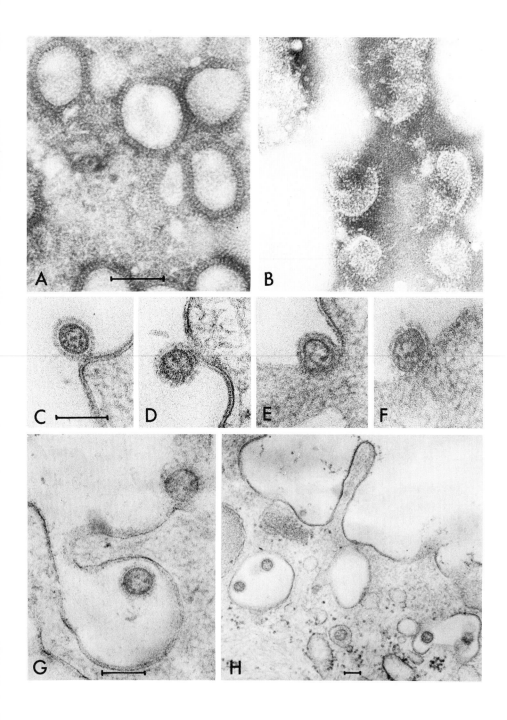

The influenza virus-infected cell is a good system for studying the factors that influence the movement of proteins within the cell. The transport of polypeptides NS and NP into the nucleus does not require continuing RNA or protein synthesis; the entry of HA into membranes is an energy-requiring process (White et al., 1970). Indeed, HA is always associated with membranes, being synthesized on membrane-associated ribosomes and moving within minutes through the smooth endoplasmic reticulum into plasma membrane (Compans, 1973; Stanley et al., 1973). Using various inhibitory sugars which have been shown to block the completion of viral glycoproteins without affecting the synthesis of other viral proteins (Gandhi et al., 1972), it has been shown that glucosamine is added to the HA almost immediately the polypeptide is synthesized and fucose is added shortly after (Stanley et al., 1973).

The HA glycoprotein is sometimes cleaved enzymatically into two smaller glycoproteins, $HA_1$ and $HA_2$, which nevertheless remain associated by disulfide bonds (Lazarowitz et al., 1971; Stanley and Haslam, 1971). Posttranslational cleavage may be fortuitous rather than essential because it occurs to different degrees with different virus–cell combinations and some systems yield infectious, hemagglutinating virions in which the HA is totally uncleaved (Stanley et al., 1973). Brief pulse labeling followed by prolonged chases have shown that all the other major virus-coded proteins are completely stable and not products of posttranslational cleavage of any high molecular weight precursor (Taylor et al., 1969).

## Replication of Viral RNA

From the third hour of infection onward, over 80% of the RNA synthesized in the presence of actinomycin D is vRNA (Duesberg and Robinson, 1967; Nayak and Baluda, 1968; Scholtissek and Rott, 1970). Clearly, some kind of regulatory mechanism must encourage the preferential synthesis of vRNA at the expense of cRNA, even though the latter is required for both translation and replication. Pons (1970) has postulated that the individual RNA molecules may remain linked during replication, perhaps held in sequence along a backbone of NP, but evidence for this is lacking. Nayak and Baluda (1968) described a structure that appeared to represent the replicative intermediate: a rapidly labeled 14 S RNA, partially RNase-resistant, and precipitable by 2 M NaCl. The replicase has not yet been identified, nor is it known whether

---

PLATE 6–1. Entry of influenza virus; fusion or engulfment? Electron micrographs of chorioallantoic membrane at intervals after infection, suggesting either fusion of viral envelope and plasma membrane (A–F) or engulfment of intact virions and subsequent intracellular uncoating (G and H) (bars = 100 nm). (A and B) Negatively stained preparations of virions attached to surface of entodermal cells at 4°C after 1 minute (A) and 10 minutes (B) at 37°C. (C to F) Thin sections showing successive stages suggesting fusion of envelope with plasma membrane. (G and H) Thin sections showing engulfment (G) and intact intravesicular virions seen 20 minutes after attachment. (A–F, from Morgan and Rose, 1968, courtesy of Dr. C. Morgan; G and H, from Dales and Choppin, 1962, courtesy of Dr. S. Dales.)

replication (or transcription for that matter) occurs in the nucleus or the cyto-plasm.

## Assembly and Release

Assembly of influenza virions coincides with their release from the cell by budding (Duc-Nguyen et al., 1966). The process has been examined in detail by Compans and Dimmock (1969). Areas of plasma membrane containing hemag-glutinin and neuraminidase develop visible projections about the same time as a new layer of protein (presumably M) is laid down on the inside of the mem-brane. Ribonucleoprotein (RNP) associates with these areas of morphologically altered membrane and spherical or filamentous virions bud off. The process will be discussed in greater detail in the section on paramyxoviruses, but two points are of particular interest here: the state of the RNP and the role of neura-minidase.

RNP isolated from infected cells occurs in short lengths corresponding to the particular class of RNA molecule each contains (Duesberg, 1969; Kingsbury and Webster, 1969; Pons et al., 1969). Structurally it resembles that isolated directly from virions (see Chapter 3; Fig. 3–6), viz., a double helix with a single-stranded loop at one end. Compans et al. (1970a) have calculated that the probability of the correct combination of five (now considered to be seven or more) RNP pieces being selected at random from the intracellular pool for incorporation into a budding virion is unrealistically low unless one assumes that the average virion contains an excess of such pieces. On the other hand, it is still possible that the natural state of RNP in infected cells and virions is a single continuous strand incorporating a complete set of RNA molecules and that the RNP fragments de-scribed are the result of breakage at weak points between the individual RNA's (Pons, 1970).

Neuraminidase moves into localized regions of the plasma membrane, often starting near one pole of cells spread out in monolayer, at about the same time as hemagglutinin (Maeno and Kilbourne, 1970). No doubt the enzyme is respon-sible for the absence of neuraminic acid in the glycoproteins and glycolipids of such areas of membrane (Klenk et al., 1970b). Yet it seems improbable that neuraminidase plays a vital role in the insertion of glycoproteins into mem-branes, peplomer formation or budding itself, because many budding viruses lack the enzyme altogether. However, neuraminidase may play a role in expedit-ing the elution of budded virions from the outside of the cell (Compans et al., 1970a) because anti-neuraminidase IgG (but not the monovalent Fab fragments obtained by pepsin digestion) inhibits the release of virus (Becht et al., 1971). Anti-hemagglutinin has no effect, but concanavalin A, which appears to bind more strongly to neuraminidase than to hemagglutinin, does block viral produc-tion (Rott et al., 1972).

## Regulation of Viral Gene Expression

The viral structural proteins are not made in anything like equimolar num-bers, or even in the same proportion as they occur in the virion itself (White et al., 1970). In particular, some regulatory mechanism must suppress the expression of the genes for NA and P proteins which are barely detectable in the infected cell, whereas NP and NS are made in very large amounts. The synthesis of these two proteins becomes detectable during the second hour after infection and rises to a plateau between 5 and 8 hours. More recently, Skehel (1972) has shown that the relative rates of synthesis of viral proteins change with time; early in the cycle NS is abundant and M scarce, whereas the converse applies later.

## Alterations of Cellular Metabolism

Cellular protein synthesis is substantially shut down with some virus–cell combinations, beginning at about 2 hours after infection when viral proteins start to be made in quantity and becoming almost total late in the cycle (White et al., 1970; Skehel, 1972). In other systems, shutdown is minimal until quite late (Lazarowitz et al., 1971). A newly synthesized viral protein appears to be responsible (Long and Burke, 1970). Since cellular RNA synthesis is not suppressed by influenza virus infection (Scholtissek et al., 1969) but if anything is stimulated during the first 3 hours (Mahy et al., 1972), the repressor protein probably acts at the level of translation; it may well be NS.

## Requirement for Cellular RNA Transcription

Influenza virus differs from most other RNA viruses in that its replication is inhibited by actinomycin D (Barry et al., 1962). If the drug is added during the first 2 hours after infection the synthesis of viral RNA, protein, and transcriptase are all blocked; if added later there is little overall inhibition of viral yield but cRNA synthesis continues to be affected, despite the fact that transcriptase synthesis is no longer blocked and actinomycin has no effect on the transcription of cRNA in vitro (Barry, 1964; White et al., 1965; Scholtissek and Rott, 1970).

Preirradiation of the host cells with ultraviolet light also inhibits influenza virus multiplication (Barry, 1964). Large doses have no effect on the capacity of cells to support the multiplication of parainfluenza virus added subsequently, but the capacity of preirradiated cells to support the growth of influenza virus is just as sensitive as is the virus–cell complex itself shortly after infection (White and Cheyne, 1966). Moreover, parainfluenza virus will grow in enucleated fragments of cytoplasm whereas influenza virus will not (Cheyne and White, 1969). All this tends to support the hypothesis, first put forward by Barry (1964) that influenza requires the expression of some cellular gene(s) early in the multiplication cycle.

Recently, it has been shown that the drug, α-amanitine, which inhibits cellular DNA-dependent RNA polymerase II (see Chapter 4) also blocks the production of influenza but not parainfluenza virus, with much the same time characteristics as actinomycin (Rott and Scholtissek, 1970). The drug does not inhibit the virion transcriptase in vitro (Penhoet et al., 1971) and has no effect on the synthesis of total RNA if added to infected cells later than 2 hours after infection (Mahy et al., 1972). However, Mahy and his colleagues found that α-amanitine inhibits the early burst of RNA synthesis detected in influenza-infected cells during the second hour. Since this RNA is transcribed at a time when they can demonstrate an increase in activity of a DNA-dependent RNA polymerase (?II) they believe it to be cellular RNA.

Rott and Scholtissek, however, do not accept that any of the data need be interpreted in terms of a nuclear requirement in influenza multiplication. They have postulated that the action of the various inhibitory drugs and of UV irradiation might be to liberate or activate a cellular RNase that destroys influenza RNA in the nucleus or to block viral RNA transport (Scholtissek *et al.*, 1969; Scholtissek and Rott, 1970). On the other hand, Dimmock (1969) has proposed that the inhibitory effects of actinomycin and UV irradiation are due to the considerable changes in the nucleoli that are morphologically apparent following treatment of normal cells with these agents. In influenza virus infection, the nucleoli are also grossly distorted (Compans and Dimmock, 1969), a nonstructural antigen is serologically demonstrable there (Dimmock, 1969), and the main nonstructural protein, NS, rapidly accumulates there from the second hour of infection onward (Taylor *et al.*, 1970; Lazarowitz *et al.*, 1971).

## Incomplete Virus

In 1946, von Magnus described a phenomenon which now bears his name (review, von Magnus, 1954). Repeated passage of influenza virus at high multiplicity led to progressive increase in the proportion of noninfectious virus particles among the progeny. These "incomplete" particles represent the prototype of what are now more generally known as DI (defective interfering) particles, produced under comparable circumstances with a wide range of viruses (Huang and Baltimore, 1970b). Incomplete influenza particles are deficient in the largest class of RNA molecule (Duesberg, 1968a; Pons and Hirst, 1969; Choppin and Pons, 1970) and in transcriptase (Chow and Simpson, 1971). It has been postulated that the smaller RNA molecules compete more successfully for the viral replicase (Blair and Duesberg, 1970; Huang and Baltimore, 1970b; Choppin and Pons, 1970) and it is quite possible that the largest RNA molecule(s) codes for a polymerase component.

In general, it is true to say that passage of influenza at high multiplicity will encourage the production of incomplete virus, but there are some interesting exceptions. Influenza A0/ strain WSN grown in MDBK cells does not produce any substantial number of noninfectious particles when passaged at high or low multiplicity in MDBK, or even in HeLa cells in which the multiplication cycle is usually abortive (Choppin, 1969); this can be viewed as an example of "host-controlled modification" (see Chapter 7). Choppin and Pons (1970) consider that the various virus–cell systems can be arranged in a sequence of decreasing productivity; at the top end would be highly productive systems such as any viral strain in MDBK cells, where the progeny are fully infectious even at high multiplicity for the first 1–2 passages; in the middle would be the classic von Magnus situation, applicable to most virus–cell combinations, where DI particle production increases only with repeated passage at high multiplicity; at the bottom end of the scale are "nonpermissive" cell lines, such as HeLa, in which the cycle is abortive (nonproductive) even at low multiplicty.

## PARAMYXOVIRUS

The most important differences between the multiplication of orthomyxoviruses and paramyxoviruses are based on the fact that the genome of the former is segmented and that the latter is not. It will be recalled that the paramyxovirus RNA occurs as a single strand of molecular weight 6.5–7.5 million daltons enclosed in a tubular nucleocapsid that also carries a transcriptase. Shorter molecules of cRNA transcribed by this enzyme serve as mRNA (reviews, Blair and Duesberg, 1970; Kingsbury, 1970). Paramyxovirus multiplication, unlike that of orthomyxoviruses, is resistant to actinomycin D or preirradiation of the cell with ultraviolet light (Barry *et al.*, 1962) and can occur in enucleated cells (Cheyne and White, 1969). Most paramyxoviruses carry in their envelope not only a hemagglutinin and neuraminidase like the orthomyxoviruses, but also a hemolysin and a factor that causes cells to fuse into multinucleate syncytia (Okada, 1962).

Paramyxoviruses grow more slowly than influenza virus, do not usually shut down host macromolecular synthesis to any substantial extent, and commonly set up persistent infections in cultured cells. They have served as the main models for the study of carrier cultures (see Chapter 12) and of the phenomena of budding (review, Choppin *et al.*, 1971) and cell fusion (see Chapter 9). More general aspects of paramyxovirus multiplication have been reviewed by Robinson and Duesberg (1968) and Compans and Choppin (1971). Most of the recent biochemical investigations have been conducted on Newcastle disease virus (NDV), Sendai (a strain of parainfluenza virus type 1), and simian virus type 5 (SV5), also known as parainfluenza virus type 5, hence the discussion that follows will be confined to an integrated synopsis of the findings obtained with these three viruses.

### Initiation of Infection

Though the adsorption of paramyxoviruses to susceptible cells has not been studied in great detail, the presence of hemagglutinin and neuraminidase in the viral envelope make it a reasonable assumption that the process is basically the same as for the orthomyxoviruses. As with other enveloped viruses, controversy surrounds the mechanisms of penetration and uncoating. Compans *et al.* (1966) presented electron micrographs showing examples of cytosis of SV5 (Plate 6–2) whereas Durand *et al.* (1970) have claimed that NDV nucleoprotein is released following fusion of the viral envelope with isolated chick cell membranes.

Homma (1971) has described an interesting example of host-controlled modification of a paramyxovirus that is relevant to the understanding of penetration. Sendai virus undergoes only a single cycle of multiplication in L cells, the progeny being noninfectious for L cells (but not for eggs) and having a greatly reduced hemolysin activity. Treatment of this L cell-grown virus with a low concentration of trypsin increases its infectivity for L cells 1000-fold and also restores much of the hemolytic activity. Similarly, a high yield of highly infectious virus with high hemolysin content is obtained by treating the L cells themselves with trypsin at the time new virus is budding. The data indicate that trypsin removes a protein from the surface of L cell-grown Sendai virus which otherwise hinders the apposition of viral envelope with plasma membrane. It would be consistent with the hypothesis that paramyxoviruses enter cells by membrane fusion aided by the hemolysin, which may correspond with the lysolecithin of the viral envelope (Barbanti-Brodano et al., 1971) or with lipid-bound peplomers (Hosaka and Shimizu, 1972).

Another isolated finding of some interest is that Sendai virus multiplication is blocked by the drug ouabain, which is known to inhibit the plasma membrane enzyme Na,K-ATPase, leading to an inability of the cell to maintain a high intracellular $K^+$ concentration (Nagai et al., 1972).

### Transcription

Most of the RNA synthesized in NDV-infected cells is single-stranded cRNA falling into two broad size classes, 18 and 35 S; both are associated with polyribosomes (Kingsbury, 1966; Bratt and Robinson, 1967). Similar findings have been obtained with Sendai virus (Blair and Robinson, 1970). It appears, therefore, that paramyxovirus mRNA consists of cRNA molecules which are shorter than the vRNA template (50–57 S) from which they are transcribed. Following transcription, polyadenylic acid totaling 4 S is covalently attached to 18 S cRNA (Pridgen and Kingsbury, 1972). Transcription of Sendai virus cRNA from parental vRNA templates occurs even in the presence of cycloheximide (Robinson, 1971b) presumably because the viral transcriptase is carried into the cell as an integral constituent of the nucleocapsid (Huang et al., 1971; Robinson, 1971a; Stone et al., 1971). Later in the cycle, cycloheximide again has no effect on the synthesis of cRNA, which is still the dominant RNA species being made, but the synthesis of vRNA (57 S) declines rapidly (Robinson, 1971b).

Portner and Kingsbury (1972) have isolated a structure from Sendai virus-infected cells which they believe to represent the transcriptive intermediate (TI). It is a rapidly labeled RNA with some base-pairing and some single-stranded tails, as evidenced by partial RNase-resistance and precipitation by 1 M NaCl, which sediments heterogeneously (28–60 S), and on melting yields a mixture of single-stranded 50 and 18 S RNA. Identification of the 50 S material as vRNA and the 18 S as cRNA by hybridization of the latter to RNA extracted from virions has to be interpreted in the face of the finding that a substantial minority of the 50–57 S RNA found in purified NDV or Sendai virions is complementary to the majority ("vRNA") species (Robinson, 1970).

Presumably some TI's must synthesize full length (50–57 S) cRNA molecules in order that they, in turn, may serve as templates for the production of more vRNA. The outstanding remaining question about the transcription of paramyxovirus cRNA is how it is regulated to ensure the synthesis of both long and short transcripts. One possibility is that all transcripts are full length but that some are subsequently cleaved, perhaps in stages, to produce monocistronic messengers. Alternatively, one would have to postulate that a transcriptase, perhaps different from the known virion transcriptase, has the capacity to "read through" the punctuation points in vRNA. Elucidation of this interesting situation will have significance extending well beyond the confines of the paramyxoviruses.

The transcriptase found associated with the nucleocapsid of the virion (Huang et al., 1971; Robinson, 1971a; Stone et al., 1971) may be the RNA-dependent RNA polymerase that is found in quantity late in the multiplication cycle in the "microsomal fraction" of the cytoplasm (Mahy et al., 1970). Both enzymes require $Mg^{2+}$ and the four ribonucleoside triphosphates, have a temperature optimum of 28°C and a pH optimum of 8, are not inhibited by actinomycin D, and produce short lengths of single-stranded cRNA in vitro.

## Translation

The five or six polypeptides found in paramyxovirions (see Chapter 3) are also detectable in electrophorograms of infected cells. Two of these contain carbohydrate, about one-half of which is glucosamine and the remainder galactose, mannose, and fucose (Klenk et al., 1970a; Choppin et al., 1971). No virus-induced nonstructural proteins have yet been described.

By immunofluorescence (Reda et al., 1964) or electron microscopy (Compans et al., 1966) large aggregations of RNP can be seen to accumulate in the cytoplasm. With some paramyxoviruses, notably the measles-rinderpest-distemper subgroup, nuclear inclusions are also detectable by these means or, indeed, by conventional staining techniques, but the protein that accumulates in the nucleus may play no vital role in multiplication (Llanes-Rodas and Liu, 1965).

## Replication of RNA

The replicative intermediate (RI) has not been positively identified. Portner and Kingsbury (1972) described a slowly labeled 24 S RNA which was almost totally RNase-resistant and on melting yielded a 50 S single-stranded RNA that hybridized with virion RNA only slightly more efficiently than vRNA hybridized with itself. It is not yet clear that this postulated RI is not RF, nor has the replicase yet been isolated.

Portner and Kingsbury (1971) have described defective (noninfectious) Sendai virions which contain only short RNA molecules and interfere with the multiplication of Sendai virus (but not other paramyxoviruses) apparently by inhibiting viral RNA replication rather than transcription. The mechanism may be similar to the auto-interference established by the "T" particles of vesicular stomatitis virus, which is discussed in some detail in the *Rhabdovirus* section.

## Assembly and Release

The assembly of paramyxoviruses has been comprehensively studied in Choppin's laboratory (reviews, Choppin et al., 1971; Compans and Choppin, 1971). First, the viral glycoproteins become incorporated into the plasma membrane; erythrocytes or ferritin-conjugated antibody are capable of adhering to such areas of membrane. Then three morphological changes became apparent in the particular areas from which budding occurs: (1) the glycoproteins assume their characteristic spikelike appearance, (2) an electron-dense layer (presumably M protein) develops on the inside of the plasma membrane beneath these spikes, and (3) nucleocapsids align themselves under the thickened membrane. Only when and where these three changes are apparent does budding occur (Plate 6–2). In addition to spherical particles, long filaments are produced, usually containing multiple nucleocapsids, only one of which is required for infectivity (Compans et al., 1966; Hosaka et al., 1966).

Though we have arbitrarily listed the morphological changes leading up to budding in a particular order, the precise sequence of events is not yet known. Much must depend on the affinity that each species of viral protein has for other proteins and for the lipids of the plasma membrane.

The membrane protein (M) appears to be the target to which RNP homes; the two proteins doubtless have an affinity for one another. Coinfection with SV5 and vesicular stomatitis virus (VSV) produces many phenotypically mixed progeny with the bullet-shaped morphology and the nucleocapsid (and therefore the genome) of VSV, plus all the VSV proteins together with the two SV5 peplomer glycoproteins but no SV5 M protein (McSharry et al., 1971). This is consistent with the hypothesis that VSV M protein is recognized only by VSV ribonucleoprotein.

Paramyxovirus nucleocapsids may accumulate in huge cytoplasmic inclusions clearly demonstrable by electron microscopy (Plate 6–2), or by light microscopy, where they are seen as the large, irregular, acidophilic inclusions that characterize the *Paramyxovirus* genus. In the virus–cell systems that have been studied, the extent of the excess accumulation of nucleocapsids seems to be inversely proportional to the yield of infectious virus (Compans et al., 1966).

## Alterations in Cellular Metabolism

In general, the paramyxoviruses do not cause a dramatic shutdown of synthesis of cellular macromolecules. Clearly, there is negligible damage in the case of the persistent infections. The most detailed studies have been confined to NDV, which does suppress the synthesis of cellular RNA, protein, and DNA (Wheelock and Tamm, 1961) following the synthesis of viral protein (Wilson, 1968). The extent of the shutdown varies greatly from strain to strain; there appears to be some correlation with virulence for the chicken but it is by no means absolute (Moore et al., 1972). Thacore and Youngner (1970) showed that there is no simple association between cell shutdown and CPE, because host protein and RNA synthesis was depressed less rapidly in chick fibroblasts than

---

PLATE 6–2. *Stages of Paramyxovirus multiplication (bars = 100 nm). Different stages of multiplication of paramyxovirus SV5 (from Compans et al., 1966). (A) Attachment. (B) Viropexis. (C) Accumulation of nucleocapsids. (D and E) Budding from the plasma membrane, with some filamentous forms. (Courtesy Dr. P. W. Choppin.)*

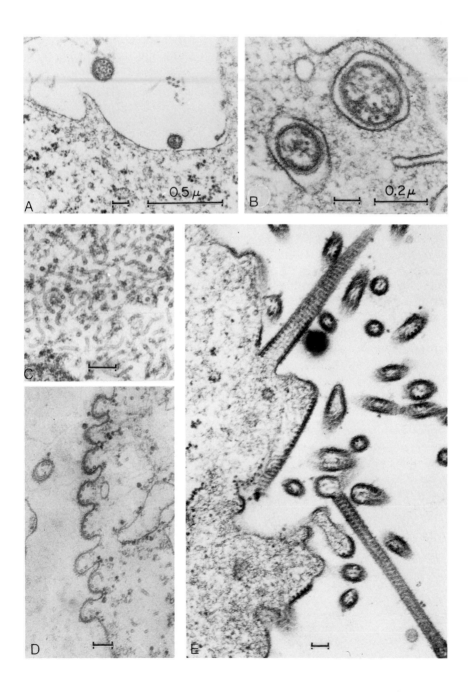

in L cells yet visible cell damage was more marked and viral yield greater in chick cells, and interferon synthesis was detectable only in L cells. NDV-induced shutdown of DNA synthesis, which is equally demonstrable at any stage in the mitotic cycle of synchronized cells, coincides with, and is probably secondary to the decline in protein synthesis (Ensminger and Tamm, 1970). Fuchs and Kohn (1971) have drawn attention to the fact that increased incorporation of radioactive thymidine into cellular DNA, such as they observed between 0.5 and 1.5 hours after infection of Sendai virus-infected cells, does not always mean increased DNA synthesis; since they found no increase in $^{32}P$ incorporation into DNA they postulated a transient block in the endogenous pathway of thymidine synthesis.

## Cell Fusion

The paramyxoviruses have a special capacity to cause cells to fuse together to form a multinucleated giant cell known as a "polykaryocyte" or "syncytium" (Okada, 1962). Sendai virus and NDV, in particular, have been extensively studied from this point of view. Two types of virus-induced fusion can be distinguished: "fusion from without" and "fusion from within" (Bratt and Gallaher, 1970; see also Chapter 9). Fusion from without probably results from the fusion of the viral envelope to the plasma membranes of adjacent cells so causing them to flow together. It has been postulated to be due to the hemolysin since both activities are destroyed under conditions that fail to destroy the hemagglutinating activity or infectivity of the virion (Barbanti-Brodano et al., 1971; Hosaka and Shimizu, 1972). Fusion from without is brought about equally well by live virus or virus whose infectivity has been inactivated by ultraviolet light (Okada, 1962) or β-propiolactone (Enders et al., 1967), and occurs very rapidly (1–3 hours or less) at high multiplicity.

Fusion from within is one of the characteristic cytopathic effects of paramyxoviruses in cell culture and in vivo, occurring to different extents with different virus strains and cell types. The effect is often most pronounced following infection at low multiplicity; it requires protein synthesis and is enhanced at high pH (8.3), the change develops only late in the cycle, and may be blocked by antiviral antiserum as late as 5.5 hours after infection (Bratt and Gallaher, 1970). The mechanism may be similar to that of fusion from without, namely fusion of the envelope of newly synthesized, extracellular or budding virus to the plasma membrane of an adjacent cell, or fusion of a virus-specific area of the plasma membrane of a cell to that of an adjacent cell. However, Poste (1970) has proposed that the necessary modification of the plasma membrane might be brought about by released lysosomal enzymes.

## CORONAVIRUS

The human coronaviruses were discovered relatively recently and at first could only be satisfactorily grown in organ cultures of human embryonic nasal, tracheal, or esophageal epithelium (Almeida and Tyrrell, 1967; review, Bradburne and Tyrrell, 1971). Consequently, nothing is yet known about the biochemistry of their replication; studies so far have been restricted to electron microscopy. W. Becker *et al.* (1967) examined WI-38 cells infected with the human coronavirus strain 229E, and chick embryo choriollantoic membrane infected with avian infectious bronchitis virus (IBV). In both cases, the virus was found to bud into cisternae of the endoplasmic reticulum and cytoplasmic vesicles, but not from the plasma membrane. Budding was preceded by a pronounced crescent-shaped protein thickening in the membrane. No helical nucleocapsids were seen at any stage during the latent period (6–12 hours depending on multiplicity) or the subsequent 12-hour period of viral production. Oshiro *et al.* (1971) confirmed these findings and also observed cytoplasmic inclusions containing virions and tubular structures within a granular matrix. The optimum temperature of growth is 33°–35°C, not 37°C, and the optimum pH near 7.0 (Bradburne and Tyrrell, 1971).

## ARENAVIRUS

Although the type species of this genus, lymphocytic choriomeningitis virus, has been studied for many years as the inducer of persistent infections in mice (see Chapters 9 and 12), little information at the cellular or molecular level is available about its multiplication, nor have other arenaviruses been adequately studied.

Virions of a great variety of shapes and sizes bud from those areas in the plasma membrane where viral peplomers have been incorporated (Murphy *et al.*, 1970; Plate 3–18). Ribosomes are regularly enclosed within the virion, and are also prominent in aggregates within the cytoplasmic inclusions demonstrable by immunofluorescence.

## BUNYAMWERA SUPERGROUP

Only recently have we come to realize that the Bunyamwera supergroup is by far the largest group of arboviruses (see Chapter 1). Virtually nothing is yet known about their multiplication, though Pettersson *et al.* (1971) have examined the structure of two typical members and shown them to be large enveloped viruses of helical symmetry. Murphy *et al.* (1968a), in an electron microscopic study of three members of the group, showed that infected mouse brain cells develop large numbers of cytoplasmic vacuoles. Virions acquire an envelope in the process of budding into Golgi vesicles and cisternae of the endoplasmic reticulum. I. Holmes (1971) confirmed and extended these findings in cultured cells. Lazdins and Holmes (1973) have demonstrated the two major viral proteins in Bunyamwera virus-infected Vero cells. Cellular protein and RNA synthesis are markedly suppressed. Most of the newly synthesized virions remain cell-associated for several hours.

## RHABDOVIRUS

Detailed studies with this genus have been focused on the relatively safe and rapidly growing vesicular stomatitis virus (VSV).

## The Multiplication Cycle

Vesicular stomatitis virus (VSV) grows extremely rapidly to high titers in cell culture, the latent period being 2.5 hours, with yields approaching $10^5$ virions per cell.

## Initiation of Infection

As with other enveloped viruses, there is evidence for entry by both phagocytosis and fusion. Simpson et al. (1969) have published electron micrographs showing cytosis. On the other hand, Heine and Schnaitman (1971) presented convincing evidence of fusion of viral envelope to cell membrane. Following adsorption at 4°C, they raised the temperature to 37°C and within 10 minutes demonstrated by ferritin-conjugated antibody labeling viral antigens distributed throughout patches of plasma membrane. Cell fractionation revealed the envelope proteins, G and M, to be in cytoplasmic membranes, whereas the ribonucleoprotein (RNP) was found free in the cytoplasmic sap. Evidence is still lacking as to whether both modes of entry are equally likely to lead to successful initiation of infection.

## Transcription

The transcriptase of the vesicular stomatitis virion (Baltimore et al., 1970) transcribes the complete viral genome in vitro into small molecules of cRNA (Bishop, 1971). Purified RNP, together with its four associated minor proteins, comprises the active transcription complex, indicating that the enzyme transcribes the RNA template while it is still firmly attached to protein N (Bishop and Roy, 1972). The enzyme does not catalyze viral RNA replication; all the in vitro product is hybridizable to vRNA, which itself shows no self-annealing (Bishop and Roy, 1971; Baltimore et al., 1971).

Transcription of mRNA from the parental genome in vivo can be studied in the absence of RNA replication by preventing translation with cycloheximide (Marcus et al., 1971). Using this inhibitor and an inoculum of purified B particles, Huang and Manders (1972) were able to show that the whole viral genome is transcribed in vivo into two main types of cRNA, one with a sedimentation coefficient of 28 S (not transcribed in vitro), and the other a heterogeneous collection of about 13–15 S. Both classes of cRNA almost certainly represent mRNA since they are labeled in brief radioactive pulses and are found in polyribosomes from which they may be released by EDTA (Stampfer et al., 1969; Huang et al., 1970; Mudd and Summers, 1970b; Petric and Prevec, 1970; Schincariol and Howatson, 1970). The 13–15 S (or 10–16 S) material can be resolved by gel electrophoresis into at least one-half dozen size classes representing a molecular weight range of about 0.25 to 1 million daltons, which would correspond quite closely with the sizes of putative monocistronic messengers for the known viral proteins (Wild, 1971; Huang and Manders, 1972). Following transcription, poly A is added to these monocistronic mRNA's (Mudd and Summers, 1970b; Huang and Manders, 1972).

The transcriptive intermediate (TI) has not been positively identified but is probably present in the partially RNase-resistant material sedimenting heterogeneously at around 30 S, which is rapidly labeled, mainly in the short cRNA molecules released on melting (Stampfer et al., 1969; Mudd and Summers, 1970b; Wild, 1971). As newly synthesized vRNA associates rapidly with protein N (Huang et al., 1970; Mudd and Summers, 1970b), it is quite likely that the nucleocapsid, rather than naked vRNA, serves as the main template for transcription late as well as early in the cycle.

When crude mixtures of B and T particles are used as inoculum the situation is much more complicated. Late in the cycle, in particular, in noncycloheximide-treated cells, besides vRNA (40 S), cRNA (13–15 and 28 S), TI, RI, and RF, a confusing range of shorter RNA's are made which doubtless represent vRNA, cRNA, TI, RI, and RF of T particle origin. The situation has been systematically clarified by Stampfer et al. and will be discussed in detail in the section on T particles.

## Translation

Viral proteins become demonstrable by gel electrophoresis as early as 1 hour after infection and are being synthesized at maximum rate by 4 hours (Mudd and Summers, 1970a; Wagner et al., 1970). There is some confusion about the precise number and nature of virus-induced proteins involved. All workers recognize the five major structural proteins known as L, G, M, N and NS (see Chapter 3), but it is unclear whether any of the minor peaks described by Mudd and Summers (1970a) correspond with the two additional minor structural proteins A and B, recently detected, together with L and NS, in RNP from the virion (Bishop and Roy, 1972). Kang and Prevec (1971) were not able to detect any nonstructural proteins even when the background of host protein synthesis was reduced by prolonged pretreatment with actinomycin D. There is no evidence of posttranslational cleavage (Mudd and Summers, 1970a; Wagner et al., 1970), a finding not unexpected in the light of the evidence that the small cRNA molecules in polysomes are probably monocistronic messengers.

However, there are two reports suggesting the existence of a nonglycosylated precursor of the peplomer glycoprotein, G. Kang and Prevec (1971) believe this to be an intracellular precursor which migrates somewhat more rapidly than the virion equivalent, while Printz and Wagner (1971) have presented evidence for the production of an unglycosylated spike glycoprotein by one of their ts mutants at nonpermissive temperature.

The glycoprotein, G, is found chiefly in smooth and plasma membranes, as is the membrane protein, M, in smaller quantities. The nucleoprotein, N, forms a pool, free in cytoplasm, which can be "chased" into plasma membrane (Wagner et al., 1970; Cohen et al., 1971a). Proteins M and L are incorporated into virions shortly after being synthesized, whereas there is a longer delay with N and G (Kang and Prevec, 1971). Cohēn *et al.* (1971a) have pointed out the importance of including appropriate controls in experiments purporting to demonstrate the location of viral proteins in membranes *in vivo*, because soluble VSV proteins, particularly M, adsorb strongly to plasma membranes isolated from normal cells.

## Replication of Viral RNA

The transcription of monocistronic cRNA molecules may be an excellent way of producing proteins without the necessity for posttranslational cleavage, but it is hard to imagine such short molecules being intermediates in RNA replication. A transcriptase that may or may not be identical with the one carried in the virion must transcribe cRNA molecules of full length which would then serve as template for the synthesis of vRNA. Little is yet known about this process, but Baltimore *et al.* (1971) have reported finding some such long cRNA strands in VSV-infected cells among a large excess of vRNA.

The replicative intermediate (RI) is probably to be found along with the TI in the rapidly labeled, heterogeneously sedimenting, partially RNase-resistant RNA already described (Stampfer *et al.*, 1969; Mudd and Summers, 1970b). Replicase activity has not been convincingly separated from transcriptase (Wilson and Bader, 1965) but the description of two distinct complementation groups of *ts* mutants defective in RNA synthesis suggests the existence of two polymerases (Wunner and Pringle, 1972).

## Assembly and Release

VSV is released by budding, mainly from the plasma membrane (Howatson and Whitmore, 1962), but with some virus–cell combinations budding occurs predominantly into cytoplasmic vesicles (Zee *et al.*, 1970). The process has been discussed in detail for paramyxoviruses, as has also the work of McSharry *et al.* (1971) demonstrating the key role of VSV membrane protein (M) in serving as a specific attachment site for the nucleocapsid of VSV rather than SV5 in phenotypically mixed rhabdovirions from doubly-infected cells. The fact that the length of T particles bears a fairly direct relationship to the length of the abbreviated RNA they contain suggests that only the area of M protein-lined membrane to which RNP is adsorbed is thrust out as a budding virion. Conversely, abnormally long particles are found to contain multiple nucleocapsids.

Most of the rhabdoviruses whose morphogenesis has been examined by electron microscopy acquire their envelopes by budding (review, Howatson, 1970). However, Hummeler (1971) has interpreted some of his findings as demonstrating *de novo* synthesis of rabiesvirus envelope in cytoplasmic inclusions (Negri bodies). Viral nucleocapsids in the unwound form become recognizable in these inclusions, then arrays of bullet-shaped particles appear in the filamentous matrix. With other strains of rabies (and VSV), however, Hummeler and his colleagues find that virions mature only by budding. Matsumoto and Kawai (1969) also described a difference between the budding behavior of rabiesvirus in different types of host tissue; in neurons viral morphogenesis was completely restricted to intracytoplasmic membranes closely associated with viral inclusions, whereas in cultured chick fibroblasts budding was from the plasma membranes.

## Regulation of Viral Gene Expression

Mudd and Summers (1970b) produced some evidence for the preferential transcription of the smallest class of VSV cRNA molecules but a definitive answer must await the clear resolution by gel electrophoresis of the three main cRNA classes into pure populations of individual molecules.

There is certainly some control, perhaps at the level of translation, over the rate of synthesis of VSV proteins, because they are not found in equimolar proportions. The nucleocapsid protein (N) is made in excess and forms a large cytoplasmic pool of nucleocapsid, while the membrane protein (M) is made in limiting amounts (Wagner et al., 1970). Moreover, the rate of synthesis of individual proteins fluctuates with time (Kang and Prevec, 1971). Whereas NS is made predominantly early in the cycle, and is therefore presumed to have an early function, the two envelope proteins (M and G) are made progressively more rapidly as infection progresses.

## Alterations in Cellular Metabolism

The extent of cell shutdown varies greatly from one rhabdovirus to another. It is minimal, for example, in the case of persistent infections by rabiesvirus, where infected cells continue to produce virus almost indefinitely. On the other hand, VSV, which as usual is the only rhabdovirus to have been at all closely examined from this point of view, does suppress the synthesis of cellular macromolecules.

At high multiplicity, infectious or UV-inactivated VSV B or T particles suppress the synthesis of host RNA and protein (Huang and Wagner, 1965; Wagner et al., 1970). Presumably this particular phenomenon results from the toxic action of proteins contained in the inoculum because the UV-irradiated particles have nonfunctional RNA. There is, however, another mechanism of suppression of cellular protein synthesis which seems to be dependent on translation of a viral protein (Wertz and Youngner, 1972). This is brought about by live virus only and does not occur in the presence of cycloheximide. Rather unexpectedly, cellular protein synthesis was found to be suppressed more rapidly by a small-plaque mutant (90% by 2 hours) than by a large-plaque mutant (90% by 8 hours). Huang et al. (1970) demonstrated the breakdown of polyribosomes by their strain of VSV during the third hour after infection and reformation of a quite different population of smaller virus-specific polysomes, reaching a peak during the fourth hour.

## Defective Interfering Particles

The T (truncated) VSV particles described by Cooper and Bellett (1959) and then by Hackett (1964) have been designated DI (defective interfering)-T particles by Stampfer *et al.* (1971) to emphasize their two main characteristics: a total lack of infectivity in the absence of infectious (B) particles, and the capacity to interfere specifically with the replication of B. The majority of DI-T particles described in the literature are one-third the length of the B particle, but it is now appreciated that with different virus strains and cell types a wide variety of sizes are possible (review, Huang and Baltimore, 1970b).

Prevec and Kang (1970) described two classes of T particles formed by their particular VSV strain, only the longer of which would interfere successfully with a heterologous strain. The length of the T particle is also characteristic of the parent strain in the case of Pringle's *ts* mutants (Reichmann *et al.*, 1971). Some of these mutants yield predominantly DI-T particles on undiluted passage at the permissive temperature. Autointerference by DI-T particles is more readily demonstrable in certain cell types than in others (Perrault and Holland, 1972b).

Cloned B particles do not produce T particles until about the third passage at high multiplicity (Stampfer *et al.*, 1971). The proportion of T particles increases on continued passage and the resulting interference leads to a proportionate drop in the yield of B. Total viral protein synthesis is suppressed (Mudd and Summers, 1970a; Wagner *et al.*, 1970) as is total viral RNA synthesis (Stampfer *et al.*, 1969). Yet transcription of B RNA is not affected when T particles are added prior to infection with B in the presence of cycloheximide (Huang and Manders, 1972).

DI-T RNA is necessary for the establishment of interference; UV-inactivated particles have no interfering capacity (Huang and Wagner, 1966a). Yet T particles lack transcriptase activity *in vitro* (Bishop and Roy, 1971) and synthesize no RNA *in vivo* (Stampfer *et al.*, 1969). Clearly, transcription of T RNA is not required in order that interference occur. This conclusion is indirectly supported by the observation that, although DI-T particles from any of Pringle's *ts* mutants will interfere with the multiplication of wild-type B particles, none of them are able to complement *ts* B mutants from other complementation groups (Reichmann *et al.*, 1971). Furthermore, Sreevalsan (1970b) has produced interference with purified RNA from T particles. The current hypothesis, therefore, is that T RNA specifically interferes with B RNA replication (but not transcription) by binding the VSV replicase with high affinity (Stampfer *et al.*, 1969; Sreevalsan, 1970b).

## LEUKOVIRUS

Leukoviruses multiply in a way that is radically different from that of all other viruses. First, the cells which support the growth of the avian and murine leukoviruses of subgenera A and B invariably carry one or more copies of the DNA provirus of the appropriate leukovirus as an integral part of the cellular genome (see Chapter 14). Infection of such cells with infectious virions is thus always a homologous superinfection. Although it is clear that under suitable conditions infectious virus can be synthesized from the information contained in the resident provirus, it is not known whether the resident viral genetic information influences the multiplication of the superinfecting virus. Second, there is good evidence that an essential step early in the multiplication of the leukoviruses involves the synthesis of DNA and there is much circumstantial data to indicate that this DNA is a copy of the incoming RNA genome. In most virus-infected as well as in uninfected cells the prevailing flow of information is from DNA to RNA. In leukovirus-infected cells at least some information travels in the reverse direction, that is from RNA to DNA. In all probability this novel transfer is mediated by the RNA-dependent DNA polymerases which are an integral part of all infectious leukoviruses. Many facets of the multiplication of leukoviruses and their effects on the host cells are described in Chapter 14; here we shall consider only the events that follow orthodox "infection" of chick cells with virions of avian leukosis or Rous sarcoma virus.

## The Multiplication Cycle

The growth rates of different strains of avian leukosis virus vary considerably (Hanafusa and Hanafusa, 1966), but as with most other viruses, attention has been focused on systems in which events are highly synchronized and relatively rapid (Hanafusa, 1969b; Bader, 1972b). The progress of infection may be measured either by titrating cell-associated and extracellular infectivity (Vogt and Rubin, 1963; Fig. 6–1) or by observing the morphological transformation of cells infected with Rous sarcoma virus (Hanafusa, 1969b).

## Initiation of Infection

The capacity of chick cells to support leukovirus infection is genetically determined (see Table 14–1). Attachment of virions is unaffected by the susceptibility of the cells in question; the specificity of blocking that is related to the envelope glycoproteins of the virion (virus-attachment interference; see Chapter 8) occurs at the penetration step (Piraino, 1967). Dales and Hanafusa (1972) have drawn attention to the existence of morphologically distinguishable attachment sites on the surface of chick embryo cells.

As with other enveloped viruses, there are proponents for the views that whole virions are engulfed (Sarkar et al., 1970; Dales and Hanafusa, 1972) or that the nucleoid enters after fusion of the viral envelope and the plasma membrane (Miyamoto and Gilden, 1971). After penetration, inoculum particles move rapidly to the perinuclear zone, and autoradiography shows that at least some of the viral RNA enters the nucleus, where most of the events that follow take place.

## Transcription

In contrast to all other RNA viruses, "transcription" in the case of leukoviruses involves two processes: the production of proviral DNA from the input viral RNA and the subsequent synthesis of mRNA on this DNA template.

**Transcription of DNA from Viral RNA.** The idea that the multiplication of leukoviruses might be different from that of other RNA viruses was first suggested by Temin (1962), on the basis of experiments which showed that the structure carrying viral information in transformed cells was inherited in a regular fashion at cell division (Temin, 1960). The chemical nature of this structure was left undefined until Temin (1964) proposed that the "provirus" was DNA, a notion that many virologists found difficult to accept at that time. However, several new discoveries, added to the earlier observations, have confirmed Temin's views.

1. The multiplication of leukoviruses is inhibited by actinomycin D, a drug which blocks specifically the transcription of DNA into RNA (Bader, 1965). This result suggests that viral RNA is made directly from DNA templates rather than from the RNA replicative intermediate familiar with other RNA viruses (Fig. 6–2). In fact, double-stranded RNA has never been found in cells infected by leukoviruses, and no RNA complementary to viral RNA can be detected at all in leukovirus-infected cells (Coffin and Temin, 1972).

2. By using specific inhibitors of DNA synthesis, it has been shown that initiation of leukovirus production requires the synthesis of a new species of DNA (Bader, 1966). The synthesis of this DNA occurs in the absence of protein synthesis (Bader, 1966, 1972a) suggesting that the enzymes responsible pre-exist in the cell or are carried in as part of the virion. In cells infected with high multiplicity of Rous sarcoma virus, the synthesis of DNA begins at about 1.5 hours and is completed by 6 or 7 hours after infection.

3. There is good kinetic evidence that at least some of the newly synthesized DNA contains viral sequences. Boettiger and Temin (1970) and Balduzzi and Morgan (1970) have shown that if 5-bromodeoxyuridine (BUdR) is present during the time of synthesis of the DNA the provirus is rendered light-sensitive. Furthermore, the presence of BUdR during provirus synthesis induces both lethal and conditional lethal mutants of Rous sarcoma virus (Bader and Bader, 1970; Bader and Brown, 1971). In all probability the proviral DNA is synthesized from the incoming viral RNA using the virion-associated RNA-dependent DNA polymerase (Baltimore, 1970; Temin and Mizutani, 1970). The virions of all infectious leukoviruses so far examined contain endogenous polymerase systems capable of synthesizing DNA copies of virion RNA (see Chapters 3 and 14), and mutants which lack these systems are noninfectious (Hanafusa and Hanafusa, 1971).

4. At least some cells transformed by RNA tumor viruses contain increased amounts of DNA that will hybridize to viral RNA (Baluda and Nayak, 1970). The total amount of proviral DNA synthesized after infection is unknown, but because resistance to DNA inhibitors increases logarithmically after infection (Bader, 1972a) it is likely that several copies are made. However, in spite of many attempts no one has yet been able to isolate the newly formed proviral DNA.

**Transcription and Translation of mRNA.**  In order to account for the genetic stability of transformed cells it is believed that at least one of the newly synthesized copies of proviral DNA becomes integrated into the cellular DNA and is transmitted to the daughter cells at mitosis. RNA is transcribed from the proviral DNA by one or other cellular DNA-dependent RNA polymerase. Presumably this RNA serves two functions; it may act as mRNA and direct the synthesis of virus-specific proteins or it may be incorporated into the genome of newly synthesized virions.

The mechanism of these processes is unknown. However, virus-specific RNA can be detected in both the nucleus and the cytoplasm of chick cells transformed by Rous sarcoma virus (Leong et al., 1972). This RNA is heterogeneous in size, with molecules longer than unit length viral RNA in the nucleus and shorter molecules in the cytoplasm. Infected cells contain about 6000 genome equivalents of viral RNA (Coffin and Temin, 1972) and as much as 1% of the total cell RNA may contain viral sequences. Presumably some of this RNA is messenger, and becomes attached to cytoplasmic ribosomes where it is translated into viral proteins. Immunofluorescence studies reveal that some of the internal antigens of the virion are assembled in the nucleus, and later migrate to the cytoplasm, whereas the glycoproteins that form the peplomers of the envelope accumulate in the plasma membrane (Vogt and Rubin, 1961; Payne et al., 1966). Lai and Duesberg (1972b) have recently reported that the glycopeptides obtained by pronase digestion of glycoproteins of avian leukoviruses released from transformed cells are larger than those of the corresponding strains released from normal cells.

It will be clear from the foregoing account that it is difficult to separate the processes leading to the formation of new viral RNA that will be incorporated into progeny virions from the synthesis of viral mRNA. Both are believed to be transcribed from integrated proviral DNA. Viral RNA enters virions about 1 hour after it has been synthesized (Bader, 1970; Baluda and Nayak, 1969; Okano and Rich, 1969).

### Assembly and Release

Leukoviruses lend themselves to electron microscopic study; most of the earlier studies of them, and many current papers, deal with the detection of characteristic "C-type particles" as they assemble at the plasma membrane and are released by budding. Differences in assembly and release that distinguish the viruses of subgenus A and subgenus B were described in Chapter 3. The leukoviruses are the only viruses known to change morphologically after leaving the cell. The chemical alterations involved in this postbudding "maturation" have only just begun to be analyzed (Cheung et al., 1972; Canaani et al., 1973). Rous sarcoma virions harvested within 3–5 minutes of budding differ from those harvested an hour or more later in a number of respects. Initially the RNA is mainly of the 30–40 S class, but with time it appears to associate with 4–12 S "primer" RNA to form 60–70 S RNA.

### Alterations in Cellular Metabolism

Leukoviruses are quite unlike other viruses in their effects on cellular metabolism. In no circumstances are they cytocidal. In the fully integrated form, the viral genome may be carried as repressed "cellular" genes, or expression may be limited to the production of one or a few antigens (see Chapters 12 and 14). After superinfection of chick cells (which already carry an integrated viral genome) with avian leukoviruses one of two results follows: with the non-transforming avian leukosis viruses, virions are released but the cells are essentially unchanged, whereas with the sarcoma viruses the cells are morphologically transformed and multiply with less restraint than normal cells (see Chapter 14).

## REOVIRUS

A remarkable amount of information has accumulated on the molecular biology of reovirus multiplication since the first edition of this book was published. Each of the ten double-stranded RNA molecules of the genome has been identified as a separate gene (see Chapter 3) and the corresponding mRNA's transcribed *in vitro* by the transcriptase present in the core of the virion have been translated in a mammalian cell-free system into eight of the ten viral proteins (McDowell *et al.*, 1972). The structural role of several of the seven virion proteins is now known (see Chapter 3) and using *in vitro* systems we can soon expect to learn something about the nonstructural proteins and the regulatory mechanisms that program the multiplication cycle. Equally promising are the *ts* mutants whose properties are reviewed in Chapter 7.

Reovirus multiplication has been recently reviewed by Shatkin (1971), Joklik and Zweerink (1971), and Millward and Graham (1971).

### The Multiplication Cycle

Most of the work on this genus has been done with reovirus type 3 growing in suspensions of L cells. After infection at high multiplicity, virus begins to appear intracellularly after a latent period of 6 hours and maximum yields of just over 1000 pfu per cell (20,000 virions) are reached by 16–18 hours (Silverstein and Dales, 1968). Since actinomycin D in moderate doses does not block multiplication it can be used to depress cellular protein synthesis (Gaunt and Graham, 1969; Shatkin, 1969). Multiplication is restricted to the cytoplasm and is often associated with spindle tubules (Rhim *et al.*, 1962; Dales, 1963b). Characteristic eosinophilic crescentic perinuclear inclusions are seen by conventional staining methods; these contain both single- and double-stranded RNA's, capsid proteins, and virions (Gomatos *et al.*, 1962; Rhim *et al.*, 1962; Dales *et al.*, 1965b; Mayor and Jordan, 1965; Silverstein and Dales, 1968).

### Initiation of Infection

The mechanisms of entry and uncoating of reovirions have been elucidated by Silverstein and Dales (1968). The sequence of events bears a striking similarity to that seen with the poxviruses, which also multiply in the cytoplasm and contain a transcriptase within the inner protein core.

Virions are taken into the cell by cytosis. The phagocytic vacuole then fuses with a lysosome to produce a "phagosome," in which intact virions can be clearly visualized by electron microscopy (Plate 6–3). The outer capsid is then

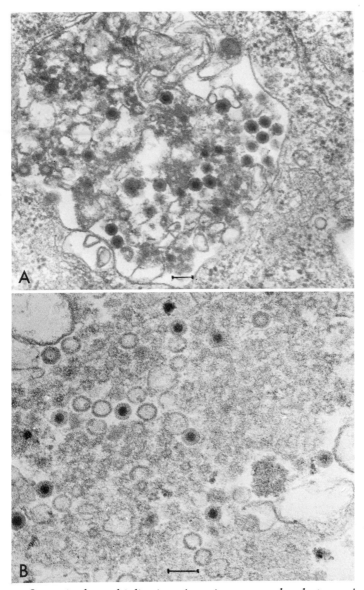

PLATE 6–3. *Stages in the multiplication of reovirus as seen by electron microscopy of thin sections of infected L cells (bars = 100 nm). (A) Lysosome from cell sampled 2 hours after infection; many inoculum particles in various stages of uncoating. (B) Assembly of virions in cytoplasmic "factory." (From Silverstein and Dales, 1968, courtesy Dr. S. Dales.)*

removed by lysosomal hydrolases, leaving a "subviral particle" of smaller diameter and higher buoyant density which has lost the outer coat proteins. The main protein of the outer capsid, $\sigma_3$, seems to be released first, then $\mu_2$ is cleaved to give a new protein of lower molecular weight and $\sigma_1$ is removed (Chang and Zweerink, 1971; Silverstein et al., 1972). Hence the subviral particle contains proteins $\lambda_1$, $\lambda_2$, $\mu_1$, $\sigma_2$, and a cleavage product of $\mu_2$; cores derived by chymotrypsin digestion of virions are similar but lack the $\mu_2$ cleavage product. This first stage of uncoating is accomplished within about an hour of infection and occurs even in the presence of cycloheximide. Indeed, it seems that most virions never become completely uncoated, but remain as subviral particles throughout the multiplication cycle (Schonberg *et al.*, 1971; Silverstein *et al.*, 1972). Removal of $\sigma_3$ is sufficient to activate the transcriptase and subviral particles produce mRNA for many hours.

## Transcription

The *in vivo* product of the transcriptase consists of ten separate and distinct classes of single-stranded RNA which can be hybridized to the ten double-stranded virion RNA's of the corresponding size classes (Shatkin *et al.*, 1968; Watanabe *et al.*, 1968b; Fig. 6–4). The precise characterization of these transcripts has been greatly facilitated by the discovery that chymotrypsin-derived cores are prolific producers of such single-stranded RNA *in vitro* (Shatkin and Sipe, 1968b).

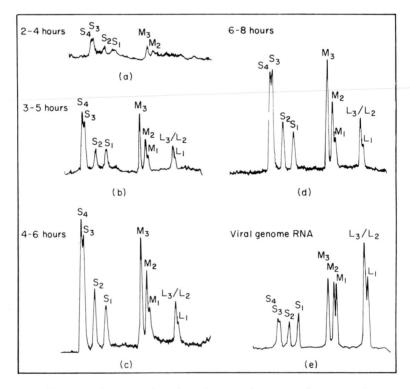

FIG. 6–4. *Tracings from acrylamide gel autoradiograms showing relative rates of formation of reovirus mRNA species during the infection cycle (a to d), as determined by hybridizing labeled RNA from the cytoplasm of infected cells to an excess of genome RNA; compared with the tracing of genome RNA derived from virions (e). (From Zweerink and Joklik, 1970.)*

The transcriptase of the core, which is activated by removal of the outer capsid, requires $Mg^{2+}$ or $Mn^{2+}$ and functions optimally at a temperature of 50°C or more, transcribing all ten single-stranded RNA's at a constant rate of at least a dozen nucleotides per second for 2 days at 37°C. There is no transcriptional control *in vitro*, i.e., the number of molecules of each class of RNA transcribed is inversely proportional to its molecular weight; the parental double-stranded RNA template is conserved (Skehel and Joklik, 1969). Similar findings were obtained by Levin *et al.* (1970b) using subviral particles from infected cells.

Gillies *et al.* (1971) have published striking electron micrographs which actually show up to ten single-stranded RNA molecules being simultaneously extruded from active chymotrypsin cores *in vitro* (Plate 6–4), the implication

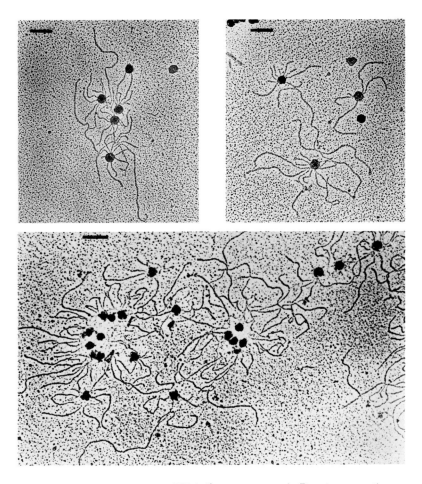

PLATE 6–4. *Reovirus messenger RNA (bars = 0.2 μm). Reovirus reaction cores prepared by chymotrypsin digestion were incubated in polymerase incubation mixture, spread by the Kleinschmidt technique, and rotary shadowed before photography (Gillies et al., 1971). (Unpublished photograph, Gillies et al., 1972, courtesy Dr. A. R. Bellamy.)*

being that each virion must contain at least ten functional transcriptase molecules. It appears that the transcripts may leave the core through the twelve
hollow capsomers at its vertices (Luftig *et al.*, 1972). Zweerink and Joklik
(1970) have suggested that the whole inner core may consist of polymeric
transcriptase molecules constructed of the polypeptides $\lambda_1$, $\lambda_2$, and $\sigma_2$.

All ten single-stranded RNA transcripts have the same polarity, as indicated
by the fact that there is no self-annealing of the product; it can be concluded
that only one strand of the double-stranded template is transcribed (Schonberg
*et al.*, 1971; Sakuma and Watanabe, 1971). Recently, McDowell *et al.* (1972)
have formally demonstrated that the transcripts represent monocistronic mRNA
molecules. In a cell-free extract from rabbit reticulocytes, single-stranded RNA's
transcribed from chymotrypsin cores *in vitro* were translated into all eight of
the known reovirus-coded polypeptides.

All classes of reovirus mRNA have (p)ppG at their 5' end (Levin *et al.*,
1970a) and cytosine at the 3' terminus (Banerjee *et al.*, 1971). Nichols *et al.*
(1972b) have determined the twenty-five nucleotides from the 5' end of mRNA
of the *s* class and found that the first nine, (p)ppGCCAUUUUU, correspond with
the sequence known to occur in the untranslated region at the 5' ends and the
intercistronic regions of certain RNA phages. They postulate that it represents
the ribosome attachment site, the polymerase recognition site, or the binding
site for a linker. The initiation codon was not present in the first twenty-five
residues, hence, must be even further in. Levin *et al.* (1972) have presented
evidence for the existence of an AUG initiator codon somewhere in reovirus
mRNA of all three size classes; Met-tRNA$_F$, the usual initiator of protein
synthesis in prokaryotic and eukaryotic systems, was found to form a complex
with reoviral mRNA at a puromycin reactive site on mouse fibroblast L-929
ribosomes *in vitro*.

Nichols *et al.* (1972a) propose that the 3000 or so oligonucleotides found in
the average reovirion may be the product of abortive transcription. The 5'-G-
terminated oligonucleotides resemble the sequence at the 5' end of reovirus
mRNA; the authors postulate that for the short time during which the subviral
particle is acquiring its outer capsid as the final step in the assembly process, the
transcriptase may continue to initiate successfully but be physically constrained
from completing transcription; they calculate that the entire oligoucleotide
complement of the virion could be transcribed in less than 3 minutes. The other
major species of oligonucleotide in the reovirion, polyadenylate (Bellamy and
Joklik, 1967; Shatkin and Sipe, 1968a), could arise by reiterative copying of the
short sequences of U known to occur near the 5' terminus of reovirus *s* mRNA
(Nichols *et al.*, 1972a, b).

## Translation

By 4 hours after infection, acrylamide gel electrophorograms of infected cells
resolve all seven structural reovirus proteins (Loh and Shatkin, 1968; R. Smith
*et al.*, 1969), one of which, $\mu_2$, is a cleavage product of $\mu_1$ (Zweerink *et al.*,
1971). In addition, Zweerink *et al.* have detected two nonstructural proteins,

$\mu_0$ and $\sigma_{2A}$, making a total of eight primary gene products (Fig. 3–10); two additional nonstructural proteins must be made in trace amounts or comigrate with others on the gel, because all ten species of mRNA are recoverable from polysomes (Ward *et al.*, 1972). The nine virus-specific proteins described are those identifiable in L cells infected at 31°C; at 37°C three others arise late in the cycle as a result of proteolytic cleavage (perhaps an artifact) of $\mu_1$ or $\mu_2$ (Zweerink *et al.*, 1971).

In vitro L cell ribosomes using endogenous reovirus mRNA make the same eight primary gene products, but there is no cleavage of $\mu_1 \rightarrow \mu_2$, or of $\mu_1$ or $\mu_2 \rightarrow$ XYZ (McDowell and Joklik, 1971).

## Replication of Viral RNA

The template for the synthesis of new double-stranded RNA *in vivo* or *in vitro* is single-stranded mRNA (Schonberg *et al.*, 1971). In other words, single-stranded "+" RNA serves as the template both for the translation of viral proteins and for the replication of viral RNA. Following a short pulse with radioactive uridine all the label is found in the "−" strand of double-stranded product. Whereas single "+" strands (mRNA) are transcribed from double-stranded RNA templates and accumulate in the cytoplasm during the first 8 hours after infection, single "−" strands are never found in a free state *in vivo* or *in vitro* but exclusively in double-stranded RNA.

Obviously an enzyme other than the virion transcriptase is required to catalyze the production of these double-stranded molecules from single-stranded mRNA. Such an enzyme ("replicase") making double-stranded RNA of all size classes, was recovered from L cells by Watanabe *et al.* (1968a); it was first detectable 4 hours after infection and reached maximum levels by 8 hours. Replicase and transcriptase activities, still template-bound, have been isolated separately from infected cells (Sakuma and Watanabe, 1971). However, the authors have pointed out that the two enzymes are not necessarily completely different; the replicase could conceivably be converted into the transcriptase when the enzyme–template complex associates with additional protein to form the subviral particle.

## Assembly

Virions are assembled in cytoplasmic "factories" in perinuclear locations often in association with spindle tubules (Dales *et al.*, 1965b; Mayor and Jordan, 1965; Silverstein and Dales, 1968), though this association is not essential because colchicine does not inhibit reovirus growth (Dales, 1963b). The inclusions are seen by electron microscopy to contain crystalline aggregates (and single particles) of virions, empty virions, and subviral particles within a matrix of fine filaments (Plate 6–3). Virions are released from the cell either singly or in such aggregates, over several hours, without budding.

Subviral particles are intermediates in the assembly of virions. Astell *et al.* (1972) have constructed virions (as identified by electron microscopy and buoyant density) by mixing subviral particles with soluble proteins from infected cells. The main outer capsid protein, $\sigma_3$, together with small amounts of $\mu$ and $\sigma_1$, were specifically the proteins selected from the pool. Transcriptase activity ceased with the attachment of $\sigma_3$ and could be activated again by removal of the same protein with chymotrypsin. Cleavage of $\mu_1 \rightarrow \mu_2$ may also represent a late step in morphogenesis (Zweerink et al., 1971).

Some of the ts mutants of Fields et al. (1971) and Ikegami and Gomatos (1972) accumulate subviral particles, whereas others accumulate empty capsids at the nonpermissive temperature. If these represent true "assembly" mutants they could provide rapid insight into the morphogenesis of the reovirion (review, Joklik and Zweerink, 1971).

Serial transfer of reovirus at high multiplicity yields noninfectious progeny after six to eight passages (Nonoyama et al,. 1970). The largest double-stranded RNA molecule is missing from these defective virions.

### Regulation of Viral Gene Expression

Although neither transcription nor translation is regulated in vitro, both appear to be controlled in vivo. Watanabe et al. (1968b) reported that if double-stranded RNA synthesis is inhibited by cycloheximide added at the time of infection, only three or four of the mRNA species are transcribed, whereas in the absence of cycloheximide, Zweerink and Joklik (1970) found all ten species of mRNA to be transcribed in vivo at all times between 2 and 8 hours after infection in a molecular ratio that remained constant throughout the cycle. There was a tendency for the number of molecules of each species transcribed to be inversely proportional to its molecular weight, mRNA of the l class being particularly scarce and s relatively plentiful, but there were several exceptions; e.g., s1 and s2 were transcribed only one-half as often as s3 and s4. Translation seemed to be independently controlled; five viral proteins were synthesized in abundance and others, notably $\sigma_1$, in very small amounts. With time the molecular ratio changed relatively little but those proteins comprising the outer capsid tended to be made in greater proportion later in the cycle.

Replication of RNA begins as early as 3 hours after infection but is predominantly confined to the later part of the cycle. Most of the RNA synthesized for the first 8 hours after infection is single-stranded (+) mRNA; virions completed after that time contain double-stranded RNA labeled mainly in the "−" strand (Acs et al., 1971).

### Alterations of Cellular Metabolism

Cellular macromolecular synthesis is not significantly affected until relatively late in the cycle. Host RNA synthesis is suppressed at high multiplicity of infection. Host protein synthesis is unaffected for the first 7 hours, and even by 9 hours has declined only 50% (Ensminger and Tamm, 1969; Zweerink and Joklik, 1970). The kinetics of shutdown of cellular DNA synthesis is very similar (Ensminger and Tamm, 1969) and results from a reduction in the number of initiation sites, not in the rate of chain growth (Hand and Tamm, 1972).

## ORBIVIRUS

Bluetongue, the prototype of this genus (reviews, Verwoerd, 1970; Howell and Verwoerd, 1971) multiplies in BHK21 or L cells with a latent period of 4 hours, reaching maximum titers after 12 hours. Unlike the reoviruses, bluetongue shuts down cell protein synthesis quite rapidly (Huismans, 1971).

Bluetongue virus shows the same correspondence of mRNA's and genome fragments as *Reovirus* (see Chapter 3). The ten genome segments are transcribed into ten species of mRNA whose molar ratios after *in vitro* or *in vivo* synthesis, and as isolated from polyribosomes, indicate that there is some specific regulation at the transcription level but probably not during the initial stages of translation (Huismans and Verwoerd, 1972). Further details of the molecular biology of viral multiplication have yet to be worked out.

Electron microscopic studies of the growth cycles of several orbiviruses in cell culture and mouse brain indicate maturation within "granular" or "reticular" cytoplasmic inclusions containing many filaments and tubules; release occurs by cell lysis, not by budding (Murphy *et al.*, 1971a).

CHAPTER 7

# Viral Genetics

## INTRODUCTION

Molecular genetics, the most illuminating branch of molecular biology, developed from and continues to be concentrated upon the intensive study of one bacterium, *Escherichia coli*, and the numerous viruses that parasitize it (Cairns *et al.*, 1966; Hayes, 1968; Stent, 1971). Genetic studies of animal viruses have lagged far behind, principally because of the much greater difficulty of successfully manipulating the diploid cells of vertebrates than the haploid cells of *E. coli*. Effective genetic studies of animal viruses date from Dulbecco's introduction of plaque assays in vertebrate cells (see Chapter 2), and they are now expanding under the triple impacts of improved cell culture technology, more versatile and precise biophysical and biochemical techniques for studying viruses and their components, and the growing use of conditional lethal mutants.

In this chapter we shall examine two facets of animal virology: mutations in viral genomes and interactions between viral genetic material or its products in mixedly infected cells. Interactions between viral and cellular genomes are discussed in Chapters 13 and 14, which are concerned with oncogenic viruses.

Interactions between viruses are not just the result of laboratory manipulations, for multiple infections of vertebrate cells are probably more common, both in the laboratory and in nature, than are single infections. Many, if not all cells, in the intact animal and in cell cultures, may be infected with noncytocidal viruses, either free or as provirus, so that superinfection of cells with different viruses (i.e., mixed infections) is very common. Likewise, spread of infection from a cell which may have been infected with a single viral particle usually involves multiple infection of the neighboring cells, in solid organs like the liver, on mucous membranes, and in cell monolayers. The effects of such multiple infections depend upon whether the same virus or different viruses or mutants are involved.

Simultaneous multiple infection with several infectious particles of the same virus may shorten the latent period and reduce asynchrony; this is important in biochemical studies of viral replication. Because of the low efficiency of plating of most animal viruses (see Chapter 2), most of the particles involved in multiple infections do not succeed in establishing infection in their own right; some of these particles are noninfectious, others are potentially infectious but, because of chance factors, fail to initiate infection. However, such particles may participate in genetic interactions, and they may reduce the yield of infective virus either by interference (see Chapter 8) or by the production of "incomplete" virus. Infection of the same cell by different viruses may be followed by (a) their independent replication, (b) interference with the growth of either or both viruses, (c) complementation or enhancement of the growth of either or both viruses, (d) genetic recombination, (e) phenotypic mixing, or (f) various combinations of these interactions.

There is as yet no agreed terminology for interactions between viruses. We shall use "recombination" as a general term to cover all cases of exchange of genetic material between viruses whether this operates at the intramolecular level or by reassortment of separate molecules of nucleic acid; and "complementation" to include all cases of enhanced yield of either or both viruses, in mixedly infected cells.

## NUMBERS OF GENES IN VIRUSES

All the DNA viruses and several groups of RNA viruses have genomes consisting of a single molecule of nucleic acid, but the genomes of some RNA viruses consist of several separate pieces of RNA (Chapter 3). If a gene is defined as the nucleotide sequence that specifies a polypeptide, it is possible to estimate the numbers of genes in different viruses from our knowledge of the amount of viral nucleic acid in the infectious particle (Table 7–1). Several assumptions of varying reliability are involved in making such estimates: (1) that the genetic code for viruses, as for organisms, is a nonoverlapping triplet code, an assumption that there is no reason to doubt, (2) that there are no substantial repetitive sequences in viral nucleic acids, and (3) that all the viral nucleic acid codes for polypeptides. [An earlier suggestion (Subak-Sharpe and Hay, 1965) that herpes simplex virus codes for some tRNA's has now been withdrawn (Bell et al., 1971)]. (4) That the "average" viral polypeptide consists of about 500 amino acids and has a molecular weight of about 50,000 daltons. [This figure accords more nearly with observation (55,000 for 70 viral polypeptides, including glycoproteins: data from Chapter 3) than the figure of 20,000 daltons previously suggested (Fenner, 1968)]. (5) That each viral polypeptide is specified by a unique nucleotide sequence. The possibility that ambiguous posttranslational cleavage may occur with some RNA viruses (Cooper et al., 1970b) and that the structural proteins of some DNA viruses may consist of precursor proteins and polypeptides derived from them (Katz and Moss, 1970a, b) has raised doubts about the universal applicability of the assumption that each polypeptide is specified by a unique nucleotide sequence. Further, there is a larger semantic problem; since picornavirus RNA, acting as mRNA, is translated as a single giant protein that is subsequently cleaved (see Chapter 6) is there only one gene in such viruses, or should the cleavage products be regarded each as the product of a separate gene? We have adopted the latter convention, which makes more sense when discussing gene function.

TABLE 7–1

*Comparison of the Numbers of Genes in Animal Viruses* [a]

| | | NUCLEIC ACID | | | |
| | | ESTIMATED MOLECULAR WEIGHT ($\times 10^6$) | TYPE AND CONFIG- URATION [b] | CALCULATED NUMBER OF GENES | POLYPEPTIDES IN VIRION [c] |
| GENUS | VIRUS | | | | |
|---|---|---|---|---|---|
| *Parvovirus* | Minute virus of mice | 1.8 | SS-D | 3 | 3 |
| *Polyomavirus* | Polyoma | 3.0 | DS-D | 3 | 3 |
| *Adenovirus* | Adenovirus type 2 | 23 | DS-D | 23 | 5–9 |
| *Herpesvirus* | Herpes simplex | 100 | DS-D | 100 | 12–24 |
| *Iridovirus* | Frog virus 3 | 130 | DS-D | 130 | 15 |
| *Poxvirus* | Vaccinia | 160 | DS-D | 160 | 30 |
| *Enterovirus* | Poliovirus | 2.6 | SS-R | 5 | 4 |
| *Alphavirus* | Sindbis | 4 | SS-R | 8 | 3 |
| *Orthomyxovirus* | Influenza A | 5 | SS-R | 10(7) [d] | 7 |
| *Paramyxovirus* | Newcastle disease | 7 | SS-R | 14 | 6 |
| *Leukovirus* | Rous sarcoma | 12 | SS-R | 24 | 10 |
| *Rhabdovirus* | Vesicular stomatitis | 4 | SS-R | 8 | 5 |
| *Reovirus* | Reovirus 3 | 15 | DS-R | 15(10) [d] | 7 |
| *Orbivirus* | Bluetongue | 15 | DS-R | 15(10) [d] | 7 |

[a] Numbers calculated from the estimated molecular weights of their nucleic acids, on the assumption of a nonoverlapping triplet code with genes containing about 1500 nucleotides or nucleotide pairs.
[b] DS, double-stranded; SS, single-stranded; D, DNA; R, RNA.
[c] Data from Chapter 3; does not include virion-associated enzymes.
[d] Figure in parentheses indicates number of separate fragments in genome.

Even allowing for the fact that gel electrophoresis fails to resolve all the polypeptides of the complex large DNA virions, it is clear that structural polypeptides account for only a minority of the genes of these viruses. The remainder presumably code for enzymes involved in DNA replication, regulating proteins, and in case of the poxviruses, perhaps even enzymes concerned with the synthesis of viral polysaccharides and lipids.

In viruses with fragmented genomes (*Orthomyxovirus* and *Reovirus*), each molecule of RNA represents a single gene (see Chapter 6). There is a reasonably close agreement between the number of genome fragments and the calculated number of genes. At least half the genome of most RNA viruses appears to specify structural proteins.

## TECHNICAL METHODS

In spite of the large amount of effort that has been devoted to animal virus genetics since 1950, little more has been achieved in the way of genetic mapping than was accomplished with bacteriophages within 2 years of the first demonstration of recombination between phage mutants. The principal reason for this slow process is the unsatisfactory nature of the systems used, a difficulty that has been to some extent self-imposed, in that the viruses first chosen for intensive study were selected because of their importance in human and veterinary medicine rather than for their suitability for genetic analysis. A large part of the difficulty is inherent, residing in the nature of animal cells. The host cells for bacteriophage studies are haploid single cells and their genetic constitution can be readily manipulated by the experimenter. Although animal cells can be cloned and the genetics of somatic animal cells is a developing branch of science, the much greater technical difficulties of handling them, and the fact that they are diploid and contain so much more genetic material than bacteria, have meant that animal cells cannot be readily manipulated genetically.

There are also technical difficulties associated with the actual initiation of infection of animal cells with viruses, which do not occur in the bacterial cell–bacteriophage systems. Elution of virus from animal cells often occurs and the low efficiency of plating (usually <0.1) means that over 90% of the physically recognizable viral particles fail to establish infection. Under conditions of multiple infection, such as are employed in experiments on genetic interactions, all cells receive several of these noninfectious particles as well as one or more infectious particles. The contribution of these noninfectious particles to interactions between viruses varies with different systems, but adds to the difficulty of accurate quantitation. With influenza virus, the infection of cells at high multiplicities leads to the von Magnus effect, i.e., the production of a large majority of particles that lack a complete genome (see Chapters 3 and 6); a rather similar phenomenon has now been recognized with several other viruses (see below). In other cases, infections at high multiplicities lead to "toxic" changes in the cells (Chapter 9), which may interfere with the uptake of virus and possibly with viral synthesis. Active or inactive viruses, and even certain viral components, may cause interference and thus alter the yields from infected cells (Schlesinger, 1959).

Nevertheless, mapping by recombination analysis has been achieved with poliovirus and herpesvirus (see below), and is technically possible with adenovirus and vaccinia virus. There are many genera for which intramolecular recombination has yet to be demonstrated. For some of these the best hopes for success in mapping probably lie with the exploitation of more direct biochemical approaches such as the use of inhibitors like pactamycin (Rekosh, 1972) and the identification of the function of each of the separate genome fragments of viruses like influenza virus and reovirus (McDowell et al., 1972).

The introduction of minicultures (Sambrook et al., 1966) and the adaptation of replica-plating techniques to such cultures (Robb and Martin, 1970) should accelerate genetic studies of viruses with which these systems can be used. Stephenson et al. (1972) have reported further improvements in both techniques. Enrichment procedures, e.g., isatin-$\beta$-thiosemicarbazone (or rifampicin) with poxviruses (Sambrook et al., 1966), and the plaque enlargement method, originally used for bacteriophage T4 (Edgar and Lielausis, 1964) and used in selecting ts mutants of poliovirus (Cooper et al., 1966) and reovirus (Fields and Joklik, 1969) are worth further investigation as ways of accelerating the selection of ts mutants, or improving the yield of "early function" ts mutants.

## HOST-CONTROLLED MODIFICATION

Mutations are undoubtedly the most important cause of phenotypic varia-tion, but phenotypic nonheritable changes in viruses (host-controlled modifica-tion) can be brought about by growing the viruses in particular hosts.

All the RNA viruses except Picornaviridae, *Reovirus*, and *Orbivirus* mature by acquiring an envelope as they bud through the cytoplasmic membranes of the host cell. Such envelopes always include host cell components, mainly lipids, and the carbohydrate moieties of cellular glycolipids and viral glycoproteins. Host-cell modification can occur by the inclusion of any of these three cellular components in the envelope. Thus, Newcastle disease virus grown in chick em-bryos contain blood group substances and Forssman antigen in the glycolipid of its envelope (Klenk *et al.*, 1972), but these are lost after a single passage of the virus in duck eggs (Rott *et al.*, 1966). Simpson and Hauser (1966) showed that host-cell modification in the susceptibility of myxoviruses to lipases was due to the nature of the lipids, which may also be responsible for differences in particle density of Newcastle disease virus when grown in different cells (Sten-back and Durand, 1963) and difference in resistance to inactivating agents (Drake and Lay, 1962).

The neutralization of chick embryo-grown influenza virus by antisera raised against the carbohydrate of normal chick allantoic cell membranes (Laver and Webster, 1966) is an example of host-controlled modification of the glycoprotein comprising the peplomers of the viral envelope. Host-controlled modification of T-even bacteriophages is related to the glucosylation of their unique base, hydroxymethylcytosine, which may be affected by mutations in certain viral genes or by genes in their bacterial hosts (review, Revel and Luria, 1970). With other bacteriophages (e.g., λ and fd), the "restriction" in host range is due to changes in the viral molecules associated with multiplication in a par-ticular host that do not alter their genetic message but render the molecules sensitive to a limited endonucleolytic attack when they enter the restrictive bacterial host (review, Boyer, 1971). Nothing comparable to this has been rec-ognized among animal viruses.

## MUTATION

### Terminology

Papers dealing with animal viruses make extensive and rather indiscriminate use of the words "strain," "type," "variant," and "mutant" to describe particular preparations of virus. The term "wild type" originated from the work of animal and plant breeders, who used it to describe the original type of whatever animal or plant they used. In genetics, it has come to have a narrower meaning, and is used to designate the parental strain from which a particular set of mutants is derived. In many circumstances, this parental strain is itself a mutant of the original "wild" virus as it occurs in nature, but the meaning of wild type in a genetical sense is usually quite clear.

The use of the terms "species," "type," "strain," and "variant" in the literature is confused by the fact that different conventions about the "rank" of serologically distinguishable viruses have developed among workers specializing in the study of different genera. For example, arbovirologists have tended to name new species every time a virus has been isolated from a new ecosystem, often to find later that it is serologically close to a known agent. On the other hand, some of the "types" of Picornaviridae share no detectable antigens and should perhaps be regarded as separate species, as should influenza virus "types" A and B. Perhaps the International Committee on the Nomenclature of Viruses will eventually attempt to bring some order to the present chaotic situation. Workers in the influenza virus field have recently rationalized methods of designating influenza A viruses (World Health Organ., 1971b).

In laboratory manipulations, variations are frequently observed, and the viruses which produce such deviations from the wild-type strain are often called "variants." In the past, these have usually been recognized only after several serial passages of the original viral stock and the accumulation of an unknown number and variety of mutations. A further difficulty arose with RNA tumor viruses, for their serial passage in mice or chickens led to their contamination with a number of related viruses, e.g., no fewer than five genetically different avian leukoviruses have been recovered from the original Bryan high titer strain of Rous sarcoma virus. Most modern work in viral genetics hinges on the use of mutants which differ from the wild type by one or a small number of defined mutational steps.

## Spontaneous Mutations

Some viruses appear to yield a very high proportion of mutants that have similar phenotypes. For example, all poxviruses of the vaccinia subgroup that produce red ($u+$) pocks on the chorioallantoic membrane were found to produce white ($u$) pocks with a frequency as high at 1% (Downie and Haddock, 1952; Fenner, 1958; Gemmell and Fenner, 1960). With rabbitpox virus, these $u$ mutants were shown by recombination tests and by biological characterization to be different in genotype as well as in phenotypic behavior in host systems other than the chorioallantoic membrane (Gemmell and Fenner, 1960). Spontaneous $ts$ mutants are quite common with some viruses (e.g., 2.3% in vesicular stomatitis virus; Flamand, 1970), but, of course, they include mutations in several different genes. Although cloning by single particle infections of single cells should reduce genetic heterogeneity it is sometimes difficult to build up high-titer genetically homogeneous stocks (Walen, 1962).

The essential character of a spontaneous mutational change in a virus is its random occurrence during viral replication, which may be demonstrated by the random distribution of mutant clones among the yields of individual cells (Luria, 1951). Analysis of plaque mutants of poliovirus (Dulbecco and Vogt, 1958), encephalomyocarditis virus (Breeze and Subak-Sharpe, 1967), and the $morph' \rightarrow morph^I$ mutation of Rous sarcoma virus (Temin, 1961) showed that in each case the mutants were clonally distributed. Calculations of the mutation rates, which have been made in only a few cases and only with riboviruses, have

## TABLE 7–2

*Mutation Rates Observed with Some Animal Viruses*

| VIRUS | CHARACTER STUDIED | MUTATION | MUTATION RATE ($-\text{LOG}_{10}$) |
|---|---|---|---|
| Poliovirus | Reduced ability to plaque in acid medium [a] | $d \longrightarrow d^+$ | 4.5–4.7 |
| | Plaque size [b] | $m \longrightarrow m^+$ | 6.5 |
| | Ability to replicate at 40°C [c] | $rct/40^- \longrightarrow rct/40^+$ | 4.7–5.7 |
| | Resistance to guanidine [c] | $g \longrightarrow g^+$ | 5.4–7.2 |
| Encephalomyocarditis virus | Plaque size [d] | $r \longrightarrow r^+$ | 5.0 |
| | | $r^+ \longrightarrow r$ | 3.7–4.0 |
| Influenza A1 and influenza B | Resistance to β-inhibition [e] | $s \longrightarrow r$ | 7.5–8.2 |

[a] Dulbecco and Vogt (1958).
[b] Takemori and Nomura (1960).
[c] Carp (1963).
[d] Breeze and Subak-Sharpe (1967).
[e] Medill-Brown and Briody (1955).

given widely different results (Table 7–2). Some of the mutation rates found (e.g., those of the mutations affecting plaque size in encephalomyocarditis virus) were extremely high for both the primary mutation and for reversion. However, it is not known in this case whether the phenotypic change is associated with mutations in a single gene.

### Induced Mutations

Mutations can be induced in viruses, or their infectious nucleic acids, by physical and chemical treatments. The mechanisms by which these act have been reviewed by Freese (1963) using data derived from experiments with bacteriophages, bacteria, and tobacco mosaic virus. Nitrous acid, hydroxylamine and ethylmethyl sulfonate act *in vitro* and have been used as mutagens for both DNA and RNA viruses. Nitrosoguanidine is an effective mutagen for RNA viruses *in vitro*, but with DNA viruses acts better, and at much lower concentrations, on replicating virus (Tegtmeyer *et al.*, 1970). 5-Bromodeoxyuridine and 5-fluorouracil act by producing base substitutions in replicating DNA or RNA, respectively. Ther use is, therefore, restricted to either DNA or RNA viruses, with the interesting exception, explicable on the basis of the DNA provirus hypothesis (see Chapter 13) that 5-bromodeoxyuridine is a mutagen for Rous sarcoma virus (Bader and Brown, 1971).

To obtain induced mutants, suspensions of virus (or sometimes their infectious nucleic acids) are exposed to the appropriate chemical mutagen, or the drugs are added to cultured cells supporting viral replication. After treatment, which results in a substantial reduction in infectivity or a reduced yield, an increased proportion of mutants is found in the survivors compared with the original population. Control experiments show that the mutant fraction cannot have arisen by selection from the original stock during exposure to the mutagen or at the time of plating.

## KINDS OF MUTANTS OF ANIMAL VIRUSES

A vast number of mutants of animal viruses has been described. Many workers, using a large variety of different viruses, have described mutant phenotypes, but few have gone beyond this stage to utilize these mutants for any sort of genetic investigation, mainly because of the technical problems already outlined. A tabulation of the types of mutants found among viruses of vertebrates was presented by Fenner and Sambrook (1964). They include plaque- and pock-type mutants, host-dependent and temperature-dependent mutants, mutants lacking the capacity to induce thymidine kinase synthesis, mutants which are resistant to (or dependent on) certain chemicals like guanidine, mutants which differ in the temperature-sensitivity of their infectivity or their enzymatic (neuraminidase) activity, mutants that differ in the antigenic properties of their coat proteins, mutants which differ in the resistance to a variety of inhibitors (in horse and bovine serum, in agar, etc.) of their capacity to produce plaques, and many others.

### Characters Used for Genetic Studies

Good phenotypic characters, which are a prerequisite for genetic studies, are provided by mutants that have phenotypic expression with complete penetrance; they should be easily scored, and they should result from single mutations and be reasonably stable.

Almost all the work on recombination up to 1959 was carried out with different natural or laboratory strains of influenza or vaccinia virus. A large number of different characters was utilized: serological behavior and thermostability of the hemagglutinin, the inhibition of hemagglutination (by heated virus) by different mucoids, and pathogenicity for the chick embryo, the mouse lung, and the mouse brain. Several of these characters are of doubtful value because they are either polygenic or covariant.

After the introduction of the plaque technique to animal virology, Dulbecco and his associates began work on the genetics of poliovirus. Although these pioneering studies were very elegant, we now recognize that they could not have achieved their purpose of providing a complete genetic map of poliovirus because most of the genetic characters of poliovirus studied were concerned with phenotypic changes associated with the coat protein, and most showed a high degree of covariation. The markers for resistance to horse and bovine serum inhibitors (*ho* and *bo*) were independent of seven others used, but variation in any one of the latter was usually but not always associated with a change in others of that group (Dulbecco, 1961). Much early work on recombination was rendered fruitless by the unwitting employment of covariant instead of independent characters, but if their nature is appreciated covariation may simplify scoring and thus be of considerable practical value, e.g., in selecting for attenuation. The unequivocal demonstration of recombination with poliovirus was not possible until noncovariant characters (*ho, bo,* and *g*) were used (Hirst, 1962; Ledinko, 1963).

Two approaches have been used in the genetic analysis of viruses by recombination or complementation. All the early work involved selection for phenotypic changes that reflected a mutation in the particular gene responsible for that function. Later work, initiated by Campbell (1961) with bacteriophage λ and by Epstein *et al.* (1963) with phage T4, made use of *conditional lethal mutations,* in which one selects for a phenotypic change of such a general nature that it could result from a mutation in any one of a wide variety of different genes. The screening tests for the two commonly used conditional lethal mutations are based on the fact that the mutation registers as lethal if one attempts to grow the mutant in a particular host (*host-dependent mutants*) or at a certain (elevated) temperature (*temperature-sensitive mutants*), but is harmless if the mutant is grown in a permissive host or at a permissive temperature.

With bacteriophages, both types of conditional lethal mutant have been used: temperature sensitivity of viral development (*ts*) and susceptibility to host-cell suppressors (*amber* and *ochre*); with animal viruses only *ts* mutants have been used. The use of conditional lethal mutants has increased greatly during recent years and they form the basis of almost all current genetic studies of animal viruses.

### Plaque Mutants

Various aspects of plaque mutants have been reviewed by Takemoto (1966). Many plaque size mutants are either resistant or sensitive to polysaccharide inhibitors in agar, and it is this property which determines the mutant phenotype (Liebhaber and Takemoto, 1963). Various other reasons for the variation in plaque morphology have been proposed. Thus, the large-plaque mutants of adenovirus (Kjellén, 1963), vesicular exanthema of swine virus (McClain and Hackett, 1959), and measles virus (Rapp, 1964) are released from the host cells more rapidly, and, in the latter two cases, more completely than the small-plaque mutants. A small-plaque mutant of Western equine encephalitis virus is more dependent upon bicarbonate and folic acid for growth and is also more thermolabile than the large-plaque type (Quersin-Thiry, 1961). The large-plaque mutant of foot-and-mouth disease virus is more acid stable (Mussgay, 1959). Some temperature-sensitive mutants are also mutant in plaque morphology when grown under permissive conditions, and vice versa (Takemoto and Martin, 1970). Likewise, most *cys* (cysteine dependence) and structural protein *ts* mutations in poliovirus are covariant (Cooper *et al.,* 1970c). If double mutation can be excluded, such covariation is useful both for the selection of mutants and for their easy recognition in mixed infection experiments.

### Cold-Adapted Mutants

Most attenuated strains of virus currently in use in human and veterinary medicine were obtained by serial passage of the wild-type virus in a novel host (formerly embryonated eggs, now cultured cells) and often at low temperatures of incubation (e.g., poliovirus, Sabin, 1961; measles virus, Schwarz, 1964; rubella virus, Kenny *et al.* 1969), for it had been observed that such procedures usually, but not always (e.g., Van Kirk *et al.*, 1971), led to attenuation of virulence for the relevant host animal. Such "cold-adapted" mutants often have a *ts* phenotype, i.e., they are unable to grow at an elevated temperature, but exceptions have been found (Cooper, 1964), and often the cold-adapted mutants grow better than wild type at reduced temperatures. Continued serial passage at high concentrations, the procedure followed in obtaining these mutants, undoubtedly selects strains that are mutated in many sites.

## Host-Dependent Conditional Lethal Mutants

So far no animal cells have been discovered which are comparable to the *su*, *su*⁺ mutants of *E. coli*, that have proved so valuable in the analysis of host-dependent conditional lethal mutants of bacteriophages. Host-dependent mutants of several animal viruses have been recognized, but they do not seem to be comparable to their bacteriophage counterparts, in which novel tRNA's found in the permissive strain of bacterium see the nonsense mutation in the phage mRNA, not as a chain termination codon, but as a codon that can be translated into a normal amino acid. The basis for the host-dependent phenotype has been recognized only in the special case of polyoma virus mutants that will undergo productive infection only in cells transformed by polyoma virus (Benjamin, 1970); otherwise such mutants have been little exploited.

**PK-Negative Mutants of Rabbitpox Virus.** Many *u* (white pock) mutants of rabbitpox virus fail to grow in pig kidney (PK) cells (McClain, 1965), and two that were studied in detail were also defective in L cells and HeLa cells (Sambrook *et al.*, 1965). A total of 34 PK-negative mutants, obtained from rabbitpox virus mutagenized with bromodeoxyuridine (Sambrook *et al.*, 1966) and from *u* mutants that had been selected for white pock phenotype on the chorioallantoic membrane (Gemmell and Fenner, 1960), were examined physiologically (Fenner and Sambrook, 1966). Like *amber* mutants, they have low and reproducible reversion rates and are relatively "non leaky." All had the *u* phenotype, although many *u* mutants are not PK-negative. Study of DNA synthesis and antigen production in the restrictive cells showed that the multiplication of different mutants was blocked at stages varying from DNA synthesis to maturation. In spite of this phenotypic diversity, they did not recombine with each other, although they recombined readily with either *ts* mutants (Padgett and Tomkins, 1968) or with *u* PK-positive mutants.

**KB-Negative Mutants of Adenovirus.** Takemori *et al.* (1968, 1969) have isolated and partially characterized a number of mutants of adenovirus type 12 that give varied results when plaqued in human embryonic kidney (HEK) and KB cells. All plaqued in HEK cells; some (kb) failed to plaque in KB cells. Mutants designated *cyt* were found to produce large clear plaques in HEK cells, instead of the small fuzzy-edged plaques characteristic of the wild-type virus (*cyt*⁺). In a particular line of KB cells, in which the parental *cyt*+ virus produced typical small *kb*+ fuzzy-edged plaques, some of the *cyt* mutants produced large clear plaques (*cyt, kb*+) whereas others failed to plaque (*cyt, kb*). *kb* mutants obtained from spontaneous *cyt* mutants reverted with low frequency to *cyt*+*kb*+; ultraviolet-induced *cyt kb* did not revert. The *cyt kb* mutants produced T antigen but no hexon antigen in KB cells, in which their growth was complemented by wild-type virus but not by other adenovirus serotypes. All except one of the *cyt* mutants was of lower tumorigenicity for baby hamsters than the parental *cyt*+ virus.

**Host Range Mutants of Polyoma Virus.**   Polyoma-transformed mouse cells have an increased but not absolute resistance to reinfection with polyoma virus, probably because of the strong selection for resistance to lytic infection that accompanies their isolation. Since such cells contain the integrated polyoma virus genome (see Chapter 13), Benjamin (1970) (review, Benjamin, 1972) reasoned that they could be used to select for mutants defective in genes supplied by the integrated genome. Four such mutants were isolated, which plaqued in polyoma-transformed 3T3 cells but not in normal 3T3 cells. The restriction was not related to attachment or uncoating, since infectious DNA extracted from the mutants behaved in a similar manner. All four mutants had also lost their ability to transform rat embryo fibroblasts and BHK cells. Presumably these mutants depend for their growth on a virus-coded product that is expressed by the viral genome(s) integrated into the cellular DNA of the transformed cells. The nature and function of the product are unknown.

**Host Range Mutants of Vesicular Stomatitis Virus.**   Vesicular stomatitis virus (VSV) has a very broad host range. Simpson and Obijeski (1973a, b) discovered a class of host-restricted (*hr*) mutants of VSV by plating mutagenized virus on a mixed indicator system of permissive chick embryo fibroblasts and nonpermissive HeLa cells. In tests on twenty-eight cell lines of diverse origin, all of which were permissive for wild-type virus, some were fully permissive for *hr*; others showed varied degrees of restriction. The *hr* mutants fall into several phenotypic clones.

## Temperature-Sensitive (*ts*) Conditional Lethal Mutants

Current genetic work with animal viruses is based almost entirely on the use of temperature-sensitive mutants, whose use has increased greatly during the last few years. They are produced by missense mutations, which alter the nucleotide sequence of the wild-type virus in such a way that the resulting protein is unable to assume or maintain its correct functional configuration at the restrictive temperature. Usually the complete polypeptide specified by the mutated gene is synthesized, but its function is defective at the restrictive temperature and normal (or nearly so) at the permissive temperature. If the defective protein is a structural component, virions produced at the permissive temperature may, but need not, be less heat stable than wild-type virions; if the defective protein is not a structural protein the virion is usually no less heat stable than is the wild type. Likewise, viral enzymes synthesized by *ts* mutants under permissive conditions may be less heat stable than the wild-type enzymes [e.g., DNA polymerases of vaccinia virus (Basilico and Joklik, 1968), or T4 bacteriophage (de Waard *et al.*, 1965)]. On the other hand, the RNA polymerases of certain *ts* mutants of Semliki Forest virus, if synthesized at the permissive temperature, were as heat stable as the wild-type enzyme, but their synthesis was temperature sensitive (Martin, 1969).

Temperature-sensitive mutants have several other properties that are worth noting. Although they may be nonfunctional, the mutant proteins may have the same immunological specificity as wild-type proteins, if the mutation does not affect the antigenic determinants or the configuration of the protein that is of antigenic significance. They are usually "leaky," i.e., some functional activity is found even at the restrictive temperature. Conversely, they rarely function as effectively as wild-type virus at the permissive temperature, since temperature sensitivity is probably a continuously variable function. In addition, it may be difficult to produce large stocks of single-step *ts* mutants since revertants to *ts*$^+$, which are likely to occur with a frequency of $10^{-5}$ or higher, may overgrow the mutant even at the permissive temperature.

**Selection and Production of Stocks of *ts* Mutants.** Since *ts* mutants now occupy such an important place in viral genetics we shall describe their selection, production, and use in some detail in the next two sections. The essential prerequisite for work with *ts* mutants is a good plaque assay system. The stock virus is cloned at the restrictive temperature to eliminate spontaneous mutants that may be present in the stock. Spontaneous *ts* mutants may then be selected from plaques grown at the permissive temperature and tested at permissive and restrictive temperatures (e.g., Flamand, 1970; Hirst and Pons, 1972). More commonly, the cloned stock is treated with a mutagen, either *in vitro* or during a cycle of growth at the permissive temperature. Most of the treated virus is inactivated by this treatment. Taking precautions to ensure that the suspensions do not contain clumps, the survivors are plated at the permissive temperature and a large number of plaques are picked off the assay plates. These are then assayed at the permissive and the restrictive temperature and putative *ts* mutants are selected. The strict temperature control required is best provided by immersing the culture flasks in trays in a water bath.

It is easy to obtain *ts* mutants; indeed, the majority of the survivors of mutagen treatment may be, to some extent, temperature-sensitive (Sambrook *et al.*, 1966). It is much more difficult to select useful mutants and to build up and maintain satisfactory working stocks of them. In most circumstances, it is desirable to work with single-step mutants (as far as temperature sensitivity is concerned); these may revert during the production of working stocks. Another restriction is imposed by the leakiness of many *ts* mutants that makes them unsuitable for use in genetic analysis or physiological experiments. A further requirement is that suspensions used for mixed infection experiments must consist of well-dispersed single virions, for clumps containing genetically different mutants may give rise to misleading results. The occurrence of heteropolyploids in the yields of some enveloped viruses is almost unavoidable (Simon, 1972; and see below) and may constitute an absolute barrier to some sorts of genetical experiment.

**Use of** *ts* **Mutants in Analysis of Gene Functions.** If the *ts* function is irreversible, temperature-shift experiments can be used to determine whether the defect occurs in a gene that functions early or late. The procedure is as follows: Replicate tubes are incubated initially at the permissive temperature (in "step-up" experiments) or the restrictive temperature (in "step-down" experiments), and at intervals tubes are transferred to the restrictive and permissive temperature, respectively. After an appropriate interval all tubes are assayed for infective virus. A defective early function, e.g., defective synthesis of the viral nucleic acid, gives a good yield in step-up experiments but a poor yield in step-down experiments. The reverse is true if the temperature-sensitive defect occurs late in the infectious cycle, for example, a defect in maturation.

The next useful test is to determine the heat stability of the virion, for this will reveal many cases in which the defect involves a structural protein, or perhaps a temperature-sensitive virion-associated polymerase. With some viral genera, mapping by genetic recombination can be used to determine where the *ts* defect is located. Then the production of new viral nucleic acid at the restrictive temperature can be determined, either by autoradiography or by studying the incorporation of radioactive precursors. Finally, studies are made of viral polypeptides, by functional tests in the case of enzymes, by fluorescent-antibody staining of cells or by other immunological tests, especially if antigen-specific as well as broad anti-virion antibodies are available, and by polyacrylamide gel electrophoresis of infected cells. With oncogenic viruses an important functional test is the capacity of the mutant to induce and to maintain the transformed state. It may also be useful to determine the virulence of *ts* mutants for experimental animals. Before embarking on these functional tests it is usual to sort out the mutants by complementation tests (see below).

In the pages that follow, we set out in summary form the data available for different groups of animal viruses. The comprehensive examination of a suitably large collection of *ts* mutants clearly calls for a continued study; the only viruses subjected to any adequate investigation thus far are poliovirus (Cooper), Sindbis virus (Pfefferkorn), and polyoma virus (di Mayorca, Eckhart). Excellent systems are available for reovirus (Joklik, Fields), vesicular stomatitis virus (Pringle, Flamand), and adenovirus (Williams), and there is certain to be a substantial expansion in the use of *ts* mutants of the oncogenic viruses, especially SV40 and the leukoviruses. The very large genomes of poxviruses, iridoviruses and herpesviruses make the comprehensive study of their gene functions a considerable but rewarding task, on which Subak-Sharpe and his colleagues have now ventured with herpes simplex virus.

**Poliovirus.** Cooper (1968, 1969) found that in a linear, additive genetic map constructed by recombination experiments, poliovirus *ts* mutants that were clustered to the left-hand side of the map were defective in viral RNA synthesis, and those spread out on the right-hand side of the map appeared to have defects in virion proteins although there were also various associated (pleiotropic) defects. Further study of the RNA-defective mutants revealed two types of defect (Cooper *et al.*, 1971). One group of mutants failed to synthesize any RNA, double- or single-stranded; two mutants produced a normal amount of double-stranded RNA plus a considerable amount of low molecular weight single-stranded RNA. The first defect is probably associated with an enzyme required to make or maintain double-stranded RNA ("replicase I"), the second with a different gene function needed for the production (or maintenance) of progeny 35 S strands ("replicase II"). Experiments with foot-and-mouth disease virus (Arlinghaus and Polatnick, 1969) also suggest that two "polymerases" are needed for the replication of single-stranded viral RNA's. The difficulty of isolating other kinds of *ts* mutant, coupled with consideration of the coding capacity of the poliovirus genome (2500 amino acids), and the proportion of the map occupied by structural protein genes (about one-half, Cooper *et al.*, 1971), makes it likely that poliovirus has only three gene functions. These are (5' → 3'): structural protein/regulator, replicase I and replicase II; each gene function, however, may involve several discrete polypeptides, not necessarily all virus-coded.

Contrary to the expectation aroused by the inhibitory effect of guanidine on viral RNA synthesis, the map locus for guanidine resistance lies in a structural protein. The phenotype of the two *ts* mutants which produced double- but not single-stranded RNA appeared to be very like that of *ts*+ grown in the presence of guanidine, and that of certain other *ts* mutants whose single defect occurred in a structural protein of the virion. Mixed infection of cells with *ts*+ and the latter *ts* mutants at the restrictive temperature, or with guanidine-sensitive and -resistant mutants in the presence of guanidine, suppressed all viral RNA synthesis (Cooper *et al.*, 1970a). Cooper suggests that changes in the secondary or tertiary structure of a particular virion protein (which has not yet been identified), induced either phenotypically by guanidine treatment, or by genotypic alteration of the primary structure by certain *ts* defects, exert a controlling effect on the replication of poliovirus RNA. Recently, Cooper *et al.* (1973) have found that the virus-induced repression of host protein synthesis is also controlled by structural protein, as judged by *ts* mutants. Proteins comprising the viral polypeptides (VP0+VP1+VP3) (see Chapter 3) and corresponding to the precursors to the structural units, are found associated with ribosome precursors and with the replication complex. Such structural protein precursors may have important regulatory functions at several stages of poliovirus multiplication (see Chapter 6).

**Sindbis and Semliki Forest Viruses.** With both these alphaviruses, *ts* mutants could be classed as RNA+ or RNA− according to whether they did or did not produce viral RNA at the restrictive temperature (Burge and Pfefferkorn, 1966a; Tan *et al.*, 1969). Temperature-shift experiments with the Semliki Forest virus RNA− mutants showed that they were defective in an early function, and most of them failed to produce any species of virus-specific RNA at the restrictive temperature. Martin (1969) found that although no RNA polymerase activity developed in cells infected with some of these RNA− mutants incubated at the restrictive temperature, polymerase produced at the permissive temperature, tested *in vitro*, was active and stable at the restrictive temperature. He concluded that in these mutants the synthesis of the polymerase, rather than the enzyme molecule itself, was temperature-sensitive.

RNA$^+$ mutants of both viruses included some that were defective in either nucleocapsid formation or envelope protein synthesis (Table 7–6), and one example of an RNA$^+$ mutant that made both nucleocapsids and membrane protein at the restrictive temperature but failed to synthesize infectious virus (Burge and Pfefferkorn, 1968; Tan *et al.*, 1969; Yin and Lockhart, 1968). The hemagglutinating ability of the virions of one Sindbis RNA$^+$ mutant was reversibly lost when the hemagglutination test was carried out at 37°C instead of 25°C (Yin, 1969). The physiological defects of three RNA$^+$ mutants of Semliki Forest virus (lack of production of nucleocapsid, or production of nucleocapsid but failure of envelopment) were reflected in the electron microscopic appearance of infected cells incubated at the restrictive temperature (Tan, 1970).

**Vesicular Stomatitis Virus (VSV).**  Pringle and his colleagues have published preliminary data on the physiological characteristics of *ts* mutants of VSV. A total of 210 mutants of VSV, Indiana serotype, belonged to four nonoverlapping complementation groups (Pringle and Duncan, 1971; see Table 7–7). Physiological findings were in accord with the grouping by complementation. The mutants of three complementation groups were defective in early functions and failed to synthesize viral proteins at the restrictive temperature (Wunner and Pringle, 1972). All mutants of two of these groups were also RNA$^-$, the third group contained one RNA$^-$ and one RNA$^+$ mutant. The fourth complementation group was RNA$^+$ and synthesized all four viral proteins at both 31° and 39°C. Flamand (1970) reported similar results with seventy-five spontaneous mutants (Table 7–7). Forty-eight mutants of VSV, New Jersey serotype, that could be allocated to six complementation groups, likewise contained a majority of RNA$^-$ mutants (Pringle *et al.*, 1971).

Deutsch and Berkaloff (1971) showed that one of Flamand's RNA$^+$ mutants had a heat-sensitive virion. The envelope disintegrated at 40°C and released intact nucleocapsid. In general, heat-sensitivity of infectivity of the virion was found to vary considerably between different mutants belonging to the same complementation group, including those belonging to one of the RNA$^-$ groups (Holloway *et al.*, 1970; Pringle and Duncan, 1971). The latter finding was explicable by the demonstration that the RNA polymerase, a virion-associated enzyme, was thermolabile in some RNA$^-$ mutants (Szilágyi and Pringle, 1972).

**Reovirus.**  Fields and Joklik (1969) defined five recombination groups (designated A–E) among thirty-five *ts* mutants of reovirus type 3, and Cross and Fields (1972) have recently defined a sixth group (F) and possibly one other. These groups can also be defined by complementation tests, and they accord with the functional behavior of the mutants. Mutants of groups A, B, and F are RNA$^+$, while C, D, and E are RNA$^-$.

The RNA$^-$ mutants have been further defined by Ito and Joklik (1972a, c) and Fields *et al.* (1972). All three groups were defective in their ability to synthesize double-stranded RNA at the restrictive temperature, but were able to synthesize all ten species of mRNA, although at a reduced rate. In temperature-shift experiments, mutants of groups C and D abruptly ceased to synthesize double-stranded RNA at 39°C; mutants of group E that had begun synthesis at 31°C were unaffected. The *ts* defect in mutants of group D appears to reside in a structural polypeptide of the cores (see Chapter 3) (probably $\mu_1$, the precursor of $\mu_2$) which must therefore be involved in the synthesis of double-stranded RNA (Ito and Joklik, 1972c). Mutants of groups C and D induced normal factories within which empty viral capsids accumulated (Fields *et al.*, 1971); the failure to detect a structural anomaly in group E mutants may have been due to the fact that the temperature shift was made too late (Ito and Joklik, 1972a).

Ikegami and Gomatos (1968, 1972), have isolated a total of fourteen *ts* mutants of reovirus type 3. All had a late defect that resulted in inhibition of continued synthesis of viral cores concomitantly with an inhibition of host and viral protein synthesis, due to inhibition of translation. Six mutants that were tested showed greatly reduced virulence for baby hamsters (Ikegami and Gomatos, 1968).

**Polyoma Virus and Simian Virus 40 (SV40).**   Genetic experiments with these viruses, and with the avian leukoviruses (see below) have been focused upon their relevance to the mechanism of cellular transformation by viruses (review, Di Mayorca and Callender, 1970; Benjamin, 1972). With poloyoma virus, *ts* mutants selected for their inability to plaque at the restrictive temperature in mouse cells were examined for their capacity to transform and induce cellular DNA synthesis in mouse cells (Eckhart, 1969; Di Mayorca *et al.*, 1969), and to transform hamster cells (Fried, 1970; Eckhart, 1971); more limited studies on the isolation and physiological properties of SV40 *ts* mutants have been reported (Tegtmeyer and Ozer, 1971; Kimura and Dulbecco, 1972).

With polyoma virus five complementation groups were distinguished (Table 7–3). Mutants of groups I and IV, which could be differentiated only by complementation tests, showed no defects at the restrictive temperature other than the failure to produce infectious virus. Mutants of complementation group II, that includes the original *ts*(a) of Fried (1965, 1970), are unable to transform cells at the restrictive temperature. The mutant *ts*3 (complementation group V) seems to be restricted in a function that is involved in the maintenance of at least some facets of the transformed cell phenotype (Dulbecco and Eckhart, 1970; see Chapter 13).

**Leukovirus.**   Until recently, difficulties of cloning precluded genetic analysis of leukoviruses. Temperature-sensitive mutants have now been recovered from the Schmidt-Ruppin strain of Rous sarcoma virus (G. Martin, 1970; Kawai and Hanafusa, 1971) and the B77 avian sarcoma virus (Toyoshima and Vogt, 1969). Analysis of the *ts* defects have revealed mutants that are temperature-sensitive

TABLE 7–3

*Temperature-Sensitive Mutants of Polyoma Virus: Functional Behavior and Complementation*

| RESTRICTED FUNCTION AT 39°C | WILD TYPE | COMPLEMENTATION GROUP | | | | |
|---|---|---|---|---|---|---|
| | | I [a] | II [b] | III | IV [a] | V [c] |
| Infectious virus | + | − | − | − | − | − |
| Viral DNA | + | + | − | − | + | − |
| Transformation | + | + | − | + | + | −(+) [d] |
| T antigen | + | + | − | − | .. [e] | .. [e] |
| Stimulation of cellular DNA synthesis | + | + | + | + | + | − |

[a] Groups I and IV differentiated by complementation tests.
[b] Includes mutant *ts*(a).
[c] Noncomplementing mutants; includes *ts* 3.
[d] Transformation is temperature-dependent.
[e] No data.

in respect to cellular transformation but not virus production (G. Martin, 1970; Kawai and Hanafusa, 1971) or show coordinate temperature sensitivity in both transformation and viral multiplication (Friis *et al.*, 1971). Further analysis of the first class of mutant suggested that a virus-coded, heat-labile protein that was not a structural component of the virion, played an essential role in cell transformation.

Using a microtiter method and automated replica testing devices Stephenson *et al.* (1972) have been able to recover nine *ts* mutants of murine leukemia virus from 1000 clonal virus-producing cultures, but have not yet examined their functional defects.

## INTERACTIONS BETWEEN VIRUSES

Interactions between viruses in mixedly infected cells may involve physical integration (temporary or permanent) of parts of the viral genomes (genetic recombination, reassortment, reactivation, and heteropolyploidy), or may simply result from one virus making "temporary" use of a protein specified by the other (complementation and phenotypic mixing). Often both processes occur simultaneously in the same cell.

### Definitions

In order to clarify the discussion we shall define the terms that are used, and illustrate their nature with a simple diagram (Fig. 7–1).

**Genetic Recombination.** This term is used in a general sense to denote the exchange of nucleic acid between different parental viruses so that the progeny, called *recombinants*, contain sequences of nucleotides derived from each parent. Two varieties can be distinguished among animal viruses: *intramolecular recombination*, which involves the rearrangement of sequences within a single

A. Intramolecular recombination

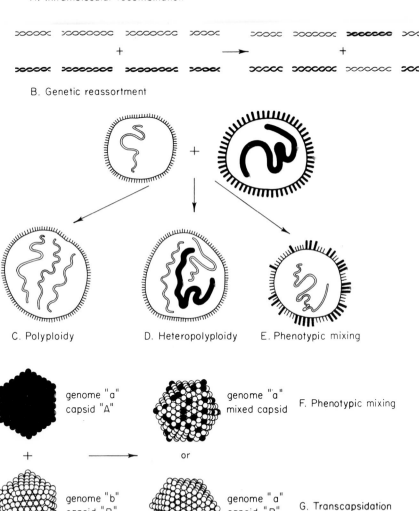

FIG. 7–1. *Genetic recombination, polyploidy, phenotypic mixing and transcapsidation. (A) and (B) Genetic recombination; (A) intramolecular recombination, according to model of Hotchkiss (1971). (B) reassortment of genome fragments, as in Reovirus and Orthomyxovirus. (C) and (D) Polyploidy; (C) polyploidy, as seen in unmixed infections with Paramyxovirus; (D) heteropolyploidy, as may occur in mixed infections with Paramyxovirus. (E)–(G) Phenotypic mixing; (E) with enveloped viruses; (F) with icosahedral capsid; (G) extreme case of transcapsidation or genomic masking.*

nucleic acid molecule (Hotchkiss, 1971), and *genetic reassortment,* in which separate molecules of nucleic acid of viruses that have fragmented genomes (influenza virus and reovirus) are exchanged, so that some of the progeny, called *reassortants,* contain genomes with nucleic acid molecules derived from each parent.

**Genetic Reactivation** (Table 7–4).   This is a special case of recombination or reassortment in which one or both of the parental viruses has been inactivated and is therefore noninfectious on its own, but contributes to the production of infectious progeny in a mixed infection. When the inactivated parental viruses are of the same strain the process is called *multiplicity reactivation;* when one parent is inactive and the other infectious the process is called *cross-reactivation* or *marker rescue.*

**Polyploidy.**   This refers to the enclosure within a single envelope of more than one complete nucleocapsid. If the enclosed nucleocapsids come from genetically different parental viruses the resulting particles are *heteropolyploids* (or heteroploids, Simon, 1972).

**Complementation** (Table 7–5).   This refers to the interaction of gene products in mixedly infected cells such that the yield of one or both parental viruses is enhanced but their genotypes are unchanged.

**Nongenetic Reactivation.**   This is a special case of complementation in which virus that has been rendered noninfectious by heating or other treatments that affect viral protein rather than nucleic acid is reactivated by complementation.

**Phenotypic Mixing.**   This refers to the production of progeny with structural proteins derived from both parental viruses, and *transcapsidation* or *genomic masking* describes the extreme case of phenotypic mixing where the genome of one parent is enclosed within a capsid that is coded for by another.

## GENETIC RECOMBINATION

Classically, three techniques are available for mapping of viral genomes (Hayes, 1968): (a) Mapping by comparison of recombination frequencies in two-factor crosses. The expectation is that, if crossing-over between the two nucleic acid molecules can occur randomly at any point, then the frequency with which a progeny virion will contain two characters derived from different parents will be proportional to the distance that separates those two loci in the nucleic acid molecule. It is thought that an endonuclease is required to open up the DNA molecules while the recombinational event occurs and a ligase subsequently repairs the break. Two-factor crosses are subject to many inherent sources of error since the results of different crosses are compared, and with most animal viruses there are still substantial unsolved technical problems involved in determining recombination frequencies reliably. (b) Mapping by deletion. Point mutations may be allocated to specific regions by analysis against a "library" of deletions whose sequence is known (Benzer, 1959). (c) Mapping by three-

TABLE 7-4

Nucleic Acid Interactions; Genetic Recombination and Reactivation[a]

| PHENOMENON | PARENT 1 | PARENT 2 | PROGENY | COMMENT |
|---|---|---|---|---|
| Intramolecular recombination | ABC | ABC | ABC, **ABC**, **ABC**, (**ABC**) | With mutants of DNA viruses and Picornaviridae |
| | ABC | **AST** | **ABT, ASC** | With different strains of vaccinia virus |
| | ABCD | XYZ | ABCZ | Loss of adenovirus genes (D), addition of SV40 genes (Z) |
| | ABC | 123 | 12ABC3 | Oncogenic viruses; integration of viral genes (ABC) into genome of cell (123) |
| Gene reassortment | A/**B**/C | A/B/**C** | A/B/C | With *Reovirus* and *Orthomyxovirus* |
| | A/B/C | A/**S**/**T** | A/B/**T**, A/**S**/C | |
| Cross-reactivation: | | | | |
| Between UV-inactivated virus and virus of a different but related strain | A̸B̸C | **AST** | A**S**C | Rescue of gene C from inactivated parent, by intramolecular recombination or gene reassortment |
| | A̸/B̸/C | A/**S**/**T** | A/**S**/C | |
| Multiplicity reactivation: | | | | |
| Between virions of same virus inactivated in different genes | ABC̸ | AB̸C | ABC | Recombination or reassortment of genes B and C to yield viable virus |
| | A/B/C̸ | A/B̸/C | A/B/C | |

[a] A, etc., active viral genes; 1, etc., active cellular genes; **B**, etc., mutant genes; A̸, etc., inactivated genes; ABC, continuous linear genome; A/B/C, segmented genome.

# TABLE 7-5

Gene Product (Protein) Interactions: Complementation and Phenotypic Mixing[a]

| PHENOMENON | PARENT 1 | PARENT 2 | PROGENY | COMMENT |
|---|---|---|---|---|
| **Complementation** | | | | |
| (a) Between conditional lethal mutants of the same virus (under restrictive conditions) | $\dfrac{ABC \downarrow\downarrow\downarrow}{a\ b\ c}$ | $\dfrac{ABC \downarrow\downarrow\downarrow}{a\ b\ c}$ | ABC, ABC, (ABC) | Reciprocal; both mutants rescued; sometimes recombination also |
| (b) Between defective virus and unrelated helper virus | $\dfrac{A\not{B}C \downarrow\downarrow\downarrow}{a\quad c}$ | $\dfrac{BYZ \downarrow\downarrow\downarrow}{byz}$ | A$\not{B}$C and BYZ | Defective virus is rescued by gene product "b" of helper BYZ |
| **Phenotypic mixing** | | | | |
| (a) Enveloped viruses | $\dfrac{ABC}{a}$ | $\dfrac{XYZ}{x}$ | $\dfrac{ABC,}{ax}\ \dfrac{XYZ}{ax}$ | Mixed peplomers in envelopes, genomes unaltered; also parental phenotypes |
| (b) Nonenveloped viruses | $\dfrac{ABC}{a}$ | $\dfrac{XYZ}{x}$ | $\dfrac{ABC,}{ax}\ \dfrac{XYZ}{ax}$ $\dfrac{ABC,}{x}\ \dfrac{XYZ}{a}$ | Mixed capsomers in capsids. Genome of one parent with capsid of the other (transcapsidation). Not always reciprocal |

[a] A, etc., active viral genes; **B**, etc., mutant genes; $\not{B}$, defective gene B; a, a, etc., product of gene A, etc., product of mutant gene **B**; ABC, ABC genome; $\overline{ax}$, proteins in envelope (or capsid).

factor reciprocal crosses. This offers an unambiguous method for determining the order of loci and has been applied to poliovirus (Cooper, 1968) and herpesvirus (Brown *et al.*, 1973).

Two genera of animal viruses present a novel problem for genetic mapping. The genomes of reovirus and influenza virus are fragmented, and consist of several separate pieces of RNA, each coding for a particular viral polypeptide (see Chapters 3 and 6). There is no evidence that the fragments are arranged in any particular order in the virion, so that "mapping" of these viruses consists of relating particular distinguishable pieces of nucleic acid to their gene products (e.g., Ito and Joklik 1972a, b, c). *In vitro* systems now available for transcription and translation of the reovirus genome (McDowell *et al.*, 1972) should make it possible to isolate "pure" populations of each molecule of reovirus RNA (=gene), transcribe mRNA from it, and synthesize the corresponding polypeptide.

Genetic recombination has been clearly demonstrated with most DNA viruses. Further there is evidence with certain members of several genera (*Herpesvirus, Adenovirus,* Papovaviridae, and, via the DNA provirus, the RNA tumor viruses) that the viral genome may be integrated with cellular chromosomes, probably by a process of recombination (see Chapters 13 and 14).

The situation is quite different for RNA viruses. No process like recombination of RNA molecules has been recognized in uninfected vertebrate cells, and recombination does not occur with RNA phages (Hayes, 1968). Over twenty years ago, however, genetic recombination was reported with influenza virus (Burnet and Lind, 1951), and high frequency recombination has been repeatedly found with this virus ever since (reviews, Kilbourne, 1963; Webster and Laver, 1971). A similar result has also been reported with reovirus (Fields and Joklik, 1969; Fields, 1971), but not with any other RNA viruses. This high frequency recombination is almost certainly due to the exchange of pieces of the viral RNA of the fragmented genomes of these two viruses, a process defined earlier as "genetic reassortment." With RNA viruses of several other genera, claims made for recombination have now been withdrawn and the results are explained by complementation, either of clumped virions, or among enveloped viruses, by complementing heteropolyploids (vesicular stomatitis virus, Wong *et al.*, 1971; respiratory syncytial virus, Wright and Chanock, 1970; Newcastle disease virus, Dahlberg and Simon, 1969b). The only instance of what appears to be intramolecular recombination among RNA viruses involves the picornaviruses, poliovirus (Cooper, 1968, 1969), and foot-and-mouth disease virus (Pringle, 1968). The evidence that this is intramolecular recombination is set out below; it would be of great interest to determine its mechanism and to define the enzymes involved.

So far, rabbitpox virus (a close relative of vaccinia virus), poliovirus, and herpes simplex virus are the only agents with which recombination has been used to assign an order to a group of mutants (Gemmell and Fenner, 1960; Fenner and Sambrook, 1966; Cooper, 1968; Brown *et al.*, 1973). Reproducible two-factor crosses have been performed with *ts* mutants of rabbitpox, and recombination has been demonstrated with *ts* mutants of adenovirus type 5.

## Examples of Recombination

**Poxvirus.** In early studies, recombination was demonstrated between two different strains of vaccinia virus, both on the chorioallantoic membrane (Fenner and Comben, 1958) and in the progeny obtained from single mixedly infected HeLa cells (Fenner, 1959). Recombination was demonstrated between a number of different members of the vaccinia subgenus, and within the myxoma subgenus, but not between poxviruses belonging to different subgenera (Woodroofe and Fenner, 1960; Bedson and Dumbell, 1964).

Recombination between eighteen *u* (white pock) mutants of rabbitpox virus allowed their ordered linear arrangement, in four groups (Gemmell and Cairns, 1959; Gemmell and Fenner, 1960). Subsequently, it was found that all PK-negative mutants of rabbitpox virus fell into one of these groups but did not recombine with each other (Fenner and Sambrook, 1966).

Padgett and Tomkins (1968) carried out two-factor crosses with twenty *ts* mutants of rabbitpox virus. Recombination frequencies with most pairs tested were between 10 and 20%, but a few were much lower (0.01—1%). These *ts* mutants have not been ordered.

**Herpesvirus.** Temperature-sensitive mutants of herpes simplex virus type 1 (HSV1) recombine readily. By making three-point crosses between nine *ts* mutants from eight complementation groups, and a ninth unselected morphological marker, Brown *et al.* (1973) have produced a linear linkage map of HSV1. Mutants of three DNA⁻ complementation groups are clustered close together on the linkage map.

Temperature-sensitive mutants of herpes simplex virus type 2 (HSV2) likewise fall into several complementation groups; mutants from different groups will recombine and recombination also occurs between certain mutants of HSV1 and HSV2 (Timbury and Subak-Sharpe, 1973). Successive progeny testing of (HSV1–HSV2) recombinant clones showed that a proportion of genomes retained the potential for segregating with respect to *ts* or plaque morphology markers, suggesting the occurrence of partial heterozygotes. One recombinant tested for envelope antigens by neutralization tests behaved as an intermediate between HSV1 and HSV2.

**Adenovirus.** Williams and Ustacelebi (1971a, b) have demonstrated recombination between *ts* mutants of adenovirus type 5, using two-point crosses.

**Polyoma Virus.** In spite of the ready recombination between polyoma viruses and cellular DNA's (see Chapter 13), it has been difficult to demonstrate recombination between different mutants of either SV40 or polyoma virus.

**Picornaviridae.** Much of the effort of Dulbecco's group between 1953 and 1958 was devoted to the genetics of poliovirus, but their efforts to demonstrate recombination were unsuccessful. Using noncovariant markers, Hirst (1962) and Ledinko (1963) subsequently provided evidence suggesting that recombination did occur, with the low frequency that might be expected from the small size of the genome. With equal inputs of the two parents, the recombination frequency was 0.4%, rising during the growth cycle from 0.2% at the time of appearance of the first progeny virus to 0.4% at the end of viral production; Cooper (1968), using one pair of *ts* mutants, confirmed that there was a small but significant increase in recombination from 0.28% at 3 hours to 0.42% at 7 hours.

Cooper (1968) examined recombination between different pairs of *ts* mutants of poliovirus. When careful attention was paid to experimental details the recombination rates in two-factor crosses were reproducible and reached a maximum of 0.85%. The recombination frequencies with selected pairs were additive, allowing them to be arranged in a linear order that corresponds with their ordering by physiological tests (Fig. 7–2). Three-factor crosses using $g^r$ confirmed many of the sequences in the genetic map, which comprises a single linkage group. The $g^r$ locus is found in the region concerned with virion proteins.

Pringle and his colleagues have reported that recombination occurs between strains of one serological subtype of foot-and-mouth disease virus, but not between serotypes (Pringle, 1968; Pringle and Slade, 1968) and with a somewhat higher frequency between *ts* mutants derived from a single parental clone (Pringle *et al.*, 1970). In experiments with subtype strains, the recombinants were sometimes obtained from single plaques that yielded one or both parental strains as well, probably because the plaques were initiated by viral aggregates (Pringle and Slade, 1968).

## The Evidence for Intramolecular Recombination with Picornaviridae

"Recombination" between RNA viruses whose genome consists of a single RNA molecule appears to be limited to the two examples just described, poliovirus and FMDV. Claims for other viruses have now been explained by complementation due to viral aggregates or heteropolyploidy. It is therefore important to examine the evidence for regarding the results obtained with these picornaviruses as being due to genetic recombination. The most compelling evidence would be the demonstration of reciprocal recombinants in the yield of single mixedly infected cells. An attempt by Slade and Pringle (1971) to demonstrate reciprocity in single-cell yields of FMDV recombinants was frustrated by the inefficient infection process, and in the poliovirus system the occurrence of 0.1 to 0.5% reciprocal (double *ts*) mutants would not stand convincingly above the background of spontaneous production of such mutants.

The conclusion that poliovirus recombination occurs at all is primarily based, as in many other systems, on the detection of double (in some cases triple) mutants in significant excess over what is expected from spontaneous mutation, as judged from the self-crosses. This excess is up to 30 ×, usually 10–20 ×, the spontaneous background, and occurs with a reproducible frequency that is characteristic of the parental mutant pair. The nature of the poliovirus genome makes it very unlikely that this is due to genetic reassortment of the type found with reovirus and influenza virus.

Trivial explanations, like clumping of virions, are excluded both by technical manipulation and by the consistency of the results obtained with particular crosses on repeated occasions. Cooper and his colleagues (P. Cooper, personal

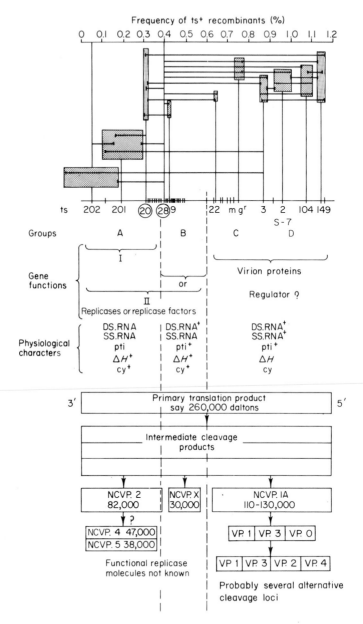

FIG. 7–2. *Genetic map of poliovirus type 1, based upon experiments with tempera-ture-sensitive (ts), guanidine-resistant (g^r), dextran sulfate (m), and S-7 resistant mu-tants (Cooper, 1968; Cooper et al., 1971). Upper figures represent the scale (percentage recombinants); lower figures are the identification numbers of some of the mutants examined. Recombination frequencies determined by two-factor crosses; the stippled boxes represent the discrepancies in the additivity of recombination frequencies. The location (to the left or to the right) of all the mutants with respect to ts 28 has been found by three-factor crosses involving ts 28 adapted to guanidine resistance (ts 28 g^r); g^r indicates the locus for its guanidine resistance. The mutants have been allocated to*

communication, 1973) have examined the possibility that poliovirus recombinants contain an abnormal amount of RNA by comparing the S value, buoyant density, and UV sensitivity of recombinants and parental virus. No differences were found, nor could any segregants be detected in serial progeny testing of some 50 recombinant clones by a technique that was sensitive enough to detect $10^{-5}$ of one parental phenotype. Finally, the results obtained by Cooper and his colleagues in numerous experiments over several years have been self-consistent; recombination frequencies are additive, and data on some forty mutants involving recombination, physiological, and inhibitor studies all give the same result for the genetic map.

The only likely explanation for these results is that poliovirus can be involved in precisely ordered molecular exchange similar to that recognized in the DNA bacteriophages and E. coli. Specific enzymes analogous to those for DNA breakage-reunion may be involved, but because of the limited base-pairing in the poliovirus replication complex (Öberg and Philipson, 1971) it remains possible that appreciable template dissociation can occur. Thus recombination of poliovirus may arise by reassortment of templates and nascent chains in vivo followed by continued replicase activity to complete the chains (a form of copy-choice), without recourse to novel RNA-combining enzymes.

## Genetic Reassortment

This term applies to the special case of high frequency recombination among viruses with fragmented genomes, due to the exchange of whole pieces of RNA, which may be regarded as being equivalent to genes. The occurrence of intramolecular recombination is not excluded among these viruses and may explain some of the observed low frequency recombination. Although many more observations have been made with influenza virus we shall discuss the case of reovirus first, since the best studies of the nature of high frequency recombination were made with this virus.

**Reovirus.** Reovirus has a fragmented genome consisting of ten separate pieces of double-stranded RNA. Mixed infection experiments with ts mutants of reovirus type 3 at the permissive temperature yielded either a high proportion of recombinants ($ts^+$) or none at all (Fields and Joklik, 1969). Five recombination (and complementation) groups were defined, and Fields et al., (1972)

---

physiological groups. Mutants of group A (characterized by ts 20) appear to be defective in a function DS·RNA$^+$ needed to make or maintain double-stranded RNA; mutants of group B (characterized by ts28) are defective in a function SS·RNA$^+$ needed to produce progeny RNA free of the double-stranded complex. All of the mutants allocated to groups C and D appear to be defective in the production of a virion protein (DH$^+_2$ thermostable; cy$^+$, cystine independent). The 5' end of the message is to the right of the map; separation of functional gene products occurs by posttranslational cleavage at many loci. NCVP·IA is the main virion protein precursor, and appears also to give rise to the regulator protein; NCVP·2 and NCVP·X may be related to the replicase proteins. (Courtesy of P.D. Cooper.)

have subsequently added one and possibly two more. The recombinants occur early and do not increase in frequency during the growth cycle, suggesting that there is an equal likelihood of reassortment at all stages of the cycle. Fields (1971) carried out a statistical analysis of the frequency of recombinants in crosses between mutants of four recombination groups. The similar relatively high frequencies (of the order of 3 to 8%) found with all pairwise combinations were strongly suggestive of reassortment of pieces of viral RNA rather than intramolecular recombination.

**Influenza Virus.**  Recombination in animal viruses was first demonstrated with influenza virus (Burnet and Lind, 1951), and a great deal of work was carried out in embryonated eggs on recombination between a variety of natural and laboratory strains of influenza A virus, from both human and animal sources. Recombination also occurs with influenza B, but not between strains of influenza A and B (review, Kilbourne, 1963).

Because of the high frequency of recombination both Burnet (1959) and Hirst (1962) suggested the possibility of reassortment, long before the structure of the genome was known. Accurate measurement of recombination frequencies became possible only when *ts* mutants were studied by plaque assay methods. With few exceptions, pairs of mutants showed either no recombination, or they recombined at rates between 3 and 10% (Simpson and Hirst, 1968), far higher than could be explained by intramolecular recombination. However, as Fields (1971) observed with reovirus, the recombination frequencies were substantially lower than would be expected if reassortment were random.

Using plaque assay in cultured cells, reassortment has now been demonstrated between influenza A strains of human origin (Kilbourne *et al.*, 1967) or human and animal origins (Kilbourne, 1968). The most common type of reassortant recognized had the ribonucleoprotein antigen of one parent and either the hemagglutinin or the neuraminidase of the other (Laver and Kilbourne, 1966; Easterday *et al.*, 1969). Indeed, such antigenic hybrids can be "made to order" (Webster, 1970). Genetic reactivation and reassortment occurs readily in the respiratory tracts of both swine and chickens (Webster *et al.*, 1971; Webster and Campbell, 1972), a finding of considerable epidemiological interest (see Chapter 17).

It is possible that intramolecular recombination also occurs with influenza viruses. For example, Staiger (1964) recorded recombination frequencies of 0.1 to 0.2% with two small-plaque mutants of fowl plague virus, and low frequency recombination is found in some crosses with *ts* mutants, although it is hard to judge its statistical significance.

**Avian Leukoviruses.**  Vogt (1971b) has presented preliminary evidence suggestive of recombination by reassortment between avian leukosis viruses with different host range marker attributes and the Prague strain of Rous sarcoma virus. These results have been confirmed and extended by Kawai and H. Hanafusa (1972), using the same principle but different sarcoma and leukosis viruses. The frequency of recombination exceeded 8%, suggesting reassortment rather than intramolecular recombination. It will be recalled (Chapter 3) that there is evidence for a subunit structure of leukovirus RNA, although the pieces are linked together by short regions of hydrogen bonding.

## INTERACTIONS OF VIRAL NUCLEIC ACID AND CELLULAR DNA

There is growing evidence that integration of viral or virus-specific DNA into the cellular genome is relatively common and may be a very important feature in malignancy (see Chapters 13 and 14) and possibly in some latent infections (see Chapter 12). With DNA viruses, integration has now been demonstrated with herpesviruses, adenoviruses, and papovaviruses. With RNA viruses, it is possible that the normal method of survival and transmission of many leukoviruses may be the integration of the DNA copy of their genome into the genome of every cell of most or all members of the species concerned. If this is true, the virus is normally represented by a totally or partially repressed set of cellular genes, and viral replication is associated with derepression either because of the activity of other genes or under the influence of extrinsic factors.

The transfer of genetic information concerned with specific enzymes to mammalian cells by viruses has been claimed with UV-irradiated herpes simplex virus, which converted thymidine kinase-less L cells to a state whereby they synthesized thymidine kinase that was probably of viral specificity (Munyon *et al.*, 1972).

## GENETIC REACTIVATION

Recombination or reassortment of the genetic material is not limited to active viruses but can occur between an active and a suitably inactivated parent (cross-reactivation or marker rescue) or between two parental viruses each inactivated by damage to different parts of their genome (multiplicity reactivation).

### Cross-Reactivation

With influenza virus in particular, use of a nonplaquing active virus and an irradiated plaquing parent has facilitated the recovery of reassortants (Simpson and Hirst, 1961; Sugiura and Kilbourne, 1966). In some systems, the variety of recombinant progeny obtained after cross-reactivation is much less than is found after recombination between two active parental viruses (McCahon and Schild, 1971). Cross-reactivation can also be demonstrated with viruses that undergo intramolecular recombination, e.g., vaccinia virus (Abel, 1962b).

### Multiplicity Reactivation

Since it is probably achieved by genetic recombination, the occurrence of multiplicity reactivation with suitably inactivated viruses would be expected to parallel the occurrence of genetic interactions between two active viruses, and such is the case. Multiplicity reactivation has been demonstrated with poxviruses, where clumping of virions in the inoculum plays an important role in increasing the frequency of reactivation, presumably by ensuring the proximity of interacting molecules of viral DNA (Abel, 1962a; review, Sharp, 1968). It can be readily demonstrated with influenza virus (Barry, 1961a, b) but not with the paramyxovirus, Newcastle disease virus (Barry, 1962). UV-irradiated "incomplete" influenza virus does not show multiplicity reactivation (Rott and Scholtissek, 1963), a finding explained by the fact that all particles lack the same piece of RNA (Pons and Hirst, 1969).

Reoviruses inactivated by UV-irradiation readily undergo multiplicity reactivation, even when the virions are deliberately dispersed before inoculation (McClain and Spendlove, 1966). Cross-multiplicity reactivation occurs between all three serotypes, suggesting a substantial degree of genetic relatedness between them (greater, for example, than between influenza A and B). UV-irradiated adenovirus 12 and UV-irradiated SV40 both undergo multiplicity reactivation (Yamamoto and Shimojo, 1971).

## HOST-CELL REACTIVATION AND PHOTOREACTIVATION

Even in the absence of multiplicity reactivation, the survival curves obtained after UV-irradiation of bacterial and animal viruses are multicomponent (Abel, 1962b). With bacteriophages the more gradual slope of the lower part of the inactivation curve is due to repair of the UV-induced lesions by host-cell enzymes, a phenomenon called host-cell reactivation (review, Rupert and Harm, 1966). In experiments conducted with two herpesviruses (pseudorabies and herpes simplex), host-cell reactivation has been demonstrated in animal cells also (Závadová and Závada, 1968; Lytle, 1971).

Repair of UV-induced damage to bacteriophage DNA is enhanced by exposure to visible light (review, Jagger, 1958), due to the splitting of UV-induced pyrimidine dimers by a light-dependent enzyme (Setlow, 1964). Cleaver (1966) sought but failed to find evidence of photoreactivation in mammalian cells, and Pfefferkorn et al. (1966) demonstrated that pseudorabies virus, which multiplies in the nucleus, and frog virus 3 (Pfefferkorn and Boyle, 1972b) which multiplies in the cytoplasm, underwent photoreactivation in chick cells but not in rabbit cells (Pfefferkorn and Coady, 1968). This is correlated with the capacity of chick but not rabbit cells to eliminate UV-induced thymine dimers when exposed to visible light, a finding which parallels the failure to demonstrate photoreactivating enzyme activity in cells of placental mammals (reviews, Cook, 1970, 1972), although it is present in avian cells.

## COMPLEMENTATION

The term complementation is used to describe cases in which a protein(s) specified by the genome of one virus enables a second virus to multiply (Table 7–5). Often an enhanced yield of one virus in a mixed infection is accompanied by a depressed yield of the other (interference).

### Incomplete and Defective Viruses

Even when considering the simplest virus–cell system, where the cell is susceptible and the virus cytocidal, virologists tend to take an oversimplified view, and ignore all particles in the inoculum and the yield except those that are able to initiate productive infection. The situation is really much more complex, for the majority of the inoculum always consists of noninfectious particles, the cell produces excessive amounts of all viral components, and most progeny particles are noninfectious. In addition, there are three complications that involve the viral genome and warrant consideration here.

**Incomplete Virus.** Under certain circumstances (commonly serial passage at high multiplicity) the infection of susceptible cells yields a very small proportion of infectious particles but many particles that are "incomplete" in that they lack a complete genome (review, Huang and Baltimore, 1970b). The situation is found among viruses with fragmented genomes, like influenza virus (review, von Magnus, 1954) and reovirus (Nonoyama et al., 1970) when the incomplete virus lacks a particular piece of RNA (see Chapters 3 and 6). It is also found with vesicular stomatitis virus where the incomplete virus (T component) is a spherical particle with a genome only one-third the full size (Huang et al., 1966). Serial passage of Sindbis virus and poliovirus also yields defective particles (S. Schlesinger et al., 1972; C. Cole et al., 1971). Among DNA viruses, serial passage at high multiplicity yields particles of adenovirus (Mak, 1971) and SV40 (Uchida et al., 1968) whose genome is smaller than normal. Huang and Baltimore (1970b) have postulated that the shorter (defective) viral nucleic acid molecules compete more effectively for replicase.

**Abortive Infection.** The incomplete virus particles just described may cause abortive infections in which some viral biosynthetic processes occur but there is no yield. However, even some virus particles that possess complete genomes and are fully infectious in one cell system may be defective and cause abortive infection in another type of cell. Examples are herpes simplex virus in dog kidney cells (Spring et al., 1968), mouse encephalomyelitis virus in HeLa cells (Sturman and Tamm, 1969), and human adenoviruses in simian cells (see below). Such viruses may be regarded as host-dependent mutants; they multiply in cells that provide a particular genetic function that they themselves lack.

In addition, there exist viruses that are absolutely defective, in all cell systems, and survive in nature only because they can be complemented by another, *helper* virus. Where there is a natural association of helper and defective virus the latter may be called a *satellite* virus of the helper virus.

The viruses that transform cells provide an even more common illustration of abortive infection, which is indeed a prerequisite for transformation (Chapters 13 and 14). Some viruses are only "conditionally defective" in that they may be rescued by a process of complementation by a "helper" virus; they are known as "helper-sensitive" defective viruses.

Two situations that will be discussed at greater length later in this chapter are (a) the complementation of adenovirus-associated viruses (adeno-satellite viruses) by adenoviruses, and (b) the defectiveness of human adenoviruses in simian cells and their complementation by SV40 and other agents.

## Complementation between Related Viruses

The gene as a unit of function can be defined by tests for complementation, that determine whether two mutants, which may or may not exhibit similar phenotypes, are defective in the same function. Complementation tests, which enable functional units to be defined genetically in the absence of any information about the chemical nature of the function, are often used to sort mutants into groups before carrying out recombination experiments (Edgar *et al.*, 1964) or physiological tests (Burge and Pfefferkorn, 1968). Unlike genetic recombination, complementation does not involve exchange of viral nucleic acid, but reflects the fact that one virus provides a gene product in which the other is defective, so enabling both to multiply in the mixedly infected cell (Table 7–5). Complementation between two defective viruses under restrictive conditions is often asymmetrical, i.e., one parental type dominates the yield.

Two types of complementation between mutant viruses have been recognized: nonallelic (intergenic) complementation in which mutants defective in different genes can assist each other's multiplication (perhaps to unequal extents) by supplying missing gene products or functions; and allelic complementation (often called intragenic), in which function of a defective gene may be restored, at least partially, by the formation of hybrid polymeric protein molecules from two defective products of the same gene which have undergone mutations at different sites (Fincham, 1966). Allelic complementation occurs between *ts* but not between *amber* mutants of T4 (or between *ts* and *amber* alleles), but has not been unequivocally recognized yet with viruses of vertebrates, although its existence was postulated as one possible explanation for complementation between certain RNA$^-$ mutants of Sindbis virus (Burge and Pfefferkorn, 1966b).

Mutants with defects anywhere in the same gene are said to fall into the same complementation group, i.e., they will not complement one another but will complement a mutant from a different group. Theoretically there should be as many complementation groups as genes but it usually happens that mutations in some genes are absolutely lethal (rather than conditionally lethal) and cannot be complemented even by the wild-type virus. Biochemical analysis of the viral macromolecules made in cells infected separately with the two complementary *ts* mutants may indicate the stage at which the multiplication cycle is blocked, and thus pinpoint the function of the defective gene. However, interpretation is often difficult because the presence of a single nonfunctional protein may have secondary effects on concurrent as well as subsequent biochemical processes.

Now that satisfactory techniques have been developed for resolving and identifying viral nucleic acid and protein molecules directly, *ts* mutants are not necessary to identify the various structural proteins of the virion. Their main value is in providing clues about the intracellular functions of structural and nonstructural proteins, e.g., identifying regulatory roles for structural proteins or the existence of two polymerases rather than one. This is especially true in the analysis of virus-induced cellular transformation (see Chapters 13 and 14).

**Poxvirus.** The host-dependent PK-negative ($p$) mutants of rabbitpox virus did not exhibit complementation when PK cells were mixedly infected with $p$ mutants that exhibited different phenotypic traits. However, mixed infection of the restrictive cells with a $p$ mutant and a $p^+$ virus (either wild type or a $u\ p^+$) resulted in the rescue of the $p$ mutants by complementation (Fenner and Sambrook, 1966). Rabbitpox $ts$ mutants showed efficient mutual complementation when inoculated together at the restrictive temperature (Padgett and Tomkins, 1968). There are so many genes in the poxvirus genome that it is not surprising that each of the eighteen $ts$ mutants tested fell into a different complementation group.

**Herpesvirus.** Complementation is readily recognized with $ts$ mutants of herpes simplex virus type 1 and type 2. Hay and Subak-Sharpe (1973) have recognized eight complementation groups in a suite of $ts$ mutants of HSV1 and Timbury (1971) at least ten groups with HSV2. Not unexpectedly, cross-complementation between HSV1 and HSV2 also occurs (Timbury and Subak-Sharpe, 1973).

**Adenovirus.** With adenovirus type 5, Williams and Ustacelebi (1971a, b) have distinguished fourteen complementation groups, only one of which is DNA$^-$. Likewise, Suzuki et al. (1972) were able to group twelve $ts$ mutants of adenovirus type 31 into eight complementation groups; physiological defects corresponded with the grouping by complementation tests.

**Polyoma Virus.** Complementation tests and tests for transformation at restrictive temperatures have been used to group $ts$ mutants of polyoma virus (Eckhart, 1969; Di Mayorca et al., 1969). Table 7-3 sets out the complementation groups and the properties of mutants belonging to each group. Kimura and Dulbecco (1972) have distinguished two complementation groups among six $ts$ mutants of SV40.

**Enterovirus.** Several reports have appeared of complementation between drug-requiring and drug-resistant enteroviruses. Ikegami et al. (1964) and Agol and Shirman (1965) showed that guanidine-sensitive and guanidine-dependent polioviruses of different immunological types could complement each other in mixed infections under conditions (presence and absence of guanidine respectively) which prevented multiplication of one of the viruses. Cords and Holland (1964b) showed that this sort of complementation also occurred with other combinations of enteroviruses, e.g., poliovirus and coxsackievirus B1. In these cases, the virus not inhibited may provide both the RNA polymerase to replicate the RNA of the inhibited virus and sometimes the capsid proteins to enclose the new viral RNA, producing "transcapsidants," with "genomic masking."

With $ts$ mutants of poliovirus, Cooper (1965) found that although complementation occurred, it was restricted to a few combinations of mutants, was highly asymmetric, and the yield was low. This may be because picornavirus

RNA is translated initially into a single polypeptide, hence, a *ts* defect in any part of this molecule may interfere with its function, or that of the proteins resulting from the first round of cleavage. Consequently, complementation has not proved useful for either genetic or physiological studies.

**Alphavirus.** Burge and Pfefferkorn (1966b) allocated 23 *ts* mutants of Sindbis virus to five complementation groups. The yield from cells mixedly infected with complementing mutants was only 1–2% of wild-type yield, but was 5 to 100 times the mutant yields, and it contained both genotypes. There was excellent correlation between the complementation groups and physiological tests on representative mutants (Table 7–6).

TABLE 7–6

*Functional Defects of ts Mutants of Sindbis Virus[a]*

| COMPLEMENTATION GROUP | RNA SYNTHESIS AT 40°C | NUCLEOCAPSID FORMATION AT 40°C | HEMADSORPTION AT 40°C | PRESUMED DEFECT IN |
|:---:|:---:|:---:|:---:|:---|
| A | 0 | 0 | 0 | An enzyme for RNA synthesis |
| B(?) [b] | 0 | 0 | 0 | Another enzyme for RNA synthesis (?) |
| C | + | 0 | + | A nucleocapsid protein |
| D | + | + | 0 | An envelope protein |
| E | + | + | + | ? |
| Fails to complement | + | + | 0 | ? |

[a] From Pfefferkorn and Burge (1968).
[b] May be due to intracistronic complementation.

**Orthomyxovirus.** Complementation has been demonstrated between *ts* mutants of influenza virus (Simpson and Hirst, 1968), but these viruses can be grouped so much better on the basis of reassortment data that grouping by complementation has not been used.

**Paramyxovirus.** Dahlberg and Simon (1968) isolated a number of nitrous acid-induced *ts* mutants of Newcastle disease virus. By taking precautions to minimize the effects of eluted inoculum by appropriate treatment of infected cells with neutral red and white light they could detect very low yields, and were able to distinguish nine nonoverlapping complementation groups among twenty-nine *ts* mutants.

**Rhabdovirus.** Complementation has proved very useful in the genetic analysis of vesicular stomatitis virus (VSV) and several interesting results have emerged. Flamand and Pringle (1971) have compared the numerous mutants

TABLE 7–7

Complementation Groups of ts Mutants of Vesicular Stomatitis Virus[a]

| | INDIANA SEROTYPE | | | | NEW JERSEY SEROTYPE | | |
|---|---|---|---|---|---|---|---|
| | INDUCED | | SPONTANEOUS [b] | | | 5-FU-INDUCED | |
| COMPLEMENTATION GROUP | NO. | RNA PHENOTYPE | NO. | RNA PHENOTYPE | COMPLEMENTATION GROUP | NO. | RNA PHENOTYPE |
| I | 177 | —[c] | 58 | — | A | 17 | — |
| II | 2 | —/+ | 2 | + | B | 21 | — |
| III | 3 | + | 4 | + | C | 4 | + |
| IV | 22 | — | 4 | — | D | 1 | + |
| V | 0 | 0[c] | 3 | + | E | 3 | —/+ |
| Unallocated | 6 | 0 | 0 | 0 | F | 2 | — |
| | | | | | Unallocated | 1 | — |

[a] From Pringle et al. (1971).
[b] Flamand (1970).
[c] + or —, Induce or fail to induce viral RNA synthesis; 0, no data.

obtained from VSV, Indiana serotype, as spontaneous mutants (Flamand, 1970) and after mutagenization (Pringle, 1970). They show a close correspondence (Table 7–7). Subsequently, Pringle et al. (1971) recovered forty-eight ts mutants of VSV, New Jersey serotype and found that all but one of them could be allocated to one or other of six complementation groups (Table 7–9). However, none of the Indiana mutants was able to complement any of the New Jersey mutants, a rather surprising result when compared with the tolerance exhibited by VSV in phenotypic mixing with a paramyxovirus of the peplomers in its envelope (Choppin and Compans, 1970). This lack of complementation may be related to the phenomenon of heterotypic interference (in the absence of homotypic interference) with these two serotypes of VSV (Cooper, 1958; see Chapter 8).

## Rescue of Defective Viruses by "Helper" Viruses

Another type of complementation between closely related viruses that has assumed considerable importance in the study of the oncogenic RNA viruses is the rescue of defective viruses. A similar phenomenon has also been reported with host-dependent mutants of other viruses.

**Leukoviruses.** The avian and murine leukoviruses which produce solid tumors, Rous sarcoma and Moloney sarcoma viruses, appear to be defective in that yields of infective virus occur only when a cell is mixedly infected with the sarcoma virus and another avian or a murine leukovirus, respectively (see Chapter 14). Some strains of avian sarcoma virus are not defective, although their yield of infectious virus can be enhanced by superinfection; Moloney (murine) sarcoma virus is defective both in its capacity to produce transformed cells and to yield infectious virus. The helper virus specifies at least one viral coat protein (Rubin, 1965; Hanafusa, 1965).

## Nongenetic Reactivation

In 1936, Berry and Dedrick (1936) set out to see whether the phenomenon of "transformation," which Griffith (1928) had demonstrated with pneumococcus, also occurred with myxoma and fibroma viruses. They showed that following mixed infection with live fibroma virus and heat-inactivated myxoma virus, rabbits died of myxomatosis. Subsequent work (Fenner *et al.*, 1959; Hanafusa *et al.*, 1959), showed that the Berry-Dedrick phenomenon was due to multiplication of the undamaged myxoma virus genome, and did not involve either transformation or genetic recombination between the viruses concerned. For this reason it was called "nongenetic reactivation" (Fenner, 1962). A similar phenomenon has now been reported with frog virus 3, an *Iridovirus*, and with adenoviruses.

**Poxvirus.**  Nongenetic reactivation (review, Fenner, 1962) occurs between poxviruses of all subgenera that have been tested. Viruses inactivated by heating, or chemicals, such as urea, ether or a number of others that destroy viral infectivity by altering the viral coat without attacking the nucleic acid, could all be reactivated. Active poxviruses, or particles whose DNA had been inactivated by nitrogen mustard (Joklik *et al.*, 1960; Rönn *et al.*, 1970) or UV irradiation (Dunlap and Patt, 1971) will act as reactivating particles. The explanation appears to be that the reactivable virus contains a functional genome but an inactivated virion transcriptase. The reactivating virus contains an undamaged transcriptase which transcribes early mRNA for an enzyme that uncoats both genomes and allows the reactivable virus to multiply.

**Iridovirus.**  Although morphologically quite different from the poxviruses, frog virus 3 (FV3) is also a large DNA virus that multiplies in the cytoplasm. Gravell and Naegele (1970) have shown that heat-inactivated FV3 can be reactivated by UV-irradiated particles, especially if the mixed particles are clumped so that cells engulf both types of particles together (Gravell and Cromeans, 1971). Uncoating of FV3 is a less complex process than that of poxvirus (see Chapter 5), and reactivation of FV3 will occur without prior protein synthesis. FV3 contains a virion-associated protein kinase, whose activity is essential for nongenetic reactivation, possibly because it may activate an inactive virion-associated DNA-dependent RNA polymerase by phosphorylation (Gravell and Cromeans, 1972).

**Adenovirus.**  Béládi *et al.* (1970) found that heat-inactivated adenovirus types 1 and 6 could be rescued if cells were simultaneously inoculated with UV-irradiated adenovirus type 8, but the mechanism of reactivation was not further elucidated.

## Complementation between Unrelated Viruses

All the examples of recombination and complementation that have been discussed so far refer to interactions between viruses of the same species or genus. The adenoviruses exhibit a striking variety of complementation reactions with other viruses (reviews, Rapp and Melnick, 1966; Rapp, 1969), acting as a defective virus in some systems, and as a helper virus in others. The rescue of adenoviruses in situations where they are defective involves both complementation and recombination; the results are so important that they will be described in a separate section (see below).

Adenoviruses act as helper viruses in relation to parvoviruses. There are two subgenera of the genus *Parvovirus*. Subgenus A consists of viruses like rat virus and H1, that are competent in a variety of cell types. In other types of cells, which are nonpermissive, their replication is complemented by coinfection of the cells with human adenovirus (Ledinko *et al.*, 1969). Subgenus B comprises the adenovirus-associated viruses (AAV) which are completely dependent upon complementation by adenoviruses for their replication, both in cultured cells and apparently in nature (Blacklow *et al.*, 1967a).

Adenoviruses from widely divergent host species (man, monkey, dog, mice, chicken) will complement AAV, but there are large differences in their helper ability, and one simian adenovirus (SA7) fails to act as a helper (Boucher *et al.*, 1969). AAV does not multiply in simian cells infected with human adenoviruses (Blacklow *et al.*, 1967b), in which adenovirus DNA and T antigen are synthesized. However, Ito and Suzuki (1970) found that a *ts* mutant of adenovirus 31 that produced some early proteins but did not synthesize adenovirus DNA at the restrictive temperature nevertheless complemented AAV as well as wild-type adenovirus 31. Further, herpes simplex virus, but none of several other viruses tested, exerts a partial helper function, in that it promotes the synthesis of AAV antigen and AAV DNA, but not complete virions (Blacklow *et al.*, 1971; Boucher *et al.*, 1971).

## ADENOVIRUSES AND SV40

### Growth of Human Adenoviruses in Simian Cells

For years, human adenoviruses were isolated and cultivated in rhesus monkey kidney cells, but it is now clear that they will grow in these cells only by virtue of complementation by SV40 or perhaps some other virus (Lewis *et al.*, 1966; Butel and Rapp, 1967). SV40 is noncytocidal for rhesus monkey kidney cells, in which it often occurs as a latent infection; it is usually assayed in African green monkey kidney cells, in which it produces a characteristic cytoplasmic vacuolation (Sweet and Hilleman, 1960). Human adenoviruses do not multiply in previously uninfected simian cells, but O'Conor *et al.* (1963) found that if these cells were infected with SV40 prior to or at the same time as their exposure to adenovirus type 12, multiplication of both viruses could be detected by thin-section electron microscopy. This observation has been abundantly confirmed and extended (Rabson *et al.*, 1964; O'Conor *et al.*, 1965a; Schell *et al.*, 1966). Complementation by SV40 has been observed with all adenoviruses tested, but herpes simplex, measles, rabbit papilloma, and human wart viruses are ineffective as "helper" viruses. However, the growth of human adenoviruses in African green monkey kidney cells is enhanced by the simian adenovirus, SV15, which can replicate in these cells (Naegele and Rapp, 1967).

As with host-dependent mutants of other viruses (herpes simplex, Roizman and Aurelian, 1965; rabbitpox, Fenner and Sambrook, 1966), even in the absence of the helper virus, adenoviruses undergo some steps in the replication cycle in the restrictive host cells. Although they do not synthesize any of the virion antigens (penton, hexon, or fiber), African green monkey kidney cells infected with adenoviruses on their own synthesize T antigen (Malmgren *et al.*, 1966) and viral DNA (Rapp *et al.*, 1966; Reich *et al.*, 1966). In the presence of SV40, the amount of adenovirus DNA synthesized may be increased slightly, and all the virion antigens and a large yield of infectious adenovirus are produced, the yield of SV40 being slightly depressed.

## Adenovirus–SV40 Hybrids

In 1964, three groups of workers independently made an observation which suggested that adenovirus type 7 and SV40 might have formed virions which contained some SV40 genetic matehial within an adenovirus capsid (Huebner *et al.*, 1964; Rowe and Baum, 1964; Rapp *et al.*, 1964). The stock of adenovirus type 7 which they used was isolated in 1955 and had been carried through twenty-two serial passages in primary rhesus monkey kidney cells. By the ninth serial passage it was recognized that the stock virus was contaminated with SV40. It was passed twice in primary African green monkey kidney cells in the presence of potent anti-SV40 serum to get rid of the contaminating virus, and all seed preparations after this were shown to be free of infectious SV40. However, when concentrated suspensions of this "purified" virus were inoculated into newborn hamsters to determine the oncogenicity of adenovirus 7, the surprising discovery was made (Huebner *et al.*, 1964) that the majority of hamsters carrying primary tumors or transplants of these developed both adenovirus 7 and SV40 T antigens. Experiments in cultured cells gave similar results (Rowe and Baum, 1964; Rapp *et al.*, 1964); at no time could SV40 capsid antigens or infectious SV40 be detected. Transformation or tumor production could be neutralized by antiserum to adenovirus 7 but not with antiserum to SV40, and heating for 10 minutes at 56°C destroyed the capacity of the virus to induce SV40 T antigen. This treatment inactivates adenovirus 7 but has little or no effect on SV40. All this evidence indicated that SV40 genetic material was enclosed within adenovirus type 7 capsids. The virus stock was given the name E46[+]. At first it was suspected that a portion of the SV40 genome had been included together with a complete adenovirus 7 genome during packing of the adenovirus capsid. However, analysis of the infectivity of the E46[+] pool showed that it contained two sorts of viruses—complete adenovirus 7 and the hybrid particles, which turned out to contain incomplete adenovirus genomes and defective SV40 genomes (Boeyé *et al.*, 1965; Rowe and Baum, 1965; Rowe *et al.*, 1965). When the stock was titrated on human embryonic kidney cells, plaques were produced in numbers proportional to dose, i.e., cooperative effects between particles were not found. However, the progeny virus (adenovirus 7) could not productively infect monkey cells and lacked the ability to induce SV40-specific T antigen. When the stock was grown on AGMK cells, plaque formation followed a two-hit curve (Rowe and Baum, 1965; Boeyé et al., 1966; see Chapter 2) and the plaques contained both adenovirus 7 and hybrid particles.

From this it was deduced that: (a) because adenovirus 7 will not grow in monkey cells, its presence in the plaques formed by the E46+ pool in AGMK cells must mean that the stock contained SV40 genes (in the hybrid particles) that had acted as helper; (b) because no plaques that appeared on HEK cells contained hybrid particles, the hybrid must be unable to grow on HEK cells and must therefore contain a defective adenovirus genome; and (c) because no plaques that appeared on AGMK cells contained only hybrid, the hybrid must also be defective on AGMK cells and must depend for its growth on coinfection with a wild-type adenovirus 7 virus. To put it another way, of the two viruses in the E46+ stock only adenovirus 7 lytically infects HEK cells; productive infection of AGMK cells depends on the presence both of adenovirus 7 and hybrid virus genomes.

The defective hybrid particles in the E46+ pool have very similar physical properties to those of adenoviruses, and so far it has proved impossible to separate the hybrid particles from the excess nonhybrid adenovirus 7 particles (Rowe and Baum, 1965; Rowe et al., 1965; Baum et al., 1970). However, the hybrid particles are differentiated from the nondefective adenovirus 7 virions because they are able to induce SV40 T antigen and transplantation antigen (Rowe and Baum, 1965; Rapp et al., 1964; Rapp et al., 1966) and they transform hamster cells in vitro with the highest efficiency yet reported for a DNA tumor virus (Duff and Rapp, 1970b; Duff et al., 1972). Further, they have a greatly increased oncogenic potential in hamsters (Duff and Rapp, 1970b; Duff and Rapp, 1971; Duff et al., 1972), producing ependymomas which are morphologically identical to those produced by SV40 (Kirschstein et al., 1965).

Three lines of evidence prove unambiguously that the adenovirus and SV40 DNA's in the hybrid particles are covalently linked. First, the SV40 DNA sequences in the hybrid have a buoyant density both in neutral and alkaline conditions which is very close to that of adenovirus 7 DNA (Baum et al., 1966). Second, heteroduplexes formed between hybrid DNA and adenovirus 7 DNA show that a DNA sequence which maps between 0.05 and 0.21 of the distance along the nonhybrid adenovirus 7 DNA is missing from the hybrid. The deleted DNA has been partially replaced by an SV40 DNA sequence equivalent in length to 75% of the complete SV40 genome (Kelly and Rose, 1971). Presumably this SV40 sequence codes for the undefined helper function required for the replication of human adenoviruses in AGMK cells. Third, the process called transcapsidation confirms that the SV40 and adenovirus DNA's in the hybrid behave as a single molecule during replication. Transcapsidation occurs when monkey cells are infected with the E46+ pool (which contains both adenovirus 7 particles and hybrid particles) together with an excess of another immunologically distinct adenovirus. The progeny of such infections contains particles which consist of the original hybrid DNA wrapped in a capsid specified by the other adenovirus (Rowe and Pugh, 1966). Transcapsidation has been reported between E46+ and adenoviruses 1, 2, 5, 6, and 12 (Rowe, 1965; Rapp et al.,

1968). After transcapsidation with adenovirus 2, transcapsidated E46$^+$ can be partially separated from adenovirus 2 by centrifugation (Baum *et al.*, 1970). A summary of the results of infection of various cell types with adenovirus 7, SV40, and the E46$^+$ stock is shown in Table 7–8.

After the discovery of E46$^+$, many other adenovirus types were found to have acquired SV40 genetic material as a result of either unwitting or deliberate passage with SV40 (Rowe, 1965; Lewis *et al.*, 1966). The hybrid populations could be classified into two main groups: those like E46$^+$ and the adenovirus 3 hybrid that did not produce free infectious SV40 virus, and those like the adenovirus 1, 2, 4, 5, and 12 hybrid populations (for refs. see Lewis *et al.*, 1969). None of the latter group is completely defined, but the best studied by far is the adenovirus-SV40 hybrid population called Ad$_2$$^{++}$. The Ad$_2$ refers to the adenovirus phenotype (capsid). The double plus ($^{++}$) indicates that some particles not only contain SV40 genomes but produce SV40 progeny as well (Lewis *et al.*, 1969). This virus stock forms plaques with one-hit efficiency on both AGMK cells and HEK cells and induces SV40-specific T antigen. Isolation of plaques from both sorts of cells shows that the Ad$_2$$^{++}$ population consists of nondefective adenovirus 2 virions, nonhybrid SV40, and several classes of hybrid particles which can be purified by cloning (Lewis and Rowe, 1970).

The following classes of hybrid particles have been recognized: (a) Ad$_2$$^{++}$ HEY, which contains a complete SV40 genome bound to a fragment of adenovirus 2 DNA of molecular weight about $13 \times 10^6$ daltons (Wiese *et al.*, 1970) and which yields infectious SV40 virus with a high efficiency; (b) Ad$_2$$^{++}$ LEY which contains a complete SV40 genome bound to an almost complete adenovirus 2 genome (Wiese *et al.*, 1970). Ad$_2$$^{++}$ LEY yields SV40 with an efficiency of $10^{-4}$ to $10^{-3}$ of the Ad$_2$$^{++}$ HEY isolate (Lewis and Rowe, 1970); and (c) nondefective adenovirus 2–SV40 hybrids.

The last group of viruses contain a complete adenovirus genome covalently linked to a fragment of SV40 DNA (Lewis *et al.*, 1969), so that these hybrids cannot yield infectious SV40, but they can replicate without the aid of helper viruses both in HEK and in AGMK cells. This means that genetically pure stocks of hybrid virus can be obtained. Several of these hybrids have been isolated—Ad$_2$$^+$ND$_1$, Ad$_2$$^+$ND$_2$, Ad$_2$$^+$ND$_3$, etc.,—but work has been published on only one of them, Ad$_2$$^+$ND$_1$. The plus indicates that the particles contain SV40 sequences but do not produce infectious SV40 progeny; "ND" indicates that the particle is nondefective, i.e., it replicates without helper (Lewis *et al.*, 1969). Biophysical measurements show that Ad$_2$$^+$ND$_1$ DNA has a buoyant density of 1.175 g/ml and a $T_m$ of 75.1°C (Crumpacker *et al.*, 1971), almost identical with adenovirus 2 DNA. Its molecular weight is $25 \times 10^6$ daltons. Hybridization with SV40 RNA made *in vitro* suggests that about 1% of the Ad$_2$$^+$ND$_1$ genome consists of SV40 sequences (Crumpacker *et al.*, 1971); a more accurate estimate has been obtained by heteroduplex mapping which shows that a piece of DNA equal in length to 15.8% of the SV40 genome is present in the hybrid genome (Sharp, 1972). This small piece of DNA (about 800 base pairs) could code for a protein of about 25,000 daltons, which must have a dual function: (a) to code for the helper function which enables the hybrid

## TABLE 7–8

### Growth Properties of Adenovirus 7, SV40, E46$^+$ and Ad$_2$$^+$ND$_1$

| VIRUS | DESCRIPTION | CELLS | |
|---|---|---|---|
| | | HUMAN EMBRYO KIDNEY (HEK) | AFRICAN GREEN MONKEY KIDNEY (AGMK) |
| Adenovirus 7 | | Productive infection; good virus yield | T antigen, +; virion antigens +; no virus yield |
| SV40 | | Partially permissive | Productive infection |
| Mixture of adenovirus 7 and SV40 | | | Normal yield of both adenovirus and SV40 |
| E46$^+$ | Capsid: adenovirus 7 DNA: 84% of adenovirus 7 (20 × 10$^6$ daltons) plus 75% of SV40 DNA (2.25 × 10$^6$ daltons) | Plaques produced by adenovirus 7; no plaques or T antigen produced by SV40. Hybrid is lost during growth on HEK. The plaques that appear are due to the infectious adenovirus 7 particles in the E46$^+$ pool | Because simultaneous infection with E46$^+$ and adenovirus 7 is required, plaque formation follows 2-hit kinetics. Both E46$^+$ and adenovirus 7 appear in the yield. |
| Ad$_2$$^+$ND$_1$ | Capsid: adenovirus 2 DNA: almost complete adenovirus 2 (22 × 10$^6$ daltons) plus 16% of SV40 DNA 0.5 × 10$^6$ daltons) | Multiplies to yield Ad$_2$$^+$ND$_1$; produces SV40 U antigen | Multiplies to yield Ad$_2$$^+$ND$_1$; produces SV40 U antigen |

to grow without assistance in AGMK cells, and (b) it probably codes for the SV40-specific U antigen found in cells infected by the hybrid (Lewis and Rowe, 1971; Lewis *et al.*, 1969).

SV40-specific RNA can be detected in cells lytically infected by the hybrid (Oxman *et al.*, 1971). This RNA comes entirely from one segment of the "E" strand of the SV40 genome (Sambrook *et al.*, 1972b) and corresponds to part of the "early" sequences of SV40 DNA (Oxman *et al.*, 1971).

Although we know nothing of the helper function supplied to adenoviruses by SV40 in monkey cells and nothing about the mechanism by which adenovirus-SV40 hybrids are generated, the nondefective hybrid viruses are becoming increasingly important as tools for studying the molecular biology of SV40. This is because they contain defined sequences of SV40 DNA which are used in all sorts of experiments, from mapping SV40 deletion mutants to *in vitro* protein-synthesizing systems and studies of *in vivo* viral RNA synthesis and processing. In all probability, the hybrids will be exploited even more extensively in the next few years especially when other adenovirus 2 nondefective hybrids already isolated by Lewis come into general use.

## PHENOTYPIC MIXING

Following mixed infection by two viruses that share certain features, such as the type of capsid or the property of maturing by budding through the plasma membrane (see Chapter 6), some of the progeny may acquire phenotypic characteristics from both parents although their genotype is unchanged (Fig. 7–1; Table 7–4). For example, when cells are mixedly infected with antigenically different strains of influenza virus (Burnet and Lind, 1953), or with influenza and Newcastle disease viruses (Granoff and Hirst, 1954), or with vesicular stomatitis virus and the simian paramyxovirus SV5 (Choppin and Compans, 1970), the envelopes of some of the progeny particles contain viral antigens characteristic of each of the parental viruses. Phenotypic mixing of enveloped viruses may be accompanied by genetic reassortment (e.g., with two strains of influenza A), but characteristically the genomes are those of one or other parent, or mixed complete genomes in heteropolyploid particles (see below). Phenotypic mixing, probably limited to the peplomers, appears to be universal in cells mixedly infected with related viruses that mature by budding through the same cellular membranes. Defective avian and murine sarcoma viruses, for example, always have the pseudotype of the helper virus used, as a phenotypic character, and, in addition to the examples already quoted, phenotypic mixing has been demonstrated with alphaviruses (Burge and Pfefferkorn, 1966c), between measles and Sendai viruses (Norrby, 1965), and with mutants of herpes simplex virus (Roizman, 1965).

Phenotypic mixing also occurs readily with some of the nonenveloped icosahedral viruses. Among enteroviruses, it has been demonstrated with different poliovirus serotypes (Ledinko and Hirst, 1961), with which phenotypic mixing was evident to a high degree in the earliest maturing virus, and in almost every virion in the final yield of mixedly infected cells. Unrelated enteroviruses also undergo phenotypic mixing, as with echovirus 7 and coxsackievirus A9 (Itoh and Melnick, 1959), poliovirus and coxsackievirus B1 (Holland and Cords, 1964), and foot-and-mouth disease virus and a bovine enterovirus (Trautman and Sutmoller, 1971). In the last two cases, a majority of progeny particles seems to have completely heterologous capsids; a situation sometimes referred to as *genomic masking* or, in the rather different circumstances that obtain with adenoviruses and adenovirus-SV40 hybrids (Rapp and Melnick, 1966; see above) as *transcapsidation*. Genomic masking of foot-and-mouth disease virus may have important epidemiological consequences. Phenotypic mixing with adenoviruses has been shown to involve heteropolymer hexon capsomers containing polypeptides derived from viruses of two different serotypes, as well as capsids containing capsomers of both parents (Norrby and Gollmar, 1971).

## HETEROZYGOSIS AND POLYPLOIDY

These two terms have well recognized meanings in the genetics of higher organisms, where polyploidy refers to the presence in one cell of multiple copies of the species' suite of chromosomes, and heterozygosis implies that diploid chromosomes differ in allelic markers. In early experiments on recombination with T-even bacteriophages, Hershey and Chase (1951) demonstrated the occurrence of heterozygotes in the yields of mixedly infected bacteria. They arose with a frequency of 1 to 2% among the progeny of a cross with respect to every locus examined. Heterozygous phage particles contain normal double-stranded DNA in which the information carried in the two strands is different in the heterozygous region, i.e., heteroduplex DNA (see Hayes, 1968). This sort of heterozygosis has now been demonstrated with herpes simplex virus (Timbury and Subak-Sharpe, 1973). Heterozygotes have also been described in the Ff bacteriophages, which are rod-shaped phages that contain single-stranded DNA, but in this case two (or more) separate viral DNA molecules are enclosed within a single elongated capsid (Salivar *et al.*, 1967). A situation like this is very common with animal viruses and it is probably more convenient to refer to these particles as being polyploid; heteropolyploid if the genomes are derived from genetically different parents (Fig. 7–1).

Polyploidy is very common among animal viruses that mature by budding from the cell membranes, several nucleocapsids being enclosed within a single envelope (review, Simon, 1972), as can be seen by electron microscopic observations of polynucleocapsids in togaviruses (Matsumura *et al.*, 1971; Higashi *et al.*, 1967) where the process is uncommon and in paramyxoviruses (Prose *et al.*, 1965; Yunis and Donnelly, 1969), where it is very common. Hosaka *et al.* (1966) with parainfluenza type 1 virus, and Dahlberg and Simon (1969a) with Newcastle disease virus, have clearly demonstrated that normal populations of virions of these paramyxoviruses include many virions with multiple complete nucleocapsids. This provides an explanation for the frequent occurrence of "heterozygotes" (heteropolyploids) in the yields of cells mixedly infected with different strains of Newcastle disease virus (Granoff, 1962; Dahlberg and Simon, 1969b), and influenza virus (Hirst, 1962). When cells are infected with recognizably different strains of Newcastle disease virus almost all the progeny is phenotypically mixed, and cells infected with single particles of this phenotypically mixed virus yield both the original parental genotypes and some phenotypically mixed progeny. Heteropolyploidy, with complementation between the genomes enclosed within a single envelope, gives experimental results that have been misinterpreted as recombination (Wright and Chanock, 1970; Dahlberg and Simon, 1969b; Pringle et al., 1971), and may constitute an almost insuperable difficulty in determining whether intramolecular recombination occurs in such systems (Simon, 1972).

## MAPPING BY SERIAL INACTIVATION AND DIRECT BIOCHEMICAL ANALYSIS

### Target Size Measurements

Serial inactivation of viral infectivity and other biological properties of viruses has been used to measure the "target size" of the genes controlling particular biological functions. Ethylene iminoquinone inactivates the capacity of influenza virus to synthesize various viral proteins in a particular order: infectious virions, hemagglutinin, neuraminidase, ribonucleoprotein, and cell shutdown protein (Scholtissek and Rott, 1964). The capacity of polyoma virus to transform is more resistant to inactivation by irradiation than infectivity (Latarjet et al., 1967). With avian sarcoma viruses a variety of results has been obtained; some radiation-damaged particles can transform but not reproduce, others can reproduce but not transform (Toyoshima et al., 1970; Hanafusa, 1970).

### Determination of Gene Order with Antibiotics

Painstaking studies over a number of years by Cooper involving recombination between ts mutants indicated that the "genes" of poliovirus could be ordered into two groups: (a) those coding for the structural proteins of the virion (NCVP1), which represented just under one-half the genome and appeared to be situated at the 5' end of the RNA molecule, (b) those coding for RNA polymerase I (NCVP2) situated at the 3' end (Cooper, 1969; Cooper et al., 1971). Now that a wide range of inhibitors of protein synthesis with accurately defined modes of action have become available, the precise "gene" sequence within such RNA molecules can be determined directly by measuring the order in which proteins are synthesized. Pactamycin, for example, specifically inhibits initiation of protein synthesis but permits completion and release of polypeptides already commenced (see Chapter 4); hence, if radioactive amino acids are added after pactamycin, proteins coded by the 3' end of the mRNA will be more heavily labeled than those coded by the 5' end. Emetine, on the other hand, "freezes" the nascent polypeptide on its ribosome, so that a very short radioactive pulse followed by emetine will permit the release of only those polypeptides labeled toward the C-terminal end, i.e., in that region coded by the 3′ end of the mRNA. Using these two inhibitors to analyze the kinetics of labeling of the various cleavage products of the polio precursor protein, Rekosh *et al.* (1970) were able to confirm Cooper's provisional map and Rekosh (1972) went on to show that the "gene order" within NCVP1 is (reading from the 5′ end of the RNA molecule): VP4, VP2, VP3, VP1. Rueckert (1972) has performed similar experiments with EMC virus, and by comparing the kinetics of "pulse-chase" with those of "progressive" labeling of the various proteins has derived an almost identical map, together with complete details of the cleavage program, which occurs in up to four successive rounds.

## ADAPTATION TO NEW HOSTS

Until the last decade or so, when molecular biological techniques began to be applied to the subject, animal virology was developed mainly as a branch of pathology, by investigators concerned with viruses as agents of disease. This still remains a prime motive for their study, although fields like tumor virology provide valuable tools for the analysis of the function of eukaryotic cells as well as the potential for understanding cancer.

The essential prerequisite for the experimental investigation of an animal virus or the disease it causes is the production in some laboratory host (which includes cultured cells as well as intact animals) of recognizable signs of infection associated specifically with the virus in question. This may happen the first time that the virus is inoculated into an experimental host, e.g., cowpox virus when it is inoculated from an infected bovine onto the chorioallantoic membrane or into the rabbit skin, and neither the nature of the lesions nor the efficiency of plating for the new host changes with serial passage. In other cases, minimal signs of infection are observed initially, yet after serial passage, sometimes prolonged, a lethal infection is regularly produced, as in the adaptation of poliovirus and dengue virus to rodents (Armstrong, 1939; Sabin and Schlesinger, 1945). Similar results are found with cultured cells, and most virological research is performed with strains of virus adapted to produce characteristic lesions, e.g., plaques, in cultured cells. A frequent byproduct of such adaptation to a new experimental host is the coincident attenuation of the virus for its original host.

The practical importance of adaptation and attenuation has resulted in an enormous volume of literature concerned with growth of animal viruses in different hosts, involving primary adaptations, secondary adaptations from one laboratory host to another or from one organ or tissue of a particular host to another, and studies of the effects of such adaptations on the pathogenic capacity of the virus in its original host. Because of the complexity of the experimental material virtually nothing is known of the mechanism of adaptation. We can postulate that adaptation usually results from selection of those viral mutants best equipped to multiply in the novel host, but it has not been possible to analyze satisfactorily even the simplest examples of adaptation to cultured cells, let alone the complexity of adaptation required to produce death in an intact animal (Fenner and Cairns, 1959).

Current research in animal virus genetics suggests two other mechanisms of adaptation that may be important in specific cases. With human adenoviruses, "adaptation" to monkey cells was essentially due to complementation by latent SV40 in those cells, followed in a few instances by the production of novel viable adenovirus–SV40 hybrids that could multiply in simian cells without other SV40 helper activity. The leukoviruses of various animals present another picture, for here virtually all cells of many species carry integrated viral genomes which may be expressed to a varying degree, and multiplication of an added leukovirus may be complemented by the activity of the endogenous virus.

## CHAPTER 8

# Interference and Interferon

## INTRODUCTION

Depression of viral yield by interference is one of the usual consequences of the infection of cells with more than one virus. The possibility that an understanding of the phenomenon will lead to practical means of preventing viral infections of human and veterinary importance has attracted the interest of epidemiologists and clinicians. The discovery of a virus-induced interfering agent, interferon, raises particularly challenging and as yet unsolved problems for virologists—challenging both from the viewpoint of understanding what interferon is and how it acts, and in the possibility of potentially controlling viral diseases by mobilizing the body's own interferon system.

Viral interference can be defined as a state, induced by an interfering virus, that is characterized by resistance of cells or tissues to infection by a challenge virus. The interfering virus does not necessarily have to multiply in order to induce interference, and the ability of the challenge virus to multiply may be partially or completely inhibited.

At least four types of viral interference can be distinguished: (a) virus-attachment interference, due to destruction of cellular receptors by interfering virus, (b) homologous, or defective particle, interference, (c) heterologous interference between unrelated viruses due to a structural component or virus-induced protein interfering with the replication of the challenge virus, and (d) interference mediated by interferon.

Our concepts of intracellular interference, as distinct from virus-attachment interference, were long dominated by the belief that interference was always associated with the presence of the interfering virus in the cell in which interference occurred (see Schlesinger, 1959). This view was radically changed by the discovery by Isaacs and Lindenmann (1957) that an "interfering" dose of influenza virus could induce cells to produce and release a nonviral protein which could modify uninfected cells and protect them against viral infection. They named this soluble inhibitor "interferon," and characterized it by its failure to affect viruses *in vitro*, its nonspecificity, in relation to virus, and its capacity to transfer resistance from one cell (or cell culture) to another in the absence of the original interfering virus.

## VIRUS-ATTACHMENT INTERFERENCE

The alteration or destruction of cellular receptors for virions is one type of interference that is clearly different from all other cases. Baluda (1959) showed that the "homologous interference" which he had previously described in chick cells treated with UV-irradiated Newcastle disease virus (Baluda, 1957) was due to destruction of the cellular receptors, probably by the viral neuraminidase.

Another example of interference due to the effects of viral infection on cellular receptors is interference with infection of chick cells by Rous sarcoma virus, first demonstrated by Rubin (1960). This interference is due to the fact that chick cells are often infected with a noncytocidal, nontransforming avian leukovirus that abrogates the cellular receptors for Rous sarcoma virus, but has no effect on the susceptibility of the cells to infection by several unrelated viruses (Rubin, 1961; Hanafusa et al., 1964). Steck and Rubin (1966a,b) have made a detailed analysis of this type of interference. It is specific for avian leukoviruses of either of the major antigenic groups and may be induced rapidly, by large doses of interfering virus, or slowly, in cells infected with small doses. In both cases, interference appears to be due to blockage of the cellular receptors, which are specific for viruses of each of the antigenic types of the avian leukoviruses.

## HOMOLOGOUS INTERFERENCE

One of the most characteristic features of interferon is its activity against a wide range of viruses. Intracellular (as distinct from virus-attachment) interference, which is active only against homologous or even only against homotypic viruses, must therefore operate by some mechanism other than interferon production. Examples are known from experiments with influenza virus, vesicular stomatitis virus, and poliovirus.

Henle and Rosenberg (1949) described a situation in which interference with active influenza virus by UV-irradiated virus was strictly homologous, i.e., UV-irradiated type A influenza virus interfered with type A but not type B influenza virus, and in which interference was still effective when UV-inactivated (interfering) virus was added to the allantoic sac up to 3 hours after the active virus.

With vesicular stomatitis virus (VSV), of which there are two serotypes (N.J. and Ind.), Cooper (1958) described a unique example in which interference was homologous but heterotypic. UV-irradiated VSV-NJ interfered with active VSV-Ind but not with active VSV-NJ, and vice versa. This interesting situation has not been further explored, and indeed the whole subject of homologous interference would now repay closer investigation.

In homologous interference with polioviruses, the characteristics of interference seem to depend upon whether the interfering virus is able to multiply or not, rather than whether the challenge virus is homotypic or heterotypic (Pohjanpelto and Cooper, 1965). With viruses that are able to multiply, Cords and Holland (1964a) found that interference between equal multiplicities of virus was not induced unless the interfering virus was allowed to replicate for an hour before challenge. Interference induced by a guanidine-sensitive virus was reversed when the cells were challenged in the presence of guanidine, so that action of endogenous interferon was unlikely. The most likely explanation for this type of interference seems to be that there is a competition for replicating sites or for some cell component needed for viral synthesis.

A familiar result in infectivity titrations carried out in experimental animals is the occurrence of a "prozone;" animals infected with very large doses may survive, whereas those infected with smaller doses die. Similar effects are obtained in nonantibody-producing systems, like embryonated eggs or cultured cells, when the yields of very large and smaller doses of virus are compared. This particular type of homologous interference has been referred to previously (Fenner, 1968) as autointerference, and may, in some cases, be due partly to the presence of interferon in the inoculum; this can be removed by partial purification of the virus or its effects avoided by using heterologous cell systems. All preparations of virus consist of a mixture of infective and noninfective particles, the latter usually greatly exceeding the former. The usual explanation given for autointerference, other than that due to interferon in the inoculum, is that with large inocula cells may receive particles of both types, and the inactive particles interfere with the replication of active particles, either by endogenous interferon production or by some other mechanism.

A special type of autointerference occurs with influenza and vesicular stomatitis viruses. When concentrated suspensions of either of these viruses are passaged serially there is severe autointerference, as far as infectious virus is concerned, but large amounts of noninfectious "virus" are produced (see Chapter 6). This "incomplete virus" (as it is called with influenza, von Magnus, 1954) or "T component" (for vesicular stomatitis virus, Cooper and Bellett, 1959) shows strong homotypic and weak heterotypic interference, but does not interfere with heterologous viruses (Huang and Wagner, 1966a).

The T component of vesicular stomatitis virus can be separated from the virions by density gradient centrifugation and it has now been well characterized. It consists of spherical particles about 65 mm in diameter (compared with rod-shaped particles measuring 65 × 180 nm for infectious VSV) with a detailed ultrastructure very like that of VSV virions (Huang et al., 1966). These particles contain only one-third as much viral RNA as the infectious virus (Huang and Wagner, 1966b). The interfering activity is dependent upon the functional activity of the RNA of the T component and must take place at some early stage in the biosynthesis of VSV, after penetration. From a number of studies on the effects of defective interfering particles at different stages of the replication of normal virions, it appears that there is no direct interference with viral transcription or protein synthesis and the defective T particles possess no transcriptase activity nor are they transcribed (Stampfer et al., 1969; Portner and Kingsbury, 1971; Huang and Manders, 1972). The most likely explanation for their inhibitory action is that the fragment of RNA from defective particles competes directly with nondefective viral RNA for the replicative enzymes.

Huang and Baltimore (1970b) have listed a number of RNA and DNA viruses that produce defective interfering (DI) particles when cells are infected with high multiplicities of virus; these DI particles interfere specifically with the intracellular replication of nondefective homologous virus. They suggest that such DI particles could be important determinants in the course of both acute, self-limiting viral infections and of persistent slowly progressing viral diseases. Lymphocytic choriomeningitis (LCM) is one persistent infection in which DI particles could well be important. Following infection before or shortly after birth, young mice become immunologically tolerant and the virus persists throughout their lifetime (Mims, 1966a; see Chapter 12). By infecting cultured mouse L cells with high multiplicities of LCM virus, Lehmann-Grube et al. (1969) established a system that produced only DI particles. Such cells exhibited intracellular interference with superinfecting normal LCM virus and it is conceivable that in a similar manner DI particle products might maintain persistent infection in adult tolerant mice. LCM virus fails to induce interferon in either infected animals or cultured cells.

## HETEROLOGOUS INTERFERENCE

An example of heterologous interference in which interferon plays no part was first described by Marcus and Carver (1965, 1967). They used the term "intrinsic interference," a term that could also be applied to other examples of heterologous interference. Cells infected with rubella virus (which is noncytocidal) and cells carrying a mouse leukemia virus or undergoing noncytopathic infection with Sindbis, West Nile, or MEF-poliovirus were found to be absolutely resistant to infection with several other viruses. Interference was induced only by active virus and it was insensitive to actinomycin but sensitive to puromycin, i.e., it resulted from the action of a protein(s) specified by the virus. Intrinsic interference was confined to the infected cells, and it was manifest at some stage after penetration, i.e., it was not due to alteration of receptors. Of several viruses tested, Marcus and Carver could demonstrate intrinsic interference only with NDV, but Wainwright and Mims (1967) observed it in mouse cells infected with lymphocytic choriomengitis virus and challenged with parainfluenza virus type 1.

Vaccinia virus (a poxvirus) and frog virus 3 (FV3; an iridovirus) are both DNA viruses that replicate in the cytoplasm (McAuslan, 1969a; Granoff, 1969). They are in no way related biologically, but some structural component of FV3 can inhibit the replication of vaccinia virus in cells coinfected with both viruses (Aubertin and Kirn, 1969; Vilaginès and McAuslan, 1970). The molecular events in poxvirus replication are known in some detail (Chapter 5). Vilaginès and McAuslan (1970), using systems in which FV3 replication was nonpermissive, showed that a structural component of FV3 could associate with and block the transcription of uncoated poxvirus DNA. FV3 had no effect on transcription initiated by coated poxvirus genomes, supporting the contention that the inhibitor combined directly with poxvirus DNA to repress transcription. By first infecting cells with vaccinia virus in the presence of a reversible specific inhibitor of DNA synthesis, one can set up conditions where all mechanisms necessary for poxvirus DNA synthesis are primed to initiate DNA synthesis upon reversal of the inhibitor (Kates and McAuslan, 1967c). Under these conditions, cycloheximide and actinomycin D, potent inhibitors of protein and RNA synthesis, cannot block initiation of poxvirus DNA synthesis when added just prior to the reversal step; in striking contrast, superinfecting FV3 particles can do so. Attempts to isolate a repressor protein from FV3, or to isolate a repressed complex of newly replicated poxvirus DNA and an FV3 polypeptide, have not been successful.

Two other examples of heterologous interference between DNA viruses, one a bacteriophage system and the other an animal virus system, provide some interesting comparisons. Hayward and Green (1965) found that transcription of bacteriophage lambda mRNA in *Escherichia coli* is inhibited by superinfection with phage T4 and Giorno and Kates (1971) studied the inhibition of vaccinia virus replication in adenovirus-infected HeLa cells. In both cases, complete inhibition of the challenged virus requires synthesis of viral proteins. In cells infected with both adenovirus and vaccinia virus, it appears that some adenovirus capsid protein(s) synthesized late in adenovirus infection blocks the association of poxvirus mRNA with ribosomes.

FV3-mediated interference differs from the rubella-NDV, λ-T4 or adenovirus-poxvirus systems in that no protein synthesis is necessary to bring about interference, suggesting that FV3 proteins are highly active or have some quite unusual affinities. These examples of heterologous interference suggest that judicious exploitation of the phenomenon might well be a useful adjunct to specific chemical inhibitors in studying the programming of molecular events in viral multiplication.


## INTERFERON-MEDIATED INTERFERENCE

Isaacs (1963) described interferon as "an antiviral substance produced by the cells of many vertebrates in response to virus infection. It appears to be of protein or polypeptide nature, it is antigenically distinct from virus, and it acts by conferring on cells resistance to the multiplication of a number of different viruses."

Our definition today would be very little different: "Interferons are cell-coded proteins induced by foreign nucleic acids; they are nontoxic for cells but are able to inhibit the multiplication of vertebrate viruses in cells of homologous species." Because estimates of molecular weights of interferon from a given source vary considerably, and since there are differences in mode of formation (i.e., release, as opposed to induced synthesis), it is usual to refer to "interferons" as a family of protein species rather than to an "interferon." It is also now recognized that a number of "nonviral" inducers of interferons are contaminated with viral nucleic acids and that other nonviral agents may cause release of preformed interferons *in vitro* and *in vivo*. The statement that interferon is of cellular origin rather than a viral protein is based on the following evidence:

(a) there are nonviral inducers of interferon, (b) interferon induced in chick cells by herpes simplex virus and purified 4500-fold shows the same physicochemical properties as that induced by influenza virus (Lampson *et al.*, 1965), and (c) actinomycin D, which blocks cellular DNA-dependent RNA polymerase, also inhibits interferon synthesis in cells infected with RNA viruses whose replication is insensitive to the drug (Heller, 1963).

## BIOLOGICAL PROPERTIES OF INTERFERONS

### Viral Specificity

Interferon produced by one cell type is biologically active against a wide variety of viruses; an interferon produced by treatment of chick cells with influenza virus, for example, may inhibit the multiplication of homologous and heterologous influenza viruses, togaviruses, vaccinia virus, etc., in chick cells.

The sensitivity of animal viruses to a given preparation of interferon varies markedly. In general, togaviruses and vesicular stomatitis virus (VSV) appear to be among the most susceptible, but there is a large variation in sensitivity that depends on the cell–interferon system used as an indicator. For example, vaccinia virus is more sensitive than VSV to chick and mouse interferons, but the reverse is found with respect to human interferons (Riley *et al.*, 1966; Ruiz-Gomez and Isaacs, 1963; Gallager and Khoobyarian, 1969). Similar situations have been found with respect to the sensitivity of vaccinia and Semliki Forest viruses to chick and bovine interferons (Riley *et al.*, 1966; Finter, 1968). Wagner *et al.* (1963) found that two variants of VSV that show the same sensitivities to chick interferon have different sensitivities to mouse interferon. Stewart *et al.* (1969) compared five different viruses for their relative sensitivity to interferons from a variety of host species. Vaccinia virus was the most sensitive to human, rabbit, and bat interferons. On the other hand, Semliki Forest virus, the least sensitive to hamster and mouse interferons, was relatively sensitive to bat and rabbit interferons. These results indicate the difficulty of comparing the potency of interferon preparations from different species of animal, and underline the importance of stating the species of interferon used when relative sensitivities of viruses are determined. It seems unlikely that the induction of interferon during the assays is responsible for the observed differences in relative sensitivities (Gifford, 1963; Stewart *et al.*, 1969).

### Cell Specificity

Interferon has a high degree of host cell species specificity. Tyrrell (1959) showed that interferons produced in calf cells and chick cells, respectively, were much more effective in cells of the homologous than of the heterologous species, and subsequent work with crude and purified interferons has confirmed this.

Earlier reports of the absence of species specificity were probably due to the virus-inhibiting activities of impurities in the interferon preparations. It is important that purified preparations of interferon should be used in such experiments; for example, reduced yields of vaccinia virus were obtained in chick cells treated with crude chick interferon or normal allantoic fluid, or mouse serum interferon or normal mouse serum (Buckler and Baron, 1966), but if the cell cultures were washed after treatment and prior to infection with challenge virus only the chick interferon exhibited antiviral activity. Fantes (1966) followed the potency of chick interferon through a rigid purification procedure and showed that as purity increased, inhibitory activity in chick cells rose and activity in heterologous cells declined. Absolute species specificity was demonstrated with highly purified chick and mouse interferons (Merigan, 1964). The species specificity of interferon is not due to the failure of heterologous cells to adsorb it (Gifford, 1963; Merigan, 1964), but rather to whether it elicits antiviral activity in the treated cell.

Interferon prepared in cells derived from different organs of the same animal show no evidence of organ or tissue specificity, although some types of cultured cells are more sensitive than others to the antiviral activity of interferons prepared in homologous and heterologous tissues (Riley and Gifford, 1967). Tumor cell lines are usually less sensitive to interferon, and often produce less interferon, than cells of the same species in primary culture (Rotem *et al.*, 1964), and different lines of HeLa (Cantell, 1961) and L cells (Wagner, 1965) differ in their sensitivity to the action of interferon on viruses.

## Immunogenicity

Interferon is a cellular protein and should therefore by immunogenic in animals of another species. However, either because of its poor immunogenicity or the small immunogenic mass employed in most experiments, the only report of consistent success in the production of antibodies against interferon is that of Paucker (1965), who established that antibodies against L cell interferon could be produced by prolonged immunization of guinea pigs and that L cell, chick, and HeLa cell interferons were antigenically different. In subsequent neutralization experiments with purified mouse interferons, Boxaca and Paucker (1967) found no antigenic differences between interferons of 30,000 and 90,000 molecular weights, elicited by viral or by nonviral inducers. This suggests that each mammalian species produces a single characteristic species of interferon.

## PHYSICOCHEMICAL PROPERTIES OF INTERFERONS

From a consideration of its physicochemical and biological properties we can set down a number of criteria which must be satisfied if a virus-inhibitory substance is to be classed as an interferon. This is important, because a number of other substances that occur in extracts of infected and uninfected cultures may under some conditions inhibit the growth of certain viruses. For example, Buckler and Baron (1966) have reported that an inhibitor for vaccinia virus found in the supernatant fluids of some virus-infected cells could be distinguished from interferon only by its inhibitory activity in heterologous as well as homologous cells, and by the fact that this activity could be removed by washing the treated cells. When "interferons" with atypical biological characteristics are found, it is important to purify them and to measure their molecular weight and charge.

Properties common to all interferons so far examined are: (a) they are non-dialyzable, (b) they are not sedimented by centrifugation at 100,000 $g$ for 4 hours, (c) they are inactivated by treatment with proteolytic enzymes (trypsin, pepsin, and chymotrypsin) but unaffected by DNase, RNase, or lipase, and (d) they are stable over a wide pH range (pH 2–10). These properties are consistent with the idea that interferons are proteins, and the procedures used to purify interferons involve standard techniques in protein chemistry including precipitation by ammonium sulfate, purification by gel filtration or electrophoresis, and separation on ion exchangers. Outlines of procedures for purifying interferons from many different sources are provided by Fantes (1966). Chick interferon is one of the most stable and easiest to purify to high titers and Merigan et al. (1965) achieved a 20,000-fold purification with a specific activity of $10^6$ units per milligram of protein.

The best preparations of interferons that have been examined chemically seem to be proteins containing some carbohydrate, including glucosamine. Their isoelectric points are in the range 6.5–7.0; they are stable over a pH range of approximately 2–10 and disulfide groups are required for antiviral activity since cysteine and $\beta$-mercaptoethanol can reduce activity markedly. The biological activity of interferons from different animal species varies greatly in heat stability (Fantes, 1966).

There have been conflicting estimates of the molecular weights of different interferons. In general, these estimates fall into three major classes, approximately 100,000, 50,000, and 25,000 daltons, respectively. Levy et al. (1970) suggested that interferons that appear early, which tend to have higher molecular weights, may be those preformed in cells, i.e., their appearance is not prevented by inhibiting cellular RNA and protein synthesis. Whether or not this is so, many of the variable results could be due to aggregation of a basic monomeric unit or to complexing with other cell products during synthesis or extraction.

Experiments by Carter (1970) with preparations of purified human and mouse interferons induced by Newcastle disease virus give strong support to the idea that interferon exists as multimers of a basic unit. Mouse interferon, purified 500-fold, when subjected to electrofocusing was found as two equally active molecular forms of 38,000 and 19,000 daltons, respectively. Similarly, human interferon purified 1500-fold existed as 24,000 and 12,000 daltons molecular weight species, showing that native molecules can exist as dimers of similar subunits. In a recent review, Colby and Morgan (1971) refer to a polynucleotide-induced human interferon isolated as a molecular weight species of 96,000 daltons that was dissociable into forms with molecular weights of 24,000 and 12,000 daltons.

If all interferons are multimers of a fundamental unit of molecular weight around 12–19,000 daltons, and have a common mode of antiviral action, it is reasonable to consider them as a homogeneous group of substances. If however, the considerable differences in molecular weights reflect major qualitative differences, or if interferons from different sources differ in their mode of antiviral action, we are clearly dealing with a heterogeneous collection of substances linked only by the important practical consideration that they are all nucleic acid-induced antiviral substances of cellular origin. The latter definition would suffice if we wish to use "interferons" as an operational term embracing all known or unknown substances with this biologically important property. The alternative is to restrict the term to substances with properties indistinguishable (except for species-specificity) from chick or mouse interferons, the first interferons to be studied in detail and still the only ones whose intracellular mode of action has been analyzed to any extent.

Lockhart (1966) has suggested the following criteria for the acceptance of a viral inhibitor as an interferon: (a) The inhibitor must be a protein and be formed as a result of addition of an inducing agent to cells or animals. (b) The antiviral effect must not result from nonspecific toxic effects on the cells. (c) The putative interferon must inhibit the growth of viruses in cells through an intracellular action involving cellular RNA and protein synthesis. (d) The inhibitor must be active against a range of unrelated viruses. (e) The supposed interferon must show a marked specificity of its antiviral activity in cells of the animal species in which it was produced.

## METHODS OF ASSAY OF INTERFERONS

So far the only way of assaying interferon is by demonstration of its biological activity. A unit of biological activity has not been standardized but in general relates to the minimum amount of interferon needed to induce significant resistance to some measurable effect of viral multiplication. Using an appropriate cell and challenge virus, one can assay resistance by inhibition of plaque formation, direct measurement of the yield of infectious virus under single-step growth conditions, or by quantitating some activity associated with virus, such as hemagglutination. Togaviruses have been widely used for interferon assays because they plaque well and are very sensitive to inhibition by interferon.

## INDUCTION OF INTERFERON

### Influence of Cell Type and State

Virtually all types of cells from all vertebrates appear to have the capacity to produce interferon, although there are some exceptions, such as VERO cells, a clone of monkey cells that appears to lack this capacity. As yet, there is no convincing evidence that interferons are produced by insect, plant, or bacterial cells. In general, human cells are comparatively poor producers of interferon, and unless conditions are established that will substantially enhance its production, it is unlikely that the expectation of controlling viral infections in man by active "interferonization" will be fulfilled. To this end, much current activity concerned with the therapeutic use of interferon is directed to finding the most potent inducers and the optimum conditions for interferon synthesis and release.

Different cell lines and different sublines of the same cell type vary in their ability to produce interferon. Cantell and Paucker (1963) found that with the same preparation of Newcastle disease virus one HeLa cell line produced ten to fifty times more interferon than another, and similar observations have been made with other lines of transformed cells (Henle and Henle, 1963).

There is some evidence that the "age" of a cell, i.e., whether it is derived from a very young or an older embryo or animal, or whether in an established culture a short or a long period has elapsed since the last mitosis, may affect its ability to produce interferon, "Younger" cells in general make less interferon (Ho and Enders, 1959; Isaacs and Baron, 1960; Heineberg *et al.*, 1964). Carver and Marcus (1967) showed that chick embryo fibroblasts aged *in vitro* exhibited an increased capacity to produce interferon and an increased sensitivity to interferon action; in an earlier study of the factors affecting the susceptibility of chick allantoic cells to influenza virus *in vitro*, White (1959) had demonstrated a progressive increase in resistance with age of embryonic development.

### Inducers of Interferon

Mobilization of the body's own interferon system ("interferonization") would be of obvious therapeutic value in protecting an animal against a variety of viral infections. With this in mind, an active search for nontoxic nonviral inducers of interferon is in progress (Colby, 1971; Colby and Morgan, 1971). One can make a broad division of interferon inducers into those effective only in intact animals and those active both *in vivo* and *in vitro*. The latter group includes DNA and RNA viruses, naturally occurring double-stranded RNA's and synthetic helical polyribonucleotides. The *in vivo* inducers are a heterogenous collection that includes those just mentioned as well as endotoxins, protozoa, bacteria, and synthetic polyanions, such as polysulfonates and polycarboxylates.

It appears that interferon inducers that are active only in intact animals might exert their effects on the membranes of cells of the reticuloendothelial system and in some way stimulate the release of preformed interferon rather than its synthesis (Ho and Kono, 1965; Ke *et al.*, 1966). Synthetic copolymers of maleic acid anhydride elicit interferon in mice (Regelson, 1967) and a variety of polycarboxylated and polysulfonated polymers stimulate interferon release in treated animals; the structural characteristics of a number of such polyanions have been discussed by Merigan (1967).

## VIRUSES AND NUCLEIC ACIDS AS INDUCERS OF INTERFERON

### Induction by Poly I·Poly C

It had been observed that when chick or mouse cells in culture were incubated with chick or mouse RNA, the RNA from heterologous species was more effective in inducing resistance of cells to challenge virus. This led Rotem *et al.* (1963) to suggest that viruses might elicit interferon by presenting a foreign nucleic acid to a cell. Implicit in their proposal was the idea that the interferon induction mechanism involves the recognition of specific nucleotide sequences.

The report of Braun and Nakano (1965) that complexes of polynucleotides stimulated antibody formation in mice prompted others (Field *et al.*, 1967) to examine the capacity of polynucleotides to induce interferon in rabbits and cultured rabbit cells. A duplex of polyriboinosinic acid and polycytidylic acid (poly I·poly C) was found to be the most active of a number of polynucleotides tested. This important finding greatly stimulated the search for a nonviral interferon inducer that could be used in the prophylaxis of viral diseases, and also provided an opportunity to study the relation between interferon induction and the primary and secondary structure of the inducer molecule. Numerous references to these studies can be found in recent reviews (Colby, 1971; Colby and Morgan, 1971).

No particular sequence of nucleotides is necessary for polynucleotides to act as interferon inducers; natural double-stranded RNA's from a variety of sources (e.g., reovirus, viruses from *Penicillium*) are good inducers, and the homopolymeric duplex poly I·poly C, the alternating copolymer poly I-C, and double-stranded polynucleotides with sequences including adenyl, uridyl, or guanyl residues are all active inducers. The secondary structure is important; single-stranded poly I, poly C, or poly A (polyriboadenylic acid) are inactive even at a concentration 10,000-fold greater than those at which poly I·poly C is active as an inducer. The polydeoxynucleotide duplexes (e.g., the duplex of polydeoxyguanylic acid and polydeoxycytidylic acid) are virtually inactive or exhibit very weak activity. The intracellular target site for interferon induction thus appears to recognize both the molecular configuration and the acid functional group in an inducer; in particular, there is a requirement for a double-stranded helical molecule and for 2' hydroxyl groups in the sugar moieties of the nucleotides.

It is established that the antiviral state induced by poly I·poly C is due to interferon production. However, the mechanism by which interferon is produced in response to synthetic polynucleotides differs in some respects from that following infection with single-stranded RNA viruses. For example, in some cultured cells, interferon appears more rapidly and its production is less sensitive to inhibition by actinomycin D when induced by synthetic polynucleotides (Finkelstein et al., 1968). Even in systems where input reovirus double-stranded RNA was the inducer, interferon production was much more sensitive to cycloheximide and actinomycin D inhibition than was induction by poly I·poly C (Long and Burke, 1971). Furthermore, interferon production in response to synthetic inducers *in vivo* and in some cultured cells is not suppressed by inhibitors of protein synthesis. In fact, in some cases, cycloheximide added after synthesis has started actually enhances interferon production, from which it has been suggested that in such systems an endogenous inhibitor of interferon is normally synthesized soon after introduction of the synthetic inducer (Vilček, 1970).

## Induction by Viral RNA

The interferon-inducing capacity of a number of fungal extracts including "statolon" (a complex polysaccharide) and "helenine" was eventually shown to be due to the presence in such preparation of viruses containing a double-stranded RNA genome (Lampson et al., 1967; Banks et al., 1968; review, Kleinschmidt, 1972). The viral RNA was an active inducer of interferon only when in its native double-stranded form; all activity was lost when it was converted to the single-stranded form. Double-stranded RNA is not unique to fungal phages; purified noninfectious double-stranded RNA from reovirus (Tytell *et al.*, 1967) and the double-stranded replicative form of MS₂ phage (Field *et al.*, 1967) are highly active inducers of interferon.

In the appropriate hosts virtually all animal viruses elicit interferon (Colby, 1971). In general, the enteroviruses are poor inducers, probably because they shut down cellular RNA synthesis rapidly and efficiently. This contention is supported by the observation that a strain of poliovirus which, unlike wild-type poliovirus, does not shut off cellular RNA synthesis, is a good inducer of interferon (Johnson and McLaren, 1965). FV3 shuts off host RNA synthesis rapidly and does not induce interferon (Gravell and Granoff, personal communication).

The togaviruses and myxoviruses are usually good inducers of interferons, but the amount induced by various members of these groups varies considerably. Newcastle disease virus is noteworthy in that it is a good inducer, but is quite insensitive to the antiviral action of interferon. In some permissive cell culture systems (chick embryo), the lytic infectious cycle is accompanied by production of a viral function that blocks induction of interferon. Interferon production is stimulated when this viral function is not expressed, either because the virus has been UV-irradiated or a host has been used in which infection is naturally abortive (L cells) (Youngner *et al.*, 1966).

There has been much controversy over whether virus-induced interferon is induced by single-stranded viral RNA, or by double-stranded RNA formed as an intermediate during viral replication. Much of the experimental data has been reviewed by Colby and Morgan (1971). In many experiments, inhibitors of protein synthesis and RNA synthesis have been used to establish that the input single-stranded RNA of the infecting virus, or some other viral component, is the inducer. Alternatively, virus whose infectivity has been inactivated by UV irradiation has been used to support the same argument. Most of these results could be explained on the basis of failure completely to prevent transcription of the input RNA to form double-stranded RNA, either by virion-associated transcriptases or by preformed host RNA polymerase. This view is supported in part by the application of temperature-sensitive mutants of Semliki Forest virus to the problem. Using mutants which were blocked in RNA synthesis at nonpermissive temperature, Lomniczi and Burke (1970) found that at low input multiplicities, RNA synthesis was a necessary prerequisite for interferon induction. However, the problem is further complicated by the observation that at high input multiplicities of infection, RNA synthesis was not necessary to elicit interferon, indicating the possibility of two types of interferon production.

Further support for the view that double-stranded RNA is the prime inducer of interferon comes from studies with synthetic polynucleotides, to be discussed below, and from a remarkable piece of scientific detective work by Colby and Duesberg (1969), who were intrigued that vaccinia, a virus with a DNA genome that was not expected to synthesize double-stranded RNA, should be able to induce interferon. Colby and Duesberg then successfully searched for a virus-specific induced double-stranded RNA in infected cells. This RNA (see Chapter 5) is produced during the normal transcription of vaccinia messenger, and is an active inducer of interferon.

### Induced Synthesis of Interferon

It is clear that interferon elicited by virus or poly I·poly C in cultured cells represent *de novo* synthesis of an induced protein. Both actinomycin D and puromycin, for example, will prevent interferon production if applied to cells prior to the inducing stimulus (Walters *et al.*, 1967; Buchan and Burke, 1966).

The idea that preformed interferon might exist in cells and be released rather than synthesized *de novo* came from observations that inhibitors of RNA and protein synthesis not only failed to prevent interferon production following injection of animals with nonviral inducers (Youngner *et al.*, 1965) but in fact yields were actually increased by cycloheximide. The enhanced level of interferon produced in the presence of low levels of cycloheximide is inhibited by puromycin, suggesting that the increase in interferon does require protein synthesis (Vilček, 1969; Ke and Ho, 1971; Tan *et al.*, 1971). Although there is not a great deal of evidence that protein synthesis in eukaryotic cells is controlled in the same manner as in bacterial cells, the data on interferon induction are compatible with the Jacob-Monod model for induced enzyme synthesis in bacteria. According to this model, the operon for interferon synthesis is under the negative control of a repressor protein; inhibition of repressor synthesis or inactivation of the repressor by the binding of inducer leads to derepression of the interferon operon and enhanced rates of interferon synthesis.

While the existence of mRNA for interferon is implied from studies with actinomycin D and puromycin (Heller, 1963; Wagner and Huang, 1965), Montagnier's group (DeMaeyer-Guignard *et al.*, 1972) conducted a remarkable experiment, that if confirmed is an unequivocal demonstration of an interferon mRNA and its translation. Messenger RNA was extracted from interferon-producing monkey cells, which, when introduced into chick cells, directed them to synthesize not chick but monkey interferon, i.e., the RNA used was indeed a specific messenger for interferon rather than a nonspecific RNA inducer of interferon. The same mRNA preparation when introduced into VERO cells, a simian line that normally cannot be induced to synthesize interferon, caused these cells to produce interferon.

Currently, considerable effort is being applied to determining if this simple model for interferon induction is tenable, to the problem of whether the primary inducer acts directly on the postulated repressor or acts indirectly through some other cell process and to establishing the optimum conditions favoring the largest possible production of interferon.

## MECHANISM OF INTERFERON ACTION

Any proposed mechanism of action of interferon must take into consideration the following points: (a) Interferon is completely nontoxic for cells, therefore, it cannot affect cellular macromolecular syntheses, and (b) interferon is equally effective against RNA and DNA viruses, so that its primary site of action is likely to be some step common to both sorts of virus such as translation of protein rather than transcription of mRNA or replication of RNA or DNA.

## Effects on Translation

Interferon does not act either directly upon virus particles or on their adsorption to and penetration of cells. However, it is not clear whether interferon acts on the cell membrane or whether it has to penetrate the cell in order to act. When cells are treated with interferon, only minute quantities are removed from the medium (Buckler and Baron, 1966). The effectiveness of interferon in inducing a resistant state is dependent on the concentration rather than the absolute amount of interferon (Baron et al., 1967), suggesting that transport of interferon into the cell is concentration-dependent. Interferon can bind to cells held at low temperature and the development of resistance in such cells can be inhibited by trypsin treatment prior to transferring cells to 37°C (Friedman, 1967). Early investigations (Taylor, 1964, 1965; Lockhart, 1964) suggested that interferon induces the synthesis of a new protein which inhibits the synthesis of viral macromolecules without affecting the normal synthetic activity of host cells. Such inhibition could occur at the level of mRNA transcription or translation. Despite the considerable effort that has gone into the study of this problem the exact mechanism has not been unequivocally established. A number of comprehensive reviews on interferon action have been published recently (Levy et al., 1970; Colby and Morgan, 1971) and rather than attempt a complete survey of the field, this section will be restricted to selected papers which provide data on the mechanism of action.

Support for interferon inhibition at the level of translation was provided by Marcus and Salb (1966), who compared the interaction of labeled viral and host cell RNA species with ribosomes from interferon-treated and control chick embryo fibroblasts (CEF). Sindbis virus RNA combined with CEF ribosomes at low temperatures. Upon raising the temperature, these ribosome–RNA complexes underwent breakdown coupled with the incorporation of amino acids into polypeptides. In contrast, when the source of ribosomes was interferon-treated cells, ribosome–RNA complexes were formed as before but these did not undergo similar breakdown. They concluded that ribosomes from interferon-treated cells could combine with but not translate viral RNA, although they retained the ability to translate host cell mRNA isolated from polysomes. Mild trypsin treatment of such ribosomes restored their ability to translate a viral RNA (Marcus and Salb, 1968). Marcus and Salb proposed that interferon induced the synthesis of a cell-coded "translation inhibitory protein" which bound to and modified the ribosomes' capacity to translate viral RNA.

Kerr and co-workers (Kerr et al., 1970; Kerr, 1971) undertook a careful reexamination of these results, using a system of cell ribosomes extracted from interferon-treated cells and viral RNA identical with that used by Marcus and Salb. In particular, they sought evidence for, but did not find, changes in the proteins of the ribosome fraction that might reflect the presence of the hypothetical inhibitory protein; they also failed to repeat Marcus and Salb's demonstration of translational inhibition. Kerr et al. (1970) found that the breakdown of ribosome–RNA complexes described by Marcus and Salb is not related to amino acid incorporation and cannot be taken as a reliable index of translation. In marked contrast to the results of Marcus and Salb, there was no difference between ribosomes from interferon-pretreated and control cells in their ability to respond to polyuridylic acid or undergo breakdown during amino acid incorporation. Nevertheless, the Marcus and Salb paper made a valuable contribution in that it clearly set out a working hypothesis regarding interferon action which stimulated investigators to test it experimentally. This model is still accepted in principle although not in detail regarding the action of the antiviral protein. It should also be kept in mind that there is no direct evidence for an antiviral protein produced in response to interferon. If there is such a protein, the genetic site of its synthesis is located on a different chromosome from that for interferon synthesis, as demonstrated by Cassingena et al. (1971), who studied interferon production and action in hybrid cell lines.

Kerr's results do not disprove the ribosome modification hypothesis but suggest that the alteration in the interferon-treated cell might involve a factor required for viral protein synthesis which is not an integral part of the ribosome but which could associate with it under certain conditions. Subsequently, Kerr (1971) reported a reduction in the frequency of translation of encephalomyocarditis (EMC) RNA with interferon-treated cell ribosomes. This difference was reproducible provided a particular fraction of the cell's microsomes was used as a source of ribosomes. The data do not indicate whether the limitation in translation is at the level of initiation, peptide bond formation, or the release of polypeptides. Further progress requires a definition of the number and nature of factors controlling the translation of viral RNA.

If interferons act against all viruses by the same mechanism, viz., the induction of an antiviral protein that interferes with translation, then presumably the antiviral protein acts at the same site for each virus. However, in such a situation, one would expect that all species of interferon would inhibit a given spectrum of viruses to the same extent, which is not the case (see above).

The idea that interferon induces, via an antiviral protein, a modification not of the ribosome but of other factors involved in translation, focuses attention on evidence for alternative mechanisms of interferon action. It is conceivable that the mechanism could depend on the virus–cell system used since the replication of both DNA and RNA viruses is inhibited in cells protected by the action of interferon. Most studies of the action of interferon on DNA viral replication have been conducted with vaccinia virus. Joklik and Merigan (1966) preincubated L cells with mouse interferon, then infected them with vaccinia virus and followed the synthesis of vaccinia virus early mRNA and its association with ribosomes. The rate of vaccinia mRNA synthesis was not inhibited but enhanced by interferon treatment. The primary defect was failure of this early messenger to combine with ribosomes, so that early viral proteins could not be synthesized. A difficulty in the interpretation of the Joklik-Merigan experiment is that the interferon preparations used led to disintegration of infected cells beginning 3 hours postinfection whereas untreated infected cells remained intact for 24 hours, so that the results obtained may not represent a general action of interferon. Similar observations are reported by Horak et al. (1971), who noted that vaccinia virus causes a drastic cytopathic effect in interferon-treated L cells that differs from the usual cytopathic effects in poxvirus-infected cells. Until this situation is clarified, the hypothesis that the primary action of interferon in poxvirus-infected cells is simply to block association of mRNA with ribosomes must be considered tentative.

It is of some interest that suppression of the induction of poxvirus-induced enzymes requires higher doses of interferon than the inhibition of synthesis of infectious virus (Bodo and Jungwirth, 1967; Levine et al., 1967).

### Effects on Transcription

Horak et al. (1970) repeated the experiments of Joklik and Merigan (1966) and extended them to the cowpox-infected chick fibroblast system. They found that early synthesis of vaccinia mRNA in mouse cells was not depressed by mouse interferon and that this mRNA did associate with ribosomes to form polysomes. Later, viral mRNA synthesis and polysome formation were reduced. In cowpox-infected chick embryo fibroblasts, interferon pretreatment reduced the incorporation of uridine into poxvirus mRNA at the earliest times studied (1 hour postinfection) and also reduced the formation of polysomes. Because of the extended times used to adsorb virus to cells and the lack of data on the time course of viral DNA synthesis in the system studied, it is not clear whether Horak and co-workers were actually measuring early or late cowpox mRNA synthesis. To ensure that the mRNA being studied is actually "early" mRNA, Bialy and Colby (1972) measured the effect of interferon pretreatment on mRNA synthesis by vaccinia virus arrested at the core stage, and concluded that one target of interferon is early mRNA synthesis by the virion-bound transcriptase. However, in similar experiments, Jungwirth et al. (1972) could not show any reduction of early vaccinia mRNA synthesis by mouse interferon. With any complex inhibitor in a system as complex as the virus-infected cell there could be more than one specific target. The problem is how to study the effect under the minimal conditions needed to block viral multiplication. In order to measure quantitatively particular steps in multiplication, it is sometimes necessary to use high inputs of virus particles in the presence of chemical inhibitors to enhance the rate of a particular step. Under these conditions, either the effects of interferon may be largely overcome or the high levels of interferon necessary to cause an inhibition may be misleading. It will be necessary to see if interferon or any protein induced by it in vivo will act on virion-bound transcriptase in vitro before any definite conclusion can be reached.

Vesicular stomatitis virus (VSV), like vaccinia virus, has an RNA polymerase associated with particles. This polymerase transcribes parental viral RNA in vivo in the presence of cycloheximide and actinomycin D, which block cellular protein and RNA synthesis. Marcus et al. (1971) have shown that in chick embryo cells pretreated with chick but not with mouse interferon, the synthesis of VSV-specific RNA is delayed and the total synthesis is reduced by about 60%. The possibility that the differences measured were associated with uridine uptake by cells rather than uridine incorporation into RNA was excluded, but the possibility remains that interferon pretreatment of cells led to an increased rate of breakdown of newly synthesized viral RNA. Since a dose of interferon that reduced VSV replication by 98% produced only a 50% reduction in viral RNA synthesis, it seems unlikely that the primary site of the inhibition of viral multiplication is RNA transcription. However, there could be two different states of virion RNA polymerase complexes within a cell. The refractory material might represent RNA produced by enzyme complexes of nonproductive particles physiologically and topographically isolated from the interferon system.

A striking example of interferon treatment blocking viral transcription is provided by the case of SV40 early mRNA synthesis. In the presence of cytosine arabinoside, which blocks DNA synthesis, SV40 infection can be limited to early viral functions and techniques are available to quantitate early viral mRNA. In primate cells pretreated with primate interferon, the synthesis of SV40-specific early mRNA was inhibited by about 90% as was the synthesis of virus-specific T antigen (Oxman and Levin, 1971). Taken at face value, the evidence for a block in transcription is convincing, but Oxman and Levin considered other possible interpretations. They speculate that the SV40 genome contains a special early gene, "proto-early," whose product could be either a sigmalike factor for modification of host RNA polymerase or a virus-specified DNA-dependent RNA polymerase which could be essential for transcription of the early SV40 genes. Interferon-induced inhibition of translation of the postulated proto-early gene would thus result in a marked reduction of total early SV40 RNA. It is significant that pretreatment with interferon inhibits the synthesis of T antigen in 3T3 cells infected with SV40 (Oxman and Black, 1966), but has no effect on the production of the T antigen in SV40-transformed 3T3 cells in which the relevant viral genetic material is presumed to be integrated into the cellular DNA (Oxman et al., 1967).

Bearing in mind that other interpretations of the data can be made, it seems that in three different virus–cell systems there is evidence that interferon could inhibit transcription of viral mRNA. The key questions are (a) what is the primary target of the interferon-mediated inhibition in vivo, and (b) is there only one site of action, or may several viral processes be directly affected?

## ENHANCEMENT

Several examples have been reported in which mixed infection of cells by unrelated viruses (at least one being non-cytocidal) may result in increased growth and/or cytopathic effects. This has usually been called "enhancement." As with interference, it is likely that different mechanisms may operate in different cases.

One of the first instances to be recognized arose from efforts to devise an in vitro assay for hog cholera virus, which is noncytocidal in swine cells. Kumagai et al. (1961) found that Newcastle disease virus (NDV), which normally does not produce cytopathic effects in swine testis cells until the seventh or eighth day after inoculation, produced severe cytopathic effects on the third or fourth day in cells which had been infected with hog cholera virus soon after cultivation. Growth curves of NDV in hog cholera-infected and normal cells showed that preinfection with hog cholera virus produced obvious enhancement of growth of NDV.

Hermodsson (1963) investigated the enhancement of NDV in another system. On its own, NDV shows strong autointerference in calf kidney cell cultures and produces a persistent infection in them. Parainfluenza type 3 is noncytocidal for these cells but produces a steady-state infection. Infection of calf kidney cells with parainfluenza type 3 virus was antagonistic to added interferon produced by a third unrelated virus, and to the production of interferon by NDV. Mixed infections led to enhancement in the growth of NDV, which Hermodsson ascribed to an anti-interferon effect of the parainfluenza virus infection. A similar situation has been described in human or chick cells infected with mumps and Sindbis viruses (Frothingham, 1965).

Experiments described by Valle and Cantell (1965) and Cantell and Valle (1965) are of interest in revealing the limitation of the phenomenon to certain combinations of viruses. The plaque count and viral yield of vesicular stomatitis virus in chicken or human cells were increased by preinfection of the monolayers with Sendai virus (parainfluenza type 1) at a multiplicity of about one, but not by preinfection with several other myxoviruses. Preinfection with Sendai virus had no effect on the plaque counts of pseudorabies, vaccinia, West Nile, or polioviruses, and depressed the plaque count of Newcastle disease virus. Further, plaque mutants of vesicular stomatitis virus showed different amounts of enhancement. Cantell and Valle (1965) excluded *in vitro* effects or changes in the efficiency of attachment as causative factors, and showed that Sendai virus (which under these conditions failed to produce interferon) had a potent antagonistic effect against added interferon added simultaneously or after infection of the monolayer with Sendai virus. All these examples of enhancement involve the use of a paramyxovirus, usually as the "enhancer" but sometimes as the enhanced virus. It is not known to what extent this is due to the fact that these viruses readily produce noncytocidal steady-state infections, which are a necessary requirement for the demonstration of enhancement.

There are at least two well documented examples of enhanced multiplication of RNA viruses due to prior infection of cells with DNA viruses. The DNA viruses, rabbit fibroma and Yaba viruses, are both members of the *Poxvirus* genus and produce benign tumors in rabbits or monkeys, respectively. Infection of the appropriate cultured cells by these viruses leads to persistent viral infection, altered cell morphology, and continued cell multiplication. Certain RNA viruses grow either very poorly or not at all in the original cell line but if cells are first infected with fibroma or Yaba viruses, the RNA viruses will replicate to high titers in such cells. The first indication of such a phenomenon was the report by Ginder and Friedewald (1951), who found that Semliki Forest virus would multiply in and cause necrosis of fibromas on domestic rabbits and that the virus would grow in cultures of minced fibromas but not in normal adult rabbit tissues. More striking examples are provided by cultured cell lines in which the replication of the challenge virus can be quantitated. Sindbis or Japanese encephalitis virus will not grow at all in monkey kidney cell lines. However, if cells are first infected with Yaba virus (Tsuchiya and Tagaya, 1970) then Sindbis and Japanese encephalitis virus will replicate extensively. Multiplication of poliovirus, echovirus, and vesicular stomatitis virus is also enhanced ten- to fiftyfold over the yields obtained in cells not preinfected with Yaba virus.

Establishment of a persistent fibroma infection of a cloned line of rabbit kidney cells enhanced the yield of vesicular stomatitis virus, Sindbis virus, or NDV normally produced in the control cell line by as much as a hundred thousand-fold (Padgett and Walker, 1970). All the RNA viruses mentioned are quite sensitive to interferon action and the enhancement effect might be explained in part by the failure of the virus to induce interferon in cells preinfected with the enhancing virus. In fact, Tsuchiya and Tagaya state that Japanese encephalitis virus induces interferon in the normal monkey cell line but not in the Yaba virus-infected line.

*CHAPTER 9*

# Pathogenesis: The Spread of Viruses through the Body

## INTRODUCTION

At the molecular and cellular levels with which we have been concerned so far, viruses behave quite differently from other infectious agents. This clear distinction disappears when we consider their effects on whole organisms and on populations; both the pathogenesis and the epidemiology of viral infections have much in common with these aspects of many bacterial and protozoal infections.

In this and the following two chapters we will consider the interactions of viruses with vertebrate organisms from three points of view: (a) the way in which different viruses spread through the body, (b) the immune response and its influence in viral infections, and (c) genetic resistance and some nonimmunological factors that affect virus–host interactions. A great variety of diseases results from such interactions, which involve several hundred viruses and many animal species. Descriptions of viral diseases in a single animal, man, can be found in Horsfall and Tamm (1965), and in abbreviated form in Fenner and White (1970). Much less work has been done on viral infections of other animals and no comprehensive modern textbook exists, hence in this chapter many of the examples will come from human viral infections.

Except in immunology, which is discussed in Chapter 10, our understanding of the interactions between animal viruses and whole animals has increased more slowly than has our knowledge of the molecular events that occur when viruses infect cells in culture. This slow rate of growth is reflected in the fact that this chapter has required a less drastic revision than most of the other chapters in this book.

## MECHANISMS OF CELL DAMAGE

Before exploring viral pathogenesis at the level of organism we shall summarize briefly what is known of the cytopathic effects of viruses at the cellular level.

As Table 9–1 indicates, a variety of types of interaction is possible between virus and cell, the outcome in any particular case being largely determined by the genetic constitution of the cell and virus concerned. These interactions can be grouped into three main types: (a) cell destruction, caused by cytocidal

TABLE 9–1

*Types of Virus–Cell Interaction*

| VIRUS | CELL | INTERACTION | EFFECT ON CELL | YIELD OF INFECTIOUS VIRUS |
|---|---|---|---|---|
| Poliovirus | HeLa | Cytocidal | Destruction | + |
| | Mouse | Nil | Nil | — |
| Poliovirus RNA | Mouse | Cytocidal | Destruction | + |
| Rabbitpox | Chick | Cytocidal | Destruction | + |
| | PK | Cytocidal | Destruction | + |
| Rabbitpox PK—[a] | Chick | Cytocidal | Destruction | + |
| | PK | Cytocidal | Destruction | — |
| Avian leukovirus | Chick | Steady-state | Nil | + |
| Avian leukovirus, RSV [b] | Chick | Noncytocidal | Transformation | — |
| Polyoma | Mouse | Cytocidal (most cells) | Destruction | + |
| | Mouse | Noncytocidal (few cells) | Transformation | — |
| | Hamster | Noncytocidal | Transformation | — |

[a] PK negative mutant of rabbitpox virus (see Chapter 7).
[b] RSV, Rous sarcoma virus—a mutant avian leukovirus (see Chapter 14).

viruses, (b) steady-state infection, caused by some noncytocidal viruses, and (c) transformation, caused in certain cells by the oncogenic viruses. The latter two types of interaction are discussed at length in Chapters 12 through 14; here we will concentrate on the mechanism of cell killing by cytocidal viruses.

## Causes of Cell Death

Cytocidal virus–cell interactions are characterized by the production of lethal cellular damage due to the effects of products of the viral genome on the cell or its regulatory mechanisms. Despite the impressive expansion in our knowledge of the molecular biology of viral multiplication over the last decade, we still are not at all certain about how viruses kill cells—or indeed, at another level, how they cause disease.

Gross biochemical changes occur in cells infected with cytocidal viruses. Early virus-coded proteins often shut down host protein and RNA synthesis, which in itself is incompatible with survival of the cell. Furthermore, large numbers of viral macromolecules accumulate late in the infectious cycle; some of these, particularly certain capsid proteins, may be directly toxic. Viral proteins or virions themselves sometimes congregate in large crystalline aggregates or inclusions that visibly distort the cell. Infected cells usually swell substantially; it would be surprising if there were not consequential changes in membrane permeability. Eventually the cell's own lysosomal enzymes must leak out and result in autolytic digestion of the cell.

Thus, there are numerous changes in the virus-infected cell which, individually or cumulatively, are lethal. In a sense, only the first lethal change is relevant, even though it may not necessarily be the one that produces the first visible "cytopathic effect," or the final dissolution of the cell. Most of the literature on the subject is concerned with identifying the event responsible for visible cyto- pathic damage. Such studies have demonstrated that cell damage may still occur when expression of the viral genome is limited, either because the virus is inac- tivated (e.g., by UV irradiation), or it is a conditional lethal mutant (see Chapter 7) or its multiplication is suppressed by chemical inhibitors (see Chapter 4). The occurrence of severe cytopathic effects in cells infected with polioviruses whose multiplication is blocked with guanidine early in the growth cycle (see Bablanian, 1972), or with host-dependent mutants of rabbitpox virus that fail to synthesize DNA or more than a few recognizable proteins (Sambrook *et al.*, 1965) provide good examples. Abortive infections of this type are, of course, not transmissible to other cells, although the products of the damaged cells may be "toxic."

## Shutdown of Cellular Macromolecular Synthesis

Most cytocidal viruses code for early proteins which shut down the synthesis of cellular RNA and protein; DNA synthesis is usually affected secondarily. The shutdown is particularly rapid and severe in infections caused by picornaviruses, poxviruses, and herpesviruses, all of which are rapidly cytopathogenic. With several other genera the shutdown is late and gradual, while with noncytocidal viruses such as leukoviruses there is, by definition, no shutdown and no cell death. The biochemical events involved were discussed in detail in Chapters 5 and 6.

## Cytopathic Effects of Viral Proteins

Viral capsid proteins in high dosage are often toxic for animal cells and may be the principal cause of cytopathic effects. This may follow the accumulation of viral proteins late in the multiplication cycle after infection at low multiplicity, or may be seen quite early as a laboratory artifact resulting from the experimental use of very large inocula. The best studied example is adenovirus.

Two of the virion proteins of adenoviruses have a direct cytopathic effect on cultured cells. Large doses of virus cause cell clumping and detachment from the substrate, without killing the cells (Pereira and Kelly, 1957; Rowe *et al.*, 1958). The same effect can be produced by purified penton antigen and is then com- pletely reversible (Valentine and Pereira, 1965). On the other hand, the fiber antigen of adenoviruses inhibits the multiplication of adenoviruses, vaccinia virus, and poliovirus (Pereira, 1960), probably because of its inhibitory effect on cellular syntheses of RNA, DNA, and protein (Levine and Ginsberg, 1967). This generalized shutdown of all viral and cellular activity is not to be confused with the more specific shutdown of cellular (but not viral) RNA and protein synthesis by virus-coded "early" proteins that is an important feature of the multiplication cycle of many cytocidal viruses.

Frog virus 3, which is unable to multiply at 37°C (see Chapter 5), is highly toxic when large doses are inoculated intraperitoneally in mice, lesions being seen in the nuclei of hepatic cells within 3 hours and death occurring within 18 to 36 hours (Kirn *et al.*, 1972). Inactivation of viral infectivity by heat or γ irradiation did not affect the toxicity, which was ascribed to a component of the virion, as yet unidentified.

## Release of Enzymes by Lysosomes

Allison and his colleagues (reviews, Allison 1967a; 1971b), have suggested that activation of lysosomal enzymes may be an important mechanism in the production of other kinds of cellular damage by viruses. They distinguish three stages of activation by staining fixed and unfixed cells for a lysosomal enzyme, acid phosphatase. In first-stage activation the membranes of lysosomes show increased permeability, but the enzymes remain within them. This state is reversible. Second-stage activation is marked by diffusion of the enzymes into the cytoplasm, often with secondary adsorption to the nucleus; the cells become rounded up. In third-stage activation, the acid phosphatase marker disappears, due to diffusion or inactivation.

Slight first-stage activation may occur in cells infected with noncytocidal viruses, and the cell detachment produced by the penton antigen of adenoviruses is associated with first-stage activation of lysosomes (Allison and Mallucci, 1965). In some viral infections, lysosomal changes do not progress beyond the first (reversible) stage, and since such lysosomes take up more neutral red than usual red plaques may be produced (see Plate 2–3), as with Thiry's (1963) red plaque mutant of Newcastle disease virus, or Mallucci's (1965) red plaques of mouse hepatitis virus. Second-stage activation is found in cells infected with cytocidal viruses; it leads to a failure of the affected cells to take up neutral red and the occurrence of the familiar unstained plaques. The final dissolution of the dead cell probably results from autolysis brought about by the liberation of enzymes from the cell's own lysosomes.

## Nonspecific Histological Changes

In addition to changes due to the specific effects of viral multiplication, most virus-infected cells also show nonspecific lesions, very like those induced by physical or chemical agents. The most common early and potentially reversible change is what histopathologists call "cloudy swelling," which is associated with changes in the permeability of the plasma membrane. Electron microscopical study of such cells reveals diffuse changes in the nucleus, plasma membrane, endoplasmic reticulum, mitochondria, and polyribosomes, prior to any changes in the lysosomes (Trump et al., 1965). Margination of the nuclear chromatin and pyknosis are late nonspecific changes that probably occur after death of the cell.

## Inclusion Bodies

The most characteristic morphological change in virus-infected cells, recognized by histologists since before the turn of the century, is the "inclusion body" —an area with altered staining behavior. Depending on the causative virus, such inclusions may be single or multiple, large or small, round or irregular in shape, intranuclear or intracytoplasmic, acidophilic or basophilic (Fig. 9–1). They develop progressively as the multiplication cycle proceeds and generally repre-

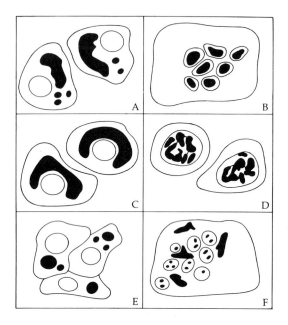

Fig. 9–1. Inclusion bodies in virus-infected cells. (A) Vaccinia virus—intracytoplasmic acidophilic inclusion (Guarnieri body: B body, see text). (B) Herpes simplex virus —intranuclear acidophilic inclusion (Cowdry type A); cell fusion produces syncytium. (C) Reovirus—perinuclear intracytoplasmic acidophilic inclusion. (D) Adenovirus—intranuclear basophilic inclusion. (E) Rabies virus—intracytoplasmic acidophilic inclusions (Negri bodies). (F) Measles virus—intranuclear and intracytoplasmic acidophilic inclusions; cell fusion produces syncytium.

sent "virus factories," i.e., foci in which viral nucleic acid or protein is being synthesized or virions assembled. The presence of such virus-specific products in these factories can sometimes be demonstrated by autoradiography and immunofluorescence (Cairns, 1960) or by electron microscopy (see Dalton and Hagenau, 1972).

On other occasions cellular organelles may be involved. For example, the bizarre crescent-shaped perinuclear inclusions of reovirus-infected cells are revealed by electron microscopy to involve the mitotic spindle fibers with which the replication of that virus is closely associated (Dales, 1963b). Cytomegaloviruses characteristically produce large intranuclear inclusion bodies. Eosinophilic inclusions are also seen in the cytoplasm; these contain large aggregates of lysosomes (McGavran and Smith, 1965).

In other situations, inclusion bodies do not appear to be directly associated with viral multiplication. For example, cells infected with poxviruses always contain cytoplasmic factory areas that may constitute large inclusion bodies, but with some virus–cell combinations other inclusion bodies are also produced in the cytoplasm, e.g., with ectomelia, cowpox, and fowlpox viruses, but not with vaccinia. Japanese investigators have distinguished the latter as "A-type" cytoplasmic inclusions, in contradistinction to the "B-type," which are viral factories (Matsumoto, 1958; Morita, 1960; Kato and Kamahora, 1962; Ishihashi et al., 1971; see Plate 5–3). The classic type A intranuclear inclusion body of herpesvirus-infected cells (Cowdry, 1934) is in a sense an artifact of fixation and staining. Herpesviruses develop in the nucleus and characteristic early changes are seen there by electron microscopy (Chapter 5), but only with difficulty in stained sections. Late in infection, after the viruses have crossed the nuclear membrane into the cytoplasm, the center of the nucleus is occupied by a homogeneous Feulgen-negative eosinophilic area which is separated by a halo from the marginated chromatin.

## Cell Fusion

A wide variety of different agents will cause cultured cells to fuse so that they form polykaryocytes or syncytia (Roizman, 1962a). Among these, viruses occupy an important place, and virus-induced cell fusion has become an important tool in experimental biology (review, Poste, 1970). Only a few kinds of virus produce cell fusion. Paramyxoviruses (including measles and respiratory syncytial virus, which derives its name from the property) and herpesviruses are the most important groups, but cell fusion has been observed with some strains of vaccinia virus (McClain, 1965), with visna virus (a leukovirus) (Harter and Choppin, 1967), and with avian infectious bronchitis virus, a coronavirus (Akers and Cunningham, 1968).

Several viruses induce rapid cell fusion, even when rendered noninfectious by UV irradiation. This property of parainfluenza virus has been widely exploited during the last few years to study problems in virology, like the rescue of integrated viral genomes (Koprowski et al., 1967) and the ability of "insusceptible" cells to support viral replication (Tegtmeyer and Enders, 1969). Even more important for general biology has been the production of heterokaryons and stable replicating cell hybrids with cells of different species, a discovery that is being vigorously exploited to study the fundamental control mechanisms operating in gene expression in differentiated cells (reviews, Harris, 1970; Watkins, 1971a, b).

Viruses that cause early cell fusion when massive doses are added to cultured cells also produce polykaryocytes several hours after monolayers are infected at low multiplicity, and a few kinds of virus that do not produce early cell fusion do induce polykaryocytosis in infected cultures (e.g., Rous sarcoma virus; Moses and Kohn, 1963). Virus-induced cell fusion plays a role in the cytopathic effects of some viruses in the intact animal, and polykaryocytosis ("giant-cell formation") is a characteristic feature of infection with measles virus (Warthin, 1931), varicella (Tyzzer, 1905), and most other viruses of the *Paramyxovirus* and *Herpesvirus* genera.

Virus-induced polykaryocytosis is conditioned by the genetic characteristics of both cell and virus. For example, herpes simplex virus produces large polykaryocytes in HeLa cells but not in human diploid fibroblasts, yet the latter cells support viral growth better than HeLa cells (Albanese et al., 1966). Conversely, the macroplaque mutant of herpes simplex virus is much more effective than the microplaque strain in producing polykaryocytes in HEp-2 cells (Roizman, 1962b).

The mechanism of virus-induced polykaryocytosis is not fully understood. Poste (1970) suggests that cell fusion requires modification of the glycoproteins in the plasma membrane by lysosomal enzymes released on to the cell surface. This is followed by fusion of the membranes, an energy-dependent process that requires fundamental reorganization of their membrane structure. Lysosomal stabilizing agents such as hydrocortisone or chloroquine may prevent cell fusion.

## Viral Hemolysins

A few of the viruses that induce polykaryocytosis also lyse cells, a phenomenon most readily demonstrable with red blood cells. The phenomenon is restricted to paramyxoviruses (including measles), and can be distinguished from the capacity of these viruses to cause cell fusion by its differential sensitivity to inactivation by heat, trypsin, and UV irradiation (Okada and Tadokoro, 1962; Neurath, 1964). The hemolysin appears to be a component of the viral envelope distinct from the neuraminidase and hemagglutinin (Norrby and Falksveden, 1964). Recent evidence about its chemical nature was discussed in Chapters 3 and 6. There is no evidence that it plays a part in the pathogenesis of the diseases produced by paramyxoviruses.

## Viral Infection and Mitosis

Synthesis of cellular RNA and protein is inhibited during mitosis (see Chapter 4). If cells arrested in metaphase by vinblastine are infected with viruses, uptake and uncoating occur but viral macromolecules are not made (Marcus and Robbins, 1963). Highly cytocidal viruses like poliovirus and Mengo virus irreversibly damage cellular macromolecular syntheses very soon after infection and thus prevent cellular division. In synchronized cells infected with Mengo virus, only those cells already in mitosis at the time of addition of virus were able to divide; all others failed to enter prophase (Tobey et al., 1965). Results in other cell–virus systems vary; with herpes simplex virus in HeLa cells mitosis is shut down as early as 1 hour after infection (Stoker and Newton, 1959), whereas in adenovirus-infected KB cells no effect is observed until 10 hours after infection (Green and Daesch, 1961). Herpes simplex virus infection of cells that are already committed to mitosis causes aberrant cleavage of nuclei, and mitotic events are often synchronized in the polykaryocytes formed by infection with measles virus (Roizman and Schluederberg, 1962). The timing of mitotic inhibition by different viruses is a direct reflection of the timing and extent of their inhibitory effect on cell protein synthesis, which is known to be vital to the cell right up to the moment mitosis begins (Robbins and Scharff, 1966).

Cellular macromolecular syntheses and cell division may be unaffected in steady-state noncytocidal infections (see Chapter 12), whereas a stimulation of cellular DNA synthesis is a characteristic early effect of the infection of resting cells with SV40 or polyoma virus (see Chapter 13).

## Chromosomal Aberrations

Chromosomal aberrations have been described in cultured cells infected with viruses of the following genera: *Polyomavirus, Adenovirus, Herpesvirus, Paramyxovirus,* and *Leukovirus,* with unconfirmed reports relating to poliovirus, vaccinia virus, and yellow fever virus (reviews, Stich and Yohn, 1970; Nichols, 1970). A variety of effects have been described; chromatid breaks and fragmentation being the most common, while "pulverization" is a rarer effect observed sometimes in cells infected with adenoviruses, herpes simplex, and measles viruses. Chromosomal changes have also been described in the circulating lymphocytes of patients suffering from measles, chickenpox, hepatitis, and other viral infections (Nichols, 1970), but their significance is difficult to evaluate. The changes are often nonrandom, i.e., are more common in certain chromosomes than in others, suggesting the possibility of some sort of specificity, but the abnormalities which follow radiation or radiomimetic chemicals tend to affect the same chromosomes. Accordingly, it is hard to accept the view that they represent a primary lesion of any relevance to viral multiplication. A visible chromosomal abnormality is a gross and late manifestation of many earlier and more basic biochemical changes, perhaps at sites quite remote from DNA itself. It is unlikely that they indicate any direct interaction between viral and cellular genomes. Allison (1967a) believes that the chromosomal aberrations result from nonspecific damage by nucleases released from lysosomes late in the infectious cycle.

Much effort has gone into the karyological analysis of virus-induced and other neoplasms. Chromosomal aberrations are common, but they do not appear to be a necessary process in neoplastic transformation. The more important and subtle changes associated with integration of viral into cellular DNA's (see Chapters 13 and 14) are not revealed as chromosomal abnormalities by ordinary karyological techniques.

## CELL DAMAGE IN THE INFECTED ANIMAL

All of the changes described in virus-infected cells in culture occur equally in the whole animal, although infected cells are often more difficult to locate in stained sections of tissue because infection may be limited by the body's defences or by lack of susceptibility of all but a few cell types in a given organ. Nevertheless, inclusion bodies, multinucleate giant cells, and so on are sufficiently in evidence to be pathognomonic of certain viral infections in animals and man, e.g., Negri bodies in rabies, Warthin's giant cells in measles, and cytomegalic cells in cytomegalovirus infection.

Destruction of infected cells by direct viral action clearly plays an important part in pathogenesis. However, sublethal changes such as cloudy swelling can also have profound physiological effects if they occur in certain cells, such as the endothelial cells of small vessels, or in solid organs like the brain. Minor functional abnormalities in myocardial cells or neurons may be of critical importance. The central nervous system shows extreme vulnerability to early damage, as evidenced, for example, by the early paresthesias in poliomyelitis in man.

Cell damage is often the result not of direct viral cytotoxicity, but of immunological responses to the products of viral infection (see Chapter 10). Further, the intact organism differs from cultured cells in having a circulatory system. Circulatory disturbances due to inflammation, or to direct or indirect viral damage to blood vessel walls, inevitably contribute to pathological changes, with production of edema, anoxia, hemorrhage, or infarction. If disturbances are severe the tissues supplied by affected vessels undergo necrosis.

## Toxic Effects

When an animal is injected with a very large dose of virus by a route that does not allow the virus access to the cells in which it is capable of undergoing productive infection, that animal may nevertheless develop a fever and other signs of "toxicity." Influenza virus injected parenterally into mice provides a classical example (review, Burnet, 1960). Mims (1960a, b) showed by studies with fluorescent antibody staining that nonadapted influenza virus undergoes a single cycle of multiplication in the brain after intracerebral inoculation and in the hepatic cells of the liver after intravenous inoculation. In the latter case, the Kupffer cells remove and destroy most of the injected virus but with an overwhelming inoculum some virus "leaks" through to the hepatic parenchymal cells where it undergoes an abortive cycle of multiplication and produces severe liver damage.

"Toxic" symptoms are often seen early in severe viral infections like smallpox. Viral antigens and early (IgM) antibodies to them are usually present in the circulation and the tissues before symptoms become evident (see Chapter 10). The formation of immune complexes and the activation of complement may initiate a series of physiological effects (reviews, Dixon, 1963; Symposium, 1971) that lead to the onset of symptoms. This common mechanism may explain the similarity of "toxic" symptoms in diseases in which the distribution of the virus in the body may be very different, e.g., smallpox and severe influenza.

## INFECTIONS OF THE SKIN

### The Histology of Skin

The skin consists of three parts (Fig. 9–2): (a) the epidermis: a layer of stratified squamous epithelium, with hair follicles and sebaceous glands as appendages; (b) the dermis, which consists of a gel-like matrix of loose connective tissue containing collagenous, reticular, and elastic fibers, a rich plexus of capillary blood vessels and small lymphatic vessels, and many nerve trunks and nerve endings; and (c) superficial fascia, which in many places contains the subcutaneous fat. The epidermis forms a protective covering whose surface consists of the stratum corneum, a tough layer of dead keratinized cells which are continually being rubbed off and replaced from below, by active multiplication of the cells of the stratum germinativum. As they are displaced toward the surface, the basal cells differentiate into flat laminated plates, that have thick cell membranes and become filled with keratin. The epidermis contains no blood or lymphatic vessels or nerve fibers, and the intact epidermis is impermeable to viruses.

A thin (35 nm) basement membrane separates the epidermis from the dermis, which contains many scattered migratory cells, both fibroblasts, which elaborate collagen, and macrophages. There is an extremely rich plexus of fine lymphatics just beneath the epidermis and scattered throughout the dermis, so that trauma that allows penetration of the dermis probably involves rupture of lymphatics and, therefore, potentially at least, passage of the inoculum into the lymphatic

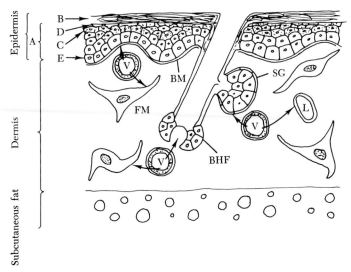

FIG. 9–2. *Diagram of skin showing the structures of importance in the pathogenesis of viral lesions. The epidermis consists of a layer several cells thick of living cells, the stratum Malpighii (A), covered by a layer of dead keratinized cells, the stratum corneum (B). The cells of the upper part of the stratum Malpighii contain increasing amounts of keratin granules; sometimes two narrow layers, the stratum granulosum (C) and the stratum lucidum (D), are distinguished. The basal layer of dividing epidermal cells is the stratum germinativum (E). The dermis contains blood vessels (V), lymphatic vessels (L), and dermal cells (fibroblasts and macrophages—FM). The ground substance contains collagenous, reticular, and elastic fibers. The hair follicle (BHF) and sebaceous gland (SG) are appendages of the epidermis.*

circulation. The macrophages also play an important part in carrying viruses from one part of the dermis to another and from the dermis to the epidermal cells; their mobility may be affected by state of the "ground substance" which is affected by physiological (especially hormonal) changes (Asboe-Hansen, 1969).

## Virus-Induced Skin Tumors

Three types of virus-induced tumor develop in the skin (see Table 9–2 and Chapter 13), each involving a different class of cells. Virus-induced papillomas are confined to the epidermal cells; viral replication begins in the basal cells but virions do not mature until the cells become keratinized. Fibromas caused by poxviruses of the myxoma subgenus are associated with extensive proliferation of the fibroblasts in the dermis; although the epidermal cells may be infected they do not proliferate. The lesions of Yaba monkey tumor poxvirus infection are associated with the infiltration and proliferation of macrophages in the dermis, i.e., they are histiocytomas. No RNA viruses cause a skin lesion at the site of entry, and such lesions are not produced by parvoviruses, adenoviruses, or herpesviruses, although herpes simplex often produces localized ulcers of the skin and mucous membranes.

TABLE 9–2

Viruses Causing Tumors in the Skin

| DISEASE | HOST | LOCAL TUMOR | GENERALIZED LESIONS | NATURAL TRANSMISSION |
|---|---|---|---|---|
| *Poxvirus* | | | | |
| Orf subgenus | | | | |
|   Milker's node | Cow, man | + | — | Mechanical—trauma |
| Myxoma subgenus | | | | |
|   Myxoma | *Sylvilagus brasiliensis* | + | — | Mechanical—arthropod vectors |
| | *Oryctolagus cuniculus* | + | + (Skin and internal organs) | Mechanical—arthropod vectors |
|   Rabbit fibroma | *Sylvilagus floridanus* | + | — | Mechanical—arthropod vectors |
| | *Oryctolagus cuniculus* | + | Rarely (skin only) | Not transmissible |
|   Hare fibroma | *Lepus europaeus* | + | Sometimes (mainly on skin) | Probably mechanical—arthropod vectors |
|   Squirrel fibroma | *Sciurus carolinensis* | + | Sometimes (mainly on skin) | Mechanical—arthropod vectors |
| Birdpox subgenus | | | | |
|   Fowlpox | Chickens | + | Sometimes (throughout organism) | Mechanical—arthropod vectors (also by respiratory route) |
| Unclassified poxviruses | | | | |
|   Yaba monkey tumor | Primates | + | — | Not known |
|   Tana virus disease | Monkeys | + | — | Probably mechanical—arthropod vectors |
| | Man | + | — | |
|   Molluscum contagiosum | Man | + | + (On skin only) | Probably mechanical—trauma |
|   Swinepox | Swine | + | + (On skin only) | Mechanical—arthropod vectors |
| *Papillomavirus* | | | | |
|   Rabbit papilloma | *Sylvilagus floridanus* | + | — | Mechanical—arthropod vectors |
|   Human wart | Man | + | — | Mechanical—trauma |
|   Bovine papilloma | Cow | + | — | Probably mechanical—arthropod vectors |

## Skin Lesions in Generalized Infections

The blood vessels of the dermis, whose structure is like that of small arterioles and venules and capillary loops elsewhere in the body, play a major role in the causation of the rashes that characterize many generalized viral diseases. The sensory nerves in the dermis are probably the sites of entry of viruses which travel by neural routes to the central nervous system (e.g., varicella-zoster virus), and in attacks of zoster the virus passes centrifugally down the nerve fibers and infects the cells of the basal layer of the epidermis. In relation to viral lesions in the skin, it is worth drawing attention to the highly artificial nature of intra-dermal or subcutaneous injections of viruses in relatively large volumes of fluid, compared with natural infection of the skin either directly through lesions pro-duced by trauma or arthropod bites, or after hematogenous or neural spread of viruses that produce rashes.

**Local Skin Lesions in Mousepox.** Mousepox is the standard experimental model of a generalized infection with rash (see below). Natural infection follows entry of virus from the contaminated environment into abrasions of the skin (Fenner, 1947; Roberts, 1962b) or the respiratory tract (Briody, 1959; Roberts, 1962a). The dose needed to infect the epidermis alone, by light scarification, is much larger than if scarification includes the dermis, probably because the macrophages and fibroblasts in the dermis are much more susceptible than the epidermal cells to virus introduced by scarification. Subsequently, virus spreads through the network of fibroblasts and the macrophages in the dermis and causes infection of the epidermis, with the production of a primary lesion in the skin. The swelling associated with the primary lesion is partly a nonspecific response to injury, and partly associated with a cell-mediated immune response; the infected cells themselves do not multiply. Free virus and probably infected macrophages enter dermal lymphatic vessels, leading to generalization of the infection.

**Viruses Causing Skin Lesions after Generalization.** A large number of vi-ruses, belonging to many different genera, cause skin lesions as part of a general-ized infection, or in particular segments of the skin during recrudescent infections (Table 9–3). There is, nevertheless, a clear association between taxonomic group-ings and the likelihood of producing rashes. The pathogenesis of rashes is de-scribed below and in Chapter 12.

## INFECTIONS OF THE RESPIRATORY TRACT

### Anatomy and Histology of the Respiratory Tract

The respiratory tract comprises a conducting system of tubes connecting the exterior of the body with the respiratory system, which consists of many small air vesicles, the alveolar sacs and alveoli, where exchange of gases between blood and air takes place. Inspired air passes in succession through the nose, pharynx, larynx, trachea, and bronchial tubes of various sizes before it reaches the alveoli. Conventionally, the upper respiratory tract consists of the conducting system

TABLE 9–3

*Viruses and Cutaneous Rashes* [a]

| VIRUS GROUP | RASH PRESENT | RASH ABSENT OR NOT RECORDED |
|---|---|---|
| DNA viruses | | |
| *Parvovirus* | Nil | — |
| Papovaviridae | Nil | — |
| *Adenovirus* | Adenoviruses (?1,2,3,7) | Adenoviruses (most) |
| *Herpesvirus* | Herpes simplex | Cytomegaloviruses |
| | Varicella-zoster | |
| | Mononucleosis | |
| *Iridovirus* | African swine fever | — |
| *Poxvirus* | Many | — |
| RNA viruses | | |
| Picornaviridae | Coxsackieviruses A9, A16, A23 | Coxsackieviruses (most) |
| | (B1, B3, B5, and others rarely) | |
| | Echoviruses 9,16 (4,6,18) | Echoviruses (others) |
| | Vesicular exanthema virus | Polioviruses |
| | Foot-and-mouth disease virus | Cardioviruses |
| Togaviridae | *Alphavirus*—Chikungunya, | *Alphavirus* (most) |
| | O'nyong-nyong, Ross River | |
| | *Flavivirus*—West Nile, | *Flavivirus* (most) |
| | dengue viruses | |
| | Ungrouped—rubella | |
| *Orthomyxovirus* | Nil | — |
| *Paramyxovirus* | Measles | Parainfluenza viruses |
| | Distemper | Mumps |
| | Rinderpest | Newcastle disease virus |
| | | Respiratory syncytial virus |
| *Coronavirus* | Nil | — |
| *Leukovirus* | Nil | — |
| *Rhabdovirus* | Vesicular stomatitis | Rabies |
| *Reovirus* | Reovirus 2 | Reoviruses (others) |
| *Orbivirus* | Nil | — |
| *Arenavirus* | Machupo, Junin | Lymphocytic choriomeningitis |
| Ungrouped | — | Hepatitis |

[a] Modified from Mims (1966b).

down to the larynx. Inside the nasal cavity the stratified squamous epithelium of the nose changes to a pseudostratified ciliated epithelium which is continued down the trachea and the bronchi into the terminal bronchioles. There are many mucus-secreting goblet cells among ciliated epithelial cells, and numerous mixed mucous glands beneath the basement membrane discharge over the epithelium and keep it moist. Throughout the subepithelial tissue of the nasopharynx and trachea there are scattered lymphocytes and collections of lymphoid tissue, the largest of which constitute the tonsils and adenoids. This lymphoid tissue probably plays an important role in local production of antibody, most of which appears to be IgA. The turbinate bones in the nasal cavity provide a complex arrangement of "baffle plates," and in the lower nasal conchae there are rich venous plexuses that warm the air passing through the nose.

From the point of view of viral infections, the important features of the conducting system are (a) the ciliated epithelial cells, which are the cells that support the multiplication of many of the respiratory viruses, (b) the diffuse and nodular lymphoid tissue which produces antibody that is released on to the epithelial surface, and (c) ciliary movement which moves the film of mucus that covers the epithelium up the respiratory tree to the pharynx, where it is removed by swallowing or expectoration. The movement of the mucus blanket by ciliary action is an important protective mechanism, and interference with this movement, either by ciliary paralysis, or by alteration of the secretion of mucus, may lower resistance to viral infections. Viruses that cause respiratory symptoms appear to spread between the epithelial cells; other viruses that initiate infection in the respiratory tract but cause generalized infection or lesions like herpangina probably move downward through the basement membrane to the subepithelial lymphoid tissues.

The terminal bronchioles pass into alveolar ducts, tortuous thin-walled tubes that are beset with outpouchings, the alveolar sacs, which in turn contain some two to four alveoli. Gaseous exchanges take place through the alveolar walls, which contain a dense network of capillaries. The alveolar wall has a very thin continuous cellular covering separated from the capillary endothelium by a thin homogeneous basement membrane. The alveolar macrophages that are found free in the alveoli appear to be derived from the blood stream.

## Infections Localized to the Respiratory Tract

Among gregarious animals like man, viruses that infect the respiratory tract are numerous and important. Some of these enter the body via the respiratory tract, where they multiply and are excreted without causing significant systemic infection. Others enter via the respiratory tract "silently" (i.e., there is no detectable local lesion) and then spread through the body to give a generalized infection, often with secondary involvement of the respiratory tract followed by excretion of virus (Table 9–4).

Three examples of viral infection of the respiratory tract will be described: mousepox, as a model for viruses that cause no initial respiratory symptoms but progress to a generalized infection; influenza, in which viral multiplication may occur throughout the upper and lower respiratory tract; and rhinovirus infections, which are strictly confined to the upper respiratory tract.

**Initial Respiratory Infection in Mousepox.** An obvious primary lesion develops when poxviruses are implanted in the skin, but no primary lesion has been found in those infections in which the virus enters via the respiratory tract, as in smallpox and rabbitpox. Possible reasons for this difference have been investigated in mousepox by Roberts (1962a), who showed by fluorescent antibody staining after inhalation of a virus aerosol that the first cells infected in the lower respiratory tract were alveolar macrophages and mucosal cells. Spread of

TABLE 9–4

*Viral Infections of the Respiratory Tract*

| | | |
|---|---|---|
| Localized upper respiratory (common cold, pharyngitis, etc.) | *Adenovirus* | human types 1-7, 14, 21 |
| | *Orthomyxovirus* | influenza A (man, birds, horses) influenza B (man) |
| | *Paramyxovirus* | parainfluenza types 1-4 (man and other animals) respiratory syncytical virus |
| | *Coronavirus* | human strains |
| | Picornaviridae | |
| | *Rhinovirus* | 90 serotypes |
| | *Enterovirus* | coxsackievirus A21, B2-B5 |
| Localized lower respiratory (croup, bronchitis, bronchiolitis, pneumonia) | *Adenovirus* | human types 1-7, 14, 21 |
| | *Herpesvirus* | infectious bovine rhinotracheitis, equine rhinopneumonitis, avian laryngotracheitis |
| | *Orthomyxovirus* | influenza A and B |
| | *Paramyxovirus* | parainfluenza types 1-3; respiratory syncytial virus |
| Generalized infection-- viral entry via respiratory tract | *Poxvirus* | smallpox, mousepox, sheep pox, etc. |
| | *Herpesvirus* | chickenpox |
| | Papovaviridae | polyoma (mice) |
| | *Orthomyxovirus* | fowl plague (influenza A) |
| | *Paramyxovirus* | Newcastle disease, measles, distemper, rinderpest, mumps |
| | Picornaviridae | foot-and-mouth disease |

poxviruses from the initially infected cells of the bronchial mucosa is slow, being delayed centrally by the basement membrane. The small local lesions that do develop are clinically and epidemiologically silent, and they may be difficult to find unless the whole lung is examined by serial sections. Virus may be transported directly to the lymph node by migration of infected alveolar macrophages, without concomitant multiplication in the lung, and thus the infection can become generalized without the development of a primary lesion. Mice infected by aerosols of variable particle size also develop fairly extensive lesions in the upper respiratory tract, especially in the olfactory mucosa over the nasal turbinates and septum. Virus expelled from the noses of mice during the incubation period of mousepox probably comes from these lesions (Roberts, 1962a).

Using aerosols of dried particles 1 μm or less in diameter, Westwood *et al.* (1966) studied the infection of rabbits with rabbitpox and vaccinia viruses, and of monkeys with variola virus. The cells initially involved (Lancaster *et al.*, 1966) were those of the bronchiolar epithelium and alveoli. Rabbitpox was the only one of these three diseases that spread to noninfected contacts; rabbits infected by aerosol did not become infective until there was an obvious ocular and nasal discharge, after the disease had become generalized.

No precise information is available on the early stages of infection of man with chickenpox, measles, mumps, rubella, or smallpox viruses. There is epidemiological evidence that infection occurs via the respiratory tract, but fails to cause symptoms or lead to secondary cases until the mouth and upper respiratory tract are involved in the lesions of the generalized disease.

**Influenza.** The pathogenesis of influenza has been more extensively investigated than that of any other respiratory disease (review, Stuart-Harris, 1965). Nearly all these studies have been carried out with influenza in experimental animals rather than in domestic animals (McQueen et al., 1968) or in man. Influenza virus produces symptoms by multiplication in the cells of the upper and lower respiratory tract; generalization via the blood stream may occur (Lehmann and Gust, 1971), but is rare and its significance in the disease process is unknown.

A great deal of work has been carried out on the production of infectious aerosols, and on the influence of droplet size and state on airborne infection (reviews by Wells, 1955; Proctor, 1966). Infection of man can be initiated by influenza viruses in the wet or dry state, in small or large aggregates, and when lodged on the upper or lower respiratory passages. Virus particles on the mucus film that covers the epithelium of the upper respiratory tract may undergo one of several fates. If the individual has previously recovered from an infection with that strain of influenza virus, antibody in the mucus secretion may combine with the virus and neutralize it. Nasal mucus also contains glycoprotein inhibitors which combine with virus and could prevent it from attaching to the specific receptors on the host cell; one possible function of the viral neuraminidase is to release virus from those inhibitors. Eventually the viral particle(s) that will cause infection come into contact with specific receptors on the surface of the cells of the respiratory mucosa and initiate infection as described in Chapter 6.

The successful initial infection of a few cells with influenza virus, and even its passage from these cells once more into the respiratory mucus and then into other cells, does not inevitably produce progressive infection, because the inflammation that follows cell injury results in an increased diffusion of plasma constituents (antibody and glycoprotein inhibitor) which may inactivate the virus and cut short the infection. Interferon, which has been demonstrated in the lungs of mice infected with influenza virus (Isaacs and Hitchcock, 1960), may also play a role in limiting viral spread.

If it does not lead to inactivation of the virus, the outpouring of fluid that follows early cellular injury may disperse the virus and thus increase the extent and severity of infection; in mice, Taylor (1941) showed that sublethal infections with influenza virus could be converted into lethal infections by instilling saline intranasally.

Although the early stage of release of virus does not cause obvious cell damage, the end result is necrosis and desquamation of the respiratory epithelium. In man, the nasopharynx, trachea, bronchi, and bronchioles are involved to a variable extent in different cases, and in influenzal pneumonia the fixed alveolar cells may show cytopathic changes (Hers et al., 1958).

Fluorescent antibody staining has been used to study the pathogenesis of influenza in ferrets (Liu, 1955), mice (Albrecht et al., 1963), and man (Hers and Mulder, 1961). In ferrets infected by the intranasal instillation of a viral suspension, antigen was initially demonstrable in a few ciliated cells on the nasal mucosa, over the turbinates, and later spread over the whole of the epithelium. This was followed by desquamation which coincided with the onset of manifest illness. Some ferrets developed pneumonia, and in these animals infected cells were found in the bronchial epithelium and in the mediastinal lymph nodes. Histological studies of human influenza have shown that ordinarily damage is confined to the epithelial cells of the upper respiratory tract, but in cases of pneumonia fluorescent antibody staining showed that there were foci of infection in the cells of the bronchi and bronchioles and in alveolar cells and alveolar macrophages. The experimental disease is greatly influenced by the method of infection. Large particle aerosols or small volumes of virus-containing fluids given intranasally infect, in the first place, the upper respiratory tract, whereas small particle aerosols or larger volumes given intranasally to anesthetized animals can straight away infect the lower respiratory tract and produce pneumonia.

Although there have been reliable reports of the occasional recovery of influenza virus from tissues other than those of the respiratory tract (e.g., Kaji et al., 1959; Naficy, 1963; Lehmann and Gust, 1971), viremia in the sense of the regular blood-borne dissemination of viable influenza virus is rare in human influenza, although irregular viremia does occur in mice (Hamre et al., 1956). However, viral antigen must be disseminated through the body, either free or perhaps after phagocytosis by alveolar macrophages as described by Liu (1955) in ferrets and by Albrecht et al. (1963) in mice. Such viral components are important in promoting the immune response, and, in turn, antigen–antibody complexes in the circulation and tissues activate complement and thus contribute to the constitutional symptoms of influenza.

Environmental factors such as atmospheric temperature, relative humidity, and atmospheric pollution may affect susceptibility to respiratory virus infections. Hope-Simpson (1958) found that the incidence of common colds increased as the atmospheric temperature decreased, but only as a result of the consequent change in relative humidity; attempts to demonstrate a direct influence of temperature have failed (Douglas et al., 1968). The evidence for an effect of air pollution on viral infections is less convincing, although the condition of patients with chronic bronchitis is closely related to air pollution levels in cities (Waller, 1971). Cigarette smoking has been associated with increased susceptibility to influenza virus infections, both in man (Finklea et al., 1969) and in experimental animals (Spurgash et al., 1968). Cigarette smoke has been shown to depress mucociliary activity in experimental animals (Falk et al., 1959; Dahlam, 1964), and goblet cells are more numerous in the bronchial epithelium of smokers (Chang, 1957), but no differences in mucociliary activity between smokers and nonsmokers could be detected by Quinlan et al. (1969).

Secondary bacterial infection is an important complication of respiratory virus infections. Studies with influenza and parainfluenza viruses in mice (Harford and Hara, 1950; Degre, 1970) have shown that the ability of the lung to remove or inactivate intranasally inoculated bacteria ("pulmonary clearance," Green, 1968) is seriously impaired several days after peak viral titers have

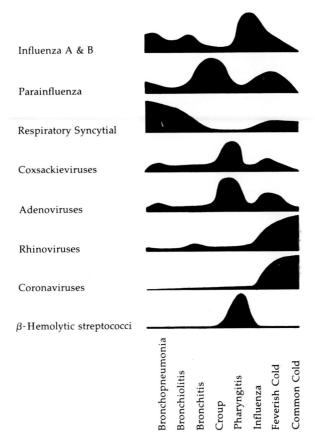

Influenza A & B

Parainfluenza

Respiratory Syncytial

Coxsackieviruses

Adenoviruses

Rhinoviruses

Coronaviruses

β-Hemolytic streptococci

Bronchopneumonia · Bronchiolitis · Bronchitis · Croup · Pharyngitis · Influenza · Feverish Cold · Common Cold

FIG. 9–3. *Diagram illustrating the frequency with which particular viruses produce disease at various levels of the human respiratory tract. (Courtesy Dr. D.A.J. Tyrrell.)*

been attained, because of defective function of alveolar macrophages. Increased susceptibility to bacteria would be expected in respiratory mucosa infected with viruses because of epithelial cell necrosis, plasma exudate, and defective muco-ciliary transport.

**Rhinovirus Infections.** Rhinoviruses, of which some ninety serological types are currently recognized (review, Jackson and Muldoon, 1973) have been recovered from the nose and throat but not from the feces or conjuctiva. Isolations have been made over a period of 4 days, but not during convalescence. Although mild respiratory illnesses indistinguishable from "common colds" can be produced by a variety of different viruses (see Fig. 9–3), the rhinoviruses and the human coronaviruses are by far the most important agents and they appear to cause no other type of disease.

The classic signs of the common cold in man are the profuse discharge of thin watery mucus from the upper respiratory tract, and the subsequent production of a thick mucopurulent secretion, which in the later stages but not initially may occasionally be associated with secondary bacterial infection of the mucous membranes of the nose and nasal sinuses (Tyrrell, 1965). Ciliary activity is retained in mild colds although it is severely inhibited in the catarrhal stage of severe colds (Tyrrell, 1965). The initial viral damage may render the mucous membrane more susceptible to invasion by the scant but varied bacterial flora which occurs normally on these surfaces, and common colds sometimes lead to purulent sinusitis, or less commonly otitis media.

No equivalent of the human common cold is known in experimental animals, although chimpanzees can be asymptomatically infected with rhinoviruses (Dick, 1968). Extensive human volunteer experiments have been carried out, both before and after the isolation of the rhinoviruses, but virtually nothing is known of the pathology or pathogenesis of these infections. The better growth of many strains at 33° rather than 37°C (Tyrrell and Parsons, 1960) may explain their localization to the mucosa of the upper respiratory tract.

## Respiratory Viruses in Organ Cultures

Hoorn and his colleagues (review, Hoorn and Tyrrell, 1969) have studied the effects of several viruses that multiply in the human respiratory tract on organ cultures obtained from human adult and fetal nasal and tracheal mucosae. Results reflected the localization of the natural diseases; rhinoviruses, for example, multiplied well in the ciliated nasal epithelium, poorly in tracheal epithelium, and not at all in esophageal or palatal epithelium. Apart from small areas of degeneration of the superficial cells there were no lesions. Organ cultures fail to provide an explanation for the profuse outpouring of watery mucus which is such a characteristic early feature of common colds; this probably depends upon intact vascular and nervous systems as well as the secreting cells of mucous glands of the nasal mucosa. Parainfluenza viruses, various strains of influenza virus, and adenovirus types 2 and 7 all grew in tracheal mucosa with destruction of ciliated epithelial cells. Polioviruses multiplied as readily in organ cultures of nonciliated epithelium (gut, palate) as in ciliated epithelium; ciliary activity was unaffected.

## INFECTIONS OF THE ALIMENTARY TRACT

### Structure and Function of the Alimentary Tract

The epithelium of the mouth is constantly being injured during the process of chewing food, and breaches of its surface act as the portals of entry of viruses, of which the more important examples in man are herpes simplex and herpangina (a coxsackievirus infection). In generalized infections with a rash, there is usually an enanthem in the mouth and esophagus; this differs from the exanthem produced on the skin because although the mouth and esophagus are lined by stratified squamous epithelium they do not have a dry cornified superficial layer. Virus from these mucosal lesions is excreted in saliva and by coughing, and may constitute an important mode of transfer of infection (see Chapter 16). The enanthem may also involve the intestine, as in severe measles.

There are many collections of nodular lymphoid tissue in the pharyngeal mucous membrane; the larger ones constitute the tonsils and adenoids. They lack the lymphatic sinuses found in lymph nodes, and the tonsils are penetrated by deep crypts covered with epithelium. Their importance in viral infections is uncertain; they are probably local sources of secretory antibodies, they may serve as portals of entry of poliovirus, for example, and they are often latently infected, especially with adenoviruses, but also with such viruses as herpes simplex and cytomegalovirus.

The stomach and duodenum are rarely affected by viruses, either primarily or secondarily. The acidic and proteolytic secretions of the stomach, and the bile which enters the duodenum, inactivate most swallowed viruses. The only viruses that survive these conditions and are able to infect cells of the small intestine are enteroviruses, reoviruses, hepatitis viruses, adenoviruses, and possibly parvoviruses. The rhinoviruses are acid labile, while most of the enveloped viruses are dissociated by bile salts.

The small intestine is a long tube with an enormous surface area, due to the formation of circular folds and microscopic villi; it is the major site for absorption of nutritive material into the blood and lymph vessels. The villi are outgrowths of the mucous membrane about 1 mm long, between the bases of which glands open. The simple columnar epithelium that covers the villi consists of columnar cells with a striated border, which is seen in the electron microscope to consist of closely packed microvilli. There is a constant shedding and replacement of the cells of the intestinal villi, the cell population being completely replaced every 7–8 days. The connective tissue of the cores of the villi and the spaces between them, the lamina propria, contains small secretory glands, a rich plexus of blood and lymphatic vessels, and free macrophages, lymphocytes, and plasma cells. There are also numerous small collections of lymphoid tissue throughout the lamina propria; in some places these are massed together to form Peyer's patches. Both the solitary and aggregated lymphocytes and plasma cells appear to be important sources of IgA. Specific antiviral antibodies produced in these cells probably play an important role in protection against reinfection by enteric viruses.

## Viruses Affecting the Alimentary Tract

Viruses of most genera never cause primary infection of the gastrointestinal tract, because they cannot survive passage through the acidic stomach contents and the bile-containing duodenum, but there can be gastrointestinal involvement secondary to systemic spread of infection. The only DNA viruses that commonly cause primary infection of the gastrointestinal tract are the adenoviruses, many of which have been recovered only from human feces (Rosen et al., 1962).

**Adenovirus Infections of the Alimentary Tract.** Clemmer (1965) has made observations on the pathogenesis of an avian adenovirus in its natural host, the chicken. After oral inoculation the virus was largely confined to the gastrointestinal tract, with highest concentrations in the esophagus, terminal ileum, and cecae. Fluorescent antibody studies of the terminal ileum 6 days after infection

TABLE 9–5

*Viral Infections of the Intestinal Tract and Liver*

| | | |
|---|---|---|
| Intestinal tract | *Adenovirus* | human and avian |
| | *Parvovirus* | subgenus A |
| | Picornaviridae | all species of *Enterovirus* |
| | *Coronavirus* | transmissible gastroenteritis of swine |
| | | mouse hepatitis |
| | *Paramyxovirus* | Newcastle disease |
| | *Reovirus* | all species |
| | Unclassified | LIVIM, EDIM (mice, see text) |
| Liver | *Poxvirus* | mousepox |
| | *Herpesvirus* | EB virus (infectious mononucleosis), |
| | | herpes simplex (human infants, mice) |
| | *Adenovirus* | infective canine hepatitis |
| | Picornaviridae | duck hepatitis |
| | *Arenavirus* | lymphocytic choriomeningitis |
| | *Coronavirus* | mouse hepatitis |
| | *Flavivirus* | yellow fever |
| | *Reovirus* | type 3 (mice) |
| | Unclassified | Rift valley fever (mice) |
| | | human hepatitis types A and B |

revealed scattered single infected cells on the surface of the villi, both columnar and goblet cells being affected (Clemmer and Ichinose, 1968). There was no lymphoid hyperplasia or inflammatory reaction. Recovered birds were refractory to challenge infection regardless of their serum antibody titer, which may reflect the overriding importance of local (IgA) antibody.

Experiments with human volunteers showed that adenoviruses administered in enteric-coated (hard gelatin) capsules produced selective inapparent enteric infections (Chanock *et al.*, 1966). Virus did not spread easily to susceptible contacts, although marital partners sometimes suffered inapparent infections (Stanley and Jackson, 1969). Vaccinated individuals developed neutralizing antibody in their serum, but not in nasal secretions (T. Smith *et al.*, 1970). However, prior enteric adenovirus type 4 infection is highly effective in preventing acute respiratory tract disease caused by adenovirus type 4 (Edmondson *et al.*, 1966).

**Transmissible Gastroenteritis of Swine.**   This severe infection of piglets, in which damage is restricted to the small intestine, is caused by a coronavirus (Tajima, 1970). The pathogenesis of the disease has been investigated by Hooper and Haelterman (1966). Infection of the small intestine follows ingestion of the virus, and symptoms occur within 24 hours. Fluorescent antibody studies (Pensaert *et al.*, 1970) showed that virus multiplies in columnar epithelial cells of the jejunum, causing rapid and extensive loss of cells so that the villi are greatly shortened. There is no inflammatory reaction and the villi are not denuded of cells, the shortened stubs being covered with immature epithelial cells. The loss of so much of the digestive and absorptive surface of the jejunum results in severe watery diarrhea, dehydration, electrolyte imbalance, and sometimes death. Viremia may occur and virus has been recovered from the kidneys and lungs, but this appears to be irrelevant as far as the symptomatology is concerned. Transmissible gastroenteritis of swine is very like cholera in man, a disease localized to the small gut and causing symptoms because of malabsorption, fluid loss, and electrolyte imbalance.

**Intestinal Infections of Suckling Mice.** There are two viral infections of suckling mice, produced by viruses of unknown taxonomic status, that appear to infect intestinal epithelial cells. Lethal intestinal virus infection of mice, or LIVIM (Kraft, 1966), produces an acute diarrheal disease in infant mice with death 48 hours after oral infection. There are cytoplasmic inclusions in swollen epithelial cells of intestinal villi, and affected cells are shed to give ulcerated mucosal areas, with high viral titers in intestinal contents. Older mice show less epithelial cell degeneration and a rapid proliferative response in intestinal crypts, and there is generally no ulceration and no disease (Biggers et al., 1964). The antigenically distinct epidemic diarrheal disease of infant mice (EDIM) gives a less severe disease with some necrosis of intestinal epithelium, viremia, and recovery. Attempts to grow EDIM and LIVIM in cell cultures have failed, but EDIM has recently been cultivated in organ cultures of mouse intestinal epithelium (Rubenstein et al., 1971).

**Enterovirus Infections.** The enteroviruses are primarily parasites of the intestinal tract. Some may cause diarrhea in man (Dolin et al., 1972; reviews by Kibrick, 1965; Wenner and Behbehani, 1968); most infections are symptomless. Some enteroviruses (poliovirus and some of the coxsackieviruses and echoviruses) may cause generalized infection after entry and initial multiplication in the gastrointestinal tract. Important though these viruses are, we know very little of the pathogenesis of infection of the gastrointestinal tract by them.

## SYSTEMIC INFECTIONS

There are two circulatory systems in the vertebrate body, the lymphatic system and the blood vessels, both of great importance in the pathogenesis of systemic viral infections.

### The Lymphatic System

The lymphatic system has a dual importance in viral infections: (a) lymphatic vessels constitute one of the principal routes by which virus passes from the exposed surfaces (skin, respiratory mucosa, or alimentary tract) into the body of the animal, and (b) the lymphoid tissue associated with the lymphatic system contains the cells which mount the immune response. Lymphatic vessels occur throughout the tissues of the body, and they constitute a closed system that discharges eventually into veins. The lymphatic capillaries are thin-walled tubular structures slightly larger than blood capillaries, with blind ends. Their walls are formed of a single layer of flat endothelial cells which abut directly against the surrounding tissues, so that viral particles in the tissues readily enter into lymphatic capillaries. The lymph passes into successively larger lymphatic vessels, then through lymph nodes, and finally enters two large trunks, the right lymphatic duct and the thoracic duct, which discharge into large veins. The lymph nodes are large accumulations of lymphoid tissue located along the course of lymphatic vessels; their structure is described in the next chapter.

## Blood Vessels

Blood vessels afford the principal route for the dissemination of viruses throughout the body, the term viremia signifying the presence of infective virus in the bloodstream. Viruses enter the blood from tissues where they have multiplied by passing through blood vessel walls, or by entering lymphatics which discharge into the blood. They may also enter the blood directly after growth in circulating blood cells or in endothelial cells lining blood vessels. To pass from the blood into any tissue virus, either free or cell-associated, must pass through a blood vessel wall. Blood vessels or the heart may themselves be infected and damaged by viruses, coxsackievirus B, for instance, causing myocarditis in man and in experimental animals. Finally, in addition to their importance in transporting and distributing viruses throughout the body, many of the important events in infected tissues, such as inflammation, necrosis and repair, are based on changes in blood vessels.

From the point of view of viral infections the most important elements in the blood vascular system are the small capillaries and venules and the sinusoids of the liver, spleen, and bone marrow. Capillaries and venules have a complete endothelial lining (Plate 9–1), which constitutes a barrier to the passage of viruses from the circulation to the cells of all other organs and tissues of the body. This barrier is obviously not impassable, for virions may be passively transferred across the endothelial cells by micropinocytosis, without multiplying, or they may multiply in the endothelial cells before being released from both sides of them, or virus may be carried into the tissues by infected leukocytes which migrate from the capillaries by diapedesis. The basement membrane differs in thickness and density in different parts of the body, and may present a further barrier to the spread of free viral particles.

In the liver, spleen, and bone marrow, arteries and veins are connected by sinusoids, whose walls are formed not by a continuous layer of endothelial cells, but by irregularly scattered phagocytic and nonphagocytic cells. In the liver, for instance, electron microscopic studies suggest that there are openings in the walls of the sinusoids, but the littoral cells appear to be so efficient in phagocytosis that viral particles rarely gain direct access to the parenchymal cells (Plate 9–2). There is no basement membrane between the cells lining the sinusoids and the parenchymal cells of the liver, merely a narrow submicroscopic space (space of Disse) into which microvilli of the parenchymal cells project. Active phagocytosis of circulating virions by the Kupffer cells is important in many viral infections.

## The Systemic Spread of Viruses

Systemic infection with generalization of the virus but no generalized signs and symptoms occurs in many infections that appear to be localized, as is evident

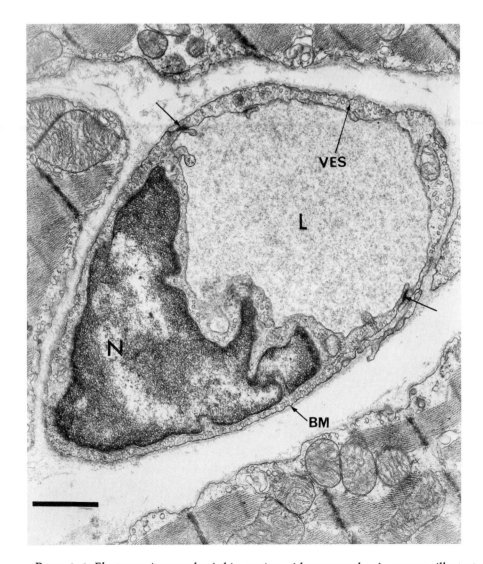

PLATE 9–1. *Electron micrograph of thin section of heart muscle of a mouse, illustrating a blood capillary (bar, 1 μm). L, lumen. Two endothelial cells are present, one of which contains a nucleus (N); there are desmosomes (arrows) at their junctions. There is a well-developed basement membrane (BM) and the cytoplasm of the endothelial cells contains many caveolae and vesicles (VES). The surrounding striated muscle cells contain many mitochondria. (Courtesy of Professor B. Morris.)*

if organs distant from the local lesion are assayed for virus. Fibroma virus infection in domestic rabbits is a good example. Ordinarily the lesions are strictly confined to the site of introduction of virus into the skin, but viral infection of the internal organs (without the production of lesions) is widespread. In occasional individuals, or under particular circumstances, disseminated skin lesions

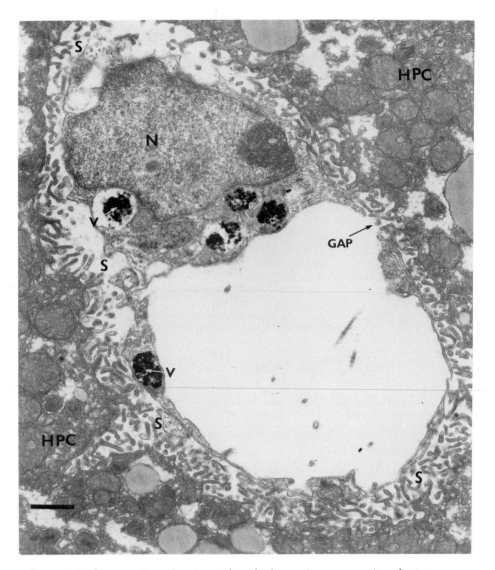

PLATE 9–2. *Cross section of a sinusoid in the liver of a mouse, after the intravenous injection of India ink (bar, 1 μm). The cavity of the sinusoid is enclosed by the cytoplasm of a Kupffer cell which is incomplete (GAP), and is very thin over much of the sinusoid. Outside the endothelial lining of the sinusoid is the space of Disse (S) into which project numerous microvilli of the hepatic parenchymal cells (HPC). Several vacuoles (V) in the Kupffer cell are filled with carbon particles which have been phagocytosed. N, nucleus of Kupffer cell. (Courtesy of Professor B. Morris.)*

may occur (Fenner and Ratcliffe, 1965). Perhaps a more interesting problem is why some localized infections do remain strictly localized, for in all viral infections of the skin or mucous membranes virions are potentially able to enter the lymphatics and thus to become generalized. The failure of rhinoviruses to replicate optimally at body temperature (37°C) probably prevents their multiplication in the internal organs, and human influenza virus does not seem to grow readily in cells other than those of the respiratory mucosa. This is not a property of all myxoviruses, since fowl plague and Newcastle disease in chickens, and mumps and measles in man, are generalized diseases. Arbovirus infections of vertebrates are always generalized, and the viremia is an obligatory stage for infection of arthropod vectors.

Generalization may occur without the production of symptoms, or it may be associated with lesions in certain target organs, such as the skin and the central nervous system; more rarely the liver, heart, or certain glands may be primarily affected. In the following pages we shall first discuss various aspects of viremia and then consider the spread of infection through the vertebrate body in two important types of generalized infection, those characterized by a rash of the skin and mucous membranes and those involving the central nervous system.

## The Nature of Viremia

Viremia occurs in most viral infections, including many of those described earlier as "localized" infections, even though it may not be readily detectable. Rhinovirus infections of the upper respiratory tract and infection of the skin with warts virus may be exceptions.

The blood consists of plasma and formed elements: platelets, leukocytes (polymorphonuclear cells, monocytes, and lymphocytes), and erythrocytes. Different viruses may be characteristically associated with particular types of leukocytes, with platelets, or with erythrocytes, or the virus may be free in the plasma. In poxvirus infections, the viremia is associated with leukocytes, mainly with lymphocytes and monocytes. Leukocyte-associated viremia is a feature of several other types of infection, including the only generalized orthomyxovirus infection, fowl plague, and infections with several members of the *Paramyxovirus* genus (e.g., rinderpest, distemper, and measles). Occasionally a proportion of the virus in the circulation is associated with erythrocytes, to which it is adsorbed superficially, as in Rift Valley fever, hog cholera, parvovirus infections, and lymphocytic choriomeningitis. In the case of Colorado tick fever, erythrocytes contain virions and viral antigen (Emmons *et al.*, 1972), presumably after infection during erythropoiesis in the bone marrow. In some cases, virus is confined to the plasma; the togaviruses and picornaviruses that cause viremia fall into this group. Finally, in some infections the viremia is mixed, i.e., the virus is partly in the plasma and partly cell-associated. Infectious polyoma virus DNA can be recovered from the blood of mice into which it has been inoculated intraperitoneally (Bendich *et al.*, 1965); it is conceivable therefore that "viremia" could in some circumstances be due to the circulation of viral nucleic acid.

Most tissues contain resident extravascular macrophages (histiocytes), and the sinusoids and the lymphatic channels through lymph nodes are lined by macrophages. Aschoff introduced the term "reticuloendothelial system" to designate all the fixed cells that are actively phagocytic, as demonstrated by the uptake of significant quantities of vital dyes or India ink when these are injected into the blood stream. Whether virus circulates free in the plasma or is cell-associated has a considerable influence on its passage from the circulation to extravascular sites where it may multiply. Free virus circulating in the blood is, as a rule, continuously removed by cells of the reticuloendothelial system. Viremia can, therefore, be maintained only if there is a continued release of virus into the blood, or if the clearance system is impaired. Infected leukocytes can initiate infection in various parts of the body by passing through the walls of small vessels, and plasma-associated virus can enter tissues by passing through or growing through vessel walls as discussed above. The circulating leukocytes could themselves constitute a source of multiplying virus. Blood leukocytes maintained in culture (from which polymorphonuclear leukocytes are rapidly lost) support limited multiplication of many viruses, but are unlikely to be very important *in vivo* (Gresser and Lang, 1966). The major sources which maintain the viremia include organs with extensive sinusoids, like the liver, spleen, and bone marrow, the endothelial cells of the blood vessels themselves, the lymphoid tissues (via the thoracic duct), and sometimes cells of the voluntary muscles and connective tissues (probably via the thoracic duct).

## Chronic Viremia

As well as acute viremia of the kind just described, chronic viremia occurs as a major feature of a number of different viral infections (see Chapter 12). Even in chronic viremias that are principally leukocyte- or erythrocyte-associated, virions also occur free in the plasma. Characteristically, little or no antibody is found in the serum of such animals, which used to be considered to be immuno- logically tolerant to the viruses they carried (Volkert and Larsen, 1965). How- ever, closer examination showed that the virus usually circulates as infectious virus–antibody complexes, the infectivity of which can be neutralized by incuba- tion with anti-immunoglobulin and complement (Notkins *et al.*, 1966, 1968). Animals with chronic viremia characteristically get immune complex disease (Symposium, 1971; see Chapter 12). Viral antigen–antibody complexes formed in antigen excess are deposited in kidney glomeruli and blood vessel walls lead- ing in the case of Aleutian disease of mink, for example, to severe glomerulone- phritis and vasculitis.

## Chronic Antigenemia

Because of the importance of viremia, and the extreme sensitivity of infec- tivity titrations, attention has been concentrated upon circulating virions rather than circulating antigens. The recognition that a vast excess of antigen (compared with virions) circulates for prolonged periods in the blood of many people who have recovered from serum hepatitis has directed attention to antigenemia in this disease and to its pathological effects (Almeida and Waterson, 1969; Gocke *et al.*, 1971). No studies have been made of antigenemia in the other chronic viremic diseases, but consideration of the size of immune complexes involved in the model immune complex disease, serum sickness (Dixon, 1963; Wilson and Dixon, 1971; Cochrane, 1971) suggests that complexes of antigen and antibody are probably of greater pathogenetic importance than virion–antibody complexes.

There are a few old records of antigenemia in the early stages of acute generalized viral diseases (smallpox in man, Downie *et al.*, 1953; myxomatosis in rabbits, Rivers and Ward, 1937, and Fenner and Woodroofe, 1953; and yellow fever in man, Hughes, 1933), but no investigations have been carried out using modern techniques. The occurrence of anticomplementary activity in the serum of rabbits infected with myxoma virus before neutralizing antibody could be detected (Fenner and Woodroofe, 1953) and in acute and early convalescent sera in serum hepatitis (Shulman and Barker, 1969) indicates that immune complexes are present in the circulation during the latter part of the incubation period. It is possible that antigenemia and the early formation of immune complexes, rather than viremia as such, provides the pathogenetic basis for the fever and symptoms that mark the end of the incubation period in many generalized viral infections. The relatively sudden onset would be consistent with the need for a precise balance to be achieved between antigen and antibody before complexes are formed and complement is activated. The occurrence of a prodromal rash in many of the exanthemata could also be explained on this basis.

### Clearance of Viruses from the Circulation

There is an extensive literature on the clearance of particulate matter from the blood stream, concerned with both inert particles and microorganisms (Rowley, 1962), the clearance of viruses being effected mainly by macrophages of the reticuloendothelial system. Mims (1964) and his colleagues have devoted particular attention to the role of these cells in viral infections. Figure 9–4 illustrates clearance rates of viruses of various sizes after their intravenous

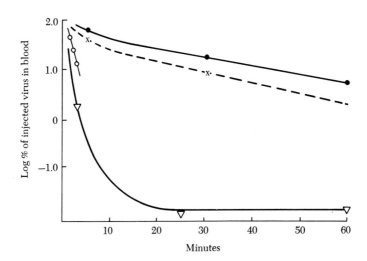

FIG. 9–4. *Clearance curves of viruses of different sizes after the intravenous inoculation of large doses into mice.* ●, *T7 bacteriophage,* ~ *30 nm diameter;* ×, *Rift Valley fever virus, 30 nm diameter;* ○, *vesicular stomatitis virus,* ~ *90 nm diameter;* ∇, *vaccinia virus,* ~ *250 nm diameter. (From Mims, 1964.)*

injection into mice. As with inert particles, large viruses like vaccinia tend to be cleared more rapidly than small ones, although the nature of the particle is also important (see below). One feature of the clearance curves of viruses is the constant presence of an "uncleared tail." This can sometimes be accounted for by the continued circulation of cell-associated virus, and in the case of plasma-associated viruses by assuming that the uptake of particles by the reticuloendothelial macrophages is partly reversible.

## Infection of the Liver

Quantitatively by far the most important cells concerned with the removal of particles from the circulation are the "fixed" macrophages of the liver, which may play an important part in viral infections of this organ (see Table 9–5). In mice, the littoral cells of the liver sinusoids constitute a functionally, although not an anatomically, complete lining, and under normal circumstances no particle in a sinusoid can enter a hepatic parenchymal cell except through a macrophage. A wide variety of different interactions of viruses and liver macrophages has been observed (Mims, 1964). Poliovirus type 1, in contrast to poliovirus types 2 and 3, is not cleared from the blood after intravenous injection into mice, but it is cleared normally in cynomolgus monkeys. Under certain conditions the liver macrophages digest the viruses which they clear from the blood stream. Using fluorescent antibody staining, Mims found that the CL (dermal) strain of vaccinia virus, influenza virus, and myxoma virus injected intravenously in the mouse were taken up by the liver macrophages, but within an hour the specific fluorescence in these cells had faded away and it never reappeared. In these cases, the hepatic macrophages were clearly protective, for when these same viruses were injected up the common bile duct and thus brought into direct contact with the hepatic cells, all of them multiplied in the hepatic parenchymal cells, as judged by the reappearance of large amounts of antigen 10–18 hours later.

Some viruses may be taken up by liver macrophages and pass passively across them to reach the hepatic cells, without growth in the macrophages. When large doses of Rift Valley fever virus are inoculated intravenously in mice the hepatic cells may show typical nuclear changes within an hour, inclusion bodies are seen in 3 hours, and mice die after 6 hours, with very extensive hepatic damage (Mims, 1957). Rift Valley fever virus does not appear to multiply in the liver macrophages (McGauran and Easterday, 1963) and clearly a large proportion of the inoculated virus must pass straight through into the hepatic cells.

Sometimes the material taken up from the blood by macrophages is excreted in the bile. This occurs with inert particles as well as with substances in solution, and it may occur with viruses. For example, type 1 poliovirus appears in the hepatic bile within an hour of being injected intravenously into cynomolgus monkeys, and T7 bacteriophage can be obtained from the gall bladder bile of mice 2 hours after an intravenous injection. Mims points out that if hepatitis virus in the liver is excreted into the bile, as is poliovirus in monkeys, there need be no other source of fecal virus in infective hepatitis.

In several generalized infections, virus is ingested by the littoral cells of the liver sinusoids and after growing in these may infect the hepatic cells. The details of this process as it occurs in mousepox are described below; it is also believed to be important in infectious canine hepatitis (Coffin *et al.*, 1953), and in yellow fever in monkeys (Tigertt *et al.*, 1960), to mention two diseases in which fluorescent antibody studies have been carried out. Infection of the liver is by no means an invariable feature of generalized viral infections, although the liver macrophages must always be exposed to large amounts of any virus that circulates in the blood. Guinea pig-adapted strains of lymphocytic choriomeningitis virus are taken up by Kupffer cells when injected intravenously into mice, and this leads to infection of the parenchymal cells of the liver. Mouse-adapted strains of virus, in contrast, are not cleared from the blood and therefore fail to infect the liver (Tosolini and Mims, 1971). Sometimes liver macrophages are regularly infected, but there is no spread of the infection to hepatic cells, as shown by fluorescent antibody studies of lactic dehydrogenase virus infection of mice (Porter *et al.*, 1969a), or Aleutian disease virus infection of mink (Porter *et al.*, 1969b).

## GENERALIZED INFECTIONS WITH RASHES

### Pathology of Skin Lesions in Rashes

Rashes are more easily seen in human beings than in furred and feathered beasts, and many of the descriptive data are therefore derived from human infections (review; Mims, 1966b). The individual lesions in generalized rashes are usually described as macules, papules, vesicles, or pustules. A lasting local dilatation of subpapillary dermal blood vessels produces a macule, which becomes a papule if there is also edema and an infiltration of cells into the area. Primary involvement of the epidermis usually results in vesiculation, ulceration, and scabbing, but prior to ulceration a vesicle may be converted to a pustule if there is a copious cellular exudate. Secondary changes in the epidermis may lead to desquamation, and more severe involvement of the dermal vessels to hemorrhagic and petechial rashes, although coagulation defects and thrombocytopenia may also be important in the genesis of such lesions. Table 9–3 summarizes current information on the association of rashes with infections due to different viruses; it is clear that some viruses (and most viruses of some groups) usually produce a rash whereas others never do so.

### The Pathogenesis of Smallpox

As a background for the discussion of generalized infections with rash we will briefly describe the sequence of events in the human infection, smallpox (Fig. 9–5). Epidemiological studies have established clearly that there is an incubation period of 10 to 12 days before symptoms of illness appear. These usually begin abruptly with fever, headache, loss of appetite, and nausea. There may be a prodromal erythematous rash and, after a few days, signs associated

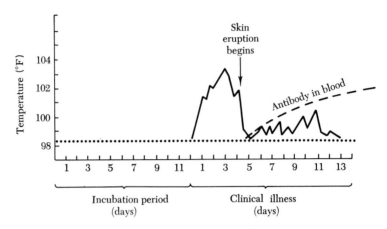

FIG. 9–5. *The stages of infection in human smallpox. (From Downie, 1963.)*

with infection of the skin and mucous membranes of the oropharynx appear. The skin lesions progress from macules through papules, vesicles, and pustules which begin to dry up after 10 to 11 days. In milder cases, fever abates soon after the rash appears, and antibody can usually be detected in the serum at about this time. Cases are not infectious during the incubation period.

In its essential features, smallpox is characteristic of exanthematous infections caused by a variety of viruses. Questions concerning pathogenesis include the following: what is happening to the virus during the incubation period and why is the patient both symptom-free and noninfectious during this period; what causes the prodromal rash and what processes determine the sudden onset of illness; why does the virus multiply especially in certain of the internal organs and why does it localize in the skin; what determines recovery; and what are the mechanisms by which viral multiplication causes disease or death? We are still far from being able to answer most of these questions, but some of them have been solved by studies of model infections in experimental animals. The models used for study of the exanthemata include mousepox in mice (Fenner, 1948b), rabbitpox in rabbits (Bedson and Duckworth, 1963; Westwood *et al.*, 1966), and smallpox in monkeys (Hahon, 1961). These poxvirus infections are satisfactory as models for smallpox and probably chickenpox, but not for diseases like measles and rubella in which the initial virus–cell interaction is noncytocidal.

## The Incubation Period

Problems concerning the portal of entry have been touched upon in the preceding sections and will be discussed in relation to the spread of infection in Chapter 16. What happens during the long incubation period of generalized viral infections was first studied with mousepox (Fenner, 1948b), which closely resembles human smallpox in many of its essential features. At regular intervals after infection with small doses of virus, infected mice were killed and their organs were assayed for infectious virus. There was a regular sequence of the steps of infection, multiplication, and release in a succession of tissues and organs (Fig. 9–6). In mousepox induced by infection of the skin, a "primary lesion" developed at the site of implantation of the virus, and its appearance (the first sign of disease) marked the end of the incubation period. Meanwhile, the virus had proceeded through the lymphatics to the local lymph node, then to the bloodstream (primary viremia), and had then localized in the liver and spleen. A

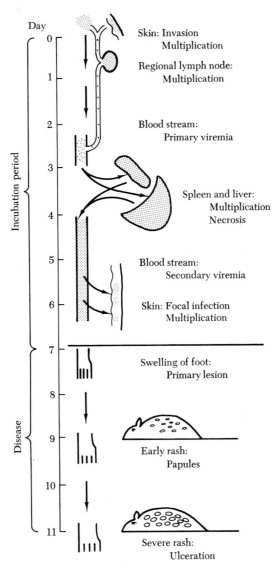

FIG. 9–6. *Scheme illustrating the possible sequence of events during the incubation period and development of signs of disease in mousepox. (From Fenner, 1948b.)*

secondary viremia occurred after the virus had multiplied extensively in the parenchymal cells of the liver and in the spleen, and this was followed by focal infection of the skin and mucous membranes, producing the rash.

Essentially the same mechanism as described here for mousepox has been demonstrated in myxomatosis (Fenner and Woodroofe, 1953), in rabbitpox (Bedson and Duckworth, 1963; Westwood et al., 1966), and somewhat less certainly in sheeppox (Plowright et al., 1959) and canarypox (Burnet, 1933). Fenner (1948b) suggested that a similar series of events occurred during the incubation period of all the exanthematous viral infections, and apart from the final site of localization it is probable that the same pattern underlies all generalized hematogenous infections. The liver and spleen are probably only rarely the "central focus," as they are in mousepox; in different viral infections the lung, the bone marrow, the lymph nodes, the muscles, or the endothelial cells of capillaries throughout the body may serve as the central site where the viral proliferation which establishes the viremia occurs.

## Localization in Organs and in the Skin

The factors that determine the secondary localization of virus in generalized viral infections are unknown. To some extent, secondary localization must be related to the nature of the viremia, i.e., whether the virus is cell associated or free in the plasma. In mousepox, it seems probable that virus is removed from the circulation at the time of primary viremia by the reticuloendothelial cells of the liver and spleen. The mode of invasion of the liver by mousepox virus was investigated by Mims (1959a, b) using fluorescent antibody staining of sections of the liver as well as infectivity assays. By inoculating mice intravenously with large doses of virus he showed that most of the injected virus was very rapidly cleared, mainly by the Kupffer cells of the liver, which contained large amounts of antigen immediately after injection. The intensity of fluorescence of the littoral cells decreased in the first few hours, and then increased again at 5 hours as new viral antigen was synthesized. Infection of the parenchymal cells of the liver was first seen 10 hours after the inoculation, and there was a stepwise increase in viral titer in the liver as virus grew first in the littoral cells and then in successive groups of parenchymal cells. By 23 to 30 hours almost all cells of the liver were infected.

In the skin, virus first infects the endothelium of the capillaries and venules in the dermis, and from there it spreads to the epidermis where typical degenerative changes occur, resulting in the formation of vesicles (Mims, 1968b). The focal eruption of the exanthemata also involves the mucous membranes of the mouth and upper respiratory tract. Since there is no keratinized squamous epithelium here, these lesions break down more rapidly than the skin lesions and are important in determining the infectiousness of the patient in the early stages. Just why virus multiplies so extensively in the skin, and what determines the peculiar and characteristic distributions of the rashes in many of the exanthemata, is unknown. One factor may be the lower temperature of the skin, for many viruses fail to multiply readily, at least in developing eggs and cultured cells, at the temperatures attained during fever (38°–41°C).

## The Production of Symptoms and the Development of the Rash

The incubation period ends with the production of symptoms and signs of infection, and in smallpox this appears to occur at about the time of the massive secondary viremia. The severity of symptoms is related to the level of the viremia, probably because this is an index of the degree of multiplication of the virus in the internal organs. The level of the viremia also affects the degree of secondary localization of the virus in the skin and mucous membranes, and thus the severity of the rash.

The general symptoms which usher in disease are common to a wide variety of infections; viral, protozoal, and bacterial. No toxins separable from the viral particle have been demonstrated in viral infections although pyrogens are produced when influenza virus interacts with leukocytes. The usual explanation for the general symptoms is that they are due to the toxic and pyrogenic effects of products of cells destroyed during viral multiplication. It is more likely that the prodromal rash and the abrupt onset of general symptoms, like the subsequent simultaneous development of the skin lesions, has an immunological basis. There are soluble antigens in the serum in the early stages of smallpox, and the early signs may be due to the interaction of these antigens and the developing antibodies, especially IgM, to produce immune complexes that activate complement and thus trigger a series of pharmacological and physiological reactions. In vaccinia infections, the typical skin lesions do not appear, in spite of the growth of virus in the skin, unless there is an immune response (Pincus and Flick, 1963).

**The Rash of Measles.** There is evidence that the rash in measles has an immunological basis (see Chapter 10); a possible scheme for its pathogenesis is as follows (Burnet, 1968). Measles virus initially infects cells of the respiratory tract, and multiplies freely in the cells first affected, but does not induce necrosis. Virus passes to the lymph nodes, perhaps by the movement of infected alveolar macrophages, as found by Roberts (1962a) after aerosol infection of mice with mousepox virus. Multiplication then proceeds slowly, the appearance of Warthin's giant cells being an index of this lymphoid infection. Necrosis is still not a feature of the infected cell, but small amounts of virus are released into the circulation and seeded out to the epithelial surfaces of the body, including the oropharynx, conjunctiva, respiratory tract, skin, and even the bladder and alimentary canal. Growth of virus does not lead to the production of lesions until the onset of a generalized immune reaction at the end of the incubation period. This leads to the breakdown of infected cells in mucosae, with formation of lesions, shedding of virus to the exterior and into the circulation, and appearance of the rash shortly afterward. This scheme would account for the pathogenesis of other exanthemata, such as rubella, in which the initial virus–cell interaction is relatively noncytocidal. There is increasing evidence that antibodies and sensitized lymphocytes can cause the destruction of infected cells *in vitro*, but our ignorance of the *in vivo* situation is illustrated by the fact that even in these very common diseases (measles and rubella) the actual site of viral multiplication in the skin remains to be demonstrated, although virus has now been isolated from punch biopsies of the rubella skin lesions (Heggie, 1971a).

## INFECTIONS OF THE NERVOUS SYSTEM

### Structure and Function of the Central Nervous System

The nervous system is immensely complicated; we can do no more in this introduction than mention some of the anatomical features that are important for an understanding of the pathogenesis of viral infections. The central nervous system (CNS) consists of the brain and the spinal cord; both contain white matter, which consists of myelinated nerve fibers, and gray matter, made up mainly of the nerve cells themselves, or neurons. The number of neurons, although very large, is exceeded by the number of other cells which maintain the physical integrity of the nervous tissue and supply its nutritional needs, the neuroglial cells and cells of the rich vascular plexus which pervades the CNS. There are no lymphatics within the CNS. Neuroglial cells react rapidly in viral infections of the CNS, whether or not they are infected, and they may play an important part in pathogenesis (Zlotnik, 1968). The neurons vary greatly in size, shape, and the number and mode of branching of their processes. They consist of a cell body and several processes (dendrites and axon). The cell body comprises the nucleus and surrounding cytoplasm, which often contains a rich endoplasmic reticulum and associated ribosomes (Nissl's bodies), mitochondria, and Golgi apparatus. The processes comprise several or many short dendrites that make contact through synapses with other nerve cells, and one axon which may extend for long distances within either the central or the peripheral nervous system. There are no ribosomes in the axon.

The brain is enveloped with a number of sheaths and contains several cavities, the ventricles, which are lined by the ependyma and are connected with the subarachnoid space which surrounds the central nervous tissue. The ventricles and subarachnoid space contain the cerebrospinal fluid, which contains an occasional mononuclear cell and has a characteristic composition different from that of the blood plasma. The brain and its sheaths are enclosed within a rigid bony skull.

**Intracerebral Injection.**   One of the common experimental procedures in virology is intracerebral inoculation, i.e., the inoculation of virus-containing fluids directly into the brain. This is a highly traumatic procedure (Mims, 1960c), since the pressure exerted during routine intracerebral inoculations in the mouse is 200–300 cm $H_2O$, i.e., twenty to thirty times the pressure of the cerebrospinal fluid. This leads to the breakdown of anatomical barriers (Fig. 9–7). The arachnoid villi, which form a barrier between cerebrospinal fluid and blood, are the first to break down and the injected material then enters the blood stream. Virus is thus deposited along the needle track and overflows into the cerebrospinal fluid, where it comes into contact with the ependymal and meningeal cells, and then into the blood stream, which rapidly distributes the inoculum throughout the body.

### Structure of Peripheral Nerves

Some viruses pass along the peripheral nerves to reach the posterior ganglion cells or the CNS; a knowledge of the structure of the peripheral nerves is there-

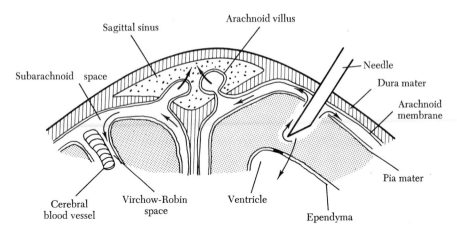

FIG. 9–7. *The fate of material injected by the intracerebral route. (From Mims, 1960a.)*

fore important for understanding the neural spread of viruses. The nerve fiber is composed of an axon, which is the long peripheral extension of the cytoplasm of a neuron, and a number of ectodermal coverings. All axons in the peripheral nervous system are surrounded by Schwann cells, which lay down a myelin sheath, forming a continuous investment for the nerve fiber from its beginning at the spinal root or in the ganglia, almost to its peripheral termination. Some peripheral nerve fibers (the unmyelinated nerves) lack a well-developed myelin sheath, but all have associated Schwann cells. Groups of nerve fibers are bound together by connective tissue to form peripheral nerve trunks that are associated with arteries, veins, and lymphatic vessels, in which the lymph flows centrifugally. At their peripheral ends the motor, secretory, and sensory nerves terminate in nerve endings, of which there are several kinds associated with different functions. The sensory nerve endings in the skin are important in centripetal and centrifugal neural spread of certain viruses; they are free nerve endings which do not penetrate further into the epidermis than the granular layer.

## Generalized Infections Involving the Central Nervous System

Disease of the central nervous system (CNS) is usually an exceptional complication rather than the normal consequence of infection, but many viruses belonging to several different groups occasionally invade the CNS (Table 9–6). Poxviruses may infect mesenchymal cells of the meninges and ependyma in mice, but they do not multiply in neurons even after intracerebral inoculation (Mims, 1960c), and postvaccinial encephalitis in man is probably not due primarily to multiplication of virus in the brain but to an immunological reaction. Several of the herpesviruses affect cells of the CNS. Herpes simplex virus is probably the most common cause of sporadic fatal encephalitis in man (Meyer *et al.*, 1960); zoster is due to the activation of latent chickenpox virus, initially in the cells of the posterior ganglia, and pseudorabies may involve the CNS and the peripheral nerves of naturally infected animals. Neural involvement is almost unknown in

TABLE 9–6

*Viral Infections of the Central Nervous System*

| Meningitis | *Poxvirus* [a] | vaccinia (mice, rabbit) |
| | | ectromelia (mice) |
| | *Herpesvirus* | herpes simplex |
| | Picornaviridae | coxsackievirus B1-B6, A7, A9, A23 |
| | | echovirus 4, 6, 9, 16, 30 |
| | | poliovirus 1-3 |
| | *Orthomyxovirus* [a] | influenza (mice) |
| | *Paramyxovirus* | mumps |
| | *Arenavirus* | lymphocytic choriomeningitis |
| Encephalitis or | *Herpesvirus* | herpes simplex, B virus, pseudorabies |
| myelitis | | Marek's disease [b] |
| | Picornaviridae | poliovirus 1-3, coxsackievirus A7 |
| | | Theiler's virus (mice), Teschen |
| | | disease (swine) |
| | Togaviridae | all species in infant mice [a] |
| | | many species in natural infections |
| | *Orthomyxovirus* | fowl plague |
| | | influenza [a] (mice) |
| | *Paramyxovirus* | mumps, Newcastle disease |
| | | distemper, measles (inc. SSPE) |
| | *Coronavirus* | JHM virus [b] |
| | *Rhabdovirus* | rabies |
| | *Leukovirus* | visna (sheep) |
| | *Reovirus* [a] | (mice) |

[a] In experimental animals only.
[b] Demyelination.

infections with adenoviruses, but a virus of the genus *Polyomavirus*, serologically related to SV40, is found in the glial cells in the rare human disease, progressive multifocal leukoencephalopathy (Padgett *et al.*, 1971; Weiner *et al.*, 1972). Although human influenza viruses remain localized in the respiratory tract, fowl plague virus sometimes causes a diffuse encephalitis. The human paramyxo-viruses that multiply in the CNS are mumps virus, which produces meningitis and sometimes encephalitis, and measles virus, which may involve the CNS during the acute disease and under rare circumstances causes a slowly develop-ing fatal disease, subacute sclerosing panencephalitis (Payne *et al.*, 1969; see Chapter 12). Among animals, distemper in dogs and Newcastle disease in chickens, two diseases caused by paramyxoviruses, are associated with encepha-litis. The viruses which most commonly cause encephalitis in man belong to the Togaviridae. Some togaviruses (e.g., dengue virus) usually produce a rash but do not involve the CNS, whereas others, like Japanese encephalitis virus, never produce a rash but involve the CNS in an appreciable proportion of cases. Many of the enteroviruses produce CNS disease as a rare complication. Echoviruses and coxsackieviruses may produce meningitis, and polioviruses characteristically infect the anterior horn cells of the CNS in a small proportion of infected human beings.

Four infectious agents that behave like viruses but have not yet been chemically characterized are incriminated in the etiology of some rare diseases of the CNS that have been collectively termed the subacute spongiform viral encephalopathies (see Chapter 12). All have a very long incubation period and the disease process progresses slowly until it kills the infected host. None of the agents is immunogenic, and the CNS lesions are noteworthy for the complete absence of an inflammatory response. Infection can occur by natural routes, probably oral and via minor lesions of the skin. Virus can be recovered from visceral organs as well as from the brain, and spread probably occurs by the hematogenous route.

## Cellular Sites of Viral Multiplication

Table 9–7 lists available evidence about the cells in which viruses may multiply in the brains of mice, derived from studies using fluorescent antibody. The neurons constitute less than 5% of the cells in the brain (Nurnberger and Gordon, 1957), and if the term "neurotropic" is limited to viruses which multiply in neurons the poxviruses and ordinary strains of influenza virus do not merit this designation. Some viruses affect a wide range of cell types (e.g., herpes simplex and influenza strain NWS), some fail to multiply in the mesenchymal cells of the ependyma and meninges but multiply widely in neurons or glial cells (e.g., many togaviruses), and multiplication of rabies is not only confined to the neurons but affects particular anatomical areas with an extraordinary specificity. Some parvoviruses, such as rat virus, have a tropism for certain types of dividing cells (Tennant et al., 1969) and infection of mitotic cerebellar cells in newborn animals leads to selective destruction of these cells causing a characteristic cerebellar hypoplasia (Margolis and Kilham, 1965). Bluetongue vaccine virus selectively infects immature subventricular cells in the developing brain of the fetal lamb (Osburn et al., 1971) and newborn mouse (Narayan and Johnson, 1972), resulting in characteristic cerebral lesions. The brain becomes insusceptible as the subventricular cells migrate and differentiate.

## Hematogenous Spread to the CNS

Study of the pathogenesis of viral infections involving the CNS provides an excellent example of the way in which use of the wrong laboratory model impeded understanding of natural infections. Hurst (1936) and Friedemann (1943) summarized the results of intensive investigations carried out in the 1930's with statements implying that there was a "blood–brain barrier" that was virtually impermeable to viruses. Reconsideration of the hematogenous route, the potential importance of which had been emphasized much earlier by Doerr and Vochting (1920), followed recognition of the importance of viremia induced in primates by feeding poliovirus (Bodian, 1955, 1956). Precise study of the cells involved in the spread of viruses to the CNS awaited the application of fluorescent antibody staining methods to these problems. Johnson's (1964a) experiments with herpes simplex virus have provided a valuable model, which illustrates the possibility of simultaneous spread of viruses to the CNS by several routes

TABLE 9–7

*Susceptibility of Cells in the CNS of the Mouse to Infection with Different Viruses as Shown by Staining with Fluorescent Antibodies* [a]

| | MENINGES AND/OR EPENDYMA | GLIA | NEURONS |
|---|---|---|---|
| *Poxvirus* | | | |
| Vaccinia | + | − | − |
| Rabbitpox | + | − | − |
| Mousepox | + | − | − |
| *Herpesvirus* | | | |
| Herpes simplex | + | + | + |
| *Togaviridae* | | | |
| *Alphavirus* | | | |
| Sindbis | + | + | + |
| Venezuelan equine encephalitis | + | − | + |
| Semliki Forest virus | − | + | + |
| Ross River virus | +[b] | − | ± |
| *Flavivirus* | | | |
| West Nile | − | + | + |
| Murray Valley encephalitis | − | + | + |
| Tick-borne encephalitis | − | + | + |
| Japanese encephalitis | − | + | + |
| *Orthomyxovirus* | | | |
| Standard strains influenza | + | − | − |
| Neuro-adapted strains (NWS and WSN) | + | + | + |
| *Paramyxovirus* | | | |
| Sendai (parainfluenza 1) | + | + | − |
| *Arenavirus* | | | |
| Lymphocytic choriomeningitis | | | |
| In mature mice | + | − | − |
| In persistently infected (tolerant) mice | + | + | + |
| *Rhabdovirus* | | | |
| Rabies | − | − | + |
| *Reovirus* | + | + | + |

[a] Modified from Johnson and Mims (1968).
[b] Ependyma only.

(hematogenous, via the peripheral nerves, and via the olfactory route). Current information about probable pathways of spread of neuropathogenic viruses to the CNS is summarized in Table 9–8; the importance of blood-borne virus is evident, and adequate viremia must be maintained, as discussed earlier.

Viruses could spread from the blood to the brain cells by several routes (Fig. 9–8). Growth through the endothelium of small cerebral vessels has been clearly demonstrated in several systems (see below), and there is suggestive evidence that virions may sometimes be passively transferred across the vascular endothelium. The production of meningitis rather than encephalitis by several viruses (coxsackieviruses, echoviruses), and the ease with which these agents are recovered from the cerebrospinal fluid, makes it likely that the virus in the blood

TABLE 9–8

*Pathways of Spread of Viruses to the CNS* [a]

| PATHWAY | IN EXPERIMENTAL MODELS | IN MAN (PROBABLE) |
|---|---|---|
| Neural | Herpes simplex, B virus, rabies, polioviruses | Rabies, B virus, polioviruses (sometimes) |
| Olfactory | Herpes simplex and some togaviruses | Possibly herpes simplex in adults |
| Hematogenous | Polioviruses, herpes simplex, lymphocytic choriomeningitis, togaviruses | Polioviruses (usually), coxsackieviruses, echoviruses, lymphocytic choriomeningitis, and mumps, herpes simplex in children; rubella and cytomegaloviruses in fetus |
| | The spongiform viral encephalopathies: scrapie, kuru, Creutzfeld-Jakob disease, and mink encephalopathy | Kuru, Creutzfeld-Jakob disease |

[a] Modified from Johnson and Mims (1968).

grows or passes through the chorioid plexus or meningeal vessels. Virus in the meninges can directly infect underlying brain cells and virus entering the cerebrospinal fluid can grow or pass through meningeal and ependymal cells and thus reach brain cells (Fig. 9–8). There is no evidence on these points in picornavirus infections, but some evidence has been provided in experiments with neuro-adapted influenza virus and Japanese encephalitis virus in mice (Mims, 1960c; Hamashima *et al.*, 1959). Tosolini (1970) showed that lymphocytic choriomeningitis virus injected intravenously in adult mice caused choriomeningitis after infecting small blood vessels in the chorioid plexus and meninges; cerebral vessels were also infected, but this did not lead to infection of glial cells or neurons. Three examples of the way in which viremia is maintained and virus passes from the blood to the CNS will be discussed in more detail.

**Herpes Simplex Virus in Mice.**   Johnson (1964a) showed that after intraperitoneal injection into suckling mice herpes simplex virus multiplied in the viscera, especially in the liver and spleen, and produced viremia. The mice developed signs of encephalitis on the fourth day and died within 24 hours. Foci of flourescent neural and glial cells were found in the brain randomly distributed deep in the cerebrum and cerebellum and unrelated to neural tracts. This distribution suggested hematogenous spread, which was clearly demonstrated by the specific fluorescence of endothelial cells in capillaries of mice whose blood had been replaced with India ink just before they were killed.

**Poliomyelitis in Primates.**   Viremia occurs in natural infections of man and after feeding virus to primates (Bodian, 1955; Sabin, 1956). Studies on human infants vaccinated orally with live poliovirus vaccines have shown that infection

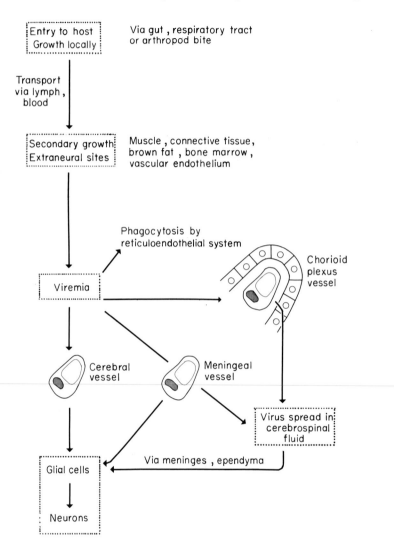

FIG. 9–8. *Steps in the hematogenous spread of virus to the central nervous system.*

may be established in the small intestine, as well as in the nasopharynx, for virus administered in hard gelatin capsules is infective (Horstmann *et al.*, 1959). Low-grade viremia, especially with type 2 poliovirus, is a frequent sequel of vaccination in children lacking antibodies to poliovirus (Horstmann *et al.*, 1964), but is not associated with any untoward effects. With the large doses used for vaccination, establishment of alimentary infection and fecal excretion of the virus occur within a few days, and viremia may be found by the third or fourth day after infection.

Fluorescent antibody staining methods were used to determine the course of infection of cynomolgus monkeys infected orally with poliovirus type 1 (Kanamitzu, 1967). Specific fluorescence was seen, first in pharyngeal epithelial cells, and later in submucosal but not epithelial cells of the oropharynx and the lower alimentary tract. Fluorescent foci were seen in the local lymph nodes, and, after viremia was established, in the macrophages of spleen and liver. Fluorescent cells were found in the CNS only in animals that suffered paralytic infections. The route of passage from blood stream to neural cells was not determined.

The overall picture in poliovirus infection is thus not greatly different from what is found in other generalized viral infections. Infection is established presumably in the mucosal cells of the throat or possibly the ileum and virus soon gains access to the local accumulations of lymphoid tissue within the mucosa, and then via the lymphatic system to the regional lymph nodes. Viremia occurs, followed by secondary seeding of virus; in experimental primates very high concentrations of virus may be found in the brown fat (Shwartzman et al., 1955). Invasion of the CNS probably occurs via the blood stream (Nathanson and Bodian, 1962) although some laboratory strains of virus (e.g., the MV strain) invade the brain by the neural route (Nathanson and Bodian, 1961).

**Togavirus Infections in Mice.** The importance of the hematogenous route in infections with these viruses is well established (review, Albrecht, 1968). Most recent studies have used immunofluorescence as well as viral assay methods, and the growth and spread of virus has been examined after subcutaneous or intravenous routes of inoculation, which are similar to the natural routes of infection. Using assay methods only, Malkova (1960) found growth of tick-borne encephalitis virus greatest in the lymph nodes and spleen, but Albrecht (1962) using fluorescent antibody staining and a different strain of virus reported infection of several extraneural sites, notably muscle, but found no evidence of involvement of vascular endothelium. Immunofluorescence studies with West Nile virus in mice (Kundin et al., 1963) also demonstrated infection of many extraneural tissues including widespread involvement of vascular endothelial cells. In Sindbis virus infections of suckling mice, infection was limited to muscle, fibroblasts, and vascular endothelial cells (Johnson, 1965a); these cells constitute the "central focus" which maintains the viremia. Study of brain sections of mice infected with Sindbis virus showed that infection of endothelial cells of the small vessels in the brain clearly preceded infection of surrounding neurons and glial cells (Johnson, 1965a). However, although Albrecht (1962) showed that tick-borne encephalitis virus in mice and chick embryos invaded the CNS from the blood, he could find no evidence that the vascular endothelial cells were infected. This virus might "leak" across the blood–brain junction, for molecules of ferritin, which are not much smaller than virions, readily do so (Bondareff, 1964). Treatments that sustain viremia thereby promote invasion of the CNS; Zisman et al. (1971) showed that when clearance of circulating yellow fever virus was impaired by the injection of infected mice with silica, there were higher viral titers in the brain and increased mortality.

## Neural Spread to the CNS

Spread from the periphery to the central nervous system is possible without generalization of viruses through the blood stream, for the peripheral nerves and the nerve fibers of the olfactory bulb offer potential direct pathways. These routes are undoubtedly followed in some natural infections, notably rabies (Table 9–8), and studies on CNS invasion in poliomyelitis and viral encephalitis were long dominated by the belief that spread via the olfactory bulbs was of major importance.

Although earlier workers thought in terms of "conduits"—the axon, the lymphatics, and the tissue spaces between nerve fibers (Wright, 1953)—the route of spread along peripheral nerves is usually by growth within endoneural cells. This was first suggested by Hurst (1933), who had found inclusion bodies within the Schwann cells of peripheral nerves in rabbits infected with pseudo-rabies virus (a herpesvirus), and was conclusively demonstrated by Johnson (1964a) with suckling mice infected subcutaneously with herpes simplex virus. Endoneural cells of small subcutaneous nerve fibers were infected within 24 hours, and over the next 4 days infection of the endoneural cells (Schwann cells and fibroblasts) could be traced by fluorescent antibody staining up the corresponding peripheral nerves, to the dorsal root ganglia, and into the appropriate segment of the spinal cord. No virus was recoverable from the blood or viscera. Electron microscopic studies have provided confirmatory evidence that herpes simplex virus multiplies in the nuclei of the Schwann cells (Rabin et al., 1968; Constantine and Mason, 1971).

However, this cannot be the only method of neural spread of viruses, for in identical experiments with "fixed" (i.e. attenuated) rabies virus, no fluorescent endoneural cells could be found (Johnson, 1965b). Disease could be prevented by cutting the appropriate peripheral nerves; and with intact nerves the appropriate dorsal root ganglion cells were the first cells showing specific staining. The rapidity of spread to the spinal cord, and the evidence that fixed rabies virus would grow only in neurons, suggested that the virus spread via some conduit.

A similar conclusion was reached from studies with "street" (i.e., virulent) rabies virus (Yamamoto et al., 1965), and also in rabbits inoculated into the sciatic nerve with herpes simplex virus (Johnson and Mims, 1968). Studies with herpes simplex virus in suckling mice suggest a neural pathway other than Schwann cells, possibly the axon itself (Kristensson et al., 1971). Centrifugal transmission along the peripheral nerves from the appropriate dorsal root ganglion cells to skin and mucous membranes appears to be the most likely mode of spread of virus in recurrent herpes simplex and in herpes zoster (see Chapter 12).

Olfactory spread of viruses has been demonstrated in experimental situations but is unlikely to occur in nature. In suckling mice inoculated intranasally with herpes simplex virus, Johnson (1964a) demonstrated that infection could spread from the olfactory mucosa to the CNS by two routes. In some mice, extensive infection of the mucosal and submucosal cells was followed by direct extension across the cribriform plate to the meninges; in others, infection of the perineural and endoneural cells of the olfactory fibers resulted in initial infections of the parenchymal cells of the olfactory bulb. In mice exposed to aerosol infection with West Nile virus, Nir et al. (1965) observed early infection of the olfactory bulbs and subsequently widespread involvement of the brain. Viremia did not occur until some days later, and there was no evidence of infection of cells of the nasal mucosa.

## Spread of Viruses Within the CNS

In the CNS, as in other organs of the body, viral infection is not necessarily synonymous with disease. Having reached the CNS by one of the pathways just described, disease does not result unless virus spreads within the CNS, and unless infection of cells of the CNS results in some direct or secondary interference with normal function. In the CNS, as elsewhere, infection of cells may depend upon the existence of suitable cellular receptors on the cytoplasmic membrane. Presence or absence of receptors may be a factor of major importance, although not the only factor, in determining the susceptibility of different cells in the CNS, and at different ages of the host animal (Kunin, 1962).

The spread of viruses within the CNS after intracerebral inoculation, which is the most widely used method of inoculation in experimental studies, has already been described (Fig. 9–7); virus enters the blood and cerebrospinal fluid and is rapidly disseminated throughout the length of the neuraxis. This is a highly artificial situation, although something like it occurs with viruses that enter the cerebrospinal fluid. In most natural infections, spread must occur through brain parenchyma, possibly along the extracellular spaces which have been described (Brightman, 1965). Although these channels appear to be only 20 nm wide they may contain viruses up to 45 nm in diameter (Blinzinger and Müller, 1971). Viruses as large as rabies (80 nm in its smallest diameter) appear to spread readily through CNS tissue after intracerebral injection and synchronously reach many neurons, some of them distant from ependymal and meningeal surfaces (Johnson and Mercer, 1964). Sometimes spread may occur between contiguous cells; this is seen with herpes simplex virus both in the CNS (Johnson, 1964a, b) and in cultured cells (Black and Melnick, 1955; Stoker, 1958). The CNS of the newborn mouse is nonmyelinated, and this may facilitate direct spread and help account for their high susceptibility to viral infections.

## Production of Disease by Viral Infection of the CNS

In cells of the CNS, as in other cells, viral infection is not necessarily cytocidal; if damage occurs it may be due to the host's immune response (Webb and Smith, 1966; and Chapter 10) rather than any direct effect of the virus. Poliovirus provides the best evidence of cytocidal effects on neuronal cells (reviews, Bodian, 1958, 1959). Chromatolysis develops in the motor neuron prior to other pathological changes, and is followed by necrosis and phagocytosis by glial cells (neuronophagia). The mononuclear cell inflammatory reaction ("perivascular round cell infiltration") that is the hallmark of viral encephalitis was thought to be a nonspecific response to cell damage, but the inflammatory reaction does not correlate well topographically with cell necrosis, either in poliovirus infections or in mumps virus encephalitis in hamsters in which there is severe vasculitis but viral multiplication appears to be confined to distant, morphologically normal neurons (Johnson, 1967). Reconstitution studies in immunosuppressed mice infected with Sindbis virus revealed that the perivascular inflammatory reaction ("cuffing") is an immunologically specific reaction mediated by sensitized lymph node and bone marrow cells, i.e., it is an expression of cell-mediated immunity (McFarland et al., 1972).

The basis of disease in other CNS infections is more obscure. Rabies virus is noncytopathogenic in cultured cells (Fernandes et al., 1965); and in man and the mouse rabies virus infection evokes none of the inflammatory reactions or cell necrosis found in the encephalitides (Du Pont and Earle, 1965; Johnson, 1965b). Yet lethal disease is produced, indicating serious impairment of cell function with minimal histological changes. In other situations, extensive infection of neurons causes no symptoms; for example, the extensive CNS infection of mice congenitally infected with lymphocytic choriomeningitis virus, readily demonstrable by fluorescent antibody staining (Mims, 1966a), has no deleterious effect. However, adoptive transfer of sensitized spleen cells to adult mice made tolerant to LCM virus infection leads to neurological symptoms and death (G. Cole et al., 1971). Still other changes are produced by some of the viruses that cause slowly progressive diseases of the CNS (Chapter 12). In visna, there is demyelination and cell infiltration in ependyma, subependyma, and meninges (Sigurdsson et al., 1957); an immunopathological mechanism for demyelination has been suggested (Gudnadóttir and Pálsson, 1965), but there is at present little evidence for this. Scrapie and the scrapielike agents cause a primary neuronal degeneration with no signs of any inflammatory or immunological response (Gibbs and Gajdusek, 1971).

Johnson et al. (1967) have described an interesting example of the production of disease (hydrocephalus) some weeks after an inapparent viral infection of the CNS. When suckling hamsters were inoculated intracerebrally with unadapted strains of mumps virus, viral multiplication, which was asymptomatic, was limited to the ependymal cells lining the ventricles. A few weeks after inoculation, when viral multiplication had ceased, severe stenosis of the aqueduct of Sylvius produced hydrocephalus in all the inoculated animals. A similar pathogenesis accounts for the hydrocephalus produced in experimental animals by reovirus type 1 (Margolis and Kilham, 1969).

## CONGENITAL INFECTIONS

In generalized infections with viremia, virus may be transmitted to the ovum or fetus. In severe acute viral infections, this usually causes fetal death and abortion (e.g., smallpox, Dixon, 1962), but the more interesting situations are those in which the embryo is not killed but sustains a nonlethal infection (review by Catalano and Sever, 1971). Although once thought to be rare, congenital infections with viruses are not uncommon (Table 9–9). The embryo may be unharmed by the virus, or it may suffer damage to developing organs insufficiently severe to cause fetal death, but severe enough to produce congenital malformations (Manshaw, 1970). In some congenital infections, the animal infected before birth acquires tolerance to the virus involved.

TABLE 9–9

*Congenital Infections with Viruses*

| VIRUS GROUP | EXAMPLE | EFFECT ON FETUS |
|---|---|---|
| *Poxvirus* | Smallpox (man) | Fetal death (abortion) |
| *Herpesvirus* | Cytomegalovirus (man) | Severe neonatal disease; cerebral damage and prolonged viruria in survivors |
| | Equine rhinopneumonitis | Abortion in mares and in guinea pigs |
| | Feline herpesvirus | Fetal death (abortion) |
| | Herpes simplex virus (type 2) | Severe neonatal disease |
| | Herpeslike virus (dogs) | Acute hemorrhagic disease |
| *Parvovirus* | H1 (hamsters) | Fetal death with developmental defects |
| | RV (rats) | Fetal death and resorption; hepatitis and cerebrellar hypoplasia in newborn |
| *Leukovirus* | Avian leukosis (chickens) | Inapparent; prolonged viremia, immunological tolerance, eventually leukemia (rarely) |
| | Murine leukemia (mice) | Inapparent; prolonged viremia, eventual leukemia (sometimes) |
| Togaviridae | Hog cholera vaccine (pigs) | Edema and limb malformation |
| | Rubella (man) | Developmental defects, prolonged viremia, no immunological tolerance |
| | Bovine viral diarrhea-mucosal disease (cow) | Cerebellar hypoplasia |
| *Arenavirus* | Lymphocytic choriomeningitis (mice) | Inapparent; prolonged viremia, immunological tolerance |
| *Orbivirus* | Bluetongue vaccine (sheep) | Lambs stillborn or with symptoms of CNS disease |
| *Reovirus* | Reovirus type 1 (mice) | Fetal death and resorption |
| | Reovirus type 2 (mice) | Inapparent |

In this chapter we will consider two aspects of congenital infection: the routes of transfer of virus from parent to progeny in mammals and in birds, and the development of congenital defects due to viral infections. Immunological tolerance and the pathogenesis of late disease in animals suffering congenital infections are discussed elsewhere (Chapters 10 and 12).

## Congenital Infections in Mammals

Recent studies on murine leukemia, avian leukosis, and mammary tumor virus infections of mice reveal that congenital transfer of the integrated viral genome occurs in these conditions and perhaps in all similar infections (review, Todaro and Huebner, 1972; see Chapter 14). In addition to such hereditary transfer of viral genetic material, viremic disease in the mother may be followed by congenital transfer of virions (Fig. 9–9).

One well studied example of congenital infection is lymphocytic choriomeningitis of mice (LCM). Immunofluorescence studies with material from early embryos to adults 8 months old (Mims, 1966a) showed that infected cells were present in every section examined, of whatever organ or age. The occurrence of infected germinal epithelium and infected ova showed that transovarial trans-

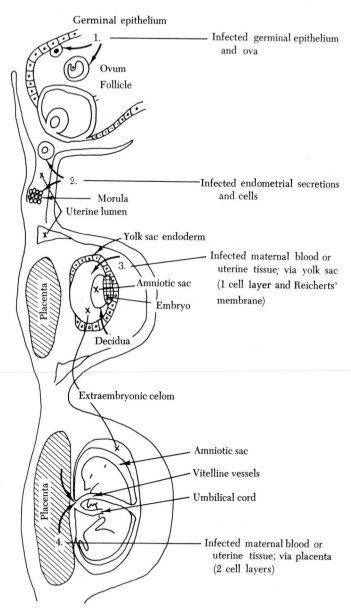

Germinal epithelium

1. ———————————— Infected germinal epithelium
and ova

Ovum

Follicle

2. ————————————— Infected endometrial secretions
and cells

Morula

Uterine lumen

Yolk sac endoderm

Infected maternal blood or
uterine tissue; via yolk sac
(1 cell layer and Reicherts'
membrane)

3.

Amniotic sac

Embryo

Placenta

Decidua

Extraembryonic celom

Amniotic sac

Vitelline vessels

Umbilical cord

Placenta

Infected maternal blood or
uterine tissue; via placenta
(2 cell layers)

4.

FIG. 9–9. *Schematic drawing of the reproductive system of the mouse to show possible routes of infection of the embryo by viruses. In other mammals the situation is similar, but (a) the yolk sac is very variable in different mammals, and (b) the placental junction (number of cell layers and their permeability) varies in different species of mammal and at different stages of pregnancy.*

mission may occur, but many ova did not contain antigen. However, all the young mice in a litter were infected, and the observation of infected cells in the blood, oviduct epithelium, endometrium, and placenta suggested that ample opportunity was available for infection to occur after ovulation. In early embryos, almost every cell contained antigen. In older embryos and in adult mice, less than 50% of cells contained antigen at any particular time, but the fact that all cells resisted superinfection with LCM virus, in the absence of interferon production, led Mims and Subrahmanyan (1966) to suggest that every cell in these mice was probably infected with the virus and remained infected for life. Infected male carrier mice can transfer LCM virus during copulation, since many of their secretions and excretions contain virus, but this does not appear to be an important way of initiating congenital infection; the spermatozoa do not carry virus although the testes are infected.

In most congenital infections, virus is transferred from maternal blood to fetal tissues across the placenta, which presents a physical barrier to the transit of viral particles. The barrier consists of between one and four layers of cells, depending on the species of animal and the stage of pregnancy. As in the case of the blood–skin and blood–brain barriers, viruses can be carried across, leak across, or grow across the barrier. In rubella, cytomegalovirus, and smallpox infections in man there are foci of placental infection, suggesting that these viruses grow across the placental barrier. Mims (1969) showed that in pregnant mice undergoing primary infection with LCM virus the fetus was infected either via the avascular yolk sac or via the placenta, depending on the stage of pregnancy; fetal infection followed placental infection.

## Congenital Infections of Birds

As indicated in Fig. 9–10, viruses can be transferred from parent birds to a fertilized egg by a variety of different routes. In addition to the hereditary transmission of integrated leukovirus genetic information, which probably occurs in all birds (Chapter 14), avian leukosis virions may be transferred congenitally (Burmester, 1962; Rubin et al., 1961, 1962). The majority of fertile eggs laid by viremic birds are infected, and virtually all cells obtained from an infected embryo are infected and yield virus. Most hens with high level persistent viremia have themselves been congenitally infected and are immunologically tolerant to the antigens of that strain of avian leukosis virus. Under conditions of close contact with infected birds, most chicks that were not congenitally infected were horizontally infected after hatching, but their viremia was usually transient and they developed antibody. Some of these birds were none the less able to transmit congenitally, perhaps because the cells of their ovaries continued to harbor and produce virus. Although most cells obtained by culturing the testes of infected roosters contained virus there was no evidence of congenital transmission of virions by the rooster, suggesting that they were lost when the cytoplasm was shed in the last stages of spermatogenesis.

In an electron microscopic study of the tissues of first- and second-generation congenitally infected chickens, Di Stefano and Dougherty (1966) found that virus multiplied throughout the reproductive system of female birds. Very high

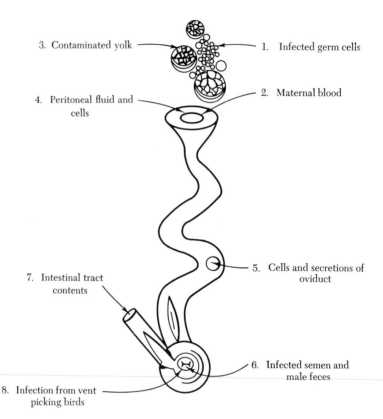

3. Contaminated yolk

1. Infected germ cells

4. Peritoneal fluid and cells

2. Maternal blood

7. Intestinal tract contents

5. Cells and secretions of oviduct

6. Infected semen and male feces

8. Infection from vent picking birds

Fig. 9–10. *Schematic drawing of hen's reproductive system showing various ways disease agents may gain entrance into eggs. (From Cottral, 1952.) (1) Infected germ cells. The female germ cells may carry disease agents into eggs. (2) Maternal blood. Some birds have a slight hemorrhage following ovulation. Thus, maternal blood which could carry disease agents is often included in the egg, forming the so-called blood-spot. (3) Contaminated yolk. The yolk material accumulates over a period of several weeks before ovulation occurs. Disease agents could be deposited in the yolk and thus carried into the egg. (4) Peritoneal fluid and cells. Peritoneal infections can be carried into the egg via the infundibulum by means of peritoneal fluid and cells. (5) Cells and secretions of oviduct. The egg spends about 25 hours in the oviduct of the hen. During this time the chalazae, the albumen, shell membranes, shell, and cuticle are added to the egg. An infection from the oviduct cells could be easily included in the egg. (6) Infected semen and male feces. The seminal fluid or spermatozoa could carry disease agents. In addition, feces from the male bird occasionally enters the female during copulation. (7) Intestinal tract contents. In many hens a physiological prolapse of the oviduct occurs at the time the egg is laid. This allows the everted wall of the oviduct to come in contact with the cloacal lining, and contamination can thus be taken into the oviduct as it is retracted. (8) Infection from vent-picking birds. Contamination on the bills of birds can be transferred to the oviduct of other birds that are the victims of cannibalism. The vent-picking habit is quite common in laying hens, especially in groups which contain hens that are slow to retract their oviduct after laying.*

concentration occurred in the oviduct, especially in the albumen-secreting region. The two most likely routes for congenital transmission of avian leukosis virions appeared to be infection of the germinal cells in the ovary and infection of the zygote in the proximal portion of the oviduct (i.e., sites 1 and 5 of Fig. 9–10).

## Congenital Defects Due to Viruses

Infection of the fetus or embryo by viruses may produce a variety of different results (Table 9–9). Most viral infections of the mother have no effect on the fetus. Sometimes abortion occurs, as in smallpox and equine rhinopneumonitis, or fetal death and resorption as in parvovirus infection of rats. At other times the embryo survives but develops congenital defects due to interference with the normal development of particular organs or tissues. Some viruses appear to affect particular organs at certain stages of fetal development, especially during morphogenesis of these organs, and sometimes virions cross the placenta more easily at certain stages of pregnancy. Cytopathogenic viruses generally kill the infected fetus, whereas relatively noncytopathogenic viruses, such as rubella, are more likely to allow fetal survival and the development of congenital abnormalities.

**Rubella in Man.** Gregg (1941) made a major contribution to human medicine when he recognized that there was an association between certain congenital abnormalities and maternal rubella contracted in the early months of pregnancy. This has since been abundantly confirmed, and a variety of abnormalities has been recognized of which the most common are cataract, deafness, and congenital heart disease (Krugman, 1965). The infected fetus is usually damaged, but some abnormalities do not become apparent until many months after birth.

As well as producing congenital defects which tend to predominate in one or a few organs (brain, heart, eye, or ear), rubella may cause very severe neonatal disease ("rubella syndrome") with hepatosplenomegaly, purpura, and jaundice. Such cases are usually fatal soon after birth, but if they survive they present as classic persistent infections (Chapter 12). Rubella virus can be regularly recovered from the nasopharynx, urine, and cerebrospinal fluid of infants with the rubella syndrome throughout pregnancy and for many weeks after birth, and from lens tissue as late as 3 years after birth (Rawls, 1968). Children who have contracted rubella *in utero* usually have high titers of neutralizing antibody, including IgM, which must be synthesized by the fetus because, unlike IgG, it cannot cross the placenta.

Little is known of the pathogenesis of the congenital abnormalities in rubella (Catalano and Sever, 1971). In the infected human fetus, small necrotic foci have been described in the heart, lens, and inner ear (Tondury and Smith, 1966), which could cause abnormal development of these organs. Lesions in blood vessels have been described in the infected placenta and infant, which may lead to general or local hypoxia and thus affect fetal development. Noncytopathic depression of mitotic activity could also be important in rapidly developing organs. Naeye and Blanc (1965) ascribed the retarded growth ("runting") found in many rubella-infected infants to the reduced numbers of cells they observed in many of their organs. Damage due to congenital infection is much more common in fetuses infected during the first trimester, and the particular organs affected tend to be those that were forming at the time of fetal infection. In the rapidly changing fetus, different mechanisms may operate in the causation of different types of lesion. The presence of maternal neutralizing antibody in the fetal circulation may prevent viremia after the first few weeks but would have no effect on clones of persistently infected cells (Rawls and Melnick, 1966).

**Cytomegalovirus Infections in Man.** The cytomegaloviruses are members of the herpesvirus group and cause widespread subclinical infection in man (Weller, 1971). Apart from activation of latent infections in older individuals and in immunosuppressed renal transplant patients, or transmission by blood transfusion in the latter, disease is almost always due to infection acquired congenitally, usually from mothers who suffer an acute (inapparent) infection during pregnancy (Hanshaw, 1970). Cytomegalic cells are present in the chorionic villi at birth (Cochard et al., 1963), and the important clinical features in neonates include hepatosplenomegaly, thrombocytopenic purpura, hepatitis and jaundice, microcephaly, mental retardation, and persistent viruria.

**Parvovirus Infections of Rats.** Kilham's "rat virus" (RV) has been shown to produce congenital infection in naturally infected rats (Kilham and Margolis, 1966) as well as cerebellar hypoplasia in hamsters and cats inoculated by the intracerebral route as newborn animals. In the latter the external germinal layer of cells in the developing cerebellum is infected. These cells are actively dividing and about to migrate and take up their position below the Purkinje cells. Their loss leads to a grossly hypoplastic cerebellum with a disordered disposition of Purkinje cells (Margolis and Kilham, 1968). Congenital infection in rats may lead to resorption of the fetus, to severe hepatitis in the newborn, or to milder infections compatible with survival. Some of the surviving progeny of infected mothers developed cerebellar hypoplasia of the type found in inoculated hamsters and cats. Histological examination of infected newborn rats revealed characteristic inclusion bodies in the endothelial cells of small capillaries; such growth through the vessel wall probably explains the passage of the virus across the placenta and into the neuronal cells of the brain.

H1 virus occurs naturally in hamsters, and if given orally to pregnant hamsters infects the fetus within 24 hours and causes mongoloidlike facial deformities in the newborn (Kilham and Margolis, 1969).

## INFECTIONS OF LYMPHOID TISSUES

Many viruses fail to infect lymphoid tissues, i.e., spleen, lymph nodes, thymus, and circulating lymphoid cells, and the recovery of virus from a lymph node may merely reflect the lymphatic drainage of virions from a peripheral site of growth. Other viruses grow in the lymphocytes, reticular cells, or macrophages of lymphoid tissues, as shown by fluorescent antibody staining or electron microscopy. Before the development of reliable diagnostic techniques there had been exhaustive and detailed studies of leukocyte responses in viral infections. In some viral diseases there is characteristically a lymphopenia, in others a neutropenia, and there is often lymphocytosis during recovery and convalescence. The hematological pattern reflects the changing course of inflammatory and immunological events in the tissues, on occasion the direct action of virus in leukocytes and the bone marrow responses. The bone marrow itself may be infected but has received little study. Less attention is now paid to blood leukocyte changes, but leukocytes may be infected or may transport viruses, they may produce interferon or be immunologically active, and they probably play an important part in pathogenesis (Gresser and Lang, 1966).

Associated with this wide range of functions, the lymphoid tissue may respond to different viruses in a variety of ways. (a) Viral multiplication may occur without a cytopathic effect, as in the case of mice infected congenitally with LCM or leukemia viruses, in which there is persistent infection with immune tolerance. (b) Viral multiplication may occur in lymphoid tissues with cytopathic effects, such as inclusions or giant cell formation, e.g., in infections with cytomegaloviruses, herpesvirus, and measles. The infected leukocytes are eventually killed, but in the meantime help to distribute virus throughout the body. Rinderpest virus produces rather more severe cytopathic effects in lymph nodes of cattle (Tajima and Ushijima, 1971). There is an accompanying hyperplastic immune cellular response which contributes to the lymphoid tissue changes and the immune responses may play some part in the production of pathological changes. (c) Typically, a lymph node drains virions and viral antigens from a peripheral site of growth and is at the same time a site for the evolution of the immune response. The reaction of immune cells or antibody with viral antigen in the node may lead to edema and sometimes necrosis, the virus itself causing a noncytopathic infection, as in LCM virus in adult mice (Mims and Tosolini, 1969; Tosolini and Mims, 1971), or no significant infection of lymphoid tissues, as in cowpox virus in mice (Wallnerova and Mims, 1971). (d) Lymphoid tissue lesions may be produced by a direct destructive action of virus. In mousepox, when such lesions are severe, a defect in the (cell-mediated) immune response results, with extensive viral growth in the liver and eventual death, whereas in animals that recover, a strong immune response is generated by uninfected cells of the lymphoid series (Mims, 1964). Venezuelan encephalitis virus (Victor et al., 1956), and the mouse thymic necrosis virus (Rowe and Capps, 1961), also appear to replicate in lymphoid tissues and destroy them. Mice infected with mouse hepatitis virus (a coronavirus) show necrosis of follicles in the spleen with local growth of virus (Biggart and Ruebner, 1970); the necrosis is prevented by cortisone, which could act either by stabilizing cell membranes or by inhibiting an immunological contribution to the lesions. (e) Finally, lymphoid tissue lesions may be caused by corticosteroid hormones. A severe infection acts as a stress, and comparable noninfectious stresses such as cold, starvation, or injury induce an increased output of corticosteroid hormones. These hormones have profound effects on lymphoid tissues, but their contribution to pathological changes has rarely been distinguished from direct viral or immunological effects on lymphoid tissues (Wallnerova and Mims, 1971).

Many viral infections cause immunosuppression, generally associated with growth of virus in lymphoid tissues, but the mechanisms are not understood (see Chapter 10).

## INFECTIONS OF MISCELLANEOUS ORGANS

In the preceding pages there has been discussion of viral infections of the skin, respiratory tract, alimentary tract, central nervous system, and the fetus. There are other important organs and tissues which may be infected and damaged, but which have been less thoroughly studied (Table 9–10).

Hepatitis (Table 9–5) is produced by a variety of different viruses (review, Piazza, 1969). The mechanisms of infection of the liver have been discussed earlier, and liver damage may be due to direct viral cytopathology or indirect inflammatory or immunopathological effects.

Viral infections of the pancreas have received little attention since the classic studies on coxsackievirus B pancreatitis in suckling mice (Pappenheimer et al., 1951), and the role of viruses in the pathogenesis of diabetes mellitus remains uncertain. A variant of encephalomyocarditis virus (M) that fails to kill adult mice has been found to produce localized lesions in the beta cells of the islets of Langerhans, often with chronic hyperglycemia and a "diabetes-like" syndrome (Münterfering et al., 1971; Craighead and Steinke, 1971). In view of the late onset of symptoms in most cases of human diabetes and the complex nature of its inheritance, viral infection could be an important environmental factor in human diabetes. Viral infections of other endocrine organs such as adrenals, pituitary, and thyroid have been described, but their relationship to endocrine disease has not been demonstrated (review, Levy and Notkins, 1971).

The salivary glands are affected by viruses such as mumps, and by a number of others causing persistent infection (see Chapter 12), e.g., rabies, herpes simplex, and cytomegaloviruses. Mumps and cytomegalovirus also infect the mammary glands, as does foot-and-mouth disease virus (Burrows et al., 1971) and certain tick-borne viruses, and can be recovered from human milk. Milk is the principal vehicle for transmission of virions of the mouse mammary tumor virus.

Ovaries and ova may be infected by vertically transmitted viruses. Infection of the testes occurs, for instance, in hamsters given attenuated strains of Venezuelan equine encephalomyelitis virus (Vestergaard and Scherer, 1971); sometimes viruses are present in semen, as with foot-and-mouth disease in bulls (Cottral et al., 1968), but this reflects the presence of virus in accessory gland secretion rather than the infection of spermatozoa. Viral orchitis is an important complication of mumps and less commonly of other viral infections (Riggs and Sanford, 1962). Viruses may be present in urine as a result of infection of epithelial cells lining kidney tubules, for instance, in polyoma virus in mice (Levinthal et al., 1962), or cytomegalovirus in man (Fetterman et al., 1968), but usually viruses isolated from kidneys are present in very small amounts, and are not associated with kidney disease. Viruses causing kidney disease generally do so by indirect immunological means (immune complex disease: see Chapter 10).

TABLE 9-10

*Viral Infections of Miscellaneous Organs*

| TARGET ORGAN | Natural Infections | | Laboratory Infections | |
|---|---|---|---|---|
| | VIRUS | HOST | VIRUS | HOST |
| Heart | Encephalomyocarditis | Rodents | | |
| | Coxsackievirus B | Man | | |
| Kidney | Polyoma | Mice | | |
| | Cytomegalovirus | Man | | |
| | Measles | Man | | |
| | Rubella | Man | | |
| Striated muscle | Coxsackieviruses | Man | Coxsackieviruses | Mice |
| | Togaviruses | Man | Reoviruses | Mice |
| | | | Foot-and-mouth disease | Mice |
| | | | Togaviruses (many) | Mice |
| Bone | Herpesvirus | Cat | Parvovirus | Hamster |
| Joints | Togaviruses (incl. rubella) | Man | | |
| Brown fat | | | Poliovirus | Monkey |
| | | | Reovirus | Mice |
| | | | Foot-and-mouth disease | Mice |
| | | | Togaviruses | Mice |
| Salivary glands | Cytomegaloviruses | Many species | | |
| | Herpes simplex | Man | | |
| | Lymphocytic choriomeningitis | Mice | | |
| | Rabies | Many species | | |
| | Mumps | Man | | |
| Pancreas | Mumps | Man | Coxsackievirus B | Mice |
| | | | Encephalomyocarditis | Mice |
| Mammary glands | Flaviviruses (tick borne) | Many species | | |
| | Mammary tumor virus | Mice | | |
| | Mumps | Man | | |
| | Foot-and-mouth disease | Cattle | | |
| Ovary | Murine leukemia | Mice | Cowpox | Mice |
| | Lymphocytic choriomeningitis | Mice | | |
| | Mumps | Man | | |
| Testis | Mumps | Man | Venezuelan equine encephalitis | Hamsters |

Myositis is a feature of human infections with many togaviruses, influenza virus, and coxsackievirus B. At the experimental level, infant mice are particularly susceptible. Infection of striated muscle has been described in newborn mice infected with orthomyxoviruses, coxsackievirus A, reoviruses, foot-and-mouth disease virus, and a variety of togaviruses. Many of these viruses and also coxsackievirus B and encephalomyocarditis virus can cause myocarditis. Coxsackievirus B is an important cause of myocarditis or pericarditis in man (Pankey, 1965).

Other tissues which are sometimes infected include bone (Heggie, 1971b; Hoover and Griesemar, 1971), brown fat (Murphy *et al.*, 1972), and adrenal glands (Boulter *et al.*, 1961).

## THE INCUBATION PERIOD AND ITS SIGNIFICANCE IN PATHOGENESIS

The incubation period of an infectious disease is the period that elapses between infection and the first clinical sign or symptom. Consideration of the mode of spread of viruses in the infected animal, set out earlier in this chapter, allows us to make some generalizations about the incubation period in natural infections. It is short (lasting 1–3 days) only in diseases in which the symptoms are due to the effects of viral multiplication at the portal of entry. Thus respiratory infections (e.g., by rhinoviruses or influenza virus) produce symptoms by virtue of their multiplication in the respiratory tract and have short incubation periods. By contrast, the incubation period is relatively long in generalized infections, in which the virus spreads in stepwise fashion through the body before reaching the target organ(s) in which symptoms are produced.

Other factors also influence the length of the incubation period. The long incubation periods of some localized infections like papilloma of rabbits and human warts are presumably due to slow multiplication of the virus and transformation of cells, or perhaps to the slow growth of a clone of transformed cells, whereas in generalized infections by togaviruses the combination of a rapidly multiplying virus and intravenous injection as the mode of infection may lead to a relatively short incubation period. When the immune response is an important factor in disease production, a period of at least a week is needed for the generation of this response and the production of disease.

This concept of the incubation period is not relevant for diseases that recur after prolonged intervals of latency (e.g., herpes zoster), nor for two important groups of diseases in which the virus–cell interaction is initially noncytocidal (see Chapter 12): slow infections like scrapie, which have a very long incubation period, and chronic infections in which immunopathological or neoplastic disease may occur after prolonged asymptomatic infection.

## VIRULENCE AT THE LEVEL OF ORGANISM

At the cellular level, cell–virus relationships can vary from a situation in which the virus kills the cell (cytocidal viruses), through situations where there appears to be a state of noncytocidal steady-state infection (as with lymphocytic choriomeningitis virus in mouse cells) to those in which the virus fails to initiate productive multiplication but causes cellular transformation, as with polyoma virus in hamster cells. Somewhat the same spectrum of responses can be seen at the level of organism, but just as the pathogenesis of infection in a vertebrate is much more complicated than the cycle of multiplication in a cultured cell, so also are the different types of host response to viral infection. Long usage has associated the term "virulence" with those properties of a virus which lead to severe symptoms, of whatever type is characteristic for infections with that virus, and what is called high "virulence" for a particular virus obviously depends to a large extent on the pathogenesis of the disease associated with it. It is possible to compare the virulence of variola major and alastrim, or influenza in two different epidemics, but we cannot compare the virulence of different strains of smallpox with the virulence of different influenza viruses, except in a crude way by comparing their lethality. With viruses that cause disease by multiplication in vulnerable target organs such as the heart, central nervous system, or liver, a major factor in our assessment of virulence is whether the virus gains access to susceptible cells, as well as the degree of multiplication and cell damage caused when they do reach these cells.

There is a spectrum of virulence. At one extreme, most viruses infecting man produce no signs or symptoms and minimal pathological changes. At the other extreme, mice injected intravenously with Rift Valley fever virus die in 6 hours, following a single cycle of viral multiplication in hepatic cells (Mims, 1957). Since virulence is a compound of the reaction of host and virus, comparative statements on the level of virulence of a particular strain of virus can be made only if all other factors are kept constant; the species and age of the host, the route of inoculation, the dose, and so on. In spite of these difficulties, the virulence of a virus is one of its two most important properties as far as man is concerned, the other being its transmissibility. Transmissibility may vary independently of virulence, representing a separate genetic attribute, as shown in mousepox (Fenner, 1949), and with influenza virus in mice (Schulman, 1967).

The production of live virus vaccines is a practical exercise in viral genetics directly concerned with virulence. So far it has been conducted in an empirical manner, but with considerable success (Chapter 15). Early genetic studies involving the direct analysis of virulence gave disappointing results (Burnet, 1955) because so many genes affect this property. However, this very fact can now be exploited by selecting temperature-sensitive conditional lethal mutants (Fenner, 1969). Since they fail to multiply well at body temperature such mutants would be expected to be attenuated, and experiments with the virulence in animals of temperature-sensitive mutants of viruses belonging to six genera have borne out this expectation (Fenner, 1972).

CHAPTER 10

# Pathogenesis: The Immune Response

## INTRODUCTION

Vertebrates differ from all other living organisms in their capacity to respond in a highly specific way to foreign macromolecules, notably proteins, by producing an immune response. Such foreign macromolecules, which are called antigens, are said to be "immunogenic" when they produce a positive response and "tolerogenic" when their administration results in immunological tolerance. The microorganisms and viruses that cause infections constitute the most important natural antigens, and the word "immunity" is used to describe the specific type of host resistance associated with the immune response.

Both virology and immunology can trace their beginnings to the investigations of Edward Jenner (1798) into the protection of man against smallpox by prior inoculation with cowpox virus, and it is common knowledge that individuals who have recovered from smallpox, measles, or chickenpox are specifically resistant to second attacks. Since immunology is now an even broader and more rapidly advancing field of biological science than virology, we cannot hope to cover its complexities in this chapter. The interested reader is referred to some of the excellent introductory texts now available (Burnet, 1969, Humphrey and White, 1970), and to recent review articles dealing with particular facets of the subject.

The specific altered responsiveness of the immune host is expressed in two ways, both mediated by cells of the lymphocyte series: The synthesis of specific immunoglobulins, the antibodies, and the development of cell-mediated immunity, one expression of which is delayed hypersensitivity.

## THE IMMUNOGLOBULINS

All antibodies belong to a single group of globular proteins that have been called the immunoglobulins (World Health Organ., 1964), and have been subdivided into five classes by immunoelectrophoresis. Each immunoglobulin molecule is a polymer consisting of "heavy" and "light" polypeptide chains, the heavy but not the light chains being different in each of the five immunoglobulin classes (Table 10–1).

## TABLE 10–1

*Immunoglobulin Classes*[a]

| PROPERTY | IgG | IgA[b] | IgM | IgD | IgE |
|---|---|---|---|---|---|
| Molecular weight | 145,000 | 385,000 (170,000) | 850,000 | N.D. | 200,000 |
| Sedimentation coefficient | 7 S | 11.4 S (6.9 S) | 19 S | 6.2–6.8 S | 8 S |
| Light chains | $\kappa\lambda$ | $\kappa\lambda$ | $\kappa\lambda$ | $\kappa\lambda$ | $\kappa\lambda$ |
| Heavy chains | $\gamma$ | $\alpha$ | $\mu$ | $\delta$ | $\epsilon$ |
| Serum concentration (mg/100 ml) | 800–1680 | (140–420) | 50–190 | 0.3–40 | 0.01–0.07 |
| Proportion in: | | | | | |
| Blood | 40% | 40% | 80% | N.D. | N.D. |
| Extracellular fluid | 60% | 0 | 20% | | |
| Secretions and mucosae | 0 | 60% | 0 | | High |
| Rate of synthesis (mg/kg/day) | 20–40 | 2.7–55 | 3.2–16.9 | 0.03–1.49 | N.D. |
| Site of synthesis | Spleen and lymph nodes | Submucosal tissues | Spleen and lymph nodes | ? | Submucosal tissues |
| Half-life (days) | 25 | 6 | 5 | 2.8 | 0.5–2.0 |
| Complement fixation | + | − | + | N.D. | — |
| Transfer to offspring | Via placenta | Via milk | No transfer | Via milk | No transfer |
| Functional significance | Major systemic immunoglobulin | Present on mucosal surfaces | Appears early in immune response | Unknown | Allergenic, responses, on epithelial surfaces |

[a] Data for man: World Health Organ (1964), Cohen and Milstein (1967).
[b] Secretory IgA; data for serum IgA in parentheses.
[c] N.D., no data.

Immunoglobulin G (IgG) is the major serum globulin and the only class of immunoglobulin that can pass across the placenta. It includes most antibodies and dominates the late stages of the antibody response. IgM (macroglobulin) is the first kind of antibody formed, both in ontogeny and during the antibody response of mature animals. Both IgG and IgM are synthesized by plasma cells of the spleen and lymph nodes, and to a lesser extent by those in the submucosal tissues.

IgA is only a minor component (about 10%) of the lighter $\gamma$-globulins in plasma, but it is the principal immunoglobulin on mucous surfaces and in colostrum, and IgA-secreting plasma cells are very numerous in the submucosal tissues of the respiratory and intestinal tracts (Tomasi and Bienenstock, 1968; Dayton et al., 1971). The biological significance of IgD is unknown, but the other minor immunoglobulin, IgE, includes the reaginic antibodies that are involved in anaphylactic reactions (Ishizaka, 1970; Bennich and Johansson, 1971). It represents no more than 0.002% of the total serum immunoglobulins, but IgE-forming plasma cells are very numerous in the respiratory and gastrointestinal mucosa.

The structure of IgG has been investigated in great detail and the amino acid sequences of IgG from several species of animal have been determined (G. Smith

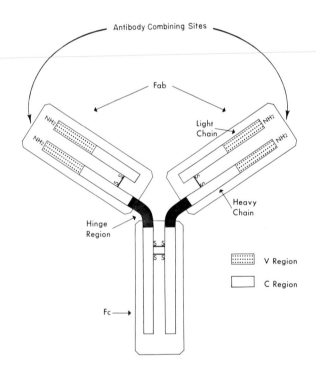

Fig. 10–1. Basic four-chain structure of immunoglobulin G molecule. Hinge region, the two arms of the molecule can open or close at this region to bridge antigenic sites. Papain cleaves the molecule here to give a single Fc and two Fab fragments. V, variable; C, constant.

*et al.*, 1971). A simplified model is illustrated in Fig. 10–1. The molecule is composed of two heavy and two light polypeptide chains, held together by disulfide bonds. For a given IgG molecule the two light chains are either κ or λ; both heavy chains are γ (Table 10–1). The light chains (molecular weight 20,000 daltons) and the heavy chains (molecular weight 50,000 daltons) are under the control of two independent sets of structural genes. Amino acid sequence studies on given molecular species of immunoglobulin (myeloma proteins) show that the amino terminal portions of light and heavy chains have a unique sequence called the variable (V) region, while the rest of the chain is identical throughout an immunoglobulin class or subclass and is called the constant (C) region. Independent C and V genes determine these sequences. The V regions of the chains are the antigen-binding sites, so that the specificity of the antibody produced by a given cell is determined by the amino acid sequence of the V regions.

Enzymatic digestion of IgG with papain splits the molecule into three parts. Two of these (Fab) are similar and carry the specific combining sites, each of which involves the variable region of neighboring parts of the heavy and light chains. The third (Fc) is devoid of combining sites but contains the species-specific antigenic determinants of γ-globulin, and carries the chemical groups that are responsible for the transport of IgG (but not IgM or IgA or most other proteins) across the placenta of some mammals. Treatment with pepsin degrades IgG by destroying fragment Fc, leaving a molecule which contains the two specific combining groups. This can be split by reagents that disrupt disulfide bonds into two almost equal and similar portions, each of which possesses a single combining group whose ability to react with the antigen is unimpaired. In viral neutralization tests, the degraded but bivalent antibody combines firmly with the viral antigens, whereas combination between univalent antibody (Fab) and viral antigen is readily reversible (Lafferty, 1963, 1965).

IgM is a polymer of five subunits, each of which has the basic four-chain structure, with two heavy (μ) and two light (κ or λ) chains, joined together by disulfide bonds. IgA obtained from serum has a sedimentation coefficient of 7 S and consists of two heavy (α) chains and two light (κ or λ) chains. Secretory IgA, which has a sedimentation coefficient of 11 S, is a dimer consisting of two 7 S components plus a secretory component (sometimes called the "secretory piece" or the "transport piece") which is a nonimmunoglobulin glycoprotein with a molecular weight of 50,000 to 60,000 (Dayton *et al.*, 1971). IgE molecules contain two light chains and two unique heavy (ε) chains.

Not only may a single antigen provoke the formation of these different classes of immunoglobulin, but the antibody molecules of each class of immunoglobulin are heterogeneous in their capacity to bind antigen. The antigen-binding capacity of each combining site on the antibody molecule is defined as affinity, whereas the term avidity refers more broadly to the ability of antibodies to bind antigen (World Health Organ., 1970). Thus, each of the ten combining sites on IgM antibody has the same affinity as the two combining sites on an IgG antibody, so that the IgM antibody exhibits greater avidity because of the greater number of combining sites per molecule. The avidity of viral antibody can be measured by determining the equilibrium constant of the virus–antibody complex after equilibrium filtration (Fazekas de St. Groth and Webster, 1961). There are changes in the avidity of antibodies formed after viral immunization, and hyperimmune antibody is generally more avid than antibody produced earlier in the immune response. This is because during immunization, after the change from IgM to IgG synthesis, there is progressive selection for antigen-sensitive cells whose progeny can form antibody of high affinity.

## COMPLEMENT

There are two aspects of *in vivo* humoral immune reactions: specific attachment of antibody to antigen, and the effector process, by which the antigen–antibody complex is disposed of by the host animal. In viral infections, neutralizing antibody may exert its protective effects by preventing virus from entering susceptible cells; or, alternatively, cells, particularly phagocytes, may ingest and destroy the virus–antibody complex more readily than the virus alone. In addition, a serum component termed complement, which interacts with many immune complexes (i.e., antibody combined with antigen), may play a role in the disposal of virus–antibody complexes. Complement is present in normal serum and is heat-sensitive. It is a complex system consisting of nine numbered components (eleven distinct proteins) whose structure and function are now being elucidated at the molecular level (Müller-Eberhard, 1968). Complement reacts with neither antigen nor antibody alone, but only with the antigen–antibody complexes formed either free in the plasma or tissues or on the surfaces of cells. As a result of the reaction chemical mediators may be liberated, causing inflammation with the migration of polymorphonuclear leukocytes from blood vessels into tissues. Complement is important in serological tests with viruses and viral components (complement-fixation tests). Because they have more combining sites, IgM antibodies are more effective than IgG, molecule for molecule, in fixing complement and in sensitizing cells to the lytic action of complement (see below). IgA antibodies do not fix complement.

The addition of fresh normal serum often enhances the neutralizing activity of specific immune serum. In a study of the neutralization of herpes simplex virus, Yoshino and Taniguchi (1965) found that although "early" and "late" rabbit antisera (both containing mainly IgG) bound equally well to the virus, early sera required complement if they were to neutralize viral infectivity by other than a very slow process. Only the "early" (C1-C4) components of complement are needed (Linscott and Levinson, 1969). Early antibodies are present in smaller quantities and have lower affinity than late antibodies and this may be why their effectiveness is increased by the action of complement, or by antiglobulins. Complement may thus play an important part in increasing the antiviral action of antibodies produced very early in infection. Although the practice of heating immune serum before carrying out neutralization tests ("inactivating" the serum) is desirable from the point of view of standardization of the test, the results thus obtained may not reflect the total neutralizing capacity of the serum *in vivo*.

One major significance of complement, however, is that it is an effector system acting on cells whose antigen-carrying membranes have been sensitized by antibody. To an increasing extent it is becoming apparent that many viral infections lead to specific antigenic alterations of the cell surface, and cytotoxicity involving such antigens, antibodies, and complement may be important in the pathogenesis of many viral diseases, notably those due to oncogenic viruses. All of the components of complement are necessary for the lysis of sensitized target cells; interaction with selected complement components leads to lesser changes in target cells such as histamine release or promotion of phagocytosis.

## CELL-MEDIATED IMMUNITY

In addition to the production of immunoglobulins, animals may respond to the inoculation of viral and other antigens by the development of cell-mediated immunity (CMI), which is classically expressed in the delayed hypersensitivity tuberculin reaction (World Health Organ., 1969b). The delayed reaction in the skin requires specific antigen for its initiation, and consists of a slowly evolving mixed cellular accumulation which includes predominantly mononuclear cells (lymphocytes and macrophages). Passive transfer of delayed hypersensitivity can be achieved with suspensions of sensitized lymphoid cells, of which the important component is the lymphocyte, but not with immune serum. Delayed hypersensitivity occurs in many viral infections (Allison, 1967b), the accelerated reaction to vaccinia virus originally described by Jenner providing the classic example.

Recent advances in immunology have given a clearer understanding of the central role of the lymphocyte in CMI. It seems likely that sensitized thymus-derived lymphocytes, i.e., those carrying on their surface receptors (antibodies) specific for the sensitizing antigen, on encountering the antigen, either free or on the surface of cells, respond by releasing a variety of chemical mediators, known as lymphokines. These induce the inflammatory cell infiltration and the mitotic or functional changes in lymphocytes and macrophages that are the hallmark of CMI reactions. A number of *in vitro* methods for studying CMI have been developed, but their relevance for the *in vivo* situation is not yet clear, and delayed hypersensitivity skin reactions remain the classic correlate of CMI in the intact individual.

Macrophages seem to play an important secondary role in CMI responses, and behave as "immune" macrophages as a result of the action of the lymphokines released when sensitized lymphocytes encounter antigen (Rocklin, 1971). After "activation" following this specific immunological reaction they develop lysosomes, increased phagocytic and digestive powers, and thus express a changed reactivity, nonspecific in nature, toward viruses or cells bearing viral antigens. If any antibody is present, this altered reactivity will appear to be immunologically specific. If cytophilic antibody is present on the cell surface of macrophages, they may show immunologically specific behavior without being activated by sensitized lymphocytes.

Homograft rejection is primarily an expression of cell-mediated immunity, although antibodies also may play a part during the later stages of the rejection process. Host lymphocytes recognize foreign antigens on the grafted cell surfaces and then initiate CMI responses leading to the destruction of the grafted cells. In the case of large grafts, vascular changes may play a part in rejection, but with single grafted cells the same mechanism operates as in immune surveillance for tumor cells (Burnet, 1970). Here, lymphocytes circulating through the body encounter small numbers of cells bearing foreign antigens and kill them, either directly or via chemical mediators. CMI against foreign cells is important in animal virology because many viruses (notably "budding" viruses and the oncogenic viruses) produce novel antigens on the surfaces of infected cells. Host responses to such cells are important in controlling both the spread of infections and the growth of virus-induced tumors. On the other hand, vigorous responses can cause damage in infected tissues and thus produce disease (see below).

## CELLS AND TISSUES INVOLVED IN THE IMMUNE RESPONSE

The operative cells in the immune response are those of lymphoid tissue: macrophage, lymphocyte, and plasma cell. Immunoglobulin production (IgG, IgM, and IgA) is a function of lymphocytes and plasma cells, and lymphocytes appear to be important in establishing and executing the specific reactions of delayed hypersensitivity and homograft rejection (Gowans and McGregor, 1965). Recent work has analyzed the immune response in terms of the roles of two types of lymphocyte, derived from the thymus (T cells) or from the bone marrow (B cells) (Parrott and de Sousa, 1971). Effective antibody responses to certain antigens depend upon an intimate interaction between T and B cells, although it is the latter cell type that develops into plasma cells and secretes the antibody. By themselves B cells initiate the antibody response less effectively. However, T cells on their own may respond to antigen by dividing and producing a clone of cells reactive in CMI responses. Macrophages seem to be necessary for the initial uptake of antigens in the form of large particles or cells, making them available, after intracellular processing, for neighboring lymphocytes. A more definitive account of these phenomena should be possible within the next few years.

During development of the individual mammal or bird there are at least three organs, the thymus, bone marrow, and bursa of Fabricius (birds) which are necessary for the full development of immune competence. These organs also play a part in the maintenance of immune competence in the adult. Since the thymus and the bursa also appear to play important roles in leukemogenesis (see Chapter 12), their ontogenic development will be briefly reviewed (Good and Gabrielson, 1964; Miller, 1966). In all vertebrates, the thymus is the organ in which lymphocytes can first be clearly identified (in the fetus or newborn), closely followed by the tonsil in mammals and the bursa of Fabricius in birds. Other tissues (lymph nodes, spleen, Peyer's patches, etc.) become lymphoidal late in fetal life or even after birth, and fail to undergo further development in germ-free animals. In normal animals, germinal centers and plasma cells develop in these organs after birth, presumably as a result of exposure to microbes in the extrauterine environment.

Thymectomy in neonatal chicks and rodents interferes drastically with the capacity of the mature animal to develop delayed hypersensitivity and to reject homografts, and in some species affects the ability of the animal to produce an antibody response to some antigens (Miller, 1965). Children born with thymic dysplasia may develop normal levels of serum globulins but lack the capacity to exhibit delayed hypersensitivity (Fulginiti et al., 1966; Good, 1968). In birds, thymectomy has no effect on the antibody response but early bursectomy prevents $\gamma$-globulin synthesis. Thus the thymus and the bursa of Fabricius appear to be primary lymphoid organs that control the development of cells found in other lymphoid tissues, by mechanisms which are still obscure. In mammals, a primary lymphoid organ necessary for the development of antibody responses and equivalent to the bursa has not yet been identified. The bone marrow supplies cells (B cells) which can differentiate into antibody-producing cells. In addition, cells of bone marrow origin are continuously seeded into the thymus where their multiplication and differentiation into competent immunocytes (T cells) occurs. It appears that the thymus also exerts a less well-defined humoral influence on lymphoid cell differentiation. Once these cells have matured, their further differentiation and multiplication is no longer dependent on the thymus but occurs within the secondary lymphoid organs after the appropriate antigenic stimulation.

The important organs in the immune response of mature animals are the lymph nodes and the spleen, but local accumulations of the appropriate lymphoid cells, e.g., in Peyer's patches or scattered through the lamina propria of the intestinal tract and the respiratory mucosa, also engage in immune responses. The lymph nodes and the spleen are in effect filters interposed in the lymph–vascular and blood–vascular systems, respectively. The tortuous blood sinuses of the spleen are lined with reticuloendothelial cells, which phagocytose circulating antigens or microorganisms. With some antigens these cells carry out a scavenger function, but they often appear to play a special role in processing antigens for antibody production by cells of the lymphocyte series.

In mammals, the lymph nodes are distributed along the major lymphatic pathways in such a way that lymph never enters the blood without passing through at least one lymph node. Antigens or microorganisms from the various tissues of the body are taken up by afferent lymphatics and enter a lymph node at its periphery (Fig. 10–2) where the afferent lymphatic discharges into the marginal lymph sinus. After percolating through the spongelike medullary sinuses, lined by macrophages, the lymph leaves the node at the hilum by a single efferent lymphatic vessel. Lymphatic vessels eventually join the thoracic lymph duct whose contents enter the great veins.

There is a constant and large-scale recirculation of lymphocytes from blood to tissues to lymph and then back to blood. As well as leaving blood vessels in the tissues and entering the afferent lymphatics, lymphocytes in the blood also enter lymph nodes directly via postcapillary venules. The lymphocytes engaged in this constant recirculation are mostly the thymus-derived cells concerned with

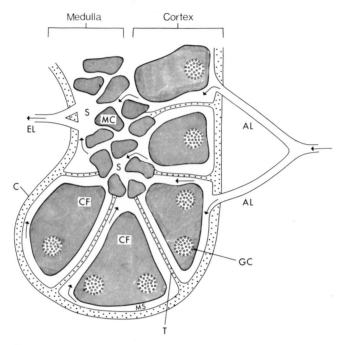

FIG. 10–2. *Diagramatic section of a lymph node. (Details differ according to species of animal, and region of the body.) CF, cortical follicle; GC, germinal center. Lymph enters the node by afferent lymphatics (AL) and leaves by an efferent lymphatic (EL). Connective tissue capsule (C) and trabeculae (T) provide structural framework of node. The marginal sinus (MS) and medullary sinuses (S) are lined by macrophages; lymph in the medulla flows between medullary cords (MC) which consist of blood vessels with surrounding lymphocytes and a macrophage sleeve. Postcapillary venules are the site of lymphocyte transit between blood and lymph. Cortico-medullary junction is not well defined, and is called the "paracortical" area when it contains areas of thymus-dependent cells.*

the primary reaction to foreign antigens. They can thus monitor both tissues and the lymph nodes draining tissues.

At different stages of ontogeny and maturation and at different phases of the immune response, cells in the lymph nodes and spleen may synthesize mainly IgM or IgG (Mellors and Korngold, 1963); lymphoid cells in the mucous surfaces of the respiratory and alimentary tracts synthesize IgA and IgE, and those in the salivary glands and mammary gland mainly IgA.

## THE IMMUNE RESPONSE IN MATURE ANIMALS

In natural viral infections, the infectious dose often consists of a few virus particles, whose "antigenic mass" is extremely small compared with the numbers of antigen molecules used by immunologists. Extensive viral multiplication must occur before there is sufficient antigen to stimulate the immune mechanism, and the viral antigens must reach the appropriate cells.

## Humoral Immunity

Differences that have been described in the antibody response to soluble proteins and particulate antigens are probably related to differences in their clearance by the phagocytic cells of the reticuloendothelial system. After intravenous inoculation, antigenic serum proteins circulate with a half-life similar to that of isologous serum proteins until the onset of antibody formation leads to their accelerated removal (immune elimination). Viruses are particulate antigens, although during all viral infections specific soluble antigens (some but not all of which are components of the virion) are also produced. After intravenous inoculation most particulate antigens are rapidly cleared from the circulation by phagocytosis (Rowley, 1962) even in the absence of specific acquired immunity. The distinction between soluble and particulate antigens is not absolute, and some viruses, e.g., poliovirus type 1 in the mouse (Mims, 1964) and bacteriophage $\phi\chi174$ in the guinea pig (Uhr et al., 1962), disappear from the circulation at a slow rate until the onset of immune elimination.

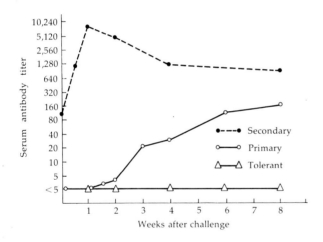

Fig. 10–3. The serum antibody titers in normal and tolerant rats after immunization with flagellin, in doses of 10 µg, showing the difference between the primary and secondary response, and the lack of production of antibody by tolerant rats. (From Nossal and Ada, 1964.)

There are several important differences in the responses of animals on their first exposure to an antigen (primary response) and on subsequent exposure (secondary response). Figure 10–3 illustrates the differences between primary and secondary responses of normal animals to a particulate antigen, and the failure of a "tolerant" animal to produce detectable antibody. With replicating antigens, the production of which extends over several days, it is almost impossible to obtain a "pure" primary response, and this is also true of large doses of nonreplicating antigens. Studies by Uhr and his colleagues and by Nossal and Ada (Ada et al., 1963; Nossal et al., 1963) have emphasized the fact that quite small doses of potent nonreplicating antigens will elicit a primary response and that large doses of such antigens may well obscure important features of this response.

In general, parenteral introduction of an antigen into a previously untreated animal is followed, after an inductive phase of variable length, by production of a limited amount of IgM antibody, and with antigens introduced in the appropriate manner, by the onset of delayed hypersensitivity. Upon reexposure to the antigen a secondary response occurs, which is characterized by (a) a shorter induction period than the primary response, (b) the production of very much larger amounts of antibody than are found after a primary response, (c) a slower rate of decline of antibody synthesis, and (d) the production of mainly IgG antibodies.

The capacity to respond in this accelerated manner probably persists for life, although in the absence of restimulation this "immunological memory" fades. A primary response in very young animals is not necessarily followed by secondary reactivity. In the case of secretory IgA antibodies, very little is known about primary and secondary responses, or immunological memory. Secretory responses are often short-lived compared with circulating antibody responses.

TABLE 10–2

*Half-Lifetimes of Immunoglobulins of Different Animals*

| SPECIES OF ANIMAL | HALF-LIFETIME (IN DAYS) | | |
|---|---|---|---|
| | IgG | IgM | IgA |
| Mouse | 2.5–5.4 | 0.5 | 1.2 |
| Guinea pig | 5.5 | 1.1 | — |
| Rabbit | 5.5 | 3 | — |
| Man | 23 | 5 | 6.0 |

The levels of immunoglobulins in the serum (and in the extravascular fluids) depends not only on their rates of synthesis, but also upon their catabolism. The half-lives of different immunoglobulins differ in a characteristic way (IgM is more rapidly catabolized than IgG), and the half-lives of all immunoglobulins are longer in large than in small animals, which have a higher metabolic rate (Table 10–2).

## Cell-Mediated Immunity

At the cellular level, little is known of the dynamics of primary and secondary responses in CMI. Work on transplantation immunity has shown that sensitized thymus-derived lymphocytes that were generated in a primary response gave rise to an accelerated but more transient response on secondary stimulation (Sprent and Miller, 1971). This suggests that a qualitative as well as a quantitative change occurs in the lymph cell population during the generation of immunological memory.

## IMMUNOLOGICAL TOLERANCE

One of the major advances in immunology during the last half century was the recognition of the phenomenon of immunological tolerance (Medawar, 1961). This consists in the specific absence, partial or complete, of the immune response of a mature animal to a particular antigen and can sometimes be produced by the administration of the appropriate antigen during fetal life. The classic examples are erythrocyte mosaicism in dizygotic cattle twins, lymphocytic choriomeningitis virus infection in mice (Traub, 1939; Volkert and Larsen, 1965), and the acceptance of homologous skin transplants by suitably prepared mice (Billingham et al., 1956). The situation is much more complex than originally thought. Tolerance can also be induced in immunologically mature animals, even with small doses of antigen (Mitchison, 1964), and conversely immunity can be induced in many neonatal or fetal animals. Tolerance to soluble antigens induced experimentally can be "broken" by the inoculation of the mature animals with serologically related antigens (Weigle, 1961), or by adoptive immunization, i.e., the passive transfer of syngeneic immunologically competent lymphoid cells.

## IMMUNITY IN THE FETUS AND NEWBORN

The productive immune response just described is found in mature animals. In view of the importance of congenital infections with viruses, and because it throws much light on the development and nature of the immune response, it is useful to consider the ontogeny of immunity (review, Miller, 1966).

Measurement of immunoglobulin synthesis in fetal animals is complicated by passive transfer of maternal immunoglobulins, but if allowance is made for this there is normally little immunoglobulin synthesis in the developing fetus. The capacity to produce antibodies may nevertheless be present. After challenge of fetal lambs (which have a gestation period of 21 weeks) with several different antigens including the bacteriophage $\phi_\chi 174$, IgM antibodies are found as early as the seventh week of gestation, and IgG appears 4–6 weeks later (Silverstein et al., 1963). The situation in other species is similar, and in humans the capacity to synthesize IgM and, to a lesser extent, IgG starts about the twentieth week of gestation (van Furth et al., 1965). The IgM concentration in the serum of normal newborn infants is very low, since this immunoglobulin cannot cross the placental barrier. Stiehm et al. (1966) have suggested that detection of IgM in cord blood by immunoelectrophoresis is presumptive evidence of intrauterine infection. Generally, the capacity to develop delayed hypersensitivity or reject homografts appears during late embryonic life, often later than the capacity to produce antibodies. In fetal lambs, skin grafts made before 75 days' gestation are completely tolerated, but at 85 days they are rejected (A. Silverstein et al., 1964).

Mammals are born at differing physiological ages, and when corrections are made for physiological maturity it appears that immunocompetence appears at about the same equivalent age in different mammals (Solomon, 1971).

In children, synthesis of IgG usually becomes appreciable at 1 to 2 months after birth and reaches the adult's level between 1 and 4 years of age. IgM begins to rise at about the same time and reaches adult concentrations quickly, especially in the presence of infection, at a time when IgG production is still defective. Experiments with germ-free animals show that the postnatal rise of serum immunoglobulin levels occurs as a result of exposure to antigens present in the extrauterine environment. Fetuses are immunocompetent before birth, but they are protected by the placenta from antigenic stimulation, except in the case of *in utero* infections. Immunocompetence may be high soon after birth but it is often masked by passively acquired maternal antibodies (Solomon, 1971). The vulnerability of newborn animals to many viral infections is attributable as much to a physiological susceptibility as to immunological deficiency.

The immunological reactivity of animals rises to a maximum during their maturity and falls somewhat during senescence (Gatti and Good, 1970). Experiments with mice show that maximum reactivity is attained at the age of about 7 months and that the fall in reactivity in old age is due to a fall in the number of reactive cells; there is no change in the response of individual reactive cells (Makinodan and Peterson, 1964; Wigzell and Stjernwärd, 1966).

## THE IMMUNE RESPONSE TO VIRAL INFECTION

Both humoral and cell-mediated immunity play important roles in viral infections. Antibodies, since they can be tested and assayed without great difficulty, have received most attention. Recent advances in immunology have brought new light to bear on cell-mediated responses, and various *in vitro* tests for this type of immunity have been developed (review, Bloom, 1971). These *in vitro* tests, although of uncertain significance in the infected host, have extended and clarified the *in vivo* observations. Attempts have been made to distinguish between the roles of antibody and cell-mediated immunity in viral infections, and useful distinctions have sometimes been possible in experiments involving various immunosuppressive treatments or cell and serum transfer (Allison, 1972a). In most experiments, however, this distinction has not been sharp, and the possible contribution of interferon has remained in doubt.

Since most of our information concerns antibody this will be dealt with first and at greater length; cell-mediated immunity will be discussed separately in a section on recovery from viral infections.

Viruses are, in general, "good" antigens, and in viral infections a minute quantity of antigen, quite insufficient on its own to provoke significant antibody production, multiplies many millionfold. In the process, several different antigens are produced. Some of these are structural components of the virion, others are virus-specific enzymes concerned with the manufacture of the viral structural proteins and the viral nucleic acid. In cells infected with some poxviruses, for instance, no fewer than twenty antigens can be distinguished by gel diffusion tests; less than one-half of these antigens are eventually incorporated into the virions. Few of the antibodies which react with these viral antigens play any role in protection against viral infection or reinfection.

It has been a common observation in studies on the serological response in viral infections that neutralizing and complement fixing antibodies are first detected at different times and persist for different lengths of time (Lennette, 1959). This is not due only to differences in the sensitivity of the tests. Different viral antigens may be involved and to some extent the temporal differences may reflect the fact that early (IgM) antibodies fix complement more effectively than IgG antibodies. Also, complement-fixing antibodies may be ephemeral because many of them are directed against soluble viral antigens produced in cells only during the acute infection, whereas virion antigens remain in the host for a longer period. In persistent infections (see Chapter 12), there is often persistent production of complement-fixing antibodies.

The intensity of the immune response, in an animal previously not exposed to the virus, depends upon a number of factors. These include the amount and potency of the antigen and its distribution to cells concerned with antibody production. Some viruses, or at least some viral components, appear to be very potent antigens (e.g., phage $\phi\chi174$ and phage f2) and a few may be relatively weak antigens [e.g., lymphocytic choriomeningitis virus (LCM) in mice], but relative antigenic potency cannot properly be assessed without consideration of the species of the host animal. Even with LCM virus in mice very high antibody titers may be produced by adoptively immunizing congenitally infected (tolerant) mice, presumably because the antigenic mass in this situation is very large (Volkert and Larsen, 1965).

Various parameters of the immune response of rabbits to parenteral inoculation with nonreplicating viral antigens have been examined with two viruses, poliovirus (Svehag and Mandel, 1964a, b; Svehag, 1964a, b) and influenza virus (Webster, 1965, 1968a, b). In both cases, the antibody response was characterized by a short inductive phase, the initial appearance of IgM antibodies, and the later appearance of IgG. Small doses of poliovirus produced only a transitory IgM response, with no detectable immunological memory; repeated small doses gave rise to repeated transitory IgM responses. Large doses were followed by the appearance of IgM and IgG and the establishment of immunological memory. The kinetics of formation of IgM and IgG were different, and X irradiation prior to exposure to the antigen abolished the IgG, but not the IgM response.

In Webster's experiments, the route of inoculation of influenza virus in rabbits had a significant effect on the serum antibody level, intravenous inoculation giving higher levels than intraperitoneal, which in turn was more effective than subcutaneous inoculation. Multiple doses of antigen, by all routes, produced more antibody molecules. Studies on the quality (avidity), quantity (number), and specificity of IgM and IgG antibodies to influenza virus showed that the IgM antibodies were of high avidity and gave high levels of cross-reactivity. The early IgG antibodies were of low avidity and failed to cross-react with related strains of influenza virus. The avidity of the IgG antibodies increased during the primary response, and on secondary stimulation the number of antibody molecules increased greatly, but there was only a slight increase in their avidity.

There are a number of factors which contribute to the observed differences in avidity, specificity, and cross-reactivity of antibodies in this experimental influenza virus system. The surface of a virus particle consists of repeating subunits, and IgM antibodies, with more reactive groups per molecule than IgG antibodies, could bind more effectively to the virus and give greater avidity. Also, during immunization, there is progressive selection for antigen-sensitive cells whose progeny form antibodies of higher affinity. Antibody populations are heterogeneous, even when produced against simple, chemically defined antigenic determinants, and may vary in "goodness of fit." Finally, viruses as antigens could provide a basis for cross-reactions in two ways: (a) there may be two or more different antigens in the viral coat, and these may be shared differently with other strains of the virus (e.g., the neuraminidase and hemagglutinin of NWS, A2, and recombinant X7 influenza virus; Laver and Kilbourne, 1966); and (b) two different strains of virus may share parts of the antigenic determinant, e.g., there may be a sharing of a particular 9, 10, or 11 out of, say 12 amino acids which make up the antigenic determinant.

## NEUTRALIZATION OF VIRUSES BY ANTIBODY

Neutralization tests are an important tool in the virologist's armamentarium (see Chapter 2). In this chapter we shall first describe what is known about the mechanism of neutralization as it occurs in such tests, which can be considered as an *in vitro* system although the final assay for infectious virus must be carried out in cultured cells or experimental animals. Then we shall analyze the ways in which neutralizing antibodies react with viruses in intact animals.

### Mechanisms of Neutralization

Viruses are not rendered irretrievably noninfectious by reaction with neutralizing antibody, nor does antibody appear to be appreciably altered by the reaction. Reversal by simple dilution can occur, and even the firm complexes formed by prolonged incubation of virus and antibody can be dissociated to yield infectious virus by a variety of physical and chemical treatments (review, Mandel, 1971b).

There is an interesting contrast between the views of bacteriologists, who regard antibody as an opsonin that promotes phagocytosis (Rowley, 1962) and virologists, who have thought of antibody as inhibiting attachment of virus to the cell and preventing engulfment of attached virions (Fazekas de St. Groth, 1962). The bacteriologist is usually concerned with the behavior of phagocytes and the virologist with epithelial surfaces or cultured cells. As far as cells are concerned, it is impossible to think of virions coated with antibody as being totally different from antibody-coated bacteria or indeed antibody-coated inert particles, and there is evidence that phagocytes and other cells are only quantitatively and not qualitatively different in their phagocytic capacity. There is now evidence that virus neutralized by antibody may attach to cells and be engulfed by them.

Although neutralizing antibody in large amounts can indeed inhibit the attachment of virions to cells, this does not seem to be its primary mode of action when concentrations akin to those encountered in the body are used. For example, Smith *et al.* (1961), using particle counts of vaccinia virus, found an 80% reduction in the attachment to L cells of virus treated with immune rabbit serum diluted 1:5, after centrifugation of virus on to the cells and subsequent thorough washing. Yet this antiserum diluted 1:250 produced "100% plaque inhibition," indicating that even the attached virions failed to initiate infection. Dales and Kajioka (1964) showed that after treatment with antiserum the surface of vaccinia virions was covered with a coating of dense filaments and often the particles were clumped together. However, some single virions and aggregates attached to L cells. A larger proportion of antibody-coated virions than of normal virus remained at the cell surface for a prolonged period, but most were engulfed and could be seen in phagocytic vesicles. A quantitative study showed that prior treatment with antibody (a) reduced the number of cell-associated virions, (b) blocked the release of virus or viral cores from the phagocytic vesicles, and (c) rendered the virus susceptible to degradation within the vesicle. Experiments with vaccinia virus whose DNA had been labeled with tritiated thymidine showed that with antibody-treated virus the viral DNA was degraded very soon after the viral coats had broken down intracellularly.

Results very like those described with vaccinia were obtained with Newcastle disease virus that had been treated with antibody (Silverstein and Marcus, 1964; Granoff, 1965). In experiments with labeled poliovirus, Mandel (1967a, b) found that the effects of neutralizing antibody on attachment depended upon the class of antibody (IgG or IgM), the ratio of antibody to virus, and how long they were allowed to interact. Most of the neutralized virus adsorbed strongly to HeLa cells, and with low concentrations of IgG allowed to react with poliovirus for a long time, attachment to cells was considerably greater than found with untreated virus. Although virus neutralized by antibody attached strongly to cells, and eluted to a lesser extent than untreated poliovirus, the viral RNA of the neutralized virus was completely degraded by the cells. It may be that the normal mode of attachment of poliovirus to cells and penetration of their genome (see Chapter 6) does not bring the particles into contact with lysosomal enzymes, whereas virions opsonized by antibody are actively phagocytosed and therefore become available to such enzymes when lysosomes fuse with phagocytic vesicles.

Investigations by Lafferty (1963, 1965) have underlined the heterogeneity of antibody molecules within a single class of immunoglobulin in their capacity to neutralize virus, and the influence of the type of host cell and its physiological state in determining the result of infectivity assays. In tests with rabbitpox virus, he showed that the neutralizing capacity of rabbit IgG from specific antiserum was increased by treatments which decreased the negative charge and decreased by treatments which increased the negative charge carried by the molecule, although none of the treatments used altered the combining power of the antibody with the virus. The same antibody preparation showed widely different neutralizing activity when assayed in chick embryo fibroblasts under different physiological conditions. Lafferty's experiments emphasize the fact that the neutralizing activity is not solely a function of the antibody molecule itself, but depends upon how the virus–antibody complex is handled by a particular cell. In different cases there may be differences in attachment, in engulfment, or in the subsequent fate of ingested virus–antibody complexes. An extreme example has been provided by Hawkes (1964; Hawkes and Lafferty, 1967), who found that chicken antibody against flaviviruses specifically enhanced the infectivity of these viruses by as much as a factor of 10 when assayed on chick embryo cells, but showed substantial neutralization when the assay was carried out in heterologous cells.

In theoretical discussions of the mechanism of neutralization of viruses by antibody, great importance has been attached to the "nonneutralizable" or "persistent" fraction, and several different interpretations of its significance have been advanced (Dulbecco et al., 1956; Fazekas de St. Groth et al., 1958). In part, the explanation may be that neutralization by antibody is a multi-hit process, for virus in the persistent fraction is rapidly neutralized by added anti-globulin (Ashe and Notkins, 1966). More important, however, are the observations by Wallis and Melnick (1967) and Wallis (1971), which suggest that the "persistent fraction" is primarily a consequence of the presence of viral aggregates in the suspensions used. Antibody molecules may be attached to such aggregates, or to isolated virions, without necessarily neutralizing their infectivity. Filtration through membrane filters to remove aggregates greatly enhances the titer of neutralizing antibodies added subsequently, and virtually eliminates the persistent fraction (Fig. 10–4). However, it is important to remember that viral aggregates, to which antibody molecules are often attached, are a normal feature of viremia (Notkins et al., 1966).

## Neutralization in the Intact Animal

The ways in which antibodies and cells involved in cell-mediated immunity affect viruses in the intact animal are still not clear. Antibacterial antibodies are usually thought to act against bacteria by opsonizing them and thus promoting their phagocytosis and subsequent intracellular digestion. This also happens in viral infection. Mims (1964) found that when mousepox virus was mixed with immune rabbit serum and injected intravenously in mice the virus–antibody complexes were taken up by macrophages in the liver. In contrast to nonneutralized virus, which multiplied in the macrophages or hepatic cells, the ingested complexes disappeared, presumably being digested by the macrophages.

Since virus within cells is protected from the action of antibody, neutralization can occur only when viruses pass from the infected cells into tissue spaces, lymph, or the blood. Some viruses, e.g., poliovirus and influenza virus, are rapidly released from the free surfaces of the cells in which they multiply and are immediately exposed to the action of neutralizing antibody in their environment, which prevents the infection of neighboring cells. Others, like vaccinia and herpes simplex virus, often pass directly from one cell to another, and expanding lesions can then occur even in the presence of a high concentration of neutralizing antibody (Black and Melnick, 1955).

Neutralizing antibodies may be potentiated or depend for their action on complement. There is electron microscopic evidence that enveloped viruses are physically destroyed by complement after reaction with antibody (Berry and Almeida, 1968; Almeida and Laurence, 1969). Rubella and avian infectious bronchitis virions treated with heated antiserum were clumped, but unheated antiserum caused immune lysis of the virion, with "complement holes" clearly visible in the envelope, resembling those seen in the surface membrane of cells following immune lysis. It is not clear whether this process plays an important

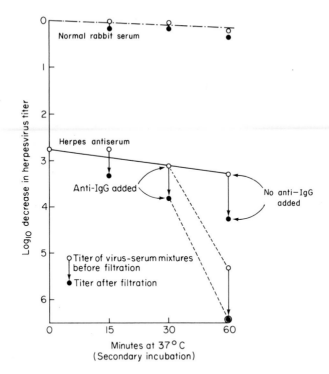

FIG. 10–4. *Removal with anti-IgG of "persistent fraction" (nonneutralized virus) from a virus–antibody mixture. Undiluted, monodisperse herpesvirus was mixed with an equal volume of 1:20 (a) normal rabbit serum or (b) herpesvirus hyperimmune rabbit antiserum, and samples were incubated at 37°C for 1 hour and then 18 hours at 4°C (primary incubation). "0" time in the figure represents the results of this primary incubation. Samples were then incubated for a further 1 hour at 37°C (secondary incubation). At several times during the secondary incubation, samples were assayed for infectivity before (○) or after (●) filtration through a membrane filter that allowed passage of single herpesvirions but not of clumps. Halfway through the secondary incubation anti-IgG was added to samples before and after filtration, and they were assayed immediately and after a further ½ hour at 37°C. (From Wallis, 1971.)*

role in the neutralization of enveloped viruses *in vivo*, but naked isosahedral virions are clearly unaffected.

As has been discussed, infectious virus can be recovered from virus–antibody complexes by several different treatments. In the body, these complexes probably circulate until the virus is thermally inactivated (Smorodintsev, 1957), or until they are ingested by macrophages and digested. It seems likely that nonneutralizing antibodies also play a part in recovery from infection, for by reacting with viral antigens in infected tissues to form immune complexes, such antibodies could initiate the inflammatory infiltrations that lead to an antiviral effect.

Neutralizing antibodies are directed against surface components of the virion, but sometimes antibodies reacting in this way with the intact virion may nevertheless fail to neutralize. Virus–antibody complexes are present in the blood in mice persistently infected with lymphocytic choriomeningitis and lactic dehydrogenase viruses, and can be neutralized or deposited by treatment with antiglobulins, but the circulating complexes are infectious (Oldstone and Dixon, 1970; Notkins et al., 1968).

Macrophages may have an important antiviral role (Silverstein, 1970), especially in relation to cell-mediated immunity, but sometimes their activity is partly attributable to antibody. Different investigators have reported conflicting results on the susceptibility to viral infection of macrophages from normal and immune animals, and no clear conclusions can yet be drawn. Roberts (1964) found that minute amounts of antibody inhibited the infection with ectromelia virus of peritoneal macrophages obtained from normal mice and from ectromelia-immune mice which were hyperimmunized with an intraperitoneal injection of ectromelia virus 8–10 days earlier. In the absence of added antibody, thoroughly washed cells from immune animals were more susceptible to infection, as judged by fluorescent antibody staining. This was ascribed to the increased size and increased nonspecific phagocytic activity of "activated" macrophages. Earlier, however, Maral (1957) had found that even after thorough washing of rabbit peritoneal macrophages, the yield of myxoma virus was much lower with cells from an immune rabbit than with cells from a normal animal. The experiments of Tompkins et al. (1970a) with vaccinia virus in rabbits gave further evidence that immune macrophages are more resistant to infection in vitro than are normal macrophages. There is still some uncertainty about the significance of these experiments, because cytophilic antibody can remain attached to immune macrophages (Heise et al., 1968). Nevertheless, under some circumstances, "activated" macrophages may express antibacterial or antiviral activity independently of antibody. Liver and spleen macrophages which had been activated during a graft-versus-host reaction or Listeria infection were more resistant to ectromelia virus infection (Blanden, 1971b), and those activated by ectromelia or LCM virus infection were more resistant to Listeria (Blanden and Mims, 1973).

## SECRETORY ANTIBODIES ON MUCOSAL SURFACES

In a series of classic experiments on the effect of cholera vaccine in guinea pigs, Burrows and co-workers (Burrows et al., 1947) showed that protection against infection was correlated with local fecal antibody (coproantibody) rather than with serum antibody. It appears that viral infections (and thus live virus vaccines) may confer a specific local immunity in addition to producing the well-recognized humoral response. This argument was advanced by Sabin (1959) when advocating the use of live poliovirus vaccines, which were said to produce a "gut immunity" that did not depend upon the presence of demonstrable antibody in the blood (see also Paul et al., 1959). The discovery of the production and local activity of IgA antibodies has important implications in understanding the role of antibodies in all infections that are localized to mucous membranes, or have an initial stage of multiplication in mucous membranes before they invade the body (Tomasi and Bienenstock, 1968). As far as the intestinal tract is concerned, Crabbé *et al.* (1965) found that the stroma of the normal human duodenal and jejunal mucosa contains large numbers of plasma cells, the majority of which were shown by fluorescent antibody staining to be synthesizing IgA, while a small number synthesized IgG and IgM. After infection with live poliovirus vaccine (Sabin), there was a sustained IgA antibody response in the nasal mucosa and the duodenum, as well as a serum IgG, IgM, and IgA response (Ogra *et al.*, 1968, and Fig. 10–5). Parenteral administration of inactivated vaccine (Salk) gave identical serum responses but no local IgA production and these subjects showed less resistance to oral infection with poliovirus.

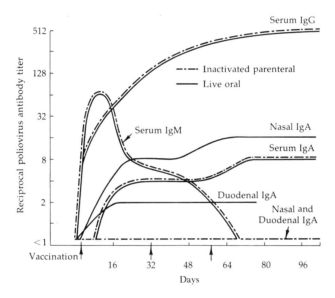

FIG. 10–5. *Comparison of antibody responses to live oral and inactivated parenteral poliovirus vaccines. Serum IgG, IgM, and IgA responses were identical; nasal and duodenal IgA were produced by oral vaccination with live vaccine, but not by parenteral injection of inactivated vaccine. (Modified from data of Ogra et al., 1968.)*

In serum and internal secretions (peritoneal, synovial, cerebrospinal fluids), there is approximately six times as much IgG as IgA, in contrast to external secretions where IgA is locally synthesized (breast, salivary, intestinal, and respiratory secretions), which contain more IgA than IgG. Serum IgA synthesized in the spleen and elsewhere, or administered passively, may be transported from the plasma across the mucous membranes by a mechanism which involves a specific "transport piece" (South *et al.*, 1966), but in some cases this has been shown not to occur (Tomasi and Bienenstock, 1968). Current investigations show that IgA found in human nasal secretions is of major importance in protection against infections with respiratory viruses (Rossen *et al.*, 1971). Surprisingly, patients with selective IgA deficiency do not necessarily show increased susceptibility to infections of the respiratory and gastrointestinal tracts. Perhaps there are compensatory increases in the concentration of other immunoglobulins in external secretions. Locally produced IgG or IgM antibodies may also be important in certain parts of the body, raising local antibody levels independently of serum levels, e.g., in the brains of sheep infected with louping-ill virus (Doherty et al., 1971).

Further, there may be antiviral factors other than immunoglobulins on mucosal surfaces. Cell-mediated immunity is demonstrable in vitro with cells taken from the respiratory tract of locally immunized animals (Henney and Waldman, 1970; Mackaness, 1971), and interferon may be important. Although lysozyme appears not to be antiviral, the possibility remains that other antiviral substances occur.

## PASSIVE IMMUNITY

Passive immunization occurs naturally when maternal antibodies are transferred to the fetus or newborn animal, and can be achieved artificially by inoculation with antisera. The survival time of isologous immunoglobulins differs according to the class of immunoglobulin and the species of animal (Table 10–2). Heterologous antisera provoke an antibody response to themselves and are eliminated more rapidly than homologous immunoglobulins (immune elimination; Talmage, et al., 1951). Much of the isotypic antigenic specificity of γ-globulin resides in the Fc fragment, which is devoid of the specific combining sites, and peptic digestion of γ-globulin may provide antibodies with reduced heterologous antigenic specificity.

Passive immunization has been of considerable theoretical importance in establishing the role of antibodies in protection against viral infectons. It is of practical importance in man as replacement therapy in the agammaglobulinemias, and some agammaglobulinemic children have been given monthly injections of human γ-globulin (IgG) for as long as 13 years without evidence of isosensitization (Janeway and Rosen, 1966). Human γ-globulin has also been employed for the prevention of several viral diseases (see Chapter 15).

### Congenital Immunity and Viral Infections

By the time they reach adult life, all animals, including man, have been exposed to a wide variety of infectious agents and have produced antibodies to many of these. The amount of antibody found in the circulation at any time is dependent upon the particular antigen involved, the intensity of the antigenic stimulus, and the interval since infection. Transfer of immunoglobulins from mother to offspring occurs in many species, by a variety of mechanisms. In birds, transmission takes place from the hen to the egg via the follicular endothelium and then to the embryo via the yolk sac. The situation in mammals is complex. Most work (Brambell, 1970) has been concerned with the transfer of immunoglobulins from the mother to the circulation of the fetus or newborn animal (Table 10–3). In some species (notably the ungulates, but also in rodents), immunoglobulins ingested in the colostrum or milk may pass undigested across the

TABLE 10–3

*Routes of Passive Transfer of Immunoglobulins from the Mother to the Extracellular Fluids of the Fetus or Newborn Animal*

| | ROUTE OF TRANSFER | | |
|---|---|---|---|
| ANIMAL | YOLK SAC | PLACENTA | COLOSTRUM AND/OR MILK |
| Birds | + | N.R. [a] | N.R. |
| Rodents (rat, mouse) | + | − | +(for 16–20 days) |
| Rabbit | + | − | [b] |
| Ungulates (cow, goat, sheep, pig) | − | − | +(for 36 hours) |
| Primates | − | + | [b] |

[a] N.R., not relevant.
[b] Immunoglobulins from milk and colostrum are present and may be active against viruses within the lumen of the gut of the newborn animal, although they may not pass into the circulation.

gut wall of the newborn animal and enter the circulation. In all mammals, IgA in the colostrum or milk may be important in providing local (gut) immunity in newborn animals. This is probably responsible for the poor results found in oral vaccination of breast-fed infants with live poliovirus vaccines (Plotkin *et al.*, 1966). In rodents and rabbits, immunoglobulins are transferred to the fetal circulation via the fetal yolk sac, and also during the first week after birth by means of colostrum and milk. Among ungulates there is no transfer of immunoglobulins *in utero*. In primates, there is a high concentration of immunoglobulins in milk but the intestinal route does not constitute an important mode of parenteral transfer; however, IgG but not IgM or IgA is readily transmitted across the placenta. These differences are related to the types of placenta found in different mammals (Brambell, 1970).

Congenital immunity may be of considerable epidemiological and evolutionary importance. It may completely protect the otherwise very susceptible young animal from infection with bacteria and viruses which occur commonly in the environment, or it may so modify infection with an otherwise lethal agent that the infected young are actively immunized without serious injury, and are then immune to subsequent reinfection. An example of this is the effect of maternal immunity on the responses of mice to infection with virulent ectromelia virus (Fenner, 1948a). In unprotected young, this infection, which is enzootic in many mouse colonies, is invariably and rapidly lethal. Protection with maternal antibody, which was shown by foster nursing experiments to be obtained mainly in the milk during the first week after birth, enables young mice to sustain a nonlethal infection which makes them immune for life.

The role of congenital immunity in viral infections of man has been described by Anderson and Hamilton (1949). In a study of the incidence of herpes simplex infections in an orphanage, they found that babies living in a heavily contaminated environment failed to become infected until they were about 11 months old. At birth, all infants had circulating antibody equivalent in titer to their mother's, but this fell until neutralizing antibody could not be detected by pock reduction tests after the seventh month. Immunity against primary infection lasted for a further 3–4 months, indicating a high efficiency of antibody in the prevention of infection. There is little doubt that congenital immunity is also important in other more serious viral infections, such as poliomyelitis, measles, and smallpox. Although infection with these agents in early childhood may be somewhat less severe than in later life (Burnet, 1952), infection in the serologically unprotected newborn infant is severe and often lethal.

Transmissible gastroenteritis of swine (see Chapter 9) provides a good example of antibody supplied in the colostrum and milk reacting locally with virus within the gut of an infected piglet. Parenterally administered antiserum is ineffective; to provide protection antibodies must be repeatedly supplied to the piglet during the first few days of life, either in the mother's milk or by the oral administration of antiserum (Haelterman, 1965).

## THE DURATION OF ACTIVE IMMUNITY

The lifelong immunity found after attacks of a number of viral diseases, e.g., smallpox, measles, and yellow fever, is proverbial, whereas most adults have suffered from several attacks of influenza and the common cold. A number of factors contribute to this situation, the most important being (a) the intensity of the antibody response in generalized viral infections and the probability that second infections never become clinically apparent, (b) the antigenic stability and hence monotypic nature of the viruses that cause many of the generalized infections, compared with the antigenic variability and multiplicity of antigenic types found with other agents, and (c) the persistence of virus or viral antigens in the body.

In generalized viral infections, reinfections would almost certainly be aborted before the stage of symptom production by the anamnestic response which occurs during the long incubation period. Repeated unrecognized subclinical infections of this sort could play a part in maintaining active immunity at a high level. In measles, for example, Stokes *et al.* (1961) recorded rises in the titers of neutralizing antibody in several measles-immune individuals who had recently been exposed to reinfection. Even in the absence of such reexposure, however, immunity may persist for very many years. The classic example is Panum's (1847) study of measles in the Faroe Islands, where successive measles epidemics were separated by an interval of 65 years. Each epidemic affected virtually all those who had not been infected previously and left untouched those who had been. In only a few instances has it been possible to test for antibodies in the absence of possible reexposure over a period of many years, but reports with yellow fever (Sawyer, 1931), poliomyelitis among the eskimos (Paul *et al.*, 1951), and Rift Valley fever (Sabin and Blümberg, 1947) indicate persistence of recognizable antibody for 75, 40, and 12 years, respectively.

Panum's evidence also shows that measles virus in 1846 was antigenically very similar (in its coat antigens) to that of 1781, and experience with immunoprophylaxis since then has confirmed the antigenic stability of measles virus. Similar considerations apply to all the acute exanthemata of man, in which second attacks are virtually unknown. Recurrent clinical attacks of the same type of disease caused by what appears to be the same virus (e.g., foot-and-mouth disease in cattle, and common colds and influenza in man) gave rise to the suspicion that several antigenically distinct types of these viruses must exist, and this has now been conclusively demonstrated. The continuing change in antigenic pattern in certain viruses (influenza virus, and probably foot-and-mouth disease virus) has been called immunological drift (Burnet, 1955), now more commonly known as antigenic drift. It is an evolutionary change in the envelope antigens of influenza virus imposed by natural selection (see Chapter 17); immunity to reinfection with exactly the same antigenic type of influenza virus is of much longer duration than the apparent freedom from "reinfection." However, even strictly homologous immunity is less durable after superficial infections of mucous membranes, which have a short incubation period, than after systemic infections, which have a long incubation period during which a secondary response can be set in train.

## CHANGES IN THE REACTIVITY OF ANTIBODIES AFTER MULTIPLE INFECTIONS

The antibody response to serial infections of animals with the same or with antigenically related viruses (or injections of inactivated vaccines prepared from them) is affected by two processes which at first sight appear to operate in competition; there is a broadening of the specificity of the antibodies produced, but a heightened reaction to the original virus rather than that used for challenge.

It was pointed out earlier that the degree of cross-reactivity of antibodies to influenza virus was related to their avidity, which was in turn related to the intensity and nature of the antigenic stimulus. Multiple infections or injections of vaccines increase avidity and therefore increase cross-reactivity of antibodies. If the antigens used for challenge are different but related there may be a further broadening of the specificity. Relevant data come from experiments with influenza viruses and togaviruses. For instance, repeated infections of guinea pigs with the same strain of influenza virus or with several serologically related strains evoked antibodies that reacted with antigenically different homotypic strains, to at least some of which the immunized animals had never been exposed (Henle and Lief, 1963). Sequential infection of man with two flaviviruses led to the production of neutralizing antibody which cross-reacted with a third flavivirus to which the subjects had never been exposed (Wisseman et al., 1962), but the existence of cross-reacting antibodies was not necessarily an index of resistance to infection (Wisseman et al., 1966). These broadly reactive antibodies pose difficulties in the interpretation of serological surveys.

From the foregoing experiments it might be thought that vaccination with a series of carefully selected antigenically related viral vaccines would produce a broad immunity which might be effective against all viruses of the subgroup concerned (e.g., alphaviruses or flaviviruses). However, this aim cannot be achieved because of a feature of immunological memory that Francis (1955) has called "original antigenic sin." A study of the response of various age groups of the human population to vaccination with different strains of influenza A showed that the serological response of a person to influenza type A is dominated throughout his life by the type of antibody produced as a result of his first experience of an influenza A virus (Francis, 1953, 1955; Davenport and Hennessy, 1956). This response is characteristic of infections of long-lived animals with any virus of which there are several cross-reacting antigenic types (e.g., togaviruses, Smithburn, 1954 and Hearn and Rainey, 1963; paramyxoviruses, van der Veen and Sonderkamp, 1965; enteroviruses, Mietens et al., 1964). It can be reproduced in laboratory animals either by successive infections with cross-reacting viruses, or successive vaccination with inactive vaccines made from such cross-reacting viruses.

Possible reasons for the dominating influence of the first infection upon subsequent reactions to related antigens have been investigated by Fazekas de St. Groth and Webster (1964). Analysis of the number and avidity of antibodies produced after secondary vaccination with homologous and heterologous (but cross-reacting) influenza type A viruses, and with influenza B virus, suggested that when an animal which had been primarily stimulated with one antigen was then given a serologically related antigen, there was stimulation of the cells that had been involved in the primary response. Larger doses of the heterologous antigen given as a secondary stimulus flooded the initially primed cells and the overflow caused a standard primary response in previously unstimulated cells. If the completely unrelated influenza B was used as the second stimulus there was no change in the antibodies to influenza A and the response to influenza B was a typical primary response.

Original antigenic sin is of practical importance in two fields, serological epidemiology (Chapter 16) and vaccination (Chapter 15). In epidemiological studies, efforts may be made to determine the pattern of previous infection by analysis of the antibodies present in a cross section of the population. Within the limits of the sensitivity of the serological assay, such surveys give unequivocal data in infections with monotypic viruses, as in myxomatosis in rabbits or measles in man. However, in human or animal infections with influenza viruses or togaviruses exact interpretation of the serological results, in terms of whether or not individuals have been exposed to particular viruses, is almost impossible. Usually the most that can be said is that a certain proportion of a particular age group in the population gives serological evidence of prior infection with some member of the relevant group, for example, a type A influenza virus or flavivirus.

If highly specific antibody is needed to give protection against a particular (new) strain of influenza A, the existence of original antigenic sin imposes barriers to obtaining such a specific response by vaccination in all those who have previously been infected (or vaccinated) with another strain of influenza A virus. This can be overcome only by the use of very large doses of antigen, which may be expensive and toxic (see Chapter 15).

## RECOVERY FROM VIRAL INFECTION

In the past, the significance of the immune response for the infected host has been largely assessed in terms of resistance to reinfection, where important roles for circulating or secretory antibodies have been established. Viral vaccines are designed to generate such antibodies. Less attention has been paid to the role of the immune response in recovery from primary viral infections. The coincidence in time between recovery and the appearance of circulating antibodies, and the protection conferred by antiserum given during the incubation period of measles, focused attention on the humoral response as being of major importance in promoting recovery. However, in some viral infections other factors play a greater role, notably cell-mediated immunity and nonimmunological factors like interferon. Indeed, Baron (1963) presented an argument that interferon, not antibody, was the most important factor in recovery. Temporal and spatial relationships are important; to be effective in natural recovery antiviral factors must be present in adequate concentrations, and in the right places, early in the pathogenesis of the infection under study.

Recent experimental studies on mechanisms of recovery have made use of various methods of suppressing the immune response, but unfortunately none of the available techniques is sufficiently selective to give an unambiguous result. Cyclophosphamide greatly reduces antibody production, and neonatal thymectomy followed by the administration of antilymphocytic serum is a powerful suppressant of cell-mediated immunity, but both treatments have other effects as well; for example, the antibody response to some viral antigens is thymus-dependent. Nevertheless, use of these techniques plus selective restoration, by the administration of immune cells or immune serum, is beginning to reveal a pattern of responses in different kinds of viral infections (reviews, Allison, 1972a, b).

### Circulating Antibody

Experiments with immunosuppression and restoration by antibody reveal that viruses that produce systemic disease with a plasma viremia are probably controlled primarily by circulating antibody. For example, if adult mice that have been inoculated with coxsackievirus or yellow fever virus are treated with cyclophosphamide, they die from infections that cause no symptoms in untreated animals. Passive immunization with specific antibody as late as 48 or 96 hours after infection greatly reduces the amounts of virus in the blood and in target organs and protects such immunosuppressed mice (Nathanson and Cole, 1970; Zisman et al., 1971). Neonatal thymectomy, on the other hand, does not increase the susceptibility of mice to enterovirus infections, suggesting that cell-mediated immunity plays a minor role in recovery. Likewise, children with severe hypogammaglobulinemia, but intact cell-mediated immunity, are more liable to develop paralytic poliomyelitis after exposure to vaccine strains than are normal children (Schur et al., 1970).

## Cell-Mediated Immunity

There is ample evidence that viruses elicit cell-mediated immunity. This is demonstrable by delayed hypersensitivity following intradermal injection of viral antigens, a reaction transferable by cells and not by serum, and also by *in vitro* tests like lymphocyte transformation and liberation of a macrophage migration factor that are correlated with cell-mediated immunity (review, Allison, 1972a). Cell-mediated immunity is largely abrogated by neonatal thymectomy and treatment with antilymphocyte serum. Such treatments greatly aggravate infections of mice due to herpesviruses and poxviruses, but have little effect on enterovirus or togavirus infections (Table 10–4).

TABLE 10–4

*The Effects of Impairment of Humoral or Cell-Mediated Immune*
*Responses on Recovery from Various Viral Infections*

| ANIMAL | IMMUNODEFICIENCY | INFECTIONS AGGRAVATED | INFECTIONS UNAFFECTED |
|--------|-------------------|------------------------|------------------------|
| Man [a] | Hypogammaglobulinemia with intact CMI | Paralytic poliomyelitis | Vaccination with vaccinia virus |
| Man [a] | Deficient CMI (with or without normal immunoglobulins) | Vaccinia Herpes simplex Varicella-zoster Cytomegalovirus Measles | |
| Mouse | Suppression of antibody by cyclophosphamide | Coxsackievirus B | |
| Mouse | Impairment of CMI by thymectomy and anti-lymphocyte serum | Herpes simplex Mousepox | Influenza Sendai Yellow fever |

[a] For references, see Allison (1972a).

Restoration experiments confirm the importance of cell-mediated immunity in recovery from systemic infections in which viremia is cell-associated. For example, Blanden (1970) showed that mice infected with sublethal doses of ectromelia virus died if they were treated with antilymphocyte serum, apparently as a result of the uncontrolled growth of virus in the liver. The antilymphocyte serum suppressed the CMI response, but not the antibody or interferon responses to ectromelia virus. Transfer of splenic lymphocytes to mice infected 1 day earlier with ectromelia virus demonstrated that immune but not normal lymphocytes caused striking inhibition of viral growth and a fall in viral titer in target organs (Blanden, 1971a). This effect was demonstrable within 24 hours of cell transfer, and was greatly reduced when transfused cells had been treated with antilymphocyte serum, anti-$\theta$ (thymus-derived cell antigen) serum, or anti-light chain serum. The recipient of the immune cells did not develop detectable antibody. When mouse hyperimmune anti-ectromelia serum was transfused, antibody was present in high titer in recipients, and there was some inhibition of viral growth, but no fall in viral titer such as was produced by transfer of immune cells. Passively transferred interferon had no effect. Blanden concluded that thymus-derived lymphocytes were the primary agents of the antiviral effect. He also showed that an infiltration of lymphocytes and monocytes into liver lesions accompanied the antiviral effect (Blanden, 1971b), and that there was a need for a radiation-sensitive host component, probably a bone marrow-derived macrophage, for the full expression of the antiviral effect of the transfused lymphocytes. Comparable experiments have shown that immune lymphocytes have antiviral activity in lymphocytic choriomeningitis virus infection of mice (Mims and Blanden, 1972).

Sensitized lymphocytes could have an important antiviral effect in several ways. By liberating lymphokines on exposure to antigens in tissues they would induce the migration and activation of macrophages. These macrophages, perhaps with the help of small amounts of antibody or locally produced interferon, could phagocytose and digest infected material. Further, sensitized lymphocytes encountering intact infected cells, which bear viral antigens on their surface, could kill such cells before virus was liberated (Allison, 1971a; Porter, 1971b). Such antigens are found on all cells infected with viruses that mature by budding from the plasma membrane, and also on other virus-infected cells, e.g., poxviruses and herpesviruses (see Chapters 5 and 6). Finally, sensitized lymphocytes may liberate interferon on exposure to antigen, and this could have a significant antiviral effect in tissues.

Thus, cell-mediated immunity plays a central role in recovery from at least some viral infections, mainly those in which plasma viremia is not important and infected cells (in the blood stream and elsewhere) have virus-specified antigens on their surfaces.

### Secretory Antibody

Restoration experiments with secretory antibody are difficult to carry out, especially in respiratory virus infections. However, since antilymphocyte serum has no detectable effect on the pathogenicity for mice of either influenza virus (Hirsch and Murphy, 1968a), or Sendai virus (Mims and Murphy, 1973), antibodies available on the epithelial surfaces (i.e., secretory antibody) probably play a role in recovery.

Although attention is usually concentrated on the effects of neutralizing antibodies, other kinds of antibody may also be important. For example, antibody directed against any viral antigen present in infected tissues could, by forming immune complexes, induce the inflammatory infiltrates that lead to an antiviral effect. Conceivably antibody, cell-mediated immunity, and interferon each play a part in recovery from all viral infections, although their relative importance may vary considerably in different situations.

## THE IMMUNE RESPONSE AS A CAUSE OF PATHOLOGICAL CHANGES

In certain viral infections it is clear that the immune response is a major factor in causing pathological changes and disease. Lymphocytic choriomeningitis

(LCM) virus is nonpathogenic for mice in the absence of an immune response, and animals infected as adults can be protected by X irradiation, neonatal thymectomy, antilymphocyte serum, and other immunosuppressive treatments (Rowe et al., 1963; Hotchin, 1962; review, Lehmann-Grube, 1971). Viral growth patterns are similar in immunosuppressed animals, but the infection is primarily noncytopathic for mouse cells and pathological changes are minimal unless there is a significant immune reaction in infected tissues. Primary importance has been attributed to cell-mediated immune responses (Hotchin, 1962), strains of mice with less vigorous responses being less susceptible (Tosolini and Mims, 1971), but the relative importance of antibody is still uncertain. Immune lymphocytes have been shown to induce lesions and disease in adult mice infected with LCM virus and made tolerant with cyclophosphamide treatment (G. Cole et al., 1971; D. Gilden et al., 1972a). Recipients of immune lymphocytes did not produce significant amounts of antibody before dying of classic LCM disease, thus suggesting that cell-mediated immunity rather than humoral factors were responsible (D. Gilden et al., 1972b).

In other situations, the humoral immune responses to LCM virus can be pathogenic. Mice infected neonatally or in utero show immune tolerance, with lifelong and widespread infection of tissues, but their tolerance has been shown to be incomplete (Oldstone and Dixon, 1967). Throughout life, small amounts of antibody are produced, which react with virus in the blood to form immune complexes, and these are precipitated out in kidney glomeruli and eventually cause glomerulonephritis. Antibody can be eluted from glomeruli, and treatments that increase or decrease antibody formation will increase or decrease the severity of glomerulonephritis. In mice persistently infected with lactic dehydrogenase virus (Porter and Porter, 1971; Oldstone and Dixon, 1971a) or murine leukemia viruses (Hirsch et al., 1968), and in mink infected with Aleutian disease virus (Porter et al., 1969b), immune complexes are deposited in glomeruli in the same way. Glomerulonephritis is of variable severity, perhaps because of differences in the rate of deposition of complexes. SWR/J strain mice persistently infected with lymphocytic choriomeningitis virus all develop glomerulonephritis, and 63.7 $\mu$g IgG could be eluted from each kidney, whereas only 4.1 $\mu$g IgG per kidney was eluted from the same strain of mice persistently infected with lactic dehydrogenase virus, and here there is a low incidence of glomerulonephritis (Oldstone and Dixon, 1971b).

When certain classes of antibodies react with viral antigens on the surface of infected cells, fixation of complement can lead to cell damage. This has been shown to occur in vitro with noncytopathic infections caused by rabies (Wiktor et al., 1968), and LCM (Oldstone and Dixon, 1969), and also with cytopathic infections caused by herpes simplex, vaccinia, influenza, and other viruses (Brier et al., 1971; Porter, 1971). There is no clear evidence about the part played by such reactions in the infected host, but activation of complement would result in the release of mediators of inflammation and thus cause pathological changes.

The immune response has been invoked as a pathogenic agent in many other viral infections, but the evidence is generally less complete than with the LCM model system. In the experiments with CMI and vaccinia virus (Pincus and

Flick, 1963), there was evidence that this immune response caused the primary skin lesion. In measles infection also, there are reasons for supposing that the CMI response causes the rash as well as having an antiviral action in the lungs. On infection with measles virus, patients with a defective CMI response caused by various lymphoreticular tumors or by thymic aplasia may show progressive growth of virus in the lungs leading to giant cell pneumonia, but no rash (Nahmias *et al.*, 1967).

Killed measles virus vaccines cause the formation of circulating antibodies rather than CMI, and on subsequent infection with live virus there are unusual lesions in the skin and lungs, presumably immunopathological in nature (Fulginiti *et al.*, 1967). It is likely that immunopathological mechanisms contribute to other viral diseases, such as dengue hemorrhagic fever (Halstead, 1969; see Chapter 17), viral diseases of the central nervous system (Webb and Smith, 1966; Nathanson and Cole, 1970), and respiratory syncytial virus pneumonia (Chanock *et al.*, 1968). The balance between the humoral and the cell-mediated immune responses to a viral infection can play an important part both in resistance and in pathogenesis. Resistance often depends on one rather than the other arm of the immune response. Further factors favoring either type of response may permit the "immunological engineering" (Parish, 1972) that will make vaccines more effective, and at the same time increase our understanding of viral immunopathology.

Finally, there is evidence that certain viral infections can induce autoimmune disease in the affected host. For instance, mice of the NZB strain and (NZB × NZW) $F_1$ hybrids develop hemolytic anemia and severe glomerulonephritis of autoimmune origin. Autoimmunity appears to be triggered by infection with murine leukemia virus, the affected strain of mice perhaps having a genetically determined predisposition to autoimmune responsiveness (Mellors, 1969; Tonietti *et al.*, 1970).

## IMMUNOSUPPRESSION BY VIRUSES

It has long been known that when tuberculin-positive individuals suffer from measles they temporarily become tuberculin-negative, and studies in Greenland have shown that measles exacerbates preexisting tuberculosis (Christensen *et al.*, 1953). *In vitro* studies show that human lymphocytes will support the growth of measles virus, and such infected cells fail to respond to tuberculin; nor do lymphocytes taken from individuals rendered temporarily tuberculin-negative by measles (Smithwick and Berkovich, 1966).

In the past 10 years, direct tests for immune function have shown that several other viral infections influence the immune responses (Notkins *et al.*, 1970). Immunosuppression is seen, for instance, in mice infected with leukemia viruses, cytomegalovirus, lymphocytic choriomeningitis virus, and in chickens with Marek's disease or avian leukosis. The reduced response is to unrelated antigens, and is thus distinct from the immunologically specific effect seen in tolerance. Most reports deal with depressed antibody responses. Cell-mediated immunity is more difficult to measure, but delayed skin-graft rejection has been described in mice infected with leukemia and lactic dehydrogenase viruses. A few viruses, such as Venezuelan equine encephalitis virus in mice and paradoxically, lactic dehydrogenase virus, act as adjuvants, increasing antibody responses.

The mechanism of immunosuppression is not understood, but it may be significant that immunosuppressive viruses generally induce strong cell-mediated immune responses. They also replicate in lymphoid tissues and macrophages, but generally cause little or no histological damage. A vigorous multiplication of responding lymphoid cells could dilute out cells reacting to unrelated antigens and make them less readily available. Also, these cells could be sequestered non-specifically in infectious foci and thus be prevented from engaging in immune responses. Immune responses are depressed, not abolished, and individual cells infected with murine leukemia virus, for instance, can at the same time produce antibody to sheep red cells (Celada et al., 1970).

Immunosuppression by infectious agents is not restricted to viruses; several nonviral infections including leprosy (Waldorf et al., 1966), malaria (Salaman et al., 1969), and leishmaniasis (Turk and Bryceson, 1971), have been shown to lead to immunosuppression.

## IMMUNOLOGICAL TOLERANCE IN VIRAL INFECTION

Under natural conditions immunological tolerance is more likely to be evoked by viral infections than by any other foreign agents or materials. However, it is now becoming clear that not all congenital viral infections render tolerant the individual exposed to them in embryonic life. If the virus is cytocidal the embryonic tissues are so highly susceptible, and the embryo so vulnerable, that fetal death is the usual result. The important natural noncytocidal viral infections which commonly produce tolerance are those due to the leukoviruses of birds (Rubin et al., 1962) and rodents (Axelrad, 1965; Klein and Klein, 1966). The classic example of immunological tolerance in a viral infection is lymphocytic choriomeningitis (LCM) in mice (Volkert and Larsen, 1965). Tolerance is subtotal, and small amounts of virus-specific antibody are produced in mice congenitally infected with LCM and murine leukemia viruses. In both instances, tolerance can be overcome by adoptive immunization (Volkert and Larsen, 1965).

Rubella virus is relatively noncytocidal in human cells, including those from embryonic tissues, and it is readily transferred to the embryo early in embryonic life. Neutralizing antibody (maternal IgG) has been detected in the fetus as early as the fifth month of gestation, and at birth the rubella-infected infant has both maternal IgG and its own IgM antibodies. The level of the passively transferred IgG falls until about 6 months after birth, when it is replaced by the child's own immunoglobulins, both IgG and often a persistent abnormally high concentration of IgM (Alford, 1965; Soothill et al., 1966; Sever et al., 1966). The absence of tolerance is perhaps surprising, since children who develop congenital abnormalities are usually infected during the first trimester of pregnancy, when the human fectus is immunologically null (van Further et al., 1965). Perhaps the development of tolerance requires that the virus should multiply for a prolonged period in the developing lymphoid system. This is ensured when infection originates in the fertilized egg at the earliest stage of development, as with LCM and leuko-viruses (Mims, 1968a). In rubella in man, not only is infection at a later stage of development, during organogenesis, but maternal antibody is present after the first 3 weeks or so. This antibody may limit infection to certain developing clones of cells in various organs of the body and thus prevent the adequate exposure of the developing antibody-synthesizing system to the rubella antigens.

## IMMUNOLOGICAL ASPECTS OF INFECTIONS WITH ONCOGENIC VIRUSES

Neoplastic cells of many types are immunogenic for the host in which they have arisen, and novel cellular antigens (tumor-specific transplantation antigens, or TSTA) are found in both chemically and virus-induced tumors (Sjögren, 1965; Habel, 1969). Every individual chemically or physically induced tumor produces a different TSTA (Klein and Klein, 1962) (Table 10–5), but TSTA's induced by

TABLE 10–5

*Tumor-Specific Transplantation Antigens (TSTA's) Capable of Inducing Rejection Responses in Isologous Hosts* [a]

|  | COMMENT |
|---|---|
| Chemical carcinogens<br>3-Methylcholanthrene<br>1,2,5,6-Dibenzanthracene<br>9,10-Dimethylbenzanthracene<br>3,4,9,10-Dibenzpyrene<br>3,4-Benzpyrene<br>p-Dimethylaminoazobenzene<br>Physical agents<br>Films<br>Millipore filter<br>Cellophane film<br>Radiation<br>UV<br>$^{90}$Sr | No cross-reactivity of TSTA's in tumors induced by any one agent, and no cross-reactivity between those induced by different agents |
| DNA viruses<br>Polyoma<br>SV40<br>Shope papilloma<br>Adenovirus 12, 18 | Complete cross-reactivity of TSTA's in tumors induced by any given virus; no cross-reactivity between those induced by different viruses |
| RNA viruses<br>Mammary tumor agent<br>Murine leukemia<br>Gross<br>Graffi<br>Moloney ⎤<br>Rauscher ⎬ Group A<br>Friend ⎦<br>Rous sarcoma (Schmidt-Ruppin<br>strain) | Complete cross-reactivity of TSTA's induced by any given virus; no cross-reactivity between those induced by different viruses except for those of Group A which cross-react |

[a] Modified from Klein (1966).

viruses are virus-specific, i.e., they are the same in histologically different tumor cells produced by a particular virus in the same or in different host animals, and different viruses produce different TSTA's.

The existence of TSTA's in cells transformed by an oncogenic virus was first demonstrated by Sjögren et al. (1961) and Habel (1961, 1962) with virus-free polyoma tumor cells. Inoculation of graded doses of such cells into preimmunized and normal syngeneic adult mice showed that the tumor cells were antigenically different from cells of the hosts from which they were derived, and that the antigenic change produced was common to all tumors induced by polyoma virus in mouse cells, either in vivo or in vitro. Other virus-specific TSTA's have been recognized in all virus-induced tumors whether induced by DNA or RNA viruses. These new antigens are located on the cell surface; the mechanism of their induction is discussed in Chapters 13 and 14. The oncogenic RNA viruses differ from the oncogenic DNA viruses in that cell transformation is accompanied by continued production of virions that mature by budding from the cell membrane: TSTA's can nevertheless be distinguished from viral envelope antigen by cross-absorption tests (Steeves, 1968), and by electron microscopic labeling with hybrid antibodies (Aoki et al., 1970).

Although transplant rejection tests offer the clearest demonstration of the antigenic novelty of virus-transformed cells, several in vitro tests have also been developed which probably involve the same antigen (Hellström and Hellström, 1969). These include membrane immunofluorescence and tests for immune cytolysis (recognized by release of $^{51}Cr$ from labeled cells) occurring when tumor cell suspensions are incubated with the sera of resistant syngeneic animals in the presence of complement, or with sensitized lymphocytes. The effects of antibodies depend upon several factors, notably the concentration of TSTA on the cell surface; if the TSTA concentration is high, as in the case of murine leukemia due to Moloney virus (Klein et al., 1966), antibodies have a substantial inhibiting effect. With lower antigenic concentrations the cells may be less sensitive. Leukemias induced in mice by the naturally occurring Gross virus are weakly antigenic, reducing the chances of rejection by antibody or immune cells (Klein, 1968). Leukemia viruses isolated from laboratory propagated tumors (e.g., Moloney) are strongly antigenic but sublines with reduced antigenicity can be isolated by immunoselection (Fenyö et al., 1968). Even "tolerant" mice naturally infected with Gross virus, however, produce antibody which reacts with the infected cell surface (Oldstone et al., 1972), and their lymphoid cells attack leukemic cells in vitro (Wahren and Metcalf, 1970). Not all types of specific antibody molecule are cytotoxic, however, even in the presence of complement. Noncytotoxic antibodies can actually cause enhancement of tumor growth by blanketing the antigenic sites on the transformed cells and thus protecting them from destruction by immune cells.

The immunological implications of TSTA in relation to oncogenesis have been studied in greatest detail in the polyoma virus–mouse system (review, Law et al., 1966). When they are infected with polyoma virus, mice suffer an acute generalized infection and develop high titer antiviral antibodies. An unknown but small proportion of virus–cell interactions leads not to the production of more infective virus, but to transformation of the cell to a tumor cell, and this is accompanied by the appearance of a new antigen on the surface of the transformed cell. In immunologically incompetent animals, like the newborn, such cells are not inhibited by the immune mechanism and they may multiply without restraint to produce tumors. In immunologically active individuals there is an immune response to this new antigen (as well as to the virion antigens), and the transformed cells may be eliminated by a cell-mediated immune response. Such animals are then resistant to challenge with larger doses of polyoma virus-induced tumor cells than are previously uninfected mice. This resistance can be passively transferred by immune lymphoid cells but not by serum, whereas resistance to infection with polyoma virus can be passively transferred by immune serum. The importance of the cellular immune response in preventing the production of tumors by polyoma virus was confirmed by experiments using neonatal thymectomy (Law and Ting, 1965), and antilymophocyte serum (Allison and Law, 1968). Allison (1971a) has reviewed evidence that both age and strain-dependent resistance against polyoma virus oncogenesis in mice is due to a cell-mediated immune response against the TSTA. The TSTA induced by polyoma virus could not be reduced in immunogenicity by serial passage in polyoma-immune mice (Sjögren, 1964b), and appears to be more stable than the TSTA induced by leukemia viruses (see above); it may be the key change responsible for the malignant behavior of the cell (Klein, 1968).

A cell-mediated immunological protective mechanism also operates with adenovirus type 12 tumors in mice; Allison et al. (1967) found that antilymphocyte serum was as effective as neonatal thymectomy in increasing the incidence of such tumors. Murine leukemia viruses induce TSTA's in infected cells, and antilymphocyte serum increases the incidence of leukemia by depressing cell-mediated immune surveillance (Hirsch and Murphy, 1968b). Neonatal thymectomy, in contrast to the result with polyoma and adenovirus 12, decreases the incidence of leukemia, probably because the leukemic process originates in thymus-derived cells (Allison and Law, 1968).

Thus immunological surveillance, operating through cellular rather than humoral mechanisms, appears to play an important role in eliminating potentially neoplastic cells bearing novel antigens. Burnet (1970) has made the suggestion that this may have been the evolutionary basis of the homograft reaction. The phenomenon of allogeneic inhibition appears to offer an alternative mechanism for surveillance, although a weaker and possibly more ancient one. In allogeneic inhibition, tumor cells are killed after contact with lymphocytes (or even hepatic cells) bearing different histocompatibility antigens under conditions where an immune response is impossible (Hellström and Hellström, 1969). Although this could provide a mechanism for tissue homeostasis whereby cells with novel surface antigens are eliminated, the generality of the phenomenon is still not clear. For instance, cells from genetically distinct strains of mice can develop normally side by side in the same fetus to form a healthy allophenic mouse (Mintz and Silvers, 1967), in apparent contradiction to the concept of allogeneic inhibition.

In addition to TSTA, transformed cells may develop fresh, but nonviral antigens on their surfaces. Heterophil (Forssman-reactive) and fetal antigens appear on cells transformed by SV40 and polyoma virus (Pearson and Freeman, 1968; Baranska *et al.*, 1970; Burger, 1971b), and TL (thymic) antigen appears on the cells of TL strains of mice after the development of viral leukemia (Old *et al.*, 1963). These are host-coded antigens which are exposed or derepressed in infected transformed cells. A fuller understanding of such changes in cell membranes may help account for the neoplastic behavior of transformed cells.

## NOVEL ANTIGENS ON THE SURFACE OF CELLS UNDERGOING CYTOCIDAL INFECTIONS

The development of novel transplantation antigens can be readily studied in noncytocidal infections, whether they are due to viruses that become integrated or to steady-state infections. Large amounts of viral antigen(s) accumulate on the surface of all cells from which enveloped viruses mature by budding from the cytoplasmic membrane, as demonstrated by immunofluorescence or hemadsorption. Novel surface antigens may also be produced by nonbudding viruses, and can be detected if the infection is not acutely cytocidal, as in the case of vaccinia, fibroma, and herpes simplex viruses (Veda *et al.*, 1969; Tompkins *et al.*, 1970a, b; Roane and Roizman, 1964).

The importance of these phenomena in pathogenesis and recovery is as follows. Virus-induced antigens on the cell surface, whether viral coat antigens or novel transplantation antigens, and even soluble antigens or virions secondarily absorbed to the cell surface, can be the target for a host immune response which is thereby directed against the cell. In noncytocidal infections, an immune response against infected host cells may be of great importance. Thus, in mice infected as adults with LCM virus, a primarily noncytopathic, nonpathogenic infection is converted into a pathological and lethal disease because of the immune response to infected cells. With RNA tumor viruses, immune responses to antigen-bearing tumor cells may control tumor development. An immune response to the host cell surface may be irrelevant if the infection is acutely cytocidal. With less acutely cytocidal infections, however, viral antigens may appear early on the infected cell surface, as in the case of vaccinia or fibroma viruses (Veda *et al.*, 1969; Tompkins *et al.*, 1970a, b) and the immune response could exert an antiviral effect by destroying infected cells before virus is produced.

## CHAPTER 11

# Pathogenesis: Nonimmunological Factors and Genetic Resistance

## INTRODUCTION

The immune response, which has been discussed in detail in the previous chapter, is a highly specific reaction that is of major importance in protecting vertebrates from viral infection, in promoting their recovery if they are infected, and as a determinant of the symtomatology of many viral diseases. Resistance is also affected by many other physiological factors, such as interferon, body temperature, nutrition, and hormones. These factors may interact with each other; they are sometimes called "nonspecific" because they do not exhibit immunological specificity. Many of these nonspecific factors, and the immune response itself, are under genetic control. This chapter is concerned with nonspecific factors in resistance to viral infection, with particular emphasis on interferon, and with genetic resistance, which may operate through effects on the immune response, on cellular receptors, or on these nonspecific factors.

## INTERFERENCE AND INTERFERON IN THE INTACT ANIMAL

Interference between viruses of vertebrates was first observed in experiments with intact animals. In experiments with two strains of yellow fever virus in the monkey, Hoskins (1935) reported the existence of a protective effect that was not due to immune reactions. Findlay and MacCallum (1937) confirmed and extended this, and made the important observation that in contrast with what was then known about interference in plants, interference could be demonstrated between unrelated animal viruses as well as between different strains of the same virus.

Intact animals were obviously not the best system in which to elucidate the mechanism of interference, and with the development of methods for growing influenza virus in the allantoic sac of the developing chick embryo a great deal of work was carried out with homologous and heterologous interference in this system. This and much other work carried out up to 1958 has been summarized in a critical review by Schlesinger (1959). In 1957, studies on interference were given a new impetus by the discovery of interferon. Details of its nature, mode of production, and mode of action have been worked out in cultured cells (see Chapter 8), but from the time of its discovery there has been a great deal of interest in the possible role of interferon in the pathogenesis of viral infections (Baron, 1963, 1970, 1973; de Clercq and Merigan, 1970a).

It is clear from the evidence presented in Chapter 8 that interferons are potent inhibitors of viral multiplication in cell cultures and chick embryos. In order to understand their role in the intact animal we need to know answers to the following questions: (a) Are they produced *in vivo*? (b) Are there cellular differences in the efficiency of production of interferon or in sensitivity to the action of interferon? (c) Is interferon active in the intact animal locally and at a distance? (d) What is the half-life of circulating interferon and what cells remove it from the circulation? (e) Is there a correlation between interferon production and virulence? (f) Are there natural or experimental situations in which other mechanisms of resistance are not operative but interferon production and activity are normal? Some of these questions can now be clearly answered, although we are still far from being able to define precisely the role of interferon in the pathogenesis of viral infections.

## Production of Interferon *in Vivo*

**Virus-Induced Interferon.**   There is now ample evidence that interferon is produced in virus-infected animals. The local occurrence of interferon in the site of multiplication of the infecting virus has been demonstrated in the rabbit skin after infection with vaccinia virus (Nagano and Kojima, 1958), in the throat washings of patients with influenza (Gresser and Dull, 1964), in the mouse lung in influenzal infections (Isaacs and Hitchcock, 1960), in the mouse brain after infection with togaviruses (Hitchcock and Porterfield, 1961), and in many other situations. In mice inoculated intracerebrally with West Nile virus, viral multiplication and interferon production are virtually confined to the brain (Subrahmanyan and Mims, 1966), but both in experimental situations, e.g., lactic dehydrogenase elevating virus in the mouse (Evans and Riley, 1968), and in natural infections, e.g., respiratory syncytial virus infection in man (Gray *et al.*, 1967), interferon is found in the blood, and can exert a protective effect at sites distant from those where the inducing virus is multiplying.

**Nonviral Inducers.**   As pointed out in Chapter 8, many nonviral substances, when injected into experimental animals, will induce the formation of interferon, the best studied being bacterial endotoxin and the synthetic double-stranded polynucleotide, polyinosinic: polycytidylic acid (poly I·poly C). The interferon-inducing capacity of certain mold products, such as statolon and helenine, is due to double-stranded RNA viruses present in them (de Clercq and Merigan, 1970b). When endotoxin or poly I·poly C is injected intravenously into experimental animals, peak levels of interferon in the serum are reached within 2–3 hours, whereas the interferon induced by the intravenous injection of Newcastle disease virus does not appear until about 6 hours. Also, endotoxin-induced interferon, unlike virus-induced interferon, appears not to be sensitive to the action of various inhibitors of RNA and protein synthesis. It was formerly believed that endotoxin and poly I·poly C induced the release of performed interferon, whereas viruses induced interferon synthesis. More recent work has made this distinction less clear, and the pattern of interferon induction may not be essentially different in the two situations (de Clercq and Merigan, 1970b).

Nonviral interferon inducers have various actions other than the induction of interferon. Endotoxin and poly I·poly C, like Newcastle disease virus or influenza virus, influence the reticuloendothelial system, act as immunogens, and induce fever in injected animals (de Clercq and Merigan, 1970b). Further, induction of interferon in animals, as in cultured cells (see Chapter 8), is followed by a refractory period before the response can be reinduced. These side effects, and the toxicity of poly I·poly C for man, present problems which have so far prevented clinical use of nonviral interferon inducers.

**Cells Involved.** In most experimental studies the interferon inducers have been given intravenously. Poly I·poly C also induces interferon when given by the intranasal route. Reticuloendothelial cells have been implicated in the uptake of the inducer and in interferon production, and peritoneal macrophages produce interferon *in vitro* (Smith and Wagner, 1967). Extirpation experiments show that the spleen is particularly active (Fruitstone *et al.*, 1966); Osborn and Walker (1969) developed an ingenious assay system which showed that the mouse spleen contained up to 30,000 interferon-producing cells after the intravenous injection of Newcastle disease virus.

Interferon induction by the intravenous injection of either nonviral or non-replicating viral inducers is quite distinct from the situation in a viral infection, during which the infecting virus multiplies, spreads through the body, and induces interferon in a variety of susceptible cells.

## The Fate of Circulating Interferon

Interferon administered to mice by intravenous injection disappears rapidly from the circulation, at the rate of over 95% per hour (Baron *et al.*, 1966a; Ho and Postic, 1967). This rapid fall in titer indicates that interferon, which is a relatively small protein, is distributed throughout the extracellular space of the body. Interferon induced in pregnant rabbits by the intravenous inoculation of Newcastle disease virus diffuses into the placenta, the fetal blood, and the fetal tissues (Ho and Postic, 1967). Assay of organs of mice snap-frozen 5 minutes after the intravenous inoculation of interferon showed that significant **amounts** could be detected in the lungs and kidney but none in the liver, spleen, brain, muscle, or intestine (Subrahmanyan and Mims, 1966). Small amounts were excreted in the urine.

In the rabbit, Ho and Postic (1967) found that different interferons were cleared from the plasma by the kidneys at very different rates; 68% of the plasma volume of virus-induced interferon was cleared in an hour, but only 3.3% of the plasma volume of endotoxin-induced interferon. Since between 0.2 and 2% of passively administered interferon is excreted in the urine of rabbits and the amount of interferon excreted in the 30 hours after induction by Newcastle disease virus was 100,000–300,000 units, Ho and Postic calculated that the total amount of interferon induced per kilogram of rabbit might be between 5 and 150 million units. It is clear that unless there is a defect in clearance, the detection of appreciable amounts of interferon in the plasma of virus-infected animals implies its rapid and continuous production and release into the blood.

## Local and Systemic Protection by Interferon

Isaacs and Westwood (1959) showed that interferon injected into the skin of the rabbit 24 hours before infection inhibited the multiplication of vaccinia virus, and similar results have been reported in man (Scientific Committee on Interferon, 1962).

Evidence about the systemic effects of interferon comes from experiments on the passive administration of interferon and on the effects of the active production of circulating interferon by a mild generalized viral infection on the production of lesions by another virus. Baron et al. (1966b) carried out protection experiments with several viruses which kill mice by multiplication in the brain. Both passively administered interferon (in mouse serum) and interferon induced by the intravenous inoculation of a large dose of Newcastle disease virus had a protective effect in mice challenged by the intracerebral inoculation of encephalomyocarditis and vesicular stomatitis viruses, or by the intraperitoneal inoculation of a togavirus. Finter (1966) reported comparable results in mice infected with Semliki Forest virus. Gresser and his associates have shown that interferon administered passively may have to be given in repeated doses for protection (Gresser et al., 1968). In all experiments in animals, the protection afforded by interferon can usually be overcome by increased doses of the challenge virus (Finter, 1964; Baron et al., 1966b).

In assessing the effectiveness of interferon in intact animals, two facts established in experiments with cultured cells (see Chapter 8) must be borne in mind: several hours' exposure of cells is needed before interferon exerts its full antiviral effects, and the amount of interferon that needs to be taken up by cells in order to produce resistance is very small.

## Interferon and Recovery from Viral Infections

There is no doubt that in cell cultures interferon can play a major role in the protection of cells against viral infection and under suitable conditions in the protection of a culture, as demonstrated in the mixed cell culture experiments of Gresser et al. (1965) and Glasgow (1965). In the intact animal, many factors are involved in recovery, and it is difficult to decide their relative importance. In different types of infection, either secretory antibody, humoral antibody, and/or cell-mediated immunity play important roles. Interferon is potentially important in several situations but since no disorders are known in which man or animals are unable to produce interferon decisive experiments are difficult.

Interferon administered passively, or actively induced by infection with an attenuated virus, can undoubtedly protect animals against otherwise lethal viral infections. Interferon is produced as soon as the first few cells are infected, without the delay required by the immune response. While it is likely that interferon produced during the course of a viral infection protects some cells from infection, both locally and in distant target organs, it has so far proved difficult to devise experiments to evaluate the importance of these effects in promoting recovery.

However, there is much circumstantial evidence that interferon is important in recovery from viral infection. Decreased interferon production caused by altered temperature, chemical inhibitors, or different viral strains has been associated with impaired recovery, but the situation is often complex with multiple determinants of virus virulence. Virulence of some viruses is associated with a weak interferon response, but in other instances there is no correlation (Baron, 1970). Hansen *et al.* (1969) have implicated the responsiveness of infected cells to interferon action as a factor. Mice of the C3HRV strain were much more resistant than C3H mice to flavivirus infections, and although both produced equal amounts of interferon, the cells of the resistant mice were more susceptible to the action of interferon, i.e., a genetic difference in interferon sensitivity rather than interferon production appeared to determine the severity of infection. Several nonspecific factors involved in resistance may operate at least in part by their effects on interferon production; this appears to be the case with body temperature, with some types of stress, and with the effects of some hormones.

Interferon is also active against leukemia viruses of mice and therefore against leukemia (Gresser *et al.*, 1968) and Gresser has shown that interferon inhibits transplanted mouse tumors which were not originally induced by viruses (Gresser *et al.*, 1969).

Despite this encouraging data, there are results that temper optimism in relation to the possible therapeutic use of interferon. Neither natural infections with viruses nor immunization with live attenuated vaccines confers substantial protection against simultaneous or subsequent heterologous viral infections elsewhere in the body. Indeed, certain attenuated viruses can be combined to produce potent multivalent vaccines (see Chapter 15). Perhaps, under natural circumstances, interferon expedites recovery because transient high concentrations occur in the immediate vicinity of infected cells. It may be impossible, in terms of antiviral therapy, to produce enough interferon in the right place at the right time.

## OTHER NONIMMUNOLOGICAL FACTORS IN RESISTANCE

Although a large number of situations have been described in which some natural product of an animal susceptible to viral infection appears to have an antiviral effect, only specific immunoglobulins, specifically altered cells which can participate in delayed hypersensitivity reactions, and interferon have been characterized well enough to attempt to evaluate their roles in protection and recovery.

As we have already seen, it is difficult to establish the precise roles of the immune response and interferon in the pathogenesis of viral infections in the intact animal. It is even more difficult to analyze the effects of the variety of other nonspecific factors that have been shown at one time or another to affect the course of viral infections.

## Body Temperature

Experimental observations on the effects of temperature on the severity of infectious diseases date back to Pasteur et al. (1878), who found that chickens lost their resistance to anthrax if their temperature was lowered. Lwoff (1959) sharpened the interest of animal virologists in the effects of temperature on the multiplication of viruses in cultured cells and on their pathogenic behavior in experimental animals, but he greatly oversimplified the situation at the cellular and especially at the organismal level (Bennett and Nicastri, 1960). Exposure of animals to abnormally low temperatures, for example, increases thyroid activity and the stress response, increases metabolism and the susceptibility of the liver to chemical injury, and affects such different physiological responses as the circulation in the extremities of the body, the immune response, and the inflammatory response. Exposure to abnormally high temperatures has a similarly wide variety of physiological effects.

Lwoff confirmed the observation of Dubes and Wenner (1957) that poliovirus cultivated in cells at low temperatures lost its capacity to grow at 37°C, and by serial passage he was able to select temperature-sensitive and temperature-resistant strains. He then established a relationship between thermoresistance and the neurovirulence of poliovirus in intact animals (Lwoff et al., 1959b; Lwoff, 1961; Sabin, 1961). Attenuated strains invariably grow poorly at high temperatures, and cultivation at a low temperature has proved a useful method both for the selection of vaccine strains and for their propagation. Pérol-Vauchez et al. (1961) established a similar relationship with encephalomyocarditis virus in mice, and showed that attenuated (thermosensitive) strains were more virulent in mice whose body temperature had been reduced, and less virulent in mice whose body temperature had been raised, compared with mice maintained at normal ambient temperatures.

All these experiments were carried out with multiple-step mutants obtained after serial passage of the wild-type virus in high concentration and at progressively lower temperatures. Such mutants have the advantage for animal experiments that they do not readily revert to wild type, nor do they show "leak" when cultured at high temperatures. In order to test the effects of single-step mutations to temperature sensitivity, Fenner (1972) studied the virulence of several ts mutants of Semliki Forest virus (Tan et al., 1969) and rabbitpox virus (Padgett and Tomkins, 1968). In both cases, the virulence of single-step ts mutants was much less than that of the wild-type viruses, and there was a close correlation between the cutoff temperature of particular mutants and the lethality of Semliki Forest virus in mice, at high, moderate, and low environmental temperatures. Neither interferon production nor antibody production showed any significant relationship with lethality.

Experiments with coxsackievirus B1 in mice illustrate other facets of the relationship between environmental temperature and resistance to viral infection. Coxsackievirus B1 produces viremia and multiplies extensively in most tissues of suckling mice, but only to any marked extent in the pancreas of adult mice, even though viremia is regularly observed. Exposure of adult mice to a cold environment, however (4°C), results in an infection with persisting viremia,

high levels of virus in the liver and other organs, and an invariably fatal outcome (Boring *et al.*, 1956). The increased susceptibility of cold mice was not
due to increased production of corticosteroid hormones (Walker and Boring,
1958). An elevated temperature (36°C), although it had an adverse effect on the
mice, caused marked inhibition of viral invasion and multiplication (Walker and
Boring, 1958). Even when mice were transferred from 25° to 36°C as late as
36 hours after infection, when viral multiplication was well advanced in the
pancreas, further multiplication was suppressed and the virus was eliminated.
Ruiz-Gomez and Sosa-Martinez (1965) confirmed the adverse effects of low
temperature on adult mice infected with coxsackievirus. Mice kept at 25°C developed much interferon but little virus in their livers whereas the reverse occurred
at 4°C; they suggested that the effects of temperature may operate through
regulation of interferon production.

Some of the earliest observations on the effect of ambient temperature on
viral disease were made with myxoma virus by Thompson (1938) and Parker and
Thompson (1942), who showed that sustained high room temperature, which
raised the internal temperature of the rabbits from 39° to 40°C, and their skin
temperatures from 33°–36°C to 38°–40°C, saved rabbits from death after infection with virulent myxoma virus, otherwise invariably lethal. Marshall (1959)
confirmed this, and using a slightly attenuated strain of myxoma virus showed
that low ambient temperatures greatly increased the severity of the disease and
the mortality rate. Measurements of viremia and of antibody production in rabbits exposed to high and to low temperatures showed that both viral multiplication and the antibody response were affected in ways which would increase the
severity of the disease at the lower temperature and diminish it at the high temperature (Fig. 11–1). Interferon production was not measured.

Other experiments involved mice and dogs. Roberts (1964) found that mice
infected with ectromelia virus were about 100 times as susceptible when kept at
2° rather than 20°C. Mice kept in the cold showed an increased metabolic rate,
and this was associated with more severe liver damage without an increase in
viral growth in the liver. Newborn pups are highly susceptible to generalized
canine herpesvirus infections, to which dogs become resistant by the age of 4 to
8 weeks. Exposure of litters of virus-infected newborn pups to high ambient
temperature, and lowering the body temperature of 4 to 8-week-old dogs, reversed both situations. Carmichael *et al.* (1969) concluded that the high susceptibility of newborn pups to disseminated canine herpesvirus infection might be
due in part to their low and poorly regulated body temperature.

Thus in different situations increased body temperatures (or the higher temperature of some parts of the body than of others) may suppress viral multiplication directly, by an effect on some step in multiplication, or indirectly by
promoting the efficiency of the host response in terms of the production of
antibody and/or interferon. Low body temperatures usually have the reverse
effects, and may, in addition, alter the susceptibility of target organs to damage.

Most of this experimental work has been concerned with the effects of different environmental temperatures. However, almost all severe viral infections in
the larger mammals are accompanied by fever (38°–41°C), usually at the end of

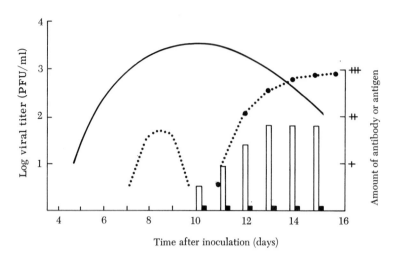

FIG. 11–1. *The effects of high and low temperatures on the response of rabbits to infection with myxoma virus. Diagram illustrates the high viremia (—) in rabbits kept in the cold room (4°C for 16 hours out of each 24) and the low viremia (···) in rabbits kept in the hot room (37°–39°C for 16 hours out of each 24). The cold-room rabbits produced no detectable circulating antibody, but large amounts of soluble antigen (white columns) in their serum; the hot-room rabbits produced high titer antibody (●··●) but no soluble antigen (black columns). Almost all the cold-room rabbits died from severe myxomatosis; only 9% of the hot-room rabbits died from myxomatosis. (From Fenner and Ratcliffe, 1965.)*

the incubation period, when viral multiplication is largely completed, and pyrogenic antigen–antibody complexes are present (Root and Wolff, 1968). All the antiviral mechanisms associated with high temperature operate during fever, but there is virtually no information on the extent to which the fever may be causally related to the cessation of viral growth. Lwoff (1959, 1969) has postulated that fever may represent a natural defence mechanism against viruses and that virulent mutants able to multiply in febrile animals are the result of natural selection. Conversely, temperature-sensitive mutants are logical contenders for safe live virus vaccines against a number of diseases (see Chapter 15).

There is one situation in which elevated body temperature promotes viral multiplication, namely the production of "fever blisters" or recurrent herpes simplex in man. Artificial fever induced by physical means results in herpes labialis in about 50% of cases (Warren *et al.*, 1940), and fever blisters are a frequent complication of some febrile diseases (malaria, influenza, streptococcal and pneumococcal infections) but are very rarely found in others (tuberculosis, smallpox, and typhoid fever).

## Hormones

Hormones play an important role in maintaining homeostasis and in regulating many of the physiological reactions in the animal body. The corticosteroids have readily demonstrable effects on resistance to viral infections both in man and in many experimental animals. Many instances of the deleterious effect of corticosteroids on viral infections have been described, of which we shall quote only a few. Adult mice treated with cortisone develop fatal infections with coxsackievirus B1, which is normally lethal only in newborn mice (Kilbourne and Horsfall, 1951), and cortisone greatly increases the susceptibility of hamsters (Shwartzman and Fisher, 1952) and monkeys (Stuart-Harris and Dickinson, 1964) to poliovirus infection. Exacerbation of otherwise mild viral infections (e.g., vaccinia and varicella) is a well recognized complication in human patients receiving corticosteroid therapy (Kass and Finland, 1958). Blindness has been known to follow ill-advised administration of cortisone to patients with herpetic keratoconjunctivitis (Kaufman, 1965b). The anti-inflammatory effects of the corticosteroids and their capacity to depress the immune response have been amply documented (reviews by Kass and Finland, 1953, 1958). The immune response is also influenced by other hormones (Ciba Foundation Study Group, 1970). Hydrocortisone inhibits the production of virus-induced interferon in chick embryos (Smart and Kilbourne, 1966), cultured cells (Reinicke, 1965), and intact animals (Rytel and Kilbourne, 1966).

Pregnancy affects the pathogenesis and the severity of several viral diseases, probably because of the hormonal changes. Smallpox is particularly severe in pregnant women and abortion is common, and in the days before widespread vaccination paralytic poliomyelitis was more frequent in pregnant women than in control groups (Siegel and Greenberg, 1955). Sprunt (1932) found that the skin lesions of myxomatosis were less pronounced in pregnant than in nonpregnant rabbits but lesions in the lungs and liver were more extensive; the mortality was very high in both groups. Another example of the association of ovarian hormone function with susceptibility to the effects of viruses is provided by the occurrence of mammary tumors in female but not male mice infected with the mammary tumor virus, unless the males are treated with estrogenic hormones (Gross, 1970).

## Nutrition

Malnutrition can interfere with any of the mechanisms that act as barriers to the multiplication or progress of viruses through the body. It has been repeatedly demonstrated that almost any severe nutritional deficiency will interfere with the production of antibodies and the activity of phagocytes, and the integrity of the skin and mucous membranes is impaired in many types of nutritional deficiency (review, Scrimshaw et al., 1968). The synergistic effects of malnutrition and infectious diseases are evident in the data on human mortality in the age period 1–4 years in technically advanced and in underdeveloped countries. Measles, for example, is rarely a cause of death in the developed countries, and this is not merely due to the availability of chemotherapy for secondary bacterial infections. However, there is a substantial mortality from measles in most underdeveloped countries (reviews, Morley, 1967; Naficy and Nategh, 1972); in different South American states, for example, the mortality rate per 100,000 population in 1959 was 105–418 times higher than in the United States of America (Scrimshaw,

1964). The higher mortality was confined to children in the lowest socioeconomic groups. Such effects are seen particularly in children aged 1–4 years because this is the period of malnutrition during the change from breast milk to an adult type diet.

In a careful study (Woodruff and Kilbourne, 1970; Woodruff, 1970), it was shown that mice which from the age of 3 weeks, received one-quarter the food intake necessary for optimal growth, developed marked atrophy of lymphoid tissue and lymphocytopenia. They were then much more susceptible than normal mice to coxsackievirus B3 infection, which produced severe and often fatal lesions in liver and heart, with a marked absence of cell infiltration. There were indications that the effect involved an inadequate host immune response to the infection. A similar depletion of the thymolymphatic system and depression of cell-mediated immunity has been observed in children with protein-calorie malnutrition (Smythe et al., 1971).

Studies of nitrogen balance in human beings have demonstrated that even mild infections, viral or bacterial, lead to substantial nitrogen losses (Beisel, 1966). These are readily sustained in well-nourished individuals but in those suffering from protein deficiencies relatively mild infections can lead to severe disease by the synergistic effects of the nutritional and the dietary stresses. Poskitt studied children in Uganda suffering from measles, and found that there were significant falls in plasma albumin levels, sometimes with the production of kwashiorkor (Poskitt, 1971). Even mild infections like smallpox vaccination may lead to lowered serum levels of vitamins A and C in individuals suffering from malnutrition (Scrimshaw, 1964).

Overfeeding increases the severity of some generalized viral infections. Newberne (1966) and Scrimshaw et al. (1968) found that dogs rendered obese by eating an excessive amount of a balanced diet were substantially more susceptible to canine distemper than were dogs given a normal or low-level caloric intake. The severe effect of obesity appeared to be related to the endocrine response to infection with distemper virus.

## Other Nonspecific Factors

Isolated reports have been published of a variety of other nonspecific factors that raise or lower the resistance of animals to viral infections.

**Stress.** "Stress" is a very general term that is used to describe almost any type of abnormal or unusual environmental influence on animal behavior; thus an animal may be subjected to nutritional stress, heat stress, psychological stress, the stresses of infection, and so on. These stresses call into play an adaptive host response involving the pituitary–adrenal axis (corticosteroid hormones) and the autonomic nervous system (adrenaline), which functions by combating the harmful effects of the stresses. Various stresses may therefore influence a viral infection by inducing an increased output of corticosteroids, whose adverse effects have been referred to earlier. Infection itself is often an important stress, and can cause changes in lymphoid tissues and liver which are attributable to the action of corticosteroids (Wallnerova and Mims, 1971). Scrimshaw (1964) has emphasized the tremendous importance of the synergism between malnutrition and the stresses of infectious diseases in human beings.

Rasmussen and his colleagues have studied the effects of psychological stress on the susceptibility of mice to herpes simplex, coxsackievirus, and vesicular stomatitis viral infections. Two types of response were observed. The cumulative response was characterized by increased susceptibility and appeared after several weeks of daily stress exposures. It was correlated with pronounced anatomical changes: adrenal hypertrophy and atrophy of spleen, thymus, and liver (Rasmussen et al., 1957; Johnsson and Rasmussen, 1965). The transitory response was characterized by periods in which mice were either more or less susceptible to viral challenge than control mice (Jensen and Rasmussen, 1963); the increased susceptibility was correlated with diminished interferon production immediately after the period of stress (Jensen, 1968). In very similar experiments, Solomon et al. (1967) described increased interferon production after stress.

**Trauma.** Most of the evidence concerning the effects of trauma on viral infections relates to poliomyelitis. To affect the severity of the disease, trauma must occur during the early stages of the incubation period, or if the effects of the "trauma" are prolonged, shortly before infection. The effects of the traumatic experience are to increase the likelihood of paralysis, its severity, and sometimes its localization. Physical exertion and nonspecific trauma increase the likelihood of paralysis (Horstmann, 1950), injections of pertussis vaccine or diphtheria or tetanus toxoids increase the likelihood of paralysis in the injected limb (McCloskey, 1950); and tonsillectomy increases the likelihood of bulbar poliomyelitis (Aycock, 1942). There is also statistical evidence to show that persons whose tonsils have been removed at any time in the past are more likely to suffer from bulbar poliomyelitis than those with intact tonsils (Anderson and Rondeau, 1954). This probably has an immunological rather than a "traumatic" basis. Ogra (1971b) showed that removal of the tonsils and adenoids reduced both preexisting IgA levels in the pharynx and the local IgA response after vaccination.

There is abundant evidence (review, Mims, 1966b) that the skin lesions in exanthematous viral diseases localize in "provoked" skin areas, probably because of the increased vascularity and the inflammatory response associated with the provoking factor.

**Concurrent Infections.** Since latent or chronic infections with bacteria, protozoa, and viruses are common in both man and animals, concurrent infections are also common. Sometimes the two infections apparently have no effect on each other; with other combinations of agents there are interactions which may either increase or decrease the severity of either infection. Veterinary virologists are particularly conscious of the potentiating effects on viral diseases of coinfection with parasites because of the universal infestation of livestock with protozoa and worms.

Infections with respiratory viruses commonly increase the susceptibility of the respiratory tract to infection with bacteria, e.g., after rhinovirus or influenza virus infections, and after measles. Measles also appears to exacerbate the severity of tuberculosis in man (see Chapter 10).

Resistance to some viruses may be affected by other parasites which affect cells of the reticuloendothelial system. Canaries experimentally infected with malaria, for example, had a lower titer of Western equine encephalitis virus in their blood than control birds, or birds with a latent infection with *Plasmodium relictum* (Barnett, 1956). On the other hand, Gledhill (1956) found that the pathogenicity of mouse hepatitis virus was greatly enhanced by concurrent infection of mice with a normally harmless blood parasite, *Eperythrozoon coccoides*. This strain of mouse hepatitis virus normally caused an inapparent infection; mice concurrently infected with *E. coccoides*, or with Friend or Moloney leukemia viruses (Gledhill, 1961), suffered from severe hepatitis. Gledhill *et al.* (1965) have suggested that these effects may operate via the phagocytic cells of the liver, for *E. coccoides* and the leukemia viruses increase the phagocytic index, as measured by the clearance of carbon particles. This may in some way increase the susceptibility of the Kupffer cells to mouse hepatitis virus so that it multiplies readily in these cells and spreads more readily to the parenchymal cells of the liver. However, since stilbestrol enhances phagocytosis but has no effect on the pathogenicity of mouse hepatitis virus, increased phagocytosis alone does not account for the changed activity of the virus.

Increased resistance to the Mengo strain of encephalomyocarditis (EMC) virus was found in mice infected with the protozoa *Toxoplasma gondii* and *Besnoitia jellisoni* (Remington and Merigan, 1969). The antiviral effect did not appear to be related to interferon production by the protozoa, and is perhaps associated with a change in the reactivity of macrophages, because Freund's complete adjuvant and BCG, the vaccine strain of tubercle bacilli, have similar effects on EMC virus infection.

## THE INFLUENCE OF AGE ON RESISTANCE TO VIRAL INFECTIONS

It is a commonplace observation in clinical medicine and in experimental virology that the response to many viruses differs greatly at different ages, the changes being particularly dramatic in the neonatal period. Both immunological and nonspecific factors are involved in these changing patterns. The significance of congenital infections is discussed in Chapters 9, 10, and 12; here we shall consider some statistical data on case mortality rates in man and some experimental data obtained in studies with viral infections of laboratory animals.

### Vital Statistics in Human Viral Infections

Most statistics on morbidity in human viral infections are difficult to interpret because of differences in the immune state of the population at various ages and variations in the opportunity for infection, but age-specific case–mortality rates give some idea of the severity of various diseases at different ages. Considering only postnatal life, and excluding the effects of antibody acquired passively from the mother, viral infections tend to be very severe in the perinatal period, moderately severe in infancy, mild during childhood, and severe in the aged.

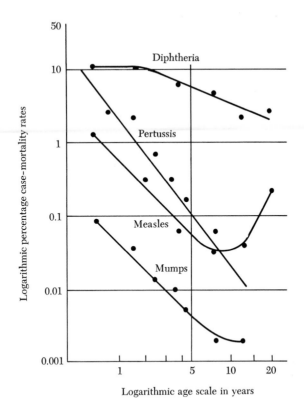

FIG. 11–2. *The effects of age on the case–mortality rates of some bacterial and viral infections of man, plotted on logarithmic scales. (From Burnet, 1952.)*

Burnet (1952) has plotted the case–mortality rates observed in a number of different infectious diseases of man on a partially logarithmic time scale which allocates the same space to infancy and childhood to the age of 12 years as to the rest of life. Some of his data are illustrated in Figs. 11–2 and 11–3. They bring out a fact we have mentioned elsewhere, namely that at the level of organism the similarities between viruses and other infectious agents are much more apparent than at the cellular level, presumably because the manifestations of infections are dominated by the host response. With most infections there is a steady fall in mortality from birth to childhood, which is probably due to the increasing effectiveness with which the homeostatic mechanisms of the body deal with the disturbances induced by the infectious agent. The rise in the mortality of infectious diseases acquired in old age is presumably due to the steadily progressing inadequacies of the homeostatic mechanisms (review, Makinodan et al., 1971).

The 1918–1919 epidemic of influenza showed a unique curve with a peak in young adult life (Fig. 11–3). This is not a general characteristic of influenza, for all previous and subsequent epidemics have the characteristic peaks of mortality in the very young and the very old. The reason for this "young adult" peak of

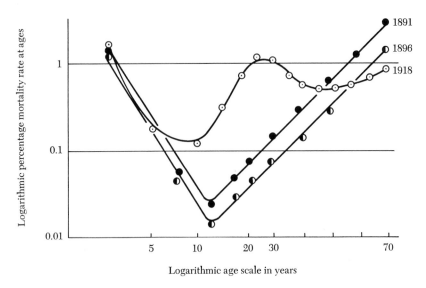

FIG. 11–3. *Mortality by age from respiratory infections in England and Wales in the influenza years 1891 and 1918 and a noninfluenza year 1896. Time scale for age changes from logarithmic to linear at 25 years. (From Burnet, 1952.)*

mortality is obscure. Burnet (1960) has suggested that it may have been due to the more intense local inflammatory reaction evoked in young adults; Francis and Maassab (1965) believe that it was a result of excess physiological and environmental stress contributing to pneumonia rather than a result of selective viral action.

## Effects of Age in Experimental Infections

The great susceptibility of newborn animals to many viral infections has been of considerable importance in laboratory studies of viruses. Thus the coxsackieviruses were discovered as a result of the use of suckling mice (Dalldorf and Sickles, 1948), and the newborn mouse is still the most sensitive host system available for the recovery of arboviruses, and still the only system for the growth in the laboratory of some serotypes of coxsackievirus A. Maternal transfer of antibody to common infectious agents derives its ecological importance from the lethality of such infections in newborn animals which are not thus protected. The ease with which oncogenic adenoviruses and papovaviruses produce tumors in suckling hamsters and other baby rodents but not in adult animals is a striking example of age-dependent susceptibility that is discussed further in Chapter 13.

In laboratory animals, the first few weeks of life are a period of very rapid physiological change. During this time, mice, for example, pass from a stage of immunological nonreactivity (to many antigens) to normal responsiveness. This change profoundly affects their reaction to viruses like lymphocytic choriomeningitis (which induces a tolerant state when inoculated into newborn mice,

Hotchin, 1962) and the oncogenic viruses. Newborn rabbits and chickens, also, may suffer from severe generalized disease when inoculated with "oncogenic" viruses (fibroma virus, Duran-Reynals, 1940b, 1945 and Allison, 1966; Rous sarcoma virus, Duran-Reynals, 1940a), whereas older animals develop tumors. There is good evidence that in both these cases the changed reactivity is due mainly to the maturation of the immune response. Prevention of generalization of fibroma virus is due primarily to humoral antibodies, the regression of established tumors to cell-mediated immunity (Allison, 1966). Since the former response develops sooner than the latter, inoculation of newborn rabbits with small doses of virus or inoculation after the first few days of life results in the development of large localized tumors.

There is some evidence that very young animals may produce less interferon than adults in response to certain inducers. Heineberg *et al.* (1964) could find little interferon in the tissues of infant mice infected with coxsackievirus B1, although adult mice, which were much less susceptible, developed substantial amounts of interferon. Sawicki (1961) found that the more susceptible suckling mice developed less interferon in the lungs than did adults at comparable intervals after the intranasal inoculation of parainfluenza type 1 virus. Most other investigators [Craighead (1966) with intranasal infections of mice with parainfluenza type 3 virus; Vilček (1964) with intracerebral inoculations of mice with Sindbis virus] have found that the more susceptible young mice produce as much interferon as older less susceptible animals. Although 2-week-old mice were much more susceptible to the lethal effects of infection with cowpox virus than were 6-week-old mice, Subrahmanyan (1968) found no age-dependent differences in the production of interferon after their inoculation by suitable routes with cowpox, West Nile, or influenza viruses. The balance of evidence suggests that differences in interferon production are not an important factor in age-dependent resistance to viral infection.

The experimental system in which the greatest efforts have been made to determine the operative mechanisms relates to the different susceptibility of adult and suckling mice to encephalitis after peripheral inoculation, both age-groups being susceptible to intracerebral inoculation. This difference has been observed with viruses belonging to many groups: herpesviruses, most of the togaviruses, vesicular stomatitis virus, coxsackieviruses, etc. As in all other systems, maturation of the immunological response may play a part (Overman and Kilham, 1953; Overman, 1954), but other factors are also important. The concept of the development of "barriers" to the ready access of virus from the periphery to the central nervous system (CNS) (Fig. 11–4) was proposed many years ago (Sabin and Olitsky, 1937a, b). In a more recent investigation of the problem, using herpes simplex virus in suckling and adult mice as the model system, Johnson (1964b) found differences in the release of virus by their macrophages. Peritoneal macrophages from newborn and adult mice, maintained *in vitro*, were equally susceptible to infection with herpes simplex virus, but with adult macrophages the virus failed to spread to other cells, whether these were HeLa cells or suckling or adult macrophages, whereas spreading infection always occurred when suckling macrophages were primarily infected. Further work

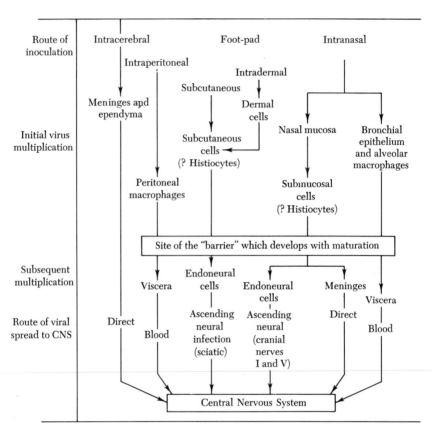

FIG. 11–4. *Diagram of the pathogenesis of herpesvirus encephalitis in mice. In young suckling mice, infection spread readily to the CNS; this spread was arrested with increasing age at the site shown by the barrier. When the barrier was overcome in young adult mice by massive intraperitoneal or intranasal inocula of virus, infection followed the same pathways to the CNS. (From Johnson, 1964b.)*

extended Johnson's observations (Hirsch *et al.*, 1970) and it was shown that adult macrophages produced more interferon than suckling mouse macrophages. Curiously, adult macrophages appeared insensitive to interferon from macrophages, although sensitive to other types of mouse interferon. The macrophages may well constitute one component of the "barrier" to spread to the CNS illustrated in Fig. 11–4. Spread from the blood to target organs such as brain or muscle is also facilitated in infant mice because viremia is prolonged as a result of slower clearance by reticuloendothelial cells. There are indications that virus-sized particles may pass more readily across capillary walls in suckling animals (Suter and Majno, 1965) and this also favors the dissemination of infection into susceptible organs.

In some other viral infections age-dependent susceptibility to extraneural inoculation may depend upon changes in the viral receptors. Group B coxsackieviruses infect suckling but not adult mouse brains and Kunin (1962) showed that cell fragments from suckling mouse brains adsorb these viruses more actively than those from adult mouse brains. Studies of age-dependent differences in cell susceptibility are difficult because the mere removal of cells from tissues for *in vitro* testing is known to lead to important changes in viral susceptibility.

## GENETIC RESISTANCE

It is a matter of common knowledge that different species of animal differ in their resistance to particular viral infections. A variety of mechanisms govern genetic resistance; some operate at the cellular level, others only at the level of the organism. Within a susceptible species the reactions of individual animals that have not been previously exposed may differ greatly. During an epidemic of poliomyelitis, for example, one child in a family may become paralyzed while one sibling develops nothing more than a headache and another escapes with an infection that is completely subclinical. We know little about the reasons for these differences. To some extent they are due to genetic differences between the animals, as can be revealed by breeding experiments, but if it could be repeatedly tested, a single animal would doubtless be found to differ from one time to another in its susceptibility to a standard infection, due to physiological changes of the kind we have just described, and to chance factors in the distribution of virus particles and their encounters with cells and antiviral substances.

### Resistance at the Cellular Level

The absence of suitable receptors on the plasma membrane is the most obvious reason for cellular resistance to animal viruses (see Chapters 5 and 6), but in some virus–cell systems attachment and penetration may occur without subsequent viral multiplication because the cell is intrinsically incapable of supporting the growth of that virus.

A few examples have been reported of cells that are restrictive for certain viruses or mutants (e.g., PK cells for some rabbitpox virus mutants, Fenner and Sambrook, 1966; and a variety of different cells for vesicular stomatitis host-restricted mutants, Simpson and Obijeski, 1973a; see Chapter 7). In these cases, entry and uncoating are normal but viral multiplication does not proceed to completion. The mechanisms are unknown. A possible reason for such cellular insusceptibility is what bacteriophage workers call "restriction" (review, Arber and Linn, 1969). In nonaccepting bacteria, the DNA of restricted bacteriophages is injected normally but is rapidly degraded to acid-soluble fragments by endonucleases. Virtually no information is available as to whether a comparable process of rapid degradation of the viral nucleic acid plays a part in determining cellular resistance to animal viruses.

Cellular susceptibility of the type shown by permissive (suppressor-positive, $su^+$) bacteria when they are infected with *amber* mutants of bacteriophage (Hayes, 1968) has yet to be demonstrated with animal viruses; neither $su+$ animal cells nor suppressible mutants of animal viruses have been recognized.

## Cellular Receptors

In a few instances, it has been shown that the susceptibility of intact animals, or some of their organs, is dependent upon the presence or absence of specific cellular receptors. Here organ specificity is a direct reflection of cellular specificity; the best examples are provided from studies with influenza, polioviruses, and the avian leukoviruses.

**Influenza.** Bacterial neuraminidase (receptor-destroying enzyme: RDE) will destroy the cellular receptors for influenza virus (see Chapter 6). Stone (1948) showed that pretreatment of mice with RDE, given intranasally, conferred substantial protection against subsequent intranasal challenge with influenza virus. The protection was quite short-lived (1–2 days) because regeneration of receptors occurred rapidly. Fazekas de St. Groth (1948b) investigated the regeneration of receptors by giving RDE intranasally and investigating the subsequent adsorption-elution behavior of influenza virus in the *in vitro* mouse lung preparations. Receptors were progressively reduced by a single effective dose of RDE for about 8 hours; a subsequent increase in receptors starting at 30 hours reached the normal level in 6 days. The capacity of the cells to regenerate receptors was not exhausted by intensive RDE treatment, for regeneration was more rapid after 14 days' consecutive treatment with RDE than after a single effective dose. A moderate degree of protection against intracerebral infection with the influenza strain NWS, which multiplies in the neurons (Mims, 1960a), followed prior intracerebral inoculation of large amounts of RDE (Cairns, 1951).

**Poliovirus.** Evidence for the importance of cellular receptors in determining susceptibility of cultured cells to enterovirus infection is summarized in Chapter 6. Susceptibility to poliovirus also affords interesting examples of the importance of cellular receptors in the intact animal. Mice are insusceptible to infection with most strains of poliovirus. If, however, they are injected intracerebrally with poliovirus nucleic acid (Holland *et al.*, 1959), or with poliovirus genomes enclosed within the capsids of a mouse-pathogenic enterovirus (e.g., coxsackievirus B1, Cords and Holland, 1964b), the cells are infected with poliovirus genetic material and support a single cycle of multiplication. Since the progeny particles have the surface properties of normal poliovirions they are unable to attach to and produce infection of other cells.

**Avian and Murine Leukoviruses.** The resistance of some lines of chickens to infection with certain strains of avian leukovirus is a direct consequence of the genetically determined absence of particular receptor substances in their cells; both the intact organism and cultured cells derived from it fail to adsorb the virions. Prince (1958a, b) proposed that a single pair of autosomal genes controlled susceptibility of the chorioallantoic membrane to Rous sarcoma virus, with susceptibility dominant to resistance. Bower *et al.* (1965) confirmed this, and Crittenden *et al.* (1964) showed that the same situation existed when tests were made for tumor production in 1-day-old chicks and focus or pock production in cultured cells or on the chorioallantoic membrane, respectively. Other work (Rubin, 1965; Vogt and Ishizaki, 1965) has shown that cellular susceptibility to the avian leukoviruses depends upon the presence or absence of the appropriate cellular receptors, and the avian leukoviruses have been grouped into five antigenic subtypes (A, B, C, D, and E) on the basis of the affinity of their envelope proteins for the receptors on various lines of chicken cells. A gene (now called "tumor virus" *tv; a, b, c, d,* or *e*) responsible for the production of cellular receptor substance could render cells susceptible regardless of whether it was present in one or two alleles, so that susceptibility and not resistance would be expected to be dominant (group B of Fig. 11–5), as was found in the genetic experiments. Expression of the *tv* gene is controlled by a dominant epistatic "inhibitor" gene "I," which may ebbe related to the *gs* gene that determines whether cells produce viral antigen (Payne *et al.,* 1971).

There is a complex genetic control of murine viral leukemogenesis (reviews, Lilly, 1972; Lilly and Pincus, 1973). In the first place, the growth of virus in cells may be genetically controlled, as in the case of the FV-1 gene controlling N (NIH strain) or B (BALB/c) type susceptibility of mouse embryo fibroblasts to leukemia virus infection (Pincus *et al.,* 1971). Factors such as the susceptibility of cells to transformation, host resistance to the transformed cells, and the behavior of possible helper viruses may also be under genetic control. Finally, there is now convincing evidence (see Chapter 14) that the genome of murine leukemia viruses is integrated into the genome of the host cell and transmitted vertically in ova and sperm, the infectious agent thus constituting one or perhaps several separate "host" genetic factors. At present little is known of this fascinating borderland between infection and heredity.

## In Vivo **and** In Vitro **Susceptibility**

It is a commonplace observation that cultured cells have a spectrum of sensitivity to viral infection much broader than that of the organ or the species of animal from which they came. Early notions of "tissue tropisms" have lost much of their validity today. An organ or tissue may appear refractory to infection in the animal merely because the virus fails to gain access to it, or because its cells fail to adsorb the virions. Cells from the same organ may prove quite susceptible when tested *in vitro.* With poliovirus, for example, Evans *et al.* (1954) showed that although monkey testicular cells were susceptible to infection with poliovirus when cultured *in vitro,* the same cells were refractory to infection *in vivo,* even when the tissues were minced *in situ* before infection. Searching for an explanation, Holland (1961) has studied a particularly suitable model system, the human amnion. Cell cultures prepared from the amnion multiply little or not at all; they are essentially preparations of surviving cells (Lehmann-Grube, 1961), and strips of the intact membrane can be maintained for long periods *in vitro.* Freshly trypsinized amnion cells, like the amnion from which they were derived, were not susceptible to infection with poliovirus and failed to adsorb it. After maintenance in culture for 7 days, the dispersed cells became susceptible to infection, in parallel with the development of the capacity of cell homogenates to adsorb virus. Strips of amnion, however, remained insusceptible to infection and failed to develop receptor substance. Gresser *et al.* (1965) reinvestigated this system, and found that amniotic membranes in organ culture supported prolonged persistent infection with poliovirus, but only about 1% of the cells were infected. Unlike the superficially similar situation in carrier cultures, this was not due to interferon. They suggested that the increased infection rate that followed cultivation of trypsinized amnion cells might be due to their increased phagocytic activity (and hence increased uptake of virus), but pointed out that this interpretation is not necessarily incompatible with Holland's idea that surface receptors for poliovirus developed after cultivation. It is possible that the susceptibility of the dispersed amnion cells depends not upon the development of receptor sites, but upon their exposure, because of the removal or disappearance of an intercellular matrix substance (Chany *et al.*, 1966).

There is now much evidence to suggest that some glycoproteins in the plasma membrane may be masked by others that can be readily removed by enzymatic treatment (Burger, 1971a; see Chapter 13). Such glycoproteins may act as viral receptors; their expression on the cell surface depends upon whether the cell has been transformed, or dedifferentiated, as often happens in cell culture.

## Genetic Resistance due to the Immune Response

All strains of mice are susceptible to infection with polyoma virus, both as newborn and adult animals. Normal adult mice of all strains are resistant to the oncogenic effects of polyoma virus, but in most strains inoculation of newborn animals is followed by the development of neoplastic growths. C57Bl and A strains of mice are resistant to the oncogenic effects of polyoma virus introduced at birth, and the evidence indicates that resistance is at the level of the organism and is probably associated with the immunological reactivity of these mice. Specific immune responses in mice and guinea pigs are known to be under genetic control (reviews, McDevitt and Benacerraf, 1969; Ríhová-Skarova & Ríha, 1972).

Schell (1960a, b) showed that the resistance of C57Bl mice to footpad inoculation with ectromelia virus was correlated with their prompt immune response. He found no differences in the susceptibility of cultured mouse embryo fibroblasts from susceptible and resistant strains of mice, and Roberts (1964) could detect no differences in the responses of their peritoneal macrophages to infection with ectromelia virus.

Tosolini and Mims (1971) investigated the resistance of C57Bl mice to the lethal action of lymphocytic choriomeningitis (LCM) virus. In this infection, the pathological changes are the result of a host immune response to infection and the resistance of C57Bl mice was associated with a weak cell-mediated immune response to LCM virus infection, as compared with susceptible strains of mice.

## Resistance among Inbred Mice

By selective breeding, Webster (1937) developed strains of mice which differed greatly in resistance to two flaviviruses, St. Louis encephalitis and louping-ill viruses. He showed that resistance to these viral infections was inherited independently of resistance to bacterial infections, and that inherent susceptibility or resistance was correlated with the level of viral multiplication, both in the brain of the intact animal (Webster and Clow, 1936) and in cultures containing minced brain tissue (Webster and Johnson, 1941).

In a few cases, it has been possible to study the genetics of resistance to viral infection. Figure 11–5 illustrates what may be expected on three hypotheses of the inheritance of resistance: (A) a single pair of alleles with resistance dominant makes an important contribution, (B) a single pair of alleles with susceptibility dominant makes an important contribution, and (C) many genes affect susceptibility. Tests made with $F_1$ hybrids and with backcrosses to the parental strains showed that the resistance of PRI mice and the susceptibility of C3H and BSVS mice to flaviviruses was under single gene control with resistance dominant (group A of Fig. 11–5) (Sabin, 1954). The demonstration that virus multiplied freely in foreign (non-PRI) tumor cells implanted in the peritoneum of resistant mice, but not in the brains of the same animals, suggested that the immune response was not responsible. In further investigations of the same system,

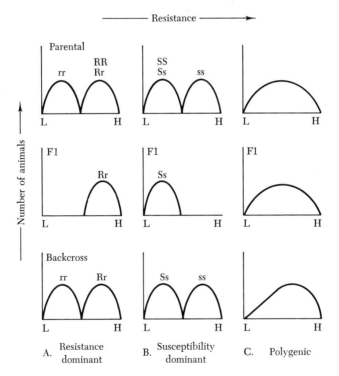

FIG. 11–5. Diagram illustrating the expected modes of inheritance of resistance against an infection on three hypotheses: (A) A single pair of alleles with resistance dominant makes an important contribution. (B) A single pair of alleles with susceptibility dominant makes an important contribution. (C) Many genes affect susceptibility. Upper curves: Parental generations, with resistance plotted on the abscissa in arbitrary units (L, low resistance; H, high resistance). Middle curves: The resistance of $F_1$ progeny of matings of highly susceptible and highly resistant animals. Lower curves: The resistance of back-crosses of $F_1$ with suitable parental types. (Modified from Allison, 1965.)

Goodman and Koprowski (1962) confirmed Sabin's findings and made further observations on the mechanism. Resistance is not due to an enhanced ability to produce interferon (Vainio *et al.*, 1961) or to more rapid or more effective production of antibody. Vainio (1963a, b) further showed that the yield of West Nile virus (a flavivirus) in cultures of splenic macrophages from susceptible mouse strains was 100–1000 times greater than in cultures from resistant strains. Differences in viral yield of lesser and varying degree was found in cultures of cells from brain, lung, and kidney. Interferon was implicated in greater resistance of C3HRV mice than C3H mice to flavivirus infection by Hansen *et al.* (1969), who found that cells of the C3HRV strain produced the same amount of interferon but were more susceptible to its action. Production of interferon has been shown to be under genetic control in mice (de Maeyer and de Maeyer-Guignard, 1969).

Breeding experiments with mice and the coronavirus, mouse hepatitis virus, provide an example of (B), a single pair of alleles with susceptibility dominant. Bang and Warwick (1960) found that PRI mice (which were resistant to flavivirus infections) were susceptible to infection with mouse hepatitis virus, whereas C3H mice were resistant. By cultivating the liver tissue in such a way that both the hepatic cells and macrophages survived, they found that in cultures made from livers of newborn PRI mice the macrophages were rapidly destroyed by mouse hepatitis virus, whereas the epithelial and fibroblastic cells were not affected. The cells (including macrophages) produced from explants of livers of resistant C3H mice were unaffected by the virus. Tests of hybrids and of their cells *in vitro* showed that susceptibility was inherited, and that segregation occurred in the backcross as predicted in Fig. 11–5, B, and also in the F2 generation. Peritoneal macrophages exhibited the same characteristics as the macrophages from cultured liver (Kantoch *et al.*, 1963). Shif and Bang (1966) used a plaque assay of MHV in peritoneal macrophages to show that macrophages from susceptible backcross C3H mice were just as susceptible as those from PRI mice. Further, virus adsorbed to and penetrated both susceptible and resistant macrophages. In the latter, it persisted for several days without being uncoated but was eventually degraded, whereas in susceptible cells eclipse and viral multiplication occurred (Shif and Bang, 1970). These results suggest that resistance to a viral disease in the intact animal may be under simple genetic control which operates by affecting the ease of infection or destruction of certain key cell types. However, Allison (1965) found that mice of the VSBS strains, of which normal weanlings and adults are genetically resistant to a virulent strain of mouse hepatitis virus, became highly susceptible after neonatal thymectomy, i.e., both the resistance gene and an intact immune response were required for mice to survive infection.

Although it has not yet been worked out completely, the genetic resistance of C57Bl mice to polyoma virus-induced tumors appears to be an example of incompletely dominant resistance depending on two or three independent genes (category C above), all of which are required for the expression of resistance (Jahkola, 1965). In both polyoma and mousepox the resistance of C57Bl mice operates only at the level of organism, not at the level of either cell or organ.

## Natural Selection for Resistance

As noted above, Webster (1937) was able to obtain mice highly resistant to certain viral infections by selective breeding, and his experience has been widely confirmed with other pathogens and with other species of animal. It is generally believed that natural selection has operated in the past in a somewhat similar way with the more lethal infections of man and animals, most diseases as we know them now being the relatively benign end result of a long period of mutual adjustment of host and parasite. A natural experiment illustrating this, myxomatosis in the Australian rabbit, is discussed in Chapter 17.

## The Host Range of Viruses

In books which deal systematically with viral infections of vertebrates (Andrewes and Pereira, 1972; Horsfall and Tamm, 1965) there usually appears a section entitled "Host Range" in the descriptions of each virus or group of viruses, in which are listed the susceptible host animals. Such lists are correct only in a positive sense, for very few viruses have been adequately tested in a large number of individual animals, over a wide range of species, and by a variety of routes of inoculation. In addition, with few exceptions the only response studied has been the production of acute or sometimes late signs of disease.

Myxomatosis in Europe provides a good example of the importance of studying a large number of animals of a particular species before deciding that the species is insusceptible. In laboratory tests, the hare (*Lepus europaeus*) appeared to be completely resistant to infection with myxoma virus (Bull and Dickinson, 1937), but when myxomatosis occurred on a continental scale in Europe after 1952 a number of cases of severe myxomatosis were reported in hares (Fenner and Ratcliffe, 1965). The initial recovery by Smith *et al.* (1933) of influenza virus in ferrets (which happened to be available then in Hampstead because they were being used for the study of canine distemper) further illustrates the point, and the demonstration by Deinhardt *et al.* (1967) and A. Holmes (1971) and London *et al.* (1972) that human hepatitis viruses can be propagated in marmosets and rhesus monkeys are more recent examples.

# INTRODUCTION

The thinking of clinicians and virologists about viral diseases has long been dominated by acute febrile diseases like smallpox, measles, poliomyelitis, and influenza. In such diseases, as we have seen in the preceding chapters, the course of the infection may be summarized as follows. The causative virus enters the body, multiplies in one or more tissues, and spreads either locally or through the blood stream. When viral multiplication has reached a critical level, after an incubation period of 2 days to 2 or 3 weeks, symptoms of disease appear, associated with localized or widespread tissue damage. Nonspecific and specific host defenses are mobilized during the incubation period, and unless the disease is fatal the host has usually eliminated the infecting agent within 2 or 3 weeks of the onset of symptoms. Virus can ordinarily be isolated from the blood or secretions only in the short period just before and just after the appearance of symptoms. Some viruses (e.g., measles and smallpox in man) almost always cause acute disease; many others produce acute *infections* in which the pathogenetic mechanisms are similar but often there is no clinical *disease*, i.e., the infection is subclinical (Table 10–4 of Fenner and White, 1970).

Quite distinct from the acute infections, however, are those in which virus persists for months or years, i.e., *persistent viral infections* (review, Hotchin, 1971). As we shall see, persistent infections are associated with a great variety of pathogenetic mechanisms and clinical manifestations, and it is difficult to classify them satisfactorily. For convenience, we shall subdivide the persistent infections of intact animals into three categories, recognizing that there is some overlap:

1. Persistent infections with intermittent acute episodes of disease between which virus is usually not demonstrable: *latent infections* (see Table 12–1).

2. Persistent infections in which virus is always demonstrable and often shed, but disease is either absent, or is associated with immunopathological disturbances: *chronic infections* (see Table 12–2).

3. Persistent infections with a long incubation period followed by slowly progressive disease that is usually lethal: *slow infections* (see Table 12–3).

The key distinctions between these three groups of persistent infections are illustrated diagrammatically in Fig. 12–1. In slow infections, the concentration of

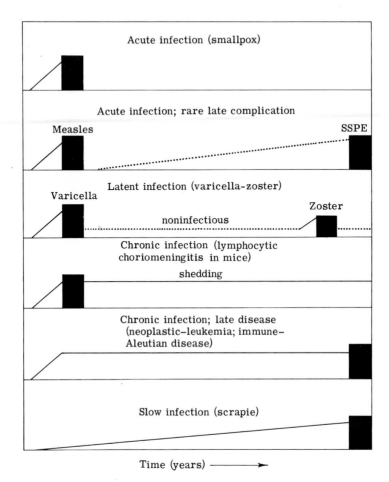

FIG. 12–1. *Diagram illustrating acute and various kinds of persistent infections. Solid line, demonstrable infectious virus; dotted line, virus not readily demonstrable; box, disease episode.*

virus in the body builds up gradually over a prolonged period until disease finally becomes manifest. Chronic infections, on the other hand, can be regarded as acute infections (clinical or subclinical) following which the host fails to reject the virus; sometimes disease supervenes late in life as a result of an immuno-pathological or neoplastic complication. The distinction between chronic and latent infections may be merely a matter of the ease of demonstration of infectious virus, or it may have a more fundamental basis, e.g., with herpesviruses between recrudescences of endogenous disease.

Before describing persistent infections of animals it will be useful to review what is known of persistent infections of cultured cells.

## PERSISTENT INFECTIONS OF CULTURED CELLS

One of the unexpected discoveries that emerged from the widespread use of cultured animal cells was the common occurrence of persistent infection of such cells by viruses that gave no overt sign of their presence (review, Smith, 1970). Historically, interest in these persistent infections was quickened by two observations: the discovery of adenoviruses in explanted adenoids and the revelation that monkey kidney cell cultures used for Salk poliovirus vaccine production were often contaminated with viruses of simian origin.

Adenoviruses were originally discovered when the cell monolayers that grew out of explanted fragments of "normal" human adenoids underwent virus-induced degeneration *in vitro* (Rowe *et al.*, 1958). The same adenoids failed to yield virus when homogenates were inoculated directly onto monolayers of susceptible cells. We now realize that adenoviruses persist for years in lymphoid tissues, and may be recovered from about one-third of all tonsils and adenoids removed during the first decade of life (Israel, 1962). The nature of the association between adenovirus and cell remains obscure; the two most likely possibilities are its persistence in an integrated state, or the continued presence and slow multiplication of very small amounts of infectious virus. The latter explanation is favored by the work of Strohl and Schlesinger (1965), who showed by infectious center assays of naturally infected adenoid tissues that less than one cell in $10^7$ was infected, although at least 0.5% of the cells were susceptible. It appears that infectious virus multiplies continuously but slowly in these infected tonsils and adenoids. Antibody and interferon may play a role in limiting the rate of spread from cell to cell.

Another situation in which cultured cells were found unexpectedly to yield virus derived from the extensive use of monkey kidney cells for the isolation of viruses and the manufacture of vaccines. Over fifty different simian viruses have been recovered from "normal" monkey kidney cell cultures. Some of these establish weakly cytocidal or steady-state associations with the cells, but often the simian viruses are accidental blood-borne contaminants of the kidneys (reviews by Hull, 1968; Hsiung, 1968; Kalter and Heberling, 1971), kept in check by antibody bathing the tissue *in vivo*, but producing cytopathic effects when the cells are washed and cultured. Similar problems have plagued Gibbs and Gajdusek in their attempts to grow kuru virus in long-term cultures of tissues from chimpanzees. Viruses recovered from such tissues included adenoviruses, reoviruses, and leukoviruses, as well as other agents not yet identified (Gajdusek *et al.*, 1969; Hooks *et al.*, 1972a, b).

In most cases, the relation of these viruses to the cells which carry them is unknown; they have been said to be "occult," a term that "appropriately surrounds our state of ignorance with an aura of mysticism." However, careful studies of a variety of types of persistent infections in carefullly controlled *in vitro* systems are beginning to shed some light on the nature of the virus–cell interaction.

The following information needs to be obtained by experiment before one can determine what is happening in persistently infected cultured cells (modified from Walker, 1968).

1. Are all cells infected and continuously producing virus, or are only a minority involved?

2. Do infected cells divide, or do they die?

3. Can the culture be "cured" of infection by cloning cells in the presence of antiviral antibody?

4. Can the balance be altered toward cell destruction by washing inhibitory substances from the medium?

Basically, the great variety of virus–cell interactions that have been described as persistent infections of cultured cells can be allocated to one of three categories:

1. The viral genome is integrated with that of all cells in the culture (tumor viruses).

2. All cells continuously produce a noncytocidal virus (*steady-state infection*).

3. A cytocidal virus is kept in check by inhibitors in the medium or by the presence in the culture of a minority of susceptible and a majority of genetically resistant cells (*carrier culture*).

Tumor viruses are described in Chapters 13 and 14. The other two situations will be discussed below.

## Steady-State Infections

Steady-state infections are not uncommon in cells infected with RNA viruses that mature by budding from the plasma membrane. In such cells, large amounts of infectious virus may be continuously released from cells whose metabolism and multiplication are scarcely affected. Such persistently infected cells can be superinfected with other viruses without noticeably affecting the multiplication of either virus. Alternatively, the yield of either the superinfecting virus or the noncytocidal virus may be enhanced (complementation) or reduced (interference) (see Chapters 7 and 8).

A typical example is the paramyxovirus SV5, a common contaminant of primary cultures of rhesus monkey kidney cells, in which it multiplies to high titer with little cytopathic effect. The infected cells survive and multiply, and produce large amounts of infectious virus for many days (Choppin, 1964). Cellular DNA, RNA, and protein synthesis are hardly affected by the virus (Holmes and Choppin, 1966); viral RNA synthesis amounts to less than 1% of the cellular RNA synthesis. Electron microscopic observations reveal little cellular damage (Compans et al., 1966), despite the fact that large numbers of virions are maturing at the plasma membrane. Infection of monkey kidney cells with SV5 does not interfere with the growth of vaccinia virus, influenza virus, vesicular stomatitis virus, or three different picornaviruses (Choppin, 1964).

Similar steady-state infections have been described with most paramyxoviruses and with several togaviruses, rhabdoviruses, and leukoviruses, which are likewise relatively noncytocidal viruses that do not shut down the metabolism of the cell and are released by budding from the plasma membrane. Most of the situations described in the literature share the following common features: (a) most or all cells in the culture are infected, (b) virus is released continuously but at a slow rate, and (c) the cultures cannot be "cured" by antibody.

Carrier cultures of rubella virus can be readily established from virtually any organ of a fetus with "rubella syndrome." Most of the cells produce virus continuously; they are not killed but multiply more slowly than normal. The cultures do not produce interferon and cannot be "cured" by antibody (Rawls and Melnick, 1966; review by Rawls, 1968). Almost exactly the same properties characterize the carrier cultures that result from experimental infection of the LLC MK2 cell line with rubella virus, except that interferon is detectable.

The avian leukosis viruses and Rous sarcoma virus produce noncytocidal steady-state infections in chick embryo fibroblasts. The avian leukosis viruses fail to affect cellular morphology and behavior, in spite of the continued release by budding from the plasma membrane of large numbers of virions for a prolonged period, but Rous sarcoma virus induces cellular transformation. Depending upon the strain of Rous sarcoma virus and the presence or absence of concurrent infection of the cells by a "helper" avian leukosis virus, Rous sarcoma cells may or may not yield infectious Rous sarcoma virus (see Chapter 14).

Productive (i.e., virus-producing) infections of cells by DNA viruses are almost all cytocidal. Unlike the enveloped RNA viruses, which bud from the plasma membrane of infected cells without necessarily destroying them, DNA virus production seems, in general, to be incompatible with cell survival, although with some DNA viruses nonproductive persistent infections, in which the viral DNA is integrated into the cell genome, are readily produced (see Chapter 13). Two examples have been described in which productive infection by a DNA virus appears to have a minimal cytocidal effect, e.g., fibroma virus in rabbit kidney cells and SV40 in rhesus monkey kidney cells.

Most naturally occurring strains of Shope fibroma virus are relatively noncytocidal when grown in serially cultured rabbit kidney cells (Hinze and Walker, 1971). Infected cells alter in morphology, lose contact inhibition, and multiply at a slightly slower rate than control cells until further increase is limited by overcrowding (Hinze and Walker, 1964). Cellular and viral DNA synthesis alternate. Experiments with $\gamma$-irradiated cultures showed that infection with fibroma virus led to alteration of cell morphology and loss of contact inhibition in the absence of cell division (Tompkins et al., 1969).

Sweet and Hilleman (1960) reported that simian virus 40 (SV40) multiplied to high titer in cells derived from rhesus or cynomolgus monkey kidneys without producing any cytopathic effect. Its presence in these cells, which had been widely used for the cultivation of many viruses for several years, was recognized only when supernatant fluids from such cultures were tested in African green monkey kidney cells, in which a characteristic cytopathic change (vacuolation of the cytoplasm) was produced. More detailed study of SV40-infected rhesus monkey kidney cells (Easton, 1964) showed that the infection was in fact cytocidal, but it progressed very slowly and obvious cytopathic changes were restricted to the nucleus. Rhesus monkey cells infected with SV40 can be superinfected with several other viruses. Interference does not occur; indeed, growth of adenoviruses is greatly enhanced (see Chapter 7). Stocks of several viruses grown in these persistently infected cells (e.g., adenoviruses, poliovirus vaccine strains) became contaminated with SV40. Largely as a result of these experiences monkey kidney cell cultures have been abandoned for vaccine production.

## Carrier Cultures

Persistent viral infections are often established in cells maintained by serial culture in the laboratory, after they have been deliberately infected with any of a variety of viruses. These *carrier cultures* (reviews, Walker, 1964, 1968) differ from the steady-state infections that we have just described in that virus-free cells can always be recovered by cloning from a carrier culture, whereas in steady-state infections all the cells are virus producers. Three types of carrier culture can be distinguished: (a) most of the cells in the culture are genetically resistant to the carried virus but the infection is perpetuated in a minority of susceptible cells (e.g., coxsackievirus A9 in HeLa cells, Takemoto and Habel, 1959), (b) the cells are genetically susceptible to the carried virus but the transfer of virus from cell to cell is limited by antiviral factors (usually antibody) in the medium (e.g., herpes simplex virus in HeLa cells, Wheeler, 1960), or (c) the cells are genetically susceptible, but most cells are made temporarily refractory by interfering factors (e.g., interferon) produced within the carrier culture (e.g., Newcastle disease virus in L cells, Henle, 1963). Often more than one mechanism of cell protection operates and sometimes the situation undergoes an evolving series of cell–virus relationships. For example, Thacore and Youngner (1969) described a situation in which a small plaque mutant of Newcastle disease virus was selected that induced more interferon than the wild-type virus. It was produced by only a small proportion of the cells, but the number of producer cells increased dramatically if the medium was changed more often.

Although the investigation of carrier cultures has demonstrated that there are a number of ways that an infecting virus can persist in a population of cells and remain relatively inconspicuous, these results have not yet been effectively applied to understanding latent infections in the intact animal.

## PERSISTENT INFECTIONS IN INTACT ANIMALS

Viruses that produce persistent infections in cultured cells may also produce persistent infections in the intact animal. As in cultured cells, such infections may be accompanied by the production of infectious virus or they may be attributable to the integration of viral nucleic acid into the cellular DNA. In addition, there are a few viruses (e.g., herpes simplex virus) that produce cytocidal infections in most cultured cell systems but persistent infections *in vivo*.

As outlined in the Introduction, we have grouped persistent infections into three categories which for convenience can be called *latent, chronic,* and *slow* infections. With the help of appropriate tables we shall first describe illustrative examples of infections thus classified, and then consider the pathogenetic mechanisms that allow such viruses to persist in the infected host.

## LATENT INFECTIONS

We use this expression for a small but important group of herpesvirus infections, of which the best understood are herpes simplex and varicella-zoster in man. These are characterized by typical acute primary infection, followed by

TABLE 12–1

Latent Infections: Virus Occult between Acute Episodes of Disease

| EXAMPLES | SYMPTOMS | SITES OF INFECTION | | VIRUS SHEDDING | ANTIBODIES |
|---|---|---|---|---|---|
| | | BETWEEN DISEASE EPISODES | DURING ACUTE ATTACK | | |
| Herpes simplex | 1. Primary stomatitis 2. Recurrent fever blisters | Occult, in cerebral or dorsal root ganglion cells | 1. Epithelial cells 2. Schwann cells, then cells of corresponding dermatome | Sporadically in saliva between attacks. Plentiful in recurrent fever blisters | + |
| Varicella-zoster | 1. Generalized primary infection: varicella 2. Late recurrent skin eruption: herpes zoster | Occult in cerebral or dorsal root ganglion cells | 1. Generalized with rash 2. Schwann cells, then cells of corresponding dematome | 1. Varicella: from throat and skin lesions 2. Zoster: from skin lesions. Contacts contract varicella | + |

apparent disappearance of the virus. Yet months or years later, acute disease recurs, perhaps more than once, and infectious virus is then readily demonstrable. There is ample evidence that the recurrent episodes of acute disease do not result from exogenous reinfection, although this can rarely occur (zoster, Thomas and Robertson, 1971; herpes simplex, Nahmias, 1971a), but from endogenous exacerbation of infections that have lain dormant for many years.

## Herpes Simplex

First infections with herpes simplex virus are usually manifest as an acute stomatitis contracted in early childhood (Burnet and Williams, 1939), although clinical illness probably occurs in only 10–15% of primary infections. Typically, at intervals of months or years after recovery from the primary infection, blisters from which the virus can be recovered appear, usually around the lips and nose. A variety of stimuli can trigger this recurrent viral activity, e.g., exposure to ultraviolet light, fever, menstruation, nerve injury, or emotional disturbances. Patients who have completely recovered from stomatitis may excrete virus intermittently in their saliva for several weeks, and recovery of virus from the saliva has been reported in 2.5% of asymptomatic adults at any given time (Buddingh et al., 1953), suggesting low-grade chronic multiplication and periodic release of infectious virus. However, the site in which the virus is multiplying is by no means clear. Kaufman et al. (1968) report intermittent secretion of virus from the conjunctiva, lacrimal, and salivary glands in patients subject to recurrent herpetic keratitis. Douglas and Couch (1970), on the other hand, detected no virus in parotid gland secretions but isolated virus from 2% of all specimens of oral secretions taken from known carriers between attacks of herpes labialis. There was no relationship between antibody levels and the frequency of either shedding or recurrent attacks.

Apart from such observations, the virus is not demonstrable between recurrent attacks of fever blisters. There are no data available to indicate with certainty either the site or the form in which it persists, but there is strong presumptive evidence implicating the nerve cells of the trigeminal nerve ganglion (Paine, 1964; Kibrick and Gooding, 1965). Attempts to recover herpes simplex virus by cultivation for as long as 5 months of skin biopsies taken from the sites of recurrent herpes lesions, between the attacks, were negative (Rustigian et al., 1966). Further, Stalder and Zurukzoglu (1936) found that when facial skin of a patient with herpes simplex was transplanted to another part of the body, recurrent eruptions occurred at the site from which the skin was removed and not at the site to which it had been transplanted.

Direct evidence for the persistence of herpes simplex virus in sensory ganglion cells has now been provided in both mice (Stevens and Cook, 1971b) and rabbits (Stevens et al., 1972). In mice that survived the acute infection, virus could be demonstrated several months later in explant cultures of spinal ganglion cells, but not in other tissues. When antiviral antibody was included in the culture medium there was electron microscopic evidence of virions only in the neurons, but not in satellite cells. Herpes simplex infection of mice is not accompanied by recurrent disease, but rabbits infected in the eye regularly get recurrent ocular herpes. In such animals, homogenized trigeminal ganglia yielded virus during the acute infection, but not subsequently. However, explanted ganglia, but no other parts of the trigeminal tract or brain, yielded herpes simplex virus up to 8 months after infection.

## Herpes Zoster

This human disease is characterized by a vesicular rash that is usually limited to an area of skin and mucous membrane served by a single sensory ganglion (Head and Campbell, 1900). Hope-Simpson (1965) reviewed his experience of the epidemiology and clinical manifestations of zoster in a rural population, and presented the following picture. It occurs, at an average annual rate of 3 to 5 per thousand, only in persons who have already had varicella. Young people are seldom attacked, and both the frequency and the severity of the disease tend to increase with age. Second attacks are as common among those already attacked as first attacks are in the general population, and they occur more commonly than expected by chance in the same dermal segment as in the first attack. The distribution of the incidence of attacks of zoster among individual ganglia roughly resembles the distribution of the varicella rash, which occurs mainly on the trunk and face. During the period of active cutaneous zoster eruption there is an extensive infiltration of round cells in the sensory ganglia, focal hemorrhages occur, and nerve cell destruction follows. Zoster may occur "spontaneously," but is especially common as an accompaniment of lymphoproliferative disorders, immunosuppression, and x irradiation of the spine (Sokol and Firat, 1965; Rifkind, 1966; Spencer and Anderson, 1970).

Epidemiological evidence (Hope-Simpson, 1965) is supported by virological and serological studies (Weller and Coons, 1954; Taylor-Robinson, 1959) which show that zoster and varicella are two clinical manifestations of the activity of a single virus. There is also very strong evidence for the view that zoster is almost always due to reactivation of virus that has remained latent since an attack of varicella, perhaps many years earlier, and that the sites of the latent virus are the cells of one or more of the dorsal root and cranial nerve (sensory) ganglia.

## Pathogenesis of Recurrent Herpes Simplex and Herpes Zoster

It is likely that the same mechanism operates in the two natural diseases and the experimental model in rabbits. All data are consistent with the view that during a primary attack of herpes simplex or varicella virus moves to the ganglia along the sensory nerves, probably by spread in Schwann cells of the nerve sheath (Johnson, 1964a). In recurrent herpes simplex or zoster the virus moves down the sensory nerves again until it reaches the skin, where it proliferates and produces vesicles (Fig. 12–2).

The state of the virus between attacks is unknown, but it cannot be recovered from presumably infected tissues. In the experimental model in rabbits, infectivity develops after explantation, initially in the neurons of the appropriate ganglion. Herpesviruses can exist in an integrated state with the cellular genome (see Chapter 13) and may conceivably persist in ganglion cells in this state.

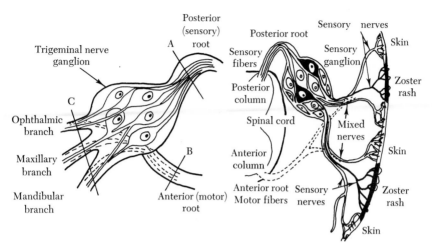

FIG. 12–2. Diagram illustrating the probable pathogenesis of recurrent herpes simplex (left) and herpes zoster (right). (Latter modified from Hope-Simpson, 1965.) Left, herpes simplex virus is presumed to be latent in the sensory nerve cells of the trigeminal nerve ganglion. Recurrent viral activity is triggered by fever, ultraviolet light, etc., and also by nerve injury. Section of posterior (sensory) root of trigeminal nerve (A) produces herpes simplex lesions in skin innervated by maxillary and mandibular branches of that nerve. Section of the motor root (B) or the branches (C) has no such effect. Right, varicella virus is presumed to become latent in sensory cells of the dorsal root ganglion. Upon activation the virus grows down the sensory nerve and infects the skin to produce the vesicles of herpes zoster.

However, it must be remembered that with other viruses integration probably does not occur without cellular DNA synthesis, which neurons do not exhibit, since they do not divide. Perhaps the viral genome persists as a repressed episome, or the virus simply remains for very long periods in a state of "suspended animation" in nondividing neurons.

## CHRONIC INFECTIONS

A wide variety of conditions fall within this category (Tables 12–2 and 12–3). We have already referred to the fact that many viruses have been recovered during the cultivation of tissues from "normal" animals. Often organs that yield viruses on explant culture do not yield infectious virus if ground up and inoculated into appropriate cultures or experimental animals.

Then there are a number of situations in which the features in common are the absence of disease but persistent or recurrent evidence of infectious virus, which may be shed or transferred by blood transfusion, so that infection of susceptible hosts can occur in the absence of apparent disease in the transmitters. It will be noted that herpes simplex virus could be correctly classified as a chronic infection as well as being classed as a latent infection, for the reasons already given.

TABLE 12-2

*Chronic Infections: Virus Always Demonstrable, Often No Disease*

| EXAMPLES | HOST | SITES OF INFECTION | VIRUS SHEDDING | ANTIBODIES | DISEASE |
|---|---|---|---|---|---|
| Various viruses: (SV40, reovirus, adenovirus, herpesvirus, paramyxoviruses) | Monkeys | Kidneys | Variable | + | None recognized. Viruses found in explanted cells |
| Cytomegalovirus | Many species | Salivary glands | Saliva | + | None recognized |
| Cytomegalovirus | Man | Salivary glands Kidney Circulating leukocytes | Saliva Urine | + | Rarely. Activated by immuno-suppression. Sometimes, transmitted by transfusion of fresh blood |
| EB virus | Man | Lymphoid tissue Circulating leukocytes | Nil | + | Usually none. Association with Burkitt's lymphoma |
| Hepatitis B | Man | Liver, serum | Feces (slight) Serum transmission | + | Rarely chronic hepatitis. Often transmitted by transfer of serum |
| Rubella | Man | Widespread | Urine Respiratory | + | Rubella syndrome → death or recovery. Teratogenic defects |

## Cytomegalovirus and EB Virus Infections

Generalized cytomegalovirus infection is occasionally seen in hospitals, and may result from two alternative sources of infection. In individuals with lymphoreticular disease, or undergoing prolonged immunosuppression, generalized cytomegalovirus infection may represent activation of an endogenous latent infection (Spencer and Anderson 1970). More commonly, perhaps, it is an exogenous infection resulting from the transmission of cytomegalovirus during the transfusion of large volumes of fresh blood (posttransfusion mononucleosis, Caul et al., 1971). Both situations reflect the widespread occurrence of healthy carriers of cytomegalovirus in the general population (Rifkind et al., 1967). There may be many similar unrecognized chronic infections in other animals.

The EB virus is another ubiquitous herpesvirus carried asymptomatically by large numbers of normal people. Henle et al. (1968) have presented persuasive evidence that it is the causative agent of infectious mononucleosis (reviews by Henle and Henle, 1972; Smith and Bausher, 1972). Two types of human cancer, Burkitt lymphoma and a nasopharyngeal carcinoma commonly found in the Chinese consistently carry several genome equivalents of EB virus DNA and synthesize EBV antigens but not infectious virus (zur Hausen et al., 1970). Lymphoblastoid cells from such patients, and indeed from many normal people, carry the viral genome in a repressed state which can be activated by drugs (Hamper et al., 1972).

## Serum Hepatitis (Hepatitis B)

Perhaps the best known chronic human infection is serum hepatitis, which is acquired by the transmission of virus in serum from a chronic carrrier. What is known of the etiological agent was described in Chapter 3, in which attention was drawn to the common occurrence of a large amount of a virus-associated antigen ("Australia" antigen or HB-Ag) in the serum. Following clinical or subclinical infection the antigen disappears quite rapidly from the serum of about one-half the cases, but in the remainder it may persist for several years. The concentration of antigen in the serum of carriers may be extremely high ($10^{12}$ to $10^{13}$ HB-Ag particles per milliliter), and although concentrations of infectious virions are very much lower (about $10^6$ infectious doses per milliliter) such sera may contain a substantial amount of infectious virus.

Sera from some acute or chronically ill individuals, but not from chronic asymptomatic carriers, contain antigen–antibody complexes (Almeida and Waterson, 1969). Immune complexes may play a part in the acute disease, perhaps in some forms of chronic hepatic disease, as well as in the glomerulonephritis and polyarteritis nodosa that are very occasionallly seen in HB-Ag carriers (Gocke et al., 1971).

## Rubella Syndrome

Babies born of mothers infected with rubella virus during the first trimester of pregnancy may suffer from congenital defects (see Chapter 9), and sometimes exhibit a wide range of signs and symptoms known collectively as the "rubella syndrome" (reviews, Rawls, 1968; Banatvala, 1970). Virus can be isolated from virtually any organ of such babies, although only a minority of cells in any given organ are infected (Rawls, 1968). Maternal antibody (IgG) may inhibit the spread of virus but fail to eliminate the virus-producing cells which continue to divide at a slower rate than normal. Usually the resulting clones of infected cells are lost about 6 months after birth, but occasionally in special sites they may persist for much longer.

There is no immunological tolerance, and large amounts of IgM are synthesized by the fetus and continue to be made after birth.

### Chronic Infections with Late Immunopathologic Disease

Finally, there are a number of chronic infections in which severe disease, of either immunopathological or neoplastic origin, appears as a late phenomenon (Table 12–3). Included are infections of mice that have long been regarded as the classic cases of persistent infections associated with immunological tolerance (see Chapter 10): lymphocytic choriomeningitis and lactic dehydrogenase virus infections.

**Lymphocytic Choriomeningitis Virus Infection of Mice (LCM).** After congenital infection (see Chapter 9), mice are born normal and appear to be normal for most of their lives, although they have persistent viremia and viruria; almost every cell in the mouse is infected with LCM virus and remains so throughout the life of the mouse (Mims and Subrahmanyan, 1966). Circulating free antibody cannot be detected, but immunological tolerance is not complete, and the virus circulates in the bloodstream as virus–IgG–complement complexes, which are infectious (Oldstone and Dixon, 1970). Late in life such mice may exhibit "late disease" due to the deposition of viral antigen–antibody complexes in the glomeruli.

There is an important genetic component in the response of mice to LCM virus infection. The strain of mice mainly used by Oldstone and Dixon produced antibody readily (even though it did not eliminate viremia) and suffered early and severe immune disease; the mice used by Mims and presumably by Traub (1939) in his early studies were at the other end of the spectrum. There is evidence (Oldstone et al., 1972) that the immune response to LCM virus and susceptibility to the immunopathological disease is associated with the H-2 locus.

Other arenaviruses also set up chronic infections in their natural hosts. For example, Machupo virus, following inoculation into the cricetine rodent *Calomys callosus* shortly after birth, establishes a chronic infection with viremia and viruria persisting for at least a year. As with LCM, neutralizing antibody is not detectable (Justines and Johnson, 1969).

**Lactic Dehydrogenase-Elevating Virus Infection (LDV).** This is widespread among laboratory mice (reviews, Notkins, 1965, 1971). The causative virus has not yet been classified; its properties are described in Chapter 3. LDV grows in mouse macrophages or mouse embryo cell cultures, without cytopathic effect. After inoculation into mice it multiplies rapidly, almost exclusively in macro-

TABLE 12-3

Chronic Infections: Virus Always Demonstrable, Late Immunopathological or Neoplastic Disease, Nonneutralizing Antibodies

| EXAMPLES | HOST | SITES OF INFECTION | NONNEUTRALIZING ANTIBODIES | DISEASE |
|---|---|---|---|---|
| Lymphocytic choriomeningitis | Mouse | Widespread, including lymphoid tissue | + | Glomerulonephritis |
| Lactic dehydrogenase virus | Mouse | Macrophages | + | Immune complexes in glomeruli, but no disease |
| Aleutian disease | Mink | Macrophages | ++ | Arteritis, glomerulonephritis, hyperglobulinemia |
| Equine infectious anemia | Horse | Macrophages | ? | Anemia, vasculitis, glomerulonephritis |
| Murine leukemia | Mouse | Widespread | + | Immune complexes in glomeruli. Leukemia occasionally |
| Avian leukosis | Chicken | Widespread | + | Leukemia occasionally. Sarcoma rarely |

phages, to achieve a high-titer viremia that persists for life. Early in infection there is impairment of the clearance of several enzymes from the blood (hence the name) and some derangement of immunological function, but later these changes diminish. Antibodies are produced but free circulating antibody is not found; the viremia is associated with circulating infectious virus–antibody complexes. Eventually immune complexes accumulate in the glomeruli and cause mild late glomerulonephritis, but, in general, LDV is a benign infection in spite of the prolonged high-titer viremia. Since many stocks of mice bear this inapparent infection, which nevertheless does produce physiological changes, it has been an unrecognized complicating factor in much experimental work, both on pathogenesis and as a contaminant of viruses that have been passaged through mice (Notkins, 1971).

**Autoimmune Hemolytic Disease and Membranous Glomerulonephritis in NZB Mice.** Although once thought to be "autoimmune" diseases, these conditions are probably late developments of congenital infection of a particular strain of mice with a murine leukemia virus (Mellors, 1969). An autoimmune hemolytic anemia is seen in NZB mice and in F1 (NZB × NZW) hybrids, both diseases probably being precipitated by the murine leukemia virus. The hybrids develop high titers of circulating anti-DNA antibodies, and deposition of antigen–antibody complexes in glomeruli leads to glomerulonephritis.

**Aleutian Disease of Mink.** This disease (reviews, Porter and Larsen, 1968; Porter, 1971a) affects mink and ferrets, but mink with the Aleutian coat color mutation are more severely affected than others, probably because the deposits of immune complexes in their glomeruli are larger than in other genotypes. The virus has not been identified with certainty, but passes through Millipore filters of 50 nm average pore size. It grows rapidly and to high titer after inoculation into mink, multiplying almost exclusively in macrophages in the liver and spleen. Viremia persists for life, death occurring after about 4 to 5 months in Aleutian mink and up to 1 year in other genotypes.

Antibody production begins within 2 weeks of inoculation and antibody titers rise to very high levels by 2 months. The circulating virus is complexed with antibody but is not neutralized. The basic lesion is a systemic plasmacytosis with overproduction of IgG, most of which is antiviral antibody. Deposits of virus–antibody–complement in the kidneys result in glomerulonephritis and renal failure and deposits in blood vessels lead to a degenerative arteritis. About 10% of mink infected for a year convert from a heterogeneous hyperglobulinemia to a monoclonal gammopathy (Porter et al., 1965). Aleutian disease of mink is thus at the opposite end of the spectrum from lymphocytic choriomeningitis in relation to antibody production, yet in both diseases there is persistent high-level viremia in which virus circulates in the plasma as virus–antibody complexes.

**Equine Infectious Anemia.** The virus that causes equine infectious anemia has not yet been classified, but it appears to be a small RNA virus (review, Hensen, 1973). It is mechanically transmitted by arthropods or surgical instruments. Some infected horses die of the acute disease; those that survive become chronically infected although viremia is demonstrable only during the recurrent febrile attacks that occur at intervals of weeks or months. In the chronic stage, neutralizing antibody is always present even during viremia, when infectious virion–antibody complexes can be demonstrated in the serum. Many of the chronic lesions probably have an immunological basis.

## Chronic Infections with Late Neoplasia

Leukemias and lymphomas induced by the RNA tumor viruses fall into this category, which is unique in animal virology in that these viruses are maintained as DNA copies of their genomes, integrated with the cellular genome and inherited genetically in all individuals of certain species, including mice and chickens (see Chapter 14). In certain strains of rodents and chickens and under conditions that vary according to the host animal, the virus is activated and infectious noncytocidal virions are then produced. These may infect neighboring cells, or if injected into young animals of the same species may infect them also. Rarely, spontaneous horizontal spread of murine RNA tumor viruses may occur; this seems to be more common with the chicken–avian leukovirus system studied by Rubin.

**Murine Leukemia.** The natural history of this infection depends upon the genetic background of the murine hosts. In a high leukemia strain of mice like AKR no free virus is found in early embryos, but it appears spontaneously when their cells are cultured (Hartley et al., 1969). In late embryonic life, infectious virus is found in most embryos, and its titer rises to a high level in the first 2 weeks postnatally (Rowe and Pincus, 1972). Viremia persists for life and large amounts of virus can be extracted from certain organs and tissues, notably bone. Early onset of viremia appears to be correlated with early onset of leukemia, certainly in different inbred lines of mice and possibly in individual mice.

It was long thought that congenital exposure to viral antigens should induce immunological tolerance (Burnet and Fenner, 1949). Partial tolerance is found in AKR mice, for viremia persists and free circulating antibody cannot be demonstrated, but the occurrence of immune complexes in the kidneys shows that some antibody is formed (Klein and Klein, 1966; Oldstone et al., 1972). The viremia consists of circulating infectious virus–antibody immune complexes (Branca et al., 1971). Thus AKR mice have a persistent infection with murine leukemia virus, acquired by activation of the integrated viral genome and resulting in late, mild "immune" disease. In addition, most AKR mice get leukemia within 10 months of birth, a process in which the thymus plays an essential role (Furth et al., 1964, 1966; Metcalf, 1966). Thymectomy prevents the spontaneous development of lymphoid leukemia caused by the leukovirus which they harbor, without inducing any diminution of immunological competence, and grafts of cultured thymic reticulum cells can restore their capacity to get leukemia. Even if the thymus is intact, however, many mice have circulating leukemia virus for months without getting leukemia. Perhaps chance mutations arising in the very large populations of virions or of lymphoid cells found in AKR mice determine the expression of leukemogenesis, the mutant being amplified by its own multiplication and accelerated multiplication of the altered leukocytes.

In low leukemia lines of mice like C57Bl and BALB/c infectious virus may not be expressed at all during the life of the mouse, i.e., the "infection" is latent rather than persistent and is transmitted genetically to succeeding generations. gs antigen is expressed in early BALB/c embryos (Heubner et al., 1970) and in a small proportion of BALB/c mice virus is expressed as the animals age; a proportion of such animals get leukemia.

**Avian Leukosis.**  As with mice and murine leukemia virus, all chickens carry avian leukosis virus genetic material integrated into the cellular genome (Baluda, 1972; Hanafusa et al., 1972). It may be expressed in the hen and congenitally transmitted to the embryo as infectious virus (Rubin et al., 1961), or it may never be expressed as infectious particles, although noninfectious particles and gs antigen may be demonstrable (Dougherty et al., 1967). "Leukosis-free" flocks are essentially those free of demonstrable infectious virus, not of integrated viral genetic material. In flocks with persistent expressed virus, Rubin et al. (1961, 1962) showed that chickens may acquire the virus either congenitally or by contact infection immediately after hatching; if they receive some maternal antibody they may be protected from contact infection until they are 3–4 weeks old. Congenitally infected chickens lack free circulating antibody and sustain high-level viremia for life. Chicks infected immediately after hatching may be rendered partially tolerant and sustain prolonged high-level viremia, but birds that are infected later usually have an inapparent infection of limited duration, with low-level transient viremia and active immunization.

Disease due to avian leukosis virus is relatively rare even in heavily infected flocks. Rubin et al. (1962) recorded 31 cases of visceral lymphomatosis and 2 cases of osteopetrosis in 744 naturally infected chickens, and these occurred between 114 and 378 days after hatching. Both cases of osteopetrosis occurred in congenitally infected birds, and visceral lymphomatosis was six times higher in this group than in birds infected by contact. Rubin's data suggest that disease occurs most commonly in birds with a persistent high-level viremia, but there is no satisfactory explanation of why disease occurs in some of the persistently infected birds and not in others.

## SLOW INFECTIONS

Three groups of diseases are listed in this category (Table 12–4), which includes the original "slow-virus infections" of Sigurdsson (1954). Group A comprises the slowly progressive infections of sheep: visna, maedi, and progressive pneumonia, caused by serologically related leukoviruses (see Chapter 1). Group B consists of four obscure infections of the central nervous system (scrapie, mink encephalopathy, kuru, and Creutzfeldt-Jakob disease) that have been designated the subacute spongiform viral encephalopathies. Group C con-

TABLE 12–4

*Slow Infections: Long Incubation Period, Slowly Progressive Fatal Disease*

| GROUP | EXAMPLES | HOST | SITES OF INFECTION | ANTIBODIES | DISEASE |
|---|---|---|---|---|---|
| A. Nonneoplastic leukovirus | Visna<br>Maedi<br>Progressive pneumonia | Sheep | Brain<br>Lung<br>Lung | + | Slowly progressive, recurrent viremia, immunopathological component |
| B. Subacute spongiform encephalopathy | Scrapie<br>Mink encephalopathy<br>Kuru<br>Creutzfeld-Jakob disease | Sheep<br>Mink<br>Man<br>Man | CNS and lymphoid tissue | − | Slowly progressive encephalopathy |
| C. Other human CNS diseases | Measles: subacute sclerosing panencephalitis | Man | CNS | ++ | Acute measles, recovery, then slowly progressive encephalitis years later |
| | Papovavirus: progressive multifocal leukoencephalopathy | Man | CNS | + | Progressive encephalopathy, following immunosuppression |

sists of two unrelated diseases of the human CNS, subacute sclerosing panencephalitis and progressive multifocal leukoencephalopathy.

Although in many respects these diseases resemble some of those described in the previous section as "chronic infections," they share a feature not found in the latter, namely that in all cases disease progresses slowly and inexorably to death.

### Group A. Slow Leukovirus Infections without Neoplasia

**Progressive Pneumonia.** This is a cosmopolitan disease of sheep that is now recognized to be the same as the Icelandic disease called "maedi" (Kennedy *et al.*, 1968; Takemoto *et al.*, 1971). The incubation period lasts 2–3 years, the onset is insidious, the course protracted, and the mortality very high. The earliest lesions appear to be scattered periarterial infiltrations of the lung with lymphocytes and mononuclear cells; later there is a diffuse mesenchymal proliferation of the lung with enlargement of the bronchial and mediastinal lymph nodes (Sigurdsson *et al.*, 1952, 1953).

**Visna.** This is a demyelinating disease of sheep that was recognized in Iceland between the years 1935 and 1951. It disappeared from the field in 1951 when all sheep in the affected area were slaughtered in an attempt to eradicate maedi. The incubation period is very long: about 2 years in both natural and experimental infections. The onset is insidious, with paresis, which progresses slowly but inevitably to total paralysis and death. Visna virus has been grown in tissue cultures of sheep cells, in which it causes cytopathic changes; a latent period of about 20 hours is followed by active multiplication for a further 24 hours—a pattern not unlike that of many viruses that cause acute infections (Thormar, 1963). The pathogenesis of the disease in sheep infected by intracerebral inoculation has been studied by Gudnadóttir and Pálsson (1965). There is an increased white cell count and increased protein in the cerebrospinal fluid from 1 to 2 months after inoculation, and virus may be found intermittently over a period of many months in the cerebrospinal fluid, in whole blood, and in the saliva. Neutralizing antibodies develop about 3 months after inoculation and slowly rise to high levels, without eliminating virus from the blood or tissues, and virus can be isolated from the blood in spite of high concentrations of neutralizing antibody.

In the intact animal, visna virus does not seem to be directly cytopathogenic. Gudnadóttir and Pálsson suggest that the characteristic demyelination may be a late immune disease, due to an antigen–antibody reaction on the surface of infected glial cells.

Intrapulmonary inoculation of visna virus into sheep produces pulmonary lesions like maedi, and CNS lesions like visna are sometimes noted after infection with maedi; visna and maedi viruses are closely related antigenically and are probably only variant strains of one virus.

### Group B. Subacute Spongiform Viral Encephalopathies

Gibbs and Gajdusek (1969) have suggested that one of the many synonyms of Creutzfeldt-Jakob disease, subacute spongiform encephalopathy, should be

TABLE 12–5

*Natural and Experimental Host Range of Subacute Spongiform Viral Encephalopathies [a,b]*

| HOST | KURU | CREUTZFELDT-JAKOB | SCRAPIE | MINK ENCEPHALOPATHY |
|---|---|---|---|---|
| Man | + | + | NT | NT |
| Chimpanzee | + | + | — | — |
| Spider monkey | + | + | NT | NT |
| Rhesus monkey | — | — | — | ± |
| Sheep | — | — | + | — |
| Goat | — | — | + | + |
| Mink | — | — | + | + |
| Mouse | — | — | + | + |
| Rat | — | — | + | — |
| Hamster | — | — | + | + |
| Gerbil | NT | NT | + | NT |
| Ferret (albino) | — | — | NT | ± |
| Incubation period | | | | |
|   Natural disease | 5–10 Years | Unknown | 3–5 Years | 8–12 Months |
|   Experimental disease | 14–38 Months | 12–14 Months | 4–6 Months | 4–8 Months |

[a] +, Clinical disease and confirmatory histopathological lesions; ±, histopathological lesions in the absence of clinical disease; —, no disease or histopathological lesions; NT, not tested.

[b] From Gibbs and Gajdusek (1971).

used as a generic name for four diseases that have strikingly similar clinico-pathological features and causative agents: namely, scrapie of sheep and goats, mink encephalopathy, and kuru and Creutzfeldt-Jakob disease in man. Scrapie has been transmitted to mice (Chandler, 1961) and other rodents, and recently, after a very long incubation period, to monkeys (Gibbs and Gajdusek, 1972); mink encephalopathy has been transmitted to ferrets, and kuru and Creutzfeldt-Jakob disease to chimpanzees and a few other primates (Table 12–5; review by Gibbs and Gajdusek, 1971). The basic neurocytological lesion in all these diseases is a progressive vacuolation in the dendritic and axonal processes of the neurons, and to a lesser extent, in astrocytes and oligodendrocytes; an extensive astroglial hypertrophy and proliferation; and finally a spongiform change in the gray matter (Lampert *et al.*, 1971a, b). At no stage is there the slightest sign of an inflammatory or a cellular immune response. Not only are the diseases in the natural hosts and the experimental animals similar in clinical and pathological features, but the etiological agents share several very unusual properties, notably a complete lack of immunogenicity and a high level of resistance to heat, formalin, and radiation (see Chapter 3). They can be distinguished from each other only by their host ranges and the incubation periods in experimental animals (Table 12–5). Since it is the best studied, we shall consider scrapie as the exemplar of the group.

**Scrapie.** Scrapie has been recognized as a natural infection of sheep since the eighteenth century. Transmission occurs with difficulty by contact, more commonly vertically from ewe to lamb (Dickinson *et al.*, 1965). Infection was widely disseminated in Britain by the inoculation of sheep with a louping-ill vaccine that was contaminated with the scrapie agent (Greig, 1939; Gordon, 1946). The incubation period in natural infections or experimentally infected sheep is very long, up to 3 years, and once symptoms have appeared the disease progresses slowly but inexorably. Early signs are intermittent, the sheep are excitable and may have convulsions. Later there is parasthesia and intense pruritis, emaciation, paralysis, and death. The neurocytopathological findings are those characteristic of the subacute spongiform viral encephalopathies (see above). Different breeds of sheep vary greatly in their susceptibility to inoculation with scrapie virus (Gordon, 1966), and it has been argued that the natural disease behaves strictly as a Mendelian recessive trait (Parry, 1962).

Scrapie can be transmitted to mice and certain other animals (Table 12–5) by the intracerebral inoculation of suspensions of brain and spinal cord of affected animals; the clinical and pathological findings are similar to the manifestations of scrapie in sheep and goats (Morris and Gajdusek, 1963). The use of mice, in which the incubation period is relatively short (4–8 months) has accelerated the rate of research in scrapie. Contact infection can apparently occur between inoculated and uninoculated mice (Morris *et al.*, 1965). Eklund *et al.* (1967) have studied the pathogenesis of scrapie in mice after subcutaneous inoculation. The pattern of spread of the virus was not unlike that found in other generalized viral infections (Chapter 9) but occurred at a much slower rate (Table 12–6). The lymphoid tissue was affected early, and high titers of virus could be recovered from the lymph nodes and spleen throughout the infection. Systemic spread occurred between the fourth and eighth week, presumably via the blood stream, although virus was never detected in the serum or blood clot. Virus was first found in the brain 16 weeks after inoculation and clinical symptoms became apparent at 28 to 32 weeks. Both in the mouse (Eklund *et al.*, 1967) and in the rat (Pattison, 1965b) there is an early astrocytosis; although conceivably the astrocyte may be the site of primary damage, and neuronal vacuolation and degeneration may be secondary changes consequent upon abnormal metabolism of the astrocyte, electron microscopic observations suggest a primary neuronal involvement (Lampert *et al.*, 1971b). In spite of the extensive involvement of the lymphoid cells, there was no indication of an immune response to scrapie virus; antibodies have not been demonstrated using a wide range of sensitive methods (Gibbs and Gajdusek, 1971). Further, neither thymectomy nor chemical immunosuppression affect the progress of the disease (McFarlin *et al.*, 1971; Worthington and Clark, 1971). No interferon is produced (Katz and Koprowski, 1968), or does inoculated interferon affect the disease (Gresser and Pattison, 1968).

Brain tissue explanted from scrapie-infected mice has been maintained in culture for 100 serial passages, during which it always contained the scrapie agent (Haig and Clarke, 1971). The titers were low and infectivity was cell associated and appeared to increase synchronously with cell multiplication. The agents of kuru and Creutzfeldt-Jakob disease (Gajdusek *et al.*, 1972) have likewise been maintained for many serial passages in explanted brain tissue.

## TABLE 12-6

### Spread of Scrapie Virus through Mice Infected by Subcutaneous Inoculation[a]

| | 1 | 2 | 4 | 8 | 12 | 16 | 20 | 24 | 28 | 29[f] | 32 | 36 | 42 |
|---|---|---|---|---|---|---|---|---|---|---|---|---|---|
| Weeks after inoculation | 1 | 2 | 4 | 8 | 12 | 16 | 20 | 24 | 28 | 29[f] | 32 | 36 | 42 |
| Percentage of surviving mice sick with scrapie | | | | | | | | 6.8 | 26 | 40 | 60 | 63 | 25 |
| Percentage of total mice dead from scrapie | | | | | | | | | 4.5 | 6.9 | 25 | 61 | 73 |
| Tissues examined[b] | | | | | | | | | | | | | |
| Spleen | 4.5[c] | | 3.5 | 5.6 | 5.6 | 6.2 | 6.2 | 5.5 | 5.5 | 5.7 | 5.6 | 5.2 | 5.5 |
| Peripheral lymph nodes | | | 3.4 | 5.6 | 4.7 | 4.7 | 5.2 | 4.5 | 4.8 | 5.4 | 5.5 | 5.6 | 4.6 |
| Thymus | | | | 4.2 | 4.8 | 5.5 | 5.0 | 5.4 | 4.5 | 0.8[d] | 5.2 | 5.6 | 4.5 |
| Submaxillary salivary gland | | | | 5.8 | 5.5 | 6.5 | 5.2 | 6.2 | 6.0 | 6.4 | 5.6 | 3.4 | 2.5 |
| Lung | | | | | 3.5 | 3.4 | 3.2 | 3.2 | 2.4 | 2.5 | | 2.2 | 3.8 |
| Intestine | | | | | 2.2 | 2.2 | 3.4 | 5.3 | 5.5 | 4.6 | 5.4 | 5.2 | 4.5 |
| Spinal cord | | | | | 1.4 | 5.6 | 4.5 | 6.6 | 6.5 | 7.4 | 7.4 | 6.7 | 6.6 |
| Brain | | | | | | 4.4 | 3.2 | 5.7 | 6.3 | 6.7 | 6.5 | 7.2 | 7.4 |
| Bone marrow (femur) | | | | | | | | 1.7 | 2.8 | 4.8 | 3.5 | 5.0 | 3.6 |
| Uterus | | | | | | | +[e] | | | + | | + | + |
| Liver | | | ←——— Not examined ———→ | | | | | | + | + | + | + | + |
| Kidney | | | ←——— Not examined ———→ | | | | | | | + | | + | + |

[a] From Eklund et al. (1967).

[b] Blood clot, serum, and testis were also examined, but virus was never detected in them.

[c] Negative log$_{10}$ of dilution of tissue suspension that contained 1 LD$_{50}$ per 0.03 ml when inoculated intracerebrally into mice. Blank spaces indicate virus was not detected in any dilution.

[d] Questionable whether thymus was removed.

[e] Virus was detected in 10$^{-1}$ dilution only and not all mice were affected.

[f] From week 29 on, only sick mice were examined.

The foregoing account reveals that scrapie behaves as a typical infectious disease, and filtration shows that the causative agent is the size of a small virus. Unusual features are the absolute absence of any sign of an immune response and the lack of sensitivity to either interferon or measures that depress the immune response. Tests on the inactivation of infectivity by a variety of physical and chemical treatments (see Chapter 3) also give results unlike those of a conventional virus. To the extent that they have been tested, these unusual biological and physicochemical properties are shared by the agents of the other three subacute spongiform encephalopathies (Gajdusek and Gibbs, 1973). It was suggested by Diener (1972a) that scrapie virus may be a small molecule of free nucleic acid protected from adverse effects by close association with cellular membranes.

**Kuru and Creutzfeldt-Jakob Disease.** Because of their significance in human medicine these two human representatives of the subacute spongiform encephalopathies will be mentioned briefly. Kuru is a disease confined in space to a group of 50,000 highland New Guineans; it was probably spread by ritualistic cannibalism and is now disappearing as that habit has disappeared (review, Gajdusek and Gibbs, 1973). Its significance lies in the fact that it was the first human degenerative disease of the central nervous system to be shown to have a viral etiology (Gajdusek et al., 1966), a finding whose importance was amplified by the demonstration that the rare but cosmopolitan presenile dementia of the Creutzfeldt-Jakob type could also be transmitted to chimpanzees (Gibbs and Gajdusek, 1969; review by Roos et al., 1972).

**Mink Encephalopathy.** This is the fourth member of the group. In the absence of either virions, purified nucleic acid, or evidence of antigenicity it is impossible to determine the relationships between these four agents. Mink encephalopathy could possibly be caused by a strain of scrapie "virus" transferred by feeding mink with infected sheep meat; and kuru could conceivably be a localized epidemic of Creutzfeldt-Jakob disease associated with cannibalism under peculiar circumstances. The possibility that both kuru and Creutzfeldt-Jakob disease may be related to scrapie cannot be ruled out, for scrapie has now been transmitted to primates, albeit so far only to a single animal, and with an incubation period of over 5 years (Gibbs and Gajdusek, 1972).

### Group C. Subacute Sclerosing Panencephalitis and Progressive Multifocal Leukoencephalopathy

We conclude with two other infections of the human brain, but they are caused by typical viruses quite unlike the bizarre agents just discussed. In many respects, both of these might more properly be designated chronic infections but they are tentatively placed alongside the slow infections of the CNS to emphasize common clinical and pathological features of this ever-growing assemblage of diseases of man.

**Measles and Subacute Sclerosing Panencephalitis (SSPE).** Measles was long thought to be an acute self-limited disease never associated with viral persistence or late disease. Most cases of this universal human ailment probably conform to this description, but recently a long-recognized but rare chronic disease of the CNS, subacute sclerosing panencephalitis (SSPE), was shown to be a late sequel to measles (Connolly *et al.*, 1967; Sever and Zeman, 1968; Horta-Barbosa *et al.*, 1969; review, Ter Meulen *et al.*, 1972). SSPE is characterized by the slow development of neuronal degeneration, and is invariably fatal. The interval between recovery from measles and the onset of frank SSPE is always several years. Although brain cells from affected individuals show serological and electron microscopic evidence of measles virus infection, infectious measles virus cannot be recovered directly, but only by cocultivation of brain tissue with indicator cells, i.e., the virus in the brain is genetically complete but normal multiplication is suppressed. Measles virus is not confined to the brain in cases of SSPE, for it has also been recovered by cocultivation from lymph node biopsies of early cases (Horta-Barbosa *et al.*, 1971a); indeed, the authors suggest that infected leukocytes may have carried the virus to the brain in the first instance.

Several investigators have demonstrated that the levels of antibody to measles virus in the serum in cases of SSPE are extremely high, and antibody is also found regularly in the cerebrospinal fluid (Sever and Zeman, 1968). The latter finding appears to be pathognomonic of SSPE. If patients are diagnosed early and antibody levels followed, they tend to rise progressively through the later stages of the disease, indicating (a) constant antigenic stimulation due to release of measles virus antigens and (b) failure of circulating antibody to arrest the progress of the disease (Horta-Barbosa *et al.*, 1971b).

Wear and Rapp (1971) have developed an interesting laboratory model for persistent infection of the brain by measles virus that may have relevance for SSPE. Human measles virus adapted to newborn hamster brain produced high viral titers and acute encephalitis. When sucklings from immune mothers were used, passively transferred antibody protected them from acute encephalitis, allowing virus to multiply and remain latent in the brain. Neurological signs could be produced 51 days after infection by treatment of such animals with the immunosuppressant cyclophosphamide, and measles virus could be recovered from the brain by cocultivation.

The critical question, yet to be answered, is why this slow complication of measles develops in certain rare individuals and not others. It does not seem to be correlated with unusual severity of the original acute infection, or with "allergic" encephalitis which is a much more frequent complication occurring within a few days of the appearance of the rash, or with the atypical measles giant-cell pneumonia without a rash that characterizes measles in children suffering from severe defects of cellular immunity (reviews by Ter Meulen *et al.*, 1972; Sever and Zeman, 1968). Burnet (1968) postulated that SSPE might reflect a specific defect in cell-mediated immunity against measles antigens resulting from viral destruction of the clone(s) of immunocytes with measles specificity. The evidence on this point that has accumulated since is conflicting, as is that on whether SSPE is caused by a particular mutant of measles virus.

**Progressive Multifocal Leukoencephalopathy (PMLE).**   This rare subacute progressive demyelinating disease has only been recognized in individuals whose immunological responsiveness has been severely depressed by malignant disease or immunosuppressive therapy (Cavanagh *et al.*, 1959; Manz *et al.*, 1971). Electron microscopic study of brain biopsies reveals intranuclear accumulations of viral particles morphologically indistinguishable from polyoma virus (Zu Rheim, 1969), and these have now been cultured (Padgett *et al.*, 1971; Weiner *et al.*, 1972). The latter isolate showed strong serological cross-reactivity with SV40, whereas Padgett's isolate showed only slight cross-reactivity with this virus. A serological survey of normal people by Shah (1972) suggested that there is a moderate incidence of human infection with a virus that cross-reacts with SV40, but is not associated with any known disease. Progressive multifocal leukoencephalopathy, which is very rare, may represent the activation of such persistent infection in immunosuppressed individuals.

## FACTORS IN THE PATHOGENESIS OF PERSISTENT INFECTIONS

It is clear from the foregoing account that a wide variety of different conditions is included within the broad designation of persistent infections. However it is worth while enquiring whether there are any common mechanisms whereby the viruses that cause these infections bypass the host defences that ensure the elimination of virus in the acute viral infections. Several mechanisms appear to be involved, but in the current state of knowledge we can only speculate about why some of these factors play a dominant role in certain persistent infections but not in others.

### Growth in Protected Sites

Herpes simplex virus and varicella virus avoid immune elimination by remaining within cells of the nervous system, in an occult form in the ganglion cells during the intervals between disease episodes, and within the Schwann cells of the nerve sheaths prior to acute recurrent episodes of disease. Likewise, other herpesviruses such as human cytomegalovirus and EB virus appear to bypass immune elimination, but in this instance they persist in lymphoid tissue and circulating leukocytes. Other viruses grow in cells in epithelial surfaces, e.g., kidney tubules, salivary gland, or mammary gland, and are persistently shed in the appropriate secretions and excretions. Most such viruses are not acutely cytopathogenic, and perhaps because they are released on the lumenal borders of cells they do not provoke an immunological inflammatory reaction. Secretory IgA, which does have access to the infected cells, does not cause complement activation and therefore fails to produce an inflammatory response.

### Integrated Genomes

RNA tumor viruses persist as DNA copies of their genomes integrated into cellular genomes. Until activated, the viral genome is only partially expressed, if at all, and it can be regarded as part of the cellular genetic material. When activated, the RNA tumor viruses multiply in lymphoid tissue and induce the production of nonneutralizing rather than neutralizing antibodies.

## Nonimmunogenicity

The unknown agents that cause the subacute spongiform encephalopathies seem to be completely nonimmunogenic, they fail to induce interferon, and they are not demonstrably susceptible to interferon action. There appears to be no mechanism whereby the host can control the multiplication and pathological effects of these agents.

## Nonneutralizing Antibodies

Viruses that cause persistent plasma-associated viremia usually multiply in lymphoid tissue and macrophages (sometimes more generally) and they characteristically induce production of nonneutralizing antibodies. These antibodies combine with viral antigens and virions in the serum to form immune complexes which may produce immune complex disease. Further, these nonneutralizing antibodies may block the action of both neutralizing antibodies and cell-mediated immunity.

## Defective Cell-Mediated Immunity

We still know relatively little about cell-mediated immunity (CMI) in viral infections (see Chapter 10), but several circumstances can be envisaged that suggest a role for CMI in viral persistence. The CMI response is antiviral (see Chapter 10), and this antiviral action can be weakened because of immuno-suppression, tolerance, the presence of blocking antibodies on target cells (Porter, 1971b), failure of immune lymphocytes to reach target cells, or finally because less viral antigen is expressed on the surface of the target cell. These factors are likely to be important in persistent infections such as visna, SSPE, PMLE and in those caused by herpesviruses. Finally we may again note the fact that many persistent viruses multiply extensively in macrophages and lymphocytes, and thus affect several parameters of the immune response. Indeed, depressed CMI may diminish the rate of destruction of infected cells and thereby prolong the release of viral antigens. The resulting protracted immunogenic stimulus would explain the very high antibody levels found in SSPE and Aleutian disease of mink, for example.

## Tolerance

Many persistent infections are associated with a very weak antibody response, especially in congenitally infected animals. Immunological tolerance is rarely complete, but there is a severe degree of specific hyporeactivity in conditions like congenital LCM, LDV infection and leukovirus infections. Tolerance to a viral antigen may be genetically determined, and the immune response to several specific antigens has been shown to be under genetic control (Mc-Devitt and Bodmer, 1972). Another factor favoring tolerance would be the presence in host tissues of antigens closely related to viral antigens (molecular mimicry), but there is no clear evidence for or against this possibility. Another kind of "tolerance" about which even less information is available, is what might be called "interferon-tolerance." Little or no interferon is produced in mice congenitally infected with LCM, LDV, or murine leukemia virus, but in each case the virus involved is sensitive to interferon. Interferon "tolerance" appears to be specific to the virus involved, possibly involving a recognition mechanism at the mRNA level. Experiments with varicella (Armstrong and Merigan, 1971) cast doubt on the suggestion that low interferon production and low sensitivity are invariably characteristics of viruses that produce persistent infections, but this would hardly be expected in such a diverse group.

## Temperature-Sensitive Mutants

Recent unpublished reports from several laboratories indicate that persistent infections may often be attributable to *ts* mutants. All strains of Newcastle disease virus isolated by J. S. Youngner from L cell carrier cultures have been found to be *ts* mutants, while V. Stollar finds that the Sindbis virus strains recovered from his carrier cultures of *Aedes albopictus* cells are also temperature-sensitive. Some persistent infections *in vivo* may have a similar explanation. Temperature-sensitive mutants have been isolated from animals carrying foot-and-mouth disease, vesicular stomatitis, and parainfluenza viruses. Of particular interest with regard to slow infections of the CNS is the recent observation that certain of B. N. Field's *ts* mutants of reovirus establish persistent infections in the brains of hamsters.

*CHAPTER 13*

# Viral Oncogenesis: DNA Viruses

About one-quarter of the 600 or so known animal viruses possess oncogenic potential; i.e., they have the capacity to initiate in animals the sort of cellular proliferation that leads to the development of tumors. Both DNA and RNA viruses are capable of causing this change in cellular behavior. However, whereas oncogenic potential among the RNA viruses seems to be restricted to one particular genus, oncogenic viruses occur among four of the six genera of the DNA viruses. On the other hand, the RNA tumor viruses are the etiological agents of natural leukemias and sarcomas in a large number of animal species, whereas no DNA viruses except the herpesviruses cause malignant tumors in their natural hosts. Nevertheless, the normal manifestation of infection by some DNA viruses is a benign tumor, and under highly artificial laboratory conditions, several other DNA viruses are potently oncogenic.

It is this last set of viruses, which includes the adenoviruses, polyoma virus, and SV40, that has been responsible for the great surge of interest in DNA tumor viruses during the last decade, because they provide model systems in which at least some of the events that lead to tumor formation in animals can be duplicated *in vitro*. The central phenomenon of the system is cell transformation. When cultured cells are infected with these viruses, a proportion of them assumes a new set of stable properties that closely resemble the properties of cells derived directly from "spontaneous" or virus-induced tumors. The model system is useful because it enables us to study in a quantitative way the interaction of the viruses with genetically homogeneous populations of cells free of interference from the host immune system, and it allows us to use the techniques of molecular biology to investigate aspects of the virus–cell interaction that are inaccessible in whole animals. Before dealing in detail with this *in vitro* system, we will first discuss the role of DNA viruses in causing tumors under natural conditions.

## TUMORS PRODUCED UNDER NATURAL CONDITIONS

### Poxvirus

Several poxviruses produce slight epithelial proliferation in their natural hosts, and fowlpox and molluscum contagiosum viruses produce quite large localized tumors in the skin of chickens and man, respectively (see Table 9–1). Two poxviruses, Yaba poxvirus and rabbit fibroma virus, produce large tumors in the skin, both in their natural hosts and in experimental infections.

**Yaba Poxvirus.** The natural reservoir of this virus, which is a typical poxvirus (Noyes, 1965; Yohn and Gallagher, 1969) is unknown; it is presumed to be an African primate. Yaba poxvirus causes large localized benign tumors of the skin in natural infections of Asian monkeys in West Africa, and after subcutaneous or intradermal inoculation of rhesus and cynomolgus monkeys, some African primates (Bearcroft and Jamieson, 1958; Niven *et al.*, 1961), and humans (Grace and Mirand, 1963). Pulmonary tumors develop in some monkeys after inhalation of aerosols containing the virus (Wolfe *et al.*, 1968); after intravenous inoculation in rhesus monkeys hundreds of small tumors may develop in the subcutaneous tissues, heart, muscles, and lungs.

The growths are typically tumors of the subcutaneous tissues, and the epithelium is rarely involved except when ulceration occurs. The tumors are histiocytomas; their pathogenesis has been described by Sproul *et al.* (1963). Within 48 hours of subcutaneous inoculation histiocytes migrate into the dermis and undergo striking morphological changes. The nuclei and nucleoli are greatly enlarged and rapid proliferation of the infected cells occurs. Cytoplasmic inclusion bodies can be seen as early as the third day but are not conspicuous until the end of the proliferative stage, in the third week. The proliferating cells resemble fibroblasts, but they never produce collagen and sequential studies reveal their origin from histiocytes. Healing is due to the disintegration of the tumor cells, which often become multinucleate before they disappear.

There is no indication that the tumors produced in mature animals ever become malignant; the numerous tumors which follow intravenous inoculation are probably due to viral infection in many sites and not to cellular metastasis. No studies have been reported on the effects of the Yaba virus on newborn animals.

The virus grows in both monkey and human cells *in vitro*, with a prolonged growth cycle and eventual cytopathic changes (Yohn *et al.*, 1966).

**Rabbit Fibroma Virus.** In their natural hosts, all four members of the myxoma subgenus of the poxviruses (myxoma virus; rabbit, hare, and squirrel fibroma viruses) produce benign fibromas (Fenner and Ratcliffe, 1965), although the most familiar disease due to viruses of this subgenus is severe generalized myxomatosis of European rabbits. The best studied poxvirus which can cause a proliferative lesion is Shope's rabbit fibroma virus, which produces fibromas in both its natural host, the cottontail (*Sylvilagus floridanus*), and in the domestic rabbit (*Oryctolagus cuniculus*).

Cottontail and domestic rabbits react differently to infection with fibroma virus, although the same peak titer of infectious virus is attained in each (Dalmat and Stanton, 1959). The first specific response consists of groups of proliferating cells that arise focally around small blood vessels and hair follicles, and soon make up the entire cell populations of the affected areas. In domestic rabbits, the lesions regress rapidly; in cottontail rabbits their longer persistence is accompanied by the appearance of glycoprotein-containing cytoplasmic inclusion bodies in the fibroma cells (Fisher, 1953), and more characteristic cytoplasmic viral inclusion bodies in the overlying epithelial cells. Newborn domestic rabbits suffer a generalized and usually lethal disease (Duran-Reynals, 1940b, 1945), and the fibromas which occur in cottontail rabbits infected at birth may become very large (Yuill and Hanson, 1964) and persist for many months (Kilham and Dalmat, 1955). If adult domestic rabbits infected with fibroma virus are treated with tar, various carcinogens, or cortisone, they develop much larger and more persistent tumors which have many of the characteristics of fibrosarcomas (Andrewes and Ahlström, 1938).

Studies on the growth of fibroma virus in cultured rabbit kidney cells, which have not been very revealing from the point of view of virus-induced transformation, are described in Chapter 12.

### Herpesvirus

At least five different herpesviruses have been associated with natural cancers of animals and man. (a) Marek's disease of chickens, a highly infectious lymphomatosis, is caused by a herpesvirus (Marek's disease herpesvirus—MDHV) (Witter et al., 1969). (b) There is strong suggestive evidence that Lucké frog kidney carcinoma is caused by a herpesvirus (Mizell et al., 1969). (c) Circumstantial and inferential evidence indicates that infection with Epstein-Barr virus (EBV), which is a causative agent of infectious mononucleosis, may contribute to the etiology of Burkitt's lymphoma of man (Henle et al., 1969; Klein, 1972) and possibly to nasopharyngeal carcinoma (Henle et al., 1970). (d) *Herpesvirus saimiri*, isolated from squirrel monkeys, causes a rapidly lethal lymphatic leukemia or reticulum cell sarcoma in owl monkeys and marmosets (Meléndez et al., 1970). (e) Finally, there have been speculations, based on sero-epidemiological evidence, that herpes simplex virus type 2 may be linked to human cervical carcinoma (Nahmias et al., 1970).

**Marek's Disease of Chickens.** Since 1907, when the disease was first recognized (Marek, 1907), chicken farms have been scourged periodically by a lymphomatosis which may kill up to 70% of closely housed birds in a single outbreak. The disease is marked by massive lymphocytic invasions of the internal organs and nerve trunks; death usually occurs within 6–8 weeks. No infectious virus can be isolated directly from the tumor cells and no viral particles can be seen in the cells by electron microscopy, and for many years the highly contagious nature of the disease remained an enigma. The paradox was resolved recently by the discovery that sick chickens shed large amounts of virus from their feather follicles (Calnek et al., 1969, 1970; review, Purchase, 1972); the virus is identical morphologically to typical herpesviruses (Epstein et al., 1968) and its injection into susceptible young chickens leads to the development of classic Marek's disease (Biggs et al., 1968; Churchill et al., 1969; Cook and Sears, 1970). It is not clear why virus production is restricted to the feather follicles; certainly the viral genome is present in other sorts of cells in diseased chickens because the disease can be transmitted artificially with virus-free, intact lymphoma cells. What prevents full viral expression in these cells is unknown.

The virus grows well in chick and duck embryo fibroblasts (Biggs and Payne, 1967; Biggs *et al.*, 1968) to produce cytopathic effects. So far there are no reports of cell transformation *in vitro* and no lines of lymphoma cells have been established.

**Lucké Carcinoma of Frogs.** Many frogs (*Rana pipiens*) in New England carry adenocarcinomas of the kidney (Lucké, 1938). The tumors are thought to be caused by a herpesvirus; first, because histological examination of the tumors of frogs collected in winter time shows the presence both of inclusion bodies and of herpeslike viruses (McKinnel and Zambernard, 1968; Mizell *et al.*, 1968), and second, because preparations of partially purified virus induce tumors in tadpoles (Mizell *et al.*, 1969). Unfortunately this evidence is not conclusive because of possible contamination of the herpesvirus preparations by some of the other viruses that are known to exist in frog adenocarcinomas (Granoff, 1972). Before it becomes unambiguously clear that herpesviruses are indeed the etiological agent of the renal tumors, it will be necessary to show that highly purified viral preparations are oncogenic. This is not easy to achieve because there are no *in vitro* systems in which the Lucké herpesvirus will multiply, and no lines of frog adenocarcinoma cells have yet been established.

Nevertheless the tumor is extremely interesting. By contrast to tumors of frogs captured in winter, those of frogs caught in the summer do not contain visible herpesvirus particles; however, the virus can be induced in them by storing the frogs for several weeks at 4°C (Mizell *et al.*, 1968). This phenomenon raises the possibility that the Lucké herpesvirus might be present in a proviral form in the multiplying tumor cells of summer frogs. Furthermore, there seems to be a reciprocal relationship between tumor formation and viral growth. Tumors develop and metastasize much more frequently when the frogs are incubated at summer temperature (23°–26°C) than when they are kept cold (Lucké, 1938), and there have been speculations that the viral maturation process which occurs at low temperature leads to death of the tumor cells (Klein, 1972). What prevents the complete expression of viral functions in the tumor cells of summer frogs is unknown.

**Epstein-Barr Virus.** One kind of cancer accounts for about one-half of all cancer in children in East Africa, namely Burkitt's lymphoma, named after Dennis Burkitt, who first recognized the abnormally high incidence of the tumors and suggested that their peculiar geographical distribution was the result of infection with an oncogenic virus transmitted by an insect (Burkitt, 1963). This idea has been a great stimulus for epidemiological research, most of which is aimed at determining whether or not Burkitt's lymphoma is in any sense an infectious disease. Evidence favoring this hypothesis has come from Pike and his co-workers (1967), who have shown that the distribution of the disease obeys certain of the mandatory principles required for lateral transmission—it occurs in time-space clusters and shows epidemic drift. Furthermore, a similiar disease found subsequently in New Guinea shows a similar type of distribution (Booth *et al.*, 1967). On the other hand, cases of Burkitt's lymphoma have been reported at low frequency in virtually every area of the world in which an adequate search has been made (O'Conor *et al.*, 1965b; Wright, 1966; Burkitt, 1967). This last finding, together with the virological evidence discussed later, persuaded Burkitt to modify his original idea and postulate that the lymphoma may be caused by a combination of a virus which is ubiquitous and an environmental factor which is restricted geographically and which may be vector-associated (Burkitt, 1969; Cook and Burkitt, 1971).

Burkitt's ideas stimulated the search for the etiological agent(s) of the tumor, and, in 1964, Epstein, Achong, and Barr reported the discovery of herpesvirus particles in a cultured line of Burkitt's lymphoma cells. Since that time the same virus, now called Epstein-Barr virus or EBV, has been isolated regularly from almost all cultures of lymphocytes obtained from Burkitt's lymphoma patients (review, Epstein, 1970). The virion is enveloped, possesses 162 capsomers, and is identical morphologically to the other members of the genus *Herpesvirus*. However, it is quite distinct antigenically (Henle and Henle, 1972) and does not multiply in any cell system other than human lymphoid cell lines (Epstein *et al.*, 1965; Rabson *et al.*, 1966).

There has been no convincing demonstration of EBV particles in fresh biopsy specimens of Burkitt tumor material, and the tumor material is usually negative for viral capsid antigens (Nadkarni *et al.*, 1970). However, EBV-specific surface antigens are present in the majority of the cells (Klein *et al.*, 1967; Henle *et al.*, 1969; review, Klein, 1971). When placed in culture the cells divide, and after 3 to 10 days 0.1–0.5% of the actively growing cells from some biopsy specimens contain viral capsid antigens (Nadkarni *et al.*, 1970) and viral particles. The number of cells that spontaneously yield virus at any one time can be increased up to fiftyfold by treatment of the cultures with 5-bromodeoxyuridine (Hampar *et al.*, 1971; Gerber, 1972), or by arginine starvation (Hinuma *et al.*, 1967). Treatment with 5-bromodeoxyuridine also induces viral synthesis in those lymphoma lines which do not produce virus spontaneously (Gerber, 1972; Hampar *et al.*, 1972). Because the percentage of cells that contains virus remains constant over many cell generations (Rabson *et al.*, 1966) and because the cells that produce virus die, it seems that viral synthesis is activated as a result of a stochastic process in a small fraction of the multiplying cell population. Direct evidence favoring this idea comes from experiments which have shown that cloning of the virogenic lines gives rise to cultures which also are virogenic (Hinuma and Grace, 1968; Maurer *et al.*, 1970). Further, nucleic acid hybridization experiments show the presence of multiple copies of EBV DNA closely associated with the chromosomes of nonvirogenic lines of cells of Burkitt's lymphoma (zur Hausen and Schulte-Holthausen, 1970; zur Hausen *et al.*, 1970; Nonoyama and Pagano, 1971; zur Hausen, 1972). The mechanism by which the viral genome is activated is unknown.

There is evidence for cell transformation *in vitro* by EBV in two systems. First, exposure to EBV of human leukocytes obtained from peripheral blood changes them from cells unable to multiply in tissue culture into lymphoblastic cells capable of continuous proliferation (Henle *et al.*, 1967; Pope *et al.*, 1968; Gerber *et al.*, 1969; reviews by Epstein, 1970; zur Hausen, 1972); the cells show loss of contact inhibition and grow in floating clumps. These transformed cell lines acquire a characteristic chromosomal abnormality—a constriction in the long arm of chromosome No. 10 (Miles and O'Neill, 1967; review, zur Hausen, 1972), which also occurs in cultured Burkitt's lymphoma cells. However, infection with EBV induces many other chromosomal changes including fragmentations and condensation effects (Whang-Peng et al., 1970; Macek et al., 1971), so the significance of the abnormality in chromosome No. 10 is unclear. Similar, although less extensive, results have been reported for the transformation of human fibroblasts by EBV (Probert and Epstein, 1972).

All African patients with Burkitt's lymphoma have antibodies against EBV (Henle et al., 1969). However, the sera of a large number of normal adults also show anti-EBV activity, albeit at lower titer (reviews, Klein, 1971, 1972). This paradox was resolved when it was discovered that serum conversion from EBV-negative to EBV-positive occurs when people contract infectious mononucleosis (Henle et al., 1968). On the basis of very extensive prospective and retrospective serum surveys (Niederman et al., 1968; Evans et al., 1968; Henle and Henle, 1972), it is now proved that the etiological agent of infectious mononucleosis, which was thought to be an undiscovered virus, is, in fact, EBV. A high proportion of infections must be subclinical. Not surprisingly, lymphoblast cells from mononucleosis patients, like those from Burkitt's lymphoma, grow well in culture and can be induced to release EBV (Gerber, 1972).

The central question remains whether EBV causes Burkitt's lymphoma. There is no doubt that (a) EBV causes infectious mononucleosis which, although it is a self-limiting disease, in other respects resembles quite closely the early stages of lymphoma; (b) EBV causes transformation of lymphoid cells *in vitro*; (c) there is a serological association between EBV and Burkitt's lymphoma; and (d) EBV antigens are detectable in the cells of all Burkitt tumors and virions can be recovered following cultivation *in vitro*. None of this circumstantial evidence is conclusive, but it has been powerful enough to persuade some people that EBV is a human tumor virus and is probably the etiological agent for Burkitt's lymphoma (see, for example, Epstein, 1971). Other workers are more skeptical. Presumably the question will be resolved when the results of a large prospective epidemiological study conducted by WHO become available in a few years' time. In any case, EBV remains a fascinating virus. The greatest obstacle to further progress with the system has been an inability to find tissue culture systems in which EBV reproduces lytically. Virus stocks are grown of necessity in lymphoid cell lines which give poor yields. Consequently, the preparation of the large amounts of virus required for biochemical studies remains a major problem and little work of significance has been reported.

*Herpesvirus saimiri.* In 1969, Melendez isolated a herpesvirus from a culture of squirrel monkey kidney cells; it causes a fatal lymphoma or reticulum cell tumor when inoculated into several species of marmosets, owl monkeys, cinnamon and ringtail monkeys (Melendez et al., 1970), and rabbits. Originally it was thought that the virus existed as a harmless passenger in its natural host species (Melendez et al., 1970), but this conclusion may need to be revised because it is now known that the majority of these monkeys have antibodies to the virus (Deinhardt *et al.*, 1971), which may have interfered with the tests for pathogenicity and tumorigenicity. Furthermore, the fact that this herpesvirus can be rescued from the peripheral white cells of most squirrel monkeys by cocultivation is reminiscent of the situation with EBV in humans. The parallel can be extended further; just as Burkitt tumors are free of infectious EBV, so tumors caused by injection of *Herpesvirus saimiri* into marmosets contain neither viral antigens nor particles. Several lymphoma cell lines have been established (Deinhardt *et al.*, 1971; Rabson *et al.*, 1971) which show a remarkable resemblance to human cells carrying EBV, both in their general growth properties and also because after a few days in culture virus-synthesizing cells can be demonstrated by immunofluorescence and electron microscopy. The natural history of *Herpesvirus saimiri* is of great interest since it provides a possible model for the interactions between EBV and humans.

**Other Possible Connections between Herpesviruses and Cancer.**   Three other situations have been described linking herpesviruses with cancer.

a. *Cervical carcinoma.* A strong association between the possession of antibodies to herpes simplex virus type 2 and cervical carcinoma of women has been reported by three groups (Nahmias *et al.*, 1970; Rawls *et al.*, 1969; Royston *et al.*, 1970), and cervical smears taken from women with carcinoma were reported to contain cells in which herpesvirus-specific antigens could be demonstrated (Royston and Aurelian, 1970). In addition, UV-inactivated herpes simplex virus type 2 has been reported to transform hamster cells in tissue culture (Duff and Rapp, 1971). At present, the circumstantial evidence indicates no more than that herpes simplex virus type 2 infection is often a covariable of cervical cancer; both diseases are correlated with a high frequency of sexual intercourse and especially with sexual promiscuity. Much more work is required before we can decide whether there is a direct link between the virus and the neoplasm.

b. *Nasopharyngeal cancer.* Serological associations have been found between EBV and nasopharyngeal cancer seen in the Chinese population in Southeast Asia (Old *et al.*, 1968; Henle *et al.*, 1970) and between EBV and Hodgkin's disease (Johansson *et al.*, 1970; Levine *et al.*, 1971); furthermore, EBV DNA has been demonstrated regularly in biopsies of nasopharyngeal carcinoma. Whether EBV is merely a passenger in these tumors or whether there is a causal relationship is unknown.

c. *Rabbit lymphoma.* A rabbit herpesvirus has been isolated that causes lymphomas in wild cottontail rabbits (Hinze, 1971).

In summary, it is clear that in a wide variety of animals, herpesviruses enter into a permanent association with leukocytes which results in little or no virus production but a great proliferation of the cells. In some cases (for example, *Herpesvirus saimiri* in marmosets, Marek's disease virus in chickens) we know the relationship is oncogenic; in others (EBV in humans) we cannot be sure. In all systems, the herpesvirus genome seems to enter into a stable relationship with that of the cell, and in every case the cells die when virus production is induced. The balance of the system is delicate and fascinating; and it is ripe for investigation by molecular biologists.

## Papillomavirus

All members of the genus *Papillomavirus* produce benign tumors in their natural hosts; sometimes these may become malignant. The other genus of the Papovaviridae, *Polyomavirus*, contains two viruses, SV40 and polyoma virus of mice, that have been intensively investigated as oncogenic viruses in experimental systems (see below), but they do not appear to produce tumors under natural conditions.

**Rabbit Papilloma Virus.** This virus, which is transmitted by biting arthropods (Dalmut, 1968) is widespread among cottontail rabbits (*Sylvilagus floridanus*) in the midwest states of the United States, causing large persistent keratinized tumors of the skin ("horns"). Histopathological examination shows that the growths are epithelial (Shope and Hurst, 1933). Domestic rabbits (*Oryctolagus cuniculus*) can be readily infected with rabbit papilloma virus and develop similar lesions, but these tumors rarely contain infectious virus (Shope, 1933), although they do contain infectious DNA (Ito, 1962).

In both species, viral DNA replication is completely repressed in the deeper proliferating epithelial cells of the tumors. It is induced after the initiation of keratinization in most cells of cottontail papillomas but in very few cells of the papillomas of domestic rabbits, followed in cottontail but not domestic rabbit papillomas by the synthesis of viral proteins and the production of virions (Noyes and Mellors, 1957; Noyes, 1959; Orth *et al.*, 1971).

Squamous cell carcinomas develop in a high proportion of persistent papillomas between 4 and 9 months after their initiation (Rous and Beard, 1935), both in experimentally produced papillomas of cottontail and domestic rabbits and in natural lesions of cottontails (Syverton *et al.*, 1950; Syverton, 1952). These carcinomas usually yield no infectious virions but they do contain nonvirion antigens (Kidd, 1942; Evans *et al.*, 1962), which are probably analogous to the T and transplantation antigens found in tumors caused by other papovaviruses. More detailed analysis of cell–virus interactions in papilloma has been hampered by the lack of a satisfactory tissue culture assay system.

With the more recent discoveries of virus-free tumors in hamsters inoculated with SV40, polyoma virus, and certain adenoviruses (see below), rabbit papillomavirus is now seen to fit into a pattern common to most oncogenic DNA viruses; yet there are certain features in which the rabbit papillomas are unique. These relate to the recovery of infectious nucleic acid from the "noninfectious" tumors and the recognition of increased arginase activity in virus-infected cells. Ito (1962) extracted infectious DNA from carcinomas which had supervened on papillomas produced in domestic rabbits by either papilloma virus or its DNA. In the latter group, positive results were rare but the resulting tumors were classical. Papilloma virus induces an arginase in the squamous epithelium of cottontail and domestic rabbits, both in intact animals (Rogers, 1959, 1962) and in cultured cells from cottontail skin (Passen and Schultz, 1965). Although Rogers suggested that this enzyme was virus-coded (Rogers, 1962; Rogers and Moore, 1963), Orth *et al.* (1967) have produced strong evidence for the view that the L-arginase produced in papillomas during infection by Shope papilloma virus is specified by the host's genome.

**Human Papilloma Virus.**   The virus which causes contagious human warts is identical morphologically to the Shope papilloma virus of rabbits (Williams *et al.*, 1961; Crawford, 1965). Wart virus occasionally induces a cytopathic effect in monkey kidney cells in culture (Mendelson and Kligman, 1961) and in murine and human fetal skin cells (Oroszlan and Rich, 1964), but there are no reliable *in vitro* infectivity assays available. The virus transforms human cells in culture at very low frequency (Noyes, 1965).

Warts seem to form as a consequence of infection of a single basal epithelial cell; the cell is stimulated to divide, and the wart finally consists of the clonal descendants of the original infected cell (Murray *et al.*, 1971). There is no evidence that there is spread of virus with recruitment of neighboring basal cells into the wart. This observation fits well with the fact that no infectious virus can be detected in the deep layers of the epidermis; viral antigens are found and viral particles mature only in the outer keratinized cells of the skin. It seems, then, that the viral genome is present in the dividing cells deep inside the wart in a proviral form, and that it is converted into infectious mature virions only when the host cells begin to keratinize. The mechanism by which all this occurs is unknown.

**Bovine Papilloma Virus.**   This virus is morphologically identical to the other two papilloma viruses. Both intact virus and isolated DNA transform bovine and murine cells with low efficiency (Black *et al.*, 1963a; Thomas *et al.*, 1963, 1964).

## TUMOR PRODUCTION BY ADENOVIRUSES

The oncogenic potential of certain human adenoviruses was first recognized when Trentin *et al.* (1962) showed that human adenovirus type 12 produced sarcomas when inoculated into newborn hamsters. Since that time intensive work in several laboratories has demonstrated that at least twelve of the thirty-one serotypes of human adenoviruses, as well as adenoviruses from several other animals, are oncogenic for various rodents (see Table 13–1). In most cases, the tumors were undifferentiated sarcomas, although occasionally they resembled malignant lymphomas (Larson *et al.*, 1965).

The capacity of the thirty-one serotypes of human adenoviruses to produce tumors in hamsters varies, as judged by the latent period and the minimum effective dose, and the viruses have been classified into three subgroups — the highly oncogenic types (12, 18, 31), the weakly oncogenic types (3, 7, 11, 14, 16, 21), and nononcogenic adenoviruses (reviews, Green, 1969, 1970; and see Table 3–2). We do not know what determines the range of oncogenicity of the different adenovirus types. At one time it was thought that there was a correlation between the base composition of the DNA of different adenoviruses and their ability to form tumors. It is true that human adenoviruses with a low G+C content are more oncogenic than those with a high G+C content, but it is difficult to attach any special significance to this observation, first, because simian adenovirus 7 does not conform to the pattern [this virus is highly oncogenic but its DNA contains 60% G+C (Piña and Green, 1968)], and second, viruses representative of each subgroup are equally efficient in producing neoplastic trans-

TABLE 13–1

*Induction of Tumors and Transformation of Cells by Adenoviruses*

| HUMAN ADENOVIRUSES [a] | | ANIMAL SPECIES IN WHICH MALIGNANT TUMORS ARE INDUCED | CELLS TRANSFORMED *in vitro* |
|---|---|---|---|
| GROUP | TYPES | | |
| A | 12, 18, 31 | Hamster, mouse, rat Mastomys | Hamster, rat, rabbit, human |
| B | 3, 7, 11, 14, 16, 21 | Hamster | Hamster, rat |
| C | 1, 2, 5, 6 | — | Hamster, rat |
| D | 9, 10, 13, 15, 17 19, 26 | — | Hamster, rat |
| OTHER ADENOVIRUSES | | | |
| Simian [b] | | Hamster, rat, mouse | Hamster |
| Bovine (type 3) [c] | | Hamster | Hamster |
| Infectious canine hepatitis [d] | | Hamster | Hamster |
| Chicken (CELO) [e] | | Hamster | Hamster |

[a] McAllister *et al.* (1969a).
[b] Hull *et al.* (1965).
[c] Gilden *et al.* (1967).
[d] Sarma *et al.* (1967).
[e] Anderson *et al.* (1969).

formation of rat cells (McAllister and Macpherson, 1968; McAllister *et al.*, 1969b). Perhaps the ability to classify adenoviruses into groups of different oncogenicity depends more on the characteristics of the assay systems than on inherent properties of the viruses.

In the adenovirus/hamster system repeated attempts to extract infectious virus from tumors have failed, even after extensive treatment of the cells with chemical and physical inducing agents (Trentin *et al.*, 1962; Huebner *et al.*, 1962; Landau *et al.*, 1966). However, Huebner and his colleagues (1965) showed that the tumors contained an antigen (T antigen) detectable by complement fixation or fluorescent antibody staining. The T antigen is not a virion antigen, but it does appear to be virus specific (see Chapter 5). Further, it is subgroup specific (see Table 3–2). Since animals bearing tumors produce antibody directed against the T antigen, these antigens could serve as a marker for possible adenovirus-induced tumors in humans. Sera from 389 patients with advanced solid tumors were screened for antibodies to T antigens of adenoviruses 12 and 27, but did not differ significantly from control sera (Gilden *et al.*, 1970; Rowe and Lewis, 1968).

A second kind of test for the presence of adenovirus genes in human or animal cancer cells is to analyze them for the presence of adenovirus-specific RNA. In contrast to adenovirus-induced hamster tumors in which viral RNA is easily demonstrable, over 200 human cancer specimens were uniformly negative [M. Green and T. Devine, quoted by McAllister *et al.* (1972b) and Green (1972)]. Even though the sensitivity of this test was not very great, when the results are taken together with the negative serological data and the failure to induce tumors into cynomolgus or rhesus monkeys by injection of adenoviruses [Manaker and Landon (1972), cited in McAllister *et al.*, 1972b)], it seems unlikely that adenoviruses are important causative agents of human tumors, if indeed they cause any at all.

## TRANSFORMATION OF CULTURED CELLS BY ADENOVIRUSES

McBride and Wiener (1964) reported *in vitro* transformation of cultures of kidney cells from newborn hamsters with adenovirus type 12. Piled-up colonies of epithelioid cells appeared after an incubation period of 8 to 10 weeks; the cells did not produce detectable virus, but they did contain adenovirus 12 T antigen. Following this finding, more rapid and reproducible transformation assays for other adenovirus serotypes were developed using several types of animal cells (Table 13–1).

The subgroups A, B, C, and D correspond to hemagglutination groups 4, 1, 3, and 2, respectively (Rosen, 1960; McAllister *et al.*, 1969a; see Table 3–2), and each subgroup possesses a specific T antigen. It is interesting that even though adenoviruses belonging to subgroups C and D do not cause tumors in rats, they transform rat cells as efficiently as the highly oncogenic adenovirus 12. About $5 \times 10^7$ infectious particles are required for a single transformation event. This reflects the very low proportion of virus–cell interactions that leads to transformation; the great majority result in cell destruction and virus production, which have thwarted attempts to establish permanent lines of adenovirus-transformed human cells. However, several lines of adenovirus-transformed rat and hamster cells have been obtained. They grow to high density in culture, form colonies in agar, and cause malignant tumors when inoculated into susceptible hamsters or rats.

### State of the Viral Genome in Transformed Cells

Cells transformed by adenoviruses, *in vitro* or *in vivo*, do not appear to contain the complete genome of the infecting virus in a state in which it can ever be activated or induced (Landau *et al.*, 1966). Several lines of evidence indicate, however, that at least some adenovirus genes persist indefinitely in transformed cells.

**Viral DNA Sequences in Transformed Cells.** There are two different hybridization techniques that have been used to quantitate the amount of viral DNA in transformed cells, both of which were first developed with the polyoma system. The first method, worked out by Westphal and Dulbecco (1968), involves synthesis *in vitro* of highly radioactive RNA from purified viral DNA using *E. coli* DNA-dependent RNA polymerase. The RNA is annealed to transformed or control cell DNA's immobilized on nitrocellulose filters, and the number of copies of viral DNA per transformed cell is calculated from reconstruction experiments using mixtures of untransformed cell DNA and known amounts of viral DNA. The second method, developed by Gelb *et al.* (1971b), takes advantage of the fact established by Britten and Kohne (1968) that the rate of reannealing of any

DNA sequence in solution is proportional to its concentration. It is possible, therefore, to determine directly the number of viral genome copies in transformed cell genomes by following the rate of reannealing of small amounts of highly radioactive DNA in the presence of DNA from transformed cells, without relying on reconstruction experiments. Both methods are very sensitive and are capable of detecting minute amounts of viral DNA. However, they give slightly different estimates of the number of viral genomes per transformed cell, that obtained by DNA–RNA filter hybridization always being the higher.

Green and his co-workers (Green *et al.*, 1970), using the DNA–RNA hybridization method, have shown that all adenovirus-transformed cell lines carry multiple copies of viral DNA; adenovirus 12-transformed hamster cells have 18–20 copies (400–460 $\times$ 10$^6$ daltons of DNA) per diploid mammalian genome (3 $\times$ 10$^{12}$ daltons of DNA), adenovirus 7 hamster tumor cells 28–30 copies, and adenovirus 2-transformed rat cells 7–10 copies. However, the same line of adenovirus 2-transformed rat cells shows only 1–2 copies of viral DNA per diploid cell genome as determined by reassociation kinetics (Green, 1972; Pettersson and Sambrook, 1973). Which of these two estimates is the more accurate is unknown. No experiments concerning the state of the viral DNA in transformed cells have been reported, although there is some evidence (Green *et al.*, 1970) that it is closely associated with the chromosomes.

**Transcription of Viral Genes in Transformed Cells.**    Adenovirus-specific RNA can be detected in cells transformed by members of subgroups A, B, and C by pulse labeling with uridine followed by extraction of the RNA and hybridization to viral DNA. However, in spite of many attempts, no virus-specific RNA has been found in cells transformed by the "nononcogenic" adenovirus subgroup D (McAllister *et al.*, 1972b) even though T antigen is present in them. Possibly only a small segment of the viral genome is transcribed or the viral RNA is more rapidly turned over in these cells than is the case in cells transformed by other adenoviruses.

The viral RNA's found in cells transformed by the different tumorigenic classes of adenoviruses are subgroup specific. The viral RNA extracted from cells transformed by the subgroup A viruses, for example, hybridizes well with DNA's of viruses of that subgroup, but very poorly with the DNA's of viruses of the other subgroups. Similiar results have been obtained with RNA isolated from cells transformed with subgroups B and C (Fujinaga and Green, 1967; Fujinaga *et al.*, 1969). This result shows that few if any viral genes common to all three subgroups are transcribed in transformed cells, suggesting that there is no particular adenovirus "cancer gene."

Up to 5% of the mRNA present in the polyribosomes of cells transformed by adenoviruses of groups A and B is virus specific (Green *et al.*, 1970), so that the viral DNA must be transcribed much more efficiently than the cellular DNA, or else the viral mRNA is much more stable than that of the cell. Competition hybridization experiments show that the sequences of viral mRNA on the polyribosomes represent not the whole viral genome, but rather a small specific subset of the "early" sequences of viral DNA (Green *et al.*, 1970). The mechanism by which this small specific segment of the viral genome is transcribed at high frequency is unknown. Finally, recent data suggests that the same RNA molecules that contain viral sequences also contain sequences homologous to those found in reiterated host DNA (Tsuei *et al.*, 1972). The most likely interpretation of this result is that transcription of adenovirus genes in transformed cells occurs from an integrated genome and that the hybrid host–viral RNA molecules exist because of "read-through" synthesis by RNA polymerase from the host DNA sequences into the viral, or vice versa.

### Virus-Specific Proteins in Transformed Cells

Although they do not contain or release infectious virus, adenovirus-transformed cells do produce virus-specific antigens. In spite of early reports suggesting that adenovirus 12 fiber antigen was produced in transformed cells (Huebner *et al.*, 1964; Berman and Rowe, 1965), it now seems likely that adenovirus-transformed cells do not synthesize virion proteins. However, two other kinds of virus-specific antigens are regularly found: transplantation antigens (TSTA) and T antigens.

**Tumor-Specific Transplantation Antigens (TSTA).** Trentin and Bryan (1966) have shown that hamsters and mice can be protected against adenovirus type 12 tumor cell transplants by preimmunization of adult animals with adenovirus type 12; the tumor cells therefore presumably contain a novel virus-specific transplantation antigen (TSTA). The specificity of adenovirus type 12 TSTA was investigated by Sjögren *et al.* (1967). Infection of mice with adenoviruses types 7, 12, or 18 produced resistance to adenovirus type 12-induced mouse tumor cells; i.e., there was at least partial cross-reactivity between the TSTA's induced by these three oncogenic adenoviruses. Infection of mice with the nononcogenic adenovirus type 5 failed to produce resistance to the adenovirus type 12 tumor cells.

**T Antigens.** The most interesting antigens associated with adenovirus-transformed and tumor cells are the T antigens (review, Huebner, 1967). These occur transiently in lytically infected cells, but immunofluorescence and complement fixation tests show that they are always present in transformed cells and persist over many generations of cultivation *in vitro*. Cells transformed by members of each subgroup contain the subgroup-specific T antigen (Huebner, 1967). Although T antigen is by far the most abundant of the viral antigens in transformed cells, it has never been purified to homogeneity, and there is disagreement about the properties of the partially purified antigen (Taritian *et al.*, 1967; Gilead and Ginsberg, 1968; Tockstein *et al.*, 1968). Its function is unknown, and there is no direct evidence that it is virus-coded, although the circumstantial evidence is strong. T antigen may consist of more than a single virus-coded early protein.

### Virus-Specific Repressors

Finally, there has been an attempt to determine whether virus-specific repressors are present in transformed cells. Strohl *et al.* (1966) have tested the susceptibility of adenovirus type 12-induced hamster tumor cells to infection with adenovirus type 2. All cells produced adenovirus type 2 virion antigens; the yield of infectious adenovirus type 2 was very low, but could be increased 100-fold by treatment with mitomycin C. They suggested that most cells showed superinfection immunity, and that the small amount of infectious virus found was yielded by cells which had been "induced" (in the sense that lysogenic bacteria can be induced), so that they yielded both adenovirus type 12 fiber antigens and complete infectious adenovirus type 2.

## Summary

In summary, adenoviruses do not seem to be the causative agents of human tumors, but under artificial laboratory circumstances some of them produce sarcomas in animals and transform cells in culture. The transformed cells contain at least part of the viral genome, probably integrated into cellular DNA. Some viral DNA sequences are transcribed at high frequency into messenger RNA which presumably codes for the virus-specific antigens that are found in transformed cells. Whether these proteins have functions which are required to maintain the cell in a transformed state is not known.

## TUMOR PRODUCTION BY POLYOMA VIRUS AND SV40

### Polyoma Virus

This virus was discovered when Gross (1953) reported that cell-free filtrates from leukemic Ak mice, when inoculated into newborn C3H mice, induced fibrosarcomas and parotid tumors as well as leukemia. It was soon apparent that at least two agents were involved, a leukemia virus and another virus (or viruses) that caused the solid tumors (Gross, 1956). It was found that the wide variety of histologically different solid tumors were in fact due to a single virus, which was isolated in tissue culture by Stewart et al. (1957) and designated polyoma virus by Stewart and Eddy (1959).

Polyoma virus is endemic in populations of wild and laboratory mice, but it has never been associated with spontaneous tumors of mice; it is generally regarded as a harmless infectious virus that produces tumors in newborn mice only under artificial laboratory circumstances (Rowe, 1961).

Stewart (1960) summarized data available up to that time on the oncogenic properties of polyoma virus; little more work has been done on this topic since then. Besides producing a wide variety of histologically distinguishable tumors when inoculated into newborn mice, polyoma virus produces tumors in several other species of animals if they are inoculated shortly after birth. Hamsters are particularly susceptible, not only when inoculated as newborn animals but also up to the age of 3 weeks. Hamster tumors (especially sarcomas of the kidney) may be recognizable as early as 7 days after inoculation, whereas in mice tumors are rarely seen before the sixth week and are usually detected much later. Sarcomas are also induced in rats, but in newborn rabbits polyoma virus produces multiple benign subcutaneous fibromas which regress within 4 months.

The tumors induced by polyoma virus do not contain any infectious virus, and no viral particles can be seen in the cells by electron microscopy. Furthermore, no infectious virus can be rescued either from polyoma virus-induced tumor cells or from cells transformed by polyoma virus *in vitro* either by co-cultivation of the neoplastic cells with permissive cells or by treatment with chemical or physical inducing agents. This finding is in sharp contrast to the situation with simian virus 40 (see below) and provides one of the major points of difference between two viruses that resemble each other closely in many other respects.

### Simian Virus 40 (SV40)

Simian virus 40 was discovered by Sweet and Hilleman (1960) as a contaminant of cultures of monkey kidney cells of the type being used to produce poliovirus vaccines. Independently, Eddy *et al.* (1961) set out to determine whether viruses found in rhesus monkey kidney cells (Hull *et al.*, 1958) could produce tumors in hamsters. They did not test isolated viruses, but merely extracts from frozen and thawed cultures of rhesus monkey kidney cells. Such extracts, from some batches of cells, produced transplantable tumors when inoculated in baby hamsters. Subsequently, it was shown (Girardi *et al.*, 1962; Eddy *et al.*, 1962) that the oncogenic agent was SV40. Discoveries about the oncogenic properties of SV40 have to some extent followed the same pattern as with polyoma virus, and some of its properties make SV40 an even better agent for studies on cellular transformation.

SV40 produces tumors in baby hamsters (Girardi *et al.*, 1962) and multimammate rats (Rabson *et al.*, 1962) but not in other species (Eddy, 1964), although cultured cells from a variety of animals can be transformed *in vitro*. The histopathology of subcutaneous fibrosarcomas in hamsters has been described by Black and Rowe (1964), Eddy (1964), and Girardi and Hilleman (1964) and of ependymomas by Gerber and Kirschstein (1962). Black *et al.* (1966) used clonal lines of cells to analyze the types of tumors that single cells could produce. The results show that the progeny of a single SV40-transformed hamster kidney cell are capable of producing tumors of several kinds in the hamster. Most tumors showed some degree of epithelial differentiation; histologically they appeared to be nephroblastomas.

Although occasional tumors release small amounts of infectious virus, most tumors produced by SV40 are virus free (Sabin and Koch, 1963a; Gerber and Kirschstein, 1962; Ashkenazi and Melnick, 1963). However, several observers have reported that very small amounts of virus could be recovered from most, but not all, virus-free tumors when tumor cells were cultivated for prolonged periods on the sensitive cells of African green monkeys (Gerber and Kirschstein, 1962). Sabin and Koch (1963a,b) used this method to show that the potential capacity to produce infectious virus probably occurred in every cell of the tumor by demonstrating that virus could be recovered from the tumors which developed in hamsters inoculated with approximately two tumor cells.

Sabin and Koch (1963b) reported that the yield of virus could be increased by several different procedures, including X irradiation of the cultured tumor cells.

Gerber (1964) confirmed this and showed that minute amounts of virus could be induced from cultured cells derived from a virus-free hamster ependymoma by treatment with UV light, $H_2O_2$, or mitomycin. Although culture fluids and cell lysates from these ependymoma cell cultures were consistently negative for SV40 infectivity, Gerber (1963) found that if the tumor cells were plated on indicator monolayers of African green monkey kidney cells, they produced small colonies; adjacent to these the African green monkey kidney cells showed characteristic vacuolation. Experiments with rabbit kidney cells (which can be infected by SV40 DNA but not by SV40 virions) and with "parabiotic" cultures of "virogenic" and indicator cells showed that this transfer of infectivity depended upon the direct contact of viable virogenic cells with the indicator cells, and did not involve the transfer of virions (Gerber, 1964, 1966).

The interpretation of these results remained in doubt until 1967, when Watkins and Dulbecco (1967) and Koprowski and his co-workers (Koprowski et al., 1967) independently demonstrated that SV40-transformed cells were indeed completely free of virus and that virus could be rescued by fusion of the transformed cells with permissive cells. These experiments showed beyond any doubt that a total set of viral genes persisted in each and every cell of a transformed line. In view of this success with SV40, it is disappointing that all attempts to rescue polyoma virus from transformed cells in a similar way have failed. Possible explanations of this paradox will be postponed until we have discussed in detail the interactions of the two viruses with cultured cells.

## INTERACTIONS OF POLYOMA VIRUS AND SV40 WITH CULTURED CELLS

Polyoma virus and SV40 interact with cells in two distinct and mutually exclusive ways; either there is a productive response which yields viral progeny and results in death of the infected cells, or an abortive or nonproductive infection during which very little or no progeny virus is produced and the cells survive. During nonproductive infection the cells change their pattern of growth and acquire properties very similar to those of tumor cells. In most cases, the change is transient so that after a few generations the cells resume their normal growth characteristics. This process is called abortive transformation. However, some of the infected cells are altered permanently, and are said to be stably transformed. They retain indefinitely the properties of tumor cells which were displayed transiently by the abortive transformants.

Whether polyoma virus and SV40 give rise to productive or nonproductive infections depends largely on the species of cells being infected (see Table 13–2). During infection of mouse cells by polyoma virus and of monkey cells by SV40, the response by the overwhelming majority of cells is productive—new virus is synthesized by cells which are fated to die. Sometimes, however, a few cells become transformed. During infection of rat and hamster cells by polyoma virus and of mouse cells by SV40, transformation occurs at much higher frequency, and no virus at all is produced. Human and hamster cells infected by SV40 show a mixed response, with some cells of the culture yielding virus and others being nonproductively infected. Even though the response appears mixed in the total

TABLE 13–2

*Response of Cells of Different Species of Mammal to Infection
by Polyoma Virus and SV40*

| VIRUS | PRODUCTIVE INFECTION | NONPRODUCTIVE INFECTION, WITH TRANSFORMATION |
|---|---|---|
| Polyoma | Mouse cells (the vast majority of cells in any culture) | Mouse cells (a minority) Hamster cells Rat cells |
| SV40 | Monkey cells | Mouse cells Hamster cells Human cells |

cell population, at the level of individual cells it is quite distinct. Either a cell supports a fully permissive response, or it does not. If it does not, it may become stably transformed as a result of exposure to the virus.

We do not know why some cells give rise to lytic and others to nonproductive infections. Basilico (1970) found that lines of hybrid cells formed by fusing permissive mouse cells and nonpermissive hamster cells could support the growth of polyoma virus. Each of the hybrid lines contained different proportions of mouse and hamster chromosomes and supported different levels of viral multiplication. The hybrid lines with the highest numbers of mouse chromosomes gave the greatest yields of virus, and the lines with small numbers of mouse chromosomes the lowest. This result is interpreted to mean that permissiveness of the hybrids is a consequence of the production by the mouse chromosomes of a factor or factors necessary for the multiplication of polyoma virus. What the factor(s) are and how they act is unknown. The result also suggests that hamster cells are nonpermissive because they lack this factor and eliminates the possibility that cells are nonpermissive because they contain specific repressors of viral synthesis.

The various interactions of polyoma virus and SV40 with cells have been surveyed in three recent extensive reviews (Green, 1970a; Benjamin, 1972; Sambrook, 1972). The productive responses have been discussed already in Chapter 5 of this book. In the rest of this chapter we will concentrate on events that occur during nonproductive infection and on the properties of stable lines of transformed cells.

## EVENTS DURING NONPRODUCTIVE INFECTION WITH POLYOMA VIRUS AND SV40

The initial stages of viral infection appear to be very similar in permissive and nonpermissive cells. The virus attaches to the cell and is uncoated. Soon after infection of mouse cells with SV40, virus-specific RNA can be detected (Martin, 1970); but because very little is made detailed analysis has been impossible, and we do not know whether the RNA corresponds to "early" or "late" viral DNA sequences, nor from which strand of viral DNA it is copied. Virus-specific T antigen appears in the nucleus of the cell, and there is an increase in the activity of some of the enzymes concerned with manufacture of precursors for DNA synthesis (see, for example, Kit et al., 1967a and Kit, 1968). The viral multiplication cycle seems to stop at this stage in nonpermissive cells; neither viral DNA nor capsid protein synthesis is detectable and no progeny virus is made. However, by 18–24 hours after infection cellular DNA synthesis is induced, and many of the infected cells in the culture divide a few times and behave for a few cell generations like transformed cells—"abortive transformants." Two observations show that this effect is a consequence of viral gene function: first, some mutants of polyoma virus fail to show the effect (Benjamin and Norkin, 1973), and second, it is prevented by interferon (Dulbecco and Johnson, 1970; Taylor-Papadimitriou and Stoker, 1971). Abortive transformants grow for a few generations under conditions that restrict the growth of normal cells—in suspension in sloppy agar or methylcellulose (Stoker, 1968) or in liquid medium depleted of serum growth factors (H. Smith et al., 1970, 1971). In addition, they show the membrane alterations that are typical of tumor cells, i.e., they become agglutinable by concanavalin A (Ben-Bassat et al., 1970) and wheat germ agglutinin (Sheppard et al., 1971), and a new set of antigens is exposed (Irlin, 1967).

In most cells of the abortively transformed population, the changes in both growth properties and membranes are evanescent, and after a few days the cells settle back into their original state. T antigen disappears (Oxman and Black, 1966) along with viral mRNA, and the lectin-mediated agglutinability decreases as the membrane returns to its normal configuration. The cells will no longer grow in suspension or in depleted medium. However, at least in the SV40–mouse system, some of the cells do not emerge completely unchanged from their encounter with the virus; after they return to an apparently normal phenotype, they continue to carry for a large number of cell generations several copies of viral DNA (H. Smith et al., 1972). No infectious virus can be rescued from these cells by fusion with permissive cells, and they are negative for SV40-specific T antigen (H. Smith et al., 1971). The mechanism by which viral DNA is preserved in the cells and its physical state are both unknown.

Not all of the surviving cells are abortively transformed cells; some remain permanently fixed in the transformed state. Why some cells give rise to stable transformants and some do not is not known. However, at least one round of cell division after infection is necessary to achieve stable "fixation" of the transformed state (Todaro and Green, 1966b). Virus-coded functions are necessary because the transforming ability of virus is inactivated by UV irradiation (Basilico and di Mayorca, 1965; Latarjet et al., 1967; Aaronson, 1970), and temperature-sensitive and host-range mutants exist which are unable to transform cells (review, Benjamin, 1972; and see Chapter 7). However, not all of the viral genes are essential for transformation because mutants in "late" viral functions (Benjamin, 1972) and some types of defective viruses that cannot replicate (Shiroki and Shimojo, 1971) have unimpaired transforming ability. Furthermore, determination of the target size for inactivation of transforming and reproductive capacity, after exposure of virus to X rays (Basilico and di Mayorca, 1965) and other agents that affect the DNA (Benjamin, 1965; Aaronson, 1970), showed that the target size for transforming ability is no more than one-half as big as that for plaque production (Latarjet et al., 1967). This can be interpreted to indicate that less than one-half as many genes are required for transformation as are necessary for productive infection. On the other hand, the target size for the induction of cellular DNA synthesis (see Chapter 5) is the same as that for plaque production (Basilico et al., 1966).

In contrast to the narrow range of species of intact animals in which SV40 is oncogenic, cultured cells from many different species have been transformed: hamster (Rabson and Kirschstein, 1962; Ashkenazi and Melnick, 1963; Black, 1966), human (Shein and Enders, 1962; Koprowski et al., 1962; Girardi et al., 1965), rabbit, mouse, pig (Black and Rowe, 1963), and bovine (Diderholm and Wesslén, 1965). Jensen et al. (1964) reported that a few human patients with terminal cancer, who were inoculated with isologous SV40-transformed human fibroblasts, developed small sarcomatous nodules which regressed. About 0.02 to 0.03% of normal human fibroblasts in culture become transformed after infection with SV40. However, cells from humans with certain chromosomal aberrations, such as Fanconi's anemia, Downs' syndrome, or Klinefelter's syndrome, seem to be up to ten times more susceptible to transformation (Todaro et al., 1966; Todaro and Martin, 1967; Potter et al., 1970; Mukerjee et al., 1970). The differences between cells are eliminated by using SV40 DNA instead of intact virions, suggesting that "normal" human cells are resistant to transformation by SV40 because of a block at an early stage of infection, such as attachment or uncoating.

## Efficiency of Transformation

The SV40–mouse cell combination is the most efficient system of transformation available with the papovaviruses, and frequencies as high as 40% have been reported by Todaro and Green (1966a). With the best polyoma virus system, only about 5% of the cells ever become transformed. The inability to transform all the cells does not seem to be a consequence of genetic inhomogeneity or physiological differences in the cell population (Macpherson and Stoker, 1962; Black, 1964; Basilico and Marin, 1966), so it seems that stable transformation is the result of a chance event which occurs with low probability in virus-infected cells. Because there is a linear relationship between the number of transformed clones and virus dose (Stoker and Abel, 1962; Macpherson and Montagnier, 1964; Todaro and Green, 1966a), it is likely that a single virus particle has the ability to cause transformation. In order to overcome the inefficiency of the process, however, cells must be infected at multiplicities of $10^3$ to $10^5$ pfu/cell in order to detect transformation.

## Transformation by Viral DNA

Transformation of hamster cells can be produced with polyoma DNA. Although both the twisted and open circular forms of the molecule (see Chapter 3) can cause transformation, the twisted molecules are about ten times more efficient (Bourgaux et al., 1965). Comparison of transformation by virus and by

DNA shows that there was about one transformation for every $10^8$ to $10^9$ DNA molecules, compared with one per $10^6$ intact virions.

SV40 DNA has been shown to carry transforming ability for human (Aaronson and Todaro, 1969) and bovine cells (Diderholm et al., 1965). However, for unknown reasons, all attempts to transform mouse cells by SV40 DNA have failed (see Pagano and Hutchison, 1971).

## PROPERTIES OF TRANSFORMED CELLS

Transformed cells grow under conditions which restrict the multiplication of normal cells. For example, transformed cells will divide when suspended in medium containing agar or methylcellulose (Macpherson and Montagnier, 1964) or in medium depleted of serum growth factors (H. Smith et al., 1971); untransformed cells will not. It is this sort of differential growth that provides the basis for assays used to select colonies of transformed cells from a background of untransformed cells (review, Sambrook and Pollack, 1972). The most commonly used assay for transformation by polyoma virus was developed by Macpherson and Montagnier (1964). Cells are infected in suspension and then plated in medium containing either 0.33% agar or 1.2% methylcellulose. The stably transformed cells grow to form large colonies which are visible to the naked eye; untransformed cells divide only once or twice. For SV40 a different assay is used, mainly for historical reasons. Monolayers of cells are infected with virus, and the cells are then plated at low density either in complete medium (Todaro and Green, 1964) or in medium which has been depleted of serum growth factors (H. Smith et al., 1970). In complete medium, both transformed and untransformed cells grow to form colonies. Those formed by the transformed cells are thicker and bigger, are composed of disorganized masses of cells, and are easy to distinguish from those formed by the untransformed cells, which are thinner and more orderly. In depleted medium, only transformed cells grow to produce large colonies. By picking and subculturing transformed clones it is possible to establish lines of transformed cells that can be used to define the differences between transformed and normal cells. These can be considered under two headings: changes in cell growth and alterations in the cell surface.

### Changes in Cell Growth

As well as the ability to grow under the selective conditions described above, transformed cells form tumors more readily than untransformed cells on injection into susceptible animals (Aaronson and Todaro, 1968; Jarrett and Macpherson, 1968; Pollack and Teebor, 1969; Marin and Macpherson, 1969), multiply to higher cell densities in culture, grow in a less oriented manner, require less serum, show higher rates of cell movement, show changes in composition of glycoproteins and glycolipids, display new surface antigens, build up chromosome aberrations, and form colonies when plated on resting cultures of nontransformed cells. Although this is only a partial list of the differences that have been reported between the two sorts of cells, it is quite clear that gene products of polyoma and SV40 cannot be directly and separately responsible for all of these changes, since there are more changes reported than there are virus-coded proteins to cause them. Clearly then most of the observed alterations must be pleiotropic or they must be secondary and tertiary effects which occur as a consequence of some central, primary event. The various changes which characterize transformed cell lines tend to behave in a covariant manner, but they can be dissociated one from another, so that any one transformed cell line will not necessarily display all the properties characteristic of transformed cells as a class.

## Alterations in Cell Surfaces

The surfaces of transformed and tumor cells differ from those of normal cells in two major respects. First, new antigens can be detected on the surface of transformed cells (review, Hellström and Hellström, 1969), and second, transformed cells react with lectins in a different way from normal cells.

**Transplantation Antigens (TSTA).** This class of antigens is demonstrated by showing that animals that have been immunized with polyoma virus or SV40 or with virus-transformed or tumor cells from another animal species more readily reject isologous cells transformed by the same viral species than do unimmunized animals (review, Melnick et al., 1970; see Chapter 10). The antigens have been demonstrated to occur on the surfaces of tumor cells and transformed cells produced by polyoma virus (Habel, 1961; Sjögren, et al., 1961; Sjögren, 1965) and SV40 (Defendi, 1963, 1968; Koch and Sabin, 1963; Khera et al., 1963; Goldner et al., 1964). They are virus specific in that animals immunized with polyoma virus-transformed cells are not protected against challenge by SV40-transformed or tumor cells and vice versa. The same transplantation antigen is induced by different strains of polyoma virus (Friedman and Rabson, 1964) and is present in the cells of all the many types of tumors produced by the virus in mice. However, some strains of the virus are more efficient than others at inducing the antigen (Hare, 1964, 1967; Hare and Godal, 1965; Jarrett, 1966). Although the structure of the TSTA's is unknown, they do not cross-react with the intranuclear T antigens or virion antigens (Sjögren, 1964a). Since they are also synthesized during the lytic cycle (Girardi and Defendi, 1970), TSTA's are possibly virus-coded, but they might well be cell-coded proteins modified or induced by a consequence of viral infection.

**Surface Antigens (S Antigens).** Cells transformed by polyoma virus or SV40 often but not invariably display on their surfaces virus-specific antigens (Levin et al., 1969; A. S. Levine et al., 1970) which can be detected by any one of a variety of in vitro immunological tests, including cytotoxicity (Tevethia and Rapp, 1965; Hellström, 1965; Tevethia et al., 1965; Hellström and Sjögren, 1965, 1966, 1967; H. Smith et al., 1970; Tevethia et al., 1968; Wright and Law, 1971), immunofluorescence (Irlin, 1967; Malmgren et al., 1968), and mixed hemagglutination (Häyry and Defendi, 1968, 1969; Metzgar and Oleinek, 1968). Some of these antigens may be identical to TSTA (review, Butel et al., 1972), but others certainly are not (Wright, 1971). At least some of the antigens detected by in vitro tests are cell-coded, since they can be exposed on the surface of normal cells by mild protease treatment (Häyry and Defendi, 1970). Furthermore, Duff and Rapp (1970a) were able to show that sera from pregnant hamsters reacted with the membrane of SV40-transformed hamster cells, and Baranska *et al.* (1970) demonstrated antigenic cross-reactions between SV40-transformed mouse cells and unfertilized mouse eggs. It seems then that some, if not all, of the antigens detected on the surface of virus-transformed cells by *in vitro* tests may be cell-coded embryonic antigens. It is not clear how the virus causes the exposure of these antigens or why antigens which are cell-coded should be so virus specific.

**Reaction with Lectins.**   Lectins is the name given to a diverse group of proteins or glycoproteins, occurring mostly in plants, which share the property of binding to carbohydrate residues (Watkins and Morgan, 1952). A vast number of different lectins is known, but most attention has been given to concanavalin A and wheat germ agglutinin, because these two lectins show an ability to distinguish between normal and malignant cells. Suspensions of many sorts of tumor cells (Aub *et al.*, 1963, 1965a, b; Liske and Franks, 1968) and chemically or virally transformed cells are agglutinated much more readily by concanavalin A and wheat germ agglutinin than are suspensions of normal cells (Burger and Goldberg, 1967; Burger, 1969; Inbar and Sachs, 1969; Inbar *et al.*, 1969; Burger, 1971). Because the agglutination is reversible by haptens [$\alpha$-methylglucose for concanavalin A (Kalb and Levitzki, 1968) and N-acetylglucosamine for wheat germ agglutinin (Burger and Goldberg, 1967)], it seems likely that the lectins react with sugar groups on the cell surfaces and cause agglutination by forming cross-links between cells. The reason for the difference in agglutinability between the two sorts of cells is unknown, but it seems to be due to extensive architectural changes in the cell membrane which accompany malignant transformation, rather than to an unequal number of concanavalin A and wheat germ agglutinin binding sites on normal and transformed cells (Cline and Livingston, 1971; Ozanne and Sambrook, 1971a; Nicholson, 1971; Jovin-Arndt and Berg, 1971). Mild treatment of normal cells with proteases can cause them to become agglutinable (Burger, 1969; Inbar and Sachs, 1969); and Burger has suggested that the agglutinin receptor sites are present on normal cells in a "cryptic" form (review, Burger, 1971a) and become unmasked by the treatment with proteolytic enzymes, or as a consequence of transformation. Singer and Nicolson (1972) have proposed an alternative explanation for the change in agglutinability based on the fluid mosaic model of membrane structure (see Fig. 4–2), in which the agglutinin binding sites are clustered in agglutinable cells and dispersed in nonagglutinable cells. Whichever explanation turns out to be correct, it is clear that the agglutinability of transformed cells is somehow affected by the expression of viral genes, since mutants of polyoma virus are known which are unable to convert cells transformed by them from the nonagglutinable to the agglutinable state (Benjamin and Burger, 1970).

Burger (1971a) has reviewed evidence which suggests that the architectural change which accompanies agglutinability may be involved in the control of cell division. However, much more work is required to resolve whether this association is causal or casual.

**Other Changes.** There are consistent reports of alterations in the carbohydrate and glycolipid composition of membranes in transformed cells (Wu *et al.*, 1969; Hakomori and Murakami, 1968; Hakomori *et al.*, 1969; Mora *et al.*, 1969; Brady and Mora, 1970). Most of these reports show that the glycolipids present in transformed cells are deficient in terminal sugars and that the activity of at least one specific sugar transferase is lower in transformed cells than normal cells (Cumar *et al.*, 1970; Mora *et al.*, 1971). However until our knowledge of membrane biochemistry is more complete, the relationship of these changes to lectin-mediated agglutinability, surface antigens, tumorigenicity, and control of cell growth will be obscure.

## Viral Genes in Transformed Cells

The presence of viral genes in transformed cells can be demonstrated by rescue experiments or by the detection of viral DNA sequences, their activity by the detection of viral mRNA's or virus-specific proteins.

**Rescue of Infectious Virus: SV40.** The great majority of cell lines transformed by SV40 carry no trace of infectious virus or infectious viral DNA (Sabin and Koch 1963b; Kit *et al.*, 1968), and most attempts to induce viral multiplication by physical and chemical methods, such as exposure to mitomycin C, proflavine, $H_2O_2$, amino acid starvation, and X irradiation have failed or are successful only in causing the release of minute amounts of virus ($10^{-5}$ to $10^{-3}$ pfu/cell) (Sabin and Koch, 1963b; Gerber, 1964; Burns and Black, 1968, 1969; Rothschild and Black, 1970; Kaplan *et al.*, 1972). However, if lines of SV40-transformed cells are mixed with permissive cells, infectious virus becomes detectable a day or two later (Gerber and Kirchstein, 1962; Sabin and Koch, 1963b). Contact between the two sorts of cells is essential (Gerber, 1966) and the yield of virus is always low. The sensitivity of the system can be increased by forcing the transformed cells and the indicator cells to fuse into heterokaryons by the use of inactivated Sendai virus (Koprowski *et al.*, 1967; Watkins and Dulbecco, 1967; Tournier *et al.*, 1967), or by treating the transformed cells with base analogs before fusion with the permissive cells (review, Watkins, 1971a). Although the frequency varies from clone to clone, virus can always be rescued from all derivatives of the same parental line (Watkins and Dulbecco, 1967; Dubbs *et al.*, 1967), suggesting that every cell of the population carries at least one complete SV40 genome. Some lines of SV40-transformed cells yield virus only rarely, and some never yield at all, even though viral DNA and T antigen are detectable in the cells; the most likely explanation of this result is that the nonyielding cells were transformed by defective viral mutants (Dubbs and Kit, 1968). Virus can be rescued as readily from cells transformed with low multiplicities of virus as from cells transformed with high multiplicities (Kit and Brown, 1969); it seems that there is no correlation between the number of viral genomes in the transformed cells and the efficiency with which they can be recovered.

The virus recovered from the transformed cells is identical in most respects to the parental transforming virus (Tournier *et al.*, 1967; Takemoto *et al.*, 1968) although there is one report which suggests that the rescued virus transforms more efficiently than the original virus (Todaro and Takemoto, 1969). The mechanism of viral rescue by cell fusion is unknown. Presumably it involves excision of the viral genome from its integration site, thus allowing multiplication of the virus in the permissive environment donated to the heterokaryon by the monkey cell.

**Rescue of Infectious Virus: Polyoma Virus.** Although a few exceptional polyoma virus-transformed cell lines exist in which viral multiplication can be induced by physical and chemical methods (Fogel and Sachs, 1969, 1970), the great majority of cell lines transformed by polyoma never yield virus, even when the manipulations that are so successful with SV40 are applied. The most plausible explanation for this difference in the behavior of the two viruses is that all the cells commonly used for transformation by polyoma virus are slightly permissive, so that if the cells are to survive, transformation must be caused by viral mutants that are defective in multiplication. Even though polyoma virus cannot be rescued, the evidence outlined below provides ample proof that viral genes persist in transformed cells.

**Detection of Viral DNA Sequences by Hybridization.** Although the weight of the DNA of polyoma virus or SV40 is only about one-millionth of that of cellular DNA, it has been possible to detect and quantitate viral DNA sequences in transformed cells, using the two different nucleic acid hybridization techniques discussed earlier in this chapter and in Chapter 2. Using RNA synthesized *in vitro* from viral DNA as a diagnostic probe for integrated viral DNA sequences, Westphal and Dulbecco (1968) found that there was always some hybridization to DNA extracted from untransformed cells, but a significantly greater amount to transformed cell DNA. Different transformed cell lines contained different numbers of viral genome equivalents ranging from about five copies in some lines of mouse cells transformed by polyoma virus to as many as fifty-eight in a hamster tumor cell line induced by SV40. As the hybridization procedure has been improved over the years, the estimate of viral genomes per cell obtained using *in vitro* synthesized viral RNA has been lowered (H. Westphal, personal communication; A. J. Levine *et al.*, 1970), and most SV40-transformed cell lines are now thought to contain about two to five copies of viral DNA per cell (A. J. Levine *et al.*, 1970). This estimate is in fairly good agreement with the value of one to three genome equivalents per cell obtained by Gelb *et al.* (1971b) using renaturation kinetics.

The state of the viral DNA in one line of SV40-transformed mouse cells was examined by Sambrook *et al.* (1968). Using the DNA–RNA hybridization technique, they found that all the detectable viral DNA sequences remained associated with very high molecular weight cellular DNA during zonal and equilibrium centrifugation in alkaline gradients. No free circular DNA molecules containing viral sequences were found. From these experiments, it seems that the SV40 DNA is covalently linked to the chromosomal DNA of the transformed cell. Nothing is known of the location of the viral sequences within the cell genome.

The physical evidence in favor of integration is supported by the elegant genetic experiments of Weiss and her colleagues (Weiss *et al.*, 1968; Weiss, 1970), which exploited the fact that following repeated cell division human chromosomes are gradually lost from mouse–human hybrid cells. It was found that hybrids prepared by fusing SV40-transformed human cells with mouse cells retained the virus-specific T antigen until virtually all the human chromosome complement had been lost. Because T antigen synthesis could not be associated with any single specific human chromosome, these results suggest either (a) that there are viral genomes associated with many different chromosomes in transformed cells, or (b) that there are a few SV40 genomes which may be located on different chromosomes in different cells of the same transformed cell line.

**Transcription of Viral Genes in Transformed Cells.** The first definite evidence that the viral DNA in transformed cells served as a template for RNA synthesis was obtained by Benjamin (1966), who demonstrated that pulse-labeled RNA from cells transformed by polyoma virus and SV40 contains base sequences homologous to the transforming virus. This finding stimulated a great amount of work by many groups aimed at defining the portions of the viral genome that were transcribed, because these should contain the sequences that code for any viral functions required to maintain the cells in the transformed state. The following results have been obtained: (a) The proportion of the total RNA in the cells that is virus specific is very small (0.01%, Benjamin, 1966). (b) The viral RNA found in transformed cells can compete in hybridization tests with about 80% of the viral sequences transcribed early during lytic infection and about 30 to 40% of the RNA sequences found late in lytic infection (Aloni *et al.*, 1968; Sauer and Kidwai, 1968; Oda and Dulbecco, 1968; Tonegawa *et al.*, 1970; Sauer, 1971). These experiments suggest that in transformed cells most of the "early" viral DNA sequences and a small fraction of the "late" sequences are transcribed, adding up to a total of about 40% of the SV40 genome. However, this result does not seem to fit with (c) saturation hybridization experiments, which show that up to 100% of the viral genome is transcribed in certain cell lines (Martin and Axelrod, 1969a, b). A possible resolution of the paradox seems to be that (d) in at least one line of transformed cells, as well as RNA sequences corresponding to all the true "early" and a small proportion of the "late" genes, some of the viral RNA that is synthesized is "anti-late" (Sambrook *et al.*, 1972). This was shown using hybridization of RNA to separated "early" and "late" strands of viral DNA. Because the "anti-late" RNA would not have been detected in the earlier competition hybridization experiments, but would have appeared in the direct saturation experiments, it probably accounts for the different results obtained with the two systems. (e) Giant RNA molecules ($4 \times 10^6$ daltons), containing covalently linked host and viral RNA sequences, have been found in the nuclei of mouse cells transformed by SV40 (Lindberg and Darnell, 1970; Tonegawa *et al.*, 1970; Wall and Darnell, 1971). The most plausible explanation for the existence of these molecules is that they result from transcription of an integrated SV40 genome, as a consequence of "read through" of RNA polymerase from cellular into viral sequences or vice versa. This is supported by the observation that interferon fails to interfere with the translation of the mRNA for T antigen in transformed cells but inhibits T antigen synthesis in productively infected cells (Oxman and Black, 1966; Oxman et al., 1967).

By contrast to the nuclear RNA, the smaller SV40-specific RNA found on the polyribosomes of transformed cells is not attached to host RNA (Wall and Darnell, 1971). There is some evidence from pulse chase experiments that the cytoplasmic viral RNA is derived from long nuclear precursors (Tonegawa et al., 1970, 1971).

**Virus-Specific Proteins in Transformed Cells.** Without exception, cells transformed by polyoma virus or SV40 contain T antigen which can be detected by complement fixation (Huebner et al., 1963; Koch and Sabin, 1963) or immunofluorescence (Rapp et al., 1964; Pope and Rowe, 1964). The antigen has never been purified completely, but it is heat labile and sensitive to trypsin (Gilden et al., 1965); estimates for its molecular weight range from 70,000–80,000 (Del Villano and Defendi, 1970) to 250,000 daltons (Gilden et al., 1965). Its biological function is unknown. Although T antigens are found both in lytically infected and transformed cells and are virus specific, there is no direct evidence to support the common belief that they are virus-coded products. They could conceivably be derepressed or modified cellular proteins.

SV40-transformed cells have been reported to contain a second virus-specific antigen called U antigen (Lewis et al., 1969; Lewis and Rowe, 1970, 1971). Nothing whatsoever is known of the structure or function of this antigen, which is located at the nuclear membrane of transformed cells. Apart from the virus-specific TSTA's and S antigens, which have been discussed earlier, no other candidates for virus-coded proteins have been identified in transformed cells. However there is good genetic evidence that virus-coded proteins are necessary, not only for the initiation, but also for the maintenance of the transformed phenotype, as follows:

1. *Host Range Mutants of Polyoma Virus.* Benjamin (1970) has isolated mutants of polyoma virus which will grow only in cells transformed by polyoma virus, not in normal mouse cells (see Chapter 7). These mutants, by definition, are defective in a function which is required for lytic growth; they grow in transformed cells because their defect is complemented, presumably by a polyoma virus-specific product, and which is coded by the viral genome resident in the transformed cell. Because the host range mutants are unable to transform rat or hamster cells, it follows that the polyoma virus product responsible for complementing the mutant defect must itself be necessary for transformation.

2. *The Temperature-Sensitive Mutant of Polyoma Virus: ts-3.* This mutant multiplies normally in permissive (mouse) cells incubated at 31° but fails to grow at 38.5°C. A line of hamster cells transformed by ts-3 has been established; when these cells are grown at 31°C they show all the attributes of polyoma virus-transformed cells. However, after the cells are shifted to 38.5°C, they behave in the lectin-mediated agglutinability test as untransformed cells, and they appear morphologically normal (Eckhart et al., 1971). If the cells are shifted down from 38.5° to 31°C, they regain the full panoply of transformation

characteristics within 24 hours. Not all of the transformed phenotype disappears after growth at 38.5°C; for instance, the cells retain the ability to form colonies in agar. These results prove that the presence of at least one virus-coded product is required for expression of at least some of the phenotypic characters of transformed cells.

The evidence concerning the existence of virus-specific repressors in transformed cells is conflicting. On the one hand, Jensen and Koprowski (1969) have suggested that because some lines of permissive cells transformed by SV40 are fully susceptible to superinfection (Barbanti-Brodano et al., 1970), they do not contain a repressor. A similar conclusion was drawn by Basilico (Basilico, 1971; Basilico and Wang, 1971) for cells transformed by polyoma virus, on the basis of experiments in which he found that hybrids of polyoma virus-transformed BKH cells and 3T3 cells were permissive to superinfection. On the other hand, Cassingena and his colleagues (Cassingena et al., 1969a,b; Suarez et al., 1971) have described the isolation of a protein from a variety of SV40-transformed cells which seems to have some of the properties expected of a viral repressor. The substance specifically inhibits plaque formation by SV40 on monolayers of monkey kidney cells. However, the maximum plaque reduction ever achieved by Cassingena's group is about 50%, which hardly seems enough to merit serious consideration.

## Reversion

Lines of transformed cells occasionally yield revertants that have returned to the normal phenotype, and several ingenious methods have been devised to apply selection pressure for these variants. For instance, Pollack et al. (1968) used 5'-fluorodeoxyuridine to kill dividing cells in dense cultures of SV40-transformed mouse cells. The survivors had the flattened morphology of untransformed cells, grew to low saturation density in culture, were less tumorigenic than the parent line (Pollack and Teebor, 1969), and showed reduced levels of lectin-mediated agglutinability (Pollack and Burger, 1969). Wyke (1971a,b) used 5'-bromodeoxyuridine to kill off the vast majority of polyoma virus-transformed BKH-21 cells which synthesize DNA when plated in agar, and he was able to obtain cells which resembled untransformed BHK cells in being unable to multiply in suspension. Ozanne and Sambrook (1971b) used the lectin concanavalin A to kill 99.99% of a population of mouse cells transformed by SV40; the survivors formed "flat" colonies and were less agglutinable by lectins than the original transformed cell line.

Even though Pollack's and Ozanne's revertants display a phenotype that is in most respects characteristic of untransformed cells, they contain unchanged amounts of viral DNA sequences as assayed by reassociation kinetics. Furthermore, viral RNA is detectable and virus-specific T antigen is synthesized (Ozanne et al., 1973). However, rescue of infectious SV40 by fusion of the revertants with permissive cells is erratic. Pollack (1970) reported originally that the virus recovered from the revertants possessed both infectivity and transforming ability. However, as the revertants have been maintained for longer periods in culture, virus has been more and more difficult to rescue. The reason for this and the mechanism(s) by which reversion occurs are unknown. We do not know for sure whether mutations are involved, let alone whether they are in viral or cellular genes. However, Renger and Basilico (1972) recently described the isolation of a line of SV40-treated mouse cells which has a transformed phenotype at low and a more normal phenotype at high temperatures. SV40 virus, which resembles wild-type virus in all respects tested, can be rescued from these cells. Thus it seems that the mutation leading to reversion in this case is in the cellular and not the viral genes. The existence of the line also provides proof that ongoing synthesis of at least one cell-coded function is required to maintain cells in the transformed state.

In addition to the methods described above, several other techniques have been used to isolate revertants. For instance, variants of polyoma virus-transformed hamster cells have been isolated which can grow on monolayers of normal cells fixed with glutaraldehyde. Why this procedure should select for or induce revertants is not clear; however, about 90% of the clones that were selected were less tumorigenic than the parent cells and behaved like normal cells *in vitro* (Rabinowitz and Sachs, 1968, 1969, 1970a,b; Inbar *et al.*, 1969) Presumably the revertants continue to carry polyoma virus genes because they contain polyoma virus-specific T antigen. By contrast, the revertant cells isolated by Marin and Littlefield (1968), Marin and Macpherson (1969), and Marin (1971) appear to have lost the viral genes. Since these revertants were obtained after forced loss of chromosomes from transformed cells, it seems likely that some of the chromosomes which were lost carried the viral genomes responsible for transformation.

The revertants display a phenotype which is essentially identical to that of untransformed cells, hence they can be used in transformation assays. However it turns out that of all of the types of revertant available, only those isolated by Marin and Littlefield are sensitive to retransformation. Why the revertants obtained by Wyke, Pollack, and Ozanne should be resistant is not clear; many explanations are possible ranging from the mundane (cells resistant to superinfection because of a surface change) to the exciting (integration site blocked by resident genome, repressors, trans-dominant effects, etc.); at the moment we do not know which (if any) is correct.

If there is a mechanism which blocks retransformation, it must be directed specifically against the original transforming virus, because several lines of hamster and mouse cells exist which have been sequentially transformed by polyoma virus and SV40 (Todaro *et al.*, 1965; Takemoto and Habel, 1966). These cells contain the T antigens and transplantation antigens of both viruses, and hybridization experiments have demonstrated the presence of both viral genomes (Benjamin, 1966; Westphal and Dulbecco, 1968; Gelb *et al.*, 1971b). These results suggest that there are different integration sites for SV40 and polyomavirus DNA's in the cellular genome.

## ADENOVIRUS-SV40 HYBRID VIRUSES

The properties of hybrid viruses containing, within adenovirus capsids, genomes made up of covalently linked DNA sequences derived from adenovirus and SV40 were described in Chapter 7. Populations of adenovirions containing such particles (E46$^+$ pool) produce tumors in hamsters that contain both SV40 and adenovirus 7 T antigens (Huebner *et al.*, 1964). After *in vitro* transformation the SV40 T antigen was found in all transformed cells, that of adenovirus 7 in none (Black and Todaro, 1965), and the ependymomas produced by intracerebral inoculation of newborn hamsters with mixed adenovirus type 7-SV40 preparations were morphologically identical with those produced by SV40 (Kirschstein *et al.*, 1965). Tests for transplantation immunity showed that the mixed virions conferred resistance comparable to that afforded by SV40 alone (Rapp *et al.*, 1966).

The SV40 component likewise dominates transformation and tumor production by certain adenovirus 2-SV40 preparations; the morphology of hamster cells transformed *in vitro*, the chromosomal aberrations, and the T antigen all resemble those found after transformation by SV40 (Black and White, 1967; Igel and Black, 1967). Adenovirus type 2 is nononcogenic, but the tumors produced in hamsters inoculated with cells transformed by defective adenovirus 2-SV40 hybrids contain areas that are histologically like adenovirus tumors, and sera from hamsters bearing these tumors produce antibodies to some adenovirus early antigens. The best studied of the nondefective hybrid viruses obtained from these populations of virions by Lewis and his colleagues (Ad2$^+$ND1), which contains only enough SV40 DNA to code for protein of molecular weight 25,000 daltons, is nononcogenic for baby hamsters (Lewis *et al.*, 1969).

## CHAPTER 14

# Viral Oncogenesis: RNA Viruses

## INTRODUCTION

All tumor viruses have one property in common—they can induce tumors when inoculated into susceptible animals. This feature apart, they are a collage of heterogeneous viruses with no taxonomic relationships. They can be arbitrarily classified into two groups: the DNA tumor viruses which have double-stranded DNA genomes of various sizes, and the RNA tumor viruses, the genomes of which are single-stranded RNA molecules of molecular weight about 10 million daltons.

As we saw in Chapter 13, the DNA tumor viruses include members of four of the six taxonomic groups of DNA viruses: Papovaviridae, *Adenovirus, Poxvirus,* and *Herpesvirus.* By contrast, only one of the eleven currently recognized groups of RNA viruses has been unequivocally associated with neoplastic disease. The oncogenic RNA viruses belong to two of the four subgenera of the genus *Leukovirus* (subgenera A and B, see Table 1–16); they have also been called oncornaviruses, C-type viruses, Rousviruses or Thylaxoviridae. They are widely distributed in normal animal populations and cause leukemias and sarcomas in a large number of animal species. Their virions are large and complex structures (see Chapter 3) whose organization is still not completely defined. Subgenus B comprises a single virus, the mouse mammary tumor virus, that differs structurally in minor ways from the C-type viruses of subgenus A, and is antigenically quite distinct from the murine C-type viruses.

Although C-type viruses (subgenus A of Table 1–16) have been isolated from many sorts of vertebrates, they are divided into two major groups, avian and mammalian, on the basis of the species from which they were isolated, members of each group containing a common internal antigen. Each group has been further classified in two ways. Historically the first subdivision was based on pathogenic potential and distinguished those viruses which cause solid tumors (the sarcoma viruses) from those that produce latent infections which may sometimes progress so as to cause neoplasia of the lymphoid or hemopoietic tissues (the avian leukosis and murine leukemia viruses). The other scheme is based on the properties of the glycoproteins in the viral envelope, which are responsible for their surface antigenic properties, their host range *in vitro,* and their ability to interfere with the growth of other viruses of the same group. In this way, the avian tumor viruses have been thoroughly classified into five subgroups—A, B, C, D, and E (Duff and Vogt, 1969; Hanafusa, 1970; Hanafusa *et al.*, 1970; Vogt, 1970). Classification of the mammalian viruses is less advanced; they are divided into subgroups on the basis of their species of origin (mouse, cat, hamster, rat, etc.) and on internal subgroup-specific antigens (Freeman *et al.*, 1970). The mouse subgroup can be further subdivided using the same criteria as the avian leukoviruses, but more work is needed to clarify the exact assignment of individual viruses (Levy *et al.*, 1969; Hartley *et al.*, 1970).

As well as these three major groups of RNA tumor viruses, leukoviruses that probably belong to subgenus A have been reported in snakes (Zeigel and Clark, 1971), monkeys (Wolfe *et al.*, 1971), pigs (Howard *et al.*, 1967), bovines (Ferrer *et al.*, 1971), and perhaps man (McAllister *et al.*, 1972a). Thus, leukoviruses are found in most mammals including primates (see Huebner *et al.*, 1970). In chickens, mice, and cats these viruses cause cancer under natural conditions; it is this reasoning which leads to expectation that RNA tumor viruses may also be associated with some human cancers.

## AVIAN RNA TUMOR VIRUSES IN CHICKENS

The positive association of viruses with malignant disease dates from 1908 when Ellerman and Bang showed that fowl leukemia could be transmitted by filtered extracts of leukemic cells and blood serum. In those days, however, leukemia was not classified as a neoplastic disease and so the discovery did not have as much impact as that of Rous (1911), who reproduced a solid tumor indefinitely by serial passage in fowls using cell-free, bacteria-free filtrates of extracts of a transplantable sarcoma that had appeared spontaneously in a Plymouth Rock hen (Rous, 1910). These two types of neoplasia in chickens were caused by viruses which are now known to be closely related both serologically and in their physical and chemical properties.

### Avian Leukosis

The avian leukoviruses may initiate a variety of neoplastic diseases in chickens. Most are related to the hemopoietic system; visceral (and ocular) lymphomatosis, erythroblastosis, myeloblastosis, and osteopetrosis; solid tumors (sarcomas, carcinomas, and endotheliomas) are rarer manifestations (Beard, 1963). Two different lines of evidence suggest that all these types of lesion may be produced by the same or closely related agents. First, it has been a common experience that what is apparently a pure preparation of a single virus may induce several types of pathological change, although in most reported instances the evidence for clonal purity is inadequate. Peterson *et al.* (1964), for example, using the East Lansing leukosis-free stock of chickens, showed that the RPL-12 strain of leukosis-inducing virus produced visceral lymphomatosis, erythroblastosis, osteopetrosis, and more rarely fibrosarcomas or endotheliomas, recalling the diversity of tumors induced in mice by polyoma virus. Second, there is a common group antigen present in all avian leukoviruses, demonstrable by complement fixation tests and fluorescent antibody staining (see Chapter 3); neutralization tests have revealed that different coat antigens are found in viruses of the major antigenic subgroups of the avian tumor viruses. The group antigen is found in viruses associated with all the different types of tumor listed above.

Infection of domestic chickens with avian leukosis virus is very common and widespread. Usually it fails to produce symptoms for much or the whole lifespan of the bird, although visceral lymphomatosis, the most common clinical form, was stated to be responsible for an annual loss to the poultry industry in the United States exceeding $60 million in 1955 (Gross, 1970). Some aspects of the pathogenesis and epidemiology of leukosis in chickens are described in Chapter 12. The well-nigh universal occurrence of the virus in domestic chickens has made experimentation very difficult, and until assay methods were developed which bypassed the use of intact birds, assay of "wild" strains of avian leukosis virus was possible only in special leukosis-free flocks, and even then it was necessary to wait 9 months for the completion of an assay. Experimental observations with avian leukosis viruses in intact birds have therefore been carried out on a large scale only with a few variant strains selected for their high virulence. Workers at Duke University (review, Beard, 1963) have carried out extensive studies with the BAI strain A of avian myeloblastosis virus, by using carefully selected plasmas of very high viral content ($10^{11}$ to $10^{12}$ particles per milliliter). By ignoring later and different pathological effects (other than myeloblastosis) due to smaller doses of virus, they were able to complete assays within 30 days.

## Rous Sarcoma

The other avian leukoviruses that have been extensively used in experimental investigations are those which cause sarcomas after subcutaneous inoculation, and of the several viruses available (Beard, 1963) Rous sarcoma virus (RSV) has been by far the most widely employed. Several sorts of RSV's are in common use; their probable relationships and derivations have been investigated by Morgan (1964) and Ishizaki and Vogt (1966). The Bryan high titer strain, which has been used for most of the studies in chickens and chick cells over the last 15 years, will produce macroscopic tumors within 5–6 days of the inoculation of large doses into young susceptible chicks. Under controlled conditions the response is qualitatively and quantitatively uniform (Bryan et al., 1954). Newly hatched chicks may respond with multiple hemorrhagic lesions (Duran-Reynals, 1940a), and Borsos and Bang (1957) have shown that counts of the hemorrhagic lesions produced in chick embryos may be employed as a method of quantitation of RSV. Counting of foci produced on the chorioallantoic membrane, first described by Keogh (1938), was extensively employed for assay purposes (Groupé et al., 1957; Rubin, 1957; Prince, 1958), but has now been completely displaced by focus counts in chick embryo fibroblast monolayers (see below).

Certain other strains of RSV (Schmidt-Ruppin, Carr-Zilber, Prague) became important when it was discovered that they (but not the Bryan high titer strain) could produce tumors in mammals. All these strains differ in other ways from the Bryan high titer strain; they produce less distinct foci in chick embryo fibroblasts, and they are not "defective" (see below).

Studies in cultured cells have been so much more effective than those made in intact animals for obtaining an understanding of the relation between virus and cell that in this section we will consider only data relating directly to malignancy at the level of the whole animal. Two aspects merit comment; the relation of *in vitro* cellular transformation to malignancy, and "noninfective tumors."

There is clear-cut evidence that all cells transformed by RSV are malignant, i.e., are capable of producing lethal metastasizing tumors in chickens (Trager and Rubin, 1966). In a study of mixed clones of transformed and nontransformed cells produced in chick embryo fibroblasts following infection with RSV, these workers found that visible transformation of a cell was correlated with its ability to grow into a tumor. Nontransformed cells derived from RSV-infected cells (by segregation) did not produce tumors.

Shortly after the demonstration that Rous sarcoma was caused by a specific virus it was recognized that the yield of infective virus from tumor to tumor was highly variable, and that some tumors contained no detectable virus at all (Rous and Murphy, 1914). Investigations over the last few years have shown that there are two classes of noninfective Rous sarcomas which have quite different origins. The first class consists of tumors that are induced by large doses of RSV in mature birds which are able to respond immunologically to the virus. In such situations, the yield of virus from the tumors decreased greatly with increasing tumor age, and most tumors become noninfective within 30 or 40 days (Carr, 1943; Rubin, 1962). Chickens rendered immunologically tolerant to RSV by infecting them as embryos with a "nonpathogenic" avian leukosis virus never develop noninfective tumors after inoculation as adult birds with large doses of RSV of the same antigenic group; indeed, the virus content of these tumors is always very high (Rubin, 1962). Histological examination of the tumor in these two types of bird showed that in nontolerant animals the tumor is extensively invaded with lymphocytes, which often appear to be attached to individual tumor cells, and the tumors usually show complete regression by the end of the fourth week. On the other hand, there are very few lymphocytes in the tumors of tolerant birds, the tumor cells are actively multiplying, and regression never occurs. Rubin (1962) concluded that the destruction of the tumor in the nontolerant birds is due to an immunological reaction to the tumor cells, which is mediated by lymphocytes. The disappearance of RSV is a consequence of the destruction of the virus-producing cells.

Low dose noninfective tumors (Bryan *et al.*, 1955; Prince, 1959) have a completely different origin, which became clear only after the discovery that RSV was apparently defective. The discovery of this apparent defectiveness and its significance are discussed below; in the present context low dose noninfective tumors can be explained as arising from infection of leukosis-free chickens with "defective" RSV, uncontaminated by any helper virus.

Although they are referred to as "avian" leukoviruses, investigations with these agents have been carried out almost exclusively with domestic chickens. When it was first recovered, Rous chicken tumor I could only be transplanted to Plymouth Rock chickens but it was eventually passed to many other breeds, and finally the virus was shown to produce tumors in a variety of birds, including pheasants, guinea fowl, ducks, Japanese quail, and turkeys. Sporadic investigations have been made of the behavior in these birds of different leukoviruses derived from domestic chickens. On ecological grounds it is unlikely that the avian leukoviruses are confined to chickens, and Morgan (1965) has demonstrated antibodies to RSV in sera from a variety of birds in East Africa. Although some of these birds could have been infected with RSV as a result of contact with domestic chickens, it seems unlikely that all of them were infected in this way; it appears reasonable to assume that these viruses occur widely in birds.

## ROUS SARCOMA VIRUS IN MAMMALS

Some strains in Rous sarcoma virus produce tumors in mammals (Svet-Moldavsky, 1958; Zilber, 1965; Ahlstrom and Forsby, 1962), positive results being most readily obtained with the Schmidt-Ruppin and Carr-Zilber strains (derived originally from Rous' chicken tumor I) and the Prague strain (derived from another isolate, by Engelbreth-Holm; Morgan, 1964), although tumors have also been produced in hamsters inoculated with the Bryan and Harris strains (Rabotti et al., 1965). The demonstration by Hanafusa and Hanafusa (1966a) that the capacity of RSV to produce tumors in hamsters depends on the envelope antigen has made it possible, by phenotypic mixing in nonproducer cells with the appropriate helper virus, to produce tumors in mammals with any strain of RSV.

Several different mammals are susceptible to the induction of tumors when injected with RSV as newborn animals, the spectrum of susceptibility being similar to that of polyoma virus. Sarcomas have been produced in rats, cotton rats, guinea pigs, mice, hamsters, and monkeys (reviews, Zilber, 1965; Svoboda, 1966, 1968). Hamsters remain susceptible when adults. In many cases, cell-free virus fails to induce tumors in mammals, although inoculation of suspensions of viable chicken tumor cells (or chicken cells transformed in tissue culture) produces tumors which consist of mammalian cells. The reason for this is unknown. It is clear from the number of successful reports of the induction of tumors in mammals with cell-free preparations (review by Vogt, 1965) that the requirement for cells is not essential. Perhaps inoculation of virus-yielding cells potentiates the formation of tumors because the cells supply large quantities of virus over a prolonged period of time or they fuse with some of the host cells, thereby increasing the chance of yielding a virus with the right envelope antigen.

Unlike cells of chicken tumors, which usually yield infectious RSV, mammalian tumors induced by RSV show several different types of response (review, Svoboda and Hlozanek, 1970). In a few cases—hamster tumors induced by the Prague strain of RSV (Svoboda and Klement, 1963; Klement and Versely, 1965), rat tumors caused by B77-RSV (Altaner and Svec, 1966), and a rabbit sarcoma produced by the Carr-Zilber strain of RSV (Kryukova, 1966)—very small amounts of infectious virus can be detected by injection of cell-free filtrates into chickens. Such virus-yielding tumors seem to be very rare and most mammalian tumors induced by RSV usually do not contain infectious RSV. However, complete infectious RSV usually can be recovered after cocultivation or Sendai virus-mediated fusion of the mammalian tumor cells with permissive chicken cells (Shevlyaghin et al., 1969; reviews, Svoboda and Hlozanek, 1970; Watkins, 1971b). The efficiency of the process seems variable, but the presence of structurally intact viable tumor cells is required. Although the mammalian cells are virus-free before cocultivation, they do contain novel virus-specific antigens. They develop a tumor-specific transplantation antigen (TSTA) (Jonssen and Sjögren, 1966; Sjögren and Jonssen, 1970) which is common to avian tumor viruses of different subgroups (Bubenik and Bauer, 1967; Bauer et al., 1969), and the avian leukovirus group-specific antigen can usually be detected in them (Huebner et al., 1964). Finally there are reports that some tumors induced after inoculation of RSV into mammals do not yield virus under any conditions. However, since these tumors do not contain detectable group-specific antigen it is not clear that they were in fact caused by RSV (Thurzo et al., 1969; Harris et al., 1969).

These experiments raise more questions than they answer. For instance, why is it that tumors induced by RSV in chickens yield virus constantly, while those caused by the same virus in mammals usually do not? Why should some mammalian tumors yield small amounts of RSV spontaneously while others never yield or require fusion with permissive cells? Temin (1971a) has suggested that there is a quantitative block to production of RSV in mammalian cells so that synthesis or processing of viral RNA from an integrated viral DNA template is inhibited.

## INTERACTIONS OF AVIAN TUMOR VIRUSES WITH CULTURED AVIAN CELLS

The most common outcome of infection of cells with nononcogenic viruses is death of the cells and production of new viral particles. It is clear that such a terminal interaction cannot take place in cells transformed by oncogenic viruses. In the case of DNA tumor viruses, transformed cells lose growth control but do not yield infectious virus. However, the RNA tumor viruses exhibit no such clear-cut distinction between productive infection and transformation. The infected cells are never killed but whether they become transformed, or yield virus, or both, depends on the interplay of many variables which include the genotype and phenotype of both the cell and the virus.

There are four general classes of interaction between a cell and an RNA tumor virus: (a) the cell is resistant to infection—there is no virus production and the cell does not become transformed, (b) the virus multiplies in a cell but does not transform it, (c) the virus multiplies in a cell and transforms it, or (d) the cell is transformed but the virus fails to multiply.

Because the outcome of infection with RNA tumor viruses is so varied, several types of *in vitro* assay have been devised (see also Chapter 2). The one most commonly used for avian sarcoma viruses is the focus-forming assay using chick embryo fibroblasts. This technique was foreshadowed by Halberstaedter *et al.* (1941) and further developed by Manaker and Groupé (1956), before being perfected by Temin and Rubin (1958). Cells are infected with dilutions of virus and after several days some of the cells (the proportion increasing linearly) form "foci," and show morphological alterations and different growth properties similar to those already described for DNA tumor viruses. The number of virus particles is directly proportional to the number of focus-forming units. By contrast, most strains of avian leukemia viruses do not transform fibroblasts and usually cause no visible sign of infection. They are usually titrated by serological techniques which detect virus-specific antigens in trans- formed cells (Vogt and Rubin, 1963), or by interference assays. Interference assays are based on the fact that cells infected with leukosis viruses will not support the growth of a superinfecting sarcoma virus of the same antigenic subgroup; the reduction of the yield of sarcoma virus can be used as a measure of the amount of leukosis virus present in the stock (Rubin, 1960).

Both of these assays are tedious and will probably be replaced by a new assay for certain avian leukosis viruses which has been recently described (Kawai and Hanafusa, 1972). Cultures of chick fibroblasts are infected with a temperature-sensitive mutant of RSV and incubated at 41°C. Under these condi- tions no transformation occurs. When the cultures are superinfected with avian leukosis viruses of antigenic subgroups B or D plaques are produced after 7 days of incubation at 41°C. It is not understood why this test works at all, or why leukosis viruses of subgroups A, C, and E do not form plaques, but the method provides an assay for viral infectivity that is simpler, more rapid, and more accurate than the earlier methods.

Although leukosis viruses are usually said to be nontransforming, in fact certain strains do cause morphological alterations in cultured cells (Baluda and Goetz, 1961; Moscovici *et al.*, 1969; Calnek, 1964; Bolognesi *et al.*, 1968) and they can transform cell types other than fibroblasts, such as lymphopoietic or hemopoietic cells (Hanafusa, 1970). It seems, therefore, that the ability of a virus to transform cells depends on the state of differentiation of the cell. Why some avian tumor viruses transform chick fibroblasts and others do not is un- known. Duesberg and his colleagues (Duesberg and Vogt, 1970; Duesberg *et al.*, 1971; Martin and Duesberg, 1972) have shown that nontransforming (leukosis) viruses contain less RNA than sarcoma viruses, and lack a specific subunit of RNA. Furthermore, Vogt (1971a) has presented evidence suggesting that non- transforming viruses are spontaneous segregants of sarcoma viruses, perhaps formed by the chance exclusion of the subunit of the sarcoma virus genome which is responsible for transformation. However, more work is required to prove this hypothesis.

## Bryan High Titer Strain of RSV

In 1962, Rubin and Vogt discovered that the Bryan high titer strain of RSV contained two viruses: (a) RSV, which transformed fibroblasts, and (b) a four- to tenfold larger amount of a nontransforming avian leukosis virus. This Rous-associated virus (RAV) was easily isolated from RSV by cloning. It was non-cytocidal, interfered with focus formation by RSV, and produced erythroblastosis when inoculated into chicken embryos, but it failed to produce sarcomas in chickens. Soon afterward, Temin (1962) and Hanafusa et al. (1963) realized that RAV played an essential role in the synthesis of infectious RSV. They found that a small proportion of single foci and Rous sarcoma cells produced by the inoculation or avian leukosis virus-free chick fibroblasts with Bryan strain RSV failed to release infectious RSV. The most important factor determining the incidence of "nonproducer" cells was the multiplicity of virus infection. At low multiplicities of infection with the Bryan high titer stock, many of the foci were nonproducers; at high multiplicites virtually all of them produced infectious RSV. Furthermore, the appearance of producer clones followed two-hit kinetics. Hanafusa et al. (1963) concluded that the nonproducer state was a consequence of solitary infection of cells with particles of RSV that were defective in that they could not on their own yield infectious progeny. The producer state occurred when cells were simultaneously infected with RSV and RAV, or when nonproducer cells were subsequently superinfected with any avian leukosis virus (Temin, 1962; Hanafusa et al., 1963). Therefore it seemed that the avian leukosis virus RAV in the Bryan high titer stock provided a helper function essential for the synthesis of infectious RSV. Only the Bryan high titer and the Bryan standard strains of RSV need helper virus in order to complete viral multiplication (Hanafusa et al., 1963); other strains of RSV transform cells as efficiently as the Bryan strains but are not defective in replication (Hanafusa, 1964; Dougherty and Rasmussen, 1964; Golde and Vigier, 1966).

## Pseudotypes of Rous Sarcoma Virus

The infectious sarcoma virus that is rescued from nonproducer cells by superinfection with an avian leukosis virus has envelope properties (antigenic make up, host range, and interference pattern) which are determined by the helper virus (Hanafusa et al., 1964; Hanafusa, 1965; Vogt, 1965); its genome, however, seems to be identical to that of the original transforming virus (Hanafusa et al., 1963). This means that rescue is not due to a recombination event between the sarcoma and helper virus genomes but suggests that envelope glycoproteins are provided to the RSV by the helper virus. Such phenotypically mixed particles, which contain one sort of viral genome in the envelope of another virus, are now known to occur frequently in many types of avian tumor viruses (Hanafusa and Hanafusa, 1966a; Vogt, 1967a). They are called pseudotypes (Rubin, 1965) and the accepted nomenclature is RSV (RAV-1) where the genotype of the virus is written first and the phenotype is shown in parenthesis.

Whether an avian leukosis virus acts as a helper virus, or interferes with RSV, depends upon two factors—the order in which cells are infected with the two agents and the antigenic subgroups to which the viruses belong. If added to uninfected cells before infection with an RSV preparation which has the same coat antigen, avian leukosis virus produces viral-attachment interference (see Chapter 8). If added to cells which are already infected with RSV, the result of superinfection with an avian leukosis virus depends upon whether the cells are yielding RSV or not and upon the antigenic subgroups of the two viruses involved In nonproducer cells, the added avian leukosis virus acts as a helper and leads to the formation of infectious RSV, the pseudotype of which will depend upon the antigenic group of the avian leukosis virus. In cells which are already yielding RSV, the leukosis virus is excluded by virus-attachment interference if it has the same envelope antigen as the RSV being produced, but (depending upon the presence of appropriate cellular receptors) it may infect the cells and can then act as a subsidiary helper virus if it belongs to a different antigenic subgroup. Stock "Rous sarcoma virus" is often mixed antigenically and in its pathogenic capacity; Hanafusa and Hanafusa (1966a) have demonstrated no fewer than three "Rous-associated viruses" in Bryan high titer strain of RSV.

## Altered Host Range Properties of Virus from "Nonproducer" Cells

When the defectiveness of the Bryan high titer strain was first discovered it was customary to measure the infectivity of RSV in chicken cells of the C/O genotype (see Table 14–1). These are the most common genetic type of chicken cells and were thought to be susceptible to almost all strains of avian tumor viruses. The first hint that this might not be so came when Dougherty and Di Stefano (1965) reported electron microscopic observations showing that small numbers of particles that were morphologically indistinguishable from the virions of RSV could be found budding from the surface of RSV-infected "nonproducer" cells. Robinson (1967) confirmed this observation by biochemical methods, which showed that the nonproducer cells do in fact yield particles with the physical and chemical properties of RSV. Shortly afterward it was reported (Robinson et al., 1967) that the particles carried biological activity, because after they were reintroduced into C/O cells by fusing cells with UV-irradiated Sendai virus, the cells became transformed and again began to yield viral particles of the same phenotype. Finally it was realized that the particles obtained from the RSV-infected nonproducer cells actually were fully infectious for many species of wild fowl (Vogt, 1967b; Weiss, 1967) and for chicken cells of certain genotypes. In other words, the virus obtained from nonproducer cells possessed a completely different host range from any of the other avian leukosis viruses. Vogt (1967b) called the virus RSV (O) because he thought it did not need helper virus to supply an envelope in the nonproducer cells.

The host range of avian tumor viruses is defined by the spectrum of types of fowl cells that the viruses can infect. The outcome in any particular virus/ cell combination is defined by the envelope antigens of the virus and by the presence or absence of specific receptors on the cell surface. The expression of these receptors in chickens is controlled by a single dominant autosomal gene (Crittenden *et al.*, 1967). Thus, the most convenient and reliable way to classify avian tumor viruses into subgroups is to determine their plating efficiency (EOP, Table 14–1) on different strains of chicken cells. Table 14–1 shows the host of

TABLE 14–1

*Host Range of Avian Tumor Virus Subgroups* [a]

| | SUBGROUP | | | |
| CELL TYPE | A | B | C | D |
|---|---|---|---|---|
| Chicken | | | | |
| C/O | S [b] | S | S | S |
| C/A | R [b] | S | S | S |
| C/B | S | R | S | S |
| C/AB | R | R | S | S |
| C/BC | S | R | R | S |
| Bobwhite quail | R | R | R | R |
| Duck | R | R | S [c] | SR to S |

[a] From Duff and Vogt (1969).
[b] S, EOP $\geq 0.1$; SR, EOP $= 0.01–0.1$; R, EOP $\leq 0.01$.
[c] Except RSV (RAV-7), which does not transform these cells.

the four major subgroups (Duff and Vogt, 1969). RSV(O) does not fit into any of the four subgroups: it is always noninfectious for over 90% of C/O chick cells and all C/B and C/AB embryos. However, RSV(O) is fully infectious for quail and C/A chick embryos, given that the virus is grown in certain sorts of chicken cells—those that contain "helper factor."

## Helper Factor and the COFAL Test

COFAL is the name given to an antigen detected by a complement fixation test (COFAL test) (Sarma *et al.*, 1964); the antigen is indistinguishable from the group-specific (*gs*) antigen of the avian leukosis viruses. The cells of some strains of normal chickens give a positive COFAL reaction, whereas other strains give a negative reaction (Dougherty and Di Stefano, 1966). Very surprisingly, the expression of the COFAL antigen in chickens is under the control of a single dominant autosomal gene (Payne and Chubb, 1968). This raises the possibility that the genes that code for this viral antigen may well be part of the genome of an integrated avian leukovirus.

Infectious RSV(O) is produced only in those cells which carry *gs* antigen (Weiss, 1969a). When RSV(O) is cloned in *gs⁻* cells, virus particles are produced which are not infectious for any known host cell. The basis for this lack of infectivity is not known; the most likely explanation seems to be that the virus is a host range variant of the Bryan strain for which no susceptible host cells have been identified. The virus is not totally incapable of biological activity since it replicates and causes transformation if its entry into cells is facilitated by Sendai virus-induced cell fusion (Weiss, 1969a; T. Hanafusa et al., 1970).

These experiments show that $gs^+$ cells contain something which modifies RSV(O) so as to allow it to infect C/A and quail cells (Weiss, 1969a; H. Hanafusa et al., 1970; Vogt and Friis, 1971). Clearly the function of the helper substance (called chick factor, chf) must be to provide to RSV(O) an envelope that is capable of attaching to the receptors on cells of those particular species.

## Chick Factor

The close relationship between chf and the gs antigen was confirmed by Weiss and Payne (1971) who showed that chf was controlled by a dominant autosomal gene at, or closely linked to, a similar gene for gs. However, the first clue concerning the nature of chick factor came when it was noticed that the outcome of infection of $gs^-$ cells with RSV(O) propagated in $gs^+$ cells was dependent on the culture conditions. As we have seen cloning of RSV(O) in $gs^-$ cells gives rise to noninfectious progeny. However, several workers (Weiss, 1969a; H. Hanafusa et al., 1970; Vogt and Friis, 1971; Weiss and Payne, 1971) showed that infection of $gs^-$ cells with RSV(O) at high multiplicity allowed propagation of infectious RSV(O). This observation strongly suggested that the chick factor could be rescued from $gs^+$ cells and transmitted to $gs^-$ cells by infection with RSV(O). The nature of the factor became clear when T. Hanafusa et al. (1970) isolated the factor from a stock of leukosis virus grown on $gs^+$ cells and showed it to be a typical avian tumor virus, containing 60–70 S RNA. The virus was called RAV-60. Because it has the same envelope properties as RSV(O) it can be propagated in quail and C/A chicken cells.

Under normal conditions, uninfected $gs^+$ cells do not release any mature virus particles (Weiss, 1969a; H. Hanafusa et al., 1970). However, Vogt and Friis (1971) have been able to identify occasional COFAL-positive chickens that release spontaneously a nontransforming virus which they called RAV-O. This virus is indistinguishable from RAV-60 in its envelope properties and its helper function for RSV(O), but for unknown reasons it infects quail cells very inefficiently (review, Weiss, 1972). RAV-60 and RAV-O are classified together in the subgroup E of avian tumor viruses. Because RSV(O) grown on $gs^+$ cells has envelope properties donated by RAV(O), its pseudotype is RSV (RAV-O). The noninfectious particles produced by cloning of RSV on $gs^-$ cells are called RSV(−) (Weiss, 1972).

Subgroup E viruses can be induced from $gs^+$ cells of gallinaceous birds of many species from all over the world. Thus, they seem to be avian tumor viruses which under most circumstances exist in an integrated state in the genome of $gs^+$ birds. Normally they do not appear in the form of free virus, but occasional embryos do produce RAV-O spontaneously, and virus production can be induced in $gs^+$ cells superinfected with other avian sarcoma and leukosis viruses.

The question remains whether $gs^-$ cells contain chf in a repressed form. Three different sorts of experiments have been reported. First, it has been shown

(Weiss, 1972; Hanafusa *et al.*, 1972) that the correlation between the presence of helper activity for RSV(O) and *gs* antigen is not absolute. Those chickens that are positive in the COFAL test always show helper activity, but some strains of chickens exist that are COFAL negative but nevertheless support the growth of RSV(O). One explanation for this finding is that the sensitivity of the COFAL test is low; another is that the *gs⁻* chickens contain *chf* in a cryptic state. To test this hypothesis, methods have been developed that are capable of detecting both spontaneous release of RAV-O and the presence of latent helper factor in cells (Weiss *et al.*, 1971; Weiss, 1971; Hanafusa *et al.*, 1972). These methods are extremely elegant, but highly technical, and the reader should consult the original papers for details of the procedures. Suffice it to say that the use of these more sensitive techniques has provided unequivocal evidence that in many *gs⁻* embryos there does exist an endogenous viral genome which can be activated by infection with other avian leukosis viruses.

Second, Varmus *et al.* (1971) prepared radioactive DNA from RAV-O using the DNA polymerase and template RNA endogenous in the virion. The DNA was used to measure the number of copies of viral genomes in DNA extracted from *gs⁺ chf⁺* and *gs⁻ chf⁻* embryos, by reassociation kinetics. No significant differences were found. This result implies that integrated viral genomes may be ubiquitous in all avian embryos, whether or not they express *gs* or contain *chf*, and fits well with other experiments which show that the DNA of normal uninfected chick cells contains sequences in common with viral RNA (Harel *et al.*, 1966; Baluda and Nayak, 1970; Yoshikawa-Fukada and Ebert, 1969; Baluda, 1972).

The third line of evidence is the most convincing. Weiss *et al.* (1971) showed that infectious avian leukosis virus can be induced both from *gs⁺ chf⁺* and *gs⁻ chf⁻* cells by treatment with ionizing radiation, chemical mutagens, or carcinogens. The rescued viruses are typical in all respects of avian leukosis viruses; they act as helpers for RSV(O), they possess RNA-dependent DNA polymerase, they are nontransforming, and with one exception they can be classified as subgroup E viruses. The mechanism of induction is unknown, but it seems that the process may occur in two steps, since nearly all *gs⁻* cells treated with 20-methylcholanthrene become *gs⁺* and acquire helper activity, but few of them release complete virus.

These experiments prove that the complete viral genome is present both in *gs⁺* and in *gs⁻* cells, and it seems that whether or not chickens are fated to develop leukemia, they carry in their genomes sequences of DNA which code for avian leukosis viruses. Clearly this result has far-reaching significance for the origin of these viruses, their mode of transmission (see Chapter 12), and the mechanism by which they cause tumors; it has led to a number of theories which unify several diverse phenomena. However, the fact that all chickens seem to harbor genetic information specified by the avian tumor virus also has implications on a much more mundane level. For instance, it is possible that the endogenous viral genomes complement defects in strains of RSV that we think of as nondefective; certainly the presence of the resident genomes complicates experiments such as the kinetics of inactivation of avian leukosis viruses by UV or X irradiation, and causes problems in the interpretation of complementation and recombination experiments with these viruses. Some of the earlier results will have to be reassessed in the light of current knowledge.

## PROPERTIES OF CELLS TRANSFORMED BY AVIAN TUMOR VIRUSES

Cells transformed by avian RNA tumor viruses display many of the same properties shown by cells transformed by DNA tumor viruses (review, Macpherson, 1970). They have altered morphologies which are controlled by the viral genome, because mutants of RSV produce distinct morphological changes (Temin, 1960), and they form malignant tumors on injection into susceptible animals (Trager and Rubin, 1966).

Transformed cells grow in conditions that restrict the multiplication of normal cells, so that they grow to higher saturation densities in liquid medium (Colby and Rubin, 1969; Hanafusa, 1969b; Temin, 1965). As in the case of cells transformed by DNA tumor viruses, this overgrowth probably results from decreased requirement of transformed cells for serum growth factors (Temin, 1966, 1967, 1968b).

The plasma membranes of the transformed cells contain new virus-specific antigens. Some of these are viral envelope antigens (Gelderblom et al., 1972; Kurth and Bauer, 1972); others are present on the surfaces of transformed cells which do not yield virus and seem to be group-specific transplantation antigens (Gelderblom et al., 1972). In all probability, the same transplantation antigen is produced on the surfaces of all species of cells transformed by the same virus.

Often transformed cells show a change in lectin-mediated agglutinability, but the extent of this change varies. Moore and Temin (1971) found that transformation of chicken, mouse, or rat cells by RSV or murine sarcoma virus resulted in small changes in agglutinability by wheat germ agglutinin or concanavalin A, but other workers report larger changes (Kapeller and Doljanski, 1972). One reason for the disparity may be the fact that some sorts of cells transformed by RNA tumor viruses produce greatly increased amounts of acid mucopolysaccharides (Hamerman et al., 1968; Van Tuyen et al., 1967) especially hyaluronic acid (Ishimoto et al., 1966 ) which may interfere with the agglutination assay (Martin and Burger, 1972). Even allowing for this, it is clear that the magnitude of the change in agglutinability of cells transformed by RNA viruses is not as great as that produced by the oncogenic DNA viruses. Possibly the glycoprotein viral peplomers that project from the plasma membrane of cells transformed by leukoviruses interfere with the exposure of agglutinin-sensitive sites.

A variety of less well-characterized alterations has been reported; for example, increased rates of glucose uptake (Hatanaka and Gilden, 1970; Hatanaka and Hanafusa, 1970; Martin et al., 1971), sensitivity to dibucaine (Rifkin and Reich, 1971), sensitivity to polysaccharide inhibitors of growth (Montagnier,

1969), and changes in the glycolipid (Hakomori *et al.*, 1971) and glycoprotein (Warren *et al.*, 1972) composition of membranes, but nothing is known of the molecular basis of these changes.

## TRANSFORMATION OF CULTURED MAMMALIAN CELLS BY AVIAN TUMOR VIRUSES

Several workers have reported that RSV will produce transformation of rat, mouse, hamster, and human cells (reviews, Hanafusa, 1969a; Temin, 1971a; Svoboda and Hlozanek, 1970). In most cases, the transformed cells are positive for avian leukosis *gs* antigen although the amount of antigen present in different transformed cell lines seems to be very variable. In some cases, its presence can be detected only by injection of the cells into test animals, when antibody directed against avian *gs* antigen is produced (Bataillon, 1969; Harris *et al.*, 1969; Thurzo *et al.*, 1969). Typically, the transformed cells are free of infectious virus, although small amounts have been detected in occasional cell lines (Klement and Vesely, 1965; Svec *et al.*, 1966, 1970). Although sarcoma virus production cannot be induced by superinfection with leukosis virus, virus can usually be rescued from transformed mammalian cells by cocultivation or fusion with "uninfected" chick cells (Svoboda *et al.*, 1967; Vigier, 1967). In some cases, almost all the heterokaryons produce virus (Machala *et al.*, 1970), but in others the efficiency of rescue seems to be much lower (Svoboda and Dourmashkin, 1969; Yamaguchi *et al.*, 1969; Altaner and Temin, 1970). The mechanism by which rescue occurs is not known. The most likely hypotheses are that the chick cells cause release of the avian leukosis provirus from the mammalian cell chromosome or that chick cells stimulate transcription of the genome into viral RNA which becomes assembled into viral particles. However, much more work is required to test the assumptions on which both of these hypotheses are based.

The virus recovered from transformed mammalian cells does not have the same properties as the original transforming sarcoma virus; it transforms mammalian cells with higher efficiency and in some cases has altered envelope properties (Kryukova *et al.*, 1970; Altaner and Temin, 1970; Obukh and Kryukova, 1969). The reason for these changes is not known but Temin (1971a) has suggested that they may be a result of recombination of the avian sarcoma virus with leukoviruses endogenous to mammalian cells.

Revertants of BHK21/13 hamster fibroblasts transformed with Schmidt-Ruppin strain of RSV have been isolated (Macpherson, 1965). They showed reduced tumorigenicity, had lost the avian leukosis *gs* antigen, and did not yield virus during heterokaryon formation with chick cells. By contrast to revertants of cells transformed by DNA tumor viruses, Macpherson's revertants were easily retransformed with the original virus. The retransformants again threw off revertants (Macpherson, 1968, 1970). The most likely explanation for these results and similar experiments of Todaro and Aaronson (1969) with murine sarcoma virus is that the viral genome has been completely lost from the revertants. This implies that the proviral DNA is not integrated so stably as the genome of DNA tumor viruses in transformed mammalian cells.

## MECHANISM OF TRANSFORMATION BY AVIAN TUMOR VIRUSES

Infection of cells with avian tumor viruses does not result in cell death. Instead, the cells survive and usually but not necessarily begin to yield virus. Many of them divide and assume a new set of stable heritable properties. Such cells are said to be transformed. The length of time between infection with virus and the appearance of morphologically transformed cells varies widely in different systems. In the case of mammalian cells infected with RSV, transformed foci only appear several weeks after infection. In the chick system, events proceed much more rapidly so that cells showing morphological evidence of transformation can be detected 3–5 days after infection with low multiplicities of virus (Nakata and Bader, 1968) and as little as 24 hours after infection with high multiplicities (Hanafusa, 1969a,b; Bader and Bader, 1970).

It is known that continued expression of at least some genes of the RNA tumor viruses is necessary for the maintenance of at least some of the phenotypic characteristics of transformed cells. This important result comes from studies of temperature-sensitive mutants of RSV. Cells transformed by some of these mutants lose their transformed phenotype when shifted from permissive to nonpermissive temperatures (Toyoshima and Vogt, 1969; Martin, 1970; Biquard and Vigier, 1972; Martin et al., 1971; Kawai and Hanafusa, 1971; Warren et al., 1972).

Two questions then arise: (a) How does a virus whose genetic material is RNA maintain itself in a population of dividing cells? and (b) What is the viral function which causes cells to maintain the transformed phenotype?

### DNA Provirus

In the early 1960's, Temin realized that the orderly transmission of genetic information by RNA in a dividing population of cells raised enormous difficulties. His response to the dilemma was to utilize the concept of provirus originally introduced to explain the phenomenon of lysogeny in bacteria (reviews, Lwoff, 1953; Bertani, 1958). At first the physical structure of the provirus was undefined, but in 1964 Temin's ideas crystallized (Temin, 1964b) and he suggested that during transformation the RNA genome of the avian tumor viruses was copied into DNA, which was integrated into the cell genome and inherited during cell division just like any other host genes.

For several years few took this proposal seriously. However circumstantial evidence in favor of the idea slowly accumulated, and in 1970, when Mizutani and Temin, and Baltimore independently discovered an RNA-dependent DNA polymerase in the particles of RNA tumor viruses, the provirus theory suddenly gained universal respectability and acceptance.

Three sorts of experiments provide evidence in favor of a DNA provirus.

**Dependence of Transformation and Viral Production on DNA Synthesis.**
Treatment of cells with any one of a variety of inhibitors of DNA synthesis
early after infection with RSV blocks subsequent production of virus (Temin
1964a, 1968a; Bader 1965, 1966; Murray and Temin, 1970; Bader, 1972a, b).
This suggests that some step involving DNA synthesis is required during the
first few hours of viral infection. The most convincing evidence that this DNA
is viral and not cellular comes from the experiments of Balduzzi and Morgan
(1970) and Boettiger and Temin (1970). They infected stationary chick cells
with RSV in the presence of 5-bromodeoxyuridine. After 24 hours the BUdR
was removed and infection was allowed to proceed. If the cultures were exposed
to light, the cells did not become infected with virus. This result suggests that
the DNA which contained BUdR was synthesized early during infection and
that this DNA was required for subsequent viral production. Although cytoplasmic DNA synthesis has been detected in cells after infection with murine
sarcoma virus (Hatanaka *et al.*, 1971) all attempts to isolate the BUdR-labeled
DNA have been unsuccessful and we do not know for sure whether it consists
of host or viral sequences. However, high multiplicity infections are considerably
more resistant to inactivation by light than low multiplicity infections (Boettiger
and Temin, 1970). This result suggests that the greater the number of infecting
genomes the greater is the amount of synthesis of BUdR-labeled DNA, and
implies that the DNA is specified by the virus, rather than the host. Furthermore, the yield of RSV from cells that have been treated with BUdR during
the first 12 hours of infection contains a larger proportion of temperature-
sensitive mutants and defective particles than virus obtained from untreated
cells (Bader and Bader, 1970; Bader and Brown, 1971).

**Detection of Proviral DNA by Hybridization.**   As we have already seen, it
is possible to detect virus-specific DNA in the genome of both normal and
stably transformed chick cells. Several workers using the filter hybridization
method have found more sequences of virus-specific DNA in cells after infection
with RSV than in uninfected cells (Baluda and Nayak, 1970; Baluda, 1972;
Rosenthal *et al.*, 1971). Presumably the increase is a consequence of the synthesis of proviral DNA copies of the incoming viral RNA. Varmus *et al.* (1971),
however, were unable to detect any difference in the number of viral DNA
copies in the two sorts of cells. The cause of this discrepancy in results is
unknown, but one possible explanation is that Varmus *et al.* used a radioactive
viral DNA synthesized *in vitro* as a probe for viral DNA sequences in the
cell genome; it is known that such synthetic viral DNA corresponds only to
about 10–20% of the sequences of the whole viral RNA, and for this reason it
is a relatively insensitive tool to detect small changes in numbers of complete
genomes.

It is clear that the presence of integrated viral DNA sequences in both
normal and transformed cells may make the detection of new proviral DNA
synthesized after infection very difficult. However, the fact that viral DNA
sequences are present at all in cell genomes provides support for the proviral
DNA theory.

**The Presence of RNA-Dependent DNA Polymerase in RNA Tumor Virus Particles.** Because the presence of puromycin or cycloheximide during the first 12 hours does not affect the final outcome of infection (Bader, 1966, 1972a), it seems that the formation of the provirus does not depend on the synthesis of new proteins. This means that the enzyme responsible for converting the incoming viral genome into its proviral form either must exist in the cells before infection, or it must be carried into the cells as part of the virus particles. Evidence in favor of the latter hypothesis came when Mizutani and Temin (1970) using RSV and Baltimore (1970a) using Rauscher murine leukemia virus, discovered RNA-directed DNA polymerase activity in the viral particles.

This finding caused great excitement and led to the rapid accumulation of a vast amount of information on the properties and distribution of the enzyme, which has been very well reviewed recently (Gallo, 1971; Temin and Baltimore, 1972). The main properties of the virion polymerase are shown in Table 14–2.

TABLE 14–2

*Properties of Virion-Associated RNA-Dependent DNA Polymerase*

1. RNA-dependent DNA polymerase activity is present in all RNA tumor viruses whether or not they are transforming. It is also found in visna virus, foamy viruses (see Table 1–16), and certain "viruslike" particles isolated from humans and animals.
2. The polymerase system is present in the viral core (see Chapter 3); for maximal activity *in vitro*, the surrounding viral envelope must be removed by treatment with detergents.
3. The enzyme activity requires the presence of all four deoxynucleotide triphosphates, a reducing agent, manganese or magnesium ions, and a pH about 8. The enzyme from mammalian viruses works best at 37°–40°C, while the optimum temperature for the avian viral enzyme is 40°–45°C.
4. The reaction is primer-dependent and works both from endogenous and exogenous templates. The products of the endogenous reaction are short pieces of DNA (10 S) some of which are single- and some double-stranded. All sequences of viral RNA are represented in the transcript but some regions of the viral RNA are copied much more frequently than others.
5. The production of double-stranded DNA is inhibited by actinomycin D; certain rifampicin compounds depress synthesis both of double- and single-stranded DNA.
6. The RNA-dependent DNA polymerase does not correspond to any of the recognized antigens in the virions. However, each group of RNA tumor viruses has an antigenically distinct enzyme.
7. Other enzymes are present in the core of the viral particles including nucleases, ligase, phosphatases, and kinases.
(For references see Gallo, 1971; Temin and Baltimore, 1972).

Even though none of the properties of RNA-dependent DNA polymerase is inconsistent with its postulated role in the synthesis of proviral DNA, we have no direct evidence concerning the activity of the enzyme in cells after infection, so that assessment of its biological function is difficult. However, two approaches have been utilized to try to provide evidence that the enzyme is involved in transformation and/or viral replication.

**RSVα.** Hanafusa and Hanafusa (1968) isolated a variant of the Bryan strain of RSV called RSVα, which is noninfectious, cannot transform chick cells, and does not contain RNA-dependent DNA polymerase activity (Hanafusa and Hanafusa, 1971; Robinson and Robinson, 1971). However, RSVα can be rescued as an infectious and transforming pseudotype [RSVα (RAV)] by superinfection of RSVα-producing cells with RAV. The defect in RSVα particles is not due to envelope properties and penetration of the viral particles into cells is normal (H. Hanafusa et al., 1970; Hanafusa and Hanafusa, 1971). Thus, it seems that RSVα is defective in RNA-dependent DNA polymerase and it is this deficiency which is responsible for the failure of the mutant to transform. Because cells transformed by the RSVα(RAV) pseudotype yield enzyme-deficient nontransforming RSVα particles, it follows that the RNA-dependent DNA polymerase is required only to establish transformation and not for maintenance of the transformed state. However, it is by no means certain that the only defect in RSVα is in the RNA-dependent DNA polymerase, and further work on a suite of such mutants will be required to establish this crucial point.

**Chemical Inhibitors.** Several compounds (the rifampicins and the streptovaricins) which inhibit RNA-dependent DNA polymerase *in vitro* (Gurgo et al., 1971; Gallo et al., 1970, 1971; Brockman et al., 1971) have been reported to block infection by RNA tumor viruses (Calvin et al., 1971; Carter et al., 1971; Rickert and Balduzzi, 1971). However, before this result can be accepted as proof that the viral polymerase is required for infection and/or transformation we would have to be sure that these inhibitors were absolutely specific and nontoxic. Unfortunately, both of these points are in doubt.

In summary then, there is good evidence that viral RNA is converted to a proviral DNA form during the early stages of infection. An enzyme exists in viral particles which catalyzes the formation of DNA *in vitro* from viral RNA template. There is no conclusive evidence to show that this is the enzyme responsible for provirus formation.

## Fate of Provirus in Cells

Because all attempts to isolate newly synthesized proviral DNA have failed, we do not know how many copies of viral DNA are made and we have no idea of its structure or fate. However, Temin believes that cell division is essential for viral production and cell transformation by the RNA tumor viruses. If stationary cells are infected with RSV they do not produce virus or undergo transformation. If serum is added to such cells, even as much as 7 days after infection, the cells are stimulated to divide. They then become transformed and viral production begins (Temin, 1968b; Murray and Temin, 1970). Bader (1972), on the other hand, using much higher input multiplicities and a different strain of RSV, found that viral production and morphological changes could be detected in infected cells even when cell division was blocked with vinblastine. The discrepancy between these two sets of results will probably not be resolved until methods become available to isolate the virus-specific DNA that is synthesized in the early stages of infection. Until then, the most reasonable way to explain these results is that at least some of the proviral DNA finishes up in an integrated form in the cell genome, and that in some circumstances the integration event may require mitosis. It is not known whether or not integration occurs at a specific site; presumably in either case proviral DNA becomes indistinguishable from a set of host genes, so that it is replicated and segregated into daughter cells in exactly the same way as typical host genes.

In all likelihood, proviral DNA is transcribed into RNA by one or other of the cellular DNA-dependent RNA polymerases. In those cells which synthesize virus, this RNA must be translated into virus-coded proteins which are assembled into progeny particles, as discussed in Chapter 6. We know from studies on temperature-sensitive mutants of avian sarcoma viruses that at least some virus-coded proteins are responsible for the establishment and maintenance of the transformed state (Toyoshima and Vogt, 1969; Martin, 1970; Kawai and Hanafusa, 1971). However, we have no idea what the functions of these proteins are, or how they interact with the cell to produce the multifaceted changes in cell behavior that occur as a consequence of transformation.

## MURINE LEUKOVIRUSES: SUBGENUS A

The development of inbred strains of mice was accompanied by their extensive use for studies on transplantable tumors. Over the years a number of viruses have been recovered from these transplanted tumors. Several of these viruses have been shown to induce leukemia (or more precisely disseminated lymphosarcomatosis) in mice (review, Moloney, 1964). Those which have been studied most intensively are the Gross Passage A (Gross, 1957), Friend (Friend, 1957), Moloney (Moloney, 1960), and Rauscher (Rauscher, 1962) viruses (see Monograph, 1966). They contain two common group-specific antigens (Geering et al., 1966; Schäfer et al., 1971) and an interspecific "mammalian" antigen which is also present in hamster, canine, feline, and rat leukoviruses, but not in avian leukoviruses.

The murine leukemia viruses are classified into subgroups on the basis of serological tests (see Table 14–3).

TABLE 14–3

*Subgroups of Murine Leukemia Viruses*

| FMR SUBGROUP | GROSS–AKR SUBGROUP |
|---|---|
| Friend | All the "wild-type" or |
| Moloney | naturally occurring mouse |
| Rauscher | leukemia viruses, and the |
|  | Kirsten virus |

Two other murine leukemia viruses, the Graafi and WM 1-B are not neutralized by anti-FMR or anti-Gross AKR antisera, and they may, therefore, form a third group, However, none of these groups is as well defined and distinct as the antigenic subgroups of the avian tumor viruses.

Table 14–4 summarizes information on the origins and behavior of the four best studied viruses which produce leukemias in mice. Nearly all of the con-

TABLE 14-4

*Biological Characteristics of Some Mouse Leukemia Viruses*

| VIRUS | ORIGIN | SENSITIVITY OF MICE[a] | | | | SUBGROUP ANTIGENS | |
| | | SUCKLING (%) | ADULT (%) | STRAINS | TYPE OF DISEASE | ENVELOPE (DETECTED BY NEUTRALIZATION)[b] | TSTA (DETECTED BY TRANSPLANT REJECTION)[e] |
|---|---|---|---|---|---|---|---|
| Gross passage A | AK, C58 mice (spontaneous leukemia) | 62–100 | 44–60 | C3H,C3Hf | Lymphoid | G+ | G+ |
| Moloney | Sarcoma | 100 | 100 | Many strains | Lymphoid | G− | G− |
| Friend | Ehrlich carcinoma cells | 100 | 100 | Swiss;DBA/2 | Erythroid | G− | G− |
| Rauscher[d] | Schwartz ascites tumor cells | 100 | 100 | Many strains | Mixed, erythroid and lymphoid[e] | G− | G− |

(Moloney and Friend envelope entries joined by brace marked [f])

[a] Moloney (1962).
[b] Hartley *et al.* (1965) and Rowe *et al.* (1966).
[c] Klein (1966a).
[d] Rauscher (1962).
[e] Pluznik *et al.* (1966).
[f] Minor differences detectable.

siderable amount of information which has been amassed on these viruses was acquired by bioassays in rats or mice, using leukemia, splenomegaly, or in a few cases spleen colony counts (Axelrad, 1965) as the endpoint. Since all strains of leukemia virus have had many serial passages in mice, many of which themselves carry other leukemia viruses, it is probable that many stocks of murine leukemia viruses are mixtures, like the Bryan high titer strain of Rous sarcoma virus.

## PATHOGENESIS OF MURINE LEUKEMIA

A vast amount has been written on the pathological response of mice infected with different strains of leukemia virus (see Monograph, 1966; Gross, 1970). We cannot attempt to survey this work in any detail. Natural infections of mice with murine leukemia viruses, like natural infections of birds with avian leukosis viruses, are usually latent infections. Except in genetically highly susceptible strains of mice symptoms of disease occur late in life, and the production of lymphoma and lymphoid leukemia, which are the usual manifestations of natural murine leukemia, depend upon the presence of an intact thymus gland (see Chapter 12). Some types of "murine leukemia," like some types of avian leukosis, differ in their pathological picture; the erythroid "splenic" disease produced by Friend virus, for example, is not thymus-dependent.

The incidence of spontaneous leukemia varies greatly in different mouse lines. This is not due only to the presence or absence of virus. Embryos of high leukemia lines of mice regularly yield virus; however, there is no correlation between the incidence of leukemia and virus production in low incidence lines, because embryos of some low leukemia lines do and some do not, yield murine leukemia viruses (Rowe et al., 1966).

### Host range in Vitro

Murine leukemia viruses can be classified into three groups on the basis of their ability to multiply in cells of NIH (Swiss) mice or BALB/c mouse embryos (Hartley et al., 1970).

1. Virus strains which grow equally well in both sorts of cells (NB-tropic).

2. Viruses which grow 100–1000 times better in NIH cells than BALB/c cells (N-tropic).

3. Viruses which grow 30–100 times better in BALB/c cells than NIH cells (B-tropic).

Because the tropism of the murine leukemia viruses is not reversed by passage in resistant cells and because it is independent of the viral pseudotype, host range appears to be a genetic rather than a phenotypic character of each virus. There is no correlation between host range and viral serotype—most of the FMR subgroup is NB-tropic but the Gross AKR subgroup contains both B- and N-tropic strains.

The sensitivity of mouse cells to infection by the leukemia viruses seems to be controlled by recessive genes, because N/B mice, which are the $F_1$ progeny of the cross between N- and B-type mice are resistant to both N and B strains of leukemia virus. However, the resistance of cells is never absolute, so it is thought to be caused by an inhibitor in the cells which partially inhibits viral multiplication. This situation is in marked contrast to that obtaining in chickens where sensitivity to infection by avian leukoviruses is dominant, and resistance depends on the failure of viral penetration because of lack of cellular receptors.

Because the murine leukemia viruses replicate in cell culture but do not produce cytopathic effects, many ingenious assay systems have been devised including complement fixation (Huebner, 1967; Huebner et al., 1964; Hartley et al., 1965 [COMUL Test]); immunodiffusion (Berman and Sarma, 1965; Fink and Cowles, 1965), fluorescent antibody techniques (Vogt, 1965; Kelloff and Vogt, 1966; Rowe et al., 1966), and detection of interference in cultures infected with closely related sarcoma viruses (Sarma et al., 1967; Kelloff et al., 1970a). However, all these earlier methods have been superseded by more recent techniques, the most commonly used of which are the XC test (Rowe et al., 1970b), and an assay based on measurement of RNA-dependent DNA polymerase activity (Kelloff et al., 1972).

The XC test depends on the fact that cells derived from a tumor originally induced by the Prague strain of Rous sarcoma virus (Svoboda, 1960) form syncytia when mixed with cells infected with murine leukemia viruses. The mechanism of fusion is unknown, but it seems to be mediated through a heat-stable lipoprotein which is present both on the surface of cells infected with murine leukemia virus and in the envelope of the virus itself (Johnson et al., 1971). The test is set up by infecting monolayers of mouse cells with dilutions of murine leukemia virus. Five to six days after infection the medium is removed, the cells are lightly irradiated with UV light, and about $10^6$ XC tumor cells are added. The medium is replaced and the XC cells begin to grow. Four days later they have formed a monolayer which is complete apart from the areas where mouse cells infected with leukemia virus have fused with the XC cells. After staining these areas appear as "plaques" which can be counted in the usual way (Rowe et al., 1970b). The procedure has been successfully applied to both laboratory and field isolates of mouse leukemia virus.

The assay for RNA-dependent DNA polymerase devised by Kelloff et al., (1972) is of general use for the measurement of any leukovirus. It is based on measuring the amount of enzyme activity in the supernatant medium of cultures infected with these viruses. The new assay is just as accurate and much more rapid than the older techniques and should be most useful as a screening procedure for mutants of the noncytopathogenic leukoviruses.

### Antigens in Infected Cells

Cells infected by murine leukemia viruses contain several new antigens, which are detected by a variety of techniques. This subject has been comprehensively reviewed (Pasternak, 1969) and we will present only a very brief outline.

**Group-Specific Antigen.**  The first approach to an in vitro assay for mouse leukemia viruses was modeled on the test for a complement-fixing antigen in avian leukoviruses, detectable with antisera from hamsters bearing tumors due to Rous sarcoma virus (Sarma *et al.*, 1964). A complement fixation test was developed which depended on the use of group-specific antibodies produced in the sera of Osborne-Mendel rats bearing progressively growing Rauscher virus-induced lymphosarcomas (Hartley *et al.*, 1965). These sera react with antigens found in tumor cells produced by Rauscher, Friend, Moloney, and AKR leukemia viruses, and with extracts of mouse embryo fibroblasts infected with all leukemia viruses so far tested (Rowe *et al.*, 1966; Huebner 1967). Like the complement fixation test for avian leukoviruses (Armstrong *et al.*, 1964), this complement fixation test reveals a group-specific antigen shared by all laboratory strains of murine leukoviruses and most naturally occurring strains as well (Table 14–4). The structure of this complement-fixing antigen, which is part of the virions of the mouse leukemia viruses, is discussed in Chapter 3.

**Soluble Antigens.** The murine leukemia viruses and cells infected by them contain soluble antigens analogous to those that are found in avian leukovirus-infected cells. This was first demonstrated by Geering *et al.* (1966) in gel diffusion tests with sera from rats infected with Gross leukemia virus and with antigens released by ether treatment of purified preparations of murine leukemia viruses. Such antigens have been found in all strains of murine leukemia virus that have been tested, and react in complement fixation tests with antisera from hamsters which have been inoculated with Moloney sarcoma virus-induced hamster tumor cells (review, Pasternak, 1969). The same antigens can be detected by immunofluorescence, and Hilgers *et al.* (1972) have been able to demonstrate: (a) that the avian and murine leukovirus soluble antigens are unrelated to one another, (b) that RNA tumor viruses from chickens, mice, hamsters, and cats contain species-specific viral antigens that are distinguishable from one another, (c) that the murine leukemia/sarcoma viruses share at least one common antigen, and (d) that the soluble antigens produced by murine mammary tumor viruses (*Leukovirus* subgenus B) are unrelated to the soluble antigens of all other RNA tumor viruses.

**Transplantation Antigens in Cells Infected with Murine Leukoviruses.** Transplantation protection experiments show that novel cell-associated antigens (TSTA's) develop in cells infected with several of the murine leukoviruses (Klein, 1966a). Moloney, Rauscher, and Friend viruses cross-react by the transplantation test. Cells infected with Gross virus develop a cellular antigen (G, Table 14–4) which does not cross-react with those produced by Moloney, Rauscher, or Friend viruses. Cells infected with mouse leukemia viruses produce infectious virus by budding from the cytoplasmic membrane, and it is possible the new cellular antigen involved in transplant rejection tests is a viral envelope protein (review by Pasternak, 1969).

Antigens which react in cytotoxicity, immunofluorescence, and fluorochromatic tests have been found on the surface of cells infected by mouse leukemia viruses (reviews by Pasternak, 1969; Aoki *et al.*, 1970). It is not known whether the antigens detected by these diverse techniques are identical to each other or to TSTA, or whether they are virus- or cell-coded.

## MURINE SARCOMA VIRUSES

The murine sarcoma viruses were discovered when Harvey (1964) and Moloney (1966), working independently, found that the inoculation of very large doses of Moloney virus into infant mice led to the rapid development of sarcomas (Perk and Moloney, 1966). These tumors contain complement-fixing antigens which react in the test with Rauscher rat sera, and they appear to contain two agents, the sarcoma-inducing agent and Moloney leukemia virus. Subsequently, Kirsten and Mayer (1967) isolated a murine sarcoma virus from the Kirsten strain of MLV that had been passed in rats.

Further investigations led Hartley and Rowe (1966) to the conclusion that the mouse sarcoma virus is a defective virus that multiplies only in the presence of related leukemia virus. They noticed initially that stocks of murine sarcoma virus produced foci of altered cells in cultures of mouse embryo fibroblasts and in a continuous line of mouse embryo cells. Moloney leukemia virus (MLV) grew well enough to produce specific complement-fixing antigens, but induced no proliferative foci. The focus-forming virus (MSV) had many of the properties of MLV, and cells transformed by MSV contained CF antigens reactive with the sera of rats carrying transplanted Rauscher or Moloney lymphosarcoma tumors and MSV-induced fibrosarcomas. The focus-forming activity was neutralized by specific rat antisera to Moloney leukemia virus and by sera of rats carrying MSV tumors. If a pool of this virus grown in cultured cells was assayed by the focus-forming test and by the CF test for viral antigen the end points were $10^{3.4}$ and $10^{5.6}$, the excess virus behaving like Moloney leukemia virus.

Murine sarcoma virus, therefore behaves in cultured cells as a transforming, focus-forming virus, generally in association with an excess of murine leukemia virus. Dual infection of mouse cells with these two viruses causes transformation and concomitant production of both murine sarcoma virus and murine leukemia virus (Hartley and Rowe, 1966; Bassin et al., 1968; O'Connor and Fischinger, 1968; Todaro and Aaronson, 1969). However, plating of cells in soft agar infected with high dilutions of the stock effectively separates the murine sarcoma and leukemia viruses, and gives rise to three sorts of cells, all of which have been transformed by murine sarcoma virus.

1. *MSV Nonproducer Cells.* These are mouse, hamster or rat cells transformed by the sarcoma virus but which lack any evidence of spontaneous viral production or viral antigens. However, the sarcoma genome can be rescued by superinfection with mouse leukemia virus (Aaronson and Rowe, 1970; Bassin et al., 1970; Aaronson et al., 1970; Rowe, 1971; Aaronson and Weaver, 1971) or by treatment with 5-bromodeoxyuridine (Aaronson, 1971c). These experiments show conclusively that the murine sarcoma genomes are not defective for transformation: earlier reports (Hartley and Rowe, 1966) that focus formation by murine sarcoma virus was a two-hit phenomenon were due to artifacts in the assay system.

2. *S$^+$L$^-$ (Sarcoma-Positive, Leukemia-Negative) Cells.* These are 3T3 cells transformed by murine sarcoma viruses which do not release infectious viral particles except on superinfection with murine leukemia viruses. Unlike nonpro-

ducer cells, however, S$^+$L$^-$ cells release small amounts of biologically inactive particles (Bassin et al., 1970, 1971) and contain murine leukemia virus antigens (Fischinger et al., 1972).

3. *S$^+$H$^-$ (Sarcoma-Positive, Helper-Negative) Cells.* These are hamster or rat cells transformed by MSV which contain mouse *gs* antigen and which release large numbers of noninfectious viral particles. They are called sarcoma-positive, helper-negative because of the absence of detectable helper virus of hamster, rat, or mouse origin. The sarcoma virus is present in virions released from S$^+$H$^-$ cells because sarcomogenic activity can be demonstrated after artificial formation of pseudotypes by the fusion of viral envelopes that occurs during cosedimentation of S$^+$H$^-$ particles with murine leukemia helper viruses (Gadzar et al., 1971). S$^+$H$^-$ particles are not fully characterized but there is a claim that by contrast to all other RNA tumor viruses, they contain RNA of size 30–40 S (Gadzar et al., 1971).

It seems that rescue of the sarcoma virus genome by superinfection of S$^+$L$^-$ and nonproducer cells is a consequence of the leukemia virus providing envelope functions to the sarcoma genome, first because induction of the sarcoma genome requires replication of the leukemia virus (Nomura et al., 1971) and, second, because the biologically inert particles produced by S$^+$L$^-$ and S$^+$H$^-$ cells do not possess RNA-dependent DNA polymerase (Peebles et al., 1972).

There are several possibilities to explain the different properties of the nonproducer, S$^+$L$^-$, and S$^+$H$^-$ cells. First, there could be genetic differences between the murine sarcoma virus genomes in the three sorts of transformants, with the S$^+$L$^-$ and S$^+$H$^-$ cells possessing viral information which the sarcoma virus genome in nonproducer cells either does not have or does not express. Second, the sarcoma virus genomes might be identical, but each of the three sorts of cells might contain different helper functions. Third, S$^+$L$^-$ and S$^+$H$^-$ cells could have arisen by infection with a complete sarcoma virus genome and a defective leukemia virus genome. However, clonal analysis of the rescued sarcoma virus from S$^+$L$^-$ cells (Aaronson et al., 1972) showed that it retained its characteristics through several passages in different cells, proving that the sarcoma virus genome in S$^+$L$^-$ cells is different from that in nonproducer cells. The reason for this is unknown; Aaronson et al. (1972) suggest that the sarcoma genome found in S$^+$L$^-$ cells may be the consequence of a recombination event between sarcoma and leukemia virus genomes.

### Growth in Primary Mouse Cells

Most of the current work with the murine leukemia/sarcoma viruses involves the use of stable lines of mouse cells, such as 3T3. However, these viruses also infect primary and secondary cultures of mouse cells, and all the early work was carried out with this system. When mouse embryo cells are infected with a mixture of murine sarcoma and leukemia viruses, some of them become transformed by the sarcoma virus genome. These transformed cells are different from all other types of transformed cells in that they grow poorly in culture (Simons et al., 1969; Parkman et al., 1970), perhaps because of an increased synthesis of toxic metabolites (Kotler, 1970). Therefore, instead of the usual formation of piled-up foci of transformed cells, "plaque foci" develop in the mouse cell monolayer because of secondary infection of contiguous cells by newly synthesized leukemia and sarcoma viruses. Although many important results have been obtained with this system, it is clear that plaque foci can form only if leukemia and sarcoma viruses are released from the transformed cells. For this reason it was difficult to detect transformed cells of the nonproducer or $S^+L^-$ classes; consequently, most workers now prefer the 3T3 system which forms the usual sort of foci when infected by MSV.

## Murine Sarcoma Virus in Cells of Other Species

Several workers have reported that growth of murine sarcoma virus in hosts where its helper is unable to replicate can result in the emergence of a sarcoma virus that has undergone striking changes in host range and antigenic composition. For instance, passage of any of several strains of murine sarcoma virus in hamsters produces a virus that has lost the ability to transform mouse cells and to induce sarcomas in mice, but has gained the ability to transform hamster cells and to cause sarcomas in hamsters (Bassin et al., 1968; Klement et al., 1970; Sarma et al., 1970; Kelloff et al., 1970b). In addition, such hamster-tropic viruses are no longer neutralized by anti-murine leukovirus antiserum. These changes in envelope properties seem to be due to the fact that the new sarcoma virus that emerges depends on hamster helper viruses for its growth instead of mouse helper viruses. A similar phenomenon has also been described for MSV grown on rat cells (Ting, 1968; Aaronson, 1971b). The most likely candidates for the new helper viruses are leukoviruses that are indigenous to hamsters or rats and which have been induced by infection with murine sarcoma virus (Gilden et al., 1971). Because there are no genetic markers available on the sarcoma virus genome, we do not know whether it too, has undergone changes. However, this possibility seems unlikely in view of the observation that cocultivation of cell lines derived from the virus-induced hamster tumors with mouse cells and mouse helper viruses yields sarcoma viruses with envelope properties indistinguishable from the original precursor murine sarcoma viruses (Klement et al., 1970; Sarma et al., 1970). Aaronson (1971), however, reached the opposite conclusion on the basis of experiments in which he followed the alteration of murine sarcoma virus during passage in human cells. He transformed a line of human fibroblasts with a clonal isolate of Kirsten murine sarcoma virus, and propagated the line for many generations. After prolonged passage, the host range of the virus altered markedly in that it transformed human fibroblasts very efficiently, but was completely inactive on mouse cells. Because the virus did not revert to its original phenotype on passage in nonhuman cells, Aaronson believes it to have undergone a genetic change rather than an envelope substitution. He postulates that the altered virus may have been created by recombination between the murine sarcoma virus genome and a latent human leukovirus. However much more work is required to establish this point.

## MAMMARY TUMOR VIRUS: LEUKOVIRUS SUBGENUS B

The announcement in 1933 by the staff of the Roscoe B. Jackson Memorial Laboratory of the discovery of a nonchromosomal influence of maternal origin which played a decisive role in the development of mouse mammary cancer caused great interest at that time. In the succeeding 30 years, the "influence" was shown to be a virus transmitted through the mother's milk (Bittner, 1936) and the complex interactions of strain susceptibility, hormonal influences, viral interference, and the mammary tumor virus on the development of mammary tumors in mice were demonstrated (review, Gross, 1970).

The virus has now been purified and its structure shown to be similar to that of the mouse leukemia viruses except that the nucleoid is placed eccentrically within the envelope (Bernhard and Bauer, 1955; see Chapter 3). Such particles are called "B particles" to distinguish them from the more symmetrical "C-type" viruses (Bernhard, 1958; Dalton, 1972). By contrast to the other leukoviruses, in which intracytoplasmic particles are not seen, immature spherical particles of MTV occur within the cytoplasm of infected cells ("intracytoplasmic A particles"). They mature by budding from the cytoplasmic membrane. Like the other murine leukoviruses MTV contains single-stranded RNA with a molecular weight of about 10 million daltons (Chapter 3) and carries RNA-dependent DNA polymerase activity, but it is serologically unrelated to all other murine tumor viruses (Hilgers et al., 1971, 1972).

MTV particles are found in very large quantities in normal breast tissue as well as in the milk of strains of mice which develop mammary tumors with high frequency. The virus is infectious when injected into young mice and assays for MTV have usually depended upon the development of mammary carcinoma as the end point. In spite of many attempts no satisfactory tissue culture assays for the virus have been described. Until these are developed it will be difficult to investigate in vitro the relationship between virus and cell which determine the malignant transformation. Despite this limitation, work on the biology of MTV took on a new twist recently, when it was discovered that Bittner's original strain contained two types of MTV called MTV-S and MTV-L (Mühlbock and Bentvelzen, 1969). MTV-S causes tumors in female mice soon after birth and its sole mode of transmission is through the mother's milk. Only strains of mice that carry a specific genetic allele $MS^e$ can be infected by and can transmit MTV-S. The other virus strain MTV-L is transmitted equally well through the germ cells of both sexes, but never through the milk. MTV-L has lower oncogenic potential than MTV-S, but the tumors it does induce contain B-type particles (Calafat and Hageman, 1968). However, these particles have never been isolated in an infectious form and it has been suggested that MTV-L is usually transmitted as a provirus (Daams et al., 1968).

A third type of MTV, called MTV-P has been isolated from certain strains of mice. Like MTV-S it is highly oncogenic but it is transmitted through both sperm and eggs as well as through milk (Mühlbock and Bentvelzen, 1969).

Timmermans et al. (1969) found that application of carcinogens to strains of mice previously believed to be free of MTV induces a highly oncogenic form of the virus. The relationship of the strain of MTV to MTV-S, MTV-P, and MTV-L is not clear, but the fact that induction of viral particles, viral antigens, or tumors occurred in seven genetically diverse strains of mice after treatment with urethan suggests that every mouse contains in its genome at least one copy of some form of MTV as a provirus (Bentvelzen *et al.*, 1972). Although this hypothesis accounts for the genetic transmission of these viruses, the mechanism by which they are induced either naturally or after treatment with carcinogens is unknown. The most likely explanation is that under normal conditions the transcription of the provirus is repressed, but addition of carcinogens causes transient or permanent derepression, allowing synthesis of virus, which can then be transmitted through the milk. Presumably when MTV indigenous to one strain of mouse is introduced into other strains it is transferred only in the mother's milk.

## HUMAN TUMOR VIRUSES

Quite clearly, mammary cancer in mice is the consequence of a complex interaction between viral and genetic factors. Recently there has been speculation that human breast cancers have a similar etiology. This idea is based on two facts: First, the familial occurrence of breast cancer in women suggests either that the disease itself or susceptibility to it is genetically transmitted (Anderson, 1972; Mühlbock, 1972), and second, human milk contains virus particles which are morphologically similar to those of mouse mammary tumor viruses (Moore *et al.*, 1971; Sarkar and Moore, 1972), contain 70 S RNA (Schlom *et al.*, 1972), and exhibit RNA-dependent DNA polymerase activity (Schlom *et al.*, 1971, 1972; Das *et al.*, 1972). Unfortunately, large quantities of this virus are not available so that hybridization experiments have not yet been undertaken. However, Spiegelman *et al.* (1972) and Axel *et al.* (1972a, b) have used DNA synthesized by murine MTV particles in the endogenous RNA-dependent DNA polymerase reaction in experiments designed to find out whether human breast cancers contain RNA with base sequences in common with those of MTV. It is found that polysomal RNA extracted from 19 out of 29 malignant human tumors showed a detectable amount of specific hybridization with MTV DNA. RNA extracted from control breast tissue did not hybridize. These results mean that base sequences homologous with those from a virus that can cause mammary tumors in mice are present in human mammary tumors. Although experiments of this sort cannot provide definitive evidence of a viral etiology for human breast cancer, they strongly suggest that this is the case. Furthermore they provide a method to search for viral involvement in other sorts of neoplastic diseases of humans. Spiegelman and his collaborators have already begun investigations with human sarcomas and leukemias (Kufe *et al.*, 1972; Hehlmann *et al.*, 1972). A large proportion of these tumors contained RNA homologous to the RNA of the Rauscher murine leukemia virus, but not to murine mammary tumor virus or avian myeloblastosis RNA's.

These experiments are certainly exciting and evocative, and they show a close connection between viruses and certain sorts of human tumors. However they do not resolve whether the relationship is causal or coincidental. Direct proof of a viral etiology for human tumors will involve at the least, consistent isolation of virus particles from the tumors in question and a demonstration that the same particles can cause similar tumors in laboratory animals. C-type particles have been isolated from human tumors from time to time, e.g., RD114 of McAllister et al. (1972a), now considered to be an irrelevant cat leukovirus. Much more work will be required to provide definite proof that any human tumor is caused by a virus.

## OTHER RNA TUMOR VIRUSES

Leukovirus particles have been demonstrated in normal and malignant tissues of so many different sorts of animals that it seems that they are universally distributed throughout all vertebrates. Most of these viruses are poorly characterized and will not be discussed here. Two of them, however, are especially interesting.

### Feline Leukemia Virus

This virus was first isolated in 1964 by Jarrett and his colleagues from cats with lymphosarcomas, which are the most common form of neoplasms in these animals. Since that time the virus (feline leukemia virus—FeLV) has been isolated so consistently from these tumors (Rickard et al., 1969; Kawakami et al., 1967; Jarrett, 1970) that it now seems possible that all cat leukemias are of viral origin. FeLV induces leukemia after inoculation into kittens. It is a typical C-type particle containing 60–70 S RNA (Jarrett et al., 1971) and RNA-dependent DNA polymerase (Spiegelman et al., 1970; Hatanaka et al., 1970). The nucleoid carries the interspecific (gs-3) antigenic determinant of the mammalian C-type viruses as well as the feline species-specific (gs-1) antigen (Gilden et al., 1971). The RNA-dependent DNA polymerase can be distinguished from the enzyme found in other classes of C-type particles by serological tests (Scolnick et al.', 1972; Parks et al., 1972). The virus can be grown in the laboratory in a variety of normal and embryonic cell lines; like the murine leukemia and avian leukosis viruses, FeLV is produced continuously by infected cells without noticeable cytopathic effect. The virus is usually assayed either by interference techniques (Sarma and Log, 1971), complement fixation (Sarma et al., 1971), or RNA-dependent DNA polymerase activity (Kelloff et al., 1972). Perhaps the most interesting finding is that FeLV can replicate in human cells (Jarrett et al., 1969), raising the possibility that some human leukemias may be caused by infection with FeLV. However, in spite of intensive epidemiological studies no connection has been established between the occurrence of leukemia in humans and the presence of cats in their households.

C-type particles have been shown to be causative agents of two sorts of solid tumors of cats—fibrosarcomas (Synder and Theilen, 1969; Gardner et al., 1970) and liposarcomas (Rickard et al., 1969). The same preparations of virus also induce sarcomas in dogs, rabbits, marmosets, and monkeys (Rickard et al., 1969; Theilen et al., 1969), and can transform human embryo (Sarma et al., 1970b) and human osteosarcoma cells (McAllister et al., 1971). Whether these virus stocks contain only sarcoma genomes or whether they are mixtures of FeLV with a feline sarcoma virus is not yet clear.

## Mason-Pfizer Monkey Tumor Virus

Chopra and Mason (1970) found that cells of a spontaneous mammary tumor in a rhesus monkey contained large numbers of type A particles. More importantly the cells released C-type particles which could multiply in a variety of monkey cells and in human leukocytes, without producing morphological transformation or cytopathic effect (Jensen *et al.*, 1970; Chopra *et al.*, 1971; Ahmed *et al.*, 1971). The virus contains RNA-dependent DNA polymerase (Schlom and Spiegelman, 1971; Gallo *et al.*, 1972).

So far there is no proof that this virus was the cause of the monkey tumor from which it was isolated. It has been injected into susceptible animals in order to see whether mammary cancers develop, but so far no positive results are available.

## INDUCTION OF MAMMALIAN LEUKOVIRUSES, ONCOGENES, PROTOVIRUSES, PROVIRUS

It is certain (a) that at least some RNA tumor viruses cause at least some of the cancers that occur in animals and (b) that production of C-type particles can be induced in cells that are destined never to be malignant. It is also probable that the information for C-type particle production is vertically transmitted as part of the genome of the host cell. A number of hypotheses have been proposed in bringing these seemingly disparate observations together.

### The Oncogene Hypothesis

In 1969, Huebner and Todaro presented a hypothesis (the oncogene hypothesis) which could explain these observations, based mainly on the following experimental result.

After long-term culture, several lines of 3T3 and 3T12 cells derived both from outbred Swiss and from inbred BALB/c mice began to shed virus with properties very similar to those of murine leukemia virus (Aaronson *et al.*, 1969). The production of virus seemed to depend largely on the conditions of cell culture, and was favored in old cultures maintained at high cell density. External contamination was eliminated as a source of the virus by a series of elegant reconstruction experiments, and Aaronson *et al.* (1969) concluded that the virus or the capacity to produce it was present in each of the original embryo cells. This led Huebner and Todaro (1969) to propose that the cells of most or all vertebrate species contain genomes of RNA tumor viruses and that the virus-specific information is vertically transmitted from parent to offspring. Every cell of a vertebrate body, therefore, would contain this information. Depending on a complex interplay between the host genotype and environmental conditions, viral production could be elicited at some stage in the life of the individual animal. Since C-type viruses carry oncogenic information as part of their genetic make up, tumor formation might sometimes ensue after induction of C-type particles.

Since 1969 no evidence has appeared to challenge this hypothesis (cynics might say that it is impossible to provide such evidence given the amorphous nature of the idea). However, a great deal of new experimental data is consistent with the oncogene hypothesis (review, Todaro and Huebner, 1972).

It has been found that synthesis of C-type particles, virus-specific RNA-dependent DNA polymerase, and/or viral antigens can be induced in clones of mouse embryo cells by treatment with carcinogens or mutagens (Table 14–5). For instance, treatment of "virus-free" clonal lines of mouse cells with low concentrations of 5-bromodeoxyuridine or 5-iododeoxyuridine elicits low levels of viral production within 3 days (Lowy et al., 1971). Every clone of mouse cells that has been tested in this way has been induced to release virus (Aaronson et al., 1971a). This result means that the information for viral production is present in every mouse cell but is not usually expressed. This conclusion is supported by hybridization data (Gelb et al., 1971a) which shows that the genome of normal mouse cells contains sequences of DNA which are homologous to those found in the double-stranded DNA synthesized in the endogenous polymerase reaction by murine RNA tumor viruses.

TABLE 14–5

Rodent Continuous Cell Lines with Clones That Have Been
Shown to Contain Inducible C-Type Viruses[a]

| ANIMAL SPECIES AND STRAIN | CELL LINE |
| --- | --- |
| Mouse (AKR) | Line 26 |
| | Line 32 |
| Mouse (BALB/c) | BALB/3T3 |
| | BALB/3T12 |
| | 3T3-SV40 |
| Mouse (random-bred Swiss) | 3T6 |
| | 3T12 |
| Rat | NRK |
| Chinese hamster | CHO |

[a] From Todaro and Huebner (1972) and Todaro (1973).

The virus that is induced is morphologically identical to other C-type particles and behaves in vitro just like any other mouse leukemia virus, except that it often grows very poorly in the original host cells, but replicates quite well in Swiss 3T3 cells. A possible explanation for this phenomenon is the N and B tropism of murine leukemia viruses discussed earlier in this chapter.

The induction of C-type particles is not restricted to mouse cells. We have already seen that a similar phenomenon has been found in chick cells (Weiss et al., 1971), and Aaronson (1971c) and Klement et al. (1971) have shown that endogenous C-type particles can be induced from rat cells (see Table 14–5). In fact, out of all the species of animal cells tested only the Syrian hamster line BHK-21 fail to yield C-type viruses (Todaro and Huebner, 1972). The reason for this exception is not known but none the less the observations are extremely striking, and they make it difficult to avoid the conclusion that the information for RNA tumor viruses is present in many, if not all vertebrate cells.

The mechanism by which the synthesis of virus is induced, either "spontaneously" or in mutagen-treated cells, is not understood. Because the viral genes are not normally expressed in vertebrate cells, it is presumed that induction requires derepression of some part of the host genome. How this is carried out, and whether the derepression is transient or permanent is not known.

Viral gs antigens and to a lesser extent C-type particles have been demonstrated in the tissues of chick embryos from "leukosis-free" flocks (Dougherty et al., 1967; Allen and Sarma, 1972) and in tissues derived from mouse embryos of many different strains, including wild *Mus musculus* (Huebner et al., 1970). Mice of "low-leukemia" strains lose these antigens in the late stages of embryogenesis and they do not reappear until very late in adult life. Huebner et al. (1970) interpret these results to mean that there is regulation of the expression of the "virogenes;" it implies that natural cancers develop when the regulation system breaks down so that the "virogene" and the "oncogene" contained in it, or even the oncogene alone, are activated. Whether the appearance of gs antigens and sometimes of C-type particles during embryogenesis is accidental or whether the viral products have an essential developmental function is, of course, not known.

The oncogene hypothesis, which predicts that spontaneous neoplasms are a consequence of events occurring within the genome of cells, fits well with epidemological data which shows that the majority of naturally occurring tumors of vertebrates do not occur in time-space clusters and do not, therefore, behave as infectious diseases. Of course, the hypothesis does not bear on the fact that some tumors are transmissible by viruses in an orthodox way, or on the artificial production of cancers by tumor viruses in the laboratory.

In summary, the data so far accumulated on the biology of the C-type viruses show a very close association between viral and cellular genomes; although hardly any of the details of this association are understood, the general observations fit quite well with the oncogene hypothesis. Perhaps the key question still to be resolved is whether the viruses induced from normal animals and normal cells are tumorigenic (i.e., whether they carry the oncogene).

## Proviruses and Protoviruses

Huebner and Todaro's hypothesis is by no means the only one that has been put forward to explain the behavior of the C-type viruses. We shall close this chapter by considering briefly two ideas put forward by Temin.

**The Provirus Hypothesis.**   The provirus is defined as the intracellular form of a tumor virus that is passed from parent to daughter cells at mitosis and contains the information for the production of new virus and for the maintenance of the cells in a transformed state (Temin, 1961, 1962). It is implied that proviral DNA is integrated into the genome of the transformed cell as a consequence of infection of the cell with a tumor virus. In this sense the provirus hypothesis is different from the oncogene hypothesis which postulates that the information for viral production is part of the genetic make up of the cell and behaves as a cellular gene. The provirus theory, then, is concerned with events which occur when viruses infect cells, and it seems to be correct only in certain limited situations, for instance, during the transformation of rodent cells by avian RNA tumor viruses where hybridization experiments show the addition of viral DNA sequences into the cell genome (Baluda, 1972). Although the proviral theory has been very useful in explaining what happens when a tumor virus meets a species of cell that it has never encountered previously, it is limited because this sort of interaction is highly artificial and probably only occurs in laboratories. The provirus theory does not address itself to what we now think is the central phenomenon of RNA tumor viruses: The vertical transmission of endogenous C-type virus genomes in vertebrate cells and the way in which the expression of these genomes is controlled.

**The Protovirus Hypothesis.**   Recently Temin (1971b) has put forward the protovirus theory, which proposes the following: a region of DNA in a cell is transcribed into RNA; the RNA is then copied back into DNA which can become integrated in the genome either of the original cell or other cells in the body, or both. The main function of this system is to provide a mechanism for gene amplification and cellular differentiation. Temin postulates that successive DNA to RNA to DNA information transfers could lead to a modification of the original information so that occasionally cells will be produced which contain sequences of nucleic acid which code for the synthesis of a C-type virus. The protovirus hypothesis and the oncogene theory differ because the oncogene theory assumes that no modification of information occurs, only derepression of existing information.

While the protovirus theory may have important things to say about differentiation, gene amplification, and the origin of viruses, it is incompatible with the existing data. The experiments of Rowe and Aaronson (1971) and Weiss *et al.* (1971) seem to show conclusively that every cell in mice and chickens has the capacity to produce C-type particles. It is very difficult to see how this situation can have arisen purely as the result of a random process; it seems much more likely that the viral information is conserved through genetic transmission and does not arise *de novo* in somatic cells.

It is worth pointing out that these three models—oncogene, protovirus, and provirus—are not mutually exclusive and it is possible that all of them are correct. For instance, it is possible that the information contained in C-type viruses is vertically transmitted and that the events that occur during induction are as follows. First, new copies of viral RNA are produced by derepression of the oncogene/virogene; some of this RNA is translated into virus-coded RNA-dependent DNA polymerase which acts on the viral RNA to generate a DNA provirus which integrates into the genome of the induced cell at a different site from the one occupied by the original oncogene/virogene. In its new site the proviral DNA is not repressed and virus-coded functions are expressed, some of which may cause cells to become transformed. The central problem in current research both with the DNA and RNA tumor viruses is the identification and characterization of the virus-coded products that are responsible both for the initiation and maintenance of the transformed state.

## EPILOGUE

The great upsurge of interest in tumor viruses in the past 10 years has come about as a consequence of the development of experimental systems which seem to mimic *in vitro* at least some of the events by which tumor viruses induce cancers in animals. For the first time the events leading to malignancy have become accessible at the molecular level. The central property of the *in vitro* systems is cellular transformation, which occurs when fibroblasts are infected with some kinds of tumor viruses.

Of the RNA tumor viruses, only the sarcomagenic types can transform fibroblasts and viral multiplication may or may not occur concomitantly. In the case of the DNA tumor viruses, however, transformation and viral synthesis are mutually exclusive. For example, transformation by polyoma virus and SV40 can occur only after infection of cells in which complete expression of viral genes is restricted or after infection of fully permissive cells with defective viral mutants. Furthermore, DNA tumor viruses cause two types of transformation: (a) abortive transformation which occurs with high frequency but lasts only for a few cell generations, and (b) stable transformation which occurs more rarely but results in permanent changes in cell phenotype. Whether the transformation is stable or abortive, and whether it is mediated by DNA or RNA viruses, virus-coded functions are required both for the establishment and for the maintenance of the transformed phenotype.

Transformation by both the DNA and the RNA tumor viruses seems to involve the stable association of viral nucleic acid sequences with the cell DNA. This occurs in the case of polyoma virus, SV40, and adenoviruses by the direct integration of viral DNA into the cell genome. With the RNA viruses the situation is not so clear-cut because most, if not all, vertebrate cells contain viral sequences as part of their normal genetic makeup. Consequently, it is not entirely clear whether transformation necessarily involves the addition of new viral sequences to the cell genome or activation of preexisting indigenous viral DNA. However, the fact that cells transformed by certain temperature-sensitive mutants of RNA tumor viruses show a temperature-sensitive phenotype indicates that the incoming viral genome plays a dominant role in determining the phenotype of the transformed cell. Because transformation is a stable heritable trait, it seems likely that the nucleic acid sequences of the viral RNA genome become attached in a permanent fashion to the cell DNA.

The mechanism by which the addition of these small amounts of viral nucleic acid causes large changes in cell behavior is still obscure and remains one of the focal points of current tumor virus research. The problem is difficult because the changes which cells undergo at transformation are extremely diverse and range from a variety of alterations in cell surface structure and function to changes in chromosome number and distribution, to the capacity to grow under conditions which restrict the growth of normal cells. The interrelationship of these alterations which often occur in a covariant manner is not understood, and the key question seems to be how to distinguish the changes which cause transformation from those which are a consequence of transformation. The main hope for the future seems to lie in the characterization of large numbers of viral and cell mutants.

We do not know whether the oncogenic RNA and DNA viruses induce tumors by the same mechanism. The fact that RNA tumor viruses possess enzymes potentially capable of synthesizing complete DNA copies of their genomes suggests that the two sorts of viruses may be fundamentally similiar in the way that they convert normal cells into malignant cells. However, even if the DNA and RNA tumor viruses turn out to be identical in this respect, the biology of the two groups is extremely different.

RNA tumor viruses have been isolated from at least three orders of vertebrates—reptiles, birds, and mammals. They all show a remarkable morphological similarity and can only be distinguished one from another by sensitive serological tests. Some are capable of causing neoplastic disease of the hemopoietic systems of their hosts; others cause solid tumors. However, the most remarkable characteristic of this class of viruses is that they can be transmitted vertically as part of the genome of the host animal. Expression of viral genes occurs during embryogenesis and this has led to speculation that viral functions are involved in development. Induction of C-type viruses occurs spontaneously both in animals and cells in culture, and its frequency can be increased by laboratory manipulation.

The DNA tumor viruses are much more varied, being drawn from several different genera. Only one of these, the herpesviruses, has been established as a cause of malignant tumors in its natural hosts. With other DNA tumor viruses the sole sign of infection is a benign tumor. Most of them, however, seem to produce no more than inapparent or innocuous infections in their natural hosts and only cause tumors under highly artificial laboratory circumstances.

So far there is no conclusive evidence for human tumor viruses. However, there is a large effort to find out whether there is any causal relationship between herpesviruses and human cancer and large sums of money are being spent in order to isolate the putative natural C-type viruses of humans.

## CHAPTER 15

# Prevention and Treatment of Viral Diseases

## INTRODUCTION

There are three main approaches to the prophylaxis and therapy of viral infections: (a) chemotherapeutic agents that inhibit viral multiplication directly; (b) agents that induce the body to synthesize interferon or to mount other types of nonimmunological defence; and (c) vaccines that elicit an immune response. All three can be used to prevent infection (chemoprophylaxis or immunoprophylaxis) or to treat established or impending disease (chemotherapy or immunotherapy), but for obvious reasons all methods are more effective the earlier they are applied and some, notably active immunization, are usually ineffective after infection has begun. A number of viral vaccines have been outstandingly successful: notably those directed against smallpox, yellow fever, poliomyelitis, measles, and rubella, all of which are generalized diseases. Vaccines against the many diseases caused by viruses localized to the respiratory tract have been much less effective, and antiviral chemotherapy is still in its infancy.

## EFFICACY OF VIRAL VACCINES

### Immunological Considerations

The immunological principles upon which vaccination is based are discussed in Chapter 10. The long-term protection that follows systemic infection with such viruses as smallpox or measles is primarily a function of humoral (antibody-based) immunity that depends upon immunoglobulin G (IgG), although cellular immunity may play a part (reviews by Allison, 1972a; Allison and Burns, 1972). Effective vaccines have been developed against many generalized viral diseases of both man and his domestic animals (see Table 15–3 and 15–4).

On the other hand, immunity after superficial infection of the respiratory mucous membrane is relatively short-lived (Chanock, 1970), and depends upon immunoglobulin A (IgA) synthesized by lymphoid cells in the epithelium and secreted into the mucus (reviews, Tomasi and Bienenstock, 1968; Ogra and Karzon, 1971; Chanock, 1971; Rossen et al., 1971). Because immunity after natural infection is relatively short-lived, most studies on vaccination against respiratory viruses have of necessity been conducted over a very short interval, often only a matter of weeks. At least some of the protection attributed to specific IgA may, in fact, be due to the nonspecific antiviral action of interferon (which tends to be induced in larger amounts by avirulent than by virulent viruses, Wheelock *et al.*, 1968), or to cross-reacting antibodies of low avidity recalled by a secondary response to the heterologous vaccine. Both of these factors may turn out to be valuable adjuncts to specific IgA in practice, and help to underline the advantages of topical administration of live intranasal vaccines, but it is likely that their influence would wane rapidly. The poor immunity that follows natural infection with most of the respiratory viruses (Chanock, 1970) does not augur well for the future of vaccines directed against them.

Attention has already been drawn (Chapter 10) to the fact that most of the viruses that cause generalized viral infections are monotypic, whereas great antigenic diversity is found among the viruses that cause localized infections of superficial mucous membranes. This antigenic diversity raises further difficulties for vaccination; however, the severe lower respiratory tract infections of young children tend to be attributable to a relatively narrow range of viral serotypes: one respiratory syncytial virus, four parainfluenza viruses, one influenza virus, and three adenoviruses. Effective vaccines against these agents, plus *Mycoplasma pneumoniae* (Chanock, 1970; Chanock *et al.*, 1971) might make an appreciable impact on the total spectrum of serious respiratory disease in children.

Antigenic drift constitutes an additional difficulty with vaccines against influenza virus; its basis is discussed in Chapters 10 and 17.

## Practical Considerations

The acceptability of a vaccine against any given disease can be determined by the public need for the vaccine and the degree of protection it affords. Where the need is clear, as in the case of lethal diseases such as smallpox or rabies, or even influenza, we have been inclined to accept lower standards of safety (in the first two cases) or efficacy (in the latter) than we would today for new vaccines directed against more trivial illnesses. This is relevant to current work on respiratory vaccines; they will only be acceptable if they are completely safe and reasonably efficacious, and if the disease is sufficiently debilitating to be worth preventing. As severe and lethal bacterial diseases progressively disappear, levels of public expectation concerning freedom from viral diseases may be expected to rise.

## SAFETY OF VIRAL VACCINES

Throughout the history of vaccine development there have been a succession of accidents, some of which were disastrous, others potentially so (Table 15–1). Most of the dangers are associated with live vaccines and will be discussed in detail under that heading; they include (a) insufficient attentuation of the vaccine strain leading to disease and sometimes spread to contacts, and/or back-mutation to high virulence, and (b) the presence of adventitious viruses in the cells used to grow the vaccine virus. The risks have now been considerably diminished by the establishment of elaborate laboratories and government authorities to test vac-

TABLE 15–1

*Some Problems Encountered during the Development
and Use of Viral Vaccines*

| VACCINE | PROBLEM |
|---|---|
| Live vaccines | |
| Poliovirus (Sabin) | Contaminating viruses (e.g., SV40) |
| | Back-mutation to virulence (type 3) |
| | Interference by endemic enteroviruses |
| Smallpox | Inadequate attenuation leading to complications |
| | Overattenuation leading to lack of protection |
| Measles (Edmonston) | Inadequate attenuation leading to fever and rash |
| Rubella (HPV-77) | Inadequate attenuation leading to arthritis in adult females |
| Yellow fever | Contaminating hepatitis B in "stabilizing" medium |
| Inactivated vaccines | |
| Poliovirus (Salk) | Residual live virulent virus |
| | Contaminating virus (SV40) resisting formaldehyde |
| Influenza | Pyrogenicity |
| Rabies (Semple) | Allergic encephalomyelitis |
| Measles⎫ Respiratory syncytial⎬ | Hypersensitivity reactions on subsequent natural infection or live vaccine booster |

cines for safety and efficacy and to evaluate reports of clinical trials in man. These and other technical details of vaccine manufacture are described in an excellent monograph from the National Cancer Institute (1968), and in reviews by Potash (1968) and Bachrach and Breese (1968).

There is one immunological problem that has arisen only with inactivated vaccines. Instead of being protected, children immunized with inactivated measles or respiratory syncytial vaccine may, when subsequently exposed to natural infection or to live vaccine, develop more serious illness than do controls (Chanock *et al.*, 1968; Isacson and Stone, 1971; Rossen *et al.*, 1971). Chanock *et al.* (1968) have drawn an analogy between the severe bronchiolitis and pneumonitis experienced by children preimmunized with respiratory syncytial virus vaccine and the severity of natural infection in very young infants with high levels of circulating maternal IgG; both may be caused by an antigen–antibody-complement reaction in the lung, similar to the mould-induced disease "farmer's lung." Isacson and Stone (1971), on the other hand, regard the postvaccination condition as an allergic reaction following prior sensitization, due to viral attachment to IgE bound to mast cells, as in asthma, while Rossen *et al.* (1971) point out that the lung is a primary shock organ in anaphylaxis. Other possible explanations include cell lysis by sensitized lymphocytes interacting with viral glycoproteins present in the plasma membrane of the infected epithelium, or complement-mediated lysis of virus-infected cells by cytotoxic antibodies, or by antibodies acting in concert with normal lymphocytes. The practical implications of these findings are that inactivated respiratory viral vaccines must be carefully evaluated before being used and that even living attenuated vaccines delivered by nasal spray could produce bronchial spasm and exudation following repeated application in an atopic individual (Rossen et al., 1971).

Isacson and Stone (1971) record hypersensitivity reactions to components of vaccines other than the virus itself. The most serious is the "autoimmune" encephalomyelitis which quite frequently follows repeated daily injections of rabies vaccines containing rabbit brain tissue. Fortunately, no renal damage has been noted in the millions of patients who have received poliovirus vaccines grown in monkey kidney cells. In the days before chick embryo-grown vaccines were administered in semipurified form, individuals sensitive to eggs, or presensitized by other chick embryo-grown vaccines, frequently developed allergic reactions to influenza, mumps, or yellow fever vaccines. Likewise, penicillin hypersensitivity used to be relatively frequent in recipients of Salk vaccine, but now vaccine viruses are grown in media containing less commonly used antibiotics.

## LIVE VACCINES

Almost all of the successful viral vaccines used in man and animals are living avirulent viruses (Tables 15–3 and 15–4). They present obvious advantages over inactivated vaccines (Table 15–2). Multiplication in the host, leading to a

TABLE 15–2

*Advantages and Disadvantages of Live and Inactivated Vaccines*

| ADVANTAGES | DISADVANTAGES |
|---|---|
| Live vaccines | |
| (1) Single dose, given by | (1) Reversion to virulence [a] |
| (2) natural route invokes | (2) Natural spread to contacts |
| (3) full range of immuno- | (3) Contaminating viruses [a] |
| logical responses, (4) local | (4) "Human cancer viruses" [b] |
| IgA as well as systemic | (5) Viral interference |
| antibody production, leading | (6) Inactivation by heat in the tropics |
| to (5) possibility of local | |
| eradication of wild-type | |
| viruses | |
| | |
| Inactivated vaccines | |
| (1) Stability | (1) Multiple doses and boosters needed, (2) given by injection, therefore (3) local IgA fails to develop. (4) High concentration of antigen needed: production difficulties |

[a] With care these difficulties have been largely overcome.

[b] Theoretical objection that has been raised against use of human diploid cell strains as substrate for vaccine production.

prolonged immunogenic stimulus of similar kind and magnitude to that oc-curing in natural subclinical infection, gives rise to a substantial immunity. In some cases, the virus can be inoculated by the natural route, i.e., orally or nasally, where it can induce the continuing synthesis of immunoglobulin A by lymphoid cells present in tonsils, Peyer's patches, and throughout the respiratory epithe-lium, and the resulting local immunity may suffice to prevent reimplantation of wild strains.

## Derivation of Vaccine Strains

A few vaccines still used in veterinary practice consist of fully virulent infec-tious virus injected in a site where its multiplication is of little consequence, e.g., contagious pustular dermatitis in sheep, where vaccination on the inside of the thigh was introduced by Glover (1928). Since immunization with live smallpox virus ("variolation") was abandoned more than a century ago this procedure has been considered too dangerous for use in man, although the ingenious technique of delivering respiratory pathogens, such as adenoviruses, to the lower reaches of the alimentary tract in enteric coated capsules (Couch et al., 1963) is an in-teresting exception.

Most vaccine strains are mutants which are sufficiently attenuated to produce no symptoms (or trivial ones) but not so avirulent that they fail to multiply extensively in the body and elicit a lasting immunity. It is not always easy to strike this balance. The process of attenuation of most current vaccines has been empirical, based on the knowledge that prolonged passage of a virus in a foreign host tends to select mutants better suited to growth in the new host than in the old one (see Chapter 7). The emergence of rare mutants is facilitated by frequent passage of large viral inocula (by contrast with the practice of infrequent passage at limit dilution used for the maintenance of laboratory stocks of wild-type viruses in their original state). To demonstrate loss of virulence the vaccine virus is extensively tested in animals, but the ultimate test of avirulence for man can only come from experimental inoculation of volunteers. The acceptable level of virulence sometimes requires a degree of compromise about the immunogenicity of a vaccine. For example, the original Edmonston strain of live measles vaccine produced quite high fever in a substantial proportion of children, and even oc-casional rashes. For a time it was recommended that human immunoglobulin be simultaneously inoculated into the opposite arm to minimize these side effects. Nowadays, however, it is considered wiser to use the more attenuated Schwarz vaccine (Schwarz, 1964), which is slightly less immunogenic, but produces negligible side effects.

Today it is no longer necessary to rely on the purely empirical process of rapid serial passage in a foreign host to produce an avirulent mutant, for our knowledge of animal virus genetics is sufficiently advanced for us to be able to apply rational procedures to the selection of candidate strains for attenuated viral vaccines (Fenner, 1969, 1972; see Chapter 7). Chanock and his colleagues have made good use of genetic principles to derive ts mutants of respiratory syncytial virus (Wright et al., 1971) and influenza virus (Mills and Chanock, 1971), which grow poorly at 37°C, the temperature of the lower respiratory tract of man, but well at 33°C, the temperature of the nose. The "cold-adapted" influenza mutants of Maassab (1969) and Beare *et al.* (1972) are somewhat different but have also been used successfully as vaccine strains in man. Much the same principle can be used to select potentially avirulent variants of poliovirus, by taking advantage of the fact that most attenuated strains are unable to grow at 40°C (Lwoff, 1969).

Recently, workers in two of these laboratories have gone further and used recombination of a *ts* mutant with a wild strain of influenza virus to produce a genetic reassortment (see Chapter 7) with the hemagglutinin of the virulent strain but the *ts* lesion of the attenuated mutant (Maassab *et al.*, 1972; Murphy *et al.*, 1972). These reassortments grow satisfactorily only in the upper respiratory tract of animals (or at 33°C in cultured cells) and one of them has been successfully used as a living intranasal vaccine to protect human volunteers against challenge with virulent influenza virus (Murphy *et al.*, 1972).

### Potential Problems

In spite of their obvious advantages, live vaccines are subject to a number of problems that do not occur with inactivated vaccines (Table 15–2). They are: (a) genetic instability, (b) contamination by adventitious viruses, (c) interference by wild viruses, and (d) heat lability. Now that the problems have been recognized, ways have been found to circumvent each of them, and most vaccines commercially available for human use today are in no way hazardous.

**Genetic Instability.** Some vaccine strains are not as avirulent as could be wished, but tests have shown that further attenuation leads to an unacceptable loss of immunogenicity or even to loss of the capacity to grow in man when administered by the desired route. Indeed, each of the human live vaccines listed in Table 15–3, with the exception of poliovirus vaccine, produces mild symptoms in a minority of recipients, and vaccinia very occasionally causes death. In spite of early apprehension, there is no evidence that measles vaccine gives rise to subacute sclerosing panencephalitis. Of more concern is the theoretical possibility that live vaccine strains may back-mutate to, or toward, the virulent wild type. Only for poliovirus type 3 have we statistical evidence that this may occur in practice, albeit very rarely (see below).

A related question is whether vaccine strains, unchanged or virulent, are capable of spreading to the vaccinee's contacts. This would be particularly dangerous if rubella vaccine, which is excreted in the throat, were teratogenic and could spread to pregnant women, but this has not occurred. Vaccinia virus is occasionally transmitted, by direct contact, to other people, or more commonly to other parts of the vaccinee, and can be dangerous if it enters the eye.

**Contaminating Viruses.** Many cells used for culture are contaminated with viruses (see Chapter 12). Because of methods of shipping and holding monkeys, monkey kidney cells are particularly liable to be contaminated, with any or several of over fifty known simian viruses. The risks are less with closed, specific-pathogen-free colonies of animals, such as the rabbits used for the production of Cendehill strain rubella vaccine, but even in this situation viruses transferred congenitally cannot be eliminated. For example, most of the older human viral vaccines were grown in chick embryos or cultured chick embryo fibroblasts

TABLE 15–3

Viral Vaccines Recommended for Use in Man [a]

| DISEASE | VACCINE STRAIN | CELL SUBSTRATE | ATTENUATION | INACTIVATION | ROUTE |
|---|---|---|---|---|---|
| Smallpox | Vaccinia | Calf or sheep skin | + | — | Intradermal |
| Yellow fever | 17D | Chick embryo | + | — | Subcutaneous |
| Poliomyelitis | Sabin 1, 2, 3 | WI-38 | + | — | Oral |
| Measles | Schwarz | Chick fibroblasts | + | — | Subcutaneous |
| Rubella | RA 27/3 | WI-38 | + | — | Subcutaneous |
| | Cendehill | Rabbit kidney | + | — | Subcutaneous |
| Mumps | Jeryl Lynn | Chick fibroblasts | + | — | Subcutaneous |
| Rabies [b] | Pitman-Moore | WI-38 | — | β-Propiolactone | Intramuscular |
| Influenza | A/HK/68 (H3N2) | Chick embryo | — | Formaldehyde and deoxycholate | Subcutaneous |
| Adenovirus | 4, 7 | WI-38 | — | Live, in enteric-coated capsules | Oral |

[a] A great variety of different viral strains and cellular substrates are used in different countries; the selection listed is not comprehensive.
[b] Experimental.

which were contaminated with avian leukosis viruses. Fortunately no ill seems to have come of it. Contaminating viruses are clearly more likely to be dangerous in live vaccines, but inactivated vaccines are not free from this risk, for the adventitious virus may be more resistant to the inactivating agent than the vaccine strain. For example, many early batches of formaldehyde-inactivated poliovaccine were contaminated by live SV40.

The discovery of adenovirus-SV40 hybrids (see Chapter 7) has added a new dimension to the problem of contaminating viruses. Such recombinants, some of which are highly oncogenic, are indistinguishable serologically from the parent adenovirus since they have identical capsids. Up to the present time, this intergeneric hybridization remains a unique laboratory artifact, but in relation to live virus vaccines it represents a risk that will have to be borne in mind.

The problem of contaminating viruses has led to the abandonment of monkey kidney cells for the preparation of vaccines. Human diploid cells like WI-38 (Hayflick, 1965) constitute a logical replacement that could be used for all human vaccine viruses. Although they offer many advantages their use in some countries (notably United States until recently) has been prohibited because of the hypothetical risk that they might carry a "human cancer virus." However, several million people in other countries have now received WI-38-grown vaccines against a number of viral diseases and no untoward complications have been reported (National Cancer Institute, 1968; World Health Organization, 1972a).

**Interference.**  If a live vaccine such as the oral poliovirus vaccine is given by the natural route of infection, preexisting enteroviral infection of the gut of the vaccinated individual may prevent the establishment of the vaccine strain by interference. This presents a problem in some underdeveloped countries where enteroviruses are present in the intestines of the majority of children at any particular time. It is obviated by the practice of giving trivalent vaccine on three occasions, 6–8 weeks apart. In spite of earlier doubts, interference between the three serotypes comprising the poliovirus vaccine itself does not occur to any appreciable extent under this regimen, but the vaccine strains may themselves interfere with the circulation of virulent strains of poliovirus, if given at the time of an epidemic (Sabin, 1957; Koprowski, 1957).

Interference between live viruses in a polyvalent vaccine does not usually occur following parenteral administration, probably because the viruses are rapidly dispersed throughout the body. For instance, an experimental live measles-mumps-rubella vaccine appears, in preliminary trials, to give satisfactory antibody responses against all three viruses (Stokes et al., 1971). However, difficulties have been encountered with polyvalent live measles-yellow fever-smallpox vaccines in trials carried out in Africa (see Wheelock et al., 1968).

**Heat Lability.**  Since a reasonable titer of active virus must be maintained, shipping and handling of live vaccines calls for good organization of public health services, including refrigeration and prompt use of the vaccine after reconstitution from the lyophilized state. The problem is greatest in tropical countries where the ambient temperature is high and health services are often inadequate.

## INACTIVATED VACCINES

Inactivated vaccines are "dead" in the sense that their infectivity has been destroyed, usually by treatment with formaldehyde. The only inactivated viral vaccine now in widespread use in man is that directed against influenza. An inactivated rabies vaccine is employed in particular emergencies; others like Salk poliovirus vaccine and inactivated measles vaccine have been displaced by attenuated live-virus vaccines.

The major difficulty with inactivated vaccines lies in producing enough virus to provide the necessary quantity of the relevant antigens and in administering the vaccine in such a way as to promote local IgA synthesis where this is relevant, as well as systemic antibody production. With influenza vaccine the large dose of virus required often causes febrile reactions, especially in young children. This difficulty can be overcome by disrupting the virions with ether or deoxycholate, and such "split vaccines" are now in use (Duxbury et al., 1968; Davenport, 1971).

At an experimental level efforts are being made to produce effective polyvalent vaccines to control the major viruses that cause respiratory disease: respiratory syncytial virus, parainfluenza types 1, 2, and 3, influenza type A2, adenovirus types 3, 4, and 7, and *Mycoplasma pneumoniae*. Inactivated vaccines may be better suited to this end than living vaccines, because of potential difficulties with mutual interference, especially if the vaccines were to be administered intranasally. In order to ensure that immunogenic amounts of each inactivated virus can be contained in a volume suitable for injection, modern techniques for concentrating and purifying viruses will need to be employed in the development of such polyvalent vaccines (Reimer et al., 1966).

### Inactivation of Viruses

Information about the inactivation of viruses by physical and chemical treatments is important both for the production of inactivated vaccines and in relation to preserving the infectivity of live viruses, whether they are required for vaccinating or for laboratory studies. The older vaccines were inactivated by phenol, ultraviolet (UV) or γ-radiation, or formalin. Most inactivated vaccines used today have been treated with 1:4000 formaldehyde for long periods at 37°C until there is no residual infectivity but antigenicity and immunogenicity are little affected (Gard, 1960). However, the whole process is empirical and we know very little about the chemistry of the inactivation of viruses by formalin or any other agent.

For inactivated vaccines an agent is needed that specifically inactivates the viral nucleic acid (which carries the infectivity) without affecting the capsid or envelope protein (which carries the immunogenicity). Formalin does not, of course, come into this category, since it denatures protein by cross-linking, as well as reacting with the amino groups of nucleotides (Bachrach, 1966). Likely candidates are β-propiolactone (Sikes et al., 1971) or oxidized spermine (Bachrach, 1972). One of the main difficulties with agents whose sole target is nucleic acid, e.g., hydroxylamine, is that mutants may result unless every nucleic acid molecule is subjected to several hits, or the action of the agent is such that mutants cannot be formed. Multiplicity reactivation, which is known to occur with several viruses (Chapter 7), is a strong contraindication to UV irradiation.

A symposium on Inactivation of Viruses published in the Annals of the New York Academy of Sciences (1960) served to underline how little was then known about the chemistry of the processes involved. Bachrach (1966) has discussed in considerable detail the mechanism of denaturation of viruses by physical and chemical agents, while other reviews have been published on inactivation by UV irradiation (Kleczkowski, 1968), ionizing radiation, and heat (Ginoza, 1968). Potash (1968) has discussed the practical details of inactivation methods with specific reference to viral vaccine production.

Some recent papers throw further light on inactivation by heat and radiation. Dimmock (1967) and Fleming (1971) have carefully examined the kinetics of heat inactivation of picornaviruses and togaviruses, respectively. An Arrhenius plot of the data reveals that two distinct processes contribute to inactivation. Above 41°C inactivation occurs mainly as a result of denaturation of protein; below that temperature something else is chiefly responsible, perhaps very slow denaturation of RNA. The overall rate of inactivation increases markedly with temperature. As a rough rule of thumb it could be said that one can measure the inactivation of viruses in terms of seconds at 100°, minutes at 60°, hours at 37°, days at 4°, and years at −70°C, the temperature at or beneath which viruses are usually stored (Ward, 1968).

The pH of the suspending medium and the ionic environment also affect the rate of heat inactivation (review, Wallis et al., 1965). Various cations render some viruses heat stable and others heat labile. Concentrated $Mg^{2+}$ in particular has a profound effect in both directions, depending on the suspending medium and the particular virus (e.g., herpes simplex virus is especially vulnerable). Protein seems to exert some protective effect against thermosensitization by high or low salt concentrations. Poliovirus is also protected against heat inactivation by L-cystine, presumably by preventing oxidation of SH groups (Pohjanpelto, 1961).

Of course, enveloped RNA viruses are readily destroyed by lipid solvents such as ether, chloroform, and deoxycholate, which are still used by the arbovirus workers as a rapid initial screening test to determine whether a new isolate is likely to fall into the Bunyamwera or togavirus groups. Detergents, such as sodium dodecyl sulfate (SDS), sarkosyl, and Non-Idet P40 are widely used in biochemical studies of viruses to dissociate not only the envelope but the whole virion. Though SDS and phenol denature protein completely they have no effect on RNA or DNA and are therefore employed to extract infectious nucleic acid from virions.

Proctor et al. (1972) have classified DNA viruses into three classes according to their "weighted sensitivity" (inactivation cross section: molecular weight of the viral DNA) to inactivation by UV radiation of wavelength 254 nm. Class I, the most sensitive, are single-stranded DNA viruses. Class II are the double-stranded DNA coliphages when assayed in "repair-deficient" bacteria. Class III, the most resistant, contains the mammalian double-stranded DNA viruses which undergo repair in competent hosts. Although photoreactivation does not occur in placental mammals, there are other enzymatic repair mechanisms (Cook, 1972).

## Subunit Vaccines

If inactivated vaccines are to be used there is a good case to be made for vaccines consisting solely of the relevant immunogenic material, i.e., the capsid proteins of nonenveloped icosahedral viruses or the peplomers of enveloped viruses. Techniques are now available that make this a possibility for some agents, although production costs would doubtless still be too high. Adenovirus hexon and fiber antigens (see Chapter 3) and influenza hemagglutinin (Brand and Skehel, 1972; Hayman *et al.*, 1973) have been purified to the point of crystallization. Experimental "subunit" vaccines have been prepared from rabies and foot-and-mouth disease and measles viruses as well as influenza virus and adenovirus (review, Neurath and Rubin, 1971).

## Adjuvants

The immunological response to antigens can be greatly enhanced if they are administered in an emulsion with adjuvants. The mechanism is complex, involving among other things stimulation of phagocytosis and other activities of the reticuloendothelial system, and a delayed release and degradation of the antigen. Adjuvants may be particularly valuable in enabling a number of highly concentrated and purified viruses to be incorporated into a single polyvalent inactivated vaccine in a volume suitable for injection. However, conventional adjuvants (e.g., Freund's) based on mineral oils are not accepted for use in man because they are nonmetabolizable and are potentially carcinogenic. Suspicions about the carcinogenicity of Arlacel A have so far prevented United States licensing of a metabolizable (peanut oil) adjuvant developed by Hilleman (1966).

## PASSIVE IMMUNIZATION

Instead of actively immunizing with viral vaccines it is possible to confer short-term protection by the administration of preformed antibody, as immune serum or concentrated immunoglobulin prepared from it (reviews by Krugman, 1963; Pollack, 1969; World Health Organization, 1966b). Human immunoglobulin is usually preferred in human medicine, because heterologous protein provokes an immune response to the foreign antigenic components. When only heterologous immune sera are available (e.g., for rabies), this antigenicity may be reduced without impairment of immunological potency by peptic digestion of the immunoglobulins. Pooled normal human immunoglobulin can be relied on to contain high titers of antibody against all the common viruses that cause systemic disease in man.

Although passive immunization should be regarded as an emergency procedure for the immediate protection of unimmunized individuals exposed to special risk, it is an important prophylactic measure in several human viral infections. Human immunoglobulin has proved to be most effective in the short-term (up to 7 months) prophylaxis of infectious hepatitis; it can be administered with advantage to contacts of cases in schools and institutions or to travellers or military personnel about to visit parts of the world where hepatitis is prevalent

(Krugman, 1963). Experience with prevention of hepatitis B using human normal immunoglobulin has on the whole not been so favorable. Incidentally, the immunoglobulin does not contain hepatitis B virus because the latter is lost during the cold ethanol precipitation step in the purification procedure.

In measles, immunoglobulin will abort the disease if given in large doses within 5–6 days of exposure, or lead to a modified attack of diminished severity if a smaller dose is given during the same period; it is still used to provide emergency protection for exposed siblings if they have a history of chronic respiratory illness and have not been immunized with live measles vaccine. It is much less effective in chickenpox, but when given to premature, debilitated, or immunodeficient children within 3 days of exposure it somewhat reduces the severity of the ensuing disease. The efficacy of $\gamma$-globulin in preventing the development of congenital defects in babies of women exposed to rubella during the first trimester is highly questionable and its use has been abandoned.

In rabies, combined passive and active immunization of persons bitten by rabid animals has been considered to give better protection than active immunization with killed vaccine alone, but recent studies by Sikes et al. (1971) using new concentrated vaccines do not support the belief. In the future, rabies prevention is likely to be based on a single injection of potent $\beta$-propiolactone-inactivated WI-38-grown virus, and conceivably poly I·poly C (see Chemotherapy).

Vaccinia-specific human immunoglobulin, obtained from recently vaccinated donors, has a place in the prevention of smallpox and the treatment of vaccination complications. Following exposure to smallpox, vaccination is, of course, essential and methisazone has a place (see Chemotherapy), but human immunoglobulin, anti-vaccinia, may also be given into the other arm. This reagent also reduces the incidence of postvaccination complications, hence should be used in situations where vaccination must be given in spite of clear contraindications such as eczema, pregnancy, or immunosuppression. It is a valuable therapeutic measure in cases of eczema vaccinatum or autoinoculation of the eye; in the latter instance, it can be instilled into the eye as well as intramuscularly.

The importance of passive immunization is diminishing now that satisfactory vaccines for such diseases as poliomyelitis, measles, mumps, and rubella are widely used. Its major role today is in the prevention of infectious hepatitis.

## THE PSYCHOLOGY OF IMMUNIZATION

Fear is probably the major factor motivating people to seek or accept immunization for themselves and their children, so that when a feared disease like poliomyelitis has almost disappeared and is no longer seen as a threat to the individual, it is difficult to maintain community-wide vaccination. This has led to a resurgence of measles in the United States and of poliomyelitis in some other countries. Disappearance of pathogenic viruses from a community as a result of widespread immunization removes the beneficial influence of repeated subclinical infections to boost immunity and leaves both the immunized, and particularly the unimmunized, unusually vulnerable should the agent ever reappear. Continuation of immunization practices after the threat appears to have vanished but the virus has not been totally eradicated calls for highly organized public health services.

An important but rather neglected aspect of this problem stems from the un-necessarily complicated immunization schedules to which mothers are required to adhere. Polyvalent vaccines, dead or alive, would be a major practical advan-tage. Greater safety would be assured by the use of presterilized plastic syringes already loaded with a single dose of vaccine. Jet guns which quickly and pain-lessly deliver a jet of vaccine through the skin under high pressure should do much to lessen patients' objections and also to speed up crash immunization pro-grams. Nevertheless, as far as members of the public are concerned, there is little doubt that oral vaccines will continue to remain the most popular.

## MEDICAL AND VETERINARY VACCINES

Space does not permit a detailed discussion of all known viral vaccines; in veterinary medicine alone there are scores in everyday use around the world. Table 15–4 shows a selection of the more important of these. Since human viral vaccines (Table 15–3) have been developed with greater attention to the require-ments of safety and efficacy, they provide most of what we now accept as the basic principles of the subject.

Certain parameters are used to judge the degree of efficacy and safety of any given vaccine. The immunological response is measured mainly in terms of serum IgG and local IgA at various times after immunization by various routes. Re-sistance to challenge, whether natural or artificial, is the most important practical criterion of the efficacy of a vaccine. Safety is assessed in terms of short- and long-term toxic side effects, virus shedding, and transmissibility.

Valuable reviews have been written by Evans (1969), Hilleman (1969) and Chanock (1970), and the Technical Reports of WHO Expert Committees and of WHO Conferences give authoritative accounts of particular vaccines (World Health Organization, 1966a, b, 1967b, 1969a, 1971a, 1972a, b; Pan American Health Organization, 1971).

## HUMAN VIRAL VACCINES

### Smallpox

It has been known from ancient times that one attack of smallpox conferred lifelong protection against subsequent attacks of the same disease. Long before the formulation of the germ theory of infectious diseases, this knowledge was exploited in China and the Middle East, and later in the Western world, to protect infants against smallpox by deliberately inoculating them with material from smallpox pustules. The practice, known as "variolation," continued up to the end of the eighteenth century (and, indeed, is still practiced in some parts of the world, see below); the mortality rate was 0.2–0.5%. In 1798, Jenner, having observed that milkmaids were immune to smallpox, published his famous paper on protection against smallpox by inoculation with cowpox. For half a

TABLE 15–4

*Some Veterinary Viral Vaccines*

| DISEASE | VIRAL GENUS | NATURE OF VACCINE |
|---|---|---|
| | Dog | |
| Canine hepatitis | Adenovirus | Live attenuated |
| Canine distemper | Paramyxovirus | Live attenuated |
| Rabies | Rhabdovirus | Live attenuated |
| | Cat | |
| Panleukopenia | Parvovirus | Formalin-inactivated, in adjuvant |
| | Cattle | |
| Infectious bovine rhinotracheitis | Herpesvirus | Live attenuated |
| Foot-and-mouth disease | Picornavirus | Formalin-inactivated |
| Bovine virus diarrhea | ? Togaviridae | Live attenuated |
| Rinderpest | Paramyxovirus | Live attenuated |
| Vesicular stomatitis | Rhabdovirus | Live attenuated |
| | Sheep | |
| Sheep pox | Poxvirus | Live virulent (special site) or live attenuated |
| Scabby mouth | Poxvirus | Live virulent (special site) |
| Louping ill | Flavivirus | Formalin-inactivated |
| Rift Valley fever | Bunyamwera | Live attenuated |
| Bluetongue | Orbivirus | Live attenuated |
| | Horse | |
| Equine rhinopneumonitis | Herpesvirus | Live attenuated |
| Venezuelan equine encephalitis | Alphavirus | Live attenuated |
| Western equine encephalitis | Alphavirus | Live attenuated |
| Infectious arteritis | ? Togaviridae | Live attenuated |
| Equine influenza | Orthomyxovirus | Live attenuated |
| African horse sickness | Orbivirus | Live attenuated |
| | Pig | |
| African swine fever | Iridovirus | Live attenuated |
| Teschen disease | Enterovirus | Live attenuated |
| Swine fever (hog cholera) | ? Togaviridae | Live attenuated, with antiserum |
| Transmissible gastroenteritis | Coronavirus | Formalin-inactivated, in adjuvant |
| | Chicken | |
| Fowlpox | Poxvirus | Live attenuated |
| Marek's disease | Herpesvirus | Live attenuated |
| Infectious laryngotracheitis | Herpesvirus | Live attenuated |
| Fowl plague | Orthomyxovirus | Formalin-inactivated |
| Newcastle disease | Paramyxovirus | Live attenuated |
| Infectious bronchitis | Coronavirus | Live attenuated |

century people continued to be "vaccinated" with material derived from cows or by arm-to-arm passage in man until commercial production of vaccinia virus in the skin of calves and sheep began. We are now uncertain whether the present strains of vaccinia virus were derived originally from cowpox or from smallpox, or perhaps from recombinants between them. The virus is still grown commercially on the skin of these same animals, and WHO regulations stipulate that the number of contaminating bacteria per milliliter of vaccine shall not exceed 500—a requirement that would never be tolerated in a "modern" vaccine (World Health Organization, 1972b). Today's techniques of inoculating the virus by multiple puncture with a bifurcated needle or a foot-operated jet injector are more sophisticated than in Jenner's day, but it is typical of the conservative approach to viral vaccines that alternative methods of producing the virus have not been widely adopted, though many have been described, including chick embryo- and cell culture-grown vaccines.

Recent surveys (e.g., Lane et al., 1969) indicate that there is about one complication in every thousand primary vaccinations, and one death in every million. Autoinoculation, especially of the eyes and lids, accounts for nearly one-half these occurrences and generalized vaccinia for about another one-quarter. The much more serious postvaccinial encephalitis has an incidence of only about one in a hundred thousand primary vaccinations and fatal progressive vaccinia (vaccinia necrosum) is even rarer. Eczema vaccinatum has decreased markedly in incidence since physicians became aware of the dangers of vaccinating children with eczema. An attenuated strain, CV1-78, with a low affinity for human skin is now available for vaccinating eczematous individuals (Kempe et al., 1968); alternatively, the standard strain may be used with anti-vaccinia immunoglobulin injected into the other arm, but it is wisest not to vaccinate such people at all. The risks of all complications are substantially lower on revaccination, and lower in children than in adults, hence the recommendation that children be vaccinated routinely during the second year of life. There is merit in this idea for children who are likely to be traveling abroad in later life, but the prevailing view is that universal smallpox vaccination should be abandoned in advanced Western countries (Dick, 1966, 1971; Lane et al., 1971).

An interesting study by Mack (1972) of the 958 cases of smallpox that have occurred in Europe during the last 20 years indicates that relatively few of the forty-nine separate episodes of importation gave rise to more than a handful of cases and that most of the dissemination occurred inside hospitals. Spread within the largely unvaccinated general community almost never went beyond the first round of immediate contacts and was relatively easily contained by prompt detection and "ring vaccination." Experiences such as these have persuaded the British and United States governments to discourage routine vaccination of the indigenous population and to abandon the previous requirement of a valid smallpox vaccination certificate from all visitors, although the latter is still demanded of travelers from areas such as the Indian subcontinent and Ethiopia which comprise the main endemic foci of smallpox today. Medical and public health personnel in all countries are potentially at risk and should be vaccinated, as, of course, should the whole population of endemic areas from which the World Health Organization is attempting to eradicate the disease.

Theoretically, prospects for global eradication of smallpox are good (see Chapter 17). The virus has no animal reservoir, no arthropod vector, and no capacity to persist as a latent infection in man. The vaccine is effective. Yet, since WHO embarked on its eradication campaign, progress has been disappointingly slow. Notifications were fairly constant at about 100,000 cases per year until 1968 when they fell steadily to 33,304 in 1970, but with civil war in Bangladesh the numbers rose again the 52,098 in 1971. Citing one example of the unusual problems faced in such a campaign it was stated that "in Afghanistan itinerant variolators have been moving from village to village leaving multiple outbreaks of smallpox in their wake" (World Health Organization, 1972b).

## Rabies

Rabies is perhaps the most uniformly lethal, and is certainly the most feared of all diseases of man. Yet it is so rare in most of the world that it would not be surprising to find that in some Western countries more deaths have resulted from rabies "prophylaxis" than from the disease itself. The commonly used human vaccines are unacceptably dangerous and in urgent need of replacement. Until a much better vaccine is readily available, anti-rabies vaccination should only be administered to those professionally at risk (veterinarians, trappers, speleologists) or those bitten in an endemic area by an animal considered likely to be infected; captured animals should be kept under surveillance, or killed and their brains examined by immunofluorescence.

Rabies vaccination practice is quite unique in that active immunization is commenced *after* an individual has been bitten by a potentially rabid animal, a procedure that is successful only because of the unusually long incubation period of the disease. The remarkable number of different rabies vaccines (Plotkin and Clark, 1971) is partly attributable to the very long history of research into this disease, going back to the days of Pasteur himself; indeed, that is the era to which some of the current vaccines belong, for one cannot conceive of them being licensed were they to be submitted for initial approval today.

The Semple vaccine is a 10% emulsion of phenol-inactivated rabiesvirus-infected rabbit brain which carries a risk of allergic encephalomyelitis variously assessed at 0.01–3%, of whom 20% die and another 30% sustain permanent sequelae. Such a vaccine is totally unacceptable even though it is said that most of the encephalitogenic material can be removed with fluorocarbon. Vaccines made from suckling mouse or rat brain, on the basis of the fact that brains from young animals contain less myelin, also cause far too many CNS reactions. The $\beta$-propiolactone-inactivated duck embryo vaccine is much less hazardous, but produces a high incidence of Arthus-type local reactions and a moderate number of generalized allergic reactions, especially in those presensitized to chick embryo tissue, e.g., by prior yellow fever immunization. Moreover, the antibody response to the duck embryo vaccine is too low during the crucial interval between 1 and 2 months after exposure to be certain of its value. The Flury-HEP living attenuated strain, derived by prolonged passage in chick embryos, is widely and successfully used in domestic animals but gives completely inadequate antibody responses in man.

Koprowski and his colleagues at the Wistar Institute have devoted many years to the development of a safer rabies vaccine. In an extensive trial, they have now compared almost a dozen separate vaccines grown in various non-neural tissues for their efficacy in monkeys (Sikes *et al.*, 1971). All gave superior immunity to that provided by the rabbit brain, mouse brain, and duck embryo vaccines which were used as reference standards. The best was a concentrated $\beta$-propiolactone-inactivated vaccine grown in human diploid fibroblasts. A single dose injected 4–6 hours after rabiesvirus had been inoculated into the neck muscles of monkeys saved the lives of the animals. This constitutes a major advance because the practice in the past has been to continue daily injections of vaccine for at least 2 weeks, a procedure that promotes the development of hypersensitivity. Meanwhile Wiktor and colleagues, and Crick and Brown at Pirbright in England, have independently developed experimental vaccines consisting of partially purified immunogenic subunits of the rabiesvirion.

## Yellow Fever

Theiler's 17D strain of yellow fever virus was attenuated by passage in tissue cultures of chick embryos from which the central nervous tissue had been removed; it was subsequently grown in embryonated eggs (World Health Organization, 1971a). A single subcutaneous injection gives a very high degree of immunity for at least 10 years. Complications (encephalitis in infants) are rare although hypersensitivity reactions occasionally occur in people with egg allergy. The World War II disaster in which several thousand United States troops contracted hepatitis B after yellow fever vaccination (Sawyer *et al.*, 1944) was traced to the human serum used to stabilize the vaccine. The early yellow fever vaccines were also almost certainly contaminated by avian leukosis viruses, but follow-up has not revealed any excess incidence of malignancy in the recipients; current yellow fever vaccines are leukosis-free. The so-called Dakar vaccine or French neurotropic strain, grown in mouse brain, should never be used again after the experience during the 1965 yellow fever epidemic in Senegal when the vaccine caused 245 cases of encephalitis in children, with 23 deaths.

Yellow fever was the first disease for which global eradication (now known to be an unrealistic objective) was proposed (see Chapter 17). Currently the aim is to eliminate urban yellow fever by mosquito control, immunize human populations in or near regions where sylvan yellow fever occurs, and attempt to prevent international distribution of the virus by airport protection ("yellow fever-free airports"), and compulsory vaccination of travellers from or passing through yellow fever endemic areas.

Experimental vaccines have been developed for a number of other togaviruses, but the only ones widely used are an inactivated Japanese encephalitis vaccine in man (in Japan, Taiwan and China) (Hammon *et al.*, 1971), and Venezuelan and Western equine encephalitis live vaccines in horses in the United States (Gillette, 1971).

## Poliomyelitis

In Western countries, where it first arose as an epidemic disease at the turn of this century (Chapter 17), poliomyelitis has now passed into history as a result of immunization. First the Salk formalin-inactivated vaccine, and then the Sabin and Koprowski living attenuated oral vaccines, were responsible for this remarkable change (Salk, 1958; Koprowski, 1957; Sabin, 1957). Sweden, the country in which paralytic poliomyelitis first became apparent in epidemic form, and which for many years continued to have the highest rate of poliomyelitis in the world, eradicated the disease by the use of Salk vaccine from 1957 onward (Gard, 1967); there has been no case of poliomyelitis in Sweden since 1967. In other parts of the world reliance has been placed on living attenuated oral vaccines, mainly the Sabin strains, and the results have been equally dramatic (Sabin, 1967b). In a massive drive against poliomyelitis during 1960, the USSR immunized 90% of the nonadult population within a matter of months with the result that not only the disease, but also the virus itself was reduced to very low levels. The United States, Canada, and Australia began with Salk vaccine in 1954–1955 but made a smooth transition to Sabin vaccine in the early 1960's and have been essentially free of the disease ever since. Only in the developing countries have there been problems, mainly of a practical nature, which are slowly being solved in the following ways.

Sabin vaccine is a "trivalent" (types 1, 2, and 3) oral vaccine which can be administered quickly and simply by unqualified personnel. Because of the presence of IgA in breast milk the vaccine is not usually given to children under the age of 6 months but in some countries women habitually breast feed their infants for up to 2 years. This difficulty can be circumvented to some extent provided the child is not breast-fed for about 6 hours before and for the same period after administration of the vaccine.

The second major problem in the developing countries of the tropics is the high incidence (sometimes up to 90%) of intercurrent infection with enteroviruses; this can interfere with the multiplication of the oral live vaccine. The solution in this case is simply to follow the routine advocated in all countries of giving three doses of the trivalent vaccine, spaced 6–8 weeks apart, but this can pose a substantial administrative problem in the remote and inaccessible villages of Africa, Asia, and South America. The tendency for the type 3 component of the vaccine to be overgrown by the other two types has been counteracted by increasing the proportion of the more slowly growing serotype.

Inactivation of this living vaccine can occur, particularly in the heat of the tropics, if inadequate precautions are taken to ensure its continued refrigeration; magnesium chloride is commonly used as a stabilizing agent (Wallis et al., 1965).

Some concern has been felt about the possibility that the type 3 vaccine strain (Leon) may be insufficiently attenuated. It is certainly less stable genetically than the other two and has been reported to back-mutate on passage in man to regain neurovirulence for the monkey. Statistical analysis of "vaccine-associated" cases of poliomyelitis (defined as those instances in which paralysis develops within a month after Sabin vaccination) indicates a higher than chance incidence of virulent type 3 isolations and suggests that this component of the vaccine may indeed be causing paralysis in something less than one in a million recipients (Melnick, 1971). It is also notable that, probably as a result of the relatively poor immunogenicity of the type 3 attenuated strain, a much higher proportion of the rare instances of paralysis that occur at any time in vaccinated individuals are attributable to this type rather than type 1, which used to be the most common serotype causing paralysis. A new type 3 strain is needed for the vaccine, perhaps the Czechoslovakian USOL-D strain (Melnick, 1971).

In many parts of the world, poliovaccine is still grown in monkey kidney cells, despite the salutary warnings provided by the Marburg, B virus, and SV40 incidents and the discovery of other cryptic simian viruses. The World Health Organization Committee on Biological Standardization (1972a) has laid down very strict recommendations on the quarantine and surveillance of monkeys in closed colonies, screening of cell cultures, and safety testing of the finished product, but it is difficult to justify the continued use of primary monkey kidney cells. Particularly, for an oral vaccine, the human diploid cell strain WI-38 would seem to be in all ways superior.

## Measles

Though a relatively harmless disease in the average child, measles is complicated by serious sequelae (encephalitis, pneumonia) sufficiently often to make it a more common cause of death than was poliomyelitis 20 years ago. Moreover, in undernourished children in many of the developing countries measles is quite a lethal disease (Morley, 1967). The development of a successful vaccine by Enders and his colleagues was, like his cultivation of poliovirus, an important innovation which led to the virtual disappearance of measles in many cities in the United States within 5 years of its licensing in 1963. The fact that there has been a slight resurgence of the disease during the last few years (Fig. 16–3) reflects the complacency of the public when a disease becomes rare.

The original (Edmonston) strain of attenuated measles vaccine virus was obtained by serial passage in cultured chick embryo cells; the currently used Schwarz strain is a more highly attenuated virus which produces fewer side effects but equally satisfactory immunity (Enders and Katz, 1967). Reactions to the Schwarz vaccine (Schwarz, 1964) consist only of occasional fever, rash, and transient electroencephalographic changes. It is administered subcutaneously in a single dose to children in their second year of life, and gives a seroconversion rate of over 95%. Hemagglutination-inhibition titers of about 100 are achieved within a month and have not dropped by more than 50% over a period of 8½ years (Krugman, 1971). It is now believed that boosters may not be necessary; when natural reinfection does occur it is always subclinical and will itself act as a booster infection.

## Mumps

Mumps does not pose an important threat to life, but serious complications (orchitis, meningoencephalitis, pancreatitis) do occur, particularly in adults. A case can therefore be made for immunizing adolescent and adult males (perhaps after a preliminary skin test for delayed hypersensitivity) upon entry into military establishments or under other conditions where they are at increased risk. A highly satisfactory living attenuated vaccine is now available derived from the

Jeryl Lynn strain by passage in chick embryos and then chick fibroblasts until it had lost virulence for man (Deinhardt and Shramek, 1969; Hilleman, 1970a). Serum neutralizing antibody levels achieved 1 month after a single subcutaneous injection are about eightfold lower than those following natural infection, but have shown no decline over the 5 years since the vaccine was licensed for use in the United States; protection exceeds 95%. Henceforth many young children will probably receive mumps immunization in the form of a trivalent measles-mumps-rubella vaccine (Stokes et al., 1971).

## Rubella

Unlike the vaccines we have discussed so far, immunization against rubella is directed not at protection of the vaccinee, but at the prevention of the teratogenic effects of wild rubella virus on the fetus.

The original vaccine strain HPV-77 was developed by Parkman and Meyer (1969) by 77 consecutive passages in vervet monkey kidney cells and has since been passed five times in duck embryo tissue culture (Hilleman et al., 1971b). Meanwhile, Huygelen et al. (1969) in Belgium developed their Cendehill-51 strain by three passages in African green monkey kidney cells followed by 51 passages in primary rabbit kidney cells at 34°C. A third vaccine, RA 27/3, was derived by Plotkin et al. (1969) by attenuation entirely in the human diploid cell strain WI-38. The three vaccines have much in common and will be discussed together. They all induce HI-antibody titers of about 50 to 200, i.e., three to ten times lower than in natural rubella infection, in over 95% of nonimmune recipients following subcutaneous injection, but RA 27/3 can also be administered intranasally with comparable results. The HI and neutralizing antibody titers are somewhat higher with RA 27/3 and immunodiffusion reveals that one of the two principal species of precipitating antibody detected following natural infection is induced by RA 27/3 but not by the other two vaccines (Le Bouvier and Plotkin, 1971). Serum antibody titers have declined less than twofold during the 5 years that have elapsed since trials began. Protection against clinical rubella is good, but recent experience has shown that subclinical infection of vaccinees occurs very commonly (Horstmann et al., 1970).

It is not known whether attenuated rubella strains are teratogenic and no certain cases of embryopathy due to vaccine strains have been reported. However, since they have been detected in surgically aborted fetuses (Vaheri et al., 1972), the prudent policy is not to immunize a pregnant woman and to advise against conception for 2 months following immunization. The risk of natural transfer of vaccine strains to pregnant women seems to be nil. Although the virus is regularly excreted for a few days in the throats of vaccinees, the titers are only about one hundredth of those found in natural infection and no well-documented instance of person-to-person transmission has been recorded. The HPV-77 strains are not satisfactory for use in adults because they give rise to a substantial number of cases of arthritis, which although transient, gives cause for some concern. All three vaccines produce occasional low fever, headache, lymphadenopathy, and rarely a rash, especially in adults.

Two strategies of vaccination have been adopted. In the United States an attempt was made initially to eradicate the virus by mass immunization of all children. This is now seen to have limited prospects of success, and several authorities suggest that the sounder policy epidemiologically is to protect the population at risk by mass immunization of girls at the age of 11 to 12 years, and immunization of mothers in the few days immediately postpartum when they are least likely to become pregnant again. The latter policy was adopted from the outset in most European countries, and in Australia.

## Vaccination against Respiratory Diseases

All the examples quoted so far are generalized infections of man in which antibody in the serum has an excellent opportunity of neutralizing virus before symptoms develop. In addition, some of these vaccines, e.g., live poliovirus vaccine, may prevent implantation of virus in the cells of the gut. In these generalized infections, the efficacy of immunization is beyond question and efforts have been concentrated upon the development of sufficiently safe and potent vaccines. With viruses that cause localized infections of the respiratory tract, however, clinical or subclinical infection does not give rise to lasting immunity, even to the same serotype. Quite apart from the additional difficulties due to antigenic drift and multiple serotypes, referred to earlier, it seems unlikely that vaccination will ever afford a really satisfactory method of prevention of most of these diseases. Nevertheless, some, notably influenza, are so important that an enormous effort has been made to develop effective vaccination procedures.

## Influenza

Formalin-inactivated influenza vaccines were developed in the early days of World War II and administered to the United States military forces. The degree of protection conferred ranges from 75% (Davenport, 1971) down to zero in some recent outbreaks. Results depend very much on the recipients' experience of the homologous and heterologous strains of influenza virus and on whether the prevalent agent has undergone some degree of antigenic drift since the vaccine was made, but the level of protection does not approach that achieved by inactivated poliovaccine, for example, or by any of the live vaccines used in generalized infections.

Inactivated vaccine made up of whole virions, inoculated intramuscularly or subcutaneously, frequently causes pyrexia and local inflammation. Reimer et al. (1966) have perfected techniques of purifying large batches of virus by continuous-flow zonal ultracentrifugation; inactivated vaccines prepared from such material are only marginally less pyrogenic but should help to eliminate the hypersensitivity reactions sometimes encountered with less highly purified vaccines in recipients allergic to eggs. Dissociation of the viral envelope with lipid solvents such as ether, deoxycholate or (experimentally), tri(N-butyl) phosphate and Tween 80 does diminish pyrogenicity considerably (Duxbury et al., 1968;

Davenport, 1971). Now that highly purified hemagglutinin is obtainable (Brand and Skehel, 1972; Hayman *et al.*, 1973) this should make a satisfactory nontoxic "subunit" vaccine. Intranasal administration of inactivated influenza vaccines does not induce satisfactory local immunity, but there may be merit in subcutaneous priming followed by an intranasal booster which has been found to produce a rapid secondary rise in intranasal IgA; similar results are obtained with primary intranasal immunization of people with preexisting serum antibodies as a result of previous natural infection (Rossen *et al.*, 1971).

Recently a number of research groups have been reexamining the value of live influenza vaccines, originally introduced experimentally by Burnet during World War II (Burnet, 1942). Such vaccines have been in general use in USSR for several years (Zhdanov, 1967). Beare *et al.* (1971) have compared the human responses to three separately derived live vaccines, administered intranasally. All produced moderate antibody responses but an unfortunately high incidence (about 50%) of afebrile coryza. Temperature-sensitive mutants provide a theoretically attractive kind of attenuated mutant that would multiply in the nose but not in the lungs (Mackenzie, 1969; Maassab, 1969). Mills and Chanock (1971) selected a stable influenza A2 *ts* mutant which was unable to replicate above 37°C in cultured calf kidney cells. When administered intranasally to hamsters this mutant grew poorly in their lungs but quite satisfactorily in the turbinates of the upper respiratory tract, and protected the animals against subsequent challenge with wild-type virus. The mutant (H2N2) was then recombined with a calf kidney-grown A2/Hong Kong/68 (H3N2) strain to yield reassortants in which the RNA molecule carrying the *ts* lesion had been transferred to the H3N2 virus (Murphy *et al.*, 1972). One of the reassortants, *ts*-1 (E), seems sufficiently attenuated to make a suitable vaccine. It grew readily in the upper respiratory tract of human volunteers, producing no symptoms in most of the men and mild coryza in the rest, but induced moderate titers of neutralizing antibody in both the nose and the serum, and gave complete protection against challenge with wild-type virus a few weeks later. The reassortant was genetically stable in man and did not spread to contacts.

A further complication in vaccination against influenza that is not encountered among the viruses causing generalized infections is the fact that the protective (envelope) antigens are constantly changing. There are two conflicting theories about the origin of so-called "pandemic" strains. Fazekas de St. Groth (1969; 1970) holds that all human strains arise by mutation from preexisting human viruses; other investigators (Kilbourne, 1968; Webster and Laver, 1971; Webster, 1972) while accepting this origin for changes attributable to antigenic drift, believe that the major "antigenic shifts" that cause pandemics reflect the emergence of genetic reassortants between animal and human influenza strains. Both hypotheses highlight the need constantly to "update" the vaccine strain so that it has the appropriate envelope antigens, notably the hemagglutinin (Pereira, 1969; Webster and Laver, 1971), although the neuraminidase may have some minor importance (Schulman *et al.*, 1968). This requirement is equally critical whether a live or inactivated vaccine is used; recombination, as developed by Kilbourne (Kilbourne, 1969; Kilbourne *et al.*, 1971), Chanock (Murphy *et al.*,

1972), Beare and Hall (1971), and Maassab (Maassab *et al.*, 1972) provides a method for rapidly selecting a strain with appropriate antigens and growth characteristics.

Fazekas (1969, 1970) has advocated that an attempt be made to anticipate future antigenic changes by preparing a "prospective" vaccine from a mutant selected by growing virus in the presence of homologous antibody and passing the progeny through a column containing antibody. It is claimed that such "senior" mutants cross-protect against all "earlier" strains, and a vaccine has been developed on these lines. It is unlikely that such a vaccine would be better than one made from the strain causing disease at any particular time, which as we have seen does not give protection in any way comparable to that obtained in generalized viral infections, nor is it likely that this procedure would anticipate the major antigenic shifts like A1(H1N1) to A2(H2N2) (see Chapter 17). Nevertheless, the results of the vaccine trials will be awaited with interest, for an effective method of coping with the antigenic changes associated with antigenic drift would be valuable.

Policy on immunization against influenza varies from country to country and from time to time, reflecting the imperfections of available vaccines. The groups particularly at risk are the aged, debilitated pulmonary and cardiac invalids, and infants. A case can also be made for immunizing schoolchildren because influenza, like most other infections, tends to spread in schools then be carried home to infect the family (Jordan *et al.*, 1958), and some studies have indicated that immunization of schoolchildren can also diminish the circulation of virus in the community as a whole (Monto *et al.*, 1970). Immunization is carried out in the autumn; an annual "booster" is required.

## Respiratory Syncytial Virus

Respiratory syncytial (RS) virus is the most important cause of serious lower respiratory tract infection in infants. Chanock *et al.* (1968) attempted to protect babies parenterally with concentrated inactivated vaccines which induced very high levels of serum antibody, but, as discussed earlier in this chapter, the vaccinees' response to subsequent challenge was alarming. Following natural exposure to RS virus immunized children suffered more serious bronchiolitis and pneumonitis than did unprotected controls.

Turning toward the alternative of a live vaccine Chanock and his colleagues reasoned that *ts* mutants might be safe because of the temperature differential between the upper and lower respiratory tract. A *ts* mutant was isolated which was unable to grow at 37°C in cultured cells but multiplied asymptomatically in the upper respiratory tract of adults and protected them against subsequent challenge with virulent RS (Wright *et al.*, 1971). Though the immunogenicity of this particular mutant was minimal the study established the feasibility of *ts* mutants as human respiratory vaccines.

## Adenovirus

A relatively small number of serotypes of adenovirus, notably types 3, 4, and 7, regularly cause substantial outbreaks of "acute respiratory disease"

(ARD) among military recruits in United States Army camps. By enclosing virus in enteric-coated capsules, Couch et al. (1963) introduced live unattenuated adenovirus 4 directly into the intestine without running any risk of the virus multiplying in the throat and causing a respiratory disease, nor of its being inactivated by the acid environment of the stomach. The resulting immunity to challenge was good and excreted virus spread only rarely to contacts, probably via the alimentary route (see Chanock, 1970). With the discovery that some adenoviruses are oncogenic for baby rodents (Trentin et al., 1962), further work on adenovirus vaccines was temporarily suspended, but has been resumed now that it is realized that adenovirus 4 is not carcinogenic for rodents and that in all probability no adenovirus is carcinogenic for man (Green, 1972). Recently, Top et al. (1971) demonstrated that adenoviruses types 4 and 7 can be administered simultaneously in enteric-coated capsules with little or no interference between the two viruses, and that the protection against subsequent challenge was of the order of 50 to 75%. Needless to say, the adenoviruses employed in these vaccines are grown in human embryonic fibroblasts, not monkey kidney, because of the danger of formation of adenovirus-SV40 hybrids (see Chapter 7).

## PROSPECTS FOR NEW VACCINES

### Hepatitis

One of the outstanding human public health problems remaining in virology is the conquest of hepatitis, for infectious hepatitis (A) and serum hepatitis (B) are increasing at an alarming rate throughout the world.

Existing information on the pathogenesis and epidemiology of both kinds of hepatitis point to the prospect that a vaccine would be as effective as in poliomyelitis, and Krugman et al. (1971) have demonstrated the practicability of active immunization by successfully protecting children against hepatitis B by two injections of a 1:10 dilution of a preparation of human serum that contained virus and Australia antigen, and had been boiled for 1 minute. At present, a satisfactory way of cultivating these viruses in cell cultures is lacking. Preliminary reports of cultivation of hepatitis B virus in organ cultures of human embryonic liver (Zuckerman et al., 1972) are encouraging but this is not a potential source of virus on a commercial scale.

### Herpesviruses

Many members of the genus Herpesvirus share the undesirable property of persisting in the body for years, perhaps for life, and giving rise sporadically to endogenous disease. The fever blisters of herpes simplex are unpleasant, while recurrent keratitis can lead to blindness. Herpes zoster (due to the reactivation of varicella virus) is an exceedingly painful condition. Cytomegalovirus is responsible for many cases of the postperfusion syndrome and is the most common cause of mental retardation due to microcephaly. The EB virus causes a common and debilitating disease of adolescence, infectious mononucleosis, and may possibly be involved in certain malignancies. Prevention of all these diseases is warranted, if feasible. All are monotypic and adequate levels of circulating antibody might be expected to prevent primary infection but to have little effect on the recurrence of established infections.

Currently there are difficulties in obtaining adequate yields of cell-free virus, and the evaluation of the safety of attenuated live virus vaccines would be extremely difficult because of the propensity of herpesviruses to produce latent infections.

## Slow Viral Infections and Oncogenic Viruses

Gajdusek *et al.* (1971) have discussed the prospects of controlling chronic degenerative diseases with vaccines, if it were firmly established that they were virus-induced. The problems in such diseases and in virus-induced tumors are considerable, not only in terms of the cultivation and production of adequate amounts of virus for a vaccine but because of the pathogenesis of such infections, and the length of time needed for clinical trials. Nevertheless, the discovery of attenuated herpesvirus strains which immunize successfully against Marek's disease of fowls (Purchase *et al.*, 1971) gives some encouragement (Hilleman, 1972; 1973).

## ANTIVIRAL CHEMOTHERAPY AND CHEMOPROPHYLAXIS

While steadily improving standards of public and personal hygiene and several highly effective vaccines have dramatically reduced the threat to human life and health imposed by the more dangerous viral pathogens, they have had no impact on the morbidity caused by the respiratory viruses. As we have already pointed out, the multiplicity of species and serotypes of such viruses rules out the feasibility of preventive immunization and calls for a different approach, which can only be through chemotherapy or chemoprophylaxis.

## Mechanisms of Action of Antiviral Chemotherapeutic Compounds

The major obstacle to effective antiviral chemotherapy is the obligatory dependence of viruses upon the metabolic pathways and organelles of the host cell. Indeed, it was once thought self-evident that any agent blocking a metabolic process essential to viral multiplication must inevitably kill the cell. Now we know that there are in fact several biochemical processes that are essential to the virus yet of no consequence to the cell; any of these processes constitutes a logical point of chemotherapeutic attack (Table 15–5).

**Attachment and Penetration.** Attachment of virus to the cell's plasma membrane often necessitates the apposition of specific complementary receptors on the surface of virus and cell respectively (see Chapters 5 and 11). Agents that interfere with either could theoretically block infection. For example, neuraminidase destroys the glycoprotein receptors in the lungs of mice and renders them refractory to influenza infection until new receptors appear some hours later. Neuraminidase is not seriously contemplated as a potential chemothera-

TABLE 15–5

*Mechanisms of Action of Antiviral Chemotherapeutic Agents*

| DRUG | PROBABLE POINT OF ACTION | PRINCIPAL VIRUSES INHIBITED |
|------|--------------------------|------------------------------|
| Amantadine | Penetration or uncoating | Influenza A2; *Arenavirus* |
| Idoxuridine ⎫ | DNA replication | *Herpesvirus* |
| Ara-A, Ara-C ⎬ | | *Poxvirus* |
| Rifamycins | ? Reverse transcriptase | *Leukovirus* |
| | Assembly | *Poxvirus* |
| Thiosemicarbazones | Translation of late viral mRNA | *Poxvirus* |
| Interferon | Transcription or translation | All viruses |
| Poly I · poly C ⎫ | Induction of interferon | Many viruses |
| Statolon ⎪ | ? Stimulation of | |
| Pyran copolymer ⎬ | reticuloendothelial | |
| COAM ⎪ | system | |
| Tilorone ⎭ | | |

peutic agent for use in man, but illustrates the principle well. One class of drugs (amantadine) that does not inhibit attachment appears to have some effect on the process of "penetration."

**Transcription.**   The transcriptase carried by most RNA viruses (see Table 6–1), as well as the poxviruses, is essential for the transcription of "early" mRNA from the partially uncoated viral core. Since all such transcriptases are virus specific, and all except that of the poxviruses are RNA-dependent, there may very well be chemicals that will inhibit these enzymes without affecting the DNA-dependent cellular polymerases. Evidence is accumulating that interferon may block transcription (see Chapter 8). The reverse transcriptase may render leukoviruses particularly vulnerable to attack at this point; some of the newer rifamycin derivatives show promise, and the search is bound to be widened now that a novel rationale for anti-cancer therapy has emerged.

Posttranscriptional cleavage of the large molecules of polycistronic mRNA transcribed from most DNA viruses and perhaps some RNA viruses, and/or the subsequent addition to poly A, may conceivably be susceptible to selective chemotherapy, though the latter process at least seems to be a regular occurrence in normal mammalian cells.

**Translation.**   Translation of proteins from viral mRNA is, perhaps somewhat unexpectedly, the process affected by at least two of the very few successful antiviral agents known so far. Methisazone, a thiosemicarbazone, appears to interact with some early product of infection to block the translation of "late" vaccinia mRNA, while interferon (or another cellular protein derepressed by interferon) may block the translation of all viral mRNA's. Since the cell must continue to translate its own messengers it may seem paradoxical that chemicals could discriminate specifically against viral mRNA. However, the latter may carry a different initiation codon, or have a ribosomal attachment site that can be blocked either directly or by substances that bind to ribosomes. We know that the opposite phenomenon is a regular feature of the multiplication of most cytocidal viruses, namely a virus-coded protein is produced that specifically blocks the translation of cellular mRNA's without adversely affecting viral protein synthesis.

**Replication of Viral Nucleic Acid.**   This process occurs very rapidly, and often in resting cells. Inhibitors of DNA synthesis, such as 5-iodo-2'-deoxyuridine, adenine arabinoside, and cytosine arabinoside, although toxic when administered systemically, have turned out to be quite useful agents against DNA viruses multiplying in the localized environment of the cornea, where relatively few cells are dividing. The RNA viruses should, in theory, be far more vulnerable at this particular point in their multiplication cycle, for they are dependent upon novel virus-specific RNA-dependent RNA polymerases (replicases) that are not present at all in the normal mammalian cell. Certain benzimidazole derivatives, such as HBB, inhibit RNA replication by picornaviruses in cultured cells, though the effect is probably not directed specifically at the replicase.

**Posttranslational Cleavage of Proteins.**   This process may be unique to viruses. All of the picornavirus (and probably togavirus) proteins arise in this way, and cleavage seems to constitute a vital step in the late stages of assembly of several other viruses. A number of toxic chemicals have been used experimentally to inhibit cleavage proteases, and rifampicin blocks the cleavage of a precursor of a vaccinia viral core protein; the evidence suggests that this may be secondary to inhibition of an early step in the maturation of the virion.

**Assembly.**   Guanidine, which disrupts hydrogen bonds, may exert its very varied effects on the multiplication of picornaviruses by so distorting the tertiary configuration of the capsid protein precursor that mature virions are not assembled.

**Regulation of Gene Expression.**   Regulation of viral RNA transcription, translation, and replication may be under the control of viral proteins. Quite subtle changes exerted by a simple drug like guanidine might act in this way to throw the whole cycle "out of gear."

### The Search for Antiviral Agents

Although recent advances in the molecular biology of viral multiplication provide us with a new rationale on which to base a logical search for antiviral agents, there has not yet been a corresponding harvest of new drugs. Only three groups of agents, all of which have been available since the early 1960's, have emerged for practical use: the pyrimidine nucleosides, methisazone, and amantadine, and they have had a negligible effect on human morbidity from viruses.

Further, it is apparent that even if effective antiviral chemotherapy does become a reality, we will have to face the same problems of drug resistance that have created such difficulties for antibacterial chemotherapy in recent years. In laboratory tests, resistant mutants to almost all the present antiviral agents emerge with great rapidity, and in the case of HBB and guanidine, drug-dependent mutants have also been obtained.

Meanwhile reports of new antiviral agents appear at the rate of more than one a day, but the great majority of such "potentially useful" substances have not even been shown to be nontoxic to cultured cells at the minimum virus-inhibitory concentration. Even fewer have been successfully tested in animals, and very few, indeed, ever get to the stage of properly controlled human trials. Most "antiviral agents" picked up by empirical screening have no more selective toxicity than substances like actinomycin D, fluorodeoxyuridine, or cyclohexi-mide, which inhibit the synthesis of viral macromolecules at concentrations no lower than those that block synthesis of the corresponding cellular macromole-cules. Although some of these substances have proved valuable tools for analyzing the viral multiplication cycle (see Chapter 4), they have no po-tential in human medicine. The task of the pharmaceutical companies is to mount a considered, logical search for agents thought likely to inhibit a known viral enzyme or process, select the most promising from *in vitro* assays, and demonstrate lack of toxicity for cultured cells, then for animals, at the minimum virostatic dose. The "chemotherapeutic index" of a compound can be defined as the ratio between the lowest effective antiviral concentration and the highest nontoxic concentration. Substances displaying an index no greater than unity in cultured cells are not worth pursuing further because, even though they may be nonlethal to animals at virostatic concentrations, it can usually be assumed that they are killing some cells, often the crucial dividing cells of the bone marrow and the liver, and therefore doing an unacceptable amount of damage. On the other hand, substances that fail to cause cytopathic effects or slow cell division in cultured cells at virostatic concentrations often turn out to be dis-appointingly ineffective *in vivo*; guanidine and HBB are good examples.

### The Prospects for Antiviral Chemotherapy

The ideal chemotherapeutic agent would have a high chemotherapeutic index, i.e., a high margin of safety in man, as well as certain other desirable phar-macological properties such as a reasonably long half-life *in vivo*, solubility, and the capacity to penetrate the target cells. Hopefully it would also be broad spectrum so that it could be used against a wide range of viruses without the delay involved in laborious laboratory diagnosis. Until rapid diagnostic tech-niques such as immunofluorescence and electron microscopy become much more highly developed and generally available than they are today, the delays (and costs) involved in establishing a definitive virological diagnosis would invalidate chemotherapy oriented toward specific viruses, except in two situations: (a) the few viral diseases caused by monotypic viruses such as herpes simplex, vari-cella-zoster, smallpox, mumps, and measles where clinical diagnoses are reliable, and (b) large-scale epidemic outbreaks of a particular disease like influenza.

Another theoretical difficulty of antiviral chemotherapy is the fact that viral multiplication may be almost over by the time the patient presents with an established illness. In the case of most respiratory infections, this may mean that antiviral agents will have to be used prophylactically rather than therapeutically, a procedure that would be practicable only for the short-term protection of family and professional contacts of the sentinel case, or perhaps more widespread protection of whole communities in the face of impending epidemics of serious viral diseases like influenza. However, the success attending the use of tetracyclines in several generalized rickettsial infections whose pathogenesis parallels that of the generalized viral infections suggests that if satisfactory drugs existed they would be useful even in established generalized diseases. Further there are many viral diseases characterized by prodromal symptoms that may last up to a couple of days and provide time for therapy to be initiated; the childhood exanthemata come into this category. Drugs could be valuable also to prevent the development of complications, such as orchitis or meningitis in mumps. Chronic diseases, such as warts or hepatitis, slow diseases, such as kuru, or recurrent endogenous diseases, such as herpes simplex and zoster, might be very suitable candidates for chemotherapy. Embryopathy caused by rubella or cytomegaloviruses might also be preventable and antiviral chemotherapy might have a major role in human medicine if viruses were found to play a continuing determinative role in human cancer.

Finally, it is appropriate to comment on the use of antibacterial agents in the treatment of viral diseases. For a variety of reasons, physicians the world over tend to overprescribe antibiotics for the treatment of infections of the respiratory and alimentary tracts. Over 90% of respiratory infections and many of gastrointestinal infections are of viral etiology and therefore totally refractory to treatment with existing antibacterial antibiotics. The only valid circumstances for the use of antibiotics in the treatment of viral infections are the following: (a) to prevent bacterial superinfection, e.g., otitis media or pneumonia in a bronchiectatic child with measles, or bronchopneumonia in an elderly cardiac or pulmonary invalid, or perhaps in any infant with respiratory syncytial virus infection of the lower respiratory tract, (b) to "play safe" in potentially serious illnesses where there is real diagnostic doubt about the possibility of a bacterial etiology until results come back from the laboratory, e.g., in meningitis or pneumonia.

In the following pages we make no attempt to review the multitude of substances for which some degree of antiviral activity *in vitro* has been claimed. Rather, we will discuss in some detail the few antiviral agents that have been clearly shown to be of real value in human medicine, then make brief mention of a selection of others which seem to show promise. Readers interested in more comprehensive listings are referred to such publications as those edited by Herrmann and Stinebring (1970), Bauer (1972) and Shugar (1972a), and to the annual volumes entitled "Antimicrobial Agents and Chemotherapy."

## INTERFERON

Superficially, human interferon suggests itself as the ideal antiviral chemotherapeutic agent for use in man. It is a natural by-product of human viral infections, completely nontoxic and nonallergenic in man, and active against a broad spectrum of viruses (Chapter 8). Yet, at the time of writing, interferon has been available for no less than 15 years during which period research workers in universities and pharmaceutical companies throughout the world have endeavored without success to convert Isaacs' discovery into a practicable proposition for human use. Interferon and its potential significance has been discussed at length in several books (Rita, 1968; Vilček, 1969; Finter, 1973) and reviews (e.g., Wheelock *et al.*, 1968; Baron, 1970; de Clercq and Merigan, 1970a; Ho, 1970; Colby and Morgan, 1971).

Soon after their initial discovery, Isaacs and Lindenmann demonstrated that exogenously administered interferon could protect animals against viral infection (Isaacs and Westwood, 1959; Isaacs and Hitchcock, 1960). There followed countless papers which showed pronounced protection; some of the more important are listed in Table 15–6.

Most of the studies listed in Table 15–6 were conducted in the mouse or the rabbit, which presented themselves as convenient models for the investigation of several systemic and localized viral diseases. Some important generalizations should be made about the results obtained. (a) Prophylaxis is invariably more successful than therapy; best results, and often the only positive result, are obtained by administering interferon some hours before viral challenge and continuing therapy for some days thereafter. Treatment of an established infection is rarely successful, and never if started late in the illness. (b) There is a clear dose-response relationship; higher doses of interferon are needed to combat higher doses of challenge virus. (c) Topical application of interferon to the eye, skin, or respiratory tract followed by challenge via the same route tend to be more successful than systemic administration for the prevention of a generalized infection.

These promising results in animals have not been followed by comparable success in man (see Table 15–7). The Scientific Committee on Interferon set up by the Medical Research Council of Great Britian has published a number of reports on its own carefully controlled human trials. The latest of these is decidedly discouraging. Almost a million "research standard" units of human interferon delivered by spray gun over a 24-hour period failed to give any protection against intranasal challenge with influenza B virus. Only when the duration of the treatment was extended to one day before and three days after challenge with rhinovirus type 4, and the dose of interferon increased to impracticable levels (the total dosage per patient being the yield from the leukocytes cultured from 15–25 liters of human blood) was a detectable reduction in severity of symptoms and virus-shedding obtained (Merigan *et al.*, 1973). This comes in the wake of a large-scale trial by Solov'ev (1969) which had seemed to show that man could be protected against natural exposure to influenza A2 by intranasal prophylaxis with human interferon. Indeed, in Moscow, human interferon for prophylactic administration by nasal spray is already available for distribution through pharmacies.

Even if future studies in man were to give cause for greater encouragement than we have had so far, there are forbidding problems in the commercial mass-production of the very large quantities of interferon that appear to be required.

## TABLE 15-6

Antiviral Protection by Interferon and Interferon Inducers in Animals

| DRUG | ROUTE[b] | CHALLENGE VIRUS[c] | ROUTE | ANIMAL | PROPHYLACTIC OR THERAPEUTIC[d] | REFERENCE |
|---|---|---|---|---|---|---|
| **Administered interferon** | | | | | | |
| Interferon | ID | Vaccinia | ID | Rabbit | P | Isaacs and Westwood (1959) |
| Interferon[a] | IP | VSV | IC | Mouse | P | Glasgow and Habel (1963) |
| Interferon | IV | EMC, VSV | IC | Mouse | P | Baron et al. (1966b) |
| Interferon | IP | Transplantable tumors | IP | Mouse | P | Gresser et al. (1968, 1971) |
| Interferon | IP | SFV | IP | Mouse | T | Finter (1964, 1966) |
| **Interferon inducers** | | | | | | |
| Influenza virus | IN | Bunyamwera | IP | Mouse | P | Isaacs and Hitchcock (1960) |
| Statolon | IP | MM | SC | Mouse | P | Kleinschmidt and Murphy (1967) |
| Pyran copolymer | IP | Mengo, EMC, VSV | IP | Mouse | P | Merigan and Regelson (1967) |
| COAM | IP | Mengo, FMDV | IP | Mouse | P | Billiau et al. (1970) |
| Tilorone | Oral | Mengo, EMC, SFV / VSV, HSV, influenza | SC / IC, IP, IN | Mouse | P | Krueger and Mayer (1970) |
| Poly I · poly C | Eye | HSV | Eye | Rabbit | T | Park and Baron (1968) |
| Poly I · poly C | IN | PVM / Col-SK, Vaccinia | IN | Mouse | T | Nemes et al. (1969) |
| Poly I · poly C | IP | VSV | IN, IV | Mouse | P | de Clercq and Merigan (1969) |
| Poly I · poly C | IP | Influenza A2 | IN | Mouse | P | Hill et al. (1969) |
| Poly I · poly C | IN | Rabies | IM | Rabbit | T | Fenje and Postic (1971) |
| Poly I · poly C | IV | SFV | SC | Mouse | T | Worthington and Baron (1971) |
| Poly I · poly C | IP | EAV | IP | Hamster | P | Lieberman et al. (1972) |

[a] Leukocytes primed to produce interferon by pretreatment with UV-irradiated vaccinia virus.

[b] Abbreviations: ID, intradermal; IC, intracerebral; IP, intraperitoneal; IV, intravenous; IM, intramuscular; IN, intranasal; SC, subcutaneous.

[c] VSV, vesicular stomatitis virus; SFV, Semliki forest virus; EMC, encephalomyocarditis virus; FMDV, foot-and-mouth disease virus; HSV, herpes simplex virus; PVM, pneumonia virus of mice; Col-SK, Columbia SK virus; EAV, equine abortion virus.

[d] P, effective prophylactically; T, effective therapeutically (some hours or days after virus inoculation) and prophylactically.

## TABLE 15-7

Antiviral Protection by Interferon and Interferon Inducers in Man[a]

| DRUG | ROUTE | VIRUS | ROUTE | PROPHYLACTIC OR THERAPEUTIC | REFERENCE |
|---|---|---|---|---|---|
| Interferon | ID | Vaccinia | ID | P | Scientific Committee on Interferon (1962) |
| Interferon | IN spray | Influenza A2 | IN | P | Solov'ev (1969) |
| Measles vaccine | SC | Vaccinia | ID | P | Petralli et al. (1965) |
| Poly I · poly C | IN | Rhinovirus 13 | IN | P | Hill et al. (1971) |

[a] Abbreviations as in Table 15-6.

There is much evidence that man (and monkey) are not nearly as readily protected by homologous interferon as the mouse and the rabbit (see Finter, 1973). The use of leukocytes, available to blood banks as a by-product of the preparation of human plasma (Cantell *et al.*, 1968), or human amnions obtained from maternity hospitals (Chany *et al.*, 1968) cannot be seriously contemplated in the long term as sources of human cells for interferon production. Diploid strains of human embryonic fibroblasts such as WI-38 (Hayflick 1965) could provide an adequate supply of cells, but precautions would still need to be taken to ensure that the virus used as interferon inducer (often Sendai or NDV), or indeed any adventitious viruses, e.g., hepatitis, were inactivated by exposure to pH 2, precipitation, or the various chromatographical procedures that have now become standard steps in the protocol for purification of interferon. Billiau *et al.* (1972) have shown that human skin fibroblasts, can be repeatedly stimulated with poly I · poly C to produce a 6-hour burst of interferon about once every 24 hours, whereas most other cells become refractory to secondary stimulation by virus or any other inducer (see Chapter 8).

The problem of the therapeutic use of interferon is solely a quantitative one, for the substance is harmless and the body can tolerate unlimited amounts. Nevertheless, present indications are that if interferon cannot be demonstrated to protect man against superficial infections of the respiratory tract under laboratory conditions designed to favor the drug, there does not appear to be much prospect for its use as a chemotherapeutic agent for treating established viral infection, systemic or superficial.

## SYNTHETIC POLYNUCLEOTIDES AND OTHER INTERFERON INDUCERS

Attention over the last few years has moved to the possibility of harnessing the body's capacity to manufacture its own interferon. Although the levels of interferon achieved in human serum following natural viral infection or artificial stimulation do not approach those found in mice ($<10^2$, compared with $>10^4$ units/ml), there is strong circumstantial evidence that these low levels are protective. For example, Petralli *et al.* (1965) demonstrated that smallpox vaccination "failed to take" in people immunized with the live attenuated Edmonston measles vaccine 9–10 days earlier, and that this coincided with the time of maximum serum interferon levels.

### Polyinosinic Acid: Polycytidylic Acid (Poly I · Poly C)

The mainstream of the current search for interferon-inducers is directed not toward the use of live attenuated viruses but to synthetic chemicals such as the polynucleotide poly I · poly C, otherwise known as (poly rI)·(poly rC) or poly rI: rC (Field *et al.*, 1967b; review, Hilleman, 1970b). Such double-stranded RNA's are powerful inducers of interferon (see Chapter 8).

Poly I · poly C has been convincingly shown to protect animals against a wide variety of viral infections (for selected examples, see Table 15–6). As with interferon itself, prophylaxis is much more successful than therapy, but there are some reports of effective treatment of established disease, e.g., herpetic keratoconjunctivitis in rabbits (Park and Baron, 1968), or progressive pneumonia of mice (Field et al., 1967b; Nemes et al., 1969), vesicular stomatitis (de Clercq and Merigan, 1969) or Semliki Forest virus encephalitis (Worthington and Baron, 1971) in mice. The long incubation period of rabies may have been an influential factor in the striking success obtained by Fenje and Postic (1971) in protecting rabbits and mice against intramuscularly injected rabiesvirus by inoculating poly I · poly C intravenously or into the same muscle as the virus even up to 3 days later.

In man, poly I · poly C administered intranasally in frequent divided dosage totaling 0.05–0.1 mg/kg/day for 1 day before and 4 days after challenge with rhinovirus type 13 inoculated by the same route produced a marginal reduction in the number and severity of common colds (Hill et al., 1971). Topical application of poly I · poly C to the respiratory tract may conceivably have some value in human medicine since high concentrations of the drug can be attained by this route with no appreciable toxic effects. However, systemic administration of this particular polynucleotide seems to be contraindicated for any but grave viral illnesses because of the unacceptable side effects. Careful human trials conducted on terminal cancer patients by two independent groups (Hill et al., 1971; Hilleman et al., 1971a) have given almost identical results: intravenous administration of 0.1 mg/kg/day over a period of 1½ to 3 weeks induced a persistent fever and some indication of reversible disturbances of hemopoiesis and liver function. But the toxic effects of poly I · poly C in man are relatively minor compared with those described in animals, where the principal acute damage is hematological and hepatic, with death from thrombosis and hemorrhage, and survivors may develop hypersensitivity, runting, perhaps due to mitotic inhibition, or embryotoxicity (see Hill et al., 1971; Hilleman et al., 1971a; de Clercq, 1972; Finter, 1973). Another major drawback of poly I · poly C as a potential antiviral agent is that, following a single dose, the animal becomes temporarily refractory to interferon-induction by a second dose (Ho et al., 1970; Hill et al., 1971). This "refractory state" or "tolerance" or "hyporeactivity" is seen with all varieties of interferon inducer (Chapter 8). Field et al. (1972) were able to raise antibodies to poly I · poly C in rabbits only after a prolonged course of intravenous injections; though this antibody depressed the production of interferon by subsequent doses of poly I · poly C it is almost certainly not the explanation of the rapidly developing but short-lasting refractory state in rabbits, or in man, where no antibodies have been detected. Hyporeactivity in vivo, as well as in cultured cells, is probably due to the synthesis of a repressor of interferon synthesis (Vilček, 1970).

In animals, depending on the poly I · poly C dose, appreciable titers of interferon become detectable in serum after 2 to 3 hours and persist for about a day before the animal enters the hyporeactive state (Field et al., 1967b). In man, the maximum serum titers of interferon induced are very much lower than those achieved in mice or rabbits. Peak concentrations are reached about 12 to 24 hours after an intravenous dose of 0.1 mg/kg of poly I · poly C and disappear by 48 hours; following intranasal administration no interferon has been detected in serum and only low levels were found in nasal washings (Hill *et al.*, 1971). There was some evidence of the development of a refractory state 2 days after a single intravenous dose. Assuming that hyporeactivity does occur in man as in other animals and cultured cells, the refractory state imposes a major, perhaps insuperable, flaw in the rationale of interferon induction, for unless much higher interferon titers than currently attainable can be achieved, it is improbable that a single dose would provide enough interferon to prevent a viral infection, let alone cure one.

Much current research on poly I · poly C is focused on attempts to increase the therapeutic ratio by enhancing the drug's antiviral potency or decreasing its toxicity for man, but there is evidence that the two properties are closely linked; pyrogenicity generally increases hand in hand with antiviral activity as the dose of a given polynucleotide is raised or as the molecule is manipulated to change its physicochemical properties. According to de Clercq (1972) the following structural characteristics favor high antiviral potency: (a) presence of the 2'-hydroxyl group (deoxyribopolynucleotides are inactive); (b) high molecular weight ($>10^5$ and ideally $>10^6$ daltons); (c) stable helical configuration with high thermal stability ($T_m>60°C$); and (d) RNase resistance (though there are exceptions). However, others do not agree that the structural requirements (c) and (d) are so strict; they find no absolute correlation with thermal stability or RNase resistance but argue for the importance of providing conditions that maximize the uptake of the molecule by the cell; normally only a minute proportion of poly I · poly C is taken up by cells, but DEAE-dextran greatly stimulates activity (Colby and Chamberlin, 1969; Colby, 1971; Pitha and Carter, 1971). Workers in some laboratories (e.g., de Clercq, 1972) are seeking double-stranded polyribonucleotides that resist the RNase for double-stranded RNA that is found in human serum (Nordlund *et al.*, 1970). Others (Carter *et al.*, 1972) have synthesized unstable poly I · poly C molecules with mismatched bases or interruptions in the C strand which have near-normal antiviral activity but are much more rapidly hydrolyzed, in the hope that this will diminish their toxic activity.

It is generally assumed that the *in vivo* protection conferred by poly I · poly C is mediated via interferon induction (Schafer and Lockart, 1970; Colby and Morgan, 1971; de Clercq, 1972), but even this hypothesis is open to question. The levels of interferon induced by poly I · poly C inoculated by any route fail to reach those achieved following natural viral infection, and even the low titers that do result do not always correlate well with the degree of protection obtained (Field *et al.*, 1967b; de Clercq and Merigan, 1970b; Colby and Morgan, 1971). For example, Lieberman *et al.* (1972) could detect no interferon in hamsters protected by poly I · poly C against lethal equine abortion virus infection. It may be that other effects of poly I · poly C such as its well-documented stimulation of phagocytosis and various other immunological and nonimmunological reticuloendothelial activities (Braun and Nakano, 1967) could play an important role in the drug's action *in vivo* if not *in vitro*. The adjuvant effect of poly I · poly C is so marked that Hilleman *et al.* (1971a) have incorporated it into an experimental influenza vaccine. Macrophages become cytotoxic following the administration of poly I · poly C to mice (Alexander and Evans, 1971), which may explain the reported action of poly I·poly C against numerous chemically or virus-induced, and "spontaneous" tumors (Hilleman *et al.*, 1971a). Interferon itself has also recently been reported to suppress the growth *in vitro* and *in vivo* of several murine transplantable tumors (Gresser *et al.*, 1971).

### Statolon

Statolon (Kleinschmidt and Murphy, 1967) is an interferon inducer derived from the fungus *Penicillium*, which turned out to be a mycophage containing double-stranded RNA. Kleinschmidt (1972) has recently compared the chemotherapeutic potential of statolon favorably with the synthetic double-stranded RNA's. For instance, a single intraperitoneal dose of statolon protects mice against MM virus challenge, and one intranasal dose protects them against influenza; in both cases, the effect lasts for many days. However, the agent is toxic, causing among other things mitotic inhibition in the regenerating liver of partially hepatectomized mice, and it is difficult to believe that an RNA derived from a fungal virus would find acceptance in human medicine.

### Pyran Copolymer

Before it was appreciated that the activity of statolon was attributable to double-stranded viral RNA it was postulated that polyanionic compounds, in general, might be expected to be effective interferon inducers (see Kleinschmidt and Murphy, 1967). Acting on this premise, Merigan and Regelson (1967) discovered that pyran, a random copolymer of maleic anhydride and divinyl ether, protected mice against EMC, mengo, VSV, and a number of other viruses. However, when inoculated into man, this synthetic anionic polycarboxylate plastic caused severe toxicity, notably fever and thrombocytopenia, and could be demonstrated to persist in the reticuloendothelial system for long periods. Hence it cannot be considered for future human use. Nevertheless, the discovery precipitated a search for other less toxic synthetic anionic polymers.

### Polyacetal Carboxylic Acids

Searching for interferon inducers that might be less toxic than polyacrylic acid by virtue of being biodegradable, Billiau *et al.* (1970) demonstrated antiviral activity in a variety of chlorite-oxidized oxypolysaccharides, of which the most active was chlorite-oxidized oxyamylose (COAM). When injected intraperitoneally COAM protected mice against lethal infections with mengovirus over a period of several weeks; activity against certain other viruses was less dramatic but real. The minimum effective antiviral dose was over 100-fold lower than the $LD_{50}$, but toxic effects (e.g., splenomegaly) were observed at all prophylactic dose levels. It is not clear whether these agents act primarily via interferon induction or via effects on the reticuloendothelial system, such as macrophage stimulation.

## Tilorone

Tilorone hydrochloride is a diamine, 2,7-bis[2-(diethylamino) ethoxy]-fluoren-9-one dihydrochloride. It is an interesting interferon inducer, not only because of its relatively low molecular weight but also because it is administered by mouth. Krueger and Mayer (1970) showed that oral tilorone protected mice against parenteral challenge with several viruses. Peak titers of interferon were obtained 24 hours after ingestion. Although de Clercq and Merigan (1971) were able to demonstrate a close correlation between tilorone dose, the resulting interferon titer, and the degree of protection of mice against VSV, they could not induce any interferon production in cultured cells, hence concluded that tilorone, which is toxic for the hemopoeitic and reticuloendothelial systems, might act like bacterial endotoxins to release "preformed" interferon.

## IDOXURIDINE AND OTHER HALOGENATED NUCLEOSIDES

### Idoxuridine

As if to deny the logic of the principles upon which the rational search for antiviral agents is based, the most successful group of drugs available so far are highly toxic to mammalian cells. The halogenated pyrimidines are potent inhibitors of cellular DNA synthesis, yet one of them, 5-iodo-2'-deoxyuridine (Idoxuridine, IUdR, IDU, see Fig. 15–1), has proved itself in practice to be a very useful chemotherapeutic agent when applied topically to the human eye in early cases of keratoconjunctivitis or dendritic ulcer due to herpes simplex virus (reviews by Kaufman, 1965a, b; Prusoff, 1967, 1972; Schabel and Montgomery, 1972). This paradoxical situation is attributable to the localized nature of the disease; herpetic ulcers are initially quite superficial lesions, readily accessible to high concentrations of idoxuridine applied topically at frequent intervals. The cornea is relatively avascular, hence the drug remains localized. Moreover, viral DNA synthesis, which is proceeding rapidly, is more vulnerable than that of the slowly proliferating corneal cells.

Idoxuridine treatment must be instigated early in the disease and pursued vigorously if it is to be successful (Kaufman, 1965a, b). Jawetz et al. (1970) concluded that most cases of unsuccessful idoxuridine therapy are attributable to "too little, too late" rather than to the emergence of drug-resistant mutants, though the latter do occur occasionally (Underwood et al., 1965).

More recently, claims have been advanced for the efficacy of idoxuridine in the therapy of other herpetic infections. MacCallum and Juel-Jensen (1966) treated herpes labialis (recurrent fever blisters) in man with a 5% solution of idoxuridine in dimethyl sulfoxide (DMSO), which greatly improves the penetration of the drug into the cells of the deeper layers of the skin. Presumably herpes genitalis should respond as well. Juel-Jensen (1970) has also presented evidence that continuous topical application of 40% idoxuridine in DMSO reduces the severity of herpes zoster in the elderly. Idoxuridine has even been used systemically as "heroic therapy" in cases of herpes simplex encephalitis, a disease with a 50% mortality (review, Illis and Merry, 1972). There is evidence to suggest that

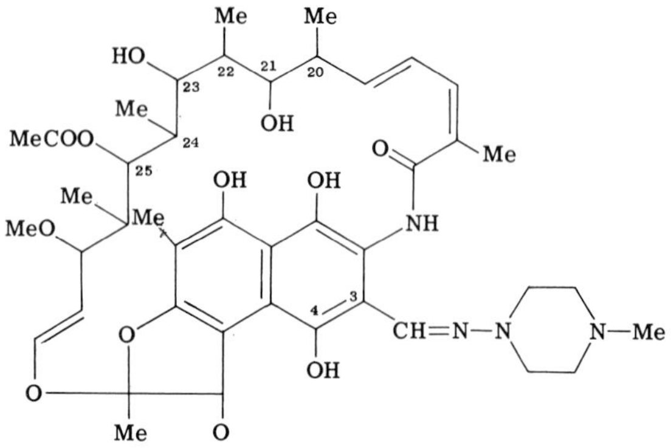

Rifampicin

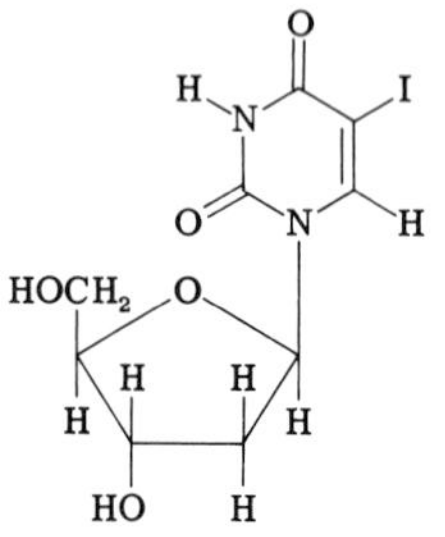

5-Iodo-2'-deoxyuridine

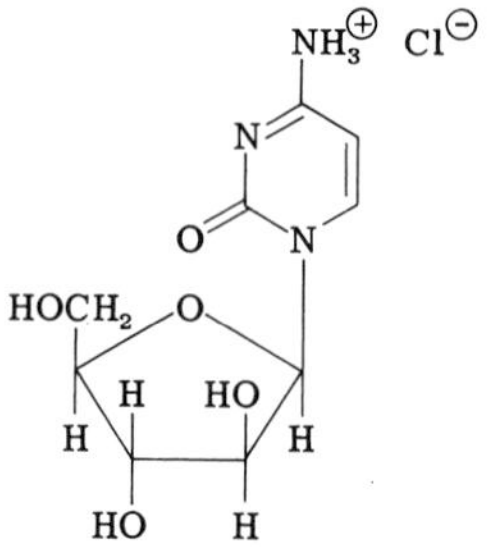

1-β-D-Arabinofuranosyl-
cytosine · HCl

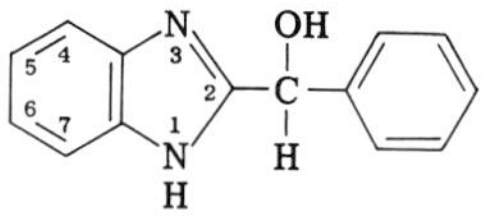

2-(α-Hydroxybenzyl)-
benzimidazole

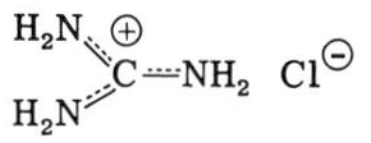

Guanidine · HCl

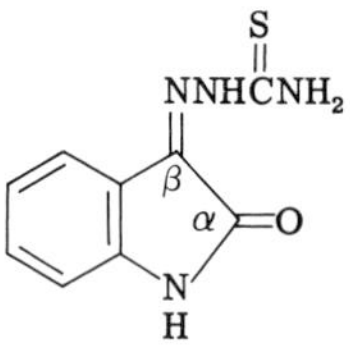

Isatin β-thiosemi-
carbazone

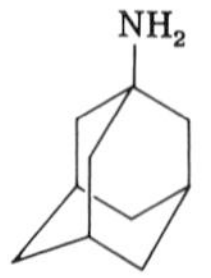

1-Amino adamantane
hydrochloride

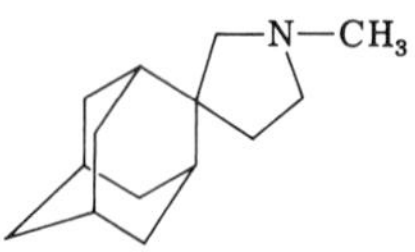

1'-Methyl spiro (adamantane-
2,3'-pyrrolidine) maleate

FIG. 15–1. *Structural formulae of some antiviral drugs. (A) Rifampicin. (B) 5-Iodo-2'-deoxyuridine (Idoxuridine). (C) 1-β-D-Arabinofuranosylcytosine · HCl (cytosine arabinoside, Ara-C). (D) 2-(α-Hydroxybenzyl)benzimidazole (HBB). (E) Guanidine · HCl (the positive charge is distributed over the whole molecule). (F) Isatin β-thiosemicarbazone. (G) Amantadine (1-amino adamantane hydrochloride). (H) 1'-Methyl spiro (adamantane-2,3'-pyrrolidine) maleate.*

the drug may increase the survival rate and lower the incidence of serious sequelae in the survivors, but the rarity of the disease makes it impossible to conduct a large-scale controlled trial (see Tomlinson and MacCallum (1970) for a critical discussion of the literature). Clearly, idoxuridine should never be administered parenterally in anything less than such desperate circumstances, because it is highly mutagenic and toxic. Doses of 100 mg/kg for as little as 4 days produce leukopenia, thrombocytopenia, alopecia, and stomatitis. On the other hand, the amounts delivered topically to the eye would produce blood levels of less than one-thousandth of this figure, even if they were totally absorbed (which they are not), therefore administration by this route is perfectly harmless. Moreover, the drug does not accumulate in the body but is rapidly degraded metabolically.

Cellular DNA synthesis is affected in three separate ways, any or all of which could also be instrumental in blocking viral multiplication: (a) IUdR, or its phosphorylated derivatives, compete with thymidine for certain enzymes involved directly or indirectly in DNA synthesis (thymidine kinase, thymidylic acid synthetase, DNA polymerase); (b) the drug also causes feedback inhibition of several enzymes, and (c) following phosphorylation, IUdR is incorporated into DNA in place of thymidine, rendering the molecule nonfunctional in transcription and/or replication (Prusoff, 1972).

As might be expected, idoxuridine is active in cell culture against most DNA viruses (Herrmann, 1961; Kaufman, 1965a, b; Prusoff, 1967) but not RNA viruses, except for the leukoviruses, in which the synthesis of DNA provirus is an essential step in multiplication (see Chapter 6). The drug is also useful therefore in topical treatment of accidental vaccinia viral infections of the eye, and has been shown to inhibit successful vaccination when administered intravenously to terminal cancer patients undergoing continuous infusion (Calabresi, 1965).

The related thymidine analog, trifluorothymidine, has recently been reported to be a significantly superior alternative to idoxuridine for the therapy of human ocular infections with herpes simplex virus (Wellings et al., 1972).

## Arabinofuranosyl Nucleosides

Though their full clinical potential has yet to be determined, there are indications that this group of compounds may be more valuable in human medicine than the halogenated nucleosides just discussed. They too interfere with both viral and cellular DNA synthesis but the chemotherapeutic index is higher.

Cytosine arabinoside (1-$\beta$-D-arabino-furanosylcytosine), abbreviated to ara-C, has a similar but not identical antiviral spectrum to idoxuridine and has been shown to be at least equally effective in treating herpetic and vaccinial infections, particularly keratitis, in man and animals (Underwood et al., 1965). There are several reports, difficult to evaluate in the absence of adequate numbers of controls, that the drug is also of benefit when administered systemically to gravely ill patients with overwhelming generalized herpes zoster or simplex infections (see Chow et al., 1971; Schabel and Montgomery, 1972). Ara-C is rapidly deaminated to the impotent derivative, arabinofuranosyluracil, but Shugar (1972b) has recently synthesized analogs that resist deamination.

Schabel (1970) and his colleagues have synthesized adenine arabinoside (9-$\beta$-D-arabinofuranosyladenine) (ara-A), which they claim has a chemotherapeutic index at least fifteen times higher than ara-C. Ara-A is active against various herpesviruses and vaccinia virus in cultured cells and animals, but has yet to be extensively tested in man. However, there are reports that it is effective against herpetic keratitis with stromal involvement, and experience in animals is encouraging. The drug protects mice against vaccinia or herpes simplex encephalitis, and hamsters, rabbits, and monkeys against herpetic keratitis (see Schabel, 1970); parenteral administration of 125 mg/kg twice daily protected hamsters against lethal infection with equine abortion virus (Lieberman *et al.*, 1972).

## THIOSEMICARBAZONES

The multiplication of poxviruses is inhibited by 1-methylisatin $\beta$-thiosemicarbazone, better known as methisazone, or the trade name Marboran (Fig. 15–1; review, Bauer, 1965). Effects of various thiosemicarbazones on a wide range of other DNA and RNA viruses in cultured cells (Bauer *et al.*, 1970) have not been reported in enough detail to judge whether there is any selectivity of action.

Appleyard *et al.* (1965) demonstrated that certain "late" vaccinia viral antigens are not made in the presence of methisazone. Woodson and Joklik (1965) examined the situation in more detail and showed that viral mRNA, early proteins, and DNA are all synthesized in the presence of the drug but that "late" mRNA is not translated into protein; polyribosomes are rapidly dissociated. It appears that methisazone interacts with some product of the first 3 hours of the cycle (perhaps a viral or cellular protein) to prevent the translation of late mRNA.

Methisazone has been successfully used for the prevention of smallpox (Bauer, 1965). In the face of serious epidemics in India and elsewhere, the drug has been administered orally to large groups of people and shown to reduce the incidence of disease by 75–95%. However, a more recent controlled trial throws doubt on whether the drug has any efficacy at all (Heiner *et al.*, 1971). Needless to say, vaccination is still by far the most important prophylactic measure against smallpox; methisazone is usually combined with vaccination in situations where the threat is so imminent that there may be insufficient time for active immunity to develop.

Though methisazone has not been shown to be effective in treating established cases of smallpox, it has been employed successfully to treat complications of smallpox vaccination (Bauer, 1965). Vaccinia gangrenosa, which is otherwise always fatal, has responded to methisazone, as have cases of the less serious but much more common complication, eczema vaccinatum.

A 3-substituted triazinoindole, known as SK & F 30097, which is an analog of methisazone but, interestingly, has structural similarities to both guanidine and the benzimidazoles, selectively inhibits the multiplication of poliovirus and rhinovirus in cell culture (Matsumoto *et al.*, 1972).

## AMANTADINE AND DERIVATIVES

A simple three-ringed symmetrical amine, 1-adamantanamine hydrochloride, or 1-amino adamantane hydrochloride, or amantadine (Symmetrel) (see Fig. 15–1) has been shown to block the multiplication of influenza A2 viruses (Davies *et al.*, 1964). It has also been claimed to have a spectrum of activity against a variety of other enveloped viruses but not against influenza B (Cochran *et al.*, 1965). The precise mode of action is still unclear but may be to block some aspect of viral penetration (Davies *et al.*, 1964) or uncoating (Kato and Eggers, 1969). The latter authors presented evidence purporting to show that amantadine does not inhibit "penetration," as measured by escape from neutralizability by antiserum, but does block "uncoating," defined as loss of photosensitivity of neutral red-treated virus, but the mechanism of action of amantadine must still be considered an open question. It has recently been reported to block the multiplication of arenaviruses, including lymphocytic choriomeningitis virus, both at the stage of penetration and also later in the cycle, perhaps exerting its effect on cellular membranes (Welsh *et al.*, 1971).

The efficacy and safety of amantadine in man have been the subject of considerable controversy. Certainly the benefits of the drug are marginal at best and the incidence of side effects upon the central nervous system is disturbing if the recommended dosage is exceeded, as well it might be in a drug freely available for a common ailment like "flu" (Sabin, 1967a). Nevertheless there have now been some very extensive double-blind clinical trials which do indicate a prophylactic effect. For example, Smorodintsev *et al.* (1970) demonstrated a twofold reduction in incidence of influenza A2 in some thousands of subjects given 100 mg of amantadine by mouth each day before and after exposure to the virus. Insomnia (in 1% of subjects) and dyspepsia (2%) were the main side effects reported. Togo *et al.* (1970) using 200 mg per day claimed success in treating established cases of human influenza. Resistance to amantadine (Cochran *et al.*, 1965) and cross-resistance to other derivatives (Oxford *et al.*, 1970) develop on passage of influenza A2 in its presence in cultured cells or the lungs of mice.

Rimantadine (α-methyl-1-adamantanethylamine hydrochloride) administered intraperitoneally to mice reduced the incidence of naturally acquired contact infection with influenza by almost one-half in an experiment where amantadine gave no protection (Schulman, 1968). More recently, Beare *et al.*, (1972) have presented evidence that 1'-methyl spiro(adamantane-2,3'-pyrrolidine) maleate (1:1) at a dosage of 35 mg orally before challenge with a partially attenuated strain of influenza A2/Hong Kong lowered the attack rate by almost 50%. The new derivative was said to have a wider spectrum than amantadine in cell culture, though it too was inactive against influenza B.

## SOME OTHER COMPOUNDS OF POTENTIAL VALUE

### Rifamycin Derivatives (Ansamycins)

Rifamycin-SV and the derivative, rifampin, otherwise known as rifampicin (Fig. 15–1) inhibit bacterial RNA synthesis by binding to DNA-dependent RNA polymerase, but have no such effect on mammalian polymerases (reviews, Wehrli and Staehelin, 1971; Thiry and Lancini, 1972). Hence, when Heller et al. (1969) and Subak-Sharpe et al. (1969) demonstrated these agents to be active against poxviruses, which had recently been shown to contain a similar DNA-based transcriptase, the assumption was that the mechanism was the same. However, the drug concentrations required were over a thousand times higher than those effective against bacteria, and it transpired that the target was not the transcriptase *per se*, but an early step in viral assembly. As has been discussed at more length in Chapter 5, rifampicin blocks the formation of the poxvirus membrane, one of the earliest steps in assembly of the virion (Grimley et al., 1970). Both "early" and "late" viral mRNA's and proteins are synthesized satisfactorily (Ben-Ishai et al., 1969; Tan and McAuslan, 1970; Moss et al., 1971b), but a particular 125,000 dalton precursor protein fails to be cleaved to give the 76,000 dalton "core" protein that is normally formed at a late stage in assembly of the virion (Katz and Moss, 1970a, b). Upon removal of the drug, the precursor membranes that have accumulated mature into the viral envelopes (Plate 5–2), the precursor protein(s) is cleaved, and proteins synthesized during the block are incorporated into virions (Moss et al., 1971b). Rifampicin has turned out to be a valuable tool for the study of poxvirus maturation, but has no potential therapeutic application to poxvirus infections in man, because it is toxic at the doses needed to maintain adequate antiviral concentrations.

Resistant mutants (Subak-Sharpe et al., 1969) can "rescue" susceptible wild-type virus in mixed infections (Moss et al., 1971c). Mutants show cross-resistance to other rifamycin derivatives (Grimley and Moss, 1971). More recently, it has been found that some rifamycin derivatives, though not rifampicin, do inhibit the vaccinia viral transcriptase *in vitro*, but this has no bearing on the antiviral action of the drugs in cultured cells because they inhibit just as strongly the transcriptase of drug-resistant viral mutants (Szilágyi and Pennington, 1971).

Rifampicin also inhibits the activity of the reverse transcriptase of the leukoviruses. Diggelmann and Weissmann (1969) reported that rifampicin prevented the development of Rous sarcoma-transformed cell foci. It is still not clear whether this is due to a specific effect on the transcription of RNA-primed DNA or some other vital step in the transformation process itself, or whether the drug selectively or nonselectively inhibits the subsequent multiplication of transformed cells (Calvin et al., 1971), perhaps by inhibiting a cellular DNA polymerase. However, Gallo et al. (1970, 1972) and Gurgo et al. (1971) have shown quite clearly that certain rifamycin derivatives will inhibit reverse transcriptases *in vitro* without affecting DNA-dependent RNA polymerase from mammalian cells. Green et al. (1972) have synthesized a wide range of rifamycin derivatives with substituted cyclic-amine side chains in position 3 of the ansa ring and shown them to be powerful inhibitors of both the RNA-dependent DNA polymerase and the DNA-dependent DNA polymerase from murine, feline, and avian leukoviruses *in vitro*. The most active compounds were 3-piperidyl derivatives of rifamycin-SV with cyclohexyl and cyclohexylalykl substituents. However, the compounds inhibited both the *in vitro* activity of cellular DNA-dependent DNA polymerase and the multiplication of normal cells at concentrations only slightly higher than the minimum doses active against *in vitro* reverse transcriptase activity and cellular transformation by murine sarcoma virus.

## Benzimidazoles

The benzimidazole derivative, 2-($\alpha$-hydroxybenzyl)-benzimidazole (Fig. 15–1), known as HBB, inhibits the multiplication of many picornaviruses, notably coxsackie B and echoviruses, at noncytocidal concentrations in cultured cells (reviews by Tamm and Eggers, 1963; Eggers and Tamm, 1966). The ribofuranosyl benzimidazoles themselves show no selectivity of action, but HBB depresses viral RNA synthesis at concentrations about tenfold lower than those that affect cellular RNA (Bucknall, 1967). O'Sullivan *et al.* (1969) have tested a wide range of derivatives of HBB and shown several of them to be active against picornaviruses but not other RNA or DNA viruses in cultured cells. The 1-propyl derivative displayed a marginal protective effect against small doses of coxsackievirus A9 in suckling mice.

The precise mechanism of action of HBB is uncertain. Studies conducted some time ago indicated that enteroviral RNA replication is blocked, perhaps as a result of interference with the synthesis or activity of the replicase (Baltimore *et al.*, 1963; Eggers *et al.*, 1963). Superficially, there are similarities with guanidine which inhibits a similar but somewhat broader spectrum of picornaviruses. Although there is no cross-resistance between HBB-resistant and guanidine-resistant mutants, guanidine does enable HBB-dependent mutants to multiply in the absence of HBB (Tamm and Eggers, 1963; Eggers and Tamm, 1966).

## Guanidine

The simple compound guanidine hydrochloride (Fig. 15–1), which is commonly used to dissociate hydrogen bonds, has a marked inhibitory effect on the multiplication of picornaviruses in cultured cells (Rightsel *et al.*, 1961; reviews, Baltimore, 1968a, 1969; Cooper *et al.*, 1971). Its action may be completely reversed by a variety of amino acids (e.g., methionine) or methylated or ethylated amino alcohols or amines (e.g., choline or dimethylethanolamine) (see Lwoff, 1965; Mosser *et al.*, 1971). Although relatively harmless to mammalian cells *in vitro* and *in vivo* (Baltimore *et al.*, 1963), guanidine has no activity *in vivo*, possibly due to the regular and rapid emergence of drug-resistant mutants (Melnick *et al.*, 1961; Sergiescu *et al.*, 1972). Nevertheless, it has been a useful tool for the investigation of poliovirus replication (Baltimore, 1968a), providing data that has led to the enunciation of an important theory on the regulation of viral multiplication (Cooper *et al.*, 1973).

Guanidine affects a baffling array of processes that seem at first sight to be quite unrelated. (a) Viral RNA replication is blocked in cultured cells (but only in the second half of the cycle) despite the fact that replicase activity is unaffected *in vitro* (Baltimore *et al.*, 1963; Eggers *et al.*, 1963). It seems that newly synthesized cRNA molecules fail to be released from the RI, giving rise to a swollen structure designated the "guanidon" (Baltimore, 1968a; Huang and Baltimore, 1970a). (b) Viral protein synthesis is little affected until later in the cycle but cleavage of VP0 → VP2 + VP4 is blocked (Jacobson and Baltimore, 1968a). (c) Assembly of virions is disturbed and empty capsids accumulate (Jacobson and Baltimore, 1968a). Cooper et al. (1973) have attempted to relate all these findings to their concept that poliovirus multiplication is regulated by the immature capsomer (VP0, VP1, VP3), which they call the "equestron." If the primary effect of guanidine were to inhibit the cleavage of VP0 → VP2 + VP4 (Jacobson and Baltimore, 1968a) it would, according to Cooper's thesis, result in continued protein synthesis, but in failure to release cRNA from vRNA templates (Baltimore, 1968a), and accumulation of empty capsids (Jacobson and Baltimore, 1968a). This would also explain why the genetic locus for guanidine resistance maps with the capsid protein precursor, VP0,1,3, and why no progeny result from mixed infections with resistant and sensitive strains in the presence of guanidine (Cooper et al., 1970c).

CHAPTER 16

**The Epidemiology of
Viral Infections**

# INTRODUCTION

In earlier chapters we have discussed the structure and chemical composition of the virion as a physical entity, the variety of interactions between virus and animal cell that occur during viral multiplication, and the pathogenesis of viral infections. Individual cells survive for variable and often quite short periods, and all animals are mortal. Animal viruses can therefore survive in nature only if they are able to pass from one animal to another, whether of the same or another species, and this is the most hazardous step in their survival. Infection of an animal results in a vast amplification of the viral population, but the overwhelming majority of these progeny particles are not transferred to another host, but are destroyed. The likelihood of survival of individual virions is largely a matter of chance, but is affected by their genotypic characters, and natural selection operates effectively at the stage of transmission.

The study of the transfer and persistence of viruses in populations of animals comprises the subject matter of the epidemiology of viral diseases, which we will regard very broadly as an aspect of ecology dealing with viral infections rather than confining our attention to cases of overt disease.

The reductionist approach and the use of physicochemical methods are highly effective in providing valuable information about the virion, and about cell–virus interactions. At the epidemiological level and in the study of the ecology of viruses we find ourselves dealing with multiple variables and the sort of complexities that characterize the social sciences. At this level we can think of virus simply as microorganism, for the epidemiology of viral infections has many parallels with that of infections of vertebrates with microorganisms. The epidemiological resemblances are particularly close for microorganisms that are obligate intracellular parasites, e.g., the rickettsias, many protozoa, and some pathogenic bacteria. Although it is impossible to achieve in epidemiology the simplification that has been so profitable at the cellular and molecular level, recent discoveries in the molecular biology of viruses have greatly illuminated our understanding of some aspects of the ecology of viral infections.

## ROUTES OF ENTRY AND EXIT

The vertebrate body presents three large surfaces to the environment: the skin, the respiratory mucosa, and the mucosa of the intestinal tract; and two lesser surfaces, the eye and the genitourinary tract. To gain entry to the body, viruses must directly infect cells on one of these surfaces, or breach a surface (by trauma, including arthropod bite), or bypass the surfaces by congenital transmission. The same considerations apply to the escape of virus from the body. Transmission may therefore be defined in terms of the routes of exit and entry (e.g., anal–oral), or more commonly just in terms of the route of entry. The respiratory and alimentary routes are self-explanatory. Direct contact transmission involves close physical contact, e.g., EB virus and herpes simplex virus type 1 appear to be spread by kissing or other modes of exchange of oral secretions, herpes simplex type 2 by sexual intercourse. "Contact transmission" is sometimes used more broadly to include short-range airborne spread (indirect contact). Other viruses are transferred by injection, either by man (hypodermic inoculation, transfusion) or by arthropods, which may act as either mechanical or propagative vectors. Finally, some viruses are transmitted congenitally, either as part of the germ plasm, across the placenta or in the egg (of reptiles, birds, and sometimes mammals). Some viruses are transmitted by a variety of routes; other species, and even some genera, are transmitted in nature exclusively by one route (Tables 16–1 and 16–2). In some large genera (e.g. *Herpesvirus*), different species may have different characteristic routes of transmission.

## THE SKIN

The tough protective covering of the skin, the stratum corneum (see Fig. 9–2), offers an effective barrier to infection, but several viruses establish infection of the skin when the surface is breached by trauma or by inoculation, either by arthropod vectors or a hypodermic needle (Table 16–3).

TABLE 16–1

*Common Routes of Transmission of DNA Viruses*

| VIRAL GROUP | ROUTES |
|---|---|
| *Parvovirus* | Respiratory |
| Papovaviridae | Contact (human wart) |
| | Injection: mechanical by arthropod (rabbit papilloma) |
| | Respiratory (polyoma) |
| *Adenovirus* | Respiratory and alimentary |
| *Herpesvirus* | Salivary (EB virus, herpes simplex type 1) |
| | Venereal (herpes simplex type 2) |
| | Respiratory (infectious bovine rhinotracheitis) |
| | Injection: transfusion (cytomegalovirus) |
| | Transplacental (cytomegalovirus) |
| *Iridovirus* | Respiratory |
| *Poxvirus* | Contact (molluscum contagiosum) |
| | Injection: mechanical by arthropod (myxoma) |
| | Respiratory (smallpox) |

TABLE 16–2

*Common Routes of Transmission of RNA Viruses*

| VIRAL GROUP | ROUTES |
| --- | --- |
| Picornaviridae | Alimentary (*Enterovirus*) |
| | Respiratory (*Rhinovirus*) |
| Togaviridae | Injection: propagative by arthropod |
| | (*Alphavirus* and *Flavivirus*) |
| | Respiratory (rubella) |
| | Transplacental (rubella) |
| Orthomyxovirus | Respiratory |
| Paramyxovirus | Respiratory |
| Coronavirus | Respiratory (human) |
| | Alimentary (transmissible gastroenteritis |
| | of swine) |
| Arenavirus | Respiratory or alimentary |
| | Congenital |
| Leukovirus | Congenital (integrated with cellular DNA) |
| | Milk (mammary tumor virus) |
| | Respiratory (foamy agents; progressive |
| | pneumonia virus) |
| Rhabdovirus | Injection: propagative by arthropod (VSV) |
| | Animal bite (rabies) |
| Reovirus | Respiratory or alimentary |
| Orbivirus | Injection: propagative by arthropod |
| | (bluetongue) |

TABLE 16–3

*Viruses that Initiate Infection Via the Skin or Oral Mucosa*

| ROUTE | VIRUS |
| --- | --- |
| Minor trauma | *Papillomavirus*: all species |
| | *Herpesvirus*: herpes simplex type 1, EB |
| | virus, pseudorabies |
| | *Poxvirus*: mousepox, cowpox, orf, milkers' |
| | nodes, molluscum contagiosum |
| Arthropod bite: | |
| Mechanical | *Papillomavirus* (rabbit papilloma) |
| | *Poxvirus*: myxoma, fibroma, fowlpox, |
| | swinepox, Tanavirus, Yabavirus |
| Propagative | Picornaviridae (Nodamura virus) |
| | *Alphavirus* (all species) |
| | *Flavivirus* (all species) |
| | Bunyamwera supergroup (all species) |
| | *Rhabdovirus* (vesicular stomatitis, |
| | ephemeral fever) |
| | *Orbivirus* (all species) |
| Bite of vertebrate | *Herpesvirus* (B virus) |
| | *Rhabdovirus* (rabies) |
| Injection | *Herpesvirus* (by transfusion: |
| | cytomegalovirus, mononucleosis) |
| | Unclassified (hepatitis B) |

## Trauma

The skin is repeatedly damaged by minor injuries and the breaches thus produced may permit entry of viruses into the susceptible cells in the deeper layers of the epidermis or the dermis. For example, in cages of mice in which some animals had mousepox the most common mode of infection was through abrasions of the skin, the infecting virus being present as an environmental contamination either on the bedding or on the scabs or fur of other mice (Fenner, 1948a).

Certain small tumors that are confined to the skin may be naturally transmitted through small abrasions, although direct evidence is difficult to obtain and mechanical transmission by arthropods may be important. The world-wide distribution of two human skin infections, molluscum contagiosum, due to a poxvirus, and human warts, caused by a papilloma virus, is incompatible with an exclusively arthropod transmission and both of these viruses are probably spread by direct inoculation through minute abrasions. Experimental infections by intradermal injection of human wart virus, obtained either directly from warts or after passage in monkey kidney cells, produced warts after incubation periods which varied between 3 and 12 months (Mendelson and Kligman, 1961; see Chapter 13).

Pseudorabies provides an example of transmission between different species by way of minute abrasions of the skin (Shope, 1935s). Pseudorabies in swine is a highly contagious infection in which the clinical manifestations are so mild that most outbreaks go unnoticed, apart from serological evidence. Cattle that are housed in common enclosures with pigs are probably infected by virus spreading from the noses of the pigs to abraded areas on the skin of the cattle, which then sustain a very severe disease that is usually fatal, but is not transmissible from cattle to other animals.

Rabies is ordinarily transmitted by the bite of infected animals (dogs, cats, wild carnivores, bats) whose saliva contains virus. Likewise, human infections with the simian herpesvirus (B virus) are usually due to the bite of a monkey with infected saliva, but infection can be initiated by rubbing virus into trivial abrasions of the skin, or by aerosols (Davidson and Hummeler, 1960). Both rabies and probably human B virus disease are associated with neural or possibly hematogenous spread to the central nervous system. Like pseudorabies in cattle, they are dead-end infections, since they are not transmitted by arthropods and no virus is released from the infected individual into the environment.

## Artificial Inoculation

Experiments in animals attest to the ease of transmitting viral infections by inoculation, either into the subcutaneous tissues or by some other route (intracerebral, intranasal, or intraperitoneal). Inoculation can be important epidemiologically in human and veterinary medicine, and its importance is increasing in modern Western society (see Chapter 17). The classic case is hepatitis B, transmitted in serum or blood transfusions, or by contaminated syringes and needles. More recently, cases have been reported of the transmission of cytomegaloviruses and infectious mononucleosis, probably as leukocyte-associated viruses, by blood transfusions. In animals, infectious equine anemia is usually transmitted from one horse to another by hypodermic injection (Dreguss and Lombard, 1954). Finally, the most venerable of all immunization procedures, namely Jennerian vaccination against smallpox, is effected by the intradermal inoculation of vaccinia virus.

## Transmission by Arthropods

By far the most important means by which the skin is breached is by the bite of an arthropod (Marshall, 1973). Transmission may be propagative, i.e., involving multiplication of the virus in the arthropod vector, or simply mechanical. Table 16–4 lists some of the important distinctions between mechanical and propagative transmission of viruses.

TABLE 16–4

*Comparison of Mechanical and Propagative Transmission by Arthropods*

| FEATURE | MECHANICAL TRANSMISSION | PROPAGATIVE TRANSMISSION |
|---|---|---|
| Viral multiplication in vector | No | Yes |
| Source of virus | Skin lesions | Blood |
| Vector specificity | Nil, except for biting habits | Relatively high |
| Extrinsic incubation period | No | Yes |
| Interrupted feeding transmits | Yes, highly effective | Very rarely |
| Viruses transmitted | *Papillomavirus* *Poxvirus* | *Alphavirus* *Flavivirus* Bunyamwera supergroup *Rhabdovirus* (some) *Orbivirus* |

Mechanical transmission of enteric viruses by contamination of food by flies might be expected to occur (Ward *et al.*, 1945; Riordan *et al.*, 1961), but there is no evidence that flies are as important as natural vectors of enteric viruses as they are of enteric bacteria. However, a number of diseases of animals that are characterized by lumps in the skin are transmitted mechanically by biting arthropods. Some, like rabbit papilloma and rabbit fibroma (Dalmat, 1958, 1959), are localized diseases; others, like fowlpox and myxomatosis in the European rabbit (*Oryctolagus cuniculus*), may be generalized infections. Mechanical transmission by arthropods is an effective method of spreading viruses within a single species, as in rabbit papilloma in *Sylvilagus floridanus*, or both within and between species, as in myxomatosis in *Sylvilagus bachmani* and *Oryctolagus cuniculus* (Fenner and Ratcliffe, 1965). The virus acquired during probing through infected cells of the skin or subcutaneous tissue, or rarely (as in infectious equine anemia) from a blood pool, is transmitted when virions are dislodged from the contaminated mouthparts into the skin or subcutaneous tissue of the susceptible host.

Propagative transmission by arthropod vectors is the principal mode of transfer of a large number of viruses, which are called the arbo- (arthropod-borne) viruses. The requirement that these viruses must multiply in their arthropod vectors imposes an additional barrier of specificity, which many different viruses, from several taxonomic groups (Table 16–3), have successfully overcome. Multiplication of the ingested virus and its spread through the insect take some time, so that an interval of several days, called the extrinsic incubation period, elapses between the acquisition and transmission feeds.

It is not enough that a virus should be able to multiply in insect tissues. Some viruses will multiply after injection into the hemocele of nonvector mosquitoes but are not able to penetrate the gut wall [e.g., Murray valley encephalitis virus in *Anopheles annulipes* (McLean, 1955)]; other insects are susceptible nontransmitters (e.g., yellow fever virus in *Taeniorhynchus fasciolata*; Whitman and Antunes, 1937), probably because the virus that infects the mosquito after feeding and multiplies in the arthropod tissues cannot enter the salivary gland.

In the case of plant viruses, arthropod or nematode transmission is of major importance in overcoming the barrier of distance, for apart from passive spread of the seeds plants are immobile. The mobility of different vertebrates varies greatly. Some have small and restricted territories beyond which they rarely move; at the other extreme man (with the help of mechanical contrivances like motor cars and airplanes) is constantly on the move and rapidly traverses great distances. Many insects are highly mobile but they may move over smaller distances than the vertebrates to which they transmit arboviruses. However, there are a number of examples of the long distance transfer of viruses by flying vectors. Such events can be recognized only when foci of a novel and obvious disease are established in a virgin population. Two examples may be quoted. Ephemeral fever is an arboviral infection of cattle, probably spread by *Culicoides*. In 1967–1968 an epidemic occurred in Australia after an absence of 10 years, and the disease spread from the north over a distance of some 3000 miles in a period of 5 months. The pattern of its entry into and spread through the southern part of the continent was accurately predicted on the basis of the prevailing wind patterns, which presumably carried infected vectors over long distances (Murray, 1970). Some years earlier, during the early spread of myxomatosis in Australian wild rabbits, several isolated outbreaks were observed hundreds of miles away from the nearest focus of the disease; the passive carriage of infected mosquitoes on favorable winds appears to be the most probable explanation of these occurrences (Fenner and Ratcliffe, 1965).

Arthropod transmission provides a very effective way for a virus to cross species barriers, since the same arthropod may bite birds, reptiles, and mammals that rarely or never come into close contact in nature. Propagative transmission by arthropods may provide a reservoir of viruses in nature that is independent of the infection of vertebrates, since some viruses, such as the tickborne flaviviruses, may be transovarially transmitted in ticks, in which they produce inapparent infections (Burgdorfer and Varma, 1967).

## Shedding

Although lesions are produced in the skin in several localized and generalized diseases (see Table 9–3), relatively few viruses are shed from the skin lesions in a way that leads to transmission. Virus is not excreted from the maculopapular skin lesions of measles or rubella (although the enanthem is an important source of virus, and viruria occurs), or from the rashes associated with some picornavirus and togavirus infections. Smallpox and herpesvirus infections, on the other hand, produce a vesiculo-pustular rash in which virus is plentiful in the fluid of the lesions. Smallpox virus is viable in dried scabs for up to 6 months at 30°–40°C and as long as 17 months at 20°–25°C (MacCallum and McDonald, 1957), and the dust and bedlinen in smallpox hospitals are heavily contaminated with infectious virus. Contamination of the air in the vicinity of smallpox patients is due to relatively large particles of infected dust from bedlinen rather than fine droplet nuclei coming from the upper respiratory tract (Downie *et al.*, 1965).

Several herpesviruses are shed from skin lesions; for example, the occasional case of varicella that arises from exposure to a patient with herpes zoster acquires infection from the zoster rash, and Marek's disease of chickens is spread by virus shed from infected feather follicles (Nazarian and Witter, 1970).

## THE RESPIRATORY TRACT

### Inhalation

In addition to air sampling at rates that vary between 24 ml/minute for a mouse to 100 liters a minute for a horse, many animals use their noses to investigate their environment and thus increase their chances of inhaling potentially infectious material. Large inhaled particles usually lodge on the mucosa of the upper respiratory tract, where the turbinates provide a complex arrangement of "baffles" which warm and humidify the entering air and prevent ready movement of large particles to the deeper parts of the respiratory tract. Smaller particles impinge on these surfaces or they are carried directly into the bronchi and then to the alveoli before they lodge against a surface. Since there is a large "dead space" within the respiratory tract particles may, of course, be inhaled and exhaled without ever coming into contact with a surface.

Much work has been carried out on the importance of particle size in determining penetration of the respiratory tract (review, Druett, 1967). In man, most particles larger than 15 $\mu$m in diameter and about one-half those 6 $\mu$m in diameter are retained in the nose, and even with mouth breathing they hardly penetrate deeper than the secondary bronchi. Most particles less than 1 $\mu$m in diameter reach the alveoli. The normal nasal mucociliary blanket rapidly removes particles that are deposited in the nose and transfers them to the pharynx. Likewise there is an upward movement of mucus from small to larger airways (reviews, Druett, 1967; Tyrrell, 1967; Kilburn, 1968). In order to reach susceptible cells, virus must first penetrate the mucous blanket, and this is more likely to happen in the distal parts of the lung than in the larger airways or the nose. Viruses that are infectious by the respiratory route must attach to cells very rapidly, since only a few particles suffice to produce infection when given as a small intranasal drop, or as an aerosol. Experimental infections of man with coxsackievirus A21 show that if given as nasal drops it produces nasal colds (Parsons *et al.*, 1960) but when even small doses are given as a fine aerosol the virus produces a respiratory disease with marked lower respiratory tract symptoms and signs such as never occur after the administration of drops (Knight *et al.*, 1963).

Man and other gregarious animals are subject to infection with a large number of viruses that rarely spread beyond the respiratory tract. These respiratory viruses belong to five genera: *Adenovirus, Orthomyxovirus, Paramyxovirus, Coronavirus,* and *Rhinovirus.* Several other viruses initiate infection after inhalation but cause generalized infections, usually without respiratory symptoms (Table 16–5).

TABLE 16–5

*Viruses that Initiate Infection of the Respiratory Tract*

---

1. With the production of local respiratory symptoms:
    *Adenovirus* (many species)
    *Herpesvirus* (infectious bovine rhinotracheitis)
    *Rhinovirus*
    *Enterovirus* (a few species)
    *Orthomyxovirus*
    *Paramyxovirus* (parainfluenza, respiratory syncytial virus)
    *Coronavirus*
2. Producing generalized disease, usually without initial respiratory symptoms:
    Papovaviridae: polyoma
    *Iridovirus:* African swine fever
    *Herpesvirus:* varicella
    *Poxvirus:* smallpox, sheep pox
    Togaviridae: rubella
    *Paramyxovirus:* mumps, measles
    *Arenavirus:* Lymphocytic choriomeningitis

---

Few of the respiratory viruses are very resistant to environmental conditions (review, Akers, 1969). Most human respiratory viruses, for example, lose infectivity if dried in air so that infected droplets or droplet nuclei, if they are to be infectious, must spread directly from one individual to another. Respiratory viruses are therefore much more important as infections of gregarious than of relatively dispersed animals. Natural infections with respiratory viruses are usually but not necessarily confined to a single host species.

## Shedding

Many different viruses that cause either localized disease of the respiratory tract or generalized infections are expelled from the respiratory tract. Much of the extensive work carried out on airborne infection with pathogenic bacteria (review, Wells, 1955) applies to viruses. Large droplets expelled from the respiratory tract fall rapidly to the ground; smaller droplets sediment slowly and because of their relatively large surface area evaporation is so rapid that before they reach the ground they are reduced to small droplet nuclei which can remain airborne indefinitely.

In man, the relative importance of coughing, sneezing, talking, and breathing in producing infective aerosols depends upon where in the respiratory tract the virus is multiplying (review, Burrows, 1972). In general, little virus is shed by normal respiration or talking, coughs are important in infections of the lower respiratory tract and sneezes in infections of the upper respiratory tract. Other animals with long nasal passages may generate aerosols during normal respiration, and foot-and-mouth disease virus, for example, can be readily recovered from the air in the immediate vicinity of animals in the early clinical stages of disease (Sellers *et al.*, 1970).

Although large amounts of all respiratory viruses are swallowed, the only ones found in the feces are adenoviruses or the occasional enterovirus that infects the respiratory tract, since enveloped respiratory viruses and rhinoviruses are inactivated by the acidic stomach contents or by the bile salts encountered in the duodenum.

Epidemiological evidence shows conclusively that viruses responsible for the acute exanthemata of man (measles, rubella, varicella) are not excreted from the respiratory tract until late in the incubation period despite the fact that this is the site of their primary implantation. However, large amounts of virus are shed from the oropharyngeal lesions of the enanthems that are present at the end of the incubation period and during the first few days of illness.

## THE ALIMENTARY TRACT

### Ingestion

The third major surface of the body exposed to material from the environment is the alimentary tract (Table 16–6), which constitutes a very large surface area of cells: stratified epithelium in the mouth and esophagus and columnar epithelium in the intestinal mucosa. In general, infection of the intestinal tract is initiated by viruses from feces, which are resistant to the acid, bile salts, and

TABLE 16–6

*Viruses that Initiate Infection of the Alimentary Tract*

---

1. Without disease, or with production of local symptoms only:
   *Adenovirus:* avian and a few human
   Picornaviridae: some enteroviruses
   Togaviridae: bovine virus diarrhea
   *Coronavirus:* transmissible gastroenteritis of swine
   *Reovirus*
   *Parvovirus*
2. Producing generalized diseases, usually without local symptoms:
   Picornaviridae: many enteroviruses
   Unclassified: hepatitis virus A

---

enzymes that occur in the gut. Enteroviruses, several adenoviruses, some parvo-viruses and reoviruses, the coronavirus responsible for transmissible gastroenteritis of swine (see Chapter 9), and hepatitis virus A are usually spread by the alimentary route.

### Excretion

Although viruses transmitted by the alimentary route may be inactivated in feces dried in sunlight, they are in general more resistant to inactivation by environmental conditions than the respiratory viruses, especially when suspended in water, e.g., in water supplies contaminated with sewage (review, Berg, 1967). Thus, unlike the respiratory viruses, which must spread directly from infected to susceptible individuals, viruses transmitted by the alimentary route can persist for some time outside the body. Prolonged excretion of enteric viruses may occur in the feces, but this does not necessarily mean that they continue to multiply in cells of the intestinal tract, for excretion could occur via the bile.

A few viruses are excreted from the alimentary tract not in the feces, but in the oral secretions, notably cytomegalovirus, herpes simplex virus type 1, and rabies virus in saliva.

### THE EYE AND THE GENITOURINARY TRACT

Neither of these mucous surfaces is of much importance as a route of entry of viruses. Adenovirus type 8 causes epidemic keratoconjunctivitis in man, the only clinical disease caused by Newcastle disease virus in man is conjunctivitis, and herpes simplex virus is one of the most common infectious causes of blindness. When conjunctivitis occurs as part of a systemic illness, as in measles, the virus has reached the eye via the blood stream.

Venereal transfer of viruses is uncommon. In man, it is restricted to herpes simplex virus type 2 (Nahmias and Dowdle, 1968) and genital warts, but several herpesviruses are transmitted venereally in animals (Table 16–7; review by Nahmias, 1971a).

The urinary tract is not known to be a portal of entry of virus but excretion in the urine is an important mode of contamination of the environment in several diseases of laboratory animals, such as lymphocytic choriomeningitis and polyoma in mice, and it may well be important in human diseases also (Table 16–7). Viruria occurs in many generalized infections, such as mumps and measles (Utz, 1964), and is often prolonged in cytomegalovirus infections and congenital rubella.

### VERTICAL TRANSMISSION

The routes of transmission described so far apply to postnatal individuals of the same or different animal species. This is sometimes called *horizontal transmission,* to differentiate it from *vertical transmission* (Gross, 1944, 1970) which refers to the transfer of virus from an individual of one generation to its off-

TABLE 16–7

*Viruses That Initiate Infection of the Eye or Genitourinary Tract*
*or Are Excreted in Urine*

| | |
|---|---|
| 1. Ocular infections | *Adenovirus:* human type 8 and several others |
| | *Herpesvirus:* herpes simplex type 1 |
| | *Poxvirus:* accidental-vaccinia |
| | *Paramyxovirus:* in man, Newcastle disease virus |
| 2. Venereal infections | *Papillomavirus:* human warts, rarely |
| | *Herpesvirus:* herpes simplex type 2, infectious pustular vulvovaginitis, coital exanthem virus, canine herpes |
| 3. Excretion in urine (viruria) | Papovaviridae: polyoma virus |
| | *Herpesvirus:* cytomegalovirus |
| | Togaviridae: rubella |
| | *Paramyxovirus:* measles, mumps |
| | *Arenavirus:* most species |
| | Unclassified: hepatitis virus B |

spring, then to the offspring's progeny and so on. The most important routes of vertical transmission are in the germ plasm, for viruses whose genomes are integrated (see Chapter 14), in the cytoplasm of the egg, across the placenta to the fetus (see Chapter 9), or to the suckling in milk (Table 16–8).

Vertical transmission via the germ plasm, although long suspected, has only recently been convincingly demonstrated (Huebner and Igol, 1971; Todaro and Huebner, 1972; Bentvelzen *et al.*, 1970). It occurs only with the leukoviruses of the leukosis-leukemia and mammary tumor virus subgenera. It will be recalled that these viruses have a reverse transcriptase that is capable of transcribing DNA from their RNA genome (Chapter 14). There is now convincing evidence for mice and chickens, and strong circumstantial evidence for other rodents, that the DNA copy of the genome of these viruses is integrated into the cellular genome and is inherited as such. It may never spontaneously evince its presence, though by suitable manipulation it can be induced to do so, or it may be expressed either partially (as "group-specific" antigen, see Chapter 14) or as complete virions.

Cells carrying unexpressed or partially expressed leukovirus genomes (*gs* antigen) can be readily superinfected with leukovirus virions, and this may also occur congenitally, or with mammary tumor virus, in the milk. Such congenital transmission of virions has been convincingly demonstrated in chickens (Rubin *et al.*, 1961, 1962; see Chapter 12). In the aggregations of chicks found in modern brooders, horizontal transmission also occurs readily, for virus is excreted by congenitally infected chickens in both saliva and feces. Noninfected chicks may be protected for a few weeks by maternal antibody, and if they are then infected the disease is self-limited and they rarely lay infected eggs. Natural horizontal transmission of mouse leukemia viruses does not seem to occur.

As Bittner (1936) first demonstrated, the expressed form of mammary tumor virus (MTV) (i.e., as virions) may be vertically transmitted in the milk, but many, possibly all mice also carry an integrated MTV genome (Chapter 14).

TABLE 16–8

*Vertical Transmission of Viruses*

---

1. As integrated genome, in the germ plasm
     *Leukovirus:* leukosis-leukemia virus subgenus: birds, mice, cats, hamsters, etc.
     Mammary tumor virus subgenus: mice
2. As viral particles
     a. In cytoplasm of egg
          *Leukovirus:* chicken leukosis virus
          *Arenavirus:* lymphocytic choriomeningitis virus
     b. Across placenta
          *Herpesvirus:* cytomegalovirus
          Togaviridae: rubella virus, hog cholera virus
          *Arenavirus:* lymphocytic choriomeningitis
          *Orbivirus:* bluetongue
     c. In milk
          *Leukovirus:* mammary tumor virus
     d. In saliva, or recurrent skin vesicles; associated with long-term recurrent infections
          *Herpesvirus:* herpes simplex type 1, varicella-zoster

---

When a strain of MTV is transferred by foster-nursing to another line of mice its genome is not integrated and further vertical transmission of that viral strain is limited to the milk.

Several other viruses are transmitted vertically via the ovum and/or placenta (lymphocytic choriomeningitis, rubella, cytomegalovirus; see Chapter 9).

Herpes simplex virus type 1 can be transmitted "vertically," for virus can spread from a mother to her susceptible infant by salivary contamination. Then, because of its long latency and periodic recurrence, progeny of the same virus may again be transferred to the next human generation. In small isolated human populations, varicella-zoster-varicella may constitute a rather similiar cycle of "vertical" transmission (see below).

## THE EPIDEMIOLOGICAL IMPORTANCE OF IMMUNITY

In the foregoing pages we have categorized the routes of infection and excretion of viruses, and seen that despite exceptions, there is a general pattern of correspondence between the taxonomic groups and the epidemiological classifications of enteric, respiratory, and arboviruses. It is more difficult to make general statements about other aspects of the epidemiology of viral infections, partly because of their varied patterns of pathogenesis (Chapters 9 through 12).

One factor that warrants particular attention is the epidemiological importance of immunity, during the discussion of which the epidemiology of several exemplary viral infections will be described. Our attention will be focused on humoral immunity, mediated by antibodies, which is much the most important arm of the immune response in relation to reinfection, as distinguished from recovery from a primary infection.

## Active Immunity in Superficial Infections

The major antibody species evoked by infections of superficial membranes is the secretory immunoglobulin, IgA (Chapter 10), which plays a major role in the immunity to reinfection that is found for a period of months or perhaps a few years after first infection. Antibody in the serum is either irrelevant or of minor importance in such diseases (enteroviruses, Bodian and Nathanson, 1960 and Ogra, 1971a; respiratory viruses, Fazekas de St. Groth, 1950, Chanock, 1971, and Waldman *et al.*, 1971), or may even exacerbate disease (Chanock *et al.*, 1970).

**Common Colds.** Viruses of several taxonomic groups cause common colds (Fig. 9–3). Considering only the Picornaviridae, a large number of serotypes of rhinoviruses and a few enteroviruses can produce superficial infections of the mucous membrane of the upper respiratory tract. The serotypes are quite distinct antigenically and show no serological cross-reactivity, or any cross-protection. The seemingly endless succession of common colds suffered by most members of urban communities reflects a series of minor epidemics, each due to one of the several sorts of virus that can produce this syndrome or to one or other of the ninety serotypes of rhinoviruses. Only one cross-reacting strain [i.e., two antigenic subtypes of a single serotype (Kapikian *et al.*, 1967)] has been detected among the rhinoviruses. It is possible that sufficient antigenic novelty to ensure survival as agents causing short-lived superficial infections, followed by local immunity lasting several months, is achieved only by the production of mutants that show no cross-reactivity with the parental strain. Understanding of the evolution of multiple distinct serotypes of rhinoviruses (and enteroviruses), in each of several species of animal, will have to await chemical analyses of their coat proteins or the determination of their nucleic acid homologies.

However, epidemiological observations on isolated human communities illustrate both the need for susceptible subjects and/or antigenically novel viral serotypes to maintain respiratory diseases in nature, and the importance of repeated (often subclinical) infections in maintaining herd immunity. Explorers, for example, are notably free of respiratory illness during their sojourn in the Arctic or in Antarctica, despite the freezing weather, but invariably contract severe colds when they again establish contact with their fellow men. Even in larger communities it is clear that colds disappear and immunity to them falls rapidly when the community is isolated from the outside world. As Fig. 16–1 illustrates, there were no colds in Longyear City throughout the long Arctic winter, when the harbor was closed, but an epidemic began as soon as the first ship arrived at the island in spring. A similiar picture has been demonstrated on the island of Tristan da Cunha, which is visited somewhat more frequently (Shibli *et al.*, 1971). However, some agents that produce superficial respiratory infection may survive in very small isolated communities. Coxsackievirus A21 was introduced artificially into a group of thirteen men isolated for 10 months in Antarctica. It caused symptoms in the four inoculated individuals and in one contact; and although no subsequent disease occurred, all except one individual showed substantial rises in serum antibody titers at various periods during the

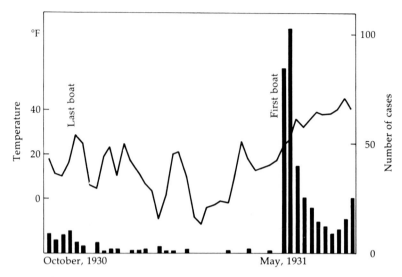

Fig. 16–1. *Acute respiratory disease in Longyear City in Spitzbergen during 1930–1931. The tracing shows the mean weekly minimum environmental temperature; the histogram shows the number of cases of respiratory disease, which virtually disappeared during the long Arctic winter, when the harbor was icebound, but reappeared in epidemic incidence shortly after the arrival of the first boat in spring. (From Paul and Freese, 1933.)*

subsequent 9 months. Five individuals showed increased titers (over those found 2 months before) on two occasions (Holmes *et al.*, 1971).

**Influenza.** In terms of the importance of secretory antibodies (Couch *et al.*, 1971) and the size and mobility of the populations affected, influenza viruses resemble the rhinoviruses. Instead of responding by the establishment in the community of a large number of distinct serotypes, both influenza A and influenza B have undergone immunoselection that has produced antigenic drift in their peplomer antigens (see Chapter 17). The strains of influenza virus circulating in any epidemic are antigenically similar, usually over the whole world, and differ from those circulating in previous epidemics, although usually they closely resemble the latter (Fig. 17–3). Exceptionally, and only with influenza A, novel serotypes appear and spread around the world, replacing the existing strains and causing major pandemics (Fig. 16–2). They probably originate in part or wholly from some animal reservoir (see Chapter 17).

**Respiratory Syncytial (RS) Virus Infections.** A third kind of epidemiological situation is found with RS virus, which causes sharp epidemics of severe bronchiolitis in very young infants every winter (Fig. 16–2). Although serological variants have been recognized using antisera produced in animals (Coates *et al.*, 1966), human sera are broadly cross-reactive to all antigenic "variants." Respiratory syncytial virus survives in nature in spite of remaining essentially constant in its antigenic make up. Secretory antibody is important for protection, but wanes rapidly, so that reinfection with viral excretion occurs relatively commonly

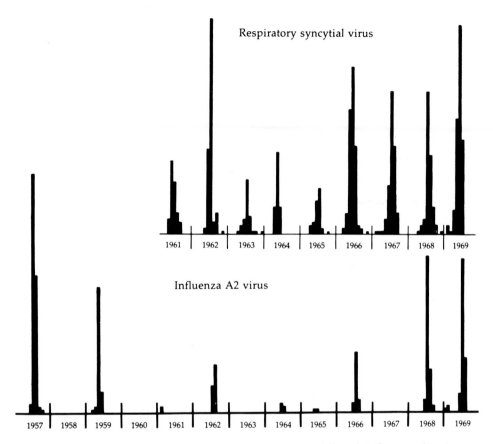

FIG. 16–2. *Epidemic occurrence of respiratory syncytial and influenza A2 viruses. The histograms show the monthly isolations of these two viruses from patients admitted to the Fairfield Hospital for Infectious Diseases, Melbourne, over the periods indicated. Compare the regular annual epidemics of respiratory syncytial virus (predominantly in infants) with the irregular occurrence of winter epidemics of influenza, which produced major peaks in 1957 (first experience of A2 virus) and again in 1968 and 1969 (Hong Kong variant of A2 virus). (The Australian winter runs from June to August.) (Courtesy Dr. A.A. Ferris.)*

in older children and in adults (Beem, 1967), who presumably bring the virus into the home and infect young infants.

The pattern of persistence of parainfluenza viruses is similar; in infections of superficial mucous membranes with agents of high infectivity, and immunity that wanes rapidly after an attack, antigenic novelty does not appear to be necessary to ensure viral survival in large and mobile host populations.

## Active Immunity in Generalized Infections

In generalized infections, except those in which the virus is introduced by inoculation or trauma, antibodies play a role in two situations; on the mucosal surface (respiratory or alimentary) where initial infection is established, and in the tissues and bloodstream during the viremic phase. The role of antibodies on the mucosae has just been discussed; in general such local immunity is relatively short-lived, whereas in generalized viral infections immunity to clinical disease is characteristically life-long. This is explicable on the basis of the effective "memory" of IgG-producing cells, which leads to a secondary response during the long incubation period (MacLeod, 1953), and the fact that serum and tissue antibodies are strategically placed to neutralize circulating virus before the secondary localization needed for symptom production can occur.

**Measles.**  Virtually all primary infections with measles virus cause disease, and the clinical picture is so distinctive that the diagnosis can be made without laboratory assistance. The virus is monotypic in time and space, and immunity persists for life. Although there is good serological evidence for recurrent inapparent reinfections (Stokes et al., 1961), Panum's experience (Chapter 10) illustrates that prolonged protection is not dependent on such reinfections.

Except for a few isolated populations, and apart from vaccination, measles is endemic in man throughout the world, its incidence rising to epidemic levels at regular intervals in a pattern that depends upon climate, season, and especially the emergence of new generations of susceptibles. Newborn infants are protected by maternal antibody for 6 to 9 months after birth, and the peak age incidence depends upon local conditions of population density and the chance of exposure (Babbott and Gordon, 1954). In urban communities, the disease occurs in epidemic proportions about every 2–3 years, exhausting most of the currently available susceptibles, and on a continental scale, in the United States of America, epidemics occur annually (Fig. 16–3). The age distribution is primarily that of children just entering school, with a peak of secondary cases (family contacts) about 2 years old. The cyclic nature of measles outbreaks is determined by several variables, including the build up of susceptibles, introduction of the virus, and environmental conditions which promote its spread. An urban population with more than 40% susceptibles can expect to experience an epidemic which will continue until the susceptibles are reduced to about 20%. In London there is a fluctuating seasonal incidence, with a peak every second winter—this is probably related more to opportunities for infection than to fluctuations in host susceptibility or changes in the virus. For example, the regular biennial periodicity observed between 1917 and 1940 was interrupted when the expected 1940 epidemic failed to appear, apparently because of the closing of the London schools due to the bombing attacks (Stocks, 1942).

When measles is introduced into a fully susceptible population it spreads rapidly, whatever the season, and attack rates are almost 100% (976 per 1000 exposed persons in southern Greenland in 1951, in an epidemic which ran through the entire population in about 40 days; Christensen et al., 1953). Such virgin-soil epidemics may have devastating consequences (review, Black et al., 1971); the severity appears more likely to be due to lack of medical care and the disruption of social life than to a higher level of genetic susceptibility. However, Black et al. (1971) found that populations of American Indians not previously exposed to measles reacted to measles vaccine with slightly higher fevers than did measles-experienced populations, possibly because of greater genetic susceptibility.

Measles virus has no animal reservoir, it is followed by permanent immunity, and recurrent excretion is unknown. The discovery that persistent infection of the brain with measles virus causes subacute sclerosing panencephalitis (Chapter 12) does not affect the epidemiological situation, for there is no evidence that cases of this disease ever excrete measles virus after their acute attack of measles. Persistence of the virus in a community depends upon a continuous supply of susceptible human beings. With an incubation period of about 12 days and maximum viral excretion early in the disease, at least thirty new susceptibles a year would be needed to maintain the virus, if the cases were evenly spaced. Obviously many more than this would be needed to maintain endemicity. Bartlett (1957, 1960) calculated that in urban areas about 2500 cases per year is the minimum needed to prevent fade-out, i.e., breaks in the continuity of transmission. He checked this estimate against data from British and American cities, and found that fade-out occurred when there were less than 4000–5000 cases per year. The minimum city population size for this number of cases was about 250,000. Island communities provide even better material to study the effects of population dispersion and reintroduction on the endemicity of measles. Black (1966) analyzed data from nineteen island communities over a period of about 15 years (Table 16–9). Measles disappeared for periods (i.e., fade-out occurred) in all islands with populations of less than half a million, and the effect of population size on the frequency of fade-out is clear from the figures. Guam and Bermuda, which form exceptions to the pattern, have large transient military or tourist populations and frequent air connections to large cities, i.e., they are insufficiently isolated to be comparable with other island populations with a similar input of susceptibles.

Widespread vaccination against measles is greatly modifying the epidemiological situation. Griffiths (1971) predicts that a partially effective vaccination campaign would have the following effects: (a) the interval between epidemics

TABLE 16–9

*Endemicity of Measles in Islands with Populations of 500,000 or Less, All of Which Had at Least Four Exposures to Measles during 1949–1964* [a]

| ISLAND | POPULATION | ANNUAL POPULATION INPUT [b] | PERCENTAGE MONTHS WITH MEASLES (1949–1964) |
|---|---|---|---|
| Hawaii | 550,000 | 16,700 | 100 |
| Fiji | 346,000 | 13,400 | 64 |
| Samoa | 118,000 | 4,400 | 28 |
| Guam | 63,000 | 2,200 | 80 |
| Tonga | 57,000 | 2,040 | 12 |
| Bermuda | 41,000 | 1,130 | 51 |

[a] Modified from F. L. Black (1966).
[b] 1956 Births less infant mortality.

would increase and, relative to the size of the (now smaller) susceptible population, epidemics would tend to be larger after vaccination than before, (b) the average age of attack would be increased, and (c) the critical community size needed to maintain endemicity would be increased dramatically; fourfold with 50% vaccination and by a factor of 100 with 90% vaccination. This would greatly enhance the possibility of measles eradication from all but the largest continental populations.

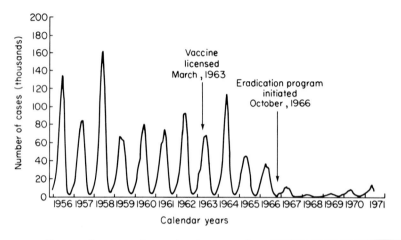

FIG. 16–3. *Reported cases of measles by 4-week periods in the United States of America, showing the effects of vaccination programs. (Center Dis. Cont., 1971.)*

Data is now available on the experience in the United States during the 8 years since live attenuated measles virus vaccine was licensed in 1963 (Fig. 16–3). The incidence of measles decreased greatly up to 1968 but has risen somewhat every year since then, probably because the levels of measles vaccination have been falling steadily since mid-1969, to levels as low as 42% in those under 4 years of age in urban poverty areas. Immigration into areas where intensive vaccination has been carried out may bring in both measles and a substantial number of nonvaccinated children. Outbreaks then occur and a few cases are found in vaccinated children (Scott, 1971), pointing up the difficulty of eradicating measles even on a national scale and in an affluent country.

Apart from reintroduction, the pattern of dispersal of the population will obviously have an effect on the endemicity of measles, since the duration of individual epidemics is correlated inversely with population density. Figure 16–4, constructed for seven island communities of comparable population size, suggests that with maximal crowding an epidemic would be spent in about 4 months in an island population with an input of 2000 to 4000 susceptibles per year. Widespread vaccination, by diminishing the number of susceptibles, would in effect increase their dispersal, and thus increase the duration of epidemics.

**Varicella.** Other monotypic human generalized infections, like rubella, mumps, and smallpox, follow a pattern similar to measles, although because of

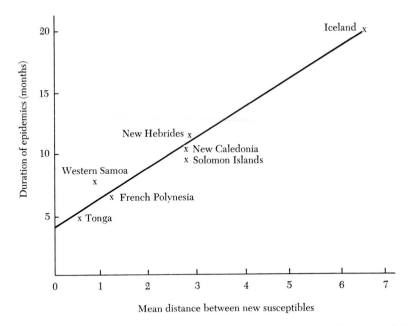

FIG. 16–4. *Relation between the average duration of measles epidemics and the degree of dispersion of populations, in island communities with about 2000–4000 new susceptible children per year. The abscissa plot*

$$\sqrt{\dfrac{1}{\dfrac{input}{kilometers^2}}}$$

*represents the mean distance between new susceptible persons. (From F.L. Black, 1966.)*

the larger proportion of subclinical infections it is impossible to obtain data of the kind available with measles. Smallpox spreads less readily than measles, and vaccination of a majority of the population has been adequate to eradicate it from all advanced countries. Varicella follows a similar pattern in relation to immunity to exogenous reinfection, but has a dramatically smaller critical community size for survival—less than 1000, compared with 250,000 to 500,000 for measles (Black, 1966). This is explained by the fact that varicella virus, after being latent for many years, may be reactivated and cause zoster (Hope-Simpson, 1965; see Chapter 12). Although it is not as infectious as chickenpox itself (attack rates of 15, compared with 70% for varicella; Gordon, 1962), zoster can, in turn, produce chickenpox in susceptible children.

**Poliomyelitis.** Poliomyelitis is a specifically human disease that may be caused by any one of three serotypes of poliovirus. Infection is usually inapparent and confined to the alimentary tract, with paralytic disease as a rare complication, so that case-to-case spread of the paralytic disease is rarely observed. The polioviruses are spread mainly by the fecal–oral route, although the pharyngeal–oropharyngeal route is probably important in older individuals in societies with a high standard of hygiene. The incubation period, as judged by the occurrence of paralysis after presumed single exposures, is said to vary between 3 and 35 days, with most cases developing after a 4- to 10-day period (Horstmann and Paul, 1947). Investigations with vaccine strains showed that multiplication and excretion can occur within 24 hours of the ingestion of virus, and contact infections can be recognized by fecal culture 3 days later (Paul et al., 1962).

Local secretory immunoglobulins play an important but relatively transient role in protecting against homotypic reinfection of the alimentary tract. Protection against paralytic disease, mediated by serum antibodies, is homotypic and prolonged. The epidemiological picture in primitive societies where infection is endemic is dominated by immunity acquired in early childhood. Epidemics in virgin-soil populations show that all age groups experience a high infection rate but paralysis is most common in adolescents and young adults. In an epidemic among Eskimos, for example, 42% of the 15–19 year age group died or were paralyzed, but there were no cases of paralysis in infants under 3 years (Adamson et al., 1949).

The appearance of paralytic poliomyelitis as an epidemic disease in the advanced Western countries, beginning about 1900, coincided with changing environmental conditions, such as increased standards of sanitation and hygiene, which postponed infection from early to late childhood (Paul, 1958). The same pattern is now being recognized in less advanced tropical countries, except where vaccination has altered the situation (Cockburn and Drozdov, 1970).

Vaccination has eradicated paralytic poliomyelitis from many parts of the world. Two types of vaccine have been used: Salk (inactivated; given by injection) and Sabin (live-attenuated, given orally) (see Chapter 15). Killed virus vaccines have virtually no effect on person-to-person circulation of polioviruses in primitive communities, although they may greatly diminish paralysis by interfering with blood-borne spread of the virus through the body. Live-virus vaccines, however, interfere with the circulation of polioviruses in the community at the time of vaccination (Hale et al., 1961); and for a prolonged period after vaccination they prevent the establishment of poliovirus infection in the gut (Paul et al., 1959; Sabin et al., 1960), probably because of local production of secretory antibodies (see Chapter 10). Mass vaccination with live-virus vaccines thus offers the possibility of interfering with the circulation of polioviruses to the point of at least local eradication, even in communities in which sanitation is poor (Sabin, 1962). The circulation of polioviruses has also been greatly reduced in advanced Western communities well vaccinated with killed virus vaccines. For instance, in Sweden, where only killed-virus vaccines have been used since 1957, paralytic poliomyelitis has disappeared since 1962 and poliovirus isolations have been very uncommon (Gard, 1967). A possible explanation is that in such communities transmission of polioviruses occurs mainly by the pharyngeal–oropharyngeal rather than the fecal-oral route. Antibody produced by Salk vaccination reduces pharyngeal infection with polioviruses and viral multiplication in that site (Wehrle et al., 1961; Howe, 1962).

## Passive Immunity

The unprotected newborn of most species of mammal are very vulnerable to almost all infectious agents, and temporary protection conferred by passively transferred antibody is vital in ensuring their survival in nature. This antibody can be transferred by several routes (see Chapter 10); it may be effective by its presence and distribution in the blood or in a few cases by its local effects within the lumen of the gut (lactogenic immunity).

**Herpes Simplex.** Facial herpes ("fever blisters") is one of the most common diseases caused by a single viral serotype. It is found in all human populations, with a pronounced familial distribution and a frequency that is related to socioeconomic conditions. The stimulus for recrudescence of viral activity varies in different individuals (see Chapter 12). Infection first occurs in early childhood, as an inapparent infection or an acute gingivostomatitis, occasionally with severe constitutional symptoms (Dodd *et al.*, 1938; Burnet and Williams, 1939). Virus is present in the saliva and may be found in feces (Buddingh *et al.*, 1953). Infants are usually infected with virus excreted by their parents during a recurrent attack of facial herpes, and the long period over which such recurrent infections occur make it possible for this virus to survive in nature even in small isolated "family unit" communities. On the other hand, where numbers of susceptible young children gather together, horizontal spread is common, and here the effects of passive immunization are clearly apparent (see Chapter 10).

Transmissible gastroenteritis of swine provides a good example of lactogenic immunity (see Chapter 10).

## Serological Surveys

The viral experience of a community can be monitored by screening tests for the appropriate antibodies. The proportion of antibody-negative individuals in different age groups gives a valuable measure of the vulnerability of the population to infection with monotypic viruses or those with a few serotypes that do not cross-react. With other viruses interpretation of the results of serological surveys must take into account a number of factors: (a) serum antibody may not reflect the level of protection, e.g., in parainfluenza or respiratory syncytial virus infections (C. Smith *et al.*, 1966), (b) a particular syndrome may be caused by several serologically different strains of a particular virus and, indeed, by viruses belonging to different groups (see Fig. 9–3), (c) the serum antibody level detectable by the technique used may fall more rapidly than the level of protection, e.g., in herpesvirus infections (Anderson and Hamilton, 1949), (d) because of cross-reactivity, the presence of "neutralizing antibody" may not necessarily indicate protection, even with viruses that have a viremic stage, and (e) original antigenic sin (see Chapter 10) may mask the serological response of older individuals to recent infections, for example, with influenza virus and togavirus infections.

It is widely recognized that in many infections complement-fixing antibody is an index of recent infection; its titer generally drops to low levels within a few months of infection whereas the titer of neutralizing antibody may remain elevated for years.

## MULTIPLE-HOST INFECTIONS: THE VIRAL ZOONOSES

The infections considered so far in this chapter have been maintained in nature within populations of a single vertebrate species. A number of viruses can spread naturally between different species of vertebrate host; the term *zoonosis* is used to describe "multiple-host" infections that are naturally transmissible from lower animals to man (Schwabe, 1969). Table 16–10 lists the viral zoonoses, of which the largest group is arboviral infection. A wide variety of animal reservoir hosts and arthropod vectors play a role in the maintenance of arboviruses in nature, and most human arbovirus infections originate from these reservoirs. Other zoonoses listed in Table 16–10 are primarily infections of domestic or wild animals transmissible only under exceptional conditions to humans engaged in particular operations involving close contact with animals.

In the zoonoses, man is the "sentinel animal;" parallel situations exist with domesticated or laboratory animals, which when appropriately exposed may acquire diseases not normally transmitted horizontally within the recipient (sentinel) species.

Most of the zoonoses have complex epidemiological patterns, often involving several vertebrates as well as the arthropod vectors. Four examples of such multiple-host infections will be briefly described: yellow fever, because it was long thought to be a single (vertebrate) host infection, but proved to be a zoonosis; tickborne flavivirus encephalitis in which there are alternate routes of human infection, by arthropod bite and by ingestion; and two nonhuman multiple-host diseases, pseudorabies and myxomatosis.

### Yellow Fever

By definition (World Health Organ., 1961) all arboviruses have at least two hosts, one a vertebrate and one an arthropod; most have several potential vertebrate and arthropod hosts. Yellow fever was once thought to be a disease peculiar to man and transmitted only by the urban mosquito, *Aedes aegypti*. This situation (Fig. 16–5) was true for urban yellow fever in South America and in Africa (Taylor, 1951). Although suspicions had existed earlier that this cycle, man-*Aedes aegypti*-man, might not constitute the only way in which yellow fever was maintained in nature, the unequivocal demonstration of jungle yellow fever in the absence of *Aedes aegypti* and of man dates from the work of Soper *et al.* (1933), and implicates monkeys as the important reservoir of the infection in nature. Intensive investigations involving the collaboration of mammalogists, ornithologists, and entomologists as well as virologists and epidemiologists were mounted in both Africa and South America (reviews, see Taylor, 1951; World Health Organ, 1971a). The ecological situation in jungle yellow fever is still not completely elucidated, but the pattern of transfer of infection everywhere appears to be the same as in the man-*Aedes*

# TABLE 16–10

*Multiple-Host Viruses, Including Those Responsible for Viral Zoonoses*

| GENUS | SPECIES | "RESERVOIR" HOST | "SENTINEL" HOST | MODE OF TRANSMISSION | |
|---|---|---|---|---|---|
| *Herpesvirus* | Pseudorabies | Swine | Cattle | Contact, through skin abrasions | |
| *Poxvirus* | Cowpox, Milkers' nodes | Cattle | Man | Contact, through skin abrasions | |
| | Orf | Sheep, goats | Man | Contact, through skin abrasions | |
| | Myxoma | *Sylvilagus* rabbits | *Oryctolagus* rabbits | Arthropod bite; mechanical | |
| *Alphavirus* | ~20 Species | Birds and mammals | Man, domestic animals | Mosquitoes | ⎫ |
| *Flavivirus* | ~30 Species | Birds and mammals | Man, domestic animals | Mosquitoes, ticks | ⎬ Arboviruses |
| *Bunyamwera* supergroup | ~80 Species | Birds and mammals | Man | Mosquitoes, phlebotomus, culicoides | ⎪ |
| *Orbivirus* | ~10 Species | Mammals | Man, domestic animals | Mosquitoes, culicoides, ticks | ⎭ |
| *Rhabdovirus* | Rabies | Canines, felines, bats, etc. | Man, cattle | Animal bite | |
| *Orthomyxovirus* | Influenza A | Horse, swine, birds | Man | Respiratory | |
| *Paramyxovirus* | Newcastle disease | Birds | Man | Contact with conjunctiva | |
| *Arenavirus* | LCM, Machupo, etc. | Rodents | Man | Respiratory | |

*aegypti* cycle, i.e., by infection of a susceptible host by mosquito bite and transfer to new mosquitoes during the short stage of viremia that occurs in vertebrate hosts.

Classic urban yellow fever has been virtually eliminated from the Americas by control of *Aedes aegypti*, but it still occurs in villages in Ethiopia and along the west coast of Africa. Jungle yellow fever, which may "overflow" to cause human disease, presents quite different epidemiological features in America and in Africa. In the large forests of tropical South and Central America, the virus is continually being moved from one area to another, so that in effect susceptible primate populations suffer recurrent epizootics. Yellow fever virus produces inapparent infections in the New World monkey *Cebus capuchinus*, which is probably the main reservoir host (Boshell, 1957), but in the forests of Central America yellow fever is associated with very high mortalities in other species of monkeys *(Ateles* and *Alouatta),* which have been virtually exterminated from those parts of the tropical American forests where *Cebus* also occurs, but persist in Guatemala and Mexico, which lie north of the habitat of *Cebus capuchinus*. Jungle yellow fever in man in Central America is mainly a disease of adult males whose work brings them into the forest, but it may occur among women and children when families are involved in crop cultivation in forest clearings, to which infected monkeys may be attracted as marauders (Causey and Maroja, 1959). The principal vectors are mosquitoes of the genus *Haemagogus*, which usually inhabit the forest canopy, although they may bite man outside and even inside houses.

In Africa, infections in nonhuman primates appear to be asymptomatic. In Uganda, jungle yellow fever (in man) involves two vector mosquitoes. *Aedes africanus* is a forest canopy mosquito which maintains an enzootic cycle in wild primates. The virus is conveyed to man by a second cycle in which monkeys raid plantations and infect local *Aedes simpsoni* mosquitoes, which then infect man and can maintain a man-mosquito-man cycle.

The common occurrence of high mortality in South American jungle primates and severe disease in South American Indians, and low mortality and mild disease in African primates (including African man), strongly supports the idea that yellow fever is an African disease which reached the Americas quite recently. Both the virus and the mosquito *Aedes aegypti* were probably transported from Africa to America on ships carrying slaves (Taylor, 1951). Since highly susceptible animal and human hosts and efficient vector mosquitoes occur in many parts of Asia, the failure of yellow fever to become established in that continent can best be explained by the supposition that it was never introduced there.

## Tick-Borne Flavivirus Encephalitis

The natural histories of most arboviruses show special characteristics, particularly in relation to overwintering (Reeves, 1961). Tick-borne flavivirus encephalitis, a disease of central Europe, illustrates two important features not found in mosquito-transmitted diseases like yellow fever (Fig. 16–5): (a) the maintenance of the virus in nature in an invertebrate host, and (b) transmission

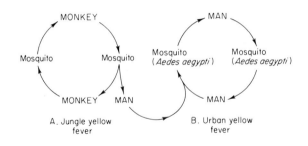

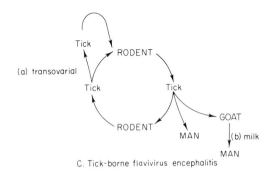

FIG. 16–5. *Cycles of arbovirus infection. (A) Jungle yellow fever, that may lead to (B) urban yellow fever. (C) Tick-transmitted flavivirus infection, with (a) transovarial transmission in ticks and (b) transmission to man in goat's milk.*

of an arbovirus from one vertebrate host to another by a mechanism other than by arthropod bite. Tick-borne encephalitis is caused by a flavivirus that is widespread in central Europe and the far east of the U.S.S.R. (Libíková, 1962). A large number of wild birds and mammals may be infected in nature; most investigators regard small rodents as the most important vertebrate reservoir. In a natural focus in Austria, Pretzmann *et al.* (1963) believed that there was a primary cycle involving wild mice and nymphal and larval ticks from which a secondary cycle developed involving nymphs and adult ticks and larger vertebrates. The secondary cycle was constantly replenished from the primary cycle and in turn provided infected larval ticks for the primary cycle, via transovarial transmission. Virus persists through the winter in ticks and in hibernating vertebrates.

Man is an incidental host for tick-borne encephalitis virus but may become involved in two ways; he may enter natural foci of the infection in pursuit of work or recreation and become infected by tick bite, or he may acquire infection from milk (Smorodintsev, 1958). Goats, cows and sheep have been shown to excrete virus in their milk (Grĕsikova and Reháček, 1959), but all milk-borne outbreaks have been associated with goat's milk. Adult and juvenile goats may acquire the virus by grazing in tick-infested pastures, and kids may be infected by drinking infected milk.

## Pseudorabies

Pseudorabies is a sporadic noncontagious disease of cattle produced by a herpesvirus. In bovines, it is regularly fatal within 2 days of the onset of illness.

Elucidation of its epidemiology by Shope (1935a,b) provides a model for the understanding of some other sporadic viral diseases. Cattle are highly susceptible but each case constitutes a blind-alley infection. Swine are also highly susceptible and sustain inapparent and highly contagious infections. The disease is recognized only when cattle are infected, which occurs if swine and cattle are maintained together, when the swine with inapparent infections transmit the virus to cattle by nuzzling and nipping. "Carrier" swine have not been recognized; the infection in these animals appears to be maintained by sequential spread. However, in view of the propensity for herpesviruses to produce recurrent disease it would not be surprising if similar recrudescent viral excretion were important in maintaining pseudorabies in swine.

### Myxomatosis in European Rabbits in America

In America, myxomatosis is maintained by mechanical arthropod transmission between certain species of *Sylvilagus* rabbits, producing a benign localized fibroma (Regnery and Marshall, 1971). Occasionally an infected arthropod bites a European (hutch) rabbit with the production of an acute, invariably lethal disease. The infected domestic rabbit may function as a source of infection for other rabbits or for arthropods (Fenner and Ratcliffe, 1965).

### SEASONS AND WEATHER

A feature of some types of viral infection is the association of cases of disease with meteorological events. In temperate climates many arbovirus infections are restricted to the summer months, when vectors are active, and the same is true of poxvirus and papilloma virus infections transferred mechanically by flying arthropods. Enterovirus infections are more prevalent in the summer, respiratory viruses in the winter. If the infectious agents are present in the community, the incidence of colds increases when the weather gets colder; a change in weather conditions is probably more important than any particular level of temperature, rainfall, or humidity. Such a change in weather may convert some inapparent infections into overt disease, and the increased excretion of virus that follows amplifies this effect by producing further infections. The crowding into restricted areas and ill-ventilated vehicles and buildings that occurs in winter may also promote the exchange of respiratory viruses. However, experiences in the Arctic (Fig. 16–1) show that cold weather alone is not enough to cause "colds," and that immunity to common colds is maintained by repeated exposure (Paul and Freese, 1933), which must often cause unrecognized infections.

Annual winter outbreaks of severe respiratory syncytial virus infections in infants are a feature of western communities in both northern and southern hemispheres, the major impact of each epidemic lasting for only 2 or 3 months (Fig. 16–2). Likewise, in spite of seeding of virus throughout the community in the summer months, epidemic influenza rarely occurs except during the winter. In contrast to the regularity of epidemics of respiratory syncytial virus infection, epidemics of influenza vary greatly in extent from year to year (Fig. 16–2). Surveys at the Great Lakes Training Center in Illinois have shown that adenoviruses spread equally well among recruits throughout the year, but twice as many recruits develop feverish illness in winter as in summer (McNamara *et al.*, 1962). Furthermore, vaccination procedures affect the incidence of respiratory disease. An experiment was conducted in which vaccination against polioviruses, adenoviruses, and influenza viruses was given to all recruits soon after their arrival at the Station, but one-half received their vaccination against smallpox, tetanus, diphtheria, and typhoid in the first 5 weeks and one-half in the second 5 weeks. Probably because of the reduced stress in those receiving staggered vaccinations, the latter showed a 20% reduction in severe respiratory illness, including pneumonia (Pierce *et al.*, 1963).

Finally, we may mention two important diseases of pigs that show a well-marked seasonal incidence, and an initial pattern inconsistent with successive spread from one animal to another, viz., swine influenza and hog cholera as they occur in the midwestern states of the United States. Once the initial cases have occurred, swine influenza is a highly contagious disease and spreads readily to other infected swine. Shope (1955, 1958, 1964) produced evidence showing that in these diseases the very different causative viruses, an orthomyxovirus and a togavirus, respectively, appeared to be widely seeded through the pig population in an occult form in swine lungworms. These small nematodes have a complex life history. The lungworm spends three of its developmental stages in its intermediate host, an earthworm, and swine acquire their lungworms by eating earthworms infected with third-stage lungworm larvae. Shope suggested that if these larvae were hatched from eggs laid by an adult lungworm while it was infecting the lung of a pig with influenza they would carry influenza virus, in an occult form. The pig acquiring the larvae (in earthworms) would then become a carrier of this occult virus, and under the influence of certain provocative stimuli the virus could be activated and produce swine influenza. In the laboratory, Shope found that multiple intramuscular injections of the bacterium *Haemophilus influenzae suis* were an effective stimulus; on the farm the appearance of disease is associated with the onset of cold, wet, inclement weather. The swine influenza virus can persist in third-stage lungworm larvae for at least 35 months (Shope, 1943), about three times as long as is necessary to account for the survival of the virus from one epizootic to the next.

## OVERWINTERING OF ARBOVIRUSES

There is no difficulty in understanding how viruses that are reactivated after periods of latency (like varicella-zoster), or that spread readily in a susceptible population, manage to survive. However, the continued survival of some viruses constitutes a major problem in the ecology of viral diseases. A case in point is the survival of mosquito-transmitted viral infections in temperate climates during the winter months—what is called "overwintering" (Reeves, 1961). Among mechanisms thought to be important are recrudescence of viremia in birds or congenital transmission in ticks (Fig. 16–5).

Hibernating vertebrates may also play a role. In cool temperate climates, bats and some small rodents, as well as snakes and frogs, hibernate during the winter months. During hibernation their body temperature falls greatly. Sulkin (1962, 1965) has reviewed reports concerning the significance of bats in the maintenance in nature of arboviruses and rabies virus. Bats can be infected with certain arboviruses, maintain an inapparent infection throughout a long period of hibernation, and become viremic when transferred to a warm environment. They may also play in important role as a natural reservoir of rabies virus, and the migratory habits of certain bats may be important in spreading the virus to wild animals resident in other geographical locations.

Snakes are potentially important as overwintering hosts for some arboviruses. Snakes naturally infected with western equine encephalitis virus have been captured in several localities in the United States (Gebhardt *et al.*, 1964) and Canada (Burton *et al.*, 1966), and both naturally and experimentally infected snakes exhibit cyclic viremia over periods of many months. The infection is inapparent, and the periodicity of viremia, which is sufficiently high to infect mosquitoes, is not strictly related to temperature. Gebhardt *et al.* (1964) found that some of the offspring of naturally infected female snakes were themselves infected. If this represents congenital transfer of the virus, it offers an alternative mode of survival in nature, independent of vectors but providing a source of virus for their infection.

Overwintering of arthropod-borne viruses is sometimes achieved by the production of persistently infectious lesions in a proportion of the individuals infected each summer. For example, in mature susceptible cottontail rabbits (*Sylvilagus floridanus*) Shope's fibroma virus transferred mechanically by mosquitoes produces tumors that appear 7–8 days after infection, reach their maximum size 2 weeks later, and may persist for 3–5 months before crusting and regressing. Infectivity of these tumors for mosquitoes is high from the fourth week until they start to regress (Dalmat, 1959). In juvenile rabbits, tumors are occasionally produced that persist in an infectious state for a very long time and could readily explain the survival of the virus in particular geographical regions throughout the winter, when vectors are inactive (Kilham and Dalmat, 1955).

## THE EPIDEMIOLOGICAL IMPORTANCE OF INAPPARENT INFECTIONS

We have repeatedly emphasized the fact that infection is not synonymous with disease. Most viral infections are inapparent. This poses insuperable difficulties, in all except a minority of diseases, in tracing case-to-case infection, even with the help of laboratory aids. Although clinical cases may be somewhat more productive sources of infectious virus than inapparent infections, the latter are much more numerous and do not restrict the movement of infected individuals, and thus provide a major source of viral dissemination. Other viral infections are latent for long periods and are either never associated with disease, produce disease only after a long period of latency, or are associated with recurrent but infrequent attacks of disease (Chapter 12). Such intermittent excretors play an important role in the perpetuation of the virus.

## EPIDEMIOLOGICAL RESEARCH

Epidemiology, the study of diseases in populations, stands at the top of the pyramid of virological research, and draws upon research at every other level. This is obvious in relation to etiology and pathogenesis, and the example of influenza virus described in Chapter 17 illustrates how molecular biology may directly illuminate epidemiological problems. In addition, there are some kinds of research that are characteristic of epidemiology itself (see Fox *et al.*, 1970, for a more complete description).

### Epidemiological Studies in Human Communities

Several different kinds of epidemiological study are carried out in human communities. In relation to attempts to determine etiology we can distinguish retrospective (case history) and prospective (cohort) studies.

**Retrospective and Prospective Studies.**   The causation of congenital defects by rubella virus provides examples of each kind of approach. Gregg (1941) was struck by the large number of cases of congenital cataract he saw in 1940–1941, and by the fact that many of the children also had cardiac defects. By interviewing the mothers he found that the great majority of them had experienced rubella early in the related pregnancy. The hypothesis that there was a causative relation between maternal rubella and congenital defects quickly received support from other retrospective studies, and prospective studies were then organized. To do this groups of pregnant women were sought who had experienced an acute exanthematous disease during pregnancy, and the occurrence of congenital defects in their children was compared with that in women who had not experienced such infections. Gregg's predictions were confirmed and the parameters defined more precisely (Swan *et al.*, 1943; Siegel and Greenberg, 1960).

Retrospective studies are valuable for providing the clues but by their nature they are not conclusive, in that they lack suitable control groups; prospective studies are much more difficult to organize but can establish etiological relationships in a more convincing way.

**Human Volunteer Studies.**   Epidemiological facets of several specifically human diseases that have not been reproduced in other animals have been studied in human volunteers, e.g., hepatitis viruses (Havens, 1946; review, Havens and Paul, 1965), rhinoviruses (reviews, Andrewes, 1965; Tyrrell, 1965), and a variety of other respiratory viruses (Knight, 1964). Many major discoveries that have led to the control of viral diseases were possible only with the use of human volunteers, but serious biological and ethical problems now arise because of an increased appreciation of the ubiquity of "occult" viruses and our ignorance of possible long-term effects of the agents being tested.

**Studies in Closed Communities.**   A special kind of epidemiological study in human beings is the long-term study of selected institutional communities, such as the Willowbrook State School in New York, in which Krugman *et al.* (1967) carried out investigations on hepatitis, and "Junior Village" in Washington (Bell *et al.*, 1961; Kapikian *et al.*, 1969) that has yielded much useful information on human respiratory and enteric viruses. These studies, conducted with careful attention to the ethical problems involved, have provided knowledge that could have been obtained in no other way (Hill, 1963; Edsall, 1969). They also provided direct and potential benefits to the participants. In Junior Village, for example, the extensive microbial surveillance provided by the investigating team was a real aid in medical management, and at Willowbrook Krugman succeeded in preventing infection and disease due to hepatitis B virus with an inactivated vaccine.

**Vaccine Trials.** Having begun with individual human volunteers and then small-scale trials in institutions, viral vaccines must be tested on a very large scale before their worth and safety can be properly evaluated. Perhaps the best-known vaccine trial was the famous "Francis Field Trial" of inactivated poliovirus vaccines, carried out in 1954 (Francis et al., 1957) but similar trials were necessary for live-virus poliovirus vaccines, for yellow fever vaccines before this, and for measles and rubella vaccines since. There is no alternative way to evaluate a new vaccine or indeed a new drug, and the design of field and clinical trials has now been developed so that they yield maximum information with minimum risk and cost (Hill, 1963).

**Long-Term Family Studies.** Another kind of investigation that provides the opportunity for both epidemiological studies and controlled trials of vaccines or therapeutic agents are long-term family studies, like the Tecumseh Study (Monto et al., 1971) and the Virus Watch program (Fox et al., 1966). Because of the present advanced state of diagnostic and serological virology, such studies now yield a much greater array of valuable data than was possible a few years ago, but they are very expensive and require long-term dedication of both personnel and money.

### Model Epidemics in Laboratory Animals

The idea of studying the epidemiology of infectious diseases by setting up model epidemics in convenient laboratory animals was developed independently by Webster (review, Webster, 1946) and Topley and Greenwood (review, Greenwood et al., 1936). It was this work which led Webster to appreciate the significance of genetic resistance in viral infections of mice (see Chapter 11), and the British workers to recognize the importance of acquired immunity in a generalized monotypic viral infection (mousepox). Sporadic attempts have been made since then to develop this approach further. Fenner (review, Fenner, 1949) used Topley's methods to investigate the natural history of mousepox, and these experiments led to his use of mousepox as a model for the investigation of the pathogenesis of acute exanthemata (see Chapter 9).

Since the respiratory viruses are of major importance among the viral infections of urban man, several attempts have been made to set up models for study of the factors that affect their transmission. Andrewes and Allison (1961) showed that Newcastle disease virus was spread between chicks mainly by large or labile airborne particles. Prolonged rather than close contact was important in transmission, as it is in human respiratory infections, in which an infrequent event that forcibly expels virus (an uncontrolled cough or sneeze) is more likely than casual contact to transmit infection. Schulman and Kilbourne (1967) studied the effects of season and relative humidity on the transmission of influenza A2 virus in mice. Even with close control of temperature and humidity in the animal rooms, transmission was somewhat better during the winter months. Using two strains of virus that were equally virulent for mice, as judged by lung lesions and virus titers, Schulman (1967) found that one of these was much more readily transmitted by airborne virus to contact mice. This difference was not due to variations in the survival of airborne virus or the amounts needed to initiate infection, but to greater shedding during the period of maximum infectiousness.

Experimental studies in epidemiology can also be carried out by computer simulation of the spread of infection in arbitrarily designed populations. For example, Elveback et al. (1968) have produced simulation models that illustrate the potential effects of live poliovirus vaccines in curtailing the spread of other enteroviruses, including heterologous strains of poliovirus.

## Experiments of Nature

"Natural experiments" have also been used for the definition and sometimes the solution of epidemiological problems. For example, the behavior of myxoma virus after its release in Australia was used by Fenner and Ratcliffe (1965) to study the changes in virus and host that occurred when an extremely virulent and readily transmitted virus spread through an almost uniformly susceptible population (see Chapter 17).

# Evolutionary Aspects of Viral Diseases

## INTRODUCTION

Viruses are ubiquitous, and have played important roles in the evolution of bacteria, plants, and animals. In the first place, infectious diseases in themselves have been powerful selective forces, whose role has been well demonstrated, for example, with malaria and the human genes for hemoglobin S and glucose-6-phosphate dehydrogenase deficiency. Speculatively, both DNA and RNA viruses may also have played a part in producing the increase in cellular DNA that has characterized evolutionary progress, by transduction or integration, directly or through the intervention of the reverse transcriptase. Viruses themselves are also subject to natural selection.

There is no fossil record of the viruses, but clues about the antiquity of some existing viruses may be obtained from biogeographical observations, or by study of the distribution of members of different viral genera throughout the animal or plant kingdoms. In a few situations it has been possible to witness evolution in action, with viruses and their vertebrate hosts.

Man's disturbances of the biosphere, and some medical innovations, have created "new" viral diseases for man himself and for his domesticated animals. Rapidly increasing urbanization, and the swift and large-scale movement of human beings over the whole world, have led to a well-nigh universal distribution of the "human" viruses, thus changing radically epidemiological patterns that evolved in small, relatively isolated human communities.

## THE ANTIQUITY OF SOME VIRUSES

All viruses depend for their replication on the energy sources, the ribosomes and some of the enzymes of their host cells. They must therefore have evolved after cells and be derived from them, either by the "parasite degeneration" of Green and Laidlaw or by the sequestration of fragments of cellular nucleic acid and its acquisition of the capacity to infect other cells (see Chapter 1). Events of this kind have probably occurred many times since the first cells developed during pre-Cambrian times, but in our short-term view of today's world, virus is always derived from preexisting virus.

There is no paleontology of viruses, and even the most daring interpretations of lesions in Egyptian mummies and in fossil bones of man or animal are confined to events of the recent past. However, parasitologists and entomologists are sometimes able to draw conclusions about evolutionary relationships in parasitic protozoa, helminths, and arthropods from a consideration of the evolutionary history of their present vertebrate hosts, and vice versa (Rothschild and Clay, 1961). Virologists may be able to use the same approach in cases where viruses are highly host specific and where recent spread of host species by man has not obscured the biogeography. The myxoma subgroup of the poxviruses provides one example (see below); with the rickettsias the occurrence of the very closely related agents of European, American, and Australian Q fever as agents of infections indigenous to three continents testifies to the antiquity and the genetic stability of *Coxiella burneti*.

## The Distribution of Viral Genera among Vertebrates

Viruses resembling those that infect vertebrates are found in plants (e.g., wound tumor virus and possibly some rhabdoviruses), and many viruses of vertebrates, belonging to several different genera, multiply in arthropods. Such situations raise unsolved problems concerning viral origins, for example, did most arboviruses arise primarily in invertebrates or in vertebrates? Other families or genera of viruses apparently occur only among vertebrate animals. If the viruses are host specific and the allocation of viral species to a particular genus is reliable, some information on the possible antiquity of the viral genus can be derived from consideration of the distribution of viral species among different orders of vertebrates.

**Herpesviruses.** Nahmias (1971a) has applied this approach to herpesviruses (Fig. 17–1). Assuming that the agents found in lower vertebrates are indeed host-specific herpesviruses, the wide distribution of the genus suggests an ancient origin. The very wide G+C range found in nucleic acids of herpesviruses is compatible with this interpretation (De Ley, 1968). Tests for genetic relatedness of viral nucleic acids might also be used to give a measure of the genetic distances. It is likely that nucleic acid hybridization is too precise a tool for such purposes, although electron microscopy of heteroduplexes may be more useful than membrane hybridization tests. The molecular weight of the nucleic acid, its base composition, and polynucleotide maps may provide more useful data.

**Leukoviruses.** The leukoviruses may also be an ancient group that has evolved with the vertebrates, as part of their genome (Todaro and Huebner, 1972). Table 17–1 sets out the serological relationships between some of the internal antigens of leukoviruses of subgenera A and B. By analogy with the orthomyxoviruses, these internal antigens, unlike the peplomers, would not be subject to selection for antigenic change. Except for the species-specific antigen $gs_1$, the cross-reactivity between the internal antigens of leukoviruses derived from a variety of vertebrate species reflects the closeness of the relationship of the host species. The mammary tumor virus (subgenus B) is seen to be antigenically as well as morphologically quite different from the leukemia-sarcoma

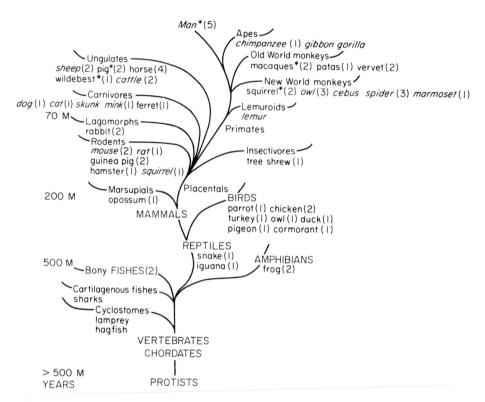

Fig. 17–1. *Evolutionary tree of vertebrates, indicating species from which herpesviruses have been recovered (from Nahmias, 1971a). Numbers in brackets, number of herpesviruses identified in each vertebrate species. Species named in italics are susceptible to herpesviruses from other species; *, species with herpesviruses that have been found to infect other species under natural conditions.*

viruses of subgenus A. Thus, as far as the investigations have been carried they are compatible with the hypothesis that the genetic material of all these viruses is maintained as part of the cellular genome. Clearly, many more data are needed on the distribution of leukoviruses among other vertebrates and their antigenic properties. Groups that would be particularly significant for such studies are the lower vertebrates, and the Australian marsupials, for Australia has been geographically isolated for almost a hundred million years. Characterization of the amino acid composition and sequence of the purified proteins might shed light on the genetic relatedness of the viruses and their hosts.

**Poxviruses of the Myxoma Subgenus.** Viruses of this subgenus (see Table 1–6; review by Fenner and Ratcliffe, 1965) were long thought to occur only as infections of a few species of wild animals in North and South America. Two serologically slightly different strains of myxoma virus occur in *Sylvilagus brasiliensis* in South America and in *Sylvilagus bachmani* in California, rabbit fibroma virus occurs naturally in *Sylvilagus floridanus* in the eastern United

TABLE 17-1

Cross-Reactions between Internal Antigens of Members of Leukovirus Genus[a]

| ANTIBODY | LEUKOVIRUS FROM | | | | | | | | |
|---|---|---|---|---|---|---|---|---|---|
| | MOUSE | RAT | HAMSTER | CAT | WOOLLY MONKEY | GIBBON APE | CHICKEN | VIPER | MOUSE MAMMARY TUMOR |
| Anti-transcriptase | | | | | | | | | |
| Mouse | ++ | ++ | ++ | + | – | – | – | – | – |
| Cat | + | ++ | ++ | ++ | – | – | – | – | – |
| Chicken | – | – | – | – | – | – | ++ | – | – |
| Anti-$g_{s3}$ (interspecific) | | | | | | | | | |
| Cat | + | + | + | ++ | + | + | – | – | – |
| Chicken | – | – | – | – | – | – | + | – | – |
| Anti-$g_{s1}$ (specific) | | | | | | | | | |
| Mouse | + | – | – | – | – | – | – | – | – |
| Cat | – | – | – | + | – | – | – | – | – |
| Chicken | – | – | – | – | – | – | + | – | – |

[a] Data from Parks et al. (1972), Scolnick et al. (1972), and Gilden and Oroszlan (1972).

States, and squirrel fibroma virus has only been found in *Sciurus carolinensis*, also in the eastern United States.

The geographical distribution of South American myxoma virus was vastly extended when it was deliberately and successfully introduced into the large population of European wild rabbits in Australia in 1950, in Europe in 1952, and in Chile in 1954. Unsuccessful attempts were made in the mid-1950's to establish rabbit fibroma virus in France in an effort to vaccinate wild rabbits against myxomatosis. In both Europe and Australia, myxoma virus has proved to be highly host specific, in spite of the fact that many species of animals must have been bitten by infected arthropods. Only the European hare has proved, on very rare occasions, to be susceptible to myxomatosis, and there is no conclusive evidence that the virus is naturally transmitted from one hare to another.

In 1959–1960, epidemics of fibromatous lesions were observed in European hares *(Lepus europaeus)* in southern France and northern Italy, and a poxvirus was recovered from the lesions. At first it was thought that this might be a strain of myxoma or fibroma virus that had become adapted to the hare. However, epizootics of the same disease were found to date back to 1908, long before the introduction into Europe of either myxoma or fibroma viruses. Serological comparison of poxviruses of the myxoma subgenus (Woodroofe and Fenner, 1965; Fenner, 1965) confirmed the position of hare fibroma virus as a previously unrecognized member of this genus. Speculation on the origins of hare fibroma virus in Europe must be tempered by the recognition of our ignorance of the mild enzootic infections of wild animals, which are only recognized when they "escape" into a domestic animal or cause disease in a game animal. However, if we assume that hare fibroma virus is the only indigenous member of the myxoma subgenus in Europe, it is not unreasonable to postulate that it must have come to Europe with the progenitors of *Lepus europaeus* at the time of the worldwide dispersal of Leporidae, probably in the early Pleistocene. It is believed that *Sylvilagus brasiliensis* also migrated from North to South America at about this time, and it probably took the prototype of the South American strain of myxoma virus with it. The differences we now see between hare fibroma virus and the other members of the myxoma subgenus may be partly due to differences existing at the time of the dispersion, and partly due to evolutionary changes since that time.

**Papilloma Viruses.**   Comparable situations may well exist in other wild animals; it would be profitable for those interested in diseases of wild life to bear these possibilities in mind when new agents are recovered from wild animals that have a restricted geographical range. The viruses that cause papillomas are highly host specific, and since virus-induced papillomas occur in many species of wild and domesticated animals these viruses may prove particularly suitable material for such investigations. Le Bouvier *et al.* (1966) found no serological cross-reactivity between the coat proteins of four mammalian papilloma viruses (human, rabbit, dog, and bovine), and Crawford (1965) failed to detect any homologies between the DNA's of human and rabbit papilloma virus. However, broader comparisons of viral proteins and nucleic acids may be needed, and the approach could profitably be extended with papilloma viruses obtained from wild animals in different parts of the world.

## CHANGES OF VIRUS AND HOST IN MYXOMATOSIS

Observations made on myxomatosis in wild European rabbits in Australia and Britain illustrate the occurrence of evolutionary changes in both the virus and its host. The history of myxomatosis has been fully documented elsewhere (review, Fenner and Ratcliffe, 1965); here we shall present the main features of the evolutionary changes.

### Myxomatosis in South America

Myxomatosis has been known since 1896 as a highly lethal "spontaneous" disease of domestic (European) rabbits in various parts of South America, and since 1954 as an enzootic disease of wild European rabbits in Chile. The natural reservoir of myxoma virus is the tropical forest rabbit of South America (*Sylvilagus brasiliensis*) in which it produces tumors identical with those produced by rabbit fibroma virus in *Sylvilagus floridanus* (see Chapter 13). Infection of domestic rabbits is due to mechanical transfer of the virus from the skin tumors in *S. brasiliensis* by infected mosquitoes. In South America, the domestic rabbit serves as a "sentinel animal" indicating the occurrence of the unobserved disease in the wild rabbits, much as cattle are the sentinel animals for pseudorabies of swine (Table 16–10). Unlike the latter, however, a sick domestic rabbit can readily transmit infection to other domestic rabbits both by direct contact and through the activity of arthropod vectors.

### Myxomatosis in Australia

The wild European rabbit was introduced into Australia in 1859 and rapidly spread over the southern half of the continent, where it became the major pest animal to the agricultural and pastoral industries. Myxoma virus was successfully introduced into Australia to help control the rabbit pest in 1950, and also spread rapidly over the southern half of the continent, causing enormous mortalities in the wild European rabbits. As had been indicated by earlier experiments, myxoma virus was highly host specific and affected only the rabbit, (apart from very rare cases in European hares) and it was spread mechanically by insect vectors, principally mosquitoes. The occurrence of this very lethal new disease offered an opportunity for the study of evolutionary changes in the virus, the host, and the disease.

**Changes in Myxoma Virus.** The virus originally liberated in Australia was extraordinarily lethal for European rabbits; detailed observations of mortality rates in excess of 99% (Myers *et al.*, 1954) reflected a situation that occurred widely over the southeastern part of Australia during the first few years after its introduction. This highly virulent virus obviously spread very readily from one animal to another during the summer, when the mosquito population was abundant, and farmers reintroduced the highly virulent virus into the wild rabbit population every spring and summer. One possible result would have been that the disease would die out after each summer, due to the disappearance of susceptible rabbits and the greatly lowered opportunity for transmission during the winter because of the scarcity of vectors. This undoubtedly occurred often in localized situations, but on a continental scale the outcome proved to be very different. The capacity to survive over the winter imposed a stringent selection favoring a less lethal disease, which would enable an infected rabbit to survive in an infectious condition for weeks instead of days. Attenuated viruses appeared within a year of the introduction of the virus, and within 3 or 4 years they became the dominant strains throughout the country (Table 17–2). Localized highly lethal outbreaks still occurred (and indeed still do occur although much more rarely) following well-timed inoculation campaigns, but in general the viruses that spread through the rabbit populations each year are the somewhat attenuated strains which offer the best opportunities for mosquito transmission. The superiority of such strains even during epizootics was attested by field experiments (Fenner *et al.*, 1957; Sobey and Myers, unpublished observations, 1964) and they obviously have advantages for overwintering. Thus the original highly lethal introduced virus has now been replaced by a heterogeneous collection of strains, all of lower virulence. A further development has recently been reported (Williams *et al.*, 1972), namely recurrent disease in recovered animals, severe enough to allow arthropod transmission to occur.

**Changes in the Genetic Resistance of Rabbits.**   With rare exceptions, rabbits that recover from myxomatosis are immune to reinfection for the rest of their lives, and immune mothers transfer passive immunity of short duration to their young. In populations of wild European rabbits in Australia, few animals have a natural lifespan of more than a year unless the rabbit density is low, so that herd immunity, due to the presence of a substantial number of actively immunized rabbits in the population, is not a very important feature in the epidemiology of the disease. However, the fact that even in the initial outbreaks some rabbits recovered suggested that selection for genetically more resistant animals might operate rapidly. Sometimes recovery was due to environmental factors, for exposure of rabbits to high environmental temperatures greatly reduces the severity of the disease (see Chapter 11). In general, however, it was found that in areas of continued annual exposure to epidemic myxomatosis the genetic resistance of the rabbits steadily increased. The early appearance of virus strains which allowed 10% of genetically unselected rabbits to recover was an important factor in allowing such genetically resistant rabbit populations to build up. Their existence was demonstrated both by tests made on uninfected wild rabbits captured in the field and tested in the laboratory, and by breeding experiments and progeny testing (Sobey, 1969). The field experiments showed that in areas of annual exposure to myxomatosis the genetic resistance changed so that the mortality rate after infection with aliquots of the same sample of virus fell from 90 to 25% within 7 years. The results of some laboratory tests are illustrated in Fig. 17–2; the trend is much the same as in the field. These

# TABLE 17-2

*Comparison of the Virulence of Naturally Occurring Strains of Myxoma Virus in Australia and Great Britain at the Time of First Epizootics and at Various Times Thereafter[a,b]*

| VIRULENCE GRADE | MEAN SURVIVAL TIME (DAYS) | CASE–MORTALITY RATE (%) | AUSTRALIA | | | | | GREAT BRITAIN | |
|---|---|---|---|---|---|---|---|---|---|
| | | | 1950–1951 | 1952–1953 | 1955–1956 | 1957–1958 | 1963–1964 | 1953 | 1962 |
| I | <13 | >99 | 100 | 4 | 0 | 3 | 0 | 100 | 4 |
| II | 14–16 | 95–99 | | 13 | 3 | 7 | 0.3 | | 18 |
| III | 17–28 | 70–95 | | 74 | 55 | 54 | 59 | | 64 |
| IV | 29–50 | 50–70 | | 9 | 25 | 22 | 31 | | 14 |
| V | — | <50 | | 0 | 17 | 14 | 9 | | 1 |

[a] Figures represent the percentages allocated to the virulence grades shown.

[b] Data from Marshall and Fenner (1960), Fenner and Chapple (1965), and Fenner and Woodroofe (1965).

experiments also show that increased resistance operates effectively against highly virulent (SS—Fig. 17–2) as well as against somewhat attenuated strains of the virus (KM13 and U).

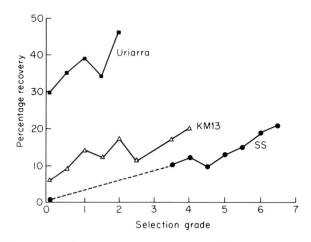

FIG. 17–2. *Changes in the genetic resistance of rabbits to myxoma virus that occurred during breeding from animals that had recovered from infection with various strains of virus. "Selection grade" indicates the parentage of rabbits under test: 0.5, one parent had recovered from myxomatosis; 1, both parents had recovered; 2, both parents and both grandparents had recovered, etc. For the initial selection two attenuated strains were used: Uriarra, from selection grade 0 to grade 2,* ■——■; *KM13, from grade 0 to grade 4,* △——△. *At grade 3.5 a selection line was begun with virulent myxoma virus, SS,* ●——●, *a strain that killed all unselected rabbits (data from Sobey, 1969).*

**The Continuing Evolution of Myxomatosis.**  In myxomatosis of the European rabbit natural selection operates to produce a type of disease which ensures that there is an adequate susceptible population, and that the virus can survive through periods of low vector density, i.e., that it can overwinter. These requirements are met, in genetically unselected rabbits, by a virus less rapidly lethal than that originally introduced. Such viruses will allow many infected rabbits to survive for prolonged periods with infectious lesions, and enough rabbits will recover and breed to ensure that there is an adequate population of new susceptible animals for the next summer epizootic.

Evolutionary changes in the virus and the host, and therefore in the disease, are still in progress, and two types of climax situation can be envisaged. It is possible that myxomatosis in wild Australian rabbits will eventually become very like the disease we know in the natural reservoir hosts in the Americas: a fibroma, with localized skin tumors in which the virus overwinters in the persistent tumors produced in juvenile animals. If this occurred it is probable that the virus obtained from such lesions would prove highly virulent for genetically unselected laboratory rabbits. Another alternative is that the climax situation may be comparable to variola major in man: a moderately severe disease with an appreciable but not a crippling mortality, for as fecund an animal as the rabbit. In this situation, as with the "benign" fibroma, it is possible that as the genetic resistance of the rabbits increases the virulence of the virus (for genetically unselected laboratory rabbits) will increase again, for too mild a generalized disease, like too severe a disease, is not conducive to successful overwintering of the virus. For example a "laboratory" mutant of virulent myxoma virus (neuromyxoma of Hurst, 1937) produces such a mild generalized disease in normal rabbits, and multiplies to such a low titer in the skin, that mosquito transmission occurs rarely and only for a brief period of time.

## INFLUENZA IN MAN

Epidemics of viral diseases usually result from some unusual ecological situation, such as the introduction of large numbers of susceptible individuals into a population where disease is endemic, a common feature of the movements of human beings during wars, or the introduction of a "new" virus into a population with no immunity, e.g., myxomatosis in the Australian rabbit or measles into human populations in Greenland. The usual pattern of established infectious diseases is endemicity with periodic fluctuations in intensity that may be seasonal or may occur at longer intervals. Among the many viruses that are associated with respiratory disease in man only three regularly cause epidemics that involve a substantial section of the population and occur at annual or longer intervals: influenza A and B and respiratory syncytial virus (Chanock et al., 1965).

Human influenza virus was first isolated in 1933 (Smith et al., 1933). Since that time viruses have been recovered from human and animal sources from many parts of the world, especially since the establishment of the Influenza Surveillance Program of the World Health Organization. Exhaustive study of their antigenic constitution has provided the opportunity for observing continuing evolutionary changes in the influenza viruses. In contrast to myxomatosis of rabbits, the mortality from influenza is too low, and the generation time of man too long, for any significant observable change to have occurred in host resistance.

Two virus-coded antigens, the hemagglutinin and neuraminidase, project from the envelope of influenza viruses (see Chapter 3). Until recently the attention of epidemiologists was concentrated almost exclusively on changes in the hemagglutinin antigen, for the antibodies to this antigen can be readily assayed by hemagglutinin-inhibition tests and they are responsible for viral neutralization. However, understanding of the ecology and evolution of the influenza viruses necessitates consideration of changes in both surface antigens.

The two well-characterized types of influenza virus, types A and B, are distinguished by differences in their internal ribonucleoprotein antigens (Chapter 1). Influenza type B appears to be an exclusively human virus. Influenza type A viruses, on the other hand, have been recovered from a number of different species of animals, including swine, horses, and several species of birds (reviews,

McQueen *et al.*, 1968; Easterday and Tumova, 1972). Because of the epidemiological importance of the envelope antigens and their independent variation, a nomenclature for viral strains has been introduced that describes the type (A or B), the time and place of isolation, and the hemagglutinin (H) and neuraminidase (N) antigens (World Health Organ., 1971b). Thus the Hong Kong virus is designated A/Hong Kong/1/68 (H3N2); avian and equine hemagglutinins and neuraminidases are designated Hav, Nav and Heq, and Neq, respectively.

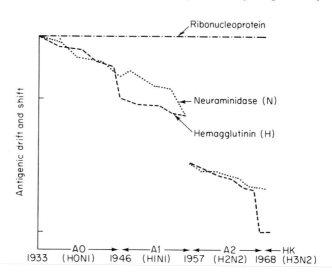

FIG. 17–3. *Diagram illustrating antigenic drift, antigenic shift, and the appearance of new subtypes of human influenza A virus. "Antigenic drift and shift" (ordinate) represents the serological relatedness of the antigens of influenza A virus recovered from man between 1933 and 1972. The internal (ribonucleoprotein) antigen (–·–·–) has not changed over the whole period. Both the hemagglutinin (– –) and the neuraminidase (·· ··) have shown antigenic drift, i.e., annual but independent changes in their antigenicity, and antigenic shift, i.e., larger changes such as the change in antigenicity of the hemagglutinin antigen of influenza A in 1946 (H0 to H1). This led to the 1947 and subsequent strains being called a new subtype, A1. There was no parallel change in the neuraminidase in 1946; it remained N1 from 1933 to 1957.*

*The appearance of Asian influenza virus (subtype A2) in 1957 was related to major changes in both the hemagglutinin and the neuraminidase antigens (antigenic shift: H1N1 → H2N2). Since 1957 these antigens of influenza A2 underwent antigenic drift until 1968, when the hemagglutinin underwent antigenic shift from H2 to H3, whereas the neuraminidase continued to show antigenic drift as N2. The new strain is designated the Hong Kong strain (HK:H3N2).*

Influenza type B causes isolated epidemics in man but it has never been implicated in pandemic influenza, whereas influenza type A has caused worldwide pandemics, those of 1889–1890, 1918–1919, 1957–1958, and 1968–1969 being the most dramatic during the past century. Between pandemics there are scattered smaller epidemics occurring every year or two, usually in winter. Two types of change in the envelope antigens of influenza viruses have been recognized. Between pandemics, minor changes in specificity occur by a process called antigenic (immunological) drift (Burnet, 1955). Antigenic drift occurs independently in the hemagglutinin and neuraminidase antigens (Robinson, 1964; Paniker, 1968), and has been observed with both influenza type A and influenza type B (Hennessy et al., 1965; Chakraverty, 1971; Fig. 17–3). By contrast, new pandemic strains of influenza A have always had hemagglutinin antigens that differed strikingly from those of previous strains (antigenic shift, Webster, 1972), while the neuraminidase antigens have sometimes been unchanged (Table 17–3).

TABLE 17–3

*Antigenic Characteristics of Pandemic Strains of Influenza Virus*

| TIME OF PANDEMIC | PREVIOUS DESIGNATION OF STRAIN | CURRENT DESIGNATION OF STRAIN | NEW PANDEMIC ASSOCIATED WITH ANTIGENIC SHIFT IN | |
|---|---|---|---|---|
| | | | H | N |
| 1918–1919 | A0 | A/London/1/35 (H0N1) | .. | .. |
| 1946 | A1 | A/FM/1/47 (HIN1) | + | − |
| 1957 | A2 | A/Singapore/1/57 (H2N2) | + | + |
| 1968 | Hong Kong | A/Hong Kong/1/68 (H3N2) | + | − |

## Evolutionary Changes in Influenza Viruses

Antigenic drift is thought to result from the selection of antigenically novel mutants during transmission of viruses through partially immune populations. Experimentally it has been shown that such mutants possess a growth advantage in the presence of antibody (Hamre et al., 1958); mutants with minor changes in the amino acid sequence of the hemagglutinin polypeptides can be obtained in the laboratory by deliberately growing the parent strain in the presence of antibodies of low avidity (Laver and Webster, 1968).

There are two main theories that seek to explain the major subtype changes [e.g., from subtype A1 (H1N1) to subtype A2 (H2N2)]: (a) new subtypes arise by direct mutation from existing human strains of influenza virus, or (b) they are derived from viruses infecting lower mammals or birds.

The weight of evidence (Webster, 1972) favors explanations involving animal as well as human strains of influenza type A. Two observations offer strong support from this view. First, the isolated and purified hemagglutinin subunits of several 1968 influenza A2 viruses (H2N2) were shown to be completely different, both antigenically and in the peptide maps of the "light" and "heavy" polypeptides, from those of several Hong Kong strains (H3N2) (Webster and Laver, 1972; Laver and Webster, 1972). Furthermore, peptide maps of the light polypeptides from the hemagglutinin subunits of the Hong Kong strain are almost the same as those of equivalent subunits from Equine 2 (A/equine /Miami/1/63–Heq2Neq2) and a duck influenza virus (A/duck/Ukraine/1 /63–Hav7Neq2) (Laver and Webster, 1973). These viruses were isolated 5 years before the Hong Kong strain appeared and could be its progenitors. Second, influenza viruses readily undergo recombination by reassortment, in eggs, mice, and cultured cells (Chapter 7). Recombination also occurs with high frequency between strains derived from different animals (birds and swine) within the respiratory tracts of either (Webster *et al.*, 1971, 1973; Webster and Campbell, 1972). The frequent recovery of "human" influenza type A viruses from other animals (pigs, Kundin, 1970; cats, Paniker and Nair, 1970; horses, Kasel and Couch, 1969) suggests that opportunities would occur in nature for recombination to take place between human and animal influenza A viruses. Several avian viruses with "human" neuraminidases may have originated in this way (N2, Webster and Pereira, 1968; N1, Schild *et al.*, 1969).

Rather than evisaging very extensive mutational changes in specifically human viruses, therefore, it seems more likely that pandemic strains of human influenza A viruses arise from animal sources. When only the hemagglutinin is changed (as in 1946 and 1968; Table 17–3) this could well have occurred by recombination; where both neuraminidase and hemagglutinin are novel (as in 1957) the new virus could be zoonotic, i.e., it could arise by the transfer of an animal virus to man, or it could represent a multiple gene recombinant.

Fazekas de St. Groth (1970) has claimed that major subtype changes (e.g., from H0N1 to H2N2) may be produced by growing H0N1 viruses under pressure of specifically selected antibody fractions from selected anti-A0 sera. This is theoretically possible, especially if only the hemagglutinin antigen is changed [as in the change from Asian (H2N2) to Hong Kong (H3N2)], but there is no evidence that such mutants have yet been produced.

An interesting and unexplained phenomenon is the rapid and virtually complete replacement, all over the world, of the strains of influenza virus current at a particular time by the strains which succeed them. For example, when subtype A1 virus spread around the world (it was first isolated in Australia in 1946, and in the following winter occurred in the northern hemisphere) viruses belonging to subtype A0 virtually disappeared, although an epidemic in Alaska in 1949 (van Rooyen *et al.*, 1949) yielded a strain of the A0 subtype. In like manner subtype A1 strains disappeared almost completely after 1958, although Isaacs *et al.* (1962) recovered an A0 strain in England in 1960. The apparent disappearance of the earlier strains is partly due to sampling, for the antigenically novel viruses are more likely to cause small epidemics that will attract the attention of public health workers, and isolated cases of disease would rarely be tested for virus. Nevertheless it is rather surprising that a situation does not arise like that commonly found with pathogenic bacteria, or with the rhinoviruses and enteroviruses, i.e., the coexistence of many different serological types rather than the regular replacement of one by another.

## ANTIGENIC DRIFT WITH OTHER VIRUSES

Burnet (1960) suggested that antigenic drift might operate as an evolutionary mechanism with the enteroviruses of man, with foot-and-mouth disease in cattle, and with dengue in man. Studies since then would tend to rule out this explanation of the sequential epidemics of dengue, of which there appear to be several stable antigenic types; immunity to any one is prolonged, but gives little protection against other serotypes. To become a source of infection for arthropods a substantial viremia must develop; this is unlikely in an immune individual if there is a substantial antigenic overlap. If something like antigenic drift operated to produce the many members of the two genera of Togaviridae, the time scale was much more prolonged than in influenza.

Seven major antigenic types of foot-and-mouth disease virus (FMDV) have been distinguished by cross-immunity tests, but all cross-react to some extent in complement fixation tests (Brooksby, 1958). There is also a wide variation in serological specificity within each type, i.e., each of the major types of FMDV appears to consist of several subtypes (Davie, 1962). Hyslop (1965) showed that serial passage of a strain of FMDV in cell cultures in the presence of increasing concentrations of specific antiserum led to the emergence of antigenic variants. The potential practical importance of antigenic drift was demonstrated by Hyslop and Fagg (1965), who obtained an antigenically changed virus from vesicles developing in cattle which had been partially immunized with a formalin-treated vaccine. Primary vesicles developed in all such partially immunized cattle on challenge with the strain used for vaccination, and secondary lesions in the majority. The antigenically altered virus was recovered after thirty-four serial passages of the virus through the vaccinated cattle. The occurrence of antigenic drift with FMDV suggests that in cattle, or in wild ungulates, the immunity that follows recovery from infection must wane rather rapidly, at least in a few animals, so that reinfection with the homologous strain can occur and produce enough virus (in the primary or secondary lesions) to infect other animals. In this respect, foot-and-mouth disease contrasts rather sharply with the viral exanthemata of man and with most togavirus infections, in which a durable immunity prevents rapid antigenic drift.

Antigenic drift is probably an important evolutionary mechanism in many viral infections of long-lived vertebrates in which most infections are limited to a mucous surface, such as the respiratory epithelium or the gut. Such situations provide an environment with small amounts of antibody on the surface of the susceptible cells of the mucous membranes, so that there is a strong selection for even minor antigenic novelty. Respiratory syncytial virus (see Chapter 16) provides an exception to this generalization, for this virus seems to have persisted as a strictly human infection of the upper respiratory tract without significant antigenic change at least since its discovery in 1956.

## NEW VIRUSES AND NEW VIRAL DISEASES

The whole pattern of public health and quarantine administration, in relation to viruses, is based upon the assumption that every virus is derived from a preexisting virus although it may have undergone mutation during the process, i.e., "new" viruses are not now developing from nonviral material, whether this be cell component or microorganism. Obviously this must have occurred in the past, but such events must be excessively rare on a human time scale. New viral diseases, on the other hand, are relatively common; most but not all are due to human intervention in some natural situation, or to changes in the social habits of man.

Man is distinguished from all other living things by the speed with which major changes have occurred in his social organization; within a few thousand years he has changed from isolated societies consisting of a few hundred hunters and food gatherers to the vast conurbations of modern man (see Table 17–4). By comparison, the social organizations of all other animals are static, except in-so-far as Western man now imposes drastic changes on populations of domestic animals. The pattern of viral infection sustained by any type of animal is greatly dependent upon its social contacts with its fellows, and its lifespan.

**Viral Diseases in Primitive Man.**  By "primitive man" we mean the nomadic hunters and food gatherers found today in a few isolated parts of the world, and characteristic of man as a species before the development of agriculture and the domestication of animals. Primitive man lived in closely knit groups, probably with little sense of hygiene, and we can assume that the viral "flora" characteristic of man as an animal was shared among all members of the group. Under these circumstances, many of the specifically human viruses of modern man could not survive. We have already discussed the basic community size needed to maintain measles virus (Chapter 16); the same situation holds for all the generalized, specifically human, viral infections in which latency and recurrent infection do not occur. Immunity due to prior infection is not as durable in many diseases of the intestinal and respiratory tract as it is in the generalized infections; even so, few agents causing such infections could survive in a long-lived animal like man in societies of a few hundred individuals. We have ample evidence of the disappearance of respiratory viruses and the diseases associated with them, influenza and the common cold, in isolated communities of modern man (see Chapter 16); this situation must have been well-nigh universal in primitive man.

Indeed, the only "specifically human" viral diseases that we could expect to survive in primitive man (and the same is true of animals with a comparable social organization and a natural lifespan exceeding a few years) are those characterized by persistent infection or by latency and recurrent disease. Most viral diseases of primitive man were presumably not "human" diseases; they were zoonoses (Table 16–10), caused by viruses of some other animal which "accidentally" infected man. The principal diseases of this type are the arbovirus infections, and their incidence in primitive man would obviously differ greatly in different parts of the world, just as it does today. A recent serological survey of an isolated Amazon tribe has now provided confirmation of this view of the viral diseases of primitive man (Black *et al.*, 1970).

**Viral Diseases in Societies Based on Agriculture.**  The development of agriculture led to a great change in human habits. No longer were men nomads, and no longer was the population size of a group limited by the availability of

TABLE 17–4

The Time Scale of Cultural Changes in Man in Relation to the Number of
Generations, World Population, and the Size of Human Communities

| NO. OF YEARS BEFORE 1973 | WORLD POPULATION (MILLIONS) | NUMBER OF GENERATIONS | CULTURAL STATE | SITE OF HUMAN COMMUNITIES |
|---|---|---|---|---|
| 500,000 | 0.1 | 25,000 | Hunter and food-gatherer | Scattered nomadic bands of <100 persons |
| 10,000 | 5 | 500 | Development of agriculture | Relatively settled villages of <300 persons |
| 6,000 | 50 | 300 | Development of irrigated agriculture | Few cities of ~100,000; mostly villages of <300 persons |
| 250 | 600 | 10 | Introduction of steam power | Some cities of ~500,000; many cities of ~100,000; many villages of ~1000 persons |
| 130 | 1,000 | 5 | Introduction of sanitary reforms | |
| 0 | 3,500 | — | Modern urbanized man | Some cities of ~5,000,000; many cities of ~500,000; fewer villages of ~1000 |

natural food. Villages developed, and the close contact between larger numbers of people led to an accelerated exchange of their gut and respiratory flora, and the development of endemic bacterial infections and perhaps some endemic enteric and respiratory viruses. When irrigation made large-scale agriculture possible, man started to live in towns and cities. The pattern of viral infections was greatly influenced by the development of these large societies, for now community size had exceeded the minimum level needed to maintain diseases like smallpox, measles, and rubella, and a large close-knit society permitted the ready spread of fecal–oral and respiratory viruses. The origins of the "proto-type" respiratory viruses is speculative. They were probably acquired from some animal source, for many animals live in large enough groups to sustain such viruses because their turnover rate (i.e., the accession of new susceptibles) is so much more rapid than in man. Smallpox virus must have been derived from a related virus in an animal host, the most likely candidate being monkey-pox virus, which even today causes sporadic nontransmissible "smallpox" among some Africans (World Health Organ., 1972b). Measles virus may well have been derived from rinderpest or distemper virus, which are serologically related (Imagawa, 1968).

Once the "prototype" respiratory viruses and enteric viruses had been successfully established in human communities, they were subjected to natural selection for survival, operating primarily at the level of transmission. Other things being equal, antigenically novel viruses had a better opportunity of becoming established and multiplying to a sufficiently high titer to be trans-mitted. This may have been accomplished initially by antigenic drift, but with the picornaviruses a large number of different noncross-reacting serotypes evolved and appear to have persisted. Sampling has not yet been extensive enough, or on a sufficiently large scale, to know whether some serotypes of these viruses actually become extinct and are replaced by new serotypes.

**Urbanization: Infectious Diseases in the Cities.**   Urbanization is a recent phenomenon, dating back only a few hundreds of years and caused initially by the recruitment of country folk to the cities. It was accompanied by intense squalor and poverty, and a greatly increased incidence of the associated infec-tious diseases. Ultimately, in more affluent countries, the dwellings improved and crowding decreased; sanitation in the form of a safe water supply and adequate disposal of excreta, and, finally, the development of effective antibacterial drugs, diminished the importance of bacterial diseases as causes of death and severe morbidity in the great conurbations. For reasons that are not clear, viral infec-tions of the gastrointestinal tract have not disappeared as rapidly as the enteric bacterial pathogens; polioviruses, reoviruses, echoviruses, and infectious hepa-titis virus still circulate widely. Perhaps this is a relative matter. We take no notice of intestinal bacteria unless they cause disease. Many enteric viruses appear to be nonpathogenic in the vast majority of cases, but because of the sort of laboratory technique required for viral isolation, all positive results are recorded. We know that there has been a great diminution in the circulation of enteric viruses in modern cities of western man compared with places like Karachi or Calcutta.

The number and variety of respiratory viral infections of man is probably in a stage of explosive expansion at the present time. As outlined earlier, man the hunter–gatherer lived in aggregations too small to maintain respiratory viruses that produced acute infection without prolonged or recurrent viral excretion. The great increase in the numbers of human beings, their crowding into ever larger cities, and the increasing communication by men between these cities all over the world, are tending to make the "human" world into a single ecological unit. The respiratory viruses of the Northern and Southern Hemispheres, and of east and west, are mingled by air travelers, so that the population of almost any large city in the world today is potentially exposed to human respiratory viruses from all over the world. With this increased exposure there is, of course, increased opportunity for evolutionary radiation, especially in the antigenicity of the viral protein coats. This may not only increase the likelihood of antigenically novel viruses with an "orthodox" pathogenic potential emerging, but may result in novel pathogenetic effects. For example, some strains of fowl plague virus (an influenza type A virus) cause severe generalized disease with high mortality in some avian species. If a human influenza virus evolved that combined such a pathogenic potential with novel coat antigens that allowed it to spread as readily as did influenza A2 (H2N2) in 1957 or influenza A2 (H3N2) in 1968, a pandemic of unequalled severity could result.

As with human viral infections, the great concentrations of livestock that characterize "industrial agriculture" in some Western countries, and the widespread shipment and aggregation of large numbers of animals for fattening, provide greatly increased opportunities for the spread of "rare" viruses through large numbers of animals and for the emergence of viral diseases of livestock.

It is not easy to see how to control the human respiratory viral infections. Vaccines are practicable and sensible for only a few relatively serious diseases; the vast majority of viral infections do not justify either the cost or the risk of vaccination. Yet in the aggregate these "minor" infections are important in producing general ill health and potentiating the adverse effects of inspired pollutants. "Air sanitation" is effective only under very special and local conditions; we can never expect to produce "sanitized" air as we do expect to produce safe drinking water, although considerable improvement should be possible in the quality of air provided to crowded public places. The most logical approach would be chemoprophylaxis. If effective specific antiviral drugs are produced, however, their use will carry all the risks associated with existing antibacterial chemotherapeutic agents, including toxic side effects (T. Smith, 1964) and the emergence of drug-resistant mutants.

**Recent Social Changes and Viral Diseases.**   Social changes and medical innovations since World War II have increased the incidence of some viral diseases, as well as decreasing the prevalence and severity of others. Perhaps the most disturbing trend is the greatly increased spread of hepatitis viruses. Hepatitis B virus was probably spread by the use of contaminated syringes in prewar venereal disease and diabetic clinics (Flaum et al., 1926), but it became a much more common disease with its unwitting inclusion in some vaccines (e.g., yellow fever vaccine, Sawyer et al., 1944) and with the greatly expanded use of human blood and blood products for therapeutic purposes (reviews, Cossart, 1972; Maycock, 1972). More recently, serum hepatitis has produced serious difficulties in the operation of hemodialysis units, affecting both patients and staff (Marmion and Tonkin, 1972). Finally, serum hepatitis has increased greatly in association with the great increase in drug abuse (Sutnik *et al.*, 1971).

Hepatitis B virus in early batches of yellow fever vaccine was due to the use of human serum as a "protective" factor. The widespread occurrence of unrecognized viruses in cell cultures, particularly primary cell cultures (see Chapter 12) furnishes another source of contamination of vaccines, highlighted by the occurrence of avian leukosis viruses in many vaccines made in chick embryos, of SV40 in poliovirus vaccines made in monkey kidney cells, and of the scrapie agent in some preparations of louping-ill vaccine made in sheep brain.

Advanced surgical procedures, notably organ transplantation, have also produced a crop of novel viral problems, due, in part, to massive blood transfusions with fresh blood (EB virus and cytomegaloviruses, see Chapter 12). The prolonged immunosuppressive therapy needed to control transplant rejection renders such patients subject to severe disease after normally trivial infections and may be associated with recurrent disease in some persistent viral infections.

## Human Colonization and Viral Diseases

Until the development of ocean-going ships in Europe in about the fourteenth century, urban and rural mankind was largely confined to the continents in which his progenitors had lived as primitive agriculturists and as nomads. The explosive period of European colonization of all the other continents had profound effects on disease patterns in the European migrants and in the indigenes. European man took his endemic diseases to new virgin populations, and there were explosive outbreaks of measles and smallpox, for example, in the native inhabitants of America and Oceania. European man likewise intruded on situations in which the indigenous inhabitants had acquired, by natural selection, considerable genetic resistance to diseases lethal to the European, e.g., falciparum malaria and yellow fever in Africa. The slave ships of the sixteenth century took the vector of urban yellow fever, *Aedes aegypti*, and the yellow fever virus to South America, where yellow fever virus became established in new jungle reservoir hosts and vectors (see Chapter 16).

The European colonists took their domesticated animals with them. Cattle, sheep, and horses, in the large herds used for pastoral purposes, encountered almost disease-free environments in Australia, where the native animal population consisted of sparse marsupials; but in Africa they were exposed to many viruses and other parasites which had evolved with the great herds of ungulates. Many of the viruses were arthropod-borne, and in the new host animals they caused devastating epizootics of diseases like bluetongue, foot-and-mouth disease, rinderpest, Rift Valley fever, Nairobi sheep disease, African horse sickness, and African swine fever. Having become established in the sheep and cattle of the colonists in Africa, some of these diseases were then transported to other continents, just as yellow fever had been; thus, in recent times, bluetongue virus became established in Europe and the United States and African horse sickness spread to India. The pastoral industries of Australia, the last continent colonized and free of serious indigenous viral diseases, have been maintained free of the great scourges of sheep, cattle, and horses by strict quarantine.

The alteration of the environment by man may also lead to other major alterations in the incidence of viral diseases. A good example of this is furnished by the history of togavirus infections in southern California (Reeves and Hammon, 1962). The floor of the San Joaquin Valley was largely a desert until man transformed it by irrigation into a rich agricultural area. Under natural conditions the togaviruses were either absent or persisted only in limited areas, but the irrigation channels and vegetation that it produced provided an almost ideal environment for mosquitoes, birds, and St. Louis and Western equine encephalitis viruses, both of which cause disease in man and the horse. The widespread construction of man-made lakes for irrigation and power production, particularly in poor heavily populated tropical regions, will lead to major ecological changes in their environs that may have important consequences in terms of arbovirus infections as well as with diseases like schistosomiasis and malaria.

The handling of exotic animals or of material derived from wild animals carries unknown risks of serious human viral disease. Two instances are well known, in both of which spread of the agent was fortunately restricted to a small number of human cases; namely Lassa fever (Frame *et al.*, 1970; Leifer *et al.*, 1970) and the Marburg agent (review, Smith, 1971).

Finally, many virologists who work with tumor viruses are becoming increasingly concerned about the safety of handling such viruses. Not only must years elapse before any cancer caused by a virus thus acquired would occur, but the risk has taken on new dimensions with the production of truly new viruses that have oncogenic potential (the nondefective adenovirus—SV40 hybrids; see Chapter 13), and because the kind of biochemical investigation now fashionable requires the handling of large amounts of concentrated virus.

## Air Travel and Viral Diseases

As long as travel from the endemic and enzootic centers of disease to disease-free areas was relatively slow, quarantine restrictions operated reasonably effectively in keeping out such diseases as smallpox from the United States and the countries of western Europe. Air travel poses much greater problems, since movement from any part of the world to another takes less than 36 hours, and the volume of air transport is increasing at a rapid rate. In 1957, pandemic influenza spread around the world mainly at the speed of ship and train travel (Stuart-Harris, 1965); in 1968, the Hong Kong virus was rapidly distributed around the world by air travel, although it did not immediately cause epidemic disease.

The threats posed by the rapid transport of viruses and their vectors from one part of the world to another have led to a series of countermeasures, which have been approved by most nations of the world. These included (until its recent relaxation by the United States and some other countries) compulsory smallpox vaccination for all international travelers, the designation of "yellow fever-free" airports, and the spraying of aircraft to kill insects. These measures, and quarantine, are reasonably effective in preventing the spread of a few serious diseases of man, and most diseases of veterinary importance. They are quite ineffective in controlling the respiratory viruses and most of the enteric viruses, although compulsory vaccination of travellers with poliovirus vaccine and eventually, if need be, with measles virus vaccine, are practicable public health measures. Neither of these viruses could cause serious epidemics in a well-vaccinated community, but without the psychological stimulus provided by the actual presence of a disease in a community considerable effort is required to maintain effective vaccination.

### "New" Viral Diseases

Apart from the transfer of viruses enzootic in one host to a novel host, in which they may produce a self-maintaining disease of some severity, several "new" viral diseases have been recognized during the past century.

**Paralytic Poliomyelitis.** Epidemic paralytic poliomyelitis appears to have been a relatively new disease, appearing at about the turn of the nineteenth century and now disappearing. Of all the enteroviruses, only the polioviruses are important causes of disease of the central nervous system in man. Many isolates of these viruses from human feces and sewage are avirulent, in the sense that they do not cause paralysis when injected intracerebrally into monkeys, and would be highly unlikely to produce poliomyelitis in man (Ramos-Alvarez and Sabin, 1954). Although there is some evidence that a disease that may have been paralytic poliomyelitis occurred in early antiquity, modern epidemic paralytic poliomyelitis first appeared in Sweden about 80 years ago, and became increasingly prevalent in Western countries as standards of sanitation improved and greater numbers of individuals first contracted infection as adolescents or young adults, since children and adolescents are much more likely to suffer paralytic disease after poliovirus infection than are infants and very young children (Burnet, 1952; Bodian and Horstmann, 1965). The increase in paralytic poliomyelitis since 1888 may not have been only a reflection of changes in the standards of hygiene and sanitation, leading to postponement of poliovirus infection until adolescence. Reliable evidence is meager, but there appear to have been no epidemics of infantile poliomyelitis in the under-developed areas of the world before 1900. Since 1940, epidemics have occurred in many such areas without there being significant changes in the mode of life. Likewise "virgin-soil" epidemics of poliomyelitis in the Eskimos and in the Pacific islands have only occurred since 1900, although there was considerable opportunity for the introduction of polioviruses with European and Asian traders before then. The picture is compatible with the occurrence and circulation of many more virulent strains of polioviruses since 1900 than before that time. Such strains, like all human viruses, are as mobile as man and have been widely distributed by military operations and by increased world-wide travel (Sabin,

1963). Since they multiply more readily in the pharynx than do more attenuated strains and probably initiate infection with smaller doses (Bodian and Horstmann, 1965), these more virulent strains spread more effectively than the natural attenuated strains.

**Influenza.** Antigenic drift in the influenza viruses leads to the periodic appearance of "new" viruses, derived by mutation of viruses previously circulating in human hosts, but the herd immunity of the human population prevents their pandemic spread. Pandemics appear to occur only when a virus appears that has a completely novel hemagglutinin (like the Hong Kong virus, in 1968), or novel hemagglutinin and neuraminidase peplomers (like the Asian virus, in 1957). Earlier, we set out persuasive evidence that these new viruses arose within nonhuman animal hosts, either by recombination with an existing human strain or by mutation of an "animal" virus and its effective spread to and among human beings.

**Dengue Shock Syndrome.** The increased mobility of modern man may lead to the development of essentially new diseases without changes in the causative viruses. This appears to be the case with hemorrhagic dengue fever. Dengue fever has long been known as a relatively mild arthropod-borne disease of man associated with a rash. The four serotypes of dengue virus have cross-reacting antigenic determinants, but the cross-reactivity does not provide lasting protection against heterologous serotypes (Whitehead et al., 1970), and sequential infections with different serotypes, with viremia, occur. Although occasional cases of hemorrhagic fever had been described in earlier outbreaks of dengue (Lumley and Taylor, 1943), a remarkable change in their incidence was noticed in the Philippines in 1954, and since then in many other countries of South and Southeast Asia (Halstead, 1966). Severe and often fatal febrile disease characterized by hemorrhage and shock has become a relatively common form of dengue especially among the local inhabitants of highly endemic areas. Epidemiological and serological observations on dengue shock syndrome suggest that this "new" disease is always associated with a second infection with a heterologous serotype (Halstead et al., 1967; Winter el al., 1969). The probable pathogenetic mechanism was described in Chapter 10; in essence it is thought that the accelerated antibody response during a heterologous viremic second infection may lead to the formation of immune complexes and complement deviation, possibly leading to the formation of the potent permeability factor C3a (review, Russell, 1971).

Of interest here are the ecological circumstances that have led to the appearance of the dengue shock syndrome, namely the increased opportunity for the infection of children with heterologous serotypes of dengue virus, a natural consequence of the increasing movement of people between villages and towns, due to better transport facilities. Movement of infected subjects could thus set up new foci, and those susceptible to second attacks move to areas where other serotypes are endemic. It is likely that as these tendencies increase hemorrhagic dengue will appear in other areas where it is at present unknown, such as coastal New Guinea.

## THE ERADICATION OF VIRAL DISEASES

Monotypic single-host viruses that do not produce recurrent transmissible disease after recovery from the initial infection can survive only in communities with a relatively high annual input of new susceptibles. Measles in man provides the best example, but smallpox and rubella are comparable; infections with rhinoviruses and enteroviruses are also probably in this situation, although because homotypic immunity in these superficial infections is less durable, the critical population size is smaller. Each of the many serotypes that cause identical symptoms constitutes a separate problem in herd immunity, so that disappearance of a particular syndrome in a small community requires the disappearance of many viruses.

Since this dependence for survival on a critical community size is due to herd immunity it should be possible to render even very large communities immune by actively immunizing a large proportion of susceptible individuals, but not necessarily all. This has led to the concept of the total eradication of infection, which may be practicable with some viral diseases, on a national or regional rather than a world-wide scale (Andrews and Langmuir, 1963). Cockburn (1961) has been one of the most vigorous advocates of global eradication, for he points out that regional eradication is basically unstable, since infection may at any time be reintroduced by carriers or vectors from outside. However, for many infectious agents the concept of global or even continental eradication is biologically unsound. Yellow fever provides an excellent example, for it was the first disease for which eradication was proposed as a deliberate program. Initially success seemed assured, but then the discovery of jungle yellow fever made it obvious that eradication of the virus was impossible (see Chapter 16). Global eradication of diseases that produce recurrent activity after long periods of quiescence, or that have an animal reservoir, constitutes an unreal goal (Dubos, 1965).

Local or regional eradication of some viral infections, however, is relatively easy and has been achieved with smallpox in most developed countries, and probably with poliomyelitis in a few places, e.g., Sweden. In view of the large critical population size needed to maintain endemicity, measles lends itself to local eradication by vaccination on a national scale, with the highly mobile population that characterizes Western society. If a virus is eradicated from a particular country a major public health problem is raised concerning the continued protection of the population. Unless eradication is global the virus will sooner or later be reintroduced by travelers. Protection could probably be maintained by continued large-scale vaccination, but the difficulties of achieving this in the apparent absence of disease have been made apparent by experience with compulsory vaccination against smallpox and more recently with poliovirus and measles vaccination. Some measure of protection of a virus-free nonvaccinated community can be achieved by quarantine (although this would be impossible with poliomyelitis, because of symptomless carriers) and by the demand for evidence of recent vaccination. The latter measure could well be extended from the earlier requirement for vaccination against smallpox, now being relaxed, to include vaccination against the polioviruses and possibly against measles virus.

The experience gained during the last decade in the WHO-sponsored campaign for the global eradication of smallpox points up the sociological difficulties of eradication. Smallpox is everywhere a severe and feared disease and there is no animal host of variola virus (although monkeypox virus may cause sporadic cases of "smallpox" in man). Variola virus is monotypic, immunity persists for many years, and vaccination is easy and effective, i.e., it has all the prerequisites for a successful eradication campaign. Yet the WHO campaign has met with repeated setbacks mainly because of social rather than technical difficulties, and the world-wide incidence of smallpox in 1966 differed little from that observed for the past 20 years. Only in the last year or so has substantial eradication been achieved in many parts of Asia and Africa, and in South America (World Health Organ., 1972b), but if complete local eradication has not been achieved social upheavals like the war in Bangladesh could clearly set back the program there for a decade.

# Bibliography

Aaronson, S. (1970). Effect of ultraviolet irradiation on the survival of simian virus 40 functions in human and mouse cells. *J. Virol.* **6**, 393.

Aaronson, S. A. (1971a). Common genetic alterations of RNA tumor viruses grown in human cells. *Nature (London)* **230**, 445.

Aaronson, S. A. (1971b). Isolation of a rat-tropic helper virus from M-MSV-(O) stocks. *Virology* **44**, 29.

Aaronson, S. A. (1971c). Chemical induction of focus-forming virus from non-producer cells transformed by murine sarcoma virus. *Proc. Nat. Acad. Sci. U.S.* **68**, 3089.

Aaronson, S. A., and Rowe, W. P. (1970). Non-producer clones of murine sarcoma virus transformed BALB/3T3 cells. *Virology* **42**, 9.

Aaronson, S., and Todaro, G. J. (1968). Basis for the acquisition of malignant potential by mouse cells cultivated *in vitro*. *Science* **162**, 1024.

Aaronson, S., and Todaro, G. J. (1969). Human diploid cell transformation by DNA extracted from the tumor virus SV40. *Science* **166**, 390.

Aaronson, S. A., and Weaver, C. A. (1971). Characterization of murine sarcoma virus (Kirsten) transformation of mouse and human cells. *J. Gen. Virol.* **13**, 245.

Aaronson, S. A., Hartley, J. W., and Todaro, G. J. (1969). Mouse leukemia virus: "spontaneous" release by mouse embryo cells after long-term *in vitro* cultivation. *Proc. Nat. Acad. Sci. U.S.* **64**, 87.

Aaronson, S. A., Jainchill, J. L., and Todaro, G. J. (1970). Murine sarcoma virus transformation of BALB/3T3 cells: lack of dependence on murine leukemia virus. *Proc. Nat. Acad. Sci. U.S.* **66**, 1236.

Aaronson, S. A., Todaro, G. J., and Scolnick, E. M. (1971a). Induction of murine C-type viruses from clonal lines of virus-free BALB/3T3 cells. *Science* **174**, 157.

Aaronson, S. A., Parks, W. P., Scolnick, E. M., and Todaro, G. J. (1971b). Antibody to the RNA-dependent DNA polymerase of mammalian C-type RNA tumor viruses. *Proc. Nat. Acad. Sci. U.S.* **68**, 920.

Aaronson, S. A., Bassin, R. H., and Weaver, C. A. (1972). Comparison of mouse sarcoma viruses in non-producer and S⁺L⁻ transformed cells. *J. Virol.* **9**, 701.

Abel, P. (1962a). Topography in vaccinia genetics. *Virology* **16**, 347.

Abel, P. (1962b). Multiplicity reactivation and marker rescue with vaccinia virus. *Virology* **17**, 511.

Abercrombie, M., and Ambrose, E. J. (1958). Interference microscope studies of cell contacts in tissue culture. *Exp. Cell Res.* **15**, 332.

Abercrombie, M., and Heaysman, J. E. (1954). Observations of the social behaviour of cells in tissue culture. II. "Monolayering" of fibroblasts. *Exp. Cell Res.* **6**, 293.

Abodeely, R. A., Lawson, L. A., and Randall, C. C. (1970). Morphology and entry of enveloped and deenveloped equine abortion (herpes) virus. *J. Virol.* **5**, 513.

Acheson, N. H., and Tamm, I. (1967). Replication of Semliki Forest virus: An electron microscopic study. *Virology* **32**, 128.

Acheson, N. H., and Tamm, I. (1970a). Purification and properties of Semliki Forest virus nucleocapsids. *Virology* **41**, 306.

Acheson, N. H., and Tamm, I. (1970b). Structural proteins of Semliki Forest virus and its nucleocapsid. *Virology* **41**, 321.

Acheson, N. H., Buetti, E., Scherrer, R., and Weil, R. (1971). Transcription of the polyoma virus genome: Synthesis and cleavage of giant late polyoma-specific RNA. *Proc. Nat. Acad. Sci. U.S.* **68**, 2231.

Acs, G., Klett, H., Schonberg, M., Christman, J., Levin, D. H., and Silverstein, S. C. (1971). Mechanism of reovirus double-stranded ribonucleic acid synthesis *in vivo* and *in vitro*. *J. Virol.* **8**, 684.

Ada, G. L., and Gottschalk, A. (1956). The component sugars of the influenza-virus particle. *Biochem. J.* **62**, 686.

Ada, G. L., and Perry, B. T. (1954). The nucleic acid content of influenza virus. *Aust. J. Exp. Biol. Med. Sci.* **32**, 453.

Ada, G. L., and Perry, B. T. (1956). Influenza virus nucleic acid: Relationship between biological characteristics of the virus particle and properties of the nucleic acid. *J. Gen. Microbiol.* **14**, 623.

Ada, G. L., Nossal, G. J. V., Pye, J., and Abbot, A. (1963). Behaviour of active bacterial antigens during the induction of the immune response. I. Properties of flagellar antigens from Salmonella. *Nature (London)* **199**, 1257.

Adamson, J. D., Moody, J. P., Peart, A. F. W., Smillie, R. A., Wilt, J. C., and Wood, W. J. (1949). Poliomyelitis in the Arctic. *Can. Med. Ass. J.* **61**, 339.

Adesnik, M. (1971). Polyacrylamide gel electrophoresis of viral RNA. *In* "Methods in Virology" (K. Maramorosch and H. Koprowski, eds.), Vol. V, p. 126. Academic Press, New York.

Agol, V. I., and Shirman, G. A. (1965). Formation of virus particles via enzyme systems and structural proteins induced by another "assistant" virus. *Vopr. Virusol.* **10**, 8. [English transl. *Fed. Proc. Fed. Amer. Soc. Exp. Biol.* **25**, T315 (1966).]

Ahlström, C. G., and Forsby, N. (1962). Sarcomas in hamsters after injection with Rous chicken tumor material. *J. Exp. Med.* **115**, 839.

Ahmed, M., Mayyasi, S. A., Chopra, H. C., Zelljadt, I., and Jensen, E. M. (1971). Mason-Pfizer monkey virus isolated from spontaneous mammary carcinoma of a female monkey. I. Detection of virus antigen by immunodiffusion immunofluorescent and virus agglutination techniques. *J. Nat. Cancer Inst.* **46**, 1325.

Akers, T. G. (1969). Survival of airborne virus, phage and other minute microbes. *In* "An Introduction to Experimental Aerobiology" (R. L. Dimmick and A. B. Akers, eds.), p. 296. Wiley (Interscience), New York.

Akers, T. G., and Cunningham, C. H. (1968). Replication and cytopathogenicity of avian infectious bronchitis virus in chicken embryo kidney cells. *Arch. Ges. Virusforsch.* **25**, 30.

Albanese, M., Bynoe, M. L., and Tyrrell, D. A. J. (1966). Studies of a strain of a herpes virus isolated from a case of upper respiratory disease. *Arch. Ges. Virusforsch.* **18**, 356.

Albrecht, P. (1962). Pathogenesis of experimental infection with tick-borne encephalitis virus. *In* "Biology of Viruses of the Tick-borne Encephalitis Complex" (H. Libíková, ed.), p. 247. Academic Press, New York.

Albrecht, P. (1968). Pathogenesis of neurotropic arbovirus infections. *Curr. Topics Microbiol. Immunol.* **43**, 44.

Albrecht, P., Blaškovič, D., Styk, B., and Koller, M. (1963). Course of A2 influenza in intranasally infected mice examined by the fluorescent antibody technique. *Acta Virol. (Prague) Engl. Ed.* **7**, 405.

Alexander, P., and Evans, R. (1971). Endotoxin and double-stranded RNA render macrophages cytotoxic. *Nature New Biol.* **232,** 76.

Al-lami, F., Ledinko, N., and Toolan, H. W. (1969). Electron microscope study of human NB and SMH cells infected with the parvovirus, H-1: Involvement of the nucleolus. *J. Gen. Virol.* **5,** 485.

Allen, D. W., and Sarma, P. S. (1972). Identification and localization of avian leukosis virus group-specific antigen within "leukosis-free" chick embryos. *Virology* **48,** 624.

Allen, D. W., Sarma, P. S., Niall, H. D., and Sauer, R. (1970). Isolation of a second group-specific antigen (gs-G) from avian myeloblastosis virus. *Proc. Nat. Acad. Sci. U.S.* **67,** 837.

Allison, A. C. (1965). Genetic factors in resistance against virus infections. *Arch. Ges. Virusforsch.* **17,** 280.

Allison, A. C. (1966). Immune responses to Shope fibroma virus in adult and newborn rabbits. *J. Nat. Cancer Inst.* **36,** 869.

Allison, A. C. (1967a). Lysosomes in virus-infected cells. *Perspect. Virol.* **5,** 29.

Allison, A. C. (1967b). Virus infections and virus-induced tumors. *Brit. Med. Bull.* **23,** 60.

Allison, A. C. (1971a). New cell antigens induced by viruses and their biological significance. *Proc. Roy. Soc.* **B 177,** 23.

Allison, A. C. (1971b). The role of membranes in the replication of animal viruses. *Intern. Rev. Exp. Pathol.* **10,** 182.

Allison, A. C. (1972a). Immunity against viruses. *In* "The Scientific Basis of Medicine." Annual Review, London, Athlone Press.

Allison, A. C. (1972b). Immune responses in persistent virus infections. *J. Clin. Pathol.* **6,** 121.

Allison, A. C., and Burns, W. H. (1972). Immunogenicity of animal viruses. *In* "Immunogenicity" (F. Borek, ed.), p. 155. North Holland, Amsterdam.

Allison, A. C., and Law, L. W. (1968). Effect of antilymphocyte serum on viral oncogenesis. *Proc. Soc. Exp. Biol. Med.* **127,** 207.

Allison, A. C., and Mallucci, L. (1965). Histochemical studies of lysosomes and lysosomal enzymes in virus-infected cell cultures. *J. Exp. Med.* **121,** 463.

Allison, A. C., and Sandelin, K. (1963). Activation of lysosomal enzymes in virus-infected cells and its possible relationship to cytopathic effects. *J. Exp. Med.* **117,** 879.

Allison, A. C., Berman, L. D., and Levey, R. H. (1967). Increased tumour production by adenovirus type 12 in thymectomized mice and mice treated with anti-lymphocyte serum. *Nature (London)* **215,** 185.

Almeida, J. D., and Lawrence, G. D. (1969). Heated and unheated antiserum on rubella virus. *Amer. J. Dis. Child.* **118,** 101.

Almeida, J. D., and Tyrrell, D. A. J. (1967). The morphology of three previously uncharacterized human respiratory viruses that grow in organ culture. *J. Gen. Virol.* **1,** 175.

Almeida, J. D., and Waterson, A. P. (1967). Some observations on the envelope of an influenza virus. *J. Gen. Microbiol.* **46,** 107.

Almeida, J. D., and Waterson, A. P. (1969). Immune complexes in hepatitis. *Lancet* **ii,** 953.

Almeida, J. D., Waterson, A. P., and Plowright, W. (1967). The morphological characteristics of African swine fever virus and its resemblance to *Tipula* iridescent virus. *Arch. Ges. Virusforsch.* **20,** 392.

Almeida, J. D., Waterson, A. P., Prydie, J., and Fletcher, E. W. L. (1968). The structure of a feline picornavirus and its relevance to cubic viruses in general. *Arch. Ges. Virusforsch.* **25,** 105.

Almeida, J. D., Waterson, A. P., Trowell, J. M., and Neale, G. (1970). The finding of virus-like particles in two Australia-antigen-positive human livers. *Microbios* **6,** 145.

Almeida, J. D., Rubenstein, D., and Stott, E. J. (1971). New antigen-antibody system in Australia-antigen-positive hepatitis. *Lancet* **ii,** 1225.

Aloni, Y., and Attardi, G. (1971). Symmetrical *in vivo* transcription of mitochondrial DNA in HeLa cells. *Proc. Nat. Acad. Sci. U.S.* **68,** 1757.

Aloni, Y., Winocour, E., and Sachs, L. (1968). Characterization of the simian virus 40-specific RNA in virus-yielding and transformed cells. *J. Mol. Biol.* **31**, 415.

Aloni, Y., Winocour, E., Sachs, L., and Torten, J. (1969). Hybridization between SV40 and cellular DNA's. *J. Mol. Biol.* **44**, 333.

Alper, T., Haig, D. A., and Clarke, M. C. (1966). The exceptionally small size of the scrapie agent. *Biochem. Biophys. Res. Commun.* **22**, 278.

Altaner, C., and Svec, F. (1966). Virus production in rat tumors induced by chicken sarcoma virus. *J. Nat. Cancer Inst.* **37**, 745.

Altaner, C., and Temin, H. M. (1970). Carcinogenesis by RNA sarcoma viruses XII. A quantitative study of infection of rat cells *in vitro* by avian sarcoma viruses. *Virology* **40**, 118.

Amako, K., and Dales, S. (1967). Cytopathology of Mengovirus infection. II. Proliferation of membranous cisternae. *Virology* **32**, 201.

Amano, Y., Katagiri, S., Ishida, N., and Watanabe, Y. (1971). Spontaneous degradation of reovirus capsid into subunits. *J. Virol.* **8**, 805.

Anderer, F. A., Schlumberger, H. D., Koch, M. A., Frank, H., and Eggers, H. J. (1967). Structure of simian virus 40. II. Symmetry and components of the virus particle. *Virology* **32**, 511.

Anderer, F. A., Koch, M. A., and Schlumberger, H. D. (1968). Structure of simian virus 40. III. Alkaline degradation of the virus particle. *Virology* **34**, 452.

Anderson, C. W., and Gesteland, R. F. (1972). Pattern of protein synthesis in monkey cells infected with simian virus 40. *J. Virol.* **9**, 758.

Anderson, D. E. (1972). A genetic study of human breast cancer. *J. Nat. Cancer Inst.* **48**, 1029.

Anderson, G. W., and Rondeau, J. L. (1954). Absence of tonsils as a factor in the development of bulbar poliomyelitis. *J. Amer. Med. Ass.* **115**, 1123.

Anderson, J., Yates, V. J., Jasty, V., and Mancini, L. O. (1969). The *in vitro* transformation by an avian adenovirus (CELO). III. Human amnion cell cultures. *J. Nat. Cancer Inst.* **43**, 575.

Anderson, N. G., Harris, W. W., Barber, A. A., Rankin, C. T., and Candler, E. L. (1966). Separation of subcellular components and viruses by combined rate- and isopycnic zonal centrifugation. *Nat. Cancer Inst. Monogr.* **21**, 253.

Anderson, S. G., and Hamilton, J. (1949). The epidemiology of primary herpes simplex infection. *Med. J. Aust.* **1**, 308.

Andrewes, C. H. (1952). Classification and nomenclature of viruses. *Annu. Rev. Microbiol.* **6**, 119.

Andrewes, C. H. (1954). Nomenclature of viruses. *Nature (London)* **173**, 620.

Andrewes, C. H. (1965). "The Common Cold." Weidenfeld & Nicolson, London.

Andrewes, C. H. (1966). Viruses and evolution. The Huxley Lecture, Univ. of Birmingham, England.

Andrewes, C. H., and Ahlström, C. G. (1938). A transplantable sarcoma occurring in a rabbit inoculated with tar and infectious fibroma virus. *J. Pathol. Bacteriol.* **47**, 87.

Andrewes, C. H., and Allison, A. C. (1961). Newcastle disease as a model for studies of experimental epidemiology. *J. Hyg.* **59**, 285.

Andrewes, C. H., and Pereira, H. G. (1972). "Viruses of Vertebrates," 3rd Ed. Ballière Tindall, London.

Andrewes, C. H., Bang, F. B., and Burnet, F. M. (1955). A short description of the *Myxovirus* group (influenza and related viruses). *Virology* **1**, 176.

Andrews, J. M., and Langmuir, A. D. (1963). The philosophy of disease eradication. *Amer. J. Public Health* **53**, 1.

Annals (1960). Inactivation of viruses. *Ann. N.Y. Acad. Sci.* **83**, 513.

Anon. (1965). Proposals and recommendations of the provisional committee for nomenclature of viruses (P.C.N.V.). *Ann. Inst. Pasteur (Paris)* **109**, 625.

Anon. (1968). Coronaviruses. *Nature (London)* **220**, 650.

Aoki, T., Boyse, E. A., Old, L. J., de Harven, E., Hammerling, U., and Wood, H. A. (1970). G (Gross) and H-2 cell-surface antigens: location on Gross leukemia cells by electron microscopy with visually labeled antibody. *Proc. Nat. Acad. Sci. U.S.* **65**, 569.

Apostolov, K., and Flewett, T. H. (1969). Further observations on the structure of influenza virus. *J. Gen. Virol.* **4**, 365.

Apostolov, K., Flewett, T. H., and Kendall, A. P. (1970). Morphology of influenza A, B, C and infectious bronchitis virus (IBV). *In* "The Biology of Large RNA Viruses" (R. D. Barry and B. W. J. Mahy, eds.), p. 3. Academic Press, London and New York.

Appleyard, G., Hume, V. B. M., and Westwood, J. C. N. (1965). The effect of thiosemicarbazones on the growth of rabbitpox virus in tissue culture. *Ann. N.Y. Acad. Sci.* **130**, 92.

Arber, W., and Linn, S. (1969). DNA modification and restriction. *Annu. Rev. Biochem.* **38**, 467.

Arif, B. M., and Faulkner, P. (1972). Genome of Sindbis virus. *J. Virol.* **9**, 102.

Arlinghaus, R. B., and Polatnick, J. (1969). The isolation of two enzyme-ribonucleic acid complexes involved in the synthesis of foot-and-mouth disease virus ribonucleic acid. *Proc. Nat. Acad. Sci. U.S.* **62**, 821.

Armstrong, C. (1939). The experimental transmission of poliomyelitis to the eastern cotton rat. *Public Health Rept. (U.S.)* **54**, 1719.

Armstrong, D., Okuyan, M., and Huebner, R. J. (1964). Complement-fixing antigens in tissue cultures of avian leucosis viruses. *Science* **144**, 1584.

Armstrong, J. A., Edmonds, M., Nakazato, H., Phillips, B. A., and Vaughan, M. H. (1972). Polyadenylic acid sequences in the virion RNA of poliovirus and eastern equine encephalitis virus. *Science* **176**, 526.

Armstrong, R. W., and Merigan, T. C. (1971). Varicella-zoster virus: Interferon production and comparative interferon sensitivity in human cell cultures. *J. Gen. Virol.* **12**, 53.

Arnott, S., Wilkins, M. H. F., Fuller, W., and Langridge, R. (1967). Molecular and crystal structures of double-helical RNA. II. Determination of diffracted intensities for the $\alpha$ and $\beta$ forms of reovirus RNA and their interpretation in terms of groups of three RNA molecules. *J. Mol. Biol.* **27**, 525.

Asboe-Hansen, G. (1969). Hormone control of connective tissue. *Brit. J. Dermatol.* **81** (Suppl. 2), 2.

Ashburner, M. (1970). Structure and function of polytene chromosomes during insect development. *Advan. Insect Physiol.* **7**, 1.

Ashe, W. K., and Notkins, A. L. (1966). Neutralization of an infectious herpes simplex virus-antibody complex by anti-$\gamma$-globulin. *Proc. Nat. Acad. Sci. U.S.* **56**, 447.

Ashkenazi, A., and Melnick, J. L. (1963). Tumorigenicity of simian papovavirus SV40 and of virus-transformed cells. *J. Nat. Cancer Inst.* **30**, 1227.

Astell, C., Silverstein, S. C., Levin, D. H., and Acs, G. (1972). Regulation of the reovirus RNA transcriptase by a viral capsomere protein. *Virology* **48**, 648.

Attardi, G., and Amaldi, F. (1970). Structure and synthesis of ribosomal RNA. *Annu. Rev. Biochem.* **39**, 183.

Attardi, G., and Ojala, D. (1971). Mitochondrial ribosomes in HeLa cells. *Nature New Biol.* **229**, 133.

Attardi, G., Parnas, H., Huang, M-L. H., and Attardi, B. (1966). Giant size rapidly labeled nuclear ribonucleic acid and cytoplasmic messenger ribonucleic acid in immature duck erythrocytes. *J. Mol. Biol.* **20**, 145.

Aub, J. C., Tieslau, C., and Lankester, A. (1963). Reactions of normal and tumor cell surfaces to enzymes: I Wheat germ lipase and associated mucopolysaccharides. *Proc. Nat. Acad. Sci. U.S.* **50**, 613.

Aub, J. C., Sanford, B. H., and Cote, M. N. (1965a). Studies on reactivity of tumor and normal cells to a wheat germ agglutinin. *Proc. Nat. Acad. Sci. U.S.* **54**, 396.

Aub, J. C., Sanford, B. H., and Wang, L. (1965b). Reactions of normal and leukemic cell surfaces to a wheat germ agglutinin. *Proc. Nat. Acad. Sci. U.S.* **54**, 400.

Aubertin, A., and Kirn, A. (1969). Interférence entre le virus 3 de la grenouille et le virus de la vaccine. Inhibition de la réplication du DNA du virus vaccinal. *C. R. Acad. Sci. (Paris)* **D268**, 2838.

Aubertin, A., and McAuslan, B. R. (1972). Virion-associated nucleases: evidence for endonuclease and exonuclease activity in rabbitpox and vaccinia viruses. *J. Virol.* **9**, 554.

Aubertin, A., Palese, P., Tan, K. B., Vilagines, R., and McAuslan, B. R. (1971). Proteins of polyhedral cytoplasmic deoxyvirus. III. Further studies on the structure of FV3 and location of the virus-associated adenosine triphosphate phosphohydrolase. *J. Virol.* **8,** 634.

Axel, R., Schlom, J., and Spiegelman, S. (1972a). Presence in human breast cancer of RNA homologous to mouse mammary tumor virus RNA. *Nature (London)* **235,** 32.

Axel, R., Schlom, J., and Spiegelman, S. (1972b). Evidence for translation of viral-specific RNA in cells of a mouse mammary carcinoma. *Proc. Nat. Acad. Sci. U.S.* **69,** 535.

Axelrad, A. (1965). Antigenic behaviour of lymphoma cell populations in mice as revealed by the spleen colony method. *Progr. Exp. Tumor Res.* **6,** 31.

Aycock, W. L. (1942). Tonsillectomy and poliomyelitis. I. Epidemiologic considerations. *Medicine* **21,** 65.

Babbott, F. L., Jr., and Gordon, J. E. (1954). Modern measles. *Amer. J. Med. Sci.* **228,** 334.

Bablanian, R. (1972). Mechanisms of virus cytopathic effects. *Symp. Soc. Gen. Microbiol.* **22,** 359.

Bablanian, R., Eggers, H. J., and Tamm, I. (1965). Studies on the mechanism of poliovirus-induced cell damage. II. The relation between poliovirus growth and virus-induced morphological changes in cells. *Virology* **26,** 114.

Bachenheimer, S. L., Keiff, E. D., Lee, L. F., and Roizman, B. (1972a). Comparative studies on the DNA of Marek's disease and herpes simplex viruses. *In* "Oncogenesis and Herpesviruses" (P. M. Biggs, G. de-Thé, and L. N. Payne, eds.) p. 105. World Health Organization, International Agency for Research on Cancer, Lyon.

Bachenheimer, S. L., Kieff, E. D., and Roizman, B. (1972b). Unpublished studies quoted by Roizman *et al.* (1973).

Bachrach, H. L. (1966). Reactivity of viruses *in vitro. Progr. Med. Virol.* **8,** 214.

Bachrach, H. L. (1968). Foot-and-mouth disease. *Annu. Rev. Microbiol.* **22,** 201.

Bachrach, H. L., and Breese, S. S. (1968). Cell cultures and pure animal virus in quantity. *In* "Methods in Virology" (K. Maramorosch and H. Koprowski, eds.), Vol. IV, p. 351. Academic Press, New York.

Bachrach, U. (1972). Antiviral action of oxidized polyamines. *In* "Virus-Cell Interactions and Viral Antimetabolites" (D. Shugar, ed.), p. 149. Academic Press, New York.

Bader, J. P. (1965). The requirement for DNA synthesis in the growth of Rous sarcoma and Rous-associated viruses. *Virology* **26,** 253.

Bader, J. P. (1966). Metabolic requirements for infection by Rous sarcoma virus. I. The transient requirement for DNA synthesis. *Virology* **29,** 444.

Bader, J. P. (1970). Synthesis of the RNA of RNA-containing tumor viruses. I. The interval between synthesis and envelopment. *Virology* **40,** 494.

Bader, J. P. (1972a). Metabolic requirements for infection by Rous sarcoma virus. III. The synthesis of viral DNA. *Virology* **48,** 485.

Bader, J. P. (1972b). Metabolic requirements for infection by Rous sarcoma virus. IV. Virus reproduction and cellular transformation without cellular division. *Virology* **48,** 494.

Bader, J. P., and Bader, A. V. (1970). Evidence for a DNA replicative genome for RNA-containing tumor viruses. *Proc. Nat. Acad. Sci. U.S.* **67,** 843.

Bader, J. P., and Brown, N. R. (1971). Induction of mutations in an RNA tumor virus by an analogue of a DNA precursor. *Nature New Biol.* **234,** 11.

Bader, J. P., Brown, N. R., and Bader, A. V. (1970). Characteristics of cores of avian leukosarcoma viruses. *Virology* **41,** 718.

Balduzzi, P., and Morgan, H. R. (1970). Mechanism of oncogenic transformation by Rous sarcoma virus. I. Intracellular inactivation of cell-transforming ability of Rous sarcoma virus by 5-bromodeoxyuridine and light. *J. Virol.* **5,** 470.

Baltimore, D. (1968a). Inhibition of poliovirus replication by guanidine. *In* "Medical and Applied Virology" (M. Sanders and E. H. Lennette, eds.), p. 340. Green, St. Louis, Missouri.

Baltimore, D. (1968b). Structure of the poliovirus replicative intermediate RNA. *J. Mol. Biol.* **32,** 359.

Baltimore, D. (1969). The replication of picornaviruses. *In* "The Biochemistry of Viruses" (H. B. Levy, ed.), p. 101. Dekker, New York.

Baltimore, D. (1970). RNA-dependent DNA polymerase in virions of RNA tumor viruses. *Nature (London)* **226**, 1209.

Baltimore, D. (1971a). Polio is not dead. *Perspect. Virol.* **7**, 1.

Baltimore, D. (1971b). Expression of animal virus genomes. *Bacteriol. Rev.* **35**, 235.

Baltimore, D., and Franklin, R. M. (1963a). Effects of puromycin and *p*-fluoro-phenylalanine on Mengovirus ribonucleic acid and protein synthesis. *Biochim. Biophys. Acta* **76**, 431.

Baltimore, D., and Franklin, R. M. (1963b). A new ribonucleic acid polymerase appearing after Mengovirus infection of L cells *J. Biol. Chem.* **238**, 3395.

Baltimore, D., and Girard, M. (1966). An intermediate in the synthesis of poliovirus RNA. *Proc. Nat. Acad. Sci. U.S.* **56**, 741.

Baltimore, D., Eggers, H. J., Franklin, R. M., and Tamm, I. (1963). Poliovirus-induced RNA polymerase and the effects of virus-specific inhibitors on its production. *Proc. Nat. Acad. Sci. U.S.* **49**, 843.

Baltimore, D., Huang, A. S., and Stampfer, M. (1970). Ribonucleic acid synthesis of vesicular stomatitis virus. II. An RNA polymerase in the virion. *Proc. Nat. Acad. Sci. U.S.* **66**, 572.

Baltimore, D., Huang, A., Manly, K. F., Rekosh, D., and Stampfer, M. (1971). The synthesis of protein by mammalian RNA viruses. *In* "Strategy of the Viral Genome" (G. E. W. Wolstenholme and M. O'Connor, eds.), p. 101. Churchill Livingstone, Edinburgh.

Baluda, M. A. (1957). Homologous interference by ultraviolet-inactivated Newcastle disease virus. *Virology* **4**, 72.

Baluda, M. A. (1959). Loss of viral receptors in homologous interference by ultraviolet-irradiated Newcastle disease virus. *Virology* **7**, 315.

Baluda, M. (1972). Widespread presence in chickens of DNA complementary to the RNA genome of avian leukosis viruses. *Proc. Nat. Acad. Sci. U.S.* **69**, 576.

Baluda, M. A., and Goetz, I. E. (1961). Morphological conversion of cell cultures by avian myeloblastosis virus. *Virology* **15**, 185.

Baluda, M. A., and Nayak, D. P. (1969). Incorporation of precursors into ribonucleic acid, protein, glycoprotein and lipoprotein of avian myeloblastosis virions. *J. Virol.* **4**, 554.

Baluda, M. A., and Nayak, D. P. (1970). DNA complementary to viral RNA in leukemic cells induced by avian myeloblastosis virus. *Proc. Nat. Acad. Sci. U.S.* **66**, 329.

Banatvala, J. E. (1970). Rubella. *Mod. Trends Med. Virol.* **1**, 116.

Banerjee, A. K., and Shatkin, A. J. (1971). Guanosine-5'-diphosphate at the 5' termini of reovirus RNA: evidence for a segmented genome within the virion. *J. Mol. Biol.* **61**, 643.

Banerjee, A. K., Ward, R., and Shatkin, A. J. (1971). Initiation of reovirus mRNA synthesis *in vitro. Nature New Biol.* **230**, 169.

Bang, F. B., and Warwick, A. (1960). Mouse macrophages as host cells for the mouse hepatitis virus and the genetic basis of their susceptibility. *Proc. Nat. Acad. Sci. U.S.* **46**, 1065.

Banks, G. T., Buck, K. W., Chain, E. B., Himmelweit, F., Marks, J. E., Tyler, J. M., Hollings, M., Last, F. T., and Stone, O. M. (1968). Viruses in fungi and interferon stimulation. *Nature (London)* **218**, 542.

Baranska, W., Koldovsky, P., and Koprowski, H. (1970). Antigenic study of unfertilized mouse eggs: cross reactivity with SV40 induced antigens. *Proc. Nat. Acad. Sci. U.S.* **67**, 193.

Barban, S., and Goor, R. S. (1971). Structural proteins of simian virus 40. *J. Virol.* **7**, 198.

Barbanti-Brodano, G., Swetly, P., and Koprowski, H. (1970). Superinfection of simian virus 40-transformed permissive cells with simian virus 40. *J. Virol.* **6**, 644.

Barbanti-Brodano, G., Possati, L., and La Placa, M. (1971). Inactivation of polykaryocytogenic and hemolytic activities of Sendai virus by phospholipase B (lysolecithinase). *J. Virol.* **8**, 796.

Barer, R., Joseph, S., and Meek, G. A. (1959). The origin of nuclear membrane. *Expt. Cell Res.* **18**, 179.

Barnard, J. E., and Elford, W. J. (1931). The causative organism in infectious ectromelia. *Proc. Roy. Soc.* **B109**, 360.

Barnett, H. C. (1956). Experimental studies of concurrent infection of canaries and of the mosquito *Culex tarsalis* with *Plasmodium relictum* and Western equine encephalitis virus. *Amer. J. Trop. Med. Hyg.* **5,** 99.

Baron, S. (1963). Mechanism of recovery from viral infection. *Advan. Virus Res.* **10,** 39.

Baron, S. (1970). The biological significance of the interferon system. *Arch. Internal Med.* **126,** 84.

Baron, S. (1973). The defensive and biological roles of the interferon system. *In* "Interferons and Interferon Inducers" (N. B. Finter, ed.), p. 267. North Holland, Amsterdam.

Baron, S., Buckler, C. E., McCloskey, R. V., and Kirschstein, R. L. (1966a). Role of interferon during viremia. I. Production of circulating interferon. *J. Immunol.* **96,** 12.

Baron, S., Buckler, C. E., Friedman, R. M., and McCloskey, R. V. (1966b). Role of interferon during viremia. II. Protective action of circulating interferon. *J. Immunol.* **96,** 17.

Baron, S., Buckler, C. E., Levy, H. B., and Friedman, R. M. (1967). Some factors affecting the interferon-induced antiviral state. *Proc. Soc. Exp. Biol. Med.* **125,** 1320.

Barron, A. L. (1971). Reactions of viruses in agar gel *In* "Methods in Virology" (K. Maramorosch and H. Koprowski, eds.), Vol. V. p. 347. Academic Press, New York.

Barry, R. D. (1961a). The multiplication of influenza virus. I. The formation of incomplete virus. *Virology* **14,** 389.

Barry, R. D. (1961b). The multiplication of influenza virus. II. Multiplicity reactivation of ultraviolet-irradiated virus. *Virology* **14,** 398.

Barry, R. D. (1962). Failure of Newcastle disease to undergo multiplicity reactivation. *Nature (London)* **193,** 96.

Barry, R. D. (1964). The effects of actinomycin D and ultraviolet irradiation on the production of fowl plague virus. *Virology* **24,** 563.

Barry, R. D., Ives, D. R., and Cruickshank, J. G. (1962). Participation of deoxyribonucleic acid in the multiplication of influenza virus. *Nature (London)* **194,** 1139.

Bartlett, M. S. (1957). Measles periodicity and community size. *J. Roy. Statist. Soc.* **120,** 48.

Bartlett, M. S. (1960). The critical community size for measles in the United States. *J. Roy. Statist. Soc.* **123,** 37.

Basilico, C. (1971). The multiplication of polyoma virus in mouse-hamster somatic hybrids *In* "Lepetit Colloquia on Biology and Medicine, Vol. 2: The Biology of Oncogenic Viruses" (L. Sylvestri, ed.), p. 12. North Holland, Amsterdam.

Basilico, C., and Burstin, S. J. (1971). Multiplication of polyoma virus in mouse-hamster somatic hybrids: A hybrid cell which produces viral particles containing predominantly host deoxyribonucleic acid. *J. Virol.* **7,** 802.

Basilico, C., and di Mayorca, G. (1965). Radiation target size of the lytic and the transforming ability of polyoma virus. *Proc. Nat. Acad. Sci. U.S.* **54,** 125.

Basilico, C., and Joklik, W. K. (1968). Studies on a temperature-sensitive mutant of vaccinia virus strain WR. *Virology* **36,** 668.

Basilico, C., and Marin, G. (1966). Susceptibility of cells in different stages of the mitotic cycle to transformation by polyoma virus. *Virology* **28,** 429.

Basilico, C., and Wang, R. (1971). Susceptibility to superinfection of hybrids between polyoma "transformed" BHK and "normal" 3T3 cells. *Nature New Biol.* **230,** 105.

Basilico, C., Marin, G., and di Mayorca, G. (1966). Requirement for the integrity of the viral genome for the induction of host DNA synthesis by polyoma virus. *Proc. Nat. Acad. Sci. U.S.* **56,** 208.

Basilico, C., Matsuya, Y., and Green, H. (1969). Origin of the thymidine kinase induced by polyoma virus in productively infected cells. *J. Virol.* **3,** 140.

Bassin, R. H., Simons, P. J., Chesterman, F. C., and Harvey, J. J. (1968). Murine sarcoma virus (Harvey); characteristics of focus formation in mouse embryo cell cultures, and virus production by hamster tumor cells. *Intern. J. Cancer* **3,** 265.

Bassin, R. H., Tuttle, N., and Fischinger, P. J. (1970). Isolation of murine sarcoma virus-transformed mouse cells which are negative for leukemia virus from agar suspension cultures. *Intern. J. Cancer* **6,** 95.

Bassin, R. H., Phillips, L. A., Kramer, M. J., Haapala, D. K., Peebles, P. T., Nomura S., and Fischinger, P. J. (1971). Transformation of mouse 3T3 cells by murine sarcoma virus:

release of virus-like particles in the absence of replicating murine leukemia helper virus. *Proc. Nat. Acad. Sci. U.S.* **68**, 1520.

Bataillon, G. (1969). Etude de l'antigène spécifique du groupe des virus oncogènes aviaires dans les clones de cellules de hamster transformées par le virus de Rous. *C. R. Acad. Sci. (Paris)* **269**, 2156.

Bauer, D. J. (1965). Clinical experience with the antiviral drug Marboran (1-methylisatin 3-thiosemicarbazone). *Ann. N.Y. Acad. Sci.* **130** (1), 110.

Bauer, D. J. (ed.) (1972). "Chemotherapy of Virus Diseases," Vol. 1, Intern. Encyclop. Pharmacol. Therap. Section 61, Pergamon, New York.

Bauer, D. J., Apostolov, K., and Selway, J. W. T. (1970). Activity of methisazone against RNA viruses. *Ann. N.Y. Acad. Sci.* **173**, 314.

Bauer, H., Bubeněk, J., Graf, T., and Allgaier, C. (1969). Induction of transplantation resistance to Rous sarcoma isograft by avian leukosis virus. *Virology* **39**, 482.

Bauer, W., and Vinograd, J. (1968). The interaction of closed circular DNA with intercalating dyes I. The superhelix density of SV40 DNA in the presence and absence of dye. *J. Mol. Biol.* **33**, 141.

Bauer, W., and Vinograd, J. (1971). Sedimentation velocity experiments in the analytical ultracentrifuge. *In* "Procedures in Nucleic Acid Research" (G. L. Cantoni and D. R. Davies, eds.), Vol. 2, p. 297. Harper and Row, New York.

Baum, S. G., Reich, P. R., Hybner, C. J., Rowe, W. P., and Weissman, S. M. (1966). Biophysical evidence for linkage of adenovirus and SV40 DNA's in adenovirus 7-SV40 hybrid particles. *Proc. Nat. Acad. Sci. U.S.* **56**, 1509.

Baum, S. G., Wiese, W. H., and Rowe, W. P. (1970). Density differences between hybrid and non-hybrid particles in two adenovirus–simian virus 40 hybrid populations. *J. Virol.* **5**, 353.

Bawden, F. C. (1941). "Plant Viruses and Virus Diseases," Chronica Botan., Waltham, Massachusetts.

Bearcroft, W. G. C., and Jamieson, M. F. (1958). An outbreak of subcutaneous tumours in rhesus monkeys. *Nature (London)* **182**, 195.

Beard, J. W. (1963). Avian virus growths and their etiologic agents. *Advan. Cancer Res.* **7**, 1.

Beare, A. S., and Hall, T. S. (1971). Recombinant influenza-A viruses as live vaccines for man. *Lancet* **ii**, 1271.

Beare, A. S., Maassab, H. F., Tyrrell, D. A. J., Slepuškin, A. N., and Hall, T. S. (1971). A comparative study of attenuated influenza viruses. *Bull. WHO* **44**, 593.

Beare, A. S., Hall, T. S., and Tyrrell, D. A. J. (1972). Protection of volunteers against challenge with A/Hong Kong/68 influenza virus by a new adamantane compound. *Lancet* **i**, 1039.

Becht, H., Hämmerling, U., and Rott, R. (1971). Undisturbed release of influenza virus in the presence of univalent antineuraminidase antibodies. *Virology* **46**, 337.

Becker, W. B., McIntosh, K., Dees, J. H., and Chanock, R. M. (1967). Morphogenesis of avian infectious bronchitis and a related human virus (strain 229E). *J. Virol.* **1**, 1019.

Becker, Y., and Olshevsky, U. (1971). The molecular composition of herpes simplex virions. *In* "Viruses Affecting Man and Animals" (M. Sanders and M. Schaeffer, eds.), p. 73. W. H. Green, St. Louis, Missouri.

Becker, Y., and Olshevsky, U. (1972). Localization of structural viral peptides in the herpes simplex virion. *In* "Oncogenesis and Herpesviruses" (P. M. Biggs, G. de-Thé and L. N. Payne, eds.), p. 420. World Health Organization, International Agency for Research on Cancer, Lyon.

Becker, Y., and Sarov, I. (1968). Electron microscopy of vaccinia virus DNA. *J. Mol. Biol.* **34**, 655.

Becker, Y., Olshevsky, U., and Levitt, J. (1967). The role of arginine in the replication of herpes simplex virus. *J. Gen. Virol.* **1**, 471.

Becker, Y., Dym, H., and Sarov, I. (1968). Herpes simplex virus DNA. *Virology* **36**, 184.

Bedson, H. S., and Duckworth, M. J. (1963). Rabbitpox: An experimental study of the pathways of infection in rabbits. *J. Pathol. Bacteriol.* **85**, 1.

Bedson, H. S., and Dumbell, K. R. (1964). Hybrids derived from the viruses of variola major and cowpox. *J. Hyg.* **62**, 147.

Beem, M. (1967). Repeated infections with respiratory syncytial virus. *J. Immunol.* **98**, 115.

Beermann, W., and Clever, U. (1964). Chromosome puffs. *Sci. Amer.* **210** (4), 50.

Beijerinck, M. W. (1899). Ueber ein Contagium vivum fluidum als Ursache der Fleckenkrankheit der Tabaksblätter. *Zentr. Bakteriol. Parasitenk. Abt. II* **5**, 27.

Beisel, W. R. (1966). Effect of infection on human protein metabolism. *Fed. Proc. Fed. Amer. Soc. Exp. Biol.* **25**, 1682.

Béládi, I., Mucsi, I., Bakay, M., and Pusztai, R. (1970). Rescue of heat-inactivated adenoviruses types 2 and 6 by ultraviolet-irradiated adenovirus type 8. *J. Gen. Virol.* **7**, 153.

Bell, D., Wilkie, N. M., and Subak-Sharpe, J. H. (1971). Studies on arginyl transfer ribonucleic acid in herpesvirus-infected baby hamster kidney cells. *J. Gen. Virol.* **13**, 463.

Bell, J. A., Huebner, R. J., Rosen, L., Rowe, W. P., Cole, R. M., Mastrota, F. M., Floyd, T. M., Chanock, R. M., and Shvedoff, R. A. (1961). Illness and microbial experiences of nursery children at Junior Village. *Amer. J. Hyg.* **74**, 267.

Bellamy, A. R., and Joklik, W. K. (1967). Studies on reovirus RNA. II. Characterization of reovirus messenger RNA and of the genome RNA segments from which it is transcribed. *J. Mol. Biol.* **29**, 19.

Bellett, A. J. D. (1967a). Preliminary classification of viruses based on quantitative comparisons of viral nucleic acids. *J. Virol.* **1**, 245.

Bellett, A. J. D. (1967b). The use of computer-based quantitative comparisons of viral nucleic acids in the taxonomy of viruses: A preliminary classification of some animal viruses. *J. Gen. Virol.* **1**, 583.

Bellett, A. J. D. (1968). The iridescent virus group. *Advan. Virus Res.* **13**, 225.

Bellett, A. J. D., and Fenner, F. (1968). Studies of base-sequence homology among some cytoplasmic deoxyriboviruses of vertebrate and invertebrate animals. *J. Virol.* **2**, 1374.

Bellett, A. J. D., and Inman, R. B. (1967). Some properties of deoxyribonucleic acid preparations from *Chilo*, *Sericesthis*, and *Tipula* iridescent viruses. *J. Mol. Biol.* **25**, 425.

Bellett, A. J. D., and Younghusband, H. B. (1972). Replication of the DNA of chick embryo lethal orphan virus. *J. Mol. Biol.*, **72**, 691.

Bellett, A. J. D., Fenner, F., and Gibbs, A. J. (1973). The viruses. *In* "Viruses and Invertebrates" (A. J. Gibbs, ed.), in press. North Holland, Amsterdam.

Ben-Bassat, H., Inbar, M., and Sachs, L. (1970). Requirement for cell replication after SV40 infection for a structural change of the cell surface membrane. *Virology* **40**, 854.

Ben-Ishai, Z., Heller, E., Goldblum, N., and Becker, Y. (1969). Rifampicin, poxvirus and trachoma agent. *Nature (London)* **224**, 29.

Benjamin, T. L. (1965). Relative target sizes for the inactivation of the transforming and reproductive abilities of polyoma virus. *Proc. Nat. Acad. Sci. U.S.* **54**, 121.

Benjamin, T. L. (1966). Virus-specific RNA in cells productively infected or transformed by polyoma virus. *J. Mol. Biol.* **16**, 359.

Benjamin, T. L. (1970). Host range mutants of polyoma virus. *Proc. Nat. Acad. Sci. U.S.* **67**, 394.

Benjamin, T. (1972). Physiological and genetic studies of polyoma virus. *Curr. Topics Microbiol. Immunol.* **59**, 107.

Benjamin, T. L., and Burger, M. M. (1970). Absence of a cell membrane alteration function in non-transforming mutants of polyoma virus. *Proc. Nat. Acad. Sci. U.S.* **67**, 929.

Benjamin, T. L., and Norkin, L. (1973). Host range mutants of polyoma virus. *In* "Molecular Studies in Viral Neoplasia," 25th Annual Symposium on Fundamental Cancer Research. In press. University of Texas M.D. Anderson Hospital and Tumor Institute, Houston, Texas.

Bennett, I. L., Jr. and Nicastri, A. (1960). Fever as a mechanism of resistance. *Bacteriol. Rev.* **24**, 16.

Bennich, H., and Johansson, S. G. O. (1971). Structure and function of human immunoglobulin E. *Advan. Immunol.* **13**, 1.

Ben-Porat, T., and Kaplan, A. S. (1963). The synthesis and fate of pseudorabies virus DNA in infected mammalian cells in the stationary phase of growth. *Virology* **20**, 310.

Ben-Porat, T., and Kaplan, A. S. (1965). Mechanism of inhibition of cellular DNA synthesis by pseudorabies virus. *Virology* **25**, 22.

Ben-Porat, T., and Kaplan, A. S. (1971). Studies on the biogenesis of herpesvirus envelope. *Nature (London)* **235**, 165.

Bentvelzen, P., Daams, J. H., Hageman, P., and Calafat, J. (1970). Genetic transmission of viruses that initiate mammary tumor in mice. *Proc. Nat. Acad. Sci. U.S.* **67**, 377.

Bentvelzen, P., Daams, J. H., Hageman, P., Calafat, J., and Timmermans, A. (1972). Interactions between viral and genetic factors in the origin of mammary tumors in mice. *J. Nat. Cancer Inst.* **48**, 1089.

Benzer, S. (1959). On the topology of the genetic fine structure. *Proc. Nat. Acad. Sci. U.S.* **45**, 1607.

Berg, G. (ed.) (1967). "Transmission of Viruses by the Water Route." Interscience, New York.

Bergoin, M., and Dales, S. (1971). Comparative observations on poxviruses of invertebrates and vertebrates. *In* "Comparative Virology" (K. Maramorosch and E. Kurstak, eds.), p. 171. Academic Press, New York.

Bergold, G. H. (1958). Viruses of insects. *In* "Handbuch der Virusforschung" (G. Hallauer and K. F. Meyer, eds.) Vol. IV, p. 60. Springer, Vienna.

Berman, L. D., and Rowe, W. P. (1965). A study of antigens involved in adenovirus 12 tumorigenesis by immunodiffusion techniques. *J. Exp. Med.* **121**, 955.

Berman, L. D., and Sarma, P. S. (1965). Demonstration of an avian leucosis group antigen by immunodiffusion. *Nature (London)* **207**, 263.

Bernhard, W. (1958). Electron microscopy of tumor cells and tumor viruses. A review. *Cancer Res.* **18**, 491.

Bernhard, W., and Bauer, A. (1955). Mise en evidence de corpuscles d'aspect virusal dans les tumeurs mammaires de la souris. Étude au microscope électronique. *C. R. Acad. Sci. (Paris)* **240**, 1380.

Berns, A. J. M., Straus, G. J. A. M., and Bloemendal, H. (1972). Heterologous *in vitro* synthesis of lens α-crystallin polypeptide. *Nature New Biol.* **236**, 7.

Berns, K. I., and Adler, S. (1972). Separation of two types of adeno-associated virus particles containing complementary polynucleotide chains. *J. Virol.* **9**, 394.

Berns, K. I., and Silverman, C. (1970). Natural occurrence of cross-linked vaccinia virus deoxyribonucleic acid. *J. Virol.* **5**, 299.

Berry, D. M., and Almeida, J. D. (1968). The morphological and biological effects of various antisera on avian infectious bronchitis virus. *J. Gen. Virol.* **3**, 97.

Berry, G. P., and Dedrick, H. M. (1936). A method for changing the virus of rabbit fibroma (Shope) into that of infectious myxomatosis (Sanarelli). *J. Bacteriol.* **31**, 50.

Bertani, G. (1958). Lysogeny. *Advan. Virus Res.* **5**, 151.

Beveridge, W. I. B., and Burnet, F. M. (1946). The cultivation of viruses and rickettsiae in the chick embryo. *Med. Res. Council Spec. Rept. Ser.* **256**.

Bialy, H. S., and Colby, C. (1972). Inhibition of early vaccinia virus ribonucleic acid synthesis in interferon-treated chick embryo fibroblasts. *J. Virol.* **9**, 286.

Biggart, J. D., and Ruebner, B. H. (1970). Lymphoid necrosis in the mouse spleen produced by mouse hepatitis virus (MHV3): an electron microscopic study. *J. Med. Microbiol.* **3**, 627.

Biggers, D. C., Kraft, L. M., and Sprinz, H. (1964). Lethal intestinal virus infection of mice (LIVIM). *Amer. J. Pathol.* **45**, 413.

Biggs, P. M., and Payne, L. N. (1967). Studies on Marek's disease. I. experimental transmission. *J. Nat. Cancer Inst.* **39**, 267.

Biggs, P. M., Churchill, A. E., Rootes, D. G., and Chubb, R. C. (1968). The etiology of Marek's disease—an oncogenic herpes-type virus. *Perspect. Virol.* **6**, 211.

Billiau, A., Desmyter, J., and de Somer, P. (1970). Antiviral activity of chlorite-oxidized oxyamylose, a polyacetal carboxylic acid. *J. Virol.* **5**, 321.

Billiau, A., van den Berghe, H., and de Somer, P. (1972). Increased interferon release and morphological alteration in human cells by repeated exposure to double-stranded RNA. *J. Gen. Virol.* **14**, 25.

Billingham, R. E., Brent, L., and Medawar, P. B. (1956). Quantitative studies on tissue transplantation immunity. III. Actively acquired tolerance. *Phil. Trans. Roy. Soc. London* **B239**, 357.

Biquard, J. M., and Vigier, P. (1972). Characteristics of a conditional mutant of Rous sarcoma virus defective in ability to transform cells at high temperature. *Virology* **47**, 444.

Birnstiel, K. L., Chipchase, M., and Spiers, J. (1971). The ribosomal RNA cistrons. *Progr. Nucleic Acid Res. Mol. Biol.* **11**, 351.

Bishop, D. H. L. (1971). Complete transcription by the transcriptase of vesicular stomatitis virus. *J. Virol.* **7**, 486.

Bishop, D. H. L., and Roy, P. (1971). Kinetics of RNA synthesis by vesicular stomatitis virus particles. *J. Mol. Biol.* **57**, 513.

Bishop, D. H. L., and Roy, P. (1972). Dissociation of vesicular stomatitis virus and the relation of the virion proteins to the viral transcriptase. *J. Virol.* **10**, 234.

Bishop, D. H. L., Obijeski, J. F., and Simpson, R. W. (1971). Transcription of the influenza ribonucleic acid genome by a virion polymerase. II. Nature of the *in vitro* polymerase product. *J. Virol.* **8**, 74.

Bishop, D. H. L., Roy, P. Bean, W. J., and Simpson, R. W. (1972). Transcription of the influenza ribonucleic acid genome by a virion polymerase. III. Completeness of the transcription process. *J. Virol.* **10**, 689.

Bishop, J. M., and Levintow, L. (1971). Replicative forms of viral RNA; structure and function. *Progr. Med. Virol.* **13**, 1.

Bishop, J. M., Koch, G., Evans, B., and Merriman, M. (1969). Poliovirus replicative intermediate: Structural basis of infectivity. *J. Mol. Biol.* **46**, 235.

Bishop, J. M., Levinson, W. E., Quintrell, N., Sullivan, D., Fanshier, L., and Jackson, J. (1970a). The low molecular weight RNAs of Rous sarcoma virus I. The 4 S RNA. *Virology* **42**, 182.

Bishop, J. M., Levinson, W. E., Sullivan, D., Fanshier, L., Quintrell, N., and Jackson, J. (1970b). The low molecular weight of Rous sarcoma virus II. The 7 S RNA. *Virology* **42**, 927.

Bishop, J. O., Pemberton, R., and Baglioni, C. (1972). Reiteration frequently of hemoglobin genes in the duck. *Nature New Biol.* **235**, 231.

Biswal, S., McCain, B., and Benyesh-Melnick, M. (1971). The DNA of murine sarcoma-leukemia virus. *Virology* **45**, 697.

Bittner, J. J. (1936). Some possible effects of nursing on the mammary gland tumor incidence in mice. *Science* **84**, 162.

Black, F. L. (1966). Measles endemicity in insular populations: Critical community size and its evolutionary implication. *J. Theoret. Biol.* **11**, 207.

Black, F. L., and Melnick, J. L. (1955). Microepidemiology of poliomyelitis and herpes-B infection. *J. Immunol.* **74**, 236.

Black, F. L., Woodall, J. P., Evans, A. S., Liebhaber, H., and Henle, G. (1970). Prevalence of antibody against viruses in the Tiriyo, an isolated Amazon tribe. *Amer. J. Epidemiol.* **91**, 430.

Black, F. L., Hierholzer, W., Woodall, J. P., and Pinhiero, F. (1971). Intensified reactions to measles vaccine in unexposed populations of American Indians. *J. Infect. Dis.* **124**, 306.

Black, P. H. (1964). Studies on the genetic susceptibility of cells to polyoma virus transformation. *Virology* **24**, 179.

Black, P. H., and Rowe, W. P. (1963). SV40-induced proliferation of tissue culture cells of rabbit, mouse, and porcine origin. *Proc. Soc. Exp. Biol. Med.* **114**, 721.

Black, P. H., and Rowe, W. P. (1964). Viral studies of SV40 tumorigenesis in hamsters. *J. Nat. Cancer Inst.* **32**, 253.

Black, P. H., and White, B. J. (1967). *In vitro* transformation by adenovirus-SV40 hybrid viruses. II. Characteristics of the transformation of hamster cells by the adeno 2-, adeno 3- and adeno 12-SV40 viruses. *J. Exp. Med.* **125**, 629.

Black, P. H., Hartley, J. W., Rowe, W. P., and Huebner, R. J. (1963a). Transformation of bovine tissue culture cells by bovine papilloma virus. *Nature (London)* **199**, 1016.

Black, P. H., Rowe, W. P., Turner, H. C., and Huebner, R. J. (1963b). A specific complement-fixing antigen present in SV40 tumor and transformed cells. *Proc. Nat. Acad. Sci. U.S.* **50**, 1148.

Blacklow, N. R., Hoggan, M. D., and Rowe, W. P. (1967a). Isolation of adenovirus-associated viruses from man. *Proc. Nat. Acad. Sci. U.S.* **58**, 1410.

Blacklow, N. R., Hoggan, M. D., and Rowe, W. P. (1967b). Immunofluorescent studies of the potentiation of an adenovirus-associated virus by adenovirus 7. *J. Exp. Med.* **125**, 755.

Blacklow, N. R., Dolin, R., and Hoggan, M. D. (1971). Studies of the enhancement of an adenovirus-associated virus by herpes simplex virus. *J. Gen. Virol.* **10**, 29.

Blackman, K. E., and Bubel, H. C. (1969). Poliovirus-induced cellular injury. *J. Virol.* **4**, 203.

Blackstein, M. E., Stanners, C. P., and Farmilio, A. J. (1969). Heterogeneity of polyoma virus DNA: isolation and characterization of non-infectious small supercoiled molecules. *J. Mol. Biol.* **42**, 301.

Blair, C. D., and Duesberg, P. H. (1970). Myxovirus ribonucleic acids. *Annu. Rev. Microbiol.* **24**, 539.

Blair, C. D., and Robinson, W. S. (1970). Replication of Sendai virus. II. Steps in virus assembly. *J. Virol.* **5**, 639.

Blanden, R. V. (1970). Mechanisms of recovery from a generalized viral infection: mousepox. I. The effects of antithymocyte serum. *J. Exp. Med.* **132**, 1035.

Blanden, R. V. (1971a). Mechanisms of recovery from a generalized viral infection: mousepox. II. Passive transfer of recovery mechanisms with immune lymphoid cells. *J. Exp. Med.* **133**, 1074.

Blanden, R. V. (1971b). Mechanisms of recovery from a generalized viral infection: mousepox. III. Regression of infectious foci. *J. Exp. Med.* **133**, 1090.

Blanden, R. V., and Mims, C. A. (1973). Macrophage activation in mice infected with ectromelia or lymphocytic choriomeningitis viruses. *Aust. J. Exp. Biol. Med. Sci.* **51**, 393.

Blaškovič, D., and Styk, B. (1967). Laboratory methods of virus transmission in multicellular organisms. *In* "Methods in Virology" (K. Maramorosch and H. Koprowski, eds.), Vol. I, p. 163. Academic Press, New York.

Blatti, S. P., Ingles, C. J., Lindell, T. J., Morris, P. W., Weaver, R. F., Weinberg, F., and Rutter, W. J. (1970). Structure and regulatory properties of eucaryotic RNA polymerase. *Cold Spring Harbor Symp. Quant. Biol.* **35**, 649.

Blinzinger, K., and Müller, W. (1971). The intercellular gaps of the neuropil as possible pathways for virus spread in viral encephalomyelitides. *Acta Neuropathol.* **17**, 37.

Bloom, B. R. (1971). *In vitro* approaches to the mechanism of cell-mediated immune reactions. *Advan. Immunol.* **13**, 102.

Blumberg, B. S., Alter, H. J., and Visnich, S. (1965). A "new" antigen in leukemia sera. *J. Amer. Med. Ass.* **191**, 541.

Bodian, D. (1955). Emerging concept of poliomyelitis infection. *Science* **122**, 105.

Bodian, D. (1956). Poliovirus in chimpanzee tissues after virus feeding. *Amer. J. Hyg.* **64**, 181.

Bodian, D. (1958). Some physiologic aspects of poliovirus infections. *Harvey Lect. Ser.* **52**, 23.

Bodian, D. (1959). Poliomyelitis: Pathogenesis and histopathology. *In* "Viral and Rickettsial Infections of Man" (T. M. Rivers and F. L. Horsfall, Jr., eds.), 3rd Ed., p. 479. Lippincott, Philadelphia, Pennsylvania.

Bodian, D., and Horstmann, D. M. (1965). Polioviruses. *In* "Viral and Rickettsial Infections of Man" (F. L. Horsfall, Jr., and I. Tamm, eds.), 4th Ed., p. 430. Lippincott, Philadelphia, Pennsylvania.

Bodian, D., and Nathanson, N. (1960). Inhibitory effects of passive antibody on virulent poliovirus excretion and on immune response in chimpanzees. *Bull. Johns Hopkins Hosp.* **107**, 143.

Bodo, G., and Jungwirth, C. (1967). Effect of interferon on deoxyribonuclease induction in chick embryo fibroblast cultures infected with cowpox cirus. *J. Virol.* **1**, 466.

Boettiger, D., and Temin, H. M. (1970). Light inactivation of focus formation by chicken embryo fibroblasts infected with avian sarcoma virus in the presence of 5-bromodeoxyuridine. *Nature (London)* **228**, 622.

Boeyé, A., Melnick, J. L., and Rapp, F. (1965). Adenovirus-SV40 "hybrids": Plaque purification into lines in which the determinant for the SV40 tumor antigen is lost or retained. *Virology* **26**, 511.

Boeyé, A., Melnick, J. L., and Rapp, F. (1966). SV40-adenovirus "hybrids": Presence of two genotypes and the requirement of their complementation for viral replication. *Virology* **28**, 56.

Bolognesi, D. P., and Bauer, H. (1970). Polypeptides of avian RNA tumor viruses I. Isolation and chemical analysis. *Virology* **42**, 1097.

Bolognesi, D. P., Langlois, A. J., Sverak, L., Bonar, R. A., and Beard, J. W. (1968). *In vitro* chick embryo cell response to strain MC29 avian leukosis virus. *J. Virol.* **2**, 576.

Bondareff, W. (1964). Distribution of ferritin in the cerebral cortex of the mouse revealed by electron microscopy. *Exp. Neurol.* **10**, 377.

Bonifas, V. H. (1967). Time course and specificity of the arginine requirement for adenovirus synthesis. *Arch. Ges. Virusforsch.* **20**, 20.

Booth, K., Burkitt, D. P., Bassett, D. J., Cooke, R. A., and Biddulph, J. (1967). Burkitt lymphoma in Papua-New Guinea, *Brit. J. Cancer* **21**, 657.

Borden, E. C., Shope, R. E., and Murphy, F. A. (1971). Physicochemical and morphological relationships of some arthropod-borne viruses to bluetongue virus—A new taxonomic group. Physicochemical and serological studies. *J. Gen. Virol.* **13**, 261.

Boring, W. D., Zu Rhein, G. M., and Walker, D. L. (1956). Factors influencing host-virus interactions. II. Alteration of Coxsackie virus infection in adult mice by cold. *Proc. Soc. Exp. Biol. Med.* **93**, 273.

Borsa, J., and Graham, A. F. (1968). Reovirus: RNA polymerase activity in purified virions. *Biochem. Biophys. Res. Commun.* **33**, 895.

Borsa, J., Grover, J., and Chapman, J. D. (1970). Presence of nucleoside triphosphate phosphohydrolase activity in purified virions of reovirus. *J. Virol.* **6**, 295.

Borsos, T., and Bang, F. B. (1957). Quantitation of hemorrhagic lesions in the chick embryo produced by the Rous tumor virus. *Virology* **4**, 385.

Borst, P. (1971). Size, structure and information content of mitochondrial DNA. *In* "Autonomy and Biogenesis of Mitochondria and Chloroplasts" (N. K. Boardman, A. W. Linnane, and R. M. Smillie, eds.), p. 260. North Holland, Amsterdam.

Borst, P., and Kroon, A. M. (1969). Mitochondrial DNA: Physiochemical properties, replication, and genetic function. *Intern. Rev. Cytol.* **26**, 107.

Boshell, J. (1957). Marche de la fièvre jaune selvatique vers les régions du nordouest de l'Amérique centrale. *Bull. WHO* **16**, 431.

Boucher, D. W., Parks, W. P., and Melnick, J. L. (1969). Failure of a replicating adenovirus to enhance adeno-associated satellite virus replication. *Virology* **39**, 932.

Boucher, D. W., Parks, W. P., and Melnick, J. L. (1970). A sensitive neutralization test for the adeno-associated satellite viruses. *J. Immunol.* **104**, 555.

Boucher, D. W., Melnick, J. L., and Mayor, H. D. (1971). Nonencapsidated infectious DNA of adeno-satellite virus in cells coinfected with herpesvirus. *Science* **173**, 1243.

Boulter, E. A., Maber, H. B., and Bowen, E. T. W. (1961). Studies on the physiological disturbances occurring in experimental rabbitpox: an approach to rational therapy. *Brit. J. Exp. Pathol.* **42**, 433.

Boulton, R. W., and Westaway, E. G. (1972). Comparisons of togaviruses: Sindbis virus (group A) and Kunjin virus (group B). *Virology* **49**, 283.

Bourgaux, O., Bourgaux-Ramoisy, D., and Stoker, M. (1965). Further studies on transformation by DNA from polyoma virus. *Virology* **25**, 364.

Bourgaux, P., Bourgaux-Ramoisy, D., and Seiler, P. (1971). The replication of the ring-shaped

DNA of polyoma virus. II. Identification of molecules at various stages of replication. *J. Mol. Biol.* **59**, 195.

Bøvre, K., Lozeron, H. A., and Szybalski, W. (1971). Techniques of RNA-RNA hybridization in solution for the study of viral transcription. *In* "Methods in Virology" (K. Maramorosch and H. Koprowski, eds.), Vol. V, p. 271. Academic Press, New York.

Bowen, E. T. W., Simpson, D. I. H., Bright, W. F., Zlotnik, I., and Howard, D. M. R. (1969). Vervet monkey disease: studies on some physical and chemical properties of the causative agent. *Brit. J. Exp. Pathol.* **50**, 400.

Bower, R. K., Gyles, N. R., and Brown, C. J. (1965). The number of genes controlling the response of chick embryo chorioallantoic membranes to tumor induction by Rous sarcoma virus. *Genetics* **51**, 739.

Bowne, J. G., and Ritchie, A. E. (1970). Some morphological features of bluetongue virus. *Virology* **40**, 903.

Boxaca, M., and Paucker, K. (1967). Neutralization of different murine interferons by antibody. *J. Immunol.* **98**, 1130.

Boyer, H. W. (1971). DNA restriction and modification mechanisms in bacteria. *Annu. Rev. Microbiol.* **25**, 153.

Brachet, J., and Mirsky, A. E., eds. (1959–1964). "The Cell," Vols. I–VI. Academic Press, New York.

Bradburne, A. F. (1970). Antigenic relationships amongst coronaviruses. *Arch. Ges. Virusforsch.* **31**, 352.

Bradburne, A. F., and Tyrrell, D. A. J. (1971). Coronaviruses of man. *Progr. Med. Virol.* **13**, 373.

Brady, R. O., and Mora, P. T. (1970). Alteration in ganglioside pattern and synthesis in SV40 and polyoma virus transformed mouse cell lines. *Biochim. Biophys. Acta* **218**, 308.

Brailovsky, C. (1966). Recherches sur le virus K du rat (*Parvovirus ratti*) 1. Une méthode de titrage par phages et sur application a l'étude du cycle de multiplication du virus. *Ann. Inst. Pasteur* (Paris) **110**, 49.

Brakke, M. K. (1967). Density-gradient centrifugation. *In* "Methods in Virology" (K. Maramorosch and H. Koprowski, eds.), Vol. II, p. 93. Academic Press, New York.

Brambell, F. W. R. (1958). The passive immunity of the young animal. *Biol. Rev.* **33**, 488.

Brambell, F. W. R. (1970). "The Transmission of Passive Immunity from Mother to Young." North Holland, Amsterdam.

Branca, M., de Petris, S., Allison, A. C., Harvey, J. J., and Hirsch, M. S. (1971). Immune complex disease. I. Pathological changes in the kidneys of Balb/c mice neonatally infected with Moloney leukaemogenic and murine sarcoma viruses. *Clin. Exp. Immunol.* **9**, 853.

Brand, C. M., and Skehel, J. J. (1972). Crystalline antigen from the influenza virus envelope. *Nature New Biol.* **238**, 145.

Brandt, C. D., Kim, H. W., Vargosko, A. J., Jeffries, B. C., Arrobia, J. O., Rindge, B., Parrott, R. H., and Chanock, R. M. (1969). Infections in 18,000 infants and children in a controlled study of respiratory tract disease. I. Adenovirus pathogenicity in relation to serologic type and illness syndrome. *Amer. J. Epidemiol.* **90**, 484.

Branton, D. (1969). Membrane structure. *Annu. Rev. Plant Physiol.* **20**, 209.

Branton, P. E., and Sheinin, R. (1968). Control of DNA synthesis in cells infected with polyoma virus. *Virology* **36**, 652.

Bratt, M. A., and Robinson, W. S. (1967). Ribonucleic acid synthesis in cells infected with Newcastle disease virus. *J. Mol. Biol.* **23**, 1.

Bratt, M. A., and Gallaher, W. R. (1970). Comparison of fusion from within and fusion from without by Newcastle disease virus. *In Vitro* **6**, 3.

Braun, W., and Nakano, M. (1965). Influence of oligodeoxyribonucleotides on early events in antibody formation. *Proc. Soc. Exp. Biol. Med.* **119**, 701.

Braun, W., and Nakano, M. (1967). Antibody formation: stimulation by polyadenylic and polycytidylic acids. *Science* **157**, 819.

Breese, S. S. (1964). An improvement for electron microscopic resolution of polyoma virus capsomers. *Virology* **24**, 125.

Breese, S. S., and De Boer, C. J. (1966). Electron microscopic observations of African swine fever virus in tissue culture cells. *Virology* **28**, 420.

Breese, S. S., and Hsu, K. C. (1971). Techniques of ferritin-tagged antibodies. *In* "Methods in Virology" (K. Maramorosch and H. Koprowski, eds.), Vol. V, p. 399. Academic Press, New York.

Breeze, D. C., and Subak-Sharpe, H. (1967). The mutability of small-plaque-forming encephalomyocarditis virus. *J. Gen. Virol.* **1**, 81.

Breindl, M. (1971). VP4, the D-reactive part of poliovirus. *Virology* **46**, 962.

Brenner, S., and Horne, R. W. (1959). A negative staining method for high resolution electron microscopy of viruses. *Biochim. Biophys. Acta* **34**, 103.

Bretscher, M. (1972). Asymmetrical lipid bilayer structure. *Nature New Biol.* **236**, 11.

Brier, A. R., Wohlenberg, C., Rosenthal, J., Mage, M., and Notkins, A. L. (1971). Inhibition or enhancement of immunological injury of virus-infected cells. *Proc. Nat. Acad. Sci. U.S.* **68**, 3073.

Brightman, N. W. (1965). The distribution within the brain of ferritin injected into cerebrospinal fluid compartments. *Amer. J. Anat.* **117**, 193.

Briody, B. A. (1959). Response of mice to ectromelia and vaccinia viruses. *Bacteriol. Rev.* **23**, 61.

Britten, R. J., and Kohne, D. E. (1968). Repeated sequences in DNA. *Science* **161**, 529.

Brockman, W. W., Carter, W., Li, Li-H., Ruesser, F., and Nichol, F. R. (1971). Streptovaricins inhibit RNA dependent DNA polymerase present in an oncogenic RNA virus. *Nature (London)* **230**, 249.

Brooksby, J. B. (1958). The virus of foot-and-mouth disease. *Advan. Virus. Res.* **5**, 1.

Brostrom, M. A., Bruening, G., and Bankowski, R. A. (1971). Comparison of neuraminidases of paramyxoviruses with immunologically dissimilar hemagglutinins. *Virology* **46**, 856.

Brown, D. D., and Weber, C. S. (1968). Gene linkage by RNA–DNA hybridization. II. Arrangement of the redundant gene sequences for 28 S and 18 S robosomal RNA. *J. Mol. Biol.* **34**, 681.

Brown, R. M. (1972). Algal viruses. *Advan. Virus Res.* **17**, 243.

Brown, S. M., Ritchie, D. A., and Subak-Sharpe, J. H. (1973). Genetic studies with herpes simplex virus type 1. The isolation of temperature-sensitive mutants, their arrangement into complementation groups and recombination analysis leading to a linkage map. *J. Gen. Virol.* **18**, 329.

Bryan, W. R., Moloney, J. B., and Calnan, D. (1954). Stable standard preparations of the Rous sarcoma virus preserved by freezing and storage at low temperatures. *J. Nat. Cancer Inst.* **15**, 315.

Bryan, W. R., Calnan, D., and Moloney, J. B. (1955). Biological studies on the Rous sarcoma virus. III. The recovery of virus from experimental tumors in relation to the initiating dose. *J. Nat. Cancer Inst.* **16**, 317.

Bryceson, A. D. M. (1970). Diffuse cutaneous leishmaniasis in Ethiopia. 3. Immunological studies. *Trans. Roy. Soc. Trop. Med. Hyg.* **64**, 380.

Bubenik, J., and Bauer, H. (1967). Antigenic characteristics of the interaction between Rous sarcoma virus and mammalian cells. *Virology* **31**, 489.

Buchan, A., and Burke, D. C. (1966). Interferon production in chick embryo cells; the effect of puromycin and *p*-fluorophenylalanine. *Biochem. J.* **98**, 530.

Buchanan, R. E., Cowan, S. T., Wikén, T., and Clark, W. A. (1958). "International Code of Nomenclature of Bacteria and Viruses." State Coll. Press, Ames, Iowa.

Bucher, D. J., and Kilbourne, E. D. (1972). A2(N2) neuraminidase of the X-7 influenza virus recombinant: determination of molecular size and subunit composition of the active unit. *J. Virol.* **10**, 60.

Buckler, C. E., and Baron, S. (1966). Antiviral action of mouse interferon in heterologous cells. *J. Bacteriol.* **91**, 231.

Buckley, S. M. (1971). Multiplication of Chikungunya and O'nyong-nyong viruses in Singh's *Aedes* cell lines. *Curr. Topics Microbiol. Immunol.* **55**, 133.

Bucknall, R. A. (1967). The effects of substituted benzimidazoles on the growth of viruses and the nucleic acid metabolism of host cells. *J. Gen. Virol.* **1**, 89.

Buddingh, G. J., Schrum, D. I., Lanier, J. C., and Guidry, D. J. (1953). Studies on the natural history of herpes simplex infections. *Pediatrics* **11**, 595.

Bull, L. B., and Dickinson, C. G. (1937). The specificity of the virus of rabbit myxomatosis. *J. Council Sci. Ind. Res.* **10**, 291.

Burdon, R. N. (1971). Ribonucleic acid maturation in animal cells. *Progr. Nucleic Acid Res. Mol. Biol.* **11**, 33.

Burgdorfer, W., and Varma, M. G. R. (1967). Trans-stadial and transovarial development of disease agents in arthropods. *Annu. Rev. Entomol.* **12**, 347.

Burge, B. W., and Huang, A. S. (1970). Comparison of membrane protein glycopeptides of Sindbis virus and vesicular stomatitis virus. *J. Virol.* **6**, 176.

Burge, B. W., and Pfefferkorn, E. R. (1966a). Isolation and characterization of conditional lethal mutants of Sindbis virus. *Virology* **30**, 204.

Burge, B. W., and Pfefferkorn, E. R. (1966b). Complementation between temperature sensitive mutants of Sindbis virus. *Virology* **30**, 214.

Burge, B. W., and Pfefferkorn, E. R. (1966c). Phenotypic mixing between group A arboviruses. *Nature (London)* **210**, 1397.

Burge, B. W., and Pfefferkorn, E. R. (1968). Functional defects of temperature-sensitive mutants of Sindbis virus. *J. Mol. Biol.* **35**, 193.

Burge, B. W., and Strauss, J. H. (1970). Glycopeptides of the membrane glycoprotein of Sindbis virus. *J. Mol. Biol.* **47**, 449.

Burger, M. M. (1969). A difference in the architecture of the surface membrane of normal and virally-transformed cells. *Proc. Nat. Acad. Sci. U.S.* **62**, 994.

Burger, M. M. (1971a). Cell surfaces in neoplastic transformation. *Curr. Topics Cell. Regul.* **3**, 135.

Burger, M. M. (1971b). Forssman antigen exposed on surface membrane after viral transformation. *Nature New Biol.* **231**, 125.

Burger, M. M., and Goldberg, A. R. (1967). Identification of a tumor specific determinant on neoplastic cell surfaces. *Proc. Nat. Acad. Sci. U.S.* **57**, 359.

Burkitt, D. (1963). A lymphoma syndrome in tropical Africa. *Intern. Rev. Exp. Pathol.* **2**, 67.

Burkitt, D. P. (1967). Burkitt's lymphoma outside the known endemic areas of Africa and New Guinea. *Intern. J. Cancer* **2**, 562.

Burkitt, D. P. (1969). Etiology of Burkitt's lymphoma—an alternative hypothesis to a vectored virus. *J. Nat. Cancer Inst.* **42**, 19.

Burlingham, B. T., and Doerfler, W. (1972). An endonuclease in cells infected with adenovirus and associated with adenovirions. *Virology* **48**, 1.

Burlingham, B. T., Doerfler, W., Pettersson, U., and Philipson, L. (1971). Adenovirus endonuclease: Association with the penton of adenovirus type 2. *J. Mol. Biol.* **60**, 45.

Burmester, B. R. (1962). The vertical and horizontal transmission of avian visceral lymphomatosis. *Cold Spring Harbor Symp. Quant. Biol.* **27**, 471.

Burnet, F. M. (1933). A virus disease of the canary of the fowl-pox group. *J. Pathol. Bacteriol.* **37**, 107.

Burnet, F. M. (1942). Influenza virus B. II. Immunization of human volunteers with living attenuated virus. *Med. J. Aust.* **1**, 673.

Burnet, F. M. (1945). "Virus as Organism." Harvard Univ. Press, Cambridge, Massachusetts.

Burnet, F. M. (1952). The pattern of disease in childhood. *Australas. Ann. Med.* **1**, 93.

Burnet, F. M. (1955). "Principles of Animal Virology." Academic Press, New York.

Burnet, F. M. (1959). Genetic interactions between animal viruses. *In* "The Viruses" (F. M. Burnet and W. M. Stanley, eds.), Vol. 3, p. 275. Academic Press, New York.

Burnet, F. M. (1960). "Principles of Animal Virology," 2nd Ed. Academic Press, New York.

Burnet, F. M. (1964). Immunological factors in the process of carcinogenesis. *Brit. Med. Bull.* **20**, 154.

Burnet, F. M. (1968). Measles as an index of immunological function. *Lancet* **ii**, 610.

Burnet, F. M. (1969). "Cellular Immunology." Melbourne Univ. Press, Melbourne.

Burnet, F. M. (1970). The concept of immunological surveillance. *Progr. Exp. Tumor Res.* **13**, 1.

Burnet, F. M., and Fenner, F. (1949). "The Production of Antibodies," 2nd Ed. Macmillan, Melbourne.

Burnet, F. M., and Lind, P. E. (1951). A genetic approach to variation in influenza viruses 4. Recombination of characters between the influenza virus A strain NWS and strains of different serological subtypes. *J. Gen. Microbiol.* **5**, 67.

Burnet, F. M., and Lind, P. E. (1953). Influenza virus recombination: Experiments using the de-embryonated egg technique. *Cold Spring Harbor Symp. Quant. Biol.* **18**, 21.

Burnet, F. M., and Williams, S. W. (1939). Herpes simplex: A new point of view. *Med. J. Aust.* **1**, 637.

Burnett, J. P., and Harrington, J. A. (1968a). Simian adenovirus SA7 DNA: chemical, physical and biological studies. *Proc. Nat. Acad. Sci. U.S.* **60**, 1023.

Burnett, J. P., and Harrington, J. A. (1968b). Infectivity associated with adenovirus type SA7 DNA. *Nature (London)* **220**, 1245.

Burns, W. H., and Black, P. H. (1968). Analysis of simian virus 40-induced transformation of hamster kidney tissue *in vitro.* V. Variability of virus recovery from cell clones inducible with mitomycin C and cell fusion. *J. Virol.* **2**, 600.

Burns, W. H., and Black, P. H. (1969). Analysis of SV40-induced transformation of hamster kidney tissue *in vitro.* VI. Characteristics of mitomycin C induction. *Virology* **39**, 625.

Burrell, C. J., Martin, E. M., and Cooper, P. D. (1970). Posttranslational cleavage of virus polypeptides in arbovirus-infected cells. *J. Gen. Virol.* **6**, 319.

Burrows, R. (1972). Early stages of virus infection: Studies *in vivo* and *in vitro. Symp. Soc. Gen. Microbiol.* **22**, 303.

Burrows, R., Mann, J. A., Greig, A., Chapman, W. G., and Goodridge, D. (1971). The growth and persistence of foot-and-mouth disease virus in the bovine mammary gland. *J. Hyg.* **69**, 307.

Burrows, W., Elliott, M. E., and Havens, I. (1947). Studies on immunity to Asiatic cholera. IV. The excretion of coproantibody in experimental enteric cholera in the guinea pig. *J. Infect. Dis.* **81**, 261.

Burton, A. N., McLintock, J., and Rempel, J. G. (1966). Western equine encephalitis virus in Saskatchewan garter snakes and leopard frogs, *Science* **154**, 1029.

Bussell, R. M., Karzon, D. T., Barron, A. L., and Hall, F. T. (1962). Hemagglutination-inhibiting, complement-fixing, and neutralizing antibody responses in ECHO 6 infection, including studies on heterotypic responses. *J. Immunol.* **88**, 47.

Butel, J. S., and Rapp, F. (1967). Complementation between a defective monkey cell-adapting component and human adenoviruses in simian cells. *Virology* **31**, 573.

Butel, J. S., Tevethia, S. S., and Melnick, J. L. (1972). Oncogenicity and cell transformation by papovavirus SV40: the role of the viral genome. *Advan. Cancer Res.* **15**, 1.

Butterworth, B. E., and Rueckert, R. R. (1972a). Gene order of encephalomyocarditis virus as determined by studies with pactamycin. *J. Virol.* **9**, 823.

Butterworth, B. E., and Rueckert, R. R. (1972b). Kinetics of synthesis and cleavage of encephalomyocarditis virus-specific proteins. *Virology* **50**, 535.

Büttner, D., Giese, H., Müller, G., and Peters, D. (1964). Die Feinstruktur reifer Elementarkörper des Ecthyma contagiosum und der Stomatitis papulosa. *Arch. Ges. Virusforsch.* **14**, 657.


Cairns, H. J. F. (1951). Protection by receptor-destroying enzyme against infection with a neurotropic variant of influenza virus. *Nature (London)* **168**, 335.

Cairns, H. J. F. (1957). The asynchrony of infection by influenza virus. *Virology* **3**, 1.

Cairns, J. (1960). The initiation of vaccinia infection. *Virology* **11**, 603.

Cairns, J. (1963). The chromosome of *Escherichia coli. Cold Spring Harbor Symp. Quant. Biol.* **28**, 43.

Cairns, J., Stent, G. S., and Watson, J. D. (eds.) (1966). "Phage and the Origins of Molecular Biology." Cold Spring Harbor Lab. Quant. Biol., Cold Spring Harbor, New York.

Calabresi, P. (1965). Clinical studies with systemic administration of antimetabolites of pyrimidine nucleosides in viral infections. *Ann. N.Y. Acad. Sci.* **130**, 192.

Calafat, J., and Hageman, P. (1968). Some remarks on the morphology of virus particles of the B type and their isolation from mammary tumors. *Virology* **36**, 308.

Calendar, R. (1970). The regulation of phage development. *Annu. Rev. Microbiol.* **24**, 241.

Caliguiri, L. A., and Tamm, I. (1970). Characterization of poliovirus-specific structures associated with cytoplasmic membranes. *Virology* **42**, 112.

Callan, H. G. (1963). The nature of lampbrush chromosomes. *Intern. Rev. Cytol.* **15**, 1.

Calnek, B. W. (1964). Morphological alteration of RIF-infected chick embryo fibroblasts. *Nat. Cancer Inst. Monogr.* **17**, 425.

Calnek, B. W., Aldinger, H. K., and Kahn, D. E. (1969). Feather follicle epithelium: a source of enveloped and infectious cell-free herpesvirus from Marek's disease. *Avian Dis.* **14**, 219.

Calnek, B. W., Ubertini, T., and Aldinger, H. K. (1970). Viral antigen, virus particles and infectivity of tissues from chickens with Marek's disease. *J. Nat. Cancer Inst.* **45**, 341.

Calvin, M., Joss, U. R., Hackett, A. J., and Owens, R. B. (1971). Effect of rifampicin and two of its derivatives on cells infected with Moloney sarcoma virus. *Proc. Nat. Acad. Sci. U.S.* **68**, 1441.

Campadelli-Fiume, G., Costanzo, F., Mannini-Palenzona, A., and La Placa, M. (1972). Impairment of host cell ribonucleic acid polymerase II after infection with Frog Virus 3. *J. Virol.* **9**, 698.

Campbell, A. (1961). Sensitive mutants of bacteriophage λ. *Virology* **14**, 22.

Canaani, E., Helm, K. V. D., and Duesberg, P. (1973). Evidence for 30-40S RNA as precursor of the 60-70S RNA of Rous sarcoma virus. *Proc. Nat. Acad. Sci. U.S.* **72**, 401.

Cantell, K. (1961). Production and action of interferon in HeLa cells. *Arch. Ges. Virusforsch.* **10**, 510.

Cantell, K., and Paucker, K. (1963). Studies on viral interference in two lines of HeLa cells. *Virology* **19**, 81.

Cantell, K., and Valle, M. (1965). The ability of Sendai virus to overcome cellular resistance to vesicular stomatitis virus. II. The possible role of interferon. *Ann. Med. Exp. Biol. Fenniae* **43**, 61.

Cantell, K., Strander, H., Hadházy, Gy., and Nevanlinna, H. R. (1968). How much interferon can be prepared in human leukocyte suspensions? *In* "The Interferons" (G. Rita, ed.), p. 223. Academic Press, New York.

Carmichael, L. E., Barnes, F. D., and Percy, D. H. (1969). Temperature as a factor in resistance of young puppies to canine herpesvirus. *J. Infect. Dis.* **120**, 669.

Carp, R. I. (1963). A study of the mutation rates of several poliovirus strains to the reproductive capacity temperature/40+ and guanidine marker characteristics. *Virology* **21**, 373.

Carp, R. I. (1967). Thymidine kinase from normal, simian virus 40-transformed and simian virus 40-lytically infected cells. *J. Virol.* **1**, 912.

Carr, J. G. (1943). The relation between age structure and agent content of Rous No. 1 sarcomas. *Brit. J. Expl. Pathol.* **24**, 133.

Carter, M. F., Biswai, N., and Rawls, W. E. (1973). Characterization of the nucleic acid of Pichinde virus. *J. Virol.* **11**, 61.

Carter, W. A. (1970). Interferon; evidence for subunit structure. *Proc. Nat. Acad. Sci. U.S.* **67**, 620.

Carter, W. A., Brockman, W. W., and Borden, E. C. (1971). Streptovaricins inhibit focus formation by MSV (MLV) complex. *Nature New Biol.* **232**, 212.

Carter, W. A., Pitha, P. M., Marshall, L. W., Tazawa, I., Tazawa, S., and Ts'O, P. O. P. (1972). Structural requirements of the $rI_n$ . $rC_n$ complex for induction of human interferon. *J. Mol. Biol.* **70**, 567.

Cartwright, B., and Brown, F. (1972). Serological relationships between different strains of vesicular stomatitis virus. *J. Gen. Virol.* **16**, 391.

Cartwright, B., Smale, C. J., and Brown, F. (1970). Dissection of vesicular stomatitis virus into the infective ribonucleoprotein and immunizing components. *J. Gen. Virol.* **7**, 19.

Cartwright, K. L., and Burke, D. C. (1970). Virus nucleic acids formed in chick embryo cells infected with Semliki Forest virus. *J. Gen. Virol.* **6**, 231.

Carver, D. H., and Marcus, P. I. (1967). Enhanced interferon production from chick embryo cells aged *in vitro*. *Virology* **32**, 247.

Casals, J. (1961). Procedures for identification of arthropod-borne viruses. *Bull. WHO* **24**, 723.

Casals, J. (1967). Immunological techniques for animal viruses. *In* "Methods in Virology" (K. Maramorosch and H. Koprowski, eds.), Vol. III, p. 113, Academic Press, New York.

Casals, J. (1971). Arboviruses : Incorporation in a general system of virus classification. *In* "Comparative Virology" (K. Maramorosch and E. Kurstak, eds.), p. 307. Academic Press, New York.

Caspar, D. L. D. (1965). Design principles in virus particle construction. *In* "Viral and Rickettsial Infections of Man" (F. L. Horsfall, Jr. and I. Tamm, eds.), 4th Ed., p. 51. Lippincott, Philadelphia, Pennsylvania.

Caspar, D. L. D., and Klug, A. (1962). Physical principles in the construction of regular viruses. *Cold Spring Harbor Symp. Quant. Biol.* **27**, 1.

Caspar, D. L. D., Dulbecco, R., Klug, A., Lwoff, A., Stoker, M. G. P., Tournier, P., and Wildy, P. (1962). Proposals. *Cold Spring Harbor Symp. Quant. Biol.* **27**, 49.

Caspersson, T., Zech, L., and Johansson, C. (1970). Analysis of human metaphase chromosome set by aid of DNA-binding fluorescent agents. *Exp. Cell Res.* **62**, 490.

Cassingena, R., Tournier, P., May, E., Estrade, S., and Bourali, M. F. (1969a). Synthese du "répresseur" du virus SV40 dans l'infection productive et abortive. *C. R. Acad. Sci. (Paris).* **268**, 2834.

Cassingena, R., Tournier, P., Estrade, S., and Bourali, M. F. (1969b). Blocage de l'action du "répresseur" du virus SV40 par un facteur constitutif des cellules permissives pour ce virus. *C. R. Acad. Sci. (Paris)* **269**, 261.

Cassingena, R., Chang, C., Vignal, M., Suarez, H., Estrade, S., and Lazar, P. (1971). Use of monkey-mouse hybrid cells for the study of the cellular regulation of interferon production and action. *Proc. Nat. Acad. Sci. U.S.* **68**, 580.

Catalano, L. W., and Sever, J. L. (1971). The role of viruses as causes of congenital defects. *Annu. Rev. Microbiol.* **25**, 255.

Caul, E. O., Clarke, S. K. R., Mott, M. G., Perham, T. G. M., and Wilson, R. S. E. (1971). Cytomegalovirus infection after open heart surgery. A prospective study. *Lancet* **i**, 777.

Causey, O. R., and Maroja, O. (1959). Isolation of yellow fever virus from man and mosquitoes in the Amazon region of Brazil. *Amer. J. Trop. Med. Hyg.* **8**, 368.

Cavanagh, J. B., Greenbaum, D., Marshall, A. A. E., and Rubenstein, L. J. (1959). Cerebral demyelination associated with disorders of the reticuloendothelial system. *Lancet* **ii**, 525.

Ceccarini, C., and Eagle, H. (1971). Induction and reversal of contact inhibition of growth by pH modification. *Nature New Biol.* **233**, 271.

Celada, F. C., Asjo, B., and Klein, G. (1970). The presence of Moloney virus-induced antigen on antibody-producing cells. *Clin. Exp. Immunol.* **6**, 317.

Center Dis. Cont. (1971). *Mortality Morbidity* **20** (No. 33), U.S. Center for Disease Control.

Chakraverty, P. (1971). Antigenic relationship between influenza B viruses. *Bull WHO* **45**, 755.

Champoux, J. J., and Dulbecco, R. (1972). An activity from mammalian cells that untwists superhelical DNA, a possible swivel for DNA replication. *Proc. Nat. Acad. Sci. U.S.* **69**, 143.

Chan, V. F., and Black, F. L. (1970). Uncoating of poliovirus by isolated plasma membranes. *J. Virol.* **5**, 309.

Chandler, R. L. (1961). Encephalopathy in mice produced by inoculation with scrapie brain material. *Lancet* **i**, 1378.

Chang, C. T., and Zweerink, H. J. (1971). Fate of parental reovirus in infected cell. *Virology* **46**, 544.

Chang, S. C. (1957). Microscopic properties of whole mounts and sections of human bronchial epithelium in smokers and non-smokers. *Cancer* **10**, 1246.

Chanock, R. M. (1970). Control of acute mycoplasmal and viral respiratory tract disease. *Science* **169**, 248.

Chanock, R. M. (1971). Local antibody and resistance to acute viral respiratory tract disease. *In* "The Secretory Immunologic System" (D. H. Dayton, P. A. Small, R. M. Chanock, H. E. Kaufman, and T. B. Tomasi, eds.), p. 83. National Inst. Child Health and Human Development, National Institutes of Health, Bethesda, Maryland.

Chanock, R. M., Mufson, M. A., and Johnson, K. M., (1965). Comparative biology and ecology of human virus and mycoplasma respiratory pathogens. *Progr. Med. Virol.* **7**, 208.

Chanock, R. M., Ludwig, W., Huebner, R. J., Cate, T. R., and Chu, L. W. (1966). Immunization by selective infection with type 4 adenovirus grown in human diploid tissue culture. I. Safety and lack of oncogenicity and tests for potency in volunteers. *J. Amer. Med. Ass.* **195**, 445.

Chanock, R. M., Parrott, R. H., Kapikian, A. Z., Kim, H. W., and Brandt, C. D. (1968). Possible role of immunological factors in pathogenesis of RS virus lower respiratory tract disease. *Perspect. Virol.* **6**, 125.

Chanock, R. M., Kapikian, A. Z., Mills, J., Kim, H. W., and Parrott, R. H. (1970). Influence of immunologic factors in RS disease of the lower respiratory tract. *Arch. Environ. Health* **21**, 347.

Chanock, R. M., Kapikian, A. Z., Perkins, J. C., and Parrott, R. H. (1971). Vaccines for nonbacterial respiratory diseases other than influenza. *In* "International Conference on the Application of Vaccines against Viral, Rickettsial and Bacterial Diseases of Man," p. 101. Sci. Publ. No. 226. Pan. Amer. Health Organ., Washington, D. C.

Chany, C., and Brailovsky, C. (1967). Stimulating interaction between viruses (stimulons). *Proc. Nat. Acad. Sci. U.S.* **57**, 87.

Chany, C., and Vignal, M. (1970). Effect of prolonged interferon treatment on mouse embryonic fibroblasts transformed by murine sarcoma virus. *J. Gen. Virol.* **7**, 203.

Chany, C., Fourier, F., and Falcoff, E. (1968). A simple system for mass production of human interferons: the human amniotic membrane. *In* Ciba Foundation Symposium "Interferon" (G. E. W. Wolstenholme and M. O'Connor, eds.), p. 64. Churchill, London.

Chany, C., Gresser, I., Vendrely, C., and Robbe-Fossat, F. (1966). Persistent polioviral infection of the intact amniotic membrane. II. Existence of a mechanical barrier to viral infection. *Proc. Soc. Exp. Biol. Med.* **123**, 960.

Chardonnet, Y., and Dales, S. (1970a). Early events in the interaction of adenoviruses with HeLa cells. I. Penetration of type 5 and intracellular release of the DNA genome. *Virology* **40**, 462.

Chardonnet, Y., and Dales, S. (1970b). Early events in the interaction of adenoviruses with HeLa cells. II. Comparative observations on the penetration of types 1, 5, 7, and 12. *Virology* **40**, 478.

Chardonnet, Y., and Dales, S. (1972). Early events in the interaction of adenoviruses with HeLa cells. III. Relationship between an ATPase activity in nuclear envelopes and transfer of core material: a hypothesis. *Virology* **48**, 342.

Chesterton, C. T., and Green, M. H. (1968). Early-late transcription switch: Isolation of a lambda DNA-RNA polymerase complex active in the synthesis of late RNA. *Biochem. Biophys. Res. Commun.* **31**, 919.

Cheung, K-S., Smith, R. E., Stone, M. P., and Joklik, W. K. (1972). Comparison of immature (rapid harvest) and mature Rous sarcoma virus particles. *Virology* **50**, 851.

Cheyne, I. M., and White, D. O. (1969). Growth of paramyxoviruses in anucleate cells. *Aust. J. Exp. Biol. Med. Sci.* **47**, 145.

Choppin, P. W. (1964). Multiplication of a myxovirus (SV5) with minimal cytopathic effects and without interference. *Virology* **23**, 224.

Choppin, P. W. (1969). Replication of influenza virus in a continuous cell line: high yield of infectious virus from cells inoculated at high multiplicity. *Virology* **39**, 130.

Choppin, P. W., and Compans, R. W. (1970). Phenotypic mixing of envelope proteins of the parainfluenza virus SV5 and vesicular stomatitis virus. *J. Virol.* **5**, 609.

Choppin, P. W., and Pons, M. W. (1970). The RNA's of infective and incomplete influenza virions grown in MDBK and HeLa cells. *Virology* **42**, 603.

Choppin, P. W., Murphy, J. S., and Stoeckenius, W. (1961). The surface structure of influenza virus filaments. *Virology* **13**, 548.

Choppin, P. W., Compans, R. W., and Holmes, K. V. (1968). Replication of the parainfluenza virus SV5 and the effects of superinfection with poliovirus. *In* "Medical and Applied Virology" (M. Sanders and E. H. Lennette, eds.), p. 16. W. Green, St. Louis, Missouri.

Choppin, P. W., Klenk, H.-D., Compans, R. W., and Caliguiri, L. A. (1971). The parainfluenza virus SV5 and its relationship to the cell membrane. *Perspect. Virol.* **7**, 127.

Chopra, H. C., and Mason, M. M. (1970). A new virus in a spontaneous mammary tumor of a rhesus monkey. *Cancer Res.* **30**, 2081.

Chopra, H. C., Zelljadt, I., Jensen, E. M., Mason, M. M., and Woodside, N. J. (1971). Infection of cell cultures by a virus isolated from a mammary carcinoma of a rhesus monkey. *J. Nat. Cancer Inst.* **46**, 127.

Chow, A. W. C., Foerster, J., and Hryniuk, W. (1971). Cytosine arabinoside therapy for herpesvirus infections. In "Antimicrobial Agents and Chemotherapy—1970" (G. L. Hobby, ed.). p. 214. Amer. Soc. Microbiol.

Chow, N.-L., and Simpson, R. W. (1971). RNA-dependent RNA polymerase activity associated with virions and subviral particles of myxoviruses. *Proc. Nat. Acad. Sci. U.S.* **68**, 752.

Christensen, P. E., Schmidt, H., Bang, H. O., Andersen, V., Jordal, B., and Jensen, O. (1953). An epidemic of measles in southern Greenland, 1951. Measles in virgin soil. III. Measles and tuberculosis. *Acta Med. Scand.* **144**, 450.

Churchill, A. E., Payne, L. N., and Chubb, R. C. (1969). Immunization against Marek's disease using a live attenuated virus. *Nature (London)* **221**, 744.

Ciba Foundation Study Group (1970). "Hormones and the Immune Response." Ciba Foundation Study Group No. 36. J. & A. Churchill, London.

Ciba Foundation Symposium, (1972). "Ontogeny of Acquired Immunity." Excerpta Medica, Amsterdam.

Clark, H. F., and Koprowski, H. (1971). Isolation of temperature-sensitive conditional lethal mutants of "fixed" rabies virus. *J. Virol.* **7**, 295.

Clarke, J. K., Gay, K. W., and Attridge, J. J. (1969). Replication of simian foamy virus in monkey kidney cells. *J. Virol.* **3**, 358.

Clarkson, S. G., and Runner, M. N. (1971). Transfer RNA changes in HeLa cells after vaccinia virus infection. *Biochim. Biophys. Acta* **238**, 498.

Clayton, D. A., and Vinograd, J. (1967). Circular dimer and catenate forms of mitochondrial DNA in human leukemic leukocytes. *Nature (London)* **216**, 652.

Clayton, D. A., and Vinograd, J. (1969). Complex mitochondrial DNA in leukemic and normal human myeloid cells. *Proc. Nat. Acad. Sci. U.S.* **62**, 1077.

Clayton, D. A., Davis, R. W., and Vinograd, J. (1970). Homology and structural relationships between the dimeric and monomeric circular forms of mitochondrial DNA from human leukemic leukocytes. *J. Mol. Biol.* **47**, 137.

Cleaver, J .E. (1966). Photoreactivation: a radiation repair mechanism absent from mammalian cells. *Biochem. Biophys. Res. Commun.* **24**, 569.

Clemmer, D. L. (1965). Experimental enteric infection of chickens with an avian adenovirus (strain 93). *Proc. Soc. Exp. Biol. Med.* **118**, 943.

Clemmer, D. L., and Ichinose, H. (1968). The cellular site of virus replication in the intestine of chicks infected with an avian adenovirus. *Arch. Ges. Virusforsch.* **25**, 277.

Cline, M. J., and Livingston, D.C. (1971). Binding of ³H-concanavalin A by normal and transformed cells. *Nature New Biol.* **232**, 154.

Coates, H. V., Alling, D. W., and Chanock, R. M. (1966). An antigenic analysis of respiratory syncytial virus isolates by a plaque reduction test. *Amer. J. Epidemiol.* **83**, 299.

Cochard, A.-M., Le-Tan-Vinh, and Lelong, M. (1963). Le placenta dans la cytomegalie congenitale. Étude anatomico-clinique de 3 observations personnelles. *Arch. Fr. Pediat.* **20**, 35.

Cochran, K. W., Maassab, H. F., Tsunoda, A., and Berlin, B. S. (1965). Studies on the antiviral activity of amantadine hydrochloride. *Ann. N.Y. Acad. Sci.* **130**, 432.

Cochrane, C. G. (1971). Mechanisms involved in the deposition of immune complexes in tissues. *J. Exp. Med.* **134**, 75s.

Cockburn, T. A. (1961). Eradication of infectious diseases. *Science* **133**, 1050.

Cockburn, W. C., and Drozdov, S. G. (1970). Poliomyelitis in the world. *Bull. WHO* **42**, 405.

Coffin, J. M., and Temin, H. M. (1972). Hybridization of Rous sarcoma virus deoxyribonucleic acid polymerase product and ribonucleic acids from chicken and rat cells infected with Rous sarcoma virus. *J. Virol.* **9**, 766.

Cohen, G. H. (1972). Ribonucleotide reductase activity of synthronized KB cells infected with herpes simplex virus. *J. Virol.* **9**, 408.

Cohen, G. H., Vaughan, R. K., and Lawrence, W. C. (1971a). Deoxyribonucleic acid synthesis in synchronized mammalian KB cells infected with herpes simplex virus. *J. Virol.* **7**, 783.

Cohen, G. H., Atkinson, P. H., and Summers, D. F. (1971b). Interactions of vesicular stomatitis virus structural proteins with HeLa plasma membranes. *Nature New Biol.* **231**, 121.

Cohen, S., and Milstein, C. (1967). Structure and biological properties of immunoglobulins. *Advan. Immunol.* **7**, 1.

Cohen, S. S. (1966). Introduction to the biochemistry of *d*-arabinosyl nucleosides. *Progr. Nucleic Acid Res. Mol. Biol.* **5**, 2.

Cohn, Z. A., and Benson, B. (1965). The *in vitro* differentiation of mononuclear phagocytes. II. The influence of serum on granule formation, hydrolase production, and pinocytosis. *J. Exp. Med.* **121**, 835.

Colby, C. (1971). The induction of interferon by natural and synthetic polynucleotides. *Progr. Nucleic Acid Res. Mol. Biol.* **11**, 1.

Colby, C., and Chamberlin, M. J. (1969). The specificity of interferon induction in chick embryo cells by helical RNA. *Proc. Nat. Acad. Sci. U.S.* **63**, 160.

Colby, C., and Duesberg, P. H. (1969). Double-stranded RNA in vaccinia virus-infected cells. *Nature (London)* **222**, 940.

Colby, C., and Morgan, M. J. (1971). Interferon induction and action. *Annu. Rev. Microbiol.* **25**, 333.

Colby, C., and Rubin, H. (1969). Growth and nucleic acid synthesis in normal cells and cells infected with Rous sarcoma virus. *J. Nat. Cancer Inst.* **43**, 437.

Colby, C., Jurale, C., and Kates, J. R. (1971). Mechanism of synthesis of vaccinia virus double-stranded ribonucleic aicd *in vivo* and *in vitro J. Virol.* **7**, 71.

Cole, C. N., Smoler, D., Wimmer, E., and Baltimore, D. (1971). Defective interfering particles of poliovirus. I. Isolation and physical properties. *J. Virol.* **7**, 478.

Cole, G. A., Gilden, D. H., Monjan, A. A., and Nathanson, N. (1971). Lymphocytic choriomeningitis virus: pathogenesis of acute central nervous system disease. *Fed. Proc. Fed. Amer. Soc. Exp. Biol.* **30**, 1831.

Compans, R. W. (1971). Location of the glycoprotein in the membrane of Sindbis virus. *Nature New Biol.* **229**, 114.

Compans, R. W. (1973). Influenza virus proteins. II. Association with components of the cytoplasm. *Virology* **51**, 56.

Compans, R. W., and Caliguiri, L. A. (1973). Isolation and properties of an RNA polymerase from influenza virus-infected cells. *J. Virol.* **11**, 441.

Compans, R. W., and Choppin, P. W. (1971). The structure and assembly of influenza and parainfluenza viruses. *In* "Comparative Virology" (K. Maramorosch and E. Kurstak, eds.), p. 407. Academic Press, New York.

Compans, R. W., and Dimmock, N. J. (1969). An electron microscopic study of single-cycle infection of chick embryo fibroblasts by influenza virus. *Virology* **39**, 499.

Compans, R. W., Holmes, K. V., Dales, S., and Choppin, P. W. (1966). An electron microscopic study of moderate and virulent virus-cell interactions of the parainfluenza virus SV5. *Virology* **30**, 411.

Compans, R. W., Dimmock, N. J., and Meier-Ewert, H. (1970a). An electron microscopic study of the influenza virus-infected cell. *In* "The Biology of Large RNA Viruses" (R. D. Barry and B. W. J. Mahy, eds.), p. 87. Academic Press, New York

Compans, R. W., Klenk, H-D., Caliguiri, L. A., and Choppin, P. A. (1970b). Influenza virus proteins. I. Analysis of polypeptides of the virion and identification of spike glycoproteins. *Virology* **42**, 880.

Compans, R. W., Content, J., and Duesberg, P. H. (1972a). Structure of the ribonucleoprotein of influenza virus. *J. Virol.* **10**, 795.

Compans, R. W., Mountcastle, W. E., and Choppin, P. W. (1972b). The sense of the helix of paramyxovirus nucleocapsids. *J. Mol. Biol.* **65**, 167.

Connolly, J. H., Allen, I. V., Hurwitz, L. J., and Millar, J. H. D. (1967). Measles-virus antibody and antigens in subacute sclerosing panencephalitis. *Lancet* **i**, 542.

Constantine, V. S., and Mason, B. H. (1972). Experimental zoster-like herpes simplex in hairless mice. *J. Invest. Dermatol.* **56**, 193.

Content, J., and Duesberg, P. H. (1971). Base sequence differences among the ribonucleic acids of influenza virus. *J. Mol. Biol.* **62**, 273.

Cook, J. S. (1970). Photoreactivation in animal cells. *In* "Photophysiology" (A. C. Giesse, ed.), Vol. 5, p. 101. Academic Press, New York.

Cook, J. S. (1972). Photoenzymatic repair in animal cells. *In* "Molecular and Cellular Repair Processes" (R. F. Beers and R. M. Herriott, eds.), p. 79. Johns Hopkins Press, Baltimore, Maryland.

Cook, J. S., and McGrath, J. R. (1967). Photoreactivating enzyme activity in metazoa. *Proc. Nat. Acad. Sci. U.S.* **58**, 1359.

Cook, M. K., and Sears, J. F. (1970). Preparation of infectious cell-free herpes-type virus associated with Marek's disease. *J. Virol.* **5**, 258.

Cook, M. K., Andrewes, B. E., Fox, H. H., Turner, H. C., James, W. D., and Chanock, R. M. (1959). Antigenic relationships among the "newer" myxoviruses (parainfluenza). *Amer. J. Hyg.* **69**, 250.

Cook, P. J., and Burkitt, D. P. (1971). Cancer in Africa. *Brit. Med. Bull.* **27**, 14.

Coons, A. H., Creech, H. J., and Jones, R. N. (1941). Immunological properties of an antibody containing a fluorescent group. *Proc. Soc. Exp. Biol. Med.* **47**, 200.

Cooper, P. D. (1958). Homotypic non-exclusion by vesicular stomatitis virus in chick cell culture. *J. Gen. Microbiol.* **19**, 350.

Cooper, P. D. (1964). The mutation of poliovirus by 5-fluorouracil. *Virology* **22**, 186.

Cooper, P. D. (1965). Rescue of one phenotype in mixed infections with heat-defective mutants of type I poliovirus. *Virology* **25**, 341.

Cooper, P. D. (1967). The plaque assay of animal viruses. *In* "Methods in Virology" (K. Maramorosch and H. Koprowski, eds.), Vol. III, p. 243, Academic Press, New York.

Cooper, P. D. (1968). A genetic map of poliovirus temperature-sensitive mutants. *Virology* **35**, 584.

Cooper, P. D. (1969). The genetic analysis of poliovirus. *In* "The Biochemistry of Viruses" (H. B. Levy, ed.), p. 177. Dekker, New York.

Cooper, P. D., and Bellett, A. J. D. (1959). A transmissible interfering component of vesicular stomatitis virus preparations. *J. Gen. Microbiol.* **21**, 485.

Cooper, P. D., Johnson, R. T., and Garwes, D. J. (1966). Physiological characterization of heat-defective (temperature-sensitive) poliovirus mutants: Preliminary classification. *Virology* **30**, 638.

Cooper, P. D., Stanček, D., and Summers, D. F. (1970a). Synthesis of double-stranded RNA by poliovirus temperature-sensitive mutants. *Virology* **40**, 971.

Cooper, P. D., Summers, D. F., and Maizel, J. V. (1970b). Evidence for ambiguity in the post translational cleavage of poliovirus proteins. *Virology* **41**, 408.

Cooper, P. D., Wentworth, B. B., and McCahon, D. (1970c). Guanidine inhibition of poliovirus: a dependence of viral RNA synthesis on the configuration of structural protein. *Virology* **40**, 486.

Cooper, P. D., Geissler, E., Scotti, P. D., and Tannock, G. A. (1971). Further characterization of the genetic map of poliovirus temperature-sensitive mutants. *In* "Strategy of the Viral Genome" (G. E. W. Wolstenholme and M. O'Connor, eds.), *Ciba Found. Symp.*, p. 75. Churchill Livingstone, Edinburgh.

Cooper, P. D., Steiner-Pryor, A., and Wright, P. J. (1973). The poliovirus equestron—a model for virus regulation. *Intervirology*, 1.

Cords, C. E., and Holland, J. J. (1964a). Interference between enteroviruses and conditions effecting its reversal. *Virology* **22**, 226.

Cords, C. E., and Holland, J. J. (1964b). Alteration of the species and tissue specificity of poliovirus by enclosure of its RNA within the protein capsid of Coxsackie B1 virus. *Virology* **24**, 492.

Cormack, D. V., Holloway, A. F., Wong, P. K. Y., and Cairns, J. E. (1971). Temperature-sensitive mutants of vesicular stomatitis virus. II. Evidence of defective polymerase. *Virology* **45**, 824.

Cornick, G., Sigler, P. B., and Ginsberg, H. S. (1971). Characterization of crystals of type 5 adenovirus hexon. *J. Mol. Biol.* **57**, 397.

Cossart, Y. E. (1972). Epidemiology of serum hepatitis. *Brit. Med. Bull.* **28**, 156.

Cottral, G. E. (1952). Endogenous viruses in the egg. *Ann. N.Y. Acad. Sci.* **55**, 221.

Cottral, G. E., Gailunas, P., and Cox, B. F. (1968). Foot-and-mouth disease virus in semen of of bulls and its transmission by artificial insemination. *Arch. Ges. Virus forsch.* **23**, 362.

Couch, R. B., Chanock, R. M., Cate, T. R., Lang, D. J., Knight, V., and Huebner, R. J. (1963). Immunization with types 4 and 7 adenovirus by selective infection of the intestinal tract. *Amer. Rev. Resp. Dis.* **88** (Part 2), 394.

Couch, R. B., Douglas, R. G., Rossen, R., and Kasel, J. A. (1971). Role of secretary antibody in influenza. *In* "The Secretory Immunologic System" (D. H. Dayton, P. A. Small, R. M. Chanock, H. E. Kaufman, and T. B. Tomasi, eds.), p. 93. National Inst. Child Health and Human Development, National Institutes of Health, Bethesda, Maryland.

Cowan, S. T. (1966). Nomenclature of viruses. *Lancet* **i**, 807.

Cowan, S. T. (1970). Heretical taxonomy for bacteriologists. *J. Gen. Microbiol.* **61**, 145

Crabbé, P. A., Carbonara, A. O., and Heremans, J. F. (1965). The normal human intestinal mucosa as a major source of plasma cells containing gamma A immunoglobulins. *Lab. Invest.* **14**, 235.

Craighead, J. E. (1966). Growth of parainfluenza type 3 virus and interferon production in infant and adult mice. *Brit. J. Exp. Pathol.* **47**, 235.

Craighead, J. E., and Steinke, J. (1971). Diabetes mellitus-like syndrome in mice infected with encephalomyocarditis virus. *Amer. J. Pathol.* **63**, 119.

Crawford, L. V. (1962). The adsorption of polyoma virus. *Virology* **18**, 177.

Crawford, L. V. (1965). A study of human papilloma virus DNA. *J. Mol. Biol.* **13**, 362.

Crawford, L. V. (1966). A minute virus of mice. *Virology* **29**, 605.

Crawford, L. V. (1969). Nucleic acids of tumor viruses. *Advan. Virus Res.* **14**, 89.

Crawford, L. V., and Black, P. H. (1964). The nucleic acid of simian virus 40. *Virology* **24**, 388.

Crawford, L. V., and Crawford, E. M. (1963). A comparative study of polyoma and papilloma viruses. *Virology* **21**, 258.

Crawford, L. V., Crawford, E. M., and Watson, D. H. (1962). The physical characteristics of polyoma virus. I. Two types of particle. *Virology* **18**, 170.

Crawford, L. V., Follett, E. A. C., Burdon, M. G., and McGeoch, D. J. (1969). The DNA of a minute virus of mice. *J. Gen. Virol.* **4**, 37.

Crick, F. H. C., and Watson, J. D. (1956). Structure of small viruses. *Nature (London)* **177**, 473.

Crittenden, L. B., Okazaki, W., and Reamer, R. S. (1964). Genetic control of responses to Rous sarcoma and strain RPL12 viruses in the cells, embryos, and chickens of two inbred lines. *Nat. Cancer Inst. Monog.* **17**, 161.

Crittenden, L. B., Stone, H. A., Reamer, R. H., and Okazaki, W. (1967). Two loci controlling genetic cellular resistance to avian leukosis-sarcoma virus. *J. Virol.* **1**, 898.

Crocker, T., Pfendt, E., and Spendlove, R. (1964). Poliovirus: Growth in non-nucleate cytoplasm. *Science* **145**, 401.

Cross, G. F., Waugh, M., and Ferris, A. A. (1971). Virus-like particles associated with a faecal antigen from hepatitis patients and with Australia antigen. *Aust. J. Exp. Biol. Med. Sci.* **49**, 1.

Cross, R. K., and Fields, B. N. (1972). Temperature-sensitive mutants of reovirus type 3: studies on synthesis of viral RNA. *Virology* **50**, 799.

Crouse, H. V., Coriell, L. L., Blank, H., and Scott, T. F. McN. (1950). Cytochemical studies on the intranuclear inclusion of herpes simplex. *J. Immunol.* **65**, 119.

Crowell, R. L., and Philipson, L. (1971). Specific alterations of coxsackievirus B3 eluted from HeLa cells. *J Virol* **8**, 509

Crumpacker, C S., Henry, P. H., Kakefuda, T., Rowe, W. P., Levin, M. J., and Lewis, A. M. (1971). Studies of non-defective adenovirus 2-simian virus 40 hybrid viruses III. Base composition, molecular weight and conformation of the Ad2+ND1 genome. *J. Virol.* **7**, 352.

Cumar, F. A., Brady, R. O., Kolodny, E. H., McFarland, V. W., and Mora, P. T. (1970). Enzymatic block in the synthesis of gangliosides in DNA virus-transformed tumorigenic mouse cell lines. *Proc. Nat. Acad. Sci. U.S.* **67**, 757.

Cuzin, F., Vogt, M., Dieckman, M., and Berg, P. (1970). Induction of virus multiplication in 3T3 cells transformed by a thermo-sensitive mutant of polyoma virus II. Formation of oligomeric polyoma DNA molecules. *J. Mol. Biol.* **47**, 317.

Daams, J. H., Timmermans, A., VanderGugten, A., and Bentvelzen, P. (1968). Genetical resistance of mouse strain C57BL to mouse mammary tumor viruses II. Resistance by means of a repressed related provirus. *Genetica* **38**, 400.

Dahl, R., ad Kates, J. R. (1970). Synthesis of vaccinia virus "early" and "late" messenger RNA *in vitro* with nucleoprotein structures isolated from infected cells. *Virology* **42**, 463.

Dahlberg, J. E., and Simon, E. H. (1968). Complementation in Newcastle disease virus. *Bacteriol Proc.*, p. 162.

Dahlberg, J. E., and Simon, E. H. (1969a). Recombination in Newcastle disease virus (NDV): The problem of complementing heterozygotes. *Virology* **38**, 490.

Dahlberg, J. E., and Simon, E. H. (1969b). Physical and genetic studies of Newcastle disease virus: Evidence for multiploid particles. *Virology* **38**, 666.

Dahlman, T. (1964). Studies on tracheal ciliary activity. *Amer. Rev. Resp. Dis.* **89**, 870.

Dales, S. (1962). An electron microscope study of the early association between two mammalian viruses and their hosts. *J. Cell Biol.* **13**, 303.

Dales, S. (1963a). The uptake and development of vaccinia virus in strain L cells followed with labeled viral deoxyribonucleic acid. *J. Cell Biol.* **18**, 51.

Dales, S. (1963b). Association between the spindle apparatus and reovirus. *Proc. Nat. Acad. Sci. U.S.* **50**, 268.

Dales, S. (1965a). Penetration of animal viruses into cells. *Progr. Med. Virol.* **7**, 1.

Dales, S. (1965b). Effects of streptovaricin A on the initial events in the replication of vaccinia and reovirus. *Proc. Nat. Acad. Sci. U.S.* **54**, 462.

Dales, S., and Choppin, P. W. (1962). Attachment and penetration of influenza virus. *Virology* **18**, 489.

Dales, S., and Hanafusa, H. (1972). Penetration and intracellular release of the genomes of avian RNA tumor viruses. *Virology* **50**, 440.

Dales, S., and Kajioka, R. (1964). The cycle of multiplication of vaccinia virus in Earle's strain L cells. I. Uptake and penetration. *Virology* **24**, 278.

Dales, S., and Mosbach, E. H. (1968). Vaccinia as a model for membrane biogenesis. *Virology* **35**, 564.

Dales, S., and Siminovitch, L. (1961). The development of vaccinia virus in Earle's strain L cells as examined by electron microscopy. *J. Biophys. Biochem. Cytol.* **10**, 475.

Dales, S., Eggers, H. J., Tamin, I., and Palade, G. E. (1965a). Electron microscopic study of the formation of poliovirus. *Virology* **26**, 379.

Dales, S., Gomatos, P. J., and Hsu, K. C. (1965b). The uptake and development of reovirus in strain L cells followed with labeled viral ribonucleic acid and ferritin-antibody conjugates. *Virology* **25**, 193.

Dalgarno, L., and Kelly, M. W. (1973). Virus replication. *In* "Viruses and Invertebrates" (A. J. Gibbs, ed.), in press. North-Holland, Amsterdam.

Dalldorf, G., and Sickles, G. M. (1948). An unidentified, filtrable agent isolated from the feces of children with paralysis. *Science* **108**, 61.

Dalmat, H. T. (1958). Arthropod transmission of rabbit papillomatosis. *J. Exp. Med.* **108**, 9.

Dalmat, H. T. (1959). Arthropod transmission of rabbit fibromatosis (Shope). *J. Hyg.* **57**, 1.

Dalmat, H. T., and Stanton, M. R. (1959). A comparative study of the Shope fibroma in rabbits in relation to transmissibility by mosquitoes. *J. Nat. Cancer Inst.* **22**, 593.

Dalton, A. J. (1972). Further analysis of the detailed structure of type B and C particles. *J. Nat. Cancer Inst.* **48**, 109.

Dalton, A. J., and Hagenau, F. (ed.) (1973). "An Atlas on Viruses." Academic Press, New York.

Dalton, A. J., Rowe, W. P., Smith, G. H., Wilsnack, R. E., and Pugh, W. E. (1968). Morphological and cytochemical studies on lymphocytic choriomeningitis virus. *J. Virol.* **2**, 1465.

Dane, D. S., Cameron, C. H., and Briggs, M. (1970). Virus-like particles in serum of patients with Australia-antigen-associated hepatitis. *Lancet* **i**, 695.

Darbyshire, J. H. (1962). Agar gel diffusion studies with a mucosal disease of cattle. II. A serological relationship between a mucosal disease and swine fever. *Res. Vet. Sci.* **3**, 125.

Darbyshire, J. H. (1966). Oncogenicity of bovine adenovirus type 3 in hamsters. *Nature (London)* **211**, 102.

Darlington, R. W., and Moss, L. H. (1969). The envelope of herpesvirus. *Progr. Med. Virol.* **11**, 16.

Darlington, R. W., Granoff, A., and Breeze, D. C. (1966). Viruses and renal carcinoma of *Rana pipiens*. II. Ultrastructural studies and sequential development of virus isolated from normal and tumor tissue. *Virology* **29**, 149.

Darnell, J. E. (1968a). Considerations of virus-controlled functions. *Symp. Soc. Gen. Microbiol.* **18**, 149.

Darnell, J. E. (1968b). Ribonucleic acids from animal cells. *Bacteriol. Rev.* **32**, 262.

Darnell, J. E., Girard, M., Baltimore, D., Summers, D. F., and Maizel, J. V. (1967). The synthesis and translation of poliovirus RNA. *In* "The Molecular Biology of Viruses" (J. S. Colter and W. Paranchych, eds.), p. 375. Academic Press, New York.

Darnell, J. E., Patagoulos, G. N. Lindberg, U., and Balint, R. (1970). Studies on the relationship of mRNA to heterogeneous nuclear RNA in mammalian cells. *Cold Spring Harbor Symp. Quant. Biol.* **35**, 555.

Darnell, J. E., Philipson, L., Wall, R., and Adesnik, M. (1971). Polyadenylic acid sequences: Role in conversion of nuclear RNA into messenger RNA. *Science* **174**, 507.

Darnell, M. B., and Plagemann, P. G. W. (1972). Physical properties of lactic dehydrogenase-elevating virus and its ribonucleic acid. *J. Virol.* **10**, 1082.

Das, M .R., Vaidya, A. B., Sirsat, S. M., and Moore, D. H. (1972). Polymerase and RNA studies on milk virions from women in the Parsi community. *J. Nat. Cancer Inst.* **48**, 1191.

Davenport, F. M. (1971). Killed influenza virus vaccines: present status, suggested use, desirable developments. *In* "International Conference on the Application of Vaccines against Viral, Rickettsial and Bacterial Diseases of Man," p. 89. Sci. Publ. No. 226. Pan Amer. Health Organ., Washington, D.C.

Davenport, F. M., and Hennessy, A. V. (1956). A serological recapitulation of past experiences with influenza A; antibody response to monovalent vaccine. *J. Exp. Med.* **104**, 85.

Davenport, F. M., Minuse, E., Hennessy, A. V., and Francis, T. (1969). Interpretations of influenza antibody patterns in man. *Bull. WHO* **41**, 453.

David, A. E. (1971). Lipid composition of Sindbis virus. *Virology* **46**, 711.

Davidson, E. H., and Hough, B. R. (1971). Genetic information in oocyte RNA. *J. Mol. Biol.* **56,** 491.

Davidson, W. L., and Hummeler, K. (1960). B virus infection in man. *Ann. N.Y. Acad. Sci.* **85,** 970.

Davie, J. (1962). The classification of subtype variants of the virus of foot-and-mouth disease. *Bull. Office Intern. Epiz.* **57,** 962.

Davies, W. L., Grunert, R. R., Haff, R. F., McGahen, J. W., Neumayer, E. M., Paulshock, M., Watts, J. C., Wood, T. R., Hermann, E. C., and Hoffman, C. E. (1964). Antiviral activity of 1-adamantanamine (Amantadine). *Science* **144,** 862.

Davis, B. D., Dulbecco, R., Eisen, H. N., Ginsberg, H. S., and Wood, W. B. (1967). "Microbiology." Harper and Row, New York.

Davis, R. W., Simon, R., and Davidson, N. (1971). Electron microscope heteroduplex methods for mapping regions of base sequence homology in nucleic acids. *In* "Methods in Enzymology" (L. Grossman and K. Moldave, eds.), Vol. 21, p. 413. Academic Press, New York.

Dayton, D. H., Small, P. A., Chanock, R. M., Kaufman, H. E., and Tomasi, T. B. (eds.) (1971). "The Secretory Immunologic System." National Institute of Child Health and Human Development, Bethesda, Maryland.

de Clercq, E. (1972). Nucleic acids as interferon inducers. *In* "Virus-Cell Interactions and Viral Antimetabolites" (D. Shugar, ed.), p. 65. Academic Press, New York.

de Clercq, E., and Merigan, T. C. (1969). Local and systemic protection by synthetic polyanionic interferon inducers in mice against intranasal vesicular stomatitis virus. *J. Gen. Virol.,* **5,** 359.

de Clercq, E., and Merigan, T. C. (1970a). Current concepts of interferon and interferon induction. *Annu. Rev. Med.* **21,** 17.

de Clercq, E., and Merigan, T. C. (1970b). Induction of interferon by non-viral agents. *Arch. Internal Med.* **126,** 94.

de Clercq, E., and Merigan, T. C. (1971). Bis-DEAE-Fluorenone: mechanism of antiviral protection and stimulation of interferon production in the mouse. *J. Infect. Dis.* **123,** 190.

de Clercq, E., Wells, R. D., and Merigan, T. C. (1972). Studies on the antiviral activity and cell interaction of synthetic double-stranded polyribo- and polydeoxyribonucleotides. *Virology* **47,** 405.

deDuve, C. (1963). General properties of lysosomes *In* "Lysosomes" (A. V. S. deReuck and M. Cameron, eds.), CIBA Found. Symp., p. 1. Churchill, London.

Defendi, V. (1963). Effect of SV40 virus immunization on growth of transplantable SV40 and polyoma virus tumors in hamsters. *Proc. Soc. Exp. Biol. Med.* **113,** 12.

Defendi, V. (1968). Studies with virally induced transplantation antigens. *Transplantation* **6,** 642.

Defendi, V., and Jensen, F. (1967). Oncogenicity by DNA tumor viruses: enhancement after ultraviolet and cobalt-60 radiations. *Science* **157,** 703.

Degré, M. (1970). Synergistic effect in viral-bacterial infection. II. Influence of viral infection on the phagocytic ability of alveolar macrophages. *Acta Pathol. Microbiol. Scand.* **B78,** 41.

d'Hérelle, F. (1917). Sur un microbe invisible antagoniste des bacilles dysentériques. *C. R. Acad. Sci.* **165,** 373.

Deinhardt, F., and Shramek, G. J. (1969). Immunization against mumps. *Progr. Med. Virol.* **11,** 126.

Deinhardt, F., Holmes, A. W., Capps, R. B., and Popper, H. (1967). Studies on the transmission of human viral hepatitis to marmoset monkeys, I. Transmission of the disease, serial passages, and description of liver lesions. *J. Exp. Med.* **125,** 673.

DeLange, R. J., and Smith, E. L. (1971). Histones: structure and function. *Annu. Rev. Biochem.* **40,** 279.

DeLay, P. D., Stones, S. S., Karzon, D. T., Katz, S., and Enders, J. (1965). Clinical and immune response of alien hosts to inoculation with measles, rinderpest, and canine distemper viruses. *Amer. J. Vet. Res.* **26,** 1359.

deLey, J. (1968). Molecular biology and bacterial phylogeny. *In* "Evolutionary Biology" (T. Dobzhansky, M. Hecht, and W. Steine, eds.), Vol. 2, p. 103. Appleton-Century Crofts, New York.

Del Villano, B., and Defendi, V. (1970). Preparation and use of immunosorbent to study the molecular composition of SV40 T antigen from hamster tumor cells. *Bacteriol. Proc.*, p. 188.

de Maeyer, E., and de Maeyer-Guignard, J. (1969). Gene with quantitative effect on circulating interferon induced by Newcastle diseas virus. *J. Virol.* **3**, 506.

de Maeyer-Guignard, J., de Maeyer, E., and Montagnier, L. (1972). Interferon messenger RNA: Translation in heterologous cells. *Proc. Nat. Acad. Sci. U.S.* **69**, 1203.

deRobertis, E. D. P., Nowinski, W. W., and Saez, F. (1970). "Cell Biology," 5th. Ed, p. 452. Saunders, Philadelphia, Pennsylvania.

de-Thé, G., and Notkins, A. L. (1965). Ultrastructure of the lactic dehydrogenase virus (LDV) and cell-virus relationships. *Virology* **26**, 512.

de-Thé, G., Ishiguro, H., Heine, U., Beard, D., and Beard, J. W. (1963). Multiplicity of cell response to BAI strain A (myeloblastosis) avian tumor virus. VI. Ultrastructural aspects of adenosine triphosphate activity of nephroblastoma cells and virus. *J. Nat. Cancer Inst.* **30**, 1267.

Deutsch, V., and Berkaloff, A. (1971). Analyse d'un mutant thermolabile du virus de la stomatité vesiculaire (VSV). *Ann. Inst. Pasteur (Paris)* **121**, 101.

de Waard, A., Paul, A. V., and Lehman, I. R. (1965). The structural gene for deoxyribonucleic acid polymerase in bacteriophages T4 and T5. *Proc. Nat. Acad. Sci. U.S.* **54**, 1241.

Diamond, L. S., Mattern, C. F. T., and Bartgis, I. L. (1972). Viruses of *Entamoeba histolytica*. I. Identification of transmissible virus-like agents. *J. Virol.* **9**, 326.

Dick, E. C. (1968). Experimental infections of chimpanzees with human rhinoviruses types 14 and 43. *Proc. Soc. Exp. Biol. Med.* **127**, 1079.

Dick, G. (1966). Smallpox: a reconsideration of public health policies. *Progr. Med. Virol.* **8**, 1.

Dick, G. (1971). Routine smallpox vaccination. *Brit. Med. J.* **3**, 163.

Dickinson, A. G., Young, G. B., Stamp, T., and Renwick, C. C. (1965). An analysis of natural scrapie in Suffolk sheep. *Heredity* **20**, 485.

Diderholm, H., Stenkvist, B., Pontén, J., and Wesslén, T. (1965). Transformation of bovine cells *in vitro* after inoculation of simian virus 40 or its nucleic acid. *Exp. Cell Res.* **37**, 452.

Diener, T. O. (1972a). Is the scrapie agent a viroid? *Nature New Biol.* **235**, 218.

Diener, T. O. (1972b). Viroids. *Advan. Virus Res.* **17**, 295.

Diggelmann, H., and Weissmann, C. (1969). Rifampicin inhibits focus formation in chick fibroblasts infected with Rous sarcoma virus. *Nature (London)* **224**, 1277.

di Mayorca, G., and Callender, J. (1970). Significance of temperature-sensitive mutants in viral oncology. *Progr. Med. Virol.* **12**, 284.

di Mayorca, G., Callender, J., Marin, G., and Giordano, R. (1969). Temperature-sensitive mutants of polyoma virus. *Virology* **38**, 126.

Dimmock, N. J. (1969). New virus-specific antigens in cells infected with influenza virus. *Virology* **39**, 224.

Dimmock, N. J., and Watson, D. H. (1969). Proteins specified by influenza virus in infected cells: analysis by polyacrylamide gel electrophoresis of antigens not present in the virus particle. *J. Gen. Virol.* **5**, 499.

Dingman, C. W., and Peacock, A. C. (1971). Polyacrylamide gel electrophoresis of RNA. *In* "Methods in Nucleic Acid Research" (G. L. Cantoni and D. R. Davies, eds.), Vol. 2, p 623. Harper and Row, New York.

Di Stefano, H. S., and Dougherty, R. M. (1966). Mechanisms for congenital transmission of avian leukosis virus. *J. Nat. Cancer Inst.* **37**, 869.

Dixon, C. W. (1962). "Smallpox." Churchill, London.

Dixon, F. J. (1963). The role of antigen-antibody complexes in disease. *Harvey Lect. Ser.* **58**, 21.

Dobos, P., Arif, B .M., and Faulkner, P. (1971). Denaturation of Sindbis virus RNA with dimethylsulphoxide. *J. Gen. Virol.* **10**, 103.

Dodd, K., Johnson, I. M., and Buddingh, G. J. (1938). Herpetic stomatitis. *J. Pediat.* **12**, 95.

Doerfler, W., Lundholm, U., and Hirsch-Kaufman, M. (1972). Intracellular forms of adenovirus DNA. I. Evidence for a deoxyribonucleic acid-protein complex in baby hamster kidney cells infected with adenovirus type 12. *J. Virol.* **9**, 297.

Doerr, R., and Vochting, K. (1920). Sur le virus de l'herpes febrile. *Rev. Gen. Ophthalmol.* **34**, 409.

Doherty, P. C., Reid, H. W., and Smith, W. (1971). Louping-ill encephalomyelitis in the sheep. IV. Nature of the perivascular inflammatory reaction. *J. Comp. Pathol.* **81**, 545.

Dolin, R., Blacklow, N. R., DuPont, H., Buscho, R. F., Wyatt, R. G., Kesel, J. A., Hornick, R., and Chanock, R. M. (1972). Biological properties of Norwalk agent of acute infectious nonbacterial gastroenteritis. *Proc. Soc. Exp. Biol. Med.* **140**, 578.

Dougherty, R M. (1964). Animal virus titration techniques. In "Techniques in Experimental Virology" (R. J. C. Harris, ed.), p. 169. Academic Press, New York.

Dougherty, R. M., and Rasmussen, R. (1964). Properties of a strain of Rous sarcoma virus that infects mammals. *Nat. Cancer Inst. Monog.* **17**, 337.

Dougherty, R. M., and DiStefano, H. S. (1966). Lack of relationship between infection with avian leukosis virus and the presence of COFAL antigen in chick embryos. *Virology* **29**, 586.

Dougherty, R. M., DiStefano, H. S., and Roth, F. K. (1967). Virus particles and viral antigens in chicken tissues free of infectious avian leukosis virus. *Proc. Nat. Acad. Sci. U.S.* **58**, 808.

Douglas, R. G., and Couch, R. B. (1970). A prospective study of chronic herpes simplex virus infection and recurrent herpes labialis in humans. *J. Immunol.* **104**, 289.

Douglas, R. G., Lindgren, K. M., and Couch, R. B. (1968). Exposure to cold environment and rhinovirus common cold. *New Engl. J. Med.* **279**, 742.

Downie, A. W. (1963). Pathogenesis of generalized virus diseases. *Vet. Rec.* **75**, 1125.

Downie, A. W., and Haddock, D. W. (1952). A variant of cowpox virus. *Lancet* **i**, 1049.

Downie, A. W., McCarthy, D., Macdonald, A., MacCallum, F. O., and Macrae, A. D. (1953). Virus and virus antigen in the blood of smallpox patients. Their significance in early diagnosis and prognosis. *Lancet* **ii**, 164.

Downie, A. W., Meiklejohn, M., St. Vincent, L., Rao, A. R., Babu, B. V. S., and Kempe, C. H. (1965). The recovery of smallpox virus from patients and their environment in a smallpox hospital. *Bull. WHO* **33**, 615.

Drake, J. W., and Lay, P. A. (1962). Host-controlled variations in NDV. *Virology* **17**, 56.

Dreguss, M. N., and Lombard, L. S. (1954). "Experimental Studies in Equine Infectious Anemia." Univ. of Pennsylvania Press, Philadelphia, Pennsylvania.

Druett, H. A. (1967). The inhalation and retention of particles in the human respiratory system. *Sym. Soc. Gen. Microbiol.* **17**, 167.

Drzeniek, R. (1972). Viral and bacterial neuraminidases. *Curr. Topics Microbiol. Immunol.* **59**, 35.

Dubbs, D. R., and Kit, S. (1968). Isolation of defective lysogens from simian virus 40-transformed mouse kidney cultures. *J. Virol.* **2**, 1272.

Dubbs, D. R., Kit, S., de Torres, R. A., and Anken, M. (1967). Virogenic properties of bromodeoxyuridine-sensitive and bromodeoxyuridine-resistant simian virus 40-transformed mouse kidney cells. *J. Virol.* **1**, 968.

Dubes, G. B., and Wenner, H. A. (1957). Virulence of polioviruses in relation to variant characteristics distinguishable on cells *in vitro*. *Virology* **4**, 275.

Dubos, R. (1965). "Man Adapting." Yale Univ. Press, New Haven, Connecticut.

Duc-Nguyen, H., Rose, H. M., and Morgan, C. (1966). An electron microscopic study of changes at the surface of influenza-infected cells as revealed by ferritin-conjugated antibodies. *Virology* **28**, 404.

Duesberg, P. H. (1968a). The RNA's of influenza virus. *Proc. Nat. Acad. Sci. U.S.* **59**, 930.

Duesberg, P. H. (1968b). Physical properties of Rous sarcoma virus RNA. *Proc. Nat. Acad. Sci. U.S.* **60**, 1511.

Duesberg, P. H. (1969). Distinct subunits of the ribonucleoprotein of influenza virus. *J. Mol. Biol.* **42**, 485.

Duesberg, P. H. (1970). On the structure of RNA tumor viruses. *Curr. Topics Microbiol. Immunol.* **51**, 79.

Duesberg, P. H., and Robinson, W. S. (1965). Isolation of the nucleic acid of Newcastle disease virus (NDV). *Proc. Nat. Acad. Sci. U.S.* **54**, 794.

Duesberg, P. H., and Robinson, W. S. (1967) One the structure and replication of influenza virus. *J. Mol. Biol.* **25**, 383.

Duesberg, P. H., and Vogt, P. K. (1970). Differences between the ribonucleic acids of transforming and nontransforming avian tumor viruses. *Proc. Nat. Acad. Sci. U.S.* **67**, 1673.

Duesberg, P. H., Martin, G. S., and Vogt, P. K. (1970). Glycoprotein components of avian and murine RNA tumor viruses. *Virology* **41**, 631.

Duesberg, P. H., Vogt, P. K., and Canaani, E. (1971). Structure and replication of avian tumor virus RNA. *In* "Lepetit Colloquia on Biology and Medicine, Vol. 2: The Biology of Oncogenic Viruses" (L. Sylvestri, ed.), p. 154. North Holland, Amsterdam.

Duff, R., and Rapp, F. (1970a). Reaction of serum from pregnant hamsters with surface of cells transformed by SV40. *J. Immunol.* **105**, 521.

Duff, R., and Rapp, F. (1970b). Quantitative characteristics of the transformation of hamster cells by PARA (defective simian virus 40)-adenovirus 7. *J. Virol.* **5**, 568.

Duff, R., and Rapp, F. (1971). Oncogenic transformation of hamster cells after exposure to herpes simplex virus type 2. *Nature New Biol.* **233**, 48.

Duff, R. G., and Vogt, P. (1969). Characteristics of two new avian tumor virus subgroups. *Virology* **39**, 18.

Duff, R., Knight, P., and Rapp, F. (1972). Variation in oncogenic and transforming potential of para (defective SV40)-adenovirus 7. *Virology* **47**, 849.

Dulbecco, R. (1952). Production of plaques in monolayer tissue cultures by single particles of an animal virus. *Proc. Nat. Acad. Sci. U.S.* **38**, 747.

Dulbecco, R. (1961). Poliovirus mutations seen from the point of view of molecular biology. *In* "Poliomyelitis," 5th Intern. Poliomyelitis Conf., p. 21. Lippincott, Philadelphia, Pennsylvania.

Dulbecco, R. (1969). Cell transformation by viruses. *Science* **166**, 962.

Dulbecco, R. (1970). Topoinhibition and serum requirement of transformed and untransformed cells. *Nature (London)* **227**, 802.

Dulbecco, R., and Eckhart, W. (1970). Temperature-dependent properties of cells transformed by a thermosensitive mutant of polyoma virus. *Proc. Nat. Acad. Sci. U.S.* **67**, 1775.

Dulbecco, R., and Johnson, T. (1970). Interferon sensitivity of the enhanced incorporation of thymidine into cellular DNA induced by polyoma virus. *Virology* **42**, 368.

Dulbecco, R., and Vogt, M. (1954). Plaque formation and isolation of pure lines with poliomyelitis viruses. *J. Exp. Med.* **99**, 167.

Dulbecco, R., and Vogt, M. (1958). Study of the mutability of *d* lines of polioviruses. *Virology* **5**, 220.

Dulbecco, R., and Vogt, M. (1963). Evidence for a ring structure of polyoma virus DNA. *Proc. Nat. Acad. Sci. U.S.* **50**, 236.

Dulbecco, R., Vogt, M., and Strickland, A. (1956). A study of the basic aspects of neutralization of two animal viruses, Western equine encephalitis virus and poliomyelitis virus. *Virology* **2**, 162.

Dulbecco, R., Hartwell, L. H., and Vogt, M. (1965). Induction of cellular DNA synthesis by polyoma virus. *Proc. Nat. Acad. Sci. U.S.* **53**, 403.

Dunlap, R. C., and Patt, J. K. (1971). Reactivation of heated vaccinia virus by UV-irradiated vaccinia virus. *Proc. Soc. Exp. Biol. Med.* **136**, 1.

Dunker, A. K., and Reuckert, R. R. (1971). Fragments generated by pH dissociation of ME-virus and their relation to the structure of the virion. *J. Mol. Biol.* **58**, 217.

Dunnebacke, T. H., and Kleinschmidt, A. K. (1967). Ribonucleic acid from reovirus as seen in protein monolayers by electron microscopy. *Z. Naturforsch.* **22b**, 159.

Dunnebacke, T. H., Levinthal, J. D., and Williams, R. C. (1969). Entry and release of poliovirus as observed by electron microscopy of cultured cells. *J. Virol.* **4**, 505.

Dupont, J. R., and Earle, K. M. (1965). Human rabies encephalitis. *Neurology* **15**, 1023.

Dupraw, E. (1970). "DNA and Chromosomes." Holt and Rinehart, New York.

Durand, D. P., Brotherton, T., Chalgren, S., and Collard, W. (1970). Uncoating of myxo-viruses. In "The Biology of Large RNA Viruses" (R. D. Barry and B. W. J. Mahy, eds.), p. 197. Academic Press, London.

Duran-Reynals, F. (1940a). A hemorrhagic disease occurring in chicks inoculated with the Rous and Fuginami viruses. Yale J. Biol. Med. **13,** 77.

Duran-Reynals, F. (1940b). Production of degenerative inflammatory or neoplastic effects in the newborn rabbit by the Shope fibroma virus. Yale J. Biol. Med. **13,** 99.

Duran-Reynals, F. (1945). Immunological factors that influence the neoplastic effects of the rabbit fibroma virus. Cancer Res. **5,** 25.

Duxbury, A. E., Hampson, A. W., and Sievers, J. G. M. (1968). Antibody response in humans to deoxycholate-treated influenza virus vaccine. J. Immunol. **101,** 62.


Eagle, H. (1959). Amino acid metabolism in mammalian cell cultures. Science **130,** 432.

Easterbrook, K. B. (1966). Controlled degradation of vaccinia virions in vitro: An electron microscopic study. J. Ultrastruct. Res. **14,** 484.

Easterday, B. C., and Tumova, B. (1972). Avian influenza viruses: in avian species and the natural history of influenza. Advan. Vet. Sci. Comp. Med. **16,** 201.

Easterday, B., Laver, W. G., Pereira, H. G., and Schild, G. C. (1969). Antigenic composition of recombinant virus strains produced from human and avian influenza A viruses. J. Gen. Virol. **5,** 83.

Easton, J. M. (1964). Cytopathic effect of simian virus 40 on primary cell cultures of rhesus monkey kidney. J. Immunol. **93,** 716.

Eaton, B. T., Donaghue, T. P., and Faulkner, P. (1972). Presence of poly (A) in the poly-ribosome-associated RNA of Sindbis-infected BHK cells. Nature New Biol. **238,** 109.

Echols, H. (1971). Lysogeny: viral repression and site-specific recombination. Annu. Rev. Biochem. **40,** 827.

Eckhart, W. (1969). Complementation and transformation by temperature-sensitive mutants of polyoma virus. Virology **38,** 120.

Eckhart, W. (1971). Induced cellular DNA synthesis by "early" and "late" temperature-sensitive mutants of polyoma virus. Proc. Roy. Soc. **B177,** 59.

Eckhart, W., Dulbecco, R., and Burger, M. M. (1971). Temperature-dependent surface changes in cells infected or transformed by a thermosensitive mutant of polyoma virus. Proc. Nat. Acad. Sci. U.S. **68,** 283.

Eddy, B. E. (1964). Simian virus 40(SV-40): an oncogenic virus. Progr. Exp. Tumor Res. **4,** 1.

Eddy, B. E., Borman, G. S., Berkeley, W. H., and Young, R. D. (1961). Tumors induced in ham-sters by rejection of rhesus monkey kidney cell extracts. Proc. Soc. Exp. Biol. Med. **107,** 191.

Eddy, B. E., Borman, G. S., Grubbs, G. E., and Young, R. D. (1962). Identification of the on-cogenic substance in rhesus monkey kidney cell cultures as simian virus 40. Virology **17,** 65.

Edgar, R. S., and Lielausis, I. (1964). Temperature-sensitive mutants of bacteriophage T4D: their isolation and genetic characterization. Genetics **49,** 649.

Edgar, R. S., Denhardt, G. H., and Epstein, R. H. (1964). A comparative genetic study of conditional lethal mutations of bacteriophage T4D. Genetics **49,** 635.

Edmondson, W. P., Purcell, R. H., Gundelfinger, B. F., Love, J. W. P., Ludwig, W., and Chanock, R. M. (1966). Immunization by selective infection with type 4 adenovirus grown in human diploid tissue culture. II. Specific protective effect against epidemic disease. J. Amer. Med. Ass. **195,** 453.

Edsall, G. (1969). A positive approach to the problem of human experimentation. Daedalus **98,** 463.

Eggen, K. L., and Shatkin, A. J. (1972). In vitro translation of cardiovirus ribonucleic acid by mammalian cell-free extracts. J. Virol. **9,** 636.

Eggers, H. J., and Tamm, I. (1966). Antiviral chemotherapy. Annu. Rev. Pharmacol. **6,** 231.

Eggers, H. J., Reich, E., and Tamm, I. (1963). The drug-requiring phase in the growth of drug dependent enteroviruses. *Proc. Nat. Acad. Sci. U.S.* **50**, 183.

Ehrenfeld, E., and Hunt, T. (1971). Double-stranded poliovirus RNA inhibits initiation of protein synthesis by reticulocyte lysates. *Proc. Nat. Acad. Sci. U.S.* **68**, 1075.

Eklund, C. M., Kennedy, R. C., and Hadlow, W. J. (1967). Pathogenesis of scrapie virus infection in the mouse. *J. Infect. Dis.* **117**, 15.

Elgin, S. C. R., Froehner, S. C., Smart, J. E., and Bonner, J. (1971). The biology and chemistry of chromosomal proteins. *Advan. Cell Mol. Biol.* **1**, 1.

Ellis, E. L., and Delbrück, M. (1939). The growth of bacteriophage. *J. Gen. Physiol.* **22**, 365.

Els, H. J., and Verwoerd, D. W. (1969). Morphology of bluetongue virus. *Virology* **38**, 213.

Elveback, L. R., Ackerman, E., Young, G., and Fox, J. P. (1968). A stochastic model for competition between viral agents in the presence of interference. I. Live virus vaccine in a randomly mixing population, Model III. *Amer. J. Epidemiol.* **87**, 373.

Emmelot, P., and Bentvelzen, P., eds. (1972). "RNA Viruses and Host Genome in Oncogenesis." North Holland, Amsterdam.

Emmons, R. W., Oshiro, L. S., Johnson, H. N., and Lennette, E. H. (1972). Intra-erythrocytic location of Colorado Tick Fever Virus. *J. Gen. Virol.* **17**, 185.

Enders, J. F., and Katz, S. L. (1967). Present status of live rubeola vaccines in the United States. *In* "First International Conference on Vaccines against Viral and Rickettsial Diseases of Man," p. 295. Sci. Publ. No. 147. Pan Amer. Health Organ., Washington, D.C.

Enders, J. F., Weller, T. H., and Robbins, F. C. (1949). Cultivation of Lansing strain of poliomyelitis virus in cultures of various human embryonic tissues. *Science* **109**, 85.

Enders, J. F., Holloway, A., and Grogan, E. A. (1967). Replication of poliovirus 1 in chick embryo and hamster cells exposed to Sendai virus. *Proc. Nat. Acad. Sci. U.S.* **57**, 637.

Ennis, H. (1968). Structure-activity studies with cycloheximide congeners. *Biochem. Pharmacol.* **17**, 1197.

Ennis, H. L., and Lubin, M. (1964). Cycloheximide: Aspects of inhibition of protein synthesis in mammalian cells. *Science* **146**, 1474.

Ensminger, W. D., and Tamm, I. (1969). Cellular DNA and protein synthesis in reovirus-infected L cells. *Virology* **39**, 357.

Ensminger, W. D., and Tamm, I. (1970). The step in cellular DNA synthesis blocked by Newcastle disease or mengovirus infection. *Virology* **40**, 152.

Epstein, M. A. (1962). Observations on the mode of release of herpes virus from infected HeLa cells. *J. Cell Biol.* **12**, 589.

Epstein, M. A. (1970). Aspects of the EB virus. *Advan. Cancer Res.* **13**, 383.

Epstein, M. A. (1971). The possible role of viruses in human cancer. *Lancet* **i**, 1344.

Epstein, M. A., and Holt, S. J. (1963). Adenosine triphosphatase activity at the surface of mature extracellular herpes virus. *Nature (London)* **198**, 509.

Epstein, M. A., Achong, B. G., and Barr, Y. M. (1964). Virus particles in cultured lymphoblasts from Burkitt's lymphoma. *Lancet* **ii**, 702.

Epstein, M. A., Henle, G., Achong, B. G., and Barr, B. M. (1965). Morphological and biological studies on a virus in cultured lymphoblasts from Burkitt's lymphoma. *J. Exp. Med.* **121**, 761.

Epstein, M. A., Achong, B. G., Churchill, A. E., and Biggs, P. M. (1968). Structure and development of the herpes-type virus of Marek's disease. *J. Nat. Cancer Inst.* **41**, 805.

Epstein, R. H., Bollé, A., Steinberg, C. M., Kellenberger, E., Boy de la Tour, E., Chevalley, R., Edgar, R. S., Susman, M., Denhardt, G. H., and Lielausis, A. (1963). Physiological studies of conditional lethal mutations of bacteriophage T4D. *Cold Spring Harbor Symp. Quant. Biol.* **28**, 375.

Erikson, E., and Erikson, R. L. (1970). Isolation of amino acid acceptor RNA from purified avian myeloblastosis virus. *J. Mol. Biol.* **52**, 387.

Erikson, E., and Erikson, R. L. (1971). Association of 4S ribonucleic acid with oncornavirus ribonucleic acids. *J. Virol.* **8**, 254.

Erikson, R. L. (1969). Studies on the RNA from avian myeloblastosis virus. *Virology* **37**, 124.

Erikson, R. L., and Erikson, E. (1972). Transfer ribonucleic acid synthetase activity associated with avian myeloblastosis virus. *J. Virol.* **9**, 231.

Estes, M. K., Huang, E-S., and Pagano, J. S. (1971). Structural polypeptides of simian virus 40. *J. Virol.* **7**, 635.

Evans, A. S. (1958). Latent adenovirus infection of the human respiratory tract. *Amer. J. Hyg.* **67**, 256.

Evans, A. S., Niederman, J. C., and McCollum, R. W. (1968). Seroepidemiologic studies of infectious mononucleosis with EB virus. *New Engl. J. Med.* **279**, 1121.

Evans, C. A., Byatt, P. H., Chambers, V. C., and Smith, W. M. (1954). Growth of neurotropic viruses in extraneural tissues. VI. Absence of *in vivo* multiplication of poliomyelitis virus, types I and II, after intratesticular inoculation of monkeys and other laboratory animals. *J. Immunol.* **72**, 348.

Evans, C. A., Weiser, R. S., and Ito, Y. (1962). Antiviral and antitumor immunologic mechanisms operative in the Shope papilloma-carcinoma system. *Cold Spring Harbor Symp. Quant. Biol.* **27**, 453.

Evans, D. G. (Ed.) (1969). Immunization against infectious diseases. *Brit. Med. Bull.* **25**, 119.

Evans, H. D. (1964). Uptake of $^3$H-thymidine and patterns of DNA replication in nuclei and chromosomes of *Vicia fava*. *Exp. Cell Res.* **35**, 381.

Evans, R., and Riley, V. (1968). Circulating interferon in mice infected with the lactic dehydrogenase-elevating virus. *J. Gen. Virol.* **3**, 449.

Everitt, E., Sundquist, B., and Philipson, I. (1971). Mechanism of arginine requirement for adenovirus synthesis. I. Synthesis of structural proteins. *J. Virol.* **8**, 742.

Everitt, E., Sundquist, B., Pettersson, U., and Philipson, L. (1973). Structural proteins of adenoviruses. X. Isolation and topography of low molecular weight antigens from the virion of adenovirus type 2. *Virology* **51**, 130.

Falk, H. L., Tremer, H. M., and Kotin, P. (1959). Effect of cigarette smoke and its constituents on ciliated mucus-secreting epithelium. *J. Nat. Cancer Inst.* **23**, 999.

Farber, J., Rovera, G., and Baserga, R. (1971). Template activity of chromatin during stimulation of cellular proliferation in human diploid fibroblasts. *Biochem. J.* **122**, 189.

Fastier, L. B. (1952). Toxic manifestations in rabbits and mice associated with the virus of western equine encephalomyelitis. *J. Immunol.* **68**, 531.

Fazekas de St. Groth, S. (1948a). Viropexis: The mechanism of influenza virus infection. *Nature (London)* **162**, 294.

Fazekas de St. Groth, S. (1948b). Regeneration of virus receptors in mouse lungs after artificial destruction. *Aust. J. Exp. Biol. Med. Sci.* **26**, 271.

Fazekas de St. Groth, S. (1950). Influenza: A study in mice. *Lancet* **i**, 1101.

Fazekas de St. Groth, S. (1962). The neutralization of viruses. *Advan. Virus Res.* **9**, 1.

Fazekas de St. Groth, S. (1969). New criteria for selection of influenza vaccine strains. *Bull. WHO* **41**, 651.

Fazekas de St. Groth, S. (1970). Evolution and hierarchy of influenza viruses. *Arch. Environ. Health* **21**, 293.

Fazekas de St. Groth, S., and Donnelly, M. (1950). Studies in experimental immunology. IV. The protective effect of active immunization. *Aust. J. Exp. Biol. Med. Sci.* **28**, 62.

Fazekas de St. Groth, S., and Webster, R. G. (1961). Methods in immuno-chemistry of viruses I. Equilibrium filtration. *Aust. J. Exp. Biol. Med. Sci.* **39**, 549.

Fazekas de St. Groth, S., and Webster, R. G. (1964). The antibody response. *Ciba Found. Symp. Cell. Biol. Myxovirus Infections*, p. 246.

Fazekas de St. Groth, S., and White, D. O. (1958). An improved assay for the infectivity of influenza viruses. *J. Hyg.* **56**, 151.

Fazekas de St. Groth, S., Watson, G. S., and Reid, A. F. (1958). The neutralization of animal viruses. I. A model of virus-antibody interaction. *J. Immunol.* **80**, 215.

Fenje, P., and Postic, B. (1971). Prophylaxis of experimental rabies with polyriboinosinic-polycytidylic acid complex. *J. Infect. Dis.* **123**, 426.

Fenner, F. (1947). Studies in infectious ectromelia of mice. II. Natural transmission: The portal of entry of the virus. *Aust. J. Exp. Biol. Med. Sci.* **25**, 275.

Fenner, F. (1948a). The epizootic behaviour of mousepox (infectious ectromelia). *Brit. J. Exp. Pathol.* **29**, 69.

Fenner, F. (1948b). The pathogenesis of the acute exanthems. *Lancet* **ii**, 915.

Fenner, F. (1949). Mousepox (infectious ectromelia of mice): A review. *J. Immunol.* **63**, 341.

Fenner, F. (1958). The biological characters of several strains of vaccinia, cowpox, and rabbit-pox viruses. *Virology* **5**, 502.

Fenner, F. (1959). Genetic studies with mammalian poxviruses. II. Recombination between two strains of vaccinia virus in single HeLa cells. *Virology* **8**, 499.

Fenner, F. (1962). The reactivation of animal viruses. *Brit. Med. J.* **2**, 135.

Fenner, F. (1965). Viruses of the myxoma-fibroma subgroup of the poxviruses. II. Comparison of soluble antigens by gel-diffusion tests, and a general discussion of the subgroup. *Aust. J. Exp. Biol. Med. Sci.* **43**, 143.

Fenner, F. (1968). "The Biology of Animal Viruses, Vol. I: Cellular and Molecular Biology." Academic Press, New York and London.

Fenner, F. (1969). Conditional lethal mutants of animal viruses. *Curr. Topics Microbiol. Immunol.* **48**, 1.

Fenner, F. (1972). The possible use of temperature-sensitive conditional lethal mutants for immunization in viral infections. *In* "Immunity in Viral and Rickettsial Diseases" (A. Kohn and M. A. Klingberg, eds.), p. 131. Plenum Press, New York.

Fenner, F., and Burnet, F. M. (1957). A short description of the Poxvirus group (vaccinia and related viruses). *Virology* **4**, 305.

Fenner, F., and Cairns, J. (1959). Variation in virulence in relation to adaptation to new hosts. *In* "The Viruses" (F. M. Burnet and W. M. Stanley, eds.), Vol. 3, p. 225. Academic Press, New York.

Fenner, F., and Chapple, P. J. (1965). Evolutionary changes in myxoma virus in Britain. An examination of 222 naturally-occurring strains obtained from 80 counties during the period October-November 1962. *J. Hyg.* **63**, 175.

Fenner, F., and Comben, B. M. (1958). Genetic studies with mammalian poxviruses. I. Demonstration of recombination between two strains of vaccinia virus. *Virology* **5**, 530.

Fenner, F., and Ratcliffe, F. N. (1965). "Myxomatosis." Cambridge Univ. Press, London and New York.

Fenner, F., and Sambrook, J. F. (1964). The genetics of animal viruses. *Annu. Rev. Microbiol.* **18**, 47.

Fenner, F., and Sambrook, J. F. (1966). Conditional lethal mutants of rabbitpox virus. II. Mutants (*p*) which fail to multiply in PK-2a cells. *Virology* **28**, 600.

Fenner, F., and White, D. O. (1970). "Medical Virology." Academic Press, New York.

Fenner, F., and Woodroofe, G. M. (1953). The pathogenesis of infectious myxomatosis: The mechanism of infection and the immunological response in the European rabbit (*Oryctolagus cuniculus*). *Brit. J. Exp. Pathol.* **34**, 400.

Fenner, F., and Woodroofe, G. M. (1965). Changes in the virulence and antigenic structure of strains of myxoma virus recovered from Australian wild rabbits between 1950 and 1964. *Aust. J. Exp. Biol. Med. Sci.* **43**, 359.

Fenner, F., Poole, W. E., Marshall, I. D., and Dyce, A. L. (1957). Studies in the epidemiology of infectious myxomatosis of rabbits. VI. The experimental introduction of the European strain of virus into Australian wild rabbit populations. *J. Hyg.* **55**, 192.

Fenner, F., Holmes, I. H., Joklik, W. K., and Woodroofe, G. M. (1959). Reactivation of heat-inactivated poxviruses: A general phenomenon which includes the fibroma-myxoma virus transformation of Berry and Dedrick. *Nature (London)* **183**, 1340.

Fenyö, E. M., Klein, E., Klein, G., and Swiech, K. (1968). Selection of an immunoresistant Moloney lymphoma subline with decreased concentration of tumor-specific surface antigens. *J. Nat. Cancer Inst.* **40**, 69.

Fernandes, M. V., Wiktor, T. J., and Koprowski, H. (1965). Endosymbiotic relationship between animal viruses and host cells. A study of rabies virus in tissue culture. *J. Exp. Med.* **120**, 1099.

Ferrer, J. F., Stock, N. D., and Lin, P. (1971). Detection of replicating C-type viruses in continuous cell cultures established from cows with leukemia; effect of the culture medium. *J. Nat. Cancer Inst.* **47**, 613.

Fetterman, G. H., Sherman, F. E., Fabrizio, N. S., and Studnicki, F. M. (1968). Generalized cytomegalic inclusion disease of the new born: localization of inclusions in the kidney. *Arch. Pathol.* **86,** 86.

Field, A. K., Lampson, G. P., Tytell, A. A., Nemes, M. M., and Hilleman, M. R. (1967a). Inducers of interferon and host resistance, IV. a double-stranded replicative form RNA (MS₂-RF-RNA) from *E. coli* infected with MS₂ coliphage. *Proc. Nat. Acad. Sci. U.S.* **58,** 2102.

Field, A. K., Tytell, A. A., Lampson, G. P., and Hilleman, M. R. (1967b). Inducers of interferon and host resistance, II. Multistranded synthetic polynucleotide complexes. *Proc. Nat. Acad. Sci. U.S.* **58,** 1004.

Field, A. K., Tytell, A. A., Lampson, G. P., and Hilleman, M. R. (1972). Antigenicity of double-stranded ribonucleic acids including poly I:C. *Proc. Soc. Exp. Biol. Med.* **139,** 1113.

Fields, B. N. (1971). Temperature-sensitive mutants of reovirus type 3: Features of genetic recombination. *Virology* **46,** 142.

Fields, B. N., and Joklik, W. K. (1969). Isolation and preliminary genetic and biochemical characterization of temperature-sensitive mutants of reovirus. *Virology* **37,** 335.

Fields, B. N., Raine, C. S., and Baum, S. G. (1971). Temperature-sensitive mutants of reovirus type 3: Defects in viral maturation as studied by immunofluorescence and electron microscopy. *Virology* **43,** 569.

Fields, B. N., Laskov, R., and Scharff, M. D. (1972). Temperature-sensitive mutants of reovirus type 3: studies on the synthesis of viral peptides. *Virology* **50,** 209.

Filshie, B. K., and Reháček, J. (1968). Studies of the morphology of Murray Valley encephalitis and Japanese encephalitis viruses growing in cultured mosquito cells. *Virology* **34,** 435.

Finch, J. T., and Gibbs, A. J. (1970). Observations on the structure of the nucleocapsids of some paramyxoviruses. *J. Gen. Virol.* **6,** 141.

Finch, J. T., and Klug, A. (1959). Structure of poliomyelitis virus. *Nature (London)* **183,** 1709.

Finch, J. T., and Klug, A. (1965). The structure of viruses of the papilloma-polyoma type. III. Structure of rabbit papilloma virus. *J. Mol. Biol.* **13,** 1.

Finch, J. T., and Klug, A. (1966). Arrangement of protein subunits and the distribution of nucleic acid in turnip yellow mosaic virus. II. Electron microscopic studies. *J. Mol. Biol.* **15,** 344.

Fincham, J. R. S. (1966). "Genetic Complementation." Benjamin, New York.

Findlay, G. M., and MacCallum, F. O. (1937). An interference phenomenon in relation to yellow fever and other viruses. *J. Pathol. Bacteriol.* **44,** 405.

Finkelstein, M. S., Bausek, G., and Merigan, T. C. (1968). Interferon inducers *in vitro*: difference in sensitivity to inhibitors of RNA and protein synthesis. *Science* **161,** 465.

Finklea, J. F., Sandifer, S. H., and Smith, D. D. (1969). Cigarette smoking and epidemic influenza. *Amer. J. Epidemiol.* **90,** 390.

Finter, N. B. (1964). Protection of mice by interferon against systemic virus infections. *Brit. Med. J.* **2,** 981.

Finter, N. B. (1966). Interferon as an antiviral agent *in vivo*: Quantitative and temporal aspects of the protection of mice against Semliki Forest virus. *Brit. J. Exp. Pathol.* **47,** 361.

Finter, N. B. (1968). Interferon assays: sensitivity and other aspects. *In* "The Interferons" (G. Rita, ed.), p. 203. Academic Press, New York, London.

Finter, N. B. (ed.) (1973). "Interferons and Interferon Inducers." North Holland, Amsterdam.

Firket, H. (1965). Cell division. *In* "Cells and Tissues in Culture" (E. N. Willmer, ed.), Vol. 1, p. 203. Academic Press, New York.

Fischinger, P. J., Schäfer, W., and Seifert, E. (1972). Detection of some murine leukemia virus antigens in virus particles derived from 3T3 cells transformed only by murine sarcoma virus. *Virology* **47,** 229.

Fisher, E. R. (1953). The nature and staining reactions of the fibroma-cell inclusions of the Shope fibroma of the rabbit. *J. Nat. Cancer Inst.* **14,** 355.

Flamand, A. (1970). Étude génétique du virus de la stomatite vésiculaire; classement de mutants thermosensible spontanés en groups de complementation. *J. Gen. Virol.* **8,** 187.

Flamand, A., and Pringle, C. R. (1971). The homologies of spontaneous and induced temperature-sensitive mutants of vesicular stomatitis virus isolated in chick embryo and BHK21 cells. *J. Gen. Virol.* **11**, 81.

Flaum, A., Malmros, H., and Persson, E. (1926). Eine nosocomiale Ikterus-Epidemie. *Acta Med. Scand. Suppl.* **16**, 544.

Fleissner, E. (1971). Chromatographic separation and antigenic analysis of proteins of the oncornaviruses. *J. Virol.* **8**, 778.

Fleming, P. (1971). Thermal inactivation of Semliki Forest virus. *J. Gen. Virol.* **13**, 385.

Fogel, M., and Sachs, L. (1969). The activation of virus synthesis in polyoma-transformed cells. *Virology* **37**, 327.

Fogel, M., and Sachs, L. (1970). Induction of virus synthesis in polyoma transformed cells by ultraviolet light and mitomycin C. *Virology* **40**, 174.

Forget, B. G., and Weissmann, S. M. (1969). The nucleotide sequence of ribosomal 5S ribonucleic acid from KB cells. *J. Biol. Chem.* **244**, 3148.

Fox, J. P., Elveback, L. R., Spigland, I., Frothingham, T. E., Stevens, D. A., and Huger, M. (1966). The Virus Watch program: A continuing surveillance of viral infections in metropolitan New York families. I. Overall plan, methods of collecting and handling information and a summary report of specimens collected and illnesses observed. *Amer. J. Epidemiol.* **83**, 389.

Fox, J. P., Hall, C. E., and Elveback, L. R. (1970). "Epidemiology: Man and Disease." Macmillan, London.

Frame, J. D., Baldwin, J. M., Gocke, D. J., and Troup, J. M. (1970). Lassa fever, a new virus disease of man from West Africa. I. Clinical description and pathological findings. *Amer. J. Trop. Med. Hyg.* **19**, 670.

Francis, T., Jr. (1953). Influenza: The newe acquayantance. *Ann. Internal Med.* **39**, 203.

Francis, T., Jr. (1955). The current status of the control of influenza. *Ann. Internal Med.* **43**, 534.

Francis, T., Jr., and Maassab, H. F. (1965). Influenza viruses. *In* "Viral and Rickettsial Infections of Man" (F. L. Horsfall, Jr. and I. Tamm, eds.), 4th Ed., p. 689. Lippincott, Philadelphia, Pennsylvania.

Francis, T., Jr., Korns, R. F., Voight, R. B., Boisen, M., Hemphill, F. M., Napier, J. A., and Tolchinsky, E. (1955). An evaluation of the 1954 poliomyelitis vaccine trials. *Amer. J. Publ. Health* **41**, part II.

Franklin, R. M. (1963). The inhibition of ribonucleic acid acid synthesis in mammalian cells by actinomycin D. *Biochim. Biophys. Acta* **72**, 555.

Franklin, R. M., and Baltimore, D. (1962). Patterns of macromolecular synthesis in normal and virus-infected mammalian cells. *Cold Spring Harbor Symp. Quant. Biol.* **27**, 175.

Franklin, R. M., Pettersson, U., Akervall, K., Strandberg, B., and Philipson, L. (1971). Structural proteins of adenovirus. V. Size and structure of the adenovirus type 2 hexon. *J. Mol. Biol.* **57**, 383.

Franklin, R. M., Harrison, S. C., Pettersson, U., Philipson, L., Brändén, C. I., and Werner, P-E. (1972). Structural studies on the adenovirus hexon. *Cold Spring Harbor Symp. Quant. Biol.* **36**, 503.

Fraser, K. B. (1969). Immunological tracing: viruses and rickettsiae. *In* "Fluorescent Protein Tracing" (R. C. Nairn, ed.), 3rd Ed., p. 192. Livingstone, Edinburgh.

Frearson, P. M., and Crawford, L. V. (1972). Polyoma virus basic proteins. *J. Gen. Virol.* **14**, 141.

Freeman, A. E., Vanderpool, E. A., Black, P. H., Turner, H. C., and Huebner, R. J. (1967). Transformation of primary rat embryo cells by a weakly oncogenic adenovirus-type 3, *Nature (London)* **216**, 171.

Freeman, A. E., Price, P. J., Igel, H. J., Young, J. C., Mavyak, J. M., Heubner, R. J. (1970). Morphological transformation of rat embryo cells induced by diethylnitrosamine and murine leukemia viruses. *J. Nat. Cancer Inst.* **44**, 65.

Freese, E. (1963). Molecular mechanisms of mutations. *In* "Molecular Genetics" (J. H. Taylor, ed.), Vol. 1, p. 207. Academic Press, New York.

French, E. L. (1952). The pyrogenic effect of the influenza-mumps group of viruses in the laboratory rabbit. *Aust. J. Exp. Biol. Med. Sci.* **30**, 479.

Frenkel, N., and Roizman, B. (1971). Herpes simplex virus: Genome size and redundancy studied by renaturation kinetics. *J. Virol.* **8**, 591.

Frenkel, N., and Roizman, B. (1972a). Ribonucleic acid synthesis in cells infected with herpes simplex virus: Controls of transcription and of RNA abundance. *Proc. Nat. Acad. Sci. U.S.* **69**, 2654.

Frenkel, N., and Roizman, B. (1972b). Separation of herpesvirus deoxyribonucleic acid duplex into unique fragments and intact strand on sedimentation in alkaline gradients. *J. Virol.* **10**, 565.

Fried, M. (1965). Cell-transforming ability of a temperature-sensitive mutant of polyoma virus. *Proc. Nat. Acad. Sci. U.S.* **53**, 486.

Fried, M. (1970). Characterization of a temperature-sensitive mutant of polyoma virus. *Virology* **40**, 605.

Fried, M., and Pitts, J. D. (1968). Replication of polyoma virus DNA I. A resting cell system for biochemical studies on polyoma virus. *Virology* **34**, 761.

Friedemann, U. (1943). Permeability of blood-brain barrier to neurotropic viruses. *A.M.A. Arch. Pathol.* **35**, 912.

Friedman, R. M. (1967). Interferon binding: The first step in establishment of antiviral activity. *Science* **156**, 1760.

Friedman, R. M. (1968). Protein synthesis directed by an arbovirus. *J. Virol.* **2**, 26.

Friedman, R. M., and Rabson, A. S. (1964). Polyoma virus strains of different oncogenicity: Transplantation immunity in mice. *Virology* **23**, 273.

Friedmann, T. (1971). *In vitro* reassembly of shell-like particles from disrupted polyoma virus. *Proc. Nat. Acad. Sci. U.S.* **68**, 2574.

Friend, C. (1957). Cell-free transmission in adult Swiss mice of a disease having the character of a leukemia. *J. Exp. Med.* **105**, 307.

Friis, R. R., Toyoshima, K., and Vogt, P. K. (1971). Conditional lethal mutants of avian sarcoma viruses. I. Physiology of *ts*75 and *ts*149. *Virology* **43**, 375.

Frothingham, T. E. (1965). Further observations on cell cultures infected concurrently with mumps and Sindbis viruses. *J. Immunol.* **94**, 521.

Fruitstone, M. J., Michaels, B. S., Rudloff, D. A. C., and Sigel, M. M. (1966). Role of the spleen in interferon production in mice. *Proc. Soc. Exp. Biol. Med.* **122**, 1008.

Frye, L. D., and Edidin, M. (1970). The rapid intermixing of cell surface antigens after formation of mouse-human heterokaryons. *J. Cell Sci.* **7**, 319.

Fuchs, P., and Kohn, A. (1971). Nature of transient inhibition of deoxyribonucleic acid synthesis in HeLa cells by parainfluenza virus 1 (Sendai). *J. Virol.* **8**, 695.

Fujinaga, K., and Green, M. (1967). Mechanism of viral carcinogenesis by deoxyribonucleic acid mammalian viruses IV. Related virus-specific ribonucleic acids in tumor cells induced by "highly" oncogenic adenovirus types 12, 18 and 31. *J. Virol.* **1**, 576.

Fujinaga, K., Pina, M., and Green, M. (1969). The mechanism of carcinogenesis by DNA mammalian viruses VI. A new class of virus-specific RNA molecules in cells transformed by group C human adenovirus. *Proc. Nat. Acad. Sci. U.S.* **64**, 255.

Fujiwara, S., and Kaplan, A. S. (1967). Site of protein synthesis in cells infected with pseudo-rabies virus. *Virology* **32**, 60.

Fulginiti, V. A., Hathaway, W. E., Pearlman, D. S., Blackburn, W. R., Reiquam, C. W., Githens, J. H., Claman, H. N., and Kempe, C. H. (1966). Dissociation of delayed hypersensitivity and antibody-synthesizing capacities in man. *Lancet* **ii**, 5.

Fulginiti, V. A., Eller, J. J., Downie, A. W., and Kempe, C. H. (1967). Altered reactivity to measles virus: atypical measles in children previously immunized with inactivated measles virus vaccines. *J. Amer. Med. Ass.* **202**, 1075.

Furlong, D., Swift, H., and Roizman, B. (1972). Arrangement of herpesvirus deoxyribonucleic acid in the core. *J. Virol.* **10**, 1071.

Furlong, N. B., and Graham, C. (1971). Cytosine arabinoside inhibits mammalian DNA polymerase. *Nature New Biol.* **233**, 212.

Furth, J. J., and Cohen, S. S. (1968). Inhibition of mammalian DNA polymerase by the 5'-triphosphate of 1-$\beta$-D-arabinofuranosylcytosine and the 5'-triphosphate of 9-$\beta$-D-arabinofuranosyladenine. *Cancer Res.* **28**, 2061.

Furth, J., Okano, H., and Kunii, A. (1964). Role of the thymus in leukemogenesis. *In* "The Thymus in Immunobiology" (R. A. Good and A. E. Gabrielson, eds.), p. 595. Harper & Row, New York.

Furth, J., Kunii, A., Ioachim, H., Sanel, F. T., and Moy, P. (1966). Parallel observations on the role of the thymus in leukaemogenesis, immunocompetence and lymphopoiesis. *Ciba Found. Symp. Thymus: Exp. Clin. Studies,* p. 288.

Gadzar, A. F., Phillips, L. A., Sarma, P. S., Peebles, P. T., and Chopra, H. C. (1971). Presence of sarcoma genome in a "non-infectious" mammalian virus. *Nature New Biol.* **234,** 69.

Gafford, L. G., and Randall, C. C. (1970). Further studies on high molecular weight fowlpox virus DNA and its hydrodynamic properties. *Virology* **40,** 298.

Gafford, L. G., Sinclair, F., and Randall, C. C. (1972). Alteration of DNA metabolism in fowlpox-infected chick embryo monolayers. *Virology* **48,** 567.

Gajdusek, D. C., and Gibbs, C. J. (1972). Transmission of kuru from man to rhesus monkey (*Macaca mulatta*) 8½ years after inoculation. *Nature (London)* **240,** 351.

Gajdusek, D. C., and Gibbs, C. J. (1973). Subacute and chronic diseases caused by atypical infections with unconventional viruses in aberrant hosts. *Perspect. Virol.* **8,** 279.

Gajdusek, D. C., Gibbs, C. J., Jr., and Alpers, M. (1965). "Slow, Latent and Temperate Virus Infections," Natl. Inst. Neurol. Diseases Blindness Monograph No. 2 U.S. Govt. Printing Office, Washington, D.C.

Gajdusek, D. C., Gibbs, C. J., Jr., and Alpers, M. (1966). Experimental transmission of a kuru-like syndrome to chimpanzees. *Nature (London)* **209,** 794.

Gajdusek, D. C., Rogers, N. C., Basnight, M., Gibbs, C. J., and Alpers, M. (1969). Transmission experiments with kuru in chimpanzees and the isolation of latent viruses from the explanted tissues of affected animals. *Ann. N.Y. Acad. Sci.* **162,** 529.

Gajdusek, D. C., Gibbs, C. J., and Lim, K. A. (1971). Prospects for the control of chronic degenerative diseases with vaccines. *Proc. Intern. Conf. Application of Vaccines against Viral, Rickettsial, and Bacterial Diseases of Man,* Pan American Health Organ. Scientific Publ. No. 226, p. 566.

Gajdusek, D. C., Gibbs, C. J., Rogers, N. G., Basnight, M., and Hooks, J. (1972). Persistence of viruses of kuru and Creutzfeldt-Jakob disease in tissue cultures of brain cells. *Nature (London)* **235,** 104.

Gall, J. D. (1967). Octagonal nuclear pores. *J. Cell Biol.* **32,** 391.

Gall, J. G., and Callan, H. G. (1962). H³ uridine incorporation in lampbrush chromosomes. *Proc. Nat. Acad. Sci. U.S.* **48,** 562.

Gall, J. G., and Pardue, M. L. (1969). Formation and detection of RNA-DNA hybrid molecules in cytological preparations. *Proc. Nat. Acad. Sci. U.S.* **63,** 378.

Gallager, J. G., and Khoobyarian, N. (1969). Adenovirus susceptibility to interferon: sensitivity of types 2, 7, and 12 to human interferon. *Proc. Soc. Exp. Biol. Med.* **130,** 137.

Gallo, R. C. (1971). Reverse transcriptase, the DNA polymerase of oncogenic RNA viruses. *Nature (London)* **234,** 194.

Gallo, R. C., Yang, S. S., and Ting, R. C. (1970). RNA dependent DNA polymerase of human acute leukaemic cells. *Nature (London)* **228,** 927.

Gallo, R. C., Yang, S. S., Smith, R. G., Herrera, F., Ting, R. C., Bobrow, S. N., Davie, C., and Fujioka, S. (1971). RNA- and DNA-dependent DNA polymerases of human and normal leukemic cells. *In* "Lepetit Colloquia on Biology and Medicine, Vol. 2: The Biology of Oncogenic Viruses" (L. Sylvestri, ed.), p. 210. North Holland, Amsterdam.

Gallo, R. C., Abrell, J. W., Robert, M. S., Yang, S. S., and Smith, R. G. (1972). Reverse transcriptase from Mason-Pfizer monkey tumor virus, avian myeloblastosis virus and Rauscher leukemia virus and its response to rifampicin derivatives. *J. Nat. Cancer Inst.* **48,** 1185.

Galper, J. B., and Darnell, J. E. (1969). The presence of N-formyl-methionyl-tRNA in HeLa cell mitochondria. *Biochem. Biophys. Res. Commun.* **34,** 205.

Gandhi, S. S., Stanley, P. M., Taylor, J. M., and White, D. O. (1972). Inhibition of influenza viral glycoprotein synthesis by sugars. *Microbios* **5,** 41.

Gantt, R. R., Stomberg, K. J., and de Oca, F. M. (1971). Specific RNA methylase associated with avian myeloblastosis virus. *Nature (London)* **234**, 35.

Gard, S. (1960). Theoretical considerations in the inactivation of viruses by chemical means. *Ann N.Y. Acad. Sci.* **83**, 638.

Gard, S. (1967). Inactivated poliomyelitis vaccine—present and future. *Proc. 1st Intern. Conf. Vacc. Viral Rickettsial Dis. Man*, Pan American Health Organ., p. 161.

Gardner, M. B., Rongey, R. W., Arnstein, P., Estes, J. D., Sarma, P., Huebner, R. J., and Rickard, C. G. (1970). Experimental transmission of feline fibrosarcoma to cats and dogs. *Nature (London)* **226**, 807.

Garon, C., and Moss, B. (1971). Glycoprotein synthesis in cells infected with vaccinia virus. II. A glycoprotein component of the virion. *Virology* **46**, 233.

Garon, C. F., Berry, K. W., and Rose, J. A. (1972). A unique form of terminal redundancy in adenovirus DNA molecules. *Proc. Nat. Acad. Sci. U.S.* **69**, 2391.

Gatti, R. A., and Good, R. A. (1970). Aging, immunity and malignancy. *Geriatrics* **25**, 158.

Gauntt, C. J., and Graham, A. F. (1969). The reoviruses *In* "The Biochemistry of Viruses" (H. B. Levy, ed.), p. 259. Dekker, New York.

Gebhardt, L. P., Stanton, G. J., Hill, D. W., and Collett, G. C. (1964). Natural overwintering hosts of the virus of Western equine encephalitis. *New Engl. J. Med.* **271**, 172.

Geering, G., Old, L. J., and Boyse, E. A. (1966). Antigens of leukemias induced by naturally occurring murine leukemia virus: Their relation to the antigens of Gross virus and other murine leukemia viruses. *J. Exp. Med.* **124**, 753.

Gelb, L. D., and Martin, M. A. (1971). The detection and quantitation of viral genomes within normal and transformed mammalian cells. *In* "Proceedings of the First Conference and Workshop on Embryonic and Fetal Antigens in Cancer" (N. G. Anderson and J. H. Coggin, eds.), p. 71. University of Tennessee Press, Oak Ridge, Tennessee.

Gelb, L., Aaronson, S. A., and Martin, M. (1971a). Heterogeneity of murine leukemia virus *in vitro* DNA; detection of viral DNA in mammalian cells. *Science* **172**, 1353.

Gelb, L. D., Kohne, D. E., and Martin, M. A. (1971b). Quantitation of simian virus 40 sequences in African green monkey, mouse and virus-transformed cell genomes. *J. Mol. Biol.* **57**, 129.

Gelderblom, H., Bauer, H., and Graf, T. (1972). Cell-surface antigens induced by avian RNA tumor viruses; detection by immunoferritin technique. *Virology* **47**, 416.

Gelderman, A. H., Rake, A. V., and Britten, R. J. (1971). Transcription of nonrepeated DNA in neonatal and fetal mice. *Proc. Nat. Acad. Sci. U.S.* **68**, 172.

Gemmell, A., and Cairns, J. (1959). Linkage in the genome of an animal virus. *Virology* **8**, 381.

Gemmell, A., and Fenner, F. (1960). Genetic studies with mammalian poxviruses. III. White (*u*) mutants of rabbitpox virus. *Virology* **11**, 219.

Gerber, P. (1963). Tumors induced in hamsters by simian virus 40: Persistent subviral infection. *Science* **140**, 889.

Gerber, P. (1964). Virogenic hamster tumor cells: Induction of virus synthesis. *Science* **145**, 833.

Gerber, P. (1966). Studies on the transfer of subviral infectivity from SV40-induced hamster tumor cells to indicator cells. *Virology* **28**, 501.

Gerber, P. (1972). Activation of Epstein-Barr virus by 5-bromodeoxyuridine in "virus-free" human cells. *Proc. Nat. Acad. Sci. U.S.* **69**, 83.

Gerber, P., and Kirschstein, R. L. (1962). SV40-induced ependymomas in newborn hamsters. I. Virus-tumor relationships. *Virology* **18**, 582.

Gerber, P., Whang-Peng, J., and Monroe, J. H. (1969). Transformation and chromosome changes induced by Epstein-Barr virus in normal human leukocyte cultures. *Proc. Nat. Acad. Sci. U.S.* **63**, 740.

Gerin, J. L., Holland, P. V., and Purcell, R. H. (1971). Australia antigen: large scale purification from human serum and biochemical studies of its proteins. *J. Virol.* **7**, 569.

Gershon, D., Sachs, L., and Winocour, E. (1966). The induction of cellular DNA synthesis by simian virus 40 in contact-inhibited and in X-irradiated cells. *Proc. Nat. Acad. Sci. U.S.* **56**, 918.

Gharpure, M. A., Wright, P. F., and Chanock, R. M. (1969). Temperature-sensitive mutants of respiratory syncytial virus. *J. Virol.* **3**, 414.

Gibbs, A. J., Harrison, B. D., Watson, D. H., and Wildy, P. (1966). What's in a virus name? *Nature (London)* **209**, 450.

Gibbs, C. J., and Gajdusek, D. C. (1969). Infection as the etiology of spongiform encephalopathy (Creutzfeldt-Jakob disease). *Science* **168**, 1023.

Gibbs, C. J., and Gajdusek, D. C. (1971). Transmission and characterization of the agents of spongiform virus encephalopathies: kuru, Creutzfeldt-Jakob disease, scrapie and mink encephalopathy. *In* "Immunological Disorders of the Nervous System." Res. Publ. Association for Research in Nervous and Mental Disease. Vol. 49, p. 383.

Gibbs, C. J., and Gajdusek, D. C. (1972). Transmission of scrapie to the cynomolgus monkey (*Macaca fascicularis*). *Nature (London)* **236**, 73.

Gibson, W., and Roizman, B. (1971). Compartmentalization of spermine and spermidine in the herpes simplex virion. *Proc. Nat. Acad. Sci. U.S.* **68**, 2818.

Gibson, W., and Roizman, B. (1973). Quoted in Furlong *et al.*, (1972).

Gifford, G. E. (1963). Studies in the specificity of interferon. *J. Gen. Microbiol.* **33**, 437.

Gilden, D. H., Cole, G. A., Morgan, A. A., and Nathanson, N. (1972a). Immunopathogenesis of acute central nervous system disease produced by lymphocytic choriomeningitis virus. I. Cyclophosphamide mediated induction of the virus-carrier state in adult mice. *J. Exp. Med.* **135**, 860.

Gilden, D. H., Cole, G. A., and Nathanson, N. (1972b). Immunopathogenesis of acute central nervous system disease produced by lymphocytic choriomeningitis virus. II. Adoptive immunization of virus carriers. *J. Exp. Med.* **135**, 874.

Gilden, R. V., and Oroszlan, S. (1972). Group-specific antigens of RNA tumor viruses as markers for subinfectious expression of the RNA virus genome. *Proc. Nat. Acad. Sci. U.S.* **69**, 1021.

Gilden, R. V., Carp, R. I., Taguchi, F., and Defendi, V. (1965). The nature and localization of the SV40-induced complement-fixing antigen. *Proc. Nat. Acad. Sci. U.S.* **53**, 684.

Gilden, R. V., Kern, J., Beddow, T. G., and Huebner, R. J. (1967). Bovine adenovirus type 3: Detection of specific tumor and T antigens. *Virology* **31**, 727.

Gilden, R. V., Kern, J., Freeman, A. E., Martin, C. E., McAllister, R. M., Turner, H. C., and Huebner, R. J. (1968). T and tumour antigens of adenovirus group C-infected and transformed cells. *Nature (London)* **219**, 517.

Gilden, R. V., Kern, J., Lee, Y. K., Rapp, F., Melnick, J. L., Riggs, J. L., Lennette, E. H., Zbar, B., Rapp, H. J., Turner, H. C., and Huebner, R. J. (1970). Serologic surveys of human cancer patients for antibody to adenovirus T antigens. *Amer. J. Epidemiol.* **91**, 500.

Gilden, R. V., Oroszlan, S., and Huebner, R. J. (1971). Antigenic differentiation of M-MSV(0) from mouse, hamster and cat C-type viruses. *Virology* **43**, 722.

Gilead, Z., and Becker, Y. (1971). Effect of emetine on ribonucleic acid biosynthesis in HeLa cells. *Europ. J. Biochem.* **23**, 143.

Gilead, Z., and Ginsberg, H. S. (1968). Characterization of a tumorlike (T) antigen induced by type 12 adenovirus. *J. Virol.* **2**, 15.

Gillespie, D. (1968). The formation and detection of DNA:RNA hybrids. *In* "Methods in Enzymology" (L. Grossman and K. Moldave, eds.), Vol. XIIB, p. 641. Academic Press, New York.

Gillette, R. (1971). VEE vaccine: fortuitous spin-off from BW research. *Science* **173**, 405.

Gillies, S., Bullivant, S., and Bellamy, A. R. (1971). Viral RNA polymerases: Electron microscopy of reovirus reaction cores. *Science* **174**, 694.

Ginder, D., and Friedewald, W. (1951). Effect of Semliki Forest virus on rabbit fibroma. *Proc. Soc. Exp. Biol. Med.* **77**, 272.

Ginoza, W. (1968). Inactivation of viruses by ionizing radiation and by heat. *In* "Methods in Virology" (K. Maramorosch and H. Koprowski, eds.), Vol. IV, p. 139. Academic Press, New York.

Ginsberg, H. S. (1969). The biochemistry of adenovirus infection. *In* "The Biochemistry of Viruses" (H. B. Levy, ed.), p. 329. Marcel Dekker, New York.

Ginsberg, H. S., Pereira, H. G., Valentine, R. C., and Wilcox, W. C. (1966). A proposed terminology for the adenovirus antigens and virion morphological subunits. *Virology* **28**, 782.

Giorno, R., and Kates, J. R. (1971). Mechanism of inhibition of vaccinia virus replication in adenovirus-infected HeLa cells. *J. Virol.* **7**, 208.

Girard, M. (1969). *In vitro* synthesis of poliovirus ribonucleic acid: Role of the replicative intermediate. *J. Virol.* **3**, 376.

Girard, M., Latham, H., Penman, S., and Darnell, J. E. (1965). Entrance of newly formed messenger RNA and ribosomes into HeLa cell cytoplasm. *J. Mol. Biol.* **11**, 187.

Girard, M., Baltimore, D., and Darnell, J. E. (1967). The poliovirus replication complex: Site for synthesis of poliovirus RNA. *J. Mol. Biol.* **24**, 59.

Girard, M., Marty, L., and Suarez, F. (1970). Capsid proteins of simian virus 40. *Biochem. Biophys. Res. Commun.* **40**, 97.

Girardi, A. J., and Defendi, V. (1970). Induction of SV40 transplantation antigen (TrAg) during the lytic cycle. *Virology* **42**, 688.

Girardi, A. J., and Hilleman, M. R. (1964). Host-virus relationships in hamsters inoculated with SV40 virus during the neonatal period. *Proc. Soc. Exp. Biol. Med.* **116**, 723.

Girardi, A. J., Sweet, B. H., Slotnick, V. B., and Hilleman, M. R. (1962). Development of tumors in hamsters inoculated in the neonatal period with vacuolating virus, SV40. *Proc. Soc. Exp. Biol. Med.* **109**, 649.

Girardi, A. J., Hilleman, M. R., and Zwickey, R. E. (1964). Tests in hamsters for oncogenic quality of ordinary viruses including adenovirus type 7. *Proc. Soc. Exp. Biol. Med.* **115**, 1141.

Glasgow, L. A. (1965). Leukocytes and interferon in the host response to viral infections. I. Mouse leukocytes and leukocyte-produced interferon in vaccinia virus infection *in vitro*. *J. Exp. Med.* **121**, 1001.

Glasgow, L. A., and Habel, K. (1963). Interferon production by mouse leukocytes *in vitro* and *in vivo*. *J. Exp. Med.* **117**, 149.

Gledhill, A. W. (1956). Quantitative aspects of the enhancing action of eperythrozoa on the pathogenicity of mouse hepatitis virus. *J. Gen. Microbiol.* **15**, 292.

Gledhill, A. W. (1961). Enhancement of the pathogenicity of mouse hepatitis virus (MHVl) by prior infection of mice with certain leukemia agents. *Brit. J. Cancer* **15**, 531.

Gledhill, A. W., Bilbey, D. L. J., and Niven, J. S. F. (1965). Effect of certain murine pathogens on phagocytic activity. *Brit. J. Exp. Pathol.* **46**, 433.

Glover, R. E. (1928). Contagious pustular dermatitis of sheep. *J. Comp. Pathol. Therap.* **41**, 318.

Gocke, D. J., Hsu, K., Morgan, C., Bombardieri, S., Lockshin, M., and Christian, C. L. (1971). Vasculitis in association with Australia antigen. *J. Exp. Med.* **134**, 330s.

Godman, G. C., Morgan, C., Breitenfeld, P. M., and Rose, H. M. (1960). A correlative study by electron and light microscopy of the development of type 5 adenovirus. *J. Exp. Med.* **112**, 383.

Goldé, A., and Vigier, P. (1966). Non-défectivité du virus de Rous de la souche de Prague. *C. R. Acad. Sci. (Paris)* **D 262**, 2793.

Goldman, R. D. (1971). The role of three cytoplasmic fibers in BHK-21 cell motility. I. Microtubules and the effects of colchicine. *J. Cell Biol.* **51**, 752.

Goldner, H., Girardi, A. J., Larson, V. M., and Hilleman, M. R. (1964). Interruption of SV40 virus tumorigenesis using irradiated homologous tumor antigen. *Proc. Soc. Exp. Biol. Med.* **177**, 851.

Goldstein, J. L., Beaudet, A. L., and Caskey, C. T. (1970). Peptide chain termination with mammalian release factor. *Proc. Nat. Acad. Sci. U.S.* **67**, 99.

Gomatos, P. J. (1968). Reovirus-specific, single-stranded RNA's synthesized *in vitro* with enzyme purified from reovirus-infected cells. *J. Mol. Biol.* **37**, 423.

Gomatos, P. J., and Tamm, I. (1963). The secondary structure of reovirus RNA. *Proc. Nat. Acad. Sci. U.S.* **49**, 707.

Gomatos, P. J., Tamm, I., Dales, S., and Franklin, R. M. (1962). Reovirus type 3: Physical characteristics and interaction with L cells. *Virology* **17**, 441.

Good, R. A. (ed.) (1968). "Immunologic deficiency diseases in man." *Birth Defects Orig. Art. Ser.* **IV,** (No. 1).

Good, R. A., and Gabrielson, A. E. (eds.) (1964). "The Thymus in Immunobiology." Harper & Row, New York.

Goodheart, C. R. (1971). DNA density of oncogenic and nononcogenic simian adenoviruses. *Virology* **44,** 645.

Goodman, G. T., and Koprowski, H. (1962). Study of the mechanism of innate resistance to viral infection. *J. Cell. Comp. Physiol.* **59,** 333.

Goodpasture, E. W., Woodruff, A. M., and Buddingh, G. J. (1932). Vaccinal infection of the chorio-allantoic membrane of the chick embryo. *Amer. J. Pathol.* **8,** 271.

Gordon, J. E. (1962). Chickenpox: An epidemiological review. *Amer. J. Med. Sci.* **244,** 362.

Gordon, W. S. (1946). Advances in veterinary research, louping ill, tickborne fever and scrapie. *Vet. Rec.* **58,** 516.

Gordon, W. S. (1966). Variations in susceptibility of sheep to scrapie and genetic implications. *In* "Report of Scrapie Seminar," p. 53. U.S. Dept. Agriculture Publ. ARS 91-53, Washington, D.C.

Gottschalk, A. (1957). Neuraminidase: The specific enzyme of influenza virus and *Vibrio cholerae. Biochim. Biophys. Acta* **23,** 645.

Gottschalk, A., Belyavin, G., and Biddle, F. (1972). Glycoproteins as influenza virus haemagglutinin inhibitors and as cellular virus inhibitors. *In* "Glycoproteins: Their Composition, Structure and Function" (A. Gottschalk, ed.), 2nd Ed., Part B, p. 1082. Elsevier, Amsterdam.

Gourlay, R. N. (1971). Mycoplasmatales virus-laidlawi, a new virus isolated from *Acholeplasma laidlawi. J. Gen. Virol.* **12,** 65.

Gowans, J. L., and McGregor, D. D. (1965). The immunological activities of lymphocytes. *Progr. Allergy* **9,** 1.

Grace, J. T., and Mirand, E. A. (1963). Human susceptibility to a simian tumor virus. *Ann. N.Y. Acad. Sci.* **108,** 1123.

Granboulan, N. (1967). Autoradiographic methods for electron microscopy. *In* "Methods in Virology" (K. Maramorosch and H. Koprowski, eds.), Vol. III, p. 618. Academic Press, New York.

Granboulan, N., and Girard, M. (1969). Molecular weight of poliovirus ribonucleic acid. *J. Virol.* **4,** 475.

Granboulan, N., Tournier, P., Wicker, R., and Bernhard, W. (1963). An electron microscope study of the development of SV40 virus. *J. Cell Biol.* **17,** 423.

Granoff, A. (1961). Studies on mixed infection with Newcastle disease virus. III. Activation of nonplaque-forming virus by plaque-forming virus. *Virology* **14,** 143.

Granoff, A. (1962). Heterozygosis and phenotypic mixing with Newcastle disease virus. *Cold Spring Harbor Symp. Quant. Biol.* **27,** 319.

Granoff, A. (1965). The interaction of Newcastle disease virus and neutralizing antibody. *Virology* **25,** 38.

Granoff, A. (1969). Viruses of amphibia. *Curr. Topics Microbiol. Immunol.* **50,** 107.

Granoff, A. (1972). Lucké tumour-associated viruses—a review. *In* "Oncogenesis and Herpesviruses" (P. M. Biggs, G. de-Thé, and L. N. Payne, eds.). p. 171. International Agency for Research on Cancer, Lyon.

Granoff, A., and Hirst, G. K. (1954). Experimental production of combination forms of virus. IV. Mixed influenza A-Newcastle disease virus infections. *Proc. Soc. Exp. Biol. Med.* **86,** 84.

Gravell, M., and Cromeans, T. L. (1971). Mechanisms involved in nongenetic reactivation of frog polyhedral cytoplasmic deoxyribovirus: evidence for an RNA polymerase in the virion. *Virology* **46,** 39.

Gravell, M., and Cromeans, T. L. (1972). Virion-associated protein kinase and its involvement in nongenetic reactivation of frog polyhedral cytoplasmic deoxyribovirus. *Virology* **48,** 847.

Gravell, M., and Naegele, R. F. (1970). Non-genetic reactivation of frog polyhedral cytoplasmic deoxyribovirus (PCDV). *Virology* **40**, 170.

Gray, C. G., Gravelle, C. R., and Chin, T. D. Y. (1967). Circulating interferon in infants and children with acute respiratory illness. *J. Pediat.* **71**, 27.

Green, E. L., and Karasaki, S. (1965). Physical and chemical properties of H-1 virus. II. Partial purification. *Proc. Soc. Exp. Biol. Med.* **119**, 918.

Green, G. M. (1968). Pulmonary clearance of infectious agents. *Annu. Rev. Med.* **19**, 315.

Green, M. (1969). Chemical composition of animal viruses. In "The Biochemistry of Viruses" (H. B. Levy, ed.), p. 1. Dekker, New York.

Green, M. (1970a). Oncogenic viruses. *Annu. Rev. Biochem.* **39**, 701.

Green, M. (1970b). Oncogenic adenoviruses. *Mod. Trends Med. Virol.* **2**, 164.

Green, M. (1972). Molecular basis for the attack on cancer. *Proc. Nat. Acad. Sci. U.S.* **69**, 1036.

Green, M., and Cartas, M. (1972). The genome of RNA tumor viruses contains polyadenylic acid residues. *Proc. Nat. Acad. Sci. U.S.* **69**, 791.

Green, M., and Daesch, G. E. (1961). Biochemical studies on adenovirus multiplication. II. Kinetics of nucleic acid and protein synthesis in suspension cultures. *Virology* **13**, 169.

Green, M., Piña, M., Kimes, R., Wensink, P., Maettattie, L., and Thomas, C. A. (1967). Adenovirus DNA. I. Molecular weight and conformation. *Proc. Nat. Acad. Sci. U.S.* **57**, 1302.

Green, M., Parson, P. T., Piña, M., Fujinaga, K., Caffier, H., and Landgraf-Lewis, I. (1970). Transcription of adenovirus genes in productively infected and in transformed cells. *Cold Spring Harbor Symp. Quant. Biol.* **35**, 803.

Green, M., Bragdon, J., and Rankin, A. (1972). 3-cyclic amine derivatives of rifamycin: strong inhibitors of the DNA polymerase activity of RNA tumor viruses. *Proc. Nat. Acad. Sci. U.S.* **69**, 1294.

Green, R. G. (1935). On the nature of filterable viruses. *Science* **82**, 443.

Greenham, L. W., and Poste, G. (1970). The role of lysosomes in virus-induced cell fusion. I. Cytochemical studies. *Microbios* **3**, 97.

Greenwood, F. C., Hunter, W. M., and Glover, J. S. (1963). The preparation of [131]I-labelled human growth hormone of high specific radioactivity. *Biochem. J.* **89**, 114.

Greenwood, M., Hill, A. B., Topley, W. W. C., and Wilson, J. (1936). Experimental epidemiology. *Med. Res. Council Spec. Rept. Ser.* **209**.

Gregg, N. McA. (1941). Congenital cataract following German measles in the mother. *Trans. Ophthalmol. Soc. Aust.* **3**, 35.

Greig, J. R. (1939). Scrapie. *Proc. 4th Intern. Cong. Comp. Pathol. (Rome)* **2**, 285.

Grešiková, M., and Reháček, J. (1959). Isolierung des Zeckenenzephalitisvirus aus Blut und Milch von Haustieren (Schaf und Kuh) nach Infektion durch Zecken der Gattung *Ixodes ricinus* L. *Arch. Ges. Virusforsch.* **9**, 360.

Gresser, I. G., and Dull, H. B. (1964). A virus inhibitor in pharyngeal washings from patients with influenza. *Proc. Soc. Exp. Biol. Med.* **115**, 192.

Gresser, I. G., and Lang, D. J. (1966). Relationships between viruses and leucocytes. *Progr Med. Virol.* **8**, 62.

Gresser, I. G., and Pattison, I. H. (1968). An attempt to modify scrapie in mice by the administration of interferon. *J. Gen. Virol.* **3**, 295.

Gresser, I., Chany, C., and Enders, J. F. (1965). Persistent polioviral infection of intact human amniotic membrane without apparent cytopathic effect. *J. Bacteriol.* **89**, 470.

Gresser, I. G., Berman, L., de-Thé, G., Brouty-Boyé, D., Coppey, J., and Falcoff, E. (1968). Interferon and murine leukemia. V. Effect of interferon preparations on the evolution of Rauscher disease in mice. *J. Nat. Cancer Inst.* **41**, 505.

Gresser, I. G., Pourali, C., Levy, J. P., Fontaine-Brouty-Boyé, D., and Thomas, M. T. (1969). Increased survival in mice inoculated with tumor cells and treated with interferon preparations. *Proc. Nat. Acad. Sci. U.S.* **63**, 51.

Gresser, I. G., Thomas, M-T., and Brouty-Boyé, D. (1971). Effect of interferon treatment of L1210 cells *in vitro* on tumour and colony formation. *Nature New Biol.* **231**, 20.

Griffith, F. (1928). Significance of pneumococcal types. *J. Hyg.* **27**, 113.

Griffiths, D. A. (1971). Measles in vaccinated communities. *Lancet* **ii,** 1423.

Grimes, W. J., and Burge, B. W. (1971). Modification of Sindbis virus glycoprotein by host-specified glycosyl transferases. *J. Virol.* **7,** 309.

Grimley, P. M., and Moss, B (1971). Similar effect of rifampicin and other rifamycin derivatives on vaccinia virus morphogenesis. *J. Virol.* **8,** 225.

Grimley, P. M., Rosenblum, E. N., Mims, S. J., and Moss, B. (1970). Interruption by rifampicin of an early stage in vaccinia virus morphogenesis: Accumulation of membranes which are precursors of virus envelopes. *J. Virol.* **6,** 519.

Grollman, A. P. (1968). Inhibitors of protein biosynthesis V. Effects of emetine on protein and nucleic acid biosynthesis in HeLa cells. *J. Biol. Chem.* **243,** 4089.

Gross, L. (1944). Is cancer a communicable disease? *Cancer Res.* **4,** 293.

Gross, L. (1956). Influence of ether, *in vitro*, on the pathogenic properties of mouse leukemia extracts. *Acta Haematol.* **15,** 273.

Gross, L. (1957). Development and serial cell-free passage of a highly potent strain of mouse leukemia virus. *Proc. Soc. Exp. Biol. Med.* **94,** 767.

Gross, L. (1970). "Oncogenic Viruses," 2nd Ed., Pergamon, Oxford.

Groupé, V., Dunkel, V. C., and Manaker, R. A. (1957). Improved pock counting for the titration of Rous sarcoma virus in embryonated eggs. *J. Bacteriol.* **74,** 409.

Gudnadóttir, M., and Pálsson, P. A. (1965). Host-virus interaction in visna infected sheep. *J. Immunol.* **95,** 1116.

Guir, J., Braunwald, J., and Kirn, A. (1971). Inhibition of host-specific DNA and RNA synthesis in KB cells following infection with frog virus 3. *J. Gen. Virol.* **12,** 293.

Gurgo, C., Ray, R. K., Thiry, L., and Green, M. (1971). Inhibitors of the RNA and DNA dependent polymerase activities of RNA tumor viruses. *Nature New Biol.* **229,** 111.

György, E., Sheehan, M. C., and Sokol, F. (1971). Release of envelope glycoprotein from rabies virions by a nonionic detergent. *J. Virol.* **8,** 649.


Habel, K. (1961). Resistance of polyoma virus immune animals to transplanted polyoma tumors. *Proc. Soc. Exp. Biol. Med.* **106,** 722.

Habel, K. (1962). Antigenic properties of cells transformed by polyoma virus. *Cold Spring Harbor Symp. Quant. Biol.* **27,** 433.

Habel, K. (1965). Specific complement-fixing antigens in polyoma tumors and transformed cells. *Virology* **25,** 55.

Habel, K. (1969a). Virus neutralization test. *In* "Fundamental Techniques in Virology" (K. Habel and N. P. Salzman, eds.), p. 288. Academic Press, New York.

Habel, K. (1969b). Antigens of virus-induced tumors. *Advan. Immunol.* **10,** 229.

Habel, K., and Eddy, B. E. (1963). Specificity of resistance to tumor challenge of polyoma and SV40 virus-immune hamsters. *Proc. Soc. Exp. Biol. Med.* **113,** 1.

Habel, K., and Salzman, N. P., (eds.) (1969). "Fundamental Techniques in Virology," Academic Press, New York.

Hackett, A. J. (1964). A possible morphologic basis for the autointerference phenomenon in vesicular stomatitis virus. *Virology* **24,** 51.

Haelterman, E. O. (1965). Lactogenic immunity to transmissible gastroenteritis of swine. *J. Amer. Vet. Med. Ass.* **147,** 1661.

Hahon, N. (1961). Smallpox and related poxvirus infections in the simian host. *Bacteriol. Rev.* **25,** 459.

Haig, D. A., and Clarke, M. C. (1971). Multiplication of the scrapie agent. *Nature (London)* **234,** 106.

Haig, D. A., Clarke, M. C., Bloom, E., and Alper, T. (1969). Further studies on the inactivation of the scrapie agent by ultraviolet light. *J. Gen. Virol.* **5,** 455.

Hakala, M. T. (1957). Prevention of toxicity of amethopterin for sarcoma-180 cells in tissue culture. *Science* **126,** 255.

Hakomori, S., and Murakami, W. T. (1968). Glycolipids of hamster fibroblasts and derived malignant transformed cell lines. *Proc. Nat. Acad. Sci. U.S.* **59,** 254.

Hakomori, S., Teather, C., and Andrews H. (1969). Organizational difference of cell surface "hematoside" in normally and virally transformed cells. *Biochem. Biophys. Res. Commun.* **33**, 563.

Hakomori, S., Saito, T., and Vogt, P. K. (1971). Transformation by Rous sarcoma virus: effects on cellular glycolipids. *Virology* **44**, 609.

Halberstaedter, L., Doljanski, L., and Tenenbaum, E. (1941). Experiments on the cancerization of cells *in vitro* by means of Rous sarcoma agent. *Brit. J. Exp. Pathol.* **22**, 179.

Hale, J. H., Lee, L. H., and Gardner, P. S. (1961). Interference patterns encountered when using attenuated poliovirus vaccines. *Brit. Med. J.* **2**, 728.

Hall, L., and Rueckert, R. R. (1971). Infection of mouse fibroblasts by cardioviruses: Premature uncoating and its prevention by elevated pH and magnesium chloride. *Virology* **43**, 152.

Halonen, P. E., Murphy, F. A., Fields, B. N., and Reese, D. R. (1968). Hemagglutinin of rabies and some other bullet-shaped viruses. *Proc. Soc. Exp. Biol. Med.* **127**, 1037.

Halstead, S. B. (1966). Mosquito-borne haemorrhagic fevers of south and South-East Asia. *Bull. WHO* **35**, 3.

Halstead, S. B. (1969). Observations related to pathogenesis of dengue hemorrhagic fever. *Yale J. Biol. Med.* **42**, 350.

Halstead, S. B., Nimmannitaya, S., Yamarat, C., and Russell, P. K. (1967). Hemorrhagic fever in Thailand: newer knowledge regarding etiology. *Jap. J. Med. Sci. Biol. Suppl.* **20**, 96.

Hamashima, Y., Kyogoku, M., Hiramatsu, S., Nakashima, Y., and Yamauchi, R. (1959). Immuno-cytological studies employing labeled active protein. III. Encephalitis Japonica. *Acta Pathol. Japan.* **9**, 89.

Hamerman, D., Barski, G., Youn, J. K., and Green, H. (1968). Increased synthesis of hyaluronic acid by a mouse cell line chronically infected with Rauscher leukemia virus. *Rev. Fr. Etud. Clin. Biol.* **13**, 800.

Hammon, W. McD., Kitaoka, M., and Downs, W. G. (1971). "Immunization for Japanese Encephalitis." Igaku Shoin, Tokyo.

Hampar, B., Derge, J. G., Martos, L. M., and Walker, J. L. (1971). Persistence of a repressed Epstein-Barr virus genome in Burkitt lymphoma cells made resistant to 5-bromodeoxyuridine. *Proc. Nat. Acad. Sci. U.S.* **68**, 3185.

Hampar, B., Derge, J. G., Martos, L. M., and Walker, J. L. (1972). Synthesis of Epstein-Barr virus after activation of the viral genome in a "virus-negative" human lymphoblastoid cell (Raji) made resistant to 5-bromodeoxyuridine. *Proc. Nat. Acad. Sci. U.S.* **69**, 78.

Hamre, D., Appel, J., and Loosli, C. G. (1956). Viremia in mice with pulmonary influenza A virus infections. *J. Lab. Clin. Med.* **47**, 182.

Hamre, D., Loosli, C. G., and Gerber, P. (1958). Antigenic variants of influenza A (PR8 strain). III. Serological relationships of a line of variants derived in squence in mice given homologous vaccine. *J. Exp. Med.* **107**, 829.

Hanafusa, H. (1964). Nature of the defectiveness of Rous sarcoma virus. *Nat. Cancer Inst. Monogr.* **17**, 543.

Hanafusa, H. (1965). Analysis of the defectiveness of Rous sarcoma virus III. Determining influence of a new helper virus on the host range and susceptibility to interference of RSV. *Virology* **25**, 248.

Hanafusa, H. (1969a). Replication of oncogenic viruses in virus-induced tumor cells—their persistence and interaction with other viruses. *Advan. Cancer Res.* **12**, 137.

Hanafusa, H. (1969b). Rapid transformation of cells by Rous sarcoma virus. *Proc. Nat. Acad. Sci. U.S.* **63**, 318.

Hanafusa, H., and Hanafusa, T. (1966a). Determining factor in the capacity of Rous sarcoma virus to induce tumors in mammals. *Proc. Nat. Acad. Sci. U.S.* **55**, 532.

Hanafusa, H., and Hanafusa, T. (1966b). Analysis of the defectiveness of Rous sarcoma virus. IV. Kinetics of RSV production. *Virology* **28**, 369.

Hanafusa, H., and Hanafusa, T. (1968). Further studies on RSV production from transformed cells. *Virology* **34**, 630.

Hanafusa, H., and Hanafusa, T. (1971). Non-infectious RSV deficient in DNA polymerase. *Virology* **43**, 313.

Hanafusa, H., Hanafusa, T., and Kamahora, J. (1959). Transformation phenomena in the pox group viruses. II. Transformation between several members of pox group. *Biken's J.* **2**, 85.

Hanafusa, H., Hanafusa, T., and Rubin, H. (1963). The defectiveness of Rous sarcoma virus. *Proc. Nat. Acad. Sci. U.S.* **49**, 572.

Hanafusa, H., Hanafusa, T., and Rubin, H. (1964). Analysis of the defectiveness of Rous sarcoma virus, II. specification of RSV antigenicity by helper virus. *Proc. Nat. Acad. Sci. U.S.* **51**, 41.

Hanafusa, H., Hanafusa, T., and Miyamoto (1970). On a chick-associated genetic factor related to the genome of avian tumor viruses. *In* "Lepetit Colloquia on Biology and Medicine, Vol. 2: The Biology of Oncogenic Viruses" (L. G. Sylvestri, ed.), p. 170. North Holland, Amsterdam.

Hanafusa, T. (1970). Virus production by Rous sarcoma cells. *Curr. Topics Microbiol. Immunol.* **51**, 114.

Hanafusa, T., Miyamoto, T., and Hanafusa, H. (1970). A type of chick embryo cell that fails to support formation of infectious RSV. *Virology* **40**, 55.

Hanafusa, T., and Hanafusa, H., Miyamoto, T., and Fleissner, E. (1972). Existence and expression of tumor virus genes in chick embryo cells. *Virology* **47**, 475.

Hand, R., and Tamm, I (1972). Rate of DNA chain growth in mammalian cells infected with cytocidal RNA viruses. *Virology* **47**, 331.

Hand, R., Ensminger, W. D., and Tamm, I. (1971). Cellular DNA replication in infections with cytocidal RNA viruses. *Virology* **44**, 527.

Hansen, B., Koprowski, H., Baron, S., and Buckler, C. E. (1969). Interferon-mediated natural resistance of mice to arbo B virus infection. *Microbios* **1B**, 51.

Hanshaw, J. B. (1970). Developmental abnormalities associated with congenital cytomegalovirus infection. *Advan. Teratol.* **4**, 64.

Hare, J. D. (1964). Transplant immunity to polyoma virus induced tumors. I. Correlations with biological properties of virus strains. *Proc. Soc. Exp. Biol. Med.* **115**, 805.

Hare, J. D. (1967). Transplant immunity to polyoma virus induced tumor cells. IV. A polyoma strain defective in transplant antigen induction. *Virology* **31**, 625.

Hare, J. D., and Godal, T. (1965). Transplant immunity to polyoma virus induced tumors. III. Evidence for heterogeneity among transplant antigens *in vivo*. *Proc. Soc. Exp. Biol. Med.* **118**, 632.

Harel, L., Harel, J., Lacour, F., and Huppert, J. (1966). Homologie entre génome du virus de la myéloblastose aviaire (AMV) et génome cellulaire. *C. R. Acad. Sci (Paris)* **263**, 616.

Harford, C. G., and Hara, M. (1950). Pulmonary edema in influenzal pneumonia of the mouse and the relation of fluid in the lung to the inception of pneumococcal pneumonia. *J. Exp. Med.* **91**, 245.

Harris, H. (1962). The labile nuclear ribonucleic acid of animal cells and its relevance to the messenger ribonucleic acid hypothesis. *Biochem. J.* **84**, 60.

Harris, H. (1970). "Cell Fusion." Clarendon Press, Oxford.

Harris, R. J. C., Chesterman, F. C., and McClelland, R. M. (1969). The properties of sarcomas induced in Wistar rats by Rous sarcoma virus (Schmidt-Ruppin). *Intern. J. Cancer* **4**, 31.

Harrison, S. C., David, A., Jumblatt, J., and Darnell, J. E. (1971). Lipid and protein organization in Sindbis virus. *J. Mol. Biol.* **60**, 523.

Harrison, V. R., Eckels, K. H., Bartelloni, P. J., and Hampton, C. (1971). Production and evaluation of a formalin-killed Chikungunya vaccine. *J. Immunol.* **107**, 643.

Harter, D. H., and Choppin, P. W. (1967). Cell-fusing activity of visna virus particles. *Virology* **31**, 279.

Hartley, J. W., and Rowe, W. P. (1966). Production of altered cell foci in tissue culture by defective Moloney sarcoma virus particles. *Proc. Nat. Acad. Sci. U.S.* **55**, 780.

Hartley, J. W., Rowe, W. P., Chanock, R. M., and Andrews, B. E. (1959). Studies of mouse polyoma virus infection. IV. Evidence for mucoprotein erythrocyte receptors in polyoma virus hemagglutination. *J. Exp. Med.* **110**, 81.

Hartley, J. W., Rowe, W. P., Capps, W. I., and Huebner, R. J. (1965). Complement fixation and tissue culture assays for mouse leukemia virus. *Proc. Nat. Acad. Sci. U.S.* **53**, 931.

Hartley, J. W., Rowe, W. P., Capps, W. I., and Huebner, R. J. (1969). Isolation of naturally occurring viruses of the murine leukemia group in tissue culture. *J. Virol.* **3**, 126.

Hartley, J. W., Rowe, W. P., and Huebner, R. J. (1970). Host range restrictions of murine leukemia viruses in mouse embryo cell culture. *J. Virol.* **5**, 221.

Harvey, J. J. (1964). An unidentified virus which causes the rapid production of tumors in mice. *Nature (London)* **204**, 1104.

Haslam, E. A., Hampson, A. W., Egan, J. A., and White, D. O. (1970a). The polypeptides of influenza virus. II. Interpretation of polyacrylamide gel electrophoresis patterns. *Virology* **42**, 555.

Haslam, E. A., Hampson, A. W., Radiskevics, I., and White, D. O. (1970b). The polypeptides of influenza virus. VI. Identification of the hemagglutinin, neuraminidase and nucleocapsid proteins. *Virology* **42**, 566.

Hatanaka, M., and Gilden, R. V. (1970). Virus-specified changes in the sugar transport kinetics of rat embryo cells infected with murine sarcoma virus. *J. Nat. Cancer Inst.* **45**, 87.

Hatanaka, M., and Hanafusa, H. (1970). Analysis of a functional change in membrane in the presence of cell transformation by Rous sarcoma virus: alteration in the characteristics of sugar transport. *Virology* **41**, 647.

Hatanaka, M., Huebner, R. J., and Gilden, R. V. (1970). DNA polymerase activity associated with RNA tumor viruses. *Proc. Nat. Acad. Sci. U.S.* **67**, 143.

Hatanaka, M., Kakefuda, T., Gilden, R. V., and Callan, E. A. O. (1971). Cytoplasmic DNA synthesis induced by RNA tumor viruses. *Proc. Nat. Acad. Sci. U.S.* **68**, 1844.

Havens, W. P. (1946). Period of infectivity of patients with experimentally induced infectious hepatitis. *J. Exp. Med.* **83**, 251.

Havens, W. P., and Paul, J. R. (1965). Infectious hepatitis and serum hepatitis. In "Viral and Rickettsial Infections of Man" (F. L. Horsfall and I. Tamm, eds.), 4 ed., p. 965. Lippincott, Philadelphia, Pennsylvania.

Hawkes, R. A. (1964). Enhancement of the infectivity of arboviruses by specific antisera produced in domestic fowls. *Aust. J. Exp. Biol. Med. Sci.* **42**, 465.

Hawkes, R. A., and Lafferty, K. J. (1967). The enhancement of virus infectivity by antibody. *Virology* **33**, 250.

Hay, J., Perera, P. A. J., Morrison, J. M., Gentry, G. A., and Subak-Sharpe, J. H. (1971). Herpes virus-specified proteins. In "Strategy of the Viral Genome" (G. E. W. Wolstenholme and M. O'Connor, eds.), p. 355, CIBA Foundation Symp., Churchill Livingstone, London.

Hayes, W. (1968). "The Genetics of Bacteria and their Viruses: Studies in Basic Genetics and Molecular Biology," 2nd Ed. Blackwell, Oxford.

Hayflick, L. (1965). The limited *in vitro* lifetime of human diploid cell strains. *Exp. Cell Res.* **37**, 614

Hayman, M. J., Skehel, J. J., and Crumpton, M. J. (1973). Purification of virus glycoproteins by affinity chromatography using *Lens culinaris* phytohaemagglutinin. *FEBS Letters* **29**, 185.

Häyry, P., and Defendi, V. (1968). Use of mixed hemagglutination technique in detection of virus-induced antigen(s) on SV40 transformed cell surface. *Virology* **36**, 317.

Häyry, P., and Defendi, V. (1969). Demonstration of specific antigen(s) on the surface of SV40 transformed cells using the mixed hemagglutination techniques. *Transplant Proc.* **1**, 119.

Häyry, P., and Defendi, V. (1970). Surface antigens of SV40 transformed tumor cells. *Virology* **41**, 22.

Hayward, W. S., and Green, M. H. (1965). Inhibition of *Escherichia* coli and bacteriophage λ messenger RNA synthesis by T4. *Proc. Nat. Acad. Sci. U.S.* **54**, 1675.

Head, H., and Campbell, A. W. (1900). The pathology of herpes zoster and its bearing on sensory localization. *Brain* **23**, 353.

Hearn, H. J., Jr., and Rainey, C. T. (1963). Cross-protection in animals infected with group A arboviruses. *J. Immunol.* **90**, 720.

Heggie, A. D. (1971a). Pathogenesis of the rubella exanthem—isolation of rubella virus from the skin. *New Engl. J. Med.* **285**, 664.

Heggie, A. D. (1971b). Pathogenesis of HI virus infection of embryonic hamster bone in organ culture. *J. Exp. Med.* **133**, 506.

Hehlmann, R., Kufe, D., and Spiegelman, S. (1972). The presence in human leukemic cells of RNA related to the RNA of a mouse leukemia virus. *Proc. Nat. Acad. Sci. U.S.* **69**, 435.

Heine, J. W., and Schnaitman, C. A. (1971). Entry of vesicular stomatitis virus into L cells. *J. Virol.* **8**, 786.

Heine, J. W., Spear, P. G., and Roizman, B. (1972). Proteins specified by herpes simplex virus. VI. Viral proteins in the plasma membrane. *J. Virol.* **9**, 431.

Heineberg, H., Gold, E., and Robbins, F. C. (1964). Differences in interferon content in tissues of mice of various ages infected with Coxsackie B1 virus. *Proc. Soc. Exp. Biol. Med.* **115**, 947.

Heiner, G. G., Fatima, N., Russell, P. K., Haase, A. T., Ahmad, N., Mohammed, N., Thomas, D. B., Mack, T. M., Khan, M. M., Knatterud, G. L., Anthony, R. L., and McCrumb, F. R. (1971). Field trials of methisazone as a prophylactic agent against smallpox. *Amer. J. Epidemiol.* **94**, 435.

Heise, E. R., Han, S., and Weiser, E. R. (1968). *In vitro* studies on the mechanism of macrophage migration inhibition in tuberculin sensitivity. *J. Immunol.* **101**, 1004.

Heller, E. (1963). Enhancement of Chikungunya virus replication and inhibition of interferon production by actinomycin D. *Virology* **21**, 652.

Heller, E., Argaman, M., Levy, H., and Goldblum, N. (1969). Selective inhibition of vaccinia virus by the antibiotic rifampicin. *Nature (London)* **222**, 273.

Hellström, I. (1965). Distinction between the effects of antiviral and anticellular polyoma antibodies on polyoma tumor cells. *Nature (London)* **208**, 652.

Hellström, I., and Sjögren, H. O. (1965). Demonstration of H-2 isoantigens and polyoma specific tumor antigens by measuring colony formation *in vitro*. *Exp. Cell. Res.* **40**, 212.

Hellström, I., and Sjögren, H. O. (1966). Demonstration of common specific antigen(s) in mouse and hamster polyoma tumors. *Intern. J. Cancer* **1**, 481.

Hellström, I., and Sjögren, H. O. (1967). *In vitro* demonstration of humoral and cell-bound immunity against common specific transplantation antigen(s) of adenovirus 12-induced mouse and hamster tumors. *J. Exp. Med.* **125**, 1105.

Hellström, K. E., and Hellström, I. (1969). Cellular immunity against tumor antigens. *Advan. Cancer Res.* **12**, 167.

Henle, G., and Henle, W. (1963). Differences in response of hamster tumor cells induced by polyoma virus to interfering virus and interferon. *J. Nat. Cancer Inst.* **31**, 143.

Henle, G., Henle, W., and Diehl, V. (1968). Relations of Burkitt's tumor-associated herpes-type virus to infectious mononucleosis. *Proc. Nat. Acad. Sci. U.S.* **59**, 94.

Henle, G., Henle, W., Clifford, P., Diehl, V., Kafuko, G. W., Kirya, B. G., Klein, G., Morrow, R. H., Munube, G. M. P., Pike, P., Tukei, P., and Ziegler, J. (1969). Antibodies to Epstein-Barr virus in Burkitt's lymphoma and control groups. *J. Nat. Cancer Inst.* **43**, 1147.

Henle, W. (1963). Interferon and interference in persistent viral infection of cell cultures. *J. Immunol.* **91**, 145.

Henle, W., and Henle, G. (1968). Effect of arginine deficient media on the herpes-type virus associated with cultured Burkitt tumor cells. *J. Virol.* **2**, 182.

Henle, W., and Henle, G. (1972). Epstein-Barr virus: The cause of infectious mononucleosis— a review. *In* "Oncogenesis and Herpesviruses" (P. M. Biggs, G. de-Thé, and L. N. Payne, eds.), International Agency for Cancer Research Publication No. 2, Lyon, France.

Henle, W., and Lief, F. W. (1963). The broadening of antibody spectra following multiple exposure to influenza viruses. *Amer. Rev. Resp. Dis. Suppl.* **88**, 379.

Henle, W., and Rosenberg, E. B. (1949). One-step growth curves of various strains of influenza A and B viruses and their inhibition by inactivated virus of the homologous type. *J. Exp. Med.* **89**, 279.

Henle, W., Diehl, V., Kohn, G., zur Hausen, H., and Henle, G. (1967). Herpes-type virus and chromosome marker in normal leukocytes after growth with irradiated Burkitt cells. *Science* **157**, 1064.

Henle, W., Henle, G., Burtin, P., Clifford, P., de Schryer, A., de-Thé, G., Diehl, V., Ho, H. C., and Klein, G. (1970). Antibodies to Epstein-Barr virus in nasopharyngeal carcinoma, other head and neck neoplasms, and control groups. *J. Nat. Cancer Inst.* **44**, 225.

Hennessy, A. V., Minuse, E., and Davenport, F. M. (1965). A twenty-one-year experience with antigenic variation in influenza B viruses. *J. Immunol.* **94**, 301.

Henney, C. S., and Waldman, R. H. (1970). Cell-mediated immunity shown by lymphocytes from the respiratory tract. *Science* **169**, 696.

Henry, C. J., Slifkin, M., and Merkow, L. (1971). Mechanism of host cell restriction in African green monkey kidney cells abortively infected with human adenovirus type 2. *Nature New Biol.* **233**, 39.

Hermodsson, S. (1963). Inhibition of interferon by an infection with parainfluenza virus type 3 (PIV-3). *Virology* **20**, 333.

Herrmann, E. C. (1961). Plaque inhibition test for detection of specific inhibitors of DNA containing viruses. *Proc. Soc. Exp. Biol. Med.* **107**, 142.

Herrmann, E. C., and Stinebring, W. R. (eds.) (1970). Second conference on antiviral substances. *Ann. N.Y. Acad. Sci.* **173** (Article 1).

Hers, J. F. P., and Mulder, J. (1961). Broad aspects of the pathology and pathogenesis of human influenza. *Amer. Rev. Resp. Dis. Suppl.* **83**, 84.

Hers, J. F. P., Masurel, N., and Mulder, J. (1958). Bacteriology and histopathology of the respiratory tract and lungs in fatal Asian influenza. *Lancet* **ii**, 1141.

Hershey, A. D. (ed.). (1971). "The Bacteriophage Lambda." Cold Spring Harbor Laboratory, New York.

Hershey, A. D., and Chase, M. (1951). Genetic recombination and heterozygosis in bacteriophage. *Cold Spring Harbor Symp. Quant. Biol.* **16**, 471.

Hershey, A. D., and Chase, M. (1952). Independent functions of viral protein and nucleic acid in growth and bacteriophage. *J. Gen. Physiol.* **36**, 39.

Hess, W. R. (1971). African swine fever virus. *Virology Monogr.* **9**, 1.

Heywood, S. M. (1969). Synthesis of myosin on heterologous ribosomes. *Cold Spring Harbor. Symp. Quant. Biol.* **34**, 799.

Hierholzer, J. C., Palmer, E. L., Whitfield, S. G., Kaye, H. S., and Dowdle, W. R. (1972). Protein composition of coronavirus OC43. *Virology* **48**, 516.

Higashi, N., Matsumoto, A., Tabata, K., and Nagatomo, Y. (1967). Electron microscope study of development of Chikungunya virus in green monkey kidney stable (VERO) cells. *Virology* **33**, 55.

Hilgers, J., Williams, W. C., Myers, B., and Dmochowski, L. (1971). Detection of antigens of the mouse mammary tumor (MTV) and murine leukemia virus (MuLV) in cells of cultures derived from mammary tumors of mice of several strains. *Virology* **45**, 470.

Hilgers, J., Nowinski, R. C., Geering, G., and Hardy, W. (1972). Detection of avian and mammalian oncogenic RNA viruses (oncornaviruses) by immunofluorescence. *Cancer Res.* **32**, 98.

Hill, A. B. (1963). Medical ethics and controlled trials. *Brit. Med. J.* **1**, 1043.

Hill, D. A., Baron, S., and Chanock, R. M. (1969). The effect of an interferon inducer on influenza virus. *Bull. WHO* **41**, 689.

Hill, D. A., Baron, S., Levy, H. B., Bellanti, J., Buckler, C. E., Cannellos, G., Carbone, P., Chanock, R. M., DeVita, V., Guggenheim, M. A., Homan, E., Kapikian, A. Z., Kirschstein, R. L., Mills, J., Perkins, J. C., Van Kirk, J. E., and Worthington, M. (1971). Clinical studies on the induction of interferon by polyinosinic-polycytidylic acid. *Perspect. Virol.* **7**, 197.

Hill, M., and Hillova, J. (1972). Recovery of a temperature-sensitive mutant of Rous sarcoma virus from chicken cells exposed to DNA extracted from hamster cells transformed by the mutant. *Virology* **49**, 309.

Hilleman, M. R. (1966). Critical appraisal of emulsified oil adjuvants applied to viral vaccines. *Prog. Med. Virol.* **8**, 131.

Hilleman, M. R. (1967). Present knowledge of the rhinovirus groups of viruses. *Curr. Topics Microbiol. Immunol.* **41**, 1.

Hilleman, M. R. (1969). Toward control of viral infections of man. *Science* **164**, 506.

Hilleman, M. R. (1970a). Mumps vaccination. *Mod. Trends Med. Virol.* **2**, 241.

Hilleman, M. R. (1970b). Double-stranded RNAs (poly I:C) in the prevention of viral infections. *Arch. Internal Med.* **126**, 109.

Hilleman, M. R. (1972). Problems and potentials for human viral cancer vaccines. *Prev. Med.* **1**, 352.

Hilleman, M. R. (1973). Viral vaccines and the control of cancer. *Perspectives Virol.* **8**, 119.

Hilleman, M. R., Lampson, G. P., Tytell, A. A., Field, A. K., Nemes, M. M., Krakoff, I. H., and Young, C. W. (1971a). Double-stranded RNA's in relation to interferon induction and adjuvant activity. *In* "Biological Effects of Polynucleotides" (R. F. Beers and W. Braun, eds.), p. 27. Springer-Verlag, New York.

Hilleman, M. R., Weibel, R. E., Villarejos, V. M., Buynak, E. B., Stokes, J., Arguedas, G. J. A., and Vargas, A. G. (1971b). Combined live virus vaccines. *In* "Proceedings of the International Conference on the Application of Vaccines against Viral, Rickettsial and Bacterial Diseases of Man," p. 397. Sci. Publ. No. 226. Pan Amer. Health Organ., Washington, D.C.

Hinuma, Y., and Grace, J. (1968). Cloning of Burkitt lymphoma cells cultured *in vitro*. *Cancer* **22**, 1089.

Hinuma, Y., Konn, M., Yamaguchi, J., Wudarski, D. J., Blakeslee, J. R., and Grace, J. T. (1967). Immunofluorescence and herpes-type virus particles in the P3HR-1 Burkitt lymphoma cell line. *J. Virol.* **1**, 1045.

Hinze, H. C. (1971). New member of the herpesvirus group isolated from wild cottontail rabbits. *Infection Immunity* **3**, 350.

Hinze, H. C., and Walker, D. L. (1964). Response of cultured rabbit cells to infection with the Shope fibroma virus. I. Proliferation and morphological alteration of the infected cells. *J. Bacteriol.* **88**, 1185.

Hinze, H. C., and Walker, D. L. (1971). Comparison of cytocidal and noncytocidal strains of Shope rabbit fibroma virus. *J. Virol.* **7**, 577.

Hirai, K., Lehman, J., and Defendi, V. (1971). Integration of SV40 DNA into the DNA of primary infected Chinese hamster cells. *J. Virol.* **8**, 708.

Hirsch, J. G., and Cohn, Z. A. (1964). Digestive and autolytic functions of lysosomes in phagocytic cells. *Fed. Proc. Fed. Amer. Soc. Exp. Biol.* **23**, 1023.

Hirsch, M. S., and Murphy, F. A. (1968a). Effects of antilymphoid sera on viral infections. *Lancet* **ii**, 37.

Hirsch, M. S., and Murphy, F. A. (1968b). Effect of antithymocyte serum on Rauscher virus infection of mice. *Nature (London)* **218**, 478.

Hirsch, M. S., Allison, A. C., and Harvey, J. J. (1968). Immune complexes in mice infected neonatally with Moloney leukaemogenic and murine sarcoma viruses. *Nature (London)* **223**, 739.

Hirsch, M. S., Zisman, B., and Allison, A. C. (1970). Macrophages and age dependent resistance to herpes simplex virus in mice. *J. Immunol.* **104**, 1160.

Hirst, G. K. (1941). The agglutination of red cells by allantoic fluid of chick embryos infected with influenza virus. *Science* **94**, 22.

Hirst, G. K. (1943). Adsorption of influenza virus on cells of the respiratory tract. *J. Exp. Med.* **78**, 99.

Hirst, G. K. (1962). Genetic recombination with Newcastle disease virus, polioviruses, and influenza. *Cold Spring Harbor Symp. Quant. Biol.* **27**, 303.

Hirst, G. K. (1965). Cell-virus attachment and the action of antibodies on viruses. *In* "Viral and Rickettsial Diseases of Man" (F. L. Horsfall and I. Tamm, eds.), 4th Ed., p. 216. Lippincott, Philadelphia, Pennsylvania.

Hirst, G. K., and Pons, M. (1972). Biological activity in ribonucleoprotein fractions of influenza virus. *Virology* **47**, 546.

Hirt, B., and Gesteland, R. (1971). Characterization of the proteins of SV40 and polyoma virus. *In* "Lepetit Colloquia on Biology and Medicine, Vol. 2, The Biology of Oncogenic Viruses" (L. G. Silvestri, ed.), p. 98. North Holland, Amsterdam.

Hitchcock, G., and Porterfield, J. S. (1961). The production of interferon in brains of mice infected with an arthropod-borne virus. *Virology* **13**, 363.

Ho, M. (1970). Factors influencing the interferon response. *Arch. Internal Med.* **126**, 135.

Ho, M., and Enders, J. F. (1959). Further studies on an inhibitor of viral activity appearing in infected cell cultures and its role in chronic viral infections. *Virology* **9**, 446.

Ho, M., and Kono, Y. (1965). Effect of actinomycin D on virus- and endotoxin-induced interferonlike inhibitors in rabbits. *Proc. Nat. Acad. Sci. U.S.* **53**, 220.

Ho, M., and Postic, B. (1967). Prospects for applying interferon to man. *Proc. 1st Intern. Conf. Vacc. Viral Rickettsial Dis. Man.* Sci. Publ. No. 147. Pan American Health Organ., p. 632.

Ho, M., Breinig, M. K., Postic, B., and Armstrong, J. A. (1970). The effect of pre-injections on the stimulation of interferon by a complexed polynucleotide, endotoxin and virus. *Ann. N.Y. Acad. Sci.* **173**, 680.

Ho, P. P. K., and Walters, G. P. (1966). Influenza virus-induced ribonucleic acid nucleotidyl transferase and the effect of actinomycin D on its formation. *Biochemistry* **5**, 231.

Hodge, L. D., and Scharff, M. D. (1969). Effect of adenovirus on host cell DNA synthesis in synchronized cells. *Virology* **37**, 554.

Hoggan, M. D. (1970). Adenovirus associated viruses. *Progr. Med. Virol.* **12**, 211.

Hoggan, M. D. (1971). Small DNA viruses. *In* "Comparative Virology" (K. Maramorosch and E. Kurstak, eds.), p. 43. Academic Press, New York.

Hoggan, M. D., Shatkin, A. J., Blacklow, N. R., Koczot, F., and Rose, J. A. (1968). Helper-dependent infectious deoxyribonucleic acid from adenovirus-associated virus *J. Virol.* **2**, 850.

Holland, J. J. (1961). Receptor affinities as major determinants of enterovirus tissue tropisms in humans. *Virology* **15**, 312.

Holland, J. J. (1964). Enterovirus entrance into specific host cells, and subsequent alterations of cell protein and nucleic acid synthesis. *Bacteriol. Rev.* **28**, 3.

Holland, J. J., and Cords, C. E. (1964). Maturation of poliovirus RNA with capsid protein coded by hetereologous enteroviruses. *Proc. Nat. Acad. Sci. U.S.* **51**, 1082.

Holland, J. J., and Hoyer, B. H .(1962). Early stages of enterovirus infection. *Cold Spring Harbor Symp. Quant. Biol.* **27**, 101.

Holland, J. J., and Kiehn, E. D. (1968). Specific cleavage of viral proteins as steps in the synthesis and maturation of enteroviruses. *Proc. Nat. Acad. Sci. U.S.* **60**, 1015.

Holland, J. J., McLaren, L. C., and Syverton, J. T. (1959). The mammalian cell virus relationship. IV. Infection of naturally insusceptible cells with enterovirus ribonucleic acid. *J. Exp. Med.* **110**, 65.

Holley, R. W., and Kiernan, J. A. (1968). "Contact inhibition" of cell division in 3T3 cells. *Proc. Nat. Acad. Sci. U.S.* **60**, 300.

Hollings, M. (1962). Viruses associated with a die-back disease of cultivated mushrooms. *Nature (London)* **196**, 962.

Holloway, A. F., Wong, P. K. Y., and Cormack, D. V. (1970). Isolation and characteristics of temperature-sensitive mutants of vesicular stomatitis virus. *Virology* **42**, 917.

Holmes, A. W. (1971). Transmission of human hepatitis to marmosets: further coded studies. *J. Infect. Dis.* **124**, 520.

Holmes, I. H. (1971). Morphological similarity of Bunyamwera supergroup viruses. *Virology* **43**, 708.

Holmes, I. H., and Warburton, F. (1967). Is rubella an arbovirus? *Lancet* **ii**, 1233.

Holmes, I. H., Wark, M. C., and Warburton, M. F. (1969). Is rubella an arbovirus? II. Ultrastructural morphology and development. *Virology* **37**, 15.

Holmes, K. V., and Choppin, P. W. (1966). On the role of the response of the cell membrane in determining virus virulence. Contrasting effects of the parainfluenza virus SV5 in two cell types. *J. Exp. Med.* **124**, 501.

Holmes, M. J., Allen, T. R., Bradburne, A. F., and Stott, E. J. (1971). Studies of respiratory viruses in personnel at an Antarctic base. *J. Hyg.* **69**, 187.

Holowczak, J. A. (1970). Glycopeptides of vaccinia virus. I. Preliminary characterization and hexosamine content. *Virology* **42**, 87.

Holowczak, J. A., and Joklik, W. K. (1967a). Studies on the structural proteins of vaccinia virus. I. Structural proteins of virions and cores. *Virology* **33**, 717.

Holowczak, J. A., and Joklik, W. K. (1967b). Studies on the structural proteins of vaccinia virus. II. Kinetics of the synthesis of individual groups of structural proteins. *Virology* **33**, 726.

Homma, M. (1971). Trypsin action on the growth of Sendai virus in tissue culture cells. I. Restoration of the infectivity for L cells by direct action of trypsin on L cell-borne Sendai virus. *J. Virol.* **8**, 619.

Hooks, J., Gibbs, C. J., Cutchins, E. C., Rogers, N. C., Lampert, P., and Gajdusek, D. C. (1972a). Characterization and distribution of two new foamy agents isolated from chimpanzees. *Arch. Ges. Virusforsch.* **38**, 38.

Hooks, J., Gibbs, C. J., Chopra, H., Lewis, M., and Gajdusek, D. C. (1972b). Spontaneous transformation of human brain cells grown *in vitro* and description of associated virus particles. *Science* **176**, 1420.

Hooper, B. E., and Haelterman, E. O. (1966). Concepts of pathogenesis and passive immunity in transmissible gastroenteritis of swine. *J. Amer. Vet. Med. Ass.* **149**, 1580.

Hoorn, B., and Tyrrell, D. A. J. (1969). Organ cultures in virology. *Progr. Med. Virol.* **11**, 408.

Hoover, E. A., and Griesemer, R. A. (1971). Bone lesions produced by feline herpesvirus. *Lab. Invest.* **25**, 457.

Hope-Simpson, R. E. (1958). Discussion on the common cold. *Proc. Roy. Soc. Med.* **51**, 267.

Hope-Simpson, R. E. (1965). The nature of herpes zoster: A long-term study and a new hypothesis. *Proc. Roy. Soc. Med.* **58**, 9.

Horak, I., Hilfenhaus, J., Siegert, W., Jungwirth, C., Bodo, G., and Palese, P. (1970). Interferon action: effect on the formation of poxvirus specific polysomes and viral RNA. *Z. Naturforsch.* **25b**, 1164.

Horak, I., Jungwirth, C., and Bodo, G. (1971). Poxvirus specific cytopathic effect in interferon treated L-cells. *Virology* **45**, 456.

Horne, R. W. (1967). Electron microscopy of isolated virus particles and their components. *In* "Methods in Virology" (K. Maramorosch and H. Koprowski, eds.), Vol. III, p. 522. Academic Press, New York.

Horne, R. W., and Nagington, J. (1959). Electron microscope studies of the development and structure of poliomyelitis virus. *J. Mol. Biol.* **1**, 333.

Horne, R. W., and Wildy, P. (1963). Virus structure revealed by negative staining. *Advan. Virus Res.* **10**, 101.

Horsfall, F. L., Jr., and Tamm, I. (eds.) (1965). "Viral and Rickettsial Infection of Man," 4th Ed. Lippincott, Philadelphia, Pennsylvania.

Horstmann, D. M. (1950). Acute poliomyelitis: Relation of physical activity at the time of onset to the course of the disease. *J. Amer. Med. Ass.* **142**, 236.

Horstmann, D. M., and Paul, J. R. (1947). The incubation period in human poliomyelitis and its implications. *J. Amer. Med. Ass.* **135**, 11.

Horstmann, D. M., Niederman, J. C., and Paul, J. R. (1959). Attenuated type 1 poliovirus vaccine. Its capacity to infect and to spread from "vaccinees" within an institutional population. *J. Amer. Med. Ass.* **170**, 1.

Horstmann, D. M., Opton, E. M., Klemperer, R., Llado, B., and Vignec, A. J. (1964). Viremia in infants vaccinated with oral poliovirus vaccine (Sabin). *Amer. J. Hyg.* **79**, 47.

Horstmann, D. M., Leibhaber, H., Le Bouvier, G. L., Rosenberg, D. A., and Halstead, S. B. (1970). Rubella: reinfection of vaccinated and naturally immune persons exposed to an epidemic. *New Engl. J. Med.* **283**, 771.

Horta-Barbosa, L., Fuccillo, D. A., Sever, J. L., and Zeman, W. (1969). Subacute sclerosing panencephalitis: isolation of measles virus from a brain biopsy. *Nature (London)* **221**, 974.

Horta-Barbosa, L., Hamilton, R., Wittig, B., Fuccillo, D. A., Sever, J. L., and Vernon, M. L. (1971a). Subacute sclerosing panencephalitis: isolation of suppressed measles virus from lymph node biopsies. *Science* **173**, 840.

Horta-Barbosa, L., Krebs, H., Ley, A., Chen, T-C., Gilkeson, M., and Sever, J. L. (1971b). Progressive increase in cerebrospinal fluid measles antibody levels in subacute sclerosing panencephalitis. *Pediatrics* **47**, 782.

Horwitz, M. S., and Scharff, M. D. (1969a). The production of antiserum against viral antigens. *In* "Fundamental Techniques in Virology" (K. Habel and N. P. Salzman, eds.), p. 253. Academic Press, New York.

Horwitz, M. S., and Scharff, M. D. (1969b). Immunological precipitation of radioactively labeled viral proteins. *In* "Fundamental Techniques in Virology" (K. Habel and N. P. Salzman, eds.), p. 297. Academic Press, New York.

Horwitz, M. S., Scharff, M. D., and Maizel, J. V. (1969). Synthesis and assembly of adenovirus 2. I. Polypeptide synthesis, assembly of capsomers and morphogenesis of the virion. *Virology* **39**, 682.

Horzinek, M., and Mussgay, M. (1969). Studies on the nucleocapsid structure of a group A arbovirus. *J. Virol.* **4**, 514.

Horzinek, M., Maess, J., and Laufs, R. (1971). Studies on the substructure of togaviruses. II. Analysis of equine arteritis, rubella, bovine viral diarrhoea and hog cholera viruses. *Arch. Ges. Virusforsch.* **33**, 306.

Hosaka, Y., and Shimizu, Y. K. (1972). Artificial assembly of envelope particles of HVJ (Sendai virus). II. Lipid components for formation of the active hemolysin. *Virology* **49**, 640.

Hosaka, Y., Kitano, H., and Ikeguchi, S. (1966). Studies on the pleomorphism of HVJ virions. *Virology* **29**, 205.

Hoskins, M. (1935). A protective action of neurotropic against viscerotropic yellow fever virus in *Macacus rhesus*. *Amer. J. Trop. Med.* **15**, 675.

Hotchin, J. (1962). The biology of lymphocytic choriomeningitis infection: Virus-induced immune disease. *Cold Spring Harbor Symp. Quant. Biol.* **27**, 479.

Hotchin, J. (1971). Persistent and slow virus infections. *Monogr. Virol.* **3**, 1.

Hotchkiss, R. D. (1971). Toward a general theory of recombination in DNA. *Advan. Genet.* **16**, 325.

Hovi, T., and Vaheri, A. (1970). Infectivity and some physicochemical characteristics of rubella virus ribonucleic acid. *Virology* **42**, 1.

Howard, A., and Pelc, S. R. (1952). Synthesis of desoxyribonucleic acid in normal and irradiated cells and its relation to chromosome breakage. *Heredity Suppl.* **6**, 261.

Howatson, A. F. (1970). Vesicular stomatitis and related viruses. *Advan. Virus Res.* **16**, 196.

Howatson, A. F., and Crawford, L. V. (1963). Direct counting of the capsomers in polyoma and papilloma viruses. *Virology* **21**, 1.

Howatson, A. F., and Whitmore, G. F. (1962). The development and structure of vesicular stomatitis virus. *Virology* **16**, 466.

Howe, C., and Lee, L. T. (1972). Virus-erythrocyte interactions. *Advan. Virus Res.* **17**, 1.

Howe, H. A. (1962). The quantitation of poliomyelitis virus in the human alimentary tract with reference to coexisting levels of homologous serum neutralizing antibody. *Amer. J. Hyg.* **75**, 1.

Howell, P. G., and Verwoerd, D. W. (1971). Bluetongue virus. *Virol. Monogr.* **9**, 35.

Hoyle, L. (1968). "The Influenza Viruses." *Virol. Monogr.* **4**, 1.

Hoyle, L., Horne, R. W., and Waterson, A. P. (1962). The structure and composition of the myxoviruses. III. The interaction of influenza virus particles with cytoplasmic particles derived from normal chorioallantoic membrane cells. *Virology* **17**, 533.

Hsiung, G. D. (1968). Latent virus infections in primate tissues with special reference to simian viruses. *Bacteriol. Rev.* **32**, 185.

Hsiung, G. D. (1972). Parainfluenza-5 virus. Infection of man and animal. *Progr. Med. Virol.* **14**, 241.

Hsu, T. C., and Benirschke, K. (1967). "An Atlas of Mammalian Chromosomes." Springer-Verlag, New York.

Huang, A. S., and Baltimore, D. (1970a). Initiation of polyribosome formation in poliovirus-infected HeLa cells. *J. Mol. Biol.* **47**, 275.

Huang, A. S., and Baltimore, D. (1970b). Defective viral particles and viral disease processes. *Nature (London)* **226**, 325.

Huang, A. S., and Manders, E. K. (1972). Ribonucleic acid synthesis of vesicular stomatitis virus. IV. Transcription by standard virus in the presence of defective interfering particles. *J. Virol.* **9**, 909.

Huang, A. S., and Wagner, R. R. (1965). Inhibition of cellular RNA synthesis by nonreplicating vesicular stomatitis virus. *Proc. Nat. Acad. Sci. U.S.* **54**, 1579.

Huang, A. S., and Wagner, R. R. (1966a). Defective T particles of vesicular stomatitis virus. II. Biologic role in homologous interference. *Virology* **30**, 173.

Huang, A. S., and Wagner, R. R. (1966b). Comparative sedimentation coefficients of RNA extracted from plaque-forming and defective particles of vesicular stomatitis virus. *J. Mol. Biol.* **22**, 381.

Huang, A. S., Greenawalt, J. W., and Wagner, R. R. (1966). Defective T particles of vesicular stomatitis virus. I. Preparation, morphology and some biologic properties. *Virology* **30**, 161.

Huang, A. S., Baltimore, D., and Stampfer, M. (1970). Ribonucleic acid synthesis of vesicular stomatitis virus. III. Multiple complementary messenger RNA molecules. *Virology* **42**, 946.

Huang, A. S., Baltimore, D., and Bratt, M. A. (1971). Ribonucleic acid polymerase in virions of Newcastle disease virus: comparison with the vesicular stomatitis virus polymerase. *J. Virol.* **7**, 389.

Huang, S. (1971). Hepatitis-associated antigen hepatitis. *Amer. J. Pathol.* **64**, 483.

Hubbell, W. L., and McConnell, H. M. (1968). Spin-label studies of the excitable membranes of nerve and muscle. *Proc. Nat. Acad. Sci. U.S.* **61**, 12.

Huberman, J. A., and Attardi, G. (1967). Studies of fractionated HeLa cell metaphase chromosomes. I. The chromosomal distribution of DNA complementary to 28s and 18s ribosomal RNA and to cytoplasmic messenger RNA. *J. Mol. Biol.* **29**, 487.

Huberman, J. A., and Riggs, A. D. (1966). Autoradiography of chromosomal DNA fibers from Chinese hamster cells. *Proc. Nat. Acad. Sci. U.S.* **55**, 599.

Hudack, S. S., and McMaster, P. D. (1932). I. The permeability of the wall of the lymphatic capillary. *J. Exp. Med.* **56**, 223.

Hudack, S. S., and McMaster, P. D. (1933). The lymphatic participation in human cutaneous phenomena. *J. Exp. Med.* **57**, 751.

Hudson, B., Uphold, W. B., Devinny, T., and Vinograd, J. (1969). The use of an ethidium analog in the dye-buoyant density procedure for the isolation of closed circular DNA: The variation of the superhelix density of mitochondrial DNA. *Proc. Nat. Acad. Sci. U.S.* **63**, 813.

Huebner, R. J. (1967). The murine leukemia-sarcoma virus complex. *Proc. Nat. Acad. Sci. U.S.* **58**, 835.

Huebner, R. J., and Igel, H. J. (1971). Immunological tolerance to *gs* antigens as evidence for vertical transmission of non-infectious C-type RNA virus genomes. *Perspect. Virol.* **7**, 55.

Huebner, R. J., and Todaro, G. J. (1969). Oncogenes of RNA tumor viruses as determinants of cancer. *Proc. Nat. Acad. Sci. U.S.* **64**, 1087.

Huebner, R. J., Rowe, W. P., and Lane, W. T. (1962). Oncogenic effects in hamsters of human adenovirus types 12 and 18. *Proc. Nat. Acad. Sci. U.S.* **48**, 2051.

Huebner, R. J., Chanock, R. M., Rubin, B. A., and Casey, M. J. (1964). Induction by adenovirus type 7 of tumors in hamsters having the antigenic characteristics of SV40 virus. *Proc. Nat. Acad. Sci. U.S.* **52**, 1333.

Huebner, R. J., Casey, M. J., Chanock, R. M., and Schell, K. (1965). Tumors induced in hamsters by a strain of adenovirus type 3: Sharing of tumor antigens and "neoantigens" with those produced by adenovirus type 7 tumors. *Proc. Nat. Acad. Sci. U.S.* **54**, 381.

Huebner, R. J., Kelloff, G. J., Sarma, P. S., Lance, E. T., Turner, H. C., Gilden, R. V., Oroszlan, S., Meier, H., Myers, D. D., and Peters, R. L. (1970). Group-specific antigen expression during embryogenesis of the genome of the C-type RNA tumor virus: implications for ontogenesis and oncogenesis. *Proc. Nat. Acad. Sci. U.S.* **67**, 366.

Hughes, T. P. (1933). A precipitin reaction in yellow fever. *J. Immunol.* **25**, 275.

Hull, R. N. (1968). The simian viruses. *Virology Monogr.* **2**, 1.

Hull, R. N., Minner, R. J., and Mascoli, C. C. (1958). New viral agents recovered from tissue cultures of monkey kidney cells. III. Recovery of additional agents both from cultures of monkey tissues and directly from tissue and excreta. *Amer. J. Hyg.* **68**, 31.

Hull, R. N., Johnson, I. S., Culbertson, C. G., Reimer, C. B., and Wright, H. F. (1965). Oncogenicity of the simian adenoviruses. *Science* **150,** 1044.

Hummeler, K. (1971). Bullet-shaped viruses. *In* "Comparative Virology" (K. Maramorosch and E. Kurstak, eds.), p. 361. Academic Press, New York.

Hummeler, K., Anderson, T. F., and Brown, R. A. (1962). Identification of poliovirus particles of different antigenicity as seen in the electron microscope. *Virology* **16,** 84.

Hummeler, K., Koprowski, H., and Wiktor, T. J. (1967). Structure and development of rabies virus in tissue culture. *J. Virol.* **1,** 152.

Hummeler, K., Tomassini, N., and Zajac, B. (1969). Early events in herpes simplex virus infection: A radioautographic study. *J. Virol.* **4,** 67.

Humphrey, J. H., and White, R. G. (1970). "Immunology for Students of Medicine," 3rd Ed. Blackwell, Oxford.

Hung, P. O., Robinson, H. L., and Robinson, W. S. (1971). Isolation and characterization of proteins from Rous sarcoma virus. *Virology* **43,** 251.

Hunter, G. D. (1972). Scrapie: A prototype slow infection. *J. Infect. Dis.* **125,** 427.

Huismans, H. (1971). Host cell protein synthesis after infection with bluetongue virus and reovirus. *Virology* **46,** 500.

Huismans, H., and Verwoerd, D. W. (1973). Control of transcription during the expression of the bluetongue virus genome. In press.

Hurst, E. W. (1933). Studies on pseudorabies (infectious bulbar paralysis, mad itch). I. Histology of the disease, with a note on the symptomatology. *J. Exp. Med.* **58,** 415.

Hurst, E. W. (1936). Infection of the Rhesus monkey (*Macaca mulatta*) and the guinea pig with the virus of equine encephalomyelitis. *J. Pathol. Bacteriol.* **42,** 271.

Hurst, E. W. (1937). Myxoma and the Shope fibroma. II. The effect of intracerebral passage on the myxoma virus. *Brit. J. Exp. Pathol.* **18,** 15.

Hurwitz, J, and Leis, J. (1972). RNA-dependent DNA polymerase activity of RNA tumor viruses. I. Directing influence of DNA in the reaction. *J. Virol.* **9,** 116.

Huygelen, C., Peetermans, J., and Prinzie, A. (1969). An attenuated rubella virus vaccine (Cendehill 51 strain) grown in primary rabbit kidney cells. *Progr. Med. Virol.* **11,** 107.

Hyde, J. M., and Peters, D. (1971). The organization of nucleoprotein within fowlpox virus. *J. Ultrastruct. Res.* **35,** 626.

Hyslop, N. St. G. (1965). Isolation of variant strains of foot-and-mouth disease virus propagated in cell cultures containing antiviral sera. *J. Gen. Microbiol.* **41,** 135.

Hyslop, N. St. G., and Fagg, R. H. (1965). Isolation of variants during passage of a strain of foot-and-mouth disease virus in partly immunized cattle. *J. Hyg.* **63,** 357.

Ichihashi, Y., and Matsumoto, S. (1966). Studies on the nature of Marchal bodies (A-type inclusion) during ectromelia virus infection. *Virology* **29,** 264.

Ichihashi, Y., Matsumoto, S., and Dales, S. (1971). Biogenesis of poxviruses: role of A-type inclusions and host cell membranes in virus dissemination. *Virology* **46,** 507.

Igel, H. J., and Black, P. H. (1967). *In vitro* transformation by adenovirus-SV40 hybrid viruses. III. Morphology of tumors induced with transformed cells. *J. Exp. Med.* **125,** 647.

Ikegami, N., and Gomatos, P. J. (1968). Temperature-sensitive conditional-lethal mutants of reovirus 3. I. Isolation and characterization. *Virology* **36,** 447.

Ikegami, N., and Gomatos, P. J. (1972). Inhibition of host and viral protein syntheses during infection at the nonpermissive temperature with *ts* mutants of reovirus 3. *Virology* **47,** 306.

Ikegami, N., Eggers, H. J., and Tamm, I. (1964). Rescue of drug-requiring and drug-inhibited enteroviruses. *Proc. Nat. Acad. Sci. U.S.* **52,** 1419.

Illis, L. S., and Merry, R. T. G. (1972). Treatment of herpes simplex encephalitis. *J. Roy. Coll. Phys. London* **7,** 34.

Imagawa, D. T. (1968). Relationships among measles, canine distemper and rinderpest viruses. *Progr. Med. Virol.* **10,** 160.

Inbar, M., and Sachs, L. (1969). Interaction of the carbohydrate-binding protein concanavalin A with normal and transformed cells. *Proc. Nat. Acad. Sci. U.S.* **63,** 1418.

Inbar, M., Rabinowitz, Z., and Sachs, L. (1969). The formation of variants with a reversion of properties of transformed cells III. Reversion of the structure of the cell surface membrane. *Intern. J. Cancer* **4**, 690.

International Enterovirus Study Group (1963). Picornavirus group. *Virology* **19**, 114.

Irlin, I. S. (1967). Immunofluorescent demonstration of a virus specific surface antigen in cells infected or transformed by polyoma virus. *Virology* **32**, 725.

Isaacs, A. (1957). Particle counts and infectivity titration for animal viruses. *Advan. Virus Res.* **4**, 111.

Isaacs, A. (1963). Interferon. *Advan. Virus Res.* **10**, 1.

Isaacs, A., and Baron, S. (1960). Antiviral action of interferon in embryonic cells. *Lancet* **ii**, 946.

Isaacs, A., and Hitchcock, G. (1960). Role of interferon in recovery from virus infections. *Lancet* **ii**, 69.

Isaacs, A., and Lindenmann, J. (1957). Virus interference. I. The interferon. *Proc. Roy. Soc.* **B147**, 258.

Isaacs, A., and Westwood, M. A. (1959). Inhibition by interferon of the growth of vaccinia virus in the rabbit skin. *Lancet* **ii**, 324.

Isaacs, A., Hart, R. J. C., and Law, V. G. (1962). Influenza viruses 1957–60. *Bull. WHO* **26**, 253.

Isacson, P., and Stone, A. (1971). Allergic reactions associated with viral vaccines. *Progr. Med. Virol.* **13**, 239.

Ishibashi, M. (1971). Temperature-sensitive conditional-lethal mutants of an avian adenovirus (CELO). *Virology* **45**, 42.

Ishimoto, N., Temin, H. M., and Strominger, J. L. (1966). Studies of carcinogenesis by avian sarcoma viruses. II. Virus-induced increase in hyaluronic acid synthetase in chicken fibroblasts. *J. Biol. Chem.* **241**, 2052.

Ishizaka, K. (1970). Human reaginic antibodies. *Annu. Rev. Med.* **21**, 187.

Ishizaki, R., and Vogt, P. K. (1966). Immunological relationships among envelope antigens of avian tumor viruses. *Virology* **30**, 375.

Israel, M. S. (1962). The viral flora of enlarged tonsils and adenoids. *J. Pathol. Bacteriol.* **84**, 169.

Ito, M., and Suzuki, E. (1970). Adeno-associated satellite virus growth supported by a temperature-sensitive mutant of human adenovirus. *J. Gen. Virol.* **9**, 243.

Ito, M., Melnick, J. L., and Mayor, H. D. (1967). An immunofluorescence assay for studying replication of adeno-satellite virus. *J. Gen. Virol.* **1**, 199.

Ito, Y. (1962). Relationship of components of papilloma virus to papilloma and carcinoma cells. *Cold Spring Harbor Symp. Quant. Biol.* **27**, 387.

Ito, Y., and Evans, C. A. (1961). Induction of tumors in domestic rabbits with nucleic acid preparations from partially purified Shope papilloma virus and from extracts of papillomas of domestic and cottontail rabbits. *J. Exp. Med.* **114**, 485.

Ito, Y., and Joklik, W. K. (1972a). Temperature-sensitive mutants of reovirus. I. Pattern of gene expression by mutants of groups C, D, and E. *Virology* **50**, 189.

Ito, Y., and Joklik, W. K. (1972b). Temperature-sensitive mutants of reovirus. II. Anomalous electrophoretic migration of certain hybrid RNA molecules composed of mutant plus strands and wild-type virus strands. *Virology* **50**, 202.

Ito, Y., and Joklik, W. K. (1972c). Temperature-sensitive mutants of reovirus. III. Evidence that mutants of group D ("RNA-negative") are structural polypeptide mutants. *Virology* **50**, 282.

Itoh, H., and Melnick, J. L. (1959). Double infections of single cells with ECHO 7 and Coxsackie A 9 viruses. *J. Exp. Med.* **109**, 393.

Izawa, M., Allfrey, V. G., and Mirsky, A. E. (1963). The relationship between RNA synthesis and loop structure in lampbrush chromosomes. *Proc. Nat. Acad. Sci. U.S.* **49**, 544.

Jackson, G. G., and Muldoon, R. L. (1973). Viruses causing common respiratory infections in man. I. Rhinoviruses. *J. Infect. Dis.* **127**, 328.

Jacobson, M. F., and Baltimore, D. (1968a). Morphogenesis of poliovirus. I. Association of the viral RNA with coat protein. *J. Mol. Biol.* **33**, 369.

Jacobson, M. F., and Baltimore, D. (1968b). Polypeptide cleavages in the formation of polio-virus proteins. *Proc. Nat. Acad. Sci. U.S.* **61**, 77.

Jacobson, M. F., Asso, J., and Baltimore, D. (1970). Further evidence on the formation of poliovirus proteins. *J. Mol. Biol.* **49**, 657.

Jacquemont, B., Grange, J., Gazzalo, L., and Richard, M. H. (1972). Composition and size of Shope fibroma virus deoxyribonucleic acid. *J. Virol.* **9**, 836.

Jaenisch, R., and Levine, A. (1971). DNA replication in SV40-infected cells. V. Circular and catenated oligomers of SV40 DNA. *Virology* **44**, 480.

Jaenisch, R., Mayer, A., and Levine, A. (1971). Replicating SV40 molecules containing closed circular template DNA strands. *Nature New Biol.* **233**, 72.

Jagger, J. (1958). Photoreactivation. *Bacteriol. Rev.* **22**, 99.

Jahkola, M. (1965). Inheritance of resistance to polyoma tumorigenesis in inbred mice. *J. Nat. Cancer Inst.* **35**, 595.

Jainchill, J. L., and Todaro, G. J. (1970). Stimulation of cell growth *in vitro* by serum with and without growth factor. *Exp. Cell Res.* **59**, 137.

Janeway, C. A., and Rosen, F. S. (1966). The gammaglobulins. IV. Therapeutic uses of gamma globulin. *New Engl. J. Med.* **275**, 826.

Jarrett, O. (1966). Different transplantation antigens in BHK21 cells transformed by four strains of polyoma virus. *Virology* **30**, 744.

Jarrett, O. (1970). Evidence for the viral etiology of leukemia in the domestic mammals. *Advan. Cancer Res.* **13**, 39.

Jarrett, O., and Macpherson, I. (1968). The basis of the tumorigenicity of BHK 21 cells. *Intern. J. Cancer* **3**, 654.

Jarrett, O., Laird, H. M., and Hay, D. (1969). Growth of feline leukemia virus in human cells. *Nature (London)* **224**, 1208.

Jarrett, O., Pitts, J. D., Whalley, J. M., Clason, A. E., and Hay, J. (1971). Isolation of the nucleic acid of feline leukemia virus. *Virology* **43**, 317.

Jarrett, W. F. H., Martin, W. B., Crighton, G. W., Dalton, R. G., and Stewart, M. F. (1964). Leukemia in the cat: transmission experiments with leukemia (lymphosarcoma). *Nature (London)* **202**, 566.

Jawetz, E., Coleman, V. R., Dawson, C. R., and Thygeson, P. (1970). The dynamics of IUDR action in herpetic keratitis and the emergence of IUDR resistance *in vivo*. *Ann. N.Y. Acad. Sci.* **173**, 282.

Jeanteur, Ph., and Attardi, G. (1969). Relationship between HeLa cell ribosomal RNA and its precursors studied by high resolution RNA-DNA hybridization. *J. Mol. Biol.* **45**. 305.

Jenner, E. (1798). An enquiry into the causes and effects of the variolae vaccinae, a disease discovered in some western counties of England, particularly Gloucestershire, and known by the name of the cow pox. Reprinted by Cassell, London, 1896.

Jensen, E. M., Zelljadt, I., Chopra, H. C., and Mason, M. M. (1970). Isolation and propagation of a virus from a spontaneous mammary carcinoma of a Rhesus monkey. *Cancer Res.* **30**, 2388.

Jensen, F. C., and Koprowski, H. (1969). Absence of repressor in SV40 transformed cells. *Virology* **37**, 687.

Jensen, F. C., Koprowski, H., Pagano, J. S., Ponten, J., and Ravdin, R. G. (1964). Autologous and homologous implantation of human cells transformed *in vitro* by simian virus 40. *J. Nat. Cancer Inst.* **32**, 917.

Jensen, M. M. (1968). Transitory impairment of interferon production in stressed mice. *J. Infect. Dis.* **118**, 230.

Jensen, M. M., and Rasmussen, A. F., Jr. (1963). Stress and susceptibility to viral infections. II. Sound stress and susceptibility to vesicular stomatitis virus. *J. Immunol.* **90**, 21.

Johansson, B., Klein, G., Henle, W., and Henle, G. (1970). Epstein-Barr virus (EBV)-associated antibody patterns in malignant lymphoma and leukemia, I. Hodgkin's disease. *Intern. J. Cancer* **6**, 450.

Johnson, F. B., Ozer, H. L., and Hoggan, M. D. (1971). Structural proteins of adenovirus-associated virus type 3. *J. Virol.* **8**, 860.

Johnson, F. B., Blacklow, N. R., and Hoggan, M. D. (1972). Immunologic reactivity of antisera prepared against the sodium dodecyl sulfate treated structural polypeptides of adenovirus-associated virus. *J. Virol.* **9,** 1007.

Johnson, G. S., Friedman, R. M., and Pastan, I. (1971). Analysis of the fusion of XC cells induced by homogenates of murine leukemia virus-infected cells and by purified murine leukemia virus. *J. Virol.* **7,** 753.

Johnson, R. T. (1964a). The pathogenesis of herpes virus encephalitis. I. Virus pathways to the nervous system of suckling mice demonstrated by fluorescent antibody staining. *J. Exp. Med.* **119,** 343.

Johnson, R. T. (1964b). The pathogenesis of herpes virus encephalitis. II. A cellular basis for the development of resistance with age. *J. Exp. Med.* **120,** 359.

Johnson, R. T. (1965a). Virus invasion of the central nervous system. A study of Sindbis virus infection in the mouse using fluorescent antibody. *Amer. J. Pathol.* **46,** 929.

Johnson, R. T. (1965b). Experimental rabies. Studies of cellular vulnerability and pathogenesis using fluorescent antibody staining. *J. Neuropathol. Exp. Neurol.* **24,** 662.

Johnson, R. T. (1967). Chronic infectious neuropathic agents: Possible mechanisms of pathogenesis. *Curr. Topics Microbiol.* **40,** 3.

Johnson, R. T., and Mercer, E. H. (1964). The development of fixed rabies virus in mouse brain. *Aust. J. Exp. Biol. Med. Sci.* **42,** 449.

Johnson, R. T., and Mims, C. A. (1968). Pathogenesis of viral infections of the nervous system. *New Engl. J. Med.* **278,** 23, 84.

Johnson, R. T., Johnson, K. P., and Edmonds, C. J. (1967). Virus-induced hydrocephalus: development of aqueductal stenosis in hamsters after mumps infection. *Science* **157,** 1066.

Johnson, T. C., and McLaren, L. C. (1965). Plaque development and induction of interferon synthesis by R.M.C. poliovirus. *J. Bacteriol.* **90,** 565.

Johnsson, T., and Rasmussen, A. F., Jr. (1965). Emotional stress and susceptibility to poliomyelitis virus infection in mice. *Arch. Ges. Virusforsch.* **17,** 392.

Joklik, W. K. (1962). Some properties of poxvirus deoxyribonucleic acid. *J. Mol. Biol.* **5,** 265.

Joklik, W. K. (1964a). The intracellular uncoating of poxvirus DNA. I. The fate of radioactivity-labeled rabbitpox virus. *J. Mol. Biol.* **8,** 263.

Joklik, W. K. (1964b). The intracellular uncoating of poxvirus DNA. II. The molecular basis of the uncoating process. *J. Mol. Biol.* **8,** 277.

Joklik, W. K. (1968). The poxviruses. *Annu. Rev. Microbiol.* **22,** 359.

Joklik, W. K. (1970). The molecular biology of reovirus. *J. Cell. Physiol.* **76,** 289.

Joklik, W. K. (1972). Studies on the effect of chymotrypsin on reovirions. *Virology* **49,** 700.

Joklik, W. K., and Becker, Y. (1964). The replication and coating of vaccinia DNA. *J. Mol. Biol.* **10,** 452.

Joklik, W. K., and Becker, Y. (1965). Studies on the genesis of polyribosomes. I. Origin and significance of the subribosomal particles. *J. Mol. Biol.* **13,** 496.

Joklik, W. K., and Darnell, J. E. (1961). The adsorption and early fate of purified poliovirus in HeLa cells. *Virology* **13,** 439.

Joklik, W. K., and Merigan, T. C. (1966). Concerning the mechanism of action of interferon. *Proc. Nat. Acad. Sci. U.S.* **56,** 558.

Joklik, W. K., and Zweerink, H. J. (1971). The morphogenesis of animal viruses. *Annu. Rev. Genet.* **5,** 297.

Joklik, W. K., Holmes, I. H., and Briggs, M. J. (1960). The reactivation of poxviruses. III. Properties of reactivable particles. *Virology* **11,** 202.

Jonsson, N., and Sjögren, H. O. (1966). Specific transplantation immunity in relation to Rous sarcoma virus tumorigenesis in mice. *J. Exp. Med.* **123,** 487.

Jordan, W. S., Denny, F. W., Badger, G. F., Curtiss, C., Dingle, J. H., Oseasohn, R., and Stevens, D. A. (1958). A study of illness in a group of Cleveland families. XVII. The occurrence of Asian influenza. *Amer. J. Hyg.* **68,** 190.

Jovin-Arndt, D., and Berg, P. (1971). Quantitative binding of [125]I-concanavalin A to normal and transformed cells. *J. Virol.* **8,** 716.

Józwiak, W., Koscielak, J., Madalinski, K., Brzosko, W. J., Nowoslawski, A., and Kloczekiak, M. (1971). The RNA of Australia antigen. *Nature New Biol.* **229,** 92.

Juel-Jensen, B. E. (1970). Results of the treatment of zoster with idoxuridine in dimethyl sulfoxide. *Ann. N.Y. Acad. Sci.* **173**, 74.

Jungwirth, C., and Joklik, W. K. (1965). Studies on "early" enzymes in HeLa cells infected with vaccinia virus. *Virology* **27**, 80.

Jungwirth, C., and Launer, J. (1968). Effect of poxvirus infection on host cell deoxyribonucleic acid synthesis. *J. Virol.* **2**, 401.

Jungwirth, C., Horak, I., Bodo, G., Lindner, J., and Schultze, B. (1972). The synthesis of poxvirus-specific RNA in interferon-treated cells. *Virology* **48**, 59.

Justines, G., and Johnson, K. M. (1969). Immune tolerance in *Calomys callosus* infected with Machupo virus. *Nature (London)* **222**, 1090.

Kääriäinen, L., and Soderlund, H. (1971). Properties of Semliki Forest virus nucleocapsid. I. Sensitivity to pancreatic ribonuclease. *Virology* **43**, 291.

Kacian, D. L., Watson, K. F., Burny, A., and Spiegelman, S. (1971). Purification of the DNA polymerase of avian myeloblastosis virus. *Biochim. Biophys. Acta* **246**, 365.

Kaji, M., Oseasohn, R., Jordan, W. S., Jr., and Dingle, J. H. (1959). Isolation of Asian virus from extrapulmonary tissues in fatal human influenza. *Proc. Soc. Exp. Biol. Med.* **100**, 272.

Kajioka, R., Siminovitch, L., and Dales, S. (1964). The cycle of multiplication of vaccinia virus in Earle's strain L cells. II. Initiation of DNA synthesis and morphogenesis. *Virology* **24**, 295.

Kalb, A. J., and Levitzki, A. (1968). Metal-binding sites of concanavalin A and their role in the binding of α-methyl D-glucopyranoside. *Biochem. J.* **109**, 669.

Kalter, S. S., and Heberling, R. L. (1971). Comparative virology of primates. *Bacteriol. Rev.* **35**, 310.

Kanamitzu, M., Kasamaki, A., Ogawa, M., Kasahara, S., and Imamura, M. (1967). Immunofluorescent study on the pathogenesis of oral infection of poliovirus in monkey. *Jap. J. Med. Sci. Biol.* **20**, 175.

Kang, C. Y., and Prevec, L. (1970). Proteins of vesicular stomatitis virus. II. Immunological comparison of viral antigens. *J. Virol.* **6**, 20.

Kang, C. Y., and Prevec, L. (1971). Proteins of vesicular stomatitis virus. III. Intracellular synthesis and extracellular appearance of virus-specific proteins. *Virology* **46**, 678.

Kang, H. S., and McAuslan, B. R. (1972). Virus-associated nucleases: the location and properties of deoxyribonucleases and ribonucleases in purified Frog Virus 3. *J. Virol.* **10**, 202.

Kantoch, M., Warwick, A., and Bang, F. B. (1963). The cellular nature of genetic susceptibility to a virus. *J. Exp. Med.* **117**, 781.

Kapeller, M., and Doljanski, F. (1972). Agglutination of normal and Rous sarcoma virus-transformed chick embryo cells by concanavalin A and wheat germ agglutinin. *Nature New Biol.* **235**, 184.

Kaper, J. M. (1968). The small RNA viruses of plants, animals and bacteria. *In* "Molecular Basis of Virology" (H. Fraenkel-Conrat, ed.), p. 1. Reinhold, New York.

Kapikian, A. Z., Conant, R. M., Hamparian, V. V., Chanock, R. M., Chapple, P. J., Dick, E. C., Fenters, J. D., Gwaltney, J. M., Jr., Hamre, D., Holper, J. C., Jordan, W. S., Jr., Lennette, E. H., Melnick, J. L., Mogabgab, W. J., Mufson, M. A., Phillips, C. A., Schieble, J. H., and Tyrrell, D. A. J. (1967). Rhinoviruses: A numbering system. *Nature (London)* **213**, 761.

Kapikian, A. Z., Mitchell, R. H., Chanock, R. M., Shvedoff, R. A., and Stewart, C. E. (1969). An epidemiologic study of altered clinical reactivity to respiratory syncytial (RS) virus infection in children previously vaccinated with an inactivated RS virus vaccine. *Amer. J. Epidemiol.* **89**, 405.

Kaplan, A. S. (1964). Studies on the replicating pool of viral DNA in cells infected with pseudorabies virus. *Virology* **24**, 19.

Kaplan, A. S. (1969). Herpes simplex and pseudorabies. *Virol. Monogr. No.* **5**, 1.

Kaplan, A. S., and Ben-Porat, T. (1970). Synthesis of proteins in cells infected with herpesvirus. VI. Characterization of the proteins of the viral membrane. *Proc. Nat. Acad. Sci. U.S.* **66**, 799.

Kaplan, A. S., Ben-Porat, T., and Coto, C. (1967). Studies on the control of the infective process in cells infected with pseudorabies virus. *In* "Molecular Biology of Viruses" (J. S. Colter, ed.), p. 527. Academic Press, New York.

Kaplan, J. C., Wilbert, S. M., and Black, P. H. (1972). Analysis of simian virus 40-induced transformation of hamster kidney tissue *in vitro*. VIII. Induction of infectious simian virus 40 from virogenic transformed hamster cells by amino acid deprivation or cyclohexamide treatment. *J. Virol.* **9**, 448.

Kapuler, A. M., and Acs, G. (1970). Properties of RNA transcriptase in reovirus subviral particles. *Proc. Nat. Acad. Sci. U.S.* **66**, 890.

Kapuler, A. M., Mendelsohn, N., Klett, H., and Acs, G. (1970). Four base-specific nucleoside 5'-triphosphatases in the subviral cores of reovirus. *Nature (London)* **225**, 1209.

Karasaki, S. (1966). Size and ultrastructure of the H-viruses as determined by the use of specific antibodies. *J. Ultrastruct. Res.* **16**, 109.

Kasahara, S. (1965). Histological and immunofluorescent studies on poliomyelitis in monkeys. II. Detection of viral antigen by fluorescent antibody technique. *Sapparo Med. J.* **27**, 237.

Kasel, J. A., and Couch, R. B. (1969). Experimental infection in men and horses with influenza A viruses. *Bull. WHO* **41**, 447.

Kass, E. H., and Finland, M. (1953). Adrenocortical hormones in infection and immunity. *Ann. Rev. Microbiol.* **7**, 361.

Kass, E. H., and Finland, M. (1958). Cortocosteroids and infections. *Advan. Internal Med.* **9**, 45.

Kass, S. J., and Knight, C. A. (1965). Purification and chemical analysis of Shope papilloma virus. *Virology* **27**, 273.

Kates, J. (1970). Transcription of the vaccinia virus genome and the occurrence of polyriboadenylic acid sequences in messenger RNA. *Cold Spring Harbor Symp. Quant. Biol.* **35**, 743.

Kates, J., and Beeson, J. (1970a). RNA synthesis in vaccinia virus I. The mechanism of synthesis and release of RNA in vaccinia cores. *J. Mol. Biol.* **50**, 1.

Kates, J., and Beeson, J. (1970b). Ribonucleic acid synthesis in vaccinia virus. II. Synthesis of polyriboadenylic acid. *J. Mol. Biol.* **50**, 19.

Kates, J. R., and McAuslan, B. R. (1967a). Messenger RNA synthesis by a "coated" viral genome. *Proc. Nat. Acad. Sci. U.S.* **57**, 314.

Kates, J. R., and McAuslan, B. R. (1967b). Poxvirus DNA-dependent RNA polymerase. *Proc. Nat. Acad. Sci. U.S.* **58**, 134.

Kates, J. R., and McAuslan, B. R. (1967c). Interrelation of protein synthesis and viral DNA synthesis. *J. Virol.* **1**, 110.

Kato, N., and Eggers, H. J. (1969). Inhibition of uncoating of fowl plague virus by 1-adamantanamine hydrochloride. *Virology* **37**, 632.

Kato, N., and Okada, A. (1961). The relation of the toxic agent to the subunits of influenza virus particles. *Brit. J. Exp. Pathol.* **42**, 253.

Kato, S., and Kamahora, J. (1962). The significance of the inclusion formation of poxvirus group and herpes simplex virus. *Symp. Cell. Chem.* **12**, 47.

Kato, S., Hara, J., Ogawa, M., Miyamoto, H., and Kamahora, J. (1963). Inclusion markers of cowpox virus and alastrim virus. *Biken J.* **6**, 233.

Katz, E., and Moss, B. (1970a). Formation of a vaccinia virus structural polypeptide from a higher molecular weight precursor: inhibition by rifampicin. *Proc. Nat. Acad. Sci. U.S.* **66**, 677.

Katz, E., and Moss, B. (1970b). Vaccinia virus structural polypeptide derived from a high molecular weight precursor. Formation and integration into virus particles. *J. Virol.* **6**, 717.

Katz, E., and Vogt, P. K. (1971). Conditional lethal mutants of avian sarcoma viruses. II. Analysis of the temperature-sensitive lesion in *ts*75. *Virology* **46**, 745.

Katz, M., and Koprowski, H. (1968). Failure to demonstrate a relationship between scrapie and the production of interferon in mice. *Nature (London)* **219**, 639.

Kaufman, H. E. (1965a). *In vivo* studies with antiviral agents. *Ann. N.Y. Acad. Sci.* **130**, 168.

Kaufman, H. E. (1965b). Problems in virus chemotherapy. *Progr. Med. Virol.* **7**, 116.

Kaufman, H. E., and Heidelberger, C. (1964). Therapeutic antiviral action of 5-trifluoromethyl-2'-deoxyuridine in herpes simplex keratitis. *Science* **145**, 585.

Kaufman, H. E., Brown, D. C., and Ellison, E. D. (1968). Herpes virus in the lacrimal gland, conjunctiva and cornea of man—a chronic infection. *Amer. J. Ophthalmol.* **65**, 32.

Kawai, S., and Hanafusa, H. (1971). The effects of reciprocal changes in temperature on the transformed state of cells infected with a Rous sarcoma virus mutant. *Virology* **46**, 470.

Kawai, S., and Hanafusa, H. (1972a). Plaque assay for some strains of avian leukosis virus. *Virology* **48**, 126.

Kawai, S., and Hanafusa, H. (1972b). Genetic recombination with avian tumor virus. *Virology* **49**, 37.

Kawakami, T. G., Theilen, G. H., Dungworth, D. L., Beall, S. G., and Munn, R. J. (1970). "C"-type viral particles in plasma of feline leukemia. *Science* **158**, 1049.

Kawamura, H., and Tsubahara, H. (1966). Common antigenicity of avian reoviruses. *Nat. Inst. Animal Health Quart. (Tokyo)* **6**, 187.

Kawamura, H., Shimizu, F., Maede, M., and Tsubahara, H. (1965). Avian reovirus: its properties and serological classification. *Nat. Inst. Animal Health Quart. (Tokyo)* **5**, 115.

Kaye, H. S., and Dowdle, W. R. (1969). Some characteristics of hemagglutination of certain strains of "IBV-like" viruses. *J. Infect. Dis.* **120**, 576.

Ke, Y., and Ho, M. (1971). Patterns of cycloheximide and puromycin effect on interferon production stimulated by virus or polyribonucleotide in different tissues. *Proc. Soc. Exp. Biol. Med.* **136**, 365.

Ke, Y., Singer, S. H., Postic, B., and Ho, M. (1966). Effect of puromycin on virus- and endotoxin-induced interferonlike inhibitors in rabbits. *Proc. Soc. Exp. Biol. Med.* **121**, 181.

Kedes, L. H., and Birnstiel, M. L. (1971). Reiteration and clustering of DNA sequences complementary to histone messenger RNA. *Nature New Biol.* **230**, 165.

Keller, J. M., Spear, P. G., and Roizman, B. (1970). Proteins specified by herpes simplex virus. III. Viruses differing in their effects on the social behavior of infected cells specify different membrane glycoproteins. *Proc. Nat. Acad. Sci. U.S.* **65**, 865.

Keller, W., and Crouch, R. (1972). Degradation of DNA RNA hybrids by ribonuclease H and DNA polymerases of cellular and viral origin. *Proc. Nat. Acad. Sci. U.S.* **69**, 3360.

Kelloff, G. J., and Vogt, P. K. (1966). Localization of avian tumor virus group-specific antigen in cell and virus. *Virology* **29**, 377.

Kelloff, G. J., Huebner, R. J., Chang, N., Lee, Y. K., and Gilden, R. V. (1970a). Envelope antigen relationships among three hamster-specific sarcoma viruses and a hamster-specific helper virus. *J. Gen. Virol.* **9**, 19.

Kelloff, G. J., Huebner, R. J., Lee, K. L., Toni, R., and Gilden, R. (1970b). Hamster-tropic sarcomagenic and nonsarcomagenic viruses derived from hamster tumors induced by the gross pseudotype of Moloney sarcoma virus. *Proc. Nat. Acad. Sci. U.S.* **65**, 310.

Kelloff, G. J., Hatanaka, M., and Gilden, R. V. (1972). Assay of C-type virus infectivity by measurement of RNA-dependent DNA polymerase activity. *Virology* **48**, 266.

Kelly, T. J., and Rose, J. A. (1971). Simian virus 40 integration site in an adenovirus 7-simian virus 40 hybrid DNA molecule. *Proc. Nat. Acad. Sci. U.S.* **68**, 1037.

Kempe, C. H., Fulginiti, V., Minamitani, M., and Shinefield, H. (1968). Smallpox vaccination of eczema patients with a strain of attenuated live vaccinia (CV1-78). *Pediatrics* **42**, 980.

Kennedy, R. C., Eklund, E. M., Lopez, C., and Hadlow, W. J. (1968). Isolation of a virus from the lungs of Montana sheep affected with progressive pneumonia. *Virology* **35**, 483.

Kenny, M. T., Schwarz, A. J., Bittle, J. L., Nolan, R. B., and Northrop, R. (1969). Rapid attenuation of rubella virus in African green monkey kidney tissue culture. *Symp. Ser. Immunobiol. Stand.* **11**, 219.

Keogh, E. V. (1938). Ectodermal lesions produced by the virus of Rous sarcoma. *Brit. J. Exp. Pathol.* **19**, 1.

Kerr, I. M. (1971). Protein synthesis in cell-free systems: an effect of interferon. *J. Virol.* **7**, 448.

Kerr, I. M., and Martin, E. M. (1971). Virus protein synthesis in animal cell-free systems:

nature of the products synthesized in response to ribonucleic acid of encephalomyocarditis virus. *J. Virol.* **7,** 438.

Kerr, I. M., Sonnabend, J. A., and Martin, E. M. (1970). Protein-synthetic activity of ribosomes from interferon-treated cells. *J. Virol.* **5,** 132.

Khera, K. S., Ashkenazi, A., Rapp, F., and Melnick, J. L. (1963). Immunity in hamsters to cells transformed *in vitro* and *in vivo* by SV40. *J. Immunol.* **91,** 604.

Khoury, G., Byrne, J. C., and Martin, M. A. (1972). Patterns of simian virus 40 DNA transcription after acute infection of permissive and non-permissive cells. *Proc. Nat. Acad. Sci. U.S.* **69,** 1925.

Kibrick, S. (1965). Current status of coxsackie and echo viruses in human disease. *Progr. Med. Virol.* **6,** 27.

Kibrick, S., and Gooding, G. W. (1965). Pathogenesis of infection with herpes simplex virus with special reference to nervous tissue. *Nat. Inst. Neurol. Dis. Blindness Monogr.* **2,** 143.

Kidd, J. G. (1942). The enduring partnership of a neoplastic virus and carcinoma cells. *J. Exp. Med.* **75,** 7.

Kieff, E. D., Bachenheimer, S. L., and Roizman, B. (1971). Size, composition, and structure of the deoxyribonucleic acid of herpes simplex subtypes 1 and 2. *J. Virol.* **8,** 125.

Kieff, E., Hoyer, B., Bachenheimer, S., and Roizman, B. (1972). Genetic relatedness of type 1 and type 2 herpes simplex viruses. *J. Virol.* **9,** 738.

Kiehn, E. D., and Holland, J. J. (1970). Synthesis and cleavage of enterovirus polypeptides in mammalian cells. *J. Virol.* **5,** 358.

Kilbourne, E. D. (1963). Influenza virus genetics. *Progr. Med. Virol.* **5,** 79.

Kilbourne, E. D. (1968). Recombination of influenza A viruses of human and animal origin. *Science* **160,** 74.

Kilbourne, E. D. (1969). Future influenza vaccines and the use of genetic recombinants. *Bull. WHO* **41,** 643.

Kilbourne, E. D., and Horsfall, F. L. (1951). Lethal infection with Coxsackie virus in adult mice given cortisone. *Proc. Soc. Exp. Biol. Med.* **77,** 135.

Kilbourne, E. D., Lief, F. S., Schulman, J. L., Jahiel, R. I., and Laver, W. G. (1967). Antigenic hybrids of influenza viruses and their implications. *Perspect. Virol.* **5,** 87.

Kilbourne, E. D., Schulman, J. L., Schild, G. C., Schloer, G., Swanson, J., and Bucher, D. (1971). Correlated studies of a recombinant influenza-virus vaccine. I. Derivation and characteristics of virus and vaccine. *J. Infect. Dis.* **124,** 449.

Kilbourne, E. D., Choppin, P. W., Schulze, I. T., Scholtissek, C., and Bucher, D. L. (1972). Influenza virus polypeptides and antigens—summary of influenza workshop. I. *J. Infect. Dis.* **125,** 447.

Kilburn, K. H. (1968). A hypothesis for pulmonary clearance and its implications. *Amer. Rev. Resp. Dis.* **98,** 449.

Kilham, L. (1961). Rat virus (RV) infections in hamsters. *Proc. Soc. Exp. Biol. Med.* **106,** 825.

Kilham, L., and Dalmat, H. T. (1955). Host-virus-mosquito relations of Shope fibroma in cottontail rabbits. *Amer. J. Hyg.* **61,** 45.

Kilham, L., and Margolis, G. (1964). Cerebella ataxia in hamsters inoculated with rat virus. *Science* **143,** 1047.

Kilham, L., and Margolis, G. (1965). Cerebellar disease in cats induced by inoculation of rat virus. *Science* **148,** 244.

Kilham, L., and Margolis, G. (1966). Spontaneous hepatitis and cerebellar "hypoplasia" in suckling rats due to congenital infections with rat virus. *Amer. J. Pathol.* **49,** 457.

Kilham, L., and Margolis, G. (1969). Transplacental infection of rats and hamsters induced by oral and parenteral inoculations of HI and rat viruses (RV). *Teratology* **2,** 111.

Kim, C. Y., and Bissell, D. M. (1971). Stability of the lipid and protein of hepatitis-associated (Australia) antigen. *J. Infect. Dis.* **123,** 470.

Kimberton, R. H., Millson, G. C., and Hunter, G. D. (1971). An experimental examination of the scrapie agent in cell membrane mixtures. III. Studies of the operational size. *J. Comp. Pathol. Therap.* **81,** 383.

Kimura, G., and Dulbecco, R. (1972). Isolation and characterization of temperature-sensitive mutants of simian virus 40. *Virology* **49**, 394.

Kingsbury, D. W. (1966). Newcastle disease virus RNA. II. Preferential synthesis of RNA complementary to parental viral RNA by chick embryo cells. *J. Mol. Biol.* **18**, 204.

Kingsbury, D. W. (1970). Replication and functions of myxovirus ribonucleic acids. *Progr. Med. Virol.* **12**, 49.

Kingsbury, D. W. (1972). Paramyxovirus replication. *Curr. Topics Microbiol. Immunol.* **59**, 1.

Kingsbury, D. W., and Webster, R. G. (1969). Some properties of influenza virus nucleocapsids. *J. Virol.* **4**, 219.

Kirkwood, J., Geering, G., and Old, L. J. (1972). Demonstration of group- and type-specific antigens of herpesvirus. *In* "Oncogenesis and Herpesviruses" (P. M. Biggs, G. de-Thé and L. N. Payne, eds.), p. 479. International Agency for Research on Cancer, Lyon.

Kirn, A., Gut, J-P., Bingen, A., and Hirth, Ch. (1972). Acute hepatitis produced by Frog Virus 3 in mice. *Arch. Ges. Virusforsch.* **36**, 394.

Kirschstein, R. L., Rabson, A. S., and O'Conor, G. T. (1965). Ependymomas produced in Syrian hamsters by adenovirus 7, Strain E46 ("hybrid" of adenovirus 7 and SV40). *Proc. Soc. Exp. Biol. Med.* **120**, 484.

Kirsten, W. H., and Mayer, L. A. (1967). Morphologic responses to a murine erythroblastosis virus. *J. Nat. Cancer Inst.* **39**, 311.

Kit, S. (1968). Viral-induced enzymes and the problem of viral oncogenesis. *Advan. Cancer Res.* **11**, 73.

Kit, S., and Brown, M. (1969). Rescue of simian virus 40 from cell lines transformed at high and at low input multiplicities by unirradiated or ultraviolet-irradiated virus. *J. Virol.* **4**, 226.

Kit, S., Piekarski, L. J., and Dubbs, D. R. (1967a). DNA polymerase induced by simian virus 40. *J. Gen. Virol.* **1**, 163.

Kit, S., de Torres, R. A., Dubbs, D. R., and Salvi, M. L. (1967b). Induction of cellular deoxyribonucleic acid synthesis by simian virus 40. *J. Virol.* **1**, 738.

Kit, S., Kurimura, T., Salvi, M. L., and Dubbs, D. R. (1968). Activation of infectious SV40 DNA synthesis in transformed cells. *Proc. Nat. Acad. Sci. U.S.* **60**, 1239. ·

Kjellén, L. E. (1963). A variant of adenovirus type 5. *Arch. Ges. Virusforsch* **13**, 482.

Klagsbrun, M. (1971). Changes in the methylation of transfer RNA in vaccinia infected HeLa cells. *Virology* **44**, 153.

Kleczkowski, A. (1968). Methods of inactivation by ultraviolet radiation. *In* "Methods in Virology" (K. Maramorosch and H. Koprowski, eds.), Vol. IV, p. 93. Academic Press, New York.

Klein, E., and Klein, G. (1966). Immunological tolerance of neonatally infected mice to the Moloney leukaemia virus. *Nature (London)* **209**, 163.

Klein, G. (1966a). Tumor antigens. *Annu. Rev. Microbiol.* **20**, 223.

Klein, G. (1966b). Lymphocytes and antibodies in relation to malignant disease. *In* "Viruses inducing Cancer, Implications for Therapy" (W. J. Burdette, ed.), p. 323. University of Utah Press, Salt Lake City, Utah.

Klein, G. (1968). Tumor-specific transplantation antigens. *Cancer Res.* **28**, 625.

Klein, G. (1971). Immunological aspects of Burkitt's lymphoma. *Advan. Immunol.* **14**, 187.

Klein, G. (1972). Herpesviruses and oncogenesis. *Proc. Nat. Acad. Sci. U.S.* **69**, 1056.

Klein, G., and Klein, E. (1962). Antigenic properties of other experimental tumors. *Cold Spring Harbor Symp. Quant. Biol.* **27**, 463.

Klein, G., Klein, E., and Haughton, G. (1966). Variation of antigenic characteristics between different mouse lymphomas induced by the Moloney virus. *J. Nat. Cancer Inst.* **36**, 607.

Klein, G., Clifford, P., Klein, E., Smith, R. T., Minowada, J., Kourilsky, F. M., and Burchenal, J. H. (1967). Membrane immunofluorescence reactions of Burkitt lymphoma cells from biopsy specimens and tissue culture. *J. Nat. Cancer Inst.* **39**, 1027.

Kleinschmidt, A. K. (1968). Monolayer techniques in electron microscopy of nucleic acid molecules. *In* "Methods in Enzymology" (L. Grossmann and K. Moldave, eds.), Vol. XII B, p. 361, Academic Press, New York.

Kleinschmidt, W. J. (1972). Biochemistry of interferon and its inducers. *Annu. Rev. Biochem.* **41,** 517.

Kleinschmidt, W. J., and Murphy, E. G. (1967). Interferon induction with statolon in the intact animal. *Bacteriol. Rev.* **31,** 132.

Klement, V., and Versely, P. (1965). Tumour induction with the Rous sarcoma virus in hamsters and production of infectious Rous sarcoma virus in an heterologous host. *Neoplasma* **12,** 147.

Klement, V., Hartley, J. W., Rowe, W. P., and Huebner, R. J. (1969). *J. Nat. Cancer Inst.* Recovery of a hamster-specific focus-forming and sarcomagenic virus from a "non-infectious" hamster tumor induced by the Kirsten mouse sarcoma virus. *J. Nat. Cancer Inst.* **43,** 925.

Klement, V., Nicolson, M. O., and Huebner, R. J. (1971). Rescue of focus forming virus genome from rat non-productive lines by 5'-bromodeoxyuridine. *Nature New Biol.* **234,** 12.

Klenk, H-D., and Choppin, P. W. (1969). Lipids of plasma membranes of monkey and hamster kidney cells and of parainfluenza virions grown in these cells. *Virology* **38,** 255.

Klenk, H-D., and Choppin, P. W. (1970). Glycosphingolipids of plasma membranes of cultured cells and an enveloped virus (SV5) grown in these cells. *Proc. Nat. Acad. Sci. U.S.* **66,** 57.

Klenk, H-D., and Choppin, P. W. (1971). Glycolipid content of vesicular stomatitis virus grown in baby hamster kidney cells. *J. Virol.* **7,** 416.

Klenk, H-D., Caliguiri, L. A., and Choppin, P. W. (1970a). The proteins of the parainfluenza virus SV5. II. The carbohydrate content and glycoproteins of the virion. *Virology* **42,** 473.

Klenk, H-D., Compans, R. W., and Choppin, P. W. (1970b). An electron microscopic study of the presence or absence of neuraminic acid in enveloped viruses. *Virology* **42,** 1158.

Klenk, H-D., Rott, R., and Becht, H. (1972). On the structure of the influenza virus envelope. *Virology* **47,** 579.

Kline, L. K., Weissman, S. M., and Soll, D. (1972). Investigation of adenovirus-directed 4S RNA. *Virology* **48,** 291.

Klug, A. (1965). Structure of viruses of the papilloma–polyoma type. II. Comments on other work. *J. Mol. Biol.* **11,** 424.

Klug, A., and Caspar, D. L. D. (1960). The structure of small viruses. *Advan. Virus Res.* **7,** 225.

Klug, A., and Finch, J. T. (1965). Structure of viruses of the papilloma–polyoma type. I. Human wart virus. *J. Mol. Biol.* **11,** 403.

Klug, A., and Finch, J. T. (1968). Structure of viruses of the papilloma–polyoma type IV. Tilting experiments and two side images. *J. Mol. Biol.* **31,** 1.

Klug, A., Longley, W., and Leberman, R. (1966). Arrangement of protein subunits and the distribution of nucleic acid in turnip yellow mosaic virus. I. X-ray diffraction studies. *J. Mol. Biol.* **15,** 315.

Knight, C. A. (1950). Amino acids of the Shope papilloma virus. *Proc. Soc. Exp. Biol. Med.* **75,** 843.

Knight, V. (1964). The use of voluteers in medical virology. *Progr. Med. Virol.* **6,** 1.

Knight, V., Gerone, P. J., Griffith, W. R., Couch, R. B., Cate, T. R., Johnson, K. M., Lang, D. J., Evans, H. E., Spickard, A., and Kasel, J A. (1963). Studies in volunteers with respiratory viral agents. *Amer. Rev. Resp. Dis.* **88,** Pt. 2, 135.

Koch, G., and Bishop, J. M. (1968). The effect of polycations on the interaction of viral RNA with mammalian cells: studies on the infectivity of single- and double-stranded poliovirus RNA. *Virology* **35,** 9.

Koch, M. A., and Sabin, A. B. (1963). Specificity of virus-induced resistance to transplantation of polyoma and SV40 tumors in adult hamsters. *Proc. Soc. Exp. Biol. Med.* **113,** 4.

Kohne, D. E., and Britten, R. J. (1971). Hydroxyapatite techniques for nucleic acid reassociation. *In* "Methods in Nucleic Acid Research" (G. L. Cantoni and D. R. Davies, eds.), Vol. 2, p. 500. Harper and Row, New York.

Koprowski, H. (1957). Discussion of properties of attenuated poliovirus and their behavior in human beings. *In* "Cellular Biology, Nucleic Acids and Viruses" (T. M. Rivers, ed.), Spec. Publ., Vol. 5, p. 128. N.Y. Acad. Sci., New York.

Koprowski, H., Ponten, J. A., Jensen, F., Ravdin, R. G., Moorhead, P., and Saksela, E. (1962). Transformation of cultures of human tissue infected with simian virus SV40. *J. Cell. Comp. Physiol.* **59,** 281.

Koprowski, H., Jensen, F. C., and Steplewski, Z. (1967). Activation of production of infection tumor virus SV40 in heterokaryon cultures. *Proc. Nat. Acad. Sci. U.S.* **58,** 127.

Kornberg, R. D., and McConnell, H. M. (1971). Lateral diffusion of phospholipids in a vesicle membrane. *Proc. Nat. Acad. Sci. U.S.* **68,** 2564.

Kotler, M. (1970). Control of multiplication of uninfected mouse embryo fibroblasts and mouse embryo fibroblasts converted by infection with murine sarcoma virus (Harvey). *Cancer Res.* **30,** 2493.

Kozak, M., and Nathans, D. (1972). Translation of the genome of a ribonucleic acid bacteriophage. *Bacteriol. Rev.* **36,** 109.

Kraft, L. M. (1966). Epizootic diarrhoea of infant mice and lethal intestinal virus infection of infant mice. *Nat. Cancer Inst. Monogr.* **20,** 55.

Kristensson, K., Lycke, E., and Sjöstrand, J. (1971). Spread of herpes simplex virus in peripheral nerves. *Acta Neuropathol.* **17,** 44.

Krueger, R. F., and Mayer, G. D. (1970). Tilorone hydrochloride: an orally active antiviral agent. *Science* **169,** 1213.

Krugman, S. (1963). The clinical use of gamma globulin. *New Engl. J. Med.* **269,** 195.

Krugman, S. (1965). Rubella Symposium. *Amer. J. Dis. Child.* **110,** 348.

Krugman, S. (1971). Present status of measles and rubella immunization in the U.S.: a medical progress report. *J. Pediat.* **78,** 1.

Krugman, S., Giles, J. P., and Hammond, J. (1967). Infectious hepatitis. Evidence for two distinctive clinical epidemiological and immunological types of infection. *J. Amer. Med. Ass.* **200,** 365.

Krugman, S., Giles, J. P., and Hammond, J. (1971). Viral hepatitis, type B (MS-2 strain). Studies on active immunization. *J. Amer. Med. Ass.* **217,** 41.

Kruse, P. F., and Miedema, E. (1965). Production and characterization of multiple-layered populations of animal cells. *J. Cell. Biol.* **27,** 273.

Kryukova, I. N. (1966). On two types of interaction of Rous sarcoma virus (Can strain) with mammalian cells *in vivo. Acta. Virol.* **10,** 440.

Kryukova, I. N., Obukh, I. B., and Tot, F. (1970). Characteristics of variants of Rous sarcoma virus (RSV) isolated from mouse RSV tumors. *J. Nat. Cancer Inst.* **45,** 49.

Kucera, L. S. (1970). Effects of temperature on frog polyhedral cytoplasmic deoxyribovirus multiplication: Thermosensitivity of initiation, replication and encapsidation of viral DNA. *Virology* **42,** 576.

Kucera, L. S., and Granoff, A. (1968). Viruses and renal carcinoma of *Rana pipiens*. VI. Interrelationships of macromolecular synthesis and infectious virus production in frog virus 3-infected BHK 21/13 cells. *Virology* **34,** 240.

Kufe, D., Hehlman, R., and Spiegelman, S. (1972). The presence in human sarcomas of RNA related to the RNA of a mouse leukemia virus. *Science* **175,** 182.

Kumagai, T., Shimizu, T., Ikeda, S., and Matumoto, M. (1961). A new *in vitro* method (END) for detection and measurement of hog cholera virus and its antibody by means of effect of HC virus on Newcastle disease virus in swine tissue cultures. I. Establishment of standard procedure. *J. Immunol.* **87,** 245.

Kundin, W. D. (1970). Hong Kong A2 influenza virus infection among swine during a human epidemic in Taiwan. *Nature (London)* **228,** 857.

Kundin, W. D., Liu, C., Hysell, P., and Hamachige, S. (1963). Studies on West Nile virus infection by means of fluorescent antibodies. I. Pathogenesis of West Nile virus infection in experimentally inoculated suckling mice. *Arch. Ges. Virusforsch.* **12,** 514.

Kunin, C. M. (1962). Virus-tissue union and the pathogenesis of enterovirus infections. *J. Immunol.* **88,** 556.

Kurstak, E. (1971). The immunoperoxidase technique: localization of viral antigens in cells.

*In* "Methods in Virology" (K. Maramorosch and H. Koprowski, eds.), Vol. V, p. 423. Academic Press, New York.

Kurth, R., and Bauer, H. (1972). Cell-surface antigens induced by avian RNA tumor viruses: detection by a cytoxic assay. *Virology* **47,** 426.

Lafferty, K. J. (1963). The interaction between virus and antibody. II. Mechanism of the reaction. *Virology* **21,** 76.

Lafferty, K. J. (1965). The relationship between molecular charge and the biological activity of antibody directed against rabbitpox virus. *Virology* **25,** 591.

Lai, M. M. C., and Duesberg, P. H. (1972a). Adenylic-acid rich sequence in RNAs of Rous sarcoma virus and Rauscher mouse leukemia virus. *Nature (London)* **235,** 383.

Lai, M. M. C., and Duesberg, P. H. (1972b). Differences between the envelope glycoproteins and glycopeptides of avian tumor viruses released from transformed and nontransformed cells. *Virology* **50,** 359.

Lampert, P. W., Gibbs, C. J., and Gajdusek, D. C. (1971a). Experimental spongiform encephalopathy (Creutzfeldt-Jakob disease) in chimpanzees. Electron microscopic studies. *J. Neuropathol. Exp. Neurol.* **30,** 24.

Lampert, P. W., Hooks, J., Gibbs, C. J., and Gajdusek, D. C. (1971b). Altered membranes in experimental scrapie. *Acta Neuropathol.* **19,** 81.

Lampson, G. P., Tytell, A. A., Nemes, M. M., and Hilleman, M. R. (1965). Characterization of chick embryo interferon induced by a DNA virus. *Proc. Soc. Exp. Biol. Med.* **118,** 441.

Lampson, G. P., Tytell, A. A., Field, A. K., Nemes, M. M., and Hilleman, M. R. (1967). Inducers of interferon and host resistance, I. double-stranded RNA from extracts of *Penicillium funiculosum. Proc. Nat. Acad. Sci. U.S.* **58,** 782.

Lancaster, M. C., Boulter, E. A., Westwood, J. C. N., and Randles, J. (1966). Experimental respiratory infection with poxviruses. II. Pathological studies. *Brit. J. Exp. Pathol.* **47,** 466.

Landau, B. J., Larson, V. M., Devers, G. A., and Hilleman, M. R. (1966). Studies on induction of virus from adenovirus and SV40 hamster tumors. I. Chemical and physical agents. *Proc. Soc. Exp. Biol. Med.* **122,** 1174.

Lando, D., and Ryhiner, M. L. (1969). Pourvoir infectieux du DN d'Herpesvirus hominis en culture cellulaire. *C. R. Acad. Sci. (Paris)* **269,** 527.

Landsberger, F. R., Lenard, J., Paxton, J., and Compans, R. W. (1971). Spin-label electron spin resonance study of the lipid-containing membrane of influenza virus. *Proc. Nat. Acad. Sci. U.S.* **68,** 2579.

Lane, J. M., Ruben, F. L., Neff, J. M., and Millar, J. D. (1969). Complications of smallpox vaccination, 1968. *New Engl. J. Med.* **281.**

Lane, J. M., Millar, J. D., and Neff, J. M. (1971). Smallpox and smallpox vaccination policy. *Annu. Rev. Med.* **22,** 251.

Larson, V. M., Girardi, A. J., Hilleman, M. R., and Zwickey, R. E. (1965). Studies of oncogenicity of adenovirus type 7 viruses in hamsters. *Proc. Soc. Exp. Biol. Med.* **118,** 15.

Latarjet, R., Cramer, R., and Montagnier, L. (1967). Inactivation by UV-, X-, and gamma-radiation of the infecting and transforming capacities of polyoma virus. *Virology* **33,** 104.

Latham, H., and Darnell, J. E. (1965). Distribution of mRNA in the cytoplasmic polysomes of HeLa cells. *J. Mol. Biol.* **14,** 1.

Laver, W. G. (1970). Isolation of an arginine-rich protein from particles of adenovirus type 2. *Virology* **41,** 488.

Laver, W. G. (1971). Separation of two polypeptide chains from the hemagglutinin subunit of influenza virus. *Virology* **45,** 275.

Laver, W. G. (1973). The polypeptides of influenza viruses. *Advan. Virus Res.,* in press.

Laver, W. G., and Kilbourne, E. D. (1966). Identification in a recombinant influenza virus of structural proteins derived from both parents. *Virology* **30,** 493.

Laver, W. G., and Valentine, R. C. (1969). Morphology of the isolated hemagglutinin and neuraminidase subunits of influenza virus. *Virology* **38,** 105.

Laver, W. G., and Webster, R. G. (1966). The structure of influenza viruses. IV. Chemical studies of the host antigen. *Virology* **30,** 104.

Laver, W. G., and Webster, R. G. (1968). Selection of antigenic mutants of influenza viruses. Isolation and peptide mapping of their hemagglutinating proteins. *Virology* **34**, 193.

Laver, W. G., and Webster, R. G. (1972). Studies on the origin of pandemic influenza. II. Peptide maps of the light and heavy polypeptide chains from the hemagglutinin subunits of A2 influenza viruses isolated before and after the appearance of Hong Kong influenza. *Virology* **48**, 445.

Laver, W. G., and Webster, R. G. (1973). Studies on the origin of pandemic influenza. III. Evidence implicating duck and equine influenza viruses as possible progenitors of the Hong Kong strain of human influenza. *Virology*, **51**, 383.

Laver, W. G., Wrigley, N. G., and Pereira, H. G. (1969). Removal of pentons from particles of adenovirus type 2. *Virology* **39**, 599.

Laver, W. G., Younghusband, H. B., and Wrigley, N. G. (1971). Purification and properties of chick embryo lethal orphan virus (an avian adenovirus). *Virology* **45**, 598.

Lavi, S., and Winocour, E. (1972). Acquisition of sequence homologous to host deoxyribonucleic acid by closed circular simian virus 40 deoxyribonucleic acid. *J. Virol.* **9**, 309.

Law, L. W., and Ting, R. C. (1965). Immunologic competence and induction of neoplasms by polyoma virus. *Proc. Soc. Exp. Biol. Med.* **119**, 823.

Law, L. W., Ting, R. C., and Leckband, E. (1966). The function of the thymus in tumour production by polyoma virus. *Ciba Found. Symp. Thymus: Exp. Clin. Stud.*, p. 214.

Lawrence, H. L., and Landy, M. (eds.) (1969). "Mediators of Cellular Immunity," Brook Lodge Symposium. Academic Press, New York.

Lawrence, W. C. (1971). Evidence for a relationship between equine abortion (Herpes) virus deoxyribonucleic acid synthesis and the S phase of the KB cell mitotic cycle. *J. Virol.* **7**, 736.

Lawrence, W. C., and Ginsberg, H. S. (1967). Intracellular uncoating of type 5 adenovirus deoxyribonucleic acid. *J. Virol.* **1**, 851.

Lazarowitz, S. G., Compans, R. W., and Choppin, P. W. (1971). Influenza virus structural and nonstructural proteins in infected cells and their plasma membranes. *Virology* **46**, 830.

Lazdins, I., and Holmes, I. H. (1973). Protein synthesis in Bunyamwera virus-infected cells. In preparation.

Lazdins, I., Haslam, E. A., and White, D. O. (1972). The polypeptides of influenza virus. VI. Composition of the neuraminidase. *Virology* **49**, 758.

Le Bouvier, G. L. (1971). The heterogeneity of Australia antigen. *J. Infect. Dis.* **123**, 671.

Le Bouvier, G. L., and Plotkin, S. A. (1971). Precipitin responses to rubella vaccine RA27/3. *J. Infect. Dis.* **123**, 220.

Le Bouvier, G. L., Sussman, M., and Crawford, L. V. (1966). Antigenic diversity of mammalian papillomaviruses. *J. Gen. Microbiol.* **45**, 497.

Ledinko, N. (1963). Genetic recombination with poliovirus type 1. Studies of crosses between a normal horse serum-resistant mutant and several guanidine-resistant mutants of the same strain. *Virology* **20**, 107.

Ledinko, N., and Hirst, G. K. (1961). Mixed infection of cells and polioviruses types 1 and 2. *Virology* **14**, 207.

Ledinko, N., and Toolan, H. W. (1970). Relationship between induction of thymidine kinase and potentiation of growth of H-1 virus by human adenovirus 12. *J. Gen. Virol.* **7**, 263.

Ledinko, N., Hopkins, S., and Toolan, H. (1969). Relationship between potentiation of H-1 growth by human adenovirus 12 and inhibition of the "helper" adenovirus by H-1. *J. Gen. Virol.* **5**, 19.

Lee, L. F., Kieff, E. D., Bachenheimer, S. L., Roizman, B., Spear, P. G., Burmester, B. R., and Nazerian, K. (1971). Size and composition of Marek's disease virus deoxyribonucleic acid. *J. Virol.* **7**, 289.

Lee, S. Y., Mendecki, J., and Brawerman, G. (1971). A polynucleotide segment rich in adenylic acide in the rapidly-labeled polyribosomal RNA component of mouse sarcoma 180 ascites cells. *Proc. Nat. Acad. Sci. U.S.* **68**, 1331.

Lehane, D. E., Clark, H. F., and Kargon, D. J. (1968). Antigenic relationships among frog viruses demonstrated by the plaque reduction and neutralization kinetics tests. *Virology* **34**, 590.

Lehmann, N. I., and Gust, I. D. (1971). Viraemia in influenza. A report of two cases. *Med. J. Aust.* **2**, 1166.

Lehmann-Grube, F. (1961). Preparation of cell cultures from human amniotic membranes. *Arch. Ges. Virusforsch.* **11**, 258.

Lehmann-Grube, F. (1971). Lymphocytic choriomeningitis virus. *Virology Monogr.* **10**, 1.

Lehmann-Grube, F., Slenczka, W., and Tees, R. (1969). A persistent and inapparent infection of L-cells with the virus of lymphocytic choriomeningitis. *J. Gen. Virol.* **5**, 63.

Leibowitz, R., and Penman, S. (1971). Regulation of protein synthesis in HeLa cells. III. Inhibition during poliovirus infection .*J. Virol.* **8**, 661.

Leifer, E., Gocke, D. J., and Bourne, H. (1970). Lassa fever, a new virus disease of man from West Africa. II. Report of a laboratory-acquired infection treated with plasma from a person recently recovered from the disease. *Amer. J. Trop. Med. Hyg.* **19**, 677.

Lennette, E. H. (1959). Serologic reactions in viral and rickettsial infections. *In* "Viral and Rickettsial Infections of Man" (T. M. Rivers and F. L. Horsfall, Jr., eds.). 3rd Ed., p. 230. Lippincott, Philadelphia, Pennsylvania.

Lennette, E. H., and Schmidt, N. J. (eds.) (1969). "Diagnostic Procedures for Viral and Rickettsial infections," 4th Ed. Amer. Public Health Ass., New York.

Leong, J-A., Garapin, A-C., Jackson, N., Fanshier, L., Levinson, W., and Bishop, J. M. (1972). Virus-specific ribonucleic acid in cells producing Rous sarcoma virus: detection and characterization. *J. Virol.* **9**, 891.

Levin, D. H., Acs, G., and Silverstein, S. C. (1970a). Chain initiation by reovirus RNA transcriptase *in vitro. Nature (London)* **227**, 603.

Levin, D. H., Mendelsohn, N., Schonberg, M., Klett, H., Silverstein, S., Kapuler, A. M., and Acs, G. (1970b). Properties of RNA transcriptase in reovirus subviral particles. *Proc. Nat. Acad. Sci. U.S.* **66**, 890.

Levin, D. H., Kyner, D., and Acs, G. (1972). Formation of a mammalian initiation complex with reovirus messenger RNA, methionyl-tRNA$_F$ and ribosomal subunits. *Proc. Nat. Acad. Sci. U.S.* **69**, 1234.

Levin, J. G., and Friedman, R. M. (1971). Analysis of arbovirus ribonucleic acid forms by polyacrylamide gel electrophoresis. *J. Virol.* **7**, 504.

Levin, M. J., Oxman, M. N., Diamandopoulos, G. T., Levine, A. S., Henry, P. H., and Enders, J. F. (1969). Virus-specific nucleic acids in SV40 exposed hamster embryo cell lines: correlation with S and T antigens. *Proc. Nat. Acad. Sci U.S.* **62**, 589.

Levin, M. J., Crumpacker, C. S., Lewis, A. M., Oxman, M. N., Henry, P. H., and Rowe, W. P. (1971). Studies of nondefective adenovirus 2-simian virus 40 deoxyribonucleic acid hybrid viruses. II. Relationship of adenovirus 2 deoxyribonucleic acid and simian virus 40 deoxyribonucleic acid in the Ad2$^+$ND1 genome. *J. Virol.* **7**, 343.

Levine, A. J., and Ginsberg, H. S. (1967). Mechanism by which fiber antigen inhibits multiplication of type 5 adenovirus. *J. Virol.* **1**, 747.

Levine, A. J., and Teresky, A. K. (1970). Deoxyribonucleic acid replication in simian virus 40-infected cells. II. Detection and characterization of simian virus 40 pseudovirions. *J. Virol.* **5**, 451.

Levine, A. J., Kang, H. S., and Billheimer, F. E. (1970). DNA replication in SV40 infected cells. I. Analysis of replicating SV40 DNA. *J. Mol. Biol.* **50**, 549.

Levine, A. S., Oxman, M. N., Henry, P. H., Levin, M. J., Diamandopoulos, G. T., and Enders, J. F. (1970). Virus specific deoxyribonucleic acid in simian-virus 40 exposed hamster cells: correlation with S and T antigens. *J. Virol.* **6**, 199.

Levine, M., Becker, Y., Boone, C. W., and Eagle, H. (1965). Contact inhibition, macromolecular synthesis and polyribosomes in cultured human diploid fibroblasts. *Proc. Nat. Acad. Sci. U.S.* **53**, 350.

Levine, P. H., Ablashi, D. V., Berard, C. W., Carbone, P. P., Waggoner, D. E., and Malan, L. (1971). Elevated antibody titers to Epstein-Barr virus in Hodgkin's disease. *Cancer* **27**, 416.

Levine, S., Magee, W. E., Hamilton, R. D., and Miller, O. V. (1967). Effect of interferon on early enzyme and viral DNA synthesis in vaccinia virus infection. *Virology* **32**, 33.

Levinson, W. E., Bishop, J. M., Quintrell, N., and Jackson, J. (1970). Presence of DNA in Rous sarcoma virus. *Nature (London)* **227**, 1023.

Levinthal, J. D., and Pederson, W. (1965). In vitro transformation and immunofluorescence with human adenovirus type 12 in rat and rabbit kidney cells. *Fed. Proc. Fed. Amer. Soc. Exp. Biol.* **24**, 174.

Levinthal, J. D., Jakobovits, M., and Eaton, M. D. (1962). Polyoma disease and tumors in mice: the distribution of viral antigen detected by immunofluorescence. *Virology* **16**, 314.

Levitt, N. H., and Crowell, R. L. (1967). Comparative studies of the regeneration of HeLa cell receptors for poliovirus T1 and coxsackievirus B3. *J. Virol.* **1**, 693.

Levy, H. B. (1963). Effect of actinomysin D in HeLa cell nuclear RNA metabolism. *Proc. Soc. Exp. Biol. Med.* **113**, 886.

Levy, H. B., and Merigan, T. C. (1966). Interferon and unnifected cells. *Proc. Soc. Exp. Biol. Med.* **121**, 53.

Levy, H. B., Baron, S., and Buckler, C. E. (1970). The biochemistry of interferon *In* "The Biochemistry of Viruses" (H. B. Levy, ed.). p. 579. Dekker, New York.

Levy, J. P., Varet, B., Oppenheim, E., and Leclerc, J. C. (1969). Neutralization of Graffi leukemia virus. *Nature (London)* **224**, 606.

Levy, N. L., and Notkins, A. L. (1971). Viral infections and diseases of the endocrine system. *J. Infect. Dis.* **124**, 94.

Lewis, A. M., and Rowe, W. P. (1971). Studies on non-defective adenovirus-simian virus 40 hybrid viruses. I. A newly-characterized simian virus 40 antigen induced by the Ad2+ ND₁ virus. *J. Virol.* **7**, 189.

Lewis, A. M., Baum, S. G., Prigge, K. O., and Rowe, W. P. (1966). Occurrence of adenovirus-SV40 hybrids among monkey kidney cell adapted strains of adenovirus. *Proc. Soc. Exp. Biol. Med.* **122**, 214.

Lewis, A. M., Levin, M. J., Weise W-H., Crumpacker, C. S., and Henry, P. H. (1969). A nondefective (competent) adenovirus SV40 hybrid isolated from the Ad2-SV40 hybrid population. *Proc. Nat. Acad. Sci. U.S.* **63**, 1128.

Li, K-K., and Seto, J. T. (1971). Electron microscope study of ribonucleic acid of myxoviruses. *J. Virol.* **7**, 524.

Libíková, H., ed. (1962). "Biology of Viruses of the Tick-borne Encephalitis Complex." Academic Press, New York.

Lieberman, M., Pascale, A., Schafer, T. W., and Came, P. E. (1972). Effect of antiviral agents in equine abortion virus-infected hamsters. *Antimicrob. Agents Chemother.* **1**, 143.

Liebhaber, H., and Takemoto, K. K. (1963). The basis for the size differences in plaques produced by variants of encephalomyocarditis virus. *Virology* **20**, 559.

Lilly, F. (1972). Mouse leukemia: a model of a multiple-gene disease. *J. Nat. Cancer Inst.* **49**, 927.

Lilly, F., and Pincus, T. (1973). Genetic control of murine viral leukemogenesis. *Adv. Cancer Res.* **17**, 231.

Lindberg, U., and Darnell, J. E. (1970). SV40 specific RNA in the nucleus and polysomes of transformed cells. *Proc. Nat. Acad. Sci. U.S.* **65**, 1089.

Lindstrom, D. M., and Dulbecco, R. (1972). Strand orientation of simian virus 40 transcription in productively infected cells. *Proc. Nat. Acad. Sci. U.S.* **69**, 1517.

Linscott, W. D., and Levinson, W. E. (1969). Complement components required for virus neutralization by early immunoglobulin antibody. *Proc. Nat. Acad. Sci. U.S.* **64**, 520.

Lipton, A., Klinger, I., Paul, D., and Holley, R. W. (1971). Migration of mouse 3T3 fibroblasts in response to a serum factor. *Proc. Nat. Acad. Sci. U.S.* **68**, 2799.

Liske, R., and Franks, D. (1968). Specificity of the agglutinin in extracts of wheat germ. *Nature (London)* **217**, 860.

Littauer, U. Z., Daniel, V., and Sarid, S. (1971). Phage specified transfer RNAs. *In* "Strategy of the Viral Genome" (G. E. Wolstenholme and M. O'Connor, eds.), p. 169. CIBA Foundation Symp. Churchill Livingstone, London.

Littlefield, J. W., and Basilico, C. (1966). Infection of thymidine-kinase deficient BHK cells with polyoma virus. *Nature (London)* **211**, 250.

Liu, C. (1955). Studies on influenza infection in ferrets by means of fluorescein-labeled antibody. I. The pathogenesis and diagnosis of the disease. *J. Exp. Med.* **101**, 665.

Llanes-Rodas, R., and Liu, C. (1965). A study of measles virus infection in tissue culture cells with particular reference to the development of intranuclear inclusion bodies. *J. Immunol.* **95**, 840.

Lockard, R. E., and Lingrel, J. B. (1969). The synthesis of mouse hemoglobin β-chains in a rabbit reticulocyte cell-free system programmed with mouse reticulocyte 9S RNA. *Biochem. Biophys. Res. Commun.* **37**, 204.

Lockart, R. Z. (1964). The necessity for cellular RNA and protein synthesis for viral inhibition resulting from interferon. *Biochem. Biophys. Res. Commun.* **15**, 513.

Lockart, R. Z. (1966). Biological properties of interferons: criteria for acceptance of a viral inhibitor as an interferon. *In* "Interferons" (N. B. Finter, ed.), p. 1. North Holland, Amsterdam.

Loeffler, F., and Frosch, P. (1898). Berichte der Kommission zur Erforschung der Maul-und Klauenseuche bei dem Institut für Infektions krankheiten in Berlin. *Zentr. Bakteriol. Parasitenk. Abt. 1 Orig.* **23**, 371.

Loening, U. W. (1969). The determination of the molecular weight of ribonucleic acid by polyacrylamide-gel electrophoresis. *Biochem. J.* **113**, 131 .

Loh, P. C., and Shatkin, A. J. (1968). Structural proteins of reoviruses. *J. Virol.* **2**, 1353.

Lomniczi, B., and Burke, D. C. (1970). Interferon production by temperature-sensitive mutants of Semliki Forest virus. *J. Gen. Virol.* **8**, 55.

London, W. T., Alter, H. J., Lauder, J., and Purcell, R. H. (1972). Serial transmission in rhesus monkeys of an agent related to hepatitis-associated antigen. *J. Infect. Dis.* **125**, 382.

Long, W. F., and Burke, D. C. (1970). The effect of infection with fowl plague virus on protein synthesis in chick embryo cells. *J. Gen. Virol.* **6**, 1.

Long, W. F., and Burke, D. C. (1971). Interferon production by double-stranded RNA: a comparison of induction by reovirus to that by a synthetic double-stranded polynucleotide. *J. Gen. Virol.* **12**, 1.

Longberg-Holm, K., and Philipson, L. (1969). Early events of virus-cell interaction in an adenovirus system. *J. Virol.* **4**, 323.

Lowy, D. R., Rowe, W. P., Teich, N., and Hartley, J. W. (1971). Murine leukemia virus: high-frequency activation *in vitro* by 5-iododeoxyuridine and 5-bromodeoxyuridine. *Science* **174**, 155.

Lucas-Lenard, J., and Lipmann, F. (1971). Protein biosynthesis. *Annu. Rev. Biochem.* **40**, 409.

Lucké, B. (1938). Carcinoma of the kidney in the leopard frog: the occurrence and significance of metastasis. *Amer. J. Cancer* **34**, 15.

Luftig, R. B., Kilham, S. S., Hay, A. J. Zweerink, H. J., and Joklik, W. K. (1972). An ultrastructural study of virions and cores of reovirus type 3. *Virology* **48**, 170.

Lumley, G. F., and Taylor, F. H. (1943). "Dengue." Service Publ., School of Public Health Trop. Med., No. 3, Commonwealth of Australia.

Luria, S. E. (1951). The frequency distribution of spontaneous bacteriophage mutants as evidence for the exponential rate of phage reproduction. *Cold Spring Harbor Symp. Quant. Biol.* **16**, 463.

Luria, S. E., and Darnell, J. E. (1967). "General Virology," 2nd Ed. Wiley, New York.

Luria, S. E., Williams, R. C., and Backus, R. C. (1951). Electron micrographic counts of bacteriophage particles. *J. Bacteriol.* **61**, 179.

Lwoff, A. (1953). Lysogeny. *Bacteriol. Rev.* **17**, 269.

Lwoff, A. (1957). The concept of virus. *J. Gen. Microbiol.* **17**, 239.

Lwoff, A. (1959). Factors influencing the evolution of viral diseases at the cellular level and in the organism. *Bacteriol. Rev.* **23**, 109.

Lwoff, A. (1961). Mutations affecting neurovirulence. *Proc. 5th Intern. Conf. Polyomyelitis*, p. 13, Lippincott, Philadelphia, Pennsylvania.

Lwoff, A. (1965). The specific effectors of viral development. *Biochem. J.* **96**, 289.

Lwoff, A. (1969). Death and transfiguration of a problem. *Bacteriol. Rev.* **33**, 390.

Lwoff, A., and Tournier, P. (1966). The classification of viruses. *Annu. Rev. Microbiol.* **20**, 45.

Lwoff, A., Anderson, T. F., and Jacob, F. (1959a). Remarques sur les caractéristiques de la particule virale infectieuse. *Ann. Inst. Pasteur (Paris)* **97**, 281.

Lwoff, A., Tournier, P., and Cartreaud, J. P. (1959b). L'influence de l'hyperthermie provoquée sur l'infection poliomyélitique de la souris. *C. R. Acad. Sci. (Paris)* **248**, 1876.

Lwoff, A., Horne, R., and Tournier, P. (1962). A system of viruses. *Cold Spring Harbor Symp. Quant. Biol.* **27**, 51.

Lytle, C. D. (1971). Host-cell reactivation in mammalian cells. I. Survival of ultra-violet-irradiated herpes virus in different cell lines. *Intern. J. Radiat. Biol.* **19**, 329.


Maassab, H. F. (1967). Adaptation and growth characteristics of influenza virus at 25°C. *Nature (London)* **213**, 612.

Maassab, H. F. (1969). Biologic and immunologic characteristics of cold-adapted influenza virus. *J. Immunol.* **102**, 728.

Maassab, H. F., Kendal, A. P., and Davenport, F. M. (1972). Hybrid formation of influenza virus at 25°. *Proc. Soc. Exp. Biol. Med.* **139**, 768.

McAllister, R. M., and Macpherson, I. (1968). Transformation of a hamster cell line of adenovirus type 12. *J. Gen. Virol.* **2**, 99.

McAllister, R. M., Nicholson, M. O., Reed, G., Kern, J., Gilden, R. V., and Huebner, R. J. (1969a). Transformation of rodent cells by adenovirus 19 and other group D adenoviruses. *J. Nat. Cancer Inst.* **43**, 917.

McAllister, R. M., Nicolson, M. O., Lewis, A. M., Jr., Macpherson, I., and Huebner, R. J. (1969b). Transformation of rat embryo cells by adenovirus type 1. *J. Gen. Virol.* **4**, 29.

McAllister, R. M., Filbert, J. E., Nicolson, M. O., Rongey, R. W., Gardner, M. B., Gilden, R. V., and Huebner, R. J. (1971). Transformation and productive infection of human osteosarcoma cells by a feline sarcoma virus. *Nature New Biol.* **230**, 279.

McAllister, R. M., Nicolson, M., Gardner, M. B., Rongey, R. W., Rasheed, S., Sarma, D. S., Huebner, R. J., Hatanaka, M., Oroszlan, S., Gilden, R. V., Kabigting, A., and Vernon, L. (1972a). C-type virus released from human cultured rhabdomyosarcoma cells. *Nature New Biol.* **235**, 3.

McAllister, R. M., Gilden, R. V., and Green, M. (1972b). Adenoviruses in human cancer. *Lancet* **i**, 831.

McAuslan, B. R. (1963). The induction and repression of thymidine kinase in the poxvirus-infected HeLa cell. *Virology* **21**, 383.

McAuslan, B. R. (1969a). The biochemistry of poxvirus replication. *In* "Biochemistry of Viruses" (H. B. Levy, ed.), p. 360. Dekker, New York.

McAuslan, B. R. (1969b). Rifampicin inhibition of vaccinia replication. *Biochem. Biophys, Res. Commun.* **37**, 289.

McAuslan, B. R. (1971). Enzymes specified by DNA-containing animal viruses. *In* "Strategy of the Viral Genome" (G. E. W. Wolstenholme and M. O'Connor, eds.), p. 25, CIBA Foundation Symp. Churchill Livingstone, London.

McAuslan, B. R., and Kates, J. (1966). Regulation of virus-induced deoxyribonucleases. *Proc. Nat. Acad. Sci. U.S.* **55**, 1581.

McAuslan, B. R., and Kates, J. R. (1967). Poxvirus-induced acid deoxyribonuclease: Regulation of synthesis; control of activity *in vivo*; purification and properties of the enzyme. *Virology* **33**, 709.

McAuslan, B. R., and Smith, W. R. (1968). Deoxyribonucleic acid synthesis in FV3 infected mammalian cells. *J. Virol.* **2**, 1006.

McBride, W. D., and Wiener, A. (1964). *In vitro* transformation of hamster kidney cells by human adenovirus type 12. *Proc. Soc. Exp. Biol. Med.* **115**, 870.

McCahon, D., and Schild, D. C. (1971). An investigation of some factors affecting cross-reactivation between influenza A viruses. *J. Gen. Virol.* **12**, 207.

MacCallum, F. O., and Juel-Jensen, B. E. (1966). Herpes simplex virus skin infection in man treated with idoxuridine in dimethyl sulphoxide. Results of double-blind controlled trial. *Brit. Med. J.* **2**, 805.

MacCallum. F. O., and McDonald, J. R. (1957). Survival of variola virus in raw cotton. *Bull. WHO* **16**, 247.

McCarthy, B. J. (1969). The evolution of base sequences in nucleic acids. *In* "Handbook of Molecular Cytology" (A. Lima-di-Faria, ed.), p. 3. North Holland, Amsterdam.

McClain, M. E. (1965). The host range and plaque morphology of rabbitpox virus (RPu⁺) and its U mutants on chick fibroblast PK-2a and L929 cells. *Aust. J. Exp. Biol. Med. Sci.* **43**, 31.

McClain, M. E., and Hackett, A. J. (1959). Biological characteristics of two plaque variants of vesicular exanthema of swine virus, type E54. *Virology* **9**, 577.

McClain, M. E., and Spendlove, R. S. (1966). Multiplicity reactivation of reovirus particles after exposure to ultraviolet light. *J. Bacteriol.* **92**, 1422.

McCloskey, B. P. (1950). The relation of prophylactic innoculations to the onset of polio-myelitis. *Lancet* **i**, 659.

McConkey, E. H., and Hopkins, J. W. (1964). The relationship of the nucleolus to the synthesis of ribosomal RNA in HeLa cells. *Proc. Nat. Acad. Sci. U.S.* **51**, 1197.

McDevitt, H. O., and Benacerraf, B. (1969). Genetic control of specific-immune response. *Advan. Immunol.* **11**, 31.

McDevitt, H. O., and Bodmer, W. F. (1972). Histocompatibility antigens, immune responsiveness and susceptibility to disease. *Amer. J. Med.* **52**, 1.

McDonnell, J. P., and Levintow, L. (1970). Kinetics of appearance of the products of polio-virus-induced RNA polymerase. *Virology* **42**, 999.

McDowell, M. J., and Joklik, W. K. (1971). An *in vitro* protein synthesizing system from mouse L fibroblasts infected with reovirus. *Virology* **45**, 724.

McDowell, M. J., Joklik, W. K., Villa-Komaroff, L., and Lodish, H. J. (1972). Translation of reovirus messenger RNAs synthesized *in vitro* into reovirus polypeptides by several mammalian cell-free-extracts. *Proc. Nat. Acad. Sci. U.S.* **69**, 2649.

Macek, M., Seidel, E. H., Lewis, R. T., Brunschwig, J. P., Wimberley, I., and Benyesh-Melnick, M. (1971). Cytogenetic studies of EB virus-positive and EB virus-negative lymphoblastoid cell lines. *Cancer Res.* **31**, 308.

McFarland, H. F., Griffin, D. E., and Johnson, R. T. (1972). Specificity of the inflammatory response in viral encephalitis. I. Adoptive immunization of immunosuppressed mice infected with Sindbis virus. *J. Exp. Med.* **136**, 216.

McFarlin, D. E., Raff, M. C., Simpson, E., and Nehlsen, S. H. (1971). Scrapie in immunologically deficient mice. *Nature (London)* **233**, 336.

McFerran, J. B., Clarke, J. K., and Connor, T. J. (1971). The size of some mammalian picornaviruses. *J. Gen. Virol.* **10**, 279.

McGavran, M. H., and Easterday, B. C. (1963). Rift valley fever virus hepatitis. Light and electron microscopic studies in the mouse. *Amer. J. Pathol.* **42**, 587.

McGavran, M. H., and Smith, M. G. (1965). Ultrastructural, cytochemical, and microchemical observations on cytomegalovirus (salivary gland virus) infection of human cells in tissue culture. *Exp. Mol. Pathol.* **4**, 1.

McGeoch, D. J., Crawford, L. V., and Follett, E. A. C. (1970). The DNAs of three parvoviruses. *J. Gen. Virol.* **6**, 33.

McGrath, C. M., Nandi, S., and Young, L. (1972). Relationship between organization of mammary tumors and the ability of tumor cells to replicate mammary tumor virus and to recognize growth-inhibitory contact signals *in vitro*. *J. Virol.* **9**, 367.

McGuire, T. C., and Henson, J. B. (1973). Equine infectious anemia—pathogenesis of persistent viral infection. *Perspectives Virol.* **8**, 229.

Machala, O., Donner, L., and Svoboda, J. (1970). A full expression of the genome of Rous sarcoma virus in heterokaryons formed after fusion of virogenic mammalian cells and chicken fibroblasts. *J. Gen. Virol.* **8**, 219.

Mack, T. M. (1972). Smallpox in Europe, 1950–1971. *J. Infect. Dis.* **125**, 161.

Mackaness, G. B. (1971). The induction and expression of cell-mediated hypersensitivity in the lung, *Amer. Rev. Resp. Dis.* **104**, 813.

Mackenzie, J. S. (1969). Virulence of temperature-sensitive mutants of influenza virus. *Brit. Med. J.* **3**, 757.

Mackenzie, J. S. (1970). Isolation of temperature-sensitive mutants and the construction of a preliminary genetic map for influenza virus. *J. Gen. Virol.* **6**, 63.

McKinnel, R. G., and Zambernard, J. (1968). Virus particles in renal tumors obtained from spring *Rana pipiens* of known geographic origin. *Cancer Res.* **28**, 684.

McLean, D. M. (1955). Multiplication of viruses in mosquitoes following feeding and injection into the body cavity. *Aust. J. Exp. Biol. Med. Sci.* **33**, 53.

MacLeod, C. M. (1953). Relation of the incubation period and the secondary immune response to lasting immunity in infectious diseases. *J. Immunol.* **70**, 421.

McNamara, M. J., Pierce, W. C., Crawford, Y. E., and Miller, L. F. (1962). Patterns of adenovirus infection in the respiratory diseases of naval recruits. *Amer. Rev. Resp. Dis.* **86**, 485.

Macpherson, I. (1965). Reversion in hamster cells transformed by Rous sarcoma virus. *Science* **148**, 1731.

Macpherson, I. (1968). Report on the workshop on virus induction by cell association. *Intern. J. Cancer* **3**, 318.

Macpherson, I. (1969). Agar suspension culture for quantitation of transformed cells. *In* "Fundamental Techniques in Virology" (K. Habel and N. P. Salzman, eds.), p. 214. Academic Press, New York.

Macpherson, I. (1970). The characteristics of animal cells transformed *in vitro*. *Advan. Cancer Res.* **13**, 169.

Macpherson, I., and Montagnier, L. (1964). Agär suspension culture for the selective assay of cells transformed by polyoma virus. *Virology* **23**, 291.

Macpherson, I. A., and Stoker, M. (1962). Polyoma transformation of hamster cell clones: an investigation of genetic factors affecting cell competence. *Virology* **16**, 147.

McQueen, J. L., Steele, J. H., and Robinson, R. Q. (1968). Influenza in animals. *Advan. Vet. Sci.* **12**, 285.

McSharry, J. J., and Wagner, R. R. (1971). Lipid composition of purified vesicular stomatitis viruses. *J. Virol.* **7**, 59.

McSharry, J. J., Compans, R. W., and Choppin, P. W. (1971). Proteins of vesicular stomatitis virus and of phenotypically mixed vesicular stomatitis virus–simian virus 5 virions. *J. Virol.* **8**, 722.

Maden, B. E. H. (1971). The structure and formation of ribosomes in animal cells. *Progr. Biophys. Mol. Biol.* **22**, 127.

Maeno, K., and Kilbourne, E. D. (1970). Developmental sequence and intracellular sites of synthesis of three structural protein antigens of influenza A₂ virus. *J. Virol.* **5**, 153.

Maes, R., and Granoff, A. (1967). Viruses and renal carcinoma of *Rana pipiens*. IV. Nucleic acid synthesis in frog virus 3-infected BHK 21/13 cells. *Virology* **33**, 491.

Mahy, B. W. J., Hutchinson, J. E., and Barry, R. D. (1970). Ribonucleic acid polymerase induced in cells infected with Sendai virus. *J. Virol.* **5**, 663.

Mahy, B. W. J., Hastie, N. D., and Armstrong, S. J. (1972). Inhibition of influenza virus replication by α-amanitin: mode of action. *Proc. Nat. Acad. Sci. U.S.* **69**, 1421.

Maizel, J. V., (1971). Polyacrylamide gel electrophoresis of viral proteins. *In* "Methods in Virology" (K. Maramorosch and H. Koprowski, eds.), Vol. V, p. 180. Academic Press, New York.

Maizel, J. V., and Summers, D. F. (1968). Evidence for differences in size and composition of the poliovirus-specific polypeptides in infected HeLa cells. *Virology* **36**, 48.

Maizel, J. V., Phillips, B. A., and Summers, D. F. (1967). Composition of artificially produced and naturally occurring empty capsids of poliovirus type 1. *Virology* **32**, 692.

Maizel, J. V., White, D. O., and Scharff, M. D. (1968a). The polypeptides of adenovirus. I. evidence for multiple protein components in the virion and a comparison of types 2, 7A and 12. *Virology* **36**, 115.

Maizel, J. V., White, D. O., and Scharff, M. D. (1968b). The polypeptides of adenovirus. II. Soluble proteins, cores, top components and the structure of the virion. *Virology* **36**, 126.

Mak, S. (1971). Defective virions in human adenovirus type 12. *J. Virol.* **7**, 426.

Makinodan, T., and Peterson, W. J. (1964). Growth and senescence of the primary antibody-forming potential of the spleen. *J. Immunol.* **93**, 886.

Makinodan, T., Perkins, E. H., and Chen, M. G. (1971). Immunologic activity of the aged. *Advan. Gerontol. Res.* **3**, 171.

Malkova, D. (1960). Participation of the lymphatic and blood circulations in the dissemination of tick-borne encephalitis virus to the organs of experimentally infected mice. *Acta Virol. (Prague) Engl. Ed.* **4**, 290.

Mallucci, L. (1965). Observations on the growth of mouse hepatitis virus (MHV-3) in mouse macrophages. *Virology* **25**, 30.

Mallucci, L. (1966). Effect of chloroquine on lysosomes and on growth of mouse hepatitis virus (MHV-3). *Virology* **28**, 355.

Malmgren, R. A., Rabson, A. S., Carney, P. G., and Paul, F. J. (1966). Immunofluorescence of green monkey kidney cells infected with adenovirus 12 and with adenovirus 12 plus simian virus 40. *J. Bacteriol.* **91**, 262.

Malmgren, R. A., Takemoto, K. K., and Carney, P. G. (1968). Immunofluorescent studies of mouse and hamster cell surface antigens induced by polyoma virus. *J. Nat. Cancer Inst.* **40**, 263.

Manaker, R. A., and Groupé, V. (1956). Discrete foci of altered chicken embryo cells associated with Rous sarcoma virus in tissue culture. *Virology* **2**, 838.

Manaker, R. A., and Landon, J. (1972). Cited in McAllister *et al.* (1972).

Mandel, B. (1967a). The interaction of neutralized poliovirus with HeLa cells. I. Adsorption. *Virology* **31**, 238.

Mandel, B. (1967b). The interaction of neutralized poliovirus with HeLa cells. II. Elution, penetration, uncoating. *Virology* **31**, 248.

Mandel, B. (1967c). The relationship between penetration and uncoating of poliovirus in HeLa cells. *Virology* **31**, 702.

Mandel, B. (1971a). Characterization of type 1 poliovirus by electrophoretic analysis. *Virology* **44**, 554.

Mandel, B. (1971b). Methods for the study of virus-antibody complex. *In* "Methods in Virology" (K. Maramorosch and H. Koprowski, eds.), Vol. V, p. 375. Academic Press, New York.

Mandel, M. (1969). New approaches to bacterial taxonomy: perspective and prospect. *Annu. Rev. Microbiol.* **23**, 239.

Manshaw, J. B. (1970). Developmental abnormalities associated with congenital cytomegalovirus infection. *Advan. Teratol.* **4**, 64.

Manz, H. G., Dinsdale, H. B., and Morrin, F. A. F. (1971) Progressive multifocal leukoencephalopathy after renal transplanation. *Ann. Internal Med.* **75**, 77.

Maral, R. (1957). Étude du development du virus de la myxomatose en culture de tissus. *Ann. Inst. Pasteur (Paris)* **92**, 742.

Maramorosch, K., and Koprowski, H., (eds.) (1967–1971). "Methods in Virology," Vols. I–V. Academic Press, New York.

Marchalonis, J. J. (1969). An enzymic method for the trace iodination of immunoglobulins and other proteins. *Biochem. J.* **113**, 299.

Marcus, P. I., and Carver, D. H. (1965). Hemadsorption-negative plaque tests: New assay for rubella virus revealing a unique interference. *Science* **149**, 986.

Marcus, P. I., and Carver, D. H. (1967). Intrinsic interference: A new type of viral interference. *J. Virol.* **1**, 334.

Marcus, P. I., and Robbins, E. (1963). Viral inhibition in the metaphase-arrest cell. *Proc. Nat. Acad. Sci. U.S.* **50**, 1156.

Marcus, P. I., and Salb, J. M. (1966). Molecular basis of interferon action: Inhibition of viral RNA translation. *Virology* **30**, 502.

Marcus, P. I., and Salb, J. M. (1968). On the translation of inhibitory protein of interferon action. In "The Interferons" (G. Rita, ed.), p. 111. Academic Press, New York.

Marcus, P. I., Engelhardt, D. L., Hunt, J. M., and Sekellick, M. J. (1971). Interferon action: inhibition of vesicular stomatitis virus RNA synthesis induced by virion-bound polymerase. *Science* **174**, 593.

Marek, J. (1907). Multiple Nervenentzündung (Polyneuritis) bei Hühnern. *Deut. Tierärztl. Wochenschr.* **15**, 417.

Margolis, G., and Kilham, L. (1965). Rat virus, an agent with an affinity for dividing cells. *Nat. Inst. Neurol. Dis. Blindness, Monogr. No.* **2**, 361.

Margolis, G., and Kilham, L. (1968). Virus-induced cerebellar hypoplasia. *Res. Publ. Ass. Nervous Mental Dis.* **44**, 113.

Margolis, G., and Kilham, L. (1969). Hydrocephalus in hamsters, ferrets, rats and mice following inoculations with reovirus type I. *Lab. Invest.* **21**, 189.

Marin, G. (1971). Segregation of morphological revertants in polyoma-transformed hybrid clones of hamster fibroblasts. *J. Cell. Sci.* **9**, 61.

Marin, G., and Littlefield, J. W. (1968). Selection of morphologically normal cell lines from polyoma-transformed BHK 21/13 hamster fibroblasts. *J. Virol.* **2**, 69.

Marin, G., and Macpherson, I. (1969). Reversion in polyoma-transformed cells; studies on retransformation, induced antigens and tumorigenicity. *J. Virol.* **3**, 146.

Mark, G. E., and Kaplan, A. A. (1971). Synthesis of proteins in cells infected with herpesvirus. VII. Lack of migration of structural viral proteins to the nucleus of arginine-deprived cells. *Virology* **45**, 53.

Marmion, B. P., and Tonkin, R. W. (1972). Control of hepatitis in dialysis units. *Brit. Med. Bull.* **28**, 169.

Marshall, I. D. (1959). The influence of ambient temperature on the course of myxomatosis in rabbits. *J. Hyg.* **57**, 484.

Marshall, I. D. (1973). Viruses and diptera. In "Viruses and Invertebrates" (A. J. Gibbs, ed.). North Holland, Amsterdam, in press.

Marshall, I. D., and Fenner, F. (1960). Studies on the epidemiology of infectious myxomatosis of rabbits. VII. The virulence of strains of myxoma virus recovered from Australian wild rabbits between 1951 and 1959. *J. Hyg.* **58**, 485.

Martin, E. M. (1969). Studies on the RNA polymerase of some temperature-sensitive mutants of Semliki Forest virus. *Virology* **39**, 107.

Martin, E. M., and Kerr, I. M. (1968). Virus-induced changes in host-cell macromolecular synthesis. *Symp. Soc. Gen. Microbiol.* **18**, 15.

Martin, E. M., and Sonnabend, J. A. (1967). Ribonucleic acid polymerase catalyzing synthesis of double-stranded arbovirus ribonucleic acid. *J. Virol.* **1**, 97.

Martin, E. M., Malec, J., Sved, S., and Work, T. S. (1961). Studies on protein and nucleic acid metabolism in virus-infected mammalian cells. I. Encephalomyocarditis virus in Krebs 11 mouse-ascites-tumour cells. *Biochem. J.* **80**, 585.

Martin, G. S. (1970). Rous sarcoma virus: a function required for the maintenance of the transformed state. *Nature (London)* **227**, 1021.

Martin, G. S., and Burger, M. M. (1972). Agglutination of cells transformed by Rous sarcoma virus by wheat germ agglutinin and concanavalin A. *Nature New Biol.* **237**, 9.

Martin, G. S., and Duesberg, P. H. (1972). The a subunit in the RNA of transforming avian tumor viruses. I. Occurrence in different virus strains. II. Spontaneous loss resulting in non-transforming variants. *Virology* **47**, 494.

Martin, G. S., Venuta, S., Weber, M., and Rubin, H. (1971). Temperature-dependent alterations in sugar transport in cells infected by a temperature-sensitive mutant of Rous sarcoma virus. *Proc. Nat. Acad. Sci. U.S.* **68**, 2739.

Martin, M. A. (1970). Characteristics of SV40 DNA transcription during lytic infection, abortive infection and in transformed mouse cells. *Cold Spring Harbor Symp. Quant. Biol.* **35**, 833.

Martin, M. A., and Axelrod, D. (1969a). Polyoma virus gene activity during lytic infection and in transformed animal cells. *Science* **164**, 68.

Martin, M. A., and Axelrod, D. (1969b). SV40 gene activity during lytic infection and in a series of SV40 transformed mouse cells. *Proc. Nat. Acad. Sci. U.S.* **64,** 1203.

Martin, T. E., and Wool, I. G. (1968). Formation of active hybrids from subunits of muscle ribosomes from normal and diabetic rats. *Proc. Nat. Acad. Sci. U.S.* **60,** 569.

Massie, H. R., Samis, H. V., and Baird, M. B. (1972). The effects of the buffer HEPES on the division potential of WI-38 cells. *In Vitro* **7,** 191.

Masurel, N. (1969). Serological characteristics of a "new" serotype of influenza A virus: the Hong Kong strain. *Bull. WHO* **41,** 461.

Mathews, M. B., and Korner, A. (1970). Mammalian cell-free protein synthesis directed by viral ribonucleic acid. *Europ. J. Biochem.* **17,** 328.

Mathews, M. B., Osborn, M., and Lingrel, J. B. (1971). Translation of globin messenger RNA in a heterologous cell free system. *Nature New Biol.* **233,** 206.

Mathews, M. B., Osborn, M., Berns, A. J. M., and Bloemendal, H. (1972). Translation of two messenger RNAs from lens in a cell-free system from Krebs II ascites cells. *Nature New Biol.* **236,** 5.

Matsumoto, S. (1958). Electron microscope studies of ectromelia replication. *Ann. Rep. Inst. Virus Res., Kyoto Univ. Ser.* **A1,** 151.

Matsumoto, S. (1970). Rabies virus. *Advan. Virus Res.* **16,** 257.

Matsumoto, S., and Kawai, A. (1969). Comparative studies on development of rabies virus in different host cells. *Virology* **39,** 449.

Matsumoto, S., Stanfield, F. J., Goore, M. Y., and Haff, R. F. (1972). The antiviral activity of a triazinoindole (SK and F 30097). *Proc. Soc. Exp. Biol. Med.* **139,** 455.

Matsumura, T., Stollar, V., and Schlesinger, R. W. (1971). Studies on the nature of dengue viruses. V. Structure and development of dengue virus in Vero cells. *Virology* **46,** 344.

Mattern, C. F. T., Takemoto, K. K., and Daniel, W. A. (1966). Replication of polyoma virus in mouse embryo cells: Electron microscopic observations. *Virology* **30,** 242.

Mattern, C. F. T., Takemoto, K. K., and DeLeva, A. M. (1967). Electron microscopic observations on multiple polyoma virus-related particles. *Virology* **32,** 378.

Maurer, B. A., Imamura, T., and Wilbert, S. M. (1970). Incidence of EB virus-containing cells in primary and secondary clones of several Burkitt lymphoma cell lines. *Cancer Res.* **30,** 2870.

Mautner, V., and Pereira, H. G. (1971). Crystallization of a second adenovirus protein (the fibre). *Nature (London)* **230,** 456.

Maycock, W. d'A, (1972). Hepatitis in transfusion services. *Brit. Med. Bull.* **28,** 163.

Mayor, H. D., and Jordan, L. E. (1965). Studies on reovirus. I. Morphologic observations on the development of reovirus in tissue culture. *Exp. Mol. Pathol.* **4,** 40.

Mayor, H. D., and Melnick, J. L. (1966). Small deoxyribonucleic acid-containing viruses (Picodnavirus group). *Nature (London)* **210,** 331.

Mayor, H. D., Jamison, R. M., Jordan, L. E., and van Mitchell, M. (1965). Reoviruses. II. Structure and composition of the virion. *J. Bacteriol.* **89,** 1548.

Mazzone, H. M. (1967). Equilibrium centrifugation. *In* "Methods in Virology" (K. Maramorosch and H. Koprowski, eds.), Vol. II, p. 41. Academic Press, New York.

Mécs, E., Sonnabend, J. A., Martin, E. M., and Fantes, E. H. (1967). The effect of interferon on the synthesis of RNA in chick cells infected with Semliki Forest virus. *J. Gen. Virol.* **1,** 25.

Medappa, K. C., McClean, C., and Rueckert, R. R. (1971). On the structure of rhinovirus IA. *Virology* **44,** 259.

Medawar, P. B. (1961). Immunological tolerance. *Science* **133,** 303.

Medill-Brown, M., and Briody, B. A. (1955). Mutation and selection pressure during adaptation of influenza virus to mice. *Virology* **1,** 301.

Medzon, E. L., and Bauer, H. (1970). Structural features of vaccinia virus revealed by negative staining, sectioning and freeze-etching. *Virology* **40,** 860.

Meléndez, L. V., Daniel, M. D., Hunt, R. D., Fraser, C. E. O., Garcia, F. G., King, N. W., and Williamson, M. E. (1970) Herpesvirus saimiri: V. Further evidence to consider this virus as the etiological agent of a lethal disease in primates which resembles a malignant lymphoma. *J. Nat. Cancer Inst.* **44,** 1175.

Melli, M., and Bishop, J. O. (1969). Hybridization between rat liver DNA and complementary RNA. *J. Mol. Biol.* **40,** 117.

Mellors, R. C. (1969). Murine leukemialike virus and the immunopathological disorders of New Zealand Black mice. *J. Infect. Dis.* **120,** 480.

Mellors, R. C., and Korngold, I. (1963). The cellular origin of human immunoglobulins (γ2, γIM, γIA), *J. Exp. Med.* **118,** 387.

Melnick, J. L. (1962). Papova virus group. *Science* **135,** 1128.

Melnick, J. L. (1971). Poliomyelitis vaccine: present status, suggested use, desirable developments. In "International Conference on the Application of Vaccines against Viral, Rickettsial and Bacterial Diseases of Man," p. 171. Sci. Publ. No. 226. Pan American Health Organ., Washington, D.C.

Melnick, J. L., Crowther, D., and Barrera-Oro, J. (1961). Rapid development of drug-resistant mutants of poliovirus. *Science.* **134,** 557.

Melnick, J. L., Butel, J. S., and Tevethia, S. S. (1971). Cell transformation by viruses. I. Significance of virus-specific antigens induced by deoxyribonucleic acid containing tumor viruses. *In Vitro* **6,** 335.

Mendelson, C. G., and Kligman, A. M. (1961). Isolation of wart virus in tissue culture. Successful reinoculation into humans. *Arch. Dermatol.* **83,** 559.

Merigan, T. C. (1964). Purified interferon: Physical properties and species specificity. *Science* **145,** 811.

Merigan, T. C. (1967). Induction of circulating interferon by synthetic polymers of known composition. *Nature (London)* **214,** 416.

Merigan, T. C. (1971). Interferon induction in mouse and man with viral and nonviral stimuli In "Ciba Foundation Symposium on Interferon" (G. E. Wolstenholme and M. O'Connor, eds.), p. 50. Churchill, London.

Merigan, T. C., and Regelson, W. (1967). Interferon induction in man by a synthetic polyanion of defined composition. *New Engl. J. Med.* **277,** 1283.

Merigan, T. C., Winget, C. A., and Dixon, C. B. (1965). Purification and characterization of vertebrate interferons. *J. Mol. Biol.* **13,** 679.

Merigan, T. C., Reed, S. E., Hall, T. S., and Tyrrell, D. A. J. (1973). Inhibition of respiratory virus infection by locally applied interferon. *Lancet* **1,** 563.

Merril, C. R., Geier, M. R., and Petricciani, J. C. (1971). Bacterial virus gene expression in human cells. *Nature (London)* **233,** 398.

Metcalf, D. (1966). "The Thymus." Springer, Berlin.

Metzgar, R. S., and Oleinick, S. R. (1968). The study of normal and malignant cell antigens by mixed agglutination. *Cancer Res.* **28,** 1366.

Meyer, H. M., Jr., Johnson, R. T., Crawford, I. P., Dascomb, H. E., and Rogers, N. G. (1960). Central nervous system syndromes of "viral" etiology. A study of 713 cases. *Amer. J. Med.* **29,** 334.

Meyer, H. M., Parkman, P. D., Hobbins, T. E., Larson, H. E., Davis, W. J., Simsarian, J. P., and Hopps, H. E. (1969). Attenuated rubella vaccines. Laboratory and clinical characteristics. *Amer. J. Dis. Child.* **118,** 155.

Michel, M. R., Hirt, B., and Weil, R. (1967). Mouse cellular DNA enclosed in polyoma viral capsids (pseudovirions). *Proc. Nat. Acad. Sci. U.S.* **58,** 1381.

Mietens, C., Hummeler, K., and Henle, W. (1964). Recall of complement-fixing antibodies to enteroviruses in guinea pigs. *J. Immunol.* **92,** 17.

Miles, C. P., and O'Neill, F. (1967). Chromosome studies of 8 *in vitro* lines of Burkitt's lymphoma. *Cancer Res.* **27,** 392.

Miller, D. L. (1970). The prophylactic and therapeutic uses of immunoglobulin in virus infections. *Mod. Trends Med. Virol.* **2,** 284.

Miller, J. F. A. P. (1965). The thymus and transplantation immunity. *Brit. Med. Bull.* **21,** 111.

Miller, J. F. A. P. (1966). Immunity in the foetus and the new-born. *Brit. Med. Bull.* **22,** 21.

Miller, M. E., and Schieken, R. M. (1967). Thymic dysplasia: a separable entity from "Swiss agammaglogulinemia." *Amer. J. Med. Sci.* **253,** 741.

Miller, O. L., Beatty, B. R., Hamkalo, B. A., and Thomas, C. A. (1970). Electron microscopic visualization of transcription. *Cold Spring Harbor Symp. Quant. Biol.* **35,** 505.

Mills, D. R., Pace, N. R., and Spiegelman, S. (1966). The *in vitro* synthesis of a non-infectious complex containing biologically active viral RNA. *Proc. Nat. Acad. Sci. U.S.* **56,** 1778.

Mills, J., and Chanock, R. M. (1971). Temperature-sensitive mutants of influenza virus. I. Behavior in tissue culture and experimental animals. *J. Infect. Dis.* **123,** 145.

Millward, S., and Graham, A. F. (1970). Structural studies on reovirus: discontinuities in the genome. *Proc. Nat. Acad. Sci. U.S.* **65,** 422.

Millward, S., and Graham, A. F. (1971). Structure and transcription of the genomes of double-stranded RNA viruses. *In* "Comparative Virology" (K. Maramorosch and E. Kurstak, eds.), p. 389. Academic Press, New York.

Mims, C. A. (1957). Rift Valley fever virus in mice. VI. Histological changes in the liver in reltion to virus multiplication. *Aust. J. Exp. Biol. Med. Sci.* **35,** 595.

Mims, C. A. (1959a). The response of mice to large intravenous injections of ectromelia virus. I. The fate of injected virus. *Brit. J. Exp. Pathol.* **40,** 533.

Mims, C. A. (1959b). The response of mice to large intravenous injections of ectromelia virus. II. The growth of virus in the liver. *Brit. J. Exp. Pathol.* **40,** 543.

Mims, C. A. (1960a). An analysis of the toxicity for mice of influenza virus. I. Intracerebral toxicity. *Brit. J. Exp. Pathol.* **41,** 586.

Mims, C. A. (1960b). An analysis of the toxicity for mice of influenza virus. II. Intravenous toxicity. *Brit. J. Exp. Pathol.* **41,** 593.

Mims, C. A. (1960c). Intracerebral injections and the growth of viruses in the mouse brain. *Brit. J. Exp. Pathol.* **41,** 52.

Mims, C. A. (1964). Aspects of the pathogenesis of virus diseases. *Bacteriol. Rev.* **28,** 30.

Mims, C. A. (1966a). Immunofluorescent study of the carrier state and mechanism of vertical transmission in lymphocytic choriomeningitis virus infection in mice. *J. Pathol. Bacteriol.* **91,** 395.

Mims, C. A. (1966b). The pathogenesis of rashes in virus diseases. *Bacteriol. Rev.* **30,** 739.

Mims, C. A. (1968a). Pathogenesis of viral infection of the fetus. *Progr. Med. Virol.* **10,** 194.

Mims, C. A. (1968b). The response of mice to the intravenous injection of cowpox virus. *Brit. J. Exp. Pathol.* **49,** 24.

Mims, C. A. (1969). Effect on the fetus of maternal infection with lymphocytic choriomeningitis (LCM) virus. *J. Infect. Dis.* **120,** 582.

Mims, C. A., and Blanden, R. V. (1972). Antiviral activity of immune lymphocytes in lymphocytic choriomeningitis virus infection of mice. *Infection Immunity* **6,** 695.

Mims, C. A., and Murphy, F. A. (1973). Parainfluenza virus Sendai infection in macrophages, ependyma, choroid plexus, vascular endothelium and respiratory tract of mice. *Amer. J. Pathol.,* **70,** 315.

Mims, C. A., and Subrahmanyan, T. P. (1966). Immunofluorescent study of the mechanism of resistance to superinfection in mice carrying the lymphocytic choriomeningitis virus. *J. Pathol. Bacteriol.* **91,** 403.

Mims, C. A., and Tosolini, F. A. (1969). Pathogenesis of lesions in lymphoid tissue of mice infected with lymphocytic choriomeningitis (LCM) virus. *Brit. J. Exp. Pathol.* **50,** 584.

Mintz, B., and Silvers, W. K. (1967). "Intrinsic" immunological tolerance in allophenic mice. *Science* **158,** 1484.

Mitchiner, M. B. (1969). The envelope of vaccinia and orf viruses: an electron-cytochemical investigation. *J. Gen. Virol.* **5,** 211.

Mitchison, N. A. (1964). Induction of immunological paralysis in two zones of dosage. *Proc. Roy. Soc.* **B161,** 275.

Miyamoto, K., and Gilden, R. V. (1971). Electron microscopic studies of tumor viruses. I. Entry of murine leukemia virus into mouse embryo fibroblasts. *J. Virol.* **7,** 395.

Miyamoto, K., and Morgan, C. (1971). Structure and development of viruses as observed in the electron microscope. XI. Entry and uncoating of herpes simplex virus. *J. Virol.* **8,** 910.

Miyamoto, T., Hinuma, Y., and Ishida, N. (1965). Intracellular transfer of hemadsorption type 2 virus antigen during persistent infection of HeLa cell cultures. *Virology* **27,** 28.

Mizell, M., Stackpole, C. W., and Helperen, S. (1968). Herpes-type virus recovery from "virus-free" frog kidney tumors. *Proc. Soc. Exp. Biol. Med.* **127,** 808.

Mizell, M., Toplin, J., and Isaacs, J. J. (1969). Tumor induction in developing frog kidneys by a zonal centrifuge purified fraction of the frog herpes-type virus *Science* **165,** 1134.

Mizutani, S., and Temin, H. M. (1970). An RNA-dependent DNA polymerase in virions of Rous sarcoma virus. *Cold Spring Harbor Symp. Quant. Biol.* **35,** 847.

Mizutani, S., and Temin, H. M. (1971). Enzymes and nucleotides in virions of Rous sarcoma virus. *J. Virol.* **8,** 409.

Mizutani, S., Boettiger, D., and Temin, H. M. (1970). A DNA-dependent DNA polymerase and a DNA endonuclease in virions of Rous sercoma virus. *Nature (London)* **228,** 424.

Mizutani, S., Temin, H. M., Kodama, M., and Wells, R. D. (1971). DNA ligase and exonuclease activities in virions of Rous sarcoma virus. *Nature New Biol.* **230,** 232.

Mölling, K., Bolognesi, D. P., and Bauer, H. (1971a). Polypeptides of avian RNA tumor viruses. III. Purification and identification of a DNA synthesizing enzyme. *Virology* **45,** 298.

Mölling, K., Bolognesi, D. P., Bauer, H., Busen, W., Plassmann, H. W., and Hausen, P. (1971b). Association of viral reverse transcriptase with an enzyme degrading the RNA moiety of RNA-DNA hybrids. *Nature New Biol.* **234,** 240.

Moloney, J. B. (1960). Biological studies on a lymphoid-leukemia virus extracted from sarcoma 137. I. Origin and introductory investigations. *J. Nat. Cancer Inst.* **24,** 933.

Moloney, J. B. (1964). The rodent leukemias: Virus-induced murine leukemias. *Annu. Rev. Med.* **15,** 383.

Moloney, J. B. (1966). A virus-induced rhabdomyosarcoma of mice. *Nat. Cancer Inst. Monogr.* **22,** 139.

Mommaerts, A. B., Eckert, E. A., Beard, D., Sharp, D. G., and Sharp, J. W. (1952). Dephosphorylation of adenosine triphosphate by concentrates of the virus of avian erythromyeloblastic leucosis. *Proc. Soc. Exp. Biol. Med.* **79,** 450.

Monograph (1966). Conference on murine leukemia. *Nat. Cancer Inst. Monogr. No.* 22.

Montagnier, L. (1968). The replication of viral RNA. *Symp. Soc. Gen. Microbiol.* **18,** 125.

Montagnier, L. (1969). Cancérologie—necessité de précurseurs puriques pour l'expression d'un caractère de la transformation chez les cellules BHK21/13. *C. R. Acad. Sci. (Paris)* **263,** 2218.

Montagnier, L., and Sanders, F. K. (1963). Replicative form of encephalomyocarditis virus ribonucleic acid. *Nature (London)* **199,** 664.

Monto, A. S., Davenport, F. M., Napier, J. A., and Francis, T. (1970). Modification of an outbreak of influenza in Tecumseh, Michigan by vaccination of school children. *J. Infect. Dis.* **122,** 16.

Monto, A. S., Napier, J. A., and Metzner, H. L. (1971). The Tecumseh study of respiratory illness. I. Plan of study and observations on syndromes of acute respiratory disease. *Amer. J. Epidemiol.* **94,** 269.

Moore, D. H., Charney, J., and Kramarsky, B., Lasfargues, E. Y., Sarkar, N. H., Brennan, M. J., Burrows, J. H., Sirsat, S. M., Paymaster, J. C., and Varidya, A. B. (1971). Search for a human breast cancer virus. *Nature (London)* **229,** 611.

Moore, E. G., and Temin, H. M. (1971). Lack of correlation between conversion by RNA tumor viruses and increased agglutinability of cells by concanavalin A and wheat germ agglutinin. *Nature (London)* **231,** 117.

Moore, N. F., Lomniczi, B., and Burke, D. C. (1972). The effect of infection with different strains of Newcastle disease virus on cellular RNA and protein synthesis. *J. Gen. Virol.* **14,** 99.

Mora, P. T., Brady, R. O., Bradley, R. M., and McFarland, V. W. (1969). Gangliosides in DNA virus-transformed and spontaneously transformed tumorigenic mouse cell lines. *Proc. Nat. Acad. Sci. U.S.* **63,** 1290.

Mora, P. T., Cumar, F. A., and Brady, R. O. (1971). A common biochemical change in SV40 and polyoma virus transformed mouse cells coupled to control of cell growth in culture. *Virology* **46,** 60.

Morgan, C., and Howe, C. (1968). Structure and development of viruses as observed in the electron microscope. IX. Entry of parainfluenza 1 (Sendai) virus. *J. Virol.* **2,** 1122.

Morgan, C., and Rose, H. M. (1967). The application of thin sectioning. In "Methods in Virology" (K. Maramorosch and H. Koprowski, eds.), Vol. III, p. 576. Academic Press, New York.

Morgan, C., and Rose, H. M. (1968). Structure and development of viruses as observed in the electron microscope. VIII. Entry of influenza virus. J. Virol. 2, 925.

Morgan, C., Rose, H. M., Holden, M., and Jones, E. P. (1959). Electron microscopic observations on the development of herpes simplex virus. J. Exp. Med. 110, 643.

Morgan, C., Howe, C., and Rose, H. M. (1961). Structure and development of viruses as observed in the electron microscope. V. Western equine encephalomyelitis virus. J. Exp. Med. 113, 219.

Morgan, C. Rosenkranz, H. S., and Mednis, B. (1969). Structure and development of viruses as observed in the electron microscope. X. Entry and uncoating of adenovirus. J. Virol. 4, 777.

Morgan, H. R. (1964). Antibodies for Rous sarcoma virus in fowl animal and human populations of East Africa. I. Antibodies in domestic chickens and wild fowl. J. Nat. Cancer Inst. 35, 1043.

Morgan, H. R. (1965). In vitro cell-virus interactions. Nat. Cancer Inst. Monogr. 17, 392.

Morita, K. (1960). Electron microscopic study of inclusions of cowpox. Biken's J. 3, 213.

Morley, D. C. (1967). Measles and measles vaccine in preindustrial countries. Mod. Trends in Med. Virol. 1, 141.

Moroni, C. (1972). Structural proteins of Rauscher leukemia virus and Harvey sarcoma virus. Virology 47, 1.

Morris, J. A., and Gajdusek, D. C. (1963). Encephalopathy in mice following inoculation of scrapie sheep brain. Nature (London) 197, 1084.

Morris, J. A., Gajdusek, D. C., and Gibbs, C. J., Jr. (1965). Spread of scrapie from inoculated to uninoculated mice. Proc. Soc. Exp. Biol. Med. 120, 108.

Morris, N. R., and Fisher, G. A. (1963). Studies concerning the inhibition of cellular reproduction by deoxyribonucleosides. I. Inhibition of the synthesis of deoxycytidine by a phosphorylated derivative of thymidine. Biochim. Biophys. Acta 68, 84.

Morrison, J. M., Keir, H. M., Subak-Sharpe, J., and Crawford, L. V. (1967). Nearest neighbour base sequence analysis of the deoxyribonucleic acids of a further three mammalian viruses; Simian Virus 40, human papilloma virus and adenovirus type 2. J. Gen. Virol. 1, 101.

Morrow, J., and Berg, P. (1972). Cleavage of simian virus 40 DNA at a unique site by a bacterial restriction enzyme. Proc. Nat. Acad. Sci. U.S. 69, 3365.

Moscovici, C., Moscovici, M. G., and Zanetti, M. (1969). Transformation of chick fibroblast cultures with avian myeloblastosis virus. J. Cell. Physiol. 73, 105.

Moses, E., and Kohn, A. (1963). Polykaryocytosis induced by Rous sarcoma virus in chick fibroblasts. Exp. Cell. Res. 32, 182.

Moss, B. (1968). Inhibition of HeLa cell protein synthesis by the vaccinia virion. J. Virol. 2, 1028.

Moss, B., and Salzman, N. P. (1968). Sequential protein synthesis following vaccinia virus infection. J. Virol. 2, 1016.

Moss, B., Katz, E., and Rosenblum, E. N. (1969). Vaccinia virus directed RNA and protein synthesis in the presence of rifampicin. Biochem. Biophys. Res. Commun. 36, 85.

Moss, B., Rosenblum, E. N., and Grimley, P. M. (1971a). Assembly of vaccinia virus particles from polypeptides made in the presence of rifampicin. Virology 45, 123.

Moss, B., Rosenblum, E. N., and Grimley, P. M. (1971b). Assembly of virus particles during mixed infection with wild-type vaccinia and a rifampicin-resistant mutant. Virology 45, 135.

Moss, B., Rosenblum, E. N., and Garon, C. F. (1971c). Glycoprotein synthesis in cells infected with vaccinia virus. I. Non-virion glycoproteins. Virology 46, 221.

Mosser, A. G., Caliguiri, L. A., and Tamm, I. (1971). Blocking of guanidine action on poliovirus multiplication. Virology 45, 653.

Mosser, A. G., Caliguiri, L. A., and Tamm, I. (1972). Incorporation of lipid precursors into cytoplasmic membranes of poliovirus-infected HeLa cells. Virology 47, 39.

Mountcastle, W. E., Compans, R. W., Caliguiri, L. A., and Choppin, P. W. (1970). Nucleo-capsid protein subunits of simian virus 5, Newcastle disease virus, and Sendai virus. *J. Virol.* **6**, 677.

Mountcastle, W. E., Compans, R. W., and Choppin, P. W. (1971). Proteins and glycoproteins of paramyxoviruses: a comparison of simian virus 5, Newcastle disease virus and Sendai virus. *J. Virol.* **7**, 47.

Mudd, J. A., and Summers, D. F. (1970a). Protein synthesis in vesicular stomatitis virus-infected HeLa cells. *Virology* **42**, 328.

Mudd, J. A., and Summers, D. F. (1970b). Polysomal ribonucleic acid of vesicular stomatitis virus-infected HeLa cells. *Virology* **42**, 958.

Mühlbock, O. (1971). The contribution of experimental cancer research to the understanding of human disease. In "RNA Viruses and Host Genome in Oncogenesis" (P. Emmelot and P. Bentvelzen, eds.), p. 339, North Holland, Amsterdam.

Mühlbock, O., and Bentvelzen, O. (1968). The transmission of the mammary tumor viruses. *Perspect. Virol.* **6**, 75.

Müller-Eberhard, H. J. (1968). Chemistry and reaction mechanisms of complement. *Advan. Immunol.* **8**, 1.

Müntefering, H., Schmidt, W. A. K., Körber, W. (1971). Zur Virusgenese des Diabetes mellitus bei der weissen Maus. *Deut. Med. Wochenschr.* **96**, 693.

Mukerjee, D., Bowen, J., and Anderson, D. E. (1970). Simian papovavirus 40 transformation of cells from cancer patient with XY/XXY mosaic Klinefelter's syndrome. *Cancer Res.* **30**, 1769.

Munyon, W., Paoletti, E., Ospina, J., and Grace, J. T. (1968). Nucleotide phosphohydrolase in purified vaccinia virus. *J. Virol.* **2**, 167.

Munyon, W., Buchsbaum, R., Paoletti, E., Mann, J., Kraiselburd, E., and Davis, D. (1971). Electrophoresis of thymidine kinase activity synthesized by cells transformed by herpes simplex virus. *Virology* **49**, 683.

Murakami, W. T., Fine, R., Harrington, M. R., and Ben Hassan, Z. (1968). Properties and amino acid composition of polyoma virus purified by zonal ultracentrifugation. *J. Mol. Biol.* **36**, 153.

Murphy, B. R., Chalhub, E. G., Nusinoff, S. R., and Chanock, R. M. (1972). Temperature-sensitive mutants of influenza virus. II. Attenuation of *ts* recombinants for man. *J. Infect. Dis.* **126**, 170.

Murphy, F. A., Harrison, A. K., and Tzianabos, T. (1968a). Electron microscopic observations of mouse brain infected with Bunyamwera group arboviruses. *J. Virol.* **2**, 1315.

Murphy, F. A., Whitfield, S. G., Coleman, P. H., Calisher, C. H., Rabin, E. R., Jenson, A. B., Melnick, J. L., Edwards, M. E., and Whitney, E. (1968b). California group arboviruses: electron microscopic studies. *Exp. Mol. Pathol.* **9**, 44.

Murphy, F. A., Webb, P. A., Johnson, K. M., and Whitfield, S. G. (1969). Morphological comparison of Machupo with lymphocytic choriomeningitis virus: Basis for a new taxonomic group. *J. Virol.* **4**, 535.

Murphy, F. A., Webb, P. A., Johnson, K. M., Whitfield, S. G., and Chappell, W. A. (1970). Arenoviruses in Vero cells; Ultrastructural studies. *J. Virol.* **6**, 507.

Murphy, F. A., Borden, E. C., Shope, R. E., and Harrison, A. (1971a). Physicochemical and morphological relationships of some arthropod-borne viruses to bluetongue virus–A new taxonomic group. Electron microscopic studies. *J. Gen. Virol.* **13**, 273.

Murphy, F. A., Simpson, D. I. H., Whitfield, S. G., Zlotnick, I., and Carter, G. B. (1971b). Marburg virus infection in monkeys. Ultrastructural studies. *Lab. Invest.* **24**, 279.

Murphy, F. A., Taylor, W. P., Mims, C. A., and Marshall, I. D. (1973). The pathogenesis of Ross River virus infection in mice. II. Muscle, heart and brown fat lesions. *J. Infect. Dis.* **127**, 129.

Murray, M. D. (1970). The spread of ephemeral fever of cattle during the 1967–68 epizootic in Australia. *Aust. Vet. J.* **46**, 77.

Murray, R. F., Hobbs, J., and Payne, B. (1971). Possible clonal origin of common warts (verruca vulgaris). *Nature (London)* **232**, 50.

Murray, R. K., and Temin, H. M. (1970). Carcinogenesis by RNA sarcoma viruses. XIV. Infection of stationery cultures with murine sarcoma virus (Harvey). *Intern. J. Cancer* **5**, 320.

Mussgay, M. (1959). Gewinnung einer pH-stabilen Variante des Virus der Maul und Klauenseuche und ihre Vermehrung in Gewebekulturen mit saurem Medium. *Zentr. Bakteriol. Parasitenk. Abt. I Orig.* **175**, 183.

Myers, K. M., Marshall, I. D., and Fenner, F. (1954). Studies in the epidemiology of infectious myxomatosis of rabbits. III. Observations on two succeeding epizootics in Australian wild rabbits on the Riverine plain of south-eastern Australia, 1951–1963. *J. Hyg.* **52**, 337.


Nadkarni, J. S., Nadkarni, J. J., Klein, G., Henle, W., Henle, G., and Clifford, P. (1970). EB viral antigens in Burkitt tumor biopsies and early cultures. *Intern. J. Cancer* **6**, 10.

Naegele, R. F., and Granoff, A. (1971). Viruses and renal carcinoma of *Rana pipiens*. XI. Isolation of frog virus 3 temperature-sensitive mutants; complementation and genetic recombination. *Virology* **44**, 286.

Naegele, R. F., and Rapp, F. (1967). Enhancement of the replication of human adenoviruses in simian cells by simian adenovirus SV15. *J. Virol.* **1**, 838.

Naeye, R. L., and Blanc, W. (1965). Pathogenesis of congenital rubella. *J. Amer. Med. Ass.* **194**, 1277.

Naficy, K. (1963). Human influenza virus infection with proved viremia. *New Engl. J. Med.* **269**, 964.

Naficy, K., and Nategh, R. (1972). Measles vaccine and its use in developing countries. *Advan. Virus Res.* **17**, 279.

Nagai, Y., Maeno, K., Iinuma, M., Yoshida, T., and Matsumoto, T. (1972). Inhibition of virus growth by ouabain: effect of ouabain on the growth of HVJ in chick embryo cells. *J. Virol.* **9**, 234.

Nagano, Y., and Kojima, Y. (1958). Inhibition de l'infection vaccinale par un facteur liquide dans le tissu infecté par le virus homologue. *C. R. Soc. Biol.* **152**, 1627.

Nagington, J., and Horne, R. W. (1962). Morphological studies of orf and vaccinia viruses. *Virology* **16**, 248.

Nagington, J., Newton, A. A., and Horne, R. W. (1964). The structure of orf virus. *Virology* **23**, 461.

Nahmias, A. J. (1971a). Exogenous reinfection with herpes-simplex virus. *New Engl. J. Med.* **285**, 236.

Nahmias, A. J. (1971b). Herpesviruses from fish to man—a search for pathobiological unity. *In* "Pathobiology" (H. Ioachim, ed.), Vol. 2, Appleton Century Croft, New York.

Nahmias, A. J., and Dowdle, W. R. (1968). Antigenic and biological differences in Herpesvirus hominis. *Progr. Med. Virol.* **10**, 110.

Nahmias, A. J., Griffith, D., Salsbury, C., and Yoshida, K. (1967). Thymic aplasia with lymphopenia, plasma cells, and normal immunoglobulins: relation to measles virus infection. *J. Amer. Med. Ass.* **201**, 792.

Nahmias, A. J., Josey, W. E., Naib, Z. M., Luce, C. F., and Guest, B. A. (1970). Antibodies to *Herpesvirus hominis* types 1 and 2 in humans. II. Women with cervical cancer. *Amer. J. Epidemiol.* **91**, 547.

Nakai, T., and Howatson, A. F. (1968). The fine structure of vesicular stomatitis virus. *Virology* **35**, 268.

Nakata, Y., and Bader, J. P. (1968). Studies on the fixation and development of cellular transformation by Rous sarcoma virus. *Virology* **36**, 401.

Nasemann, T. H., and Bauer, E. (1957). Elektronmikroskopische untersuchungen am paravaccine virus (v. Pirquet). *Klin. Wochenschr.* **35**, 62.

Nass, M. M. K. (1969a). Mitochondrial DNA: Advances, problems, and goals. *Science* **165**, 25.

Nass, M. M. K., (1969b). Mitochondrial DNA. II. Structure and physicochemical properties of isolated DNA. *J. Mol. Biol.* **42**, 529.

Nass, M. M. K., and Buck, C. A. (1970). Studies on mitochondrial tRNA from animal cells. II. Hybridization of aminoacyl-tRNA from rat liver mitochondria with heavy and light complementary strands of mitochondrial DNA. *J. Mol. Biol.* **54**, 187.

Nathanson, N., and Bodian, D. (1961). Experimental poliomyelitis following intramuscular virus infection. I. The effect of neural block on a neurotropic and a pantropic strain. *Bull. Johns Hopkins Hosp.* **108**, 308.

Nathanson, N., and Bodian, D. (1962). Experimental poliomyelitis following intramuscular virus infection. II. The effect of passive antibody on paralysis and viremia. *Bull. Johns Hopkins Hosp.* **111**, 198.

Nathanson, N., and Cole, G. A. (1970). Immunosuppression and experimental virus infection of the nervous system. *Advan. Virus Res.* **16**, 397.

National Cancer Institute (1968). Cell cultures for virus vaccine production. *Nat. Cancer Inst. Monogr. No.* **29**.

Nayak, D. P., and Baluda, M. A. (1968). An intermediate in the replication of influenza virus RNA. *Proc. Nat. Acad. Sci. U.S.* **59**, 184.

Nazarian, K., and Witter, R. L. (1970). Cell-free transmission and *in vivo* replication of Marek's disease virus. *J. Virol.* **5**, 388.

Neefe, J. S., Stokes, J., and Reinhold, J. G. (1945). Oral administration to volunteers of feces from patients with homologous serum hepatitis and infectious (epidemic) hepatitis. *Amer. J. Med. Sci.* **210**, 29.

Neff, J. M., and Enders, J. F. (1968). Further observations on replication of type 1 poliovirus in naturally resistant fused cell cultures. *Perspect. Virol.* **6**, 39.

Nemes, M. M., Tytell, A. A., Lampson, G. P., Field, A. K., and Hilleman, M. R. (1969). Inducers of interferon and host resistance. VI. Antiviral efficacy of poly I:C in animal models. *Proc. Soc. Exp. Biol. Med.* **132**, 776.

Nermut, M. V., and Frank, H. (1971) Fine structure of influenza A2 (Singapore) as revealed by negative staining, freeze-drying and freeze-etching. *J. Gen. Virol.* **10**, 37.

Nermut, M. V., Frank, H., and Schäfer, W. (1972). Properties of mouse leukemia viruses III. Electron microscopical appearance as revealed after conventional preparation techniques as well as freeze drying and freeze etching. *Virology* **49**, 345.

Neurath, A. R. (1964). Separation of a haemolysin from myxoviruses and its possible relationship to normal chorioallantoic membrane cells. *Acta Virol. (Prague) Engl. Ed.* **8**, 154.

Neurath, A. R., and Rubin, B. A. (1971). Viral structural components as immunogens of prophylactic value. *Monogr. Virol.* **4**, 1.

Neutra, M., and Leblond, C. P. (1969). The Golgi apparatus. *Sci. Amer.* **220**, 100.

Newberne, P. M. (1966). Overnutrition on resistance of dogs to distemper virus. *Fed. Proc. Fed. Amer. Soc. Exp. Biol.* **25**, 1701.

Newman, J. F. E., Rowlands, D. J., and Brown, F. (1973). A physico-chemical sub-grouping of the mammalian picornaviruses. *J. Gen. Virol.* **18**, 171.

Newton, L. G., and Wheatley, C. H. (1970). The occurrence and spread of ephemeral fever of cattle in Queensland. *Aust. Vet. J.* **46**, 561.

Nichols, J. L., Bellamy, A. R., and Joklik, W. K. (1972a). Identification of the nucleotide sequences of the oligonucleotides present in reovirions. *Virology* **49**, 562.

Nichols, J. L., Hay, A. J., and Joklik, W. K. (1972b). 5'-Terminal nucleotide sequence of reovirus mRNA synthesized *in vitro*. *Nature New Biol.* **235**, 105.

Nichols, W. W. (1970). Virus-induced chromosome abnormalities. *Annu. Rev. Microbiol.* **24**, 479.

Nicholson, M. A., and McAllister, R. M. (1972). Infectivity of human adenovirus-1 DNA. *Virology* **48**, 14.

Nicolson, G. L. (1971). Difference in topology of normal and tumor cell membranes shown by different surface distribution of ferritin-conjugated concanavalin A. *Nature New Biol.* **233**, 244.

Niederman, J. C., McCollum, R. W., Henle, G., and Henle, W. (1968). Infectious mononucleosis. Clinical manifestations in relation to EB virus antibodies. *J. Amer. Med. Ass.* **203**, 205.

Nir, Y., Beemer, A., and Goldwasser, R. A. (1965). West Nile virus infection in mice following exposure to a virus aerosol. *Brit. J. Exp. Pathol.* **46**, 443.

Niven, J. S. F., Armstrong, J. A., Andrewes, C. H., Pereira, H. G., and Valentine, R. C. (1961). Subcutaneous "growths" in monkeys produced by a poxvirus. *J. Pathol. Bacteriol.* **81**, 1.

Nomura, S., Bassin, R. H., Turner, W., Haapala, D. K., and Fischinger, P. J. (1972). Ultraviolet inactivation of Moloney leukaemia virus: relative target size required for virus replication and rescue of "defective" murine sarcoma virus. *J. Gen. Virol.* **14**, 213.

Nonoyama, M., and Pagano, J. (1971). Detection of Epstein-Barr viral genome in non-productive cells. *Nature New Biol.* **233**, 103.

Nonoyama, M., Watanabe, Y., and Graham, A. F. (1970). Defective virions of reovirus. *J. Virol.* **6**, 226.

Nordlund, J. J., Wolff, S. M., and Levy, H. B. (1970). Inhibition of biologic activity of poly I:poly C by human plasma. *Proc. Soc. Exp. Biol. Med.* **133**, 439.

Norrby, E. C. J. (1965). Characteristics of the progeny derived from multiplication of Sendai virus in a measles virus carrier cell line. *Arch. Ges. Virusforsch.* **17**, 436.

Norrby, E. (1969). The structural and functional diversity of adenovirus capsid components. *J. Gen. Virol.* **5**, 221.

Norrby, E. (1971). Adenoviruses. *In* "Comparative Virology" (K. Maramorosch and E. Kurstak, eds.), p. 105. Academic Press, New York.

Norrby, E. C. J., and Falksveden, L. G. (1964). Some general properties of the measles virus hemolysin. *Arch. Ges. Virusforsch.* **14**, 474.

Norrby, E., and Gollmer, Y. (1971). Mosaics of capsid components produced by cocultivation of certain human adenoviruses *in vitro*. *Virology* **44**, 383.

Nossal, G. J. V., and Ada, G. L. (1964). Recognition of foreignness in immune and tolerant animals. *Nature (London)* **201**, 580.

Nossal, G. J. V., Ada, G. L., and Austin, C. M. (1963). Behaviour of active bacterial antigens during the induction of the immune response. II. Cellular distribution of flagellar antigens labelled with iodine-131. *Nature (London)* **199**, 1259.

Notkins, A. L. (1965). Lactic dehydrogenase virus. *Bacteriol. Rev.* **29**, 143.

Notkins, A. L. (1971). Enzymatic and immunologic alterations in mice infected with lactic dehydrogenase virus. *Amer. J. Pathol.* **64**, 733.

Notkins, A. L., Mahar, S., Scheele, C., and Goffman, J, (1966). Infectious virus-antibody complex in the blood of chronically infected mice. *J. Exp. Med.* **124**, 81.

Notkins, A. L., Mage, M., Ashe, W. K., and Mahar, S. (1968). Neutralization of sensitized lactic dehydrogenase virus by anti-γ-globulin. *J. Immunol.* **100**, 314.

Notkins, A. L., Mergenhagen, S. E., and Howard, R. J. (1970). Effect of virus infections on the function of the immune system. *Annu. Rev. Microbiol.* **24**, 525.

Nowinski, R. C., Old, L. J., Sarkar, N. H., and Moore, D. H. (1970). Common properties of the oncogenic RNA viruses (oncornaviruses). *Virology* **42**, 1152.

Nowinski, R. C., Fleissner, E., Sarkar, N. H., and Aoki, T. (1972). Chromatographic separation and antigenic analysis of proteins of oncornaviruses. II. Mammalian leukemia-sarcoma viruses. *J. Virol.* **9**, 359.

Nowinski, R. C., Fleissner, E., and Sarkar, N. H. (1973). Structural and serological aspects of the oncornaviruses. *Perspect. Virol.* **8**, 31.

Noyes, W. F. (1959). Studies on the Shope rabbit papilloma virus. II. The location of infective virus in papillomas of the cottontail rabbit. *J. Exp. Med.* **109**, 423.

Noyes, W. F. (1965). Studies on the human wart virus II. Changes in primary human cell cultures. *Virology* **25**, 358.

Noyes, W. F., and Mellors, R. C. (1957). Fluorescent antibody detection of the antigens of the Shope papilloma virus in papillomas of the wild and domestic rabbit. *J. Exp. Med.* **106**, 555.

Nurnberger, J. I., and Gordon, M. W. (1957). The cell density of neural tissue: Direct counting method and possible applications as a biologic reference. *In* "Ultrastructure and Cellular Chemistry of Neural Tissue" (H. Waelsch, ed.), p. 100. Harper & Row (Hoeber), New York.


Öberg, B., and Philipson, L. (1971). Replicative structures of poliovirus RNA *in vivo J. Mol. Biol.* **58**, 725.

Obert, G., Tripier, F., and Guir, J. (1971). Arginine requirement for late mRNA transcription of vaccinia virus in KB cells. *Biochem. Biophys. Res. Commun.* **44**, 362.

Obukh, I. B., and Kryukova, I. N. (1969). Malignant and transforming activity of Rous sarcoma virus. II. The study of variants of Rous sarcoma virus isolated from mouse tumors. *Intern. J. Cancer* **4**, 809.

O'Callaghan, D. J., Cheevers, W. P., Gentry, G. A., and Randall, C. C. (1968). Kinetics of cellular and viral DNA synthesis in equine abortion (herpes) virus infection of L-M cells. *Virology* **36**, 104.

O'Connor, T. E., and Fischinger, P. J. (1968). Titration patterns of a murine sarcome-leukemia virus complex; evidence for existence of competent sarcoma virions. *Science* **159**, 325.

O'Conor, G. T., Rabson, A. S., Berezesky, I. K., and Paul, F. J. (1963). Mixed infection with simian virus 40 and adenovirus 12. *J. Nat. Cancer Inst.* **31**, 903.

O'Conor, G. T., Rabson, A. S., Malmgren, R. A., Berezesky, I. K., and Paul, F. J. (1965a). Morphologic observations of green monkey kidney cells after single and double infection with adenovirus 12 and simian virus 40. *J. Nat. Cancer Inst.* **34**, 679.

O'Conor, G. T., Rappaport, H., and Smith, E. B. (1965b). Childhood lymphoma resembling "Burkitt tumor" in the United States. *Cancer* **18**, 411.

Oda, K., and Dulbecco, R. (1968). Regulation of transcription of the SV40 DNA in productively infected and transformed cells. *Proc. Nat. Acad. Sci. U.S.* **60**, 525.

Oda, K-I., and Joklik, W. K. (1967). Hybridization and sedimentation studies on "early" and "late" vaccinia messenger RNA. *J. Mol. Biol.* **27**, 395.

Ogawa, M. (1965). Experimental study on alimentary infection with poliovirus and local immunity of intestines. I. Immunofluorescent study on the sites of primary multiplication of poliovirus following alimentary infection in monkeys. *Sapparo Med. J.* **28**, 300.

Oglesby, A. S., Schaffer, F. L., and Madin, S. H. (1971). Biochemical and biophysical properties of vesicular exanthem of swine virus. *Virology* **44**, 329.

Ogra, P. L. (1971a). The secretory immunoglobulin system of the gastrointestinal tract. *In* "The Secretory Immunologic System" (D. H. Dayton, P. A. Small, R. M. Chanock, H. E. Kaufman, and T. B. Tomasi, eds.), p. 259. National Inst. Child Health and Human Development, National Institutes of Health, Bethesda, Maryland.

Ogra, P. L. (1971b). Effect of tonsillectomy and adenoidectomy on nasopharyngeal antibody response to poliovirus. *New Engl. J. Med.* **284**, 59.

Ogra, P. L., and Karzon, D. T. (1971). Formation and function of poliovirus antibody in different tissues. *Progr. Med. Virol.* **13**, 156.

Ogra, P. L., Karzon, D. T., Rightland, F., and MacGillivray, M. (1968). Immunoglobulin response in serum and secretions after immunization with live and inactivated poliovaccine and natural infection. *New Engl. J. Med.* **270**, 893.

Ohno, S., and Nozima, T. (1966). Mode of action of interferon on the early stages of vaccinia virus multiplication in chick embryo fibroblasts. *Acta Virol.* **10**, 310.

Okada, Y. (1962). Analysis of giant polynuclear cell formation caused by HVJ virus from Ehrlich's ascites tumor cells. I. Microscopic observation of giant polynuclear cell formation. *Exp. Cell Res.* **26**, 98.

Okada, Y., and Tadokoro, J. (1962). Analysis of giant polynuclear cell formation caused by HVJ virus from Ehrlich's ascites tumor cells. II. Quantitative analysis of giant polynuclear cell formation. *Exp. Cell. Res.* **26**, 168.

Okano, H., and Rich, M. A. (1969). Time cycle for intracellular synthesis of murine leukaemia virus. *Nature (London)* **224**, 77.

Oki, T., Fujiwara, Y., and Heidelberger, C. (1971). Utilization of host-cell DNA by vaccinia virus replicating in HeLa cells irradiated intranuclearly with tritium. *J. Gen. Virol.* **13**, 401.

Old, L. J., Boyse, E. A., and Stockert, E. (1963). Antigen properties of experimental leukemias. I. Serological studies in vitro with spontaneous and radiation-induced leukemias. *J. Nat. Cancer Inst.* **31**, 977.

Old, L. J., Boyse, E. A., Geering, G., and Oettgen, H. F. (1968). Serologic approaches to the study of cancer in animals and in man. *Cancer Res.* **28**, 1288.

Oldstone, M. B. A., and Dixon, F. J. (1967). Lymphocytic choriomeningitis: production of antibody by "tolerant" infected mice. *Science* **158**, 1193.

Oldstone, M. B. A., and Dixon, F. J. (1969). Pathogenesis of chronic disease associated with persistent lymphocytic choriomeningitis viral infection. I. Relationship to antibody production in neonatally infected mice. *J. Exp. Med.* **129**, 483.

Oldstone, M. B. A., and Dixon, F. J. (1970). Persistent lymphocytic choriomeningitis viral infection. III. Virus-antiviral antibody complexes and associated chronic disease following transplacental infection. *J. Immunol.* **105**, 829.

Oldstone, M. B. A., and Dixon, F. J. (1971a). Lactic dehydrogenase virus-induced immune complex type of glomerulonephritis. *J. Immunol.* **106**, 1260.

Oldstone, M. B. A., and Dixon, F. J. (1971b). Immune complex disease in chronic viral infections. *J. Exp. Med.* **134**, 32s.

Oldstone, M. A., Aoki, T., and Dixon, F. J. (1972). The antibody response of mice to murine leukemia virus in spontaneous infection: Absence of classical immunological tolerance. *Proc. Nat. Acad. Sci. U.S.* **69**, 134.

Olshevsky, U., and Becker, Y. (1970). Herpes simplex virus structural proteins. *Virology* **40**, 948.

Olshevsky, U., Levitt, J., and Becker, Y. (1967). Studies on the synthesis of herpes simplex virions. *Virology* **33**, 323.

Oroszlan, S., and Rich, M. A. (1964). Human wart virus: *in vitro* cultivation. *Science* **146**, 531.

Oroszlan, S., Hatanaka, M., Gilden, R. V., and Huebner, R. J. (1971a). Specific inhibition of mammalian C-type virus deoxyribonucleic acid polymerases by rat antisera. *J. Virol.* **8**, 816.

Oroszlan, S., Foreman, C., Kelloff, G., and Gilden, R. V. (1971b). The group-specific antigen and other structural proteins of hamster and mouse C-type viruses. *Virology* **43**, 665.

Orth, G., Vielle, F., and Chanjeux, J. P. (1967). On the arginase of the Shope papilloma. *Virology* **31**, 729.

Orth, G., Jeanteur, P., and Croissant, O. (1971). Evidence for and localization of vegetative viral DNA replication by autoradiographic detection of RNA-DNA hybrids in sections of tumors induced by Shope papilloma virus. *Proc. Nat. Acad. Sci. U.S.* **68**, 1876.

Osborn, J. E., and Walker, D. L. (1969). The role of individual spleen cells in the interferon response of the intact mouse. *Proc. Nat. Acad. Sci. U.S.* **62**, 1038.

Osburn, B. I., Johnson, R. T., Silverstein, A. M., Prendergast, R. A., Jochim, M. M., and Levy, S. E. (1971). Experimental viral-induced congenital encephalopathies. II. The pathogenesis of bluetongue vaccine virus infection in fetal lambs. *Lab. Invest.* **25**, 206.

Oshiro, L. S., Schieble, J. H., and Lennette, E. H. (1971). Electron microscopic studies of Corona-virus. *J. Gen. Virol.* **12**, 161.

O'Sullivan, D. G., Pantic, D., Dane, D. S., and Briggs, M. (1969). Protective action of benzimidazole derivatives against virus infections in tissue culture and *in vivo*. *Lancet* **i**, 446.

Ouchterlony, O. (1964). Gel diffusion techniques. In "Symposium on Immunological Methods" (J. F. Ackroyd, ed.), p. 55. Blackwell, Oxford.

Overman, J. R. (1954). Antibody response of suckling mice to mumps virus. II. Relation of onset of antibody production to susceptibility to mumps virus infection. *J. Immunol.* **73**, 249.

Overman, J. R., and Kilham, L. (1953). The inter-relation of age, immune response, and susceptibility to mumps virus in hamsters. *J. Immunol.* **71**, 352.

Oxford, J. S., Logan, I. S., and Potter, C. W. (1970). *In vivo* selection of an influenza A2 strain resistant to amantadine. *Nature (London)* **226**, 82.

Oxman, M. N., and Black, P. H. (1966). Inhibition of SV40 T antigen formation by interferon. *Proc. Nat. Acad. Sci. U.S.* **55**, 1133.

Oxman, M. N., and Levin, M. J. (1971). Interferon and transcription of early virus-specific RNA in cells infected with simian virus 40. *Proc. Nat. Acad. Sci. U.S.* **68**, 299.

Oxman, M. N., Baron, S., Black, P. H., Takemoto, K. K., Habel, K., and Rowe, W. P. (1967). The effect of interferon on SV40 T antigen production in SV40 transformed cells. *Virology* **32**, 122.

Oxman, M. N., Levine, A. S., Crumpacker, C. S., Levin, M. J., Henry, P. H., and Lewis, A. M. (1971). Studies of nondefective adenovirus 2-simian virus 40 hybrid viruses. IV. Characterization of the simian virus 40 ribonucleic acid induced by wild-type simian virus 40 and by the nondefective hybrid virus Ad2⁺ND1. *J. Virol.* **8**, 215.

Ozanne, B., and Sambrook, J. (1971a). Binding of radioactively labelled concanavalin A and wheat germ agglutinin to normal and virus-transformed cells. *Nature New Biol.* **232**, 156.

Ozanne, B., and Sambrook, J. (1971b). Isolation of lines of cells resistant to agglutination by concanavalin A from 3T3 cells transformed by SV40. *In* "Lepetit Colloquia on Biology and Medicine, Vol. 2: The Biology of Oncogenic Viruses" (L. Sylvestri, ed.), p. 248. North Holland, Amsterdam.

Ozanne, B., Sharp, P. A., and Sambrook, J. (1973). Transcription of simian virus 40. II. Viral RNA in different lines of transformed cells. *J. Virol.* **12**, in press.

Ozer, H. L., Takemoto, K. K., Kirschstein, R. L., and Axelrod, D. (1969). Immunochemical characterization of plaque mutants of simian virus 40. *J. Virol.* **3**, 17.

Padgett, B. L., and Tomkins, J. K. (1968). Conditional lethal mutants of rabbitpox virus. 3. Temperature-sensitive (*ts*) mutants; physiological properties, complementation and recombination. *Virology* **36**, 161.

Padgett, B. L., and Walker, D. L. (1970). Effect of persistent fibroma virus infection on susceptibility of cells to other viruses. *J. Virol.* **5**, 199.

Padgett, B. L., Walker, D. L., Zu Rheim, G. M., Eckroade, R. J., and Dessel, B. H. (1971). Cultivation of papova-like virus from human brain with progressive multifocal leukoencephalopathy. *Lancet* **i**, 1257.

Pagano, J. S. (1970). Biologic activity of isolated viral nucleic acids. *Progr. Med. Virol.* **12**, 1.

Pagano, J., and Hutchison, C. A. III (1971). Small, circular, viral DNA: Preparation and analysis of SV40 and ØX 174 DNA. *In* "Methods in Virology" (K. Maramorosch and H. Koprowski), Vol. V, p. 79. Academic Press, New York.

Paine, T. F. (1964). Latent herpes simplex infection in man. *Bacteriol. Rev.* **28**, 472.

Pan American Health Organization (1971). *Proc. Intern. Conf. Application Vaccines Against Viral Rickettsial Bacterial Dis. Man, Sci. Publ. No. 226.* Washington, D.C.

Paniker, C. K. J. (1968). Serological relationships between the neuraminidases of influenza viruses. *J. Gen. Virol.* **2**, 385.

Paniker, C. K., and Nair, C. M. (1970). Infection with A2 Hong Kong influenza virus in domestic cats. *Bull. WHO* **43**, 859.

Pankey, G. A. (1965). Effects of viruses on the cardiovascular system. *Amer. J. Med. Sci.* **250**, 193.

Panum, P. L. (1847). Beobachtungen über das Maserncontagium. *Arch. Pathol. Anat. Physiol. Virchows* **1**, 492.

Pappenheimer, A. M., Kunz, L. J., and Richardson, S. (1951). Passage of coxsackie virus (Connecticut-S strain) in adult mice with production of pancreatic disease. *J. Exp. Med.* **94**, 45.

Pardue, M. L., and Gall, J. G. (1970). Chromosomal localization of mouse satellite DNA. *Science* **168**, 1358.

Parish, C. R. (1972). The relationship between humoral and cell-mediated immunity. *Transplant. Rev.* **13**, 35.

Park, J. H., and Baron, S. (1968). Herpetic keratoconjunctivitis: therapy with synthetic double-stranded RNA. *Science* **162**, 811.

Parker, R. F., and Thompson, R. L. (1942). The effect of external temperature on the course of infectious myxomatosis of rabbits. *J. Exp. Med.* **75**, 567.

Parkman, P. D., and Meyer, H. M. (1969). Prospects for a rubella virus vaccine. *Progr. Med. Virol.* **11**, 80.

Parkman, P. D., Buescher, E. L., and Artenstein, M. S. (1962). Recovery of rubella virus from army recruits. *Proc. Soc. Exp. Biol. Med.* **111**, 225.

Parkman, R., Levy, J. A., and Ting, R. C. (1970). Murine sarcoma virus: the question of defectiveness. *Science* **168**, 387.

Parks, W. P., and Todaro, G. J. (1972). Biological properties of syncytium-forming ("foamy") viruses. *Virology* **47,** 673.

Parks, W. P., Melnick, J. L., Rongey, R., and Mayor, H. D. (1967). Physical assay and growth cycle studies of a defective adeno-satellite virus. *J. Virol.* **1,** 171.

Parks, W. P., Scolnick, E. M., Ross, J., Todaro, G. J., and Aaronson, S. A. (1972). Immunological relationships of reverse transcriptases from ribonucleic acid tumor viruses. *J. Virol.* **9,** 110.

Parrott, D. M. V., and de Sousa, M. (1971). Thymus-dependent populations and thymus-independent populations: origin, migratory patterns and lifespan. *Clin. Exp. Immunol.* **8,** 663.

Parry, H. B. (1962). Scrapie: a transmissible and hereditary disease of sheep. *Heredity* **17,** 75.

Parsons, J. T., and Green, M. (1971). Biochemical studies on adenovirus multiplication. XVIII. Resolution of early virus-specific RNA species in Ad 2 infected and transformed cells. *Virology* **45,** 154.

Parsons, J. T., Gardner, J., and Green, M. (1971). Biochemical studies on adenovirus multiplication. XIX. Resolution of late viral RNA species in the nucleus and cytoplasm. *Proc. Nat. Acad. Sci. U.S.* **68,** 557.

Parsons, R., Bynoe, M. L., Pereira, M. S., and Tyrrell, D. A. J. (1960). Inoculation of human volunteers with strains of Coe virus isolated in Britain. *Brit. Med. J.* **1,** 1776.

Passen, S., and Schultz, R. B. (1965). Use of the Shope papilloma virus-induced arginase as a biochemical marker *in vitro*. *Virology* **26,** 122.

Pasternak, G. (1969). Antigens induced by the mouse leukemia viruses. *Advan. Cancer Res.* **12,** 1.

Pasteur, L., Joubert, and Chamberland (1878). Sur le charbon des poules. *C. R. Acad. Sci. (Paris)* **87,** 47.

Patil, S. R., Merrick, S., and Lubs, H. A. (1971). Identification of each human chromosome with a modified Giemsa strain. *Science* **173,** 821.

Pattison, I. H. (1965a). Resistance of the scrapie agent to formalin. *J. Comp. Pathol. Therap.* **75,** 159.

Pattison, I. H. (1965b). Experiments with scrapie with special reference to the nature of the agent and the pathology of the disease. *Nat. Inst. Neurol. Dis. Blindness Monogr. No.* **2,** 249.

Paucker, K. (1965). The serological specificity of interferon. *J. Immunol.* **94,** 371.

Paul, J. (1972). "Cell and Tissue Culture," 4th Ed. E. & S. Livingstone, Edinburgh.

Paul, J., and Gilmour, R. S. (1968). Organ-specific restriction of transcription in mammalian chromatin. *J. Mol. Biol.* **34,** 305.

Paul, J. H., and Freese, H. L. (1933). An epidemiological and bacteriological study of the "common cold" in an isolated Arctic community (Spitsbergen). *Amer. J. Hyg.* **17,** 517.

Paul, J. R. (1958). Endemic and epidemic trends of poliomyelitis in Central and South America. *Bull. WHO* **19,** 747.

Paul, J. R., Riordan, J. T., and Melnick, J. L. (1951). Antibodies to three different antigenic types of poliomyelitis virus in sera from North Alaskan eskimos. *Amer. J. Hyg.* **54,** 275.

Paul, J. R., Horstmann, D. M., and Niederman, J. C. (1959). Immunity in poliomyelitis infection: Observations in experimental epidemiology. *In* "Immunity and Virus Infection" V. A. Najjar, ed.), p. 233. Wiley, New York.

Paul, J. R., Horstmann, D. M., Riordan, J. T., Opton, E. M., Niederman, J. C., Isacson, E. P., and Green, R. H. (1962). An oral poliomyelitis vaccine trial in Costa Rica. *Bull. WHO* **26,** 311.

Payne, F. E., Solomon, J. J., and Purchase, H. G. (1966). Immunofluorescent studies of group-specific antigen of the avian sarcoma-leukosis viruses. *Proc. Nat. Acad. Sci. U.S.* **55,** 341.

Payne, F. E., Baublis, J. V., and Itabashi, H. H. (1969). Isolation of measles virus from cell cultures of brain from a patient with subacute sclerosing panencephalitis. *New Engl. J. Med.* **281,** 585.

Payne, L. N., and Chubb, R. C. (1968). Studies on the nature and genetic control of an antigen in normal chick embryo which reacts in the COFAL test. *J. Gen. Virol.* **3,** 379.

Payne, L. N., Pani, P. K., and Weiss, R. A. (1971). A dominant epistatic gene which inhibits cellular susceptibility to RSV (RAV-0). *J. Gen. Virol.* **13**, 455.

Pearson, G., and Freeman, G. (1968). Evidence suggesting a relationship between polyoma-virus-induced transplantation antigen and normal embryonic antigen. *Cancer Res.* **28**, 1665.

Pearson, G. D., and Hanawalt, P. C. (1971). Isolation of DNA replication complexes from uninfected and adenovirus-infected HeLa cells. *J. Mol. Biol.* **62**, 65.

Pedersen, I. R. (1970). Density gradient centrifugation studies on lymphocytic choriomeningitis virus and on viral ribonucleic acid. *J. Virol.* **6**, 414.

Pedersen, I. R. (1971). Lymphocytic choriomeningitis virus RNAs. *Nature New Biol.* **234**, 112.

Peebles, P. T., Haapala, D. K., and Gadzar, A. F. (1972). Deficiency of ribonucleic acid dependent deoxyribonucleic acid polymerase in non-infectious virus-like particles released from murine sarcoma virus transformed hamster cells. *J. Virol.* **9**, 488.

Pene, J. J., Knight, E., and Darnell, J. E. (1968). Characterization of a new low molecular weight RNA in HeLa cell ribosomes. *J. Mol. Biol.* **33**, 609.

Penhoet, E., Miller, H., Doyle, M., and Blatti, S. (1971). RNA-dependent RNA polymerase activity in influenza virions. *Proc. Nat. Acad. Sci. U.S.* **68**, 1369.

Penman, S. (1969). Preparation of purified nuclei and nucleoli from mammalian cells *In* "Fundamental Techniques in Virology" (K. Habel and N. P. Salzman, eds.), p. 35. Academic Press, New York.

Penman, S., and Summers, D. (1965). Effects on host-cell metabolism following synchronous infection with poliovirus. *Virology* **27**, 614.

Penman, S., Scherrer, K., Becker, Y., and Darnell, J. E. (1963). Polyribosomes in normal and poliovirus-infected HeLa cells and their relationship to messenger-RNA. *Proc. Nat. Acad. Sci. U.S.* **49**, 654.

Penman, S., Becker, Y., and Darnell, J. E. (1964). A cytoplasmic structure involved in the synthesis and assembly of poliovirus components. *J. Mol. Biol.* **8**, 541.

Penman, S., Wesco, C., and Penman, M. (1968). Localization and kinetics of formation of nuclear heterodisperse RNA, cytoplasmic heterodisperse RNA and polysome-associated messenger RNA in HeLa cells. *J. Mol. Biol.* **34**, 49.

Pennington, T. H., Follett, E. A. C., and Szilagyi, J. F. (1970). Events in vaccinia virus-infected cells following the reversal of the antiviral action of rifampicin. *J. Gen. Virol.* **9**, 225.

Pensaert, M. B., Burnstein, T., and Haelterman, E. O. (1970). Cell culture adapted SH strain of transmissible gastroenteritis virus of pigs: *in vivo* and *in vitro* studies. *Amer. J. Vet. Res.* **31**, 771.

Pereira, H. G. (1960). A virus inhibitor produced in HeLa cells infected with adenovirus. *Virology* **11**, 590.

Pereira, H. G. (1969). Influenza: antigenic spectrum. *Progr. Med. Virol.* **11**, 46.

Pereira, H. G., and Kelly, B. (1957). Dose-response curves of toxic and infective actions of adenovirus in HeLa cell cultures. *J. Gen. Microbiol.* **17**, 517.

Pereira, H. G., and Skehel, J. J. (1971). Spontaneous and tryptic digestion of virus particles and structural components of adenoviruses. *J. Gen. Virol.* **12**, 13.

Pereira, H. G., Huebner, R. J., Ginsberg, H. S., and Van Der Veen, J. (1963). A short description of the adenovirus group. *Virology* **20**, 613.

Pereira, H. G., Valentine, R. C., and Russell, W. C. (1968). Crystallization of an adenovirus protein (the hexon). *Nature (London)* **219**, 946.

Pereira, M. S., Pereira, H. G., and Clarke, S. K. R. (1965). Human adenovirus type 31. A new serotype with oncogenic properties. *Lancet* **i**, 21.

Pérol-Vauchez, Y., Tournier, P., and Lwoff, M. (1961). Atténuation de la virulence du virus de l'encéphalomyocardite de la souris par culture à basse température. Influence de l'hypo- et l'hyperthermie sur l'évolution de la maladie expérimentale. *C. R. Acad. Sci. (Paris)* **253**, 2164.

Perrault, J., and Holland, J. J. (1972a). Absence of transcriptase activity or transcription-inhibiting ability in defective interfering particles of vesicular stomatitis virus. *Virology* **50,** 159.

Perrault, J., and Holland, J. (1972b). Variability of vesicular stomatitis autointerference with different host cells and virus serotypes. *Virology* **50,** 148.

Perry, R. P. (1963). Selective effects of actinomycin D in the intracellular distribution of RNA synthesis in tissue culture cells. *Exp. Cell Res.* **29,** 400.

Perry, R. P. (1965). The nucleolus and the synthesis of ribosomes. *Nat. Cancer Inst. Monogr.* **18,** 325.

Perry, R. P. (1969). Nucleoli: the cellular sites of ribosome production *In* "Handbook of Molecular Cytology" (A. Lima-de-Faria, ed.) p. 620. North Holland, Amsterdam.

Pestka, S. (1971). Inhibitors or ribosome functions. *Annu. Rev. Microbiol.* **25,** 487.

Peters, D. H. A., and Müller, G. (1963). The fine structure of the DNA-containing core of vaccinia virus. *Virology* **21,** 266.

Peterson, R. D. A., Burmester, B. R., Frederickson, T. N., Purchase, H. G., and Good, R. A. (1964). Effect of bursectomy and thymectomy on the development of visceral lymphomatosis in the chicken. *J. Nat. Cancer Inst.* **32,** 1343.

Petralli, J. K., Merigan, T. C., and Wilbur, J. R. (1965). Action of endogenous interferon against vaccinia infection in children. *Lancet* **ii,** 401.

Petric, M., and Prevec, L. (1970). Vesicular stomatitis virus—a new interfering particle, intracellular structures, and virus-specific RNA. *Virology* **41,** 615.

Pettersson, U., and Höglund, S. (1969). Structural proteins of adenoviruses. III. Purification and characterization of adenovirus type 2 penton antigen. *Virology* **39,** 90.

Pettersson, U., and Sambrook, J. (1973). The amount of viral DNA in the cells transformed by genome of adenovirus type 2. *J. Mod. Biol.* **73,** 125.

Pettersson, U., Philipson, L., and Höglund, S. (1967). Structural proteins of adenoviruses. I. Purification and characterization of the adenovirus type 2 hexon antigen. *Virology* **33,** 575.

Pettersson, R., Kääriäinen, L., von Bonsdorff, C-H., and Oker-Blom, N. (1971). Structural components of Unkuniemi virus: a non-cubical tick-borne arbovirus. *Virology* **46,** 721.

Pfefferkorn, E. R., and Boyle, M. K. (1972a). Selective inhibition of the synthesis of Sindbis virion proteins by an inhibitor of chymotrypsin. *J. Virol.* **9,** 187.

Pfefferkorn, E. R., and Boyle, M. K. (1972b). Photoreactivation of a cytoplasmic virus. *J. Virol.* **9,** 474.

Pfefferkorn, E. R., and Burge, B. W. (1968). Morphogenic defects in the growth of *ts* mutants of Sindbis virus. *Perspect. Virol.* **6,** 1.

Pfefferkorn, E. R., and Coady, H. M. (1968). Mechanism of photoreactivation of pseudorabies virus. *J. Virol.* **2,** 474.

Pfefferkorn, E. R., and Burge, B. W., and Coady, H. M. (1966). Characteristics of the photoreactivation of pseudorabies virus. *J. Bacteriol.* **92,** 856.

Pfeiffer, S. E., and Tolmach, L. J. (1967). Selecting synchronous populations of mammalian cells. *Nature (London)* **213,** 139.

Philipson, L., Lonberg-Holm, K., and Pettersson, U. (1968). Virus-receptor interaction in an adenovirus system. *J. Virol.* **2,** 1064.

Philipson, L., Wall, R., Glickman, R., and Darnell, J. E. (1971). Post-transcriptional addition of polyadenylic acid sequences to virus-specific RNA during adenovirus growth. *Proc. Nat. Acad. Sci. U.S.* **68,** 2806.

Phillips, B. A. (1971). *In vitro* assembly of poliovirus. II. Evidence for the self-assembly of 14S particles into empty capsids. *Virology* **44,** 307.

Phillips, B. A., and Sydiskis, R. J. (1971). Synthesis of viral products by cell-free extracts. *Progr. Med. Virol.* **13,** 83.

Phillips, B. A., Summers, D. F., and Maizel, J. V. (1968). *In vitro* assembly of poliovirus-related particles. *Virology* **35,** 216.

Piazza, M. (1969). "Experimental Viral Hepatitis." Thomas, Springfield, Illinois.

Pierce, W. E., Stille, W. T., and Miller, L. F. (1963). A preliminary report on effects of routine military inoculations on respiratory illness. *Proc. Soc. Exp. Biol. Med.* **114,** 369.

Pike, M. C., Williams, E. H., and Wright, B. (1967). Burkitt's tumor in the west Nile district of Uganda. *Brit. Med. J.* **2**, 395.

Piña, M., and Green, M. (1965). Biochemical studies on adenovirus multiplication. IX. Chemical and base composition analysis of 28 human adenoviruses. *Proc. Nat. Acad. Sci. U.S.* **54**, 547.

Piña, M., and Green, M. (1968). Base composition of the DNA of oncogenic simian adenovirus SA7 and homology with human adenovirus DNA. *Virology* **36**, 321.

Pincus, T., Rowe, W. P., and Lilly, F. (1971). A major genetic locus affecting resistance to infection with murine leukemia virus. II. Apparent identity to a major locus described for resistance to Friend murine leukemia virus. *J. Exp. Med.* **133**, 1234.

Pincus, W. B., and Flick, J. A. (1963). Inhibition of the primary vaccinial lesion and of delayed hypersensitivity by an antimononuclear cell serum. *J. Infect. Dis.* **113**, 15.

Piraino, F. (1967). The mechanism of genetic resistance of chick embryo cells to infection by Rous sarcoma virus—Bryan strain (BS-RSV). *Virology* **32**, 700.

Pitha, P. M., and Carter, W. A. (1971). The DEAE dextran: polyriboinosinate-polyribocytidylate complex: physical properties and interferon induction. *Virology* **45**, 777.

Pitkanen, A., McAuslan, B., Hedgpeth, J., and Woodson, B. (1968). Induction of poxvirus ribonucleic acid polymerases. *J. Virol.* **2**, 1363.

Plotkin, S. A., and Clark, H. F. (1971). Prevention of rabies in man. *J. Infect. Dis.* **123**, 227.

Plotkin, S. A., Boué, A., and Boué, J. G. (1965). The *in vitro* growth of rubella virus in human embryo cells. *Amer. J. Epidemiol.* **81**, 71.

Plotkin, S. A., Katz, M., Brown, R. E., and Pagano, J. S. (1966). Oral poliovirus vaccination in newborn African infants. *Amer. J. Dis. Child.* **111**, 27.

Plotkin, S. A., Farquhar, J. D., Katz, M., and Buser, F. (1969). Attenuation of RA 27/3 rubella virus in WI-38 human diploid cells. *Amer. J. Dis. Child.* **118**, 178.

Plowright, W., MacLeod, W. G., and Ferris, R. D. (1959). The pathogenesis of sheep pox in the skin of sheep. *J. Comp. Pathol. Therap.* **69**, 400.

Plowright, W., Perry, C. T., Pierce, M. A., and Parker, J. (1970). Experimental infection of the argasid tick, *Ornithodorus moubata porcinus*, with African swine fever virus. *Arch. Ges. Virusforsch.* **31**, 33.

Plummer, G., Goodheart, C. R., Henson, D., and Bowling, C. P. (1969). A comparative study of the DNA density and behavior in tissue cultures of fourteen different herpesviruses. *Virology* **39**, 134.

Plus, N., and Atanasiu, P. (1966). Selection d'un mutant du virus rabique adapté à un Insecte: *Drosophila melanogaster*. *C. R. Acad. Sci. (Paris)* **263**, 89.

Pluznik, D. H., Sachs, L., and Resnitzky, P. (1966). The mechanism of leukemogenesis by the Rauscher leukemia virus. *Nat. Cancer Inst. Monogr.* **22**, 3.

Pogo, B. G. T., and Dales, S. (1969). Two deoxyribonuclease activities within purified vaccinia virus. *Proc. Nat. Acad. Sci. U.S.* **63**, 820.

Pohjanpelto, P. (1961). Response of enteroviruses to cystine. *Virology* **15**, 225.

Pohjanpelto, P., and Cooper, P. D. (1965). Interference between polioviruses induced by strains which cannot multiply. *Virology* **25**, 350.

Pollack, R. E. (1970). Cellular and viral contributions to maintenance of the SV40-transformed state. *In Vitro* **6**, 58.

Pollack, R. E., and Burger, M. M. (1969). Surface-specific characteristics of a contact-inhibited cell line containing the SV40 genome. *Proc. Nat. Acad. Sci. U.S.* **62**, 1074.

Pollack, R. E., and Teebor, G. W. (1969). Relationship of contact inhibition to tumor transplantability, morphology and growth rate. *Cancer Res.* **29**, 1770.

Pollack, R. E., Green, H., and Todaro, G. J. (1968). Growth control in cultured cells: selection of sublines with increased sensitivity to contact inhibition and decreased tumor producing ability. *Proc. Nat. Acad. Sci. U.S.* **60**, 126.

Pollock, T. M. (1969). Human immunoglobulin in prophylaxis. *Brit. Med. Bull.* **25**, 202.

Pons, M. W. (1970). On the nature of the influenza virus genome. *Curr. Topics Microbiol. Immunol.* **52**, 142.

Pons, M. W. (1971). Isolation of influenza virus ribonucleoprotein from infected cells. Demonstration of the presence of negative-stranded RNA in viral RNP. *Virology* **46**, 149.

Pons, M. W. (1972). Studies on the replication of influenza virus RNA. *Virology* **47**, 823.

Pons, M. W., and Hirst, G. K. (1969). The single- and double-stranded RNA's and the proteins of incomplete influenza virus. *Virology* **38**, 68.

Pons, M. W., Schulze, I. T., and Hirst, G. K. (1969). Isolation and characterization of the ribonucleoprotein of influenza virus. *Virology* **39**, 250.

Pope, J. H., and Rowe, W. P. (1964). Detection of specific antigen in SV40-transformed cells by immunofluorescence. *J. Exp. Med.* **120**, 121.

Pope, J. H., Horne, M. K., and Scott, W. (1968). Transformation of foetal human leukocytes in vitro by filtrates of a human leukaemic cell line containing herpes-like virus. *Intern. J. Cancer* **3**, 857.

Porter, D. D. (1971a). A quantitative view of the slow virus landscape. *Progr. Med. Virol.* **13**, 339.

Porter, D. D. (1971b). Destruction of virus-infected cells by immunological mechanisms. *Annu. Rev. Microbiol.* **25**, 283.

Porter, D. D., and Larsen, A. E. (1968). Virus-host interactions in Aleutian disease of mink. *Perspect. Virol.* **6**, 173.

Porter, D. D., and Porter, H. G. (1971). Deposition of immune complexes in the kidneys of mice infected with lactic dehydrogenase virus. *J. Immunol.* **106**, 1246.

Porter, D. D., Dixon, F. J., and Larsen, A. E. (1965). The development of a myeloma-like condition in mink with Aleutian disease. *Blood* **25**, 736.

Porter, D. D., Porter, H. G., and Deerhake, B. B. (1969a). Immunofluorescence assay for antigen and antibody in lactic dehydrogenase virus infection of mice. *J. Immunol.* **102**, 431.

Porter, D. D., Larsen, A. E., and Porter, H. G. (1969b). The pathogenesis of Aleutian disease of mink. I. *In vivo* viral replication and the host antibody response to viral antigen. *J. Exp. Med.* **130**, 575.

Portner, A., and Kingsbury, D. W. (1971). Homologous interference by incomplete Sendai virus particles: changes in virus-specific ribonucleic acid synthesis. *J. Virol.* **8**, 388.

Portner, A., and Kingsbury, D. W. (1972). Identification of transcriptive and replicative intermediates in Sendai virus-infected cells. *Virology* **47**, 711.

Poskitt, E. M. E. (1971). Effect of measles on plasma albumin levels in Ugandan village children. *Lancet* **ii**, 68.

Poste, G. (1970). Virus-induced polykaryocytosis and the mechanism of cell fusion. *Advan. Virus Res.* **16**, 303.

Potash, L. (1968). Methods in human virus vaccine preparation. *In* "Methods in Virology" (K. Maramorosch and H. Koprowski, eds.), Vol. IV, p. 372. Academic Press, New York.

Potter, C. W., Potter, A. M., and Oxford, J. S. (1970). Comparison of transformation and T antigen induction in human cell lines. *J. Virol.* **5**, 293.

Prage, L., and Pettersson, U. (1971). Structural proteins of adenoviruses. VII. Purification and properties of an arginine-rich core protein from adenovirus type 2 and type 3. *Virology* **45**, 364.

Prage, L., Pettersson, U., Höglund, S., Lenberg-Holm, K., and Philipson, L. (1970). Structural proteins of adenoviruses. IV. Sequential degradation of the adenovirus type 2 virion. *Virology* **42**, 341.

Prescott, D. M., and Bender, M. A. (1962). Synthesis of RNA and protein during mitosis in mammalian tissue culture cells. *Exp. Cell Res.* **26**, 260.

Prescott, D. M., Kates, J., and Kirkpatrick, J. B. (1971). Replication of vaccinia virus DNA in enucleated L-cells. *J. Mol. Biol.* **59**, 505.

Pretzmann, G., Loew, J., and Radda, A. (1963). Untersuchungen in einem Naturherd der Frühsommer-Meningoencephalitis (FSME) in Niederösterreich. *Zentr. Bakteriol. Parasitenk. Abt. I Orig.* **190**, 299.

Prevec, L., and Kang, C. Y. (1970). Homotypic and heterotypic interference by defective particles of vesicular stomatitis virus. *Nature (London)* **228**, 25.

Price, R., and Penman, S. (1972). Transcription of the adenovirus genome by an α-amanitin-sensitive ribonucleic acid polymerase in HeLa cells. *J. Virol.* **9**, 621.

Pridgen, C., and Kingsbury, D. W. (1972). Adenylate-rich sequences in Sendai virus transcripts from infected cells. *J. Virol.* **10**, 314.

Prince, A. M. (1958). Quantitative studies on Rous sarcoma virus. I. The titration of Rous sarcoma virus on the chorioallantoic membrane of the chick embryo. *J. Nat. Cancer Inst.* **20**, 147.

Prince, A. M. (1959). Quantitative studies on Rous sarcoma virus. IV. An investigation of the nature of "non-infective" tumors induced by low doses of virus. *J. Nat. Cancer Inst.* **23**, 1361.

Pringle, C. R. (1968). Recombination between conditional lethal mutations within a strain of foot-and-mouth disease virus. *J. Gen. Virol.* **2**, 199.

Pringle, C. R. (1970). Genetic characteristics of conditional lethal mutants of vesicular stomatitis virus induced by 5-fluorouracil, 5-azacytidine and ethyl methane sulfonate. *J. Virol.* **5**, 559.

Pringle, C. R., and Duncan, I. B. (1971). Preliminary physiological characterization of temperature-sensitive mutants of vesicular stomatitis virus. *J. Virol.* **8**, 56.

Pringle, C. R., and Slade, W. R. (1968). The origin of hybrid variants derived from subtype strains of foot-and-mouth disease virus. *J. Gen. Virol.* **2**, 319.

Pringle, C. R., Slade, W. R., Elworthy, P., and O'Sullivan, M. (1970). Properties of temperature-sensitive mutants of the Kenya 3/57 strain of foot-and-mouth disease virus. *J. Gen. Virol.* **6**, 213.

Pringle, C. R., Duncan, I. B., and Stevenson, M. (1971). Isolation and characterization of temperature-sensitive mutants of vesicular stomatitis virus, New Jersey serotype. *J. Virol.* **8**, 836.

Printz, P., and Wagner, R. R. (1971). Temperature-sensitive mutants of vesicular stomatitis virus: Synthesis of virus-specific proteins. *J. Virol.* **7**, 651.

Probert, M., and Epstein, M. A. (1972). Morphological transformation *in vitro* of human fibroblasts by Epstein-Barr virus: preliminary observations. *Science* **175**, 202.

Proctor, D. F. (1966). Airborne disease and the upper respiratory tract. *Bacteriol. Rev.* **30**, 498.

Proctor, W. R., Cook, J. S., and Tennant, R. W. (1972). Ultraviolet photobiology of Kilham rat virus and the absolute ultraviolet sensitivities of other animal viruses: influence of DNA strandedness, molecular weight, and host-cell repair. *Virology* **49**, 368.

Prose, P. H., Balk, S. D., Liebhaber, H., and Krugman, S. (1965). Studies with a myxovirus recovered from patients with infectious hepatitis. II. Fine structure and electron microscopic demonstration of intracytoplasmic internal component and viral filament formation. *J. Exp. Med.* **122**, 1151.

Prusoff, W. H. (1967). Recent advances in chemotherapy of viral diseases. *Pharmacol. Rev.* **19**, 209.

Prusoff, W. H. (1972). Viral and host cell interactions with 5-iodo-2'-deoxyuridine (idoxuridine). *In* "Virus-Cell Interactions and Viral Antimetabolites" (D. Shugar, ed.), p. 135, FEBS Symp, Vol. 22. Academic Press, New York.

Puck, T. T. (1964). Studies on the life cycle of mammalian cells. *Cold Spring Harbor Symp. Quant. Biol.* **29**, 167.

Purchase, H. G. (1972). Recent advances in the knowledge of Marek's disease. *Advan. Vet. Sci. Comp. Med.* **16**, 223.

Purchase, H. G., Witter, R. L., Okazaki, W., and Burmester, B. R. (1971). Vaccination against Marek's disease. *Perspect. Virol.* **7**, 91.


Quersin-Thiry, L. (1961). Nutritive requirements of a small plaque mutant of western equine encephalitis virus. *Brit. J. Exp. Pathol.* **42**, 511.

Quigley, J. P., Rifkin, D. B., and Reich, E. (1971). Phospholipid composition of Rous sarcoma virus host cell membranes and other enveloped RNA viruses. *Virology* **46**, 106.

Quinlan, M. F., Salman, S. D., Swift, D. L., Wagner, H. N., and Proctor, D. F. (1969). Measurement of mucociliary function in man. *Amer. Rev. Resp. Dis.* **99**, 13.


Rabin, E. R., Jenson, A. B., and Melnick, J. L. (1968). Herpes simplex virus in mice: electron microscopy of neural spread. *Science* **162**, 126.

Rabinowitz, M., and Fisher, J. M. (1962). A dissociative effect of puromycin on the pathway of protein synthesis by Ehrlich ascites tumor cells. *J. Biol. Chem.* **237**, 477.

Rabinowitz, Z., and Sachs, L. (1968). Reversion of properties in cells transformed by polyoma virus. *Nature (London)* **220**, 1203.

Rabinowitz, Z., and Sachs, L. (1969). The formation of variants with a reversion of properties of transformed cells. II. *In vitro* formation of variants from polyoma virus-transformed cells. *Virology* **38**, 343.

Rabinowitz, Z., and Sachs, L. (1970a). Control of the reversion properties in transformed cells. *Nature (London)* **225**, 136.

Rabinowitz, Z., and Sachs, L (1970b). The formation of variants with a reversion of properties of transformed cells. IV. Loss of detectable polyoma transplantation antigen. *Virology* **40**, 193.

Rabotti, G. F., Raine, W. A., and Sellers, R. L. (1965). Brain tumors (gliomas) induced in hamsters by Bryan's strain of Rous sarcoma virus. *Science* **147**, 504.

Rabson, A. S., and Kirschstein, R. L. (1962). Induction of malignancy *in vitro* in newborn hamster kidney tissue infected with simian vacuolating virus (SV40). *Proc. Soc. Exp. Biol. Med.* **111**, 323.

Rabson, A. S., O'Conor, G. T., Kirschstein, R. L., and Branigan, W. J. (1962). Papillary ependymomas produced in *Rattus (Mastomys) natalensis* inoculated with vacuolating virus (SV40). *J. Nat. Cancer Inst.* **29**, 765.

Rabson, A. S., O'Conor, G. T., Berezesky, I. K., and Paul, F. J. (1964). Enhancement of adenovirus growth in African green monkey kidney cell cultures by SV40. *Proc. Soc. Exp. Biol. Med.* **116**, 187.

Rabson, A. S., O'Conor, G. T., Baron, S., Whang, J. J., and Legallais, F. Y. (1966). Morphologic, cytogenetic and virologic studies *in vitro* of a malignant lymphoma from an African child. *Intern. J. Cancer* **1**, 89.

Rabson, A. S., O'Conor, G. T., Lorenz, D. E., Kirschstein, R. L., Legallais, F. Y., and Tralka, T. S. (1971). Lymphoid cell-culture line derived from lymph node of marmoset infected with *Herpesvirus saimiri*—preliminary report. *J. Nat. Cancer Inst.* **46**, 1099.

Radloff, R., Bauer, W., and Vinograd, J. (1967). A dye buoyant density method for the detection and isolation of closed circular duplex DNA in HeLa cells. *Proc. Nat. Acad. Sci. U.S.* **57**, 1514.

Rakusanova, T., Ben-Porat, T., Himeno, M., and Kaplan, A. S. (1971). Early functions of the genome of herpesvirus: I. Characterization of the RNA synthesized in cycloheximide-treated, infected cells. *Virology* **46**, 877.

Ralph, R. K., and Colter, J. S. (1972). Evidence for the integration of polyoma virus DNA in a lytic system. *Virology* **48**, 49.

Ramos, B. A., Courtney, R. J., and Rawls, W. E. (1972). Structural proteins of Pichinde virus. *J. Virol.* **10**, 661.

Ramos-Alvarez, M., and Sabin, A. B. (1954). Characteristics of poliomyelitis and other enteric viruses recovered in tissue culture from healthy American children. *Proc. Soc. Exp. Biol. Med.* **87**, 655.

Randall, C. C., and Gafford, L. G. (1962). Histochemical and biochemical studies on isolated viral inclusions. *Amer. J. Pathol.* **40**, 51.

Rapp, F. (1964). Plaque differentiation and replication of virulent and attenuated strains of measles virus. *J. Bacteriol.* **88**, 1448.

Rapp, F. (1969). Defective DNA animal viruses. *Annu. Rev. Microbiol.* **23**, 293.

Rapp, F., and Melnick, J. L. (1966). Papovavirus SV40, adenoviruses and their hybrids: Transformation, complementation, and transcapsidation. *Progr. Med. Virol.* **8**, 349.

Rapp, F., Melnick, J. L., Butel, J. S., and Kitahara, T. (1964). The incorporation of SV40 genetic material into adenovirus 7 as measured by intranuclear synthesis of SV40 tumor antigen. *Proc. Nat. Acad. Sci. U.S.* **52,** 1348.

Rapp, F., Butel, J. S., Feldman, L. A., Kitahara, T., and Melnick, J. L. (1965). Differential effects of inhibitors on the steps leading to the formation of SV40 tumor and virus antigens. *J. Exp. Med.* **121,** 935.

Rapp, F., Feldman, L. A., and Mandel, M. (1966). Synthesis of virus deoxyribonucleic acid during abortive infection of simian cells by human adenoviruses. *J. Bacteriol.* **92,** 931.

Rapp, F., Jerkofsky, M., Melnick, J. L., and Levy, B. (1968). Variation in the oncogenic potential of human adenoviruses carrying a defective SV40 genome (PARA). *J. Exp. Med.* **127,** 77.

Raska, K., and Strohl, W. A. (1972). The response of BHK 21 cells to infection with type 12 adenovirus. VI. Synthesis of virus-specific RNA. *Virology* **47,** 734.

Raska, K., Prage, L., and Schlesinger, R. W. (1972). Effects of arginine starvation on macromolecular synthesis in infection with type 2 adenovirus. II. Synthesis of virus-specific RNA and DNA. *Virology* **48,** 472.

Raskas, H. (1971). Release of adenovirus messenger RNA from isolated nuclei. *Nature New Biol.* **233,** 134.

Raskas, H., and Green, M. (1971). DNA:RNA and DNA:DNA hybridization in virus research. *In* "Methods in Virology" (K. Maramorosch and H. Koprowski, eds.), Vol. V, p. 293. Academic Press, New York.

Raskas, H., and Okubo, C. (1971). Transport of RNA in KB cells infected with adenovirus type 2. *J. Cell Biol.* **49,** 438.

Rasmussen, A. F., Marsh, J. T., and Brill, N. O. (1957). Increased susceptibility to herpes simplex in mice subjected to avoidance-learning stress or restraint. *Proc. Soc. Exp. Biol. Med.* **96,** 183.

Rauscher, F. J. (1962). A virus-induced disease of mice characterized by erythropoiesis and lymphoid leukemia. *J. Nat. Cancer Inst.* **29,** 515.

Rawls, W. E. (1968). Congenital rubella: the significance of virus persistence. *Progr. Med. Virol.* **10,** 238.

Rawls, W. E., and Melnick, J. L. (1966). Rubella virus carrier cultures derived from congenitally infected infants. *J. Exp. Med.* **123,** 795.

Rawls, W. E., Tomkins, W. A. F., and Melnick, J. L. (1969). The association of herpesvirus type 2 and carcinoma of the uterine cervix. *Amer. J. Epidemiol.* **89,** 547.

Razin, S. (1969). Structure and function in mycoplasma. *Annu. Rev. Microbiol.* **23,** 317.

Reda, I. M., Rott, R., and Schäfer, W. (1964). Fluorescent antibody studies with NDV-infected cell systems. *Virology* **22,** 422.

Reed, S. (1967). Transformation of hamster cells *in vitro* by adenovirus type 12. *J. Gen. Virol.* **1,** 405.

Reeves, W. C., (1961). Overwintering of arthropod-borne viruses. *Progr. Med. Virol.* **3,** 59.

Reeves, W. C., and Hammon, W. McD. (1962). Epidemiology of the arthropod-borne viral encephalitides in Kern County, California. *Univ. Calif. (Berkeley) Publ. Public Health* **4.**

Regelson, W. (1967). Prevention and treatment of Friend leukemia virus (FLV) infection by interferon-inducing synthetic polyanions. *In* "The Reticuloendothelial System and Atherosclerosis" (N. R. Di Luzia and R. Paoletti, eds.), p. 315. Plenum, New York.

Regnery, D. C., and Marshall, I. D. (1971). Studies on the epidemiology of myxomatosis in California. IV. The susceptibility of six leporid species to Californian myxoma virus and the relative infectivity of their tumors for mosquitoes. *Amer. J. Epidemiol.* **94,** 508.

Reháček, J., Dolan, T., Thompson, K., Fischer, R. G., Rehaçek, Z.; and Johnson, H. (1971). Cultivation of oncogenic viruses in mosquito cells *in vitro*. *Curr. Topics Microbiol. Immunol.* **55,** 161.

Reich, E. (1966). Binding to DNA and inhibition of DNA functions by actinomycins. *Symp. Soc. Gen. Microbiol.* **16,** 266.

Reich, E., and Goldberg, I. H. (1964). Actinomycin and nucleic acid function. *Progr. Nucleic Acid Res. Mol. Biol.* **3,** 183.

Reich, E., Franklin, R. M., Shatkin, A. J., and Tatum, E. L. (1962). Action of actinomycin D on animal cells and viruses. *Proc. Nat. Acad. Sci. U.S.* **48**, 1238.

Reich, P. R., Baum, S. G., Rose, J. A., Rowe, W. P., and Weissman, S. M. (1966). Nucleic acid homology studies of adenovirus type 7-SV40 interactions. *Proc. Nat. Acad. Sci. U.S.* **55**, 336.

Reichmann, M. E., Pringle, C. R., and Follett, E. A. C. (1971). Defective particles in BHK cells infected with temperature-sensitive mutants of vesicular stomatitis virus. *J. Virol.* **8**, 154.

Reimer, C. B., Baker, R. S., Newlin, T. E., Havens, M. L., van Frank, R. M., Storvick, W. O., and Miller, R. P. (1966). Comparison of techniques for influenza virus purification. *J. Bacteriol.* **92**, 1271.

Reinicke, V. (1965). The influence of steroid hormones and growth hormones on the production of influenza virus and interferon in tissue culture. II. The influence of metandienonum, d-aldosterone, testosterone, oestradiol and growth hormones. *Acta Pathol. Microbiol. Scand.* **64**, 553.

Rekosh, D. (1972). Gene order of the poliovirus capsid proteins. *J. Virol.* **9**, 479.

Rekosh, D. M., Lodish, H. F., and Baltimore, D. (1970). Protein synthesis in *Escherichia coli* extracts programmed by poliovirus RNA. *J. Mol. Biol.* **54**, 327.

Remington, J. S., and Merigan, T. C. (1969). Resistance to virus challenge in mice infected with protozoa and bacteria. *Proc. Soc. Exp. Biol. Med.* **131**, 1184.

Renger, H. M., and Basilico, C. (1972). Mutation causing temperature-sensitive expression of cell transformation by a tumor virus. *Proc. Nat. Acad. Sci. U.S.* **69**, 109.

Renkonen, O., Kääräinen, L., Simons, K., and Gahmberb, C. G. (1971). The lipid class composition of Semliki Forest virus and of plasma membranes of host cells. *Virology* **46**, 318.

Revel, H. R., and Luria, S. E. (1970). DNA-glucosylation in T-even phage: genetic determination and role in phage-host interaction. *Annu. Rev. Genet.* **4**, 177.

Rhim, J. S., Jordan, L. E., and Mayor, H. D. (1962). Cytochemical fluorescent antibody and electron microscopic studies on the growth of reovirus (ECHO 10) in tissue culture. *Virology* **17**, 342.

Rickard, C. G., Post, J. E., Noronha, F., and Barr, L. M. (1969). A transmissible virus-induced lymphatic leukemia of the cat. *J. Nat. Cancer Inst.* **42**, 937.

Rickert, N. J., and Balduzzi, P. (1971). Mechanism of oncogenic transformation by Rous sarcoma virus II. Effect of rifampin on Rous sarcoma virus infection. *J. Virol.* **8**, 62.

Rifkin, D. B., and Compans, R. W. (1971). Identification of spike proteins of Rous sarcoma virus. *Virology* **46**, 485.

Rifkin, D. B., and Reich, E. (1971). Selective lysis of cells transformed by Rous sarcoma virus. *Virology* **45**, 172.

Rifkind, D. (1966). The activation of varicella-zoster infections by immuno-suppressive therapy. *J. Lab. Clin. Med.* **68**, 463.

Rifkind, D., Goodman, N., and Hill, R. B. (1967). The clinical significance of cytomegalovirus infection in renal transplant recipients. *Ann. Internal. Med.* **66**, 1116.

Riggs, S., and Sanford, J. P. (1962). Viral orchitis. *New Engl. J. Med.* **266**, 990.

Rightsel, W. A., Dice, J. R., McAlpine, R. J., Timm, E. A., McLean, I. W., Dixon, G. J., and Schabel, F. M. (1961). Antiviral effect of guanidine. *Science* **134**, 558.

Říhova-Škárová, B., and Říhá, I. (1972). Host genotype and antibody formation. *Curr. Topics Microbiol. Immunol.* **57**, 159.

Riley, B. P., and Gifford, G. E. (1967). Studies on tissue specificity of interferon. *J. Gen. Microbiol.* **46**, 293.

Riley, B. P., Troy, S. T., and Gifford, G. E. (1966). Relative effects of interferon on virus plaque formation. *Proc. Soc. Exp. Biol. Med.* **122**, 1142.

Riman, J., and Beaudreau, G. S. (1970). Viral DNA-dependent DNA polymerase and the properties of thymidine labeled material in the virions of an oncogenic RNA virus. *Nature (London)* **228**, 427.

Riordan, J. T., Paul, J. R., Yoshioka, L., and Horstmann, D. M. (1961). The detection of poliovirus and other enteric viruses in flies. Results of tests carried out during an oral poliovirus vaccine trial. *Amer. J. Hyg.* **74**, 123.

Rita, G. (ed.) (1968). "The Interferons." Academic Press, New York.

Ritzi, E., and Levine, A. J. (1970). Deoxyribonucleic acid replication in simian virus 40 in-fected cells. III. Comparison of simian virus 40 lytic infection in three different monkey kidney cell lines. *J. Virol.* **5,** 686.

Rivers, T. M., and Ward, S. M. (1937). Infectious myxomatosis of rabbits. *J. Exp. Med.* **66,** 1.

Roane, P. R., and Roizman, B. (1964). Studies of the determinant antigens of viable cells. II. Demonstration of altered antigenic reactivity of HEp-2 cells infected with herpes simplex virus. *Virology* **22,** 1.

Robb, J. A., and Martin, R. G. (1970). Genetic analysis of simian virus 40. I. Description of microtitration and replica-plating techniques for virus. *Virology* **41,** 751.

Robberson, D. L., Kasamatsu, H., and Vinograd, J. (1972). Replication of mitochondrial DNA. Circular replicative intermediates in mouse L cells. *Proc. Nat. Acad. Sci. U.S.* **69,** 737.

Robbins, E., and Scharff, M. D. (1966). Some macromolecular characteristics of synchronized HeLa cells. *In* "Cell Synchrony Studies in Biosynthetic Regulation" (I. L. Cameron and G. M. Padilla, eds.), p. 353. Academic Press, New York.

Roberts, J. A. (1962a). Histopathogenesis of mousepox. I. Respiratory infection. *Brit. J. Exp. Pathol.* **43,** 451.

Roberts, J. A. (1962b). Histopathogenesis of mousepox. II. Cutaneous infection. *Brit. J. Exp. Pathol.* **43,** 462.

Roberts, J. A. (1964). Growth of virulent and attenuated ectromelia virus in cultured macrophages from normal and ectromelia-immune mice. *J. Immunol.* **92,** 837.

Robertson, J. D. (1959). The ultrastructure of cell membranes and their derivatives. *Biochem. Soc. Symp.* **16,** 3.

Robinson, D. J., and Watson, D. H. (1971). Structural proteins of herpes simplex virus. *J. Gen. Virol.* **10,** 163.

Robinson, H. L. (1967). Isolation of noninfectious particles containing Rous sarcoma virus RNA from the medium of Rous sarcoma virus-transformed nonproducer cells. *Proc. Nat. Acad. Sci. U.S.* **57,** 1655.

Robinson, R. Q. (1964). Natural history of influenza since the introduction of the A2 strain. *Progr. Med. Virol.* **6,** 82.

Robinson, W. S. (1970). Self annealing of subgroup 2 myxovirus RNAs. *Nature (London).* **225,** 944.

Robinson, W. S. (1971a). Ribonucleic acid polymerase activity in Sendai virions and nucleocapsid. *J. Virol.* **8,** 81.

Robinson, W. S. (1971b). Sendai virus RNA synthesis and nucleocapsid formation in the presence of cycloheximide. *Virology* **44,** 494.

Robinson, W. S., and Duesberg, P. H. (1968). The myxoviruses. *In* "Molecular Basis of Virology" (H. Fraenkel-Conrat, ed.), p. 255, Reinhold, New York.

Robinson, W. S., and Robinson, H. L. (1971). DNA polymerase in defective Rous sarcoma virus. *Virology* **44,** 457.

Robinson, W. S., Robinson, H. L., and Duesberg, P. H. (1967). Tumor virus RNA's *Proc. Nat. Acad. Sci. U.S.* **58,** 825.

Roblin, R., Harle, E., and Dulbecco, R. (1971). Polyoma virus proteins. I. Multiple virion components. *Virology* **45,** 555.

Rocklin, R. E. (1971). Mediators associated with delayed hypersensitivity. *J. Reticuloendothel. Soc.* **10,** 50.

Roeder, R. G., and Rutter, W. J. (1969). Multiple forms of DNA-dependent RNA polymerase in eukaryotic organisms. *Nature (London)* **224,** 234.

Roeder, R. G., and Rutter, W. J. (1970). Specific nucleoplasmic RNA polymerases. *Proc. Nat. Acad. Sci. U.S.* **65,** 675.

Rogers, S. (1959). Induction of arginase in rabbit epithelium by the Shope rabbit papilloma virus. *Nature (London)* **183,** 1815.

Rogers, S. (1962). Certain relations between the Shope virus-induced arginase, the virus and the tumor cells. *In* "The Molecular Basis of Neoplasia," M. D. Anderson Symp., p. 483. Univ. of Texas Press, Austin, Texas.

Rogers, S., and Moore, M. (1963). Studies on the mechanism of action of the Shope rabbit papilloma virus. I. Concerning the nature of the arginase in the infected cell. *J. Exp. Med.* **117**, 521.

Roizman, B. (1962a). Polykaryocytosis. *Cold Spring Harbor Symp. Quant. Biol.* **27**, 327.

Roizman, B. (1962b). Polykaryocytosis induced by viruses. *Proc. Nat. Acad. Sci. U.S.* **48**, 228.

Roizman, B. (1965). Abortive infection of canine cells by herpes simplex virus. III. The interference of conditional lethal virus with an extended host rang mutant. *Virology* **27**, 113.

Roizman, B. (1969a). The herpesviruses—a biochemical definition of the group. *Curr. Topics Microbiol. Immunol.* **49**, 1.

Roizman, B. (1969b). Herpesviruses. *In* "The Biochemistry of Virus Replication" (H. B. Levy, ed.), p. 415. Dekker, New York.

Roizman, B. (1970). Herpesviruses, membranes and the social behavior of infected cells. *Proc. 3rd Intern. Symp. Appl. Med. Virol. Fort Lauderdale, Florida*, p. 37. W. Green, St. Louis, Missouri.

Roizman, B., and Aurelian, L., (1965). Abortive infection of canine cells by herpes simplex virus. I. Characterization of viral progeny from cooperative infection with mutants differing in capacity to multiply in canine cells. *J. Mol. Biol.* **11**, 528.

Roizman, B., and de-Thé, G. (1972). The classification of herpesviruses: a proposal. *Bull. WHO.* **46**, 547.

Roizman, B., and Schluederberg, A. E. (1962). Virus infection of cells in mitosis. III. Cytology of mitotic and amitotic HEp-2 cells infected with measles virus. *J. Nat. Cancer Inst.* **28**, 35.

Roizman, B., and Spear, P. G. (1971). Herpesviruses: current information on the composition and structure. *In* "Comparative Virology" (K. Maramorosch and E. Kurstak, eds.), p. 135, Academic Press, New York.

Roizman, B., and Spear, P. G. (1973). Herpesviruses. *In* "An Atlas on Viruses" (A. J. Dalton and F. Hagenau, eds.), p. 83. Academic Press, New York.

Roizman, B., Spring, S. B., and Schwartz, J. (1969). The herpesvirion and its precursors made in productively and abortively infected cells. *Fed. Proc. Fed. Amer. Soc. Exp. Biol.* **28**, 1890.

Roizman, B., Bachenheimer, A. L., Wagner, E. K., and Savage, T. (1970). Synthesis and transport of RNA in herpesvirus infected mammalian cells. *Cold Spring Harbor Symp. Quant. Biol.* **35**, 753.

Roizman, B., Spear, P. G., and Kieff, E. D. (1973). Herpes simplex viruses I and II: a biochemical definition. *Perspect. Virol.* **8**, 129.

Rokutanda, M., Rokutanda, H., Green, M., Fujinaga, K., Ray, R. K., and Gurgo, G. (1970). Formation of viral RNA-DNA hybrid molecules by the DNA polymerase of sarcoma-leukemia viruses. *Nature (London)* **227**, 1027.

Rönn, O., Inglot, A. D., and Lucke, E. (1970). Non-genetic reactivation of differently inactivated vaccinia virus. *Arch. Ges. Virusforsch.* **32**, 291.

Roos, R., Gajdusek, D. C., and Gibbs, C. J. (1973). The clinical characteristics of transmissible Creutzfeldt-Jakob disease. *Brain* **96**, 1.

Root, R. K., and Wolff, S. (1968). Pathogenic mechanisms in experimental immune fever. *J. Exp. Med.* **128**, 309.

Rose, G. G. (1970). "Atlas of Vertebrate Cells in Tissue Culture," Academic Press, New York.

Rose, J. A., and Koczot, F. (1971). Adenovirus-associated virus multiplication. VI. Base composition of the deoxyribonucleic acid strand species and strand-specific *in vivo* transcription. *J. Virol.* **8**, 771.

Rose, J. A., Hoggan, M. D., and Shatkin, A. J. (1966). Nucleic acid from an adeno-associated virus: Chemical and physical studies. *Proc. Nat. Acad. Sci. U.S.* **56**, 86.

Rose, J. A., Hoggan, M. D., Koczot, F., and Shatkin, A. J. (1968). Genetic relatedness studies with adenovirus-associated viruses. *J. Virol.* **2**, 999   .

Rose, J. A., Berns, K. I., Hoggan, M. D., and Koczot, F. (1969). Evidence for a single-stranded adenovirus-associated virus genome: formation of a DNA density hybrid on release of viral DNA. *Proc. Nat. Acad. Sci. U.S.* **64**, 863.

Rose, J. A., Maizel, J. V., Inman, J. K., and Shatkin, A. J. (1971). Structural proteins of adenovirus-associated viruses. *J. Virol.* **8**, 766.

Rosen, L. (1960). A hemagglutination-inhibition technique for typing adenoviruses. *Amer. J. Hyg.* **71**, 120.

Rosen, L. (1964). Hemagglutination. *In* "Techniques in Experimental Virology" (R. J. C. Harris, ed.), p. 257. Academic Press, New York.

Rosen, L. (1969). Hemagglutination with animal viruses. *In* "Fundamental Techniques in Virology" (K. Habel and N. P. Salzman, eds.), p. 276. Academic Press, New York.

Rosen, L., Hovis, J. F., and Bell, J. A. (1962). Observations on some little-known adenoviruses. *Proc. Soc. Exp. Biol. Med.* **111**, 166.

Rosen, L., Melnick, J. L., Schmidt, N. J., and Wenner, H. A. (1970). Subclassification of enteroviruses and ECHO virus type 34. *Arch. Ges. Virusforsch.* **30**, 89.

Rosenbergova, M., Lacour, F., and Huppert, J. (1965). Mise en evidence d'une activité nucléasique associée an virus de la myéloblastose aviaire, lors de tentatives de purification de ce virus et de son acide ribonucléique *C. R. Acad. Sci. (Paris)* **260**, 5145.

Rosenkranz, H. S., Rose, H. M., Morgan, C., and Hsu, K. C. (1966). The effect of hydroxyurea on virus development. II. Vaccinia virus. *Virology* **28**, 510.

Rosenthal, P. N., Robinson, H. L., Robinson, W. S., Hanafusa, T., and Hanafusa, H. (1971). DNA in uninfected and virus-infected cells complementary to avian tumor virus RNA. *Proc. Nat. Acad. Sci. U.S.* **68**, 2336.

Rossen, R. D., Kasel, J. A., and Couch, R. B. (1971). The secretory immune system: its relation to respiratory viral infection. *Progr. Med. Virol.* **13**, 194.

Rotem, Z., Cox, R. A., and Isaacs, A. (1963). Inhibition of virus multiplication by foreign nucleic acid. *Nature (London)* **197**, 564.

Rotem, Z., Berwald, Y., and Sachs, L. (1964). Inhibition of interferon production in hamster cells transformed *in vitro* with carcinogenic hydrocarbons. *Virology* **24**, 483.

Rothschild, H., and Black, P. H. (1970). Analysis of SV40-induced transformation of hamster kidney tissue *in vitro*. VII. Induction of SV40 virus from transformed hamster cell clones by various agents. *Virology* **42**, 251.

Rothschild, M., and Clay, T. (1961). "Fleas, Flukes and Cuckoos: a Study of Bird Parasites." Arrow Books, London.

Rott, R., and Scholtissek, C. (1963). Investigations about the formation of incomplete forms of fowl plague virus. *J. Gen. Microbiol.* **33**, 303.

Rott, R., and Scholtissek, C. (1970). Specific inhibition of influenza replication by α-amanitin. *Nature (London)* **228**, 56.

Rott, R., Drzeniek, R., Saber, M. S., and Reichert, E. (1966). Blood group substances, Forssman and mononucleosis antigens in lipid-containing RNA viruses. *Arch. Ges. Virusforsch.* **19**, 273.

Rott, R., Becht, H., Klenk, H. -D., and Scholtissek, C. (1972). Interactions of concanavalin A with the membrane of influenza virus infected cells and with envelope components of the virus particle. Z. *Naturforsch.* **27b**, 227.

Roumiantzeff, M., Summers, D. F., and Maizel, J. V. (1971). *In vitro* protein synthetic activity of membrane-bound poliovirus polyribosomes. *Virology* **44**, 249.

Rous, P. (1910). A transmissible avian neoplasm (sarcoma of the common fowl). *J. Exp. Med.* **12**, 696.

Rous, P. (1911). Transmission of a malignant new growth by means of a cell-free filtrate. *J. Amer. Med. Assoc.* **56**, 198.

Rous, P., and Beard, J. W. (1935). The progression to carcinoma of virus-induced rabbit papilloma (Shope). *J. Exp. Med.* **62**, 523.

Rous, P., and Murphy, J. B. (1914). On the causation by filterable agents of three distinct chicken tumors. *J. Exp. Med.* **19**, 52.

Rouse, H. C., and Schlesinger, R. W. (1972). The effects of arginine starvation on macromolecular synthesis in infection with type 2 adenovirus. I. Synthesis and utilization of structural proteins. *Virology* **48**, 463.

Rovera, G., Baserga, R., and Defendi, V. (1972). Early increase in nuclear acidic protein synthesis after SV40 infection. *Nature New Biol.* **237**, 240.

Rowe, W. P. (1961). The epidemiology of mouse polyoma virus infection. *Bacteriol. Rev.* **25**, 18.

Rowe, W. P. (1965). Studies of adenovirus-SV40 hybrid viruses. III. Transfer of SV40 gene between adenovirus types. *Proc. Nat. Acad. Sci. U.S.* **54**, 711.

Rowe, W. P. (1967). Some interactions of defective animal viruses. *Perspect. Virol.* **5**, 123.

Rowe, W. P. (1971). The kinetics of rescue of the murine sarcoma virus genome from a non-producer line of transformed mouse cells. *Virology* **46**, 369.

Rowe, W. P., and Baum, S. G. (1964). Evidence for a possible genetic hybrid between adenovirus type 7 and SV40 viruses. *Proc. Nat. Acad. Sci. U.S.* **52**, 1340.

Rowe, W. P., and Baum, S. G. (1965). Studies on adenovirus SV40 hybrid viruses. II. Defectiveness of the hybrid particles. *J. Exp. Med.* **122**, 955.

Rowe, W. P., and Capps, W. I. (1961). A new mouse virus causing necrosis of the thymus in new born mice. *J. Exp. Med.* **113**, 831.

Rowe, W. P., and Lewis, A. M. (1968). Serologic surveys for viral antibodies in cancer patients. *Cancer Res.* **28**, 13.

Rowe, W. P. and Pincus, T. (1972). Quantitative studies of naturally occurring murine leukemia virus infection of AKR mice. *J. Exp. Med.* **135**, 429.

Rowe, W. P., and Pugh, W. E. (1966). Studies of adenovirus-SV40 hybrid viruses. V. Evidence for linkage between adenovirus and SV40 genetic materials. *Proc. Nat. Acad. Sci. U.S.* **55**, 1126.

Rowe, W. P., Huebner, R. J., Gilmore, L. K., Parrott, R. H., and Ward, T. G. (1953). Isolation of a cytopathogenic agent from human adenoids undergoing spontaneous degeneration in tissue culture. *Proc. Soc. Exp. Biol. Med.* **84**, 570.

Rowe, W. P., Hartley, J. W., Roizman, B., and Levy, H. B. (1958). Characterization of a factor formed in the course of adenovirus infection of tissue cultures causing detachment of cells from glass. *J. Exp. Med.* **108**, 713.

Rowe, W. P., Black, P. H., and Levey, R. H. (1963). Protective effect of neonatal thymectomy on mouse LCM infection. *Proc. Soc. Exp. Biol. Med.* **114**, 248.

Rowe, W. P., Baum, S. G., Pugh, W. E., and Hoggan, M. D. (1965). Studies of adenovirus SV40 hybrid viruses. I. Assay system and further evidence of hybridization. *J. Exp. Med.* **122**, 943.

Rowe, W. P., Hartley, J. W., and Capps, W. I. (1966). Tissue culture and serologic studies of mouse leukemia viruses. *Nat. Cancer Inst. Monogr.* **22**, 15.

Rowe, W. P., Murphy, F. A., Bergold, G. H., Casals, J., Hotchin, J., Johnson, K. M., Lehmann-Grube, F., Mims, C. A., Traub, E., and Webb, P. A. (1970a). Arenoviruses: a proposed name for a newly defined group. *J. Virol.* **5**, 651.

Rowe, W. P., Pugh, W. E., and Hartley, J. W. (1970b). Plaque assay techniques for murine leukemia viruses. *Virology* **42**, 1136.

Rowe, W. P., Pugh, W. E., Webb, P. A., and Peters, C. J. (1970c). Serological relationship of the Tacaribe complex of viruses to lymphocytic choriomeningitis virus. *J. Virol.* **5**, 289.

Rowe, W. P., Hartley, J. W., Lander, M. R., Pugh, W. E., and Teich, N. (1971). Non-infectious AKR mouse embryo cell lines in which each cell has the capacity to be activated to produce infectious murine leukemia virus. *Virology* **46**, 866.

Rowley, D. (1962). Phagocytosis. *Advan. Immunol.* **2**, 241.

Royston, I., and Aurelian, L. (1970). Immunofluorescent detection of herpesvirus antigens in exfoliated cells from human cervical carcinoma. *Proc. Nat. Acad. Sci. U.S.* **67**, 204.

Royston, I., Aurelian, L., and Davies, H. J. (1970). Genital herpesvirus findings in relation to cervical neoplasia. *J. Reprod. Med.* **4**, 109.

Rubenstein, D., Milne, R. G., Buckland, R., and Tyrrell, D. A. J. (1971). The growth of the virus of epidemic diarrhoea of infant mice (EDIM) in organ cultures of intestinal epithelium. *Brit. J. Exp. Pathol.* **52**, 442.

Rubin, H. (1957). The production of virus by Rous sarcoma cells. *Ann. N. Y. Acad. Sci.* **68**, 459.

Rubin, H. (1960). A virus in chick embryos which induces resistance *in vitro* to infection with Rous sarcoma virus. *Proc. Nat. Acad. Sci. U.S.* **46**, 1105.

Rubin, H. (1961). The nature of a virus-induced cellular resistance to Rous sarcoma virus. *Virology* **13**, 200.

Rubin, H. (1962). The immunological basis for non-infective Rous sarcomas. *Cold Spring Harbor Symp. Quant. Biol.* **27**, 441.

Rubin, H. (1965). Genetic control of cellular susceptibility to pseudotypes of Rous sarcoma virus. *Virology* **26**, 270.

Rubin, H., Cornelius, A., and Fanshier, L. (1961). The pattern of congenital transmission of an avian leukosis virus. *Proc. Nat. Acad. Sci. U.S.* **47**, 1058.

Rubin, H., Fanshier, L., Cornelius, A., and Hughes, W. F. (1962). Tolerance and immunity after congenital and contact infection with an avian leukosis virus. *Virology* **17**, 143.

Rueckert, R. R. (1971). Picornaviral architecture. *In* "Comparative Virology" (K. Maramorosch and E. Kurstak, eds.), p. 256. Academic Press, New York.

Rueckert, R. R., Dunker, A. K., and Stoltzfors, C. M. (1969). The structure of mouse-Elberfeld virus: a model. *Proc. Nat. Acad. Sci. U.S.* **62**, 912

Ruiz-Gomez, J., and Isaacs, A. (1963). Interferon production by different viruses. *Virology* **19**, 8.

Ruiz-Gomez, J., and Sosa-Martinez, J. (1965). Virus multiplication and interferon production at different temperatures in adult mice infected with Coxsackie B1 virus. *Arch. Ges. Virusforsch.* **17**, 295.

Rupert, C. S., and Harm, W. (1966). Reactivation after photobiological damage. *Advan. Radiat. Biol.* **2**, 1.

Russell, P. K. (1971). Pathogenesis of the dengue shock syndrome: Evidence for an immunologic mechanism. *In* "Immunopathology" (P. A. Miescher, ed.), p. 426. Grune & Stratton, New York.

Russell, W. C., and Becker, Y. (1968). A maturation factor for adenovirus. *Virology* **35**, 18.

Russell, W. C., and Skehel, J. J. (1972). The polypeptides of adenovirus-infected cells. *J. Gen. Virol.* **15**, 45.

Russell, W. C., Hayashi, K., Sanderson, P. J., and Pereira, H. G. (1967). Adenovirus antigens— a study of their properties and sequential development in infection. *J. Gen. Virol.* **1**, 495.

Russell, W. C., McIntosh, K., and Skehel, J. J. (1971). The preparation and properties of adenovirus cores. *J. Gen. Virol.* **11**, 35.

Rustigian, R., Smurlow, J. B., Tye, M., Gibson, W. A., and Shindell, E. (1966). Studies on latent infection of skin and oral mucosa in individuals with recurrent herpes simplex. *J. Invest. Dermatol.* **47**, 218.

Rytel, M. W., and Kilbourne, E. D. (1966). The influence of cortisone on experimental viral infection. VIII. Suppression by cortisone of interferon production in mice injected with Newcastle disease virus. *J. Exp. Med.* **123**, 767.

Sabin, A. B. (1954). Genetic factors affecting susceptibility and resistance to virus disease of the nervous system. *Res. Publ. Ass. Res. Nervous Mental Dis.* **33**, 57.

Sabin, A. B. (1956). Pathogenesis of poliomyelitis (reappraisal in light of new data). *Science* **123**, 1151.

Sabin, A. B. (1957). Properties of attenuated polioviruses and their behavior in human beings. *In* "Cellular Biology, Nucleic Acids and Viruses" (T. M. Rivers, ed). Spec. Publ. Vol. 5. p. 113. N. Y. Acad. Sci., New York.

Sabin, A. B. (1959a). Reoviruses. *Science* **130**, 1387.

Sabin, A. B. (1959b). Characteristics of naturally acquired immunity in poliomyelitis and of immunity induced by killed and live-virus vaccine. *In* "Immunity and Virus Infection" (V. A. Najjar, ed.), p. 211. Wiley, New York.

Sabin, A. B. (1961). Reproductive capacity of polioviruses of diverse origins at various temperatures. *Perspec. Virol.* **2**, 90.

Sabin, A. B. (1962). Oral poliovirus vaccine, recent results and recommendations for optimum use. *Roy. Soc. Health J.* **82**, 51.

Sabin, A. B. (1963). Poliomyelitis in the tropics. Increasing incidence and prospects for control. *Trop. Geogr. Med.* **15**, 38.

Sabin, A. B. (1967a). Amantadine hydrochloride: analysis of data related to its proposed use for prevention of A2 influenza virus disease in human beings. *J. Amer. Med. Ass.* **200**, 943.

Sabin, A. B. (1967b). Poliomyelitis: accomplishments of live virus vaccine. *In* First International Conference on Vaccines against Viral and Rickettsial Diseases of Man," p. 171, Sci. Publ. No. 147. Pan American Health. Organ., Washington, D.C.

Sabin, A. B., and Blumberg, R. W. (1947). Human infection with Rift Valley fever virus and immunity twelve years after single attack. *Proc. Soc. Exp. Biol. Med.* **64**, 385.

Sabin, A. B., and Koch, M. A. (1963a). Evidence of continuous transmission of noninfectious SV40 viral genome in most or all of SV40 hamster tumor cells. *Proc. Nat. Acad. Sci. U.S.* **49**, 304.

Sabin, A. B., and Koch, M. A. (1963b). Behavior of noninfectious SV40 viral genome in hamster tumor cells: Induction of synthesis of infectious virus. *Proc. Nat. Acad. Sci. U.S.* **50**, 407.

Sabin, A. B., and Olitsky, P. K. (1937a). Influence of host factors on neuroinvasiveness of vesicular stomatitis virus. I. Effect of age on the invasion of the brain by virus instilled in the nose. *J. Exp. Med.* **66**, 15.

Sabin, A. B., and Olitsky, P. K. (1937b). Influence of host factors on neuroinvasiveness of vesicular stomatitis virus. II. Effect of age on the invasion of the peripheral and central nervous systems by virus injected into the leg muscles or the eye. *J. Exp. Med.* **66**, 35.

Sabin, A. B., and Schlesinger, R. W. (1945). Production of immunity to dengue with virus modified by propagation in mice. *Science* **101**, 640.

Sabin, A. B., Ramos-Alvarez, M., Alvarez-Amezquita, J., Pelon, W., Michaels, R. H., Spigland, I., Koch, M. A., Barnes, J. M., and Rhim, J. S. (1960). Live, orally given poliovirus vaccine. Effects of rapid mass immunization on population under conditions of massive enteric infection with other viruses. *J. Amer. Med. Ass.* **173**, 1521.

Saikku, P., von Bonsdorff, C-H, and Oker-Blom, N. (1970). The structure of Uukuniemi virus. *Acta Virol.* **14**, 103.

Saikku, P., von Bonsdorff, C-H., Brummer-Korvenkontio, M., and Valieri, A. (1971). Isolation of non-cubic ribonucleoprotein from Inkoo virus, a Bunyamwera supergroup arbovirus. *J. Gen. Virol.* **13**, 335.

Sakuma, S., and Watanabe, Y. (1971). Unilateral synthesis of reovirus double-stranded ribonucleic acid by a cell-free replicase system. *J. Virol.* **8**, 190.

Salaman, M. H., Wedderburn, N., and Bruce-Chwatt, L. J. (1969). The immunodepressive effect of a murine plasmodium and its interaction with murine oncogenic viruses. *J. Gen. Microbiol.* **59**, 383.

Salb, J. M., and Marcus, P. I. (1965). Translational inhibition in mitotic HeLa cells. *Proc. Nat. Acad. Sci. U.S.* **54**, 1353.

Salivar, W. O., Henry, T. J., and Pratt, D. (1967). Purification and properties of diploid particles of coliphage M13. *Virology* **32**, 41.

Salk, J. E. (1958). Basic principles underlying immunization against poliomyelitis with a noninfectious vaccine. *In* "Poliomyelitis: Papers and Discussions Presented at the Fourth International Poliomyelitis Conference," p. 66. Lippincott, Philadelphia, Pennsylvania.

Salser, W., Bolle, A., and Epstein, R. (1970). Transcription during bacteriophage T4 development: A demonstration that distinct subclasses of the "early" RNA appear at different times and that some are "turned off" at late times. *J. Mol. Biol.* **49**, 271.

Salzman, L. A. (1971). DNA polymerase activity associated with purified Kilham rat virus. *Nature New Biol.* **231**, 174.

Salzman, L. A., and White, W. L. (1970). The structural proteins of Kilham rat virus. *Biochem. Biophys. Res. Commun.* **41**, 1511.

Salzman, N. P., and Moss, B. (1969). Analysis of radioactively labeled proteins by immunodiffusion. *In* "Fundamental Techniques in Virology" (K. Habel and N. P. Salzman, eds.), p. 327. Academic Press, New York.

Salzman, N. P., and Sebring, E. D. (1967). Sequential formation of vaccinia virus proteins and viral deoxyribonucleic acid. *J. Virol.* **1**, 16.

Sambrook, J. (1972). Transformation by polyoma virus and simian virus 40. *Advan. Cancer Res.* **16**, 14.

Sambrook, J., and Pollack, R. (1973). Isolation and culture of cells—basic methodology for cell transformation, *In* "Methods in Enzymology," Vol. 29, in press. Academic Press, New York.

Sambrook, J., and Shatkin, A. J. (1969). Polynucleotide ligase activity in cells infected with simian virus 40, polyoma virus, or vaccinia virus. *J. Virol.* **4**, 719.

Sambrook, J. F., McClain, M. E., Easterbrook, K. B., and McAuslan, B. R. (1965). A mutant of rabbitpox virus defective at different stages of its multiplication in three cell types. *Virology* **26**, 738.

Sambrook, J. F., Padgett, B. L., and Tomkins, J. N. (1966). Conditional lethal mutants of rabbitpox virus. I. Isolation of host cell-dependent and temperature-dependent mutants. *Virology* **28**, 592.

Sambrook, J. F., Westphal, H., Srinivasan, P. R., and Dulbecco, R. (1968). The integrated state of viral DNA in SV40-transformed cells. *Proc. Nat. Acad. Sci. U.S.* **60**, 1288.

Sambrook, J., Sharp, P. A., and Keller, W. (1972). Transcription of SV40 DNA I. Strand separation and hybridization of the separated strands to RNA's isolated from lytically infected and transformed cells. *J. Mol. Biol.* **70**, 57.

Sanger, J. W., and Holtzer, H. (1972). Cytochalasin B: Effects on cell morphology, cell adhesion, and mucopolysaccharide synthesis. *Proc. Nat. Acad. Sci. U.S.* **69**, 253.

Sarkar, N. H., and Moore, D. H. (1972). On the possibility of a human breast cancer virus. *Nature (London)* **236**, 103.

Sarkar, N. H., Lasfargues, E. Y., and Moore, D. H. (1970). Attachment and penetration of mouse mammary tumor virus in mouse embryo cells. *J. Microsc. (Paris)* **9**, 477.

Sarma, P. S., and Log, T. (1971). Viral interference in feline leukemia sarcoma complex. *Virology* **44**, 352.

Sarma, P. S., Turner, H. C., and Huebner, R. J. (1964). An avian leukosis group-specific complement fixation reaction. Application for the detection and assay of noncytopathogenic leucosis viruses. *Virology* **23**, 313.

Sarma, P. S., Cheong, M., Hartley, J. W., and Huebner, R. J. (1967a). A viral interference test for mouse leukemia viruses. *Virology* **33**, 180.

Sarma, P. S., Vass, W., Huebner, R. J., Igel, H., Lane, W. T., and Turner, H. C. (1967b). Induction of tumours in hamsters with infectious canine hepatitis virus. *Nature (London)* **215**, 293.

Sarma, P. S., Huebner, R. J., and Lane, W. T. (1968). Induction of tumors in hamsters with an avian adenovirus (CELO). *Science* **149**, 1108.

Sarma, P. S., Log, T., and Gilden, R. V. (1970). Studies of hamster-specific oncogenic virus derived from hamster tumors induced by Kirsten murine sarcoma virus (34551). *Proc. Soc. Exp. Biol. Med.* **133**, 718.

Sarma, P. S., Gilden, R. V., and Huebner, R. J. (1971). Complement fixation test for the feline leukemia and sarcoma viruses. *Virology* **44**, 137.

Sarov, I., and Joklik, W. K. (1972a). Studies on the nature and location of capsid polypeptides of vaccinia virions. *Virology* **50**, 579.

Sarov, I., and Joklik, W. K. (1972b). Characterization of intermediates in the uncoating of vaccinia virus DNA. *Virology* **50**, 593.

Sauer, G. (1971). Apparent differences in transcriptional control in cells productively infected and transformed by SV40. *Nature New Biol.* **231**, 135.

Sauer, G., and Kidwai, J. R. (1968). The transcription of the SV40 genome in productively infected and transformed cells. *Proc. Nat. Acad. Sci. U.S.* **61**, 1256.

Sauer, G., Koprowski, H., and Defendi, V. (1967). The genetic heterogencity of simian virus 40. *Proc. Nat. Acad. Sci. U.S.* **58**, 599.

Saunders, G. F., Shirakawa, S., Saunders, P. P., Arrigni, F. E., and Hsu, T. C. (1972). Populations of repeated DNA sequences in the human genome. *J. Mol. Biol.* **63**, 323.

Sawicki, L. (1961). Influence of age of mice on the recovery from experimental Sendai virus infection. *Nature (London)* **192,** 1258.

Sawyer, W. A. (1931). Persistence of yellow fever immunity. *J. Prevent. Med.* **5,** 413.

Sawyer, W. A., Meyer, K. F., Eaton, M. D., Bauer, J. H., Putnam, P., and Schwenkter, F. F. (1944). Jaundice in Army personnel in the western region of the United States and its relation to vaccination against yellow fever. Part I. *Amer. J. Hyg.* **39,** 337; Parts II, III and IV, *ibid.* **40,** 35.

Sbarra, A. J., and Karnovsky, M. L. (1959). The biochemical basis of phagocytosis. I. Metabolic changes during the ingestion of particles by polymorphonuclear leukocytes. *J. Biol. Chem.* **234,** 1355.

Schabel, F. M. (1970). Purine and pyrimidine nucleosides as antiviral agents—recent developments. *Ann. N. Y. Acad. Sci.* **173,** 215.

Schabel, F. M., and Montgomery, J. A. (1972). Purines and pyrimidines. *In* "Chemotherapy of Virus Diseases" (D. J. Bauer, ed.), Vol. 1, p. 231. Intern. Encyclop. Pharmacol., Therapeutics Section 61., Pergamon, Oxford and New York.

Schäfer, W., deNoronha, Lange, J., and Bolognesi, D. P. (1971). Comparative studies on group specific antigens of leukemia viruses. *In* "Lepetit Colloquia on Biology and Medicine, Vol. 2: The Biology of Tumor Viruses" (L. G. Sylvestri, ed.) p. 116. North Holland, Amsterdam.

Schäfer, W., Lange, J., Fischinger, P. J., Frank, H., Bolognesi, D., and Pister, L. (1972). Properties of mouse leukemia viruses. II. Isolation of viral components. *Virology* **47,** 210.

Schafer, T. W., and Lockart, R. Z. (1970). Interferon required for viral resistance induced by poly I·poly C. *Nature (London)* **226,** 449.

Schaffer, P., Vonka, V., Leurs, R., and Benyesh-Melnick, M. (1971). Temperature-sensitive mutants of herpes simplex virus. *Virology* **42,** 1144.

Scharff, M. D., and Robbins, E. (1966). Polyribosome disaggregation during metaphase. *Science* **151,** 992.

Scharff, M. D., Shatkin, A. J., and Levintow, L. (1963). Association of newly formed viral protein with specific polyribosomes. *Proc. Nat. Acad. Sci. U.S.* **50,** 686.

Scheele, C. M., and Pfefferkorn, E. R. (1969). Inhibition of interjacent ribonucleic acid (26S) synthesis in cells infected by Sindbis virus. *J. Virol.* **4,** 117.

Scheid, A., Caliguiri, L. A., Compans, R. W., and Choppin, P. W. (1972). Isolation of paramyxovirus glycoproteins. Association of both hemagglutinating and neuraminidase activities with the larger SV5 glycoprotein. *Virology* **50,** 640.

Schell, K. (1960a). Studies on the innate resistance of mice to infection with mousepox. I. Resistance and antibody production. *Aust. J. Exp. Biol. Med. Sci.* **38,** 271.

Schell, K. (1960b). Studies on the innate resistance of mice to infection with mousepox. II. Route of inoculation and resistance; and some observations on the inheritance of resistance. *Aust. J. Exp. Biol. Med. Sci.* **38,** 289.

Schell, K., Lane, W. T., Casey, M. J., and Huebner, R. J. (1966). Potentiation of oncogenicity of adenovirus type 12 grown in Africa green monkey kidney cell cultures preinfected with SV40 virus: Persistence of both T antigens in the tumors and evidence for possible hybridization. *Proc. Nat. Acad. Sci. U.S.* **55,** 81.

Scherrer, K., and Marcaud, L. (1965). Remarques sur les ARN messagers polycistroniques dans les cellules animales. *Bull. Soc. Chim. Biol.* **47,** 1697.

Scherrer, K., Latham, H., and Darnell, J. E. (1963). Demonstration of an unstable RNA and of a precursor to ribosomal RNA in HeLa cells. *Proc. Nat. Acad. Sci. U.S.* **49,** 240.

Schild, G. C. (1972). Evidence for a new type-specific structural antigen of the influenza virus particle. *J. Gen. Virol.* **15,** 99.

Schild, G. C., Pereira, H. G., and Schettler, C. H. (1969). Neuraminidase in avian influenza A virus antigenically related to that of human A0 and A1 subtypes. *Nature (London)* **222,** 1299.

Schild, G. C., Henry-Aymard, M., and Pereira, H. G. (1972). A quantitative, single-radial-diffusion test for immunological studies with influenza virus. *J. Gen. Virol.* **16,** 231.

Schincariol, A. L., and Howatson, A. F. (1970). Replication of vesicular stomatitis virus. 1. Viral specific RNA and nucleoprotein in infected L cells. *Virology* **42**, 732.

Schlesinger, M. J., Schlesinger, S., and Burge, B. W. (1972). Identification of a second glycoprotein in Sindbis virus. *Virology* **47**, 539.

Schlesinger, R. W. (1959). Interference between animal viruses. *In* "The Viruses" (F. M. Burnet and W. M. Stanley, eds.), Vol. 3, p. 157. Academic Press, New York.

Schlesinger, R. W. (1969). Adenoviruses: The nature of the virion and of controlling factors in productive or abortive infection and tumorigenesis. *Advan. Virus Res.* **14**, 1.

Schlesinger, S., Schlesinger, M., and Burge, B. W. (1972). Defective virus particles from Sindbis virus. *Virology* **48**, 615.

Schlom, J., and Spiegelman, S. (1971). DNA polymerase activities and nucleic acid components of virions isolated from a spontaneous mammary carcinoma from a Rhesus monkey. *Proc. Nat. Acad. Sci. U.S.* **68**, 1613.

Schlom, J., Spiegelman, S., and Moore, D. H. (1971). RNA-dependent DNA polymerase activity in virus-like particles isolated from human milk. *Nature (London)* **231**, 97.

Schlom, J., Spiegelman, S., and Moore, D. H. (1972). Reverse transcriptase and high molecular weight RNA in particles from mouse and human milk. *J. Nat. Cancer. Inst.* **48**, 1197.

Schmidt, N. J. (1969a). Tissue culture methods and procedures for diagnostic virology. *In* "Diagnostic Procedures for Viral and Rickettsial Infections", (E. H. Lennette and N. J. Schmidt, eds.), 4th Ed. p. 79. Amer. Public Health Ass., New York.

Schmidt, N. J. (1969b). Complement fixation technique for assay of viral antigens and antibodies. *In* "Fundamental Techniques in Virology" (K. Habel and N. P. Salzman, eds.), p. 263. Academic Press, New York.

Schmidt, N. J., Lennette, E. H., and Magoffin, R. L. (1969). Immunological relationship between herpes simplex and varicella-zoster viruses demonstrated by complement-fixation, neutralization and fluorescent antibody tests. *J. Gen. Virol.* **4**, 321.

Schneider, I. R., Diener, T. O., and Safferman, R. S. (1964). Blue-green algal virus LPP-1: Purification and partial characterization. *Science* **144**, 1127.

Scholtissek, C. (1969). Synthesis *in vitro* of RNA complementary to parental viral RNA by RNA polymerase induced by influenza virus. *Biochim. Biophys. Acta* **179**, 389.

Scholtissek, C., and Rott, R. (1964). Behavior of virus specific activities in tissue cultures infected with myxoviruses after chemical changes of the viral ribonucleic acid. *Virology* **22**, 169.

Scholtissek, C., and Rott, R. (1970). Synthesis *in vivo* of influenza virus plus and minus strand RNA and its preferential inhibition by antibiotics. *Virology* **40**, 989.

Scholtissek, C., Drzeniek, R., and Rott, R. (1969). Myxoviruses. *In* "The Biochemistry of Viruses" (H. B. Levy, ed), p. 219. Dekker, New York.

Schonberg, M., Silverstein, S. C., Levin, D. H., and Acs, G. (1971). Asynchronous synthesis of the complementary strands of the reovirus genome. *Proc. Nat. Acad. Sci. U.S.* **68**, 505.

Schulman, J. L. (1967). Experimental transmission of influenza virus infection in mice. IV. Relationship of transmissibility of different strains of virus and recovery of airborne virus in the environment of infector mice. *J. Exp. Med.* **125**, 479.

Schulman, J. L. (1968). Effect of 1-amantanamine hydrochloride (Amantadine HCl) and methyl-1-adamantanethylamine hydrochloride (Rimantadine HCl) on transmission on influenza virus infection in mice. *Proc. Soc. Exp. Biol. Med.* **128**, 1173.

Schulman, J. L., and Kilbourne, E. (1967). Seasonal variations in transmission of influenza virus infection in mice. *In* "Biometerology II (S. W. Tromp and W. H. Weihe, eds.), p. 83. Macmillan (Pergamon), New York.

Schulman, J. L., Khakpour, M., and Kilbourne, E. D. (1968). Protective effects of specific immunity to viral neuraminidase on influenza virus infection of mice. *J. Virol.* **2**, 778.

Schulze, I. T. (1970). The structure of influenza virus. I. The polypeptides of the virion. *Virology* **42**, 890.

Schulze, I. T. (1972). The structure of influenza virus. II. A model based on the morphology and composition of subviral particles. *Virology* **47**, 181.

Schur, P. H., Borel, H., Gelfand, E. W., Heper, C. A., and Rosen, F. (1970). Selective gamma-G globulin deficiencies in patients with recurrent pyogenic infections. *New Engl. J. Med.* **283**, 631.

Schwabe, C. W. (1969). "Veterinary Medicine and Human Health" 2nd Ed., p. 229. Williams & Wilkins, Baltimore, Maryland.

Schwartz, J., and Dales, S (1971). Biogenesis of poxviruses: identification of four enzyme activities within purified Yaba tumor virus. *Virology* **45**, 797.

Schwarz, A. J. F. (1964). Immunization against measles: Development and evaluation of a highly attenuated live measles vaccine. *Ann. Paediat.* **202**, 241.

Scientific Committee on Interferon (1962). Effect on interferon on vaccination in volunteers. *Lancet* **1**, 873.

Scolnick, E. M., Parks, W. P., Todaro, G. J., and Aaronson, S. A. (1972). Immunological characterization of primate C-type virus reverse transcriptases. *Nature New Biol.* **235**, 35.

Scott, H. D. (1971). The elusiveness of measles eradication: insights gained from three years of intensive surveillance in Rhode Island. *Amer. J. Epidemiol.* **94**, 37.

Scrimshaw, N. S. (1964). Nutrition and stress. *Ciba Found. Study Group* **17**, 40.

Scrimshaw, N. S., Taylor, C. E. and Gordon, J. E. (1968). Interactions of nutrition and infection. *WHO Monogr. Ser. No.* **57**.

Sebring, E. D., and Salzman, N. P. (1967). Metabolic properties of early and late vaccinia virus messenger ribonucleic acid. *J. Virol.* **1**, 550.

Sebring, E. D., Kelly, T. J., Thoren, M. M., and Salzman, N. P. (1971). Structure of replicating simian virus 40 deoxyribonucleic acid molecules. *J. Virol.* **8**, 478.

Sedwick, W. D., and Sokol, F. (1970). Nucleic acid of rubella virus and its replication in hamster kidney cells. *J. Virol.* **5**, 478.

Sellers, R. F., Donaldson, A. I., and Herniman, K. A. J. (1970). Inhalation, persistence and dispersal of foot-and-mouth disease virus by man. *J. Hyg.* **68**, 565.

Sergiescu, D., Horodniceanu, F., and Aubert-Combiescu, A. (1972). The use of inhibitors in the study of picornavirus genetics. *Progr. Med. Virol.* **14**, 123.

Setlow, J. K. (1964). Effects of U. V. on DNA: correlations among biological changes, physical changes and repair mechanisms. *Photochem. Photobiol.* **3**, 405.

Sever, J. L., and Zeman, W. (eds.) (1968). Conference on measles virus and subacute sclerosing panencephalitis. *Neurology* **18**, part 2.

Sever, J. L., Huebner, R. J., Fabiyi, A., Monif, G. R., Castellano, G., Cusumano, C. L., Traub, R. G., Ley, A. C., Gilkeson, M. R., and Roberts, J. M. (1966). Antibody responses in acute and chronic rubella. *Proc. Soc. Exp. Biol. Med.* **122**, 513.

Shah, K. V. (1972). Evidence for an SV40-related papovavirus infection of man. *Amer. J. Epidemiol.* **95**, 199.

Shall, S., and McClelland, A. J. (1971). Synchronization of mouse fibroblast LS cells grown in suspension culture. *Nature New Biol.* **229**, 59.

Shanmugam, G., Vecchio, G., Attardi, D., and Green, M. (1972). Immunological studies on viral polypeptide synthesis in cells replicating murine sarcoma-leukemia virus. *J. Virol.* **10**, 447.

Shapiro, D., Brandt, W. E., Cardiff, R. D., and Russell, P. K. (1971). The proteins of Japanese encephalitis virus. *Virology* **44**, 108.

Sharp, D. G. (1965). Quantitative use of the electron microscope in virus research. Methods and recent results of particle counting. *Lab. Invest.* **14**, 831.

Sharp, D. G. (1968). Multiplicity reactivation of animal viruses. *Progr. Med. Virol.* **10**, 64.

Sharp, P. A., Sugden, W., and Sambrook, J. (1973). Detection of two restriction endonucleases in *Hemophilus parainfluenzae* using analytical ethidium bromide electrophoresis. *Biochemistry*, in press.

Shatkin, A. J. (1968). Viruses containing double-stranded RNA. *In* "Molecular Basis of Virology" (H. Fraenkel-Conrat, ed.), p. 351. Reinholt, New York.

Shatkin, A. J. (1969). Replication of reovirus. *Advan. Virus Res.* **14**, 63.

Shatkin, A. J. (1971). Viruses with segmented ribonucleic acid genomes: multiplication of influenza versus reovirus. *Bacteriol. Rev.* **35**, 250.

Shatkin, A. J., and Sipe, J. D. (1968a). Single-stranded, adenine-rich RNA from purified reoviruses. *Proc. Nat. Acad. Sci. U.S.* **59**, 246.

Shatkin, A. J., and Sipe, J. D. (1968b). RNA polymerase activity in purified reoviruses. *Proc. Nat. Acad. Sci. U.S.* **61**, 1462.

Shatkin, A. J., Sipe, J. D., and Loh, P. (1968). Separation of ten reovirus genome segments by polyacrylamide gel electrophoresis. *J. Virol.* **2**, 986.

Shea, M. A., and Plagemann, P. G. W. (1971) Effects of elevated temperatures on Mengovirus ribonucleic acid synthesis and virus production in Novikoff rat hepatoma cells. *J. Virol.* **7**, 144.

Shein, H., and Enders, J. F. (1962). Transformation induced by simian virus 40 in human renal cell cultures. I. Morphology and growth characteristics. *Proc. Nat. Acad. Sci. U.S.* **48**, 1164.

Shelokov, A., Vogel, J. E., and Chi, L. (1958). Hemadsorption (adsorption-hemagglutination) test for viral agents in tissue culture with special reference to influenza. *Proc. Soc. Exp. Biol. Med.* **97**, 802.

Shelton, E., and Kuff, E. L. (1966). Substructure and configuration of ribosomes isolated from mammalian cells. *J. Mol. Biol.* **22**, 23.

Sheppard, J. R., Levine, A. J., and Burger, M. (1971). Cell surface changes after infection with oncogenic viruses: requirement for synthesis of host DNA. *Science* **172**, 1345.

Shevliaghyn, V. J., and Karazas, N. V. (1970). Transformation of human cells by polyoma and Rous sarcoma viruses mediated by inactivated Sendai virus. *Intern J. Cancer* **6**, 234.

Shevliaghyn, V. J., Biryulina, T. I., Tikhonova, Z. N., and Karazas, N. V. (1969). Activation of Rous virus in the transplanted golden hamster tumor with the aid of artificial heterokaryon formation. *Intern. J. Cancer* **4**, 42.

Shibli, M., Good, S., Lewis, H. E., and Tyrrell, D. A. J. (1971). Common colds on Tristan da Cunha. *J. Hyg.* **69**, 255.

Shif, I., and Bang, F. B. (1966). Plaque assay for mouse hepatitis virus (MHV-2) on primary macrophage cell cultures. *Proc. Soc. Exp. Biol. Med.* **121**, 829.

Shif, I., and Bang, F. B. (1970). In vitro interaction of mouse hepatitis virus and macrophages from genetically resistant mice. I. Adsorption of virus and growth curves. *J. Exp. Med.* **131**, 843.

Shiroki, K., and Shimojo, H. (1971). Transformation of green monkey kidney cells by SV40 genome: the establishment of transformed cell lines and the replication of human adenoviruses and SV40 in transformed cells. *Virology* **45**, 163.

Shope, R. E. (1933). Infectious papillomatosis of rabbits. *J. Exp. Med.* **58**, 607.

Shope, R. E. (1935a). Experiments on the epidemiology of pseudorabies. I. Model of transmission in swine and their possible role in its spread to cattle. *J. Exp. Med.* **62**, 85.

Shope, R. E. (1935b). Experiments on the epidemiology of pseudorabies. II. Prevalence of the disease among middle western swine and the possible role of rats in herd-to-herd infections. *J. Exp. Med.* **62**, 101.

Shope, R. E. (1943). The swine lungworm as a reservoir and intermediate host for swine influenza virus. III. Factors influencing transmission of the virus and the provocation of influenza. *J. Exp. Med.* **77**, 111.

Shope, R. E. (1955). The swine lungworm as a reservoir and intermediate host for swine influenza virus. V. Provocation of swine influenza by exposure of prepared swine to adverse weather. *J. Exp. Med.* **102**, 567.

Shope, R. E. (1958). The swine lungworm as a reservoir and intermediate host for hog cholera virus. I. The provocation of masked hog cholera virus in lungworm-infected swine by ascaris larvae. *J. Exp. Med.* **107**, 609.

Shope, R. E. (1964). The birth of a new disease. *In* "Newcastle Disease Virus: An Evolving Pathogen" (R. P. Hanson, ed.), p. 3. Univ. of Wisconsin Press, Madison, Wisconsin.

Shope, R. E., and Hurst, E. W. (1933). Infectious papillomatosis of rabbits. *J. Exp. Med.* **58**, 607.

Shope, R. E., Murphy, F. A., Harrison, A. K., Causey, O. R., Kemp, G. E., Simpson, D. I. H., and Moore, D. L. (1970). Two African viruses serologically and morphologically related to rabies virus. *J. Virol.* **6**, 690.

Shugar, D. (ed.) (1972a). "Virus-Cell Interactions and Viral Antimetabolites." FEBS Symposium, Vol. 22. Academic Press, New York.

Shugar, D. (1972b). Ankylated pyrimidine nucleosides and (poly)nucleotides as potential antiviral agents. In "Virus-Cell Interactions and Viral Antimetabolites" (D. Shugar, ed.) p. 193, FEBS Symposium, Vol. 22. Academic Press, New York.

Shulman, N. R., and Barker, L. F. (1969). Virus-like antigen, antibody, and antigen-antibody complexes in hepatitis measured by complement fixation. *Science* **165**, 304.

Shwartzman, G., and Fisher, A. (1952). Alteration of experimental poliomyelitis infection in the Syrian hamster with the aid of cortisone. *J. Exp. Med.* **95**, 347.

Schwartzman, G., Aronson, S. M., Teodoru, C. V., Adler, M., and Jahiel, R. (1955). Endocrinological aspects of the pathogenesis of experimental poliomyelitis. *Ann. N. Y. Acad. Sci.* **61**, 869.

Sigurdsson, B. (1954). Observations on three slow infections of sheep. *Brit. Vet. J.* **110**, 255, 307, 341.

Sigurdsson, B., Grunsson, H., and Pálsson, P. A. (1952). Maedi, a chronic progressive infection of sheep's lungs. *J. Infect. Dis.* **90**, 233.

Sigurdsson, B., Pálsson, P. A., and Trygovadóttir, A. (1953). Transmission experiments with maedi. *J. Infect. Dis.* **93**, 166.

Sigurdsson, B., Pálsson, P. A., and Grimsson, H. (1957). Visna, a demyelinating disease of sheep. *J. Neuropathol. Exp. Neurol.* **16**, 389.

Siegel, M., and Greenberg, M. (1955). Incidence of poliomyelitis in pregnancy. Its relation to maternal age, parity and gestational period. *New Engl. J. Med.* **253**, 841.

Siegel, M., and Greenberg, M. (1960). Fetal death, malformation, and prematurity after maternal rubella—results of a prospective study 1949–1958. *New Engl. J. Med.* **262**, 389.

Siegert, R. (1972). Marburg virus. *Virol. Monogr.* **11**, 97.

Siegert, R., and Braune, P. (1964a). The pyrogens of myxoviruses. I. Induction of hyperthermia and its tolerance. *Virology* **24**, 209.

Siegert, R., and Braune, P. (1964b). The pyrogens of myxoviruses. II. Resistance of influenza A pyrogens to heat, ultraviolet, and chemical treatment. *Virology* **24**, 218.

Siegert, R., Betz, E., and Schmidt, G. (1963). Zur Problematik des Pyrogenbegriffs, dargestellt am Beispiel der Viruspyrogene. Festschr. 100-Jahr-Feier Farbwerke Hoechst, p. 255 [(Quoted in Siegert and Braune (1964a).]

Sikes, R. K., Cleary, W. F., Koprowski, H., Wiktor, T. J., and Kaplan, M. M. (1971). Effective protection of monkeys against death from street virus by post-exposure administration of tissue-culture rabies vaccine. *Bull. WHO* **45**, 1.

Silverstein, A. M., Uhr, J., Kraner, K. L., and Lukes, R. J. (1963). Fetal response to antigenic stimulus. II. Antibody production by the fetal lamb. *J. Exp. Med.* **117**, 799.

Silverstein, A. M., Prendergast, R. A., and Kraner, K. L. (1964). Fetal response to antigenic stimulus. IV. Rejection of skin homografts by the fetal lamb. *J. Exp. Med.* **119**, 955.

Silverstein, S. (1970). Macrophages and viral immunity. *Semin. Haematol.* **7**, 185.

Silverstein, S. C., and Dales, S. (1968). The penetration of reovirus RNA and initiation of its genetic function in L strain fibroblasts. *J. Cell Biol.* **36**, 197.

Silverstein, S. C., and Marcus, P. I. (1964). Early stages Newcastle disease virus-HeLa cell interaction: An electron microscopic study. *Virology* **23**, 370.

Silverstein, S. C., Astell, C., Levin, D. H., Schonberg, M., and Acs, G. (1972). The mechanisms of reovirus uncoating and gene activation *in vivo. Virology* **47**, 797.

Simmons, D. T., and Strauss, J. H. (1972a). Replication of Sindbis virus. I. Relative size and genetic content of 26S and 49S RNA. *J. Mol. Biol.* **71**, 599.

Simmons, D. T., and Strauss, J. H. (1972b). Replication of Sindbis virus. II. Multiple forms of double-stranded RNA isolated from infected cells. *J. Mol. Biol.* **71**, 615.

Simon, E. H. (1961). Evidence for the non-participation of DNA in viral RNA synthesis. *Virology* **13**, 105.

Simon, E. H. (1972). The distribution and significance of multiploid virus particles. *Progr. Med. Virol.* **14**, 36.

Simons, P. J., Pepper, S. S., and Baker, R. E. U. (1969). Different cell culture characteristics of two strains of murine sarcoma virus. *Proc. Soc. Exp. Biol. Med.* **131,** 454.

Simpson, R. W., and Hauser, R. E. (1966a). Structural components of vesicular stomatitis virus. *Virology* **29,** 654.

Simpson, R. W., and Hauser, R. E. (1966b). Influence of lipids on the viral phenotype. I. Interactions of myxoviruses and their lipid constituents with phospholipases. *Virology* **30,** 684.

Simpson, R. W., and Hauser, R. E. (1968). Basic structure of group A arbovirus strains Middleburg, Sindbis and Semliki Forest examined by negative staining. *Virology* **34,** 358.

Simpson, R. W., and Hirst, G. K. (1961). Genetic recombination among influenza viruses. I. Cross reactivation of the plaque-forming capacity as a method for selecting recombinants from the progeny of crosses between influenza A strains. *Virology* **15,** 436.

Simpson, R. W., and Hirst, G. K. (1968). Temperature-sensitive mutants of influenza A virus: isolation of mutants and preliminary observations on genetic recombination and complementation. *Virology* **35,** 41.

Simpson, R. W., and Obijeski, J. F. (1973a). Conditional lethal mutants of vesicular stomatitis virus: I. Phenotypic characterization of single and double mutants exhibiting host restriction and temperature sensitivity. In preparation.

Simpson, R. W., and Obijeski, J. F. (1973b). Conditional lethal mutants of vesicular stomatitis virus: II. Virus-specified macromolecular synthesis in nonpermissive cells infected with host-restricted mutants. In preparation.

Simpson, R. W., and Obijeski, J. F. (1973c). Conditional lethal mutants of vesicular stomatitis virus: III. Permissiveness of various euploid and heteroploid cell lines for productive infections with host-restricted mutants. In preparation.

Simpson, R. W., Hauser, R. E., and Dales, S. (1969). Viropexis of vesicular stomatitis virus by L cells. *Virology* **37,** 285.

Singer, S. J., and Nicolson, G. L. (1972). The fluid mosaic model of the structure of cell membranes. *Science* **175,** 720.

Sisken, J. E., and Kinosita, R. (1961). Timing of DNA synthesis in the mitotic cycle *in vitro. J. Biophys. Biochem. Cytol.* **9,** 509.

Sjögren, H. O. (1964a). Studies on specific transplantation resistance to polyoma virus-induced tumors. III. Transplantation resistance to genetically compatible polyoma tumors induced by polyoma tumor homografts. *J. Nat. Cancer Inst.* **32,** 645.

Sjögren, H. O. (1964b). Studies on specific transplantation resistance to polyoma virus-induced tumors. IV. Stability of the polyoma cell antigen. *J. Nat. Cancer Inst.* **32,** 661.

Sjögren, H. O. (1965). Transplantation methods as a tool for detection of tumor-specific antigens. *Progr. Exp. Tumor Res.* **6,** 289.

Sjögren, H. O., and Jonsson, N. (1970). Cellular immunity to Rous sarcoma in tumor-bearing chickens. *Cancer Res.* **30,** 2434.

Sjögren, H. O., Hellström, I., and Klein, G. (1961). Transplantation of polyoma virus-induced tumors in mice. *Cancer Res.* **21,** 329.

Sjögren, H. O., Minowada, J., and Ankerst, J. (1967). Specific transplantation antigens of mouse sarcomas induced by adenovirus type 12. *J. Exp. Med.* **125,** 689.

Skehel, J. J. (1971). RNA-dependent RNA polymerase activity of the influenza virus. *Virology* **45,** 793.

Skehel, J. J. (1972). Polypeptide synthesis in influenza virus-infected cells. *Virology* **49,** 23.

Skehel, J. J., and Joklik, W. K. (1969). Studies on the *in vitro* transcription of reovirus RNA catalysed by reovirus cores. *Virology* **39,** 822.

Skehel, J. J., and Schild, G. C. (1971). The polypeptide composition of influenza A viruses. *Virology* **44,** 396.

Slade, W. R., and Pringle, C. R. (1971). Genetic characteristics of clones from individual cells multiply-infected with different strains of foot-and-mouth disease virus. *J. Gen. Virol.* **12,** 335.

Slifkin, M., Merkow, L., and Rapoza, N. P. (1968). Tumor induction by simian adenovirus 30 and establishment of tumor cell lines. *Cancer Res.* **28,** 1173.

Smart, K. M., and Kilbourne, E. D. (1966). The influence of cortisone on experimental viral infection. VI. Inhibition by hydrocortisone of interferon synthesis in the chick embryo. *J. Exp. Med.* **123**, 299.

Smith, A. E., Marcker, K. A., and Mathews, M. B. (1970). Translation of RNA from encephalomyocarditis virus in a mammalian cell-free system. *Nature (London)* **225**, 184.

Smith, C. B., Purcell, R. H., Bellanti, J. A., and Chanock, R. M. (1966). Protective effective effect of antibody to parainfluenza type-1 virus. *New Engl. J. Med.* **275**, 1145.

Smith, C. E. G. (1971). Lessons from Marburg disease. *In* "The Scientific Basis of Medicine," Annu. Rev. p. 58. Athlone, London

Smith, G. P., Hood, L., and Fitch, W. M. (1971). Antibody diversity. *Annu. Rev. Biochem.* **40**, 969.

Smith, H. S., Scher, C. D., and Todaro, G. J. (1970). Abortive infection of Balb/3T3 by simian virus 40. *Bacteriol. Proc.* p. 187.

Smith, H. S., Scher, C. D., and Todaro, G. J. (1971). Induction of cell division in medium lacking serum growth factor by SV40. *Virology* **44**, 359.

Smith, H. S., Gelb, L. D., Martin, M. A. (1972). Detection and quantitation of simian virus 40 genetic material in abortively infected Balb/3T3 clones. *Proc. Nat. Acad. Sci. U.S.* **69**, 152.

Smith, J. D., Freeman, G., Vogt, M., and Dulbecco, R. (1960). The nucleic acid of polyoma virus. *Virology* **12**, 185.

Smith, K. O. (1970). Adventitious viruses in cell cultures. *Progr. Med. Virol.* **12**, 302.

Smith, K. O., Galasso, G., and Sharp, D. G. (1961). Effect of antiserum on adsorption of vaccinia virus to Earle's L cells. *Proc. Soc. Exp. Biol. Med.* **106**, 669.

Smith, K. O., Gehle, W. D., and Thiel, J. F. (1966). Properties of a small virus associated with adenovirus type 4. *J. Immunol.* **97**, 754.

Smith, R. E., Zweerink, H. J., and Joklik, W. K. (1969). Polypeptide components of virions, top component and cores of reovirus type 3. *Virology* **39**, 791.

Smith, R. T., and Bausher, J. C. (1972). Epstein-Barr virus infection in relation to infectious mononucleosis and Burkitt's lymphoma. *Annu. Rev. Med.* **23**, 39.

Smith, R. W., Morganroth, J., and Mora, P. T. (1970). SV40 virus-induced tumor specific transplantation antigen in cultured mouse cells. *Nature (London)* **227**, 141.

Smith, T. J. (1964). Antibiotic-induced disease. *In* "Diseases of Medical Progress" (R. H. Moser, ed.), 2nd Ed., p. 3. Thomas, Springfield, Illinois.

Smith, T. J., and Wagner, R. R. (1967). Rabbit macrophage interferons. I. Conditions for biosynthesis by virus infected and uninfected cells. *J. Exp. Med.* **125**, 559.

Smith, T. J., Buescher, E. L., Top, F. H., Altemeier, W. A., and McCowan, J. M. (1970). Experimental respiratory infection with type 4 adenovirus vaccine in volunteers; clinical and immunological responses. *J. Infect. Dis.* **122**, 239.

Smith, W., Andrewes, C. H., and Laidlaw, P. P. (1933). A virus obtained from influenza patients. *Lancet* **ii**, 66.

Smith, W. R., and McAuslan, B. R. (1969). Biophysical properties of Frog Virus and its deoxyribonucleic acid: Fate of radioactive virus in the early stage of infection. *J. Virol.* **4**, 339.

Smithburn, K. C. (1954). Antigenic relationships among certain arthropod-borne viruses as revealed by neutralization tests. *J. Immunol.* **72**, 376.

Smithwick, E. M., and Berkovich, S. (1966). *In vitro* suppression of the lymphocyte response to tuberculin by live measles virus. *Proc. Soc. Exp. Biol. Med.* **123**, 276.

Smorodintsev, A. A. (1957). Factors of natural resistance and specific immunity to viruses. *Virology* **3**, 299.

Smorodintsev, A. A. (1958). Tick borne spring-summer encephalitis. *Progr. Med. Virol.* **1**, 210.

Smorodintsev, A. A., Zlydnikov, D. M., Kiseleva, A. M., Romanov, J. A., Kazantsev, A. P. and Rumovsky, V. I. (1970). Evaluation of Amantadine in artificially induced A2 and B influenza. *J. Amer. Med. Ass.* **213**, 1448.

Smythe, P. M., Schonland, M., Brereton-Stiles, G. G., Goovadia, H. M., Grace, H. J., Loening,

W. K., Mafoyane, A., Parent, M. A., and Vos, G. H. (1971). Thymolymphatic deficiency and depression of cell-mediated immunity in protein-calorie malnutrition. *Lancet* **ii**, 939.

Sneath, P. H. A. (1964). New approaches to bacterial taxonomy: Use of computers. *Annu. Rev. Microbiol.* **18**, 335.

Snyder, L., and Geiduschek, E. P. (1968). *In vitro* synthesis of T4 late messenger RNA. *Proc. Nat. Acad. Sci. U.S.* **59**, 459.

Snyder, S. P., and Theilen, G. H. (1969). Transmissible feline lymphosarcoma. *Nature (London)* **221**, 1074.

Sobey, W. R. (1969). Selection for resistance to myxomatosis in domestic rabbits. *(Oryctolagus cuniculus). J. Hyg.* **67**, 743.

Sokal, J. E., and Firat, D. (1965). Varicella-zoster infection in Hodgkin's disease. *Amer. J. Med.* **30**, 452.

Sokol, F., Schlumberger, D., Wiktor, T. J., and Koprowski, H. (1969). Biochemical and biophysical studies on the nucleocapsid and on the RNA of rabies virus. *Virology* **38**, 651.

Sokol, F., Stançek, D., and Koprowski, H. (1971). Structural proteins of rabies virus. *J. Virol.* **7**, 241.

Solomon, J. B. (1971). "Foetal and Neonatal Immunology." North Holland, Amsterdam.

Solomon, G. F., Merigan, T., and Levine, S. (1967). Variation in adrenal cortical hormones within physiologic ranges; stress and interferon production in mice. *Proc. Soc. Exp. Biol. Med.* **126**, 74.

Solov'ev, V. D. (1969). The results of controlled observations on the prophylaxis of influenza with interferon. *Bull. WHO* **41**, 683.

Sonnabend, J. A., Martin, E. M., and Mécs, E. (1967). Viral specific RNA's in infected cells. *Nature (London)* **213**, 365.

Soothill, J. F., Hayes, K., and Dudgeon, J. A. (1966). The immunoglobulins in congenital rubella. *Lancet* **i**, 1385.

Soper, F. L., Penna, H. A., Cardoso, E., Serafim, J., Jr., Frobisher, M., Jr., and Pinheiro, J. (1933). Yellow fever without *Aedes aegypti*: Study of rural epidemic in Valle do Chanaan, Espirito Santo, Brazil, 1932. *Amer. J. Hyg.* **18**, 555.

South, M. A., Cooper, M. D., Wollheim, F. A., Hong, R., and Good, R. A. (1966). The IgA system. I. Studies of the transport and immunochemistry of IgA in the saliva. *J. Exp. Med.* **123**, 615.

Spear, P. G., and Roizman, B. (1970). Proteins specified by herpes simplex virus. IV. Site of glycosylation and accumulation of viral membrane proteins. *Proc. Nat. Acad. Sci. U.S.* **66**, 730.

Spear, P. G., and Roizman, B. (1972). Proteins specified by herpes simplex xirus. V. Purification and structural proteins of the virion. *J. Virol.* **9**, 143.

Spencer, E. S., and Anderson, H. K. (1970). Clinically evident, non-terminal infections with herpesviruses and the wart virus in immunosuppressed renal allograft recipients. *Brit. Med. J.* **3**, 251.

Spendlove, R. S., McClain, M. E., and Lennette, E. H. (1970). Enhancement of reovirus infectivity by extracellular removal or alteration of the virus capsid by proteolytic enzymes. *J. Gen. Virol.* **8**, 83.

Spiegelman, S., Pace, N. R., Mills, D. R., Levisohn, R., Eikhom, T. S., Taylor, M. M., Peterson, R. L., and Bishop, D. H. L. (1968). The mechanism of RNA replication. *Cold Spring Harbor Symp. Quant. Biol.* **33**, 101.

Spiegelman, S., Burny, A., Das, M. R., Keydar, J., Schlom, J., Travnicek, M., and Watson, K. (1970). Characterization of the products of RNA-directed DNA polymerases in oncogenic RNA viruses. *Nature (London)* **227**, 563.

Spiegelman, S., Axel, R., and Schlom, J. (1972). Virus-related RNA in human and mouse mammary tumors. *J. Nat. Cancer Inst.* **48**, 1205.

Spirin, A. S., and Gavrilova, L. P. (1969). "The Ribosome." Springer-Verlag, New York.

Sprent, J., and Miller, J. F. A. P. (1971). Activation of thymus cells by histocompatibility antigens. *Nature New Biol.* **234**, 195.

Spring, S. B., and Roizman, B. (1968). Herpes simplex virus products in productive and abortive infection. III. Differentiation of infectious virus derived from nucleus and cytoplasm with respect to stability and size. *J. Virol.* **2**, 979.

Spring, S. B., Roizman, B., and Schwartz, J. (1968). Herpes simplex virus products in productive and abortive infection. II. Electron microscopic and immunological evidence for failure of virus envelopment as a cause of abortive infection. *J. Virol.* **2**, 384.

Sproul, E. E., Metzgar, R. S., and Grace, J. T., Jr. (1963). The pathogenesis of Yaba virus-induced histiocytomas in primates. *Cancer Res.* **23**, 671.

Sprunt, D. H. (1932). Infectious myxomatosis (Sanarelli) in pregnant rabbits. *J. Exp. Med.* **56**, 601.

Spurgash, A., Ehrlich, R., and Petzold, R. (1968). Effect of cigarette smoking on resistance to respiratory infection. *Arch. Environ. Health* **16**, 385.

Sreevalsan, T. (1970a). Association of viral ribonucleic acid with cellular membranes in chick embryo cells infected with Sindbis virus. *J. Virol.* **6**, 438.

Sreevalsan, T. (1970b). Homologous viral interference: Induction by RNA from defective particles of vesicular stomatitis virus. *Science* **169**, 991.

Sreevalsan, T., and Yin, F. H. (1969). Sindbis virus-induced viral ribonucleic acid polymerase. *J. Virol.* **3**, 599.

Stackpole, C. W., and Mizell, M. (1968). Electron microscopic observations of herpes-type virus-related structures in frog renal adenocarcinoma. *Virology* **36**, 63.

Staiger, H. R. (1964). Plaque-type recombinaiton in fowl plague virus. *Virology* **22**, 419.

Stalder, W., and Zurukzoglu, St. (1936). Experimentelle Untersuchungen über Herpes. Transplantation herpesinfizierter Haustellen, Reaktivierung von abgeheilten, künstlich infizierten Hautstellen, Herpesbehandlung. VI. Mitteilung. *Zentr. Bakteriol. Parisentk. Abt. I, Orig.* **136**, 94.

Stampfer, M., Baltimore, D., and Huang, A. S., (1969). Ribonucleic acid synthesis of vesicular stomatitis virus. I. Species of ribonucleic acid found in Chinese hamster ovary cells infected with plaque-forming and defective particles. *J. Virol.* **4**, 154.

Stampfer, M., Baltimore, D., and Huang, A. S. (1971). Absence of interference during high-multiplicity infection by clonally purified vesicular stomatitis virus. *J. Virol.* **7**, 409.

Stanley, E. D., and Jackson, G. G. (1969). Spread of enteric live adenovirus type 4 vaccine in married couples. *J. Infect. Dis.* **119**, 51.

Stanley, P. M., and Haslam, E. A. (1971). The polypeptides of influenza virus V. Localization of polypeptides in the virion by iodination techniques. *Virology* **46**, 764.

Stanley, P. M., Gandhi, S. S., and White, D. O. (1973). The polypeptides of influenza virus. VII. Synthesis of the hemagglutinin. *Virology* **53**, 92.

Stanley, W. M. (1935). Isolation of a crystalline protein possessing the properties of tobacco mosaic virus. *Science* **81**, 644.

Starnezer, J., and Huang, C. C. (1971). Synthesis of a mouse immunoglobulin light chain in a rabbit reticulocyte cell-free system. *Nature New Biol.* **230**, 172.

Stavis, R. L., and August, J. T. (1970). The biochemistry of RNA bacteriophage replication. *Annu. Rev. Biochem.* **39**, 527.

Steck, F. T., and Rubin, H. (1966a). The mechanism of interference between an avian leukosis virus and Rous sarcoma virus. I. Establishment of interference. *Virology* **29**, 628.

Steck, F. T., and Rubin, H. (1966b). The mechanism of interference between an avian leukosis virus and Rous sarcoma virus. II. Early steps of infection by RSV of cells under conditions of interference. *Virology* **29**, 642.

Steele, W. J. (1968). Localization of deoxyribonucleic acid complementary to ribosomal ribonucleic acid and preribosomal ribonucleic acid in the nucleolus of rat liver. *J. Biol. Chem.* **243**, 3333.

Steeves, R. A. (1968). Cellular antigen of Friend virus-induced leukemias. *Cancer Res.* **28**, 338.

Stenback, W. A., and Durand, D. P. (1963). Host influence on the density of Newcastle disease virus (NDV). *Virology* **20**, 545.

Stent, G. S. (1971). "Molecular Genetics: an Introductory Narrative." Freeman, California.

Stephenson, J. R., Reynolds, R. K., and Aaronson, S. A. (1972). Isolation of temperature-sensitive mutants of murine leukemia virus. *Virology* **48**, 749.

Stern, R., and Friedman, R. M. (1969). Chromatography of arbovirus ribonucleic acid forms on columns of benzoylated-diethylaminoethyl cellulose. *J. Virol.* **4**, 356.

Stevens, J. G., and Cook, M. L. (1971a). Restriction of herpes simplex virus by macrophages. An analysis of the cell–virus interaction. *J. Exp. Med.* **133**, 19.

Stevens, J. G., and Cook, M. L. (1971b). Latent herpes simplex virus in spinal ganglia of mice. *Science* **173**, 843.

Stevens, J. G., Nesburn, A. B., and Cook, A. L. (1972). Latent herpes simplex virus from trigeminal ganglia of rabbits with recurrent eye infection. *Nature New Biol.* **235**, 216.

Stewart, S. E. (1960). The polyoma virus. *Advan. Virus Res.* **7**, 61.

Stewart, S. E., and Eddy, B. E. (1959). Properties of a tumor-inducing virus recovered from mouse neoplasms. *Perspect. Virol.* **1**, 245.

Stewart, S. E., Eddy, B. E., Gochenour, A. M., Borgese, N. G., and Grubbs, G. E. (1957). The induction of neoplasms with a substance released from mouse tumors by tissue culture. *Virology* **3**, 380.

Stewart, W. E., Scott, W. D., and Sulkin, S. E. (1969). Relative sensitivities of viruses to different species of interferon. *J. Virol.* **4**, 147.

Stich, H. F., and Yohn, D. S. (1970). Viruses and chromosomes. *Progr. Med. Virol.* **12**, 78.

Stiehm, E. R., Ammann, A. J., and Cherry, J. D. (1966). Elevated cord macroglobulins in the diagnosis of intrauterine infections. *New Engl. J. Med.* **275**, 971.

Stocks, S. (1942). Measles and whooping-cough incidence before and during the dispersal of 1939–41. *J. Roy. Statist. Soc.* **105**, 259.

Stoker, M. G. P. (1958). Mode of intercellular transfer of herpes virus. *Nature (London)* **182**, 1525.

Stoker, M. (1968). Abortive transformation by polyoma virus. *Nature (London)* **218**, 234.

Stoker, M. G. P., and Abel, P. (1962). Conditions affecting transformation by polyoma virus. *Cold Spring Harbor Symp. Quant. Biol.* **27**, 375.

Stoker, M. G. P., and Macpherson, I. (1967). Transformation assays. *In* "Methods in Virology" (K. Maramorosch and H. Koprowski, eds.), Vol. III, p. 313. Academic Press, New York.

Stoker, M. G. P., and Newton, A. (1959). The effect of herpes virus on HeLa cells dividing parasynchronously. *Virology* **7**, 438.

Stoker, M. G. P., and Rubin, H. (1967). Density dependent inhibition of cell growth in culture. *Nature (London)* **215**, 171.

Stokes, J., Jr., Reilly, C. M., Buynak, E. B., and Hilleman, M. R. (1961). Immunologic studies of measles. *Amer. J. Hyg.* **74**, 293.

Stokes, J., Weibel, R. E., Villarejos, V. M., Arguedas, J. A., Buynak, E. B., and Hilleman, M. R. (1971). Trivalent combined measles-mumps-rubella vaccine. *J. Amer. Med. Ass.* **218**, 57.

Stollar, V. (1969). Studies on the nature of dengue viruses. IV. The structural proteins of type 2 dengue virus. *Virology* **39**, 426.

Stollar, V., Stevens, T. M., and Schlesinger, R. W. (1966). Studies on the nature of dengue virus. II. Characterization of viral RNA and effects of inhibitors of RNA synthesis. *Virology* **30**, 303.

Stollar, V., Schlesinger, R. W., and Stevens, T. M. (1967). Studies on the nature of dengue viruses. III. RNA synthesis in cells infected with type 2 dengue virus. *Virology* **33**, 650.

Stollar, V., Shenk, T. E., and Stollar, B. D. (1972). Double-stranded RNA in hamster, chick, and mosquito cells infected with Sindbis virus. *Virology* **47**, 122.

Stoltzfus, C. M., and Banerjee, A. K. (1972). Two oligonucleotide classes of single-stranded ribopolymers in reovirus A-rich RNA. *Arch. Biochem. Biophys.* **152**, 733.

Stone, H. O., Portner, A., and Kingsbury, D. W. (1971). Ribonucleic acid transcriptases in Sendai virions and infected cells. *J. Virol.* **8**, 174.

Stone, J. D. (1948). Prevention of virus infection with enzyme of *V. cholerae*. II. Studies with influenza virus in mice. *Aust. J. Exp. Biol. Med. Sci.* **26**, 287.

Stone, L. B., Takemoto, K. K., and Martin, M. A. (1971). Physical and biochemical properties of progressive pneumonia virus. *J. Virol.* **8**, 573.

Stone, R. S., Shope, R. E., and Moore, D. H. (1959). Electron microscope study of the development of the papilloma virus in the skin of the rabbit. *J. Exp. Med.* **10**, 543.

Stott, E. J., and Killington, R. A. (1972). Haemagglutination by rhinoviruses. *Lancet* **i**, 1369.

Strand, M., and August, J. T. (1971). Protein kinase and phosphate acceptor proteins in Rauscher leukemia virus. *Nature New Biol.* **233**, 137.

Strauss, J. H., Burge, B. W., Pfefferkorn, E. R., and Darnell, J. E. (1968a). Identification of the membrane protein and "core" protein of Sindbis virus. *Proc. Nat. Acad. Sci. U.S.* **59**, 533.

Strauss, J. H., Kelly, R. B., and Sinsheimer, R. L. (1968b). Denaturation of RNA with dimethylsulfoxide biopolymers. *Biopolymers* **6**, 793.

Strauss, J. H., Burge, B. W., and Darnell, J. E. (1969). Sindbis virus infection of chick and hamster cells: synthesis of virus-specific proteins. *Virology* **37**, 367.

Strauss, M. J., Shaw, E. W., Bunting, H., and Melnick, J. L. (1949). "Crystalline" virus-like particles from skin papillomas characterized by intranuclear inclusion bodies. *Proc. Soc. Exp. Biol. Med.* **72**, 46.

Strohl, W. A., and Schlesinger, R. W. (1965). Quantitative studies of natural and experimental adenovirus infections of human cells. II. Primary cultures and the possible role of asynchronous viral multiplication in the maintenance of infection. *Virology* **26**, 208.

Strohl, W. A., Rouse, H. C., and Schlesinger, R. W. (1966). Properties of cells derived from adenovirus-induced hamster tumors by long-term *in vitro* cultivation. II. Nature of the restricted response to type 2 adenovirus. *Virology* **28**, 645.

Stuart-Harris, C. H. (1965). "Influenza and Other Virus Infections of the Respiratory Tract," 2nd Ed. Arnold, London.

Stuart-Harris, C. H., and Dickinson, L. (1964). "The Background to Chemotherapy of Virus Diseases." Thomas, Springfield, Illinois.

Stubblefield, E., Klevecz, R., and Deaven, L. (1967). Synchronized mammalian cell cultures I. Cell replication cycle and macromolecular synthesis following brief colcemid arrest of mitosis. *J. Cell Physiol.* **69**, 345.

Sturman, L. S., and Tamm, I. (1969). Formation of viral ribonucleic acid and virus in cells that are permissive or non-permissive for murine encephalomyelitis virus. (GDVII). *J. Virol.* **3**, 8.

Suarez, H. G., Sonenshein, G. E., Cassingena, R., and Tournier, P. (1971). Mode of inhibition by SV40 "repressor" In Lepetit Colloquia on Biology and Medicine, Vol. 2: The Biology of Oncogenic Viruses" (L. Sylvestri, ed.) p. 1. North Holland, Amsterdam.

Subak-Sharpe, J. H. (1969). The doublet pattern of the nucleic acid in relation to the origin of viruses. *In* "Handbook of Molecular Cytology" (A. Lima-di-Faria, ed.) p. 67. North Holland, Amsterdam.

Subak-Sharpe, H., and Hay, J. (1965). An animal virus with DNA of high guanine + cytosine content which codes for sRNA. *J. Mol. Biol.* **12**, 924.

Subak-Sharpe, H., Shepherd, W. M., and Hay, J. (1966a). Studies on sRNA coded by herpes virus. *Cold Spring Harbor Symp. Quant. Biol.* **31**, 583.

Subak-Sharpe, H., Bürk, R. R., Crawford, L. V., Morrison, J. M., Hay, J., and Keir, H. M. (1966b). An approach to evolutionary relationships of mammalian DNA viruses through analysis of the pattern of nearest neighbor base sequences. *Cold Spring Harbor Symp. Quant. Biol.* **31**, 737.

Subak-Sharpe, J. H., Timbury, M. C., and Williams, J. F. (1969). Rifampicin inhibits the growth of some mammalian viruses. *Nature (London)* **222**, 341.

Subrahmanyan, T. P. (1968). A study of the possible basis of age-dependent resistance of mice to poxvirus diseases. *Aust. J. Exp. Biol. Med. Sci.* **46**, 251.

Subrahmanyan, T. P., and Mims, C. A. (1966). Fate of intravenously administered interferon and the distribution of interferon during virus infections in mice. *Brit. J. Exp. Pathol.* **47**, 168.

Sugiura, A., and Kilbourne, E. D. (1966). Genetic studies of influenza viruses. III. Production of plaque-type recombinants with $A_0$ and $A_1$ strains. *Virology* **29**, 84.

Sugiyama, T., Korant, B. D., and Lonberg-Holm, K. K. (1972). RNA virus gene expression and its control. *Annu. Rev. Microbiol.* **26**, 467.

Sulkin, S. E. (1962). The bat as a reservoir of viruses in nature. *Progr. Med. Virol.* **4**, 157.

Sulkin, S. E. (1965). Bats in relation to arthropod-borne viruses: An experimental approach with speculations. *Amer. J. Public Health* **55**, 1376.

Summers, D. F., and Maizel, J. V. (1968). Evidence for large precursor proteins in poliovirus synthesis. *Proc. Nat. Acad. Sci. U.S.* **59**, 966.

Summers, D. F., Maizel, J. V., Jr., and Darnell, J. E., (1965). Evidence for virus-specific non-capsid proteins in poliovirus-infected HeLa cells. *Proc. Nat. Acad. Sci. U.S.* **54**, 505.

Summers, D. F., Roumiantzeff, M., and Maizel, J. V. (1971). The translation and processing of poliovirus proteins. *In* "Strategy of the Viral Genome" (G. E. W. Wolstenholme and M. O'Connor, eds.), p. 111. Churchill Livingstone, Edinburgh.

Sumner, J. B. (1919). The globulins of the Jack Bean, *Canavalia ensigormis. J. Biol. Chem.* **37**, 137.

Sussenbach, J. S. (1967). Early events in the infection process of adenovirus type 5 in HeLa cells. *Virology* **33**, 567.

Sussenbach, J. S., Van der Vliet, P. C., Ellens, D. J., and Jansz, H. S. (1972). Linear intermediates in the replication of adenovirus DNA. *Nature New Biol.* **239**, 47.

Suter, E. R., and Majno, G. (1965). Passage of lipid across vascular endothelium in newborn rats. *J. Cell. Biol.* **27**, 163.

Sutnik, A. I., Cerda, J. J., Toskes, P. O., London, W. T., and Blumberg, B. S. (1971). Australia antigen and viral hepatitis in drug abusers. *Arch. Internal Med.* **127**, 939.

Suzuki, E., and Shimojo, H. (1971). A temperature-sensitive mutant of adenovirus 31, defective in viral deoxyribonucleic acid replication. *Virology* **43**, 488.

Suzuki, E., Shimojo, H., and Moritsugu, Y. (1972). Isolation and preliminary characterization of temperature-sensitive mutants of adenovirus 31. *Virology* **49**, 426.

Svec, F., Altaner, C., and Hlavay, E. (1966). Pathogenicity for rats of a strain of chicken sarcoma virus. *J. Nat. Cancer Inst.* **36**, 389.

Svec, J., Svec, F., Simkovic, D., and Thurzo, V. (1970). Induction of tumors in hamsters by rat Rous sarcoma RBI, producing chick sarcoma virus. I. Continuous production of virus oncogenic for chicks by hamster sarcoma cells. *J. Nat. Cancer Inst.* **44**, 521.

Svehag, S.-E. (1964a). The formation and properties of poliovirus-neutralizing antibody. III. Sequential changes in electrophoretic mobility of 19 S and 7 S antibodies synthesized by rabbits after a single virus injection. *J. Exp. Med.* **119**, 225.

Svehag, S.-E. (1964b). The formation and properties of poliovirus-neutralizing antibody. IV. Normal antibody and early immune antibody of rabbit origin: A comparison of biological and physicochemical properties. *J. Exp. Med.* **119**, 517.

Svehag, S.-E., and Mandel, B. (1964a). The formation and properties of poliovirus-neutralizing antibody. I. 19 S and 7 S antibody formation: Differences in kinetics and antigen dose requirement for induction. *J. Exp. Med.* **119**, 1.

Svehag, S.-E., and Mandel, B. (1964b). The formation and properties of poliovirus-neutralizing antibody. II. 19 S and 7 S antibody formation: Differences in antigen dose requirement for sustained synthesis, anamnesis and sensitivity to X-irradiation. *J. Exp. Med.* **119**, 21.

Svet-Moldavsky, G. J. (1958). Sarcoma in albino rats treated during the embryonic stage with Rous virus. *Nature (London)* **182**, 1452.

Svoboda, J. (1960). Presence of chicken tumor virus in sarcoma of the adult rat inoculated after birth with Rous sarcoma tissue. *Nature (London)* **186**, 980.

Svoboda, J. (1966). Basic aspects of the interaction of oncogenic viruses with heterologous cells. *Intern. Rev. Exp. Pathol.* **5**, 25.

Svoboda, J. (1968). Dependence among RNA-containing animal viruses. *Symp. Soc. Gen. Microbiol.* **18**, 249.

Svoboda, J., and Dourmashkin, R. (1969). Rescue of Rous sarcoma virus from virogenic mammalian cells associated with chicken cells and treated with Sendai virus. *J. Gen. Virol.* **4,** 523.

Svoboda, J., and Klement, V. (1963). Formation of delayed tumors in hamsters inoculated with Rous virus after birth and finding of infectious Rous virus in induced tumor P1. *Folia Biol.* **9,** 403.

Svoboda, J., and Hlozanek, I. (1970). Cell association in virus infection and rescue. *Advan. Cancer Res.* **13,** 217.

Svoboda, J., Machala, O., and Hlozanek, I. (1967). Influence of Sendai virus on RSV formation in mixed culture of virogenic mammalian cells and chicken fibroblasts. *Folia Biol.* **13,** 155.

Swan, C., Tostevin, A. L., Moore, B., Mayo, H., and Black, G. H. B. (1943). Congenital defects in infants following infectious diseases during pregnancy. *Med. J. Aust.* **2,** 201.

Sweet, B. H., and Hilleman, M. R. (1960). The vacuolating virus, SV40. *Proc. Soc. Exp. Biol. Med.* **105,** 420.

Swift, H., and Wolstenholme, D. R. (1969). Mitochondria and chloroplasts: nucleic acids and the problem of biogenesis (genetics and biology). *In* Handbook of Molecular Cytology" (A. Lima-di-Faria, ed.), p. 972. North Holland, Amsterdam.

Sydiskis, R. J., and Roizman, B. (1966). Polysomes and protein synthesis in cells infected with a DNA virus. *Science* **153,** 76.

Sydiskis, R. J., and Roizman, B. (1967). The disaggregation of host polyribosomes in productive and abortive infection with herpes simplex virus. *Virology* **32,** 678.

Symposium (1971). Immune Complexes and Disease. *J. Exp. Med.* **134,** 1s.

Syverton, J. T. (1952). The pathogenesis of the rabbit papilloma-to-carcinoma sequence. *Ann. N. Y. Acad. Sci.* **54,** 1126.

Syverton, J. T., Dascomb, H. E., Wells, E. B., Koomen, J., and Berry, G. P. (1950). The virus-induced papilloma-to-carcinoma sequence. II. Carcinomas in the natural host, the cottontail rabbit. *Cancer Res.* **10,** 440.

Szilágyi, J. F., and Pennington, T. H. (1971). Effect of rifamycins and related antibiotics and the deoxyribonucleic acid-dependent ribonucleic acid polymerase of vaccinia virus particles. *J. Virol.* **8,** 133.

Szilágyi, J. F., and Pringle, C. R. (1972). Effect of temperature-sensitive mutations on the virion-associated RNA transcriptase of vesicular stomatitis virus. *J. Mol. Biol.* **71,** 281.

Szilágyi, J. F., and Uryvayev, L. (1973). Isolation of an infectious ribonucleoprotein from vesicular stomatitis virus containing active RNA transcriptase. *J. Virol.* **11,** 279.

Szybalska, E. H., and Szybalski, W. (1962). Genetics of human cell lines. IV. DNA-mediated heritable transformation of a biochemical trait. *Proc. Nat. Acad. Sci. U.S.* **48,** 2026.

Szybalski, W., and Szybalska, E. (1971). Equilibrium density gradient centrifugatoin. *In* "Procedures in Nucleic Acid Research" (G. L. Cantoni and D. R. Davies, eds.), Vol. 2, p. 311. Harper and Row, New York.


Taber, R., Rekosh, D., and Baltimore, D. (1971). Effect of pactamycin on synthesis of poliovirus proteins: A method for genetic mapping. *J. Virol.* **8,** 395.

Tai, H. T., Smith, C. A., Sharp, P. A., and Vinograd, J. (1972). Sequence heterogeneity in closed simian virus 40 deoxyribonucleic acid. *J. Virol.* **9,** 317.

Tajima, M. (1970). Morphology of transmissible gastroenteritis virus of pigs. *Arch. Ges. Virusforsch.* **29,** 105.

Tajima, M., and Ushijima, T. (1971). The pathogenesis of rinderpest in the lymph nodes of cattle. *Amer. J. Pathol.* **62,** 221.

Takehara, M., and Schwerdt, C. E. (1967). Infective subviral particles from cell cultures infected with myxoma and fibroma viruses. *Virology* **31,** 163.

Takemori, N., and Nomura, S. (1960). Mutation of polioviruses with respect to size of plaque. II. Reverse mutation of a minute plaque mutant. *Virology* **12,** 171.

Takemori, N., Riggs, J. L., and Aldrich, C. (1968). Genetic studies with tumorigenic adenoviruses. I. Isolation of cytocidal (*cyt*) mutants of adenovirus type 12. *Virology* **36,** 575.

Takemori, N., Riggs, J. L., and Aldrich, C. D. (1969). Genetic studies with tumorigenic adenoviruses. II. Heterogeneity of *cyt* mutants of adenovirus type 12. *Virology* **38**, 8.

Takemoto, K. K. (1966). Plaque mutants of animal viruses. *Progr. Med. Virol.* **8**, 314.

Takemoto, K. K., and Habel, K. (1959). Virus-cell relationship in a carrier culture of HeLa cells and Coxsackie A9 virus. *Virology* **7**, 28.

Takemoto, K. K., and Habel, K. (1966). Hamster tumor cells doubly transformed by SV40 and polyoma viruses. *Virology* **30**, 20.

Takemoto, K. K., and Martin, M. A. (1970). SV40 thermosensitive mutant: synthesis of viral DNA virus-induced proteins at nonpermissive temperature. *Virology* **42**, 938.

Takemoto, K. K., and Stone, L. B. (1971). Transformation of murine cells by two "slow viruses," visna and progressive pneumonia virus. *J. Virol.* **7**, 770.

Takemoto, K. K., Todaro, G. J., and Habel, K. (1968). Recovery of SV40 virus with genetic markers of original inducing virus from SV40-transformed mouse cells. *Virology* **35**, 1.

Takemoto, K. K., Mattern, C. F. T., Stone, L. B., Coe, J. E., and Lavelle, G. (1971). Antigenic and morphologic similarities of progressive pneumonia virus, a recently isolated "slow virus" of sheep, to visna and maedi viruses. *J. Virol.* **7**, 301.

Talbot, P., and Brown, F. (1972). A model for foot-and-mouth disease virus. *J. Gen. Virol.* **15**, 163.

Talmage, D. W., Dixon, F. J., Bukantz, S. C., and Dammin, G. J. (1951). Antigen elimination from the blood as an early manifestation of the immune response. *J. Immunol.* **67**, 243.

Tamm, I., and Eggers, H. J. (1963). Specific inhibition of replication of animal viruses. *Science* **142**, 24.

Tan, K. B. (1970). Electron microscopy of cells infected with Semliki Forest virus temperature-sensitive mutants; correlation of ultrastructural and physiological observations. *J. Virol.* **5**, 632.

Tan, K. B., and McAuslan, B. R. (1970). Effect of rifampicin on poxvirus protein synthesis. *J. Virol.* **6**, 326.

Tan, K. B., and McAuslan, B. R. (1971). Proteins of polyhedral cytoplasmic deoxyviruses. I. The structural polypeptides of FV3. *Virology* **45**, 200.

Tan, K. B., and McAuslan, B. R. (1972). Binding of deoxyribonucleic acid-dependent deoxyribonucleic acid polymerase to poxvirus. *J. Virol.* **9**, 70.

Tan, K. B., Sambrook, J. F., and Bellett, A. J. D. (1969). Semliki Forest virus temperature-sensitive mutants: isolation and characterization. *Virology* **38**, 427.

Tan, Y. H., Armstrong, J. A., and Ho, M. (1971). Accentuation of interferon production by metabolic inhibitors and its dependence on protein synthesis. *Virology* **44**, 503.

Tannock, G. A., Gibbs, A. J., and Cooper, P. D. (1970). A reexamination of the molecular weight of poliovirus RNA. *Biochem. Biophys. Res. Commun.* **38**, 298.

Tattersall, P. (1972). Replication of the parvovirus MVM. I. Dependence of virus multiplication and plaque formation on cell growth. *J. Virol.* **10**, 586.

Tavitian, A., Peries, J., Chuat, J., and Boiron, M. (1967). Estimation of the molecular weight of adenovirus 12 CF antigen by rate-zonal centrifugation. *Virology* **31**, 719.

Taylor, J. (1964). Inhibition of interferon action by actinomycin. *Biochem. Biophys. Res. Commun.* **14**, 447.

Taylor, J. (1965). Studies on the mechanism of action of interferon. I. Interferon action and RNA synthesis in chick embryo fibroblasts infected with Semliki Forest virus. *Virology* **25**, 340.

Taylor, J. H. (1963). The replication and organization of DNA in chromosomes *In* "Molecular Genetics" (J. H. Taylor, ed.), Part I, p. 65. Academic Press, New York.

Taylor, J. M., Hampson, A. W., and White, D. O. (1969). The polypeptides of influenza virus. I. Cytoplasmic synthesis and nuclear accumulation. *Virology* **39**, 419.

Taylor, J. M., Hampson, A. W., Layton, J. E., and White, D. O. (1970). The polypeptides of influenza virus. IV. An analysis of nuclear accumulation. *Virology* **42**, 744.

Taylor, M. W., Cordell, B., Souhadra, M., and Prather, S. (1971). Viruses as an aid to cancer therapy: regression of soid and ascites tumors in rodents after treatment with a bovine enterovirus. *Proc. Nat. Acad. Sci. U. S.* **68**, 836.

Taylor, R. B., Duffus, P. H., Raff, M. C., and dePetris, S. (1971). Redistribution and pinocytosis of lymphocyte surface immunoglobulin molecules induced by anti-immunoglobulin antibody. *Nature New Biol.* **223**, 225.

Taylor, R. M. (1941). Experimental infection with influenza A virus in mice. *J. Exp. Med.* **73**, 43.

Taylor, R. M. (1951). Epidemiology. *In* "Yellow Fever" (G. K. Strode, ed.), p. 442. McGraw-Hill, New York.

Taylor-Papadimitriou, J., and Stoker, M. G. R. (1971). Effect of interferon on some aspects of transformation by polyoma virus. *Nature New Biol.* **230**, 114.

Taylor-Robinson, D. (1959). Chickenpox and herpes zoster. III. Tissue culture studies. *Brit. J. Exp. Pathol.* **40**, 521.

Tegtmeyer, P., and Enders, J. F. (1969). Feline herpesvirus infection in fused cultures of naturally resistant human cells. *J. Virol.* **3**, 469.

Tegtmeyer, P., and Ozer, H. L. (1971). Temperature-sensitive mutants of simian virus 40: Infection of permissive cells. *J. Virol.* **8**, 516.

Tegtmeyer, P., Dohan, C., and Reznikoff, C. (1970). Inactivating and mutagenic effects of nitrosoguanidine on simian virus 40. *Proc. Nat. Acad. Sci. U. S.* **66**, 745.

Temin, H. M. (1960). The control of cellular morphology in embryonic cells infected with Rous sarcoma virus *in vitro. Virology* **10**, 182.

Temin, H. M. (1961). Mixed infection with two types of Rous sarcoma virus. *Virology* **13**, 158.

Temin, H. (1962). Separation of morphological conversion and virus production in Rous sarcoma virus infection. *Cold Spring Harbor Symp. Quant. Biol.* **27**, 407.

Temin, H. M. (1964). The nature of the provirus of Rous sarcoma. *Proc. Nat. Acad. Sci. U. S.* **52**, 323.

Temin, H. M. (1965). The mechanism of carcinogenesis by avian sarcoma viruses. I Cell multiplication and differentiation. *J. Nat. Cancer Inst.* **35**, 679.

Temin, H. M. (1966). Studies on carcinogenesis by avian sarcoma viruses. 3. The differential effect of serum and polyoma virus on multiplication of uninfected and converted cells. *J. Nat. Cancer Inst.* **37**, 167.

Temin, H. M. (1967). Studies on carcinogenesis by avian sarcoma viruses. VI. Differential multiplication of uninfected and of converted cells in response to insulin. *J. Cell Physiol.* **69**, 377.

Temin, H. M. (1968a). Carcinogenesis by avian sarcoma viruses. *Cancer Res.* **28**, 1835.

Temin, H. M. (1968b). Studies on carcinogenesis by avian sarcoma viruses. VIII. Glycolysis and cell multiplication. *Intern. J. Cancer* **3**, 273.

Temin, H. M. (1971a). Cell transformation by RNA tumor viruses. *Annu. Rev. Microbiol.* **25**, 609.

Temin, H. M. (1971b). The protovirus hypothesis: speculations on the significance of RNA-directed DNA synthesis for normal development and for carcinogenesis. *J. Nat. Cancer Inst.* **46**, 216.

Temin, H. M. (1972). The RNA tumor viruses—background and foreground. *Proc. Nat. Acad. Sci. U. S.* **69**, 1016.

Temin, H. M., and Baltimore, D. (1972). RNA-directed DNA synthesis and RNA tumor viruses. *Advan. Virus Res.* **17**, 129.

Temin, H. M., and Mizutani, S. (1970). RNA-dependent DNA polymerase in virions of Rous sarcoma virus. *Nature (London)* **226**, 1211.

Temin, H. M., and Rubin, H. (1958). Characteristics of an assay for Rous sarcoma virus and Rous sarcoma cells in tissue culture. *Virology* **6**, 669.

Tennant, R. W., and Hand, R. E. (1970). Requirement of cellular synthesis for Kilham rat virus replication. *Virology* **42**, 1054.

Tennant, R. W., Layman, K. R., and Hand, R. E. (1969). Effect of cell physiological state on infection by rat virus. *J. Virol.* **4**, 872.

Ter Meulen, V., Katz, M., and Müller, D. (1972). Subacute sclerosing panencephalitis: a review. *Curr. Topics Microbiol. Immunol.* **57**, 1.

Tessman, E. S. (1965). Complementation groups of phage S13. *Virology* **25**, 303.

Tevethia, S. S., and Rapp, F. (1965). Demonstration of new surface antigens in cells transformed by papovavirus 40 by cytotoxic tests. *Proc. Soc. Exp. Biol. Med.* **120**, 455.

Tevethia, S. S., Katz, M., and Rapp, F. (1965). New surface antigen in cells transformed by simian papovavirus SV40. *Proc. Soc. Exp. Biol. Med.* **119**, 896.

Tevethia, S. S., Couvillion, L. A., and Rapp, F. (1968). Development in hamsters of antibodies against surface antigens present in cells transformed by papovavirus SV40. *J. Immunol.* **100**, 358.

Tevethia, S. S., Lowry, S., Rawls, W. E., Melnick, J. L., and McMillan, V. (1972). Detection of early cell surface changes in herpes simplex virus infected cells by agglutination with concanavalin A. *J. Gen. Virol.* **15**, 93.

Thacore, H. and Youngner, J. S. (1969). Cells persistently infected with Newcastle Disease virus. I. Properties of mutants isolated from persistently infected L cells. *J. Virol.* **4**, 244.

Thacore, H., and Youngner, J. S. (1970). Cells persistently infected with Newcastle disease virus. II. Ribonucleic acid and protein synthesis in cells infected with mutants isolated from persistently infected L cells. *J. Virol.* **6**, 42.

Theilen, G. H., Snyder, S. P., Wolfe, L. G., and Landon, J. C. (1969). Biological studies with viral-induced fibrosarcomas in cats, dogs, rabbits and non-human primates. *Comp. Leuk. Res.* Symposium XVIII, *Bibliotheca Haematologica* **36**, 393. S. Karger, Basel.

Theiler, M. (1957). Action of sodium deoxycholate on arthropod-borne viruses. *Proc. Soc. Exp. Biol. Med.* **96**, 380.

Thiry, L. (1963). Chemical mutagenesis of Newcastle disease virus. *Virology* **19**, 225.

Thiry, L., and Lancini, G. (1972). Mode of action of rifamycin and aminopiperazine derivatives on animal viruses and cells. *In* "Virus-Cell Interactions and Viral Antimetabolites" (D. Shugar, ed.), p. 177, FEBS Symposium, Vol. 22. Academic Press, New York.

Thomas, D. C., and Green, M. (1966). Biochemical studies on adenovirus multiplication. XI. Evidence of a cytoplasmic site for the synthesis of viral-coded proteins. *Proc. Nat. Acad. Sci. U.S.* **56**, 243.

Thomas, M., and Robertson, W. J. (1971). Dermal transmission of virus as a cause of shingles. *Lancet* **ii**, 1349.

Thomas, M., Levy, J. P., Tanzer, J., Boiron, M., and Bernard, J. (1963). Transformation *in vitro* de cellules de peau de veau embryonnaire sous l' action d'extraits accellulaires de papillomes bovins. *C. R. Acad. Sci. (Paris)* **257**, 2155.

Thomas, M., Boiron, M., Tanzer, J., Levy, J. P. and Bernard, J. (1964). *In vitro* transformation of mouse cells by bovine papilloma virus. *Nature (London)* **202**, 709.

Thompson, R. L. (1938). The influence of temperature upon proliferation of infectious fibroma and infectious myxoma *in vivo*. *J. Infect. Dis.* **62**, 307.

Thormar, H. (1963). The growth cycle of visna in monolayer cultures of sheep cells. *Virology* **19**, 273.

Thorne, H. V., Evans, J., and Warden, D. (1968). Detection of biologically defective molecules in component I of polyoma virus DNA. *Nature (London)* **219**, 728.

Thurzo, V., Simkovicova, M., and Simkovic, D. (1969). Studies on group-specific antigens in tumours induced with avian tumour viruses in rats. *Intern. J. Cancer* **4**, 852.

Tiffany, J. M., and Blough, H. A. (1970). Estimation of the number of surface projections on myxo- and paramyxoviruses. *Virology* **41**, 392.

Tigertt, W. D., Berge, T. O., Gouchenour, W. S., Gleiser, C. A., Eveland, W. C., Vorderbruegge, C., and Smetana, H. F. (1960). Experimental yellow fever. *Trans. N.Y. Acad. Sci.* **22**, 323.

Tikchonenko, T. I. (1969). Conformation of viral nucleic acids *in situ*. *Advan. Virus Res.* **15**, 201.

Timbury, M. C. (1971). Temperature-sensitive mutants of herpes simplex virus type 2. *J. Gen. Virol.* **13**, 373.

Timbury, M. C., and Subak-Sharpe, J. H. (1973). Genetic interactions between temperature-sensitive mutants of types 1 and 2 herpes simplex viruses. *J. Gen. Virol.* **18**, 347.

Timmermans, A., Bentvelzen, P., Hageman, C., and Calafat, J. (1969). Activation of a mammary tumor virus in O20 strain mice by X-irradiation and urethan. *J. Gen. Virol.* **4**, 619.

Ting, R. C. (1968). Biological and serological properties of viral particles from a non-producer rat neoplasm induced by murine sarcoma virus. *J. Virol.* **2**, 865.

Tobey, R. A., Petersen, D. F., and Anderson, E. C. (1965). Mengovirus replication. IV. Inhibition of Chinese hamster ovary cell division as a result of infection with Mengovirus. *Virology* **27**, 17.

Tobia, A. M., Schildkraut, C. L., and Maio, J. J. (1970). Deoxyribonucleic acid replication in synchronized cultured mammalian cells. I. Time of synthesis of molecules of different average guanine + cytosine content. *J. Mol. Biol.* **54**, 499.

Tockstein, G., Plarsa, H., Pina, M., and Green, M. (1968). A simple purification procedure for adenovirus 12 T and tumor antigens and some of their properties. *Virology* **36**, 377.

Todaro, G. J. (1973). Detection and characterization of RNA tumor viruses in normal and transformed cells. *Perspec. Virol.* **8**, 81.

Todaro, G. J., and Aaronson, S. A. (1968). Human cell strains susceptible to focus formation by human adenovirus type 12. *Proc. Nat. Acad. Sci. U.S.* **61**, 1272.

Todaro, G. J., and Aaronson, S. A. (1969). Properties of clonal lines of murine sarcoma virus-transformed BALB/3T3 cells. *Virology* **38**, 174.

Todaro, G. J., and Green, H. (1964). An assay for cellular transformation by SV40. *Virology* **23**, 117.

Todaro, G. J., and Green, H. (1966a). High frequency of SV40 transformation of mouse cell line 3T3. *Virology* **28**, 756.

Todaro, G. J., and Green, H. (1966b). Cell growth and the initiation of transformation by SV40. *Proc. Nat. Acad. Sci. U.S.* **55**, 302.

Todaro, G. J., and Huebner, R. J. (1972). The viral oncogene hypothesis: new evidence. *Proc. Nat. Acad. Sci. U.S.* **69**, 1009.

Todaro, G. J., and Martin, G. M. (1967). Increased susceptibility of Down's syndrome fibroblasts to transformation by SV40. *Proc. Soc. Exp. Biol. Med.* **124**, 1232.

Todaro, G. J., and Takemoto, K. K. (1969). "Rescued" SV40: Increased transforming efficiency in mouse and human cells. *Proc. Nat. Acad. Sci. U.S.* **62**, 1031.

Todaro, G. J., Lazar, G. K., and Green, H. (1965). The initiation of cell division in a contact-inhibited mammalian cell line. *J. Cell Comp. Physiol.* **66**, 325.

Todaro, G. J., Green, H., and Swift, M. R. (1966). Susceptibility of human diploid fibroblast strains to transformation by SV40 virus. *Science* **153**, 1252.

Togo, Y., Hornick, R. B., Felitti, V. J., Kaufman, M. L., Dawkins, A. T., Kilpe, V. E., and Claghorn, J. L. (1970). Evaluation of therapeutic efficiency of amantadine in patients with naturally occurring A2 influenza. *J. Amer. Med. Ass.* **211**, 1149.

Tomasi, T. B., Jr., and Bienenstock, J. (1968). Secretory immunoglobulins. *Advan. Immunol.* **9**, 1.

Tomkins, G. M., Gelehrter, T. D., Granner, D., Martin, D., Samuels, H. H., and Thompson, E. B. (1969). Control of specific gene expression in higher organisms. *Science* **166**, 1474.

Tomlinson, A. H., and MacCallum, F. O. (1970). The effect of iodo-deoxyuridine on herpes simplex virus encephalitis in animals and man. *Ann. N.Y. Acad. Sci.* **173**, 20.

Tompkins, W. A. F., Walker, D. L., and Hinze, H. C. (1969). Cellular deoxyribonucleic acid synthesis and loss of contact inhibition in irradiated and contact-inhibited cell cultures infected with fibroma virus. *J. Virol.* **4**, 603.

Tompkins, W. A. F., Zarling, J. M., and Rawls, W. E. (1970a). *In vitro* assessment of cellular immunity to vaccinia virus: contribution of lymphocytes and macrophages. *Infection Immunity* **2**, 783.

Tompkins, W. A. F., Crouch, N. A., Tevethia, S. S., and Rawls, W. E. (1970b). Characterization of surface antigen on cells infected by fibroma virus. *J. Immunol.* **105**, 1181.

Tondury, G., and Smith, D. W. (1966). Fetal rubella pathology. *J. Pediat.* **68**, 867.

Tonegawa, S., Walter, G., Bernardini, A., and Dulbecco, R. (1970). Transcription of the SV40 genome in transformed cells and during lytic infection. *Cold Spring Harbor Symp. Quant. Biol.* **35**, 823.

Tonegawa, S., Walker, G., and Dulbecco, R. (1971). Transcription of SV40 genome in transformed and lytically infected cells. *In* "Lepetit Colloquia on Biology and Medicine, Vol. 2, The Biology of Oncogenic Viruses" (L. Sylvestri, ed.), p. 65. North Holland, Amsterdam.

Tonietti, G., Oldstone, M. B. A., and Dixon, F. J. (1970). The effect of induced chronic viral infections on the immunologic diseases of New Zealand mice. *J. Exp. Med.* **132,** 89.

Toolan, H. W. (1968). The picodnaviruses H, RV and AAV. *Intern. Rev. Exp. Pathol.* **6,** 135.

Tooze, J. (ed.) (1973). "Animal Cells and their Tumor Viruses." Cold Spring Harbor Laboratory, New York.

Top, F. H., Grossman, R. A., Bartelloni, P. J., Segal, H. E., Dudding, B. A., Russell, P. K., and Buescher, E. L. (1971). Immunization with live types 7 and 4 adenovirus vaccines. I. Safety, infectivity, antigenicity, and potency of adenovirus type 7 vaccine in humans. *J. Infect. Dis.* **124,** 148.

Tosolini, F. A. (1970). The response of mice to the intravenous injection of lymphocytic choriomeningitis virus. *Aust. J. Exp. Biol. Med. Sci.* **48,** 445.

Tosolini, F. A., and Mims, C. A. (1971). Effect of murine strain and viral strain on the pathogenesis of lymphocytic choriomeningitis infection and a study of footpad responses. *J. Infect. Dis.* **123,** 134.

Tournier, P., Cassingena, R., Wickert, R., Coppey, J., and Suarez, H. (1967). Étude de méchanisme de l'induction chez de cellules de hamster Syrien transformées par le virus SV40. I. Propriétés d'une lignée cellulaire clonale. *Intern. J. Cancer* **2,** 117.

Toyoshima, K., and Vogt, P. K. (1969). Temperature-sensitive mutants of an avian sarcoma virus. *Virology* **39,** 930.

Toyoshima, K., Friis, R. R., and Vogt, P. K. (1970). The reproductive and cell-transforming capacities of avian sarcoma virus B77: inactivation with UV light. *Virology* **42,** 163.

Trager, G. W., and Rubin, H. (1966). Mixed clones produced following infection of chick embryo cell cultures with Rous sarcoma virus. *Virology* **30,** 275.

Traub, E. (1939). Epidemiology of lymphocytic choriomeningitis in a mouse stock observed for four years. *J. Exp. Med.* **69,** 101.

Trautman, R., and Sutmoller, P. (1971). Detection and properties of a genomic masked viral particle consisting of foot-and-mouth disease virus nucleic acid in bovine enterovirus protein capsid. *Virology* **44,** 537.

Travnicek, M. (1968). RNA with amino acid acceptor activity isolated from an oncogenic virus. *Biochem. Biophys. Acta* **166,** 757.

Trent, D. W., Swensen, C. C., and Qureshi, A. A. (1969). Synthesis of Saint Louis encephalitis virus ribonucleic acid in BHK-21/13 cells. *J. Virol.* **3,** 385.

Trentin, J. J., and Bryan, E. (1966). Virus-induced transplantation immunity to human adenovirus type 12 tumors of the hamster and mouse. *Proc. Soc. Exp. Biol. Med.* **121,** 1216.

Trentin, J. J., Yabe, Y., and Taylor, G. (1962). The quest for human cancer viruses. *Science* **137,** 835.

Trentin, J. J., Van Hoosier, G. L., Jr., and Samper, L. (1968). The oncogenicity of human adenoviruses in hamsters. *Proc. Soc. Exp. Biol. Med.* **127,** 683.

Trilling, D. M., and Axelrod, D. (1972). Analysis of the three components of SV40: pseudo, mature and defective viruses. *Virology* **47,** 360.

Tromans, W. J., and Horne, R. W. (1961). The structure of bacteriophage ØX174. *Virology* **15,** 1.

Trump, B. F., Goldblatt, P. J., and Stowell, R. E. (1965). Studies of necrosis *in vitro* of mouse hepatic parenchymal cells. *Lab. Invest.* **14,** 1946.

Tsuchiya, Y., and Tagaya, I. (1970). Enhanced or inhibited plaque formation of superinfecting viruses in Yaba virus-infected cells. *J. Gen. Virol.* **7,** 71.

Tsuei, D., Fujinaga, K., and Green, M. (1972). The mechanism of viral carcinogenesis by DNA mammalian viruses: RNA transcripts containing viral and highly reiterated cellular base sequences in adenovirus-transformed cells. *Proc. Nat. Acad. Sci. U.S.* **69,** 427.

Turk, J. L., and Bryceson, A. D. M. (1971). Immunological phenomena in leprosy and related diseases. *Advan. Immunol.* **13,** 209.

Twort, F. W. (1915). An investigation on the nature of ultramicroscopic viruses. *Lancet* **ii,** 1241.

Tyrrell, D. A. J. (1959). Interferon produced by cultures of calf-kidney cells. *Nature (London)* **184,** 452.

Tyrrell, D. A. J. (1965). "Common Colds and Related Diseases." Arnold, London.

Tyrrell, D. A. J. (1967). The spread of viruses of the respiratory tract by the airborne route. *Symp. Soc. Gen. Microbiol.* **17**, 286.

Tyrrell, D. A. J., and Parsons, R. (1960). Some virus isolations from common colds. III. Cytopathic effects in tissue cultures. *Lancet* **i**, 239.

Tytell, A. A., Lampson, G. P., Field, A. K., and Hilleman, M. R. (1967). Inducers of interferon and host resistance, III. double-stranded RNA from reovirus type 3 virions (Reo 3-RNA). *Proc. Nat. Acad. Sci. U.S.* **58**, 719.

Tyzzer, E. E. (1905). The histology of the skin lesions in varicella. *J. Med. Res.* **14**, 361.

Uchida, S., and Watanabe, S. (1968). Tumorigenicity of the antigen forming defective virions of simian virus 40. *Virology* **35**, 166.

Uchida, S., Watanabe, S., and Kato, M. (1966). Incomplete growth of simian virus 40 in African green monkey kidney culture induced by serial undiluted passages. *Virology* **28**, 135.

Uchida, S., Yoshiike, K., Watanabe, S., and Furuno, A. (1968). Antigen-forming defective viruses of simian virus 40. *Virology* **34**, 1.

Uhr, J. W., Finkelstein, M. S., and Baumann, J. B. (1962). Antibody formation. III. The primary and secondary antibody response to bacteriophage $\phi$X 174 in guinea pigs. *J. Exp. Med.* **115**, 655.

Underwood, G. E., Elliott, G. A., and Bathula, D. A. (1965). Herpes keratitis in rabbits: pathogenesis and effect of antiviral nucleosides. *Ann. N.Y. Acad. Sci.* **130**, 151.

Utz, J. P. (1964). Viruria in man. *Progr. Med. Virol.* **6**, 71.

Vaheri, A., and Hovi, T. (1972). Structural proteins and subunits of rubella virus. *J. Virol.* **9**, 10.

Vaheri, A., and Pagano, J. S. (1965). Infectious poliovirus RNA: a sensitive method of assay. *Virology* **27**, 435.

Vaheri, A., Vesikari, T., Oker-Blom, N., Seppala, M., Parkman, P. D., Veronelli, J., and Robbins, F. C. (1972). Isolation of attenuated rubella-vaccine virus from human products of conception and uterine cervix. *New Engl. J. Med.* **286**, 1071.

Vainio, T. (1963a). Virus and hereditary resistance *in vitro*. I. Behaviour of West Nile (E-101) virus in the cultures prepared from genetically resistant and susceptible strains of mice. *Ann. Med. Exp. Biol. Fenniae.* **41** (Suppl. 1), 1.

Vainio, T. (1963b). Virus and hereditary resistance *in vitro*. II. Behaviour of West Nile (E-101) virus in cultures prepared from challenged backcross and non-challenged susceptible mice. *Ann. Med. Exp. Biol. Fenniae* **41** (Suppl. 1), 25.

Vainio, T., Gwatkin, R., and Koprowski, H. (1961). Production of interferon by brains of genetically resistant and susceptible mice infected with West Nile virus. *Virology* **14**, 385.

Valentine, R. C., and Periera, H. G. (1965). Antigens and structure of the adenovirus. *J. Mol. Biol.* **13**, 13.

Valle, M., and Cantell, K. (1965). The ability of Sendai virus to overcome cellular resistance to vesicular stomatitis virus. I. General characteristics of the system. *Ann. Med. Exp. Biol. Fenniae* **43**, 57.

van der Veen, J., and Sonderkamp, H. J. A. (1965). Secondary antibody response of guinea pigs to parainfluenza and mumps viruses. *Arch. Ges. Virusforsch.* **15**, 721.

Van de Woude, G. F., Polatnick, J., and Ascione, R. (1970). Foot-and-mouth disease virus-induced alterations of baby hamster kidney cell macromolecular biosynthesis: Inhibition of ribonucleic acid methylation and stimulation of ribonucleic acid synthesis. *J. Virol.* **5**, 458.

van Furth, R., Schuit, H. R. E., and Hijmans, W. (1965). The immunological development of the human fetus. *J. Exp. Med.* **122**, 1173.

Van Kirk, J. E., Mills, J., and Chanock, R. M. (1971). Evaluation of low temperature grown influenza A2/Hong Kong virus in volunteers. *Proc. Soc. Exp. Biol. Med.* **136**, 34.

van Rooyen, C. E., McClelland, L., and Campbell, E. K. (1949). Influenza in Canada during 1949. *Can. J. Public Health* **40**, 447.

Van Tuyen, Vu., Maunoury, R., Febvre, H. (1967). Secretion *in vitro* de l'acide hyaluronique et du collagene par les cellules embryonnaires humaines transformées en culture par le virus du sarcome de Rous. *C. R. Acad. Sci. (Paris)* **265**, 1345.

Varmus, H., Levinson, W. E., and Bishop, J. M. (1971). Extent of transcription by the RNA-dependent DNA polymerase of Rous sarcoma virus. *Nature New Biol.* **233**, 19.

Varmus, H. E., Weiss, R. A., Friis, R. R., Levinson, W., and Bishop, J. M. (1972). Detection of avian tumor virus specific nucleotide sequences in avian cell DNAs. *Proc. Nat. Acad. Sci. U.S.* **69**, 20.

Vasquez, C., and Brailovsky, C. (1965). Purification and fine structure of Kilham's rat virus. *Exp. Mol. Pathol.* **4**, 130.

Vasquez, C., and Tourier, P. (1962). The morphology of reovirus. *Virology* **17**, 503.

Vasquez, C., and Tournier, P. (1964). New interpretation of reovirus structure. *Virology* **24**, 128.

Veda, Y., Ito, M., and Tagaya, I. (1969). A specific surface antigen induced by poxvirus. *Virology* **38**, 180.

Verwoerd, D. W. (1970). Diplornaviruses: a newly recognized group of double-stranded RNA viruses. *Progr. Med. Virol.* **12**, 192.

Verwoerd, D. W., Louw, H., and Oellermann, R. A. (1970). Characterization of bluetongue virus ribonucleic acid. *J. Virol.* **5**, 1.

Verwoerd, D. W., Els, H. J., De Villiers, E-M., and Huismans, H. (1972). Structure of the bluetongue virus capsid. *J. Virol.* **10**, 783.

Vestergaard, B. F., and Scherer, W. F. (1971). An RNA viral infection of hamster testes and uteri resulting in orchitis and effects on fertility and reproduction. *Amer. J. Pathol.* **64**, 541.

Victor, J., Smith, D. G., and Pollack, A. D. (1956). The pathology of Venezuelan equine encephalomyelitis. *J. Infect. Dis.* **98**, 55.

Vigier, P. (1967). Persistance du génome du virus de Rous dans des cellules du hamster converties *in vitro*, et action du virus Sendai inactivé sur sa transmission aux cellules de poule. *C. R. Acad. Sci. (Paris)* **264**, 422.

Vigier, P. (1970). Effect of agar, calf embryo extract, and polyanions on production of foci of transformed cells by Rous sarcoma virus. *Virology* **40**, 179.

Vilaginès, R., and McAuslan, B. R. (1970a). Restricted replication of frog virus 3 in selected variants of BHK cells. *J. Virol.* **6**, 303.

Vilaginès, R., and McAuslan, B. R. (1970b). Interference with viral messenger RNA and DNA synthesis by superinfection with a heterologous deoxyvirus. *Virology* **42**, 1043.

Vilaginès, R., and McAuslan, B. R. (1971). Proteins of polyhedral cycloplasmic deoxyvirus. II. Nucleotide phosphohydrolase activity associated with Frog Virus 3. *J. Virol.* **7**, 619.

Vilček, J. (1964). Production of interferon by newborn and adult mice infected with Sindbis virus. *Virology* **22**, 651.

Vilček, J. (1969). "Interferon," *Virol. Monogr.* **6**, 1.

Vilček, J. (1970). Metabolic determinants of the induction of interferon by a synthetic double-stranded polynucleotide in rabbit kidney cells. *Ann. N.Y. Acad. Sci.* **173**, 390.

Vinograd, J., and Lebowitz, J. (1966). Physical and topological properties of circular DNA. *J. Gen. Physiol.* **49**, 103.

Vinograd, J., Lebowitz, J., Radloff, R., Watson, R., and Laipis, P. (1965). The twisted circular form of polyoma viral DNA. *Proc. Nat. Acad. Sci. U.S.* **53**, 1104.

Vogt, M., and Dulbecco, R. (1958). Properties of a HeLa cell culture with increased resistance to poliomyelitis virus. *Virology* **5**, 425.

Vogt, M., Dulbecco, R., and Smith, B. (1966). Induction of cellular DNA synthesis by polyoma virus. III. Induction in productively infected cells. *Proc. Nat. Acad. Sci. U.S.* **55**, 956.

Vogt, P. K. (1965) A heterogeneity of Rous sarcoma virus revealed by selectively resistant chick embryo cells. *Virology* **25**, 237.

Vogt, P. K. (1967a). Phenotypic mixing in the avian tumor virus group. *Virology* **32**, 708.

Vogt, P. K. (1967b). A virus released by "non-producing" Rous sarcoma cells. *Proc. Nat. Acad. Sci. U.S.* **58**, 801.

Vogt, P. K. (1969a). Focus assay of Rous sarcoma virus. In "Fundamental Techniques in Virology" (K. Habel and N. P. Salzman, eds.), p. 198. Academic Press, New York.

Vogt, P. K. (1969b). Immunofluorescent detection of viral antigens. In "Fundamental Techniques in Virology" (K. Habel and N. P. Salzman, eds.), p. 316. Academic Press, New York.

Vogt, P. K. (1970). Envelope classification of avian RNA tumor viruses. Proc. 4th Intern. Symp. Comp. Leukemia Res. Symposium XVIII. Bibliotheca Haematologica 36, 153. S. Karger, Basel.

Vogt, P. K. (1971a). Spontaneous segregation or non-transforming viruses from cloned sarcoma viruses. Virology 46, 939.

Vogt, P. K. (1971b). Genetically stable reassortment of markers during mixed infection with avian tumor viruses. Virology 46, 947.

Vogt, P. K., and Friis, R. R. (1971). An avian leukosis virus related to RSV(0): properties and evidence for helper activity. Virology 43, 223.

Vogt, P. K., and Ishizaki, R. (1965). Reciprocal patterns of genetic resistances to avian tumor viruses in two lines of chickens. Virology 26, 664.

Vogt, P. K., and Rubin, H. (1961). Localization of infectious virus and viral antigen in chick fibroblasts during successive stages of infection with Rous sarcoma virus. Virology 13, 528.

Vogt, P. K., and Rubin, H. (1963). Studies on the assay and multiplication of avian myeloblastosis virus. Virology 19, 92.

Volkert, M., and Larsen, J. H. (1965). Immunological tolerance to viruses. Progr. Med. Virol. 7, 160.

von Bonsdorff, C-H., Saikku, P., and Oker-Blom, N. (1969). The inner structure of Uukuniemi and two Bunyamwera supergroup viruses. Virology 39, 342.

von Magnus, P. (1954). Incomplete forms of influenza virus. Advan. Virus Res. 2, 59.


Wadell, G. (1970). Structural and biological properties of capsid components of human adenoviruses. Karolinska Institutet, Stockholm.

Wagner, E. K. (1972). Evidence for transcriptional control of the herpes simplex virus genome in infected human cells. Virology 47, 502.

Wagner, E. K., and Roizman, B. (1969). RNA synthesis in cells infected with herpes simplex virus. II. Evidence that a class of viral mRNA is derived from a high molecular weight precursor synthesized in the nucleus. Proc. Nat. Acad. Sci. U.S. 64, 626.

Wagner, R. R. (1965). Interferon: A review and analysis of recent observations. Amer. J. Med. 38, 726.

Wagner, R. R., and Huang, A. S. (1965). Reversible inhibition of interferon synthesis by puromycin: Evidence for an interferon-specific messenger RNA. Proc. Nat. Acad. Sci. U.S. 54, 1112.

Wagner, R. R., Levy, A. H., Snyder, R. M., Ratcliff, G. A., and Hyatt, D. F. (1963). Biologic properties of two plaque variants of vesicular stomatitis virus (Indiana serotype). J. Immunol. 91, 112.

Wagner, R. R., Schnaitman, T. C., Snyder, R. C., Schnaitman, C. A. (1969). Protein composition of structural components of vesicular stomatitis virus. J. Virol. 3, 611.

Wagner, R. R., Snyder, R. M., and Yamazaki, S. (1970). Proteins of vesicular stomatitis virus: kinetics and cellular sites of synthesis. J. Virol. 5, 548.

Wagner, R. R., Kiley, M. P., Snyder, R. M., and Schnaitman, C. A. (1972a). Cytoplasmic compartmentalization of the protein and ribonucleic acid species of vesicular stomatitis virus. J. Virol. 9, 672.

Wagner, R. R., Prevec, L., Brown, F., Summers, D. F., Sokol, F., and MacLeod, R. (1972b). Classification of rhabdovirus proteins: a proposal. J. Virol. 10, 1228.

Wahren, B., and Metcalf, D. (1970). Tolerance breakdown and leukaemogenesis in AKR mice. In "Immunity and Tolerance in Virus Oncogenesis" (J. Dale, ed.), p. 959. Perugia Univ. Press.

Wainwright, S., and Mims, C. A. (1967). Plaque assay for lymphocytic choriomeningitis based on hemadsorption interference. J. Virol. 1, 1091.

Waite, M. R. F., and Pfefferkorn, E. R. (1970). Inhibition of Sindbis virus production by media of low ionic strength: Intracellular events and requirements for reversal. *J. Virol.* **5**, 60.

Waldman, R. H., Small, P. A., and Rowe, D. S. (1971). Utilization of the secretory immunologic system for protection against disease. *In* "The Secretory Immunologic System" (D. H. Dayton, P. A. Small, R. M. Chanock, H. E. Kaufman, and T. B. Tomasi, eds.), p. 129. National Inst. Child Health and Human Development, National Institutes of Health, Bethesda, Maryland.

Waldorf, D. S., Sheagren, J. N., Trautman, J. R., and Block, J. B. (1966). Impaired delayed hypersensitivity in patients with lepromatous leprosy. *Lancet* **ii**, 773.

Walen, K. H. (1962). Demonstration of inapparent heterogeneity in a population of an animal virus by single-burst analysis. *Virology* **20**, 230.

Walen, K. H. (1971). Nuclear involvement in poxvirus infection. *Proc. Nat. Acad. Sci. U.S.* **68**, 165.

Walker, D. L. (1964). The viral carrier state in animal cell cultures. *Progr. Med. Virol.* **6**, 111.

Walker, D. L. (1968). Persistent viral infection in cell cultures. *In* "Medical and Applied Virology" (M. Sanders and E. H. Lennette, eds.), p. 99. W. Green, St. Louis, Missouri.

Walker, D. L., and Boring, W. D. (1958). Factors influencing host-virus interactions. III. Further studies on the alteration of Coxsackie virus infection in adult mice by environmental temperature. *J. Immunol.* **80**, 39.

Walker, D. L., and Hinze, H. C. (1962). A carrier state of mumps virus in human conjunctiva cells. II. Observations on intercellular transfer of virus and virus release. *J. Exp. Med.* **116**, 751.

Walker, P. M. B. (1971). Repetitive DNA in higher organisms. *Progr. Biophys. Mol. Biol.* **23**, 145.

Wall, R., and Darnell, J. E. (1971). Presence of cell and virus specific sequences in the same molecules of nuclear RNA from virus-transformed cells. *Nature New Biol.* **232**, 73.

Wallace, R. D., and Kates, J. (1972). State of adenovirus 2 deoxyribonucleic acid in the nucleus and its mode of transcription: Studies with isolated viral deoxyribonucleic acid-protein complexes and isolated nuclei. *J. Virol.* **9**, 627.

Waller, R. E. (1971). Air pollution and community health. *J. Roy. Coll. Phys. London* **5**, 362.

Wallis, C. (1971). The role of antibody, complement, and anti-IgG in the persistent fraction of herpes virus. *In* "Viruses affecting Man and Animals" (M. Sanders and M. Schaeffer, eds.), p. 102. W. H. Green, St. Louis, Missouri.

Wallis, C., and Melnick, J. L. (1967). Virus aggregation as the cause of the non-neutralizable persistent fraction. *J. Virol.* **1**, 478.

Wallis, C., Melnick, J. L., and Rapp, F. (1965). Different effects of MgCl$_2$ and Mg SO$_4$ on the thermostability of viruses. *Virology* **26**, 694.

Wallnerova, Z., and Mims, C. A. (1971). Thoracic lymph duct cannulation of mice infected with lymphocytic choriomeningitis (L.C.M.) and ectromelia viruses. *Arch. Ges. Virusforsch.* **35**, 152.

Walter, G., Roblin, R., and Dulbecco, R. (1972). Protein synthesis in simian virus 40-infected monkey cells. *Proc. Nat. Acad. Sci. U.S.* **69**, 921.

Walters, S., Burke, D. C., and Skehel, J. J. (1967). Interferon production and RNA inhibitors. *J. Gen. Virol.* **1**, 349.

Ward, R., Melnick, J. L., and Horstmann, D. (1945). Poliomyelitis virus in fly-contaminated food collected at an epidemic. *Science* **101**, 491.

Ward, R., Banerjee, A. K., La Fiandra, A., and Shatkin, A. J. (1972). Reovirus-specific ribonucleic acid from polysomes of infected L cells. *J. Virol.* **9**, 61.

Ward, T. G. (1968). Methods of storage and preservation of animal viruses. *In* "Methods in Virology" (K. Maramorosch and H. Koprowski, eds.), Vol. IV, p. 481. Academic Press, New York.

Warner, J., Madden, M. J., and Darnell, J. E. (1963). The interaction of poliovirus RNA with *Escherichia coli* ribosomes. *Virology* **19**, 393.

Warner, J. R., Girard, M., Latham, H., and Darnell, J. E. (1966). Ribosome formation in HeLa cells in the absence of protein synthesis. *J. Mol. Biol.* **19**, 373.

Warren, L., Critchley, D., and MacPherson, I. (1972). Surface glycoproteins and glycolipids of chicken embryo cells transformed by a temperature-sensitive mutant of Rous sarcoma virus. *Nature (London)* **235**, 275.

Warren, S. L., Carpenter, C. M., and Boak, R. A. (1940). Symptomatic herpes, a sequela of artificially produced fever. *J. Exp. Med.* **71**, 155.

Warthin, A. S. (1931). Occurrence of numerous large giant cells in the tonsils and pharyngeal mucosa in the prodromal stages of measles. *A.M.A. Arch. Pathol.* **11**, 864.

Watanabe, Y., and Graham, A. F. (1967). Structural units of reovirus ribonucleic acid and their possible functional significance. *J. Virol.* **1**, 665.

Watanabe, Y., Prevec, L., and Graham, A. F. (1967). Specificity in transcription of the reovirus genome. *Proc. Nat. Acad. Sci. U.S.* **58**, 1040.

Watanabe, Y., Gauntt, C. J., and Graham, A. F. (1968a). Reovirus-induced ribonucleic acid polymerase. *J. Virol.* **2**, 869.

Watanabe, Y., Millward, S., and Graham, A. F. (1968b). Regulation of transcription of the reovirus genome. *J. Mol. Biol.* **36**, 107.

Watkins, J. F. (1964). Adsorption of sensitized sheep erythrocytes to HeLa cells infected with herpes simplex virus. *Nature (London)* **202**, 1364.

Watkins, J. F. (1971a). Fusion of cells for virus studies and production of cell hybrids. *In* "Methods in Virology" (K. Maramorosch and H. Koprowski, eds.), Vol. 5, p. 1. Academic Press, New York.

Watkins, J. F. (1971b). Cell fusion in the study of tumor cells. *Intern. Rev. Exp. Pathol.* **10**, 115.

Watkins, J. F., and Dulbecco, R. (1967). Production of SV40 in heterokaryons of transformed and susceptible cells. *Proc. Nat. Acad. Sci. U.S.* **58**, 1396.

Watkins, W. M., and Morgan, W. J. (1952). Neutralization of the anti-H agglutinin in eel serum by simple sugars. *Nature (London)* **169**, 825.

Watson, D. H. (1968). The structure of animal viruses in relation to their biological functions. *Symp. Soc. Gen. Microbiol.* **18**, 207.

Watson, D. H., and Wildy, P. (1963). Some serological properties of herpes virus particles studied with the electron microscope. *Virology* **21**, 100.

Watson, D. H., Wildy, P., and Russell, W. C. (1964). Quantitative electron microscope studies on the growth of herpes virus using the techniques of negative staining and ultramicrotomy. *Virology* **24**, 523.

Watson, D. H., Shedden, W. I. H., Elliot, A., Tetsuka, H., and Wildy, P. (1966). Virus specific antigens in mammalian cells infected with herpes simplex virus. *Immunology* **11**, 399.

Watson, D. H., Wildy, P., Harvey, B. A. M., and Shedden, W. I. H. (1967). Serological relationships among viruses of the herpes group. *J. Gen. Virol.* **1**, 139.

Watson, J. D. (1970). "The Molecular Biology of the Gene," 2nd Ed. Benjamin, New York.

Watson, J. D., and Littlefield, J. W. (1960). Some properties of DNA from Shope papilloma virus. *J. Mol. Biol.* **2**, 161.

Watson, M. L. (1959). Further observations of the nuclear envelope of the animal cell. *J. Biophys. Biochem.* **6**, 147.

Waubke, R., zur Hausen, H., and Henle, W. (1968). Chromosomal and autoradiographic studies of cells infected with herpes simplex virus. *J. Virol.* **2**, 1047.

Wear, D. J., and Rapp, F. (1971). Latent measles infection of the hamster central nervous system. *J. Immunol.* **107**, 1593.

Webb, H. E., and Smith, C. E. G. (1966). Relation of immune response to development of central nervous system lesions in virus infections of man. *Brit. Med. J.* **2**, 1179.

Webster, L. T. (1937). Inheritance of resistance of mice to enteric bacterial and neurotropic virus infections. *J. Exp. Med.* **65**, 261.

Webster, L. T. (1946). Experimental epidemiology. *Medicine* **25**, 77.

Webster, L. T., and Clow, A. D. (1936). Experimental encephalitis (St. Louis type) in mice with high inborn resistance. *J. Exp. Med.* **63**, 827.

Webster, L. T., and Johnson, M. S. (1941). Comparative virulence of St. Louis encephalitis virus cultured with brain tissue from innately susceptible and innately resistant mice. *J. Exp. Med.* **74**, 489.

Webster, R. G. (1965). The immune response to influenza virus. I. Effect of the route and schedule of vaccination on the time course of the immune response, as measured by three serological methods. *Immunology* **9**, 501.

Webster, R. G. (1968a). The immune response to influenza virus. II. Effect of route and schedule of vaccination on the quantity and avidity of antibodies. *Immunology* **14**, 29.

Webster, R. G. (1968b). The immune response to influenza virus. III. Changes in the avidity and specificity of early IgM and IgG antibodies. *Immunology* **14**, 39.

Webster, R. G. (1970). Antigenic hybrids of influenza A viruses with surface antigens to order. *Virology* **42**, 633.

Webster, R. G. (1972). On the origin of pandemic influenza viruses. *Curr. Topics Microbiol. Immunol.* **59**, 75.

Webster, R. G., and Campbell, C. H. (1972). The "*in vivo*" production of "new" influenza A viruses. II. *In vivo* isolation of "new" viruses. *Virology* **48**, 528.

Webster, R. G., and Laver, W. G. (1971). Antigenic variation in influenza virus. Biology and chemistry. *Progr. Med. Virol.* **13**, 271.

Webster, R. G., and Laver, W. G. (1972). Studies on the origin of pandemic influenza. I. Antigenic analysis of A2 influenza viruses before and after the appearance of Hong Kong influenza using antisera to the isolated hemagglutinin subunits. *Virology* **48**, 433.

Webster, R. G., and Pereira, H. G. (1968). A common surface antigen in influenza viruses from human and avian sources. *J. Gen. Virol.* **3**, 201.

Webster, R. G., Laver, W. G., and Kilbourne, E. D. (1968). Reactions of antibodies with surface antigens of influenza virus. *J. Gen. Virol.* **3**, 315.

Webster, R. G., Campbell, C. H., and Granoff, A. (1971). The "*in vivo*" production of "new" influenza A viruses. I. Genetic recombination between avian and mammalian influenza viruses. *Virology* **44**, 317.

Webster, R. G., Campbell, C. H., and Granoff, A. (1973). The "*in vivo*" production of "new" influenza viruses. III. Isolation of recombinant influenza viruses under simulated conditions of natural transmission. *Virology* **51**, 149.

Wecker, E., and Schonne, E. (1961). Inhibition of viral RNA synthesis by parafluorophenylalanine. *Proc. Nat. Acad. Sci. U.S.* **47**, 278.

Wehrle, P. F., Carbonaro, O., Day, P. A., Whalen, J. P., Reichert, R., and Pretroy, B. (1961). Transmission of polioviruses. III. Prevalence of polioviruses in pharyngeal secretions of infected household contacts of patients with clinical disease. *Pediatrics* **27**, 762.

Wehrli, W., and Staehelin, M. (1971). Actions of the rifamycins. *Bacteriol. Rev.* **35**, 290.

Weigle, W. O. (1961). The immune response of rabbits tolerant to bovine serum albumin to the injection of other heterologous serum albumins. *J. Exp. Med.* **114**, 111.

Weil, R., and Vinograd, J. (1963). The cyclic helix and cyclic coil forms of polyoma viral DNA. *Proc. Nat. Acad. Sci. U.S.* **50**, 730.

Weil, R., Michel, M. R., and Ruschmann, G. K. (1965). Induction of cellular DNA synthesis by polyoma virus. *Proc. Nat. Acad. Sci. U.S.* **53**, 1468.

Weinberg, A., and Becker, Y. (1969). Studies on EB virus of Burkitt's lymphoblasts. *Virology* **39**, 312.

Weinberg, R. A., Waarnaar, S. O., and Winocour, E. (1972a). Isolation and characterization of simian virus 40 ribonucleic acid. *J. Virol.* **10**, 193.

Weinberg, R. A., Ben-Ishai, Z., and Newbold, J. E. (1972b). Poly A associated with SV40 messenger RNA. *Nature New Biol.* **238**, 111.

Weiner, L. P., Herndon, R. M., Narayan, O., Johnson, R. T., Shah, K., Rubenstein, L. J., Preziosi, T. J., and Canley, F. K. (1972). Isolation of virus related to SV40 from patients with progressive multifocal leukoencephalopathy. *New Engl. J. Med.* **286**, 385.

Weinstein, A. J., Gazdar, A. F., Sims, H. L., and Levy, H. B. (1971). Lack of correlation between interferon production and antitumour effect of poly I·poly C. *Nature New Biol.* **231**, 53.

Weiss, E. (ed.) (1971). Arthropod cell cultures and their application to the study of viruses. *Curr. Topics Microbiol. Immunol.* **55**, 1.

Weiss, M. C. (1970). Further studies on loss of T antigen from somatic hybrids between mouse cells and SV40-transformed human cells. *Proc. Nat. Acad. Sci. U.S.* **66**, 79.

Weiss, M. C., Ephrussi, B., and Scaletta, L. J. (1968). Loss of T antigen from somatic hybrids between mouse cells and SV40 transformed human cells. *Proc. Nat. Acad. Sci. U.S.* **59**, 1132.

Weiss, R. A. (1967). Spontaneous virus production from "non-virus-producing" Rous sarcoma cells. *Virology* **32**, 719.

Weiss, R. A. (1969a). The host range of Bryan strain Rous sarcoma virus synthesized in the absence of helper virus. *J. Gen. Virol.* **5**, 511.

Weiss, R. A. (1969b). Interference and neutralization studies with Bryan strain Rous sarcoma virus synthesized in the absence of helper virus. *J. Gen. Virol.* **5**, 529.

Weiss, R. A. (1971). Cell transformation induced by Rous sarcoma virus: Analysis of density dependence. *Virology* **46**, 209.

Weiss, R. A. (1972). Helper viruses and helper cells. *In* "Symposium on RNA Viruses and Host Genome in Oncogenesis" (P. Emmelot and P. Bentvelzen, eds.), p. 117. North Holland, Amsterdam.

Weiss, R. A., and Payne, L. N. (1971). The heritable nature of the factor in chicken cells which acts as a helper for Rous sarcoma virus. *Virology* **45**, 508.

Weiss, R. A., Friis, R. R., Katz, E., and Vogt, P. K. (1971). Induction of avian tumor viruses in normal cells by physical and chemical carcinogens. *Virology* **46**, 920.

Weissbach, A., Schlabach, A., and Fridlender, B. (1971). DNA polymerases from human cells. *Nature New Biol.* **231**, 167.

Weissmann, C., Feix, G., and Slor, H. (1968). *In vitro* synthesis of phage RNA: The nature of the intermediates. *Cold Spring Harbor Symp. Quant. Biol.* **33**, 83.

Weller, T. H. (1971). The cytomegaloviruses: ubiquitous agents with protein clinical manifestations. *New Engl. J. Med.* **285**, 203.

Weller, T. H., and Coons, A. H. (1954). Fluorescent antibody studies with agents of varicella and herpes zoster propagated *in vitro*. *Proc. Soc. Exp. Biol. Med.* **86**, 789.

Wellings, P. C., Awdry, P. N., Bors, F. H., Jones, B. R., Brown, D. C., and Kaufman, H. E. (1972). Clinical evaluation of trifluorothymidine in the treatment of herpes simplex corneal ulcers. *Amer. J. Ophthalmol.* **73**, 932.

Wells, W. F. (1955). "Airborne Contagion and Air Hygiene." Harvard Univ. Press, Cambridge, Massachusetts.

Welsh, R. M., Trowbridge, R. S., Kowalski, J. B., O'Connell, C. M., and Pfau, C. F. (1971). Amantadine hydrochloride inhibition of early and late stages of lymphocytic choriomeningitis virus-cell interactions. *Virology* **45**, 679.

Wenner, H. A., and Behbehani, A. M. (1968). Echoviruses. *Virology Monogr.* **1**, 1.

Wertz, G. W., and Youngner, J. S. (1970). Interferon production and inhibition of host synthesis in cells infected with vesicular stomatitis virus. *J. Virol.* **6**, 476.

Wertz, G. W., and Youngner, J. S. (1972). Inhibition of protein synthesis in L cells infected with vesicular stomatitis virus. *J. Virol.* **9**, 85.

Westaway, E. G., and Reedman, B. M. (1969). Proteins of the group B arbovirus Kunjin. *J. Virol.* **4**, 688.

Westphal, H., and Dulbecco, R. (1968). Viral DNA in polyoma- and SV40-transformed cell lines. *Proc. Nat. Acad. Sci. U.S.* **59**, 1158.

Westwood, J. C. N., Harris, W. J., Zwartouw, H. T., Titmuss, D. H. J., and Appleyard, G. (1964). Studies on the structure of vaccinia virus. *J. Gen. Microbiol.* **34**, 67.

Westwood, J. C. N., Zwartouw, H. T., Appleyard, G., and Titmuss, D. H. J. (1965). Comparison of the soluble antigens and virus particle antigens of vaccinia virus. *J. Gen. Microbiol.* **38**, 47.

Westwood, J. C. N., Boulter, E. A., Bowen, E. T. W., and Maber, H. B. (1966). Experimental respiratory infection with poxviruses. I. Clinical, virological and epidemiological studies. *Brit. J. Exp. Pathol.* **47**, 453.

Wetmur, J. G., and Davidson, N. (1968). Kinetics of renaturation of DNA. *J. Mol. Biol.* **31**, 349.

Wettstein, F. O., Noll, H., and Penman, S. (1964). Effect of cycloheximide on ribosomal aggregates engaged in protein synthesis *in vitro*. *Biochim. Biophys. Acta* **87**, 525.

Whang-Peng, J., Gerber, P., and Knutsen, T. (1970). So-called C marker chromosome and Epstein-Barr Virus. *J. Nat. Cancer Inst.* **45**, 831.

Wheeler, C. E. (1960). Further studies on the effect of neutralizing antibody upon the course of herpes simplex infections in tissue culture. *J. Immunol.* **84**, 394.

Wheelock, E. F., and Tamm, I. (1961). Biochemical basis for alterations in structure and function of HeLa cells infected with Newcastle disease virus. *J. Exp. Med.* **114**, 617.

Wheelock, E. F., Larke, R. P. B., and Caroline, N. L. (1968). Interference in human viral infections: present status and prospects for the future. *Progr. Med. Virol.* **10**, 286.

White, D. O. (1959). The mechanism of natural resistance of the allantois to influenza infection. *Virology* **9**, 680.

White, D. O. (1973). Influenza viral proteins: identification and synthesis. *Current Topics Microbiol. Immunol.* **63**, in press.

White, D. O., and Cheyne, I. M. (1966). Early events in the eclipse phase of influenza and parainfluenza virus infection. *Virology* **29**, 49.

White, D. O., Oliphant, H. M., and Batchelder, E. (1962). Haemadsorption to single cells in suspension. *Nature (London)* **196**, 792.

White, D. O., Day, H. M., Batchelder, E. J., Cheyne, I. M., and Wansbrough, A. J. (1965). Delay in the multiplication of influenza virus. *Virology* **25**, 289.

White, D. O., Scharff, M. D., and Maizel, J. V. (1969). The polypeptides of adenovirus. III. Synthesis in infected cells. *Virology* **38**, 395.

White, D. O., Taylor, J. M., Haslam, E. A., and Hampson, A. W. (1970). The polypeptides of influenza virus and their biosynthesis. *In* "The Biology of Large RNA Viruses" (R. D. Barry and B. W. J. Mahy, eds.), p. 602. Academic Press, London and New York.

Whitehead, R. H., Chaicumpa, V., Olson, L. C., and Russell, P. K. (1970). Sequential dengue virus infections in the white-handed gibbon (*Hylobates lar*). *Amer. J. Trop. Med. Hyg.* **19**, 94.

Whitman, L., and Antunes, P. C. A. (1937). Studies on capacity of various Brazilian mosquitoes, representing the genera *Psorophora*, *Aedes*, *Mansonia* and *Culex*, to transmit yellow fever. *Amer. J. Trop. Med.* **18**, 135.

Wiese, W. H., Lewis, A. M., and Rowe, W. P. (1970). Equilibrium density gradient studies on simian virus 40-yielding variants of the adenovirus 2-simian virus 40 hybrid population. *J. Virol.* **5**, 421.

Wigzell, H., and Stjernswärd, J. (1966). Age-dependent rise and fall of immunological reactivity in the CBA mouse. *J. Nat. Cancer Inst.* **37**, 513.

Wiktor, T. J., Kuwert, E., and Koprowski, H. (1968). Immune lysis of rabiesvirus-infected cells. *J. Immunol.* **101**, 1271.

Wilcox, W. C., and Cohen, G. H. (1967). Soluble antigens of vaccinia-infected mammalian cells. II. Time course of synthesis of soluble antigens and viral structural proteins. *J. Virol.* **1**, 500.

Wilcox, W. C., and Cohen, G. H. (1969). The poxvirus antigens. *Curr. Topics Microbiol. Immunol.* **47**, 1.

Wild, T. F. (1971). Replication of vesicular stomatitis virus: Characterization of the virus-induced RNA. *J. Gen. Virol.* **13**, 295.

Wild, T. F., and Brown, F. (1970). Replication of foot-and-mouth disease virus ribonucleic acid. *J. Gen. Virol.* **7**, 1.

Wild, T. F., Burroughs, J. N., and Brown, F. (1969). Surface structure of foot-and-mouth disease virus. *J. Gen. Virol.* **4**, 313.

Wildy, P. (1971). Classification and nomenclature of viruses: first report of the International Committee on Nomenclature of Viruses. *Monogr. Virol.* **5**, 1.

Wildy, P. (1972). Herpesvirus and antigens. *In* "Oncogenesis and Herpesviruses" (P. M. Biggs, G. de-Thé and L. N. Payne, eds.), p. 409. World Health Organization International Agency for Research on Cancer, Lyon.

Wildy, P., Stoker, M. G. P., Macpherson, I. A., and Horne, R. W. (1960a). The fine structure of polyoma virus. *Virology* **11**, 444.

Wildy, P., Russell, W. C., and Horne, R. W. (1960b). The morphology of herpes virus. *Virology* **12**, 204.

Williams, J. F., and Ustacelebi, S. (1971a). Temperature-restricted mutants of human adeno-virus type 5. In "Strategy of the Viral Genome" (G. E. W. Wolstenholme and M. O'Connor, eds.), p. 275, Ciba Found. Symp. Churchill Livingstone, Edinburgh.

Williams, J. F., and Ustacelebi, S. (1971b). Complementation and recombination with temperature-sensitive mutants of adenovirus type 5. J. Gen. Virol. 13, 345.

Williams, J. F., Gharpure, M., Ustacelebi, S., and McDonald, S. (1971). Isolation of temperature-sensitive mutants of adenovirus type 5. J. Gen. Virol. 11, 95.

Williams, M. G., Howatson, D. F., and Almeida, J. D. (1961). Morphological characterization of the virus of the human common wart (verruca vulgaris). Nature (London) 189, 895.

Williams, R. T., Dunsmore, J. T., and Parer, I. (1972). Evidence for the existence of latent myxoma virus in rabbits [Oryctolagus cuniculus (L.)]. Nature (London) 238, 99.

Williamson, R., and Brownlee, G. G. (1969). The sequence of 5S ribosomal RNA from two mouse cell lines. FEBS Letters 3, 301.

Willmer, E. N., ed. (1965–1966). "Cells and Tissues in Culture," 3 Vols. Academic Press, New York.

Wilson, C. B., and Dixon, F. J. (1971). Quantitation of acute and chronic serum sickness in the rabbit. J. Exp. Med. 134, 7s.

Wilson, D. E. (1968). Inhibition of host-cell protein and ribonucleic acid synthesis by Newcastle disease virus. J. Virol. 2, 1.

Wilson, R. G., and Bader, J. P. (1965). Viral ribonucleic acid polymerase: Chick-embryo cells infected with vesicular stomatitis virus or Rous-associated virus. Biochim. Biophys. Acta 103, 549.

Wilson, V. W., and Rafelson, M. E. (1967). Studies on neuraminidases of influenza virus. III. Stimulation of activity by bivalent cations. Biochim. Biophys. Acta 146, 160.

Winocour, E. (1965). Attempts to detect an integrated polyoma genome by nucleic acid hybridization I. "Reconstruction" experiments and complementary tests between synthetic polyoma RNA and polyoma tumor DNA. Virology 25, 276.

Winocour, E. (1968). Further studies on the incorporation of cell DNA into polyoma-related particles. Virology 34, 571.

Winocour, E. (1969). Some aspects of the interaction between polyoma virus and cell DNA. Advan. Virus Res. 14, 153.

Winter, P. E., Nantapanich, S., Nisalak, A., Udomsakdi, S., Dewey, R. W., and Russell, P.K. (1969). Recurrence of epidemic dengue hemorrhagic fever in an insular setting. Amer. J. Trop. Med. Hyg. 18, 573.

Winters, W. D., and Russell, W. C. (1971). Studies on assembly of adenovirus in vitro. J. Gen. Virol. 10, 181.

Wisseman, C. L., Jr., Sweet, B. H., Kitaoka, M., and Tamiya, T. (1962). Immunological studies with group B arthropod-borne viruses. I. Broadened neutralizing antibody spectrum induced by 17D yellow fever vaccine in human subjects previously infected with Japanese encephalitis virus. Amer. J. Trop. Med. Hyg. 11, 550.

Wisseman, C. L., Jr., Kitaoka, M., and Tamiya, T. (1966). Immunological studies with group B arthropod-borne viruses. V. Evaluation of cross-immunity against type 1 dengue fever in human subjects convalescent from subclinical natural Japanese encephalitis virus infection and vaccinated with 17D strain yellow fever vaccine. Amer. J. Trop. Med. Hyg. 15, 588.

Wistar, R. (1968). Serum antibody to Salmonella flagellar antigens. I. Methods of antibody assay. Aust. J. Exp. Biol. Med. Sci. 46, 769.

Witter, R. L., Burgoyne, G. H., and Solomon, J. J. (1969). Evidence for a herpesvirus as an etiologic agent of Marek's disease. Avian Dis. 13, 171.

Wolfe, L. G., Griesemer, R. A., and Farrell, R. L. (1968). Experimental aerosol transmission of Yaba virus in monkeys. J. Nat. Cancer Inst. 41, 1175.

Wolfe, L. G., Deinhardt, F., Theilen, G. H., Rubin, H., Kawakami, T., and Bustad, L. K. (1971). Induction of tumors in marmoset monkeys by simian sarcoma virus, type 1 (Lagothrix): A preliminary report. J. Nat. Cancer Inst. 47, 1115.

Wolfson, J., and Dressler, D. (1972). Adenovirus-2 DNA contains an inverted terminal repetition. *Proc. Nat. Acad. Sci. U.S.* **69**, 3054.

Wolstenholme, G. E. W., and Knight, J. (eds.) (1970). Hormones and the immune response. *Ciba Found. Study Group No.* **36**.

Wong, P. K. Y., Holloway, A. F., and Cormack, D. V. (1971). A search for recombination between temperature-sensitive mutants of vesicular stomatitis virus. *J. Gen. Virol.* **13**, 477.

Woodroofe, G. M., and Fenner, F. (1960). Genetic studies with mammalian poxviruses. IV. Hybridization between several different poxviruses. *Virology* **12**, 272.

Woodroofe, G. M., and Fenner, F. (1962). Serological relationships within the poxvirus group : an antigen common to all members of the group. *Virology* **16**, 334.

Woodroofe, G. M., and Fenner, F. (1965). Viruses of the myxoma-fibroma subgroup of the poxviruses. I. Plaque production in cultured cells, plaque-reduction tests, and cross-protection tests in rabbits. *Aust. J. Exp. Biol. Med. Sci.* **43**, 123.

Woodruff, J. F. (1970). The influence of quantitated post-weaning undernutrition on coxsackievirus B3 infection of adult mice. II. Alterations of host defense mechanisms. *J. Infect. Dis.* **121**, 164.

Woodruff, J. F., and Kilbourne, E. D. (1970). The influence of quantitated post-weaning undernutrition on coxsackievirus B3 infection of adult mice. I. Viral persistence and increased severity of lesions. *J. Infect. Dis.* **121**, 136.

Woodson, B., and Joklik, W. K. (1965). Inhibition of vaccinia virus multiplication by isatin-β-thiosemicarbazone. *Proc. Nat. Acad. Sci. U.S.* **54**, 946.

World Health Organ. (1961). Arthropod-borne Viruses. Report of a Study Group. *WHO Tech. Rept. Ser. No.* **219**.

World Health Organ. (1964). Committee on Nomenclature of Human Immunoglobulins. *Bull. WHO* **30**, 447.

World Health Organ. (1966a). Report of Scientific Group on Human Viral and Rickettsial Vaccines. *WHO Tech. Rept. Ser. No.* **325**.

World Health Organ. (1966b). The use of human immunoglobulin. *WHO Tech. Rept. Ser. No.* **327**.

World Health Organ. (1967). Arboviruses and human disease. *WHO Tech. Rept. Ser. No.* **369**.

World Health Organ. (1969a). Report of Scientific Group on Respiratory Viruses. *WHO Tech. Rept. Ser. No.* **408**.

World Health Organ. (1969b). Cell-mediated immune responses; report of a W.H.O. Scientific Group. *WHO Tech. Rep. Ser. No.* **423**.

World Health Organ. (1970). Factors regulating the immune response: report of a WHO. Scientific Group. *WHO Tech. Rep. Ser. No.* **448**.

World Health Organ. (1971a). WHO Expert Committee on Yellow Fever, Third Report. *WHO Tech. Rep. Ser. No.* **479**.

World Health Organ. (1971b). A revised system of nomenclature for influenza viruses. *Bull. WHO* **45**, 119.

World Health Organ. (1972a). WHO Expert Committee on Biological Standardization, 24th Report, *WHO Tech. Rep. Ser. No.* **486.**

World Health Organ. (1972b). WHO Expert Committee on Smallpox Eradication Second Report, *WHO Tech. Rep. Ser. No.* **493.**

Worthington, M., and Baron, S. (1971). Late therapy with an interferon stimulator in an arbovirus encephalitis in mice. *Proc. Soc. Exp. Biol. Med.* **136**, 323.

Worthington, M., and Clark, R. (1971). Lack of effect of immunosuppression on scrapie infection in mice. *J. Gen. Virol.* **13**, 349.

Wright, D. H. (1966). Burkitt's tumour in England: a comparison with childhood lymphosarcoma. *Intern. J. Cancer* **1**, 503.

Wright, G. P. (1953). Nerve trunks as pathways of infection. *Proc. Roy. Soc. Med.* **46**, 319.

Wright, P. F., and Chanock, R. M. (1970). Complementation between temperature-sensitive mutants of respiratory syncytial virus. *Bacteriol. Proc.*, p. 193.

Wright, P. F., Mills, J., and Chanock, R. M. (1971). Evaluation of a temperature-sensitive mutant of respiratory syncytial virus in adults. *J. Infect. Dis.* **124**, 505.

Wright, P. W. (1971). *In vitro* relationship of SV40 tumor specific surface antigen to other SV40 antigens. *Nature New Biol.* **233**, 18.

Wright, P. W., and Law, L. W. (1971). Quantative *in vitro* measurements of simian virus 40 tumor specific antigens. *Proc. Nat. Acad. Sci. U.S.* **68**, 973.

Wrigley, N. G. (1969). An electron microscope study of the structure of *Sericesthis* iridescent virus. *J. Gen. Virol.* **5**, 123.

Wrigley, N. G. (1970). An electron microscope study of the structure of *Tipula* iridescent virus. *J. Gen. Virol.* **6**, 169.

Wu, H. C., Meezan, E., Black, P. H., and Robbins, P. W. (1969). Comparative studies on the carbohydrate containing membrane components of normal and virus transformed mouse fibroblasts. I. Glucosamine-labeling patterns in 3T3, spontaneously transformed 3T3 and SV40-transformed 3T3 cells. *Biochemistry* **8**, 2509.

Wunner, W. H., and Pringle, C. R. (1972). Protein synthesis in BHK21 cells infected with vesicular stomatitis virus. I. *ts* mutants of the Indiana serotype. *Virology* **48**, 104.

Wyke, J. (1971a). A method of isolating cells incapable of multiplication in suspension culture. *Exp. Cell Res.* **66**, 203.

Wyke, J. (1971b). Phenotypic variation and its control in polyoma-transformed BHK 21 cells. *Exp. Cell Res.* **66**, 209.


Xeros, N. (1962). Deoxyriboside control and synchronization of mitosis. *Nature (London)* **194**, 682.


Yagi, Y. (1971). Identification of multiple antibody components by radioimmunoelectrophoresis and radioimmunodiffusion. *In* "Methods in Immunology and Immunochemistry." (C. A. Williams and M. W. Clare, eds.), Vol. 3, p. 463. Academic Press, New York and London.

Yamaguchi, N., Takeuchi, M., and Yamamoto, T. (1969). Rous sarcoma virus production in mixed cultures of mammalian Rous sarcoma cells and chick embryo cells. *Intern. J. Cancer* **4**, 678.

Yamamoto, H., and Shimojo, H. (1971). Multiplicity reactivation of human adenovirus type 12 and simian virus 40 irradiated by ultraviolet light. *Virology* **45**, 529.

Yamamoto, T., Otani, S., and Shiraki, H. (1965). A study of the evolution of viral infection in experimental herpes simplex encephalitis and rabies by means of fluorescent antibody. *Acta Neuropathol.* **5**, 288.

Yamane, I., and Kusano, T. (1967). *In vitro* transformation of cells of hamster brain by adenovirus type 12. *Nature (London)* **213**, 187.

Yin, F. H. (1969). Temperature-sensitive behavior of the hemagglutinin in a temperature-sensitive mutant virion of Sindbis. *J. Virol.* **4**, 547.

Yin, F. H., and Lockart, R. Z. (1968). Maturation defects in temperature sensitive mutants of Sindbis virus. *J. Virol.* **2**, 728.

Yogo, Y., and Wimmer, E. (1972). Polyadenylic acid at the 3'-terminus of poliovirus RNA. *Proc. Nat. Acad. Sci. U.S.* **69**, 1877.

Yohn, D. S., and Gallagher, J. F. (1969). Some physiochemical properties of Yaba poxvirus deoxyribonucleic acid. *J. Virol.* **3**, 114.

Yohn, D. S., Haendiges, V. A., and de Harven, E. (1966). Yaba tumor poxvirus synthesis in vitro. III. Growth kinetics. *J. Bacteriol.* **91**, 1986.

Yohn, D. S., Marmol, F. R., and Olsen, R. G. (1970). Growth kinetics of Yaba tumor poxvirus after *in vitro* adaptation to Cercopithecus kidney cells. *J. Virol.* **5**, 205.

Yoshikama-Fukada, M., and Ebert, J. D. (1969). Hybridization of RNA from Rous sarcoma virus with cellular and viral DNAs. *Proc. Nat. Acad. Sci. U.S.* **64**, 870.

Yoshike, K. (1968). Studies on DNA from low-density particles of SV40: I. Heterogeneous defective virions produced by successive undiluted passages. *Virology* **34**, 391.

Yoshino, K., and Taniguchi, S. (1965). Studies on the neutralization of herpes simplex virus. I. Appearance of neutralizing antibodies having different grades of complement requirement. *Virology* **26**, 44.

Young, N. A., Hoyer, B. H., and Martin, M. A. (1968). Polynucleotide sequence homologies among polioviruses. *Proc. Nat. Acad. Sci. U.S.* **61**, 548.

Younghusband, H. B., and Bellett, A. J. D. (1971). Mature form of the deoxyribonucleic acid from chick embryo lethal orphan virus. *J. Virol.* **8**, 265.

Youngner, J. S., Stinebring, W. R., and Taube, S. E. (1965). Influence of inhibitors of protein synthesis on interferon formation in mice. *Virology* **27**, 541.

Youngner, J. S., Scott, A. W., Hallum, J. V., and Stinebring, W. R. (1966). Interferon production by inactivated Newcastle disease virus in cell cultures and in mice. *J. Bacteriol.* **92**, 862.

Yuill, T. M., and Hanson, R. P. (1964). Infection of suckling cottontail rabbits with Shope's fibroma virus. *Proc. Soc. Exp. Biol. Med.* **117**, 376.

Yunis, E. J., and Donelly, W. H. (1969). The ultrastructure of the replicating Newcastle disease virus in the chick embryo chorioallantoic membrane. *Virology* **39**, 352.

Zajac, B. A., and Hummeler, K. (1970). Morphogenesis of the nucleoprotein of vesicular stomatitis virus. *J. Virol.* **6**, 243.

Zarling, J. M., and Tevethia, S. S. (1971). Expression of concanavalin A binding sites in rabbit kidney cells infected with vaccinia virus. *Virology* **45**, 313.

Závadová, Z., and Závada, J. (1968). Host-cell repair of ultraviolet-irradiated pseudorabies virus in chick embryo cells. *Acta Virol.* **12**, 507.

Zee, Y. C., Hackett, A. J., and Talens, L. (1970). Vesicular stomatitis virus maturation sites in six different host cells. *J. Gen. Virol.* **7**, 95.

Zeigel, R. F., and Clark, H. F. (1971). Histologic and electron microscopic observations on a tumor-bearing viper: establishment of a "C"-type virus-producing cell line. *J. Nat. Cancer Inst.* **46**, 309.

Zhdanov, V. M. (1967). Present status of live influenza virus vaccines. *In* "First International Conference on Vaccines against Viral and Rickettsial Diseases of Man," p. 9. Pan Amer. Health Organ. Sci. Publ. No. 147. Washington, D.C.

Zilber, L. A. (1965). Pathogenicity and oncogenicity of Rous sarcoma virus for mammals. *Progr. Exp. Tumor Res.* **7**, 1.

Zimmerman, E. F., Heeter, M., and Darnell, J. E. (1963). RNA synthesis in poliovirus-infected cells. *Virology* **19**, 400.

Zisman, B., Wheelock, E. F., and Allison, A. C. (1971). Role of macrophages and antibody in resistance of mice against yellow fever virus. *J. Immunol.* **107**, 236.

Zlotnik, I. (1968). The reaction of astrocytes to acute virus infections of the central nervous system. *Brit. J. Exp. Pathol.* **49**, 555.

Zuckerman, A. J. (1972). "Hepatitis-Associated Antigen and Viruses." North Holland, Amsterdam.

Zuckerman, A. J., Baines, P. M., and Almeida, J. D. (1972). Australia antigen as a marker of propagation of the serum hepatitis virus in liver cultures. *Nature (London)* **236**, 78.

zur Hausen, H., and Schulte-Holthausen, H. (1970). Presence of EB virus nucleic acid homology in a "virus-free" line of Burkitt tumour cells. *Nature (London)* **227**, 245.

zur Hausen, H., Schulte-Holthausen, H., Klein, G., Henle, G., Henle, W., Clifford, P., and Samtesson, L. (1970). EBV DNA in biopsies of Burkitt tumours and anaplastic carcinomas of the nasopharynx. *Nature (London)* **228**, 1056.

Zu Rhein, G. M. (1969). Association of papova-virions with a human demyelinating disease (progressive multifocal leukoencephalopthy). *Progr. Med. Virol.* **11**, 185.

Zweerink, H. J., and Joklik, W. K. (1970). Studies on the intracellular synthesis of reovirus-specified proteins. *Virology* **41**, 501.

Zweerink, H. J., McDowell, M. J., and Joklik, W. K. (1971). Essential and non-essential non-capsid reovirus proteins. *Virology* **45**, 716.

Zwillenberg, L. O., and Bürki, F. (1966). On the capsid structure of some small feline and bovine RNA viruses. *Arch. Ges. Virusforsch.* **19**, 373.

# Subject Index*

## A

*Unless otherwise defined, a primary entry always refers to a property or structure of a virus.

B 5
C 6
D 7
E 8
F 9
G 0
H 1
I 2
J 3